TECHNOLOGICAL DICTIONARY

Vol. 1 : mechanics - metallurgy - hydraulics
and related industries

ENGLISH - FRENCH - GERMAN
with FRENCH and GERMAN indexes

DICTIONNAIRE TECHNOLOGIQUE

Tome 1 : mécanique - métallurgie - hydraulique
et industries connexes

ANGLAIS - FRANÇAIS - ALLEMAND
avec index FRANÇAIS et ALLEMAND

TECHNOLOGISCHES WÖRTERBUCH

1. Band : Mechanik - Metallurgie - Hydraulik
mit Randgebieten

ENGLISCH - FRANZÖSISCH - DEUTSCH
mit FRANZÖSISCHEM und DEUTSCHEM Indexen

Compilé par **Michel FEUTRY** Editeur-Conseil en Librairie

Avec la collaboration de

Robert MERTZ de MERTZENFELD & **Agnès DOLLINGER**
Traducteur technique   *Traductrice technique*
Membre de la S.F.T.   *Diplômée E.S.I.T. (Paris)*
                                Membre de la S.F.T

LA MAISON DU DICTIONNAIRE

95 bis, Rue Legendre, 75017 PARIS - Tél. 229.48.36

The first volume of the TECHNOLOGICAL DICTIONARY is devoted to the MECHANICAL, METALLURGICAL and HYDRAULIC industries. In addition it covers the following subjects : Automobile, boiler-plate work, electricity (automobile), mechanical tests, foundry, forging, machine tools, powder metallurgy, mounting refractory ware, strength of materials, cocks and valves, steelworks, welding, worksite terms, thermal treatment of metals, surface finishes.

Other Volumes in preparation.

This dictionary has been created by means of a computer and type set by an advanced system : high-speed PHOTOCOMPOSITION.

Although the use of data-processing techniques has resulted in a considerable time saving in placing the entries in alphabetical order, on the other hand it has obliged us to be more concise in data acquisition. The conventional grammatical indications will therefore not be found in this first issue, the use of brackets being reserved for qualificatives displaced with respect to the German substantives and the comma being employed for distinguishing and extracting synonyms.

The work includes nearly 15 000 entries to which the various synonyms have been added : this results in 20 000 terms for the English part, while the French index includes close to 25 000 terms and the German index about 30.000. The difference in these figures results from the English word which can give rise to 2, 3 (or more) French or german terms.

We hope this dictionary will provide all the assistance which its users will come to expect from it.

LA MAISON DU DICTIONNAIRE PARIS.

Le premier volume de la série DICTIONNAIRE TECHNOLOGIQUE est consacré aux industries de la MECANIQUE, de la METALLURGIE et de l'HYDRAULIQUE. Il aborde en plus les sujets suivants : Automobile, Chaudronnerie, Électricité (automobile), essais mécaniques, Fonderie, Forge, Machines-outils, Métallurgie des poudres , Montage, Réfractaires, Résistance des matériaux, Robinetterie, Sidérurgie, Soudage, Termes de Chantiers, Traitement Thermique des métaux, Traitements de Surface.

D'autres volumes sont en préparation.

Ce dictionnaire a été mis en oeuvre avec l'aide d'un ordinateur et réalisé par un procédé récent : LA PHOTOCOMPOSITION PROGRAMMEE.

Si l'informatique nous a permis de gagner un temps considérable pour l'alphabétisation, par contre, elle nous a contraint à plus de concision dans la saisie des données : le lecteur ne trouvera donc pas dans cette première édition les indications grammaticales classiques, l'usage des parenthèses ayant été réservé aux qualificatifs déplacés par rapport aux substantifs allemands, celui de la virgule servant à la distinction et l'extraction des synonymes.

L'ouvrage comporte près 15 000 entrées auxquelles il faut ajouter les différentes acceptions, soit un total de 20 000 termes pour la partie anglaise, tandis que l'index français compte près de 25 000 termes et l'index allemand environ 30.000. La différence de ces chiffres provient du mot anglais qui peut donner lieu à 2, 3 (ou plus) synonymes français ou allemands.

Nous espérons que ce dictionnaire offrira à ses utilisateurs tous les services qu'ils peuvent en attendre.

LA MAISON DU DICTIONNAIRE PARIS.

Ich darf Ihnen hiermit ein Werk vorstellen, das von einem Dokumentenberater und zwei Fachübersetzern erarbeitet worden ist. Es umfasst das weite Gebiet der mechanischen Industrien, der Metallurgie und der Hydraulik. Ein ungemeiner Arbeitsaufwand war erforderlich, um ein dreisprachiges Wörterbuch : Englisch - Französisch - Deutsch zu schaffen, das den Bedürfnissen zahlreicher Übersetzer gerecht wird, die in der Industrie und als unentbehrliche Vermittler zwischen den Wissenschaftlern und Technikern zu beiden Seiten des Rheins, des Armelkanals und des Atlantiks tätig sind.

Die Arbeit des Lexikographen ist mühselig und langwierig. Zunächst einmal müssen alle Begriffe, die mit dem Wörterbuch im Zusammenhang stehen, ausgewählt werden, und nur die Lektüre zahlreicher Werke und Fachzeitschriften kann eine mehr oder weniger erschöpfende Terminologie liefern. Sodann muss der Kontext, in welchem der gewählte Begriff verwendet worden ist, untersucht werden. Es ist bekannt, dass das gleiche Wort durch ein vor-oder nachgestelltes Adjektiv oder Verb seine Bedeutung ändern kann. Uberdies kann der gleiche Ausdruck in den einzelnen Industriebereichen einen anderen Sinn haben oder manchmal in eigenartiger, bildhafter Weise im Jargon der Techniker gebraucht werden.

Diese beiden Verfahren - Auswahl und Studium des Kontextes- müssen so oft wiederholt werden wie das Wörterbuch Sprachen enthält. Hier tritt eine neue Schwierigkeit auf, die mit der Besonderheit der beiden verwendeten Sprachen verbunden ist : zum einen handelt es sich um die Schwierigkeit der Wortkonstruktion. Bekanntlich neigt die deutsche Sprache dazu, die Substantive mit ihren Adjektiven zusammenzuziehen oder komplexe Verben zu schmieden. Der englischen Sprache ist es ebenfalls eigen, zusammengesetzte Substantive oder Adjektive zu formen. Der französische Ubersetzer muss daher in der Lage sein, diese Wörter aufzuteilen, um den Begriff unter dem ersten Buchstaben des Substantives im Wörterbuch zu finden.

Zum andern stellen die Neuschöpfungen eine schwierige Klippe dar. Die Erweiterung des Wortschatzes Hand in Hand mit neuen wissenschaftlichen oder technischen Erkenntnissen, durch welche die Autoren gezwungen werden, neue Wörter zu schaffen, und da diese Neuschöpfungen keiner festen Regel unterliegen, werden sie der oft von eigenartigen Einfällen geleiteten Vorstellung der Autoren überlassen, so dass der Lexikograph eine wahre Entschlüsselungsarbeit leisten muss.

Das sind die Schwierigkeiten, die die Autoren dieses Wörterbuches zu überwinden hatten. Nur eine lange Praxis in der Ubersetzung gestattete es ihnen, diese Aufgabe zum guten Ende zu bringen.

Deshalb gebührt den Autoren Dank, ihren Kollegen ein unerlässliches Hilfsmittel zur Verfügung zu stellen, das oft über die verschiedenen technologischen und industriellen Bereiche hinausgeht, und das als erstes dreisprachiges technologisches Wörterbuch in Frankreich herausgegeben wird. Hier muss auch dem Verleger gedankt werden, der sich der Mühe unterzogen hat, ein Werk vorzulegen, das übersichtlich und leicht nachzuschlagen ist.

Der Humanist Scaliger sagte im 16. Jahrhundert : «Wenn Du einen Verurteilten bestrafen willst, schicke ihn nicht in die Eisenbergwerke noch in die Folterkammer, lasse ihn ein Wörterbuch zusammenstellen...» Ohne diese düstere Auffassung zu teilen, sei es mir gestattet, die drei Autoren dieses Werkes zu ihrer Leistung zu beglückwünschen. Möge ihnen der verdiente Erfolg beschieden sein.

Dr. A. SLIOSBERG,

Präsident der Société Française des Traducteurs

L'ouvrage que j'ai le redoutable honneur de présenter est l'oeuvre d'un conseiller en documentation et de deux traducteurs spécialistes en la matière. Il couvre le très vaste domaine des industries de la mécanique, de la métallurgie et de l'hydraulique ; c'est dire la somme de travail qu'il a fallu fournir pour mettre sur pied un ouvrage trilingue - Anglais, Français, Allemand - qui doit satisfaire aux exigences des nombreux traducteurs qui oeuvrent dans l'industrie et qui servent aussi d'intermédiaires irremplaçables entre les scientifiques et les techniciens de part et d'autre du Rhin, de la Manche et de l'Atlantique.

Le travail du lexicographe exige un effort continu et pénible : il comporte tout d'abord la sélection des termes se rapportant à la matière du Dictionnaire, et seule la lecture de nombreux ouvrages et périodiques peut fournir un ensemble terminologique plus ou moins exhaustif ; on doit ensuite étudier le contexte dans lequel le terme choisi est utilisé, et l'on sait que le même mot peut changer de signification selon qu'il est précédé ou suivi de tel qualificatif ou de tel verbe : en outre ce même terme est susceptible d'avoir des sens différents dans divers domaines industriels ou d'être utilisé d'une manière parfois singulière, voire pittoresque dans le jargon des techniciens.

Ces deux démarches - choix et étude du contexte - doivent être répétées autant de fois que le dictionnaire comporte de langues. Il s'agit là d'une nouvelle difficulté inhérente aux deux idiomes utilisés : d'une part la construction des mots ; on sait que l'allemand a tendance à accoler les substantifs et leurs qualificatifs ou forger des verbes complexes ; les Anglophones ont également l'habitude de créer des substantifs ou des adjectifs composés, que le traducteur français doit savoir scinder pour leur assigner une place alphabétique appropriée en tenant compte de la première lettre du substantif.

Les néologismes représentent un autre écueil : l'enrichissement du vocabulaire qui va de pair avec l'acquisition de nouvelles notions scientifiques ou techniques pousse les auteurs à créer des mots nouveaux, et comme aucune règle précise ne préside à cette création, elle est laissée à leur imagination souvent fantaisiste, et demande au lexicographe un véritable effort de décodage.

Telles sont les difficultés que les auteurs de ce dictionnaire avaient à surmonter et seule une longue pratique de la traduction leur ont permis de mener à bonne fin cette tâche ardue.

On ne peut que les remercier d'avoir mis à la disposition de leurs confrères un outil indispensable qui déborde largement sur différents domaines technologiques et industriels et qui semble être le premier et unique dictionnaire technologique trilingue édité en France : à ce titre on doit également des louanges à l'éditeur qui a pris soin de présenter un ouvrage clair et facile à consulter.

L'humaniste Scaliger disait au XVIe siècle : «Si tu veux envoyer un condamné au supplice, ne l'envoie pas aux mines de fer, ni au tortionnaire, fais lui faire un dictionnaire ...» Sans prendre à la lettre cette sombre façon de concevoir la lexicographie, qu'il me soit permis de féliciter les trois auteurs de cet ouvrage pour leur performance et de leur souhaiter tout le succès qu'ils méritent.

Dr A. SLIOSBERG,

Président de la Société Française des Traducteurs

FOREWORD

The work it is my great privilege to present, is the product of a documentation consultant and of two translators, authorities in their particular fields. It covers the very vast domain of the mechanical, metallurgical and hydraulic industries, and this fact alone gives one an idea of the immense efforts that had to be expended to compile a three-language work of such a magnitude ; this dictionary must in fact meet the requirements of a large number of translators employed in industry, and who serve as irreplaceable intermediaries between scientists and technicians working on both sides of the Rhine, the channel and the Atlantic.

The work of a lexicographer entails continuous and laborious efforts : it comprises first of all the selection of the terms relevant to the contents of the dictionary, and only the reading of large numbers of works and periodicals can assure a more or less exhaustive vocabulary of words and terms. Next one must analyze the context in which the chosen term has been used, and in this connection it should be noted that the same word may have a different meaning, depending on whether it is preceded or followed by a certain adjective or verb. Moreover this same term is liable to mean different things in different industries, or to be used in a sometimes peculiar or even picturesque manner, in the slang of technicians.

These two steps - choice and analysis of contents - must be repeated as many times as the number of languages in the dictionary. This further difficulty is in fact inherent in the two languages used ; for instance the manner in which the words are constructed is different. In German nouns and their adjectives tend to be strung together, or complex verbs built up. In English there is also a tendency to form composite nouns or adjectives, and the French translator must know how to split these up in order to assign them an appropriate alphabetic position, taking into account the first letter of the noun.

Neologisms represent another stumbling block : the enriching of the vocabulary, which goes hand in hand with the acquisition of scientific and technical know-how, impels writers to coin new words ; there are no precise rules governing this coining of new words, and thus full play is given to an at times fanciful imagination. The lexicographer is thus faced with a veritable decoding task.

Outlined above are some of the difficulties that the compilers of this dictionary had to overcome, and it was only their great experience in translating that enabled them successfully to conclude this arduous task.

We must therefore not fail to thank them for having placed at the disposal of their colleagues an essential tool which extends well beyond the various technological and industrial domains, and which appears to be the first and sole three-language technological dictionary published in France. Similarly credit must also be given to the editor for having taken great care to present a work that is both clear and easy to consult.

The classical scholar Scaliger said in the 16th century : «If you wish to punish a condemned man, do not send him to the iron mines, or to the torturer, instead make him compile a dictionary ...» Without taking literally this gloomy manner of imagining lexicography, may I be allowed to congratulate the three authors of this work for their fine performance, and to wish them all the success they deserve.

Dr. A. SLIOSBERG,
President of the French Translators' Association

AN DIE LESER

Der erste Band der Reihe TECHNOLOGISCHE WORTERBUCHER ist den Industrien der MECHANIK, der METALLURGIE und der HYDRAULIK gewidmet. Uberdies werden folgende Themen gestreif : Automobilindustrie, Kesselbau, Elektrizität (für Kraftfahrzeuge), mechanische Prüfungen, Giesserei, Schmiedeindustrie, Werkzeugmaschinen, Eisenhüttenkunde, Montage, feuerfeste Stoffe, Festigkeit der Werkstoffe, Armaturen, Siderurgie, Schweisstechnik, Bauwesen, Wärmebehandlung der Metalle, Oberflächenbehandlung.

Weitere Bände werden vorbereitet.

Dieses Wörterbuch ist mit Hilfe eines Computers erarbeitet und durch ein fortschrittliches Verfahren, den programmgesteuerten Photolichtsatz, fertiggestellt worden.
Die elektronische Datenverarbeitung hat uns zwar gestattet, viel Zeit beim alphabetischen Sortieren zu gewinnen, hat uns aber gezwungen, uns kürzer bei der Datenerfassung zu fassen. So wird der Leser in dieser ersten Ausgabe nicht die üblichen grammatikalischen Erklärungen finden, zumal die Klammern für nachgestellte Ergänzungen zu den deutschen Substantiven und das Komma zur Trennung und zum Selektieren von Synonymen reserviert wurden.

Der englische Teil dieses Wörterbuchs umfasst etwa 15 000 Stichwörter, zu denen noch die verschiedenen Synonyme hinzukommen, daraus, ergeben Sich im Ganzen 20 000 Wörter, während der französische Index etwa 20 000 und der deutsche Index etwa 30 000 Wörter enthält. Der Unterschied zwischen diesen Zahlen beruht auf der Tatsache, dass ein englischer Ausdruck 2, 3 (oder mehr) französische oder deutsche Synonyme haben kann.

Wir hoffen, dass dieses Wörterbuch seinen Benutzern alle gewünschten Auskünfte liefern kann.

LA MAISON DU DICTIONNAIRE, PARIS.

	English	French	German
1	A.C ALTERNATING CURRENT	COURANT ALTERNATIF	WECHSELSTROM
2	A.C.-WELDING MACHINE	POSTE DE SOUDAGE A COURANT ALTERNATIF	WECHSELSTROMSCHWEISSMASCHINE
3	A.C.H.F. ALTERNATING CURRENT WITH HIGH FREQUENCY	COURANT ALTERNATIF HAUTE FREQUENCE	WECHSELSTROM (HOCHFREQUENZER)
4	ABATING	REDUCTION DE TREMPE	HÄRTUNGSMINDERUNG
5	ABERRATION OF LIGHT	ABERRATION	ABWEICHUNG, ABERRATION DES LICHTES
6	ABILITY OF BEING DECOMPOSED	DESTRUCTIBILITE	ZERSETZLICHKEIT
7	ABNORMAL GRAIN GROWTH	CROISSANCE ANORMALE DES CRISTAUX	KRISTALLWACHSTUM (ABNORMALES)
8	ABNORMAL STEEL	ACIER ANORMAL	STAHL (ANORMALER)
9	ABNORMAL WROUGHT IRON	FER MALLEABLE ANORMAL	SCHMIEDEEISEN (ANORMALES)
10	ABRADE (TO)	ENLEVER (OU USER) EN EMOULANT	ABSCHLEIFEN
11	ABRADING AGENT	MATIERE A POLIR	POLIERMITTEL
12	ABRAMSEN STRAIGHTENER	REDRESSEUR (DE TUBES) ABRAMSEN	ABRAMSEN-ROHRRICHTMASCHINE
13	ABRASiON	ABRASION, ENLEVEMENT PAR EMOULAGE	ABSCHLEIFEN, ABRIEB
14	ABRASION HARDNESS	DURETE DE MEULAGE	SCHLEIFHÄRTE
15	ABRASION MARKS	MARQUES DE MEULAGE	SCHLEIFSPUREN
16	ABRASION TESTS	ESSAIS D'USURE	ABSCHLEIFVERSUCH
17	ABRASIVE	ABRASIF	SCHLEIFMITTEL
18	ABRASIVE BELT	BANDE ARRASIVE, RUBAN D'EMERI	SCHLEIFBAND
19	ABRASIVE BRICK	BRIQUE ABRASIVE	SCHLEIFSTEIN
20	ABRASIVE DISK	MEULE D'EMERI	SCHLEIFSCHMIRGELSCHEIBE
21	ABRASIVE GRAIN	GRAIN ABRASIF	SCHLEIFKORN
22	ABRASIVE MATERIAL	MATIERE A POLIR	POLIERMITTEL, SCHLEIFMITTEL
23	ABRASIVE PARTICLE SIZE	GROSSEUR DES GRAINS D'ABRASIF	SCHLEIFMATERIALKORNGRÖSSE
24	ABRASIVE POINTS	POINTES ABRASIVES	SCHLEIFSPITZEN
25	ABRASIVE SHOT	GRENAILLE	GRANALIE
26	ABRASIVE WHEELS	DISQUES ABRASIFS, MEULES	SCHLEIFKÖRPER, SCHLEIFSCHMIRGELSCHEIBEN
27	ABSCISSA	ABSCISSE	ABSZISSE
28	ABSOLUTE ALCOHOL	ALCOOL ABSOLU, ALCOOL ANHYDRE	ALKOHOL (WASSERFREIER), ALKOHOL (ABSOLUTER)
29	ABSOLUTE DIMENSIONING	COTATION ABSOLUE	BEZUGSMASSSYSTEM
30	ABSOLUTE HUMIDITY OF THE AIR, MOISTURE CONTENT OF THE AIR	HUMIDITE ABSOLUE DE L'AIR	FEUCHTIGKEITSGEHALT, FEUCHTIGKEIT ABSOLUTE DER LUFT
31	ABSOLUTE MEASURING SYSTEM	SYSTEME DE MESURE ABSOLUE	MESSWERTERFASSUNG (ABSOLUTE)
32	ABSOLUTE MOTION	MOUVEMENT ABSOLU	BEWEGUNG (ABSOLUTE)
33	ABSOLUTE PRESSURE	PRESSION ABSOLUE	DRUCK (ABSOLUTER)
34	ABSOLUTE REFERENCE POINT	POINT DE REFERENCE ABSOLU	BEZUGSPUNKT (ABSOLUTER)
35	ABSOLUTE TEMPERATURE	TEMPERATURE ABSOLUE	TEMPERATUR (ABSOLUTE)
36	ABSOLUTE ZERO POINT	ZERO ABSOLU	NULLPUNKT (ABSOLUTER)
37	ABSORB (TO), SUCK UP (TO)	ABSORBER	AUFNEHMEN, AUFSAUGEN, ABSORBIEREN
38	ABSORBABLE	ABSORBABLE	ABSORBIERBAR
39	ABSORBENT	ABSORBANT	ABSORPTIONSMITTEL
40	ABSORBER	ABSORBEUR	ABSORBATOR, ABSORBER
41	ABSORBING CAPACITY FOR HEAT	CAPACITE CALORIFIQUE	WÄRMEAUFNAHMEFÄHIGKET
42	ABSORBING POWER	POUVOIR ABSORBANT, PUISSANCE ABSORBANTE	AUFSAUGEFÄHIGKEIT, ABSORPTIONSFÄHIGKEIT, ABSORPTIONSVERMÖGEN, ABSORBIERBARKEIT

43	ABSORPTIOMETER	ABSORPTIOMETRE	ABSORPTIOMETER
44	ABSORPTION	ABSORPTION	AUFSAUGEN, AUFSAUGUNG, ABSORPTION
45	ABSORPTION BAND	BANDE D'ABSORPTION	ABSORPTIONSSTREIFEN
46	ABSORPTION COEFFICIENT	COEFFICIENT D'ABSORPTION APPARENT	ABSORPTIONSKOEFFIZIENT
47	ABSORPTION CONSTANT	CONSTANTE D'ABSORPTION	ABSORPTIONSKONSTANTE
48	ABSORPTION DYNAMOMETER	DYNAMOMETRE D'ABSORPTION	ABSORPTIONSDYNAMOMETER
49	ABSORPTION EDGE	BORD OU FLANC DE LA BANDE D'ABSORPTION	ABSORPTIONSKANTE
50	ABSORPTION LIMIT	LIMITE D'ABSORPTION	ABSORPTIONSGRENZE
51	ABSORPTION OF HEAT	ABSORPTION DE CHALEUR	WÄRMEAUFNAHME
52	ABSORPTION OF WATER	ABSORPTION D'EAU	WASSERAUFNAHME
53	ABSORPTION RATIO	RAPPORT D'ABSORPTION	ABSORPTIONSVERHÄLTNIS
54	ABSORPTION SPECTRUM	SPECTRE D'ABSORPTION	ABSORPTIONSSPEKTRUM
55	ABSORPTION, BRAKE DYNAMOMETER	DYNAMOMETRE D'ABSORPTION, DYNAMOMETRE-FREIN	BREMSDYNAMOMETER, ABSORPTIONSDYNAMOMETER
56	ABSORPTIVITY	ABSORPTIVITE	ABSORPTIONSVERMÖGEN
57	ABUTMENT	SUPPORT, APPUI CALE, BUTEE, CULEE	WIDERLAGER
58	ABYSSINIAN GOLD	BRONZE D'ALUMINIUM	GOLD (ABESSINISCHES)
59	AC/DC WELDING MACHINE	POSTE DE SOUDAGE A COURANTS ALTERNATIF ET CONTINU	ALLSTROMSCHWEISSMASCHINE
60	ACCELERATED MOTION	MOUVEMENT ACCELERE	BEWEGUNG (BESCHLEUNIGTE)
61	ACCELERATING PUMP	POMPE DE REPRISE	MEMBRANPUMPE
62	ACCELERATION	ACCELERATION	BESCHLEUNIGUNG
63	ACCELERATION OF A FALLING BODY	ACCELERATION DE VITESSE D'UN CORPS TOMBANT	FALLBESCHLEUNIGUNG
64	ACCELERATION OF GRAVITY	ACCELERATION DE LA PESANTEUR	SCHWEREBESCHLEUNIGUNG
65	ACCELERATION OF GRAVITY, GRAVITY ACCELERATION	ACCELERATION DE LA PESANTEUR	SCHWEREBESCHLEUNIGUNG
66	ACCELERATOR	ACCELERATEUR	BESCHLEUNIGER
67	ACCEPTANCE	RECEPTION	ABNAHME
68	ACCEPTANCE (OF MACHINES)	RECEPTION (DE MACHINES)	ABNAHME (VON MASCHINEN)
69	ACCEPTANCE BOUNDARY	LIMITE D'ACCEPTATION	ABNAHMEGRENZE
70	ACCEPTANCE CERTIFICATE	CERTIFICAT DE RECEPTION	ABNAHMEBESCHEINIGUNG
71	ACCEPTANCE TEST	ESSAI DE RECEPTION	ABNAHMEPRÜFUNG, ABNAHMEVERSUCH
72	ACCESSIBILITY	ACCESSIBILITE	ZUGÄNGLICHKEIT
73	ACCESSIBLE	ACCESSIBLE	ZUGÄNGLICH
74	ACCESSORIES	ACCESSOIRES	ZUBEHÖR, ZUBEHÖRTEILE
75	ACCIDENT	ACCIDENT DU TRAVAIL	UNFALL, BETRIEBSUNFALL
76	ACCUMULATOR	ACCUMULATEUR	SAMMLER, SPEICHER, AKKUMULATOR
77	ACCUMULATOR BATTERY	BATTERIE D'ACCUMULATEURS	SAMMLERBATTERIE, AKKUMULATORENBATTERIE
78	ACCUMULATOR CELL, SECONDARY CELL	ELEMENT D'ACCUMULATEUR	AKKUMULATORZELLE
79	ACCUMULATOR LOCOMOTIVE	LOCOMOTIVE ELECTRIQUE A ACCUMULATEURS	SAMMLERLOKOMOTIVE, AKKUMULATORENLOKOMOTIVE
80	ACCUMULATOR METAL	METAL D'ACCUMULATEUR	AKKUMULATORENMETALL
81	ACCUMULATOR PLATE	PLAQUE D ACCUMULATEUR, LAME D ACCUMULATEUR	AKKUMULATORPLATTE
82	ACCURACY	PRECISION	GENAUIGKEIT

83	ACCURACY OF WORK	PRECISION D'USINAGE	GENAUIGKEIT DER AUSFÜHRUNG
84	ACETALDEHYDE	ALDEHYDE ACETIQUE, ALDEHYDE ETHYLIQUE, ETHYLAL, HYDRURE D ACETYLE, OXYDE D ETHYLIDENE, HYDRATE DE VINYLE, ACIDE ALDEHYDIQUE	ALDEHYD, AZETALDEHYD, ÄTHYLALDEHYD
85	ACETATE	ACETATE	ESSIGSAURES SALZ, AZETAT
86	ACETIC ACID	ACIDE ACETIQUE, VINAIGRE RADICAL	ESSIGSÄURE
87	ACETIC ANHYDRIDE	ANHYDRIDE ACETIQUE	ESSIGSÄUREANHYDRID
88	ACETIC ETHER, ETHYL ACETATE	ETHER ACETIQUE, ACETATE D'ETHYLE	ESSIGSÄUREÄTHER, ESSIGSÄUREÄTHYLESTER, ÄTHYLAZETAT
89	ACETIMETER, ACETOMETER, ACIDIMETER	ACIDIMETRE, PESE-ACIDES	SÄUREMESSER
90	ACETONE, DIMETHYL KETONE	ACETONE, DIMETHYLCETONE	AZETON, ESSIGGEIST, BRENZESSIGGEIST, DIMETHYLKETON
91	ACETYL CELLULOSE, CELLULOSE ACETATE	ACETATE, ACETYLE DE CELLULOSE	AZETYLZELLULOSE, ZELLULOSEAZETAT
92	ACETYL CHLORIDE	CHLORURE D'ACETYLE	AZETYLCHLORID
93	ACETYLENE	ACETYLENE	AZETYLEN
94	ACETYLENE CUTTING	COUPAGE A L'ACETYLENE	AZETYLENSCHNEIDVERFAHREN
95	ACETYLENE LAMP	LAMPE A ACETYLENE	AZETYLENLAMPE
96	ACETYLENE WELDING	SOUDAGE A L'ACETYLENE	AZETYLENGASSCHWEISSUNG
97	ACHESON FURNACE	FOURNEAU D'ACHESON	ACHESON-OFEN
98	ACHROMATIC	ACHROMATIQUE	ACHROMATISCH
99	ACHROMATISATION	ACHROMATISATION	ACHROMATISIEREN
100	ACHROMATISE (TO)	ACHROMATISER	ACHROMATISIEREN
101	ACHROMATISM	ACHROMATISME	ACHROMASIE, ACHROMATISMUS
102	ACICULAR	ACICULAIRE	NADEL- (KRISTALL-)FÖRMIG, NADELIG, MIT NADELSTRUKTUR
103	ACICULAR MARTENSITE	MARTENSITE ACICULAIRE	MARTENSIT (NADELFÖRMIGER)
104	ACID	ACIDE	SÄURE
105	ACID BESSEMER PIG	FONTE BRUTE BESSEMER	BESSEMER-ROHEISEN
106	ACID BESSEMER PROCESS	PROCEDE BESSEMER AU CONVERTISSEUR A GARNISSAGE ACIDE	BESSEMER-VERFAHREN
107	ACID BESSEMER STEEL	ACIER BESSEMER	BESSEMER-STAHL
108	ACID BOTTOM	SOLE ACIDE	SOHLE (SAURE)
109	ACID BRITTLENESS	FRAGILITE DE DECAPAGE	BEIZSPRÖDIGKEIT
110	ACID CORE SOLDER	ETAIN A SOUDER A FONDANT ACIDE	LOT (SAURES)
111	ACID DIP	BAIN DE DECAPAGE	DEKAPIERBAD
112	ACID DIPPING	DECAPAGE AU BAIN ACIDULE	BEIZBEHANDLUNG (SAURE)
113	ACID ELECTRIC STEEL	ACIER ELECTRIQUE ACIDE	ELEKTROSTAHL (SAURER)
114	ACID FLUX	FONDANT ACIDE	FLUSSMITTEL (SAURES)
115	ACID FORMING ELEMENT	ELEMENT ACIDIFICATEUR	SÄUREBILDNER
116	ACID OPEN HEARTH STEEL	ACIER MARTIN PAR LE PROCEDE ACIDE	SIEMENS MARTINSTAHL (SAURER)
117	ACID PIG	FONTE BESSEMER	BESSEMER-ROHEISEN
118	ACID POTASSIUM OXALATE	OXALATE ACIDE DE POTASSIUM	KALI (SAURES), KALI (OXALSAURES), KALIUMBIOXALAT
119	ACID POTASSIUM TARTRATE, CREAM OF TARTAR	TARTRE, BITARTRATE DE POTASSIUM, PIERRE DE VIN, CRISTAUX, CREME DE TARTRE	WEINSTEIN (SAURER), WEINSAURES KALI, KALIUMBITARTRAT
120	ACID PROCESS	PROCEDE ACIDE	VERFAHREN (SAURES)

121	ACID PUMP, GLASS BARREL PUMP	POMPE A ACIDE	SÄUREPUMPE
122	ACID REACTION	REACTION ACIDE	REAKTION (SAURE)
123	ACID REFRACTORIES	GARNISSAGE REFRACTAIRE ACIDE	OFENFUTTER (SAURES)
124	ACID RESISTANT	ANTI-ACIDE RESISTANT AUX ACIDES	SÄUREFEST, SÄUREBESTÄNDIG
125	ACID RESISTING ALLOYS	ALLIAGES RESISTANT AUX ACIDES	SÄUREFESTE LEGIERUNGEN
126	ACID SODIUM CARBONATE, SODIUM BICARBONATE	CARBONATE ACIDE DE SODIUM, BICARBONATE DE SOUDE	NATRON (DOPPELTKOHLENSAURES), NATRIUMBIKARBONAT
127	ACID VAPOURS	VAPEURS ACIDES	SÄUREDÄMPFE
128	ACID-PROOF	INATTAQUABLE AUX ACIDES	SÄUREBESTÄNDIG, SÄUREFEST
129	ACIDIC	ACIDE	ACIDISCH
130	ACIDIFIANT, ACIDIFIC	ACIDIFIANT	SÄUREBILDEND
131	ACIDIMETRY	ACIDIMETRIE	SÄUREGEHALTSBESTIMMUNG, AZIDIMETRIE
132	ACIDS	ACIDES	SÄUREN
133	ACIDULATE (TO)	ACIDULER	ANSÄUERN
134	ACIDULATED WATER	EAU ACIDULEE	WASSER (ANGESÄUERTES)
135	ACIERAGE	ACIERAGE	VERSTAHLEN, STAHLUNG
136	ACOUSTIC	ACOUSTIQUE	AKUSTISCH
137	ACOUSTICS	ACOUSTIQUE	SCHALLEHRE, AKUSTIK
138	ACROSS THE GRAIN FIBRES	A CONTREFIL	SENKRECHT ZUR FASER, QUER ZUR FASERRICHTUNG
139	ACTINIC	ACTINIQUE	AKTINISCH
140	ACTINIC RAYS	RAYONS ACTINIQUES	STRAHLEN (ARTINISCHE)
141	ACTINIUM	ACTINIUM	AKTINIUM
142	ACTINOMETER	ACTINOMETRE	AKTINOMETER, STRAHLENMESSER
143	ACTION OF A FORCE	ACTION D'UNE FORCE	KRAFTWIRKUNG, EIN WIRKUNG (EINER KRAFT)
144	ACTIVATION	ACTIVATION	STEIGERUNG DER REAKTIONSFÄHIGKEIT, AKTIVIERUNG
145	ACTIVATION - ACIERATION	ADDITION DE CARBONE	STAHLBILDUNG, VERWANDLUNG IN STAHL, KOHLENSTOFFZUGABE
146	ACTIVATION AGENT	ACTIVATEUR	AKTIVATOR
147	ACTIVATION ENERGY	ENERGIE D'ACTIVATION	AKTIVIERUNGSENERGIE
148	ACTIVATOR	SUBSTANCE ACTIVATRICE	AKTIVIERUNGSMITTEL, AKTIVATOR, BESCHLEUNIGER
149	ACTIVE DEPOSIT	DEPOT ACTIF	NIEDERSCHLAG (RADIOAKTIVER)
150	ACTIVE HYDROGEN	HYDROGENE ATOMIQUE	WASSERSTOFF (ATOMARER)
151	ACTUAL EFFECTIVE BRAKE HORSE POWER (B.H.P.)	CHEVAL EFFECTIF, PUISSANCE EFFECTIVE EN CHEVAUX, PUISSANCE AU FREIN EN CHEVAUX	PFERDESTÄRKE (NUTZBARE), PFERDESTÄRKE (GEBREMSTE), BREMSPFERDESTÄRKE (EFFEKTIVE), PSE
152	ACTUAL SIZE	DIMENSION REELLE	GRÖSSE (WIRKLICHE)
153	ACTUAL, REAL VALUE	VALEUR REELLE	ISTWERT
154	ACUTANGULAR, ACUTEANGLED	ACUTANGLE	SPITZWINKLIG
155	ACUTE ANGLE	ANGLE AIGU, ANGLE POINTU, ANGLE VIF	WINKEL (SPITZER)
156	ACUTE BISECTRIX	BISECTRICE AIGUE	BISEKTRIX (SPITZE)
157	ACUTE-ANGLED TRIANGLE	TRIANGLE ACUTANGLE	DREIECK (SPITZWINKLIGES)
158	ADAMANTINE LUSTRE	ECLAT ADAMANTIN	DIAMANTGLANZ
159	ADAPT PROGRAMMING LANGUAGE	LANGAGE DE PROGRAMMATION ADAPT	ADAPT-PROGRAMMIERSPRACHE
160	ADAPTER	ADAPTEUR	ADAPTOR, PASSSTÜCK
161	ADAPTIVE CONTROL	COMMANDE ADAPTIVE	STEUERUNG (ADAPTIVE)

3

162	ADD (TO), SUM (TO), ADMIX (TO)	ADDITIONNER (MATH.), AJOUTER	ZUSAMMENZÄHLEN, ADDIEREN, SUMMIEREN, BEIMISCHEN, BEIMENGEN
163	ADD WATER (TO)	ADDITIONNER DE L'EAU	VERSETZEN (MIT WASSER), ZUSETZEN (WASSER)
164	ADDENDUM	HAUTEUR DE FACE, HAUTEUR DE LA TETE D'UNE DENT, SAILLIE SUR LE PRIMITIF	ZAHNKOPFLÄNGE
165	ADDENDUM ANGLE	ANGLE DE SAILLIE	KOPFWINKEL
166	ADDENDUM CIRCLE	CERCLE DE TETE, CERCLE EXTERIEUR, CIRCONFERENCE D'ECHANFREINEMENT	KOPFKREIS, KRONENKREIS
167	ADDENDUM FLANK	CERCLE DE TETE, FLANC DE SAILLIE	KOPFKREIS, KOPFFLANKE
168	ADDITION AGENT	ELEMENT D'ADDITION, ADDITIF	ZUSATZELEMENT, BADZUSATZ
169	ADDITION, SUMMATION	ADJUVANT, ADDITION	ZUSATZ, ZUSAMMENZÄHLEN, HINZUFÜGEN, ADDITION, SUMMATION
170	ADDITIONAL LOSS	PERTE ADDITIONNELLE	VERLUST (ZUSÄTZLICHER), ZUSATZVERLUST
171	ADDRESS	ADRESSE	ADRESSE
172	ADHERENCE TEST	CONTROLE D'ADHERENCE	HAFTFÄHIGKEITSVERSUCH
173	ADHESION	ADHERENCE	HAFTFESTIGKEIT
174	ADHESION, ADHESIVE POWER	ADHESION, ADHERENCE	ADHÄSIONSKRAFT
175	ADHESIVE	ADHESIF, ADHERENT	HAFTEND, HAFT-
176	ADHESIVE PROPERTY	POUVOIR ADHESIF	HAFTVERMÖGEN
177	ADHESIVE SUBSTANCE	SUBSTANCE ADHESIVE	KLEBEMITTEL, KLEBSTOFF
178	ADIABATIC	ADIABATIQUE	ADIABATISCH
179	ADIABATIC CURVE	LIGNE, COURBE ADIABATIQUE, ADIABATIQUE	ADIABATE
180	ADJACENT ANGLE	ANGLE ADJACENT	NEBENWINKEL
181	ADJACENT, CONTIGUOUS	ADJACENT, CONTIGU	ANLIEGEND, ANGRENZEND, ANSTOSSEND
182	ADJUST (TO), REGULATE (TO)	REGLER	EINSTELLEN, REGELN, REGULIEREN, ADJUSTIEREN
183	ADJUSTABLE	REGLABLE, AJUSTABLE	VERSTELLBAR, NACHSTELLBAR
184	ADJUSTABLE HAND REAMER	ALESOIR A MAIN REGLABLE	HANDREIBAHLE (EINSTELLBARE)
185	ADJUSTABLE KEY	COIN, CLAVETTE DE SERRAGE, COIN POUR LE RATTRAPAGE DU JEU, COIN D'EPAISSEUR, CALE DE RATTRAPAGE	STELLKEIL, SPANNKEIL, NACHSTELLKEIL
186	ADJUSTABLE SNAP-GAUGE LIMIT TIPE	JAUGE-MACHOIRE REGLABLE	GRENZ-RACHENLEHRE (EINSTELLBARE)
187	ADJUSTABLE STOP	BUTEE REGLABLE	VOLLAST-EINSTELLSCHRAUBE
188	ADJUSTABLE, EXPANDING REAMER	ALESOIR EXTENSIBLE, A LAMES MOBILES	REIBAHLE (VERSTELLBARE), EXPANSIONSREIBAHLE
189	ADJUSTANCE DIES	FILIERES	DRAHTZIEHSTEIN
190	ADJUSTEMENT BY SCREW	REGLAGE PAR VIS	VERSTELLUNG DURCH SCHRAUBE
191	ADJUSTING NUT	ECROU DE REGLAGE	STELLMUTTER, NACHSTELLMUTTER
192	ADJUSTING SCREW	VIS DE REGLAGE	STELLSCHRAUBE, EINSTELLSCHRAUBE, ADJUSTIERSCHRAUBE
193	ADJUSTING SCREW	VIS DE PRESSION	DRUCKSCHRAUBE
194	ADJUSTING SLEEVE	DOUILLE DE REGLAGE	EINSTELLHÜLSE
195	ADJUSTMENT, REGULATION	REGLAGE	EINSTELLEN, EINSTELLUNG, REGELUNG, REGULIERUNG, ADJUSTIEREN
196	ADJUTAGE, AJUTAGE, DISCHARGING TUBE, MOUTHPIECE	AJUTAGE D'ECOULEMENT, EMBOUCHURE, GICLEUR	ANSATZROHR, AUSFLUSSSTUTZEN, MUNDSTÜCK, AUSFLUSSDÜSE
197	ADMIRALTY GUN METAL	BRONZE DE CANONS	GESCHÜTZBRONZE

198	ADMIRALTY METAL	LAITON DE MARINE	ADMIRALITÄTSMETALL
199	ADMISSION LINE	LIGNE D'ADMISSION DE LA VAPEUR	DAMPFEINSTRÖMLINIE, EINSTRÖMLINIE, VOLLDRUCKLINIE
200	ADMISSION PORT, STEAM PORT	LUMIERE D'ADMISSION	EINLASSSCHLITZ
201	ADMISSION VALVE, (STREAM) INLET VALVE	SOUPAPE D'ADMISSION	EINLASSVENTIL
202	ADSORB (TO)	ADSORBER	ADSORBIEREN
203	ADSORPTION	ADSORPTION	ADSORPTION
204	ADULTERATE (TO)	ADULTERER, FALSIFIER	VERFÄLSCHEN
205	ADULTERATION	ADULTERATION, FALSIFICATION, SOPHISTICATION, FRAUDE	FÄLSCHUNG, VERFÄLSCHUNG
206	ADVANCE	AVANCE	FRÜHEINSTELLUNG
207	ADZE	HERMINETTE	DÄCHSEL, DECHSEL, TEXEL, KRUMMAXT
208	AERATE (TO)	AERER UN SABLE	AUFLOCKERN (DEN SAND)
209	AERATION CELL	DIVISEUR-AERATEUR	BELÜFTUNGSELEMENT, SANDSCHLEUDER
210	AERATION OF WATER	AERATION DE L'EAU	BELÜFTUNG DES WASSERS
211	AERATOR	AERATEUR	ENTLÜFTER
212	AERIAL LINE, CONDUCTOR, OVERHEAD LINE	LIGNE AERIENNE	OBERIRDISCHE LEITUNG, FREILEITUNG, LUFTLEITUNG
213	AERIAL ROPEWAY, AERIAL CABLEWAY	TRANSPORTEUR AERIEN, TRANSPORTEUR PAR CABLE	LUFTSEILBAHN, DRAHTSEILBAHN
214	AERODYNAMIC	AERODYNAMIQUE	AERODYNAMISCH
215	AERODYNAMICS	AERODYNAMIQUE	LUFTBEWEGUNGSLEHRE, AERODYNAMIK
216	AEROSE	BRONZE (DE)	BRONZEN
217	AEROSTATICAL	AEROSTATIQUE	AEROSTATISCH
218	AEROSTATICS	AEROSTATIQUE	LUFTDRUCKLEHRE, AEROSTATIK
219	AERUGINOUS	ERUGINEUX	GRÜNSPAN, GRÜNSPANÄHNLICH
220	AERUGO, VERDIGRIS	VERT-DE-GRIS	EDELROST, GRÜNSPAN, PATINA
221	AFTER-FLOW	REVENU-FLUAGE POSTERIEUR	NACHGLÜHEN, NACHFLIESSEN
222	AGATE	AGATE	ACHAT
223	AGE HARDENING	DURCISSEMENT PAR VIEILLISSEMENT	AUSHÄRTUNG, VERGÜTUNG BEI NORMALER TEMPERATUR
224	AGE-HARDENING	DURCISSEMENT STRUCTURAL	KALT-ODER WARM-AUSHÄRTUNG
225	AGEING, AGING	VIEILLISSEMENT	ALTERUNG
226	AGEING, ARTIFICIAL	VIEILLISSEMENT ARTIFICIEL	ALTERUNG (KÜNSTLICHE)
227	AGGREGATE	AGREGAT	AGGREGAT, GEHÄUSE
228	AGGREGATE	MATIERE DE MELANGE	ZUSCHLAGSTOFF, FÜLLSTOFF (BAUW.)
229	AGGREGATION	AGREGATION	ZUSAMMENSETZUNG
230	AGING RANGE	TEMPERATURES DE VIEILLISSEMENT	ALTERUNGSBEREICH
231	AGING, CRITICAL	VIEILLISSEMENT COMPLET	ALTERUNG (VOLLSTÄNDIGE)
232	AGING, INTERRUPTED	VIEILLISSEMENT ECHELONNE	ALTERUNG (STUFENWEISE)
233	AGING, NATURAL	VIEILLISSEMENT NATUREL	ALTERUNG (NATÜRLICHE)
234	AGING, PROGRESSIVE	VIEILLISSEMENT PROGRESSIF	ALTERUNG (PROGRESSIVE)
235	AGING, STRAIN	VIEILLISSEMENT PAR TRAVAIL A FROID	ALTERUNG DURCH KALTBEARBEITUNG
236	AGITATOR	AGITATEUR	RÜHRWERK, RÜHRAPPARAT, RÜHRER
237	AGITATOR RECESS	NICHE D'AGITATEUR	RÜHRWERKNISCHE
238	AGITATOR, STIRRER	AGITATEUR MECANIQUE	RÜHRWERK
239	AILANTHUS	VERNIS DU JAPON	GÖTTERBAUM, AILANTHUSBAUM
240	AIR	AIR ATMOSPHERIQUE	LUFT (ATMOSPHÄRISCHE)

241	AIR BELT	CHAMBRE DE VENT	WINDKAMMER
242	AIR BLAST	JET D'AIR, SOUFFLERIE	LUFTSTRAHL, GEBLÄSE, LUFTSTOSS
243	AIR BLAST, WIND	AIR DE LA SOUFFLERIE	LUFT, GEBLÄSEWIND, WIND
244	AIR BLASTING	NETTOYAGE PAR JETS D'AIR	ABBLASEN MIT PRESSLUFT
245	AIR BRICK	BRIQUE CREUSE	LOCHSTEIN, LOCHZIEGEL, HOHLZIEGEL
246	AIR CHAMBER	ANTIBELIER	LUFTKAMMER
247	AIR CHANNELS	CANAUX DE VENT	WETTERLUTTE, WETTERFANG, LUTTE, WETTERLEITUNG, WINDKANÄLE
248	AIR CLASSIFICATION	TRIAGE PAR COURANT GAZEUX	WINDSICHTUNG
249	AIR COMPRESSOR	COMPRESSEUR D'AIR, POMPE D'AIR	LUFTVERDICHTER, LUFTKOMPRESSOR
250	AIR CONDITIONER	CLIMATISEUR	KLIMAANLAGE
251	AIR COOLED ENGINE	MOTEUR REFROIDI PAR AIR	MOTOR (LUFTGEKÜHLTER)
252	AIR COOLING	REFROIDISSEMENT A L'AIR	LUFTKÜHLUNG
253	AIR CURRENT	COURANT D'AIR	LUFTSTROM
254	AIR CUSHION	MATELAS, COUCHE, COUSSIN D'AIR	LUFTKISSEN, LUFTPOLSTER
255	AIR DAMPING	AMORTISSEMENT PNEUMATIQUE	LUFTDÄMPFUNG
256	AIR DASHPOT	DASHPOT A AIR, AMORTISSEUR PNEUMATIQUE, COUSSIN PNEUMATIQUE	LUFTPUFFER
257	AIR DRAUGHT	TIRAGE D'AIR	ZUG, LUFTZUG
258	AIR DRILL	MARTEAU-PIQUEUR	BOHRHAMMER
259	AIR ESCAPE VALVE	SOUPAPE A AIR	LUFTVENTIL, ENTLÜFTUNGSVENTIL
260	AIR FILTER	FILTRE A AIR	LUFTFILTER
261	AIR FURNAGE	FOUR REVERBERE	FLAMMOFEN
262	AIR GAP	INTERSTICE, ENTREFER	LUFTZWISCHENRAUM, LUFTSPALT, EISENSPALT, LUFTSTRECKE, LUFTABSTAND
263	AIR GAP OF A MAGNET	ENTREFER D'UN AIMANT	LUFTSPALT EINES MAGNETEN
264	AIR GAS	GAZ A L'AIR	LUFTGAS
265	AIR GAS THERMOMETER	THERMOMETRE A GAZ	LUFTTHERMOMETER, GASTHERMOMETER
266	AIR GATE	TRAINEE D'AIR	ENTLÜFTUNGSNUT, ENTLÜFTUNGSRILLE
267	AIR HARDENING	TREMPE A L'AIR	LUFTHÄRTUNG
268	AIR HEATER, AIR HEATING	CHAUFFAGE DE L'AIR, CHAUFFAGE A AIR CHAUD	LUFTHEIZUNG
269	AIR HOLE	EVENT	LUFTLOCH, STEIGER
270	AIR INTAKE	PRISE D'AIR	LUFTEINLASS
271	AIR KNOCK-OUT	EJECTEUR PNEUMATIQUE	AUSWERFER (PNEUMATISCHER)
272	AIR LANCE	SOUFFLETTE	ABBLASHAHN, ZERSTÄUBER
273	AIR LIFT PUMP	EMULSEUR	LUFTDRUCKPUMPE, DRUCKLUFTPUMPE
274	AIR LOCK	SAS A AIR	LUFTSCHLEUSE
275	AIR NOZZLE	BEC DE SOUFFLAGE	PRESSLUFTDÜSE
276	AIR NOZZLE, AIR GUN	SOUFFLETTE A AIR	LUFTPISTOLE, LUFTVENTIL
277	AIR OF COMBUSTION	AIR COMBURANT, COMBURANT	VERBRENNUNGSLUFT
278	AIR OPENING IN THE GRATE BARS, AIR SPACE BETWEEN GRATE BARS	INTERVALLE LIBRE, VIDE ENTRE LES BARREAUX DE LA GRILLE	ROSTSPALT, ROSTFUGE
279	AIR OPERATED CHUCK	MANDRIN PNEUMATIQUE	DRUCKLUFTFUTTER
280	AIR PATENTING	PATENTAGE A L'AIR	LUFTPATENTIEREN
281	AIR PIPE LINE	CONDUITE, CANALISATION, DISTRIBUTION D'AIR	LUFTLEITUNG
282	AIR PISTON, PNEUMATIC PISTON	PISTON A AIR	LUFTKOLBEN

283	AIR PRESSURE GAUGE	JAUGE PNEUMATIQUE	MESSER (PNEUMATISCHER)
284	AIR PUMP, VACUUM PUMP	POMPE A AIR, POMPE A VIDE, POMPE PNEUMATIQUE	LUFTPUMPE, VAKUUMPUMPE
285	AIR QUENCHING	TREMPE A L'AIR	LUFTHÄRTUNG
286	AIR RAMMER	FOULOIR PNEUMATIQUE	PRESSLUFTSTAMPFER
287	AIR REHEATER	RECHAUFFEUR D'AIR	LUFTERHITZER
288	AIR SCALE	ECAILLE	SCHUPPE
289	AIR SCOOP	PRISE D'AIR	LUFTEINLASS
290	AIR SEPARATION	SEPARATION PNEUMATIQUE	WINDSICHTUNG
291	AIR SHAFT	PUITS D'AERATION	LUFTSCHACHT
292	AIR STRANGLER OU AIR CHOKE	VOLET D'AIR	STARTERKLAPPE
293	AIR TAP, COCK	ROBINET D'AIR	LUFTHAHN
294	AIR VESSEL, AIR CHAMBER	RESERVOIR A AIR	WINDKESSEL
295	AIR-ACETYLENE WELDING	SOUDAGE AERO-ACETYLENIQUE	AZETYLEN-LUFTSCHWEISSEN
296	AIR-COOLED SURFACE CONDENSER	CONDENSEUR PAR SURFACE AU MOYEN DE L'AIR	OBERFLÄCHENKONDENSATOR MIT LUFTKÜHLUNG
297	AIR-DRY WOOD, SEASONED TIMBER	BOIS SECHE A L'AIR LIBRE	HOLZ (LUFTTROCKENES), HOLZ (GELAGERTES)
298	AIR-HARDENNING, SELF HARDENING STEEL	ACIER AUTO-TREMPANT	SELBSTHÄRTERSTAHL
299	AIR-INJECTION MACHINE	MACHINE A INJECTION PNEUMATIQUE	LUFTEINSPRITZUNGMASCHINE
300	AIR-REFINING PROCESS	AFFINAGE AU VENT	WINDFRISCHEN
301	AIR-SETTING BINDER	LIANT DURCISSANT A L'AIR	BINDEMITTEL (LUFTHÄRTENDES)
302	AIR-SETTING CEMENT	CIMENT DURCISSANT A L'AIR	ZEMENT (LUFTHÄRTENDER)
303	AIR-TIGHT	ETANCHE A L'AIR, HERMETIQUE	LUFTDICHT
304	ALABASTER	ALBATRE	ALABASTER
305	ALARM	APPAREIL D'ALARME	MELDEVORRICHTUNG, ALARMVORRICHTUNG
306	ALARM SIGNAL	SIGNAL D'ALARME	WARNZEICHEN, ALARMZEICHEN, ALARMSIGNAL
307	ALARM WHISTLE	SIFFLET D'ALARME, SIFFLET AVERTISSEUR	WARNPFEIFE, ALARMPFEIFE, SIGNALPFEIFE
308	ALBEDO	ALBEDO	ALBEDO
309	ALBERT LAY WIRE ROPE	CABLE A CABLAGE ALBERT	DRAHTSEIL IM GLEICHSCHLAG
310	ALBUMIN	ALBUMINE	EIWEISS, ALBUMIN
311	ALBUMINISED PAPER	PAPIER ALBUMINE	EIWEISSPAPIER, ALBUMINPAPIER
312	ALCOHOL ENGINE	MOTEUR A ALCOOL	SPIRITUSMOTOR
313	ALCOHOL, SPIRIT, SPIRITS OF WINE, ETHYL ALCOHOL	ALCOOL ORDINAIRE, ALCOOL VINIQUE, ALCOOL ETHYLIQUE, HYDRATE D'ETHYLE	WEINGEIST, SPIRITUS, ALKOHOL, ÄTHYLALKOHOL
314	ALCOHOLIC	ALCOOLIQUE	ALKOHOLISCH
315	ALCOHOLIC SOLUTION	SOLUTION ALCOOLIQUE	LÖSUNG (ALKOHOLISCHE)
316	ALCOHOLOMETER	ALCOOMETRE, PESE-ALCOOLS	ALKOHOLMETER
317	ALDEHYDE	ALDEHYDE	ALDEHYD
318	ALDER	AUNE, AULNE	ERLE
319	ALFAMETER	ALFAMETRE	ALFAMETER
320	ALGEBRA	ALGEBRE	ALGEBRA
321	ALGEBRAIC	ALGEBRIQUE	ALGEBRAISCH
322	ALGEBRAIC EQUATION	EQUATION ALGEBRIQUE	GLEICHUNG (ALGEBRAISCHE)
323	ALIGN (TO)	ALIGNER	AUSRICHTEN
324	ALIGNMENT	ALIGNEMENT, LIGNE DE FUITE	AUSRICHTEN, INRICHTUNGBRINGEN, FLUCHTLINIE

325	ALINEMENT	ALIGNEMENT	ABFLUCHTUNG
326	ALIPHATIC COMPOUND	COMBINAISON ALIPHATIQUE	VERBINDUNG (ALIPHATISHE)
327	ALIZARIN	ALIZARINE	ALIZARIN
328	ALKALI	ALCALI	ALKALI
329	ALKALI-METALS	METAUX ALCALINS	ALKALIMETALLE
330	ALKALIFY (TO), ALKALISE (TO)	ALCALINISER, RENDRE ALCALIN	ALKALISCH MACHEN, ALKALISIEREN
331	ALKALIMETER	ALCALIMETRE	ALKALIMETER
332	ALKALIMETRIC	ALCALIMETRIQUE	ALKALIMETRISCH
333	ALKALIMETRY	ALCALIMETRIE	ALKALIMETRIE
334	ALKALINE CLEANING	NETTOYAGE ALCALIN	REINIGUNG (ALKALISCHE)
335	ALKALINE EARTH METALS	METAUX ALCALIN-TERREUX	ERDALKALIMETALLE
336	ALKALINE EARTHS, ALKALINE EARTH METALS	TERRES ALCALINES	ERDEN (ALKALISCHE), ERDALKALIEN
337	ALKALINE REACTION	REACTION ALCALINE	REAKTION (ALKALISCHE), REAKTION (BASISCHE)
338	ALKALINE SALT	SEL ALCALIN	ALKALISALZ, SALZ (ALKALISCHES)
339	ALKALINE SOLUTION	SOLUTION ALCALINE	ALKALILAUGE
340	ALKALINTY	ALCALINITE	ALKALINITÄT, ALKALISCHE BESCHAFFENHEIT, BASIZITÄT
341	ALKALOID	ALCALOIDE	ALKALOID
342	ALKANNIN PAPER	PAPIER D'ORCANETINE, PAPIER D'ANCHUSINE	ALKANNAROTPAPIER
343	ALL-ALUMINIUM CONDUCTOR	CABLE CONDUCTEUR TOUT-ALUMINIUM	GANZ-ALUMINIUMLEITER
344	ALL-FLOTATION	FLOTTATION	SCHWIMMVERFAHREN
345	ALL-IRON	TOUT EN METAUX FERREUX	GANZEISEN
346	ALL-MINE PIG	FER VIRGINAL	FRISCHEISEN
347	ALL-WELD-METAL TEST SPECIMEN	EPROUVETTE DU METAL DEPOSE PUR	SCHWEISSGUT (REINES)
348	ALLIGATORING	PEAU DE CROCODILE	KROKODILNARBUNG, LÄNGSABBLATTERUNG
349	ALLOMERIC	ALLOMERIQUE	ALLOMER
350	ALLOMERISM	ALLOMERIE	ALLOMERISMUS
351	ALLOMORPHISM	ALLOMORPHIE	ALLOMORPHIE
352	ALLOMORPHOUS	ALLOMORPHE	ALLOMORPH
353	ALLONABLE STRESS	CONTRAINTE ADMISSIBLE	BEANSPRUCHUNG (ZULÄSSIGE)
354	ALLOTRIOMORPHIC CRYSTAL	ALLOTRIOMORPHE	ALLOTRIOMORPH
355	ALLOTROPIC TRANSFORMATION	TRANSFORMATION ALLOTROPIQUE	PHASENUMWANDLUNG
356	ALLOTROPICALL	ALLOTROPIQUE	ALLOTROP
357	ALLOTROPY	ALLOTROPIE, ISOMERIE	ALLOTROPIE, ISOMERISMUS
358	ALLOWANCE	JEU, TOLERANCE, SUREPAISSEUR D'USINAGE, JEU DE COIFFAGE	SPIEL, TOLERANZ, BEARBEITUNGSZUGABE, ABMASS SPIELRAUM
359	ALLOWANCE, PERMISSIBLE TOLERANCE, MARGIN	TOLERANCE D'USINAGE, TOLERANCE ADMISE	SPIELRAUM, ZULÄSSIGE ABWEICHUNG, TOLERANZ
360	ALLOY	ALLIAGE	LEGIERUNG
361	ALLOY (TO)	ALLIER	LEGIEREN
362	ALLOY BALANCE	BASCULE POUR ALLIAGE	LEGIERWAAGSCHALE
363	ALLOY CASTING	FONTE ALLIEE	GUSSEISEN (LEGIERTES)
364	ALLOY COATING	REVETEMENT D'ALLIAGE	LEGIERUNGSÜBERZUG, LEGIERUNGSABSCHEIDUNG
365	ALLOY CONTAMINATION	CONTAMINATION D'UN ALLIAGE	LEGIERUNGSKONTAMINATIONVERSEUCHUNG
366	ALLOY PLATE	GALVANISER	GALVANISIEREN
367	ALLOY POWDER	POUDRE ALLIEE	PULVER (LEGIERTES)

368	ALLOY STEEL	ACIER ALLIE	STAHL (LEGIERTER), SONDERSTAHL
369	ALLOY STEEL ANGLES	CORNIERES EN ACIER ALLIE	WINKELSTAHL (LEGIERTER)
370	ALLOY STEEL BILLETS	BILLETTES EN ACIER ALLIE	KNÜPPEL AUS LEGIERTEN STÄHLEN
371	ALLOY STEEL BUNDLING STRIP	FEUILLARDS EN ACIER ALLIE	BANDSTAHL (WARMGEWALZTER), LEGIERT
372	ALLOY STEEL CASTINGS	MOULAGES D'ACIER ALLIE	STAHLFORMGUSS AUS LEGIERTEN STÄHLEN
373	ALLOY STEEL FLATS	PLATS EN ACIER ALLIE	FLACHEISEN AUS LEGIERTEN STÄHLEN
374	ALLOY STEEL HEXAGONS	HEXAGONES EN ACIER ALLIE	SECHSKANTEISEN AUS LEGIERTEN STÄHLEN
375	ALLOY STEEL OCTAGONS	BARRES HUIT-PANS EN ACIER ALLIE	ACHTKANTEISEN AUS LEGIERTEN STÄHLEN
376	ALLOY STEEL ROD WIRE	FIL MACHINE EN ACIER ALLIE	WALZDRAHT AUS LEGIERTEN STÄHLEN
377	ALLOY STEEL ROUNDS	RONDS EN ACIER ALLIE	RUNDEISEN AUS LEGIERTEN STÄHLEN
378	ALLOY STEEL SHEET BARS	LARGETS EN ACIER ALLIE	PLATINEN AUS LEGIERTEN STÄHLEN
379	ALLOY STEEL SQUARES	CARRES EN ACIER ALLIE	VIERKANTSTAHL, LEGIERT
380	ALLOY STEEL THIN SHEETS	TOLES MINCES EN ACIER ALLIE	BLECHE AUS LEGIERTEN STÄHLEN
381	ALLOY SYSTEM	SYSTEME DES ALLIAGES	LEGIERUNGSSYSTEM
382	ALLOY-TREATED STEEL	ACIER FAIBLEMENT ALLIE	STAHL (NIEDRIGLEGIERTER)
383	ALLOY, ABRASION-RESISTANT	ALLIAGE RESISTANT A L'USURE	LEGIERUNG (VERSCHLEISSFESTE)
384	ALLOY, ACID RESISTANT	ALLIAGE RESISTANT AUX ACIDES	LEGIERUNG (SÄUREBESTÄNDIGE)
385	ALLOY, BEARING	ALLIAGE POUR COUSSINETS, ALLIAGE ANTI-FRICTION	LAGERLEGIERUNG
386	ALLOY, BRAZING	ALLIAGE DE BRASAGE	HARTLÖTLEGIERUNG
387	ALLOY, CORROSION RESISTANT	ALLIAGE INOXYDABLE, ALLIAGE RESISTANT A LA CORROSION	LEGIERUNG (KORROSIONSBESTÄNDIGE)
388	ALLOY, DIE CASTING	ALLIAGE POUR COULEE SOUS PRESSION	DRUCKGUSSLEGIERUNG
389	ALLOY, FUSIBLE	ALLIAGE FUSIBLE	SCHMELZLEGIERUNG
390	ALLOY, HEAT AND CORROSION RESISTANT	ALLIAGE RESISTANT A LA CHALEUR ET A LA CORROSION	LEGIERUNG (HITZE-UND KORROSIONSBESTÄNDIGE)
391	ALLOY, HEAT RESISTANT	ALLIAGE RESISTANT A LA CHALEUR	LEGIERUNG (HITZEBESTÄNDIGE)
392	ALLOY, MAGNET	ALLIAGE MAGNETIQUE	LEGIERUNG (MAGNETISCHE)
393	ALLOY, REFRACTORY	ALLIAGE REFRACTAIRE	LEGIERUNG (FEUERFISTE)
394	ALLOYING	ADDITION	ZUSATZ
395	ALLOYING ELEMENTS	ELEMENTS D'ADDITION	ZUSATZELEMENTE
396	ALLOYS, ABRASION & CORROSION RESISTANT	ALLIAGES RESISTANTS A L'ABRASION ET A LA CORROSION	LEGIERUNGEN (ABRIED-UND KORROSIONFESTE)
397	ALMALGAM	AMALGAME	AMALGAM
398	ALNICO	ALLIAGE MAGNETIQUE 'ALNICO'	ALNICO
399	ALPHA BRASS	LAITON ALPHA	ALPHA-MESSING
400	ALPHA PARTICLE	PARTICULE ALPHA	ALPHA-TEILCHEN
401	ALPHA RADIATOR	RADIATEUR ALPHA	ALPHA-STRAHLER, ALPHA-STRAHLENQUELLE
402	ALPHA RAYS	RAYONS ALPHA	ALPHA-STRAHLUNG
403	ALPHA-BETA BRASS	LAITON ALPHA-BETA	ALPHA-BETA-MESSING
404	ALTERNATE CONES	CONE ET CONTRE-CONE, PAIRE DE CONES LISSES	RIEMENKEGELTRIEB

405	ALTERNATE-IMMERSION TEST	ESSAI PAR IMMERSIONS ET EMERSIONS ALTERNEES	WECHSELTAUCHVERSUCH-PRÜFUNG
406	ALTERNATING ALTERNATE CURRENT (A.C., A.C.)	COURANT ALTERNATIF	WECHSELSTROM
407	ALTERNATING CURRENT GENERATOR, ALTERNATOR	ALTERNATEUR, DYNAMO, GENERATRICE, MACHINE A COURANT ALTERNATIF	WECHSELSTROMMASCHINE, WECHSELSTROMDYNAMO, WECHSELSTROMGENERATOR, ALTERNATOR
408	ALTERNATING CURRENT MOTOR	MOTEUR A COURANT ALTERNATIF, ALTERNO-MOTEUR	WECHSELSTROMMOTOR
409	ALTERNATIVE	ALTERNATIVE	ALTERNATIVE
410	ALTERNATIVE SOLUTION	VARIANTE	VARIANTE
411	ALTERNATOR	ALTERNATEUR	WECHSELSTROMGENERATOR
412	ALUMINA	ALUMINE	ALUMINIUMOXYD, TONERDE
413	ALUMINA, ALUMINIUM OXIDE	ALUMINE, OXYDE D'ALUMINIUM	TONERDE, ALUMINIUMOXYD
414	ALUMINATE	ALUMINATE	ALUMINAT
415	ALUMINIFEROUS	ALUMINIFERE	ALUMINIUMHALTIG
416	ALUMINIUM	ALUMINIUM	ALUMINIUM
417	ALUMINIUM ACETATE	ACETATE D'ALUMINIUM	TONERDE (ESSIGSAURE), ALUMINIUMAZETAT
418	ALUMINIUM ALLOY	ALLIAGE A BASE D'ALUMINIUM	ALUMINIUMLEGIERUNG
419	ALUMINIUM BRASS	LAITON D'ALUMINIUM	ALUMINIUMMESSING
420	ALUMINIUM BRONZE	BRONZE D'ALUMINIUM	ALUMINIUMBRONZE
421	ALUMINIUM CHLORIDE	CHLORURE D'ALUMINIUM, CHLORALUM	ALUMINIUMCHLORID, CHLORALUMINIUM
422	ALUMINIUM FILE	LIME A ALUMINIUM	ALUMINIUMFEILE
423	ALUMINIUM HYDROXIDE	ALUMINE HYDRATEE	TONERDEHYDRAT, ALUMINIUMHYDROXYD
424	ALUMINIUM INGOT METAL	ALUMINIUN EN LINGOT	ALUMINIUMROHBLOCK
425	ALUMINIUM SILICATE	SILICATE D'ALUMINE	TONERDE (KIESELSAURE), TONERDESILIKAT, ALUMINIUMSILIKAT
426	ALUMINIUM SULPHATE	SULFATE D'ALUMINIUM	TONERDE (SCHWEFELSAURE), ALUMINUIMSULFAT
427	ALUMINIUM TUBE	TUBE EN ALUMINIUM	ALUMINIUMROHR
428	ALUMINIUM WIRE	FIL D'ALUMINIUM	ALUMINIUMDRAHT
429	ALUMINIZING, ALUMINIZE	ALUMINATION	ALUMINISIEREN, VERALUMINIEREN
430	ALUMINO	ALUMINO	ALUMINO
431	ALUMINOTHERMICS	ALUMINOTHERMIE	ALUMINOTHERMIE, THERMIT-VERFAHREN
432	ALUMINUM	ALUMINIUM	ALUMINIUM
433	ALUMINUM ALLOY, WROUGHT	ALLIAGE D'ALUMINIUM FORGEABLE	KNETALUMINIUM-LEGIERUNG
434	ALUMINUM BRONZE	BRONZE D'ALUMINIUM CUPRO-ALUMINIUM	ALUMINIUMBRONZE
435	ALUMINUM CASTING ALLOY	ALLIAGE D'ALUMINIUM POUR MOULAGE	GUSSALUMINIUM
436	ALUMINUM FOIL	FEUILLE D'ALUMINIUM	ALUMINIUMFOLIE
437	ALUMINUM FORGING ALLOY	ALLIAGE D'ALUMUNIUM FORGEABLE	ALUMINIUM-SCHMIEDE-LEGIERUNG
438	ALUMINUM OXIDE	ALUMINE	TONERDE
439	ALUMINUM-BASE ALLOY	ALLIAGE A BASE D'ALUMINIUM	LEGIERUNG MIT ALUMINIUM ALS GRUNDMETALL
440	ALUMINUM-BERYLLIUM ALLOY	ALLIAGE D'ALUMINIUM-BERYLLIUM	ALUMINIUM-BERYLLIUM-LEGIERUNG
441	ALUMINUM-COATED SHEET	TOLE (D'ACIER) ALUMINEE	ALITIERTES (STAHL-)BLECH
442	ALUMINUM-KILLED STEEL	ACIER CALME A L'ALUMINIUM	STAHL (ALUMINIUM-BERUHIGTER)
443	ALUMINUM-SILICATE REFRACTORY CEMENT	CIMENT REFRACTAIRE AU SILICATE D'ALUMINIUM	MÖRTEL (ALUMINIUM-SILIKAT-FEUERFESTER)
444	ALUMINUM, ALUMINIUM	ALUMINIUM	ALUMINIUM

445	**AMALGAM**	AMALGAME	AMALGAM
446	**AMALGAMATE (TO)**	AMALGAMER	AMALGAMIEREN
447	**AMALGAMATION**	AMALGAMATION	AMALGAMIERUNG, AMALGAMATION, AMALGAMIEREN
448	**AMALGAMATION PROCESS**	PROCEDE D'AMALGAMATION	AMALGAMIERUNGSVERFAHREN
449	**AMBER**	AMBRE JAUNE, ARBRE SUCCIN	BERNSTEIN
450	**AMBIENT AIR**	AIR AMBIANT	AUSSENLUFT
451	**AMERICAN STANDARD THREAD, SELLERS THREAD**	PAS SYSTEME SELLERS	SELLERSGEWINDE
452	**AMMETER, AMPEREMETER**	AMPEREMETRE, COULOMBMETRE	STROMMESSER, STROMZEIGER, AMPEREMETER
453	**AMMONIA**	GAZ AMMONIAC, AMMONIAQUE	AMMONIAKGAS
454	**AMMONIA**	AMMONIAC	AMMONIAK
455	**AMMONIA SODA, SOLVAY SODA**	SOUDE A L'AMMONIAQUE, SOUDE SOLVAY, SEL SOLVAY	AMMONIAKSODA, SOLVAYSODA
456	**AMMONIACAL LIQUOR, GAS LIQUOR**	EAU AMMONIACALE DU GAZ	GASWASSER, AMMONIAKWASSER
457	**AMMONIACAL SOLUTION**	SOLUTION AMMONIACALE	LÖSUNG (AMMONIAKALISCHE)
458	**AMMONIUM ACETATE**	ACETATE D'AMMONIUM	AMMONIAK (ESSIGSAURES), AMMONIUMAZETAT
459	**AMMONIUM BICARBONATE**	BICARBONATE D'AMMONIUM	AMMONIAK (DOPPELTKOHLENSAURES), AMMONIUMBIKARBONAT
460	**AMMONIUM BISULPHITE**	BISULFITE D'AMMONIAQUE	AMMONIAK (SAURES), AMMONIAK (SCHWEFLIGSAURES), AMMONIUMBISULFIT
461	**AMMONIUM CARBONATE, SAL VOLATILE**	CARBONATE D'AMMONIUM, SEL VOLATIL D'ANGLETERRE	HIRSCHHORNSALZ, AMMONIUMKARBONAT, AMMONIAK (KOHLENSAURES)
462	**AMMONIUM CHLORIDE, SAL AMMONIAC**	CHLORURE D'AMMONIUM, HYDROCHLORATE, CHLORHYDRATE D'AMMONIAQUE, SEL AMMONIAC, MURIATE D'AMMONIAQUE	SALMIAK, AMMONIUMCHLORID, CHLORAMMONIUM
463	**AMMONIUM FLUORIDE**	FLUORURE D'AMMONIUM	FLUORAMMONIUM, AMMONIUMFLUORID
464	**AMMONIUM HYDROSULPHIDE, SULPHHYDRATE OF AMMONIUM**	SULFHYDRATE D'AMMONIUM	AMMONIUMSULFHYDRAT, AMMONIUMHYDROSULFID
465	**AMMONIUM NITRATE**	NITRATE D'AMMONIUM, NITRUM FLAMMANS	AMMONIAK (SALPETERSAURES), AMMONIUMNITRAT, AMMONIAKSALPETER, FLAMMENDER SALPETER
466	**AMMONIUM NITRITE**	NITRITE D'AMMONIUM	AMMONIAK (SALPETRIGSAURES), AMMONIUMNITRIT
467	**AMMONIUM OXALATE**	OXALATE D'AMMONIAQUE	AMMONIAK (OXALSAURES), AMMONIUMOXALAT
468	**AMMONIUM PERSULPHATE**	PERSULFATE D'AMMONIAQUE	AMMONIAK (ÜBERSCHWEFELSAURES), AMMONIUMPERSULFAT
469	**AMMONIUM PHOSPHATE**	PHOSPHATE D'AMMONIUM	AMMONIAK (PHOSPHORSAURES), AMMONIUMPHOSPHAT
470	**AMMONIUM STANNIC CHLORIDE, PINK SALT**	CHLORURE DOUBLE D'ETAIN ET D'AMMONIUM	PINKSALZ, AMMONIUMZINNCHLORID
471	**AMMONIUM SULPHATE**	SULFATE D'AMMONIUM	AMMONIAK (SCHWEFELSAURES), AMMONIUMSULFAT
472	**AMMONIUM SULPHIDE**	SULFURE D'AMMONIUM	AMMONIUMSULFID, SCHWEFELAMMONIUM
473	**AMMONIUM TARTRATE**	TARTRATE D'AMMONIAQUE	AMMONIAK (WEINSAURES), AMMONIUMTARTRAT
474	**AMMUNITION**	MUNITION	MUNITION
475	**AMORPHOUS**	AMORPHE	AMORPH

/

476	AMORPHOUS SULPHUR	SOUFRE AMORPHE	SCHWEFEL (AMORPHER)
477	AMOUNT OF CONTRACTION	RETRAIT, COEFFICIENT DE RETRAIT	SCHWINDMASS, UNTERMASS, SCHRUMPFMASS,
478	AMPERAGE	NOMBRE D'AMPERES, AMPERAGE, INTENSITE DU COURANT	AMPEREZAHL, STROMSTÄRKE
479	AMPERE	AMPERE	AMPERE
480	AMPERE TURN	AMPERE-TOUR	AMPEREWINDUNG
481	AMPERE-HOUR (AMP.HR)	AMPERE-HEURE, A-H	AMPERESTUNDE
482	AMPERE-MINUTE	AMPERE-MINUTE	AMPEREMINUTE
483	AMPERE-SECOND	AMPERE-SECONDE	AMPERESEKUNDE
484	AMPHOTERIC METAL	AMPHOTERE	AMPHOTER
485	AMPLIFIER	AMPLIFICATEUR	VERSTÄRKER
486	AMPLITUDE	AMPLITUDE	SCHWINGUNGSWEITE, AUSSCHLAG, AMPLITUDE
487	AMYL ACETATE	ACETATE D'AMYLE, ESSENCE DE POIRE	AMYLAZETAT
488	AMYL ACOHOL	ALCOOL AMYLIQUE	AMYLALKOHOL, PENTHYLALKOHOL, AMYLOXYDHYDRAT
489	ANAEROBIC	ANAEROBIQUE	ANAEROB
490	ANALOG	ANALOGIQUE	ANALOG
491	ANALYSE (TO)	ANALYSER	ANALYSIEREN
492	ANALYSER	ANALYSEUR	ANALYSATOR
493	ANALYSIS	ANALYSE	ANALYSE
494	ANALYTICAL	ANALYTIQUE	ANALYTISCH
495	ANALYTICAL CHEMISTRY	CHIMIE ANALYTIQUE	CHEMIE (ANALYTISCHE)
496	ANALYTICAL DETERMINATION	DETERMINATION ANALYTIQUE	BESTIMMUNG (ANALYTISCHE)
497	ANASTIGMAT	ANASTIGMAT	ANASTIGMAT
498	ANATOMICAL ALLOY	ALLIAGE OSTEOPLASTIQUE	OSTEOPLASTIKLEGIERUNG
499	ANCHERING	ANCRAGE	ANKERUNG
500	ANCHOR	ANCRE (CONSTR.)	MAUERANKER, ANKER
501	ANCHOR BOLT	BOULON D'ANCRAGE	ANKERSCHRAUBE, FUNDAMENTANKER
502	ANCHOR PLATE	CONTREPLAQUE (SCELLEE DANS LE SOL)	ANKERPLATTE (IM FUNDAMENT)
503	ANEMOMETER	ANEMOMETRE	WINDMESSER, ANEMOMETER
504	ANEROID BAROMETER	BAROMETRE ANEROIDE, BAROMETRE METALLIQUE, ANEROIDE	FEDERBAROMETER, METALLBAROMETER, DOSENBAROMETER, ANEROID BAROMETER
505	ANGLE	ANGLE (GEOM.)	WINKEL (GEOM.)
506	ANGLE (IRON)	CORNIERE	WINKELEISEN, WINKELPROFIL
507	ANGLE AT THE CENTRE	ANGLE AU CENTRE	ZENTRIWINKEL
508	ANGLE BEARING	PALIER AVEC PLAN DE SEPARATION INCLINE	LAGER (SCHRÄG), LAGER (SCHIEF GESCHNITTENES), SCHRÄGLAGER
509	ANGLE BOX SPANNER	CLEF TUBULAIRE COURBE	HÜLSENSCHLÜSSEL, AUFSTECKSCHLÜSSEL, SCHLÜSSEL (GEKRÖPFTER)
510	ANGLE BRACKET	EQUERRE D'ASSEMBLAGE	ECKWINKEL
511	ANGLE COCK	ROBINET D'ANGLE	WINKELHAHN
512	ANGLE FLANGE	BRIDE ANGULAIRE	WINKELFLANSCH
513	ANGLE GAUGE	CALIBRE D'ANGLES	WINKELLEHRE
514	ANGLE IRON	FER CORNIERE, CORNIERE, FER EN L, EQUERRE	WINKELEISEN, L-EISEN
515	ANGLE OF ADVANCE, ANGULAR ADVANCE, ANGLE OF LEAD	ANGLE D'AVANCE, ANGLE DE DECALAGE EN AVANCE	VOREILWINKEL

4

516	ANGLE OF CHAMFER	CHANFREIN D'ENTREE	ABSCHRÄGWINKEL
517	ANGLE OF CONTACT	ANGLE D'ENROULEMENT, ANGLE DE CONTACT	ANSCHMIEGUNGSWINKEL, UMSCHLINGUNGSWINKEL, GREIFWINKEL
518	ANGLE OF DEFLECTION	ANGLE DE DEVIATION	AUSSCHLAGWINKEL
519	ANGLE OF FLEXURE	ANGLE DE FLEXION	BIEGEWINKEL
520	ANGLE OF INCIDENCE	ANGLE D'INCIDENCE	EINFALLWINKEL
521	ANGLE OF INCLINATION	ANGLE D'INCLINAISON	NEIGUNGSWINKEL
522	ANGLE OF LAG, ANGULAR LAG	ANGLE DE RETARD, ANGLE DE DECALAGE EN ARRIERE	NACHEILWINKEL, VERZÖGERUNGSWINKEL
523	ANGLE OF PHASE LAG	ANGLE DE DEPHASAGE EN ARRIERE	PHASENNACHEILWINKEL
524	ANGLE OF PHASE LEAD	ANGLE DE DEPHASAGE EN AVANCE	PHASENVOREILWINKEL
525	ANGLE OF REFLECTION	ANGLE DE REFLEXION	REFLEXIONSWINKEL
526	ANGLE OF REFRACTION	ANGLE DE REFRACTION	BRECHUNGSWINKEL
527	ANGLE OF REPOSE, ANGLE OF FRICTION, LIMITING ANGLE OF RESISTANCE	ANGLE DE FROTTEMENT	REIBUNGSWINKEL
528	ANGLE OF ROTATION	ANGLE DE ROTATION	DREHUNGSWINKEL
529	ANGLE OF TAPER OF KEY	ANGLE DE CALAGE	AUFKEILWINKEL
530	ANGLE OF THREAD	ANGLE DU FILET	FLANKENWINKEL DES GEWINDES
531	ANGLE OF THREAD	ANGLE DE L'INCLINAISON DU FILET	STEIGUNGSWINKEL EINER SCHRAUBE
532	ANGLE OF TWIST	ANGLE DE TORSION	VERDREHUNGSWINKEL, DRILLUNG
533	ANGLE PLATE	EQUERRE D'ABLOCAGE	WINKELPLATTE
534	ANGLE VALVE	ROBINET D'EQUERRE, SOUPAPE D'EQUERRE	ECKVENTIL
535	ANGLE VISE	ETAU INCLINABLE	SINUS-SCHRAUBSTROCK
536	ANGSTROM UNIT	ANGSTROM	ANGSTRÖM
537	ANGULAR ACCELERATION	ACCELERATION ANGULAIRE	WINKELBESCHLEUNIGUNG, DREHBESCHLEUNIGUNG
538	ANGULAR BALL BEARING	ROULEMENT A BILLES A CHARGE RADIALE ET AXIALE COMBINEE, ROULEMENT A BILLES A POUSSEE AXIALE ET LATERALE COMBINEE	KUGELSCHULTERLAGER, KUGELDRUCK- UND TRAGLAGER
539	ANGULAR CUT	COUPE ANGULAIRE	WINKELSCHNITT
540	ANGULAR CUTTERS	FRAISES CONIQUES	WINKELFRÄSER
541	ANGULAR DISPLACEMENT	DEPLACEMENT ANGULAIRE, DECALAGE	WINKELVERSCHIEBUNG
542	ANGULAR DISTORTION	RETRAIT ANGULAIRE	WINKELSCHRUMPFUNG
543	ANGULAR MILLING CUTTER	FRAISE ANGULAIRE, FRAISE CONIQUE	WINKELFRÄSER, KEGELFRÄSER
544	ANGULAR NOTION	MOUVEMENT ANGULAIRE	WINKELBEWEGUNG
545	ANGULAR TRIANGULAR THREAD, VEE-SHAPED THREAD, V-THREAD	FILET TRIANGULAIRE	GEWINDE (SCHARFES), GEWINDE (SCHARFGÄNGIGES), DREIECK GEWINDE
546	ANGULAR VELOCITY	VITESSE ANGULAIRE	WINKELGESCHWINDIGKEIT, KREISFREQUENZ
547	ANGULAR-THREADED TRIANGULAR-THREADED SCREW, VEE-THREADED V-THREADED SCREW	VIS A FILET TRIANGULAIRE	SCHRAUBE MIT DREIECKGEWINDE, SCHRAUBE (SCHARFGÄNGIGE)
548	ANHYDRIDE	ANHYDRIDE	ANHYDRID
549	ANHYDRITE, ANHYDROUS SULPHATE OF CALCIUM	ANHYDRIT, SULFATE ANHYDRE DE CHAUX	ANHYDRIT, KALK (WASSERFREIER SCHWEFELSAURER), WASSERFREIES KALZIUMSULFAT
550	ANHYDROUS	ANHYDRE	WASSERFREI (CHEM.)

551	ANHYDROUS SODIUM CARBONATE, SODA ASH	SEL DE SOUDE, CARBONATE DE SOUDE ANHYDRE	SODA (GEGLÜHTE) (KALZINIERTE), SODASALZ (KALZINIERTES), NATRIUMKARBONAT (WASSERFREIES)
552	ANILINE	ANILINE, PHENYLAMINE	ANILIN, AMIDOBENZOL, PHENYLAMIN
553	ANILINE OIL	ANILINE DU COMMERCE	ANILINÖL, ROHANILIN
554	ANIMAL CHARCOAL, BONE CHARCOAL	CHARBON ANIMAL	KNOCHENKOHLE
555	ANIMAL OIL	HUILE ANIMALE	ÖL (TIERISCHES)
556	ANION	ANION	SAUERSTOFFION, ANION
557	ANIONIC FLOTATION	FLOTTATION ANIONIQUE	FLOTATION (ANIONISCHE), SCHWIMMAUFBEREITUNG (ANIONISCHE)
558	ANISOTROPIC	ANISOTROPE	ANISOTROP
559	ANISOTROPY, ANISOTROPISM	ANISOTROPIE	ANISOTROPIE
560	ANNEAL (TO), NORMALISE THE STEEL (TO)	RECUIRE L'ACIER	STAHL AUSGLÜHEN, STAHL NORMALISIBREN
561	ANNEALED BLACK WIRE	FIL RECUIT	DRAHT (GEGLÜHTER)
562	ANNEALING	RECUIT	AUSGLÜHEN, GLÜHFRISCHEN, TEMPERN, GLÜHEN
563	ANNEALING BOX	CAISSE DE RECUIT	GLÜHKISTE
564	ANNEALING FURNACE	FOUR A RECUIRE, FOUR DE RECUIT, FOUR DE REVENU	GLÜHOFEN
565	ANNEALING PLANT	INSTALLATION DE RECUIT	GLÜHANLAGE
566	ANNEALING POT	POT A RECUIRE	GLÜHTOPF
567	ANNEALING POTS	CAISSE DE RECUIT	GLÜHKISTE
568	ANNEALING PROCESS	TRAITEMENT DE RECUIT	GLÜHEN, TEMPERN
569	ANNEALING RESISTANCE	RESISTANCE A L'ETAT DE RECUIT	GLÜHFESTIGKEIT
570	ANNEALING TWINS	BANDE A JUMEAUX	ZWILLINGSBAND
571	ANNEALING, BLACK	RECUIT EN NOIR	SCHWARZGLÜHEN
572	ANNEALING, BLUE	BLEUISSAGE	BLAUGLÜHEN
573	ANNEALING, BRIGHT	RECUIT BLANC	BLANKGLÜHEN
574	ANNEALING, CONTINUOUS	RECUIT CONTINU	DURCHLAUFBANDGLÜHEN
575	ANNEALING, FLAME	RECUIT AU CHALUMEAU	OBERFLÄCHENGLÜHUNG
576	ANNEALING, FULL	RECUIT COMPLET	AUSGLÜHEN (VOLLSTÄNDIGES)
577	ANNEALING, INTERMEDIATE	RECUIT INTERMEDIAIRE	ZWISCHENGLÜHEN
578	ANNEALING, INVERSE	RECUIT INVERSE	GLÜHFRISCHEN (INVERSES)
579	ANNEALING, ISOTHERMAL	RECUIT ISOTHERMIQUE	GLÜHUNG (ISOTHERME)
580	ANNEALING, LOCAL	RECUIT SELECTIF	SELEKTIVES FRISCHGLÜHEN
581	ANNEALING, NORMALISING THE STEEL	RECUIT DE L'ACIER	AUSGLÜHEN, NORMALISIEREN DES STAHLS
582	ANNEALING, PERIODIC	RECUIT PERIODIQUE	KREISLAUFGLÜHUNG
583	ANNEALING, RELIEF	RECUIT DE RELAXATION	GLÜHEN (ENTSPANNENDES)
584	ANNEALING, STRESS-RELIEF	RECUIT DE DETENTE	GLÜHEN (SPANNUNGSFREIES)
585	ANNEX, EXTENSION	ANNEXE D'UN BATIMENT, BATIMENT ANNEXE	ANBAU
586	ANNUAL RING	CERCLE ANNUEL, CRUE, COUCHE ANNUELLE	JAHRESRING
587	ANNULAR CROSS SECTION	SECTION ANNULAIRE	QUERSCHNITT (RINGFÖRMIGER), QUERSCHNITT (KREISRINGFÖRMIGER)
588	ANNULAR, RING-SHAPED	ANNULAIRE	RINGFÖRMIG
589	ANNULUS	ANNEAU, COURONNE CIRCULAIRE	KREISRING
590	ANODE	ANODE	ANODE
591	ANODE BUTT	BOUT ANODIQUE	ANODENBRENNFLECK
592	ANODE CLEANING	PURIFICATION ANODIQUE	REINIGUNG (ANODISCHE)

593	ANODE COPPER	CUIVRE ANODIQUE	ANODENKUPFER
594	ANODE CORROSION EFFICIENCY	COEFFICIENT DE CORROSION ANODIQUE	KORROSIONSKOEFFIZIENT (ANODISCHE)
595	ANODE DROP	CHUTE ANODIQUE	ANODENFALL, SPANNUNGSABFALL AN DER ANODE
596	ANODE EFFECT	EFFET D'ANODE	ANODENEFFEKT
597	ANODE EFFICIENCY	RENDEMENT ANODIQUE	ANODENWIRKUNGSGRAD
598	ANODE INSOLUBLE	ANODE INSOLUBLE	ANODE (UNLÖSLICHE)
599	ANODE LAYER	COUCHE (DE PROTECTION) ANODIQUE, REVETEMENT (ENDUIT) ANODIQUE	ÜBERZUG (ANODISCHER)
600	ANODE MUD, ANODE SLIME	BOUE D'ANODE	ANODENRÜCKSTAND, ANODENSCHLAMM
601	ANODE PICKLING	DECAPAGE ANODIQUE	ANODENBEIZUNG
602	ANODE, POSITIVE ELECTRODE	ELECTRODE POSITIVE, ANODE	ELEKTRODE (POSITIVE), ANODE
603	ANODIC	ANODIQUE	ANODISCH
604	ANODIC POLARIZATION	POLARISATION ANODIQUE	ANODENPOLARISATION
605	ANODIZING	OXYDATION ANODIQUE TRAITEMENT ANODIQUE	ELOXIEREN
606	ANOLYTE	ANOLYTE, SOLUTION ANODIQUE	ANOLYT, ANODENFLÜSSIGKEIT
607	ANTHRACENE	ANTHRACENE	ANTHRAZEN
608	ANTHRACENE OIL	HUILE ENTHRACENIQUE, HUILE A ANTHRACENE	ANTHRAZENÖL
609	ANTHRACITE	ANTHRACITE, HOUILLE ECLATANTE	ANTHRAZIT
610	ANTI-AIRCRAFT GUN	CANON ANTI-AERIEN	FLAK-GESCHÜTZ
611	ANTI-CATHODE	ANTI CATHODE	ANTIKATHODE
612	ANTI-FREEZE SOLUTION	ANTIGEL	FROSTSCHUTZMITTEL
613	ANTI-KNOCK FUEL	CARBURANT ANTI-DETONNANT	KRAFTSTOFF (KLOPFFESTER)
614	ANTI-MAGNETIC	ANTIMAGNETIQUE	ANTIMAGNETISCH
615	ANTI-PIPING COMPOUND	COUVERTE	ABDECKMITTEL
616	ANTI-PITTING AGENT	AGENT ANTI-PIQURE	PORENVERMUTUNGSMITTEL
617	ANTI-SEIZURE PROPERTIES	PROPRIETES ANTI-FRICTION	ANTIFRIKTIONSWIRKUNG
618	ANTIFRICTION BEARING	ROULEMENT A BILLES, ROULEMENT A GALETS	WÄLZLAGER
619	ANTIFRICTION METAL	ALLIAGE, METAL ANTIFRICTION, ANTIFRICTION	ANTIFRIKTIONSMETALL
620	ANTILOGARITHM	ANTILOGARITHME	GEGENLOGARITHMUS, ANTILOGARITHMUS
621	ANTIMONIAL LEAD	PLOMB AIGRE, PLOMB ANTIMONIAL	ANTIMONBLEI
622	ANTIMONITE, STIBNITE	STIBINE	ANTIMONGLANZ, ANTIMONIT, GRAUSPIESSGLANZERZ
623	ANTIMONY	ANTIMOINE	ANTIMON, SPIESSGLANZ
624	ANTIMONY PENTACHLORIDE	PENTACHLORURE D'ANTIMOINE	ANTIMONPENTACHLORID, ANTIMONSUPERCHLORID
625	ANTIMONY PENTASULPHIDE	SULFURE DOREE, SOUFFRE DORE D'ANTIMOINE, PENTASULFURE D'ANTIMOINE	FÜNFFACH-SCHWEFELANTIMON, GOLDSCHWEFEL, ANTIMONPENTASULFID, SULFURAURAT
626	ANTIMONY TRICHLORIDE, BUTTER OF ANTIMONY	TRICHLORURE D'ANTIMOINE, BEURRE D'ANTIMOINE LIQUIDE	ANTIMONTRICHLORID, ANTIMONCHLORÜR, CHLORANTIMON, SPLESSGLANZBUTTER, ANTIMONBUTTER
627	ANTIMONY TRISULPHIDE	PROTOSULFURE, TRISULFURE D'ANTIMOINE	DREIFACH-SCHWEFELANTIMON, ANTIMONTRISULFID, ANTIMONSULFÜR
628	ANTISEPTIC	ANTISEPTIQUE, ASEPSISANT, ANTIPUTRIDE	FÄULNISWIDRIG, ANTISEPTISCH

3

629	ANTISEPTIC	ANTISEPTIQUE, ANTIPUTRIDE, AGENT ANTIPUTREFIANT	FÄULNISVERHINDERNDES MITTEL, ABTÖTENDES MITTEL, ANTISEPTISCHES MITTEL, ANTISEPTIKUM
630	ANVIL	ENCLUME	AMBOSS
631	ANVIL CUTTER, ANVIL CHISEL	TRANCHET D'ENCLUME, CASSE-FER	ABSCHROT, ABSCHRÖTER
632	ANVIL STAKE	TAS, TASSEAU	SCHLAGSTÖCKCHEN, STÖCKEL, AMBOSSSTÖCKEL , POLIERSTOCK
633	ANVIL STAND BLOCK	BILLOT, CHABOTTE D'ENCLUME	AMBOSSSTOCK, AMBOSSUNTERSATZ
634	APERIODIC	APERIODIQUE	APERIODISCH
635	APERIODIC DEAD BEAT INSTRUMENT, CRITICALLY DAMPED INSTRUMENT	INSTRUMENT APERIODIQUE	INSTRUMENT (GEDÄMPFTES), INSTRUMENT (APERIODISCHES)
636	APERIODIC MOTION	MOUVEMENT APERIODIQUE	BEWEGUNG (APERIODISCHE)
637	APERIODICITY	APERIODICITE	APERIODIZITÄT
638	APEX, VERTEX	SOMMET, APEX	SCHEITEL, SPITZE (GEOM.)
639	APLANAT	APLANAT	APLANAT
640	APOCHROMATIC	APOCHROMATIQUE	APOCHROMATISCH
641	APPARATUS	APPAREIL	GERÄT, APPARAT
642	APPARENT DENSITY	DENSITE APPARENTE	DURCHSCHNITTLICHE DICHTE, SCHEINDICHTE
643	APPARENT RADIATION CONSTANT	CONSTANTE DE RADIATION APPARENTE	STRAHLUNGSKONSTANTE (SCHEINBARE)
644	APPEARANCE OF CRACKS	FORMATION DE FISSURES	RISSBILDUNG
645	APPLICANT FOR A PATENT	DEMANDEUR D'UN BREVET	PATENTBEWERBER, BEWERBER UM EIN PATENT, PATENTANMELDER
646	APPLICATION FOR A PATENT	DEMANDE, REQUETE DE BREVET	PATENTANMELDUNG, EINREICHUNG EINES PATENTGESUCHES
647	APPLICATION OF A FORCE	APPLICATION D'UNE FORCE	ANGREIFEN, ANGRIFF EINER KRAFT
648	APPLIED CHEMISTRY	CHIMIE APPLIQUEE	CHEMIE (ANGEWANDTE)
649	APPLIED RESEARCH	RECHERCHE APPLIQUEE	FORSCHUNG (ANGEWANDTE)
650	APPLY A BRAKE (TO)	SERRER, APPLIQUER UN FREIN	BREMSE ANZIEHEN (EINE)
651	APPLY FOR A PATENT (TO)	DEMANDER UN BREVET	EIN PATENT ANMELDEN, EIN PATENTGESUCH EINREICHEN
652	APPROXIMATE CALCULATION	CALCUL APPROXIMATIF, APPROXIMATION	ANNÄHERUNGSRECHNUNG
653	APPROXIMATE VALUE	VALEUR APPROCHEE	NÄHERUNGSWERT, ANGENÄHERTER WERT
654	APPROXIMATIVE FORMULA	FORMULE APPROXIMATIVE, D'APPROXIMATION	NÄHERUNGSFORMEL
655	APRON	TABLIER	BLECHSCHUTZ
656	APT PROGRAMMING LANGUAGE	LANGAGE DE PROGRAMMATION APT	APT-PROGRAMMIERSPRACHE
657	AQUA FORTIS	EAU FORTE	ÄTZE, GELBBRENNSÄURE
658	AQUA REGIA, NITROHYDROCHLORIC NITROMURITICACID	EAU REGALE	KÖNIGSWASSER, SALPETERSALZSÄURE
659	AQUEOUS ALCOHOL	ALCOOL AQUEUX	ALKOHOL (WÄSSERIGER)
660	AQUEOUS AMMONIA SOLUTION	SOLUTION AMMONIACALE, ALCALI VOLATIL, AMMONIAC	SALMIAKGEIST, AMMONIAKFLÜSSIGKEIT, ÄTZAMMONIAK, SALMIAKSPIRITUS
661	AQUEOUS SOLUTION	SOLUTION AQUEUSE	LÖSUNG (WÄSSRIGE)
662	ARBOR	SUPPORT METALLIQUE DE NOYAU	METALLKERNSTÜTZE
663	ARBOR TYPE CUTTERS	FRAISES ARBREES	AUFSTECKFRÄSER
664	ARC	ARC D'UNE COURBE	BOGEN
665	ARC BLOW	SOUFFLAGE MAGNETIQUE DE L'ARC	BLASWIRKUNG (MAGNETISCHE)
666	ARC BRAZING	BRASAGE FORT A L'ARC	LICHTBOGEN-HARTLÖTEN
667	ARC DIRECT-ARC FURNACE	FOUR ELECTRIQUE A ARC DIRECT	LICHTBOGENOFEN (DIREKTER)

668	**ARC FURNACE**	FOUR A ARC ELECTRIQUE	LICHTBOGENOFEN
669	**ARC FURNACE INDIRECT**	FOUR A ARC INDIRECT	LICHTBOGENOFEN (INDIREKTER)
670	**ARC FURNACE, CARBON**	FOUR A ARC ELECTRIQUE A ELECTRODES DE CHARBON	KOHLEELEKTRODEN-LICHTBOGENOFEN
671	**ARC FURNACE, SMOOTHERED ARC**	FOUR ELECTRIQUE A ARC DIRECT	LICHTBOGENOFEN (DIREKTER)
672	**ARC LAMP**	LAMPE A ARC	BOGENLAMPE
673	**ARC LIGHT**	LUMIERE DE L'ARC VOLTAIQUE	BOGENLICHT
674	**ARC OF CIRCLE**	ARC DE CERCLE, ARC CIRCULAIRE	KREISBOGEN
675	**ARC OF CONTACT**	ARC D'ENROULEMENT, COURBE DE CONTACT	UMSPANNUNGSBOGEN, UMSCHLINGUNGSBOGEN, EINGRIFFSBOGEN
676	**ARC STREAM VOLTAGE**	TENSION (VOLTAGE) DE L'ARC	LICHTBOGENSPANNUNG
677	**ARC STRIKE**	COUP DE SOUDURE POINT D'AMORCAGE DE L'ARC	ZÜNDSTELLE
678	**ARC THICKNESS**	EPAISSEUR A L'ARC	ZAHNSTÄRKE (IM ROLLKREIS)
679	**ARC TRUE VOLTAGE**	TENSION DE SERVICE DE L'ARC	LICHTBOGEN-ARBEITSSPANNUNG
680	**ARC WELDER**	MACHINE DE SOUDAGE A L'ARC	LICHTBOGEN-SCHWEISSMASCHINE
681	**ARC WELDING**	SOUDURE A L'ARC ELECTRIQUE	LICHTBOGENSCHWEISSUNG
682	**ARC WELDING CONTACT**	SOUDAGE AU CONTACT	KONTAKT SCHWEISSEN, LICHTBOGENSCHWEISSEN
683	**ARC WELDING INERT, GAS SHIELDED**	SOUDAGE A L'ARC EN ATMOSPHERE INERTE	SCHUTZGAS SCHWEISSEN, LICHTBOGENSCHWEISSEN
684	**ARC WELDING SUBMERGED**	SOUDAGE A L'ARC SUBMERGE, PROCEDE 'UNIONMELT'	SCHWEISSEN MIT VERDECKTEN LICHTBOGEN, UNTERPULVER-SCHWEISSEN
685	**ARC WELDING, ALTERNATING CURRENT**	SOUDAGE A L'ARC A COURANT ALTERNATIF	LICHTBOGENSCHWEISSEN MIT WECHSELSTROM
686	**ARC WELDING, DIRECT CURRENT**	SOUDAGE A L'ARC A COURANT CONTINU	GLEICHSTROM-LICHTBOGENSCHWEISSEN
687	**ARCH BRICK**	VOUSSOIR, VOUSSEAU	GEWÖLBESTEIN
688	**ARCH-BUTTRESS**	ARC-BOUTANT	STREBEBOGEN
689	**ARCHIMEDEAN DRILL, BRACE, PERSIAN DRILL**	DRILLE	DRILLBOHRER
690	**ARCHIMEDEAN SPIRAL, SPIRAL OF ARCHIMEDES**	SPIRALE ARCHIMEDIENNE, SPIRALE D'ARCHIMEDE	SPIRALE ARCHIMEDISCHE
691	**ARCHITECT**	ARCHITECTE	ARCHITEKT
692	**ARCHITECTURAL BRONZE**	BRONZE ARCHITECTURAL	BAUBRONZE
693	**ARCHITECTURAL DRAWING**	DESSIN D'ARCHITECTURE	BAUZEICHNUNG
694	**AREA**	SURFACE	FLÄCHE
695	**AREA OF CIRCLE**	AIRE DU CERCLE	KREISFLÄCHE, KREISINHALT
696	**AREA OF ENGAGEMENT**	SURFACE D'ENGRENEMENT	EINGRIFFSFLÄCHE
697	**AREA OF SURFACE**	AIRE	INHALT EINER FLÄCHE, FLÄCHENINHALT
698	**AREAL COORDINATES**	COORDONNEES PLANES	KOORDINATEN-EBENE
699	**ARGILLACEOUS MARL**	MARNE ARGILEUSE	TONMERGEL
700	**ARGILLITE**	ARGILLITE	TONSCHIEFER, ARGILLIT
701	**ARGON**	ARGON	ARGON
702	**ARITHMETIC**	ARITHMETIQUE	ARITHMETIK
703	**ARITHMETICAL MEAN**	MOYENNE ARITHMETIQUE	MITTEL (ARITHMETISCHES)
704	**ARITHMETICAL PROGRESSION**	PROGRESSION ARITHMETIQUE	REIHE (ARITHMETISCHE)
705	**ARM OF A FLYWHEEL, OF A PULLEY**	BRAS D'UN VOLANT, BRAS D'UNE POULIE	ARM (EINES SCHWUNGRADES), ARM (EINER RIEMENSCHEIBE)
706	**ARM OF LEVER, LEVER ARM**	BRAS DE LEVIER DE LA FORCE	HEBELARM, KRAFTARM
707	**ARMATURE**	INDUIT, ARMATURE	ANKER EINER GLEICHSTROMMASCHINE

708	ARMATURE CORE PLATE	TOLE POUR LES INDUITS DES DYNAMOS	ANKERBLECH, DYNAMOBLECH
709	ARMATURE WINDING MACHINE	MACHINE A BOBINER LES INDUITS	ANKERWICKELMASCHINE
710	ARMATURE, KEEPER OF A MAGNET	ARMATURE, ARMURE, CONTACT D'UN AIMANT	ANKER EINES MAGNETEN, MAGNETANKER
711	ARMCO INGOT IRON	FER ARMCO	ARMCO-LISEN
712	ARMOUR PLATE	PLAQUE DE BLINDAGE	PANZERPLATTE
713	ARMOURED CABLE	CABLE ARME	PANZERKABEL
714	ARMOURED HOSE	TUBE FLEXIBLE CERCLE EN ACIER, TUBE FLEXIBLE AVEC ARMATURE EN FIL DE FER	SCHLAUCH (GEPANZERTER)
715	ARMOURING OF A CABLE	ARMATURE, REVETEMENT D'UN CABLE	BEWEHRUNG, ARMATUR EINES KABELS, KABELARMATUR
716	AROMATIC COMPOUNDS	COMPOSES AROMATIQUES	VERBINDUNGEN (AROMATISCHE)
717	ARRANGEMENT OF RIVETS	DISTRIBUTION DES RIVETS	GRUPPIERUNG DER NIETE, NIETVERTEILUNG
718	ARREST POINT	POINT DE TRANSFORMATION	UMWANDLUNGSPUNKT
719	ARROW	FLECHE	RICHTUNGSPFEIL
720	ARSENIATE	ARSENIATE	ARSENSÄURESALZ, ARSENIAT
721	ARSENIC	ARSENIC	ARSEN
722	ARSENIC ACID	ACIDE ARSENIQUE	ARSENSÄURE
723	ARSENIC PENTOXIDE	ANHYDRIDE ARSENIQUE	ARSENPENTOXYD, ARSENSÄUREANHYDRID
724	ARSENIC TRICHLORIDE	TRICHLORURE D'ARSENIQUE	ARSENTRICHLORID, CHLORARSEN
725	ARSENICAL COPPER	CUIVRE ARSENICAL	ARSENKUPFER
726	ARSENIDE	ARSENIURE	ARSENMETALL, ARSENID
727	ARSENIOUS OXIDE, WHITE ARSENIC	ARSENIQUE BLANC, ACIDE, ANHYDRIDE ARSENIEUX	ARSENIK (WEISSER), SÄURE (ARSENIGE), ARSENIGS ÄUREANHYDRID, ARSENTRIOXYD
728	ARSENIOUS SULPHIDE	TRISULFURE D'ARSENIC, ORPIMENT	ARSENTRISULFID, ARSENSUPERSULFÜR, AURIPIGMENT, OPERMENT, RAUSCHGELB, ARSENIK (GELBES)
729	ARSENITE	ARSENITE	ARSENIGSÄURESALZ, ARSENIT
730	ARSENIURETTED HYDROGEN	HYDROGENE ARSENIE, ARSENIURE D'HYDROGENE	ARSENWASSERSTOFF
731	ARTIFICIAL AGING	VIEILLISSEMENT ARTIFICIEL	ALTERUNG (KÜNSTLICHE)
732	ARTIFICIAL FUEL	COMBUSTIBLE ARTIFICIEL	BRENNSTOFF (KÜNSTLICHER)
733	ARTIFICIAL ILLUMINATION LIGHTING	ECLAIRAGE ARTIFICIEL	BELEUCHTUNG (KÜNSTLICHE)
734	ARTIFICIAL LIGHT	LUMIERE ARTIFICIELLE	LICHT (KÜNSTLICHES)
735	ARTIFICIAL MAGNET	AIMANT ARTIFICIEL	MAGNET (KÜNSTLICHER)
736	ARTIFICIAL SILK	SOIE ARTIFICIELLE	KUNSTSEIDE
737	ARTIFICIAL STONE	PIERRE ARTIFICIELLE	STEIN (KÜNSTLICHER), KUNSTSTEIN
738	ARTIFICIAL VENTILATION	VENTILATION ARTIFICIELLE	LÜFTUNG (KÜNSTLICHE)
739	AS CAST	BRUT DE COULEE	ROHGEGOSSEN
740	AS CONDITION	ETAT BRUT	ROH
741	AS DELIVERED	A L'ETAT DE LIVRAISON	IM LIEFERZUSTAND
742	AS FORGED	BRUT DE FORGE	ROHGESCHMIEDET
743	AS QUENCHED	BRUT DE TREMPE	IM ABGESCHRECKTEN ZUSTAND
744	AS ROLLED	BRUT DE LAMINAGE	IM GEWALZTEN ZUSTAND
745	AS WELDED	A L'ETAT DE SOUDAGE	IM SCHWEISSZUSTAND
746	ASBESTOS	AMIANTE	ASBEST
747	ASBESTOS CARD, ASBESTOS MILLBOARD	CARTON D'AMIANTE	ASBESTPAPPE, PRESSASBEST
748	ASBESTOS CLOTH	TOILE, TISSU D'AMIANTE	ASBESTGEWEBE

749	ASBESTOS CORD	FIL, BOUDIN D'AMIANTE	ASBESTSCHNUR
750	ASBESTOS FELT	FEUTRE D'AMIANTE	ASBESTFILZ
751	ASBESTOS GASKET	TRESSE, CORDON D'AMIANTE, AMIANTE TRESSE EN CORDELETTE	ASBESTZOPF
752	ASBESTOS PAPER	PAPIER D'AMIANTE	ASBESTPAPIER
753	ASBESTOS RING	ANNEAU D'AMIANTE	ASBESTRING
754	ASBESTOS STRIP	BANDE D'AMIANTE	ASBESTSTREIFEN
755	ASBESTOS, AMIANTHUS	AMIANTE, ASBESTE	ASBEST, AMIANT
756	ASCENDING PIPE LINE	CONDUITE ASCENDANTE	STEIGENDE LEITUNG
757	ASCENT, ACCLIVITY, RISING GRADIENT	RAMPE	STEIGUNG, ANSTIEG
758	ASH CONTENT	TENEUR EN CENDRES, POURCENTAGE DE CENDRES	ASCHENGEHALT
759	ASH PIT	CENDRIER	ASCHENRAUM, ASCHENFALL
760	ASHES	CENDRE, CENDRES	ASCHE
761	ASHLAR	PIERRE DE TAILLE	WERKSTEIN, HAUSTEIN, QUADER
762	ASHLAR WORK	MACONNERIE EN PIERRES DE TAILLE	QUADERMAUERWERK
763	ASPHALT (TO)	ASPHALTER, BITUMER	ASPHALTIEREN
764	ASPHALT MASTIC, ROCK ASPHALT MASTIC	MASTIC D'ASPHALTE	ASPHALTMASTIX
765	ASPHALTED FELT	FEUTRE ASPHALTE	ASPHALTFILZ
766	ASPHALTED TUBE	TUBE PROTEGE PAR UN RECOUVREMENT DE JUTE ASPHALTE	ASPHALTIERTES ROHR
767	ASPHALTING	ASPHALTAGE	ASPHALTIEREN
768	ASPHALTUM	ASPHALTE, BITUME SOLIDE	ASPHALT, ERDPECH
769	ASPIRATOR	ASPIRATEUR	ASPIRATOR
770	ASSEMBLING BAY	HALLE DE MONTAGE	MONTAGEHALLE
771	ASSEMBLING BOLT	BOULON D'ASSEMBLAGE	VERBINDUNGSBOLZEN, VERBINDUNGSSCHRAUBE
772	ASSEMBLING DRAWING	PLAN DE MONTAGE	MONTAGEZEICHNUNG
773	ASSEMBLING SHOP	ATELIER, HALLE DE MONTAGE	MONTAGEHALLE
774	ASTATIC	ASTATIQUE	ASTATISCH
775	ASTATIC GOVERNOR	REGULATEUR ASTATIQUE	REGLER (ASTATISCHER)
776	ASTERISM	ASTERISME	ASTERISMUS
777	ASYMMETRICAL CELL	ELEMENT ASYMETRIQUE	ELEMENT (ASYMMETRISCHES)
778	ASYMMETRICAL, DISSYMMETRICAL	ASYMETRIQUE, DISSYMETRIQUE	UNSYMMETRISCH, ASYMMETRISCH, NICHT SPIEGELGLEICH
779	ASYMMETRY, DISSYMMETRY	ASYMETRIE, DISSYMETRIE	ASYMMETRIE
780	ASYMPTOTE	ASYMPTOTE	ASYMPTOTE
781	ASYMPTOTIC	ASYMPTOTE, ASYMPTOTIQUE	ASYMPTOTISCH
782	ASYNCHRONISM	ASYNCHRONISME	ASYNCHRONISMUS
783	ASYNCHRONOUS MOTOR	MOTEUR ASYNCHRONE	ASYNCHRONMOTOR
784	ASYNCHRONOUS, NONSYNCHRONOUS	ASYNCHRONE	ASYNCHRON
785	ATHERMANCY	ATHERMANEITE	UNDURCHLÄSSIGKEIT FÜR WÄRMESTRAHLEN
786	ATHERMANOUS	ATHERMANE	UNDURCHLÄSSIG FÜR WÄRMESTRAHLEN, ATHERMAN
787	ATMOSPHERE	ATMOSPHERE	ATMOSPHÄRE
788	ATMOSPHERE (ATM.)	ATMOSPHERE (UNITE DE PRESSION)	ATMOSPHÄRE (MASSEINHEIT)
789	ATMOSPHERE PREPARED	ATMOSPHERE ARTIFICIELLE	ATMOSPHÄRE (PRÄPARIERTE)
790	ATMOSPHERE PROTECTIVE	ATMOSPHERE PROTECTRICE	SCHUTZATMOSPHÄRE

791	ATMOSPHERE SPECIAL PURPOSE	ATMOSPHERE A USAGE SPECIAL	SONDERZWECKATMOSPHÄRE
792	ATMOSPHERIC ACTION	ACTIONS ATMOSPHERIQUES	WITTERUNGSEINFLÜSSE
793	ATMOSPHERIC AGENTS	AGENTS ATMOSPHERIQUES	ATMOSPHÄRILIEN
794	ATMOSPHERIC BAROMETRIC PRESSURE, PRESSURE OF THE ATMOSPHERE	PRESSION ATMOSPHERIQUE, PRESSION BAROMETRIQUE, PRESSION DE X CM DE MERCURE	LUFTDRUCK, DRUCK (ATMOSPHÄRISCHER)
795	ATMOSPHERIC CORROSION	CORROSION ATMOSPHERIQUE	KORROSION (ATMOSPHÄRISCHE)
796	ATMOSPHERIC ELECTRICITY	ELECTRICITE ATMOSPHERIQUE	LUFTELEKTRIZITÄT, ELEKTRIZITÄT (ATMOSPHÄRISCHE)
797	ATMOSPHERIC ESCAPE VALVE	SOUPAPE D'ECHAPPEMENT	AUSPUFFVENTIL
798	ATMOSPHERIC LINE	LIGNE ATMOSPHERIQUE	LINIE (ATMOSPHÄRISCHE)
799	ATMOSPHERIC PRESSURE HEAD	MASSELOTTE A NOYAU	
800	ATOM	ATOME	ATOM
801	ATOMIC (HYDROGEN) ARC WELDING	SOUDAGE ARCATOM (A L'ARC PROTEGE, A L'HYDROGENE ATOMIQUE)	SCHUTZGAS-LICHTBOGEN- SCHWEISSEN (ATOMARES)
802	ATOMIC HEAT	CHALEUR ATOMIQUE	ATOMWÄRME
803	ATOMIC NUMBER	NOMBRE ATOMIQUE	ATOMZAHL
804	ATOMIC PLANE	PLAN RETICULAIRE	NETZEBENE, GITTEREBENE
805	ATOMIC THEORY	THEORIE ATOMIQUE	ATOMTHEORIE
806	ATOMIC VOLUME	VOLUME ATOMIQUE	ATOMVOLUMEN
807	ATOMIC WEIGHT	POIDS ATOMIQUE	ATOMGEWICHT
808	ATOMISE A LIQUID (TO)	PULVERISER (UN LIQUIDE)	ZERSTÄUBEN (EINE FLÜSSIGKEIT)
809	ATOMISER	PULVERISATEUR, DISPERSEUR DE LIQUIDES	ZERSTÄUBER
810	ATOMISING A LIQUID	PULVERISATION D'UN LIQUIDE	ZERSTÄUBEN EINER FLÜSSIGKEIT
811	ATOMIZATION	ATOMISATION, PULVERISATION	ZERSTÄUBUNG
812	ATTACHMENT	ELEMENT D'APPOINT, ACCESSOIRE	ZUSATZEINRICHTUNG
813	ATTACK	ATTAQUE	ANGRIFF
814	ATTENDANCE	CONDUITE (D'UNE MACHINE)	WARTUNG, BEDIENUNG
815	ATTRACTION	ATTRACTION	ANZIEHUNG
816	AUDIBLE SIGNAL	SIGNAL ACOUSTIQUE	HÖRBARES ZEICHEN, HÖRSIGNAL, SIGNAL (AKUSTISCHES)
817	AUER METAL	ALLIAGE D'AUER	AUER METALL, MISCHMETALL
818	AUGER	TARIERE	ZIMMERMANNBOHRER
819	AUGITE	AUGITE	AUGIT
820	AUGITE SYENITE	AUGITE-SYENITE	AUGITSYENIT
821	AURIC CHLORIDE, CHLORIDE OF GOLD, GOLD TRICHLORIDE	TRICHLORURE D'OR	GOLDTRICHLORID, CHLORGOLD, GOLDCHLORID, AURICHLORID
822	AURIFEROUS	AURIFERE	GOLDHALTIG
823	AUROUS CHLORIDE	CHLORURE D'OR	GOLDMONOCHLORID, GOLDCHLORÜR, AUROCHLORID
824	AUSTEMPERING	TREMPE ETAGEE BAINITIQUE, TREMPE EN ETAPES	ZWISCHENSTUFENVERGÜTUNG, AUSTEMPERUNG, ZWISCHENSTUFENHÄRTUNG
825	AUSTENITE	AUSTENITE	AUSTENIT
826	AUSTENITIC STEEL	ACIER AUSTENITIQUE	STAHL (AUSTENITISCHER)
827	AUSTENITIZING	AUSTENITISATION	AUSTENITISIERUNG
828	AUTO-STARTER	DEMARREUR AUTOMATIQUE	AUTOSTARTER
829	AUTOCLAVE, DIGESTER	AUTOCLAVE, DIGESTEUR, MARMITE AUTOCLAVE, MARMITE DE PAPIN	DRUCKKESSEL, DAMPFFASS, DAMPFKOCHTOPF, PAPINSCHER TOPF, DIGESTOR, AUTOKLAV
830	AUTOFRETTAGE	SOUMISSION A UNE PRE-TENSION	KALTRECKEN, VORSPANNEN
831	AUTOGEN WELDING	SOUDURE AUTOGENE	AUTOGEN-SCHWEISSUNG

3

832	AUTOGENOUS CUTTING	DECOUPAGE AUTOGENE	SCHNEIDEN (AUTOGENES)
833	AUTOGENOUS CUTTING MACHINE	MACHINE A DECOUPER AUTOGENE	SCHNEIDMASCHINE, SAUERSTOFFSCHNEID-MASCHINE (AUTOGENE)
834	AUTOGENOUS SOLDERING WELDING	SOUDURE AUTOGENE, AUTOSOUDURE	SCHWEISSUNG (AUTOGENE)
835	AUTOGENOUS WELDING	SOUDAGE AUTOGENE	AUTOGENSCHWEISSEN
836	AUTOMATIC ACCELERATION	ACCELERATION AUTOMATIQUE	BESCHLEUNIGUNG (AUTOMATISCHE)
837	AUTOMATIC ACTION MOVEMENT	AUTOMATICITE	ARBEITEN (SELBSTTÄTIGES), SELBSTTÄTIGKEIT
838	AUTOMATIC CHOKE	STARTER AUTOMATIQUE	LUFTKAPPENVERSTELLUNG (AUTOMATISCHE)
839	AUTOMATIC CYCLE	CYCLE AUTOMATIQUE	ARBEITSABLAUF (SELBSTTÄTIGER)
840	AUTOMATIC CYCLE LATH	TOUR A CYCLES AUTOMATIQUES	DREHMASCHINE MIT AUTOMATISCHEN ARBEITSABLAUFEN
841	AUTOMATIC DECELERATION	DECELERATION AUTOMATIQUE	VERZÖGERUNG (AUTOMATISCHE)
842	AUTOMATIC FEED LEVER	LEVIER D'AVANCE AUTOMATIQUE	VORSCHUBHEBEL (SELBSTTÄTIGER)
843	AUTOMATIC GOVERNOR	REGULATEUR AUTOMATIQUE	REGLER (SELBSTTÄTIGER)
844	AUTOMATIC IGNITION	ALLUMAGE AUTOMATIQUE	SELBSTZÜNDUNG
845	AUTOMATIC ISOLATING VALVE	ROBINET D'ISOLEMENT	SELBSTSCHLUSSVENTIL, ROHRBRUCHVENTIL
846	AUTOMATIC LATHE, AUTOMATIC SCREW MACHINE	TOUR AUTOMATIQUE	DREHBANK (SELBSTTÄTIGE), AUTOMAT, DREHAUTOMAT
847	AUTOMATIC LEVEL CONTROL	STABILISATEUR AUTOMATIQUE	SELBSTSTABILISIERUNGSSYSTEM
848	AUTOMATIC LUBRICATION	GRAISSAGE MECANIQUE, GRAISSAGE AUTOMATIQUE	SCHMIERUNG (SELBSTTÄTIGE)
849	AUTOMATIC MACHINE WELDING	SOUDAGE AUTOMATIQUE	AUTOMATENSCHWEISSEN
850	AUTOMATIC PROGRAMMING	PROGRAMMATION AUTOMATIQUE	PROGRAMMIEREN (MASCHINELLES)
851	AUTOMATIC SELF-ACTING LUBRICATOR	GRAISSEUR AUTOMATIQUE	SELBSTÖLER, SCHMIERVORRICHTUNG (SELBSTTÄTIGE)
852	AUTOMATIC VALVE	SOUPAPE AUTOMATIQUE, SOUPAPE DESMODROMIQUE	VENTIL (UNGESTEUERTES), VENTIL (SELBSTTÄTIGES)
853	AUXILIARY ENGINE	MACHINE AUXILIAIRE	HILFSMASCHINE
854	AUXILIARY FUNCTION	FONCTION AUXILIAIRE	HILFSFUNKTION
855	AVERAGE MEAN TEMPERATURE	TEMPERATURE MOYENNE	TEMPERATUR (MITTLERE), DURCHSCHNITTSTEMPERATUR
856	AVIATION SNIPS	CISAILLES ARTICULEES	FAUSTSCHERE
857	AXE	HACHE	AXT
858	AXES AT RIGHT ANGLES INTERSECTING	AXES D'EQUERRE	ACHSEN (SICH RECHTWINKLIG SCHNEIDENDE)
859	AXIAL COMPONENT	COMPOSANTE AXIALE	AXIALKOMPONENTE
860	AXIAL DISPLACEMENT	DEPLACEMENT AXIAL, DEPLACEMENT LONGITUDINAL, DEPLACEMENT PARALLELE A L'AXE	LÄNGSVERSCHIEBUNG, VERSCHIEBUNG (AXIALE)
861	AXIAL FLOW PUMP	POMPE HELICOIDALE	SCHRAUBENPUMPE
862	AXIAL FLOW TURBINE	TURBINE AXIALE, TURBINE PARALLELE	AXIALTURBINE
863	AXIAL RATIO IN CRYSTALS	AXE DE SYMETRIE DES CRISTAUX	KRISTALLSYMMETRIEACHSE
864	AXIS	AXE	KOORDINATENACHSE
865	AXIS OF A CRYSTAL	AXE D'UN CRISTAL	KRISTALLACHSE
866	AXIS OF A PIPE TUBE	AXE D'UN TUYAU	ROHRACHSE
867	AXIS OF OSCILLATION	AXE D'OSCILLATION	SCHWINGUNGSACHSE
868	AXIS OF ROTATION	AXE DE ROTATION	DREHACHSE, UMDREHUNGSACHSE
869	AXIS OF SYMMETRY	AXE DE SYMETRIE	SCHWERPUNKTACHSE, SYMMETRIBACHSE
870	AXIS OF THE WELD	AXE DE SOUDURE	NAHTACHSE

871	AXIS OF THE WELD BEAD	AXE DU CORDON DE SOUDURE	SCHWEISSRAUPENACHSE
872	AXIS OF X	AXE DES ABCISSES, AXE DES X	ABSZISSENACHSE, X-ACHSE
873	AXIS OF Y	AXE DES ORDONNEES, AXE DES Y	ORDINATENACHSE, Y-ACHSE
874	AXIS, CENTRE LINE	AXE GEOMETRIQUE	ACHSE, MITTELLINIE, ZENTRALLINIE
875	AXLE	ESSIEU OU ARBRE	ACHSE, WELLE
876	AXLE BEARING	BOITE D'ESSIEU	ACHSLAGER
877	AXLE BOX	BOITE D'ESSIEU, BOITE A GRAISSE, BOITE A HUILE, BOITE DE GRAISSAGE	ACHSLAGER, ACHSBÜCHSE
878	AXLE BOX DRILLING MACHINE	PERCEUSE POUR BOITE D'ESSIEU	ACHSLAGERBOHRMASCHINE
879	AXLE GREASE	GRAISSE DE VOITURE, CAMBOUIS	WAGENFETT, WAGENSCHMIERE
880	AXLE JOURNAL LATHE	TOUR POUR FUSEE D'ESSIEU	ACHSSCHENKELDREHBANK
881	AXLE LATHE	TOUR A ESSIEUX	ACHSENDREHBANK
882	AXLE NECK	FUSEE D'ESSIEU	ACHSSCHENKEL
883	AXLE WRENCH	CLEF DE CHAPEAU, CLEF D'ESSIEU	ACHSKAPPENSCHLÜSSEL, RADMUTTERSCHLÜSSEL
884	AXLE, AXLE TREE	ESSIEU	ACHSE, RADACHSE
885	AXONOMETRIC PERSPECTIVE	PERSPECTIVE AXONOMETRIQUE	PERSPEKTIVE (AXONOMETRISCHE)
886	AZURITE, BLUE CARBONATE OF COPPER	AZURITE	KUPFERLASUR, BERGBLAU
887	B.S.I. (BRITSH STANDARDS INSTITUTION)	INSTITUT BRITANNIQUE DE NORMALISATION	NORMENINSTITUT (ENGLISCHES)
888	BABBITT ('S METAL)	METAL ANTIFRICTION, METAL BLANC	WEISSMETALL
889	BABBITT METAL	REGULE, ANTIFRICTION	LAGER-WEISSMETAL
890	BACK	DOS	RÜCKEN
891	BACK DOOR	HAYON, PORTE, PORTIERE ARRIERE	HINTERTÜR
892	BACK FIRING	RETOUR DE LA FLAMME	ZURÜCKSCHLAGEN DER FLAMME
893	BACK OFF (TO), RELIEVE (TO)	DEPOUILLER, DEGAGER	HINTERDREHEN
894	BACK PRESSURE OPERATION	MARCHE EN CONTRE-PRESSION	GEGENDRUCKBETRIEB
895	BACK PRESSURE VALVE	VALVE DE CONTREPRESSION	GEGENDRUCKVENTIL
896	BACK REFLECTION	REFLEXION EN RETOUR	RÜCKSTRAHLUNG
897	BACK SAW	SCIE A DOS	RÜCKENSÄGE
898	BACK STOP	BUTEE	GEGENHALTER
899	BACK WINDOW	LUNETTE ARRIERE	RÜCKFENSTER
900	BACK-FILLING	REMBLAI, REMBLAYAGE	HINTERFÜLLUNG
901	BACK-STEP SEQUENCE	SOUDURE A PAS DE PELERIN	PILGERSCHRITTSCHWEISSVERFAHREN
902	BACK-STRIP, BACKING-STRIP	FEUILLARD-SUPPORT, PLAT-SUPPORT	KUPFERFLACHPROFIL, BANDSTAHL
903	BACK-UP ROLLS	CYLINDRES DE SUPPORT	STÜTZWALZEN
904	BACK-UP WELD	CORDON SUPPORT (A L'ENVERS)	STÜTZRAUPE, WURZELSEITIGE GEGENNAHT
905	BACKED OFF CUTTER, RELIEVED TOOTH MILLING CUTTER	FRAISE A DENTS DEPOUILLEES, FRAISE A DENTS DEGAGEES, FRAISE A PROFIL CONSTANT	FRÄSER MIT HINTERDREHEN ZÄHNEN
906	BACKFIRE	RETOUR DE FLAMME	FLAMMENRÜCKSCHLAG, RÜCKSCHLAG, RÜCKZÜNDUNG
907	BACKHAND WELDING	SOUDAGE A DROITE	RECHTSSCHWEISSUNG
908	BACKING LIGHTS	FEUX DE RECUL	RÜCKLICHT
909	BACKING MATERAIL	METAL SUPPORT	GRUNDWERKSTOFF
910	BACKING OFF, RELIEVING	DEPOUILLE, DEGAGEMENT	HINTERDREHEN
911	BACKING RING	CONTRE-JOINT	STÜTZRING
912	BACKING SAND	SABLE DE COUVERTURE	FÜLLSAND

913	BACKING-OFF LATHE, RELIEVING LATHE	TOUR A DEPOUILLER, A DEGAGER	HINTERDREHBANK
914	BACKLASH	JEU ENTRE DENTS	FLANKENSPIELRAUM
915	BACKLASH ELIMINATOR	MECANISME DE REPRISE DES JEUX	SPIELVERRINGERUNGSEINRICHTUNG
916	BACKLASH, LOST MOTION	JEU INUTILE, JEU PERNICIEUX, JEU NUISIBLE	GANG (TOTER)
917	BACKREST	DOSSIER	RÜCKENLEHNE
918	BACKSTEP WELDING	SOUDAGE EN PAS DE PELERIN	GEGENSCHRITTSCHWEISSEN
919	BACKWARD MOVEMENT	MOUVEMENT DANS LE SENS RETROGRADE, MOUVEMENT DE RECUL, RECUL	RÜCKWÄRTSBEWEGUNG, BEWEGUNG (RÜCKLÄUFIGE)
920	BAD WELD	SOUDAGE MAL FAIT, SOUDURE MAL FAITE	SCHWEISSVERBINDUG (SCHLECHTE)
921	BAFFLE	DEFLECTEUR, CHICANE, TOLE DEFLECTRICE	LEITBLECH, ABLEITBLECH
922	BAFFLE PLATE, BAFFLER, DEFLECTOR	CHICANE	LEITBLECH, VERTEILUNGSWAND
923	BAG MOULDING PROCESS	PROCEDE DE MOULAGE AU MOYEN DU SAC EN CAOUTCHOUC	GUMMISACKVERFAHREN
924	BAGASSE	BAGASSE	BAGASSE
925	BAINITE	BAINITE	BAINITE, ZWISCHENSTUFENGEFÜGE
926	BAKED CLAY	ARGILE CUITE	TON (GEBRANNTER)
927	BAKED CORE	NOYAU ETUVE	TROCKENKERN
928	BAKING	SECHAGE	TROCKNUNG
929	BALANCE	EQUILIBRE	AUSGLEICH
930	BALANCE (TO), COUNTERBALANCE (TO)	EQUILIBRER, CONTRE-BALANCER, COMPENSER	AUSWUCHTEN, AUSGLEICHEN, AUSBALANCIEREN
931	BALANCE, SCALES	BALANCE, BASCULE	WAAGE
932	BALANCED ECONOMY	ECONOMIE SAINE, ECONOMIE EN EQUILIBRE	WIRTSCHAFT (GESUNDE)
933	BALANCED FILTERS	FILTRES SYMETRIQUES	FILTER (SYMMETRISCHE)
934	BALANCED VALVE, EQUILIBRIUM VALVE	SOUPAPE EQUILIBREE, VALVE EQUILIBREE	VENTIL (ENTLASTETES), VENTIL (AUSBALANCIERTES)
935	BALANCING	EQUILIBRAGE	BALANCIEREN
936	BALANCING A VALVE	EQUILIBRAGE D'UNE SOUPAPE	ENTLASTUNG EINES VENTILS
937	BALANCING APPARATUS	APPAREIL D'EQUILIBRAGE	AUSWUCHTMASCHINE
938	BALANCING CYLINDER	CYLINDRE D'EQUILIBRAGE	AUSGLEICHZYLINDER
939	BALANCING DRUM	CYLINDRE D'EQUILIBRAGE	AUSGLEICHZYLINDER
940	BALANCING MACHINE	MACHINE A EQUILIBRER	AUSWUCHTMASCHINE
941	BALANCING THE MASSES	EQUILIBRAGE DES MASSES	MASSENAUSGLEICH
942	BALANCING, COUNTERBALANCING, EQUILIBRATION	EQUILIBRAGE, EQUILIBRATION	AUSWUCHTUNG, AUSGLEICHUNG, AUSBALANCIERUNG
943	BALATA	BALATA	BALATAHARZ, BALATA
944	BALATA BELT	COURROIE EN BALATA	BALATARIEMEN
945	BALE	BALLE (DE MARCHANDISE)	BALLEN
946	BALE PICK-UP	RAMASSE-BALLES	HEUBÜNDLER
947	BALE TRUCK	DIABLE, CABROUET	SACKKARREN, BALLENKARREN, STECHKARREN
948	BALERS, HIGH, LOW DENSITY, STATIONARY AND PICK-UP	BOTTELEUSES MECANIQUES A TOUTE DENSITE	BINDERMASCHINEN ALLER ARTEN
949	BALK, JOIST	POUTRE EN BOIS	BALKEN (HÖLZERNER)
950	BALL	BILLE	KUGEL
951	BALL (OR LUMP) OF STEEL	LOUPE, MASSE D'ACIER, GUEUSE	STAHLDEUL
952	BALL AND SOCKET BEARING	PALIER A TOURILLON SPHERIQUE	KUGELZAPFENLAGER

ꝺ

953	BALL AND SOCKET JOINT, SPHERICAL JOINT	JOINT SPHERIQUE, GENOU, JOINT A ROTULE	KUGELGELENK
954	BALL BEARING	PALIER A ROULEMENT A BILLES, ROULEMENT A BILLES, COUSSINET A BILLES, ROULEMENT	KUGELLAGER
955	BALL BEARING CUP	CAGE DE ROULEMENT A BILLES	KUGELLAGERSCHALE
956	BALL BURNISHING	POLISSAGE A BILLES	KUGELPOLIEREN
957	BALL CAGE	CAGE A BILLES	KÄFIG EINES KUGELLAGERS, KUGELKÄFIG
958	BALL JOINT SUSPENSION	SUSPENSION A ROTULE	KUGELGELENKAUFHÄNGUNG
959	BALL MILL	BROYEUR A BOULETS	KUGELMÜHLE
960	BALL PANE HAMMER	MARTEAU A PANNE BOMBEE, MARTEAU ARRONDI	HAMMER MIT KUGELFINNE
961	BALL PEEN	PANNE RONDE	KUGELFINNE
962	BALL PEEN HAMMER	MARTEAU A PANNE RONDE	HAMMER MIT KUGELFINNE
963	BALL PLANE	PANNE BOMBEE, PANNE SPHERIQUE	KUGELFINNE
964	BALL RACE	ANNEAU DE ROULEMENT, BAGUE DE ROULEMENT	KUGELRING, LAUFRING
965	BALL RACE WAY	CHEMIN DE ROULEMENT DES BILLES	KUGELBAHN, KUGELSPUR (EINES LAGERS)
966	BALL THRUST BEARING	BUTEE A BILLES	KUGELDRUCKLAGER
967	BALL THRUST TESTING MACHINE	APPAREIL A BILLER	KUGELDRUCKPRÜFAPPARAT
968	BALL VALVE	VANNE A BOISSEAU SPHERIQUE, ROBINET A BOISSEAU SPHERIQUE (A BOULE), SOUPAPE A BOULET, SOUPAPE-BILLE	KUGELVENTIL, KUGELHAHN
969	BALLAST	BALLAST	SCHÜTTUNG, MASSE (AUFGBSCHÜTTETE)
970	BALLAST UNIT	CLAPET	VENTIL
971	BALLASTING	BALLASTAGE	SCHÜTTUNG, ANSCHÜTTEN
972	BAMBOO CANE	BAMBOU	BAMBUSROHR
973	BANCA TIN	ETAIN DE BANCA	BANKAZINN
974	BAND	BANDE, RUBAN	BAND
975	BAND PULLEY	POULIE A CORDE	SCHNURSCHEIBE
976	BAND TENSION INDICATOR	INDICATEUR DE TENSION DU RUBAN	BANDSPANNUNGSANZEIGER
977	BAND-SAWING MACHINE	SCIE A RUBAN	BANDSÄGEMASCHINE
978	BAND, ENDLESS, RIBBON SAW	SCIE A RUBAN, SCIE A LAME SANS FIN	BANDSÄGE
979	BANDED STRUCTURE	STRUCTURE ZONALE	ZEILENSTRUKTUR
980	BANDS	BANDES, RUBANS, FEUILLARDS	BÄNDER, STREIFEN
981	BAR	BARRE	STAB
982	BAR AND TUBE DRAWING MACHINE	MACHINE A ETIRER LES BARRES ET LES TUBES	STANGEN-UND ROHRZIEHMASCHINE
983	BAR AUTOMATICS	TOUR AUTOMATIQUE TRAVAILLANT EN BARRE	STANGENDREHAUTOMAT
984	BAR IRON, STEEL BAR	BARREAUX VERGES DE FER, FER EN BARREAUX EN VERGES, ACIER EN BARRES	STABEISEN
985	BAR MAGNET	AIMANT EN FORME DE BARREAU, BARREAU AIMANTE	STABMAGNET
986	BAR, MEMBER (OF LATTICEWORK)	BARRE, POUTRELLE	STAB (EINES FACHWERKES)
987	BARBED WIRE	RONCE ARTIFICIELLE, FIL BARBELE	STACHELDRAHT
988	BARBED WIRE MACHINE	MACHINE A FIL BARBELE	STACHELDRAHTMASCHINE
989	BARE ELECTRODE	ELECTRODE NUE	ELEKTRODE (NACKTE)

4

990	**BARE WIRE**	FIL NU	DRAHT (NACKTER), DRAHT (BLANKER), DRAHT (NICHT ISOLIERTER ELEKTR.)
991	**BARGE**	BARGE, PENICHE	LASTKAHN
992	**BARIUM**	BARYUM	BARIUM
993	**BARIUM ACETATE**	ACETATE DE BARYUM	BARYT, BARIUMAZETAT (ESSIGSAURES)
994	**BARIUM ALUMINATE**	ALUMINATE DE BARYTE	BARIUMALUMINAT
995	**BARIUM CARBIDE**	CARBURE DE BARYUM	BARIUMKARBID
996	**BARIUM CARBONATE**	CARBONATE DE BARYUM	BARIUM (KOHLENSAURES), BARIUMKARBONAT
997	**BARIUM CHLORIDE**	CHLORURE DE BARYUM	CHLORBARIUM, BARIUMCHLORID
998	**BARIUM DIOXIDE**	BIOXYDE DE BARYUM	BARIUMSUPEROXYD, BARIUMHYPEROXYD, BARIUMDIOXYD
999	**BARIUM FLUORIDE**	FLUORURE DE BARYUM	BARUIMFLUORID, FLUOBARIUM
1000	**BARIUM HYDROXIDE**	HYDRATE DE BARYTE, HYDRATE DE BARYUM	BARIUMHYDROXYD, BARIUMOXYDHYDRAT, BARYTHYDRAT, BARYT (KAUSTISCHER), ÄTZBARYT
1001	**BARIUM NITRATE**	AZOTATE DE BARYUM	BARYT (SALPETERSAURER), BARIUMNITRAT
1002	**BARIUM PLATINOCYANIDE**	PLATINOCYANURE DE BARYUM	BARIUMPLATINZYANÜR
1003	**BARIUM SULPHATE, PERMANENT WHITE, BLANC FIXE**	SULFATE DE BARYUM, BLANC FIXE, BLANC DE BARYTE	BARIUM (SCHWEFELSAURES), BARIUMSULFAT, BARYTWEISS, NEUWEISS, MINERALWEISS, PERMANENTWEISS, SCHNEEWEISS
1004	**BARIUM SULPHIDE**	SULFURE DE BARYUM	SCHWEFELBARIUM, BARIUMSULFID
1005	**BARK**	COUCHE INTERMEDIAIRE DECARBUREE	ZWISCHENSCHICHT (ENTKOHLTE)
1006	**BARN**	UNITE DE SURFACE NUCLEAIRE, BARN	BARN
1007	**BAROGRAPH**	BAROMETRE ENREGISTREUR	BAROGRAPH
1008	**BAROMETER**	BAROMETRE	BAROMETER
1009	**BAROMETRIC**	BAROMETRIQUE	BAROMETRISCH
1010	**BAROMETRIC CONDENSER**	CONDENSEUR BAROMETRIQUE	KONDENSATOR (BAROMETRISCHER)
1011	**BAROSCOPE**	BAROSCOPE	BAROSKOP
1012	**BARREL BURNISHING**	POLISSAGE AU TAMBOUR	FASSPOLIEREN
1013	**BARREL DRILLING MACHINE**	MACHINE A ALESER EN CREUX	HOHLBOHRMASCHINE
1014	**BARREL FINISHING**	FINISSAGE AU TONNEAU	TROMMELN
1015	**BARREL OF A PUMP, PUMP BARREL**	CYLINDRE, CORPS DE POMPE	PUMPENZYLINDER
1016	**BARREL PLATING**	REVETEMENT GALVANIQUE AU TAMBOUR, GALVANOPLASTIE AU TONNEAU	ÜBERZUG IN FÄSSER (GALVANISCHER), TROMMELGALVANISIERUNG
1017	**BARREL-SHAPED ROLLER**	ROULEAU RENFLE	ROLLE (TONNENFÖRMIGE)
1018	**BARRING ENGINE**	SERVOMOTEUR DE LANCEMENT, SERVOMOTEUR DE DEMARRAGE	ANLASSMASCHINE, DREHMASCHINE, SCHALTMASCHINE
1019	**BARYTA WATER**	EAU DE BARYTE	BARYTWASSER
1020	**BARYTA, BARIUM MONOXIDE**	BARYTE, PROTOXYDE DE BARYUM	BARYT, BARIUMMONOXYD, SCHWERERDE
1021	**BARYTES, HEAVY SPAR**	BARYTINE, BARYTITE, SPATH PESANT	SCHWERSPAT, BARYT (SCHWEFELSAURER)
1022	**BASALT**	BASALTE	BASALT
1023	**BASALTIC TUFF**	BASALTIQUE	BASALTTUFF
1024	**BASE**	BASE (CHIM.)	BASE (CHEM.)
1025	**BASE**	BASE (D'UN SOLIDE)	GRUNDFLÄCHE
1026	**BASE**	BASE, SOCLE, FACE D'APPUI	BASE, UNTERBAU, GRUNDPLATTE, BERÜHRUNGSFLÄCHE

1027	**BASE**	BASE	GRUNDLINIE, BASIS (GEOM.)
1028	**BASE BOTTOM OF THREAD**	FOND, RACINE, BASE D'UN FILET	GRUND, BASIS EINES GEWINDES, GEWINDEBASIS, GEWINDEGRUND
1029	**BASE BULLION**	PLOMB IMPUR, PLOMB NON RAFFINE	BLEI (UNREINLICHES), BLEI UNRAFFINIERTES)
1030	**BASE CIRCLE, FUNDAMENTAL CIRCLE**	CERCLE DE BASE, CERCLE PRIMITIF	GRUNDKREIS
1031	**BASE METAL**	METAL NON PRECIEUX	METALL (UNEDLES)
1032	**BASE METAL**	METAL DE BASE	GRUNDMETALL
1033	**BASE OF A LOGARITHM**	BASE D'UN LOGARITHME	GRUNDZAHL, BASIS EINES LOGARITHMUS
1034	**BASE OF BEARING**	PATIN D'UN PALIER	LAGERFUSS
1035	**BASE OF COLUMN**	SOUBASSEMENT D'UNE COLONNE	FUSS EINER SÄULE, SÄULENFUSS
1036	**BASE OF TOOTH**	RACINE DE LA DENT	ZAHNWURZEL
1037	**BASE PLATE, BED PLATE, FOUNDATION PLATE**	PLAQUE DE FOND, PLAQUE DE FONDATION, PLAQUE D'ASSISE, PLAQUE DE BASE, SOCLE EN FONTE	GRUNDPLATTE EINER MASCHINE
1038	**BASE-FORMING ELEMENT**	ELEMENT BASIQUE	ELEMENT (BASENBILDENDES)
1039	**BASE-METAL TEST SPECIMEN**	EPROUVETTE DE METAL DE BASE	GRUNDMETALLPROBESTÜCK, PRÜFSTÜCK
1040	**BASE-PLATE**	EMBASE	UNTERPLATTE
1041	**BASE, FOOT**	PIED, SUPPORT	FUSSGESTELL
1042	**BASIC**	BASIQUE	BASISCH
1043	**BASIC BESSEMER PROCESS**	PROCEDE BASIQUE, PROCEDE THOMAS	THOMAS-VERFAHREN
1044	**BASIC BESSEMER STEEL**	ACIER THOMAS	THOMASSTAHL
1045	**BASIC BOTTOM & LINING**	SOLE ET GARNISSAGE BASIQUES	BODENSTEIN (BASISCHER) UND FUTTER (BASISCHES)
1046	**BASIC FLUX**	FLUX BASIQUE	FLUSS (BASISCHER)
1047	**BASIC OPEN HEARTH STEEL**	ACIER MARTIN PAR LE PROCEDE BASIQUE	SIEMENSMARTINSTAHL (BASISCHER)
1048	**BASIC OPEN-HEARTH STEEL**	ACIER SUR SOLE BASIQUE, ACIER MARTIN	SIEMENS-MARTIN-STAHL (BASISCHER)
1049	**BASIC PIG**	FONTE THOMAS	THOMASROHEISEN, ROHEISEN (BASISCHES)
1050	**BASIC PROCESS**	PROCEDE BASIQUE	VERFAHREN (BASISCHES)
1051	**BASIC REFRACTORIES**	REFRACTAIRES BASIQUES	OFENFUTTER (BASISCHES)
1052	**BASIC RESEARCH**	RECHERCHE PURE	GRUNDLAGENFORSCHUNG
1053	**BASIC SIZE**	COTE DE BASE	GRUNDMASS
1054	**BASIC SLAG**	SCORIES BASIQUES	THOMASSCHLACKE
1055	**BASICITY**	BASICITE	BASIZITÄT
1056	**BAST, PHLOEM**	LIBER	BAST
1057	**BASTARD FILE**	LIME A TAILLE BATARDE	BASTARDFEILE, VORFEILE
1058	**BASTARD SAWING OF TIMBER**	SCIAGE PARALLELE, SCIAGE EN LONG, DEBIT A LA SCIE DE LONG, GRAND DEBIT DU BOIS	SEHNENSCHNITT, FLADERSCHNITT, TANGENTIATLSCHNITT DES HOLZES
1059	**BATCH**	LOT DE COULEE	SATZ, GICHT, EINZELLOS
1060	**BATCH FURNACE**	FOUR A CHARGER, FOUR NON CONTINU	CHARGENOFEN, EÏNSATZOFEN
1061	**BATCHED GEAR**	ROCHET	KLINGWERK
1062	**BATH**	BAIN	BAD
1063	**BATH MIXING TAP**	ROBINET DE BAIGNOIRE	MISCHBATTERIE FÜR BADEWANNEN
1064	**BATH VOLTAGE**	COURANT DU BAIN	BADSTROM
1065	**BATTER**	INCLINAISON, OBLIQUITE	SCHRÄGE, NEIGUNG

1066	BATTERY	BATTERIE D'ACCUMULATEURS	BATTERIE
1067	BATTERY BOX	BAC DE BATTERIE	BATTERIEKASTEN
1068	BATTERY MAIN SWITCH	ROBINET DE BATTERIE	BATTERIESCHALTER
1069	BATTERY OF BOILERS	BATTERIE DE CHAUDIERES	KESSELBATTERIE
1070	BAUSCHINGER EFFECT	EFFET BAUSCHINGER	BAUSCHINGER-EFFEKT
1071	BAUXITE	BAUXITE	BAUXIT
1072	BAY	HALLE	HALLE
1073	BAYER PROCESS	PROCEDE BAYER	BAYER-VERFAHREN
1074	BAYONET JOINT	EMMANCHEMENT, FERMETURE A BAIONNETTE, BAIONNETTE	BAJONETTVERSCHLUSS
1075	BE IN COMPRESSION (TO), TENSION	TRAVAILLER A LA COMPRESSION, TRAVAILLER A LA TRACTION	BEANSPRUCHT WERDEN (AUF DRUCK), BEANSPRUCHT WERDEN (AUF ZUG)
1076	BE IN EQUILIBRIUM (TO)	EQUILIBRER (S')	AUFHEBEN (SICH), GLEICHGEWICHT SEIN (MIT)
1077	BE IN GEAR (TO), GEAR TOGETHER (TO), GEAR WITH (TO)...	ENGRENER ENSEMBLE, S'ENGRENER	INEINANDERGREIFEN, IM EINGRIFF STEHEN
1078	BE IN SHEAR (TO), BE SUBJECT TO SHEARING STRESS (TO)	CISAILLER, ETRE SOUMIS A UN EFFORT DE CISAILLEMENT, TRAVAILLER AU CISAILLEMENT	SCHUBBEANSPRUCHTWERDEN
1079	BE ON FULL LOAD (TO)	MARCHER A PLEINE CHARGE	VOLLBELASTET LAUFEN
1080	BE ON LIGHT LOAD (TO)	MARCHER A CHARGE INCOMPLETE	UNTERBELASTET LAUFEN
1081	BE ON NO-LOAD (TO)	MARCHER A VIDE	UNBELASTET, LEERLAUFEN
1082	BEAD	BOUDIN, CORDON DE SOUDURE	WULST, SCHWEISSRAUPE
1083	BEAD	BORD RABATTU, TOMBE	BÖRDEL, KREMPE
1084	BEAD (TO)	RABATTRE, TOMBER, RETROUSSER LES BORDS DES TOLES, RABATTRE LA COLLERETTE DES TUBES	KREMPEN, UMBÖRDELN
1085	BEAD WELD	SOUDURE LINEAIRE DE FRACTION	STRICHNAHT, ZUGNAHT
1086	BEADED IRON	FER POUR CLOTURE	GELÄNDER-EISEN
1087	BEADED TUBE	TUBE A COLLET RABATTU, TUBE A BORD RABATTU	BÖRDELROHR
1088	BEADING	RABATTEMENT, TOMBAGE, RETROUSSEMENT DES BORDS DES TOLES, RABATTEMENT DE LA COLLERETTE DES TUBES, BORDELAGE	KREMPEN, BÖRDELN, UMBÖRDELN, UMBÖRDELUNG
1089	BEADING MACHINE	MACHINE A FAIRE LES BOURRELETS	WULSTMASCHINE
1090	BEAK HORN	BIGORNE, CORNE D'ENCLUME	AMBOSSHORN
1091	BEAK IRON, BECK IRON	BIGORNE	SPERRHORN
1092	BEAKER	BECHER	BECHERGLAS
1093	BEAM	FAISCEAU, FAISCEAU ELECTRONIQUE, POUTRELLE, POUTRE	ELEKTRONENBÜNDEL, TRÄGER, BÜNDEL
1094	BEAM AND CRANK MECHANISM	MECANISME A MANIVELLE, DISPOSITIF BIELLE ET MANIVELLE	KURBELGETRIEBE
1095	BEAM BALANCE	BALANCE A FLEAU	BALKENWAAGE
1096	BEAM COMPASSES, TRAMMELS	COMPAS A VERGE	STANGENZIRKEL
1097	BEAM FIXED AT ONE END, FREELY SUPPORTED AT OTHER	POUTRE ENCASTREE A UNE EXTREMITE ET REPOSANT A L'AUTRE SUR UN APPUI	TRÄGER (HALBEINGESPANNTER, EINSEITIG EINGESPANNTER)
1098	BEAM SUPPORTED AT BOTH ENDS	POUTRE REPOSANT SOUTENUE LIBREMENT SUR DEUX APPUIS	TRÄGER (FREI AUFLIEGENDER)
1099	BEAM, GIRDER	POUTRE	TRÄGER (BAUW.)

4

1100	BEARING	COUSSINET, PALIER	LAGERSCHALE, LAGER, KOLBENSTANGENLAGER
1101	BEARING AREA	SURFACE PORTANTE	LAGEFLÄCHE, AUFLAGEFLÄCHE
1102	BEARING BALL	BILLE POUR ROULEMENTS	KUGEL EINES KUGELLAGERS
1103	BEARING BRACKET	CHAISE DE PALIER	LAGERBOCK
1104	BEARING CAP	CHAPEAU DE PALIER	LAGERDECKEL
1105	BEARING FRICTION	FROTTEMENT DANS LES PALIERS	LAGERREIBUNG
1106	BEARING METAL, BOX METAL	METAL POUR COUSSINET, COMPOSITION POUR COUSSINET, ALLIAGE POUR COUSSINET	LAGERMETALL
1107	BEARING NECK	PORTEE D'UN ARBRE	LAGERHALS EINER WELLE
1108	BEARING NEEDLES	AIGUILLES POUR ROULEMENTS	ROLLENLAGERNADELN
1109	BEARING PLATE, STRUCTURAL	TOLE DE CONSTRUCTION	BAUSTAHLBLECH
1110	BEARING PRESSURE, REACTION	PRESSION SUR LES SURFACES D'APPUI, REACTION DES APPUIS	AUFLAGERDRUCK, FLÄCHENDRUCK, STÜTZDRUCK. LAGERDRUCK, AUFLAGEPRESSUNG ; ACHSDRUCK
1111	BEARING RACE	CHEMIN DE ROULEMENT	LAUFRING
1112	BEARING ROLLER	ROULEAU, GALET POUR ROULEMENTS	ROLLE EINES ROLLENLAGERS
1113	BEARING SHELL	COQUILLE DE COUSSINET	LAGERSCHALE
1114	BEARING SPRING	RESSORT DE SUSPENSION	TRAGFEDER
1115	BEARING SURFACE	SURFACE PORTANTE, SURFACE D'APPUI	FLÄCHE (TRAGENDE), TRAGFLÄCHE, AUFLAGEFLÄCHE
1116	BEARING SURFACE FOR ROTATING SHAFT	SURFACE DE PORTEE D'UN TOURILLON, PORTEE D'UN MANETON	LAUFFLÄCHE, AUFLAGERFLÄCHE EINES ZAPFENS, ZAPFENLAUFFLÄCHE
1117	BEARING, SHAFT BEARING	PALIER, ROULEMENT	LAGER, WELLENLAGER, ZAPFENLAGER
1118	BEARING, SUPPORT	APPUI, SUPPORT (CONSTR.)	LAGER, LAGERUNG, AUFLAGER
1119	BEAT	BATTEMENT	SCHWEBUNG
1120	BEATING	BATTEMENT DES METAUX	METALLHÄMMERN
1121	BEAVER, THREADER	FILIERE A TUBES	GEWINDEKLUPPE
1122	BECOME CONCENTRATED (TO)	CONCENTRER (SE)	ANREICHERN (SICH)
1123	BECOME DEMAGNETISED (TO), RETURN TO THE UNMAGNETISED STATE (TO)	DESAIMANTER (SE)	UNMAGNETISCH WERDEN
1124	BECOME ELECTRIC (TO)	ELECTRISER (S')	ELEKTRISCH WERDEN
1125	BECOME LENGTHENED (TO), STRETCH (TO)	ALLONGER (S')	LÄNGEN (SICH)
1126	BECOME MAGNETIC (TO)	AIMANTER (S')	MAGNETISCH WERDEN
1127	BED PLATE	PLAQUE D'APPUI, SEMELLE	AUFLAGERPLATTE (BAUW.), GRUNDPLATTE
1128	BEEF TALLOW	SUIF DES BOEUFS	RINDSTALG
1129	BEESWAX	CIRE D'ABEILLES	BIENENWACHS
1130	BEILBY LAYER	COUCHE DE BEILBY	BEILBY-SCHICHT
1131	BELL CRANK LEVER, BENT LEVER	LEVIER COUDE, LEVIER A SONNETTE, EQUERRE	WINKELHEBEL
1132	BELL FURNAGE	FOUR A CLOCHE	HAUBENOFEN
1133	BELL HOUSING	CARTER D'EMBRAYAGE	KUPPLUNGSGEHÄUSE
1134	BELL METAL	BRONZE DE CLOCHE	GLOCKENMETALL, GLOCKENSPEISE, GLOCKENBRONZE
1135	BELL TYPE FURNACE	FOUR A CLOCHE	HAUBENOFEN
1136	BELL VALVE	SOUPAPE A CLOCHE, SOUPAPE DE CORNOUAILLES	GLOCKENVENTIL, KRONENVENTIL
1137	BELL, ALARM BELL	SONNERIE, TIMBRE-AVERTISSEUR	KLINGELVORRICHTUNG, LÄUTWERK

1138	BELLOW VALVE	ROBINET A SOUFFLET	VENTIL MIT FALTENBALG-ABDICHTUNG, BLASEBALGHAHN
1139	BELLOWS	SOUFFLET, SOUFFLET DE FORGE	FALTENBALG, BLASEBALG
1140	BELT	COURROIE	RIEMEN
1141	BELT CONVEYOR LOADING	CHARGEMENT PAR CONVOYEUR A BANDE	FÖRDERBANDBELADUNG
1142	BELT COUPLING, UNITING THE JOINTS OF BELTS	JONCTIONNEMENT DES COURROIES, LIAISONS DE COURROIE, REUNION DES COURROIE ATTACHE DES COURROIES	RIEMENVERBINDUNG, VERBINDEN DER RIEMENENDEN
1143	BELT CUT ALONG THE SPINE	OUTILLAGE EN ECHINE	RIEMEN AUS WIRBELBAHNEN
1144	BELT DRIVE OR GEARING	TRANSMISSION PAR COURROIES, DISPOSITIF POULIE ET COURROIE	RIEMENTRIEB, RIEMENANTRIEB
1145	BELT DRIVEN MACHINE	MACHINE A COMMANDE PAR COURROIE	MASCHINE FÜR RIEMENANTRIEB
1146	BELT FASTENER	AGRAFE POUR COURROIES	RIEMENVERBINDER, RIEMENSCHLOSS, RIEMENKLAMMER
1147	BELT FLYWHEEL, FLYWHEEL PULLEY	VOLANT-POULIE, POULIE-VOLANT	RIEMENSCHEIBENSCHWUNGRAD
1148	BELT FURNACE	FOUR A BANDE	FÖRDERBANDOFEN
1149	BELT GEARING FOR SKEW, NON-PARALLEL SHAFTS	RENVOI D'ANGLE	WINKELRIEMENGETRIEBE
1150	BELT GUIDE	GUIDE-COURROIE	RIEMENFÜHRER
1151	BELT JOINT	JONCTION, REUNION, ATTACHE DES COURROIES	RIEMENVERBINDUNG
1152	BELT PULLEY	POULIE A COURROIE	RIEMENSCHEIBE
1153	BELT PUNCH	EMPORTE-PIECE POUR COURROIE	RIEMENLOCHZANGE
1154	BELT REVERSING GEAR, REVERSING GEAR OPERATED WITH BELTS	DISPOSITIF DE CHANGEMENT DE SENS DE MARCHE PAR COURROIE	RIEMENWENDEGETRIEBE
1155	BELT RIVET	RIVET POUR COURROIES	RIEMENNIET
1156	BELT SCREW	BOULON POUR COURROIES	RIEMENSCHRAUBE
1157	BELT SHIFTER, BELT STRIKING GEAR	MECANISME DE DEBRAYAGE DE LA COURROIE	RIEMENAUSRÜCKER, RIEMENSCHALTER
1158	BELT SHIFTING	DEPLACEMENT DE LA COURROIE, PASSAGE DE LA COURROIE D'UNE POULIE SUR L'AUTRE	RIEMENVERSCHIEBUNG, RIEMENSCHALTUNG
1159	BELT SHIPPER	PASSE-COURROIE, MONTE-COURROIE, PORTE-COURROIE	RIEMENAUFLEGER
1160	BELT SHIPPING	MISE EN PLACE, MONTAGE DE LA COURROIE	AUFLEGEN DES RIEMENS
1161	BELT STRETCHER	TENDEUR DE COURROIES	RIEMENSPANNER, SPANNROLLE
1162	BELT TENSION	TENSION DE LA COURROIE	RIEMENSPANNUNG
1163	BELT; STRAP	COURROIE MOTRICE, DE COMMANDE	RIEMEN, TREIBRIEMEN
1164	BELT, BAND CONVEYOR	TRANSPORTEUR A COURROIE, TRANSPORTEUR A TAPIS ROULANT, TOILE TRANSPORTEUSE	GURTFÖRDERER, FÖRDERGURT, FÖRDERBAND, LAUFENDES BAND, TRANSPORTBAND, BANDTRANSPORTEUR, TRAINEUR
1165	BELTING (THICKNESS OF)	COURROIE (EPAISSEUR D'UNE)	RIEMENS (SCHICHT EINES)
1166	BELTING OF SO MANY PLIES	COURROIE A X PLIS, COURROIE EN X EPAISSEURS	RIEMEN (X-FACHER), RIEMEN AUS X LAGEN
1167	BENCH	ETABLI, BANC	BANK, WERKBANK
1168	BENCH DRILLING MACHINE	PERCEUSE D'ETABLI	TISCHBOHRMASCHINE, WERKBANKBOHRMASCHINE, BANKBOHRMASCHINE
1169	BENCH MOLDING	MOULAGE A LA TABLE	BANKFORMUNG
1170	BENCH TEST	ESSAI AU BANC	PRÜFSTANDVERSUCH
1171	BENCH VISE	ETAU D'ETABLI	BANKSCHRAUBSTOCK

/

1172 **BEND**	COUDE	ROHRBOGEN
1173 **BEND (TO)**	PLIER, CINTRER, COURBER	BIEGEN
1174 **BEND (TO), DEFLECT (TO)**	FLECHIR	DURCHBIEGEN (SICH)
1175 **BEND A PIPE (TO)**	CINTRER UN TUYAU, COUDER UN TUYAU, COURBER UN TUYAU	BIEGEN (EIN ROHR)
1176 **BEND RADIUS**	RAYON DE COURBURE	BIEGERADIUS
1177 **BEND TEST**	ESSAI DE FLEXION ESSAI DE PLIAGE	BIEGEPROBE
1178 **BEND TEST, FACE**	ESSAI DE FLEXION DE LA FACE DE LA SOUDURE	BIEGEPROBE AN DER OBERFLÄCHE DER SCHWEISSTELLE
1179 **BEND TEST, ROOT**	ESSAI DE PLIAGE A L'ENVERS	BIEGEVERSUCH MIT DER WURZEL IN DER ZUGZONE
1180 **BEND, PIPE BEND**	COUDE ROND	KRÜMMER, ROHRKRÜMMER, BOGENSTÜCK, BOGENROHR
1181 **BENDER**	PLIEUSE	BIEGEGLIED
1182 **BENDING**	PLIAGE, CINTRAGE, COURBAGE, FLEXION	BIEGEN, BIEGUNG
1183 **BENDING BRAKE**	JOUE DE CINTRAGE	BIEGEBACKE
1184 **BENDING MACHINE**	MACHINE A CINTRER	BIEGEMASCHINE, BIEGEPRESSE
1185 **BENDING MOMENT**	MOMENT FLECHISSANT, MOMENT DE FLEXION	BIEGEMOMENT, BIEGUNGSMOMENT
1186 **BENDING ROLLS**	MACHINE A ROULER, CYLINDRE A CINTRER, CYLINDRE DE CINTRAGE	WALZENBIEGEMASCHINE, BIEGEWALZWERK, RUNDMASCHINE, BIEGEWALZE, BIEGUNGSWALZE
1187 **BENDING STRAIN**	EFFORT DE FLEXION	BIEGEBEANSPRUCHUNG
1188 **BENDING STRENGTH**	RESISTANCE AU PLIAGE, RESISTANCE AU CINTRAGE	BIEGEFESTIGKEIT
1189 **BENDING TEST**	ESSAI A LA FLEXION, ESSAI DE PLIAGE, ESSAI DE CINTRAGE	BIEGEVERSUCH, BIEGEPROBE
1190 **BENDING TEST IMPACT TEST ON NOTCHED BAR, NICK BEND TEST**	ESSAI DE FLEXION PAR CHOC SUR BARREAUX ENTAILLES, ESSAI DE CHOC SUR ENTAILLE	EINKERBBIEGEPROBE, KERBSCHLAGPROBE, SCHLAGBIEGEPROBE MIT EINGEKERBTEN PROBESTÜCKEN
1191 **BENDING TRANSVERSE STRAIN**	EFFORT DE FLEXION, EFFORT TRANSVERSAL, TRAVAIL A LA FLEXION	BEANSPRUCHUNG AUF BIEGUNG, BIEGEBEANSPRUCHUNG
1192 **BENDING TRANSVERSE STRENGTH**	RESISTANCE A LA FLEXION TRANSVERSALE	BIEGEFESTIGKEIT
1193 **BENDING TRANVERSE STRESS**	EFFORT DE FLEXION PAR UNITE DE SECTION	BIEGESPANNUNG
1194 **BENT PIPE**	TUYAU CINTRE, COUDE, COL DE CYGNE	ROHR (GEKRÜMMTES)
1195 **BENT SPANNER, SKEW SPANNER**	CLEF COUDEE	WENDESCHLÜSSEL, SCHRAUBENSCHLÜSSEL MIT SCHRÄGEM MAUL
1196 **BENT WOOD**	BOIS BOUGE	HOLZ (GEBOGENES)
1197 **BENT, OUT OF TRUTH**	FAUSSE	VERBOGEN
1198 **BENZALDEHYDE, OIL OF BITTER ALMONDS**	BENZALDEHYDE, ALDEHYDE BENZYLIQUE, ESSENCE D'AMANDES AMERES	BENZALDEHYD, BENZOYLWASSERSTOFF, BITTERMANDELÖL
1199 **BENZENE**	BENZINE, BENZENE	BENZOL, STEINKOHLENBENZIN
1200 **BENZENE SULPHONIC ACID**	ACIDE BENZOSULFONIQUE, ACIDE PHENYLSULFUREUX	BENZOLSULFOSÄURE
1201 **BENZIDINE (PARADIAMIDODIPHENYL)**	BENZIDINE	BENZIDIN, P-DIAMIDODIPHENYL
1202 **BENZOIC ACID**	ACIDE BENZOIQUE	BENZOESÄURE
1203 **BENZOIN**	BENJOIN	BENZOEHARZ
1204 **BENZOLE**	BENZOL	ROHBENZOL
1205 **BENZOPHENONE**	BENZOPHENONE	BENZOPHENON

1206	BERLIN BLUE, PRUSSIAN BLUE, FERRIC FERROCYANIDE	BLEU DE PRUSSE, BLEU DE BERLIN	BLAU (BERLINER), PREUSSISCHBLAU, FERRIFERROZYANID
1207	BERYLLIUM	BERYLLIUM	BERYLLIUM
1208	BERYLLIUM BRONZE	BRONZE AU BERYLLIUM	BERYLLIUMBRONZE
1209	BERYLLIUM COPPER	CUPRO-BERYLLIUM	BERYLLIUMBRONZE
1210	BERYLLIUM, GLUCINUM	GLUCINIUM	BERYLLIUM
1211	BESSEMER AFTER-BLOW	SURSOUFFLAGE BESSEMER	BESSEMER-NACHBLASEN
1212	BESSEMER CONVERTER	CONVERTISSEUR BESSEMER	BESSEMER-BIRNE, BESSEMER-KONVERTER, BESSEMER-OFEN
1213	BESSEMER PIG	FONTE BESSEMER	BESSEMER-ROHEISEN
1214	BESSEMER PROCESS	PROCEDE BESSEMER	BESSEMER-VERFAHREN
1215	BESSEMER STEEL	ACIER BESSEMER	BESSEMER-STAHL
1216	BETA BRASS	LAITON BETA	BETAMESSING
1217	BETA PARTICLE	PARTICULE BETA	BETATEILCHEN
1218	BETA RAYS	RAYONS BETA	BETASTRAHLEN
1219	BETA STRUCTURE	STRUCTURE BETA	BETASTRUKTUR
1220	BETTS PROCESS	PROCEDE BETTS	BETTS-VERFAHREN
1221	BEVEL	BIAIS, BISEAU	ABSCHRÄGUNG
1222	BEVEL (TO)	BISEAUTER, TAILLER EN BIAIS, TAILLER EN SIFFLET	ABSCHRÄGEN
1223	BEVEL ANGLE	ANGLE DU CHANFREIN, BIAIS	KANTENABSCHRÄGWINKEL, ABSCHRÄGUNGSWINKEL
1224	BEVEL FRICTION GEAR	TRANSMISSION PAR CONES DE FRICTION	REIBKEGELGETRIEBE
1225	BEVEL FRICTION GEAR WHEEL	ROUE DE FRICTION CONIQUE, CONE DE FRICTION	REIBKEGELRAD
1226	BEVEL GEAR	PIGNON CONIQUE	KEGELRAD
1227	BEVEL GEARING	ENGRENAGE CONIQUE, ENGRENAGE D'ANGLE	KEGELRADGETRIEBE, WINKELRÄDERGETRIEBE
1228	BEVEL SQUARE, ANGLE BEVEL	FAUSSE EQUERRE, SAUTERELLE, ANGLOIR	SCHMIEGE, SCHRÄGMASS, SCHRÄGMESSER, SCHRÄGWINKEL, STELLWINKEL
1229	BEVEL WHEEL	ROUE CONIQUE, ROUE D'ANGLE	KEGELRAD, RAD (KONISCHES), WINKELRAD
1230	BEVEL WHEEL DRIVE	COMMANDE PAR ENGRENAGE CONIQUE	KEGELRADANTRIEB
1231	BEVELLING	CHANFREINAGE, BISEAUTAGE	ABSCHRÄGUNG, ABSCHRÄGEN
1232	BEZEL	BAGUE PERIPHERIQUE	DECKRING
1233	BI-POLAR ELECTRODE	ELECTRODE BIPOLAIRE	ELEKTRODE (ZWEIPOLIGE)
1234	BIB COCK	ROBINET A BEC COURBE	HAHN MIT ABLAUF, UNTERLAUFHAHN, ZAPFHAHN
1235	BIB COCK, FAUCET	ROBINET DE PUISAGE	AUSLAUFVENTIL
1236	BICHROMATE DIP FINISH	TRAITEMENT AU BICHROMATE	BICHROMATBEHANDLUNG
1237	BICONCAVE DOUBLE-CONCAVE LENS, CONCAVO-CONCAVE LENS	LENTILLE BICONCAVE	LINSE (BIKONKAVE)
1238	BICONVEX DOUBLE-CONVEX LENS, CONVEXO-CONVEX LENS	LENTILLE BICONVEXE	LINSE (BIKONVEXE)
1239	BICYCLE CHAIN	CHAINE DE BICYCLETTE	FAHRRADKETTE
1240	BIFILAR SUSPENSION	SUSPENSION BIFILAIRE	AUFHÄNGUNG (BIFILARE)
1241	BIFURCATED RIVETS	RIVETS	ZWEIKNOPFNIETSTIFTE
1242	BIG END BRASSES, CRANK PIN BEARING	COUSSINET DE TETE DE BIELLE	KURBELZAPFENLAGER
1243	BILLET	BILLE, BILLETTE, BUCHE	KNÜPPEL, SCHEIT, HOLZSCHEIT
1244	BILLET AND SHEET SHEARING MACHINES	CISAILLES POUR BILLETTES ET LARGETS	KNÜPPEL U. PLATINENSCHERE

1245	BILLET MILL	LAMINOIR A BILLETTES	KNÜPPELWALZWERK
1246	BILLET SHEARS	CISAILLES POUR BILLETTES	KNÜPPELSCHERE
1247	BINARY ALLOY	ALLIAGE BINAIRE	LEGIERUNG (BINÄRE)
1248	BINARY DECIMALCODE	CODE BINAIRE DECIMAL	BINÄR-DEZIMALCODE
1249	BINARY DIGIT OR BIT	BIT, POSITION BINAIRE	BIT, BINÄRSTELLE, BINÄRZEICHEN
1250	BINDER	LIANT, LUBRIFIANT	PRESSMITTEL, PRESSZUSATZ, BINDEMITTEL
1251	BINDERS AND HARVESTERS	MOISSONNEUSES ET M. LIEUSES	MÄHBINDER UND ERNTEMASCHINEN
1252	BINDING HOOP	FRETTE	SCHRUMPFRING
1253	BINDING MATERIAL, CEMENT	SUBSTANCE AGGLUTINANTE, MATIERE AGGLUTINANTE, AGGLOMERANT, LIANT, LEIN (GEOL.): CIMENT	BINDEMITTEL
1254	BINDING POST	BORNE DE RACCORDEMENT	ANSCHLUSSKLEMME
1255	BINDING WIRE	FIL DE LIGATURE	BINDEDRAHT, WICKELDRAHT
1256	BINOCULAR MICROSCOPE	MICROSCOPE BINOCULAIRE	MIKROSKOP (BINOKULARES)
1257	BINOMIAL COEFFICIENT	COEFFICIENT BINOMIAL	BINOMIALKOEFFIZIENT
1258	BINOMIAL DISTRIBUTION	LOI BINOMIALE	BINOMIALVERTEILUNG
1259	BINOMIAL EQUATION	EQUATION BINOME	GLEICHUNG (ZWEIGLIEDRIGE), GLEICHUNG (BINOMISCHE)
1260	BINOMIAL SERIES	SERIE BINOMIALE	REIHE (BINOMISCHE), BINOMIALREIHE
1261	BIPLANE	BIPLAN	DOPPELDECKER
1262	BIQUADRATIC EQUATION	EQUATION BIQUADRATIQUE, EQUATION BICARREE, EQUATION DU QUATRIEME DEGRE	GLEICHUNG VIERTEN GRADES, GLEICHUNG (BIQUADRATISCHE)
1263	BIRD-EYE	MICROFISSURE	MIKRORISS, MIKROBRUCH
1264	BIRD'S EVE VIEW	PERSPECTIVE, VUE A VOL D'OISEAU	VOGELPERSPEKTIVE
1265	BIREFRINGENT	BIREFRINGENT	DOPPELBRECHEND
1266	BISECT (TO)	PARTAGER, DIVISER EN DEUX, BISSECTER	HALBIEREN (MATH.)
1267	BISECTION	BISSECTION	HALBIEREN (MATH.)
1268	BISECTRIX	BISSECTRICE	WINKELHALBIERENDE, HALBIERUNGSLINIE EINES WINKELS
1269	BISMUTH	BISMUTH	WISMUT
1270	BISMUTH CHLORIDE	TRICHLORURE DE BISMUTH	WISMUTCHLORID, CHLORWISMUT
1271	BISMUTH NITRATE	AZOTATE, NITRATE DE BISMUTH	WISMUT (SALPETERSAURES), WISMUTNITRAT
1272	BISMUTH OXIDE	OXYDE DE BISMUTH	WISMUTSESQUIOXYD
1273	BISMUTH OXYNITRATE	AZOTATE DE BISMUTHYLE	WISMUTNITRAT (BASISCHES), WISMUTSUBNITRAT
1274	BISMUTH SOLDER	SOUDURE AU BISMUTH	WISMUTLOT
1275	BISMUTHINE, BISMUTH TRISULPHIDE	BISMUTHINE	WISMUTGLANZ, BISMUTIN
1276	BISQUE, BISCUIT	BISCUIT	BISKUIT, BISKUITGUT
1277	BITE ANGLE	ANGLE D'ATTAQUE	GREIFWINKEL
1278	BITE OF THE ROPE	COINCEMENT DE LA CORDE DANS LA GORGE	KLEMMEN DES SEILES IN DER RILLE
1279	BITTINESS	GRANULATION, GRUMELAGE	KORNBINDUNG
1280	BITUMEN	BRAI GRAS NATUREL	ASPHALT-TEER, ASPHALT-GOUDRON, BITUMEN
1281	BITUMINOUS	BITUMINEUX	BITUMENHALTIG, BITUMINÖS
1282	BITUMINOUS COAL, SOFT FAT COAL	HOUILLE GRASSE	FETTKOHLE
1283	BITUMINOUS PAINT VARNISH, TAR VARNISH	VERNIS AU BITUME, VERNIS JAPON	ASPHALTLACK

1284	BLACK ANNEALED IRON WIRE	FIL DE FER RECUIT NOIR	EISENDRAHT (SCHWARZGEGLÜHTER)
1285	BLACK BODY	CORPS NOIR	KÖRPER (SCHWARZER)
1286	BLACK BOLT	BOULON BRUT	BOLZEN (ROHER), SCHRAUBE (UNBEARBEITETE), SCHRAUBE SCHWARZE
1287	BLACK COPPER	CUIVRE NOIR BRUT	SCHWARZKUPFER, ROHKUPFER
1288	BLACK HEAT (OF A)	PORTE AU ROUGE NAISSANT	SCHWARZROTGLÜHEND
1289	BLACK HOOP AND STRIP	BANDES NOIRES	SCHWARZBLECHE
1290	BLACK NUT	ECROU BRUT DE FORGE	MUTTER (ROHE), MUTTER (SCHWARZE), MUTTER (UNBEARBEITETE)
1291	BLACK OIL	HUILE MINERALE BRUNE	MINERALÖL UNGEREINIGTES
1292	BLACK PIG IRON	FONTE GRAPHITIQUE	ROHEISEN (SCHWARZES)
1293	BLACK PINE	PIN NOIR D'AUTRICHE	SCHWARZKIEFER
1294	BLACK PLATE	TOLE NOIRE	BLECH (SCHWARZES)
1295	BLACK RED HEAT	ROUGE NAISSANT	SCHWARZROTGLUT
1296	BLACK SHEET	TOLE NOIRE, TOLE TERNE, FER NOIR	SCHWARZBLECH, BLECH (GLATTES)
1297	BLACKING	NOIR DE FONDERIE	FORMSCHWÄRZE
1298	BLACKSMITH'S TONGS, SMITH'S TONGS, PLIERS	TENAILLE, PINCE DE FORGERON	SCHMIEDEZANGE
1299	BLADE	LAME	KLINGE, BLATT
1300	BLADE OF A CUTTING TOOL	LAME D'UN OUTIL TRANCHANT, LAME TRANCHANTE	KLINGE EINES SCHNEIDWERKZEUGS
1301	BLANK	LOPIN, PIECE BRUTE, EBAUCHE	BLANKETT, ROHLING
1302	BLANK BOLT	BOULON NON FILETE	BOLZEN OHNE GEWINDE
1303	BLANK END OF PIPE	BOUT DE TUBE FERME	ROHRENDE (BLINDES)
1304	BLANK FLANGE	BRIDE DE RECOUVREMENT, BRIDE PLEINE, BRIDE OBTURATRICE, FAUSSE BRIDE	BLINDFLANSCH, DECKELFLANSCH, FLANSCHENDECKEL
1305	BLANK HOLDER	SUPPORT DE PIECE A ESTAMPER	GEGENHALTER
1306	BLANK LINER	COLONNE PERDUE NON-PERFOREE	ROHR (NICHTGELOCHTES), ROHR (VERLORENES)
1307	BLANK NUT	ECROU CARRE A SOUDER	VIERKANTMUTTER (SCHWEISSBARE)
1308	BLANK NUT, NUT BLANK	ECROU NON TARAUDE	MUTTER OHNE GEWINDE
1309	BLANKING	COUPE	SCHNEIDEARBEIT
1310	BLANKING DIE	MATRICE A DISQUE	PLATINENSCHNITT
1311	BLAST FURNACE	HAUT FOURNEAU	HOCHOFEN
1312	BLAST FURNACE SLAG, CINDER, SCORIA	SCORIE DES HAUTS-FOURNEAUX, LAITIER	HOCHOFENSCHLACKE
1313	BLAST FURNACE WASTE GAS	GAZ DE HAUT FOURNEAU	GICHTGAS, HOCHOFENGAS
1314	BLAST MAIN	CONDUITE DE VENT	WINDLEITUNG
1315	BLAST, AIR BLAST	COURANT D'AIR FORCE, VENT	GEBLÄSEWIND, WIND
1316	BLASTING	SABLAGE	SANDSTRAHLEN
1317	BLEACHING POWDER, CHLORIDE OF LIME	CHLORURE DE CHAUX	CHLORKALK, BLEICHKALK, BLEICHPULVER
1318	BLEEDER	PURGEUR	ABLASSHAHN
1319	BLEEDER SCREW	VIS DE PURGE	ENTLÜFTUNGSSCHRAUBE
1320	BLEEDER VENT	EVENT AUTOMATIQUE	LUFTABZUG (AUTOMATISCHER)
1321	BLEEDING	MIGRATION, DEGORGEMENT, SAIGNEE	AUSBLUTEN, DURCHSCHLAGEN
1322	BLEEDING OF THE MOLM	RETRAIT AU MOULE	FORMVERSATZ
1323	BLEND (TO)	MELANGER, DOSER	MISCHEN
1324	BLENDE, ZINC BLENDE, SPHALERITE	BLENDE, SPHALERITE, FAUSSE GALENE	ZINKBLENDE, BLENDE, SPHALERIT

1325	**BLENDING**	MELANGE, DOSAGE, MELANGE DE DIVERSES FRACTIONS DE POUDRES D'UNE MEME SUBSTANCE	MISCHEN, VERSCHNEIDEN, MISCHUNG, BLENDUNG
1326	**BLIND FLANGE**	BRIDE PLEINE, TAMPON OBTURATEUR	BLINDFLANSCH
1327	**BLISTER**	POQUETTE, PUSTULE	WANZE
1328	**BLISTER BAR**	BARRE EN ACIER CEMENTE	ZEMENTSTAHLSTAB
1329	**BLISTER COPPER**	CUIVRE A SOUFFLURES	ROHKUPFER (KUPFER, BLASIGES)
1330	**BLISTERING**	CLOQUAGE, SOUFFLURES, PUSTULES, BOURSOUFLURES	BLASENBILDUNG, BLASEN
1331	**BLOATING**	GONFLEMENT, MOUSSAGE	AUFBLÄHEN
1332	**BLOCK**	BLOC (D'INFORMATIONS)	BLOCK, SATZ,
1333	**BLOCK**	CHAPE DE POULIE	FLASCHE, KLOBEN, ROLLENKLOBEN, SCHERE
1334	**BLOCK ADDRESS FORMAT**	FORMAT A ADRESSES DE BLOCS	SATZADRESSEEINGABEFORMAT
1335	**BLOCK BRAKE**	FREIN A SABOT	KLOTZBREMSE
1336	**BLOCK CHAIN**	CHAINE PLATE	BLOCKKETTE
1337	**BLOCK DELETE**	SAUT DE BLOC, SUPPRESSION	SATZUNTERDRÜCKUNG
1338			
1339	**BLOCK TIN**	ETAIN EN SAUMONS	BLOCKZINN
1340	**BLOCK, CLICHE**	CLICHE TYPOGRAPHIQUE	DRUCKSTOCK, KLISCHEE
1341	**BLOCKING**	EBAUCHAGE, PREFORMAGE	VORFORMUNG
1342	**BLOCKING CONDENSER**	CONDENSATEUR DE BLOCAGE, CONDENSATEUR D'ARRET	VERBLOCKUNGSKONDENSATOR, SPERRKONDENSATOR
1343	**BLOCKING DIE**	EBAUCHE DE FORGEAGE PREALABLE	VORSCHMIEDEGESENK
1344	**BLONDIN**	BLONDIN	KABELKRAN
1345	**BLOOM**	LOUPE (MET.)	LUPPE, PUDDELLUPPE
1346	**BLOOM**	BLOOM	VORBLOCK, VORGEWALZTER BLOCK
1347	**BLOOMING**	VOILE, LOUCHE	HAUCHBILDUNG, SCHLEIERBILDUNG
1348	**BLOOMING MILL**	TRAIN EBAUCHEUR, LAMINOIR A BLOOMS	VORWALZWERK, VORBLOCKWALZWERK
1349	**BLOOMS**	EBAUCHES	ROHLINGE, LUPPEN
1350	**BLOTTING PAPER**	PAPIER BUVARD, BUVARD	LÖSCHPAPIER, FLIESSPAPIER
1351	**BLOW**	SOUFFLURE	BLASE, OBERFLÄCHENPORE
1352	**BLOW (TO), FUSE (TO)**	FONDRE (LE COUPE-CIRCUIT FOND)	DURCHBRENNEN (SICHERUNG)
1353	**BLOW HOLE**	SOUFFLURE, POCHE DE RETRAITE	GASBLASE, BLASIGE STELLE IM GUSS, GUSSBLASE, GASEINSCHLUSS, BLASE
1354	**BLOW OFF**	DECOLLEMENT	ABLÖSUNG
1355	**BLOW PIPE**	CHALUMEAU	LÖTROHR
1356	**BLOW PIPE TEST ANALYSIS**	ANALYSE AU CHALUMEAU	LÖTROHRPROBE
1357	**BLOW-OFF COCK, SLUDGE COCK**	ROBINET DE VIDANGE	ABBLASEHAHN, AUSBLASEHAHN, AUSLAUFHAHN, SCHLAMMHAHN
1358	**BLOW-OFF VALVE**	SOUPAPE DE VIDANGE, SOUPAPE DE DECHARGE, SOUPAPE D'EVACUATION	ABBLASEVENTIL, AUSBLASEVENTIL, ABLASSVENTIL, SCHLAMMVENTIL
1359	**BLOW-OFF VALVE**	VANNE D'EXTRACTION, ROBINET D'EXTRACTION	ZAHNSTANGENSCHIEBER, ABSCHLAMMVENTIL
1360	**BLOW-OUT PREVENTER**	VANNE D'ERUPTION	ERUPTIONSABSPERRVORRICHTUNG
1361	**BLOWER**	SOUFFLERIE, APPAREIL SOUFFLANT	GEBLÄSE
1362	**BLOWHOLES**	SOUFFLURES	BLASEN, LUNKER
1363	**BLOWING ENGINE**	MACHINE SOUFFLANTE	GEBLÄSEMASCHINE
1364	**BLOWING OFF A BOILER**	VIDANGE D'UNE CHAUDIERE	ABBLASEN EINES KESSELS

1365	**BLOWN GLASS**	VERRE SOUFFLE	GLAS (GEBLASENES)
1366	**BLOWN METAL**	METAL PREAFFINE	VORMETALL
1367	**BLOWPIPE**	CONDUITE DE VENT	WINDLEITUNG
1368	**BLUE (TO) THE STEEL**	BLEUIR L'ACIER	STAHL (DEN) BLAU ANLAUFENLASSEN
1369	**BLUE DIP**	BAIN BLEU, BAIN DE CHLORURE DE MERCURE	BLAUBRENNE, QUECKSILBERCHLORIDBAD
1370	**BLUE HEAT**	BLEU (CHAUDE)	BLAUGLUT
1371	**BLUE HEAT (OF A)**	CHAUFFE AU BLEU	BLAUWARM
1372	**BLUE HEAT TEST**	ESSAI DE RESISTANCE AU CHAUD BLEU	BLAUBRUCHPROBE
1373	**BLUE LINE, BLACK LINE PHOTOTYPE**	REPRODUCTION AU FERROPRUSSIATE EN BLEU SUR FOND BLANC	WEISSPAUSE, LICHTPAUSE (POSITIVE)
1374	**BLUE POWDER**	POUDRE DE ZINC OXYDE	ZINKSTAUB
1375	**BLUE PRINT, WHITE LINE PHOTOTYPE, CYANOTYPE**	BLEU, REPRODUCTION AU FERRO- PRUSSIATE EN BLANC SUR FOND BLEU	BLAUPAUSE, LICHTPAUSE (NEGATIVE)
1376	**BLUE VITRIOL, CHALCANTHITE**	VITRIOL BLEU, BLEU DE CHYPRE, COUPEROSE BLEUE	KUPFERVITRIOL, VITRIOL (BLAUER)
1377	**BLUING**	BLEUISSAGE	BLAUUNG
1378	**BLUSHING**	VOILE, LOUCHE	WEISSANLAUFEN
1379	**BOARD**	PLANCHE MINCE	BRETT
1380	**BOARD DROP HAMMER**	MARTEAU-PELIN A PLANCHE, MOUTON A PLANCHE	BRETTFALLHAMMER
1381	**BOARD, MILLBOARD**	CARTON	PAPPE, PAPPDECKEL
1382	**BOAT NAIL**	CLOU A BATEAU	SCHIFFSNAGEL
1383	**BOATSWAIN CHAIR**	NACELLE, SIEGE SUSPENDU	HÄNGESITZ
1384	**BOBBIN**	BOBINE	SPULE
1385	**BODY**	CORPS, CARROSSERIE	KÖRPER, KAROSSERIE, GEHÄUSE
1386	**BODY AT REST**	CORPS AU REPOS	KÖRPER (RUHENDER)
1387	**BODY CLEARANCE**	DETALONNAGE	HINTERFRÄSEN
1388	**BODY CLEARANCE ANGLE**	ANGLE DE DETALONNAGE DU CORPS	HINTERFRÄSWINKEL
1389	**BODY COLOUR, OPAQUE COLOUR**	COULEUR OPAQUE	FARBE (DECKENDE), DECKFARBE
1390	**BODY FILE**	LIME-FRAISE	FRÄSVORRICHTUNG
1391	**BODY FORMING MACHINE**	MACHINE A COURBER	RUNDMASCHINE
1392	**BODY IN MOTION, MOVING BODY**	CORPS EN MOUVEMENT, CORPS MOBILE	KÖRPER (BEWEGTER)
1393	**BODY OF A PIGMENT**	PROPRIETE COUVRANTE, PROPRIETE DE COUVRIR D'UNE COULEUR	DECKKRAFT EINER FARBE
1394	**BODY OF A SCREW**	NOYAU D'UNE VIS	KERN EINER SCHRAUBE, SCHRAUBENKERN
1395	**BODY OF AN OIL**	CONSISTANCE D'UNE HUILE	KONSISTENZ EINES ÖLES
1396	**BODY OF UNIFORM STRENGTH**	PRISME D'EGALE RESISTANCE	KÖRPER VON GLEICHEM WIDERSTAND
1397	**BODY SEAT RING**	SIEGE DE CORPS DE VANNE	VENTILKÖRPERSITZ
1398	**BODY SHOP**	ATELIER DE CARROSSERIE, ATELIER DE TOLERIE	KAROSSERIEWERKSTATT
1399	**BODY-CENTERED**	CENTRE CUBIQUE	KUBISCHRAUMZENTRIERT
1400	**BOHEMIAN GLASS**	VERRE BLANC DE BOHEME	GLAS (BÖHMISCHES), KALIKALKGLAS
1401	**BOIL (TO)**	BOUILLIR, ETRE EN EBULLITION	KOCHEN, SIEDEN
1402	**BOIL-OFF RATE**	TAUX D'EVAPORATION	VERDUNSTUNGSHÖHE
1403	**BOIL-OFF TEST**	ESSAI DE DEPERDITION THERMIQUE	WÄRMEVERLUSTPRÜFUNG

/

1404	**BOIL-OUT**	VAPEUR D'EBULLITION, BUEE	SIEDEDAMPF, DUNST
1405	**BOIL-OUT NOZZLE**	EVACUATION DE LA VAPEUR D'EBULLITION	SIEDEDAMPFABLASSDÜSE
1406	**BOILER**	CHAUDIERE	KOCHKESSEL, KOCHER
1407	**BOILER BODY OF A STILL**	CUCURBITE	DESTILLIERBLASE
1408	**BOILER COATING**	MATIERE CALORIFUGE POUR CHAUDIERES	KESSELABDECKUNGSMATERIAL, KESSELISOLIERMATERIAL
1409	**BOILER DRILLING MACHINE**	PERCEUSE DE CHAUDIERES	KESSELBOHRMASCHINE
1410	**BOILER EXPLOSION, BURSTING OF A BOILER**	EXPLOSION D'UNE CHAUDIERE	BERSTEN, EXPLOSION EINES KESSELS, KESSELEXPLOSION
1411	**BOILER FITTINGS, BOILER MOUNTINGS**	GARNITURES DE CHAUDIERES, APPAREILS ACCESSOIRES, ACCESSOIRES DE CHAUDIERES	KESSELAUSRÜSTUNG, KESSELZUBEHÖR, KESSELARMATUR
1412	**BOILER FLUE**	TUBE FOYER, CARNEAU INTERIEUR D'UNE CHAUDIERE	FLAMMROHR
1413	**BOILER HOUSE**	BATIMENT DES CHAUDIERES, BATIMENT DE GENERATEUR	KESSELHAUS
1414	**BOILER INSPECTION**	INSPECTION, VISITE DE LA CHAUDIERE	KESSELUNTERSUCHUNG
1415	**BOILER MAKER, BOILER SMITH**	CHAUDRONNIER EN TOLE, CHAUDRONNIER EN FER	KESSELSCHMIED
1416	**BOILER PLATE**	TOLE DE CHAUDIERE	KESSELBLECH
1417	**BOILER PRESSURE**	PRESSION DANS UNE CHAUDIERE	KESSELDRUCK
1418	**BOILER ROOM**	SALLE DES CHAUDIERES, SALLE DE CHAUFFERIE, CHAUFFERIE	KESSELRAUM
1419	**BOILER SHELL**	CORPS CYLINDRIQUE D'UNE CHAUDIERE	KESSELMANTEL
1420	**BOILER SHOP SMITHY**	ATELIER DE CHAUDRONNERIE EN FER	KESSELSCHMIEDE
1421	**BOILER TEST, PRESSURE TEST OF BOILER**	ESSAI DE LA CHAUDIERE SOUS PRESSION	KESSELDRUCKPROBE
1422	**BOILER TUBE**	TUBE DE CHAUDIERE	DAMPFKESSELROHR, KESSELROHR
1423	**BOILER WORKS**	GROSSE CHAUDRONNERIE, CHAUDRONNERIE EN FER	KESSELFABRIK
1424	**BOILING POINT**	POINT D'EBULLITION	SIEDEPUNKT
1425	**BOILING POINT (WITH A HIGT)**	EBULLITION ELEVE (A POINT D')	SCHWERSIEDEND
1426	**BOILING POINT (WITH A LOW)**	EBULLITION BAS (A POINT D')	LEICHTSIEDEND
1427	**BOILING TEMPERATURE**	TEMPERATURE D'EBULLITION	SIEDEHITZE, SIEDETEMPERATUR
1428	**BOILING WATER**	EAU BOUILLANTE	WASSER (KOCHENDES), WASSER (SIEDENDES)
1429	**BOILING, EBULLITION**	EBULLITION	SIEDEN, KOCHEN
1430	**BOLE**	BOL	BOLUS
1431	**BOLOMETER**	BOLOMETRE	BOLOMETER, STRAHLENMESSER
1432	**BOLT**	VERROU, BOULON	RIEGEL, QUERRIEGEL, SCHRAUBE UND MUTTER, BOLZEN
1433	**BOLT CIRCLE**	CERCLE DES TROUS DE BOULONS	LOCHKREIS, SCHRAUBENKREIS
1434	**BOLT DIAMETER**	DIAMETRE DU BOULON	BOLZENDURCHMESSER
1435	**BOLT DOWN (TO)**	BOULONNER	VERBOLZEN
1436	**BOLT HEAD, SCREW HEAD**	TETE D'UNE VIS, TETE D'UN BOULON	KOPF EINER SCHRAUBE, SCHRAUBENKOPF
1437	**BOLT HOLE**	TROU DE BOULON	SCHRAUBENLOCH
1438	**BOLT-CIRCLE DIAMETER**	DIAMETRE DE PERCAGE	LOCHKREUZDURCHMESSER
1439	**BOLT-DRILLING DIAMETER**	BRIDE-DIAMETRE DE PERCAGE	BOHRDURCHMESSER
1440	**BOLT-HOLE DIAMETER**	DIAMETRE DU TROU DE BOULON	BOLZENBOHRUNGSDURCHMESSER

1441	BOLT, SCREW BOLT, BOLT AND NUT	BOULON LIBRE	MUTTERSCHRAUBE, DURCHSTECKSCHRAUBE,SCHRAUBE (DURCHGEHENDE)
1442	BOLTED CONNECTION	ASSEMBLAGE PAR VIS, RACCORD A ECROUS	SCHRAUBENVERBINDUNG
1443	BOLTING	BOULONNERIE, BOULONNAGE	VERBOLZUNG
1444	BOND	APPAREIL DE CONSTRUCTON	MAUERVERBAND, VERBAND
1445	BOND	AGGLUTINANT, LIANT	GRÜNSANDBINDER, BINDEMITTEL
1446	BONDED TENDONS	CABLES INJECTES	GLIEDER (VORGERSPANNTE)
1447	BONE BLACK	NOIR ANIMAL, NOIR D'IVOIRE	BEINSCHWARZ
1448	BONE DUST, MEAL, CRUSHED BONE	OS PULVERISES, POUDRE D'OS	KNOCHENMEHL
1449	BONE GLUE	COLLE FORTE DES OS, OSSEINE	KNOCHENLEIM
1450	BONE OIL; NEAT'S FOOT OIL	HUILE DE PIED DE BOEUF, HUILE DE PIED DE MOUTON	KLAUENFETT, KLAUENÖL ; KNOCHENÖL
1451	BONNET	CAPOT, CAPOT-MOTEUR, CHAPEAU	MOTORHAUBE, HAUBE
1452	BONNET FLANGE	BRIDE DE CHAPEAU	ÜBERWURFFLANSCH, HAUBENFLANSCH
1453	BONNET STUD BOLT	TIGE FILETEE DE CHAPEAU	HAUBENGEWINDEBOLZEN
1454	BONNET, CAP	CHAPEAU	AUFSATZ
1455	BONNET, HOOD	HOTTE	RAUCHFANG
1456	BOOSTER	SURPRESSEUR, SURCOMPRESSEUR, SURVOLTEUR, AMORCE	ÜBERVERDICHTER, ZUSATZDYNAMO, ZUSATZMASCHINE
1457	BOOSTER BRAKE	SERVO-FREIN	SERVO-BREMSE
1458	BOOT	COFFRE, MALLE	KOFFERRAUM
1459	BORAX, SODIUM BIBORATE, TINCAL	BORAX, BIBORATE DE SODIUM, BORATE DE SOUDE ANHYDRE	BORAX, NATRON (BORSAURES), NATRIUMBIBORAT
1460	BORDER CURVE	COURBE LIMITE	GRENZKURVE
1461	BORE	ALESAGE, ALESAGE DE CYLINDRE	BOHRUNG, ZYLINDER
1462	BORE (TO)	ALESER	AUSBOHREN
1463	BORE HOLE	TROU DE SONDAGE	BOHRLOCH (IM GESTEIN)
1464	BORE INTERNAL DIAMETER OF A PIPE	DIAMETRE INTERIEUR D'UN TUYAU	LICHTE WEITE EINES ROHRES, ROHRWEITE
1465	BORE LINER	CHEMISE DE CYLINDRE	ZYLINDERLAUFBÜCHSE
1466	BORE ON THE LATHE (TO)	ALESER AU TOUR	AUSDREHEN
1467	BORIC (BORACIC) ACID	ACIDE BORIQUE	BORSÄURE, BORAXSÄURE
1468	BORING	FORAGE, PERCAGE, ALESAGE	BOHREN, AUSBOHREN
1469	BORING AND TURNING MILL, VERTICAL LATHE	TOUR EN L'AIR A PLATEAU HORIZONTAL	PLANDREHBANK MIT WAAGERECHTER PLANSCHEIBE, KARUSSELLDREHBANK
1470	BORING BARS	BARRES DE FORAGE	BOHRSTANGEN
1471	BORING MACHINE	MACHINE A ALESER, ALESEUSE	AUSBOHRMASCHINE, BOHRMASCHINE
1472	BORING MILL	PERCEUSE	BOHRWERK
1473	BORING ON THE LATHE	ALESAGE AU TOUR	AUSDREHEN
1474	BORING RESTS	LUNETTES D'ALESAGE	BOHR-LÜNETTEN
1475	BORING TEST, DRILL TEST	ESSAI DE FORAGE, ESSAI DE PERCAGE	BOHRVERSUCH
1476	BORINGS	COPEAUX DE FORAGE	BOHRSPÄNE
1477	BORON	BORE	BOR
1478	BORON ALLOYS	ALLIAGE AU BORE	BORLEGIERUNG
1479	BORON CARBIDE	CARBURE DE BORE	BORKARBID
1480	BORT	BORT	BORT
1481	BOSS, NAVE, HUB	MOYEU	NABE
1482	BOSSING	BATTRE EN FORME	KLOPFEN (IN DIE FORM)

1483	**BOTT**	TAMPON BOUCHON	STOPFEN
1484	**BOTTLE GLASS**	VERRE A BOUTEILLES	FLASCHENGLAS
1485	**BOTTLE JACK**	VERIN A BOUTEILLE	FLASCHENWINDE
1486	**BOTTOM BOARD**	PLAQUE DE FOND	GRUNDPLATTE
1487	**BOTTOM BRASS**	DEMI-COUSSINET INFERIEUR, COQUILLE INFERIEURE	LAGERSCHALE (UNTERE), UNTERSCHALE
1488	**BOTTOM CASTING**	COULEE EN SOURCE	GIESSEN (STEIGENDES)
1489	**BOTTOM CLEARANCE**	JEU A FOND DES DENTS, JEU DU FOND DE LA DENT	KOPFSPIEL EINES ZAHNRADES
1490	**BOTTOM CURB ANGLE**	CORNIERE DE PIED DE BAC	KESSELBODENWINKELEISEN
1491	**BOTTOM DEAD CENTER**	POINT MORT BAS	TOTPUNKT (UNTER)
1492	**BOTTOM DRAIN**	BONDE DE FOND	BODENABLAUF
1493	**BOTTOM DRAIN VALVE**	ROBINET DE FOND DE CUVE	BODENVENTIL
1494	**BOTTOM END OF CYLINDER, CYLINDER BOTTOM**	FOND DU CYLINDRE	ZYLINDERBODEN
1495	**BOTTOM NOZZLE**	TUBULURE DE FOND	BODENDÜSE
1496	**BOTTOM POURING LADLE**	POCHE A QUENOUILLE	STOPFENPFANNE
1497	**BOTTOM SWAGE**	ETAMPE, MATRICE DE DESSOUS	UNTERGESENK
1498	**BOTTOM VIEW**	VUE DE BAS EN HAUT, VUE PAR-DESSOUS	FROSCHPERSPEKTIVE
1499	**BOTTOMING TAP**	TARAUD FINISSEUR	FERTIGSCHNEIDER
1500	**BOUNDARY**	JOINT, CONTOUR DE GRAIN	KRONGRENZE
1501	**BOURDON PRESSURE GAUGE**	MANOMETRE A TUBE, MANOMETRE BOURDON	RÖHRENFEDERMANOMETER
1502	**BOURNONITE, ENDELLIONITE, WHEEL ORE**	BOURNONITE	RÄDELERZ, BOURNONIT
1503	**BOW COMPASSES, SPRING BOWS, COMPASSES**	COMPAS-BALUSTRE	FEDERZIRKEL
1504	**BOW DRILL, FIDDLE DRILL**	ARCHET	ROLLENBOHRER
1505	**BOW SAW, TURNING SAW**	SCIE A CHANTOURNER	SCHWEIFSÄGE
1506	**BOWING**	MANQUE DE PLANEITE	BALLIGKEIT
1507	**BOX**	BUIS	BUCHSBAUM
1508	**BOX CASTING**	MOULAGE EN CHASSIS	KASTENGUSS
1509	**BOX CONNECTING ROD END**	TETE DE BIELLE A CAGE FERMEE	SCHUBSTANGENKOPF (GESCHLOSSENER)
1510	**BOX COUPLING, MUFF COUPLING, BUTT COUPLING**	ACCOUPLEMENT A MANCHON	MUFFENKUPPLUNG
1511	**BOX FRAME**	CHASSIS-CAISSON	KASTENRAHMEN
1512	**BOX NUT, CAP NUT**	ECROU A CHAPEAU, ECROU BORGNE, ECROU CREUX	MUTTER (GESCHLOSSENE), ÜBERWURFMUTTER, BÜCHSENMUTTER, KAPPENMUTTER
1513	**BOX PIN**	GOUJON DE CENTRAGE	ZENTRIERSTIFT, FÜHRUNGSSTIFT
1514	**BOX PISTON**	PISTON EVIDE	HOHLKOLBEN
1515	**BOX SPANNER**	CLEF A DOUILLE, CLEF TUBULAIRE, CLEF EN BOUT	STECKSCHLÜSSEL
1516	**BRACE**	RENFORT, ENTRETOISE (CHARPENTE), BRACON (ELEMENT DE CHARPENTE METALLIQUE)	QUERVERSTREBUNG, VERSTREBUNG, STREBE
1517	**BRACE (TO)**	METTRE DES CONTREFICHES	VERSTREBEN
1518	**BRACE WITH SCREW**	VILEBREQUIN A VIS DE PRESSION	BOHRKURBEL
1519	**BRACE, BREAST BRACE**	VILEBREQUIN (OUTIL A PERCER)	BRUSTLEIER, BOHRWINDE, FAUSTLEIER
1520	**BRACES**	TIRANTS (DE SPHERE), CROISILLONS DE CHARPENTE, BRACONS DE CHARPENTE	KREUZSTREBEN, VERANKERUNG

4

1521	**BRACING**	MISE DE CONTREFICHES	VERSTREBEN
1522	**BRACKET**	ATTACHE, CONSOLE, SUPPORT	HALTERUNG, KONSOLE, TRÄGER, STÜTZE, TRAGSTÜTZE
1523	**BRACKET PEDESTAL, WALL BEARING, WALL HANGER**	PALIER MURAL, PALIER CONSOLE	WANDLAGER, KONSOLLAGER
1524	**BRACKISH WATER**	EAU SAUMATRE	BRACKWASSER
1525	**BRAGG'S METHOD**	METHODE DE BRAGG, LOI DE BRAGG	GESETZ (BRAGGSCHES), GESETZ VON BRAGG
1526	**BRAIDED COVERING**	GARNITURE EN TRESSE	GEFLECHTÜBERZUG
1527	**BRAIDED ROPE**	CORDE TRESSEE	SEIL (GEFLOCHTENES)
1528	**BRAILED WIRE**	FIL TRESSE	DRAHT (GEFLOCHTENER), DRAHT (UMKLÖPPELTER), DRAHT UMFLOCHTENER
1529	**BRAKE**	FREIN, JOUE DE CINTRAGE	BREMSE, BIEGEBACKE
1530	**BRAKE (TO)**	FREINER	BREMSEN
1531	**BRAKE BLOCK**	SABOT DE FREIN	BREMSKLOTZ
1532	**BRAKE CYLINDER, BRAKE DRUM**	TAMBOUR, CYLINDRE A FREIN	BREMSZYLINDER, BREMSTROMMEL
1533	**BRAKE DRUM**	TAMBOUR DE FREIN	BREMSTROMMEL
1534	**BRAKE FLUID TANK**	RESERVOIR DE LIQUIDE DE FREIN	BREMSFLÜSSIGKEITSBEHÄLTER
1535	**BRAKE HORSE POWER**	PUISSANCE AU FREIN	BREMSLEISTUNG
1536	**BRAKE JAW**	MACHOIRE DE FREIN	BREMSBACKE
1537	**BRAKE LEVER**	LEVIER DE FREIN	BREMSHEBEL
1538	**BRAKE LINING**	GARNITURE DE FREIN	BREMSBELAG
1539	**BRAKE LINKAGE**	TRINGLERIE DE FREIN	BREMSGESTÄNGE
1540	**BRAKE POWER**	PUISSANCE EFFECTIVE, PUISSANCE AU FREIN	WIRKLICHLEISTUNG, BREMSLEISTUNG
1541	**BRAKE RING**	BAGUE DE FREINAGE	BREMSRING
1542	**BRAKE SHOE**	SEGMENT DE FREIN	BREMSBACKE
1543	**BRAKE SHOE, BRAKE BLOCK**	SABOT DE FREIN	BREMSKLOTZ, BREMSBACKE
1544	**BRAKE STRAP**	BANDE DE FREIN, RUBAN DE FREIN, COLLIER DE FREIN	BREMSBAND
1545	**BRAKE SYSTEM**	CIRCUIT DE FREINAGE	BREMSKREIS
1546	**BRAKE TEST**	ESSAI AU FREIN	BREMSVERSUCH
1547	**BRAKE WEIGHT**	CONTREPOIDS DU FREIN	BREMSGEWICHT
1548	**BRAKE WHEEL**	POULIE DE FREIN	BREMSSCHEIBE
1549	**BRAKING**	FREINAGE, SERRAGE DU FREIN, MANOEUVRE DU FREIN	BREMSEN
1550	**BRAKING ACTION**	ACTION DU FREIN	BREMSWIRKUNG
1551	**BRAKING FORCE**	PUISSANCE DE FREINAGE, EFFORT TANGENTIEL DU FREIN	BREMSKRAFT
1552	**BRAKING RESISTANCE**	RESISTANCE AU FREINAGE	BREMSWIDERSTAND
1553	**BRALE**	PRESSE BRINELL	BRINELL-PRESSE, BRINELL-HÄRTEPRÜFER
1554	**BRANCH CONNECTION**	DERIVATION, BRANCHEMENT	ROHRANSCHLUSS
1555	**BRANCH OF A CURVE**	BRANCHE D'UNE COURBE	AST, ZWEIG, ZUG EINER KURVE
1556	**BRANCH OFF (TO)**	SE BRANCHER	ABZWEIGEN
1557	**BRANCH PIECE**	TUBULURE DE BRANCHEMENT	STUTZEN, ROHRSTUTZEN, ABZWEIGSTUTZEN, RÖHRANSATZ, ANSCHLUSSSTUTZEN
1558	**BRANCH PIPE**	TUBE DE DERIVATION	ABZWEIGROHR, ABZWEIG
1559	**BRANCH TEE**	RACCORD T	T-STUCK, T-VERSCHRAUBUNG
1560	**BRAND, BRANDING LETTER, STAMP**	ALPHABET A FRAPPER, CHIFFRES ET LETTRES A CHAUD/FROID	STEMPEL, SCHRIFTSTEMPEL
1561	**BRASILIAN SPLITTING TEST**	ESSAI D'ECRASEMENT SUR CYLINDRE DE BETON MASSIF	ZYLINDERSPALTVERSUCH

1562	BRASS	LAITON, CUIVRE JAUNE	MESSING, GELBKUPFER, GELBGUSS
1563	BRASS BILLET	BILLETTE DE LAITON	MESSINGKNÜPPEL
1564	BRASS BRIGHT DIP	BAIN DE BRILLANTAGE	GLÄNZBAD (CHEMISCHES)
1565	BRASS BUSH OF STUFFING BOX	BAGUE DE FOND	GRUNDBÜCHSE
1566	BRASS FOUNDRY	FONDERIE DE CUIVRE, ROBINETTERIE	MESSINGGIESSEREI
1567	BRASS PLATING	PLACAGE AU LAITON	VERMESSINGUNG
1568	BRASS ROD WIRE	FIL LAMINE DE LAITON, FIL MACHINE DE LAITON	MESSINGWALZDRAHT
1569	BRASS SHEET, STRIP	TOLE DE LAITON	MESSINGBLECH
1570	BRASS TUBE	TUBE EN LAITON	MESSINGROHR
1571	BRASS WIRE	FIL DE LAITON, FIL D'ARCHAL	MESSINGDRAHT
1572	BRASSES	COUSSINET EN DEUX PIECES, COUSSINET EN COQUILLES	LAGERSCHALE (GETEILTE)
1573	BRAZE (TO)	SOUDER FORT, BRASER	HARTLÖTEN
1574	BRAZED-ON TIPS	MISES BRASEES	SPITZEN (AUFGELÖTETE)
1575	BRAZIERS' COPPER	CUIVRE DE BRASAGE	HARTLÖTKUPFER
1576	BRAZING	BRASAGE, BRASEMENT, SOUDURE FORTE, BRASURE	HARTLÖTEN, HARTLÖTUNG
1577	BRAZING FURNAGE	FOURNEAU A BRASER	HARTLÖTOFEN
1578	BRAZING INSERTS	INSERTIONS DE BRASAGE	HARTLÖTUNGSEINLAGEN
1579	BREADTH OF TOOTH	LONGUEUR/LARGEUR DE LA DENT	ZAHNBREITE
1580	BREAK (TO), CRUSH (TO), STAMP (TO)	CONCASSER	ZERKLEINERN
1581	BREAKDOWN	PERTURBATION, DERANGEMENT DANS LE SERVICE, PANNE	BETRIEBSSTÖRUNG, PANNE, STÖRUNG
1582	BREAKDOWN VOLTAGE	TENSION DE CLAQUAGE	DURCHBRUCHSPANNUNG
1583	BREAKING DOWN	EBAUCHAGE	VORARBEIT, VORWALZEN
1584	BREAKING LENGTH	LONGUEUR DE RUPTURE	REISSLÄNGE
1585	BREAKING LOAD	CHARGE DE RUPTURE	BRUCHLAST
1586	BREAKING WEIGHT LOAD	CHARGE-LIMITE D'ELASTICITE	BRUCHBELASTUNG
1587	BREAKING-IN	RODAGE (D'UNE VOITURE)	EINFAHRZEIT
1588	BREAKING, CRUSHING, STAMPING	CONCASSAGE	ZERKLEINERN, ZERKLEINERUNG
1589	BREAKOUT	PERCEE	DURCHBRUCH
1590	BREAST PLATE	PLAQUE DE CONSCIENCE (DE VILLEBREQUIN), PLASTRON	BOHRBRETT, BRUSTBRETT, DRILLBRETT, BRUSTSCHEIBE
1591	BREATHER	RENIFLARD	ENTLÜFTERROHR
1592	BREATHER TANK	RESERVOIR TAMPON	ZWISCHENLAGERBEHÄLTER
1593	BREATHER VALVE, RELIEF VALVE, SAFETY VALVE	SOUPAPE DE SECURITE, RESPIRATION RENIFLARD	SICHERHEITSVENTIL, SCHNÜFFELVENTIL
1594	BREATHING LOSS	PERTE PAR RESPIRATION	ATMUNGSVERLUST
1595	BREECHES PIPE	CULOTTE	HOSENROHR, GABELROHR
1596	BRICK	BRIQUE	BACKSTEIN, MAUERSTEIN, ZIEGELSTEIN, MAUERZIEGEL
1597	BRICK (TO)	MACONNER	MAUERN
1598	BRICK EARTH	TERRE A BRIQUES	ZIEGELERDE
1599	BRICK FOUNDATION	FONDATION EN BRIQUES, FONDATION EN MACONNERIE	ZIEGELSTEINFUNDAMENT
1600	BRICKED IN, SET IN MASONRY	ENGAGE DANS UNE MACONNERIE	EINGEMAUERT
1601	BRICKLAYING	MACONNERIE	MAUERN
1602	BRICKWORK	MACONNERIE EN BRIQUES, BRIQUETAGE	ZIEGELMAUERWERK
1603	BRIDGE APPROACH	APPONTEMENT	ZUFAHRT

1604	**BRIDGING**	COURONNEMENT, FORMATION DE PONT	SCHLACKENKRANZBILDUNG, BRÜCKENBILDUNG
1605	**BRIGG'S PIPE THREAD, AMERICAN STANDARD PIPE THREAD**	PAS SYSTEME BRIGGS POUR TUBES	BRIGGS'SCHES GEWINDE
1606	**BRIGHT ANNEALED WIRE**	FIL RECUIT BLANC	BLANKGLÜHDRAHT
1607	**BRIGHT BOLT**	BOULON DECOLLETE, VIS DECOLLETEE	SCHRAUBENBOLZEN (BLANKER), SCHRAUBENBOLZEN (BEARBEITETER)
1608	**BRIGHT DIP**	BAIN DE BRILLANTAGE	GELBBRENNE
1609	**BRIGHT DIP FINISH**	BAIN DE BRILLANTAGE	GLANZBRENNE
1610	**BRIGHT HARD DRAWN WIRE**	FIL DE FER CLAIR CRU	(BLANKGEZOGENER) EISENDRAHT
1611	**BRIGHT METALLIC SURFACE**	SURFACE METALLIQUE POLIE	METALLFLÄCHE (BLANKE)
1612	**BRIGHT NUT**	ECROU DECOLLETE	MUTTER (BLANKE), MUTTER (BEARBEITETE)
1613	**BRIGHT RED HEAT**	ROUGE CLAIR, ROUGE VIF	HELLROTGLUT
1614	**BRIGHT RED HOT**	CHAUFFE AU ROUGE CLAIR	HELLROTGLÜHEND
1615	**BRIGHTENER**	AGENT DE BRILLANTAGE	GLANZMITTEL, GLANZZUSATZ
1616	**BRIGHTNESS, BRILLIANCY**	ECLAT LUMINEUX	HELLIGKEIT
1617	**BRINE**	EAU SALEE, SAUMURE	SOLE, SALZSOLE
1618	**BRINELL BALL TEST**	ESSAI PAR EMPREINTE DE BILLE, ESSAI PAR LA BILLE DE BRINELL, BILLAGE	KUGELDRUCKVERSUCH NACH BRINELL, EINDRUCKVERSUCH, BRINELLPROBE
1619	**BRINELL HARDNESS NUMBER**	CHIFFRE DE DURETE BRINELL, DURETE BRINELL	KUGELDRUCKHÄRTE, BRINELLHÄRTE
1620	**BRINELL TEST**	ESSAI BRINELL	KUGELDRUCKVERSUCH, BRINELL'SCHER
1621	**BRINGING A TOOL INTO POSITION**	MISE AU POINT D'UN OUTIL	ANSETZEN, ANSTELLEN EINES WERKZEUGES
1622	**BRIQUETTE**	BRIQUETTE DE FERRO ALLIAGE	FERROLEGIERUNGSBRIKETT
1623	**BRIQUETTE, PATENT FUEL**	AGGLOMERE, BRIQUETTE	BRIKETT, PRESSLING
1624	**BRITANNIA METAL**	METAL BRITANNIQUE	BRITANNIAMETALL
1625	**BRITISH THERMAL UNIT**	CALORIE ANGLAISE	KALORIE (ENGLISCHE)
1626	**BRITISH THERMAL UNIT (B.TH.U.)**	UNITE THERMIQUE ANGLAISE	WÄRMEEINHEIT (BRITISCHE)
1627	**BRITTLE**	CASSANT	BRÜCHIG, SPRÖDE
1628	**BRITTLENESS**	FRAGILITE, MANQUE DE SOUPLESSE	ZERBRECHLICHKEIT, SPRÖDIGKEIT
1629	**BRITTLENESS OF IRON**	FRAGILITE DU FER	SPRÖDIGKEIT DES EISENS
1630	**BRITTLENESS, ACID**	FRAGILITE PAR DECAPAGE	BEIZUNGSVERSPRÖDIGKEIT
1631	**BRITTLENESS, BLUE**	FRAGILITE AU BLEU	BLAUBRÜCHIGKEIT
1632	**BRITTLENESS, NOTCH**	FRAGILITE D'ENTAILLE	KERBSPRÖDIGKEIT
1633	**BRITTLENESS, RHEOTROPIC**	FRAGILITE RHEOTROPIQUE	RHEOTROPE SPRÖDIGKEIT
1634	**BRITTLENESS, TEMPER**	FRAGILITE DE REVENU	ANLASSPRÖDIGKEIT
1635	**BROACH (TO), REAM (TO), OPEN OUT THE RIVET HOLES (TO)**	ALESER LES TROUS DE RIVET	AUFREISSEN (DIE NIETLÖCHER)
1636	**BROACH, REAMER, RYMER, RIMER, RHYMER**	ALESOIR, EQUARRISSOIR	REIBAHLE, RÄUMER, AUSREIBER, AUFREIBER
1637	**BROACHING**	BROCHAGE	RÄUMEN
1638	**BROACHING, REAMING, OPENING OUT THE RIVET HOLES**	ALESAGE DES TROUS DE RIVET	AUFREIBEN DER NIETLÖCHER
1639	**BROAD FLANGE GIRDER**	POUTRE A LARGES AILES	BREITFLANSCHTRÄGER
1640	**BROAD FLANGE TEE IRON**	FER EN T A LARGES AILES	T-EISEN (BREITFÜSSIGES)
1641	**BROAD FLANGED BEAMS**	POUTRELLES A LARGES AILES	BREITFLANSCHTRÄGER
1642	**BROADSIDE MILL**	LAMINOIR TRANSVERSAL	QUERWALZWERK
1643	**BROADSIDE ROLLS**	CYLINDRES TRANSVERSAUX	QUERWALZEN
1644	**BROKEN STONE**	PIERRES CONCASSEES, PIERRAILLE	SCHOTTER, STEINSCHLAG, KLEINSCHLAG

1645	BROMARGYRITE, SILVER BROMIDE	BROMARGYRITE, BROMITE, BROMARGURE, BROMYRITE, ARGENT BROMURE, ARGENT VERT	BROMSILBER, SILBERBROMID, BROMIT, BROMARGYRIT
1646	BROMATE	BROMATE	BROMAT
1647	BROMIC ACID	ACIDE BROMIQUE	BROMSÄURE
1648	BROMIDE	BROMURE	BROMMETALL, BROMID
1649	BROMINE	BROME	BROM
1650	BRONZE	BRONZE	BRONZE
1651	BRONZE (TO)	BRONZER	BRONZIEREN
1652	BRONZE POWDER	POUDRE DE BRONZE, BRONZE EN POUDRE	BRONZEPULVER
1653	BRONZE TUBE	TUBE EN BRONZE	BRONZEROHR
1654	BRONZE WIRE	FIL DE BRONZE	BRONZEDRAHT
1655	BRONZE, ACID	BRONZE RESISTANT AUX ACIDES	BRONZE (SÄUREBESTÄNDIGE)
1656	BRONZE, ALPHA	BRONZE ALPHA	ALPHA-BRONZE
1657	BRONZE, COMMERCIAL	BRONZE DE QUALITE COMMERCIALE, BRONZE DU COMMERCE	HANDELSGÜTEBRONZE
1658	BRONZE, GUN METAL	BRONZE	ROTGUSS
1659	BRONZE, HARDWARE	BRONZE POUR VISSERIE	SCHRAUBENBRONZE
1660	BRONZE, PLASTIC	BRONZE PLASTIQUE, BRONZE DE COUSSINET	LAGERBRONZE
1661	BRONZE, SPRING	BRONZE DE RESSORT	FEDERBRONZE
1662	BRONZE, STATUARY	BRONZE STATUAIRE	BILDHAVERBRONZE
1663	BRONZE, TRIM	BRONZE ORNEMENTAL	ZIERBRONZE
1664	BRONZING	BRONZAGE	BRONZIEREN
1665	BROWN COAL	LIGNITE PARFAIT	BRAUNKOHLE
1666	BROWN COAL BRIQUETTE	AGGLOMERE DE LIGNITE	BRAUNKOHLENBRIKETT
1667	BROWN COAL TAR	GOUDRON DE LIGNITE	BRAUNKOHLENTEER
1668	BROWN COAL TAR OIL	HUILE DE GOUDRON DE LIGNITE	BRAUNKOHLENTEERÖL, PARAFFINGASÖL, DEUTSCHES GASÖL, ROTÖL, GELBÖL
1669	BRUSH COPPER	CUIVRE A BALAI	BÜRSTENKUPFERBLECH
1670	BUBBLE OF AIR, AIR BUBBLE	BULLE D'AIR	LUFTBLASE
1671	BUBBLE OF GAS, GAS BUBBLE	BULLE DE GAZ	GASBLASE
1672	BUBBLING	BULLAGE	BLASENBILDUNG
1673	BUCKET	PISTON A CLAPET, PISTON ELEVATOIRE	VENTILKOLBEN
1674	BUCKET	SEAU, GODET	EIMER
1675	BUCKET CONVEYOR	CONVOYEUR, TRANSPORTEUR A GODETS	FÖRDERKETTE, BECHERKETTE, BECHERKABEL, KONVEYOR
1676	BUCKET ELEVATOR	ELEVATEUR A GODETS, NORIA	BECHERWERK, SCHÖPFWERK, PATERNOSTERWERK
1677	BUCKET OF A WATER WHEEL	AUBE, PALETTE D'UNE ROUE HYDRAULIQUE	SCHAUFEL EINES WASSERRADES
1678	BUCKET SEAT	SIEGE BAQUET	KÜBELSITZ
1679	BUCKLED PLATE	TOLE BOMBEE, TOLE EMBOUTIE	BUCKELPLATTE, TROGBLECH
1680	BUCKLES	PLIS	FALTEN
1681	BUCKLING	LAMBAGE (DE TOLES)	KNICKUNG
1682	BUCKLING STRENGTH	RESISTANCE AU FLAMBAGE	KNICKFESTIGKEIT
1683	BUCKRAKES	RATEAUX-AMMEULONNEURS	SCHOBERRECHEN
1684	BUFF WHEEL	DISQUE EN BUFFLE	LEDERPOLIERSCHEIBE
1685	BUFFALO HIDE	PEAU DE BUFFLE, BUFFLE	BÜFFELLEDER

1686	BUFFER	TAMPON, BUTOIR, POLISSOIR	PUFFER, BUFFER, POLIERMASCHINE
1687	BUFFER BATTERY	BATTERIE-TAMPON	PUFFERBATTERIE
1688	BUFFER SPRING	RESSORT AMORTISSEUR DES CHOCS	PUFFERFEDER
1689	BUFFER STORAGE	MEMOIRE INTERMEDIAIRE	PUFFER-SPEICHER
1690	BUFFERS	SELS DE POLISSAGE	SCHWABBELSALZE
1691	BUFFING	EMEULAGE	SCHWABBELN
1692	BUFFING COMPOUND	COMPOSITION DE POLISSAGE A LA MEULE	SCHWABBELMITTEL
1693	BUFFING WHEEL	DISQUE POLISSEUR	POLIERSCHEIBE, SCHWABBELSCHEIBE
1694	BUILD (TO)	CONSTRUIRE	BAUEN
1695	BUILD-UP	SUREPAISSEUR	AUFMASS, ÜBERDICKE
1696	BUILD-UP SEQUENCE	PROCESSUS DE RECHARGEMENT	AUFTRAGSPROZESS
1697	BUILDING BLOCKS	PARPAINGS, BRIQUES CREUSES	BAUSTEINE
1698	BUILDING CONTRACTOR	ENTREPRENEUR DE BATIMENT	BAUUNTERNEHMER
1699	BUILDING SITE	EMPLACEMENT, TERRAIN A BATIR	BAUGRUND
1700	BUILDING STONE	PIERRE A BATIR	BAUSTEIN
1701	BUILDING UP	EPAISSISSEMENT	VERSTÄRKUNG
1702	BUILT-IN BEAM	POUTRE ENCASTREE	EINBAUTRÄGER
1703	BUILT-IN BEAM, BEAM WITH ENDS FIXED	POUTRE ENCASTREE A SES DEUX EXTREMITES	BEIDERSEITIG EINGESPANNTER TRÄGER
1704	BUILT-IN MEMBER	ELEMENT DE CHARPENTE INCORPORE	BAUTEIL (EINGEBAUTES)
1705	BUILT-UP	RAPPORTE, ACCUMULATION	ZUGEBAUT, SPEICHERUNG
1706	BUILT-UP CRANK	MANIVELLE EN PLUSIEURS PIECES	KURBEL (GEBAUTE), ZUSAMMENGEBAUTE
1707	BUILT-UP EDGE	COPEAU ADHERENT	AUFBAUSCHNEIDE
1708	BUILT-UP FLANGE	BRIDE RAPPORTEE	FLANSCHRING
1709	BUILT-UP PISTON	PISTON DE DEUX PIECES	KOLBEN (GETEILTER), ZWEITEILIGER
1710	BUILT-UP PLATE	PLAQUE MODELE	MODELLPLATTE
1711	BULB	PLONGEUR (DE THERMOMETRE), BULBE	FÜHLER, TAUCHROHR
1712	BULB ANGLE IRON	FER CORNIERE A BOUDIN	WULSTWINKEL, WINKELWULSTEISEN
1713	BULB ANGLES	CORNIERES A BOUDIN	WINKELWULSTEISEN
1714	BULB BAR	FER PLAT A BOUDIN	FLACHWULSTEISEN
1715	BULB HEAD RAILS	RAILS A BOUDIN	DOPPELKOPFSCHIENEN
1716	BULB IRON	FER A BOUDIN	WULSTEISEN
1717	BULB OF A THERMOMETER	AMPOULE DU THERMOMETRE	THERMOMETERKUGEL
1718	BULB PIPETTE	PIPETTE A CYLINDRE	VOLLPIPETTE
1719	BULB TEE	FER EN T A BOUDIN, FER A BOUDIN A PATIN	T-WULST, FLANSCHWULSTEISEN
1720	BULGE	BOMBEMENT, BOMBE, CONVEXITE, GONFLEMENT, BOSSE	AUSBAUCHUNG, AUSBEULUNG, WÖLBUNG, BEULE
1721	BULGE (TO)	SE GONFLER	AUSBAUCHEN (SICH)
1722	BULGED	BOMBE	AUSGEBAUCHT
1723	BULGING	BOMBAGE, ELARGISSEMENT	AUFWEITUNG, AUFWÖLBUNG, AUFWULSTUNG, AUSBAUCHUNG
1724	BULK DENSITY	DENSITE AU REMPLISSAGE	FÜLLDICHTE
1725	BULK SPECIFIC GRAVITY	DENSITE APPARENTE	SCHEINDICHTE
1726	BULKHEAD	CLOISON	SCHOTT, TRENNWAND
1727	BULL BLOCK	.BANC A TREFILER LES GROS FILS	GROBZUG
1728	BULL LADLE	POCHE DE GRUE DE COULEE	KRANPFANNE, KRANGIESSPFANNE

1729	**BULLION**	METAL NOBLE EN BARRES	EDELMETALLBARREN
1730	**BULLION BAR**	BARRE DE METAL NOBLE	EDELMETALLBAR
1731	**BULLOCK GEAR, HORSE GEAR, CATTLE GEAR**	MANEGE, BARITEL	GÖPEL, ROSSWERK
1732	**BUMPER**	PILETTE MECANIQUE, PARE-CHOC	PLASTSTAMPFER (MECHANISCHER), STOSSFÄNGER
1733	**BUMPING SCREEN**	CRIBLE A SECOUSSES	STOSSSIEB
1734	**BUMPING TOOL**	OUTIL A MAIN DE MARTELAGE	SCHLICHTHAMMER
1735	**BUND**	CAVALIER	TANKWALL
1736	**BUNDLE**	BOTTE (DE TUBES OU PROFILES)	ROHR-ODER FORMSTAHLBÜNDEL
1737	**BUNDLE (TO)**	BOTTELER	BÜNDELN
1738	**BUNDLE OF HOOP IRON**	BOTTE DE FEUILLARD	BUND VON BANDEISEN
1739	**BUNKER BED**	COUCHETTE, LIT-PLACARD	SCHLAFKOJE
1740	**BUNKER, COAL BUNKER**	RESERVOIR, SOUTE A CHARBON	BUNKER, KOHLENBUNKER
1741	**BUNSEN BURNER**	BRULEUR BUNSEN, BEC BUNSEN	BUNSENBRENNER
1742	**BUOYANCY**	POUSEE DE BAS EN HAUT, FORCE ASCENSIONNELLE, FLOTTABILITE	AUFTRIEB, SCHWIMMFÄHIGKEIT
1743	**BURETTE**	BURETTE	BÜRETTE
1744	**BURIED PIPE-LINES**	BRANCHEMENTS SOUTERRAINS	FERNROHRLEITUNG (EINGEERDETE)
1745	**BURN (TO)**	BRULER	VERBRENNEN, BRENNEN
1746	**BURN (TO), OVERHEAT THE IRON (TO)**	BRULER LE FER	EISEN (DAS) ÜBERHITZEN, VERBRENNEN (DAS)
1747	**BURN-UP**	COMBUSTION NUCLEAIRE	ABBRAND
1748	**BURNED INGOT**	LINGOT BRULE	VERBRANNTER BLOCK
1749	**BURNER**	BRULEUR	BRENNER
1750	**BURNING**	BRULURE	VERBRENNEN
1751	**BURNING FIRING GASEOUS FUELS**	CHAUFFAGE AU GAZ	GASFEUERUNG, VERFEUERN GASFÖRMIGER BRENNSTOFFE
1752	**BURNING FIRING OIL FUEL, OIL FIRING**	CHAUFFAGE AU PETROLE, CHAUFFAGE AU NAPHTE, CHAUFFAGE A HUILE LOURDE	ÖLFEUERUNG, VERFEUERN VON ÖL
1753	**BURNING FIRING PULVERISED COAL**	CHAUFFAGE AU CHARBON PULVERISE	KOHLENSTAUBFEUERUNG, VERFEUERN VON KOHLENSTAUB
1754	**BURNING OVERHEATING THE IRON**	BRULEMENT DU FER	ÜBERHITZUNG, VERBRENNEN DES EISENS
1755	**BURNISH**	POLI PARFAIT, BRUNI, BRUNISSURE	HOCHGLANZ
1756	**BURNISH (TO)**	BRILLANTER, POLIR BRILLANT, BRUNIR	HOCHGLANZPOLIEREN
1757	**BURNISHER**	BRUNISSOIR	POLIEREISEN, POLIERSTAHL, GLÄTTZAHN
1758	**BURNISHING**	BRUNISSAGE, GALETAGE	BRÄUNUNG, BRÜNIERUNG, PRÄGEPOLIEREN, HOCHGLANZPOLIEREN
1759	**BURNISHING BALLS**	BILLES A POLIR, BILLES POUR BRUNISSAGE	POLIERKUGELN
1760	**BURNT BRICK**	BRIQUE CUITE	ZIEGEL (GEBRANNTER)
1761	**BURNT DEPOSIT**	DEPOT BRULE	NIEDERSCHLAG (ANGEBRANNTER)
1762	**BURNT IRON**	FER BRULE	EISEN (VERBRANNTES)
1763	**BURNT STEEL**	ACIER BRULE	STAHL (VERBRANNTER)
1764	**BURR**	BAVURE, EBARBURE, EBARBE, BARBE, BARBURE	GRAT, BART
1765	**BURR REMOVER FOR TUBES, TUBE BURR REMOVER**	FRAISE EBARBEUSE	ROHRFRÄSER
1766	**BURRING**	EBAVURAGE	ENTGRATEN, ABGRATEN
1767	**BURSTING**	ECLATEMENT	PLATZEN

1768	**BURSTING OF A PIPE**	ECLATEMENT, RUPTURE D'UN TUYAU	AUFPLATZEN EINES ROHRÈS
1769	**BURSTING OF A PULLEY, OF A FLYWHEEL**	ECLATEMENT D'UNE POULIE, ECLATEMENT D'UN VOLANT	BERSTEN, EXPLOSION EINER SCHEIBE, EXPLOSION EINES SCHWUNGRADES
1770	**BUS BAR**	BARRE COLLECTRICE, RAIL DE CONTACT, BARRE OMNIBUS	STROMSCHIENE, SAMMELSCHIENE
1771	**BUSH OF BEARING**	COUSSINET EN UNE SEULE PIECE	LAGERBÜCHSE, LAGERHÜLSE, BÜCHSE, LAGERSCHALE (UNGETEILTE)
1772	**BUSHELLING**	EMPAQUETAGE, MISE EN BOTTES	PAKETIEREN
1773	**BUSHING**	BAGUE, REDUCTION MALE-FEMELLE, DOUILLE, MANCHON	LAGERSCHALE, LAGERBÜCHSE, REDUZIERSTÜCK, HÜLSE
1774	**BUSTLE PIPE**	CONDUIT ANNULAIRE DE VENT CHAUD	HEISSINDLEITUNG, RINGLEITUNG, WINDRING
1775	**BUTANE**	BUTANE, HYDRURE DE BUTYLE	BUTAN
1776	**BUTT JOINT**	JOINT BOUT A BOUT	STOSSVERBINDUNG, STUMPF, STUMPFSTOSSVERBINDUNG
1777	**BUTT JOINT, DOUBLE WELDED**	JOINT ABOUTE SOUDE DES DEUX COTES	STUMPFSTOSS (BEIDERSEITIG GESCHWEISSTER)
1778	**BUTT JOINT, JUMP JOINT**	ASSEMBLAGE, JONCTION BOUT A BOUT	STOSSVERBINDUNG
1779	**BUTT JOINT, SINGLE WELDER**	JOINT ABOUTE SOUDE D'UN SEUL COTE	STUMPFSTOSS (EINSEITIG GESCHWEISSTER)
1780	**BUTT OF TANNED HIDE**	CUIR DU CROUPON	KERNLEDER
1781	**BUTT RIVETED JOINT**	RIVURE A BANDE DE RECOUVREMENT	LASCHENNIETUNG, BANDNIETUNG
1782	**BUTT STRAP JOINT, FISHED JOINT, WELDED JOINT**	ASSEMBLAGE A COUVRE-JOINTS, JOINT A ECLISSES	ÜBERLASCHUNG, STOSS (MIT LASCHE), ÜBERLASCHTER STOSS
1783	**BUTT STRIP STRAP, COVERING STRIP, WELT, FISHPLATE, COVER STRAP**	BANDE DE RECOUVREMENT COUVRE-JOINT, ECLISSE	LASCHE, STOSSLASCHE, STOSSPLATTE
1784	**BUTT WELD**	SOUDURE BOUT A BOUT, SOUDURE PAR RAPPROCHEMENT	STOSSNAHT, STUMPF-(SCHWEISS-)NAHT
1785	**BUTT WELD (TO)**	SOUDER PAR CONTACT, SOUDER PAR ENCOLLAGE	STUMPFSCHWEISSEN
1786	**BUTT WELD, - THROAT OF**	EPAISSEUR D'UNE SOUDURE BOUT A BOUT	STUMPFNAHTDICKE
1787	**BUTT WELD, CLOSED DOUBLE BEVEL**	SOUDURE EN K SANS ECARTEMENT	K-NAHT OHNE LUFTSPALT
1788	**BUTT WELD, CLOSED DOUBLE JOINTS**	SOUDURE DOUBLE FERMEE	DOPPELNAHT OHNE LUFTSPALT
1789	**BUTT WELD, CLOSED SINGLE BEVEL**	SOUDURE EN DEMI-V SANS ECARTEMENT	HALB-V-NAHT OHNE LUFTSPALT
1790	**BUTT WELD, CLOSED SINGLE JOINT**	JOINT DE SOUDURE EN J FERMEE	J-NAHTVERBINDUNG
1791	**BUTT WELD, CLOSED SQUARE**	SOUDURE EN I SANS ECARTEMENT DES BORDS	I-NAHTVERBINDUNG (I-STOSS) OHNE LUFTSPALT
1792	**BUTT WELD, CONCAVE**	SOUDURE CONCAVE	HOHLNAHT (KONCAVE), SCHWEISSNAHT (LEICHTE)
1793	**BUTT WELD, DOUBLE BEVEL**	CHANFREIN (SOUDURE) EN K	K-NAHT
1794	**BUTT WELD, FLASH**	SOUDURE PAR ETINCELAGE	ABBRENNSTUMPFSCHWEISSUNG
1795	**BUTT WELD, JUMP**	SOUDURE (ASSEMBLAGE) EN T	T-NAHT
1796	**BUTT WELD, OPEN DOUBLE BEVEL**	SOUDURE EN K AVEC ECARTEMENT	K-NAHT MIT LUFTSPALT
1797	**BUTT WELD, OPEN SINGLE BEVEL**	SOUDURE EN DEMI-V AVEC ECARTEMENT	HALB-V-NAHT MIT LUFTSPALT
1798	**BUTT WELD, OPEN SINGLE JOINT**	SOUDURE EN J AVEC ECARTEMENT	J-NAHT MIT LUFTSPALT
1799	**BUTT WELD, RESISTANCE**	SOUDURE BOUT A BOUT PAR RESISTANCE	WIDERSTANDSSTUMPFNAHT
1800	**BUTT WELD, SINGLE BEVEL**	CHANFREIN EN DEMI-V	HALB-V-NAHT
1801	**BUTT WELD, TEE**	SOUDURE (ASSEMBLAGE) EN T	T-NAHT

1802	**BUTT WELDED JOINT, JUMP WELD**	SOUDURE PAR RAPPROCHEMENT, SOUDURE PAR ENCOLLAGE	SCHWEISSUNG (STUMPFE)
1803	**BUTT WELDED PIPE**	TUBE SOUDE BOUT A BOUT	ROHRVERBINDUNG (STUMPFGESCHWEISSTE)
1804	**BUTT WELDED TUBE**	TUBE SOUDE PAR RAPPROCHEMENT, TUBE A RAPPROCHEMENT	ROHR (STUMPFGESCHWEISSTES)
1805	**BUTT WELDING**	SOUDAGE PAR RAPPROCHEMENT, SOUDAGE BOUT A BOUT (PAR RESISTANCE)	STUMPFSCHWEISSEN
1806	**BUTT WELDING, FLASH**	SOUDAGE EN BOUT PAR ETINCELAGE	ABBRENNSTUMPFSCHWEISSEN
1807	**BUTT WELDING, RESISTANCE**	SOUDAGE EN BOUT PAR RESISTANCE	WIDERSTANDSSTUMPFSCHWEISSEN
1808			
1809	**BUTT-WELDED JOINT**	SOUDURE BOUT A BOUT	STUMPFNAHT
1810	**BUTT-WELDING ENDS, SOCKET-WELDING ENDS**	ORIFICES A SOUDER (EN BOUT, A L'INTERIEUR)	VORSCHWEISSENDEN, EINSCHWEISSENDEN
1811	**BUTTERFLY VALVE**	VANNE A PAPILLON	DROSSELKLAPPE
1812	**BUTTON**	FOND DE POCHE	PFANNENBÄR
1813	**BUTTON HEAD, ROUND HEAD, CUP HEAD, SPHERICAL HEAD, SEGMENTAL HEAD OF SCREW BOLT**	TETE RONDE D'UNE VIS	RUNDSCHRAUBENKOPF
1814	**BUTTRESS THREAD**	FILET TRAPEZOIDAL	TRAPEZFÖRMIGES, GEWINDE (HALBIERTES), TRAPEZGEWINDE
1815	**BUTTRESS THREADED SCREW**	VIS A FILET TRAPEZOIDAL	SCHRAUBE MIT TRAPEZGEWINDE
1816	**BUTYLENE**	BUTYLENE	BUTYLEN
1817	**BUTYRIC ACID**	ACIDE BUTYRIQUE	BUTTERSÄURE
1818	**BY-PASS**	CONDUITE DE DERIVATION, BY-PASS	NEBENSCHLUSS, UMFÜHRUNG
1819	**BY-PASS VALVE, AUXILIARY VALVE**	SOUPAPE DE DERIVATION	UMFÜHRUNGSVENTIL, HILFSVENTIL, ENTLASTUNGSVENTIL, ZUSATZVENTIL, UMLAUFVENTIL, UMGEHUNGSVENTIL, ZIRKULATIONSVENTIL
1820	**BY-PRODUCT**	SOUS-PRODUIT, PRODUIT SECONDAIRE	NEBENERZEUGNIS, NEBENPRODUKT
1821	**BYPASS**	DERIVATION, BY-PASS	UMGEHUNG, NEBENSCHLUSS
1822	**C-SPANNER**	CLEF A CROCHET	HAKENSCHLÜSSEL, NUTENSCHLÜSSEL
1823	**C-SPRING, CEE SPRING, COACH SPRING**	RESSORT EN C	BÜGELFEDER
1824	**CAB OVER ENGINE**	CABINE AVANCEE	FRONT FÜHRERHAUS
1825	**CABLE**	CABLE ELECTRIQUE	LEITUNGSKABEL
1826	**CABLE RAILWAY**	CHEMIN DE FER FUNICULAIRE, FUNICULAIRE	SEILBAHN, GLEISSEILBAHN
1827	**CABLE TERMINAL**	COSSE	LEITUNGSSCHUH
1828	**CABLE TWISTING**	CABLAGE	KABELVERSEILUNG
1829	**CABLE, CABLE-LAID ROPE**	GRELIN	KABEL, TAU
1830	**CABLE, WIRE**	TELEGRAMME	TELEGRAMM
1831	**CADMIUM**	CADMIUM	KADMIUM
1832	**CADMIUM PLATING**	CADMIAGE	KADMIEREN
1833	**CAESIUM**	CESIUM, CAESIUM	ZÄSIUM
1834	**CAKE**	LINGOT DE DEPART	AUSGANGSBLOCK
1835	**CAKING COAL**	HOUILLE COLLANTE	KOHLE (BACKENDE)
1836	**CALAMINE, SMITHSONITE**	CALAMINE, SIMITHSONITE	ZINKSPAT, GALMEI (EDLER), KALAMIN, SMITHSONI
1837	**CALC SINTER**	CONCRETION CALCAIRE	KALKSINTER

1838	CALCAREOUS TUFA	TUF CALCAIRE	TUFFSTEIN, KALKTUFF
1839	CALCINATION	CALCINATION, GRILLAGE	KALZINIEREN, RÖSTUNG, KALZINIERUNG
1840	CALCINATION OF ORES	CALCINATION DES MINERAIS	GLÜHEN, KALZINIEREN VON ERZEN
1841	CALCINE ORES (TO)	CALCINER LES MINERAIS	ERZE GLÜHEN, KALZINIEREN
1842	CALCINED MAGNESIA	MAGNESIE CALCINEE	MAGNESIA (GEBRANNTE)
1843	CALCINING FURNACE, KILN	FOUR DE CALCINATION	KALZINIEROFEN
1844	CALCITE	CALCITE	KALZIT
1845	CALCITE, CALCAREOUS SPAR	SPATH CALCAIRE, CALCITE	KALKSPAT, KALZIT
1846	CALCIUM	CALCIUM	KALZIUM
1847	CALCIUM BICARBONATE, CALCIUM ACID CARBONATE	BICARBONATE DE CALCIUM	KALZIUM (DOPPELTKOHLENSAURES), KALZIUMBIKARBONAT
1848	CALCIUM CARBIDE	CARBURE, ACETYLURE DE CALCIUM	KALZIUMKARBID, KARBID
1849	CALCIUM CARBONATE, CARBONATE OF LIME	CARBONATE DE CHAUX	KALK (KOHLENSAURER), KALZIUM (KOHLENSAURES), KALZIUMKARBONAT
1850	CALCIUM CHLORIDE	CHLORURE DE CALCIUM	CHLORKALZIUM, KALZIUMCHLORID
1851	CALCIUM HYDROXIDE, SLAKED LIME	CHAUX ETEINTE, CHAUX HYDRATEE, HYDRATE DE CHAUX	KALK (GELÖSCHTER), KALZIUMHYDROXYD, KALKHYDRAT
1852	CALCIUM MONO SULPHIDE, CANTON'S PHOSPHORUS	MONOSULFURE DE CALCIUM	SCHWEFELKALZIUM (EINFACH), KALZIUMSULFID, CANTONSPHOSPHOR
1853	CALCIUM OXALATE	OXALATE DE CALCIUM	KALK (OXALSAURER), KALZIUMOXALAT
1854	CALCIUM OXIDE, QUICK LIME	CHAUX VIVE, CHAUX ANHYDRE, OXYDE DE CALCIUM	ÄTZKALK, KALK (GEBRANNTER), KALZIUMOXYD
1855	CALCIUM PHOSPHATE	PHOSPHATE DE CALCIUM, PHOSPHATE TRICALCIQUE	KALK (PHOSPHORSAURER), KALZIUMPHOSPHAT
1856	CALCIUM SULFATE	SULFATE DE CALCIUM	KALZIUMSULFAT
1857	CALCULATE (TO)	CALCULER	BERECHNEN
1858	CALCULATED THEORETICAL VALUE	VALEUR THEORIQUE, VALEUR CALCULEE	WERT (ERRECHNETER), WERT (RECHNERISCHER), SOLLWERT
1859	CALCULATING MACHINE	MACHINE ARITHMETIQUE, MACHINE A CALCULER	RECHENMASCHINE
1860	CALCULATION	CALCUL	BERECHNUNG
1861	CALIBRATE (TO)	CALIBRER, JAUGER	KALIBRIEREN
1862	CALIBRATED CHAINS	CHAINES CALIBREES	KETTEN (KALIBRIERTE)
1863	CALIBRATING (OF A TANK)	ETALONNAGE (D'UN RESERVOIR)	EICHUNG (EINES BEHÄLTERS)
1864	CALIBRATING AND DIE-FORGING PRESS	PRESSE MECANIQUE A CALIBRER ET A MATRICER	KALIBRIER-UND GESENKSCHMIEDE-PRESSE
1865	CALIBRATING MACHINE	MACHINE A CALIBRER	KALIBRIERMASCHINE
1866	CALIBRATING SPRING	RESSORT TARE	KALIBRIERTE FEDER
1867	CALIBRATION	ETALONNAGE, CALIBRAGE, JAUGEAGE	EICHUNG, KALIBRIERUNG, KALIBRIEREN
1868	CALIBRATION CAPACITOR	CONDENSATEUR-ETALON	EICHKONDENSATOR
1869	CALIBRATION ERROR	ERREUR D'ETALONNAGE	EICHFEHLER
1870	CALIPER	MAITRE A DANSER	GREIFZIRKEL
1871	CALIPER GAUGE WITH DEPTH GAUGE	CALIBRE A COULISSE AVEC REGLE DE PROFONDEUR	SCHUBLEHRE MIT TIEFENMASSSTAB
1872	CALKING HAMMER	MARTEAU A MATER	STEMMHAMMER
1873	CALLIPER	COMPAS D'EPAISSEUR	KALIBER
1874	CALLIPERS	COMPAS DE CALIBRE	TASTER, TASTZIRKEL, GREIFZIRKEL
1875	CALLIPERS	MICROMETRE, PALMER	MIKROMETERLEHRE
1876	CALOMEL HALFICELL	ELECTRODE AU CALOMEL	KALOMELELEKTRODE
1877	CALORIMETER	CALORIMETRE	KALORIMETER
1878	CALORIMETRIC	CALORIMETRIQUE	KALORIMETRISCH

1879	CALORIMETRIC BOMB	BOMBE CALORIMETRIQUE	BOMBE (KALORIMETRISCHE), BOMBENKALORIMETER
1880	CALORIMETRIC DETERMINATION	ANALYSE CALORIMETRIQUE	BESTIMMUNG (KALORIMETRISCHE)
1881	CALORIMETRY	CALORIMETRIE	KALORIMETRIE
1882	CALORIZING	CALORISATION	KALORISIERUNG
1883	CAM	CAME EXCENTRIQUE BOSSE, BOSSAGE	EXZENTER (UNRUNDE SCHEIBE, HUBSCHEIBE),NOCKE, MITTNEHMER, NOCKEN, DAUMEN, KNAGGE
1884	CAM ACTION	LEVEE DE CAME	NOCKENHUB
1885	CAM FOLLOWER	GALET DE POUSSOIR	STÖSSEL
1886	CAM GEAR MOTION	MECANISME A CAME	KURVENTRIEB, EXZENTERTRIEBWERK
1887	CAM PIN	DOIGT INCLINE	NOCKENANLAUFSCHRÄGER
1888	CAM SHAFT	ARBRE A CAME S, ARBRE PORTE-CAMES	DAUMENWELLE
1889	CAMBER	BOMBAGE, CAMBRAGE, CARROSSAGE, ANGLE DE CARROSSAGE	RADSTURZ, STURZ (WINKEL), WÖLBUNG, BOMBIERUNG
1890	CAMEL HAIR	POIL, POILS DE CHAMEAU	KAMELHAAR
1891	CAMEL HAIR BELT	COURROIE EN POILS DE CHAMEAU	KAMELHAARRIEMEN
1892	CAMEL-BACK	BANDE DE RECHAPAGE	ROHLAUFSTREIFEN
1893	CAMPHOR	CAMPHRE	KAMPFER
1894	CAMPHOR OIL	HUILE DE CAMPHRE	KAMPFERÖL
1895	CAMROLLER, FOLLOWER	GALET, MOLETTE DE LA CAME	NOCKENROLLE, EXZENTERROLLE
1896	CAMSHAFT	ARBRE A CAMES	NOCKENWELLE
1897	CAMSHAFT LATHE	TOUR A ARBRE A CAMES	NOCKENWELLENDREHBANK
1898	CANADIAN ASBESTOS	AMIANTE, FIBRE DU CANADA	KANADAFASER
1899	CANDLE	CHANDELLE, BOUGIE	KERZE
1900	CANDLE POWER	INTENSITE EN BOUGIES	KERZENSTÄRKE
1901	CANNEL COAL	CANNEL-COAL	CANNELKOHLE, KENNELKOHLE, KANNELKOHLE, GASKOHLE (ECHTE)
1902	CANT	CHANFREIN	ABSCHRÄGUNG
1903	CANT FILE	LIME A BISEAU, LIME A BARRETTES	BARETTFEILE
1904	CANTILEVER	POUTRE ENCASTREE A UNE EXTREMITE, POUTRE EN PORTE-A-FAUX	FREITRÄGER
1905	CANTILEVER BEAM	POUTRE EN PORTE-A-FAUX, POUTRE NON ENTRETOISEE	FREITRÄGER
1906	CANVAS	TOILE A VOILE	SEGELTUCH
1907	CANVAS BELT	COURROIE EN CHANVRE	HANFRIEMEN
1908	CANVAS HOSE	TUYAU EN TOILE DE CHANVRE	HANFSCHLAUCH
1909	CAOT (TO)	ENDUIRE	ÜBERZIEHEN
1910	CAP	BOUCHON FEMELLE, BOUCHON, CAPUCHON, CHAPEAU	KAPPE, DECKEL, VERSCHLUSSDECKEL
1911	CAP OF A SPHERICAL TANK	CALOTTE (D'UNE SPHERE)	KUGELKAPPE, KUGELHAUBE, KUGELSCHALE, KUGELKALOTTE
1912	CAP OF BEARING	CHAPEAU, COUVERCLE DU PALIER	LAGERDECKEL
1913	CAP SCREW	VIS A TETE	KOPFSCHRAUBE
1914	CAPABLE OF BEING CAST	SUSCEPTIBLE D'ETRE COULE, QUI PEUT SE COULER	GIESSBAR
1915	CAPABLE OF BEING ROLLED	SUSCEPTIBLE D'ETRE LAMINE; QUI SE LAMINE FACILEMENT	WALZBAR
1916	CAPACITOR	CONDENSATEUR, CONDENSEUR (DE VAPEUR)	KONDENSATOR

1917	CAPACITY	PUISSANCE, CAPACITE (ELECTR.), CAPACITE, CONTENANCE	KAPAZITÄT (ELEKTR.), FASSUNGSVERMÖGEN, AUFNAHMEFÄHIGKEIT, KAPAZITÄT
1918	CAPACITY TO RESIST SHOCKS, RESISTANCE TO SHOCK	RESISTANCE AU CHOC, RESILIENCE	STOSSFESTIGKEIT, WIDERSTANDSFÄHIGKEIT GEGENSTOSS
1919	CAPE CHISEL	BEDANE	KREUZMEISSEL
1920	CAPILLARITY, CAPILLARY ACTION	CAPILLARITE	HAARRÖHRCHENWIRKUNG, KAPILLARITÄT
1921	CAPILLARY	TUBE CAPILLAIRE	HAARROHR, KAPILLARROHR
1922	CAPILLARY ACTION	ASPIRATION PAR CAPILLARITE	KAPILLARWIRKUNG
1923	CAPILLARY CONSTANT	CONSTANTE CAPILLAIRE	KAPILLARKONSTANTE
1924	CAPILLARY DEPRESSION	DEPRESSION CAPILLAIRE	KAPILLARDEPRESSION
1925	CAPILLARY ELEVATION	ASCENSION, ELEVATION CAPILLAIRE	KAPILLARELEVATION
1926	CAPITAL OF COLUMN	CHAPITEAU D'UNE COLONNE	KAPITÄL EINER SÄULE, SÄULENKAPITÄL
1927	CAPPING	FORMATION D'UNE CROUTE	DECKELBILDUNG
1928	CAPPING STRIPS	BANDES DE METAL DE RECOUVREMENT	ABDECKLEISTEN
1929	CAPSTAN	CABESTAN	SPILL
1930	CAPSTAN HEADED SCREW, TOMMY SCREW	VIS A BROCHE, VIS A LEVIER	KNEBELSCHRAUBE
1931	CAPSTAN LATHE	TOUR A REVOLVER	REVOLVERDREHBANK
1932	CAR TIPPING GEAR	BASCULEUR DE WAGONS	KIPPER, WAGENKIPPER
1933	CARBIDE	CARBURE	KARBID
1934	CARBIDE PRECIPITATION	PRECIPITATION DE CARBURE	KARBIDAUSSCHEIDUNG
1935	CARBIDE TIPPED REAMER	ALESOIR A MISE DE CARBURE	REIBAHLE (HARTMETALLBESTÜCKTE)
1936	CARBIDE-TIPPED DRILL	FORET A MISE EN CARBURE	BOHRER (HARTMETALLBESTÜCKTER)
1937	CARBOHYDRATE	HYDRATE DE CARBONE	KOHLEHYDRAT
1938	CARBOLIC ACID, PHENOL	PHENOL, ACIDE PHENIQUE, ACIDE CARBOLIQUE	KARBOLSÄURE, PHENOL, BENZOPHENOL, PHENYLSÄURE, PHENYLALKOHOL, MONOXYBENZOL, STEINKOHLENKREOSOT
1939	CARBOLINEUM	CARBOLINEUM, CARBONYLE	KARBOLINEUM
1940	CARBON	CARBONE	KOHLENSTOFF
1941	CARBON ARC	ARC AVEC ELECTRODE DE CARBONE	KOHLELICHTBOGEN
1942	CARBON ARC CUTTING	COUPAGE A L'ARC AU CARBONE	KOHLELICHTBOGENSCHNEIDEN
1943	CARBON ARC WELDING	SOUDAGE A L'ARC AVEC ELECTRODE AU CARBONE	KOHLE(LICHTBOGEN)SCHWEISSEN
1944	CARBON ARC, UNSHIELDED	ARC NON PROTEGE AVEC ELECTRODE AU CHARBON	UNGESCHÜTZTER KOHLELICHTBOGEN
1945	CARBON ARC,SHIELDED	ARC PROTEGE AVEC ELECTRODE AU CHARBON	SCHUTZGASLICHTBOGEN
1946	CARBON BRUSH	BALAI AU CHARBON	KOHLEBÜRSTE
1947	CARBON CONTENT	TENEUR EN CARBONE, DOSAGE DU CARBONE, PROPORTION DE CARBONE	KOHLENSTOFFGEHALT
1948	CARBON DIOXIDE, CARBONIC ACID GAS, CARBONIC ANHYDRIDE	ANHYDRIDE CARBONIQUE, GAZ CARBONIQUE	KOHLENSÄURE, KOHLENDIOXYD
1949	CARBON DIOXYDE AND HYDROGEN	ACIDE CARBONIQUE	KOHLENSÄURE
1950	CARBON ELECTRODE	ELECTRODE DE CHARBON	KOHLELEKTRODE
1951	CARBON FILAMENT LAMP, CARBON INCANDESCENT LAMP	LAMPE A FILAMENT DE CARBONE	KOHLEFADENGLÜHLAMPE
1952	CARBON MONOXIDE, CARBONIC OXIDE	OXYDE DE CARBONE	KOHLENMONOXYD

1953	**CARBON RESTORATION**	RECARBURATION, REGENERATION DES PIECES DECARBURISEES	AUFKOHLUNG, RÜCKOHLUNG
1954	**CARBON STEEL**	ACIER AU CARBONE	KOHLENSTOFFSTAHL, STAHL (UNLEGIERTER)
1955	**CARBON STEEL BLOOMS**	BLOOMS EN ACIER AU CARBONE	BLÖCKE AUS KOHLENSTOFFSTAHL
1956	**CARBON STEEL SHEET BARS**	LARGETS EN ACIER AU CARBONE	PLATINEN AUS LEGIERTEN STÄHLEN
1957	**CARBON STEEL, HIGH**	ACIER DUR A HAUTE TENEUR EN CARBONE	HARTSTAHL, STAHL (KOHLENSTOFFEICHER)
1958	**CARBON STEEL, LOW**	ACIER DOUX	FLUSSTAHL, STAHL (WEICHER)
1959	**CARBON STEEL, MEDIUM**	ACIER DEMI-DOUX	STAHL (HALBHARTER)
1960	**CARBON TETRACHLORIDE**	TETRACHLORURE DE CARBONE	TETRACHLORKOHLENSTOFF
1961	**CARBON-FREE**	EXEMPT DE CARBONE	KOHLENSTOFFREI
1962	**CARBON, COMBINED**	CARBONE COMBINE	KOHLENSTOFF (GEBUNDENER)
1963	**CARBON, DISSOLVED**	CARBONE DISSOUS	KOHLENSTOFF (GELÖSTER)
1964	**CARBON, DISULPHIDE**	SULFURE DE CARBONE	SCHWEFELKOHLENSTOFF, KOHLENDISULFID, SCHWEFELALKOHOL
1965	**CARBON, TEMPER**	CARBONE DE REVENU, POURCENTAGE DE CHARBON	AUSLASSKOHLENSTOFF, KOHLENSTOFFGEHALT
1966	**CARBONACEOUS**	CARBONE	KOHLENSTOFFARTIG KOHLENSTOFFHALTIG
1967	**CARBONADO, BLACK DIAMOND**	CARBONADO	KARBONAT, CARBONADO, DIAMANT (SCHWARZER)
1968	**CARBONATE**	CARBONATE	SALZ (KOHLENSAURES), KARBONAT
1969	**CARBONIFEROUS LIMESTONE**	CALCAIRE CARBONIFERE	KOHLENKALK
1970	**CARBONISATION**	CARBONISATION	VERKOHLUNG, KARBONISATION
1971	**CARBONISE (TO)**	CARBONISER, CHARBONNER	VERKOHLEN, KARBONISIEREN
1972	**CARBONITRIDING**	CARBONITRURATION	KARBONITRIEREN
1973	**CARBONITRIDING ATMOSPHERE**	ATMOSPHERE DE CARBONITRURATION	KARBONITRIERUNGSATMOSPHÄRE
1974	**CARBONIZATION**	CARBONISATION	KARBONISIERUNG, VERKOHLUNG
1975	**CARBONIZING**	CARBONISATION, COKEFACTION	VERKOHLUNG, VERKOKUNG
1976	**CARBONYL**	CARBONYLE	KARBONYL
1977	**CARBONYL IRON**	FER DE CARBONYLE	KARBONYLEISEN
1978	**CARBONYL NICKEL**	NICKEL AU CARBONYLE	KARBONYLNICKEL
1979	**CARBONYL POWDER**	POUDRE DE CARBONYLE	KARBONYLPULVER
1980	**CARBORUNDUM, CARBON SILICIDE**	CARBORUNDUN, CARBURE DE SILICUM, SILICIURE DE CARBONE	KARBORUND, KARBORUNDUM, SILIZIUMKARBID
1981	**CARBOY**	TOURIE, DAME-JEANNE, BONBONNE	KORBFLASCHE, BALLON, DEMIJOHN
1982	**CARBURATOR (U.S), CARBURETTOR (G.B)**	CARBURATEUR	VERGASER
1983	**CARBURETTED ALCOHOL**	ALCOOL CARBURE	KRAFTSPIRITUS, MOTORENSPIRITUS, ALKOHOL (KARBURIERTER)
1984	**CARBURETTER, CARBURETOR**	CARBURATEUR	VERGASER, KARBURATOR
1985	**CARBURISATION OF IRON**	CARBURATION DU FER	KOHLUNG DES EISENS
1986	**CARBURISE THE IRON (TO)**	CARBURER LE FER	EISEN (DAS) KOHLEN
1987	**CARBURIZING**	CEMENTATION PAR LE CARBONE	ZEMENTIEREN (MIT KOHLENSTOFF), AUFKOHLEN, EINSATZHÄRTUNG
1988	**CARBURIZING COMPOUNDS**	CARBURANTS	AUFKOHLUNGSMITTEL, EINSATZMITTEL
1989	**CARBURIZING CONTAINERS**	BOITES DE CEMENTATION	EINSATZKASTEN
1990	**CARBURIZING FLAME**	FLAMME CARBURANTE	FLAMME (KARBURIERENDE), FLAMME (AUFKOHLENDE)
1991	**CARBURIZING FURNACE**	FOUR DE CARBURISATION	KARBURIEROFEN
1992	**CARBURIZING, GAS**	CEMENTATION PAR GAZ	ZEMENTATION IN GASATMOSPHÄRE

/

1993	CARBURIZING, HOMOGENEOUS	CEMENTATION HOMOGENE	ZEMENTIERUNG (GLEICHARTIGE)
1994	CARBURIZING, LIQUID	BAIN DE CEMENTATION	EINSATZBAD
1995	CARBURIZING, SELECTIVE	CEMENTATION SELECTIVE	ZEMENTIERUNG (SELEKTIVE)
1996	CARBURIZING, BOX	CEMENTATION EN CAISSE	KASTENZEMENTIERUNG, KISTENZEMENTIERUNG
1997	CARDAN COUPLING	CARDAN	KARDANKUPPLUNG
1998	CARDAN'S SUSPENSION	SUSPENSION A LA CARDAN	AUFHÄNGUNG (KARDANISCHE)
1999	CARDBOARD	CARTON LEGER, CARTE	KARTON, FEINPAPPE
2000	CARDINAL STRAIN	DEFORMATION PRINCIPALE	HAUPTVERFORMUNG
2001	CARDIOID	CARDIOIDE	HERZKURVE, KARDIOIDE
2002	CARMINE PAPER	PAPIER CARMIN	KARMINPAPIER
2003	CARNAUBA WAX	CIRE DE CARNAUBA	CARNAUBAWACHS
2004	CARNOT'S CYCLE	CYCLE DE CARNOT	KREISPROZESS (CARNOTSCHER)
2005	CARPENTER	CHARPENTIER	ZIMMERMANN
2006	CARPENTRY	CHARPENTERIE	ZIMMEREI
2007	CARRIAGE PLANING MACHINE	MACHINE A RABOTER A TABLE MOBILE	TISCHHOBELMASCHINE
2008	CARRIAGE SPRING	RESSORT DE VOITURE	WAGENFEDER
2009	CARROT WEDGE	BROCHE CONIQUE	KEGELSPINDEL, KONUSSTIFT
2010	CARRYING AXLE	ESSIEU PORTEUR	TRAGACHSE
2011	CARRYING CAPACITY	CAPACITE DE CHARGE	TRAGFÄHIGKEIT
2012	CARRYING PLATE	PLATEAU DE DEMOTTAGE	TROCKENPLATTE
2013	CART	CHARRETTE	KARREN, KARRE
2014	CARTESIAN COORDINATES	COORDONNEES CARTESIENNES	KOORDINATE (KARTESISCHE)
2015	CARTRIDGE	CARTOUCHE	PATRONE
2016	CARTRIDGE BRASS	LAITON POUR CARTOUCHES	PATRONENMESSING
2017	CARTS, FARM	TOMBEREAUX	ACKERWAGEN
2018	CASCADE	CASCADE	KASKADE
2019	CASE	COUCHE SUPERFICIELLE, SURFACE	OBERFLÄCHENSCHICHT, OBERFLÄCHE
2020	CASE DEPTH	PROFONDEUR DE LA COUCHE CEMENTEE	DICKE DER ZEMENTIERSCHICHT
2021	CASE HARDENING	TREMPE DE SURFACE, CEMENTATION	EINSETZEN, EINSATZHÄRTUNG, OBERFLÄCHENARTUNG
2022	CASE HARDENING POWDER	CEMENT	EINSATZPULVER
2023	CASE HARDENING, SURFACE HARDENING	CEMENTATION DES PIECES EN ACIER DOUX , CEMENTATION PARTIELLE, TREMPE EN SURFACE	EINSATZHÄRTUNG, OBERFLÄCHENHÄRTUNG
2024	CASEHARDENING MATERIALS	CARBURANTS, CEMENTS	AUFKOHLUNGSMITTEL, EINSATZMITTEL
2025	CASEIN	CASEINE	KASEIN
2026	CASING	CARTER	EINKAPSELUNG, GEHÄUSE (EINES MASCHINENTEILS)
2027	CASING TESTER	CALIBRE POUR TUBES	ROHRKALIBER
2028	CASSETTE	CASSETTE	KASSETTE
2029	CASSINIAN OVAL	CASSINOIDE, LEMNISCATE, ELLIPSE DE CASSINI	KURVE (CASSINISCHE), ELLIPSE
2030	CAST (TO)	FONDRE, COULER, MOULER	GIESSEN, VERGIESSEN
2031	CAST BRASS	LAITON COULE	GUSSMESSING
2032	CAST COATING	REVETEMENT FONDU	GUSSPLATTIERUNG
2033	CAST HOLE	TROU VENU DE FONTE	LOCH (GEGOSSENES)
2034	CAST IRON	FONTE DE FER, FONTE	GUSSEISEN, GUSS, GRAUGUSS
2035	CAST IRON PIPE	TUYAU EN FONTE	ROHR (GUSSEISERNES), GUSSEISENROHR

2036	CAST IRON THERMIT	THERMITE DE FONTE	GUSSEISENTHERMIT
2037	CAST IRON, ALLOY	FONTE ALLIEE	GUSSEISEN (LEGIERTES)
2038	CAST IRON, CHILLED	FONTE DE COQUILLE, FONTE TREMPEE	HARTGUSS, KOKILLENGUSS, SCHALENHARTGUSS
2039	CAST IRON, CORROSION RESISTANT	FONTE RESISTANTE A LA CORROSION	GUSSEISEN (KORROSIONSBESTÄNDIGES)
2040	CAST IRON, DUCTILE	FONTE DUCTILE	GUSSEISEN MIT KUGELGRAPHIT
2041	CAST IRON, GRAY	FONTE GRISE	GRAUGUSS, GUSSEISEN MIT LAMELLENGRAPHIT
2042	CAST IRON, HIGH STRENTCH	FONTE A GRANDE RESISTANCE	FESTIGKEITSGUSS, QUALITÄTSGUSS
2043	CAST IRON, MALLEABLE	FONTE MALLEABLE	TEMPERGUSS
2044	CAST IRON, MOTTLED	FONTE TRUITEE	GUSSEISEN (MELIERTES)
2045	CAST IRON, NODULAR	FONTE DUCTILE, FONTE A GRAPHITE SPHEROIDAL	GUSSEISEN MIT KUGELGRAPHIT
2046	CAST IRON, PEARLITIC	FONTE PERLITIQUE	GUSSEISEN (PERLITISCHES)
2047	CAST IRON, PEARLITIC MALLEABLE	FONTE MALLEABLE PERLITIQUE	SCHMIEDEISEN (PERLITISCHES)
2048	CAST IRON, WEAR RESISTANT	FONTE RESISTANTE A L'USURE	GUSSEISEN (VERSCHLEISSFESTES)
2049	CAST IRON, WHITE	FONTE BLANCHE	GUSSEISEN (WEISSES)
2050	CAST METAL	METAL COULE	GUSSMETALL
2051	CAST NAIL	CLOU FONDU	NAGEL (GEGOSSENER)
2052	CAST PLATE	PLAQUE-MODELE	MODELLGUSSPLATTE
2053	CAST SHELL PROCESS	ETIRAGE A FROID DE TUBES COULES SANS SOUDURE	KALTZIEHEN VON NAHTLOSGEGOSSENEN RÖHREN
2054	CAST SOLID, CAST WHOLE	VENU DE FONTE, VENU DE FONDERIE, VENUE A LA COULEE	ZUSAMMENGEGOSSEN, IN EINEM STÜCK GEGOSSEN
2055	CAST STEEL	ACIER MOULE, FONTE D'ACIER	STAHLGUSS
2056	CAST STEEL WHEEL CENTER	CENTRE DE ROUE EN ACIER FONDU	STAHLGUSSRADSTERN
2057	CAST STEEL, CRUCIBLE CAST STEEL	ACIER FONDU AU CREUSET	GUSSSTAHL, TIEGELGUSSSTAHL
2058	CAST STRUCTURE	STRUCTURE DES ALLIAGES	GEFÜGE DER LEGIERUNGEN
2059	CAST TOOTH	DENT BRUTE DE FONTE	ZAHN (UNBEARBEITETER)
2060	CAST WELD ASSEMBLY	ASSEMBLAGE PAR SOUDURE DE PIECES MOULEES	GUSSTÜCKENZUSAMMENSCHWEISSEN
2061	CAST-IRON GROWTH	GONFLEMENT DE LA FONTE	GUSSEISENQUELLUNG
2062	CAST-IRON PIPE	CONDUITE EN FONTE	GUSSEISEN-ROHRLEITUNG
2063	CAST-ON FLANGE	BRIDE VENUE DE FONTE	FLANSCH (ANGEGOSSENER)
2064	CASTABILITY	COULABILITE	VERGIESSBARKEIT, FLIESSVERMÖGEN
2065	CASTABILITY TEST	ESSAI DE COULABILITE	VERGIESSBARKEITSVERSUCH
2066	CASTER	CHASSE	NACHLAUF
2067	CASTER (ANGLE)	CHASSE (ANGLE DE)	NACHLAUF (WINKEL)
2068	CASTING	COULEE, COULAGE, FONTE, MOULAGE	GIESSEN, VERGIESSEN, GUSS
2069	CASTING	LINGOT, PIECE MOULEE	GUSSSTÜCK
2070	CASTING (OPERATION)	COULEE (OPERATION)	GIESSEN
2071	CASTING BAY	CHANTIER DE COULEE, HALLE DE COULEE	GIESSHALLE, GIESSEREI
2072	CASTING CARS	WAGONS CIGARES/POCHES	GIESSPFANNENWAGEN
2073	CASTING COPPER	CUIVRE AFFINE	GUSSKUPFER
2074	CASTING FREE FROM INTERNAL STRESSES	FONTE SANS TENSIONS INTERNES	GUSS (SPANNUNGSFREIER)
2075	CASTING FREE FROM PIPES	FONTE SANS RETASSURES	GUSS (LUNKERFREIER)
2076	CASTING HEAT BATCH	COULEE (LOT DE...)	SCHMELZE, GICHT
2077	CASTING HOUSE OR BAY	HALLE DE COULEE	GIESSHALLE

2078	CASTING STRAINS	TENSION DE COULAGE	GUSSSPANNUNG
2079	CASTING, CAST WORK, FOUNDER'S WORK	PIECE DE FONTE MOULEE, PIECE DE FONDERIE	GUSSSTÜCK
2080	CASTINGS	PIECES COULEES (OU MOULEES)	GUSSTÜCKE
2081	CASTINGS STRAINS	TENSION DE COULEE	GUSSSPANNUNG
2082	CASTINGS, CORROSION RESISTANT	PIECES MOULEES(OU COULEES) RESISTANT A LA CORROSION	GUSSTÜCKE (KORROSIONSBESTÄNDIGE)
2083	CASTINGS, GRAY IRON	PIECES MOULEES EN FONTE GRISE	GRAUGUSSSTÜCKE
2084	CASTINGS, HEAT-RESISTING	PIECES MOULEES RESISTANTES AUX TEMPERATURES ELEVEES	GUSSSTÜCKE (HITZEBESTÄNDIGE)
2085	CASTLE NUT, CASTELLATED NUT	ECROU CRENELE, ECROU A CRENEAUX, ECROU A ENTAILLES, ECROU A FENETRES	KRONENMUTTER
2086	CASTOR OIL	HUILE DE RICIN	RIZINUSÖL, KASTORÖL
2087	CAT-HEAD	MANCHON DE CENTRAGE	ZENTRIERRING
2088	CAT'S-EYES	OEIL DE CHAT	KATZENAUGE
2089	CATACAUSTIC CURVE	CAUSTIQUE PAR REFLEXION	BRENNLINIE DURCH REFLEXION, KATAKAUSTISCHE LINIE
2090	CATALYSIS, CATALYTIC ACTION	CATALYSE	KATALYSE
2091	CATALYST	CATALYSEUR	KATALYSATOR
2092	CATALYTIC	CATALYTIQUE	KATALYTISCH
2093	CATCH	SAILLIE	NASE
2094	CATCH, DETEND, STOP PAWL, PAWL	CLIQUET D'ARRET, DOIGT D'ENCLIQUETAGE, DOIGT DE RETENUE, CHIEN	SPERRKLINKE, SPERRKEGEL, SPERRHAKEN
2095	CATENARY	CHAINETTE (GEOM.)	KETTENLINIE
2096	CATHETOMETER	CATHETOMETRE	KATHETOMETER
2097	CATHODE	CATHODE	KATODE
2098	CATHODE CLEANING	NETTOYAGE CATHODIQUE	KATODISCHE REINIGUNG
2099	CATHODE COPPER	CUIVRE ELECTROLYTIQUE	ELEKTROLYTKUPFER
2100	CATHODE DROP	CHUTE (DE TENSION) CATHODIQUE	KATODENSPANNUNGSABFALL
2101	CATHODE EFFICIENCY	RENDEMENT CATHODIQUE	KATODENWIRKUNGSGRAD
2102	CATHODE LAYER	REVETEMENT CATHODIQUE	ÜBERZUG (KATODISCHER)
2103	CATHODE PICKLING	DECAPAGE CATHODIQUE	KATODENBEIZUNG
2104	CATHODE RAYS	RAYONS CATHODIQUES	ELEKTRONENSTRAHLEN, KATODENSTRAHLEN
2105	CATHODE-ARY TUBE	TUBE CATHODIQUE, TUBE A RAYONS CATHODIQUES	ELEKTRONENSTRAHLRÖHRE, KATODENSTRAHLRÖHRE
2106	CATHODE, KATHODE, NEGATIVE ELECTRODE	ELECTRODE NEGATIVE, CATHODE	ELEKTRODE (NEGATIVE), KATHODE
2107	CATHODIC CORROSION	CORROSION CATHODIQUE	KORROSION, (KATODISCHE)
2108	CATHODIC POLARIZATION	POLARISATION CATHODIQUE	KATODENPOLARISATION
2109	CATHODIC PROTECTION	PROTECTION CATHODIQUE	KORROSIONSSCHUTZ (GALVANISCHER), SCHUTZ (KATODISCHER)
2110	CATHODIC VACUUM ETCHING	GRAVURE CATHODIQUE	ÄTZEN (KATODISCHES), GLIMMEN (KATODISCHES)
2111	CATHOLYTE	CATHOLYTE	KATOLYT
2112	CATION, KATHION	CATION, ION HYDROGENE	WASSERSTOFFION, KATION
2113	CAULK (TO), FULLER (TO)	MATER	VERSTEMMEN
2114	CAULKING TOOL, FULLERING TOOL	MATOIR	STEMMEISSEL
2115	CAULKING, FULLERING	MATAGE	VERSTEMMEN, VERSTEMMUNG
2116	CAUSTIC CURVE	CAUSTIQUE (OPT.)	BRENNLINIE
2117	CAUSTIC DIP	BAIN D'HYDROXYDE DE SODIUM	NATRIUMHYDROXYDBAD
2118	CAUSTIC EMBRITTLEMENT	FRAGILITE CAUSTIQUE	LAUGENSPRÖDIGKEIT

2119	**CAUSTIC LIQUOR, CAUSTIC SOLUTION**	LIQUIDE CAUSTIQUE	ÄTZLAUGE
2120	**CAUSTIC SODA lYE, CAUSTIC SODA SOLUTION**	LESSIVE, SOLUTION DE SOUDE CAUSTIQUE	NATRONLAUGE, ÄTZNATRONLAUGE
2121	**CAUSTIC SURFACE**	SURFACE CAUSTIQUE	BRENNFLÄCHE
2122	**CAVETTO**	CAVET, GORGE	HOHLKEHLE
2123	**CAVITATION**	CAVITATION	HOHLRAUMBILDUNG, HOHLSOGBILDUNG, KAVITATION
2124	**CAVITY**	COUCHE D'AIR, RETASSURE	LUFTSCHICHT, LUNKER
2125	**CAVITY, HOLLOW SPACE**	ESPACE CREUX, CREUX, CAVITE	HOHLRAUM
2126	**CELL**	PILE ELECTRIQUE	ZELLE (ELEKTR.)
2127	**CELL CAVITY**	CAVITE DE CELLULE	ELEMENTHOHLRAUM
2128	**CELL CONSTANT**	CONSTANTE D'UNE CELLULE ELECTROLYTIQUE	ELEKTROLYTZELLENKONSTANTE
2129	**CELLULAR RADIATOR**	RADIATEUR NID D'ABEILLES	ZELLENKÜHLER
2130	**CELLULOID**	CELLULOID	ZELLULOID, ZELLHORN
2131	**CELLULOSE**	CELLULOSE	ZELLSTOFF, ZELLULOSE
2132	**CELLULOSE ACETATE ACETYL CELLULOSE**	ACETATE DE CELLULOSE	AZETYLZELLULOSE, ZELLULOSEAZETAT
2133	**CEMENT**	CIMENT	ZEMENT
2134	**CEMENT (TO)**	CIMENTER (COUVRIR D'UNE COUCHE DE CIMENT), CEMENTER, ACIERER (MET.), MASTIQUER, LUTER	VERPUTZEN MIT ZEMENT, ZEMENTIEREN, VERKITTEN
2135	**CEMENT CONCRETE**	BETON DE CIMENT	ZEMENTBETON
2136	**CEMENT COPPER**	CUIVRE CEMENTATOIRE, CUIVRE DE CEMENT, CUIVRE REGENERE, CEMENT DE CUIVRE	ZEMENTKUPFER
2137	**CEMENT MORTAR**	MORTIER DE CIMENT	ZEMENTMÖRTEL
2138	**CEMENT STEEL, BLISTER STEEL**	ACIER CEMENTE, ACIER DE CEMENTATION, ACIER DE CARBURATION	ZEMENTSTAHL, BLASENSTAHL, EINSATZSTAHL
2139	**CEMENT, CASE HARDENING COMPOUND,**	CEMENT, AGENT DE CEMENTATION, POUDRE CARBURANTE	HÄRTEPULVER
2140	**CEMENT, GLUE FOR LEATHER**	COLLE POUR COURROIES	LEDERLEIM
2141	**CEMENTATION**	CEMENTATION	EINSATZHÄRTEN, ZEMENTIEREN, ZEMENTIERUNG
2142	**CEMENTATION, CONVERTING**	CEMENTATION (DES BARRES DE FER), ACIERATION, ACIERAGE	ZEMENTSTAHLHERSTELLUNG, ZEMENTATION, ZEMENTIEREN
2143	**CEMENTED BELT JOINT**	JONCTION DES COURROIES PAR COLLAGE	RIEMENVERBINDUNG (GELEIMTE)
2144	**CEMENTED CARBIDE CUTTING FOOLS**	OUTILS A MISES EN CARBURE	HARTMETALLSCHNEIDWERKZEUG
2145	**CEMENTED GLUED BELT**	COURROIE COLLEE	RIEMEN (GELEIMTER)
2146	**CEMENTING**	MASTICAGE, CIMENTATION (RECOUVREMENT D'UNE COUCHE DE CIMENT)	KITTEN, VERKITTEN; VERPUTZEN MIT ZEMENT, ZEMENTIEREN
2147	**CEMENTITE**	CEMENTITE	ZEMENTIT
2148	**CENTER**	CENTRE	MITTELPUNKT
2149	**CENTER BOLT**	BOULON ETOQUIAU	FEDERBOLZEN
2150	**CENTER LATHE**	TOUR A POINTE	SPITZENDREHBANK
2151	**CENTER LINE**	FIBRE NEUTRE, AXE	NEUTRALE FASER, MITTELLINIE, ACHSE
2152	**CENTER REST**	APPUI MEDIAN	MITTELAUFLAGER
2153	**CENTERING PIN**	PION DE CENTRAGE	ZENTRIERSTIFT
2154	**CENTERING RING**	BAGUE DE CENTRAGE	ZENTRIERRING
2155	**CENTERLESS CYLINDRICAL GRINDING MACHINE**	RECTIFIEUSE POUR SURFACES DE REVOLUTION SANS CENTRE	AUSSEN-RUND SCHLEIF-MASCHINE (SPITZENLOSE)

3

2156	CENTERLESS GRINDING	RECTIFICATION SANS POINTE(S)	SCHLEIFEN (SPITZENLOSES)
2157	CENTIGRADE SCALE, CELSIUS SCALE	ECHELLE CENTIGRADE, ECHELLE DE CELSIUS	CELSIUSSKALA
2158	CENTIMETRE	CENTIMETRE	ZENTIMETER
2159	CENTIMETRE-GRAMME-SECOND SYSTEM, C.G.S. UNITS	SYSTEME CENTIMETRE-GRAMME-SECONDE, SYSTEME C.G.S.	MASSSYSTEM (ABSOLUTES), GRAMM-ZENTIMETER-SEKUNDE-SYSTEM, G.C.S.-SYSTEM, ZENTIMETER-GRAMM-SEKUNDE-SYSTEM
2160	CENTRAL HEATING	CHAUFFAGE CENTRAL	ZENTRALHEIZUNG
2161	CENTRAL LOAD	CHARGE CENTRALE	BELASTUNG (ZENTRISCHE), BELASTUNG (ZENTRALE), BELASTUNG MITTIGE
2162	CENTRE	CENTRE	MITTELPUNKT (GEOM.)
2163	CENTRE (TO)	CENTRER	EINMITTELN, MITTEN, ZENTRIEREN
2164	CENTRE BIT	FORET A CENTRE, FORET A TETON, MECHE A CENTRE, MECHE A TROIS POINTES	ZENTRUMBOHRER
2165	CENTRE DISTANCE	ENTRAXE	ACHSABSTAND
2166	CENTRE HEAD	INSTRUMENT A CENTRER	ZENTRIERVORRICHTUNG
2167	CENTRE OF CURVATURE	CENTRE DE COURBURE	KRÜMMUNGSMITTELPUNKT
2168	CENTRE OF GRAVITY	CENTRE DE GRAVITE	SCHWERPUNKT, MASSENMITTELPUNKT
2169	CENTRE OF MOTION	CENTRE DE ROTATION	DREHPUNKT
2170	CENTRE OF OSCILLATION	CENTRE D'OSCILLATION	SCHWINGUNGSMITTELPUNKT
2171	CENTRE OF RIVET	CENTRE DU RIVET	NIETMITTE
2172	CENTRE PUNCH	POINTEAU DE CENTRAGE	KÖRNER, ANKÖRNER
2173	CENTRE SQUARE, RADIUS FLINDER	EQUERRE A TRACER LES CENTRES	MITTELPUNKTSUCHER, ZENTRIERWINKEL
2174	CENTRELESS GRINDER	MACHINE A MEULER SANS POINTES	SCHLEIFMASCHINE (SPITZENLOSE)
2175	CENTRIFUGAL BRAKE	FREIN CENTRIFUGE	SCHLEUDERBREMSE, ZENTRIFUGALBREMSE
2176	CENTRIFUGAL CASTING	COULEE CENTRIFUGE	SCHLEUDERGUSS
2177	CENTRIFUGAL CLUTCH	EMBRAYAGE CENTRIFUGE	FLIEHKRAFTKUPPLUNG
2178	CENTRIFUGAL COMPRESSOR, TURBO-COMPRESSOR	TURBO-COMPRESSEUR, COMPRESSEUR CENTRIFUGE	KREISELVERSICHTER, SCHAUFELVERDICHTER, TURBOKOMPRESSOR
2179	CENTRIFUGAL FAN	VENTILATEUR A FORCE CENTRIFUGE	SCHLEUDERGEBLÄSE, ZENTRIFUGGALGEBLÄSE
2180	CENTRIFUGAL FORCE	FORCE CENTRIFUGE	FLIEHKRAFT, ZENTRIFUGALKRAFT
2181	CENTRIFUGAL GOVERNOR	REGULATEUR CENTRIFUGE	FLIEHKRAFTREGLER
2182	CENTRIFUGAL LUBRICATOR	GRAISSEUR CENTRIFUGE	SCHMIERGEFÄSS (UMLAUFENDES), ZENTRIFUGALÖLER
2183	CENTRIFUGAL MOMENT	MOMENT CENTRIFUGE	ZENTRIFUGALMOMENT
2184	CENTRIFUGAL PUMP	POMPE CENTRIFUGE	KREISELPUMPE, SCHLEUDERPUMPE, ZENTRIFUGALPUMPE
2185	CENTRIFUGAL TYPE GOVERNOR, BALL GOVERNOR	REGULATEUR A FORCE CENTRIFUGE, REGULATEUR DE WATT, REGULATEUR A BOULES	FLIEHKRAFTREGLER, ZENTRIFUGALREGLER
2186	CENTRIGRADE CELSIUS THERMOMETER	THERMOMETRE CENTIGRADE	THERMOMETER (HUNDERTTEILIGES), CELESIUSTHERMOMETER
2187	CENTRING	CENTRAGE	EINMITTELN, ZENTRIEREN
2188	CENTRING AND SPOT-FACING MACHINE	MACHINE A CENTRER ET A DRESSER	ZENTRIER-U. PLANDREHMASCHINE
2189	CENTRIPETAL ACCELERATION	ACCELERATION CENTRIPETE	ZENTRIPETALBESCHLEUNIGUNG
2190	CENTRIPETAL FORCE	FORCE CENTRIPETE	ZENTRIPETALKRAFT
2191	CENTROID	CENTRE DE SURFACE	FLÄCHENMITTELPUNKT

2192	**CERAMIC METAL**	CERMET, MELANGE DE CARBURES FRITTES	CERMET, METALL-KERAMIKMISCHUNG (GESINTERTE)
2193	**CERAMICS**	CERAMIQUE	KERAMIK
2194	**CERESINE, CERASIN**	CIRE MINERALE, CERESINE	ZERESIN, OZOZEROTIN
2195	**CERIUM**	CERIUM	ZERIUM, ZER, CER
2196	**CERUSSITE, CARBONATE OF LEAD**	CERUSSITE	BLEISPAT, BLEIKARBONAT, WEISSBLEIERZ, ZERUSSIT
2197	**CESIUM**	CESIUM	CESIUM, ZÄSIUM
2198	**CETANE NUMBER**	INDICE DE CETANE	CETANZAHL
2199	**CHAFF CUTTERS**	HACHE PAILLES	STROHHÄCKSLER
2200	**CHAFING FATIGUE**	FATIGUE PAR CONTACTS DE FROTTEMENT	REIBUNGERMÜDUNG
2201	**CHAIN**	CHAINE	KETTE
2202	**CHAIN BARREL**	TAMBOUR A CHAINE, TAMBOUR POUR CHAINE-CABLE	KETTENTROMMEL
2203	**CHAIN BRAKE**	FREIN A CHAINE	KETTENBREMSE
2204	**CHAIN CASE**	CARTER DE CHAINES	KETTENKÄSTEN
2205	**CHAIN DOTTED LINE**	LIGNE UN TRAIT	LINIE (STRICHPUNKTIERTE)
2206	**CHAIN DRIVE**	COMMANDE PAR CHAINE	KETTENANTRIEB
2207	**CHAIN DRIVE, CHAIN GEARING**	TRANSMISSION PAR CHAINES	KETTENTRIEB
2208	**CHAIN DRIVE, DRIVING, DRIVE BY CHAINS**	COMMANDE PAR CHAINE	KETTENRADANTRIEB
2209	**CHAIN GRATE**	GRILLE A CHAINE	KETTENROST
2210	**CHAIN INTERMITTENT FILLET WELD**	SOUDURE D'ANGLE A RANGEES ALTERNEES SYMETRIQUES	KEHLNÄHTE (SYMMETRISCH VERSETZTE)
2211	**CHAIN LINK**	MAILLON DE CHAINE	KETTENGLIED
2212	**CHAIN LUBRICATION**	GRAISSAGE A CHAINETTE	KETTENSCHMIERUNG
2213	**CHAIN PULLEY BLOCK**	PALAN A CHAINE	KETTENFLASCHENZUG
2214	**CHAIN RIVET**	RIVET DE CHAINE	KETTENNIET
2215	**CHAIN RIVETED JOINT**	RIVURE EN CHAINE	PARALLELNIETUNG, KETTENNIETUNG
2216	**CHAIN SAW**	SCIE A CHAINETTE	KETTENSÄGE
2217	**CHAIN SHEAVE, SHEAVE WHEEL (FOR CHAIN)**	POULIE A CHAINE, POULIE A EMPREINTES, ROUE A EMPREINTES, BARBOTIN	KETTENROLLE
2218	**CHAIN TENSION ADJUSTER**	TENDEUR DE CHAINE	KETTENSPANNER
2219	**CHAIN TIGHTENER**	TENDEUR DE CHAINE	KETTENSPANNER
2220	**CHAIN WHEEL VALVE**	VOLANT A CHAINE ADAPTABLE	KETTENRADSCHIEBER
2221	**CHAIN-TYPE SULFIDE**	SULFURE EN CHAINE	SULFID (KETTENFÖRMIGES)
2222	**CHAIN, SERIES OF OPERATIONS, PROCESSES IN A FACTORY**	SUITE, GAMME DES OPERATIONS	ARBEITSGANG, FABRIKATIONSGANG
2223	**CHAINWHEEL**	VOLANT A CHAINE	KETTENRAD
2224	**CHALK**	CRAFE	KREIDE
2225	**CHALK MARL**	MARNE CALCAIRE	KALKMERGEL
2226	**CHALKING**	FARINAGE, PULVERISATION	ABKREIDEN, VERPULVERUNG
2227	**CHALKY WATER**	EAU CALCAIRE	WASSER (KALKHALTIGES)
2228	**CHAMBER ACID**	ACIDE DES CHAMBRES	KAMMERSÄURE
2229	**CHAMBER CHEST OF SLIDE VALVE**	BOITE, CHAMBRE DE DISTRIBUTION, BOITE A SOUPAPE, CHAPELLE	GEHÄUSE EINES SCHIEBERS, GEHÄUSE EINES VENTILS, VENTILKASTEN, VENTILKAMMER
2230	**CHAMFER (TO)**	CHANFREINER, ABATTRE UN ANGLE	ABFASEN, ABKANTEN, ABRÄNDERN
2231	**CHAMFERED EDGE**	CHANFREIN	KANTE (ABGEFASTE), ABFASUNG
2232	**CHAMFERED SET HAMMER**	CHASSE A BISEAU	SCHRÄGER SETZHAMMER, BALLHAMMER

2233	**CHAMFERING**	CHANFREINAGE	ABFASEN, ABKANTEN, ABSCHRÄGUNG
2234	**CHAMOIS LEATHER, WASH LEATHER**	PEAU DE NETTOYAGE, PEAU DE CHAMOIS	PUTZLEDER, WASCHLEDER, SÄMISCHLEDER
2235	**CHAMOTTE**	CHAMOTTE	SCHAMOTTE
2236	**CHANGE OF DIRECTION**	CHANGEMENT DE DIRECTION	RICHTUNGSÄNDERUNG, RICHTUNGSWECHSEL
2237	**CHANGE OF PRESSURE**	CHANGEMENT DE PRESSION	DRUCKÄNDERUNG
2238	**CHANGE OF STATE**	CHANGEMENT D'ETAT	ZUSTANDSÄNDERUNG
2239	**CHANGE OF TEMPERATURE**	CHANGEMENT DE TEMPERATURE	TEMPERATURWECHSEL, TEMPERATURÄNDERUNG
2240	**CHANGE OF VOLUME**	CHANGEMENT DE VOLUME	VOLUMENÄNDERUNG
2241	**CHANGE OVER (TO)**	INVERTIR, CHANGER LE SENS DU COURANT, COMMUTER	UMSCHALTEN (ELEKTR.)
2242	**CHANGE SPEED GEAR**	MECANISME DE CHANGEMENT DE VITESSE	WECHSELGETRIEBE
2243	**CHANGE WHEEL, SPARE WHEEL**	ROUE DE RECHANGE, ROUE DE SECOURS	WECHSELRAD, RESERVERAD, ERSATZRAD
2244	**CHANGE-OVER SWITCH, THROW-OVER SWITCH**	COMMUTATEUR	UMSCHALTER (ELEKTR.)
2245	**CHANGE, ALTERATION OF SPEED**	CHANGEMENT DE VITESSE, CHANGEMENT DE MULTIPLICATION	GESCHWINDIGKEITSÄNDERUNG
2246	**CHANGING OVER**	COMMUTATION, CHANGEMENT DU SENS D'UN COURANT	UMSCHALTEN (ELEKTR.)
2247	**CHANGING ROOM**	BARAQUE DE VESTIAIRE	UMKLEIDEBUDE
2248	**CHANNEL**	FER EN U, FER A GORGE, CANAL	RILLENEISEN, U-EISEN, KANAL
2249	**CHANNEL BEAM/IRON**	POUTRELLE EN U	U-TRÄGER
2250	**CHANNEL IRON**	FER EN E, FER EN U, FER A BRANCARDS	E-EISEN, U-EISEN
2251	**CHANNEL; (HYDR.:), FLUME, RACE**	RIGOLE, CANIVEAU, BIER	GERINNE, RINNE
2252	**CHANNELED PLATE**	TOLE STRIEE, TOLE CANNELEE	KANALBLECH
2253	**CHAPLET**	SUPPORT D'AME SUPPORT DE NOYAU	KERNSTÜTZE
2254	**CHAPMANIZING**	PROCEDE CHAPMAN	CHAPMAN-VERFAHREN
2255	**CHARACTER**	CARACTERE	ZEICHEN
2256	**CHARACTERISTIC CURVE**	CARACTERISTIQUE	KENNLINIE, CHARAKTERISTIK
2257	**CHARACTERISTIC INDEX OF A LOGARITHM**	CARACTERISTIQUE D'UN LOGARITHME	KENNZIFFER EINES LOGARITHMUS
2258	**CHARCOAL PIG IRON**	FONTE AU CHARBON DE BOIS	HOLZKOHLENROHEISEN
2259	**CHARGE**	CHARGE, CHARGE ELECTRIQUE	BESCHICKUNG, LADUNG, (ELEKTR.)
2260	**CHARGE A METALLURGICAL FURNACE (TO)**	CHARGER UN FOUR METALLURGIQUE	BESCHICKEN (EINEN METALLURGISCHEN OFFEN)
2261	**CHARGE AN ACCUMULATOR (TO)**	CHARGER UN ACCUMULATEUR	LADEN (EINEN AKKHUMULATOR)
2262	**CHARGE, VOLUME OF CHARGE**	CYLINDREE	LADUNG, ZYLINDERINHALT
2263	**CHARGING**	ALIMENTATION, CHARGEMENT	BESCHICKUNG
2264	**CHARGING A METALLURGICAL FURNACE**	CHARGEMENT D'UN FOUR METALLURGIQUE	BESCHICKUNG EINES METALLURGISCHEN OFENS
2265	**CHARGING AN ACCUMULATOR**	CHARGEMENT D'UN ACCUMULATEUR	LADEN EINES AKKUMULATORS
2266	**CHARGING CURRENT**	COURANT DE CHARGE	LADESTROM
2267	**CHARGING INDICATEUR**	INDICATEUR DE CHARGE	LADEKONTROLLAMPE
2268	**CHARGING MACHINE, FURNACE**	MACHINE A CHARGER LES FOURS	OFENBESCHICKUNGSMASCHINE
2269	**CHARGING, FILLING**	REMPLISSAGE, CHARGEMENT	FÜLLEN, FÜLLUNG
2270	**CHARPY IMPACT TEST**	ESSAI DE CHOC DE CHARPY (SUR EPROUVETTE ENTAILLEE)	KERBSCHLAGVERSUCH, CHARPY SCHLAGPROBE

2271	**CHARPY MACHINE**	MOUTON-PENDULE CHARPY	PENDELHAMMER FÜR SCHLAGVERSUCHE
2272	**CHARRED LEATHER**	CUIR BRULE, CALCINE, CARBONISE	LEDER (GERÖSTETES)
2273	**CHART**	TABLEAU, ABAQUE	TABELLE
2274	**CHASER, COMB TOOL**	PEIGNE POUR LES PAS DE VIS	STRÄHLER, GEWINDESTRÄHLER, GEWINDESTAHL, SCHRAUBSTAHL
2275	**CHASSIS**	CHASSIS	FAHRGESTELL
2276	**CHATTER MARKS**	MARQUES DE VIBRATION	RATTERMARKEN
2277	**CHATTERING OF THE VALVE**	VIBRATIONS, AFFOLEMENT DE LA SOUPAPE	FLATTERN, KLAPPERN DES VENTILS
2278	**CHECK**	CRIQUE, FELURE, FISSURE	RISS
2279	**CHECK A MEASUREMENT MADE (TO)**	VERIFIER LES MESURES	NACHMESSEN
2280	**CHECK ANALYSIS**	ANALYSE DE CONTROLE, VERIFICATION DE LA COMPOSITION CHIMIQUE, CONTRE-ANALYSE	KONTROLLANALYSE, PRÜFUNGSANALYSE
2281	**CHECK BOLT**	BOULON DE BLOCAGE	GEGENSCHRAUBE
2282	**CHECK MARKS**	MARQUES SUPERFICIELLES EN FORME DE V	ZIEHSPUREN (V-FÖRMIGE)
2283	**CHECK VALVE**	CLAPET DE RETENUE	RÜCKSCHLAGKLAPPE
2284	**CHECK VALVE, BACK PRESSURE VALVE, RETAINING VALVE, NON-RETURN VALVE**	SOUPAPE, CLAPET DE RETENUE	RÜCKSCHLAGVENTIL, RÜCKSCHLAGKLAPPE
2285	**CHECKERED PLATE**	TOLE GAUFREE, TOLE STRIEE	RIFFELBLECH
2286	**CHECKING**	FAIENCAGE, FORMATION DE CIRQUES	SPRUNGBILDUNG, RISSBILDUNG
2287	**CHEEK**	PART CENTRALE	TEIL (ZENTRALER)
2288	**CHEESE CLOTH**	MOUSSELINE	MULL
2289	**CHELATING AGENT**	AGENT DE CHELATION	CHELATBILDUNGSMITTEL
2290	**CHEMICAL**	CHIMIQUE	CHEMISCH
2291	**CHEMICAL ACTION**	PROCEDE CHIMIQUE	VORGANG (CHEMISCHER), PROZESS
2292	**CHEMICAL AFFINITY**	AFFINITE CHIMIQUE	VERWANDTSCHAFT, AFFINITÄT (CHEMISCHE)
2293	**CHEMICAL ANALYSIS**	ANALYSE CHIMIQUE	ANALYSE (CHEMISCHE)
2294	**CHEMICAL BALANCE**	BALANCE D'ANALYSE, TREBUCHET POUR ANALYSES	WAAGE (CHEMISCHE), ANALYSENWAAGE
2295	**CHEMICAL BEHAVIOUR**	PROPRIETES, CARACTERISTIQUES CHIMIQUES	VERHALTEN (CHEMISCHES)
2296	**CHEMICAL CHANGE**	TRANSFORMATION CHIMIQUE, ALTERATION CHIMIQUE	UMWANDLUNG (CHEMISCHE), VERÄNDERUNG (CHEMISCHE)
2297	**CHEMICAL COATING**	REVETEMENT CHIMIQUE	ÜBERZUG (CHEMISCHER)
2298	**CHEMICAL COMPONENT**	CONSTITUANT CHIMIQUE	KOMPONENTE (CHEMISCHE), BESTANDTEIL
2299	**CHEMICAL COMPOSITION**	COMPOSITION CHIMIQUE	ZUSAMMENSETZUNG (CHEMISCHE)
2300	**CHEMICAL COMPOUND**	COMBINAISON CHIMIQUE	VERBINDUNG (CHEMISCHE)
2301	**CHEMICAL CONSTITUTION**	CONSTITUTION CHIMIQUE	AUFBAU (CHEMISCHER), KONSTITUTION (CHEMISCHE)
2302	**CHEMICAL CORROSION**	CORROSION CHIMIQUE	KORROSION (CHEMISCHE)
2303	**CHEMICAL CUTTING FLUIDS**	FLUIDES DE COUPE CHIMIQUES, FLUIDES DE COUPE SYNTHETIQUES	SCHNEIDFLÜSSIGKEITEN (CHEMISCHE)
2304	**CHEMICAL DIP BRAZING**	BRASAGE AU TREMPE BRASAGE PAR IMMERSION	TAUCHLÖTEN
2305	**CHEMICAL ENERGY**	ENERGIE CHIMIQUE	ENERGIE (CHEMISCHE), REAKTIONSENERGIE
2306	**CHEMICAL ENGINEER**	INGENIEUR-CHIMISTE	INGENIEURCHEMIKER

3

2307	CHEMICAL EQUILIBRIUM	EQUILIBRE CHIMIQUE	GLEICHGEWISCHT (CHEMISCHES)
2308	CHEMICAL EQUIVALENT	EQUIVALENT CHIMIQUE	ÄQUIVALENT (CHEMISCHES), ÄQUIVALENTGEWICHT
2309	CHEMICAL FINISHING PROCESS	PROCEDE DE FINITION CHIMIQUE DE SURFACE	OBERFLÄCHENENDBEARBEITUNG (CHEMISCHE)
2310	CHEMICAL FORMULA	FORMULE CHIMIQUE	FORMEL (CHEMISCHE)
2311	CHEMICAL LEAD	PLOMB PUR	BLEI (REINES)
2312	CHEMICAL REACTION	REACTION CHIMIQUE	EINWIRKUNG (CHEMISCHE), REAKTION (CHEMISCHE)
2313	CHEMICAL STABILITY	STABILITE (CHIM.)	BESTÄNDIGKEIT (CHEM.)
2314	CHEMICAL SURFACE TREATING PROCESSES	PROCEDES CHIMIQUES DE TRAITEMENT DE SURFACE	OBERFLÄCHEN-BEHANDLUNGSVERFAHREN (CHEMISCHES)
2315	CHEMICAL SYMBOL	SYMBOLE (CHIM.)	ZEICHEN (CHEMISCHES), SYMBOL (CHEMISCHES)
2316	CHEMICAL WOOD PULP	PATE CHIMIQUE DE BOIS	HOLZZELLSTOFF
2317	CHEMICALLY COMBINED	CHIMIQUEMENT COMBINE	CHEMISCH GEBUNDEN
2318	CHEMICALLY COMBINED WATER, WATER OF CONSTITUTION	EAU DE CONSTITUTION, EAU CHIMIQUE	WASSER (GEBUNDENES), KONSTITUTIONSWASSER
2319	CHEMICALLY PURE	CHIMIQUEMENT PUR	CHEMISCH REIN
2320	CHEMICALS	PRODUITS CHIMIQUES	CHEMIKALIEN
2321	CHEMIST	CHIMISTE	CHEMIKER
2322	CHEMISTRY	CHIMIE	CHEMIE
2323	CHEMONUCLEAR REACTOR	REACTEUR DE RADIOCHIMIE	RADIOCHEMIEREAKTOR
2324	CHEMPUR TIN	ETAIN CHIMIQUEMENT PUR	ZINN (CHEMISCH REINES)
2325	CHEQUERED PLATE	TOLE STRIEE	RIFFELBLECH, BLECH (GERIFFELTES), BLECH (GERIPPTES)
2326	CHERRY RED HEAT	CHAUDE ROUGE-CERISE, ROUGE CERISE, COULEUR CERISE	KIRSCHROTGLUT
2327	CHERRY RED HOT	CHAUFFE, PORTE AU ROUGE CERISE	KIRSCHROTGLÜHEND
2328	CHESTNUT COAL, NUTS	GAILLETINS, TETES DE MOINEAU, NOISETTES	NUSSKOHLE
2329	CHIEF DRAUGHTSMAN	INGENIEUR-CONSTRUCTEUR EN CHEF	CHEFKONSTRUKTEUR
2330	CHILI SALPETRE, CALICHE, SODIUM NITRATE	SALPETRE, NITRATE DU CHILI, AZOTATE DE SODIUM, CALICHE	CHILISALPETER, PERUSALPETER, NATRONSALPETER, SALPETERSAURES NATRON, NATRIUMNITRAT
2331	CHILL	COQUILLE	KOKILLE, SCHALE
2332	CHILL (TO)	TREMPER (FONTE), REFROIDIR, TREMPER	ABSCHRECKEN, ABKÜHLEN
2333	CHILL CAST PIG	FER BRUT COULE EN COQUILLE	ROHEISEN IN KOKILLENGUSS
2334	CHILL CASTING	COULEE EN COQUILLE, FONTE EN COQUILLE	SCHALENGUSS, KOKILLENGUSS
2335	CHILLED CASTING	FONTE EN COQUILLE	HARTGUSS, SCHALENGUSS, KAPSELGUSS, KOKILLENGUSS
2336	CHILLED IRON	FONTE (DURCIE)	ROHEISEN (HARTGEGOSSEN)
2337	CHILLED STIFT METAL	METAL FONDU EN COQUILLE SANS DEFORMATION	METALL (VERFORMUNGSFREIES, IN METALLFORM GEGOSSENES
2338	CHILLING	TREMPE	ABSCHRECKEN, ABSCHRECKHÄRTUNG
2339	CHIMNEY STACK, FUNNEL	CHEMINEE	KAMIN, ESSE, SCHLOT, SCHORNSTEIN
2340	CHINA CLAY, KAOLIN	KAOLIN, CAOLIN	PORZELLANTON, PORZELLANERDE, KAOLIN
2341	CHINESE SCRIPT	ECRITURE CHINOISE	SCHRIFT (CHINESISCHE)
2342	CHINESE VEGETABLE TALLOW	SUIF VEGETAL DE CHINE	CHINESISCHER TALG
2343	CHIP	COPEAU, COPEAU DE BURINAGE	SPAN, MEISSELSPAN

2344	CHIP (TO), CHISEL (TO)	BURINER	MEISSELN
2345	CHIP TEST	ANALYSE DE COPEAUX	SPANANALYSE
2346	CHIP, CHIPPING, SHAVING	COPEAU	SPAN
2347	CHIPPING	BURINAGE, PIQUAGE AU MARTEAU	AUSHAUEN, ABKLOPPEN, MEISSELN
2348	CHIPS OF METAL, METAL SHAVINGS	COPEAUX METALLIQUES	METALLSPÄNE
2349	CHISEL	BURIN, CISEAU	MEISSEL
2350	CHISEL-SHAPED SOLDERING IRON	FER A SOUDER A TETE CARREE	HAMMERLÖTKOLBEN
2351	CHLORATE	CHLORATE	CHLORSÄURESALZ, CHLORAT
2352	CHLORIDE	CHLORURE	CHLORMETALL, CHLORID
2353	CHLORINATION	CHLORURATION	CHLORIERUNG
2354	CHLORINE	CHLORE	CHLOR
2355	CHLORINE MONOXIDE	ANHYDRIDE HYPOCHLOREUX	CHLORMONOXYD, UNTERCHLORIGSÄUREANHYDRID
2356	CHLOROFORM, TRICHLORMETHANE	CHLOROFORME	CHLOROFORM
2357	CHOKE	STARTER	DROSSELKLAPPE
2358	CHOKE (TO)	BOUCHER (SE), OBSTRUER (S'), ENGORGER (S')	VERSTOPFEN (SICH)
2359	CHOKE TUBE	DIFFUSEUR OU BUSE	LUFTTRICHTER
2360	CHOKED FILTER SURFACE	SURFACE FILTRANTE SATUREE	FILTERFLÄCHE (ÜBERSÄTTIGTE)
2361	CHOKED PIPE	TUYAU BOUCHE, TUYAU ENGORGE	ROHR (VERSTOPFTES)
2362	CHOKING	OBSTRUCTION, ENGORGEMENT	VERSTOPFEN, VERSTOPFUNG
2363	CHONITE, HARD RUBBER, VULCANITE	CAOUTCHOUC DURCI, EBONITE	HARTGUMMI, EBONIT, VULKANIT
2364	CHORD OF CIRCLE	CORDE D'UN CERCLE	KREISSEHNE
2365	CHORDAL TOOTH THICKNESS	EPAISSEUR DE DENT A LA CORDE	ZAHNSTÄRKE
2366	CHROMATE	CHROMATE	CHROMSÄURESALZ, CHROMAT
2367	CHROMATE DIP	CHROMATATION	CHROMATIEREN
2368	CHROMATIC	CHROMATIQUE	CHROMATISCH
2369	CHROMATIC ABERRATION	ABERRATION CHROMATIQUE, ABERRATION DE REFRANGIBILITE	ABWEICHUNG (CHROMATISCHE), ABERRATION (CHROMATISCHE)
2370	CHROME ALUM	ALUN DE CHROME	CHROMALAUN, KALICHROMALAUN, SCHWEFELSAURES CHROMOXYDKALL
2371	CHROME BRICK	BRIQUE DE CHROMITE	CHROMERZSTEIN, CHROMITSTEIN
2372	CHROME GREEN	VERT EMERAUDE	CHROMGRÜN
2373	CHROME ORE	CHROMITE	CHROMIT, CHROMEISENSTEIN
2374	CHROME RED	ROUGE DE CHROME	CHROMROT
2375	CHROME YELLOW, LEAD CHROMATE	CHROMATE DE PLOMB, JAUNE DE CHROME	CHROMGELB, BLEI (CHROMSAURES), BLEICHROMAT
2376	CHROME-NICKEL STEEL	ACIER CHROME-NICKEL	CHROMNICKELSTAHL
2377	CHROME-TANNED LEATHER	CUIR CHROME, TANNE AU CHROME	LEDER (CHROMGARES)
2378	CHROME-TUNGSTEN STEEL	ACIER AU CHROME-TUNGSTENE	CHROM-WOLFRAMSTAHL
2379	CHROMIC ACID	ACIDE CHROMIQUE	CHROMSÄURE
2380	CHROMIC ACID ANODIZING	OXIDATION ANODIQUE A L'ACIDE CHROMIQUE	OXYDATION (CHROMSAURE ANODISCHE)
2381	CHROMIC CHLORIDE	SESQUICHLORURE DE CHROME, CHLORURE CHROMIQUE	CHROMCHLORID
2382	CHROMIC OXYDE	SESQUIOXYDE DE CHROME	CHROMOXYD
2383	CHROMIC TRIOXIDE	ANHYDRIDE CHROMIQUE	CHROMSÄUREANHYDRID, CHROMTRIOXYD
2384	CHROMITE	CHROMITE	CHROMEISENSTEIN, CHROMIT
2385	CHROMIUM	CHROME	CHROM
2386	CHROMIUM NICKEL STEEL	ACIER AU NICKEL-CHROME	CHROMNICKELSTAHL
2387	CHROMIUM PAINT	MATIERE COLORANTE A BASE DE CHROME	CHROMFARBE

2388	CHROMIUM PLATING	CHROMAGE ELECTROLYTIQUE	VERCHROMUNG (ELEKTROLYTISCHE)
2389	CHROMIUM STEEL, CHROME STEEL	ACIER CHROME	CHROMSTAHL
2390	CHROMIZING	CEMENTATION PAR LE CHROME	INCHROMIEREN
2391	CHROMOUS CHLORIDE	CHLORURE CHROMEUX	CHROMCHLORÜR
2392	CHRYSOLITE	CHRYSOLITHE	CHRYSOLITH
2393	CHUCK	MANDRIN DE TOUR, MANDRIN	DREHBANKFUTTER, SPANNFUTTER
2394	CHUCK JAWS	MORS DOUX POUR MANDRINS	DREHFUTTER-BACKEN
2395	CHUTE	COULOIR, GLISSIERE, GOUTTIERE, GOULOTTE, DECHARGE	RUTSCHE, FÖRDERRINNE, BAHN (ABSCHÜSSIGE)
2396	CIGARETTE LIGHTER	ALLUME-CIGARETTE	ZIGARETTENANZÜNDER
2397	CINDER	CENDRE, LAITIER, SCORIE	ASCHE, SCHLACKE
2398	CINDER COOLER	REFRIGERANT DE CENDRES	ASCHENKÜHLER
2399	CINDER NOTCH	BEC DE PASSAGE DU LAITIER	SCHLACKENLOCH, SCHLACKENFORM
2400	CINNABAR	CINABRE	ZINNOBER, BERGZINNOBER, MERKURBLENDE, ZINNABARIT
2401	CIRCLE	CERCLE	KREIS
2402	CIRCLE OF CURVATURE	CERCLE DE COURBURE	KRÜMMUNGSKREIS
2403	CIRCLE SHEARS	CISAILLE CIRCULAIRE	KREISSCHERE
2404	CIRCLE, CIRCUMFERENCE PERIPHERY OF CIRCLE	CIRCONFERENCE, PERIPHERIE DU CERCLE	KREIS, KREISLINIE, KREISUMFANG, PERIPHERIE
2405	CIRCLIP F. GUDGEON PIN	ARRET D'AXE DE PISTON	SICHERUNGSRING F. KOLBENBOLZEN
2406	CIRCUIT	CIRCUIT ELECTRIQUE	STROMKREIS
2407	CIRCUIT BREAKER	DISJONCTEUR	AUSLÖSER, AUSSCHALTER
2408	CIRCULAR CHART RECORDER	ENREGISTREUR A DIAGRAMME CIRCULAIRE	KREISBLATTSCHREIBER
2409	CIRCULAR CONE	CONE CIRCULAIRE, CONE A BASE CIRCULAIRE	KREISKEGEL
2410	CIRCULAR CROSS SECTION	SECTION CIRCULAIRE	QUERSCHNITT (KREISFÖRMIGER)
2411	CIRCULAR CYLINDER	CYLINDRE A BASE CIRCULAIRE	KREISZYLINDER
2412	CIRCULAR ELECTRODE	MOLETTE (OU GALET) DE SOUDAGE	ELEKTRODENROLLE, SCHWEISSROLLE
2413	CIRCULAR FUNCTION	FONCTION CIRCULAIRE, FONCTION CYCLIQUE	KREISFUNKTION
2414	CIRCULAR GLASS	GLACE RONDE, VOYANT ROND	SCHAUGLAS (RUNDES)
2415	CIRCULAR GROOVE	RAINURE ANNULAIRE, RAINURE CIRCULAIRE	RINGNUT
2416	CIRCULAR INTERPOLATION	INTERPOLATION CIRCULAIRE	INTERPOLATION (ZIRKULARE)
2417	CIRCULAR KNIFE	COUTEAU CIRCULAIRE	KREISMESSER, SCHEIBENMESSER
2418	CIRCULAR MAIN, RING MAIN	CONDUITE CIRCULAIRE	RINGLEITUNG
2419	CIRCULAR MEASURE	MESURE DE L'ARC INTERCEPTE	BOGENMASS
2420	CIRCULAR MILLING MACHINE	FRAISEUSE CIRCULAIRE	RUNDFRÄSMASCHINE, KREISFRÄSMASCHINE
2421	CIRCULAR MOTION	MOUVEMENT CIRCULAIRE	BEWEGUNG (KREISFÖRMIGE), KREISSEHNE
2422	CIRCULAR PENDULUM	PENDULE CIRCULAIRE	KREISPENDEL
2423	CIRCULAR PITCH, CIRCUMFERENTIAL PITCH	PAS CIRCONFERENTIEL	UMFANGSTEILUNG, ZAHNTEILUNG, TEILKREIS
2424	CIRCULAR PLATES	PLATEAUX CIRCULAIRES	RUNDTISCHE
2425	CIRCULAR POLISHING BRUSH	BROSSE CIRCULAIRE A POLIR	BÜRSTENSCHEIBE
2426	CIRCULAR PROTRACTOR	RAPPORTEUR CERCLE ENTIER	VOLLKREISTRANSPORTEUR
2427	CIRCULAR SAW	SCIE CIRCULAIRE, DISQUE-SCIE	KREISSÄGE
2428	CIRCULAR SAWING MACHINE	SCIE CIRCULAIRE	KREISSÄGEMASCHINE
2429	CIRCULAR SHEARS	CISAILLE CIRCULAIRE	KREISSCHERE

2430	**CIRCULAR SPIRIT LEVEL, BOX SPIRIT LEVEL**	NIVEAU SPHERIQUE	DOSENLIBELLE
2431	**CIRCULAR TOOTH THICKNESS**	EPAISSEUR CIRCULAIRE DE LA DENT	ZAHNSTÄRKE IM ROLLKREIS
2432	**CIRCULATE (TO)**	CIRCULER	UMLAUFEN, ZIRKULIEREN
2433	**CIRCULATION**	CIRCULATION	UMLAUF, ZIRKULATION
2434	**CIRCULATION OF WATER**	CIRCULATION D'EAU	WASSERUMLAUF, WASSERZIRKULATION
2435	**CIRCUM-CIRCLE, CIRCUMSCRIBED CIRCLE**	CERCLE CIRCONSCRIT, CIRCONFERENCE CIRCONSCRITE	KREIS (UMGESCHRIEBENER), KREIS (UMLIEGENDER), UMKREIS
2436	**CIRCUMSCRIBED POLYGON**	POLYGONE CIRCONSCRIT	POLYGON (UMSCHRIEBENES), TANGENTENPOLYGON
2437	**CISSING**	RETRAIT, RETRACTION, RAMPAGE	PERLEN, BLASENBILDUNG
2438	**CISSOID**	CISSOIDE	ZISSOIDE, EFEUBLATTKURVE
2439	**CISTERN**	CITERNE	ZISTERNE
2440	**CISTERN BAROMETER**	BAROMETRE A CUVETTE	GEFÄSSBAROMETER
2441	**CITRIC ACID**	ACIDE CITRIQUE	ZITRONENSÄURE
2442	**CIVIL ARCHITECTURE**	ARCHITECTURE	HOCHBAU, ARCHITEKTUR
2443	**CIVIL ENGINEER**	INGENIEUR-CONSTRUCTEUR, INGENIEUR DES CONSTRUCTIONS CIVILES	BAUINGENIEUR
2444	**CIVIL ENGINEERING**	GENIE CIVIL, CONSTRUCTIONS CIVILES, TRAVAUX PUBLICS	TIEFBAU, BAUINGENIEURWESEN
2445	**CLAD PLATE**	TOLE PLAQUEE	BLECH (PLATTIERTES)
2446	**CLAD STEEL**	ACIER PLAQUE	STAHL (PLATTIERTER)
2447	**CLADDING**	DOUBLAGE, PLACAGE	PLATTIERUNG
2448	**CLAMP**	COLLIER	SPANNSCHELLE
2449	**CLAMP COUPLING, SPLIT COMPRESSION COUPLING**	MANCHON D'ACCOUPLEMENT CYLINDRIQUE, MANCHON A BOULONS NOYES, MANCHON A COQUILLES	SCHALENKUPPLUNG
2450	**CLAMP, CLIP, U-BOLT**	ETRIER	BÜGEL
2451	**CLAMP, CRAMP FOR JOINERS, SCREW CLAMP FOR WOOD WORKERS**	SERRE-JOINTS, PRESSE DE MENUISIER, PRESSE A COLLER	LEIMZWINGE
2452	**CLAMP, CRAMP, SCREW CLAMP**	SERRE-JOINTS, SERGENT, PRESSE A VIS	SCHRAUBZWINGE
2453	**CLAMPING SCREW**	VIS D'ARRET, VIS, ECROU DE SERRAGE	KLEMMSCHRAUBE, SPANNSCHRAUBE
2454	**CLAPPER BOX**	CHAPE DU BATTANT PORTE-OUTIL	MEISSELKLAPPENHALTER
2455	**CLARIFICATION, CLARIFYING**	CLARIFICATION	KLÄREN
2456	**CLARIFY (TO)**	CLARIFIER	KLÄREN
2457	**CLASP NAIL**	CLOU A CROCHET	HAKENNAGEL
2458	**CLASS OF IRON, BRAND OF IRON**	ESPECE DE FER, CATEGORIE DE FER	EISENSORTE
2459	**CLASS, VARIETY OF STEEL**	CATEGORIE SORTE D'ACIER	STAHLSORTE
2460	**CLASSIFICATION**	CLASSEMENT	KLASSIERUNG
2461	**CLAW CLUTCH, CLAW COUPLING, JAW COUPLING, DOG CLUTCH**	EMBRAYAGE, ACCOUPLEMENT A GRIFFES, ACCOUPLEMENT A DENTS	ZAHNKUPPLUNG, KLAUENKUPPLUNG
2462	**CLAW HAMMER**	MARTEAU A PANNE FENDUE, MARTEAU A DENT	KLAUENHAMMER, HAMMER MIT GESPALTENER FINNE
2463	**CLAW OF A HAMMER**	PANNE FENDUE, PANNE A PIED DE BICHE	FINNE (GESPALTENE), KLAUE EINES HAMMERS
2464	**CLAW OF DOG CLUTCH**	GRIFFE, DENT	KLAUE, KUPPLUNGSKLAUE
2465	**CLAY**	ARGILE	TON, PELIT, LEHM

2466	CLAY GROUND	TERRAIN ARGILEUX	LEHMBODEN
2467	CLAY IRONSTONE BAND, ARGILLACEOUS IRON ORE	LIMONITE ARGILEUSE, OCRE JAUNE	BRAUNEISENSTEIN (TONIGER), TONEISENSTEIN
2468	CLAY RING	BAGUE, BOURRELET EN TERRE GLAISE	TONRING
2469	CLEAN (TO), TRIM (TO), FETTLE CASTINGS (TO)	DESABLER, EBARBER, ECROUTER LES PIECES COULEES	PUTZEN (GUSSSTÜCKE)
2470	CLEAN METALLIC SURFACE	SURFACE METALLIQUE DECAPEE	OBERFLÄCHE (METALLREINE)
2471	CLEANER	PRODUIT DE NETTOYAGE	REINIGUNGSMITTEL
2472	CLEANER, ALKALINE	NETTOYANT ALCALIN	REINIGER (ALKALISCHER)
2473	CLEANER, SOLVENT	DEGRAISSEUR AU SOLVANT	FETTLÖSUNGSMITTEL, REINIGUNGSMITTEL
2474	CLEANING	NETTOYAGE DU METAL, DEGRAISSAGE	METALLREINIGUNG, ENTFETTUNG
2475	CLEANING AND POLISHING COMPOUNDS	PRODUITS DE NETTOYAGE ET DE POLISSAGE	REINIGUNGS- UND POLIERMITTEL
2476	CLEANING DIP	BAIN DE DECAPAGE PRELIMINAIRE	VORBRENNE
2477	CLEANING OIL	HUILE DE NETTOYAGE	PUTZÖL
2478	CLEANING, TRIMMING, FETTLING CASTINGS	EBARBAGE, DESABLAGE, ECROUTAGE DES PIECES COULEES	PUTZEN DER GUSSSTÜCKE
2479	CLEAR CUT SHARP IMAGE	IMAGE NETTE	BILD (SCHARFES)
2480	CLEAR OIL, PALE OIL	HUILE MINERALE BLONDE	MINERALÖL (GEREINIGTES), MINERALÖL (RAFFINNIERTES)
2481	CLEAR SPACE	VIDE LIBRE	ABSTAND (LICHTER)
2482	CLEARANCE	ESPACE NUISIBLE, ESPACE MORT, ESPACE NEUTRE	RAUM (TOTER), RAUM (SCHÄDLICHER), TOTRAUM
2483	CLEARANCE	JEU, JEU DE COIFFAGE, TOLERANCE	SPIEL, SPIELRAUM DER KERNMARKE, TOLERANZ
2484	CLEARANCE ANGLE, ANGLE OF CLEARANCE, ANGLE OF RELIEF	ANGLE D'INCIDENCE (D'UN OUTIL)	ANSTELLUNGSWINKEL, RÜCKENWINKEL
2485	CLEAT	TAQUET	TREIBKEIL, KNAGGE, ANSCHLAG
2486	CLEAVABLE ROCK	ROCHE CLIVABLE	GESTEIN (SPALTBARES)
2487	CLEAVAGE	CLIVAGE	SPALTUNG, SPALTFLÄCHE
2488	CLEAVAGE PLANE	PLAN DE CLIVAGE	SPALTFLÄCHE
2489	CLENCH THE RIVET HEADS TO	ECRASER LES RIVETS, RABBATRE L'EXCES DE LA TIGE DU RIVET	NIETE STAUCHEN
2490	CLENCHING THE RIVET HEADS	ECRASEMENT DES RIVETS	STAUCHEN DER NIETE
2491	CLEVIS	JUMELLE	SCHÄKEL
2492	CLEVIS PIN	GOUPILLE A TETE	GABELBOLZEN
2493	CLICK	DOIGT D'ENCLIQUETAGE	SCHALTKLINKE
2494	CLICK AND DETENT MOTION	ENCLIQUETAGE DOUBLE	DOPPELKLINKENGETRIEBE
2495	CLICK-WHEEL	ROUE A ROCHET	SPERRAD
2496	CLICK, RATCHET PAWL	CLIQUET	SCHALTLINKE, SCHLUBKLINKE, SCHIEBEKLINKE, SCHUBFALLE
2497	CLIMB MILLING	FRAISAGE EN AVALANT	FRÄSEN (GLEICHLÄUFIGES)
2498	CLIMBING ABILITY	TENUE DE ROUTE EN COTE	STEIGVERMÖGEN
2499	CLIMBING OF THE BELT	MONTEE DE LA COURROIE	KLETTERN DES RIEMENS
2500	CLINKER	MACHEFER (FOND), BRIQUE DURE ET TRES CUITE	KOHLENSCHLACKE, HERDSCHLACKE, KESSELSCHLACKE; KLINKER
2501	CLINOGRAPHIC PARALLEL PROJECTION	PROJECTION OBLIQUE	PARALLELPROJEKTION (SCHIEFE), PARALLELPROJEKTION (KLINOGRAPHISCHE)
2502	CLIP	ATTACHE	KLAMMER
2503	CLIP PULLEY	POULIE A GRIFFES	GREIFERSCHEIBE
2504	CLIPPING	EBARBAGE	ENTGRATUNG, ENTGRATEN

2505	CLOCK-BRASS	LAITON POUR HORLOGES	UHRENMESSING
2506	CLOCKWISE	DANS LE SENS DES AIGUILLES D'UNE MONTRE	RECHTSDREHEND, IM UHRZEIGERSINN
2507	CLOCKWISE ROTATION	ROTATION A DROITE	RECHTSDREHUNG
2508	CLOCKWISE WOUND, RIGHT-HANDRED WOUND, DEXTRORSE	ENROULE A DROITE, DEXTRORSUM	RECHTSGEWUNDEN, RECHTSGEWICKELT
2509	CLOCKWORK	MOUVEMENT D'HORLOGERIE	UHRWERK
2510	CLOGGING	COLMATAGE, ENCRASSEMENT	ABLAGERUNG, VERSCHMUTZEN, ANSETZEN VON SCHMUTZ
2511	CLOSE GRAINED IRON	FER A GRAIN FIN, FER A GRAIN SERRE	FEINKORNEISEN, EISEN (KLEINLUCKIGES)
2512	CLOSE JOINTED TUBES	TUBES RAPPROCHES	ROHRE MIT OFFENER NAHT
2513	CLOSE LAID WIRE ROPE	CABLE CLOS	SEIL (VERSCHLOSSENES), SEIL (VOLLSCHLÄCHTIGES)
2514	CLOSE PACKED	STRUCTURE COMPACTE (A)	DICHTGEPACKT
2515	CLOSE THE CIRCUIT TO	FERMER LE CIRCUIT ELECTRIQUE	STROMKREIS (DEN) SCHLIESSEN
2516	CLOSE UP THE RIVETS TO	FACONNER, FORMER LA TETE DU RIVET, RIVER	SCHLIESSKOFF (DEN) BILDEN
2517	CLOSED CELL FOAM	MOUSSE A CELLULES FERMEES	SCHAUMSTOFF (GESCHLOSSENZELLIGER)
2518	CLOSED CHAIN CLOSED RING CYCLIC HYDROCARBONS	HYDROCARBURES DE LA SERIE AROMATIQUE	KOHLENWASSERSTOFFE DER KARBOREIHE, ISOZYKLISCHE KOHLENWASSERSTOFFE
2519	CLOSED CORNER JOINT	ASSEMBLAGE EN ANGLE	ECKVERBAND
2520	CLOSED CURVE	COURBE FERMEE	KURVE (GESCHLOSSENE)
2521	CLOSED DIES	ESTAMPES FERMEES	STEMPEL (GESCHLOSSENE)
2522	CLOSED DOUBLE-BEVEL BUTT WELD	SOUDURE EN K SANS ECARTEMENT	ECKNAHTVERBINDUNG OHNE LUFTSPALT
2523	CLOSED DOUBLE-J BUTT WELD	SOUDURE DOUBLE J FERMEE	DOPPEL-J-NAHT OHNE LUFTSPALT
2524	CLOSED DOUBLE-U GROOVE WELD	SOUDURE DE BORD EN DOUBLE U FERMEE	DOPPEL-U-FUGENNAHT OHNE LUFTSPALT
2525	CLOSED DOUBLE-V GROOVE WELD	SOUDURE DE BORD EN DOUBLE-V SANS ECARTEMENT	X-FUGENNAHT OHNE LUFTSPALT
2526	CLOSED PASS	CANNELURE EMBOITEE, CANNELURE FERMEE	FLACHKABBER (GESCHLOSSENES), KABBER, KASTENKABBER
2527	CLOSED PASSES	CANNELURES FERMEES	WALZNUTEN (GESCHLOSSENE)
2528	CLOSED SINGLE-BEVEL BUTT WELD	SOUDURE EN DEMI-V SANS ECARTEMENT	HALB-V-NAHT OHNE LUFTSPALT
2529	CLOSED SINGLE-J BUTT WELD	SOUDURE EN J SANS ECARTEMENT	J-NAHT OHNE LUFTSPALT
2530	CLOSED SINGLE-U GROOVE WELD	SOUDURE DE BORD EN U FERMEE	U-FUGENNAHT OHNE LUFTSPALT
2531	CLOSED SINGLE-V GROOVE WELD	SOUDURE DE BORD EN V SANS ECARTEMENT	V-FUGENNAHT OHNE LUFTSPALT
2532	CLOSED SQUARE BUTT WELD	SOUDURE EN I SANS ECARTEMENT	I-NAHT OHNE LUFTSPALT
2533	CLOSED-LOOP SYSTEM	SYSTEME A BOUCLE DE RETOUR	RÜCKFÜHRUNGSREGELSYSTEM
2534	CLOSED, TUBE PRESSURE GAUGE	MANOMETRE A AIR COMPRIME	HEBERMANOMETER
2535	CLOSING CYLINDER	PISTON DE FERMETURE DU MOULE	SCHLIESSKOLBEN (FORM)
2536	CLOSING LINE SIDE	LIGNE DE FERMETURE	SCHLUSSLINIE
2537	CLOSING OF THE VALVE	FERMETURE DE LA SOUPAPE	VENTILSCHLUSS
2538	CLOTH MOP	MECHE EN LISIERE DE DRAP	SCHWABBEL
2539	CLOTH ROLLS	ROULEAUX DE POLISSAGE DRAPES	POLIERROLLEN (STOFFBEKLEIDETE)
2540	CLOUDBURST TREATMENT	TRAITEMENT AU JET DE SABLE	STRAHLSANDBLASEN
2541	CLOUDING	TROUBLE, LOUCHISSEMENT	TRÜBUNG, SCHLEIERBILDUNG
2542	CLUSTER	AMAS DE GUINIER-PRESTON	CLUSTER
2543	CLUSTER MILL	LAMINOIR A SIX CYLINDRES	SECHSROLLENWALZWERK
2544	CLUTCH	EMBRAYAGE	KUPPLUNG

2545	CLUTCH FACING	GARNITURE D'EMBRAYAGE	KUPPLUNGSBELAG
2546	CLUTCH HOUSING	CARTER D'EMBRAYAGE	KUPPLUNGSGEHÄUSE
2547	CLUTCH PEDAL	PEDALE D'EMBRAYAGE	KUPPLUNGSFUSSHEBEL
2548	CLUTCH RELEASE	DEBRAYAGE	AUSRÜCKEN
2549	CLUTCH THRUST BEARING	BUTEE DE DEBRAYAGE	KUPPLUNGSDRUCKLAGER
2550	CLUTRIATE (TO)	LAVER (LES MINERAIS)	SCHLÄMMEN
2551	CLYBURN SPANNER	CLEF A MOLETTE	ROLLGABELSCHLÜSSEL
2552	COACH SCREW	VIS A BOIS A TETE CARREE	WAGENSCHRAUBE
2553	COACH, CARRIAGE BOLT	BOULON DE CHARRONNAGE, BOULON DE CARROSSERIE	SCHLOSSSCHRAUBE
2554	COACHWORK NAILS	CLOUS DE CARROSSERIE	KAROSSERIENAGEL
2555	COAGULATE (TO)	SE COAGULER	GERINNEN, KOAGULIEREN
2556	COAGULATION	COAGULATION	GERINNEN, KOAGULIEREN
2557	COAGULUM	COAGULUM	GERINNSEL, KOAGULUM
2558	COAL	CHARBON, HOUILLE, CHARBON DE TERRE, CHARBON FOSSILE	KOHLE, STEINKOHLE
2559	COAL BREAKER	CONCASSEUR A CHARBON	KOHLENBRECHER
2560	COAL BRIQUETTE	AGGLOMERE DE HOUILLE	STEINKOHLENBRIKETT
2561	COAL BUNKER	SOUTE A CHARBON	KOHLENBUNKER
2562	COAL DISTRICT	REGION HOUILLERE	KOHLENGEBIET
2563	COAL DUST	POUSSIER DE HOUILLE	STAUBKOHLE
2564	COAL FIRING, FIRING, BURNING COAL	CHAUFFAGE AU CHARBON	KOHLENFEUERUNG, VERFEUERN VON KOHLEN
2565	COAL FURNACE, FURNACE FOR COAL	FOYER A CHARBON, FOYER A HOUILLE	KOHLENFEUERUNG (ANLAGE)
2566	COAL MINE, COAL PIT, COLLIERY	MINE DE HOUILLE, HOUILLERE, CHARBONNAGE	KOHLENBERGWERK, KOHLENGRUBE, KOHLENZECHE
2567	COAL TAR DYE	MATIERE COLORANTE DERIVEE DU GOUDRON DE HOUILLE	TEERFARBSTOFF
2568	COAL TAR OIL, SOLVENT NAPHTHA	HUILE DE GOUDRON DE HOUILLE	STEINKOHLENTEERÖL
2569	COAL TAR PITCH	BRAI SEC MINERAL	STEINKOHLENTEERPECH
2570	COAL TAR, GAS TAR	GOUDRON DE HOUILLE, GOUDRON DE GAZ, BRAI LIQUIDE	STEINKOHLENTEER, GASTEER
2571	COALESCED COPPER	CUIVRE ELECTROLYTIQUE COALESCE	KOALESZIERTES ELEKTROLYTKUPFER
2572	COALESCENCE	COALESCENCE	VERWACHSUNG, ZUSAMMENBALLUNG, KOALESZENZ
2573	COARSE GRAIN	GROS GRAIN	GROBKORN
2574	COARSE GRAINED SAND	SABLE A GROS GRAIN	SAND (GROBKÖRNIGER)
2575	COARSE PITCH THREAD	FILET A GRAND DIAMETRE	GROBGEWINDE, GEWINDE (GROBES)
2576	COARSE THREAD	GROS FILET	GROBGEWINDE
2577	COARSE-GRAINED STEEL	ACIER A GROS GRAIN	STAHL (GROBKÖRNIGER)
2578	COARSE-MESHED	A GRANDES MAILLES	WEITMASCHIG, GROBMASCHIG
2579	COARSE-STRAND WIRE ROPE	CABLE METALLIQUE EN GROS FILS	SEIL (GROBDRÄHTIGES)
2580	COARSELY GROUND POWDER	POUDRE GROSSIEREMENT BROYEE, GROSSE POUDRE	PULVER (GROBGEMAHLENES)
2581	COAT OF LACQUER	COUCHE DE LAQUE	LACKANSTRICH, LACKIERUNG
2582	COAT OF OIL PAINT	COUCHE DE COULEUR A L'HUILE	ÖLFARBENANSTRICH
2583	COAT OF PAINT	PEINTURE, ENDUIT	FARBANSTRICH, FARBÜBERZUG
2584	COATED ELECTRODE	ELECTRODE ENROBEE	MANTELELEKTRODE, ELEKTRODE (UMHÜLLTE)
2585	COATED PARTICLES	PARTICULES ENROBEES	TEILCHEN (UMHÜLLTE)
2586	COATED SHEETS	TOLES ENROBEES	BLECHE (UMHÜLLTE)
2587	COATING	REVETEMENT, ENROBAGE	AUFTRAG, ÜBERZUG, UMHÜLLUNG

2588	COATING, ANODIZED	REVETEMENT ANODIQUE	ÜBERZUG (ANODISCHER)
2589	COATING, ORGANIC	PRODUIT DE REVETEMENT	ANSTRICHMITTEL
2590	COATINGS	COUCHES, RECOUVREMENT, REVETEMENT	BESCHICHTUNG, SCHICHT, ÜBERZUG
2591	COATINGS, CHROMATE	COUCHE DE CHROMATE	CHROMATSCHICHT
2592	COATINGS, DIPPED	REVETEMENTS PAR TREMPE	TAUCHÜBERZUG
2593	COATINGS, ELECTROPLATED	REVETEMENTS ELECTROLYTIQUES OU GALVANOPLASTIQUES	ELEKTROPLATTIERUNG
2594	COATINGS, ENAMELED	REVETEMENTS EMAILLES	EMAILLEÜBERZUG
2595	COATINGS, LACQUER	REVETEMENT PAR PEINTURE	LACKÜBERZUG
2596	COATINGS, METALLIC	REVETEMENT METALLIQUE	METALLÜBERZUG
2597	COATINGS, NON-METALLIC	REVETEMENT NON-METALLIQUE	ÜBERZUG (NICHTMETALLISCHE)
2598	COATINGS, OXIDE	REVETEMENT D'OXYDE	OXYDBELAG
2599	COATINGS, PAINT	PEINTURES	FARBAUFTRAG
2600	COATINGS, PHOSPHATE	PHOSPHATATION	PHOSPHATISIERUNG
2601	COATINGS, ROST-INHIBITING	REVETEMENT ANTI-ROUILLE	ROSTSCHUTZMITTEL
2602	COATINGS, TIN	REVETEMENT D'ETAIN	ZINNÜBERZUG
2603	COATINGS, VITREOUS	REVETEMENT VITREUX	ÜBERGLASUNG
2604	COATINGS, ZINC	ZINGAGE	VERZINKUNG
2605	COAXIAL, COAXAL	COAXIAL	GLEICHACHSIG, KOAXIAL
2606	COB COAL, COBBLES	GAILLETTE, GAILLETTERIE	WÜRFELKOHLE
2607	COBALT	COBALT	KOBALT
2608	COBALT CARBONYL	COBALT-CARBONYLE	KOBALT-KARBONYL
2609	COBALT CHLORIDE	CHLORURE DE COBALT	KOBALTCHLORÜR, KOBALTOCHLORID, CHLORKOBALT
2610	COBALT NITRATE	AZOTATE DE COBALT	KOBALT (SALPETERSAURER), KOBALTNITRAT
2611	COBALT OXIDE	SESQUIOXYDE DE COBALT	KOBALTOXYD
2612	COBALT STEEL	ACIER AU COBALT	KOBALTSTAHL
2613	COBALT-CHROME STEEL	ACIER AU COBALT-CHROME	KOBALT-CHROMSTAHL
2614	COBALTITE, COBALT GLANCE	COBALTINE	KOBALTGLANZ, GLANZKOBALT
2615	COBALTOUS SULPHATE	SULFATE DE COBALT	KOBALTOXYDUL (SCHWEFELSAURES), KOBALTOXYDULSULFAT, KOBALTOSULFAT, KOBALTVITRIOL
2616	COBWEBBING	TOILE D'ARAIGNEE	FADENZIEHEN
2617	COCK	SOUPAPE, CLAPET, VANNE, ROBINET	KUGELHAHN, VENTIL, KLAPPE, HAHN
2618	COCK LEVER, PLUG VALVE WRENCH	CLEF DE MANOEUVRE (POUR ROBINET A BOISSEAU)	HAHNSCHLÜSSEL
2619	COCKLES	ONDULATIONS	WELLIGGRAT, WÖLBUNG
2620	COCONUT OIL	HUILE DE COCO, BEURRE DE COCO	KOKOSNUSSÖL, KOKOSFETT, KOKOSTALG
2621	COEFFICIENT	COEFFICIENT	BEIWERT, KENNZAHL, KOEFFIZIENT
2622	COEFFICIENT OF CONDUCTIVITY FOR HEAT	COEFFICIENT DE CONDUCTIBILITE CALORIFIQUE	WÄRMELEITZAHL
2623	COEFFICIENT OF CONTRACTION	COEFFICIENT DE CONTRACTION	KONTRAKTIONSZAHL
2624	COEFFICIENT OF CORROSION	COEFFICIENT DE CORROSION	KORROSIONSKOEFFIZIENT
2625	COEFFICIENT OF CUBICAL EXPANSION	COEFFICIENT DE DILATATION CUBIQUE	RAUMAUSDEHNUNGSZAHL
2626	COEFFICIENT OF DISCHARGE	COEFFICIENT D'ECOULEMENT	AUSFLUSSZAHL
2627	COEFFICIENT OF ELECTRICAL RESISTIVITY	COEFFICIENT DE RESISTIVITE ELECTRIQUE	WIDERSTANDS-BEIWERT (ELEKTRISCHER)
2628	COEFFICIENT OF EXPANSION	COEFFICIENT DE DILATATION	AUSDEHNUNGSZAHL
2629	COEFFICIENT OF FRICTION	COEFFICIENT DE FROTTEMENT	REIBUNGSKOEFFIZIENT

2630	COEFFICIENT OF LINEAR EXPANSION	COEFFICIENT DE DILATATION LINEAIRE	LÄNGENAUSDEHNUNGSZAHL
2631	COEFFICIENT OF PROPORTIONALITY	COEFFICIENT DE PROPORTIONNALITE	PROPORTIONALITÄTSFAKTOR
2632	COEFFICIENT OF RESISTANCE	COEFFICIENT DE RESISTANCE	WIDERSTANDSZAHL
2633	COEFFICIENT OF ROLLING FRICTION	COEFFICIENT DE ROULEMENT	REIBUNGSKOEFFIZIENT FÜR ROLLENDE REIBUNG
2634	COEFFICIENT OF SLIDING FRICTION	COEFFICIENT DE GLISSEMENT	REIBUNGSKOEFFIZIENT FÜR GLEITENDE REIBUNG
2635	COEFFICIENT OF SUPERFICIAL EXPANSION	COEFFICIENT DE DILATATION SUPERFICIELLE	FLÄCHENAUSDEHNUNGSZAHL
2636	COEFFICIENT OF THERMAL EXPANSION	COEFFICIENT DE DILATATION	AUSDEHNUNGSKOEFFIZIENT
2637	COEFFICIENT OF THERMOELECTRIC EFFECT	COEFFICIENT D'EFFET THERMOELECTRIQUE COUPLE THERMOELECTRIQUE	EFFEKT-KOEFFIZIENT (THERMOELEKTRISCHER)
2638	COEFFICIENT OF VELOCITY	COEFFICIENT DE VITESSE	GESCHWINDIGKEITSZAHL
2639	COERCITIVE FORCE	FORCE COERCITIVE	KOERZITIVEFELD, KOERZITIVKRAFT
2640	COERCIVE	COERCITIF	KOERZITIV
2641	COERCIVITY	FORCE COERCITIVE	KOERZITIVKRAFT
2642	COGGING	LAMINAGE DE LINGOTS	BLOCKWALZEN
2643	COHESION	COHERENCE	STANDFESTIGKEIT
2644	COHESION, COHESIVE POWER	COHESION, FORCE DE COHESION	KOHÄSION, KOHÄSTONSKRAFT, KOHÄRENZ
2645	COIL	ROULEAU, BOBINE, SERPENTIN	SPULE, ROHRSCHLANGE, BUND, ROLLE
2646	COIL OF A SPIRAL	SPIRE, TOUR DE SPIRE	WINDUNG EINER SPIRALE
2647	COIL SPRING	RESSORT HELICOIDAL	SCHRAUBENFEDER
2648	COIL WINDING	BOBINAGE	SPULENWICKLUNG
2649	COILED SPRING SUBJECTED TO BENDING	RESSORT DE FLEXION A ENROULEMENT	BIEGEFEDER (GEWUNDENE)
2650	COILING	ENROULEMENT	AUFROLLEN
2651	COINING	ESTAMPAGE, MATRICAGE, MONNAYAGE, CALIBRAGE	PRÄGEN, MÜNZEN, KALIBRIEREN
2652	COINING DIE	MATRICE D'ESTAMPAGE	PRÄGESTEMPEL
2653	COINING PRESS	PRESSE MONETAIRE	PRÄGEPRESSE
2654	COIR	FIBRE DE COCO, BOURRE DE COCO	KOKOSFASER
2655	COKE	COKE	KOKS
2656	COKE (TO)	COKEFIER, TRANSFORMER EN COKE	VERKOKEN
2657	COKE BLAST FURNACE PIG IRON	FONTE AU COKE (DE HAUT-FOURNEAU)	HOCHOFEN-ROHEISEN
2658	COKE BREAKER	CONCASSEUR A COKE, CASSE-COKE	KOKSBRECHER
2659	COKE BREEZE, DUST COKE	POUSSIER DE COKE, ESCARBILLE, BRAISETTE	LÖSCHE, KOKSLÖSCHE, KOKSABRIEB
2660	COKE FILTER	FILTRE A COKE	KOKSFILTER
2661	COKE OVEN	FOUR A COKE	KOKSOFEN
2662	COKE OVEN COKE, HARD COKE	COKE METALLURGIQUE, COKE DE FOUR	ZECHENKOKS, HÜTTENKOKS, GIESSEREIKOKS, SCHMELZKOKS
2663	COKE OVEN GAS	GAZ DE FOUR A COKE	KOKEREIGAS, KOKSOFENGAS
2664	COKE PIG IRON	FONTE AU COKE BOIS	KOKSROHEISEN
2665	COKE WORKS	USINE DE CARBONISATION DE LA HOUILLE	KOKEREI, VERKOKUNGSANSTALT
2666	COKING	COKEFACTION, COKEFICATION, TRANSFORMATION EN COKE	VERKOKUNG

2667	**COLD BEND TEST**	ESSAI DE CINTRAGE A FROID, ESSAI DE PLIAGE A FROID	KALTBIEGEPROBE
2668	**COLD BENDING**	PLIAGE A FROID, CINTRAGE A FROID	KALTBIEGUNG
2669	**COLD BLAST IRON**	FONTE A L'AIR FROID	ROHEISEN (KALT ERBLASENES)
2670	**COLD CHAMBER MACHINE**	MACHINE A CHAMBRE FROIDE	KALTKAMMER-DRUCKGIESSMASCHINE
2671	**COLD CHAMBER PRESSURE CASTING**	FONTE EN CHAMBRE FROIDE	KALTKAMMER- DRUCKGIESSEN
2672	**COLD CHISEL**	CISEAU A FROID	FLACHMEISSEL
2673	**COLD COINING**	ESTAMPAGE A FROID	KALTPRESSEN
2674	**COLD CRACK**	FELURE A FROID	KALTRISS, SPANNUNGSKALTRISS
2675	**COLD DRAW (TO)**	ETIRER A FROID	KALTZIEHEN
2676	**COLD DRAWING**	ETIRAGE A FROID	KALTZIEHEN
2677	**COLD DRAWN**	ETIRE A FROID	KALT GEZOGEN
2678	**COLD DRAWN TUBES**	TUBES ETIRES A FROID	RÖHRE (KALTGEZOGENE)
2679	**COLD FINISHING**	FINISSAGE A FROID	KALTNACHPRESSEN
2680	**COLD FLOW**	FLUAGE A FROID	KALTFLIESSEN
2681	**COLD FORGING**	FORGEAGE A FROID	KALTSCHMIEDEN
2682	**COLD FORMING**	PROFILAGE A FROID	KALTFORMUNG, KALTPROFILIEREN
2683	**COLD GALVANIZING**	GALVANISATION A FROID	VERZINKUNG (ELEKTROLYTISCHE)
2684	**COLD HAMMER (TO)**	MARTELER A FROID, ECROUIR	KALTHÄMMERN
2685	**COLD HAMMERING**	MARTELAGE A FROID, ECROUISSAGE	KALTHÄMMERN
2686	**COLD HEADING**	REFOULEMENT A FROID, FACONNEMENT DES TETES A FROID	KALTSTAUCHEN, KALTANKÖPFEN
2687	**COLD INSPECTION**	EXAMEN A FROID	KALTPRÜFUNG
2688	**COLD IRON SAW**	SCIE A FROID	KALTSÄGE
2689	**COLD PIERCING**	PENETRATION A FROID	KALTDURCHBOHRUNG
2690	**COLD PRESSING**	PRESSAGE A FROID	KALTPRESSEN
2691	**COLD REDUCING**	DEGAUCHISSAGE A FROID	HERUNTERWALZEN
2692	**COLD RIVET (TO)**	RIVER A FROID	KALTNIETEN
2693	**COLD RIVETING**	RIVURE A FROID	KALTNIETEN
2694	**COLD ROLL (TO)**	LAMINER A FROID	KALTWALZEN
2695	**COLD ROLLED BARS**	BARRES LAMINEES A FROID	KALTGEWALZTER STABSTAHL
2696	**COLD ROLLED PLATE**	TOLE LAMINEE A FROID	BLECH (KALTGEWALZTES)
2697	**COLD ROLLED STEEL**	ACIER ECROUI	STAHL (KALTGERECKTER)
2698	**COLD ROLLING**	LAMINAGE A FROID	KALTWALZEN
2699	**COLD SAWING**	SCIAGE A FROID	KALTSÄGEN
2700	**COLD SERVICE INSULATION**	ISOLATION A FROID	KÄLTEISOLIERUNGSMITTEL
2701	**COLD SET**	TRANCHE A FROID	KALTSCHROTMEISSEL
2702	**COLD SHORT IRON**	FER CASSANT A FROID, FER AIGRE	EISEN (KALTBRÜCHIGES)
2703	**COLD SHORTNESS**	FRAGILITE A FROID	KALTBRÜCHIGKEIT
2704	**COLD SHORTNESS OF IRON**	AIGREUR DU FER	KALTBRÜCHIGKEIT DES EISENS
2705	**COLD SHUT**	BRASURE A FROID	KALTLÖTSTELLE, KALTSCHWEISSE
2706	**COLD SHUT**	GOUTTE FROIDE	KALTGUSS
2707	**COLD SPRUING**	DECAPAGE A FROID	KALTABSPRITZEN
2708	**COLD STRIP MILL**	TRAIN DE LAMINAGE A FROID	KALTWALZWERK
2709	**COLD TEST**	ESSAI A FROID	KALTVERSUCH
2710	**COLD TREATMENT OF METALS**	TRAITEMENT DES METAUX A FROID	KALTBEHANDLUNG
2711	**COLD TRIMMING**	EBARBAGE A FROID	KALTABGRATEN
2712	**COLD WELDING**	SOUDURE A FROID, SOUDAGE A FROID	KALTSCHWEISSEN

2713	COLD WORKING	ECROUISSAGE, USINAGE A FROID, TRAVAIL A FROID	KALTBEARBEITUNG, KALTVERFORMUNG
2714	COLD-DRAWN TUBING	TUBE ETIRE A FROID	ROHR (KALTGEZOGENES)
2715	COLD-DRAWN WIRE	FIL ETIRE A FROID	DRAHT (KALTGEZOGENER)
2716	COLD-ROLLED	LAMINE A FROID	KALTGEWALZT
2717	COLD, CHIPPING CHISEL	CISEAU A FROID	KALTMEISSEL, BANKMEISSEL
2718	COLLAPSIBILITY	APTITUDE AU DEBOURRAGE	AUSSTOSSFÄHIGKEIT
2719	COLLAPSIBLE TOP	CAPOTE PLIANTE	KLAPPVERDECK
2720	COLLAR	ETRIER, COLLIER	BÜGEL, KLEMMSCHELLE
2721	COLLAR GAUGE, RING GAUGE	LUNETTE, BAGUE DE CALIBRE, LUNETTE VERIFICATRICE, TAMPON FILETE FEMELLE	KALIBERRING
2722	COLLAR STEP BEARING	CRAPAUDINE A PIVOT ANNULAIRE	RINGSPURLAGER
2723	COLLAR SURFACE OF THRUST BEARING	JOUE DU COUSSINET D'UNE BUTEE	RINGFLÄCHE EINES KAMMLAGERS
2724	COLLAR, FAST COLLAR	EMBASE, RONDELLE D'UN ARBRE	BUNDRING, WELLENBUND
2725	COLLECTING PIPE	TUYAU COLLECTEUR	SAMMELROHR
2726	COLLECTOR PLATES	PLAQUES COLLECTIVES	SAMMELPLATTEN
2727	COLLET	MANDRIN	SPANNDORN
2728	COLLET CHUCK	DOUILLE DE SERRAGE	ZANGENFUTTER
2729	COLLETS	MANDRIN ET PINCES PORTE-FRAISE	ZANGENFUTTER
2730	COLLIMATE (TO)	COLLIMATER	RICHTEN, EINSTELLEN
2731	COLLIMATION	COLLIMATION	KOLLIMATION
2732	COLLIMATOR	COLLIMATEUR	SPALTROHR, KOLLIMATOR
2733	COLLODION	COLLODION	KOLLODIUM, KLEBÄTHER
2734	COLLOID	COLLOIDE	KOLLOID
2735	COLLOIDAL	COLLOIDAL	KOLLOIDAL
2736	COLLOIDAL PARTICLES	PARTICULES COLLOIDALES	KOLLOIDTEILCHEN
2737	COLOPHONY	COLOPHANE, POIX SECHE, BRAI SEC	KOLOPHONIUM
2738	COLOR METALLOGRAPHY	METALLOGRAPHIE EN COULEURS	FARBMETALLOGRAPHIE
2739	COLORIMETRIC SPECTROMETRY	SPECTROSCOPIE COLORIMETRIQUE	ANALYSE (KOLORIMETRISCHE)
2740	COLORING	COLORATION	FÄRBUNG
2741	COLOUR	COULEUR (PHYS.)	FARBE (PHYS.)
2742	COLOUR A DRAWING (TO)	LAVER UN DESSIN	ZEICHNUNG (EINE) AUSTUSCHEN, ZEICHNUNG (EINE) ANLEGEN
2743	COLOUR BRUSH	PINCEAU	PINSEL, TUSCHPINSEL
2744	COLOUR OF INCANDESCENCE	TEMPERATURE DE FORGEAGE, CHALEUR DE FORGE	GLÜHFARBE
2745	COLOURED PENCIL	CRAYON DE COULEUR, PASTEL	FARBSTIFT
2746	COLOURLESS GLASS	VERRE INCOLORE	GLAS (FARBLOSES)
2747	COLUMBIUM	NIOBIUM	NIOB
2748	COLUMN	POTEAU, COLONNE	SÄULE, PFEILER
2749	COLUMN HEAD	TETE DE POTEAU	SÄULENKOPF
2750	COLUMN OF LIQUID, LIQUID COLUMN	COLONNE LIQUIDE	FLÜSSIGKEITSSÄULE
2751	COLUMN OF MERCURY	COLONNE DE MERCURE, COLONNE BAROMETRIQUE	QUECKSILBERSÄULE
2752	COLUMN OF WATER	COLONNE D'EAU	WASSERSÄULE
2753	COLUMN PIPE	COLONNE MONTANTE, CONDUITE VERTICALE	STEIGLEITUNG
2754	COLUMN SUPPORTED FRAME	CHARPENTE A POTEAUX	STÄNDERTRAGWERK
2755	COLUMN TYPE DRILLING MACHINE	PERCEUSE SUR BATI OU MONTANT, PERCEUSE A COLONNE	STÄNDERBOHRMASCHINE

2756	COLUMN WITH BOTH ENDS HINGED	POTEAU AVEC LES DEUX EXTREMITES GUIDEES	TRÄGER (BEIDERSEITSEINGESPANNTER)
2757	COLUMNAR CRYSTALS	CRISTAL BASALTIQUE	STENGELKRISTALL
2758	COLUMNAR STRUCTURE	STRUCTURE BASALTIQUE	STENGELGEFÜGE
2759	COLUMNED FRAME	CHARPENTE A POTEAUX	SÄULENGERÜST
2760	COMB GAUGE	JAUGE DE FILETAGE	GEWINDEKONTROLLEHRE
2761	COMBINATION	COMBINAISON (MATH.)	KOMBINATION (MATH.)
2762	COMBINATION DIE	MOULE A EMPREINTES MULTIPLES	KOMPOUNDSCHNITT
2763	COMBINATION DRILL AND COUNTERSINK	FORET A CENTRER	ZENTRIERBOHRER (KOMBINIERTER)
2764	COMBINATION PLIERS	PINCE MOTORISTE	MAULRINGSCHLÜSSEL
2765	COMBINE DIAGRAMS (TO)	RANKINISER, TOTALISER LES DIAGRAMMES	DIAGRAMME RANKINISIEREN
2766	COMBINE HARVESTERS AND ACCESSORY EQUIPMENT	MOISSONNEUSES-BATTEUSES	MÄHDRESCHER UND ZUSÄTZLICHE ANLAGE
2767	COMBINED CARBON	CARBONE COMBINE AU FER	KOHLENSTOFF (CHEMISCH GEBUNDENER IM EISEN)
2768	COMBINED SUCTION AND PRESSURE FAN	VENTILATEUR ASPIRANT ET SOUFFLANT	SAUG- UND DRUCKLÜFTER, SAUG- UND DRUCKVENTILATOR
2769	COMBINING MIXING CONE NOZZLE	TUYERE CONVERGENTE, AJUTAGE CONVERGENT, CONVERGENT (D'UNE TUYERE)	MISCHDÜSE
2770	COMBINING THE DIAGRAMS	RANKINISATION DES DIAGRAMMES	RANKINISIEREN VON DIAGRAMMEN
2771	COMBINING WEIGHT PROPORTION	PROPORTIONS DEFINIES	VERBINDUNGSGEWICHT
2772	COMBUSTIBLE	COMBUSTIBLE	BRENNBAR
2773	COMBUSTIBLENESS, COMBUSTIBILITY	COMBUSTIBILITE	BRENNBARKEIT, VERBRENNLICHKEIT
2774	COMBUSTION	COMBUSTION	VERBRENNUNG
2775	COMBUSTION AND EXPANSION STROKE	COURSE DE DETENTE, COURSE DE COMBUSTION	AUSDEHNUNGSHUB, EXPANSIONSHUB
2776	COMBUSTION CHAMBER	CHAMBRE DE COMBUSTION	VERBRENNUNGSRAUM, KAMMER
2777	COMBUSTION CHAMBER OF A FURNACE	CHAMBRE DE COMBUSTION D'UN FOYER	VERBRENNUNGSRAUM EINER FEUERUNG
2778	COMBUSTION CHAMBER OF AN INTERNAL COMBUSTION ENGINE, EXPLOSION CHAMBER, FIRING CHAMBER, LIGHTING CHAMBER	CHAMBRE D'EXPLOSIONS D'UN MOTEUR	VERBRENNUNGSRAUM EINES MOTORS
2779	COMBUSTION, COMPLETE	COMBUSTION COMPLETE	VERBRENNUNG (VOLLSTÄNDIGE)
2780	COMMAND	ORDRE	BEFEHL, KOMMANDO
2781	COMMERCIAL EFFICIENECY	RENDEMENT INDUSTRIEL, RENDEMENT COMMERCIAL, RENDEMENT ECONOMIQUE	WIRKUNGSGRAD (WIRTSCHAFTLICHER), WIRKUNGSGRAD (KOMMERZIELLER)
2782	COMMERCIAL POWER LINE	SECTEUR	NETZANSCHLUSS
2783	COMMERCIAL UTILISATION OF AN INVENTION	APPLICATION INDUSTRIELLE D'UNE INVENTION	VERWERTUNG (GEWERBLICHE) EINER ERFINDUNG
2784	COMMERCIAL VEHICLE	VEHICULE UTILITAIRE	NUTZFAHRZEUG
2785	COMMERCIAL ZINC, SPELTER	ZINC DU COMMERCE, ZINC BRUT	ROHZINK, HÜTTENZINK, HANDELSZINK
2786	COMMINUTION	PULVERISATION, CONCASSAGE	FEINZERKLEINERUNG
2787	COMMON ALUM, POTASH CRYSTAL ALUM	ALUN ORDINAIRE, ALUN POTASSIQUE	ALAUN, KALIALAUN, KALIUMALUMINIUMSULFAT
2788	COMMON BRIGGIAN LOGARITHM	LOGARITHME VULGAIRE, LOGARITHME DECIMAL	LOGARITHMUS (GEMEINER), LOGARITHMUS (BRIGG'SSCHER)
2789	COMMON DENOMINATOR	DENOMINATEUR COMMUN	NENNER (GEMEINSCHAFTLICHER), HAUPTNENNER, GENERALNENNER
2790	COMMON SALT, SODIUM CHLORIDE	SEL COMMUN, SEL MARIN, CHLORURE DE SODIUM	KOCHSALZ, SIEDESALZ, NATRIUMCHLORID, CHLORNATRIUM

2791	COMPACT	COMPRIME, AGGLOMERE	PRESSKÖRPER, PRESSLING
2792	COMPACT COKE	COKE DENSE	KOKS (KOMPAKTER)
2793	COMPANION DIMENSIONS	COTES DE RACCORDEMENT	ANSCHLUSSMASSE
2794	COMPARATOR	COMPARATEUR	KOMPARATOR
2795	COMPARATORS	COMPARATEURS	VERGLEICHER
2796	COMPARISON BLOCKS	CALES DE COMPARAISON	DICKENVERGLEICHER
2797	COMPARISON MEASUREMENT	MESURE COMPARATIVE	VERGLEICHSMESSUNG
2798	COMPARTMENT	COMPARTIMENT, CASE, CAISSON	ABTEIL, ABTEILUNG, FACH, RAUM, KASTEN
2799	COMPASS	BOUSSOLE	KOMPASS
2800	COMPASS CARD	ROSE DES VENTS	WINDROSE
2801	COMPASS PLANE	RABOT CINTRE	SCHIFFHOBEL
2802	COMPASS SAW, KEYHOLE SAW, PAD SAW, LOCK SAW	SCIE A GUICHET	STICHSÄGE, SPITZSÄGE, LOCHSÄGE
2803	COMPASSES WITH INK POINT, PEN POINT	COMPAS A TIRE-LIGNE	ZIEHFEDERZIRKEL
2804	COMPASSES WITH PENCIL POINT	COMPAS A PORTE-CRAYON	BLEISTIFTZIRKEL
2805	COMPASSES WITH REMOVABLE LEGS	COMPAS A PIECES DE RECHANGE	EINSATZZIRKEL
2806	COMPATIBILITY	COMPATIBILITE	VERTRÄGLICKEIT (GEGENSEITIGE)
2807	COMPENSATE (TO)	COMPENSER	AUSGLEICHEN, KOMPENSIEREN
2808	COMPENSATED PENDULUM	PENDULE COMPENSATEUR, PENDULE COMPENSE, COMPENSATEUR	KOMPENSATIONSPENDEL
2809	COMPENSATION	COMPENSATION	AUSGLEICH, AUSGLEICHUNG, KOMPENSIERUNG, KOMPENSATION
2810	COMPLEMENT OF AN ANGLE	COMPLEMENT D'UN ANGLE	KOMPLEMENT EINES WINKELS
2811	COMPLEMENTARY ANGLE	ANGLE COMPLEMENTAIRE	KOMPLEMENTWINKEL
2812	COMPLEMENTARY COLOUR	COULEUR COMPLEMENTAIRE	ERGÄNZUNGSFARBE, KOMPLEMENTÄRFARBE
2813	COMPLEX NUMBER	NOMBRE COMPLEXE	ZAHL (KOMPLEXE)
2814	COMPLEX VARIABLE	VARIABLE COMPLEXE	KOMPLEXE VERÄNDERLICHE
2815	COMPONENT	COMPOSANT	KOMPONENTE
2816	COMPONENT FORCE	FORCE COMPOSANTE	TEILKRAFT
2817	COMPONENT OF A FORCE	COMPOSANTE	SEITENKRAFT, TEILKRAFT, KOMPONENTE
2818	COMPOSITE DIE	MATRICE COMPOSITE	GESENK (ZWEITEILIGES)
2819	COMPOSITE ELECTRODE	ELECTRODE COMPOSITE	VERBUNDELEKTRODE
2820	COMPOSITE NUMBER	NOMBRE COMPOSE	ZAHL (ZUSAMMENGESETZTE)
2821	COMPOSITE PLATE	DEPOT ELECTROLYTIQUE A PLUSIEURS COUCHES	AUFLAGE (MEHRSCHICHT-ELEKTROLYTISCHE)
2822	COMPOSITE STRENGTH	RESISTANCE COMPLEXE, RESISTANCE COMPOSEE	FESTIGKEIT (ZUSAMMENGESETZTE)
2823	COMPOSITION	COMPOSITION	ZUSAMMENSETZUNG, KOMPOSITION
2824	COMPOSITION BRASS	LAITON ROUGE	ROTMESSING
2825	COMPOSITION OF FORCES	COMPOSITION DES FORCES	ZUSAMMENSETZUNG VON KRÄFTEN
2826	COMPOSITION PLANE	PLAN D'ACCOLEMENT (MACLE)	ZWILLINGSFLÄCHE
2827	COMPOUND	PATE A POLIR, COMPOSITION CHIMIQUE	MASSE, POLIERPASTE, VERBINDUNG (CHEMISCHE)
2828	COMPOUND COMPACT	EBAUCHE A PLUSIEURS CONSTITUANTS	MEHRSTOFFPRESSLING
2829	COMPOUND ENGINE	MACHINE COMPOUND	VERBUNDDAMPFMASCHINE, COMPOUNDMASCHINE
2830	COMPOUND GAUGE	MANO-VACUOMETRE	MANO-VAKUUMMETER
2831	COMPOUND GIRDER	POUTRE MIXTE OU COMPOSEE	VERBUNDTRÄGER, VERSTEIFTER WALZTRÄGER

2832	**COMPOUND OIL, MIXED OIL**	HUILE MIXTE, HUILE COMPOUND	MISCHÖL, COMPOUNDÖL
2833	**COMPOUND SCREW, DIFFERENTIAL SCREW**	VIS DIFFERENTIELLE	DIFFERENTIALSCHRAUBE
2834	**COMPOUND SPRING, LAMINATED SPRING**	RESSORT A PLUSIEURS LAMES, RESSORT A LAMES ETAGEES	FEDER (ZUSAMMENGESETZTE), FEDER (GESCHICHTETE), BLATTFEDERWERK
2835	**COMPOUND WOUND MOTOR**	MOTEUR COMPOUND	VERBUNDMOTOR, DOPPELSCHLUSSMOTOR, COMPOUNDMOTOR
2836	**COMPOUND, MULTI- STAGE COMPRESSOR**	COMPRESSEUR COMPOUND, COMPRESSEUR ETAGE	VERBUNDVERDICHTER, STUFENVERDICHTER, COMPOUNDVERDICHTER
2837	**COMPRESS (TO)**	COMPRIMER	VERDICHTEN, KOMPRIMIEREN
2838	**COMPRESS A SPRING (TO)**	COMPRIMER UN RESSORT	FEDER (EINE) ZUSAMMENDRÜCKEN
2839	**COMPRESSED AIR**	AIR COMPRIME	LUFT (VERDICHTETE) (KOMPRIMIERTE), DRUCKLUFT, PRESSLUFT
2840	**COMPRESSED AIR BRAKE**	FREIN A AIR COMPRIME	LUFTDRUCKBREMSE, DRUCKLUFTBREMSE
2841	**COMPRESSED AIR CYLINDER**	CYLINDRE A AIR	DRUCKLUFTZYLINDER
2842	**COMPRESSED AIR DRIVE**	COMMANDE PNEUMATIQUE	DRUCKLUFTANTRIEB, ANTRIEB MIT DRUCKLUFT
2843	**COMPRESSED AIR ENGINE**	MOTEUR A AIR COMPRIME, AERO- MOTEUR	PRESSLUFTMOTOR, DRUCKLUFTMOTOR
2844	**COMPRESSED AIR LOCOMOTIVE**	LOCOMOTIVE A AIR COMPRIME	DRUCKLUFTLOKOMOTIVE, PRESSLUFTLOKOMOTIVE
2845	**COMPRESSED AIR PIPING**	CONDUITE, TUYAUTERIE, DISTRIBUTION D'AIR COMPRIME	DRUCKLUFTLEITUNG, PRESSLUFTLEITUNG
2846	**COMPRESSED AIR TEST**	EPREUVE A L'AIR COMPRIME	DRUCKLUFTPROBE
2847	**COMPRESSED GAS**	GAZ COMPRIME	GAS (VERDICHTETES), GAS (KOMPRIMIERTES)
2848	**COMPRESSED OXYGEN BLOW PIPE**	CHALUMEAU A OXYGENE COMPRIME	BRENNER, LÖTBRENNER FÜR VERDICHTETEN SAUERSTOFF
2849	**COMPRESSED STEEL**	ACIER COMPRIME	PRESSSTAHL, STAHL (KOMPRIMIERTER)
2850	**COMPRESSED STEEL SHAFT**	ARBRE EN ACIER COMPRIME	WELLE (KOMPRIMIERTE)
2851	**COMPRESSIBILITY**	COMPRESSIBILITE	VERDICHTBARKEIT
2852	**COMPRESSIBILITY, ELASTICITY OF BULK OF VOLUME**	COMPRESSIBILITE	ZUSAMMENDRÜCKBARKEIT, KOMPRESSIBILITÄT, VOLUMENELASTIZITÄT
2853	**COMPRESSIBLE**	COMPRESSIBLE	ZUSAMMENDRÜCKBAR, KOMPRIMIERBAR
2854	**COMPRESSION**	COMPRESSION	VERDICHTUNG, KOMPRESSION
2855	**COMPRESSION BAR**	BARRE TRAVAILLANT A LA COMPRESSION	DRUCKSTAB, GEDRÜCKTER STAB
2856	**COMPRESSION CHAMBER**	CHAMBRE DE COMBUSTION	VERBRENNUNGSRAUM
2857	**COMPRESSION LINE**	LIGNE DE COMPRESSION	KOMPRESSIONSLINIE
2858	**COMPRESSION OF THE AIR**	COMPRESSION DE L'AIR	LUFTVERDICHTUNG
2859	**COMPRESSION PRESSURE**	PRESSION DE COMPRESSION	KOMPRESSIONSDRUCK
2860	**COMPRESSION RATIO**	TAUX DE COMPRESSION, RAPPORT DE COMPRESSION, RAPPORT VOLUMETRIQUE	VERDICHTUNGSVERHÄLTNIS
2861	**COMPRESSION RING**	SEGMENT DE COMPRESSION	KOLBENRING, KOMPRESSIONSRING
2862	**COMPRESSION SPRING, SPRING FOR COMPRESSION**	RESSORT DE COMPRESSION	DRUCKFEDER
2863	**COMPRESSION STRAIN**	EFFORT DE COMPRESSION	DRUCKBEANSPRUCHUNG
2864	**COMPRESSION STROKE**	TEMPS DE COMPRESSION, COURSE DE COMPRESSION	VERDICHTUNGSHUB
2865	**COMPRESSION TEST**	ESSAI DE COMPRESSION	DRUCKPROBE
2866	**COMPRESSION-IGNITION ENGINE**	MOTEUR DIESEL	SELBSTENTZÜNDUNGSMOTOR
2867	**COMPRESSION-RING**	SEGMENT D'ETANCHEITE	VERDICHTUNGSRING

2868	COMPRESSIVE FORCE	FORCE DE COMPRESSION	DRUCKKRAFT
2869	COMPRESSIVE STRAIN	EFFORT DE COMPRESSION, TRAVAIL A LA COMPRESSION	BEANSPRUCHUNG AUF DRUCK, DRUCKBEANSPRUCHUNG
2870	COMPRESSIVE STRENGTH	RESISTANCE A LA COMPRESSION	DRUCKFESTIGKEIT
2871	COMPRESSIVE STRESS	EFFORT DE COMPRESSION PAR UNITE DE SECTION	DRUCKSPANNUNG
2872	COMPRESSIVE TEST	ESSAI A LA COMPRESSION	DRUCKVERSUCH
2873	COMPRESSIVE YIELD STRENGTH	LIMITE D'ECRASEMENT	QUETSCHGRENZE
2874	COMPRESSOR	COMPRESSEUR	VERDICHTER, KOMPRESSOR
2875	COMPUTER DIRECTED NUMERICAL CONTROL	COMMANDE DIRECTE PAR CALCULATEUR	RECHNERSTEUERUNG (DIREKTE)
2876	COMSUMPTION OF ENERGY	DEPENSE D'ENERGIE DE FORCE MOTRICE, CONSOMMATION D'ENERGIE	KRAFTAUFWAND, ARBEITSAUFWAND, KRAFTVERBRAUCH, ENERGIEVERBRAUCH
2877	CONCAVE	CONCAVE	KONKAV
2878	CONCAVE FILLET WELD	SOUDURE D'ANGLE CONCAVE, SOUDURE EN CONGE	HOHLKEHLNAHT
2879	CONCAVE MIRROR	MIROIR CONCAVE	HOHLSPIEGEL, KONKAVSPIEGEL
2880	CONCAVITY	CONCAVITE	HOHLWÖLBUNG
2881	CONCENTRATE	CONCENTRE	KONZENTRAT
2882	CONCENTRATE (TO)	CONCENTRER	ANREICHERN
2883	CONCENTRATED STRONG SOLUTION	SOLUTION CONCENTREE, SOLUTION FORTE	LÖSUNG (KONZENTRIERTE), LÖSUNG (ANGEREICHERTE)
2884	CONCENTRATED SULPHURIC ACID	ACIDE SULFURIQUE CONCENTRE	SCHWEFELSÄURE (KONZENTRIERTE), PFANNENSAURE
2885	CONCENTRATING A SOLUTION BY VAPORISATION	CONCENTRATION D'UNE SOLUTION PAR EVAPORATION	EINDAMPFEN EINER LÖSUNG
2886	CONCENTRATION	CONCENTRATION	ANREICHERUNG, KONZENTRATION
2887	CONCENTRATION CELL	ELEMENT A DEUX LIQUIDES	KONZENTRATIONSELEMENT
2888	CONCENTRATION CELL CORROSION	CORROSION DE LA PILE DE CONCENTRATION	KONZENTRATIONSELEMENT-KORROSION
2889	CONCENTRATION POLARIZATION	POLARISATION (D'UN ELECTRODE) PAR CHUTE DE CONCENTRATION	KONZENTRATIONS POLARISATION (EINER ELEKTRODE)
2890	CONCENTRIC CIRCLES	CERCLES CONCENTRIQUES	KREISE (KONZENTRISCHE)
2891	CONCENTRICITY	CONCENTRICITE	KONZENTRIZITÄT
2892	CONCHOID	CONCHOIDE	KONCHOIDE, MUSCHELLINIE
2893	CONCHOIDAL FRACTURE	CASSURE CONCHOIDALE	BRUCHFLÄCHE (MUSCHELIGE), BRUCH (MUSCHELIGER)
2894	CONCRETE	BETON	BETON, STEINMÖRTEL, GROBMÖRTEL
2895	CONCRETE (TO), LAY CONCRETE (TO)	BETONNER	BETONIEREN
2896	CONCRETE FOUNDATION	FONDATION, ASSISE, MASSIF EN BETON	BETONFUNDAMENT
2897	CONCRETE IRON SHEARS	CISAILLE POUR FERS A BETON	BETONEISENSCHERE
2898	CONCRETE PIPE	TUYAU EN CIMENT	BETONROHR, ZEMENTROHR
2899	CONCRETE PRE-STRESSING	PRECONTRAINTE DU BETON	BETONVORSPANNUNG
2900	CONCRETE SLAB	DALLE DE BETON (SUPPORT D'UN RESERVOIR)	BETON-(DECKEN-)PLATTE, SOHLE, FUNDAMENTPLATTE
2901	CONCRETE WORKS	OUVRAGES EN BETON	BETONARBEITEN
2902	CONCRETING	BETONNAGE	BETONIEREN
2903	CONCURRENT FORCES	FORCES CONCOURANTES	KRÄFTE MIT GEMEINSAMEM ANGRIFFSPUNKT
2904	CONCURRENT HEATING	CHAUFFAGE SUPPLEMENTAIRE	HEIZUNG (ZUSÄTZLICHE)
2905	CONDENSATE	CONDENSAT	KONDENSAT
2906	CONDENSATION	CONDENSATION	KONDENSATION

2907	**CONDENSE (TO)**	SE CONDENSER	NIEDERSCHLAGEN, KONDENSIEREN (SICH)
2908	**CONDENSER**	CONDENSATEUR (ELECTR.), CONDENSEUR (CHIM.), CONDENSEUR (OPT.)	KONDENSATOR (ELEKTR.), VERFLÜSSIGER, KONDENSATOR (CHEM.), KONDENSOR (OPT.)
2909	**CONDENSER LENS**	LENTILLE CONVERGENTE	SAMMELLINSE
2910	**CONDENSING CONVERGING LENS, CONVEX LENS**	LENTILLE CONVERGENTE, LENTILLE A BORD MINCE	SAMMELLINSE, LINSE (KONVEXE)
2911	**CONDENSING STEAM ENGINE**	MACHINE AVEC CONDENSATION	KONDENSATIONSDAMPFMASCHINE
2912	**CONDITIONING**	ENLEVEMENT DE LA COUCHE SUPERFICIELLE	BESEITIGUNG DER OBERFLÄCHENSCHICHT
2913	**CONDUCT (TO) (ELECTRICITY, HEAT)**	CONDUIRE (L'ELECTRICITE, LA CHALEUR)	LEITEN (ELEKTRIZITÄT, WÄRME)
2914	**CONDUCTING MEDIUM**	MILIEU CONDUCTEUR	MEDIUM (LEITENDES)
2915	**CONDUCTING SALTS**	SELS AUGMENTANT LA CONDUCTIBILITE D'UNE SOLUTION	LEITSALZE
2916	**CONDUCTION**	CONDUCTION	ÜBERTRAGUNG
2917	**CONDUCTION OF HEAT**	CONDUCTION DE LA CHALEUR	WÄRMELEITUNG
2918	**CONDUCTIVE**	CONDUCTEUR (DE LA CHALEUR DE L'ELECTRICITE)	LEITEND (WÄRME, ELEKTRIZITÄT)
2919	**CONDUCTIVITY**	CONDUCTIBILITE, CONDUCTIVITE	LEITFÄHIGKEIT
2920	**CONDUCTIVITY FOR HEAT, THERMAL CONDUCTIVITY**	CONDUCTIBILITE THERMIQUE CALORIFIQUE	WÄRMELEITFÄHIGKEIT, WÄRMELEITVERMÖGEN
2921	**CONDUCTOR**	CONDUCTEUR (ELECTR.)	LEITER (ELEKTR.)
2922	**CONDUCTOR WIRE**	FIL CONDUCTEUR, CONDUCTEUR D'ELECRICITE, FIL ELECTRIQUE	LEITUNGSDRAHT
2923	**CONE**	DARD DE LA FLAMME	FLAMMENKEGEL
2924	**CONE**	CONE (GEOM.)	KEGEL, KONUS (GEOM.)
2925	**CONE BELT**	COURROIE TRAPEZOIDALE	KEILRIEMEN
2926	**CONE BRAKE**	FREIN A CONE	KEGELBREMSE
2927	**CONE CLUTCH**	EMBRAYAGE A CONE	KEGELKUPPLUNG
2928	**CONE COUPLING**	ACCOUPLEMENT PAR CONE	KEGELKUPPLUNG
2929	**CONE EXTRACTORS**	EXTRACTEURS DE CONES	KEGELAUSZIEHER
2930	**CONE FRICTION CLUTCH**	EMBRAYAGE A CONES	KEGELREIBUNGSKUPPLUNG
2931	**CONE KEY**	FRETTE	RINGKEIL, HÜLSENKEIL, HÜLSENKEGEL, SPANNHÜLSE
2932	**CONE OF RAYS, LUMINOUS CONE**	CONE LUMINEUX	LICHTKEGEL
2933	**CONFOCAL**	CONFOCAL	KONFOKAL
2934	**CONGLOMERATE**	CONGLOMERAT	KONGLOMERAT
2935	**CONGO RED**	ROUGE CONGO	KONGOROT
2936	**CONGO RED PAPER**	PAPIER AU ROUGE CONGO	KONGOPAPIER
2937	**CONGRUENCE**	CONGRUENCE	KONGRUENZ (GEOM.)
2938	**CONGRUENT MELTING**	FUSION CONGRUENTE	SCHMELZEN (KONGRUENTES)
2939	**CONGRUENT TRANSFORMATION**	TRANSFORMATION CONGRUENTE	ÜBERGANG (KONGRUENTER)
2940	**CONIC SECTION**	CONIQUE, SECTION CONIQUE	KEGELSCHNITT
2941	**CONICAL GUIDE**	CONE GALOPIN	KEGELTROMMEL
2942	**CONICAL HELICAL SPRING**	RESSORT CONIQUE	KEGELFEDER, SCHRAUBENFEDER (KEGELIGE)
2943	**CONICAL PENDULUM GOVERNOR**	REGULATEUR CONIQUE	KEGELPENDELREGLER
2944	**CONICAL PISTON, MARINE TYPE PISTON**	PISTON CONIQUE	TRICHTERKOLBEN, KOLBEN (KONISCHER), MARINEKOLBEN
2945	**CONICAL RING**	BAGUE CONIQUE	KEILRING
2946	**CONICAL RIVET, RIVET WITH CONICAL HEAD WITH STAFF POINT HAMMERED POINT**	RIVET A TETE CONIQUE, RIVET A TETE CHANFREINEE	NIET MIT DREIECKPROFILKOPF, NIET MIT GEHÄMMERTEM KOPF

4

2947	CONICAL ROOF	TOIT CONIQUE	KEGELDACH
2948	CONICAL SHAPE	CONICITE	KEGELFORM
2949	CONICAL SURFACE	SURFACE LATERALE D'UN CONE	MANTELFLÄCHE EINES KEGELS
2950	CONICAL TRANSITION SECTION	ELEMENTS DE REDUCTION TRONCONIQUES	REDUZIER-ELEMENTE (KEGELSTUMPFE)
2951	CONICAL VALVE, MITRE VALVE	SOUPAPE A SIEGE CONIQUE	KEGELSITZVENTIL
2952	CONJUGATE	CONJUGUE	KONJUGIERT (MATH.)
2953	CONJUGATE AXIS OF HYPERBOLA	AXE NON TRANSVERSE D'UNE HYPERBOLE	NEBENACHSE DER HYPERBEL
2954	CONJUGATE DIAMETER	DIAMETRE CONJUGUE	DURCHMESSER (ZUGEORDNETER), DURCHMESSER (KONJUGIERTER)
2955	CONNECT (TO)	RACCORDER, RELIER (ELECTR.)	ANSCHLIESSEN (ELEKTR.)
2956	CONNECT IN PARALLEL (TO)	MONTER EN PARALLELE, GROUPER EN QUANTITE	SCHALTEN NEBENEINANDER, SCHALTEN PARALLEL
2957	CONNECT IN SERIES (TO)	MONTER EN TENSION, GROUPER EN SERTIE	SCHALTEN IN REIHE, SCHALTEN HINTEREINANDER
2958	CONNECTING	RACCORDEMENT (ELECTR.)	ANSCHLIESSEN (ELEKTR.)
2959	CONNECTING BOLT	BOULON DE LIAISON	VERBINDUNGSSCHRAUBE
2960	CONNECTING PIPE ·	TUYAU DE COMMUNICATION, TUYAU DE RACCORDEMENT	ANSCHLUSSROHR, VERBINDUNGSROHR
2961	CONNECTING RING	BAGUE DE RACCORD	ZWISCHENRING
2962	CONNECTING ROD	BIELLE, TRIANGLE DE LIAISON, BARRE DE CONNEXION, TIGE DE RACCORDEMENT	PLEUELSTANGE, KOLBENSTANGE, VERBINDUNGSSTANGE, SCHUBSTANGE, TREIBSTANGE
2963	CONNECTING ROD END	TETE DE BIELLE	SCHUBSTANGENKOPF, PLEUELKOPF
2964	CONNECTING WALKWAY	PASSERELLE DE LIAISON	VERBINDUNGSSTEG
2965	CONNECTION	RACCORD	VERBINDUNG
2966	CONNECTION	RACCORD (ELECTR.)	ANSCHLUSS (ELEKTR.)
2967	CONSERVATION OF ENERGY	CONSERVATION DE L'ENERGIE	ERHALTUNG DER ENERGIE
2968	CONSISTENCY	CONSISTANCE	KONSISTENZ
2969	CONSTANT	CONSTANTE	KONSTANTE, FESTWERT
2970	CONSTANT CURRENT WELDING SOURCE	GENERATRICE DE SOUDAGE POUR COURANT CONSTANT	SCHWEISSGENERATOR FÜR KONSTANTEN STROM
2971	CONSTANT CUTTING SPEED	VITESSE DE COUPE CONSTANTE	SCHNITTGESCHWINDIGKEIT (KONSTANTE)
2972	CONSTANT DEFLECTION TEST	ESSAI DE FLEXION CONSTANTE	DURCHBIEGUNGS PRÜFUNG (KONSTANTE)
2973	CONSTANT FORCE	FORCE CONSTANTE	KRAFT (GLEICHBLEIBENDE), KRAFT (KONTINUIERLICHE)
2974	CONSTANT LOAD TEST	ESSAI DE CHARGE CONSTANTE	DAUERLASTPRÜFUNG
2975	CONSTANT OF GRAVITATION	INTENSITE DE LA PESANTEUR	GRAVITATIONSKONSTANTE
2976	CONSTANT PRESSURE	PRESSION CONSTANTE	DRUCK (GLEICHBLEIBENDER), DRUCK (KONSTANTER)
2977	CONSTANT TEMPERATURE	TEMPERATURE CONSTANTE	TEMPERATUR (GLEICHBLEIBENDE), TEMPERATUR (KONSTANTE)
2978	CONSTANT VOLTAGE	TENSION CONSTANTE	DAUERSPANNUNG
2979	CONSTANT VOLTAGE WELDING SOURCE	SOURCE DE COURANT A TENSION CONSTANTE (POUR SOUDAGE)	KONSTANTSPANNUNGSSTROMQUELLE
2980	CONSTANTAN	CONSTANTAN	KONSTANTAN
2981	CONSTITUENT	CONSTITUANT, COMPOSANT	ELEMENTARBESTANDTEIL
2982	CONSTITUENT METAL COMPONENT OF AN ALLOY	CONSTITUANT, COMPOSANT D'UN ALLIAGE, METAL CONSTITUANT	BESTANDTEIL EINER LEGIERUNG
2983	CONSTITUENT, COMPONENT	COMPOSANT (CHIM.)	BESTANDTEIL, KOMPONENTE (CHEM.)
2984	CONSTITUTION DIAGRAM	DIAGRAMME DE PHASES	ZUSTANDSDIAGRAMM

10

2985	**CONSTRAINED MOTION, DEFINITELY, FULLY CONTROLLED MOTION, POSITIVE MOTION**	MOUVEMENT A COMMANDE POSITIVE, MOUVEMENT A COMMANDE MECANIQUE	BEWEGUNG (ZWANGSLÄUFIGE)
2986	**CONSTRUCTION**	CONSTRUCTION, MONTAGE	KONSTRUKTION, ERBAUUNG
2987	**CONSTRUCTION HUT**	BARAQUE DE CHANTIER	BAUSTELLENBARACKE
2988	**CONSTRUCTIONAL METALS, COMMON**	METAUX DE CONSTRUCTION (ORDINAIRES)	BAUMETALLE (GEWÖHNLICHE)
2989	**CONSULTING ENGINEER**	INGENIEUR-CONSEIL	INGENIEUR (BERATENDER)
2990	**CONSUMABLE ELECTRODE, CONSUTRODE**	ELECTRODE FUSIBLE	ELEKTRODE (SCHMELZBARE), CONSUTRODE
2991	**CONSUMABLE INSERT RING**	JOINT FUSIBLE	ABSCHMELZVERBINDUNGSRING
2992	**CONSUMED WEIGHT**	POIDS CONSOMME	GEWICHT (VERBRAUCHTES)
2993	**CONSUMPTION OF CURRENT**	CONSOMMATION DE COURANT	STROMVERBRAUCH
2994	**CONSUMPTION OF FUEL**	CONSOMMATION DE COMBUSTIBLE	BRENNSTOFFVERBRAUCH
2995	**CONSUMPTION OF STEAM, STEAM CONSUMPTION DEMAND**	CONSOMMATION, DEPENSE DE VAPEUR	DAMPFVERBRAUCH
2996	**CONTACT**	CONTACT A LA MASSE (ELECT.), CONTACT (GEN.)	KONTAKT (ELEKTR.) BERÜHRUNG (ALLG.)
2997	**CONTACT ARC**	ARC DE CONTACT	GREIFBOGEN
2998	**CONTACT AREA**	SURFACE DE CONTACT	GREIFOBERFLÄCHE
2999	**CONTACT BAR**	BARRE DE CONTACT	KONTAKTBACKE
3000	**CONTACT BREAKER**	GRAINS DE CONTACT (RUPTEUR)	KONTAKTUNTERBRECHER
3001	**CONTACT CONDUCTOR**	ELECTRODE A CONTACT	KONTAKTELEKTRODE
3002	**CONTACT CORROSION**	CORROSION PAR CONTACT	BERÜHRUNGSKORROSION
3003	**CONTACT JAW**	JOUE DE CONTACT	KONTAKTBLOCK
3004	**CONTACT PLATING**	DEPOT PAR CONTACT	KONTAKTPLATTIERUNG
3005	**CONTACT POINT**	POINT DE CONTACT	BERÜHRUNGSPUNKT
3006	**CONTACT POINT INSERT**	POINTE DE RECHANGE	AUFSATZSPITZE
3007	**CONTACT POTENTIAL**	POTENTIEL DE CONTACT	KONTAKTPOTENTIAL
3008	**CONTACT ROLLER**	GALET DE CONTACT	KONTAKTROLLE
3009	**CONTACTS POINT**	PLOT	UNTERBRECHERKONTAKT
3010	**CONTAINER, RECELVER, TANK, DRUM**	RECIPIENT	BEHÄLTER
3011	**CONTENTS**	TENEUR, PROPORTION, DOSAGE	GEHALT, MENGENVERHÄLTNIS
3012	**CONTINOUS LUBRICATION**	GRAISSAGE CONTINU	SCHMIERUNG (BESTÄNDIGE)
3013	**CONTINOUS PAPER**	PAPIER CONTINU	PAPIER (ENDLOSES)
3014	**CONTINOUS SINTERING**	FRITTAGE CONTINU	SINTERN (KONTINUIERLICHES)
3015	**CONTINOUS SYSTEM OF ROPE DRIVING**	TRANSMISSION SUR POULIES MULTIPLES	KREISSEILTRIEB
3016	**CONTINOUS WELD**	SOUDAGE CONTINU	DURCHLAUFSCHWEISSEN
3017	**CONTINUITY**	CONTINUITE	KONTINUITÄT, STETIGER VERLAUF, STETIGKEIT
3018	**CONTINUOUS BEAM**	POUTRE CONTINUE	TRÄGER (DURCHGEHENDER), TRÄGER (DURCHLAUFENDER)
3019	**CONTINUOUS CASTING**	COULEE CONTINUE	GIESSEN (KONTINUIERLICHES)
3020	**CONTINUOUS CHAIN PASSENGER LIFT**	ASCENSEUR CONTINU	PATERNOSTERAUFZUG
3021	**CONTINUOUS CURRENT GENERATOR, DYNAMO**	DYNAMO, GENERATRICE, MACHINE A COURANT CONTINU	GLEICHSTROMMASCHINE, GLEICHSTROMDYNAMO, GLEICHSTROMGENERATOR
3022	**CONTINUOUS CURRENT MOTOR**	MOTEUR A COURANT CONTINU	GLEICHSTROMMOTOR
3023	**CONTINUOUS DIRECT CURRENT (C.C., D.C.)**	COURANT CONTINU	GLEICHSTROM
3024	**CONTINUOUS DISTILLATION**	DISTILLATION CONTINUE	DESTILLATION (STETIGE)
3025	**CONTINUOUS FUNCTION**	FONCTION CONTINUE	FUNKTION (STETIGE)

3026	**CONTINUOUS FURNACE**	FOUR CONTINU	DURCHLAUFOFEN, OFEN (KONTINUIERLICHER)
3027	**CONTINUOUS GALVANIZING**	GALVANISATION EN CONTINU	VERZINKEN (KONTINUIERLICHES)
3028	**CONTINUOUS HOT DIP**	TREMPE CONTINUE A CHAUD	EINTAUCH-VERFAHREN (KONTINUIERLICHES)
3029	**CONTINUOUS INGOT CASTING**	COULEE CONTINUE	STRANGGIESSEN
3030	**CONTINUOUS MILL**	TRAIN CONTINU (DE LAMINOIR)	WALZSTRASSE (KONTINUIERLICHE)
3031	**CONTINUOUS MOTION**	MOUVEMENT CONTINU	BEWEGUNG (STETIGE)
3032	**CONTINUOUS PATH**	COMMANDE CONTINUE	STETIGBAHNSTEUERUNG
3033	**CONTINUOUS PHASE**	PHASE CONTINUE, PHASE DISPERSIVE	PHASE (DISPERSIVE), PHASE (KONTINUIERLICHE)
3034	**CONTINUOUS PRODUCTION, PROGRESSIVE MANUFACTURE, MANUFACTURE ON THE FLOW PRINCIPE, MASS PRODUCTION WITH CONSECUTIVE OPERATIONS**	TRAVAIL A LA CHAINE, OPERATIONS A LA CHAINE	FLIESSARBEIT
3035	**CONTINUOUS ROLLING**	LAMINAGE CONTINU	WALZEN (KONTINUIERLICHES)
3036	**CONTINUOUS SPECTRUM**	SPECTRE CONTINU	SPEKTRUM (KONTINUIERLICHES)
3037	**CONTINUOUS WELD/WELDING**	SOUDURE CONTINUE	NAHT (DURCHLAUFENDE)
3038	**CONTINUOUS WORKING, CONTINUOUS RUNNING**	FONCTIONNEMENT CONTINU, SERVICE CONTINU	DAUERBETRIEB, BETRIEB (UNUNTERBROCHENER)
3039	**CONTOUR CONTROL SYSTEM**	COMMANDE DE CONTOURNAGE	BAHNSTEUERUNG
3040	**CONTOUR GRINDING**	FRAISAGE DES CONTOURS	UMRISSFRÄSEN
3041	**CONTOUR SHAPING**	FORMATION DU MODELE	MODELLFORMUNG
3042	**CONTRACTED**	RETRECI	VERENGT, EINGESCHNÜRT
3043	**CONTRACTION**	RETRAIT, STRICTION	SCHRUMPFUNG
3044	**CONTRACTION OF A JET, STREAM OF LIQUID**	CONTRACTION DE LA VEINE FLUIDE, ETRANGLEMENT DE LA VEINE LIQUIDE	EINSCHNÜRUNG, KONTRAKTION EINES FLÜSSIGKEITSSTRAHLES
3045	**CONTRACTION OF AREA**	CONTRACTION, RETRECISSEMENT, STRICTION	EINSCHNÜRUNG, VERENGERUNG, KONTRAKTION
3046	**CONTRACTION OF RIVET SHANK**	CONTRACTION DE LA TIGE DU RIVET	ZUSAMMENZIEHEN, SCHRUMPFEN DES NIETSCHAFTES
3047	**CONTRACTION RULE**	REGLE A RETRAIT	SCHWINDMASSSTAB
3048	**CONTROL**	CONTROLE, COMMANDE	KONTROLL, STEUERUNG
3049	**CONTROL (TO)**	MANOEUVRER, COMMANDER	STEUERN
3050	**CONTROL ARM**	BRAS DE SUSPENSION	SCHWINGARM
3051	**CONTROL CABLE**	CABLE DE COMMANDE	STEUERSEIL
3052	**CONTROL DEVICE**	DISPOSITIF DE CONTROLE	REGELVORRICHTUNG
3053	**CONTROL GAUGE**	JAUGE DE CONTROLE	LEHRGERÄT
3054	**CONTROL PANEL**	TABLEAU DE COMMANDE	BEDIENUNGSTAFEL
3055	**CONTROLLED ATMOSPHERE FURNACE**	FOUR A ATMOSPHERE CONTROLEE	SCHUTZGASOFEN
3056	**CONTROLLED COOLING**	REFROIDISSEMENT COMMANDE	KÜHLUNG (GESTEUERTE)
3057	**CONTROLLER**	ORGANE DE COMMANDE	STEUERUNG
3058	**CONTRUCTION CLIP ANGLE**	GOUSSET, CORNIERE SUPPORT	BEFESTIGUNGSWINKEL
3059	**CONVECTION**	CONVECTION (CHALEUR RAYONNANTE)	KONVEKTION
3060	**CONVECTION OF HEAT**	CONVECTION, CONVEXION	FORTFÜHRUNG DER WARME, KONVEKTION DER WÄRME
3061	**CONVENTIONAL MILLING**	FRAISAGE CLASSIQUE	FRÄSEN, (GEGENLÄUFIGES)
3062	**CONVERGE (TO)**	CONVERGER	ZUSAMMENLAUFEN, KONVERGIEREN
3063	**CONVERGENCE**	CONVERGENCE	KONVERGENZ
3064	**CONVERGENT**	CONVERGENT	ZUSAMMENLAUFEND, KONVERGENT
3065	**CONVERGENT SERIES**	SERIE CONVERGENTE	REIHE (KONVERGENTE)

15

3066	CONVERGING MOUTHPIECE	AJUTAGE CONVERGENT, EMBOUCHURE CONVERGENTE	AUSFLUSSSTUTZEN (SICH VERENGERNDER)
3067	CONVERSION	TRANSFORMATION, CONVERSION	UMRECHNUNG
3068	CONVERSION OF IRON INTO STEEL	TRANSFORMATION DU FER EN ACIER	EISEN-STAHLUMWANDLUG
3069	CONVERSION TABLE	TABLE DE TRANSFORMATION, DE CONVERSION	UMRECHNUNGSTAFEL
3070	CONVERSION TRANSFORMATION OF ENERGY	TRANSFORMATION D'ENERGIE	UMWANDLUNG VON ENERGIE, ENERGIEUMWANDLUNG
3071	CONVERTER	CONVERTISSEUR, COMMUTATRICE	UMFORMER, STROMUMFORMER, KONVERTERBIRNE
3072	CONVERTER MOUTH/NOSE	BEC DU CONVERTISSEUR	KONVERTERHUT
3073	CONVERTIBLE (CAR)	VOITURE DECAPOTABLE	KABRIOLETT
3074	CONVERTING CEMENTATION FURNACE	FOUR DE CEMENTATION	ZEMENTIEROFEN
3075	CONVERTING POT	CAISSE DE CEMENTATION	GLÜHKISTE, ZEMENTIERKISTE
3076	CONVEX	CONVEXE	KONVEX
3077	CONVEX FILLET WELD	SOUDURE D'ANGLE CONVEXE	VOLLKEHLNAHT, WÖLBKEHLNAHT
3078	CONVEX MIRROR	MIROIR CONVEXE	SPIEGEL (ERHABENER), KONVEXSPIEGEL
3079	CONVEXITY RATIO	RAPPORT DE CONVEXITE	KONVEXITÄTSVERHÄLTNIS WÖLBUNGSVERHÄLTNIS
3080	CONVEYOR, CONVEYING MACHINERY, HANDLING APPLIANCE	APPAREIL DE MANUTENTION, TRANSPORTEUR	FÖRDEREINRICHTUNG
3081	COOL (TO)	REFROIDIR (SE), REFRIGERER	ABKÜHLEN (SICH)
3082	COOL TIME	TEMPS DE REFROIDISSEMENT	ABKÜHLUNGSZEIT
3083	COOLANT	REFRIGERANT, FLUIDE DE REFROIDISSEMENT	KÜHLMITTEL
3084	COOLED PISTON	PISTON REFROIDI	KOLBEN (GEKÜHLTER)
3085	COOLING	REFROIDISSEMENT, REFRIGERATION	KÜHLEN, ABKÜHLEN, KÜHLUNG
3086	COOLING AGENT	AGENT FRIGORIFIQUE, AGENT DE REFRIGERATION	KÜHLMITTEL, KÄLTEMITTEL
3087	COOLING CURVE	COURBE DE REFROIDISSEMENT	ABKÜHLUNGSKURVE
3088	COOLING FAN	VENTILATEUR DE REFROIDISSEMENT	KÜHLLUFTGEBLASE
3089	COOLING PLANT FOR WATER OF CONDENSATION	INSTALLATION POUR LE REFROIDISSEMENT DES EAUX DE CONDENSATION, REFRIGERANT POUR LES EAUX DE CONDENSATION	RÜCKKÜHLANLAGE
3090	COOLING POND	BASSIN REFROIDISSANT	KÜHLTEICH
3091	COOLING STRESSES	EFFORTS DE REFROIDISSEMENT	ABKÜHLUNGSSPANNUNGEN
3092	COOLING SURFACE	SURFACE DE REFROIDISSEMENT, DE REFRIGERATION, SURFACE REFROIDISSANTE	KÜHLFLÄCHE
3093	COOLING SYSTEM	REFROIDISSEMENT, CIRCUIT DE REFROIDISSEMENT	KÜHLUNG, KÜHLSYSTEM
3094	COOLING TOWER	TOUR DE REFROIDISSEMENT	KÜHLTURM
3095	COOLING WATER	EAU DE REFROIDISSEMENT, EAU DE REFRIGERATION	KÜHLWASSER
3096	COOLING-DOWN	RETOUR A LA TEMPERATURE AMBIANTE	RÜCKKÜHLUNG ZU UMGEBUNGSTEMPERATUR
3097	COORDINATE DIMENSIONING	COTATION ABSOLUE	BEZUGSMASSSYSTEM
3098	COORDINATE DRILLING AND BORING MACHINE	PERCEUSE-ALESEUSE A COORDONNEES	KOORDINATEN-BOHR-U. AUSBOHR-MASCHINE
3099	COORDINATE LOCATING	REPERAGE PAR COORDONNEES	KOORDINATENSYSTEM
3100	COORDINATES	COORDONNEES	KOORDINATEN
3101	COPAIBA	BAUME DE COPAHU, COPAHU	KOPAIVABALSAM

3102	COPAL GUM	COPAL, COPALE	KOPAL
3103	COPAL VARNISH	VERNIS AU COPAL	KOPALLACK
3104	COPE	PARTIE DE DESSUS	FORMOBERTEIL
3105	COPE AND DRAG PATTERN	MODELE EN DEUX	MODELL (ZWEITEILIGES)
3106	COPER WIRE	FIL DE CUIVRE	KUPFERDRAHT
3107	COPING	GRUGEAGE	AUSKLINKEN
3108	COPING SAW	SCIE A GRUGER	BOGENSÄGE
3109	COPOLA SPARK ARRESTOR	PARE-ETINCELLES	FUNKENFÄNGER
3110	COPPER	CUIVRE (CUIVRE ROUGE)	KUPFER
3111	COPPER ALLOY	ALLIAGE A BASE DE CUIVRE	KUPFERLEGIERUNG
3112	COPPER ANODE	ANODE DE CUIVRE	KUPFERANODE
3113	COPPER BAR	BARRE DE CUIVRE	KUPFERSTAB
3114	COPPER COINAGE	MONNAIES DE CUIVRE	KUPFERMÜNZEN
3115	COPPER CYANIDE	CYANURE DE CUIVRE	ZYANKUPFER, KUPFERZYANID
3116	COPPER FLAT WIRE	FIL PLAT EN CUIVRE, FIL DE CUIVRE PLAT	KUPFERFLACHDRAHT
3117	COPPER FOIL	FEUILLE MINCE DE CUIVRE	KUPFERFOLIE
3118	COPPER GLANCE, VITREOUS COPPER ORE, CHALCOCITE, REDRUTHITE	CHALCOSINE, CUIVRE VITREUX, CUIVRE SUFURE GRIS	KUPFERGLANZ
3119	COPPER HAMMER	MASSE, MASSETTE EN CUIVRE	KUPFERHAMMER
3120	COPPER INGOT	LINGOT DE CUIVRE	KUPFERBARREN, KUPFERBLOCK
3121	COPPER MATT, REGULUS OF COPPER	MATTE DE CUIVRE	KUPFERSTEIN, KUPFERLECH
3122	COPPER ORE	MINERAI DE CUIVRE	KUPFERERZ
3123	COPPER PLATE	PLAQUE DE CUIVRE	KUPFERPLATTE
3124	COPPER PLATE (TO)	CUIVRER	VERKUPFERN
3125	COPPER PLATING	CUIVRAGE	VERKUPFERUNG
3126	COPPER PYRITES, CHALCOPYRITES	PYRITE CUIVREUSE, CUIVRE PYRITEUX, CHALCOPYRITE	KUPFERKIES, CHALKOPYRIT
3127	COPPER SHEET	TOLE DE CUIVRE	KUPFERBLECH
3128	COPPER SHOT	GRENAILLE DE CUIVRE	KUPFERSCHROT
3129	COPPER SLAB	LINGOT PLAT DE CUIVRE	KUPFERFLACHBLOCK
3130	COPPER STEEL	ACIER AU CUIVRE	KUPFERSTAHL
3131	COPPER STRIP	LAME DE CUIVRE, BANDE DE CUIVRE	KUPFERSTREIFEN
3132	COPPER TUBE	TUYAU EN CUIVRE	KUPFERROHR
3133	COPPER WIRE	FIL DE CUIVRE	KUPFERDRAHT
3134	COPPER-ASBESTOS GASKET	JOINT METALLOPLASTIQUE	KUPFERASBESTDICHTUNG
3135	COPPERSMITH	CHAUDRONNIER EN CUIVRE	KUPFERSCHMIED
3136	COPPERSMITHING	CHAUDRONNERIE	KUPFERSCHMIEDEN
3137	COPPERWELD	COPPERWELD	COPPERWELD
3138	COPYING LATHE	TOUR A COPIER, TOUR A REPRODUIRE	KOPIERDREHMASCHINE, FORMDREHBANK, KOPIERDREHBANK, SCHABLONENDREHBANK, FASSONDREHBANK
3139	COPYING MACHINE	MACHINE A REPRODUIRE	KOPIERMASCHINE
3140	CORBEL	CORBEAU	KRAGSTÜCK, KONSOLE
3141	CORD DRIVE, BAND DRIVE	TRANSMISSION COURROIES DE CHASSE	SCHNURTRIEB
3142	CORE	CAROTTE, NOYAU, AME, NOYAU MAGNETIQUE	MASSEKERN, BOHRKERN, KERN, ADER, SEELE
3143	CORE BAR	ARMATURE DE NOYAU	KERNSTANGE
3144	CORE BEARING BLOCK	POINT DE GUIDAGE DES NOYAUX	KERNFÜHRUNGSBLOCK
3145	CORE BINDER	LIANT DE NOYAUTAGE	TROCKENSANDBINDER, KERNBINDER

3146	**CORE BLOWING-MACHINE**	MACHINE A SOUFFLER LES NOYAUX	KERNBLASMASCHINE
3147	**CORE BOX**	BOITE A NOYAUX	KERNKASTEN
3148	**CORE DRIER**	COQUILLE DE SECHAGE	TROCKENSCHALE
3149	**CORE DRILL**	FORET ALESEUR	SPIRALSENKER
3150	**CORE GRINDER**	MACHINE A MEULER LES NOYAUX	KERNSCHLEIFMASCHINE
3151	**CORE GUM**	COLLE A NOYAUX	KERNKLEBEMITTEL
3152	**CORE HOOK**	CROCHET A NOYAUX	KERNHAKEN
3153	**CORE KNOCK-OUT MACHINE**	MACHINE A DENOYAUTER, MACHINE A DEBOURRER	AUSSCHLAGRÜTTLER
3154	**CORE MAKING MACHINE**	MACHINE A MOULER LES NOYAUX	KERNFORMMASCHINE
3155	**CORE OF A PACKING**	NOYAU D'UNE GARNITURE	EINLAGE EINER PACKUNG
3156	**CORE OF CROSS SECTION**	NOYAU DE LA SECTION	KERN EINES QUERSCHNITTES
3157	**CORE OIL**	HUILE A NOYAUX	KERNÖL
3158	**CORE OVEN**	ETUVE	KERNTROCKENOFEN
3159	**CORE PACKING**	JOINT A INSERTION	PACKUNG MIT EINLAGE
3160	**CORE PLATE**	SEGMENT DE TOLE (D'INDUIT)	ANKERSCHNITT, TRAFOBLECH, ROTOR PLATTE
3161	**CORE PRINT**	PORTEE DE MODELE	KERNMARKE
3162	**CORE ROD**	POINCON, BROCHE (FRITTAGE)	DORN, KERN
3163	**CORE SAND**	SABLE A NOYAUX	KERNSAND
3164	**CORE STRAINER**	NOYAU FILTRE	SIEBKERN
3165	**CORE STRUCTURE**	STRUCTURE DU COEUR	KERNGEFÜGE
3166	**CORE VENTS**	TIRAGE D'AIR DES NOYAUX	LUFTSTECHEN, ENTLÜFTUNG
3167	**CORE, CENTRE OF A ROPE**	AME D'UN CABLE	SEELE EINES SEILES
3168	**CORED BAR**	BARRE A NOYAU FUSIBLE	STAB MIT SCHMELZKERN
3169	**CORED CRYSTAL**	CRISTAL INHOMOGENE	KRISTALL (INHOMOGENER)
3170	**CORED STRUCTURE**	HETEROGENEITE	UNGLEICHARTIGKEIT
3171	**COREWASH**	ENDUIT POUR NOYAUX	SCHLICHTE
3172	**CORING**	SEGREGATION, MICROSEGREGATION	SEIGERUNG, ENTMISCHUNG, MIKROSEIGERUNG
3173	**CORK**	LIEGE, BOUCHON	KORK
3174	**CORK BRICK**	BRIQUE EN LIEGE	KORKSTEIN
3175	**CORK SLAB**	PLAQUE DE LIEGE	KORKPLATTE
3176	**CORK STOPPLE**	BOUCHON DE LIEGE	PFROPFEN, KORKSTOPFEN, KORK
3177	**CORN AND SEED DRESSING AND CLEANING MACHINERY**	TARARES CRIBLEURS T. TRIEURS	SAATGUT UND GETREIDEREINIGUNGS-MASCHINEN
3178	**CORNER DRILLING MACHINE**	MACHINE A PERCER LES TROUS A ANGLE	ECKBOHRMASCHINE
3179	**CORNER JOINT**	CONTRE-FICHE, JOINT EN ANGLE EXTERIEUR	GEGENSTREBE, WINKELSTOSSVERBINDUNG
3180	**CORNERING STABILITY**	TENUE DE ROUTE EN VIRAGE	KURVENFESTIGKEIT
3181	**CORRECTION**	CORRECTION	BERICHTIGUNG
3182	**CORRODE (TO)**	ATTAQUER, CORRODER	ANFRESSEN, ANGREIFEN
3183	**CORROSION**	CORROSION	VERROTTUNG, ANFRESSUNG, KORROSION
3184	**CORROSION AND HEAT RESISTING STEELS**	ACIERS RESISTANT A LA CORROSION ET A LA CHALEUR	STÄHLE (KORROSIONS- UNDWÄRMEBESTÄNDIGE)
3185	**CORROSION CREVICE**	CORROSION FISSURANTE	SPALTKORROSION
3186	**CORROSION EMBRITTLEMENT**	FRAGILITE PAR CORROSION	KORROSIONSSPRÖDIGKEIT
3187	**CORROSION FATIGUE**	FATIGUE PAR CORROSION	KORROSIONSERMÜDUNG
3188	**CORROSION FATIGUE LIMIT**	LIMITE DE FATIGUE PAR CORROSION	KORROSIONSERMÜDUNGSGRENZE
3189	**CORROSION INHIBITOR**	PRODUIT ANTI-ROUILLE	ROSTSCHUTZMITTEL

26

3190	CORROSION PIT	PIQURE DE CORROSION	KORROSIONSNARBE
3191	CORROSION PREVENTION	PROTECTION CONTRE LA CORROSION	KORROSIONSSCHUTZ, KORROSIONSVERHÜTUNG
3192	CORROSION PRODUCT SOLVENTS	SOLVANTS POUR CORROSIFS	LÖSUNGSMITTEL FÜR KORROSION
3193	CORROSION RATE	VITESSE DE CORROSION	KORROSIONSGESCHWINDIGKEIT
3194	CORROSION RESISTANCE	RESISTANCE A LA CORROSION	KORROSIONSBESTÄNDIGKEIT
3195	CORROSION STRESS	CORROSION SOUS TENSION	SPANNUNGSKORROSION
3196	CORROSIVE	CORROSIF, MORDANT	ÄTZEND
3197	CORROSIVE SUBSTANCE	SUBSTANCE CORROSIVE, CORROSIF, CORRODANT	ÄTZMITTEL
3198	CORRUGATED SHEET IRON	TOLE ONDULEE	WELLBLECH
3199	CORRUGATED TUBE	TUBE PLISSE	WELLROHR
3200	CORRUGATED WIRE NETTING	GRILLAGE ONDULE	SCHLANGENROST
3201	CORUNDUM	CORINDON	KORUND
3202	COSECANT	COSECANTE	KOSEKANTE
3203	COSINE	COSINUS	KOSINUS
3204	COST OF HANDLING	FRAIS DE MANUTENTION	FÖRDERKOSTEN
3205	COST OF PACKING, PACKING COSTS	FRAIS D'EMBALLAGE	VERPACKUNGSKOSTEN
3206	COST OF POWER	FRAIS DE FORCE MOTRICE	KRAFTBEDARFSKOSTEN
3207	COST OF PRODUCTION	FRAIS DE PRODUCTION, FRAIS DE FABRICATION	SELBSTKOSTEN, GESTEHUNGSKOSTEN, HERSTELLUNGSKOSTEN, PRODUKTIONSKOSTEN
3208	COST OF REPAIRS	FRAIS DE REPARATION	AUSBESSERUNGSKOSTEN, WIEDERHERSTELLUNGSKOSTEN, REPARATURKOSTEN
3209	COST OF TRANSPORTATION, FREIGHT CHARGES	FRAIS DE TRANSPORT	BEFÖRDERUNGSKOSTEN, VERSANDKOSTEN, FRACHTKOSTEN, FRACHT, TRANSPORTKOSTEN
3210	COST PRICE	PRIX DE REVIENT	SELBSTKOSTENPREIS, GESTEHUNGSPREIS
3211	COTANGENT	COTANGENTE	KOTANGENTE
3212	COTTER	CLE, CLEF, CLAVETTE-CLEF	VORSTECKER, VORSTECKKEIL, VORSTECKSTIFT,
3213	COTTER PIN	CLAVETTE D'ARRET	SETZKEIL, SPLINT
3214	COTTER WAY	TROU DE GOUPILLE	KEILLOCH
3215	COTTER, COTTAR	CLAVETTE EN COIN, CLAVETTE TRANSVERSALE, CHEVILLE D'ASSEMBLAGE	QUERKEIL
3216	COTTING BLOWPIPE	CHALUMEAU A DECOUPER	SCHNEIDBRENNER
3217	COTTON	COTON	BAUMWOLLE
3218	COTTON BELT	COURROIE EN COTON	BAUMWOLLRIEMEN
3219	COTTON COVERED WIRE, SILK COVERED WIRE	FIL GUIPE	UMSPONNENER DRAHT
3220	COTTON GASKET	TRESSE EN COTON	BAUMWOLLZOPF
3221	COTTON ROPE	CABLE EN COTON	BAUMWOLLSEIL
3222	COTTON SEED OIL	HUILE DE COTON	BAUMWOLLSAMENÖL, COTTONÖL, NIGGERÖL
3223	COTTON WASTE	COTON A NETTOYER	PUTZWOLLE
3224	COTTRELL BARRIER	BARRIERE DE COTTRELL	COTTRELL-SCHRANKE
3225	COULOMB	COULOMB	COULOMB
3226	COULOMB MODULUS	MODULE DE COULOMB	SCHUBMODUL
3227	COULOMETER	VOLTAMETRE	COULOMETER, VOLTAMETER
3228	COUMPOUND WOUND GENERATOR DYNAMO	DYNAMO COMPOUND	VERBUNDDYNAMO, DOPPELSCHLUSSDYNAMO, COMPOUNDDYNAMO
3229	COUNTER	COMPTEUR	ZÄHLVORRICHTUNG, ZÄHLER

3230	**COUNTER CURRENT**	COURANTS DE SENS CONTRAIRE, CONTRE-COURANT	GEGENSTROM
3231	**COUNTER ELECTROMOTIVE FORCE**	FORCE CONTRE-ELECTROMOTRICE	KRAFT (GEGENELEKTROMOTORISCHE), GEGEN-EMK
3232	**COUNTER PRESSURE**	CONTRE-PRESSION	GEGENDRUCK
3233	**COUNTER SHAFT**	ARBRE DE RENVOI	VORGELEGEWELLE, GEGENWELLE
3234	**COUNTER SINKING**	FRAISAGE, CHANFREINAGE	VERSENKEN
3235	**COUNTER WEIGHTED CRANKSHAFT**	VILEBREQUIN A CONTREPOIDS	KURBELWELLE MIT GEGENGEWICHT
3236	**COUNTER-BORING**	CHAMBRAGE	VERSENKEN
3237	**COUNTER-CLOCKWISE**	EN SENS INVERSE DES AIGUILLES D'UNE MONTRE	LINKSDREHEND, ENTGEGENGESETZT DEM UHRZEIGERSINN
3238	**COUNTER-CLOCKWISE ROTATION**	ROTATION A GAUCHE	LINKSDREHUNG
3239	**COUNTER-CLOCKWISE WOUND, LEFT-HANDED WOUND, SINISTRORSE**	ENROULE A GAUCHE, SINISTRORSUM, SENESTRORSUM	LINKSGEWUNDEN, LINKSGEWICKELT
3240	**COUNTER-SHAFT**	ARBRE DE RENVOI	VORGELEGEWELLE
3241	**COUNTERBORE**	FORET ALESEUR	ZYLINDERSENKER
3242	**COUNTERBORING**	CHAMBRAGE	VERSENKEN
3243	**COUNTERMOVEMENT, MOVEMENT IN THE OPPOSITE DIRECTION**	MOUVEMENT EN SENS CONTRAIRE	BEWEGUNG (GEGENLÄUFIGE), GEGENBEWEGUNG
3244	**COUNTERSHAFT AND ACCESSORIES , INTERMEDIATE GEARING**	TRANSMISSION INTERMEDIAIRE, RENVOI DE MOUVEMENT	VORGELEGE
3245	**COUNTERSINK**	FORET A FRAISER, FRAISE POUR LOGEMENT DE TETES DE VIS, FORET CHAMPIGNON	VERSENKBOHRER, VERSENKER, KRAUSKOPF
3246	**COUNTERSINK A RIVET HEAD, A SCREW HEAD (TO)**	NOYER UNE TETE DE RIVET, NOYER UNE TETE DE VIS, FRAISER UN TROU DE RIVET, FRAISER UN TROU DE VIS	VERSENKEN (EINEN NIETKOPF), VERSENKEN (EINEN SCHRAUBENKOPF), EINLASSEN (EINEN SCHRAUBENKOPF)
3247	**COUNTERSINKING A RIVET HEAD, A SCREW HEAD**	FRAISAGE D'UN TROU DE RIVET, FRAISSAGE D'UN TROU A VIS	VERSENKEN (EINLASSEN EINES NIETKOPFES, EINES SCHRAUBENKOPFES), EINLASSEN (EINES NIETKOPFES, EINE SCHRAUBENKOPFES)
3248	**COUNTERSUNK HEAD SCREW**	VIS A TETE NOYEE	SENKSCHRAUBE
3249	**COUNTERSUNK RIVET, RIVET WITH COUNTERSUNK POINT**	RIVET A TETE FRAISEE AVEC BOMBE	HALBVERSENKNIET, NIET MIT HALBVERSENKTEM KOPF
3250	**COUNTERSUNK SCREW HEAD**	TETE FRAISEE, TETE NOYEE D'UNE VIS	VERSENKTER SCHRAUBENKOPF
3251	**COUNTERWEIGHT, COUNTERPOISE WEIGHT, BALANCE WEIGHT**	CONTREPOIDS	GEGENGEWICHT, AUSGLEICHGEWICHT, BELASTUNGSGEWICHT
3252	**COUPLE**	COUPLE, PAIRE	PAAR
3253	**COUPLE (TO)**	ACCOUPLER	KUPPELN
3254	**COUPLING**	ACCOUPLEMENT, EMMANCHEMENT, MANCHONNAGE, ASSEMBLAGE, JONCTION, REUNION, RACCORD, EMBRAYAGE, COUPLAGE	KUPPELN, VERBINDEN, VERBINDUNG, VERSCHRAUBUNG, KUPPLUNG
3255	**COUPLING FLANGE**	PLATEAU D'ACCOUPLEMENT, BRIDE D'ACCOUPLEMENT	KUPPLUNSFLANSCH, KUPPLUNGSSCHEIBE
3256	**COUPLING ROD**	BIELLE D'ACCOUPLEMENT	KUPPELSTANGE
3257	**COUPLING SLEEVE**	MANCHON D'ACCOUPLEMENT	MUFFE, KUPPLUNGSMUFFE
3258	**COUPLING, SOCKET**	MANCHON	MUFFE
3259	**COURSE OF BRICKS**	ASSISE DE BRIQUES	ZIEGESTEINSCHICHT
3260	**COURSE RING OF A BOILER SHELL**	VIROLE D'UNE CHAUDIERE	SCHUSS EINES KESSELS, KESSELSCHUSS
3261	**COVARIANCE**	COVARIANCE	MITVERÄNDERUNG
3262	**COVER (PLATE)**	TAMPON (COUVERCLE D'UN TAMPON)	VERSCHLUSS (EINES MANNLOCHS)

3263	COVER (TO), LAG (TO), CLOTH A PIPE (TO)	ENVELOPPER UN TUYAU, ENTOURER UN TUYAU	VERKLEIDEN (EIN ROHR), UMHÜLLEN (EIN ROHR)
3264	COVER COAT	ENDUIT DE FINITION, GLACURE	AUSSENPUTZ
3265	COVER GLASS, COVER SLIP	COUVRE-OBJET	DECKGLAS
3266	COVER OF SLIDE VALVE CHEST, VALVE COVER	COUVERCLE D'UN ROBINET, COUVERCLE D'UNE SOUPAPE	DECKEL, HAUBE EINES SCHIEBERS, HAUBE EINES VENTILS
3267	COVER PLATE	PLAQUE DE RECOUVREMENT, TAMPON	DECKPLATTE, ABDECKPLATTE, FLANGEABDECKUNG
3268	COVER STRIP	RUBAN METALLIQUE DE PROTECTION	SCHUTZSTREIFEN AUS METALL
3269	COVER, LID, CAP	COUVERCLE	DECKEL
3270	COVERED CAM	CAME A RAINURE	EXZENTER (GESCHLOSSENES)
3271	COVERED ELECTRODE	ELECTRODE ENROBEE	MANTELELEKTRODE
3272	COVERED WIRE	FIL RECOUVERT	DRAHT (UMMANTELTER)
3273	COVERING POWER	POUVOIR COUVRANT	DECKFÄHIGKEIT
3274	COVERING, CLEADING	ENVELOPPE, REVETEMENT EXTERIEUR, CHEMISE	VERKLEIDUNG, UMHÜLLUNG
3275	COWHIDE	CUIR DE BOEUF, CUIR DE VACHE	RINDLEDER
3276	COWL	AUVENT	HAUBE, WINDFANG
3277	COWSTALL EQUIPMENT	INSTALLATIONS D'ETABLES	KUHSTALL-AUSSTATTUNGEN
3278	CRACK	FENTE, CRIQUE, FISSURE, RUPTURE, GERCE, COUPURE, CREVASSE, DECHIRURE, TAPURE	RISS, SPRUNG, SPALTE, BRUCH
3279	CRACK (TO)	FENDILLER (SE)	RISSIG WERDEN
3280	CRACK AND FLAW DETECTOR	DETECTEUR DE CRIQUES ET FELURES	RISS- UND FEHLERDETEKTOR
3281	CRACK AT THE EDGE	FENDILLEMENT SUR LES BORDS	KANTENRISS, QUERRISS
3282	CRACK IN HARDENED STEEL	GERCURE, GERCE, TAPURE DE L'ACIER TREMPE	HÄRTERISS
3283	CRACK/MICROCRACK	PAILLE, FISSURE (MICROFISSURE)	RISS
3284	CRACKING	CRAQUAGE OU CRACKING, FISSURATION, FISSURAGE, CRIQUAGE	KRACKEN, RISSBILDUNG, RISS
3285	CRACKING PROCESS	DISTILLATION AVEC CRACKING, CRACKING	KRACKUNG, KRACKVERFAHREN, CRACKINGDESTILLATION
3286	CRACKING ROUND THE RIVET HOLES	FENDILLEMENT DES TROUS DE RIVET	RISSIGWERDEN DER NIETLÖCHER
3287	CRACKING TOWER	TOUR DE CRAQUAGE	KRACKTURM
3288	CRADDLE	BERCEAU, SUPPORT	SATTELBEFESTIGUNG
3289	CRAFTSMAN, SKILLED WORKMAN	OUVRIER PROFESSIONNEL, OUVRIER EXPERIMENTE	GELERNTER ARBEITER
3290	CRANE	GRUE	KRAN
3291	CRANE JIB	FLECHE DE GRUE	KRANAUSLEGER
3292	CRANE LADLE	POCHE DE GRUE DE COULEE	KRANGIESSPFANNE
3293	CRANE RAIL	RAIL DE PONT ROULANT	LAUFSCHIENE, KRANSCHIENE
3294	CRANES	GRUES AGRICOLES	ACKERKRÄNE
3295	CRANK	MANIVELLE	KURBEL
3296	CRANK ARM	BRAS DE MANIVELLE	KURBELARM
3297	CRANK BOSS, BOSS OF CRANK	MOYEU DE LA MANIVELLE	KURBELNABE, KURBELAUGE
3298	CRANK CIRCLE PATH	CERCLE DECRIT PAR LA MANIVELLE	KURBELKREIS
3299	CRANK END, LARGE END, BIG END OF CONNECTING ROD	GROSSE TETE DE BIELLE	KURBELENDE DER SCHUBSTANGE
3300	CRANK HANDLE	MANETON, MANCHON DE MANIVELLE A BRAS	KURBELGRIFF

3301	CRANK LEVER	BRAS DE MANIVELLE	KURBELARM
3302	CRANK PIN	BOUTON DE MANIVELLE, MANETON	KURBELZAPFEN
3303	CRANK SHAFT BEARING, MAIN BEARING	PALIER DE L'ARBRE MANIVELLE, PALIER DE L'ARBRE DE COUCHE, PALIER DE VILEBREQUIN	KURBELWELLENLAGER, HAUPTLAGER
3304	CRANK WEB	FLASQUE D'UN ARBRE COUDE	KURBELWANGE
3305	CRANK WITH LOOSE SLEEVE, HANDLE	MANIVELLE A MANCHON	HEFTKURBEL
3306	CRANKCASE	CARTER-MOTEUR	KURBELGEHAÜSE
3307	CRANKCASE BREATHER	RENIFLARD	KURBELGEHÄUSE ENTLÜFTUNG
3308	CRANKCASE SLUDGE	CAMBOUIS	SCHMIERE (ALTE)
3309	CRANKED SHAFT	ARBRE COUDE, ARBRE A VILEBREQUIN, ARBRE-MANIVELLE	KURBELWELLE, WELLE (GEKRÖPFTE)
3310	CRANKING MOTOR	DEMARREUR	ANLASSER
3311	CRANKPIN	MANETON DE VILEBREQUIN, MANETON	KURBELWELLENZAPFEN
3312	CRANKPIN TURNING RINGS	ANNEAUX A TOURILLONNER	ZAPFENDREHRINGE
3313	CRANKSHAFT	VILEBREQUIN	KURBELWELLE
3314	CRANKSHAFT BEARING	PALIER DE VILEBREQUIN	KURBELWELLENLAGER
3315	CRANKSHAFT GEAR	PIGNON DE VILLEBREQUIN	KURBELWELLENZAHNRAD
3316	CRANKSHAFT LATHE	TOUR A ARBRE COUDE	KURBELWELLENDREHBANK
3317	CRATER	CRATERE	KRATER
3318	CRAZE	FISSURE CAPILLAIRE, MICROCRIQUE	HAARRISS
3319	CRAZING	CRAQUELURE SUPERFICIELLE	OBERFLÄCHENHAARRISSE
3320	CREASE	PLI	FALTE
3321	CREASING HAMMER	MARTEAU A SUAGE	SICKENHAMMER, STEKENHAMMER
3322	CREASING IRON	SUAGE	SICKENSTOCK, SIEKENSTOCK
3323	CREASING MACHINE	MACHINE A BORDER	SICKENMASCHINE
3324	CREEP (ING)	FLUAGE (METAUX)	KRIECHEN
3325	CREEP LIMIT	LIMITE (CONVENTIONNELLE) DE FLUAGE	DAUERSTANDGRENZE, KRIECHGRENZE (KONVENTIONELLE)
3326	CREEP RATE	VITESSE DE FLUAGE	KRIECHGESCHWINDIGKEIT
3327	CREEP STRENGHT DEPENDING ON TIME	RESISTANCE AU FLUAGE POUR UNE DUREE FINIE	ZEITSTANDFESTIGKEIT
3328	CREEP STRENGTH	RESISTANCE AU FLUAGE	ZEITSTANDFESTIGKEIT
3329	CREEPING IN BELTS	CONTRACTION DE LA COURROIE	EINKRIECHEN DES RIEMENS
3330	CREEPING OF A SOLUTION	GRIMPEMENT D'UNE SOLUTION	KRIECHEN, ÜBERKRIECHEN EINER LÖSUNG
3331	CREOSOTE OIL	CREOSOTE	KREOSOT, KREOSOTÖL
3332	CREOSOTING	CREOSOTAGE	KREOSOTIEREN
3333	CREST	SOMMET DU FILET	GEWINDESPITZE
3334	CREST OF WAVE, WAVE CREST	CRETE DE L'ONDE	WELLENBERG
3335	CRESTCLEARANCE	JEU A LA CRETE	SPITZENSPIEL
3336	CRIMP (TO)	ONDULER, PLISSER	FALTEN VERFORMEN
3337	CRIPPLING	DEFORMATION PERMANENTE	FORMÄNDERUNG (BLEIBENDE)
3338	CRIPPLING BUCKLING LOAD	CHARGE DE RUPTURE AU FLAMBAGE	KNICKLAST, KNICKBELASTUNG
3339	CRIPPLING BUCKLING STRAIN	EFFORT DE FLAMBAGE, EFFORT DE COMPRESSION SUR PIECES LONGUES	BEANSPRUCHUNG AUF KNICKUNG, KNICKBEANSPRUCHUNG
3340	CRIPPLING BUCKLING STRENGTH	RESISTANCE AU FLAMBAGE	KNICKFESTIGKEIT
3341	CRIPPLING BUCKLING STRESS	EFFORT DE FLAMBAGE PAR UNITE DE SECTION	KNICKSPANNUNG
3342	CRIPPLING BUCKLING TEST	ESSAI AU FLAMBAGE	KNICKVERSUCH

3343	**CRIPPLING, BUCKLING**	FLAMBAGE, FLAMBEMENT	KNICKUNG, KNICKEN
3344	**CRITICAL AGING**	VIEILLISSEMENT CRITIQUE	ALTERUNG (KRITISCHE)
3345	**CRITICAL ANGLE**	ANGLE LIMITE	GRENZWINKEL (OPT.)
3346	**CRITICAL HUMIDITY**	HUMIDITE CRITIQUE	FEUCHTIGKEIT (KRITISCHE)
3347	**CRITICAL POINT**	POINT CRITIQUE, POINT DE TRANSFORMATION	PUNKT (KRITISCHER), HALTEPUNKT, UMWANDLUNGSPUNKT
3348	**CRITICAL PRESSURE**	PRESSION CRITIQUE	DRUCK (KRITISCHER)
3349	**CRITICAL STATE**	ETAT CRITIQUE	ZUSTAND (KRITISCHER), GRENZZUSTAND
3350	**CRITICAL STRAIN**	DEFORMATION CRITIQUE	SPANNUNG (KRITISCHE)
3351	**CRITICAL STRESS**	TENSION CRITIQUE	BEANSPRUCHUNG (KRITISCHE)
3352	**CRITICAL TEMPERATURE**	TEMPERATURE CRITIQUE	TEMPERATUR (KRITISCHE)
3353	**CRITICAL VELOCITY**	VITESSE CRITIQUE	GESCHWINDIGKEIT (KRITISCHE)
3354	**CRITICAL VOLUME**	VOLUME CRITIQUE	VOLUMEN (KRITISCHES)
3355	**CROCUS**	ROUGE D'ANGLETERRE	ROT (PARISER)
3356	**CROP**	CHUTE	ABFALLENDE, SCHOPPENDE
3357	**CROPPING**	EBOUTAGE	SCHOPFEN
3358	**CROPPING SHEARS**	CISAILLE A DECOUPER LA CHUTE	SCHOPFSCHERE
3359	**CROSS**	CROIX	KREUZSTÜCK
3360	**CROSS ARCHES FRAME**	CHARPENTE A CROISILLONS	KREUZSTREBENTRAGWERK
3361	**CROSS BAR**	TRAVERSE	QUERBALKEN, QUERRIEGEL
3362	**CROSS COUNTRY VEHICLE**	VEHICULE TOUS TERRAINS	GELÄNDEWAGEN
3363	**CROSS CURRENT**	COURANTS CROISES	KREUZSTROM
3364	**CROSS CUT CHISEL**	BEDANE, BEC-D'ANE	KREUZMEISSEL
3365	**CROSS CUT SAW**	SCIE A TRONCONNER, SCIE PASSE-PARTOUT, PASSE-PARTOUT	SCHROTSÄGE, TRECKSÄGE, BAUCHSÄGE, QUERSÄGE
3366	**CROSS HATCHED COATINGS**	APPLICATION EN PASSES CROISEES	KREUZAUFTRAGUNG
3367	**CROSS MEMBER**	TRAVERSE	QUERTRÄGER
3368	**CROSS PIPE**	CROIX A QUATRE DIRECTIONS	KREUZROHR, KREUZSTÜCK, VIERWEGESTÜCK
3369	**CROSS PLANE**	PANNE EN TRAVERS	QUERFINNE
3370	**CROSS PRESS**	PRESSE DE COULEE	SPANNVORRICHTUNG
3371	**CROSS SECTION**	SECTION EFFICACE	WIRKUNGSQUERSCHNITT
3372	**CROSS SECTION IRON**	FER EN CROIX	KREUZEISEN
3373	**CROSS SECTION OF FRACTURE**	SECTION DE RUPTURE, PLAN DE RUPTURE	BRUCHQUERSCHNITT
3374	**CROSS SECTION OF PASSAGE**	SECTION DE PASSAGE, SECTION LIBRE	DURCHFLUSSQUERSCHNITT, DURCHGANGSQUERSCHNITT, DURCHTRITTSQUERSCHNITT
3375	**CROSS SECTION UNDER COMPRESSION**	SECTION SOUMISE A UN EFFORT DE COMPRESSION	QUERSCHNITT (GEDRÜCKTER), DRUCKQUERSCHNITT
3376	**CROSS SECTION UNDER TENSION**	SECTION SOUMISE A UN EFFORT DE TRACTION	QUERSCHNITT (GEZOGENER), ZUGQUERSCHNITT
3377	**CROSS SECTION, TRANSVERSE SECTION**	COUPE TRANSVERSALE, PROFIL TRANSVERSAL	QUERPROFIL
3378	**CROSS SECTION, TRANSVERSE SECTION**	SECTION TRANSVERSALE	QUERSCHNITT
3379	**CROSS SLIDE**	CHARIOT TRANSVERSAL	QUERSCHLITTEN
3380	**CROSS SLIP**	DEVIATION	QUERGLEITUNG
3381	**CROSS STEEL**	FER EN CROIX	KREUZPROFILEISEN
3382	**CROSS STRIP**	CORDON D'ECRASEMENT	STAUCHSTREIFEN
3383	**CROSS WIRES, SPIDER LINES**	RETICULE	FADENKREUZ
3384	**CROSS-ROLLING**	LAMINAGE TRANSVERSAL	QUERWALZEN

3385	**CROSS-WIRE WELDING**	SOUDAGE DE FILS EN CROIX	KREUZDRAHTSCHWEISSEN
3386	**CROSSED ARM GOVERNOR, PARABOLIC GOVERNOR, PARABOLIC GOVERNOR**	REGULATEUR A BRAS CROISES, REGULATEUR FARCOT	REGLER MIT GEKREUZTEN STANGEN
3387	**CROSSED BELT**	COURROIE CROISEE, COURROIE RENVERSEE	RIEMENTRIEB (GEKREUZTER), RIEMENTRIEB (GESCHRÄNKTER), KREUZTRIEB
3388	**CROSSHEAD**	ENTRETOISÈ, CROSSETTE	QUERHAUPT, KOPFPLATTE, KREUZKOPF
3389	**CROSSHEAD END SMALL END OF CONNECTING ROD**	PIED DE BIELLE, EXTREMITE DE LA BIELLE ARTICULEE AU PISTON	KREUZKOPFENDE DER SCHUBSTANGE
3390	**CROSSHEAD PIN BRASSES**	COUSSINET DE PIED DE BIELLE	KREUZKOPFZAPFENLAGER
3391	**CROSSHEAD PIN, CROSSHEAD GUDGEON, GUDGEON PIN**	TOURILLON DE CROSSE, TOURILLON A FOURCHETTE, TOURILLON-AXE POUR CHAPE-FOURCHETTE	KREUZKOPFZAPFEN
3392	**CROSSING FILE, TUMBLER FILE**	LIME FEUILLE DE SAUGE	VOGELZUNGE
3393	**CROSSING POINT**	POINT DE CROISEMENT	KREUZUNGSPUNKT
3394	**CROW BAR, PINCH BAR**	GRAND LEVIER, PINCE	BRECHSTANGE
3395	**CROW'S FEET, TICKS, ARROW HEADS**	FLECHE DE COTE	MASSPFEIL
3396	**CROWN**	GALBE, CONVEXITE, VOUSSURE, CERCLE DE TETE	BALLIGKEIT, WÖLBUNG, KRANZ
3397	**CROWN GLASS**	CROWN-GLASS	KRONGLAS, CROWNGLAS
3398	**CROWN OF PULLEY**	BOMBEMENT DE LA JANTE D'UNE POULIE	WÖLBUNG EINER RIEMENSCHEIBE, SCHEIBENWÖLBUNG
3399	**CROWN WHEEL**	COURONNE DENTEE, ROUE DE CHAMP	KRONRAD
3400	**CROWN-FACED PULLEY, PULLEY WITH CROWN FACE CONVEX RIM**	POULIE BOMBEE	RIEMENSCHEIBE (BALLIGE), RIEMENSCHEIBE (GEWÖLBTE)
3401	**CRUCIBLE**	CREUSET (DE HAUT FOURNEAU)	TIEGEL, HERD, SCHMEIZTIEGEL
3402	**CRUCIBLE FURNACE**	FOURNEAU A CALEBASSE, FOURNEAU A CREUSET	TIEGELOFEN
3403	**CRUCIBLE LIFTER**	BRANCARD DE CREUSET	TRAGSCHERE, TIEGELSCHERE
3404	**CRUCIBLE PROCESS**	PROCEDE AU CREUSET	TIEGELSCHMELZVERFAHREN
3405	**CRUCIBLE TONGS**	PINCE A CREUSET	SCHMELZTIEGELZANGE, TIEGELZANGE
3406	**CRUDE**	BRUT	ROHERDÖL
3407	**CRUDE (OIL) TANK**	RESERVOIR A PETROLE BRUT	ROHÖLTANK
3408	**CRUDE BENZINE**	ESSENCE MINERALE BRUTE	ROHBENZIN
3409	**CRUDE LEAD**	PLOMB D'OEUVRE	WERKBLEL
3410	**CRUDE METAL**	METAL BRUT	TIEGELGUSSSSTAHL
3411	**CRUDE OIL**	PETROLE BRUT	ROHERDÖL
3412	**CRUDE PETROLEUM, CRUDE OIL**	PETROLE BRUT, HUILE BRUTE DE PETROLE	ROHPETROLEUM, ROHÖL
3413	**CRUDE RUBBER**	CAOUTCHOUC BRUT, VIERGE	ROHGUMMI
3414	**CRUDE STEEL**	ACIER BRUT	ROHSTAHL
3415	**CRUDE TURPENTINE**	TEREBENTHINE	TERPENTIN
3416	**CRUISING RANGE**	AUTONOMIE	FAHRBEREICH
3417	**CRUISING SPEED**	VITESSE DE CROISIERE	REISEGESCHWINDIGKEIT
3418	**CRUMBLING, DESINTEGRATION**	EFFRITEMENT, DESINTEGRATION	ZERSTÄUBUNG, ZERFALL
3419	**CRUSHER**	MACHINE A BROYER, BROYEUR, MACHINE A CONCASSER, CONCASSEUR, MACHINE A ECRASER	ZERKLEINERUNGSMASCHINE, BRECHMASCHINE
3420	**CRUSHING**	ECRASEMENT	STAUCHUNG
3421	**CRUST**	CROUTE	SALZKRUSTE

3422	CRYOHYDRATE	CRYOHYDRATE	KRYOHYDRAT
3423	CRYOLITHE	CRYOLITHE	EISTEIN, GRÖNTARNSPAT, KRYOLITH
3424	CRYSTAL	CRISTAL (MIN.)	KRISTALL
3425	CRYSTAL ANALYSIS	ANALYSE (DETERMINATION) DE STRUCTURE CRISTALLINE	KRISTALLSTRUKTUR-ANALYSE
3426	CRYSTAL ELONGATION	ALLONGEMENT DES CRISTAUX	KRISTALLDEHNUNG
3427	CRYSTAL FACE	PLAN DE CRISTAL, FACE CRISTALLINE	KRISTALLFLÄCHE
3428	CRYSTAL GLASS	CRISTAL (VERRE)	KRISTALLGLAS
3429	CRYSTAL SPOTS	TACHE PAR CRISTAUX DE SULFURE DE CUIVRE	METALLSULFITFLECK
3430	CRYSTAL STRUCTURE	STRUCTURE CRISTALLINE	KRISTALLSTRUKTUR
3431	CRYSTAL SYSTEM	SYSTEME DES CRISTAUX	KRISTALLSYSTEM, SYNGONIE
3432	CRYSTAL UNIT	CRISTAL OSCILLATEUR	SCHWINGQUARZ
3433	CRYSTALLINE	CRISTALLIN	KRISTALLINISCH
3434	CRYSTALLINE STRUCTURE, CRYSTALLINE TEXTURE	TEXTURE CRISTALLINE	GEFÜGE (KRISTALLINISCHES)
3435	CRYSTALLINE SULPHUR	SOUFRE CRISTALLISE	SCHWEFEL (KRISTALLINISCHER)
3436	CRYSTALLISABILITY	CRISTALLISABILITE	KRISTALLISIERBARKEIT
3437	CRYSTALLISABLE	CRISTALLISABLE	KRISTALLISIERBAR
3438	CRYSTALLISATION	CRISTALLISATION	KRISTALLISATION, KRISTALLBILDUNG
3439	CRYSTALLISE (TO)	CRISTALLISER	KRISTALLISIEREN
3440	CRYSTALLITE	CRISTALLITE	KRISTALLIT
3441	CRYSTALLOGRAM	CRISTALLOGRAMME, DIAGRAMME DE DIFFRACTION A RAYONS X	KRISTALLOGRAMM, RÖNTGENBEUGUNGSBILD
3442	CRYSTALLOGRAPHY	CRISTALLOGRAPHIE	KRISTALLOGRAFIE
3443	CRYSTALLOID	CRISTALLOIDE	KRISTALLOID
3444	CUBATURE	CUBAGE (EVALUATION EN UNITES CUBES)	INHALTSBERECHNUNG EINES KÖRPERS, RAUMINHALTSBERECHNUNG, KÖRPERINHALTSBERECHNUNG, KUBATUR
3445	CUBE	CUBE (GEOM.)	WÜRFEL, KUBUS, DRITTE POTENZ
3446	CUBE (TO), FIND THE CUBIC CONTENTS (TO)	CUBER	KÖRPERINHALT (DEN) BERECHNEN, KUBIEREN
3447	CUBE ROOT	RACINE CUBIQUE	KUBIKWURZEL, DRITTE WURZEL
3448	CUBIC CAPACITY	CYLINDREE	ZYLINDERINHALT
3449	CUBIC CENTIMETRE	CENTIMETRE CUBE	KUBIKZENTIMETER
3450	CUBIC CRYSTALS	CRISTAL CUBIQUE	KRISTALL (ISOMETRISCHER), KRISTALL (KUBISCHER), KRISTALL (REGULÄRER)
3451	CUBIC DECIMETRE	DECIMETRE CUBE	KUBIKDEZIMETER
3452	CUBIC EQUATION	EQUATION CUBIQUE, EQUATION DU TROISIEME DEGRE	GLEICHUNG DRITTEN GRADES, GLEICHUNG (KUBISCHE)
3453	CUBIC FOOT	PIED CUBE ANGLAIS	KUBIKFUSS (ENGLISCHER)
3454	CUBIC INCH	POUCE CUBE ANGLAIS	KUBIKZOLL (ENGLISCHER)
3455	CUBIC METRE	METRE CUBE	KUBIKMETER
3456	CUBIC MILLIMETRE	MILLIMETRE CUBE	KUBIKMILLIMETER
3457	CULTIVATORS, RIGID TINE, SPRING-LOADED TINE, SPRING TINE	CULTIVATEURS, VIBROCULTEURS	BODENBEARBEITUNGSGERÄTE, WÜHLGRUBBER, FEDERZINKENEGGEN
3458	CUP LEATHER	CUIR EMBOUTI	LEDERSTULP, LEDERMANSCHETTE
3459	CUP LEATHER PACKING	GARNITURE EN CUIR EMBOUTI	LEDERSTULPDICHTUNG, MANSCHETTENDICHTUNG
3460	CUP SHAKE IN TIMBER	ROULURE DU BOIS	RINGKLUFT, KERNSCHÄLE DES HOLZES

3461	CUP SPRING	RONDELLE BELLEVILLE	BELLEVILLE FEDER
3462	CUP TEST MACHINE	APPAREIL D'ESSAI DE DUCTILITE	DUKTILITÄTSPRÜFMASCHINE
3463	CUPELLATION	COUPELLATION	KUPELLATION
3464	CUPOLA	CUBILOT	KUPOL.,
3465	CUPOLA CONTROL EQUIPEMENT	REGULATEUR DE DEBIT DE VENT	LUFTSTROMREGLER
3466	CUPOLA FURNACE	CUBILOT, FOURNEAU A LA WILKINSON	KUPPELOFEN, KUPOLOFEN
3467	CUPOLA MALLEABLE IRON	FONTE MALLEABLE DE CUBILOT	KUPOLOFENTEMPERGUSS
3468	CUPOLA METAL	METAL DE CUBILOT	KUPOLOFENMETALL
3469	CUPPING	EMBOUTISSAGE EN COUPE DE FILS, EMBOUTISSAGE PROFOND	TIEFZIEHEN
3470	CUPRIC CARBONATE	CARBONATE DE CUIVRE	KUPFER (KOHLENSAURES), KUPFERKARBONAT
3471	CUPRIC HYDROXYDE	OXYDE CUIVRIQUE HYDRATE	KUPFERHYDROXYD, KUPRIHYDROXYD, KUPFEROXYDHYDRAT
3472	CUPRIC NITRATE	NITRATE DE CUIVRE	KUPFER (SALPETERSAURES), KUPFERNITRAT, KUPRINITRAT
3473	CUPRIC NITRITE	NITRITE DE CUIVRE	KUPFER (SALPETRIGSAURES), KUPFERNITRIT, KUPRONITRAT
3474	CUPRIC OXIDE	OXYDE CUIVRIQUE	KUPFEROXYD
3475	CUPRIC SULPHATE	SULFATE DE CUIVRE	KUPFER (SCHWEFELSAURES), KUPFERSULFAT
3476	CUPRITE, RED OXIDE OF COPPER	CUIVRE OXYDULE	ROTKUPFERERZ
3477	CUPRO-NICKEL	CUPRO-NICKEL	KUPFERNICKEL
3478	CUPROUS CHLORIDE	CHLORURE CUIVREUX	KUPFERCHLORÜR, KUPROCHLORID
3479	CUPROUS OXIDE .	OXYDE CUIVREUX	KUPFEROXYDUL
3480	CURB WEIGHT	POIDS EN ORDRE DE MARCHE	LEISTUNGSGEWICHT (FAHRFERTIG)
3481	CURCUMIN	CURCUMINE	KURKUMAGELB, KURKUMIN
3482	CURD SOAP	SAVON BLANC ORDINAIRE	KERNSEIFE
3483	CURE	PRISE, DURCISSEMENT	AUSHÄRTEN
3484	CURING	PRISE (DU CIMENT), VULCANISATION	ZEMENTAUSHÄRTUNG, VULKANISIEREN
3485	CURLING DIE	OUTIL DE ROULAGE	ROLLWERKZEUG
3486	CURRENT	COURANT	STROM
3487	CURRENT COORDINATES	COORDONNEES COURANTES	KOORDINATEN (LAUFENDE)
3488	CURRENT DENSITY	DENSITE DU COURANT	STROMDICHTE
3489	CURRENT EFFICIENCY	RENDEMENT EN COURANT	STROMAUSBEUTE
3490	CURRENT METER	INSTRUMENT POUR LA DETERMINATION DE LA VITESSE D'UN COURANT	STRÖMUNGSMESSER
3491	CURRENT REGULATOR	REGULATEUR DE COURANT, REGULATEUR D'INTENSITE	STROMREGLER
3492	CURTAINING	DRAPERIES/COULURES EN FESTONS	VORHÄNGEBILDUNG
3493	CURTATE CYCLOID	CYCLOIDE RACCOURCIE	ZYKLOIDE (VERKÜRZTE)
3494	CURVATURE	COURBURE, CINTRE	KRÜMMUNG
3495	CURVE OF DIAGRAM, GRAPH	COURBE D'UN DIAGRAMME	SCHAULINIE, DIAGRAMMLINIE, DIAGRAMMKURVE
3496	CURVE, SWEEP	COURBE	KURVE, KRUMME LINIE
3497	CURVED ARM OF PULLEY	BRAS PARABOLIQUE D'UNE POULIE	SCHEIBENARM (GESCHWUNGENER)
3498	CURVED SURFACE, CURVED FACE	SURFACE COURBE	FLÄCHE (GEKRÜMMTE), FLÄCHE (KRUMME)
3499	CURVILINEAR	CURVILIGNE	KRUMMLINIG
3500	CURVILINEAR MOTION	MOUVEMENT CURVILIGNE	BEWEGUNG (KRUMMLINIGE)
3501	CURVOMETER	CURVIMETRE	MESSRAD, KURVIMETER

3502	CUSP OF A CURVE	POINT DE REBROUSSEMENT	SPITZE EINER KURVE
3503	CUT	COUPE	SCHNITT
3504	CUT (TO)	COUPER (GEOM.), TAILLER	DURCHSCHNEIDEN (GEOM.), SCHNEIDEN
3505	CUT BY THE AUTOGENOUS PROCESS (TO)	DECOUPER A L'AUTOGENE, DECOUPER AU CHALUMEAU	AUTOGENSCHNEIDEN
3506	CUT EDGE	RIVE CISAILLEE	KANTE (GESCHNITTENE)
3507	CUT FILES (TO)	TAILLER LES LIMES	FEILEN HAUEN
3508	CUT GEAR WHEEL	ROUE A DENTS TAILLEES, ENGRENAGE TAILLE	ZAHNRAD (BEARBEITETES), ZAHNRAD (GESCHNITTENES)
3509	CUT INTO THE REQUISITE SIZE (TO)	DEBITER AUX DIMENSIONS VOULUES	ABLÄNGEN
3510	CUT OFF (TO)	DECOLLETER, TRONCONNER, DECOUPER	ABSTECHEN
3511	CUT OFF THE RIVET HEADS (TO)	DERIVER, CHASSER LES RIVETS	ENTNIETEN, ABNIETEN, NIETE (DIE) HERAUSSCHLAGEN
3512	CUT OUTSIDE SCREW THREADS (TO)	FILETER	AUSSENGEWINDESCHNEIDEN
3513	CUT PLATES (TO)	DECOUPER DES TOLES	SCHNEIDEN (BLECHE)
3514	CUT TOOTH	DENT TAILLEE	ZAHN (BEARBEITETER)
3515	CUT WITH THE ANVIL CHISEL (TO)	COUPER AU TRANCHET, TRANCHER	ABSCHROTEN
3516	CUT-OUT RELAY	DISJONCTEUR	SCHALTSCHÜTZ
3517	CUT, KERF OF A SAW, SAW KERF, SAW CUT	TRAIT DE SCIE	SÄGESCHNITT, SÄGEEINSCHNITT, SÄGESCHNITTFUGE, SCHNITTFUGE EINER SÄGE
3518	CUT, TEETH OF A FILE	TAILLE, DENTS, DENTURE D'UNE LIME	FEILENHIEB, HIEB EINER FEILE
3519	CUTTER	FRAISE	FRÄSER
3520	CUTTER ARBOR	ARBRE PORTE-FRAISE	FRÄSERSPINDEL
3521	CUTTER COMPENSATION	CORRECTION D'OUTIL	WERKZEUGKORREKTUR
3522	CUTTERS, BLOWERS	ENSILEUSES	SILIERMASCHINEN
3523	CUTTING	COUPE, COUPAGE, TAILLE, TAILLAGE	SCHNEIDEN
3524	CUTTING ANGLE	ANGLE D'ATTAQUE, ANGLE DE COUPE	SCHNEIDWINKEL
3525	CUTTING ATTACHMENT	DISPOSITIF DE COUPE	SCHNEIDEINSATZ
3526	CUTTING DEPTH	PROFONDEUR DE COUPE	SCHNITTIEFE
3527	CUTTING DOWN	POLISSAGE DE LA SURFACE	OBERFLÄCHENPOLIERUNG
3528	CUTTING EDGE OF A TOOL	TRANCHANT D'UN OUTIL, ARETE COUPANTE	SCHNEIDE, SCHNEIDKANTE EINES WERKZEUGS
3529	CUTTING FILES	TAILLAGE DES LIMES	FEILENHAUEN
3530	CUTTING FLUID	FLUIDE DE COUPE	SCHNEIDFLÜSSIGKEIT
3531	CUTTING HEAD	TETE DE FRAISAGE (OU DE COUPE)	FRÄSKOPF
3532	CUTTING NIPPERS, NIPPING PLIERS	PINCE COUPANTE, TENAILLE A MORS COUPANTS	BEISSZANGE, KNEIPZANGE, KNEIFZANGE, ZWICKZANGE, DRAHTZWICKZANGE
3533	CUTTING OFF	DECOLLETAGE, TRONCONNAGE, TRONCONNEMENT, DECOUPAGE	ABSTECHEN
3534	CUTTING OIL	HUILE DE COUPE	SCHNEIDÖL
3535	CUTTING OUTSIDE SCREW THREADS	FILETAGE	SCHNEIDEN EINES AUSSENGEWINDES
3536	CUTTING STROKE	COURSE DE TRAVAIL, COURSE UTILE (D'UNE MACHINE-OUTIL)	ARBEITSGANG, SCHNITTGANG (EINER WERKZEUGSMASCHINE)
3537	CUTTING TIP	BUSE DE COUPE	SCHNEIBRENNERDÜSE
3538	CUTTING TOOL	OUTIL TRANCHANT, OUTIL COUPANT, OUTIL A TAILLANT	SCHNEIDWERKZEUG

3539	**CUTTING TORCH**	CHALUMEAU	SCHNEIDBRENNER
3540	**CUTTING-OFF LATHE**	TOUR A DECOLLETER	ABSTECHBANK
3541	**CUTTING-OFF MACHINE**	MACHINE A TRONCONNER, TRONCONNEUSE	ABSTECHMASCHINE
3542	**CUTTING-OFF WHEEL**	DISQUE A COUPER, DISQUE DE COUPER	SCHNEIDSCHEIBE, TRENNSCHEIBE
3543	**CUTTINGS**	COPEAUX	SPÄNE
3544	**CYANATE**	CYANATE	ZYANSÄURESALZ, ZYANAT
3545	**CYANIC ACID**	ACIDE CYANIQUE	ZYANSÄURE
3546	**CYANIDATION**	EXTRACTION DE L'OR PAR CYANURATION	CYANIDLAUGEREI, CYANIDLAUGUNGSVERFAHREN
3547	**CYANIDE**	CYANURE	ZYANMETALL, ZYANID, CYANID
3548	**CYANIDE SLIMES**	PARTICULES DE METAUX NOBLES	EDELMETALLTEILCHEN
3549	**CYANIDING**	CIRCULATION AU CYANURE, ENDURCISSEMENT AU CYANURE	ZYANSALZBARHÄRTUNG
3550	**CYANIDING FURNACE**	FOUR A CYANURATION	ZYANIDVERFAHRENOFEN
3551	**CYANOGEN**	CYANOGENE	ZYAN
3552	**CYANURIC ACID**	ACIDE CYANURIQUE	ZYANURSÄURE, TRIZYANSÄURE
3553	**CYCLE**	CYCLE, CYCLE DE TRAVAIL	PERIODE, ZYCLUS
3554	**CYCLE OF OPERATIONS , CYCLIC PROCESS**	CYCLE	KREISPROZESS
3555	**CYCLIC CURVE**	CYCLIQUE	KURVE (ZYKLISCHE)
3556	**CYCLIC PERMUTATION**	PERMUTATION CYCLIQUE	VERTAUSCHUNG (ZYKLISCHE)
3557	**CYCLOID**	CYCLOIDE	RADLINIE, ZYKLOIDE
3558	**CYCLOIDAL GEAT TEETH**	ENGRENAGE CYCLOIDAL	ZYKLOIDENVERZAHNUNG
3559	**CYCLOMETER DIAL COUNTER**	COMPTEUR A CHIFFRES, COMPTEURS A FENETRES	ZÄHLWERK MIT SPRINGENDEN ZAHLEN, SPRINGENDES ZÄHLWERK
3560	**CYCLONE**	CYCLONE PULVERISATEUR	ZYKLON
3561	**CYLINDER**	CYLINDRE (GEOM.)	ZYLINDER, WALZE (GEOM.)
3562	**CYLINDER BARREL**	FUT DE CYLINDRE	ZYLINDERLAUFBÜCHSE
3563	**CYLINDER BIT, HALF ROUND BIT**	FORET OU MECHE A CANON	KANONENBOHRER
3564	**CYLINDER BLOCK**	BLOC-CYLINDRES	ZYLINDERBLOCK
3565	**CYLINDER BORE**	ALESAGE DU CYLINDRE	ZYLINDERBOHRUNG
3566	**CYLINDER CAPACITY**	CYLINDREE	ZYLINDERINHALT
3567	**CYLINDER COVER**	COUVRE CULBUTEURS, COUVERCLE DU CYLINDRE	ZYLINDERKKOPFDECKEL, ZYLINDERDECKEL
3568	**CYLINDER GRINDER**	MACHINE A RECTIFIER L'INTERIEUR DES CYLINDRES	ZYLINDERSCHLEIFMASCHINE
3569	**CYLINDER HEAD**	TETE DU CYLINDRE, CULASSE	ZYLINDERKOPF
3570	**CYLINDER HEAD GASKET**	JOINT DE CULASSE	ZYLINDERKOPFDICHTUNG
3571	**CYLINDER LINER**	GARNITURE INTERIEURE DU CYLINDRE, CHEMISE DU CYLINDRE	EINSATZZYLINDER, ZYLINDERLAUFBÜCHSE, ZYLINDEREINSATZ, ZYLINDERFUTTER
3572	**CYLINDER OIL**	HUILE POUR CYLINDRES	ZYLINDERÖL
3573	**CYLINDER RECORDER**	ENREGISTREUR A TAMBOUR	TROMMELSCHREIBER
3574	**CYLINDER SCREW HEAD, CHEESE SCREW HEAD**	TETE CYLINDRIQUE D'UNE VIS	ZYLINDRISCHER SCHRAUBENKOPF
3575	**CYLINDER VALVE FOR COMPRESSED GAS**	ROBINET DE BOUTEILLE (DE GAZ COMPRIME)	ABSPERRVENTIL FÜR GASFLASCHEN
3576	**CYLINDRAL MOUTHPIECE**	AJUTAGE CYLINDRIQUE	ZYLINDRISCHER AUSFLUSSSTUTZEN
3577	**CYLINDRICAL**	CYLINDRIQUE	ZYLINDRISCH
3578	**CYLINDRICAL BOILER**	CHAUDIERE CYLINDRIQUE	WALZENKESSEL
3579	**CYLINDRICAL CAM, DRUM CAM**	TAMBOUR, CYLINDRE A RAINURE	SCHLITZTROMMEL, NUTENTROMMEL
3580	**CYLINDRICAL COORDINATES**	COORDONNEES CYLINDRIQUES	ZYLINDERKOORDINATEN

3581	CYLINDRICAL GAUGE, PLUG AND COLLAR GAUGE, PLUG AND RING GAUGE	TAMPON ET BAGUE, TAMPON ET LUNETTE DE CALIBRE, CALIBRE MALE ET FEMELLE	LEHRBOLZEN, LEHRDORN UND LOCHLEHRE, KALIBER UND KALIBERRING
3582	CYLINDRICAL GRADUATED MEASURE	CYLINDRE GRADUE	MESSZYLINDER
3583	CYLINDRICAL GRINDER	MACHINE A RECTIFIER LES PIECES CYLINDRIQUES	RUNDSCHLEIFMASCHINE
3584	CYLINDRICAL GUIDE	GLISSIERE CYLINDRIQUE	RUNDFÜHRUNG
3585	CYLINDRICAL HELICAL SPRING	RESSORT A BOUDIN CYLINDRIQUE	SCHRAUBENFEDER (ZYLINDRISCHE)
3586	CYLINDRICAL PIPE	TUYAU CYLINDRIQUE	ROHR (ZYLINDRISCHES)
3587	CYLINDRICAL PLUG GAUGES	CALIBRES MALES CYLINDRIQUES	MESSDORN (ZYLINDRISCHER)
3588	CYLINDRICAL RING	ANNEAU CYLINDRIQUE, TORE CIRCULAIRE	RING (ZYLINDRISCHER)
3589	CYLINDRICAL ROLLER	GALET CYLINDRIQUE	ZYLINDERROLLE
3590	CYLINDRICAL SURFACE	SURFACE LATERALE D'UN CYLINDRE	MANTELFLÄCHE EINES ZYLINDERS
3591	CYLINDRICAL WHEEL	ROUE CYLINDRIQUE, ROUE DROITE	RAD (ZYLINDRISCHES)
3592	D-SLIDE VALVE	TIROIR A COQUILLE	MUSCHELSCHIEBER, D-SCHIEBER, DREIWEGSCHIEBER
3593	D'COMER BAR	BARRE DE DEVERSOIR	FALLROHR
3594	DAILY WAGES	SALAIRE, PRIX A LA JOURNEE	TAGELOHN
3595	DAIRY MACHINERY, APPLIANCES, PLANT AND EQUIPMENT	MACHINES DE LAITERIE	MOLKEREIMASCHINEN UND ZUSATZGERÄTE
3596	DAM PLATE	PLAQUE DE DAME	WALLPLATTE, SCHLACKENBLECH
3597	DAM, BARRAGE	BARRAGE D'UNE VALLEE	TALSPERRE
3598	DAMMAR	GOMME DAMMAR, DAMMAR	DAMMARHARZ
3599	DAMP AIR, MOIST AIR	AIR HUMIDE	FEUCHTE LUFT
3600	DAMPED OSCILLATION	VIBRATION AMORTIE	SCHWINGUNG (GEDÄMPFTE)
3601	DAMPENER	AMORTISSEUR	DÄMPFER
3602	DAMPER	DAMPER, AMORTISSEUR (ELECTR.)	DÄMPFER, SCHWINGUNGSDÄMPFER
3603	DAMPER, REGISTER	REGISTRE DE TIRAGE, REGISTRE DE FUMEE	LUFTSCHIEBER, RAUCHSCHIEBER, ZUGREGLER, ESSENKLAPPE, SCHORNSTEINKLAPPE
3604	DAMPING	ATTENUATION, AMORTISSEMENT	DÄMPFUNG
3605	DAMPING CAPACITY	CAPACITE D'AMORTISSEMENT	DÄMPFUNGSVERMÖGEN, DÄMPFUNGSFÄHIGKEIT
3606	DAMPING OF OSCILLATIONS	AMORTISSEMENT DE VIBRATIONS	DÄMPFUNG, BERUHIGUNG VON SCHWINGUNGEN
3607	DANDY (US)	FOUR DE PREMIER ALLIAGE	FRISCHEREIÖFEN
3608	DANGER OF FIRE	DANGER, RISQUE, CHANCES D'INCENDIE	FEUERSGEFAHR
3609	DANGEROUS SECTION	SECTION DANGEREUSE	QUERSCHNITT (GEFÄHRLICHER)
3610	DASH BOARD	TABLEAU DE BORD	INSTRUMENTENBRETT
3611	DASHPOT	CATARACTE, DASHPOT	PUFFERVORRICHTUNG, ZEITREGLER, KATARAKT
3612	DATA	DONNEES	DATEN
3613	DATA BOOK	RECUEIL DE DONNEES	DATENREGISTER
3614	DATA REPORTS	DOSSIER TECHNIQUE	BERICHT (TECHNISCHER)
3615	DATA SHEETS	FICHES TECHNIQUES	KENNBLATT (TECHNISCHES)
3616	DATE OF APPLICATION FOR A PATENT	JOUR, DATE DU DEPOT DE LA DEMANDE DE BREVET	TAG DER ANMELDUNG, ANMELDUNGSDATUM EINER ERFINDUNG
3617	DATE OF SEALING THE PATENT	JOUR DE LA SIGNATURE DU BREVET, DATE DE L'ACCORD DU BREVET	TAG DER ERTEILUNG EINES PATENTES
3618	DAVIT	POTENCE	SCHWENKKRAN
3619	DAVIT ARM	BRAS DE POTENCE	SCHWENKKRANARM
3620	DAY SHIFT	TRAVAIL DE JOUR	TAGSCHICHT

3621	DAY WORK	TRAVAIL A LA JOURNEE	LOHNARBEIT
3622	DAYLIGHT	LUMIERE DU JOUR	TAGESLICHT
3623	DE-AERATION OF WATER	DEGAZAGE DE L'EAU	ENTLÜFTUNG DES WASSERS
3624	DEAD AXLE	ESSIEU FIXE	ACHSE (FESTSTEHENDE)
3625	DEAD CENTER	ARETE TERMINALE	TOTPUNKT
3626	DEAD CENTRE POINT	POINT MORT	TOTER PUNKT, TOTPUNKT
3627	DEAD CONDUCTOR	CONDUCTEUR SANS COURANT, CONDUCTEUR SANS TENSION	LEITER (STROMLOSER), LEITER (SPANNUNGSLOSER)
3628	DEAD DIP	DECAPAGE MAT	MATTBEIZEN, MATTBRENNEN
3629	DEAD HEAD, HEAD METAL	MASSELOTTE	KOPF (VERLORENER)
3630	DEAD HOLE	TROU EN CUL DE SAC, TROU BORGNE	LOCH (BLINDES), LOCH (NICHT DURCHGEHENDES), BLINDLOCH
3631	DEAD LOAD	CHARGE MORTE	LAST (TOTE)
3632	DEAD LOAD (SAFETY) VALVE, COWBURN VALVE	SOUPAPE DE SURETE A CHARGE DIRECTE	VENTIL MIT UNMITTELBARER GEWICHTSBELASTUNG
3633	DEAD LOAD, PERMANENT CONSTANT LOAD, STEADY QUIESCENT LOAD	CHARGE PERMANENTE	BELASTUNG (RUHENDE), BELASTUNG (STÄNDIGE), BELASTUNG (STETIGE), DAUERBELASTUNG, LAST (TOTE)
3634	DEAD MELTING	CHAUFFAGE AU-DESSUS DU POINT DE FUSION	ÜBERSCHMELZEN
3635	DEAD POINT	POINT MORT	TOTPUNKT
3636	DEAD ROASTING	GRILLAGE TOTAL	TOTALRÖSTUNG
3637	DEAD WEIGHT	POIDS PROPRE, POIDS MORT	GEWICHT (TOTES), EIGENGEWICHT, EIGENLAST
3638	DEAD WEIGHT BRAKE	FREIN A LEVIER ET CONTREPOIDS	LÜFTUNGSBREMSE, LÖSUNGSBREMSE, GEWICHTSHEBELBREMSE
3639	DEAD-BURNED MAGNESITE	MAGNESITE MORTE	MAGNESIT (TOTGEBRANNTER)
3640	DEADEN A SHOCK (TO)	AMORTIR UN CHOC	DÄMPFEN (EINEN STOSS)
3641	DEAFENING	MATIERE IMPERMEABLE AU SON, MATIERE ETOUFFANT LE BRUIT	SCHALLDÄMPFENDES MITTEL, SCHALLDICHTER STOFF
3642	DEBURRING	EBAVURAGE, EBARBAGE	ABGRATEN
3643	DEBYE-SCHERRER METHOD	METHODE DE DEBYE-SCHERRER	DEBYE-SCHERRER METHODE, PULVERMETHODE
3644	DECAGON	DECAGONE	ZEHNECK
3645	DECAGRAMME	DECAGRAMME	DEKAGRAMM
3646	DECALESCENCE	DECALESCENCE	DEKALESZENZ
3647	DECALITRE	DECALITRE	DEKALITER
3648	DECAMETRE	DECAMETRE	DEKAMETER
3649	DECANT (TO), POUR OFF (TO)	DECANTER	ABGIESSEN, ABKLÄREN, DEKANTIEREN
3650	DECANTING, POURING OFF	DECANTATION	ABKLÄREN, ABGIESSEN, DEKANTIEREN
3651	DECARBONISATION	DECARBONISATION	ENTKOHLEN, ENTKOHLUNG
3652	DECARBONISE (TO)	DECARBONISER	ENTKOHLEN
3653	DECARBURISATION OF IRON	DECARBURATION DU FER	ENTKOHLEN DES EISENS
3654	DECARBURISE THE IRON (TO)	DECARBURER LE FER	ENTKOHLEN (EISEN)
3655	DECARBURIZATION	DECARBURATION	ENTKOHLUNG
3656	DECAY	DECROISSANCE	ABNAHME
3657	DECIGRAMME	DECIGRAMME	DEZIGRAMM
3658	DECILITRE	DECILITRE	DEZILITER
3659	DECIMAL	DECIMALE	DEZIMALE, DEZIMALSTELLE
3660	DECIMAL BALANCE, DECIMAL WEIGHING MACHINE	BALANCE, BASCULE, BASCULE AU DIXIEME (AU 10)	DEZIMALWAAGE
3661	DECIMAL CODE	CODE DECIMAL	DEZIMALKODE

3662	DECIMAL FRACTION	FRACTION DECIMALE	DEZIMALBRUCH
3663	DECIMAL INTERNATIONAL CANDIE, BOUGIE DECIMALE	BOUGIE DECIMALE	DEZIMALKERZE, KERZE (INTERNATIONALE), STANDARDKERZE, NORMKERZE
3664	DECIMAL NUMBER	NOMBRE DECIMAL	DEZIMALZAHL
3665	DECIMAL SYSTEM	SYSTEME DECIMAL	DEZIMALSYSTEM
3666	DECIMETRE	DECIMETRE	DEZIMETER
3667	DECK	TOIT, VOILE, PONT	DECKE, DACH
3668	DECLINATION, DEVIATION OF THE MAGNETIC NEEDLE	DECLINAISON MAGNETIQUE	ABWEICHUNG DER MAGNETNADEL, DEKLINATION
3669	DECOCTION	DECOCTION	ABKOCHUNG, ABSUD
3670	DECODER	DECODEUR	DEKODIERER
3671	DECOMPOSABLE	DECOMPOSABLE	ZERLEGBAR, ZERSETZBAR
3672	DECOMPOSE (TO)	DECOMPOSER	ZERLEGEN (CHEM.)
3673	DECOMPOSE (TO)	DECOMPOSER (SE)	ZERSETZEN (SICH)
3674	DECOMPOSITION	DECOMPOSITION (CHIM.)	ZERSETZUNG, ZERLEGUNG (CHEM.)
3675	DECOMPOSITION POTENTIAL	TENSION DE DECOMPOSITION	ZERSETZUNGSSPANNUNG
3676	DECOMPOSITION, SPLITTING UP OF LIGHT	DECOMPOSITION DE LA LUMIERE	ZERLEGUNG DES LICHTS
3677	DECREASE OF SPEED	DIMINUTION, CHUTE DE VITESSE	GESCHWINDIGKEITSABNAHME
3678	DECREASE THE VELOCITY (TO)	REDUIRE LA VITESSE	GESCHWINDIGKEIT VERRINGERN, GESCHWINDIGKEITVERMINDERN
3679	DECREASING SPEED	VITESSE DECROISSANTE	GESCHWINDIGKEIT (ABNEHMENDE)
3680	DECTECTIVE MATERIAL	VICE DE MATIERE	MATERIALFEHLER
3681	DEDENDUM	HAUTEUR DU FLANC, HAUTEUR DU PIED DE DENT, CREUX SOUS LE PRIMITIF	ZAHNFUSSLÄNGE, ZAHNFUSSHÖHE
3682	DEDUSTING	DEPOUSSIERAGE	ENTSTAUBEN
3683	DEEP DRAWING	EMBOUTISSAGE PROFOND	TIEFZIEHEN
3684	DEEP ETCH TEST	ESSAI DE MACRO-ATTAQUE	TIEFBEIZPROBE
3685	DEEP ETCHING	ATTAQUE PROFONDE	TIEFBEIZEN
3686	DEEP HOLE DRILL	FORET POUR PERCAGE PROFOND	TIEFLOCHBOHRER
3687	DEEP HOLE DRILLING AND BORING MACHINE	MACHINE A FORER ET ALESER LES TROUS PROFONDS	TIEFLOCH-BOHRMASCHINE
3688	DEEP STAMPING STRIPS	FEUILLARDS D'ACIER POUR EMBOUTISSAGE PROFOND	TIEFZIEHBANDSTAHL
3689	DEEP WELL PUMP	POMPE POUR PUITS PROFONDS	TIEFBRUNNENPUMPE
3690	DEEP-DRAWING BRASS	LAITON A QUALITE D'EMBOUTISSAGE	TIEFZIEHQUALITÄTSMESSING
3691	DEFECT	DEFAUT	FEHLER
3692	DEFECT FLAW IN THE MATERIAL	VICE, DEFAUT DE MATIERE	MATERIALFEHLER, FEHLER, FEHLERHAFTE STELLE IM MATERIAL
3693	DEFECT IN MANUFACTURE	DEFAUT DE FABRICATION, VICE DE CONSTRUCTION, FAUTE D'EXECUTION	HERSTELLUNGSFEHLER, FABRIKATIONSFEHLER
3694	DEFECTIVE DESIGN	VICE DE CONCEPTION	ENTWURFSMANGEL
3695	DEFECTIVE OPERATION	VICE DE FONCTIONNEMENT	FEHLERHAFTER BETRIEB
3696	DEFECTIVE, FAULTY, CONDEMNED WORK	REBUTS DE FABRICATION, PIECES DEFECTUEUSES	AUSSCHUSS, AUSSCHUSSWARE
3697	DEFINITE INTEGRAL	INTEGRALE DEFINIE	INTEGRAL (BESTIMMTES)
3698	DEFLECTION	DEFORMATION (FLECHE), FLEXION, FLECHE	VERFORMUNG, DURCHBIEGUNG, BIEGUNG
3699	DEFLECTION OF A POINTER	DEVIATION, DEFLECTION, ELONGATION DE L'AIGUILLE	ZEIGERAUSSCHLAG

3700	**DEFLECTION, BENDING, FLEXURE**	FLEXION TRANSVERSALE, FLECHISSEMENT	BIEGUNG, DURCHBIEGUNG
3701	**DEFLECTION, SET**	FLECHE (MEC.)	DURCHBIEGUNG (MASS)
3702	**DEFLECTOR**	DEFLECTEUR	ABLEITBLECH
3703	**DEFLOCCULATED GRAPHITE**	GRAPHITE DEFLOCULE	GRAPHIT (AUSGEFLOCKTER), FLOCKENGRAPHIT
3704	**DEFOGGING**	DESEMBUAGE	ENTNEBELUNG
3705	**DEFORMABILITY**	DEFORMABILITE, CAPACITE DE DEFORMATION, PLASTICITE	VERFORMBARKEIT, FORMÄNDERUNGSVERMÖGEN
3706	**DEFORMATION BAND**	BANDE DE DEFORMATION	VERFORMUNGSBAND, DEFORMATIONSBAND
3707	**DEFORMATION, CHANGE OF FORM SHAPE**	DEFORMATION, CHANGEMENT DE FORME	FORMÄNDERUNG
3708	**DEFROSTER**	DEGIVREUR	ENTFROSTER
3709	**DEGASIFYING ALLOY**	ALLIAGE DEGAZEUR	ENTGASUNGSLEGIERUNG
3710	**DEGASSING**	DEGAZAGE	ENTGASUNG
3711	**DEGRAS**	DEGRAS, GRAISSE DES TANNEURS	DEGRAS
3712	**DEGREASING**	DEGRAISSAGE	ENTFETTUNG
3713	**DEGREASING COMPOUND**	COMPOSITION DE DEGRAISSAGE	ENTFETTUNGSMITTEL
3714	**DEGREE**	DEGRE	GRAD
3715	**DEGREE BAUME**	DEGRE BAUME	GRAD BAUME
3716	**DEGREE CENTIGRADE**	DEGRE CENTIGRADE	GRAD CELSIUS, CELSIUSGRAD
3717	**DEGREE ENGLER**	DEGRE ENGLER	ENGLERGRAD
3718	**DEGREE FAHRENHEIT**	DEGRE FAHRENHEIT	GRAD FAHRENHEIT, FAHRENHEITGRAD
3719	**DEGREE OF CYCLIC IRREGULARITY**	COEFFICIENT DE REGULARITE, REGULARITE CYCLIQUE	GLEICHFÖRMIGKEITSGRAD, UNGLEICHFÖRMIGKEITSGRAD
3720	**DEGREE OF HARDNESS**	DEGRE DE DURETE	HÄRTEGRAD, HÄRTESTUFE, HÄRTEMASS
3721	**DEGREE OF PRECISION**	DEGRE DE PRECISION, DEGRE D'EXACTITUDE	GENAUIGKEITSGRAD
3722	**DEGREE OF SATURATION**	DEGRE DE SATURATION	SÄTTIGUNGSGRAD
3723	**DEGREE OF SENSITIVENESS**	COEFFICIENT DE SENSIBILITE	EMPFINDLICHKEITSGRAD, UNEMPFLINDLICHKEITSGRAD
3724	**DEGREE REAUMUR**	DEGRE REAUMUR	GRAD REAUMUR, REAUMURGRAD
3725	**DEGREE TWADDELL**	DEGRE TWADDELL	GRAD TWADDELL
3726	**DEGREES OF FREEDOM**	DEGRES DE LIBERTE	FREIHEITSGRADE
3727	**DEHYDRATE (TO)**	DESHYDRATER, DESHYDRATER (SE)	HYDRATWASSER ENTZIEHEN (DAS), HYDRATWASSER VERLIEREN (DAS)
3728	**DEHYDRATION**	DESHYDRATATION	WASSERENTZIEHUNG (CHEM.)
3729	**DEIONIZATION**	DESIONISATION	ENTIONISIERUNG
3730	**DELAY**	RETARD	VERSPÄTUNG
3731	**DELAY IN DELIVERY**	RETARD DE LIVRAISON	LIEFERUNGSVERZUG
3732	**DELIQUESCE (TO)**	TOMBER EN DELIQUESCENCE	ZERFLIESSEN, DELIQUESZIEREN
3733	**DELIQUESCENCE**	DELIQUESCENCE	ZERFLIESSLICHKEIT, DELIQUESZENZ
3734	**DELIQUESCENT**	DELIQUESCENT	ZERFLIESSEND, ZERFLIESSLICH
3735	**DELIVERY OF HEAT**	CESSION DE CHALEUR	WÄRMEABGABE, WÄRMEABFUHR
3736	**DELIVERY PIPES**	TUYAUTERIE, TUYAU A PRESSION, TUYAU DE REFOULEMENT, TUBULURE DE DEVERSEMENT	DRUCKLEITUNG, DRUCKROHRLEITUNG
3737	**DELIVERY TERMS**	CONDITIONS DE LIVRAISON	LIEFERUNGSBEDINGUNGEN
3738	**DELIVERY TUBE, DELIVERY NOZZLE, DELIVERY CONE**	TUYERE DIVERGENTE, AJUTAGE DIVERGENT, TUYERE DE REFOULEMENT	DRUCKDÜSE

3739	**DELIVERY VALVE, DISCHARGE VALVE**	SOUPAPE DE REFOULEMENT, CLAPET DE REFOULEMENT	DRUCKVENTIL, DRUCKKLAPPE
3740	**DELTA IRON**	FER DELTA	DELTA-EISEN
3741	**DELTA METAL**	METAL ALLIAGE DELTA	DELTAMETALL
3742	**DEMAGNETISATION**	DESAIMANTATION	ENTMAGNETISIEREN
3743	**DEMAGNETISE (TO)**	DESAIMANTER	ENTMAGNETISIEREN
3744	**DEMAGNETIZERS**	DEMAGNETISEURS	ENTMAGNETISIERUNGSAPPARATE
3745	**DEMAND OF FUEL**	COMBUSTIBLE NECESSAIRE	BRENNSTOFFBEDARF
3746	**DEMAND OF POWER**	ENERGIE, FORCE NECESSAIRE, PUISSANCE REQUISE	KRAFTBEDARF
3747	**DENATURING AGENT**	DENATURANT	VERGÄLLUNGSMITTEL, DENATURIERUNGSMITTEL
3748	**DENDRITE**	DENDRITE	DENDRIT, TANNENBAUMKRISTALL
3749	**DENDRITIC POWDER**	POUDRE DENTRIDIQUE	PULVER (DENDRITISCHES)
3750	**DENDRITIC STRUCTURE**	STRUCTURE DENDRITIQUE	STRUKTUR (DENDRITISCHE)
3751	**DENICKELIFICATION**	DENICKELAGE	ENTNICKELUNG
3752	**DENOMINATOR**	DENOMINATEUR	NENNER
3753	**DENSIFICATION**	DENSIFICATION	VERDICHTUNG
3754	**DENSIFIER**	MOYEN D'HOMOGENEISATION	HOMOGENISIERUNGSMITTEL
3755	**DENSIMETER**	DENSIMETRE	DENSIMETER
3756	**DENSITY**	DENSITE	DICHTE, GEWICHT (SPEZIFISCHES)
3757	**DENSITY OF A GAS**	DENSITE D'UN GAZ	GASDICHTE
3758	**DENSITY OF A LIQUID**	DENSITE D'UN LIQUIDE	FLÜSSIGKEITSDICHTE
3759	**DENSITY RATIO**	RAPPORT DE DENSITE	DICHTEVERHÄLNIS
3760	**DENT**	BOSSE RENTRANTE, DEFONCEMENT	EINBEULUNG
3761	**DENT AND INDENT**	ADENT	ZAHN UND EINZAHNUNG
3762	**DEOXIDATION**	DESOXYDATION, DESOXYGENATION	SAUERSTOFFENTZIEHUNG, DESOXYDATION
3763	**DEOXIDISE (TO)**	DESOXYDER, DESOXYGENER	SAUERSTOFF ENTZIEHEN (DEN), DESOXYDIEREN
3764	**DEOXIDIZED COPPER**	CUIVRE DESOXYDE	KUPFER (DESOXYDIERTES)
3765	**DEOXIDIZER**	DESOXYDANT	DESOXYDATIONSMITTEL
3766	**DEOXIDIZING**	DESOXYDATION	DESOXYDATION
3767	**DEPLETION**	ABAISSEMENT DE CONCENTRATION	KONZENTRATIONSHERABSETZUNG
3768	**DEPOLARISATION**	DEPOLARISATION	DEPOLARISATION
3769	**DEPOLARISE (TO)**	DEPOLARISER	DEPOLARISIEREN
3770	**DEPOLARISER**	DEPOLARISANT	DEPOLARISATOR
3771	**DEPOLARIZATION**	DEPOLARISATION	DEPOLARISATION
3772	**DEPOLARIZER**	DEPOLARISANT, DEPOLARISATEUR	DEPOLARISATOR
3773	**DEPOSIT**	SEDIMENT	ABLAGERUNG (GEOL.)
3774	**DEPOSIT ATTACK**	CORROSION DUE A DES DEPOTS	BELAGKORROSION
3775	**DEPOSIT METAL**	METAL D'APPORT, METAL DE RECHARGE	AUFTRAGSMETALL
3776	**DEPOSITED METAL**	METAL FONDU APPLIQUE	SCHWEISSGUT (EINGEBRACHTES)
3777	**DEPOSITION EFFICIENCY**	RAPPORT DE DISTRIBUTION DU METAL	NIEDERSCHLAGSVERTEILUNGSVERHÄLTNIS
3778	**DEPRESSION**	DEPRESSION	UNTERDRUCK
3779	**DEPTH**	PROFONDEUR	TIEFE
3780	**DEPTH DRILLING**	FORAGE EN PROFONDEUR	TIEFBOHRUNG
3781	**DEPTH GAUGE**	CALIBRE DE PROFONDEUR, PIED A PROFONDEUR	TIEFENMASS, TIEFENLEHRE, AUSDREHWINKEL, LOCHWINKEL, SCHUBWINKEL
3782	**DEPTH OF CUT**	PROFONDEUR DE COUPE	SCHNITTIEFE

3783	**DEPTH OF GAP**	PROFONDEUR DU COL DE CYGNE	AUSLADUNGSTIEFE
3784	**DEPTH OF THREAD**	PROFONDEUR DU FILET	GANGTIEFE EINER SCHRAUBE, GEWINDETIEFE
3785	**DEPTH OF TOOTH**	HAUTEUR TOTALE DE LA DENT	ZAHNLÄNGE
3786	**DERIVATIVE**	DERIVE	DERIVAT
3787	**DERRICK BARGES**	BARGES DE MANUTENTION	BOHRTURM PONTON
3788	**DERRICK CRANE**	GRUE DERRICK, DERRICK	DERRICKKRAN
3789	**DESACTIVATION**	DESACTIVATION	ENTAKTIVIERUNG
3790	**DESCALING DIP**	BAIN DE DECALAMINAGE	ENTZUNDERUNGSBAD
3791	**DESCENDING PIPE LINE**	CONDUITE DESCENDANTE	LEITUNG (FALLENDE)
3792	**DESCENT, DECLIVITY, FALLING GRADIENT**	PENTE, DECLIVITE	GEFÄLLE (NEIGUNG)
3793	**DESEAMING**	ELIMINATION DES DEFAUTS SUPERFICIELS	ENTFERNUNG DER OBERFLÄCHENFEHLER
3794	**DESIGN (OF A PROJECT)**	TRACE, CONCEPTION, PLAN, PROJET, CALCUL, ETUDE	ENTWURF, KONZEPTION, PLANUNG, GESTALTUNG, BAUART, AUFBAU, ENTWICKELN
3795	**DESIGN (TO)**	ETUDIER, TRACER, CALCULER	ENTWERFEN, KONSTRUIEREN, AUSRECHNEN
3796	**DESIGN OFFICE/ENGINEERING OFFICE**	BUREAU D'ETUDE	KONSTRUKTIONSBÜRO
3797	**DESIGN SKETCH**	CROQUIS DE PRINCIPE	ENTWURFSZEICHNUNG
3798	**DESIGN TEMPERATURE**	TEMPERATURE DE CALCUL, TEMPERATURE D'ETUDE	TEMPERATUR (GERECHNETE)
3799	**DESIGNING**	ETUDE	ENTWERFEN, KONSTRUIEREN, KONSTRUKTION
3800	**DESIGNING DEPARTMENT, OFFICE**	BUREAU D'ETUDES	KONSTRUKTIONSBÜRO
3801	**DESILVERISE (TO)**	DESARGENTER	ENTSILBERN
3802	**DESILVERISING**	DESARGENTAGE	ENTSILBERUNG
3803	**DESLAGGING**	EVACUATION DU LAITIER	SCHLACKENABZUG
3804	**DESSICATOR**	DESSICCATEUR	EXSIKKATOR
3805	**DESTRUCTIVE DISTILLATION**	DISTILLATION AVEC DECOMPOSITION	ZERSETZUNGSDESTILLATION, SPALTUNGSDESTILLATION, DESTRUKTIVE DESTILLATION
3806	**DESTRUCTIVE TESTING**	ESSAI DESTRUCTIF	PRÜFUNG (ZERSTÖRENDE)
3807	**DESURFACING**	ENLEVEMENT DE COUCHES SUPERFICIELLES	OBERFLÄCHENGLÄTTUNG
3808	**DETACHABLE, REMOVABLE**	AMOVIBLE	ABNEHMBAR
3809	**DETAIL DRAWING**	DESSIN DE DETAIL	EINZELZEICHNUNG, STÜCKZEICHNUNG, DETAILZEICHNUNG
3810	**DETAIL OF DESIGN**	DETAIL DE CONSTRUCTION	KONSTRUKTIONSEINZELHEIT
3811	**DETAIL PART, COMPONENT**	PIECE DETACHEE	EINZELTEIL
3812	**DETAILED DRAWING**	PLAN DETAILLE	AÜSFÜHRUNGSZEICHNUNG
3813	**DETERGENT**	DETERGENT	REINIGUNGSMITTEL
3814	**DETERMINANT**	DETERMINANT	DETERMINANTE
3815	**DEVELOP (TO)**	DEVELOPPER (GEOM.)	ABWICKELN (GEOM.)
3816	**DEVELOPABLE**	DEVELOPPABLE	ABWICKELBAR (GEOM.)
3817	**DEVELOPABLE SURFACE, TORSE**	SURFACE DEVALOPPABLE, DEVELOPPABLE	FLÄCHE (ABWICKELBARE)
3818	**DEVELOPMENT OF GAS**	DEGAGEMENT DE GAZ	GASENTWICKLUNG
3819	**DEVIATION**	ECART, DEVIATION	ABWEICHUNG, REGELABWEICHUNG, ABLENKUNG (PHYS.)
3820	**DEVICE, CONTRIVANCE, ARRANGEMENT, APPLIANCE**	DISPOSITIF	VORRICHTUNG
3821	**DEVIL**	POT A FEU, BRASERO	FEUERTOPF

3822	**DEVIL'S CLAW**	CROCHET DE CHAINE	KROPFEISEN,
3823	**DEVITRIFICATION**	DEVITRIFICATION	ENTGLASUNG
3824	**DEW POINT**	POINT DE CONDENSATION	TAUPUNKT
3825	**DEXTRINE, BRITISH GUM**	DEXTRINE	STÄRKEGUMMI, DEXTRIN
3826	**DEXTRO-TARTARIC ACID**	ACIDE TARTRIQUE DEXTROGYRE	RECHTSWEINSÄURE
3827	**DEXTROROTATORY, DEXTROGYRATE**	DEXTROGYRE	RECHTSDREHEND
3828	**DIABASE**	DIABASE	DIABAS, GRÜNSTEIN
3829	**DIACAUSTIC CURVE**	CAUSTIQUE PAR REFRACTION, DIACAUSTIQUE	BRENNLINIE DURCH REFRAKTION, DIAKAUSTISCHE LINIE
3830	**DIAGNOSTIC ROUTINE**	PROGRAMME DIAGNOSTIQUE	DIAGNOSEPROGRAMM, FEHLERSUCHPROGRAMM
3831	**DIAGONAL**	DIAGONAL, DIAGONALE	DIAGONAL, DIAGONALE, ECKENLINIE
3832	**DIAGONAL CUTTING PLIERS**	PINCE COUPANTE DIAGONALE	SEITENSCHNEIDER
3833	**DIAGONAL SCALE**	ECHELLE DE PROPORTION	TRANSVERSALMASSSTAB
3834	**DIAGRAM**	DIAGRAMME, GRAPHIQUE, ABAQUE	SCHAUBILD, DIAGRAMM
3835	**DIAGRAMM OF FORCES**	DIAGRAMME DES FORCES	KRÄFTEPLAN
3836	**DIAGRAMMATIC DRAWING**	CROQUIS SCHEMATIQUE, SCHEMA	ZEICHNUNG (SCHEMATISCHE)
3837	**DIAL**	DISQUE, INDICATEUR, CADRAN DIVISE	ZIFFERBLATT, RAHMEN
3838	**DIAL BORE GAUGE**	COMPARATEUR A CADRAN POUR ALESAGES	BOHRUNGSMESSGERÄT MIT SKALA
3839	**DIAL FEED MECHANICAL PRESS**	PRESSE MECANIQUE A PLATEAU REVOLVER	REVOLVERPRESSEN (MECHANISCHE)
3840	**DIAL INDICATEUR**	INDICATEUR A CADRAN	RINGSKALA-ANZEIGER
3841	**DIAL SCALE**	ECHELLE ANNULAIRE	RINGSKALA
3842	**DIAL TEST INDICATOR**	PALPEUR A CADRAN	MESSUHR
3843	**DIAL TRAIN**	MINUTERIE	ZEIGERWERK
3844	**DIALYSE (TO)**	DIALYSER	DIALYSIEREN
3845	**DIALYSER**	DIALYSEUR	DIALYSATOR
3846	**DIALYSIS**	DIALYSE	DIALYSE
3847	**DIAMAGNETIC**	DIAMAGNETIQUE	DIAMAGNETISCH
3848	**DIAMAGNETISM**	DIAGMAGNETISME	DIAMAGNETISMUS
3849	**DIAMETER**	DIAMETRE	DURCHMESSER
3850	**DIAMETER AT BOTTOM OF THREAD, INTERNAL DIAMETER OF THREAD**	DIAMETRE INTERIEUR, PETIT DIAMETRE DU FILET, DIAMETRE DU NOYAU, DIAMETRE A FOND DE FILET	INNERER GEWINDEDURCHMESSER, KERNDURCHMESSER EINER SCHRAUBE
3851	**DIAMETER INCREMENT**	COMPLEMENT DIAMETRAL	DICKENWACHSTUM
3852	**DIAMETER OF A BOLT, OF A SCREW; EXTERNAL DIAMETER OF THREAD**	DIAMETRE EXTERIEUR, GRAND DIAMETRE DU FILET	GEWINDEDURCHMESSER (ÄUSSERER)
3853	**DIAMETER OF BOLT CIRCLE**	DIAMETRE (DU CERCLE) DE PERCAGE	LOCHKREISDURCHMESSER
3854	**DIAMETER OF BOLT HOLE**	DIAMETRE DE TROU DE BOULON	SCHRAUBENLOCHDURCHMESSER
3855	**DIAMETER OF BORE**	DIAMETRE D'ALESAGE, ALESAGE	BOHRUNG, BOHRUNGSDURCHMESSER
3856	**DIAMETER OF CHAIN LINK BAR**	DIAMETRE DE LA BARRE DE FER D'UN MAILLON	KETTENEISENSTÄRKE
3857	**DIAMETER OF HOLE**	DIAMETRE DU TROU	LOCHWEITE
3858	**DIAMETER OF PIPE**	DIAMETRE D'UN TUYAU	ROHRDURCHMESSER
3859	**DIAMETER OF RIVET**	DIAMETRE DU RIVET	NIETSTÄRKE
3860	**DIAMETER OF WIRE, SIZE OF WIRE**	DIAMETRE D'UN FIL, EPAISSEUR D'UN FIL	DRAHTSTÄRKE

3861	**DIAMETRAL PITCH**	PAS DIAMETRAL, DIAMETRE PRIMITIF, MODULE DE DENTURE	DURCHMESSERTEILUNG, VERHÄLTNISZÄHNEZAHL, DURCHMESSER DES TEILKREISES
3862	**DIAMOND**	DIAMANT	DIAMANT
3863	**DIAMOND DUST**	POUDRE DE DIAMANT, EGRISE	DIAMANTSTAUB
3864	**DIAMOND GRINDING WHEEL**	MEULE-DIAMANT	DIAMANTSCHLEIFSCHEIBE
3865	**DIAMOND HARDNESS**	ESSAI DE DURETE	DIAMANTHÄRTEPROBE
3866	**DIAMOND TOOL**	OUTIL DIAMANTE	DIAMANTWERKZEUG, SPITZSTAHL
3867	**DIAMOND WHEEL**	MEULE DIAMANTEE	DIAMANTSCHLEIFSCHEIBE
3868	**DIAPHANEITY**	DIAPHANEITE	LICHTDURCHLÄSSIGKEIT, DIAPHANITÄT
3869	**DIAPHANOUS**	DIAPHANE	LICHTDURCHLÄSSIG, DIAPHAN
3870	**DIAPHRAGM**	MEMBRANE, DIAPHRAGME, CLOISON	SCHWINGPLATTE, MEMBRAN, DIAPHRAGMA, SCHEIDEWAND
3871	**DIAPHRAGM**	DIAPHRAGME (OPT.)	BLENDE, DIAPHRAGMA
3872	**DIAPHRAGM OPERATED CONTROL VALVE**	ROBINET REGULATEUR A MEMBRANE, VANNE DE CONTROLE	MEMBRAN-REGELVENTIL
3873	**DIAPHRAGM PRESSURE GAUGE**	MANOMETRE A PLAQUE	PLATTENFEDERMANOMETER
3874	**DIAPHRAGM PUMP**	POMPE A DIAPHRAGME, A MEMBRANE	MEMBRANPUMPE
3875	**DIAPHRAGM SPRING**	RESSORT DIAPHRAGME	MEMBRANFEDER
3876	**DIAPHRAGM VALVE**	ROBINET A MEMBRANE	MEMBRAN-VENTIL
3877	**DIAPHRAHM (TO)**	DIAPHRAGMER	ABBLENDEN
3878	**DIATHERMANCY**	DIATHERMANEITE, DIATHERMANSIE	DURCHLÄSSIGKEIT FÜR WÄRMESTRAHLEN, DIATHERMANITÄT, DIATHERMANSIE
3879	**DIATHERMANOUS**	DIATHERMANE, DIATHERMIQUE	DURCHLÄSSIG FÜR WÄRMESTRAHLEN, DIATHERMAN
3880	**DIATOMACEOUS EARTH**	TERRE DE DIATOMEES	DIATOMEENERDE, KIESELGUR
3881	**DIATOMIC, DIVALENT**	DIATOMIQUE, DIVALENT, BIATOMIQUE	ZWEIWERTIG, ZWEIATOMIG
3882	**DIBASIC**	DIBASIQUE	ZWEIBASISCH
3883	**DIE**	MOULE, FILIERE, ESTAMPE, MATRICE, COQUILLE, MOULE METALLIQUE	PRÄGESTEMPEL, MATRIZE, KOKILLE, PRESSRING, FORM
3884	**DIE BLOCK**	PORTE-ESTAMPE	STEMPELBLOCK
3885	**DIE BODY**	PARTIE FIXE, CORPS DE L'ESTAMPE	FESTER STEMPELTEIL
3886	**DIE CASTING**	COULEE SOUS PRESSION	DRUCKGIESSVERFAHREN, DRUCKGUSS
3887	**DIE CUSHIONS**	SERRE-FLANS	NIEDERHALTER
3888	**DIE LINES**	MARQUE DE L'ESTAMPE	MATRIZENSPUR, TEILFUGE
3889	**DIE METAL**	METAL POUR MATRICES	GESENKMETALL
3890	**DIE SET**	MONTURE D'ESTAMPE A GUIDAGE A COLONNES	SÄULEFÜHRUNGSGESTELL
3891	**DIE SHIFT**	DEPLACEMENT DE L'ESTAMPE	STEMPELVERSCHIEBUNG
3892	**DIE SINKING**	FRAISAGE DES MATRICES	GESENKFRÄSEN
3893	**DIE STAMP**	POINCON (D'INSPECTION)	PRÜFSTEMPEL, KONTROLLSTEMPEL
3894	**DIE STAMPING**	MATRICAGE (DEFAUT DE SURFACE DE TOLE), MARQUAGE DES TOLES AU POINCON, POINCONNAGE	PRÄGUNG, GESENKSCHMIEDEN
3895	**DIE STOCK, SCREW STOCK**	MONTURE D'UNE FILIERE A COUSSINETS	KLUPPENHALTER
3896	**DIE, SCREW DIES, SCREWING DIES**	COUSSINETS DE FILIERE	SCHNEIDBACKEN, GEWINDESCHNEIDBACKEN

3897	DIE, SWAGE	MATRICE ETAMPE ESTAMPE DE FORGERON	GESENK
3898	DIELECTRIC	DIELECTRIQUE, ISOLANT, NON CONDUCTEUR	NICHTLEITER, DIELEKTRISCH, NICHTLEITEND, DIELEKTRIKUM
3899	DIELECTRIC COEFFICIENT CONSTANT, INDUCTIVE CAPACITY, INDUCTIVITY, PERMITTIVITY	CONSTANTE DIELECTRIQUE	DIELEKTRIZITÄTSKONSTANTE
3900	DIELECTRIC STRENGHT	RESISTANCE DIELECTRIQUE OU DISRUPTIVE	DURCHSCHLAGSFESTIGKEIT
3901	DIESEL ENGINE	MOTEUR DIESEL, DIESEL	DIESELMASCHINE, DIESELMOTOR
3902	DIFFERENCE	DIFFERENCE	DIFFERENZ (MATH.)
3903	DIFFERENCE GAUGE	JAUGE LIMITE DE TOLERANCE	TOLERANZLEHRE
3904	DIFFERENCE OF TEMPERATURE	DIFFERENCE DE TEMPERATURE	TEMPERATURUNTERSCHIED
3905	DIFFERENCE, LIMIT GAUGE	CALIBRE DE TOLERANCE	GRENZLEHRE, DIFFENRENZLEHRE, TOLERANZLEHRE, DOPPELKALIBER, ZWIESELKALIBER, TOLERANZKALIBER
3906	DIFFERENTIAL	DIFFERENTIEL	DIFFERENTIAL
3907	DIFFERENTIAL BRAKE	FREIN DIFFERENTIEL	DIFFERENTIALBREMSE
3908	DIFFERENTIAL CALCULUS	CALCUL DIFFERENTIEL	DIFFERENTIALRECHUNG
3909	DIFFERENTIAL COEFFICIENT	COEFFICIENT DIFFERENTIEL	DIFFERENTIALKOEFFIZIENT
3910	DIFFERENTIAL CROWNWHEEL	GRANDE COURONNE DE DIFFERENTIEL	DIFFERENTIALANTRIEBS-KEGELRAD
3911	DIFFERENTIAL DYNAMOMETER	DYNAMOMETRE DIFFERENTIEL	DIFFERENTIALDYNAMOMETER
3912	DIFFERENTIAL EQUATION	EQUATION DIFFERENTIELLE	DIFFERENTIALGLEICHUNG
3913	DIFFERENTIAL GEAR	MOUVEMENT, MECANISME, TRAIN DIFFERENTIEL, DIFFERENTIEL, ENGRENAGE DIFFERENTIEL	DIFFERENFIALGETRIEBE
3914	DIFFERENTIAL HARDENING	TREMPE DIFFERENTIELLE, TREMPE LOCALISEE	ABSCHRECKEN (ÖRTLICH BEGRENZTES)
3915	DIFFERENTIAL HEATING	CHAUFFAGE SELECTIF	SELEKTIVE ERHITZUNG
3916	DIFFERENTIAL QUOTIENT	QUOTIENT DIFFERENTIEL	DIFFERENTIALQUOTIENT
3917	DIFFERENTIAL SIDE GEAR WHEEL	ROUE DE DIFFERENTIEL	HINTERACHSWELLENRAD
3918	DIFFERENTIAL SPIDER PINION	SATELLITE DE DIFFERENTIEL	AUSGLEICHSKEGELRAD
3919	DIFFERENTIAL THERMAL ANALYSIS	ANALYSE THERMIQUE SELECTIVE	ANALYSE (SELEKTIVE THERMISCHE)
3920	DIFFERENTIAL THERMOMETER	THERMOMETRE DIFFERENTIEL	DIFFERENTIALTHERMOMETER
3921	DIFFERENTIAL U-TUBE GAUGE	MANOMETRE DIFFERENTIEL	DIFFERENTIALMANOMETER
3922	DIFFERENTIATE (TO)	DIFFERENCIER	DIFFERENZIEREN (MATH.)
3923	DIFFERENTIATION	DIFFERENCIATION	DIFFERENTIATION (MATH.)
3924	DIFFRACTION	DIFFRACTION	DIFFRAKTION, BRECHUNG, BEUGUNG
3925	DIFFRACTION GRATING	RESEAU DE DIFFRACTION	BEUGUNGSGITTER
3926	DIFFRACTION PATTERN	DIAGRAMME DE DIFFRACTION RX	RÖNTGEN-REFLEXE
3927	DIFFUSED LIGHT	LUMIERE DIFFUSE	LICHT (ZERSTREUTES), LICHT (DIFFUSES)
3928	DIFFUSER	DIFFUSEUR	AUSWURFKEGEL, DIFFUSER
3929	DIFFUSION	DIFFUSION	DIFFUSION
3930	DIFFUSION COATINGS	REVETEMENTS DE DIFFUSION	DIFFUSIONSÜBERZÜGE
3931	DIFFUSION COEFFICIENT	COEFFICIENT DE DIFFUSION	DIFFUSIONSKOEFFIZIENT
3932	DIFFUSION OF GASES	DIFFUSION DES GAZ	DIFFUSION VON GASEN
3933	DIFFUSION OF LIGHT	DIFFUSION DE LA LUMIERE	STREUUNG, DIFFUSION DES LICHTES
3934	DIFFUSION ZONE	ZONE DE DIFFUSION	DIFFUSIONGEBIET
3935	DIFFUSIVE SURFACE	SURFACE DIFFUSIVE	FLÄCHE (LICHTZERSTREUENDE)
3936	DIFFUSIVITY	DIFFUSIBILITE	DIFFUSIONSVERMÖGEN
3937	DIGEST (TO)	FAIRE DIGERER, METTRE A DIGERER	DIGERIEREN

3938	DIGESTER	LESSIVEUR	DAMPFKOCHTOPF
3939	DIGESTION	DIGESTION (CHIM.)	DIGERIEREN, DIGESTION
3940	DIGIT	CHIFFRE, POSITION	ZEICHEN, ZAHL, STELLE
3941	DIGITAL	NUMERIQUE	DIGITAL, NUMERISCH
3942	DIGITAL INPUT DATA	INFORMATION D'ENTREE NUMERIQUE	DATEN (NUMERISCH EINGEGEBENE)
3943	DIGITIZER	CODEUR NUMERIQUE	CODIERER
3944	DIHEDRAL ANGLE	DIEDRE, ANGLE DIEDRE	EBENENWINKEL
3945	DIKE	LEVEE DE TERRE	DEICH
3946	DIKES	CAVALIERS	ERDAUFSCHÜTTUNG
3947	DILATABILITY, EXPANSIBILITY	DILATABILITE	AUSDEHNUNGSFÄHIGKEIT
3948	DILATABLE, EXPANSIBLE	DILATABLE	AUSDEHNUNGSFÄHIG
3949	DILATATION	DILATATION	DILATATION, DEHNUNG
3950	DILATOMETER	DILATOMETRE	DILATOMETER
3951	DILUENT	DILUANT	VERDÜNNUNGSMITTEL
3952	DILUTE A SOLUTION (TO)	DILUER, ETENDRE UNE SOLUTION	VERDÜNNEN (EINE LÖSUNG)
3953	DILUTE SOLUTION	SOLUTION DILUEE, ETENDUE	VERDÜNNTE LÖSUNG
3954	DILUTION OF A SOLUTION	DILUTION D'UNE SOLUTION	VERDÜNNEN, VERDÜNNUNG EINER LÖSUNG
3955	DIM LIGHT	LUMIERE DIFFUSE	STREULICHT
3956	DIMENSION FIGURED	COTE	MASSZAHL, MASSBEZEICHNUNG, EINGESCHRIEBENES MASS
3957	DIMENSION LINE	LIGNE DE COTE	MASSLINIE
3958	DIMENSIONAL STABILITY	STABILITE DIMENSIONNELLE	FORMBESTÄNDIGKEIT
3959	DIMENSIONED DRAWING, DRAWING WITH DIMENSIONS	DESSIN COTE	MASSZEICHNUNG, ZEICHNUNG MIT EINGESCHRIEBENEN MASSEN
3960	DIMENSIONS, MEASUREMENTS	DIMENSIONS	ABMESSUNGEN, MASSE, AUSMASSE, DIMENSIONEN
3961	DIMETRIC PROJECTION	PROJECTION DIMETRIQUE	PROJEKTION (DIMETRISCHE)
3962	DIMINISHING REDUCING SOCKET	MANCHON DE REDUCTION D'ALESAGE	ÜBERGANGSMUFFE, ABSATZMUFFE, REDUKTIONSMUFFE
3963	DIMMER SWITCH	INVERSEUR-PHARE-CODE	DUNKELSCHALTER
3964	DIODE	DIODE	DIODE
3965	DIODE LAMP	LAMPE DIODE	ZWEIELEKTRODENLAMPE
3966	DIORITE	DIORITE	DIORIT
3967	DIP	IMMERSION, BAIN	BAD
3968	DIP (TO), IMMERGE (TO), IMMERSE (TO)	PLONGER, IMMERGER	EINTAUCHEN
3969	DIP BRAZING	SOUDURE FORTE PAR IMMERSION	TAUCHHARTLÖTUNG
3970	DIP COATING	APPLICATION D'UN REVETEMENT PAR TREMPAGE	TAUCHAUFTRAG
3971	DIP HATCH	TROU DE JAUGE	MESSSTABLOCH
3972	DIP HATCH AND VENT COMBINED	TROU DE JAUGE COMBINE AVEC EVENT	MESSSTAB-UND LUFTLOCH
3973	DIP TANK	CUVE DE TREMPAGE	TAUCHBEHÄLTER, TANK
3974	DIPHENYLAMINE	DIPHENYLAMINE	DIPHENYLAMIN
3975	DIPPING	IMMERSION, TREMPE	IMMERSION, TAUCHEN
3976	DIPPING BASKET	CORBEILLE DE TREMPAGE	TAUCHKORB
3977	DIPPING PROCESS	METHODE AU TREMPE	EINTAUCHVERFAHREN
3978	DIPPING, IMMERSION	IMMERSION, TREMPE	EINTAUCHEN
3979	DIPSTICK	JAUGE D'HUILE	ÖLMESSSTAB
3980	DIRECT ACTING	A ACTION DIRECTE	UNMITTELBAR, DIREKT WIRKEND

3981	**DIRECT COMPUTER CONTROL**	COMMANDE DIRECTE PAR CALCULATEUR	RECHNERSTEUERUNG (DIREKTE)
3982	**DIRECT COUPLED MACHINE**	MACHINE A ACCOUPLEMENT DIRECT, MACHINE A MANCHONNAGE DIRECT	MASCHINE (DIREKTGEKUPPELTE)
3983	**DIRECT DRIVE**	PRISE DIRECTE	GANG (DIREKTER)
3984	**DIRECT EXTRUSION**	FILAGE DIRECT	VORWÄRTSFLIESSPRESSEN
3985	**DIRECT NUMERICAL CONTROL**	COMMANDE DIRECTE PAR CALCULATEUR	RECHNERSTEUERUNG, (DIREKTE)
3986	**DIRECT READING DIAL CALIPER**	PIED A COULISSE A CADRAN	SCHUBLEHRE MIT DIREKTER ABLESUNG
3987	**DIRECT SPRING LOADED (SAFETY) VALVE**	SOUPAPE DE SURETE A CHARGE DIRECTE A RESSORT	VENTIL MIT UNMITTELBARER FEDERBELASTUNG
3988	**DIRECT-CONTACT CONDENSER**	CONDENSEUR PAR MELANGE	MISCHKONDENSATOR
3989	**DIRECTION INDICATOR LIGHTS**	CLIGNOTANTS, FEUX DE DIRECTION	FAHRTRICHTUNGSANZEIGER, BLINKLICHTER
3990	**DIRECTION OF A FORCE**	DIRECTION D'UNE FORCE	KRAFTRICHTUNG, RICHTUNG EINER KRAFT
3991	**DIRECTION OF MOTION**	DIRECTION DE MOUVEMENT	BEWEGUNGSRICHTUNG
3992	**DIRECTION OF ROTATION**	SENS DE ROTATION	DREHRICHTUNG, DREHSINN
3993	**DIRECTION OF THE FIBRES**	DIRECTION DES FIBRES	FASERRICHTUNG
3994	**DIRECTIONAL PROPERTIES**	ORIENTATION DES CRISTAUX	KRISTALLORIENTIERUNG
3995	**DIRECTOR**	DIRECTEUR, INTERPOLATEUR	STEUERDIREKTOR, INTERPOLATOR
3996	**DIRECTRIX**	DIRECTRICE	LEITLINIE, DIREKTRIX
3997	**DIRT COLLECTION**	PRISE DE POUSSIERES	SCHMUTZANHAFTUNG
3998	**DIRT SAND INCLUSION**	GRAINS CONTENUS DANS LA FONTE	SANDEINSCHLUSS IM GUSS
3999	**DISAGGREGATE (TO)**	DESINTEGRER, DESAGREGER	AUFLOCKERN, AUFSCHLIESSEN
4000	**DISAGGREGATION**	DESINTEGRATION, DESAGREGATION	AUFLOCKERUNG, AUFSCHLIESSUNG
4001	**DISC BRAKE**	FREIN A DISQUE	SCHEIBENBREMSE
4002	**DISC CLUTCH**	EMBRAYAGE A DISQUE	LAMELLENKUPPLUNG
4003	**DISC CRANK, CRANK DISC, CRANK PLATE**	PLATEAU-MANIVELLE	KURBELSCHEIBE
4004	**DISC FLYWHEEL**	VOLANT EN DISQUE, DISQUE-VOLANT	SCHWUNGSCHEIBE
4005	**DISC FRICTION GEAR**	TRANSMISSION PAR PLATEAUX	PLANSCHEIBENGETRIEBE
4006	**DISC GRINDER**	PONCEUSE	SCHLEIFSCHEIBE
4007	**DISC HARROWS**	PULVERISEURS A DISQUES	SCHEIBENEGGEN
4008	**DISC HOLDER**	PORTE-CLAPET	KEGELHALTERUNG
4009	**DISC LOCKNUT**	ECROU DE CLAPET	KEGELMUTTER
4010	**DISC SANDER**	PONCEUSE	SCHLEIFSCHEIBE
4011	**DISC VALVE**	SOUPAPE A SIEGE PLAN, SOUPAPE A DISQUE	EBENES VENTIL, TELLERVENTIL, SCHEIBENVENTIL, PLATTENVENTIL
4012	**DISC WHEEL**	ROUE A CENTRE PLEIN, ROUE A VOILE, ROUE A DISQUE	SCHEIBENRAD
4013	**DISC WHEEL CENTRE**	TOILE D'UNE ROUE, D'UN VOLANT	RADSCHEIBE
4014	**DISC, COMPOSITION DISC**	DISQUE (DE CLAPET)	DICHTSCHEIBE
4015	**DISC, DISK**	DISQUE	SCHEIBE
4016	**DISC, PLUG**	CLAPET (DE ROBINET)	KEGEL
4017	**DISCHARGE**	DECHARGE	ENTLADUNG
4018	**DISCHARGE (TO), EMPTY (TO)**	VIDER, DECHARGER	ENTLEEREN
4019	**DISCHARGE AN ACCUMULATOR (TO)**	DECHARGER UN ACCUMULATEUR	AKKUMULATOR ENTLADEN (EINEN)
4020	**DISCHARGE CROSS SECTION, CROSS SECTION APERTURE OF DISCHARGE ORIFICE**	SECTION DE L'ORIFICE D'ECOULEMENT	AUSFLUSSQUERSCHNITT

4021	DISCHARGE HEAD, DELIVERY HEAD OF A PUMP	HAUTEUR DE REFOULEMENT D'UNE POMPE	DRUCKHÖHE EINER PUMPE
4022	DISCHARGE PIPE-LINE	CONDUITE DE REFOULEMENT	ABLASSROHRLEITUNG
4023	DISCHARGE, DELIVERY, QUANTITY DISCHARGED	VOLUME QUI PASSE AU TRAVERS D'UN ORIFICE PAR SECONDE	AUSFLUSSMENGE, ABFLUSSMENGE
4024	DISCHARGE, EMPTYING	VIDANGE, DECHARGE	ENTLEERUNG
4025	DISCHARGING AN ACCUMULATOR	DECHARGE D'UN ACCUMULATEUR	ENTLADUNG EINES AKKUMULATORS
4026	DISCOLORATION	ALTERATION DE LA TEINTE	VERFÄRBUNG
4027	DISCONNECTABLE	DEMONTABLE	AUSEINANDERNEHMBAR, ZERLEGBAR
4028	DISCONNECTION	DEBRAYAGE	AUSRÜCKUNG
4029	DISCRIMINANT	DISCRIMINANT	DISKRIMINANTE
4030	DISENGAGE THE CATCH (TO)	DECLIQUETER, DECLENCHER	AUSKLINKEN
4031	DISENGAGEMENT RELEASE OF A CATCH	DECLENCHEMENT, DECLIQUETAGE	AUSKLINKEN
4032	DISENGAGING COUPLING	ACCOUPLEMENT DEBRAYAGE	AUSLÖSUNGSKUPPLUNG
4033	DISENGAGING COUPLING CLUTCH	ACCOUPLEMENT A DEBRAYAGE	EIN-UND AUSRÜCKKUPPLUNG, LÖSBARE KUPPLUNG, VERSCHIEBBARE KUPPLUNG, AUSRÜCKBARE KUPPLUNG
4034	DISENGAGING GEAR	MECANISME DE DEBRAYAGE, MECANISME D'EMBRAYAGE	AUSRÜCKVORRICHTUNG, EINRÜCKVORRICHTUNG
4035	DISENGAGING LEVER	LEVIER DE DEBRAYAGE, LEVIER DE DESEMBRAYAGE	AUSRÜCKHEBEL, ABSTELLHEBEL
4036	DISHED BOTTOM	FOND BOMBE	BODEN (GEWÖLBTER)
4037	DISHED DISC WHEEL	ROUE A VOILE BOMBE	SCHEIBENRAD (GEWÖLBTES)
4038	DISHED HEAD	FOND EMBOUTI	KESSELBODEN (GEWÖLBTER), KLOPPERBODEN
4039	DISHED ROOF	TOIT BOMBE	BOGENDACH
4040	DISHING	EMBOUTISSAGE	KÜMPELARBEIT
4041	DISINCRUSTANT, SCALE SOLVENT	DESINCRUSTANT CURATIF, TARTRIFUGE, ANTI-TARTRE	KESSELSTEINLÖSUNGSMITTEL
4042	DISINFECTANT	DESINFECTANT	DESINFEKTIONSMITTEL
4043	DISINTEGRATE (TO), CRUMBLE (TO)	EFFRITER (S'), DESAGREGER (SE)	ZERSTÄUBEN, ZERFALLEN
4044	DISINTEGRATION	DESINTEGRATION	ZERBRÖCKLUNG, ZERFALLEN
4045	DISINTEGRATOR	CONCASSEUR	SCHLEUDERMÜHLE, DESINTEGRATOR
4046	DISK CLUTCH	EMBRAYAGE A DISQUES	LAMELLENKUPPLUNG
4047	DISLOCATION	DISLOCATION	DISLOKATION, VERSETZUNG
4048	DISMOUNTABLE GAUGE	JAUGE DEMONTABLE	LEHRE (ZUSAMMENSTELLBARE)
4049	DISPERSED PHASE	PHASE DISPERSEE	DISPERSIONSPHASE
4050	DISPERSING AGENT	DISPERSANT, AGENT DE DISPERSION	DISPERGIERUNGSMITTEL, DISPERSIONSMITTEL
4051	DISPERSION	DISPERSION	ZERSTREUUNG, DISPERSION
4052	DISPLACEABLE	DEPLACABLE	VERSCHIEBBAR
4053	DISPLACEMENT (CEN);	DEPLACEMENT (CEN), VOLUME DEPLACE (PHYS.)	VERSCHIEBUNG, FORTRÜCKUNG (ALLG.), VOLUMEN (VERDRÄNGTES) (PHYS.)
4054	DISPLACEMENT OF A LIQUID	DEPLACEMENT D'UN LIQUIDE	VERDRÄNGUNG EINER FLÜSSIGKEIT
4055	DISPLACEMENT OF THE SHAFT CENTER LINES OF THE AXES OF THE SHAFTS	DESAXAGE	ACHSENVERSCHIEBUNG
4056	DISPOSITION, ARRANGEMENT	DISPOSITION	ANORDNUNG
4057	DISRUPTIVE STRENGTH	RESISTANCE DISRUPTIVE	ZERREISSFESTIGKEIT, DURCHSCHLAGSFESTIGKEIT
4058	DISSOCIATION	DISSOCIATION	SPALTUNG, DISSOZIATION
4059	DISSOLUTION	DISSOLUTION (ACTION) (CHIM.)	LÖSEN, AUFLÖSEN (CHEM.)
4060	DISSOLUTION	DISSOLUTION	AUFLÖSUNG

4061	DISSOLVE (TO)	DISSOUDRE, FONDRE (CHIM.), DISSOUDRE (SE)	AUFLÖSEN (SICH) (CHEM.), LÖSEN (SICH)
4062	DISSOLVING POWER	POUVOIR DISSOLVANT	LÖSUNGSVERMÖGEN, AUFLÖSUNGSVERMÖGEN
4063	DISTANCE	CHEMIN, DISTANCE, PARCOURS, TRAJET; ECARTEMENT, ESPACEMENT	WEG (MECH.); ENTFERNUNG, ABSTAND, DISTANZ
4064	DISTANCE BETWEEN BEARINGS	ECARTEMENT DES PALIERS	LAGERENTFERNUNG
4065	DISTANCE BETWEEN CENTERS	DISTANCE ENTRE POINTES, ENTRAXE, DISTANCE D'AXE EN AXE	SPITZENWEITE, ACHSABSTAND
4066	DISTANCE FROM EDGE OF PLATE TO CENTRE OF RIVET	DISTANCE DU CENTRE DU RIVET AU BORD DE LA TOLE	RANDENTFERNUNG, RANDABSTAND DER NIETE, ABSTAND DER NIETE VOM BLECHRAND
4067	DISTANCE PIECE	ENTRETOISE	ABSTANDHÜLSE, STÜCK
4068	DISTANCE PIECE	PIECE D'ECARTEMENT, ENTRETOISE	ABSTANDHALTER, DISTANZSTÜCK, ABSTANDHÜLSE
4069	DISTANT READING THERMOMETER	TELE-THERMOMETRE	FERNTHERMOMETER
4070	DISTIL (TO)	DISTILLER	DESTILLIEREN
4071	DISTILLATE	PRODUIT DE DISTILLATION, DISTILLATION	DESTILLAT, DESTILLATIONSERZEUGNIS
4072	DISTILLATION	DISTILLATION (ACTION)	DESTILLATION
4073	DISTILLATION IN STEAM	DISTILLATION A LA VAPEUR D'EAU	DESTILLATION MIT WASSERDAMPF
4074	DISTILLATION OF COAL	DISTILLATION DE LA HOUILLE	ENTGASUNG DER KOHLE
4075	DISTILLATION UNDER REDUCED PRESSURE	DISTILLATION DANS LE VIDE, VASE CLOS	DESTILLATION IM VAKUUM, VAKUUMDESTILLATION
4076	DISTILLED WATER	EAU DISTILLEE	WASSER (DESTILLIERTES)
4077	DISTORTION	ABERRATION, DISTORSION	ABWEICHUNG, VERFORMUNG, VERZERRUNG
4078	DISTRIBUTED LOAD	CHARGE REPARTIE	LAST (VERTEILTE)
4079	DISTRIBUTING MAIN	CONDUCTEUR DE RESEAU, CABLE DE DISTRIBUTION	VERTEILUNGSLEITUNG
4080	DISTRIBUTING NETWORK	RESEAU DE CONDUCTEURS ELECTRIQUES, RESEAU DE DISTRIBUTION ELECTRIQUE	LEITUNGSNETZ, STROMLEITUNGSNETZ, STROMVERTEILUNGSNETZ
4081	DISTRIBUTION OF LOAD	REPARTITION D'UNE CHARGE	LASTVERTEILUNG
4082	DISTRIBUTION OF MASSES	REPARTITION DES MASSES	MASSENVERTEILUNG
4083	DISTRIBUTION OF TEMPERATURE	DISTRIBUTION DE LA TEMPERATURE	TEMPERATURVERTEILUNG
4084	DISTRIBUTOR	DISTRIBUTEUR	VERTEILER
4085	DISTRIBUTOR ARM	DOIGT D'ALLUMEUR	VERTEILERLÄUFER
4086	DISTRIBUTOR HEAD	BOITIER DE DISTRIBUTION, DISTRIBUTEUR	VERTEILERKOPF, VERTEILERKASTEN
4087	DISTRIBUTORS : FERTILISER	DISTRIBUTEURS D'ENGRAIS	DÜNGERSTREUER
4088	DISTRIBUTORS : TRAILER AND LORRY ATTACHMENTS	DISTRIBUTEURS VEHICULES POUR ATTACHEMENT	ANHÄNGEVORRICHTUNGEN FÜR WAGEN
4089	DISTRIBUTORS AND SPREADERS, LIME	DISTRIBUTEURS DE CHAUX	KALKSTREUER
4090	DITCHING AND DRAINING PLANT AND EQUIPMENT	SOUS-SOLEUSES ET CHARRUES RIGOLEUSES	ENTWÄSSERUNGSGERÄTE UND GRABENPFLÜGE
4091	DIVERGE (TO)	DIVERGER	AUSEINANDERLAUFEN, DIVERGIEREN
4092	DIVERGENCE, DIVERGENCY	DIVERGENCE	DIVERGENZ
4093	DIVERGENT	DIVERGENT	AUSEINANDERLAUFEND, DIVERGENT
4094	DIVERGENT SERIES	SERIE DIVERGENTE	REIHE (DIVERGENTE)
4095	DIVERGING LENS, CONCAVE LENS	LENTILLE DIVERGENTE, LENTILLE A BORD EPAIS	HOHLLINSE, ZERSTREUUNGSLINSE, LINSE (KOKAVE), VERKLEINERUNGSGLAS

4096	DIVERGING MOUTHPIECE	AJUTAGE DIVERGENT, EMBOUCHURE DIVERGENTE	AUSFLUSSSTUTZEN (SICH ERWEITERNDER)
4097	DIVIDE (TO)	DIVISER (MATH.)	TEILEN, DIVIDIEREN
4098	DIVIDED JOURNAL BEARING	PALIER EN DEUX PARTIES, PALIER EN COQUILLES	LAGER (GETEILTES)
4099	DIVIDEND	DIVIDENDE	DIVIDEND
4100	DIVIDER	DIVISEUR, COMPAS	STAHLLINEAL MIT TEILUNG, TEILGERÄT
4101	DIVIDERS	COMPAS DROIT A POINTES, COMPAS A POINTES SECHES	SCHARNIERZIRKEL (GERADER), SPITZZIRKEL
4102	DIVIDING ENGINE	MACHINE A DIVISER	TEILMASCHINE
4103	DIVIDING ENGINE FOR CIRCLES	MACHINE A DIVISER LES CERCLES	KREISTEILMASCHINE
4104	DIVIDING ENGINE FOR STRAIGHT LINES	MACHINE A DIVISER LES LIGNES DROITES	LÄNGENTEILMASCHINE
4105	DIVIDING HEAD	APPAREIL DIVISEUR	TEILSCHEIBE
4106	DIVIDING PLATE	PLATEAU DIVISEUR	TEILSCHEIBE
4107	DIVISION	DIVISION (MATH.)	TEILUNG, DIVISION
4108	DIVISION OF A CIRCLE	DIVISION DU CERCLE	KREISTEILUNG
4109	DIVISION PLATE	PLATEAU DIVISEUR, PLATEAU DIVISE, A DIVISIONS, PLATE-FORME A DIVISER	TEILSCHEIBE
4110	DIVISION WALL, PARTITION WALL	CLOISON, DIAPHRAGME	ZWISCHENWAND, TRENNUNGSWAND, SCHEIDEWAND
4111	DIVISOR	DIVISEUR	DIVISOR
4112	DODECAHEDRON	DODECAEDRE	ZWÖLFFLACH, ZWÖLFFLÄCHERN, DODBKAEDER
4113	DOG CLUTCH	EMBRAYAGE A GRIFFES	KLAUENKUPPLUNG
4114	DOLERITE	DOLERITE	DOLERIT
4115	DOLLY	CONTRE-BOUTEROLLE	VORHALTER, GEGENHALTER, NIETPFANNE
4116	DOLLY	TAS, CHARIOT, DISQUE EN LISIERE DE TISSU, MEULE FLEXIBLE	SCHWABBELSCHEIBE, PLANIERKOLBEN, MONTAGEGESTELL, TRAGGESTELL
4117	DOLOMITE, MAGNESIAN LIMESTONE, CALCIUM MAGNESIUM CARBONATE	DOLOMIE, DOLOMITE	DOLOMIT
4118	DOME CAP NUT	ECROU BORGNE	HUTMUTTER
4119	DOME LIGHT	PLAFONNIER	DECKENLEUCHTE
4120	DOME ROOF	TOIT BOMBE	KUPPELDACH
4121	DOME-ROOF TANK	RESERVOIR A TOIT BOMBE	TANK MIT GEWÖLBTEM DECKEL
4122	DOOR AND WINDOW IRONS	FERS D'HUISSERIE	BESCHLÄGE FÜR TÜREN UND FENSTER
4123	DOOR LOCK	SERRURE DE PORTIERE	TÜRSCHLOSS
4124	DORMANT SCRAP	FERRAILLES DE PROTECTION PROPRE	HÜTTENSCHROTT
4125	DOTTED BROKEN LINE	LIGNE POINTILLEE, LIGNE PONCTUEE, TRAIT PONCTUE	LINIE (GESTRICHELTE), LINIE (PUNKTIERTE)
4126	DOTTING PEN, WHEEL PEN	TIRE-LIGNE A POINTILLER	PUNKTIERFEDER
4127	DOUBLE ARMED PULLEY	POULIE A BRASSURE DOUBLE	RIEMENSCHEIBE MIT DOPPELSPEICHEN
4128	DOUBLE BELT	COURROIE DOUBLE	DOPPELRIEMEN
4129	DOUBLE BLOCK BRAKE	FREIN A DEUX SABOTS, FREIN A DEUX MACHOIRES	DOPPELKLOTZBREMSE, DOPPELBACKENBREMSE
4130	DOUBLE BUTT STRAP RIVETED JOINT, BUTT RIVETED JOINT WITH TWO WELTS	RIVURE A DEUX COUVRE-JOINTS, RIVURE A DOUBLE COUVRE-JOINT	LASCHENNIETUNG (ZWEISEITIGE), LASCHENNIETUNG (DOPPELSEITIGE), DOPPELLASCHENNIETUNG
4131	DOUBLE CRANKED LEVER, T-LEVER	LEVIER A TROIS BRAS	HEBEL (DREIARMIGER)

4132	DOUBLE CUT	TAILLE CROISEE	KREUZHIEB
4133	DOUBLE CUT FILE, CROSS CUT FILE	LIME A DOUBLE TAILLE, LIME A TAILLE CROISEE	FEILE ZWEIHIEBIGE, DOPPELSCHNITTFEILE
4134	DOUBLE CUT, CROSS CUT OF A FILE	TAILLE CROISEE, DOUBLE TAILLE D'UNE LIME	KREUZHIEB, ZWEIFACHER HIEB EINER FELLE
4135	DOUBLE DECIMETRE	DOUBLE DECIMETRE	ANLEGEMASSSTAB
4136	DOUBLE DECOMPOSITION	DOUBLE DECOMPOSITION	WECHSELZERSETZUNG, UMSETZUNG (CHEMISCHE)
4137	DOUBLE ENDED PLUG LIMIT GAUGE	JAUGE-TAMPON DOUBLE A LIMITES	GRENZLEHRDORN
4138	DOUBLE FEEDING	ALIMENTATION DOUBLE	ZUFUHR (DOPPELTE)
4139	DOUBLE FLANGE	DOUBLE BRIDE	DOPPELFLANSCH
4140	DOUBLE FRAME HAMMER	MARTEAU DE FORGE DOUBLE CORPS	DOPPELSAÜLESCHMIEDEHAMMER
4141	DOUBLE FULL FILLET JOINT	JOINT SOUDE A DOUBLE CLIN	DOPPELKEHLNAHT
4142	DOUBLE GEAR	DOUBLE HARNAIS, DOUBLE TRAIN D'ENGRENAGES	DOPPELGETRIEBE
4143	DOUBLE HELICAL SPUR WHEEL, HERRINGBONE GEAR WHEEL	ROUE A CHEVRONS	WINKELRAD, PFEILRAD
4144	DOUBLE HELICAL TOOTH, HERRINGBONE TOOTH	DENT A CHEVRON, CHEVRON	WINKELZAHN, PFEILZAHN
4145	DOUBLE INTEGRAL	INTEGRALE DOUBLE	DOPPELINTEGRAL
4146	DOUBLE IRON PLANE, PLANE WITH TOP, BACK IRON	RABOT A CONTRE-FER	DOPPELHOBEL
4147	DOUBLE LEVER	LEVIER DOUBLE, A DEUX BRAS	HEBEL (DOPPELARMIGER), HEBEL ZWEIARMIGER, DOPPELHE
4148	DOUBLE MEMBER SNAP GAUGE	JAUGE-FOURCHE A TOLERANCES	GRENZ-RACHENLEHRE (DOPPELSEITIGE)
4149	DOUBLE NIPPLE	MAMELON A DEUX FILETS	DOPPELNIPPEL
4150	DOUBLE RACK IN FRAME	CADRE A CREMAILLERES, CADRE DENTE	KULISSENZAHNSTANGE
4151	DOUBLE REFRACTION	DOUBLE REFRACTION	DOPPELBRECHUNG
4152	DOUBLE RIVETED JOINT	RIVURE DOUBLE, RIVURE A DEUX RANGS	NIETUNG (ZWEIREIHIGE), NIETUNG (DOPPELTE)
4153	DOUBLE SALT	SEL DOUBLE	DOPPELSALZ
4154	DOUBLE SEATED VALVE	ROBINET A SOUPAPE DOUBLE	DOPPELSITZVENTIL
4155	DOUBLE SHEAR RIVETED JOINT	RIVURE A DEUX COUPES	NIETUNG (ZWEISCHNITTIGE)
4156	DOUBLE SHEAR STEEL	ACIER, FER DOUBLE CORROYE, FER FORT SUPERIEUR	STAHL (ZWEIMAL GEGÄRBTER), DOPPELGÄRBSTAHL
4157	DOUBLE SIDED MECHANICAL PRESS	PRESSE MECANIQUE A ARCADE	DOPPELSTÄNDER PRESSE (MECHANISCHE)
4158	DOUBLE SOCKET	MANCHON POUR RACCORDER DEUX TUYAUX COUPES	ÜBERSCHIEBER, ÜBERSCHIEBMUFFE, DOPPELMUFFE
4159	DOUBLE STANDARD	MONTANT DOUBLE, MONTANT JUMELE	DOPPELSTÄNDER
4160	DOUBLE TWISTED AUGER	MECHE TORSE, MECHE A COUTEAUX RENVERSES	SCHLANGENBOHRER
4161	DOUBLE WALL TANK	RESERVOIR A DOUBLE PAROI	TANK, DOPPELWANDIGER
4162	DOUBLE WEDGE DISC	OPERCULE DOUBLE	ZWEITEILIGER KEIL
4163	DOUBLE-ACTING	DOUBLE EFFET (A)	DOPPELTWIRKEND
4164	DOUBLE-ACTING PISTON	PISTON A DOUBLE EFFET	KOLBEN (DOPPELTWIRKENDER)
4165	DOUBLE-ACTING STEAM ENGINE	MACHINE A DOUBLE EFFET	DAMPFMASCHINE (DOPPELTWIRKENDE)
4166	DOUBLE-BEAT VALVE, CORNISH VALVE	SOUPAPE A DOUBLE SIEGE	VENTIL (DOPPELSITZIGES), DOPPELSITZVENTIL
4167	DOUBLE-ENDED SPANNER	CLEF A FOURCHE DOUBLE, CLEF DE CALIBRE DOUBLE	SCHRAUBENSCHLÜSSEL (DOPPELMAULIGER), DOPPELSCHLÜSSEL

4168	DOUBLE-EXPANSION ENGINE	MACHINE A DOUBLE EXPANSION	ZWEIFACH-EXPANSIONSMASCHINE
4169	DOUBLE-FACED HAMMER	MARTEAU A DEUX TETES	FLÄCHENHAMMER, HAMMER MIT ZWEI BAHNEN
4170	DOUBLE-WALLED	DOUBLE PAROI (A)	DOPPELWANDIG
4171	DOUBLE-WELDED LAP JOINT	JOINT A RECOUVREMENT SOUDE DES DEUX COTES	BEIDERSEITIG GESCHWEISSTE ÜBERLAPPUNGSVERBINDUNG
4172	DOUBLET	DOUBLET	DUBLETT, DOUBLET
4173	DOUBLING-OVER TEST	ESSAI DE PLIAGE	FALTVERSUCH
4174	DOUBLING-PLATE	PLATINE (PETITE PLAQUE METALLIQUE)	ZWISCHENPLATTE, PLATINE
4175	DOVETAIL	QUEUE D'ARONDE, QUEUE D'HIRONDE, TENON D'AGRAFAGE	SCHWALBENSCHWANZ
4176	DOVETAIL CUTTERS	FRAISES POUR QUEUE D'ARONDE	SCHWALBENSCHWANZFRÄSER
4177	DOVETAILED	QUEUE D'ARONDE (A)	SCHWALBENSCHWANZFÖRMIG
4178	DOVETAILED JOINT	ASSEMBLAGE A QUEUE D'ARONDE	SCHWALBENSCHWANZVERBINDUNG
4179	DOVETAILING MACHINE	MACHINE A FAIRE LES TENONS EN QUEUE D'ARONDE	ZINKENSCHNEIDMASCHINE
4180	DOWEL	GOUJON EN BOIS, CHEVILLE	DÜBEL, BOLZEN
4181	DOWEL PIN	GOUJON	PASSSTIFT, PASSBOLZEN
4182	DOWN MILLING	FRAISAGE EN DESCENDANT	GLEICHLÄUFIGES FRÄSEN
4183	DOWN PIPE	TUYAU, TUBE DE DESCENTE	FALLROHR
4184	DOWN STREAM	EN AVAL, D'AVAL	STROMABWÄRTS, MIT DEM STROM
4185	DOWN-COMER BAR	BARRE DE DEVERSOIR	FALLROHR
4186	DOWN-GATE	DESCENTE DE COULEE	EINGUSSKANAL, EINLAUF
4187	DOWN-STROKE DESCENT OF PISTON	DESCENTE DU PISTON, COURSE DESCENDANTE DU PISTON	KOLBENNIEDERGANG
4188	DOWNCOMERS BELOW TRAYS	DEVERSOIR DES PLATEAUX	FALLROHR UNTER SCHEIBEN
4189	DOWNDRAFT CARBURETTOR	CARBURATEUR INVERSE	FALLSTROMVERGASER
4190	DOWNHAND POSITION	SOUDURE EN POSITION A PLAT	SCHWEISSUNG (WAAGERECHTE)
4191	DOWNTIME	TEMPS MORT	STILLSTANDZEIT
4192	DOWNWARD VERTICAL POSITION	SOUDURE EN POSITION VERTICALE DESCENDANTE	ABWÄRTSSCHWEISSUNG
4193	DOWNWARDS PULLED DIE	MATRICE A DESCENTE COMMANDEE	ABZUGSMANTELMATRIZE
4194	DOWSON GAS, MIXED SEMI-WATER GAS	GAZ PAUVRE	MISCHGAS, KRAFTGAS, DOWSONGAS, SAUGGAS, HALBWASSERGAS
4195	DRAFT (TO)	DESSINER, TRACER	ZEICHNEN
4196	DRAFT (U.S) - DRAUGHT (GB)	TIRAGE, REDUCTION CONICITE, DEPOUILLE	ZUG, ANZUG, SCHRÄGE
4197	DRAFT ANGLE	ANGLE DE RETRAIT	AUSZIEHWINKEL
4198	DRAFTING OFFICE	BUREAU DE DESSIN	ZEICHENBÜRO
4199	DRAG	PARTIE DE DESSOUS (DE MOULE)	FORMUNTERTEIL
4200	DRAG COEFFICIENT	COEFFICIENT DE TRAINEE	WIDERSTANDSBEIWERT (LUFT)
4201	DRAG-IN	SOLUTION ADHERENTE	EINGESCHLEPPTE LOSÜNG, EÏNTRAG
4202	DRAG-OUT	SOLUTION ENTRAINEE	HERAUSGESCHLEPPTE LÖSUNG, AUSTRAG
4203	DRAGON'S BLOOD	SANG-DRAGON	DRACHENBLUT
4204	DRAIN (TO)	VIDER, VIDANGER, DRAINER, PURGER	ENTWÄSSERN
4205	DRAIN COCK	ROBINET PURGEUR, ROBINET DE VIDANGE	ABLASSVENTIL, ABLASSHAHN
4206	DRAIN PIPE	PURGE	ENTLEERUNGSROHR
4207	DRAIN PIPE, DISCHARGE PIPE, WASTE PIPE	TUYAU D'ECOULEMENT, TUYAU D'EVACUATION, TUYAU DE DECHARGE, TUYAU DE DEBIT	ABFLUSSROHR, ABLEITUNGSROHR

4208	**DRAIN PLUG**	BOUCHON DE VIDANGE	ABLASSSCHRAUBE
4209	**DRAIN THE SOIL (TO)**	DRAINER UNE TERRE HUMIDE	BODEN ENTWÄSSERN
4210	**DRAINAGE**	DRAINAGE, ASSECHEMENT, ECOULEMENT	ENTWÄSSERN, TROCKENLEGUNG
4211	**DRAUGHT GAUGE**	MANOMETRE POUR DETERMINATIONS ANEMOMETRIQUES	ZUGMESSER (FÜR SCHORNSTEINE UND GEBLÄSE)
4212	**DRAUGHT OF A CHIMNEY, CHIMNEY DRAUGHT**	TIRAGE D'UNE CHEMINEE	ZUG EINES SCHORNSTEINS, SCHORNSTEINZUG, KAMINZUG
4213	**DRAUGHTSMAN**	DESSINATEUR	ZEICHNER
4214	**DRAUGHTSMAN, DESIGNER**	MECANICIEN-CONSTRUCTEUR	KONSTRUKTEUR, KONSTRUKTIONSINGENIEUR
4215	**DRAW (TO)**	TIRER	ZIEHEN
4216	**DRAW A PERPENDICULAR (TO)**	ABAISSER UNE PERPENDICULAIRE, ABAISSER UNE VERTICALE	FÄLLEN (EIN LOT)
4217	**DRAW BAR**	BARRE, CROCHET D'ATTELAGE	ZUGHAKEN, ZUGSTANGE
4218	**DRAW BENCH**	BANC D'ETIRAGE	ZIEHBANK, DRAHTZIEHBANK
4219	**DRAW DOWN (TO)**	ECROUIR	RECKEN, STRECKEN (EISEN)
4220	**DRAW INTO WIRE(TO)**	TREFILER, ETIRER EN FIL	DRAHTZIEHEN, AUSZIEHEN (ZU DRAHT)
4221	**DRAW KNIFE**	PLANE	ZIEHMESSER, ZIEHKLINGE, SCHNITZMESSER
4222	**DRAW ON (TO), PULL ON A WHEEL (TO), DRAW ON A PULLEY (TO)**	SERRER UNE ROUE, SERRER UNE POULIE SUR L'ARBRE	AUFPRESSEN (EIN RAD), AUFPRESSEN (EINE RIEMENSCHEIBE), AUFZIEHEN (EIN RAD), AUFZIEHEN (EINE RIEMENSCHEIBE)
4223	**DRAW PLATE**	FILIERE, FILIERE DE TREFILAGE	DRAHTZIEHEISEN, ZIEHEISEN
4224	**DRAW TUBES (TO)**	ETIRER LES TUBES	ROHRE ZIEHEN
4225	**DRAW-OFF**	SOUTIRAGE-VIDANGE	ENTNAHME
4226	**DRAW-OFF NOZZLE WITH PERIPHERICAL SUMP**	TUBULURE DE PURGE AVEC CUVETTE PERIPHERIQUE	ENTLEERUNGSSTUTZEN MIT UMFANGSWANNE
4227	**DRAW-OFF PAN**	CAISSON DE SOUTIRAGE	ANZAPFUNGSTANK
4228	**DRAWABILITY**	ETIRABILITE	ZIEHBARKEIT
4229	**DRAWING**	DESSIN, PLAN	ZEICHNUNG
4230	**DRAWING**	TREFILAGE, ETIRAGE, ETIRAGE A FROID, FILAGE, RETIRURE, REFROIDISSEMENT LENT, REVENU	ZIEHEN, DRAHTZIEHEN,SCHWINDUNG, LANGSAMES ABKÜHLEN, ABLASSEN, TEMPERN
4231	**DRAWING BOARD**	PLANCHE A DESSIN	REISSBRETT, ZEICHENBRETT
4232	**DRAWING BRASS**	LAITON D'ETIRAGE	ZIEHMESSING
4233	**DRAWING COMPOUND**	GRAISSE D'ETIRAGE	ZIEHFETT
4234	**DRAWING DIE**	ANNEAU, MATRICE A ETIRER	ZIEHWERKZEUG, ZIEHRING
4235	**DRAWING DOWN**	ECROUISSAGE	RECKEN, STRECKEN DES EISENS
4236	**DRAWING INSTRUMENTS**	OUTILS ET INSTRUMENTS DE DESSIN, OUTILLAGE DU DESSINATEUR	ZEICHENGERÄT, ZEICHENUTENSILIEN
4237	**DRAWING MACHINE**	DEMOULEUSE	ABHEBEFORMMASCHINE
4238	**DRAWING OFFICE**	BUREAU DE DESSIN	ZEICHENSAAL, ZEICHENBÜRO
4239	**DRAWING OUT DEVICE**	DISPOSITIF A RETIRER	AUSZIEHVORRICHTUNG
4240	**DRAWING PAPER**	PAPIER A DESSIN	ZEICHENPAPIER
4241	**DRAWING PEN**	TIRE-LIGNE	REISSFEDER, ZIEHFEDER
4242	**DRAWING PIN**	PUNAISE	REISSZWECKE, HEFTZWEKE, REISSNAGEL, HEFTSTIFT
4243	**DRAWING TABLE OR DESK**	TABLE A DESSIN, TABLE DE DESSINATEUR	ZEICHENTISCH
4244	**DRAWING TOOL**	OUTIL A ETIRER, OUTIL A TREFILER	ZIEHWERKZEUG

4245	DRAWING, DRAFT	DESSIN (REPRESENTATION GRAPHIQUE)	ZEICHNUNG
4246	DRAWING, DRAFTING	DESSIN (ART DU DESSINATEUR)	ZEICHNEN
4247	DRAWN	ETIRE	GESCHRECKT
4248	DRAWN BARS	BARRES ETIREES	STABSTAHL (GEZOGENER)
4249	DRAWN HALF ROUNDS	DEMI-RONDS TREFILES	HALBRUNDSTAHL (GEZOGENER)
4250	DRAWN SHEETS	TOLES ETIREES	BLECHE (GESTRECKTE)
4251	DRAWN STEEL	ACIER ETIRE	STAHL (GEZOGENER)
4252	DRAWN WIRE	FIL ETIRE, FIL TREFILE	DRAHT (GEZOGENER), STRECKDRAHT
4253	DRESS ORES (TO)	TRAITER MECANIQUEMENT LES MINERAIS	ERZE AUFBEREITEN
4254	DRESSER	EBARBEUR	GUSSPUTZER
4255	DRIED SAND	SABLE SEC	SAND (TROCKENER)
4256	DRIER	ETUVE, SECHOIR, SICCATIF	TROCKENSCHRANK, SIKKATIV
4257	DRIER, SICCATIVE	SICCATIF	TROCKENSTOFF, SIKKATIV
4258	DRIERS : GRAIN	SECHOIRS A GRAINS	GETREIDETROCKNER
4259	DRIERS : GRASS AND GREEN CROP	SECHOIRS A FOURRAGES	GRASTROCKNER
4260	DRIFT	MANDRIN, BROCHE	DORN, EINTREIBDORN
4261	DRIFT (OR TAPER) PUNCH	BROCHE D'ASSEMBLAGE	DURCHTREIBER
4262	DRIFT A RIVET HOLE (TO)	ELARGIR AU MANDRIN, MANDRINER UN TROU DE RIVET	NIETLOCH AUFDORNEN (EIN)
4263	DRILL	FORET, MECHE	BOHRER
4264	DRILL (TO)	PERCER, FORER	BOHREN
4265	DRILL BIT	TREPAN	BOHRSTANGE, BOHRKRONE
4266	DRILL CHUCK	MANCHON PORTE-MECHE	BOHRFUTTER
4267	DRILL GAUGE	CALIBRE DE PERCAGE, GABARIT DE PERCAGE	BOHRLEHRE, BOHRERLEHRE
4268	DRILL JIGS	GABARITS DE PERCAGE	BOHRUNGSSTEUERVORRICHTUNG
4269	DRILL POINT SIZE	DIAMETRE DE LA POINTE DU FORET	BOHRERSPITZENDURCHMESSER
4270	DRILL STEELS FOR ROCK DRILLS	ACIERS DE TARIERES POUR FORAGES	BOHRSTÄHLE FÜR GESTEINS-BOHRUNGEN
4271	DRILL, BIT	FORET, MECHE	BOHRER
4272	DRILLABLE LINER	COLONNE PERDUE FORABLE	PERFORIERBARES VERLORENES ROHR
4273	DRILLED HOLE	TROU FORE, TROU ALESE	LOCH (GEBOHRTES), BOHRLOCH
4274	DRILLED PLATE	TOLE A TROUS FORES	BLECH MIT GEBOHRTEN LÖCHERN
4275	DRILLED RIVET HOLE	TROU DE RIVET PERCE, FORE	NIETLOCH GEBOHRTES
4276	DRILLER	PERCEUR -MECANICIEN, FOREUR	BOHRER (ARBEITER)
4277	DRILLING	PERCAGE, TARAUDAGE, FORAGE	BOHREN, BOHRUNG
4278	DRILLING HEAD STOCK	POUPEE DE PERCAGE	BOHRSPINDELSTOCK
4279	DRILLING MACHINE	MACHINE A FORER, FOREUSE, MACHINE A PERCER, PERCEUSE, ALESEUSE	BOHRMASCHINE
4280	DRILLING OIL	HUILE POUR FORER	BOHRÖL
4281	DRINKING WATER	EAU POTABLE, EAU DE BOISSON	TRINKWASSER
4282	DRIP	GOUTTE	TROPFEN
4283	DRIP CUP, DRIP PAN, OIL COLLECTOR, OIL TRAY	CUVETTE D'EGOUTTAGE	TROPFSCHALE, ÖLSCHALE, AUFFANGSCHALE, ÖLSCHIFF, ÖLFÄNGER
4284	DRIP FEED LUBRICATION, DROP FEED LUBRICATION	GRAISSAGE PAR COMPTE-GOUTTES	TROPFSCHMIERUNG
4285	DRIP MOULDING	GOUTTIERE	TROPFRINNE
4286	DRIP MOULDING PLIERS	PINCE DE JET D'EAU	REGENRINNENZANGE

4287	**DRIP OIL FEED**	GRAISSAGE COMPTE-GOUTTES	TROPFÖLSCHMIERUNG
4288	**DRIVE**	CARRE D'ENTRAINEMENT	MITNEHMERSTANGENEINSATZ
4289	**DRIVE (TO)**	COMMANDER, ACTIONNER, METTRE EN MOUVEMENT	ANTREIBEN
4290	**DRIVE AXLE**	ESSIEU MOTEUR	TREIBACHSE
4291	**DRIVE BY FRICTION (TO)**	ENTRAINER PAR FROTTEMENT	REIBUNG MITNEHMEN (DURCH)
4292	**DRIVE IN (TO), HAMMER IN A NAIL (TO)**	ENFONCER UN CLOU	EINTREIBEN (EINEN NAGEL), EINSCHLAGEN (EINEN NAGEL)
4293	**DRIVE IN A KEY (TO)**	ENFONCER UN COIN, FORCER UNE CLAVETTE DANS SA RAINURE	EINTREIBEN (EINEN KEIL)
4294	**DRIVE LINE**	TRANSMISSION	KRAFTÜBERTRAGUNG
4295	**DRIVE OUT A KEY (TO)**	CHASSER UNE CLAVETTE, DECALER	AUSTREIBEN (EINEN KEIL)
4296	**DRIVE SHAFT**	ARBRE DE TRANSMISSION	ANTRIEBSWELLE
4297	**DRIVE, DRIVING**	COMMANDE, ATTAQUE	ANTREIBEN, ANTRIEB
4298	**DRIVEN PULLEY, FOLLOWING PULLEY**	POULIE COMMANDEE, POULIE CONDUITE, POULIE MENEE, POULIE ENTRAINEE, POULIE RECEPTRICE	GETRIEBENE SCHEIBE, ANGETRIEBENE SCHEIBE
4299	**DRIVEN SHAFT**	ARBRE MENE, CONDUIT, ARBRE RECEPTEUR	WELLE (ANGETRIEBENE)
4300	**DRIVEN WHEEL, FOLLOWER**	ROUE MENEE, CONDUITE, ROUE RECEPTRICE	RAD (ANGETRIEBENES)
4301	**DRIVER**	TAQUET, SAILLANT, ENTRAINEUR, DOIGT, TOC D'ENTRAINEMENT	MITNEHMER
4302	**DRIVER'S LICENCE**	PERMIS DE CONDUIRE	FÜHRERSCHEIN
4303	**DRIVING**	ENTRAINEMENT	ANTRIEB
4304	**DRIVING BY FRICTION**	ENTRAINEMENT (PAR FROTTEMENT)	MITNEHMEN (DURCH REIBUNG)
4305	**DRIVING CHAIN**	CHAINE DE TRANSMISSION	TRIEBKETTE
4306	**DRIVING CHAIN, DRIVE CHAIN**	CHAINE DE TRANSMISSION, CHAINE MOTRICE	TRIEBKETTE, TREIBKETTE
4307	**DRIVING FIT**	EMMANCHEMENT, AJUSTAGE, MONTAGE BLOQUE	TREIBSITZ
4308	**DRIVING FLASK**	FLACON DESSECHANT, FLACON SECHEUR	TROCKENFLASCHE, TROCKENGLAS
4309	**DRIVING GEAR**	MECANISME DE COMMANDE, COMMANDE, ORGANE DE COMMANDE	TRIEB, ANTRIEBVORRICHTUNG, ANTRIEBMECHANISMUS
4310	**DRIVING IN OF A KEY**	ENCASTREMENT, CALAGE D'UNE CLAVETTE	EINTREIBEN EINES KEILS
4311	**DRIVING MECHANISM**	MECANISME DE COMMANDE	TRIEBWERK
4312	**DRIVING OUT A KEY**	DECALAGE (ACTION D'OTER LES CALES)	AUSTREIBEN EINES KEILS
4313	**DRIVING PULLEY**	POULIE MOTRICE, POULIE CONDUCTRICE, POULIE MENANTE, POULIE DE COMMANDE, POULIE D'ATTAQUE	TREIBENDE SCHEIBE, ANTRIEBSCHEIBE
4314	**DRIVING ROPE, TRANSMISION ROPE**	CABLE DE TRANSMISSION, CORDE-COURROIE	ANTRIEBSEIL, TRIEBWERKSEIL, TRANSMISSIONSSEIL, TREIBSEIL
4315	**DRIVING SHAFT**	ARBRE MOTEUR, ARBRE DE COMMANDE	TRIEBWELLE, WELLE (TREIBENDE), ANTRIEBWELLE, STEUERWELLE
4316	**DRIVING SIDE LEADING SIDE TIGHT SIDE OF A BELT, ROPE ETC.**	BRIN CONDUCTEUR, BRIN MENANT, MENEUR, BRIN MOTEUR, BRIN TENDU (D'UNE COURROIE SANS FIN)	TRUMM (ZIEHENDES), TRUMM (AUFLAUFENDES)
4317	**DRIVING SPINDLE**	BIELLE DE COMMANDE	ANTRIEBSPINDEL

4318	**DRIVING WHEEL, DRIVER**	ROUE MOTRICE, ROUE DE COMMANDE, ROUE MENANTE, ROUE CONDUCTRICE	TRIEBRAD, TREIBENDES RAD
4319	**DRIVINGS, DRIVE BY BELTS**	COMMANDE PAR COURROIE	RIEMENANTRIEB, RIEMENBETRIEB
4320	**DROP (TO), FALL INTO GEAR (TO)**	RETOMBER (CLIQUET)	EINFALLEN (KLINKE)
4321	**DROP BOTTLE**	COMPTE-GOUTTES	TROPFFLASCHE
4322	**DROP CENTER RIM**	JANTE A BASE CREUSE	TIEFBETTFELGE
4323	**DROP FEED LUBRICATOR, DRIP FEED LUBRICATOR**	GRAISSEUR COMPTE-GOUTTES	TROPFÖLER, ÖLTROPFAPPARAT
4324	**DROP FORGING**	PIECE MATRICEE	GESENKSCHMIEDESTÜCK
4325	**DROP FORGING HAMMER**	MARTEAU-PILON, MOUTON	FALLHAMMER
4326	**DROP HAMMER**	MOUTON, MARTEAU-PILON	FALLHAMMER, FALLWERK, GLEISHAMMER, PRISMENHAMMER, RAHMENHAMMER, PARALLELHAMMER, STEMPELHAMMER
4327	**DROP STAMPER**	MARTEAU-PILON, MOUTON A CHUTE LIBRE	FALLHAMMER
4328	**DROP TEST**	ESSAI DE CHOC (CHUTE), ESSAI DE FLEXION PAR CHOC, ESSAI DE CHOC SUR BARREAUX NON ENTAILLES	SCHLAGBIEGEPROBE, SCHLAGVERSUCH, FALLVERSUCH
4329	**DROP WEIGHT**	POIDS DU MOUTON	FALLGEWICHT
4330	**DROP WEIGHT TEST**	ESSAI DE CHUTE DE POIDS	GEWICHTSFALLVERSUCH
4331	**DROSS**	ECUME, CRASSE	SCHAUM, KRÄTZE, GEKRÄTZ
4332	**DROSS HOLE**	TROU A CRASSE	SCHLACKENLOCH
4333	**DRUM**	POULIE-TAMBOUR, TAMBOUR	RIEMENTROMMEL, TROMMEL
4334	**DRUM BRAKE**	FREIN A TAMBOUR	TROMMELBREMSE
4335	**DRUM CLEANING**	NETTOYAGE AU TAMBOUR	TROMMELN
4336	**DRUM LADLE**	POCHE-TONNEAU	TROMMELPFANNE, GIESSTROMMEL
4337	**DRUNKEN SAW**	SCIE OSCILLANTE	TAUMELSÄGE
4338	**DRY (TO), DESICCATE (TO), GROW DRY (TO)**	SECHER, DESSECHER	TROCKNEN
4339	**DRY AIR**	AIR SEC	LUFT (TROCKNE)
4340	**DRY AIR PUMP**	POMPE A AIR SEC	LUFTPUMPE (TROCKNE)
4341	**DRY ANALYSIS**	ANALYSE PAR VOIE SECHE	ANALYSE AUF TROCKENEM WEGE
4342	**DRY BINDER**	LIANT SEC	TROCKENBINDER
4343	**DRY CELL**	PILE SECHE	TROCKENELEMENT
4344	**DRY COAL**	HOUILLE SECHE A LONGUE FLAMME	KOHLE (TROCKNE), SANDKOHLE, SINTERKOHLE
4345	**DRY COMPRESSOR**	COMPRESSEUR SEC	VERDICHTER (TROCKENER)
4346	**DRY DISTILLATION**	DISTILLATION SECHE	DESTILLATION (TROCKENE)
4347	**DRY DRAWING**	ETIRAGE BRILLANT	GLANZZIEHEN
4348	**DRY LINER**	CHEMISE SECHE	LAUFBÜCHSE (TROCKENE)
4349	**DRY METALLURGY**	THERMOMETALLURGIE	PYROMETALLURGIE
4350	**DRY OXIDATION**	OXYDATION (OU CORROSION) SECHE	OXIDATION (ODER KORROSION) TROCKENE
4351	**DRY STEAM**	VAPEUR SECHE	DAMPFTROCKNER
4352	**DRY STRENGTH**	STABILITE A SEC	TROCKENFESTIGKEIT
4353	**DRY-SAND CASTING**	COULEE EN SABLE SEC, ETUVE	TROCKENGUSS
4354	**DRY-SAND MOULDING**	MOULAGE ETUVE, MOULAGE EN SABLE ETUVE, MOULAGE EN SABLE SEC	TROCKENSANDFORMEN
4355	**DRYING APPARATUS, DRIER**	SECHOIR	TROCKNER, TROCKENVORRICHTUNG
4356	**DRYING CHAMBER**	SECHEUR, APPAREIL SECHEUR	TROCKENKAMMER
4357	**DRYING OIL**	HUILE SICCATIVE	ÖL (TROCKNENDES), TROCKENÖL

4358	**DRYING STOVE**	ETUVE	TROCKENSCHRANK, TROCKENKASTEN, TROCKENOFEN
4359	**DRYING, DESICCATION**	SECHAGE, DESSICCATION	TROCKNEN
4360	**DUAL BARREL CARBURETTOR**	CARBURATEUR A DOUBLE CORPS	DOPPELVERGASER
4361	**DUCTILE**	DUCTILE	DEHNBAR, STRECKBAR, ZIEHBAR, SCHMIEGSAM-STRECKBAR
4362	**DUCTILE CAST IRON**	FONTE A GRAPHITE SPEROIDAL	KUGELGRAPHIT-GUSSEISEN
4363	**DUCTILITY**	DUCTILITE	DUKTILITÄT, DEHNBARKEIT, STRECKBARKEIT, ZIEHBARKEIT
4364	**DUG PEAT**	TOURBE MOTTIERE	HANDSTICHTORF
4365	**DULL BLUNT CUTTING EDGE**	ARETE TRANCHANTE EMOUSSEE	SCHNEIDE (STUMPFE)
4366	**DULL CHROMIUM PLATE**	CHOMAGE MAT	MATTVERCHROMUNG
4367	**DULL RED HEAT**	ROUGE SOMBRE	DUNKELROTGLUT
4368	**DULL RED HOT**	CHAUFFE AU ROUGE SOMBRE	DUNKELROTGLÜHEND
4369	**DULLING**	TERNISSEMENT	GLANZVERLUST
4370	**DUMMY**	PRE-MODELE, MAQUETTE	VORMODELL
4371	**DUMMY BLOCK**	BILLETTE	PRESSBLOCK (IN DER STRANGPRESSE)
4372	**DUMMY PASS**	CANNELURE, PASSE A VIDE	BLINDSTICH, BLINDKALIBER
4373	**DUMMY ROLL**	CYLINDRE A VIDE	BLINDWALZE
4374	**DUMP TEST**	ESSAI D'ECRASEMENT	STAUCHVERSUCH
4375	**DUMP TRUCK**	CAMION A BENNE BASCULANTE	KIPPER
4376	**DUMPING**	DEVERSEMENT	KIPPEN
4377	**DUNT**	FISSURE	HAARRISS
4378	**DUPLEX ALLOY**	ALLIAGE DUPLEX, ALLIAGE BINAIRE	BIMETALLLEGIERUNG
4379	**DUPLEX MANUFACTURING MILLING MACHINE**	FRAISEUSE DUPLEX A GRAND RENDEMENT	HOCHLEISTUNGSDUPLEXFRÄSMASCHINE
4380	**DUPLEX PRACTICE**	PROCEDE DUPLEX	DUPLEXVERFAHREN
4381	**DUPLEX PUMP**	POMPE DUPLEX	DUPLEXPUMPE
4382	**DUPLEX SPOT WELDER**	MACHINE DE SOUDAGE DOUBLE POINT	DOPPELPUNKTSCHWEISSMASCHINE
4383	**DUPLICATE MOLDING**	SURMOULAGE	FORMEN NACH DEM GUSSSTÜCK
4384	**DUPLICATE TEST**	CONTRE-ESSAI, CONTRE-EPROUVETTE	GEGENPROE
4385	**DURABILITY**	CONSERVABILITE, STABILITE, DURABILITE	HALTBARKEIT, DAUERHAFTIGKEIT
4386	**DURABLE**	DURABLE	DAUERHAFT
4387	**DURALUMIN**	DURALUMIN	DURALUMIN
4388	**DURVILLE POURING**	COULEE TRANQUILLE	GIESSEN (WIRBELFREIES)
4389	**DUST**	POUSSIERE	STAUB
4390	**DUST ARRESTOR**	ASPIRATEUR DE POUSSIERES	STAUBFÄNGER, GICHTSTAUBSAMMLER
4391	**DUST COLLECTING UNITS**	APPAREILS DE DEPOUSSIERAGE	ENTSTAUBUNGSAPPARATE
4392	**DUST COLLECTION BY EXHAUST VENTILATION**	CAPTAGE DU GAZ ET DEPOUSSIERAGE	GASABZUG UND STAUBABSAUGUNG
4393	**DUST PROOF**	A L'ABRI DE LA POUSSIERE	STAUBDICHT
4394	**DUSTING**	POUDRAGE AU SULFURE	EINSTAUBEN MIT SCHWEFEL, SCHWEFELBESTÄUBUNG
4395	**DUTY CYCLE**	REGIME D'UTILISATION	AUSLASTUNGSGRAD
4396	**DWELL**	ARRET TEMPORISE	VERWEILZEIT
4397	**DYE**	COLORANT	FARBSTOFF
4398	**DYNAMIC STRESS**	CONTRAINTE DYNAMIQUE	BEANSPRUCHUNG (DYNAMISCHE)
4399	**DYNAMICAL**	DYNAMIQUE	DYNAMISCH

4400	**DYNAMICS**	DYNAMIQUE	DYNAMIK
4401	**DYNAMITE**	DYNAMITE	DYNAMIT
4402	**DYNAMO (GB)**	DYNAMO	LICHTMASCHINE
4403	**DYNAMOMETER**	DYNAMOMETRE	FEDERWAAGE, LEISTUNGSMESSER, KRAFTMESSER, DYNAMOMETER
4404	**DYNAMOMETRICAL**	DYNAMOMETRIQUE	DYNAMOMETRISCH
4405	**DYNE**	DYNE	DYN, DYNE
4406	**DYNODE**	DYNODE	DYNODE
4407	**DYSPROSIUM**	DYSPROSIUM	DYSPROSIUM
4408	**EAR**	CORNE D'EMBOUTISSAGE, PLI	ECKE, FALTE
4409	**EARTH**	TERRE, MASSE (TERRE)	ERDE, MASSE
4410	**EARTH (TO), GROUND (TO)**	METTRE A LA TERRE, RELIER A LA TERRE (ELECTR.)	ERDEN
4411	**EARTH CABLE**	CABLE DE MASSE	MASSEKABEL
4412	**EARTH COLOUR, MINERAL DYE**	MATIERE COLORANTE MINERALE	ERDFARBE, MINERALFARBE
4413	**EARTH WAX, MINERAL WAX, OZOKERITE**	CIRE FOSSILE, OZOKERITE	ERDWACHS, BERGWACHS, BERGTALG, OZOKERIT
4414	**EARTH-MOVING EQUIPMENT**	EQUIPEMENT POUR NIVELER LE SOL	BAGGER, SCHÜRFRAUPE
4415	**EARTH, EARTH CONNECTION ; GROUND**	TERRE (ELECTR.), CONTACT A LA TERRE	ERDE, ERDLEITUNG, ERDUNG, ERDSCHLUSS
4416	**EARTHENWARE PIPE**	TUYAU EN POTERIE, EN TERRE CUITE	TONROHR
4417	**EARTHING**	MISE A LA TERRE	ERDUNG
4418	**EARTHING BOSS**	BOSSAGE DE MISE A LA TERRE	ERDKONTAKT
4419	**EARTHING LUG**	TAQUET DE MISE A LA TERRE	ERDUNGSKLEMME
4420	**EARTHWORK**	TERRASSEMENTS	ERDBAUTEN, ERDARBEITEN
4421	**EARTHY FRACTURE**	CASSURE TERREUSE	BRUCH (ERDIGER)
4422	**EAU DE LABARRAQUE (SODIUM HYPOCHLORITE SOLUTION)**	EAU DE LABARRAQUE (SOLUTION DE CHLORURE DE SOUDE)	LABARRAQUESCHE LAUGE, EAU DE LABARRAQUE (NATRIUMHYPOCHLORITLÖSUNG, CHLORSODALÖSUNG)
4423	**EBONY**	EBENE	EBENHOLZ
4424	**ECCENTRIC**	EXCENTRIQUE, EXCENTRE	AUSSERMITTIG, EXZENTRISCH
4425	**ECCENTRIC (ADJ.)**	EXCENTRIQUE	EXZENTER
4426	**ECCENTRIC CIRCLES**	CERCLES EXCENTRIQUES	KREISE (EXZENTRISCHE)
4427	**ECCENTRIC CRANK MOTION**	MECANISME A MANIVELLE EXCENTRIQUE	KURBELTRIEB (GESCHRÄNKTER), KURBELTRIEB (EXZENTRISCHER)
4428	**ECCENTRIC LOAD**	CHARGE EXCENTRIQUE	BELASTUNG (EXZENTRISCHE), BELASTUNG (AUSSERMITTIGE)
4429	**ECCENTRIC PISTON RING**	SEGMENT DE PISTON EXCENTRE	KOLBENRING (EXZENTRISCHER)
4430	**ECCENTRIC PRESS**	PRESSE A EXCENTRIQUE	EXZENTERPRESSE
4431	**ECCENTRIC ROD**	TIGE, BARRE D'EXCENTRIQUE	EXZENTERSTANGE
4432	**ECCENTRIC ROD END**	TETE DE LA TIGE D'EXCENTRIQUE	EXZENTERSTANGENKOPF
4433	**ECCENTRIC SHEAVE**	EXCENTRIQUE, EXCENTRIQUE CIRCULAIRE A COLLIER	EXZENTER, KREISEXZENTER, SCHEIBENKURBEL
4434	**ECCENTRIC SHEAVE PULLEY**	DISQUE D' EXCENTRIQUE, POULIE D' EXCENTRIQUE	EXZENTERSCHEIBE
4435	**ECCENTRIC STRAP HOOP**	COLLIER, BAGUE D'EXCENTRIQUE	EXZENTERBÜGEL, EXZENTERRING
4436	**ECCENTRICITY**	EXCENTRICITE	EXZENTRIZITÄT
4437	**ECONOMIZER**	ECONOMISEUR (DE CARBURANT)	KRAFTSTOFFSPARER, RAUCHGASVORWÄRMER, EKONOMISER
4438	**EDDY CURRENT**	COURANT DE FOUCAULT	WIRBELSTROM
4439	**EDDY CURRENT BRAKE**	FREIN A COURANTS DE FOUCAULT	WIRBELSTROMBREMSE

4440	EDDY CURRENTS, FOUCAULT CURRENTS	COURANTS DE FOUCAULT	WIRBELSTRÖME, FOUCAULTSCHE STRÖME
4441	EDGE	BORD, ARETE, RIVE, REBORD	KANTE, RAND, BLECHRAND
4442	EDGE CAM, PLATE CAM	CAME, EXCENTRIQUE A ONDES	UNRUNDE SCHEIBE, KURVENSCHEIBE
4443	EDGE DISLOCATION	DISLOCATION-COIN	STUFENVERSETZUNG
4444	EDGE FINDER	DISPOSITIF DE REPERAGE DES RIVES	KANTENDETEKTIONSVORRICHTUNG
4445	EDGE JOINT	JOINT SUR TRANCHES	ECKVERBAND
4446	EDGE MILLING CUTTER	FRAISE AXIALE	PLANFRÄSER, AXIALFRÄSER
4447	EDGE OF HOLE	ARETE D'UN TROU	LOCHKANTE, LOCHRAND
4448	EDGE OF REGRESSION	ARETE DE REBROUSSEMENT	RÜCKKEHRKANTE
4449	EDGE PREPARATION	PREPARATION DES BORDS	KANTENVORBEREITUNG
4450	EDGE RUNNER	MEULE VERTICALE, MOULIN A MEULES VERTICALES	KOLLERGANG, ROLLQUESTSCHE
4451	EDGER	MATRICE RONDE ET DIVISEE CAGE REFOULEUSE	RUNDGESENK, VERTEILGESENK, STAUCHGERÜST
4452	EDGING	REFOULEMENT	STAUCHEN
4453	EDGING MACHINES	PLIEUSES	ABKANTMASCHINEN
4454	EDGING MILL	CAGE REFOULEUSE	STAUCHGERÜST
4455	EDGING PASS	CANNELURE, PASSE REFOULEUSE	STAUCHKALIBER, STAUCHSTICH
4456	EDGING ROLL	CAGE A CYLINDRES VERTICAUX, CYLINDRE DE REFOULEMENT	VERTIKALGERÜST, STAUCHWALZE
4457	EFFECTIVE HEAT	CHALEUR UTILE, CHALEUR EFFECTIVE	NUTZWÄRME
4458	EFFECTIVE POWER	PUISSANCE EFFECTIVE, PUISSANCE AU FREIN	LEISTUNG (TATSÄCHLICHE), NUTZLEISTUNG, BREMSLEISTUNG
4459	EFFECTIVE PRESSURE	PRESSION REELLE, PRESSION ABSOLUE	ARBEITSDRUCK, DRUCK (ABSOLUTER)
4460	EFFECTIVE SECTION	SECTION CHARGEE	QUERSCHNITT (WIRKSAMER)
4461	EFFERVESCENCE	EFFERVESCENCE, BOUILLONNEMENT, MOUSSAGE	AUFBRAUSEN, SCHÄUMEN, AUFWALLEN
4462	EFFERVESCING STEEL	ACIER EFFERVESCENT, ACIER NON CALME	STAHL (UNBERUHIGTER)
4463	EFFICIENCY	RENDEMENT, EFFET UTILE	LEISTUNG, WIRKUNGSGRAD, GÜTEVERHÄLTNIS, NUTZEFFEKT
4464	EFFLORESCENCE	EFFLORESCENCE	AUSBLÜHUNG, AUSWITTERUNG, EFFLORESZENZ
4465	EFFUSION	EFFUSION	EFFUSION, AUSSTRÖMUNG
4466	EFFUSION OF GASES	EFFUSION DES GAZ	AUSFLUSS, EFFUSION VON GASEN
4467	EIGHTH BEND	COUDE AU 1/2	ACHTELKRÜMMER
4468	EJECT	EJECTER, DEMOULER	AUSWERFEN, AUSSTOSSEN
4469	EJECTOR	EJECTEUR	EJEKTOR, AUSWERFER
4470	EJECTOR CONDENSER	CONDENSEUR PAR EJECTION, EJECTO-CONDENSEUR	STRAHLKONDENSATOR
4471	EJECTOR/KNOCK-OUT	DEFOURNEMENT PAR EXTRACTEUR A BRAS	ARMAUSWURF
4472	EJECTOR, EXHAUSTER	EJECTEUR, ELEVATEUR	EJEKTOR
4473	ELASTIC	ELASTIQUE	FEDERND, ELASTISCH
4474	ELASTIC AFTEREFFECT	ELASTICITE DE SUITE	NACHWIRKUNG (ELASTISCHE)
4475	ELASTIC CONSTANTS	CONSTANTES D'ELASTICITE	ELASTIZITÄTSKONSTANTEN
4476	ELASTIC COUPLING	ACCOUPLEMENT ELASTIQUE, MANCHON ELASTIQUE	KUPPLUNG (ELASTISCHE)
4477	ELASTIC DEFLECTION	FLEXION ELASTIQUE	DURCHBIEGUNG (ELASTISCHE), DURCHFEDERUNG
4478	ELASTIC DEFORMATION	DEFORMATION ELASTIQUE	FORMÄNDERUNG (ELASTISCHE)

4479	ELASTIC DIAPHGRAM	DIAPHRAGME ELASTIQUE, MEMBRANE	FEDERMEMBRAN
4480	ELASTIC FORCE OF A GAS	FORCE ELASTIQUE D'UN GAZ	SPANNKRAFT EINES GASES
4481	ELASTIC LIMIT, LIMIT OF ELASTICITY	LIMITE ELASTIQUE, LIMITE D'ELASTICITE	BRUCHGRENZE, ELASTIZITÄTSGRENZE
4482	ELASTIC METALLIC PACKING	GARNITURE, JOINT PLASTIQUE	METALLPACKUNG (NACHGIEBIGE)
4483	ELASTIC MODULUS	MODULE D'ELASTICITE	ELASTIZITÄTSMODUL
4484	ELASTIC SOLID	SOLIDE ELASTIQUE	FESTSTOFF (ELASTISCHER)
4485	ELASTIC STRAIN	DEFORMATION ELASTIQUE	FORMÄNDERUNG (ELASTISCHE)
4486	ELASTIC WASHER	RONDELLE ELASTIQUE	FEDERRING, UNTERLEGRING (FEDERNDER)
4487	ELASTICITY	ELASTICITE	ELASTIZITÄT, FEDERKRAFT, FEDERWIRKUNG, EIGENFEDERUNG
4488	ELASTICITY OF COMPRESSION	ELASTICITE DE COMPRESSION	DRUCKELASTIZITÄT
4489	ELASTICITY OF FLEXURE	ELASTICITE DE FLEXION	BIEGUNGSELASTIZITÄT
4490	ELASTICITY OF TORSION	ELASTICITE DE TORSION	TORSIONSELASTIZITÄT
4491	ELASTICY OF ELONGATION	ELASTICITE DE TRACTION	ZUGELASTIZITÄT
4492	ELATED TANK	RESERVOIR SURELEVE	HOCHBEHÄLTER, HOCHRESERVOIR
4493	ELBOW	COUDE	KNIESTÜCK ·
4494	ELECTRIC ARC	ARC VOLTAIQUE	LICHTBOGEN (ELEKTR.)
4495	ELECTRIC ARC FURNACE	FOUR A ARC	ELEKTROLICHTBOGENOFEN, LICHBOGEN(ELEKTRO)OFEN
4496	ELECTRIC BLOW PIPE	CHALUMEAU ELECTRIQUE	LÖTROHR (ELEKTRISCHES)
4497	ELECTRIC BOILER	CHAUDIERE ELECTRIQUE	ELEKTRODAMPFKESSEL, DAMPFKESSEL (ELEKTRISCH GEHEIZTER)
4498	ELECTRIC BRAZING	BRASAGE DUR ELECTRIQUE	ELEKTROHARTLÖTEN
4499	ELECTRIC COIL	BOBINE	SPULE
4500	ELECTRIC CURRENT	COURANT ELECTRIQUE	STROM (ELEKTRISCHER)
4501	ELECTRIC DENSITY, CURRENT DENSITY	DENSITE ELECTRIQUE, DENSITE DE COURANT	DICHTE (ELEKTRISCHE), STROMDICHTE
4502	ELECTRIC DISCHARGE	DECHARGE ELECTRIQUE	ENTLADUNG (ELEKTRISCHE)
4503	ELECTRIC DRIVING	COMMAMDE ELECTRIQUE	ANTRIEB (ELEKTRISCHER)
4504	ELECTRIC ELECTROSTATIC FIELD	CHAMP ELECTRIQUE	FELD (ELECTRISCHES)
4505	ELECTRIC FIELD	CHAMP ELECTRIQUE	FELD (ELEKTRISCHES), SPANNUNGSFELD
4506	ELECTRIC FIELD INTENSITY	INTENSITE DE CHAMP	FELDSTÄRKE
4507	ELECTRIC FURNACE	FOUR ELECTRIQUE	OFEN (ELEKTRISCHER), ELEKTROOFEN, ELEKTROSTAHLOFEN
4508	ELECTRIC FURNACE ANNEALING	RECUIT AU FOUR ELECTRIQUE	GLÜHEN IN DEM ELEKTROOFEN
4509	ELECTRIC FURNACE IRON	FONTE ELECTRIQUE	ELEKTROGUSSEISEN
4510	ELECTRIC FURNACE PIG IRON	FONTE AU FOUR ELECTRIQUE	ELEKTROGUSSEISEN
4511	ELECTRIC FURNACE STEEL	ACIER ELABORE AU FOUR ELECTRIQUE, ACIER ELECTRIQUE	ELEKTROSTAHL
4512	ELECTRIC GENERATING CENTRAL STATION, POWER HOUSE, POWER STATION; ELECTRICITY WORKS	USINE CENTRALE ELECTRIQUE, CENTRALE ELECTRIQUE	ELEKTRIZITÄTSWERK, ELEKTRISCHE ZENTRALE
4513	ELECTRIC HEATING	CHAUFFAGE ELECTRIQUE	HEIZUNG (ELEKTRISCHE)
4514	ELECTRIC IGNITION	ALLUMAGE ELECTRIQUE	ZÜNDUNG (ELEKTRISCHE)
4515	ELECTRIC INDUCTION FURNACE	FOUR ELECTRIQUE A INDUCTION	INDUKTIONSOFEN
4516	ELECTRIC LAMP	LAMPE ELECTRIQUE	LAMPE (ELEKTRISCHE)
4517	ELECTRIC LIGHT	LUMIERE ELECTRIQUE	LICHT (ELEKTRISCHES)
4518	ELECTRIC LIGHTING	ECLAIRAGE ELECTRIQUE	BELEUCHTUNG (ELEKTRISCHE)
4519	ELECTRIC LIGHTING MAINS	LIGNE D'ECLAIRAGE	LICHTLEITUNG (ELEKTRISCHE)
4520	ELECTRIC LOCOMOTIVE	LOCOMOTIVE ELECTRIQUE	LOKOMOTIVE (ELEKTRISCHE)

4521	ELECTRIC LONG-DISTANCE LINE MAIN S	LIGNE DE TRANSPORT A GRANDE DISTANCE	FERNLEITUNG (ELEKTRISCHE)
4522	ELECTRIC MACHINE	MACHINE ELECTRIQUE	MASCHINE (ELEKTRISCHE)
4523	ELECTRIC MAINS	LIGNE, CANALISATION ELECTRIQUE, LIGNE CONDUCTRICE	LEITUNG (ELEKTRISCHE)
4524	ELECTRIC MOTORS, GENERATING EQUIPMENT, WELDING SETS AND LIGHTING PLANT.	MOTEURS ELECTRIQUES, GROUPES ELECTROGENES	ELEKTROMOTOREN, MASCHINENSÄTZE, SCHWEISS- UND BELEUCHTUNGSANLAGEN
4525	ELECTRIC POWER	COURANT ELECTRIQUE ENERGIE	STROM (ELEKTRISCHER)
4526	ELECTRIC POWER MAINS, TRANSMISSION LINE	LIGNE DE TRANSPORT DE FORCE	KRAFTLEITUNG (ELEKTRISCHE)
4527	ELECTRIC REGULATOR	REGULATEUR ELECTRIQUE	REGLER (ELEKTRISCHER)
4528	ELECTRIC RESISTANCE	RESISTANCE ELECTRIQUE	WIDERSTAND (ELEKTRISCHER)
4529	ELECTRIC SHOCK	COMMOTION, CHOC, SECOUSSE ELECTRIQUE	SCHLAG (ELEKTRISCHER)
4530	ELECTRIC SOLDERING IRON	FER A SOUDER ELECTRIQUE	LÖTKOLBEN-(ELEKTRISCHER)
4531	ELECTRIC SPARK MACHINING	ELECTROEROSION, ETINCELAGE ELECTRIQUE	FUNKENEROSION
4532	ELECTRIC STEEL	ACIER AU FOUR ELECTRIQUE	ELEKTROSTAHL
4533	ELECTRIC WELDING	SOUDAGE (A L'ARC) ELECTRIQUE	ELEKTROSCHWEISSEN, SCHWEISSEN (ELEKTRISCHES), SCHWEISSUNG (ELEKTRISCHE)
4534	ELECTRICAL ACCUMULATOR, STORAGE BATTERY, SECONDARY BATTERY	ACCUMULATEUR ELECTRIQUE, PILE SECONDAIRE, BATTERIE	ELEKTRISCHER SAMMLER, AKKUMULATOR, SEKUNDÄRELEMENT, BATTERIE
4535	ELECTRICAL COMPARATOR	COMPARATEUR ELECTRIQUE	VERGLEICHER (ELEKTRISCHER)
4536	ELECTRICAL CONDUCTIVITY	CONDUCTIBILITE ELECTRIQUE	LEITFÄHIGKEIT (ELEKTRISCHE)
4537	ELECTRICAL CONDUCTOR	CONDUCTEUR ELECTRIQUE	LEITER (ELEKTRISCHER)
4538	ELECTRICAL CONTACT METALS	METAUX UTILISES COMME CONTACT ELECTRIQUE	KONTAKTMETALLE (ELEKTRISCHE)
4539	ELECTRICAL ENERGY	ENERGIE, TRAVAIL ELECTRIQUE	ENERGIE (ELEKTRISCHE)
4540	ELECTRICAL ENGINEER	INGENIEUR-ELECTRICIEN	ELEKTROINGENIEUR
4541	ELECTRICAL ENGINEERING	ELECTRICITE INDUSTRIELLE	ELEKTROINGENIEURWESEN
4542	ELECTRICAL EQUIPMENT	EQUIPEMENT ELECTRIQUE	ANLAGE (ELEKTRISCHE)
4543	ELECTRICAL FITTER	MONTEUR-ELECTRICIEN	ELEKTROMONTEUR
4544	ELECTRICAL GRADE SHEET	TOLE ELECTRIQUE, TOLE DE TRANSFORMATEUR	ELEKTROBLECH, TRANSFORMATORENBLECH
4545	ELECTRICAL INSULATION	ISOLATION ELECTRIQUE	ISOLATION, ISOLIERUNG
4546	ELECTRICAL LOAD(ING)	PUISSANCE/ALIMENTATION	LAST (ELEKTRISCHE)
4547	ELECTRICAL OSCILLATION	OSCILLATION ELECTRIQUE	SCHWINGUNG (ELEKTRISCHE)
4548	ELECTRICAL RESISTIVITY	RESISTANCE SPECIFIQUE, RESISTIVITE	WIDERSTAND (SPEZIFISCHER), WIDERSTANDSFÄHIGKEIT
4549	ELECTRICAL VALVE	VALVE ELECTRIQUE, ELEMENT REDRESSEUR	VENTIL (ELEKTRISCHES)
4550	ELECTRICALLY MADE PIG IRON	FONTE ELECTRIQUE	ELEKTROROHEISEN
4551	ELECTRICIAN	ELECTRICIEN	ELEKTROTECHNIKER
4552	ELECTRICITY	ELECTRICITE	ELEKTRIZITÄT
4553	ELECTRICTY METER, SUPPLY METER	COMPTEUR D'ELECTRICITE	STROMZÄHLER, ELEKTRIZITÄTSZÄHLER
4554	ELECTRIFICATION	ELECTRIFICATION	ELEKTRIFIZIERUNG
4555	ELECTRIFY (TO)	ELECTRIFIER	ELEKTRIFIZIEREN
4556	ELECTRO DRILL	FLEURET ELECTRIQUE, FORET ELECTRIQUE	ELEKTROBOHRMASCHINE
4557	ELECTRO FORMING	ELECTRO-FORMAGE	GALVANOPLASTIK

4558	**ELECTRO REFINING**	AFFINAGE ELECTROLYTIQUE	VEREDELUNG (ELEKTROLYTISCHE), VERGÜTUNG (ELEKTROLYTISCHE)
4559	**ELECTRO-BAND MACHINING**	COUPE PAR ELECTRO-EROSION	FUNKENEROSIONSCHNITT
4560	**ELECTRO-CHEMICAL**	ELECTROCHIMIQUE	ELEKTROCHEMISCH
4561	**ELECTRO-CHEMISTRY**	ELECTROCHIMIE	ELEKTROCHEMIE
4562	**ELECTRO-DEPOSITION**	RECOUVREMENT D'UN METAL PAR VOIE ELECTRO-CHIMIQUE	GALVANISIERUNG
4563	**ELECTRO-DYNAMIC**	ELECTRODYNAMIQUE	ELEKTRODYNAMISCH
4564	**ELECTRO-DYNAMOMETER**	ELECTRODYNAMOMETRE	ELEKTRODYNAMOMETER
4565	**ELECTRO-EROSION**	ELECTRO-EROSION	ELEKTROEROSION
4566	**ELECTRO-GALVANISING**	ZINGAGE ELECTROCHIMIQUE ELECTROLYTIQUE	VERZINKUNG (KALTE), VERZINKUNG (ELEKTROLYTISCHE), VERZINKUNG GALVANISCHE
4567	**ELECTRO-MAGNET**	ELECTRO-AIMANT	ELEKTROMAGNET
4568	**ELECTRO-STATIC**	ELECTROSTATIQUE	ELEKTROSTATISCH
4569	**ELECTRO-TYPING**	ELECTROTYPIE	ELEKTROTYPIE
4570	**ELECTROANALYSIS**	ELECTROANALYSE	ELEKTROANALYSE, ANALYSE (ELEKTROLYTISCHE)
4571	**ELECTROCHEMICAL EQUIVALENT**	EQUIVALENT ELECTROCHIMIQUE	ÄQUIVALENT (ELEKTROCHEMISCHES)
4572	**ELECTROCHEMICAL VALVE**	REDRESSEUR ELECTROCHIMIQUE	GLEICHRICHTER (ELEKTROCHEMISCHER)
4573	**ELECTROCHEMISTRY**	ELECTRO-CHIMIE	ELEKTROCHEMIE
4574	**ELECTRODE**	ELECTRODE	ELEKTRODE
4575	**ELECTRODE HOLDER**	PORTE-ELECTRODE	ELEKTRODENHALTER, ELEKTRODENZANGE
4576	**ELECTRODE METAL**	METAL DE L'ELECTRODE	ELEKTRODENMETALL
4577	**ELECTRODE POTENTIAL**	POTENTIEL D'ELECTRODE	ELEKTRODENPOTENTIAL
4578	**ELECTRODE STUB**	BOUT D'ELECTRODE	ELEKTRODENREST
4579	**ELECTRODE TIP**	POINTE DE L'ELECTRODE	ELEKTRODENSPITZE, ELEKTRODENARBEITSFLÄCHE
4580	**ELECTRODEPOSITION**	DEPOT ELECTROLYTIQUE	ABSCHEIDUNG (ELEKTROLYTISCHE)
4581	**ELECTRODISSOLUTION**	DISSOLUTION ELECTROLYTIQUE	AUFLÖSUNG (ELEKTROLYSTISCHE)
4582	**ELECTROEXTRACTION**	EXTRACTION ELECTROLYTIQUE	EXTRAKTION (ELEKTROLYTISCHE)
4583	**ELECTROGALVANIZING**	GALVANISATION (OU ZINGAGE) ELECTROLYTIQUE	VERSINKUNG (GALVANISCHE)
4584	**ELECTROLIMIT GAGE**	CALIBRE LIMITE (OU DE TOLERANCE)	ELEKTROGRENZLEHRE
4585	**ELECTROLYSE (TO)**	ELECTROLYSER	ELEKTROLYSIEREN
4586	**ELECTROLYSIS**	ELECTROLYSE	ELEKTROLYSE
4587	**ELECTROLYTE**	ELECTROLYTE	ELEKTROLYT
4588	**ELECTROLYTIC CELL**	CELLULE (OU CUVE) ELECTROLYTIQUE	BAD (ELEKTROLYTISCHES), ELEKTROLYSEUR
4589	**ELECTROLYTIC CLEANING**	DEGRAISSAGE ELECTROLYTIQUE	REINIGUNG (ELEKTROLYTISCHE)
4590	**ELECTROLYTIC CONDENSER**	CONDENSATEUR ELECTROLYTIQUE	ELEKTROLYTKONDENSATOR
4591	**ELECTROLYTIC COPPER**	CUIVRE ELECTROLYTIQUE	ELEKTROLYTKUPFER
4592	**ELECTROLYTIC DEPOSIT**	DEPOT GALVANIQUE	NIEDERSCHLAG (GALVANISCHER)
4593	**ELECTROLYTIC DEPOSITION**	DEPOT ELECTROLYTIQUE, DEPOSITION ELECTROLYTIQUE	ABSCHEIDUNG (ELEKTROLYTICHE), FÄLLUNG (ELEKTROLYTISCHE)
4594	**ELECTROLYTIC DETERMINATION**	ANALYSE ELECTROLYTIQUE	BESTIMMUNG (ELEKTROLYTISCHE)
4595	**ELECTROLYTIC DISSOCIATION**	DISSOCIATION ELECTROLYTIQUE	DISSOZIATION (ELEKTROLYTISCHE)
4596	**ELECTROLYTIC DISSOCIATION, IONISATION**	DECOMPOSITION ELECTROLYTIQUE, IONISATION	DISSOZIATION (ELEKTROLYTISCHE), IONISATION
4597	**ELECTROLYTIC FURNACE**	FOUR A CREUSET	TIEGELOFEN
4598	**ELECTROLYTIC GOLD**	OR ELECTROLYTIQUE	ELEKTROLYTGOLD

4599	ELECTROLYTIC MANGANESE	MANGANESE ELECTROLYTIQUE	MANGAN (ELEKTROLYTISCHES)
4600	ELECTROLYTIC NICKEL	NICKEL ELECTROLYTIQUE	ELEKTROLYTNICKEL, KATHODENNICKEL
4601	ELECTROLYTIC OXIDATION	OXYDATION ELECTROLYTIQUE	OXYDATION (ELEKTROCHEMISCHE)
4602	ELECTROLYTIC PARTING	SEPARATION ELECTROLYTIQUE	SCHEIDUNG (ELEKTROLYTISCHE)
4603	ELECTROLYTIC PICKLING	DECAPAGE ELECTROLYTIQUE	BEIZUNG (ELEKTROLYTISCHE)
4604	ELECTROLYTIC POLARIZATION	POLARISATION ELECTROLYTIQUE	POLARISATION (ELEKTROLYTISCHE)
4605	ELECTROLYTIC POLISHING	POLISSAGE ELECTROLYTIQUE	POLIEREN (ELEKTROLYTISCHES), ELEKTROPOLIEREN
4606	ELECTROLYTIC RECTIFIER	REDRESSEUR (OU DETECTEUR) ELECTROLYTIQUE	ELEKTROLYTGLEICHRICHTER
4607	ELECTROLYTIC REDUCTION	REDUCTION ELECTROLYTIQUE, NEGATIVATION	REDUKTION (ELEKTROLYTISCHE)
4608	ELECTROLYTIC SILVER	ARGENT ELECTROLYTIQUE	ELEKTROLYTSILBER
4609	ELECTROLYTIC SOLUTION TENSION	TENSION DE SOLUTION ELECTROLYTIQUE	LÖSUNGSSPANNUNG
4610	ELECTROLYTIC TOUGH PITCH COPPER	CUIVRE ELECTROLYTIQUE A OXYDE CUIVREUX	CUPROOXYD ELEKTROLYT (KUPFER ENTHALTENDES)
4611	ELECTROMAGNET	ELECTRO-AIMANT	ELEKTROMAGNET
4612	ELECTROMAGNETIC	ELECTROMAGNETIQUE	ELEKTROMAGNETISCH
4613	ELECTROMAGNETIC BRAKE	FREIN ELECTROMAGNETIQUE	BREMSE (ELEKTRISCHE), BREMSE (ELEKTROMAGNETISCHE)
4614	ELECTROMAGNETIC HARDNESS ANALYZER	APPAREIL ELECTRO MAGNETIQUE POUR ESSAI DE DURETE	HÄRTEPRÜFER (ELEKTROMAGNETISCHER)
4615	ELECTROMAGNETIC INDUCTION	INDUCTION MAGNETOELECTRIQUE	INDUKTION (ELEKTROMAGNETISCHE)
4616	ELECTROMAGNETIC LEAKAGE	DISPERSION ELECTROMAGNETIQUE	STREUUNG (ELEKTROMAGNETISCHE)
4617	ELECTROMAGNETIC PERCUSSIVE WELDING	SOUDAGE PAR PERCUSSION ELECTROMAGNETIQUE	SCHLAGSCHWEISSEN (ELEKTROMAGNETISCHES)
4618	ELECTROMAGNETIC SEPARATION	SEPARATION ELECTROMAGNETIQUE	SCHEIDUNG (ELEKTROMAGNETISCHE)
4619	ELECTROMAGNETIC WELDING	SOUDAGE PAR INDUCTION	INDUKTIONSSCHWEISSEN
4620	ELECTROMAGNETISM	ELECTROMAGNETISME	ELEKTROMAGNETISMUS
4621	ELECTROMETALLURGY	ELECTRO-METALLURGIE	ELEKTROMETALLURGIE
4622	ELECTROMOTIVE FORCE (E.M.F., EMF)	FORCE ELECTROMOTRICE, F E M	KRAFT (ELEKTROMOTORISCHE), EMK
4623	ELECTROMOTOR, ELECTRIC MOTOR	MOTEUR ELECTRIQUE, ELECTROMOTEUR	ELEKTROMOTOR
4624	ELECTRON	ELECTRON	ELEKTRON
4625	ELECTRON BEAM	FAISCEAU D'ELECTRONS	ELEKTRONENSTRAHL
4626	ELECTRON BEAM FOCUSING LENS	LENTILLE DE CONCENTRATION DES ELECTRONS	ELEKTRONENSTRAHLKONZENTRATIONS-LINSE
4627	ELECTRON BEAM WELDING (EBW)	SOUDAGE PAR BOMBARDEMENT ELECTRONIQUE	ELEKTRONENSTRAHLSCHWEISSEN
4628	ELECTRON GUN	CANON A ELECTRONS	ELEKTRONENSTRAHLER, ELEKTRONENKANONE
4629	ELECTRON MICROSCOPE	MICROSCOPE ELECTRONIQUE	ELEKTRONENMIKROSKOP
4630	ELECTRON-BEAM WELDING	SOUDAGE PAR BOMBARDEMENT D'ELECTRONS	ELEKTRONENSTRAHLSCHWEISSEN
4631	ELECTRONEGATIVE	ELECTRONEGATIF	NEGATIVELEKTRISCH, ELEKTRONEGATIV
4632	ELECTRONIC COMPARATOR	COMPARATEUR ELECTRONIQUE	VERGLEICHER (ELEKTRONISCHER)
4633	ELECTRONIC IGNITION	ALLUMAGE ELECTRONIQUE	ELEKTRONIKZÜNDUNG
4634	ELECTRONIC RECTIFIER	REDRESSEUR ELECTRONIQUE	GLEICHRICHTER (ELEKTRONISCHER)
4635	ELECTRONIC REGULATOR	REGULATEUR ELECTRONIQUE	REGLER (ELEKTRONISCHER)
4636	ELECTRONIC TUBES	TUBES ELECTRONIQUES	ELEKTRONENRÖHREN
4637	ELECTRONICS	ELECTRONIQUE	ELEKTRONIK

4638	ELECTROOSMOSIS	ELECTRO-OSMOSE	ELEKTRO-OSMOSIS
4639	ELECTROPHORESIS, CATAPHORESIS	ELECTROPHORESE, CATAPHORESE	ELEKTROPHORESE, KATAPHORESE
4640	ELECTROPLATING	GALVANOSTEGIE, ELECTROCHIMIE, GALVANOPLASTIE	GALVANOSTEGIE, ELEKTROPLATTIERUNG
4641	ELECTROPOSITIVE	ELECTROPOSITIF	POSITIVELEKTRISCH, ELEKTROPOSITIV
4642	ELECTROSTATIC FORCE	FORCE ELECTROSTATIQUE	KRAFT (ELEKTROSTATISCHE)
4643	ELECTROSTATIC GENERATOR	GENERATEUR ELECTROSTATIQUE	GENERATOR (ELEKTROSTATISCHER)
4644	ELECTROSTATIC INDUCTION	INDUCTION ELECTROSIATIQUE	INDUKTION (ELEKTROSTATISCHE)
4645	ELECTROSTATIC PERCUSSIVE WELDING	SOUDAGE PAR PERCUSSION A CONDENSATEUR	KONDENSATORSTOSSENTLADUNGS-SCHWEISSEN
4646	ELECTROSTATIC SEPARATION	SEPARATION ELECTROSTATIQUE	TRENNUNG (ELEKTROSTATISCHE)
4647	ELECTROTECHNICAL	ELECTROTECHNIQUE	ELEKTROTECHNISCH
4648	ELECTROTECHNICS, ELECTROTECHNOLOGY	ELECTROTECHNIQUE	ELEKTROTECHNIK
4649	ELECTROTHERMAL EFFICIENCY	RENDEMENT ELECTROTHERMIQUE	WIRKUNGSGRAD (ELEKTROTHERMISCHER)
4650	ELECTROTHERMICS	ELECTROTHERMIQUE	ELEKTROTHERMIE, ELEKTROWÄRMELEHRE
4651	ELECTROTYPING	GALVANOPLASTIE	GAVALNOPLASTIK
4652	ELECTROWINNING	EXTRACTION ELECTROLYTIQUE	EXTRAKTION (ELEKTROLYTISCHE)
4653	ELECTRUM	ELECTRUM, ELECTRE	ELEKTRUM
4654	ELEMENT	ELEMENT, CORPS SIMPLE	ELEMENT, GRUNDSTOFF, ELEMENT (CHEMISCHES)
4655	ELEMENTARY ANALYSIS	ANALYSE ELEMENTAIRE	ELEMENTARANALYSE
4656	ELEMI	ELEMI, RESINE ELEMI	ELEMIHARZ
4657	ELETRO-ANALYSIS	ELECTROANALYSE	ELEKTROANALYSE
4658	ELEVATED TEMPERATURE	TEMPERATURE ELEVEE	TEMPERATUR (ERHÖHTE)
4659	ELEVATION, VERTICAL SECTION	ELEVATION, PROJECTION VERTICALE	AUFRISS, VERTIKALPROJEKTION
4660	ELEVATOR	ELEVATEUR (U.S. = ASCENSEUR)	ELEVATOR (U.S. = AUFZUG, LIFT)
4661	ELEVATORS AND LOADERS : HAY AND STRAW	ELEVATEURS DE FOIN ET DE PAILLE	FÖRDERUNGSANLAGE (HEU UND STROH)
4662	ELIMINATE (TO)	ELIMINER (MATH.)	ELIMINIEREN (MATH.)
4663	ELIMINATE HEAT (TO)	ELIMINER, ENLEVER DE LA CHALEUR	WÄRME ABFÜHREN
4664	ELIMINATION	ELIMINATION (MATH.)	ELIMINATION (MATH.)
4665	ELIMINATION OF AIR FROM A PIPE LINE	EVACUATION DE L'AIR D'UNE CONDUITE	ENTLÜFTUNG EINER ROHRLEITUNG
4666	ELIMINATION OF HEAT	SOUSTRACTION DE LA CHALEUR	ABLEITUNG VON WÄRME, WÄRMEABLEITUNG
4667	ELIMINATION, REMOVAL	ELIMINATION, ENLEVEMENT	ENTFERNUNG, ENTFERNEN, BESEITIGUNG
4668	ELINVAR	ELINVAR	ELINVARLEGIERUNG
4669	ELLIPSE	ELLIPSE, OVALE	ELLIPSE
4670	ELLIPSOGRAPH, ELLIPTIC TRAMMEL	ELLIPSOGRAPHE	ELLIPSENZIRKEL, ELLIPSOGRAPH
4671	ELLIPSOID	ELLIPSOIDE	ELLIPSOID
4672	ELLIPSOID OF REVOLUTION	ELLIPSOIDE DE ROTATION, ELLIPSOIDE DE REVOLUTION	DREHUNGSELLIPSOID, UMDREHUNGSELLIPSOID, ROTATIONSELLIPSOID
4673	ELLIPSOIDAL HEAD	TETE GOUTTE DE SUIF D'UNE VIS	KORBBOGENKOPF
4674	ELLIPTIC CROSS SECTION	SECTION ELLIPTIQUE	QUERSCHNITT (ELLIPTISCHER)
4675	ELLIPTIC CYLINDER	CYLINDRE A BASE ELLIPTIQUE	ZYLINDER (ELLIPTISCHER)
4676	ELLIPTIC INTEGRAL	INTEGRALE ELLIPTIQUE	INTEGRAL (ELLIPTISCHES)

4677	ELLIPTICAL OVAL GEAR WHEEL	ROUE ELLIPTIQUE, ENGRENAGE ELLIPTIQUE	ELLIPSENRAD
4678	ELLIPTICAL, OVAL WIRE	FIL OVALE, FIL ELLIPTIQUE	DRAHT (EIRUNDER), DRAHT (OVALER), DRAHT (ELLIPTISCHER)
4679	ELONGATION	ALLONGEMENT	DEHNUNG, AUSDEHNUNG
4680	ELONGATION, LONGITUDINAL EXTENSION	ALLONGEMENT	LÄNGSDEHNUNG
4681	ELUTRIATION	DECANTATION, LAVAGE (DU SABLE), TRIAGE PAR COURANT GAZEUX, LEVIGATION	SCHLÄMMUNG, SCHLÄMMEN
4682	EMBED IN CONCRETE (TO)	ENCASTRER DANS LE BETON	EINBETONIEREN
4683	EMBEDDABILITY	DEGRE D'ENCROUTEMENT	EINDRINGUNGSGRAD, EINSCHLIESSUNGSGRAD
4684	EMBOSSING	BOSSELAGE	RELIEFARBEIT
4685	EMBRITTLEMENT	FRAGILITE, ACCROISSEMENT DE LA FRAGILITE	SPRÖDIGKEIT, BRÜCHIGWERDEN, SPRÖDIGWERDEN, VERSPRÖDUNG.
4686	EMBRYO	EMBRYON, GERME	KEIM
4687	EMERGENCY BRAKE	FREIN DE SECOURS	NOTBREMSE
4688	EMERGENCY LIGHTING	ECLAIRAGE PROVISOIRE, ECLAIRAGE DE SECOURS	HILFSBELEUCHTUNG, NOTBELEUCHTUNG
4689	EMERGENCY LIGHTS	ECLAIRAGE DE SECOURS	NOTBELEUCHTUNG
4690	EMERGENCY LINE	LIGNE DE SECOURS, CANALISATION DE SECOURS	NOTLEITUNG (ELEKTR.)
4691	EMERGENT RAY	RAYON EMERGENT	STRAHL (AUSTRETENDER)
4692	EMERY ABRASIVE	EMERI	SCHMIRGEL
4693	EMERY CLOTH	TOILE EMERI	SCHMIRGELLEINEN, SCHMIRGELLEINWAND
4694	EMERY GRIND (TO)	POLIR A L'EMERI	ABSCHMIRGELN
4695	EMERY GRINDING	POLISSAGE A L'EMERI	SCHMIRGELN, ABSCHMIRGELN
4696	EMERY PAPER	PAPIER D' EMERI	SCHMIRGELPAPIER
4697	EMERY POWDER	POUDRE D'EMERI	SCHMIRGELPULVER
4698	EMERY WHEEL, EMERY GRINDER, EMERY BUFF	MEULE D'EMERI	SCHMIRGELSCHEIBE
4699	EMERY, CORUNDUM	EMERI, CORINDON	SCHMIRGEL, KORUND
4700	EMISSION	EMISSION	EMISSION
4701	EMISSIVE POWER, EMISSIVITY	POUVOIR EMISSIF	AUSSTRAHLUNGSVERMÖGEN, EMISSIONSVERMÖGEN
4702	EMISSIVITY	POUVOIR RAYONNANT	STRAHLUNGSVERMÖGEN
4703	EMIT RAYS (TO)	EMETTRE DES RAYONS	STRAHLEN AUSSENDEN, AUSSTRAHLEN
4704	EMPIRICAL FORMULA	FORMULE BRUTE	ERFAHRUNGSFORMEL, EMPIRISCHE FORMEL
4705	EMPTYING/OUTLET/SUCTION NOZZLE	TUBULURE DE SORTIE/ D'ASPIRATION	AUSGANGS-ROHRSTUTZEN
4706	EMULSIFIED OIL	HUILE EMULSIONNEE	EMULSIONSÖL
4707	EMULSIFY (TO), EMULSIONISE (TO)	EMULSIONNER	EMULGIEREN
4708	EMULSION	EMULSION	EMULSION
4709	ENAMEL	EMAIL	GLASFLUSS, SCHMELZ, EMAIL, EMAILLE
4710	ENAMEL (TO)	EMAILLER	EMAILLIEREN
4711	ENAMEL PAINT	PEINTURE VERNISSANTE	SCHMELZFARBE, EMAILFARBE
4712	ENAMELLING	EMAILLAGE	EMAILLIEREN, EMAILLIERUNG
4713	ENANTIOTROPIC	ENANTIOTROPIQUE	ENANTIOTROPISCH
4714	ENCASED BEAM	POUTRE ENROBEE	EINBETONIERTER TRÄGER

4715	**ENCLOSED MOTOR**	MOTEUR A CARTER, MOTEUR BLINDE OU CUIRASSE	MOTOR (GEKAPSELTER), MOTOR (GESCHLOSSENER)
4716	**ENCODING**	CODAGE	KODIERUNG
4717	**END ELEVATION, END VIEW**	VUE DE FACE POSTERIEURE	RÜCKANSICHT
4718	**END GRAIN SAWING OF TIMBER**	SCIAGE VERTICAL, SCIAGE CONTRE FIL, SCIAGE EN TRAVERS, TRONCONNAGE DU BOIS	HIRNSCHNITT, QUERSCHNITT DES HOLZES
4719	**END GRAIN TIMBER, CROSS GRAIN TIMBER, TIMBER CUT A CROSS THE GRAIN**	BOIS TAILLE CONTRE LE FIL	HIRNHOLZ
4720	**END JOURNAL BEARING, END OUTER BEARING**	PALIER D'EXTREMITE	STIRNLAGER
4721	**END JOURNAL, JOURNAL AT END OF SHAFT**	TOURILLON FRONTAL, TOURILLON D EXTREMITE	STIRNZAPFEN
4722	**END MILLING CUTTER**	FRAISE EN BOUT, FRAISE RADIALE, FRAISE FRONTALE, FRAISE DE FACE, FRAISE A SURFACER	STIRNFRÄSER, RADIALFRÄSER
4723	**END MILLS**	FRAISES EN BOUT	SCHAFTFRÄSER
4724	**END OF PROGRAM**	FIN DE PROGRAMME	PROGRAMMENDE
4725	**END OF TAPE (EOT)**	FIN DE BANDE	LOCHSTREIFENENDE
4726	**END PEDESTAL BRACKET**	CHAISE EN BOUT	WINKELARM, WINKELKONSOLE, QUERKONSOLE
4727	**END PLAY**	JEU AXIAL	SPIELRAUM (AXIALER)
4728	**END WALL**	PAROI FRONTALE	STIRNWAND, KOPFWAND
4729	**END-CENTERED**	A FACES CENTREES	FLÄCHENZENTRIERT
4730	**END-OF-BLOCK (EOB)**	FIN DE BLOC	BLOCKENDE
4731	**END-OF-LINE**	FIN DE LIGNE	LINIENENDE
4732	**ENDLESS CHAIN**	CHAINE SANS FIN	KETTE (ENDLOSE)
4733	**ENDLESS ROPE**	CABLE SANS FIN	SEIL (ENDLOSES)
4734	**ENDLONG THRUST PRESSURE**	POUSSEE AXIALE, POUSSEE LONGITUDINALE	LÄNGSSCHUB, AXIALSCHUB, AXIALDRUCK
4735	**ENDLONG, SIDE PLAY IN THE SHAFT, LATERAL FLOAT OF THE SHAFT**	JEU LATERAL DES ARBRES	SPIEL (SEITLICHES) DER WELLE
4736	**ENDOSMOSE**	ENDOSMOSE	ENDOSMOSE
4737	**ENDOTHERMIC**	ENDOTHERMIQUE	WÄRMEVERZEHREND, ENDOTHERMISCH, ENDOTHERM
4738	**ENDURANCE**	ENDURANCE	DAUERHAFTIGKEIT
4739	**ENDURANCE CRACK**	FISSURE DE FATIGUE / D'ENDURANCE	DAUERRISS, ERMÜDUNGSRISS
4740	**ENDURANCE FAILURE, ENDURANCE FRACTURE**	RUPTURE DE FATIGUE / D'ENDURANCE	DAUERBRUCH
4741	**ENDURANCE LIMIT**	LIMITE D'ENDURANCE	DAUERFESTIGKEIT
4742	**ENDURANCE RATIO**	RAPPORT : LIMITE DE FATIGUE, RESISTANCE DE RUPTURE PAR FRACTION	DAUERFESTIGKEITSVERHÄLTNIS, SCHWELLFESTIGKEITSVERHÄLTNIS
4743	**ENDURANCE TEST**	ESSAI DE FATIGUE / D'ENDURANCE	DAUERVERSUCH, ERMÜDUNGSVERSUCH
4744	**ENERGIZER**	SUBSTANCE ACTIVATRICE	SUBSTANZ (AKTIVIERENDE)
4745	**ENERGY**	ENERGIE	ARBEITSVERMÖGEN, ENERGIE
4746	**ENERGY EFFICIENCY**	EFFICACITE ENERGETIQUE	WIRKUNGSGRAD (ENERGETISCHER)
4747	**ENERGY INPUT, ENERGY PUT IN**	ENERGIE, PUISSANCE ABSORBEE, PUISSANCE RECUEILLIE, TRAVAIL CONSOMME	AUFGEWANDTE ENERGIE, ZUGEFÜHRTE LEISTUNG, ARBEIT, AUFGENOMMENE LEISTUNG, AUFNAHME
4748	**ENERGY OF DISCHARGE**	ENERGIE D'ECOULEMENT	AUSSTRÖMUNGSENERGIE
4749	**ENGAGE THE CATCH (TO)**	ENCLIQUETER, ENCLENCHER	EINKLINKEN
4750	**ENGAGE WITH (TO)**	ENGAGER, ENGRENER	EINGREIFEN

4751	ENGAGEMENT OF A CATCH PAWL	ENCLANCHEMENT, ENCLENCHEMENT	EINKLINKEN
4752	ENGAGING	ENGRENEMENT	EINGRIFF
4753	ENGINE HOUSE, POWER HOUSE	BATIMENT DES MACHINES	MASCHINENGEBÄUDE, MASCHINENHAUS
4754	ENGINE MOUNTING	SUPPORT MOTEUR	MOTORAUFHÄNGUNG
4755	ENGINE OIL, MACHINE OIL	HUILE POUR MACHINES	MASCHINENÖL
4756	ENGINE PART	ORGANE DE MACHINE	TRIEBWERK
4757	ENGINE ROOM	SALLE DES MACHINES, COMPARTIMENT MOTEUR	MASCHINENHALLE, MOTORRAUM, MASCHINENRAUM
4758	ENGINE SHAFT	ARBRE DE COUCHE, ARBRE PREMIER MOTEUR	MASCHINENWELLE
4759	ENGINE WORKED WITH SUPERHEATED STEAM	MACHINE A VAPEUR SURCHAUFFEE	HEISSDAMPFMASCHINE
4760	ENGINE-BRAKE	FREIN-MOTEUR	MOTOR-BREMSE
4761	ENGINEER	INGENIEUR	INGENIEUR
4762	ENGINEER-IN-CHIEF, CHIEF ENGINEER	INGENIEUR EN CHEF	OBERINGENIEUR
4763	ENGINEERING CAST IRON	FONTE MECANIQUE	MASCHINENGUSS, GUSS FÜR DEN MASCHINENBAU
4764	ENGINEERING WORKS FACTORY	ATELIER, ETABLISSEMENT DE CONSTRUCTON DE MACHINES, USINE DE CONSTRUCTON MECANIQUE	MASCHINENBAUANSTALT, MASCHINENFABRIK
4765	ENGINEMAN	MACHINISTE, CONDUCTEUR DE MACHINE, MECANICIEN	MASCHINIST, MASCHINENFÜHRER, MASCHINENWÄRTER
4766	ENGINES (PETROL, VAPORISING OIL, DIESEL)	MOTEURS FIXES ET MOBILES	MOTOREN (BENZIN, KEROSIN UND DIESEL)
4767	ENGRAVERS' COPPER	CUIVRE POUR GRAVURE	GRAVIERKUPFER
4768	ENLARGEMENT (ENLARGED PORTION) OF A PIPE	EPANOUISSEMENT D'UN TUBE	ERWEITERUNG, ERWEITERTER TEIL (EINES ROHRES)
4769	ENLARGEMENT OF A DRAWING	AGRANDISSEMENT D'UN DESSIN	VERGRÖSSERUNG EINER ZEICHNUNG
4770	ENRICHED BLAST	VENT ENRICHI D'OXYGENE	WIND (SAUERSTOFFANGEREICHERTER)
4771	ENTRAINED AIR	AIR PARASITE, RENTREE D'AIR	FALSCHLUFT
4772	ENTRAPPED SLAG	INCLUSION DE SCORIE	SCHLACKENEINSCHLUSS
4773	ENTROPY	ENTROPIE	ENTROPIE
4774	ENTRY SIDE	COTE D'ENTREE, COTE D INTRODUCTION	EINSTECKSEITE, EINTRITTSEITE
4775	ENTRY TABLE	LIGNE DE ROULEAUX D'AMENEE	ZUFUHRROLLGANG
4776	ENVELOPE	ENVELOPPE (GEOM.)	EINHÜLLENDE, HÜLLKURVE, UMHÜLLUNGSKURVE, ENVELOPPE
4777	ENVELOPE FLAME	PANACHE	BEIFLAMME
4778	EPICYCLIC GEAR TRAIN	TRAIN EPICYCLOIDAL	UMLAUFVORGELEGE, EPIZYKELVORGELEGE
4779	EPICYCLOID	EPICYCLOIDE	EPIZYKLOIDE
4780	EPICYCLOIDAL GEAR	DENTS A FLANCS EPICYCLOIDAUX, ENGRENAGE EPICYCLOIDAL	EPIZYKLOIDENVERZAHNUNG
4781	EQUAL SIDED ANGLE IRON	CORNIERE A AILES EGALES	WINKELEISEN (GLEICHSCHENKLIGES)
4782	EQUATION	EQUATION	GLEICHUNG
4783	EQUATION OF STATE	EQUATION D'ETAT	ZUSTANDSGLEICHUNG
4784	EQUI-AXED CRYSTALS	CRISTAUX EQUIAXES	KRISTALLE (GLEICHGERICTETE)
4785	EQUILATERAL TRIANGLE	TRIANGLE EQUILATERAL	DREIECK GLEICHSEITIGES
4786	EQUILIBRIUM	EQUILIBRE	GLEICHGEWICHT
4787	EQUILIBRIUM CONSTANT	CONSTANTE D'EQUILIBRE	GLEICHGEWICHTSKONSTANTE
4788	EQUILIBRIUM DIAGRAM	DIAGRAMME D'EQUILIBRE	GLEICHGEWICHTSDIAGRAMM

4789	**EQUILIBRIUM ELECTRODE POTENTIAL**	TENSION D'EQUILIBRE D'UNE ELECTRODE	GLEICHGEWICHTSPOTENTIAL EINER ELEKTRODE
4790	**EQUILIBRIUM REACTION POTENTIAL**	TENSION D'EQUILIBRE D'UNE REACTION	STATISCHES GLEICHGEWICHTSPOTENTIAL EINER REAKTION
4791	**EQUILIBRIUM TEMPERATURE**	TEMPERATURE D'EQUILIBRE	GLEICHGEWICHTSTEMPERATUR
4792	**EQUILIBRIUM VALUE**	VALEUR D'EQUILIBRE	GLEICHGEWICHTSWERT
4793	**EQUILIBRIUM, BALANCE**	EQUILIBRE	GLEICHGEWICHT
4794	**EQUIPMENT**	OUTILLAGE	WERKSGERÄT
4795	**EQUIVALENCE**	EQUIVALENCE	GLEICHWERTIGKEIT
4796	**EQUIVALENT**	EQUIVALENT	GLEICHWERTIG, AEQUIVALENT
4797	**EQUIVALENT CONDUCTIVITY**	CONDUCTIBILITE EQUIVALENTE	ÄQUIVALENTE LEITFÄHIGKEIT
4798	**EQUIVALENT RESISTIVITY**	RESISTIVITE EQUIVALENTE	ÄQUIVALENTE WIDERSTANDSFÄHIGKEIT
4799	**ERASE (TO), RUB OUT (TO)**	GRATTER, EFFACER	AUSKRATZEN, RADIEREN, AUSRADIEREN, WEGRADIEREN
4800	**ERASER, ERASING KNIFE**	GRATTOIR DE BUREAU, GRATTOIR DE DESSINATEUR	RADIERMESSER
4801	**ERASING RUBBER**	GOMME-GRATTOIR	RADIERGUMMI
4802	**ERASING, RUBBING OUT**	GRATTAGE	RADIEREN
4803	**ERASURE**	GRATTAGE (RESULTAT)	RADIERSTELLE, RASUR
4804	**ERBIUM**	ERBIUM	ERBIUM
4805	**ERECT (TO)**	MONTER	AUFSTELLEN, ZUSAMMENBAUEN, MONTIEREN
4806	**ERECT A PERPENDICULAR (TO)**	ELEVER UNE VERTICALE, ELEVER UNE PERPENDICULAIRE	SENKRECHTE ERRICHTEN (EINE)
4807	**ERECTED CONCURRENTLY OR CONSEQUENTLY**	MONTE SIMULTANEMENT OU EN CONTINUITE	GLEICHZEITIG UND FORTLAUFEND MONTIERT
4808	**ERECTING**	MONTAGE	AUFSTELLUNG, MONTAGE
4809	**ERECTING EQUIPMENT**	MATERIEL DE MONTAGE	MONTAGEMATERIAL
4810	**ERECTING MATERIAL**	MATERIEL DE MONTAGE	MONTAGEAUSRÜSTUNG
4811	**ERECTING SHOP**	ATELIER DE MONTAGE	MONTAGEHALLE, MONTIERWERKSTATT
4812	**ERECTING TOOLS**	OUTILLAGE DE MONTAGE	MONTAGEWERKZEUG
4813	**ERECTION**	MONTAGE	MONTAGE, AUFBAU
4814	**ERECTION DRAWING**	PLAN DE MONTAGE	MONTAGEZEICHNUNG
4815	**ERECTION WORK**	MONTAGE	MONTAGE
4816	**ERECTOR**	MONTEUR	AUFSTELLER, MONTEUR
4817	**ERECTOR'S TOOLS**	OUTILLAGE DU MONTEUR	MONTAGEWERKZEUG
4818	**ERG**	ERG, DYNE-CENTIMETRE	ERG
4819	**ERICHSEN CUP-TEST MACHINE**	MACHINE D'ESSAI D'EMBOUTISSAGE D'ERICHSEN	ERICHSEN-TIEFZIEHVERSUCHSMASCHINE
4820	**EROSION**	EROSION, USURE	VERSCHLEISS, EROSION
4821	**ERRATIC, BOULDER**	BLOC ERRATIQUE	FINDLING, BLOCK ERRATISCHER
4822	**ERROR**	ERREUR	FEHLER
4823	**ERROR COUNTER**	COMPTEUR D'ERREURS	FEHLERZÄHLER
4824	**ERROR IN CALCULATION**	ERREUR DE CALCUL	RECHENFEHLER
4825	**ERROR IN DESIGN**	ERREUR DE CONCEPTION DU CONSTRUCTEUR	KONSTRUKTIONSFEHLER
4826	**ERROR IN READING**	ERREUR DE LECTURE	ABLESEFEHLER
4827	**ERROR IN VALUATION**	ERREUR D'APPRECIATION	SCHÄTZUNGSFEHLER
4828	**ERROR OF COLLIMATION**	ERREUR DE COLLIMATION	KOLLIMATIONSFEHLER
4829	**ERROR OF OBSERVATION**	ERREUR D'OBSERVATION	BEOBACHTUNGSFEHLER
4830	**ERROR REGISTER**	COMPTEUR D'ERREURS	FEHLERZÄHLER

4831	ERYTHROSIN, IODEOSIN	ERYTHROSINE, TETRAIOD-FLUORESCEINE, PRIMEROSE SOLUBLE	TETRAJODFLUORESZEIN, BLAUSTICHIGES EOSIN, ERYTHROSIN, DIANTHIN, JODEOSIN
4832	ESCAPE (TO)	ECOULER (S'), ECHAPPER (S')	AUSSTRÖMEN, ENTWEICHEN
4833	ESCAPE OF STEAM, OF GASES	ECHAPPEMENT DE VAPEUR, ECHAPPEMENT DE GAZ	ENTWEICHEN, AUSSTRÖMEN VON DAMPF, VON GASEN
4834	ESCRIBED CIRCLE	CERCLE EXINSCRIT	ANGESCHRIEBENER KREIS, ANKREIS
4835	ESCUTCHEON PLATE	RONDELLE PROTECTRICE, CACHE-ENTREE	SCHLÜSSELLOCHDECKEL
4836	ESSENTIAL VOLATILE OIL	HUILE VOLATILE, HUILE ESSENTIELLE	ÖL (FLÜCHTIGES)
4837	ESTER	ETHER COMPOSE	ESTER, ZUSAMMENGESETZTER ÄTHER, SÄUREÄTHER
4838	ESTIMATE OF THE COST	DEVIS, ESTIMATION	KOSTENANSCHLAG, VORANSCHLAG
4839	ETCH (TO)	ATTAQUER A L'ACIDE	ÄTZEN
4840	ETCH BANDS	BANDES REVELEES PAR ATTAQUE CHIMIQUE	ÄTZUNG HERVORTRETENDE BÄNDER (DURCH)
4841	ETCH CRACKS	CRIQUES D'ATTAQUE CHIMIQUE	ÄTZRISSE
4842	ETCH FIGURES	FIGURES D'ATTAQUE A L'ACIDE	ÄTZFIGUREN
4843	ETCH PITS	PIQURES DE CORROSION	STOCKFLECKEN
4844	ETCHANT	REACTIF	ÄTZMITTEL
4845	ETCHED FIGURE, CORROSION FIGURE	FIGURE DE CORROSION	ÄTZFIGUR, KRAFTWIRKUNGSFIGUR
4846	ETCHING	ATTAQUE (COR.), ATTAQUE A L ACIDE	ÄTZEN, ÄTZUNG
4847	ETCHING REAGENT	CAUSTIQUE, CORROSIF, REACTIF D'ATTAQUE A L'ACIDE	ÄTZFLÜSSIGKEIT, ÄTZMITTEL
4848	ETHANE, DIMETHYL	ETHANE, HYDRURE D'ETHYLE, DIMETHYLE	ÄTHAN, ÄTHYLWASSERSTOFF, DIMETHYL
4849	ETHYL CHLORIDE	CHLORURE D'ETHYLE, ETHER CHLORHYDRIQUE	ÄTHYLCHLORID, CHLORÄTHYL, MONOCHLORÄTHAN, CHLORWASSERSTOFFÄTHER, SALZÄTHER
4850	ETHYL NITRATE	AZOTATE, NITRATE D'ETHYLE, ETHER NITRIQUE, ETHER AZOTIQUE	SALPETERSÄUREÄTHER, ÄTHYLNITRAT
4851	ETHYLENE, OLEFIANT GAS	ETHYLENE, HYDROGENE BICARBONE, ETHENE, GAZ OLEFIANT	SCHWERES KOHLENWASSERSTOFFGAS, ÄTHYLEN, ÖLBILDENDES GAS
4852	EUDIOMETER	EUDIOMETRE	GASPRÜFER, EUDIOMETER
4853	EUPRIC CHLORIDE	CHLORURE CUIVRIQUE	KUPFERCHLORID
4854	EUROPIUM	EUROPIUM	EUROPIUM
4855	EUTECTIC	EUTECTIQUE	EUTEKTIKUM, EUTEKTISCH
4856	EUTECTIC MELTING	FUSION EUTECTIQUE	SCHMELZUNG (EUTEKTISCHE)
4857	EUTECTIC MIXTURE	MELANGE EUTECTIQUE	MISCHUNG EUTEKTISCHE
4858	EUTECTIC TEMPERATURE	TEMPERATURE EUTECTIQUE	TEMPERATUR (EUTEKTISCHE)
4859	EUTECTOID	EUTECTOIDE	EUTEKTOID
4860	EUTECTOID REACTION	REACTION EUTECTOIDE	EUTEKTOIDE REAKTION
4861	EVAPORABLE, VAPORISABLE	EVAPORABLE	VERDAMPFBAR
4862	EVAPORATE (TO), VAPORISE (TO)	EVAPORER, VAPORISER, EVAPORER (S')	VERDAMPFEN
4863	EVAPORATE IN THE OPEN AIR (TO)	EVAPORER (S') A L'AIR LIBRE	VERDUNSTEN
4864	EVAPORATING DISH	CAPSULE	ABDAMPFSCHALE
4865	EVAPORATING TEMPERATURE	TEMPERATURE DE VAPORISATION	VERDAMPFUNGSTEMPERATUR
4866	EVAPORATION LOSS	PERTE PAR EVAPORATION	VERDAMPFUNGSVERLUST
4867	EVAPORATION, VAPORISATION	EVAPORATION, VAPORISATION	VERDAMPFEN

4868	**EVAPORATIVE CONDENSER**	CONDENSEUR A EVAPORATION	VERDUNSTUNGSKONDENSATOR, VERDAMPFUNGSKONDENSATOR
4869	**EVAPORATOR**	APPAREIL EVAPORATEUR	ABDAMPFVORRICHTUNG, VERDAMPFER
4870	**EVAPORATOR BODY**	CAISSE OU CORPS D'UN EVAPORATEUR	VERDAMPFERKÖRPER
4871	**EVEN FRACTURE**	CASSURE UNIE, RABOTEUSE	BRUCH (EBENER)
4872	**EVEN NUMBER**	NOMBRE PAIR	ZAHL (GERADE)
4873	**EVOLUTE**	DEVELOPPEE, EVOLUTE	EVOLUTE
4874	**EVOLUTION OF HEAT**	DEGAGEMENT DE CHALEUR	WÄRMEENTWICKLUNG
4875	**EXACT MEASUREMENT**	MESURAGE EXACT, PRECIS	FEINMESSUNG
4876	**EXAMINATION**	EXAMEN, CONTROLE	ÜBERPRÜFUNG
4877	**EXAMINATION BY SECTIONING**	INSPECTION PAR TREPANAGE	ROHRKONTROLLE DURCH TRENNUNG
4878	**EXCESS**	EXCES	ÜBERSCHUSS
4879	**EXCESS FLOW VALVE**	LIMITEUR OU REGULATEUR DE DEBIT	STROMBEGRENZUNGSVENTIL
4880	**EXCESS OF WORK**	EXCES DE TRAVAIL	MEHRARBEIT
4881	**EXCESS VOLTAGE**	SURVOLTAGE	ÜBERSPANNUNG (ELEKTR.)
4882	**EXCESS-ACETYLENE FLAME**	FLAMME REDUCTRICE	REDUKTIONSFLAMME
4883	**EXCESS-OXYGEN FLAME**	FLAMME OXYDANTE	OXYDATIONSFLAMME
4884	**EXCHANGE OF HEAT**	ECHANGE DE CHALEUR	WÄRMEAUSTAUSCH
4885	**EXCITATION**	EXCITATION	ERREGUNG (PHYS.)
4886	**EXCITATION ANODE**	ANODE D'EXCITATION	ERREGERANODE
4887	**EXCITE (TO)**	EXCITER	ERREGEN
4888	**EXCITER**	EXCITATRICE	ERREGERMASCHINE
4889	**EXFOLIATION**	AFFOUILLEMENT, ECAILLEMENT	ABBLÄTTERUNG, HÄUTUNG
4890	**EXHAUST**	ECHAPPEMENT (DES GAZ D'UN MOTEUR)	AUSPUFF, AUSLASS
4891	**EXHAUST GAS**	GAZ D'ECHAPPEMENT	AUSPUFFGAS
4892	**EXHAUST LINE**	LIGNE D'ECHAPPEMENT DE LA VAPEUR	DAMPFAUSSTRÖMLINIE, AUSSTRÖMLINIE
4893	**EXHAUST MANIFOLD**	COLLECTEUR D'ECHAPPEMENT	AUSPUFFSAMMELROHR
4894	**EXHAUST PIPE**	TUYAU D'ECHAPPEMENT, TUBULURE D ECHAPPEMENT	AUSPUFFROHR
4895	**EXHAUST PORT, EDUCTION PORT**	LUMIERE D'ECHAPPEMENT	AUSLASSSCHLITZ
4896	**EXHAUST STEAM**	VAPEUR D'ECHAPPEMENT	ABDAMPF
4897	**EXHAUST STEAM PIPE**	TUYAU D'ECHAPPEMENT DE VAPEUR	ABDAMPFROHR, DAMPFAUSLASSROHR, DAMPFABLEITUNGSROHR
4898	**EXHAUST STROKE**	COURSE D'ECHAPPEMENT	AUSPUFFHUB
4899	**EXHAUST VALVE**	SOUPAPE D'ECHAPPEMENT, SOUPAPE D EMISSION	AUSLASSVENTIL
4900	**EXHAUSTER**	EXTRACTEUR, EXHAUSTEUR	ABSAUGER, EXHAUSTOR
4901	**EXOTHERMIC**	EXOTHERMIQUE	WÄRMEGEBEND, EXOTHERMISCH, EXOTHERM
4902	**EXPAND (TO)**	DETENDRE (SE)	AUSDEHNEN (SICH), EXPANDIEREN
4903	**EXPAND A TUBE (TO)**	ELARGIR UN TUBE AU MANDRIN, MANDRINER UN TUBE	AUFWEITEN (EIN ROHR), AUFTREIBEN (EIN ROHR), AUFDORNEN (EIN ROHR)
4904	**EXPANDED METAL**	METAL DEPLOYE	STRECKMETALL
4905	**EXPANDED TUBE**	TUBE DUDGEONNE	ROHR (EINGEWALZTES)
4906	**EXPANDING CHUCKS**	MANDRINS EXPANSIBLES	SPREIZBARE DORNE
4907	**EXPANDING MANDREL**	MANDRIN DE MONTAGE EXPANSIBLE	SPREIZDORN

4908	**EXPANDING OF TUBES**	DUDGEONNAGE, SERTISSAGE DES TUYAUX AU DUDGEON	EINWALZEN VON ROHREN
4909	**EXPANDING PULLEY**	POULIE EXTENSIBLE, POULIE A EXPANSION	EXPANSIONSRIEMENSCHEIBE
4910	**EXPANDING ROLLER**	GALET TENDEUR	SPANNROLLE
4911	**EXPANDING TEST**	ESSAI DE MANDRINAGE	AUFTREIBEPROBE, AUFWEITEPROBE
4912	**EXPANSION**	AUGMENTATION DE VOLUME	VOLUMEN-AUFNAHME
4913	**EXPANSION BEND**	TUBE COMPENSATEUR, COMPENSATEUR DE DILATATION, COUDE, LYRE DE COMPENSATION, CINTRE DE DILATATION, LYRE DE DILATATION	AUSGLEICHROHR, DEHNUNGSROHR, FEDERROHR, KOMPENSATIONSROHR, LYRA, ROHRBOGENAUSGLEICHER
4914	**EXPANSION COEFFICIENT**	COEFFICIENT DE DILATION	AUSDEHNUNGSKOEFFIZIENT
4915	**EXPANSION COUPLING**	MANCHON DE DILATATION	AUSDEHNUNGSKUPPLUNG, KUPPLUNG (LÄNGSBEWEGLICHE)
4916	**EXPANSION CURVE, CURVE OF EXPANSION**	COURBE DE DETENTE	EXPANSIONSKURVE
4917	**EXPANSION HAND REAMER**	ALESOIR A MAIN EXPANSIBLE	HAND (NACHSTELLBARE)
4918	**EXPANSION JOINT**	JOINT DE DILATATION	AUSDEHNUNGSFUGE, AUSGLEICHFUGE, TEMPERATURFUGE, DEHNUNGSDICHTUNG, DEHNUNGSFUGE
4919	**EXPANSION OF A METAL**	DILATATION D'UN METAL	METALLAUSDEHNUNG
4920	**EXPANSION OF GAS**	DETENTE DES GAZ	GASEXPANSION
4921	**EXPANSION OF STEAM, OF A GAS**	DETENTE DE LA VAPEUR, DETENTE D UN GAZ	EXPANSION, AUSDEHNUNG VON DAMPF ODER GAS
4922	**EXPANSION PIECE**	PIECE DE COMPENSATION	AUSDEHNUNGSSTÜCK
4923	**EXPANSION REAMER**	ALESOIR EXPANSIBLE	REIBAHLE (NACHSTELLBARE)
4924	**EXPANSION STEAM ENGINE, ENGINE**	MACHINE A DETENTE, MACHINE A EXPANSION	EXPANSIONSDAMPFMASCHINE
4925	**EXPANSION TURBINE**	TURBINE A DETENTE	EXPANSIONSTURBINE
4926	**EXPANSION VALVE**	SOUPAPE DE DETENTE	EXPANSIONSVENTIL
4927	**EXPECTED VALUE**	ESPERANCE MATHEMATIQUE	ERWARTUNGSWERT
4928	**EXPERIMENT (TO)**	FAIRE DES ESSAIS	VERSUCHE MACHEN, EXPERIMENTIEREN
4929	**EXPERIMENTAL DETERMINATION**	DETERMINATION EXPERIMENTALE	BESTIMMUNG (EXPERIMENTELLE)
4930	**EXPERT**	EXPERT	SACHVERSTÄNDIGER, GUTACHTER
4931	**EXPERT'S REPORT**	EXPERTISE	GUTACHTEN
4932	**EXPIRATION OF A PATENT**	EXPIRATION D'UN BREVET	ABLAUF EINES PATENTES, PATENTABLAUF
4933	**EXPLODE (TO)**	EXPLOSER	BERSTEN, ZERKNALLEN, EXPLODIEREN
4934	**EXPLODED VIEW**	VUE ECLATEE	ANSICHT (EXPLODIERTE)
4935	**EXPLOSION**	EXPLOSION	EXPLOSION, BERSTEN, ZERKNALL
4936	**EXPLOSION PRESSURE**	PRESSION EXPLOSIVE	VERPUFFUNGSDRUCK, VERBRENNUNGSDRUCK, EXPLOSIONSDRUCK
4937	**EXPLOSIVE**	EXPLOSIF	EXPLOSIV, SPRENGSTOFF, EXPLOSIVSTOFF
4938	**EXPLOSIVE COMBUSTIBLE MIXTURE**	MELANGE EXPLOSIF, MELANGE TONNANT, MELANGE DETONANT	GEMISCH (BRENNBARES), GEMISCH (ZÜNDFÄHIGES), GEMISCH (ENTZÜNDLICHES)
4939	**EXPONENT**	EXPOSANT	EXPONENT
4940	**EXPONENT OF DISCHARGE**	EXPOSANT D'ECOULEMENT	AUSFLUSSEXPONENT
4941	**EXPONENTIAL DISTRIBUTION**	LOI EXPONENTIELLE	EXPONENTIALVERTEILUNG
4942	**EXPONENTIAL FUNCTION**	FONCTION EXPONENTIELLE	EXPONENTIALFUNKTION
4943	**EXPONENTIAL SERIES**	SERIE EXPONENTIELLE	EXPONENTIALREIHE
4944	**EXPOSURE**	EXPOSITION, TEMPS DE POSE	BELICHTUNG, AUFNAHME

4945	**EXTENDING HANDLE**	RALLONGE DE MANCHE	TELESKOPSTEIL
4946	**EXTENSION**	EXTENSION •	DEHNUNG
4947	**EXTENSION BAR**	RALLONGE	VERLÄNGERUNGSTAB
4948	**EXTENSION BY HEAT, EXPANSION DUE TO HEAT**	DILATATION SOUS L'INFLUENCE DE LA CHALEUR	WÄRMEAUSDEHNUNG, AUSDEHNUNG DURCH WÄRME
4949	**EXTENSOMETER**	EXTENSOMETRE	DEHNUNGSMESSER
4950	**EXTERNAL ANGLE**	ANGLE EXTERNE	AUSSENWINKEL
4951	**EXTERNAL BURR REMOVER FOR TUBES**	FRAISE EBARBEUSE FEMELLE, FRAISE D'EXTERIEUR POUR TUBES	ROHRFRÄSER ZUM AUSSENFRÄSEN
4952	**EXTERNAL CHASER**	PEIGNE POUR L'EXTERIEUR, PEIGNE A FILETER	AUSSENSTRÄHLER
4953	**EXTERNAL CYLINDRICAL GRINDING MACHINE**	RECTIFIEUSE POUR SURFACES DE REVOLUTION EXTERIEURES	AUSSEN-RUNDSCHLEIF-MASCHINE
4954	**EXTERNAL FORCE**	FORCE EXTERIEURE, EFFORT EXTERIEUR, CHARGE	ÄUSSERE KRAFT
4955	**EXTERNAL GAUGE**	JAUGE D'EXTERIEUR	AUSSENLEHRE
4956	**EXTERNAL GEARING**	ENGRENAGE EXTERIEUR	GETRIEBE MIT AUSSENVERZAHNUNG, AUSSENGETRIEBE
4957	**EXTERNAL GRINDING**	RECTIFICATION EXTERIEURE	AUSSENSCHLIFF
4958	**EXTERNAL OUTSIDE DIAMETER**	DIAMETRE EXTERIEUR	DURCHMESSER (ÄUSSERER)
4959	**EXTERNAL TEETH**	DENTURE EXTERIEURE	AUSSENVERZAHNUNG
4960	**EXTERNAL THREAD**	FILET EXTERIEUR	AUSSENGEWINDE
4961	**EXTERNAL THREADING**	FILETAGE EXTERIEUR	AUSSENGEWINDE
4962	**EXTERNAL WORK**	TRAVAIL EXTERIEUR	ÄUSSERE ARBEIT
4963	**EXTERNALLY RIBBED TUBE**	TUBE A AILETTES EXTERIEURES	RIPPENROHR MIT AUSSENRIPPEN
4964	**EXTRACT**	EXTRAIT	AUSZUG, EXTRAKT
4965	**EXTRACTION LIQUOR**	LIQUIDE D'EXTRACTION	AUSLAUGEFLÜSSIGKEIT
4966	**EXTRACTION OF METALS**	EXTRACTION DES METAUX	GEWINNUNG VON METALLEN
4967	**EXTRACTOR**	EXTRACTEUR (CHIM.)	AUSLAUGER, EXTRAKTIONSAPPARAT
4968	**EXTREME VALUE DISTRIBUTION**	LOI DES VALEURS EXTREMES	GRENZWERTVERTEILUNG
4969	**EXTRUDED MOULDINGS**	MOULURES FILEES	ZIERLEISTEN (STRANGGEPRESSTE)
4970	**EXTRUSION**	FILAGE, EXTRUSION, FILAGE A FROID	ZIEHEN, FLIESSPRESSEN, KALTSPRITZEN, STRANGPRESSEN
4971	**EXTRUSION BILLET**	EBAUCHE POUR PRESSE A FILER	STRANGPRESSENROHLING
4972	**EXTRUSION PRESS**	PRESSE A FILER	FLIESSPRESSE
4973	**EYE BOLT**	PITON	RINGBOLZEN, ÖSENSCHRAUBE
4974	**EYE HOOK**	CROC, CROCHET A OEIL	ÖSENHAKEN
4975	**EYE OF A HAMMER**	OEIL D'UN MARTEAU	AUGE EINES HAMMERS, ÖHR EINES HAMMERS, ÖSE EINES HAMMERS
4976	**EYE OF CONNECTING ROD**	OEIL DE LA BIELLE	SCHUBSTANGENAUGE
4977	**EYE-PIECE, OCULAR**	SYSTEME OCULAIRE, OCULAIRE	OKULAR
4978	**EYE, BORE OF A WHEEL, OF A PULLEY**	ALESAGE D'UNE ROUE, ALESAGE D'UNE POULIE	BOHRUNG EINES RADES, BOHRUNG EINER RIEMENSCHEIBE
4979	**EYED RIVET SNAP, EYED RIVETING SET**	BOUTEROLLE A MANCHE OU A OEIL	SCHELLHAMMER
4980	**EYELET**	OEIL, OEILLET	ÖSE, AUGE
4981	**FABRIC BELT**	COURROIE EN FIBRES TEXTILES	FASERSTOFFRIEMEN, TEXTILRIEMEN
4982	**FABRICATED VESSEL**	APPAREIL CHAUDRONNE	BEHÄLTER (TIEFGEZOGENER)
4983	**FABRICATION**	FACONNAGE, USINAGE	ERZEUGUNG, HERSTELLUNG
4984	**FABRICATION DRAWING**	PLAN D'EXECUTION	AUSFÜHRUNGSZEICHNUNG
4985	**FABRICATION JIG**	MONTAGE D'USINAGE BATI MANNEQUIN	WERKSTÜCKAUFNAHME
4986	**FABRICATION NUMBER**	NUMERO DE CONSTRUCTION	FABRIKNUMMER

4987	**FABRICATION REQUIREMENTS**	EXIGENCES D'USINAGE	VERARBEITUNGSANFORDERUNGEN
4988	**FABRICATION TOLERANCES**	TOLERANCES D'USINAGE	VERARBEITUNGSTOLERANZEN
4989	**FACE**	FACE D'ATTAQUE	STIRNFLÄCHE
4990	**FACE ANGLE**	ANGLE DE TETE	KOPFKEGELWINKEL
4991	**FACE CENTRED CUBIC**	CUBIQUE A FACE CENTREE	KUBISCH-FLÄCHEN-ZENTRIERT
4992	**FACE MILLING COTTERS**	FRAISES A DEUX TAILLES EN BOUT	STIRNFRÄSER
4993	**FACE OF A HAMMER**	TETE D'UN MARTEAU, PLANCHE DU MARTEAU	HAMMERBAHN, BAHN EINES HAMMERS
4994	**FACE OF A NUT**	PAN D'UN ECROU	SEITENFLÄCHE EINER MUTTER
4995	**FACE OF ANVIL, ANVIL FACE**	TABLE D'ENCLUME	AMBOSSBAHN
4996	**FACE OF SLIDE VALVE**	GLACE DU TIROIR	SCHIEBERSPIEGEL
4997	**FACE OF TOOTH**	FACE DE LA DENT	ZAHNFLANKE (ÜBER DEM TEILKREIS)
4998	**FACE OF WELD**	SUPERFICIE DE LA SOUDURE	SCHWEISSNAHTOBERFLÄCHE
4999	**FACE PLATE**	PLAQUE CIRCULAIRE, PLAQUE FRONTALE	PLANSCHEIBE
5000	**FACE SHIELD**	ECRAN DE SOUDAGE	GESICHTSSCHUTZMASKE
5001	**FACE SURFACING LATHE**	TOUR EN L'AIR	PLANDREHBANK, KOPFDREHBANK, SCHEIBENDREHBANK
5002	**FACE-CENTERED.**	A FACES CENTREES	FLÄCHENZENTRIERT
5003	**FACED FLANGE**	BRIDE TOURNEE	FLANSCH (BEARBEITETER)
5004	**FACING BRICK**	PIERRE DE PAREMENT, BRIQUE DE PAREMENT	VERBLENDER, VERBLENDSTEIN
5005	**FACING LATHE**	TOUR EN L'AIR	PLANDREHBANK, KOPFDREHBANK
5006	**FACING SAND**	SABLE DE CONTACT	MODELLSAND
5007	**FACING STRIP**	PORTEE D'ASSEMBLAGE DRESSEE (D'UNE PIECE DE FONDERIE)	ARBEITSLEISTE
5008	**FACTOR**	FACTEUR	FAKTOR (MATH.)
5009	**FACTOR OF SAFETY, MARGIN OF SAFETY**	COEFFICIENT DE SECURITE, FACTEUR DE SECURITE	SICHERHEITSGRAD, SICHERHEITSZUGABE, SICHERHEITSKOEFFIZIENT, SICHERHEITSFAKTOR
5010	**FACTORY**	USINE	WERK
5011	**FACTORY ACCEPTANCE GAUGE**	JAUGE DE REVISION	REVISIONSLEHRE, WERKSTATTABNAHMELEHRE
5012	**FACTORY BUILDING**	BATIMENT D'USINE	FABRIKGEBÄUDE
5013	**FACTORY, MANUFACTORY, WORKS, MILL**	FABRIQUE, MANUFACTURE, USINE, ETABLISSEMENT DE FABRICATION	FABRIK
5014	**FADING**	DECOLORATION	VERBLASSEN
5015	**FAGOT**	PAQUET DE FER A SOUDER	SCHWEISSPAKET
5016	**FAGOT SCRAP**	MITRAILLES PAQUETEES	PAKETIERSCHROTT
5017	**FAGOT, PILE**	PAQUET, TROUSSE (METALLURGIE)	SCHWEISSPAKET, PAKET
5018	**FAGOTING**	PAQUETAGE	PAKETIEREN
5019	**FAHRENHEIT SCALE**	ECHELLE FAHRENHEIT	FAHRENHEITSKALA
5020	**FAHRENHEIT THERMOMETER**	THERMOMETRE FAHRENHEIT	FAHRENHEITTHERMOMETER
5021	**FAILURE OF AN ENGINE**	RATE D'UN MOTEUR	VERSAGEN EINES MOTORS
5022	**FALL DIMINUTION OF TEMPERATURE**	ABAISSEMENT DE LA TEMPERATURE, BAISSE DE TEMPERATURE	TEMPERATURABNAHME, TEMPERATURERNIEDRIGUNG, TEMPERATURRÜCKGANG
5023	**FALL, LOOSE END (IN LIFTING TACKLE)**	EXTREMITE LIBRE, BRIN GARANT DU CABLE, GARANT (D'UN PALAN)	TRUMM (FREIES), ZUGTRUMM (EINES FLASCHENZUGS)
5024	**FALLING FREELY**	A CHUTE LIBRE	FREIFALLEND

5025	**FALLING TEMPERATURE**	TEMPERATURE ABAISSANTE	TEMPERATUR (ABNEHMENDE), TEMPERATUR (SINKENDE), TEMPERATUR (FALLENDE)
5026	**FALSE BOTTOM**	DOUBLE FOND, FAUX FOND	BODEN (FALSCHER), DOPPELBODEN, BLINDBODEN, ZWISCHENBODEN
5027	**FALSE CORE**	NOYAU EXTERIEUR, FAUX NOYAU	KERNSTÜCK
5028	**FAMILY OF CURVES**	FAISCEAU DE COURBES	SCHAR VON KURVEN, KURVENSCHAR
5029	**FAN BELT**	COURROIE DE VENTILATEUR	VENTILATORRIEMEN
5030	**FAN BLADES**	PALES DE VENTILATEUR	VENTILATORFLÜGELBLATT
5031	**FAN BRAKE**	FREIN, MOULINET A PALETTES, MOULINET DYNAMOMETRIQUE	FLÜGELBREMSE
5032	**FAN PULLEY**	POULIE DE VENTILATEUR	VENTILATORRIEMENSCHEIBE
5033	**FAN, VENTILATOR**	VENTILATEUR	LÜFTER, VENTILATOR, WINDFLÜGEL
5034	**FANTAIL**	CARNEAU DE RACCORDEMENT	VERBINDUNGSKANAL
5035	**FARAD**	FARAD	FARAD
5036	**FARADAY CAGE**	CAGE DE FARADAY	FARADAYSCHER KÄFIG
5037	**FAST FIXED RIGID PERMANENT COUPLING**	ACCOUPLEMENT FIXE, ACCOUPLEMENT RIGIDE	KUPPLUNG (FESTE)
5038	**FAST IDLE**	RALENTI ACCELERE	LEERLAUF (SCHNELLER), LEERLAUF (ERHÖHTER)
5039	**FAST PIPE FLANGE**	BRIDE FIXE	FLANSCH (FESTER)
5040	**FAST PULLEY**	POULIE FIXE	RIEMENSCHEIBE (FESTE), FESTSCHEIBE, LASTSCHEIBE, VOLLSCHEIBE
5041	**FASTEN (TO), FIX (TO)**	FIXER	BEFESTIGEN
5042	**FASTENER**	ORGANE DE FIXATION	BEFESTIGUNGSELEMENT
5043	**FASTENING SCREW**	VIS OU BOULON DE FIXATION, VIS OU BOULON DE MONTAGE	BEFESTIGUNGSSCHRAUBE
5044	**FASTENING, FIXING**	FIXAGE, FIXATION	BEFESTIGUNG
5045	**FAT**	GRAISSE	FETT (CHEM.)
5046	**FAT RICH LIME**	CHAUX GRASSE	KALK (FETTER), FETTKALK
5047	**FATIGUE**	FATIGUE	ERMÜDUNG
5048	**FATIGUE CRACK**	FISSURE PAR FATIGUE	ERMÜDUNGSRISS
5049	**FATIGUE EXPERIMENT, TEST FOR FATIGUE (OF A MATERIAL) ; RELIABILITY TEST (OF A MACHINE OR ENGINE)**	ESSAI AUX CHOCS REPETES, ESSAI D'ENDURANCE	DAUERVERSUCH
5050	**FATIGUE LIMIT**	LIMITE D'ENDURANCE OU DE FATIGUE	DAUERGRENZE
5051	**FATIGUE OF MATERIALS**	FATIGUE D'UN MATERIEL	ERMÜDUNG DES MATERIALS
5052	**FATIGUE RESISTANCE**	RESISTANCE A LA FATIGUE	DAUERFESTIGKEIT
5053	**FATIGUE STRENGTH**	LIMITE DE FATIGUE, RESISTANCE-LIMITE D'ENDURANCE	DAUERSCHWINGFESTIGKEIT
5054	**FATIGUE TESTING**	ESSAI DE RESISTANCE A LA FATIGUE	ERMÜDUNGSVERSUCH
5055	**FATTY ACID**	ACIDE GRAS	FETTSÄURE
5056	**FAULT, FLAW IN A CASTING**	DEFAUT DE COULAGE, PAILLE	GUSSFEHLER
5057	**FAULTY, DEFECTIVE, DETERIORATED, DAMAGED**	DEFECTUEUX, ENDOMMAGE, DETERIORE	SCHADHAFT
5058	**FAYING SURFACE**	SURFACE D'AFFLEUREMENT	ANLAGEFLÄCHE, PASSFLÄCHE
5059	**FEATHER KEY**	LANGUETTE	FEDER, LÄNGSFEDER, FEDERKEIL, LÄNGSNASE
5060	**FEATHER WING RIB OF VALVE**	AILETTE DE LA SOUPAPE	FÜHRUNGSRIPPE, FÜHRUNGSLEISTE, FLÜGEL EINES VENTILS
5061	**FEATHER.**	STRIE	SCHLIERE

5062	FEATHERED BOLT, LIP BOLT, SNUG HEAD BOLT	BOULON A ERGOT	NASENSCHRAUBE
5063	FEED	ALIMENTATION, AVANCE	ZUSCHIEBUNG, VORSCHUB, SPEISUNG, ZUFÜHRUNG
5064	FEED (TO)	ALIMENTER	SPEISEN
5065	FEED CHANGE LEVER	LEVIER SELECTEUR DES AVANCES	VORSCHUBWÄHLHEBEL
5066	FEED COCK	ROBINET D'ALIMENTATION	FÜLLHAHN, SPEISEHAHN,
5067	FEED GEAR MECHANISM	MECANISME DES AVANCEMENTS	VORSCHUBMECHANISMUS
5068	FEED OF A TOOL	AVANCE D'UN OUTIL	VORSCHUB EINES WERKZEUGS
5069	FEED PER TOOTH	AVANCE PAR DENT	VORSCHUB JE ZAHN
5070	FEED PIPE	TUYAU D'ALIMENTATION	SPEISEROHR, SPEISELEITUNG
5071	FEED PUMP	POMPE D'ALIMENTATION	SPEISEPUMPE
5072	FEED REGULATOR	ALIMENTATEUR AUTOMATIQUE	SPEISEREGLER
5073	FEED REVERSE LEVER	LEVIER D'INVERSION DE L'AVANCE	VORSCHUBUMSTEUERHEBEL
5074	FEED ROLLER	CYLINDRE D'ALIMENTATION, CYLINDRE ENTRAINEUR	SPEISEWALZE, LIEFERWALZE, ZUFÜHRWALZE
5075	FEED TANK	BAC D'ALIMENTATION	SPEISEBEHÄLTER
5076	FEED VALVE	SOUPAPE D'ALIMENTATION	SPEISEVENTIL
5077	FEED WATER	EAU D'ALIMENTATION	SPEISEWASSER
5078	FEED WATER HEATER	RECHAUFFEUR D'EAU D'ALIMENTATION	SPEISEWASSER-VORWÄRMER
5079	FEED-RATE OVERRIDE	CORRECTION DE VITESSE D'AVANCE	VORSCHUBGESCHWINDIGKEITSKORREKTUR
5080	FEEDBACK LOOP	BOUCLE DE RETOUR	RÜCKFÜHRKREIS
5081	FEEDER	ARTERE, CABLE PRINCIPAL, FEEDER	SPEISELEITUNG, HAUPTLEITUNG (ELEKTR.)
5082	FEEDING	ALIMENTATION	BRENNSTOFFZUFÜHRUNG
5083	FEEDING ARRANGEMENT, FEED APPARATUS	APPAREIL D'ALIMENTATION	SPEISEVORRICHTUNG
5084	FEEDING NOZZLE	TUBULURE D'ALIMENTATION	SPEISEDÜSE
5085	FEEDRATE NUMBER (FRN)	CODE DE VITESSE D'AVANCE	VORSCHUBZAHL
5086	FEELER GAUGE	CALIBRE D'EPAISSEUR	EINSTELLLEHRE
5087	FELDSPAR	FELDSPATH	FELDSPAT
5088	FELLOE	JANTE D'UNE ROUE	FELGE, RADFELGE
5089	FELSPAR	FELDSPATH	FELDSPAT
5090	FELT	FEUTRE	FILZ
5091	FEMALE GAUGE	JAUGE D'EXTERIEUR	AUSSENLEHRE
5092	FEMALE INTERNAL SCREW THREAD	FILET, FILETAGE INTERIEUR, FILETAGE FEMELLE	INNENGEWINDE, MUTTERGEWINDE
5093	FENCE OF A PLANE	JOUE D'UN RABOT	ANSCHLAG EINES HOBELS
5094	FENCING	CLOTURE	ZAUN
5095	FENDER	AILE	KOTFLÜGEL
5096	FENDER WASHER	RONDELLE D'AILE	KOTFLÜGELSCHEIBE
5097	FERMENT	FERMENT	GÄRSTOFF, GÄRUNGSSTOFF
5098	FERMENTATION	FERMENTATION	GÄRUNG, FERMENTATION
5099	FERRIC ACETATE	ACETATE FERRIQUE	EISENOXYD (ESSIGSAURES), FERRIAZETAT
5100	FERRIC CHLORIDE	CHLORURE FERRIQUE	EISENCHLORID, FERRICHLORID
5101	FERRIC HYDROXIDE	PEROXYDE HYDRATE DE FER	EISENHYDROXYD, FERRIHYDROXYD, FERRIHYDRAT, EISENOXYDHYDRAT
5102	FERRIC OXIDE	OXYDE DE FER	EISENOXYD
5103	FERRIC SULFATE	SULFATE FERRIQUE	FERRISULFAT
5104	FERRITE	FERRITE	FERRIT

5105	**FERRITE GHOST**	, BANDE DE FERRITE LIBRE	FERRITBAND (FREIES)
5106	**FERRO-ALUMINIUM**	FERRO-ALUMINIUM	ALUMINIUMEISEN, FERROALUMINIUM
5107	**FERRO-BORON**	FERRO-BORE	BOREISEN, FERROBOR
5108	**FERRO-CHROME**	FERRO-CHROME	CHROMEISEN, FERROCHROM
5109	**FERRO-CONCRETE PIPE**	TUYAU EN CIMENT ARME	ZEMENTROHR MIT EISENEINLAGE
5110	**FERRO-MANGANESE**	FERRO-MANGANESE	MANGANEISEN, EISENMANGAN, FERROMANGAN
5111	**FERRO-MOLYBDENUM**	FERRO-MOLYBDENE	MOLYBDÄNEISEN, FERROMOLYBDÄN
5112	**FERRO-NICKEL**	FERRO-NICKEL	NICKELEISEN, FERRONICKEL
5113	**FERRO-SILICON, SILICEOUS IRON**	FERRO-SILICIUM	SILIZIUMEISEN, FERROSILIZIUM
5114	**FERRO-TITANIUM**	FERRO-TITANE	TITANEISEN, FERROTITAN
5115	**FERRO-TUNGSTEN**	FERRO-TUNGSTENE	WOLFRAMEISEN, FERROWOLFRAM
5116	**FERRO-VANADIUM**	FERRO-VANADIUM	VANADIUMEISEN, FERROVANADIUM
5117	**FERROALLOYS**	FERROALLIAGES	EISENLEGIERUNGEN
5118	**FERROMAGNETIC**	FERROMAGNETIQUE	FERROMAGNETISCH
5119	**FERROPRUSSIATE PAPER**	PAPIER AU FERRO-PRUSSIATE, PAPIER PRUSSIATE, PAPIER CYANOFER, CYANOTYPE, PAPIER BLEU POUR PHOTOCALQUE	BLAUDRUCKPAPIER, EISENBLAUPAPIER, EISENZYANPAPIER
5120	**FERROUS ACETATE**	ACETATE FERREUX	ESSIGSAURES EISENOXYDUL, FERROAZETAT
5121	**FERROUS CHLORIDE**	CHLORURE FERREUX	EISENCHLORÜR, FERROCHLORID, EINFACHCHLOREISEN
5122	**FERROUS MANGANESE ORES**	MINERAIS DE MANGANESE FERREUX	EISENHALTIGES MANGANERZ
5123	**FERROUS METALLURGY**	METALLURGIE DU FER	METALLURGIE DES EISENS
5124	**FERROUS OXALATE**	OXALATE FERREUX	EISENOXALAT, FERROOXALAT, EISEN (OXALSAURES), OXYDUL
5125	**FERROUS OXYDE**	OXYDE FERREUX, PROTOXYDE DE FER	EISENOXYDUL, FERROOXYD
5126	**FERROUS SCRAP**	FERRAILLE	EISENSCHROTT
5127	**FERRUGINOUS**	FERRUGINEUX	EISENHALTIG
5128	**FERRULE**	VIROLE	ZWINGE, SPERRING
5129	**FETTLING**	EBARBAGE	ENTGRATEN, PUTZEN
5130	**FETTLING, CLEANING SHOP**	ATELIER DE DESABLAGE, ATELIER D'EBARBAGE, ATELIER D'ECROUTAGE	GUSSPUTZEREI, PUTZEREI
5131	**FIBER**	FIBRE	FASER
5132	**FIBER DIAGRAM (X-R.)**	AXE DE FIBRE	FASERACHSE
5133	**FIBER DIRECTION**	DIRECTION DES FIBRES	FASERRICHTUNG
5134	**FIBER STRESS**	EFFORT DANS LA FIBRE	FASERSPANNUNG
5135	**FIBERING**	STRUCTURE FIBREUSE	FASERSTRUKTUR
5136	**FIBRE**	FIBRE	FASER
5137	**FIBROUS**	FIBREUX	FASERIG
5138	**FIBROUS FRACTURE**	CASSURE FIBREUSE	BRUCH (FASERIGER), BRUCH (SEHNIGER)
5139	**FIBROUS IRON**	FER NERVEUX, FER A NERF	EISEN (SEHNIGES)
5140	**FIBROUS MATERIAL**	MATIERE FIBREUSE	FASERSTOFF
5141	**FIBROUS STRUCTURE, TEXTURE**	TEXTURE FIBREUSE	GEFÜGE (FASERIGES)
5142	**FIELD GUN CARRIAGE**	AFFUT DE CAMPAGNE	FELDLAFETTE
5143	**FIELD OF VIEW**	CHAMP VISUEL, CHAMP DE VISION	GESICHTSFELD, SEHFELD
5144	**FIELD STRENGTH INTENSITY**	INTENSITE DU CHAMP	FELDSTÄRKE, FELDDICHTE, FELDINTENSITÄT
5145	**FIELD SUPERINTENDENT**	CHEF DE CHANTIER	BAULEITER

5146	FIELD WELD	SOUDURE EXECUTEE SUR CHANTIER	BAUSTELLENSCHWEISSNAHT
5147	FIFTH ROOT	RACINE CINQUIEME	WURZEL (FÜNFTE)
5148	FILE	LIME	FEILE
5149	FILE (TO)	LIMER	FEILEN
5150	FILE CLEANING CARD	CARDE, BROSSE A LIMES	FEILENBÜRSTE
5151	FILE CUTTER	TAILLEUR DE LIMES	FEILENHAUER
5152	FILE HARDNESS TESTING	ESSAI DE DURETE A LA LIME	FEILENHÄRTEPROBE
5153	FILING	LIMAGE	FEILEN
5154	FILING MACHINE	MACHINE A LIMER	FEILMASCHINE
5155	FILINGS	LIMAILLE	FEILSPÄNE, FEILICHT
5156	FILL (TO), CHARGE (TO)	REMPLIR, CHARGER	FÜLLEN
5157	FILLER	CHARGE CHIMIQUE	FÜLLSTOFF
5158	FILLER CAP	BOUCHON DE REMPLISSAGE	EINFÜLLVERSCHLUSS
5159	FILLER METAL	METAL, ALLIAGE D'APPORT	FÜLLMETALL, ZUSATZWERKSTOFF
5160	FILLER PLATE SLIDE	COULISSEAU D'OBTURATION	ABSPERRUNGSSTÖSSEL
5161	FILLET	CONGE	HOHLKEHLE
5162	FILLET WELD	SOUDAGE EN ANGLE, SOUDAGE EN CONGE	KEHLNAHTSCHWEISSEN
5163	FILLING LOSS	PERTE PAR REMPLISSAGE	FÜLLVERLUST
5164	FILLING MATERIAL	MATIERE DE REMPLISSAGE	FÜLLSTOFF, AUSFÜLLMASSE
5165	FILLISTER HEAD SCREW	VIS A TETE CYLINDRIQUE	ZYLINDERSCHRAUBE
5166	FILM	COUCHE, FILM, PELLICULE	SCHICHT, FILM
5167	FILM OF OIL	PELLICULE GRASSE, MINCE COUCHE D'HUILE DE GRAISSAGE	SCHMIERSCHICHT
5168	FILM OF OXIDE	PELLICULE D'OXYDE	OXYDHAUT
5169	FILM OF WATER	PELLICULE D'EAU	WASSERHAUT
5170	FILM THICKNESS	EPAISSEUR DE FILM	FILMDICKE
5171	FILM, THIN LAYER	PELLICULE	HAUT, SCHICHT (DÜNNE)
5172	FILTER	FILTRE	FILTER
5173	FILTER (TO), STRAIN (TO)	FILTRER	FILTERN, FILTRIEREN
5174	FILTER AID	ADJUVANT DE FILTRATION	FILTERHILFSSTOFF
5175	FILTER CARTRIDGE	CARTOUCHE FILTRANTE	FILTERSIEB, FILTEREINSATZ
5176	FILTER CLOTH	ETOFFE FILTRANTE	FILTERTUCH
5177	FILTER PAPER	PAPIER A FILTRER	FILTERPAPIER, FILTRIERPAPIER
5178	FILTER TANK	BASSIN DE FILTRATION	FILTERBECKEN
5179	FILTERING MATERIAL	MATIERE FILTRANTE	FILTERSTOFF, FILTRIERMATERIAL
5180	FILTERPRESS	FILTRE-PRESSE	FILTERPRESSE
5181	FILTRATE	FILTRAT, LIQUIDE FILTRE	FILTRAT, DURCHLAUF
5182	FILTRATION, FILTERING	FILTRATION, FILTRAGE	FILTERN, FILTRIEREN, FILTRATION
5183	FIN	AILETTE, AILERON	RIPPE
5184	FIN, BURR	BAVURE (D'UNE PIECE MOULEE)	GUSSNAHT
5185	FINAL DRAWING	PLAN DEFINITIF	ZEICHNUNG (ENDGÜLTIGE)
5186	FINAL DRIVE	TRANSMISSION AUX ROUES	RADANTRIEB
5187	FINAL POSITION	POSITION FINALE	ENDLAGE, ENDSTELLUNG
5188	FINAL PRESSURE, TERMINAL PRESSURE	PRESSION FINALE	ENDDRUCK
5189	FINAL PRODUCT	PRODUIT FINAL	ENDERZEUGNIS, ENDPRODUKT
5190	FINAL STATE	ETAT FINAL	ENDZUSTAND
5191	FINAL TEMPERATURE	TEMPERATURE FINALE	ENDTEMPERATUR
5192	FINAL VALUE	VALEUR FINALE	ENDWERT
5193	FINAL VELOCITY	VITESSE FINALE	ENDGESCHWINDIGKEIT

5194	**FINAL YIELD**	RENDEMENT FINAL	ENDAUSBEUTE
5195	**FINE**	AMENDE	GELDSTRAFE
5196	**FINE BORING MACHINE**	ALESEUSE DE PRECISION	FEINBOHRMASCHINE
5197	**FINE FLUTED REAMER**	ALESOIR A FINES RAINURES	REIBAHLE (GERIFFELTE)
5198	**FINE GOLD**	OR FIN	FEINGOLD
5199	**FINE GRAINED SAND**	SABLE A GRAIN FIN	SAND (FEINKÖRNIGER)
5200	**FINE GRAINED STRUCTURE**	STRUCTURE A GRAINS FINS	GEFÜGE (FEINKÖRNIGES)
5201	**FINE PITCH THREAD**	FILET A PAS FIN	FEINGEWINDE, FEINES GEWINDE
5202	**FINE SILVER**	ARGENT FIN	FEINSILBER
5203	**FINE STRUCTURE**	STRUCTURE FINE (R.X.)	FEINSTRUKTUR
5204	**FINE THREAD**	FILET FIN	FEINGEWINDE
5205	**FINE WIRE**	FIL FIN, FIL MINCE	FEINDRAHT
5206	**FINE-MESHED**	A MAILLES SERREES	ENGMASCHIG, FEINMASCHIG
5207	**FINE-STRAND WIRE ROPE**	CABLE METALLIQUE EN FILS FINS	SEIL (FEINDRÄHTIGES)
5208	**FINENESS**	FINESSE	FEINHEIT
5209	**FINENESS OF AN ALLOY**	DEGRE DE FIN D'UN ALLIAGE	FEINGEHALT, REINGEHALT EINER LEGIERUNG
5210	**FING STRUCTURE**	STRUCTURE FINE	FEINSTRUKTUR
5211	**FINGER BAR**	CLAVETTE, BARRETTE, DOIGT	KEIL, BEFESTIGUNGSKEIL
5212	**FINGERNAILING**	DEFORMATIONS EN DEMI-LUNE DU CORDON DE SOUDURE	HALBMONDFÖRMIGE SCHWEISSNAHT-VERZERRUNGEN
5213	**FINISH**	ETAT DE SURFACE, FINISSAGE	OBERFLÄCHENBESCHAFFENHEIT
5214	**FINISH (TO)**	PARACHEVER, FINIR, RETOUCHER	NACHBEARBEITEN, SCHLICHTEN
5215	**FINISH ROLLING**	LAMINAGE FINISSEUR	FERTIGWALZEN, AUSWALZEN
5216	**FINISH WELDING**	SOUDAGE-FINITION	FERTIGSCHWEISSEN
5217	**FINISH, LAST COAT OF PAINT**	COUCHE SUPERFICIELLE (DE PEINTURE)	DECKANSTRICH
5218	**FINISHED PRODUCT**	PRODUIT FINI / PLAT / LONG	FERTIGERZEUGNIS
5219	**FINISHER**	MATRICE FINISSEUSE, CAGE FINISSEUSE	FERTIGGESENK, FERTIGGERÜST
5220	**FINISHING AND CLEANING**	PARACHEVEMENT	FERTIGBEARBEITUNG
5221	**FINISHING DIE**	DERNIERE FILIERE	ENDSTEIN
5222	**FINISHING MILL**	LAMINOIR FINISSEUR	FERTIGWALZWERK
5223	**FINISHING OPERATION**	FINISSAGE	FERTIGUNGSARBEIT
5224	**FINISHING POLISH**	FINISSAGE A LA MEULE	FERTIGSCHLEIFEN
5225	**FINISHING ROLLERS**	CAGE FINISSEURS	FERTIGWALZE
5226	**FINISHING TAPER REAMER**	ALESOIR FINISSEUR CONIQUE	KEGELSCHLICHTBOHRER
5227	**FINISHING TEMPERATURE**	TEMPERATURE DE FINISSAGE	FERTIGUNGSTEMPERATUR
5228	**FINISHING TOOL**	OUTIL A FINIR	SCHLICHTSTAHL
5229	**FINISHING TRAIN**	TRAIN FINISSEUR	FERTIGSTRASSE
5230	**FINISHING WASHER**	RONDELLE DECORATIVE	ZIERRING
5231	**FINITE DECIMAL FRACTION**	FRACTION DECIMALE TERMINEE	DEZIMALBRUCH (ENDLICHER)
5232	**FINITE SERIES**	SERIE FINIE	REIHE (ENDLICHE)
5233	**FINNED RADIATOR**	RADIATEUR A AILETTES	RIPPENROHRKÜHLER
5234	**FINNED TUBE**	TUYAU A AILETTES	RIPPENROHR
5235	**FIRE (TO)**	CHAUFFER	HEIZEN, FEUERN
5236	**FIRE (TO), START UP A BOILER (TO)**	ALLUMER LE FOYER D'UNE CHAUDIERE	ANHEIZEN, ANFEUERN (EINEN KESSEL)
5237	**FIRE BAR, GRATE BAR, FURNACE BAR**	BARREAU DE GRILLE	ROSTSTAB
5238	**FIRE BOX**	BOITE A FEU	FEUERBÜCHSE

5239	FIRE BRICK	BRIQUE REFRACTAIRE	STEIN (FEUERFESTER), SCHAMOTTESTEIN
5240	FIRE CHECK	CRIQUE A CHAUD	BRANDRISS, WARMRISS
5241	FIRE CLAY	TERRE REFRACTAIRE	TON (FEUERFESTER), SCHAMOTTE
5242	FIRE DOOR	PORTE DU FOYER	FEUERTÜR, HEIZTÜR
5243	FIRE GILDING	DORURE AU FEU	FEUERVERGOLDUNG
5244	FIRE HYDRANT	BOUCHE D'INCENDIE	UNTERFLURHYDRANT
5245	FIRE REFINING	AFFINAGE AU FEU	VEREDELUNG (THERMISCHE), VERGÜTUNG (THERMISCHE)
5246	FIRE SCALE	COUCHE D'OXYDE DE CUIVRE	KUPFEROXIDSCHICHT
5247	FIRE TEST (OIL)	POINT DE COMBUSTION, POINT DE FEU (HUILE)	BRENNPUNKT (EINES ÖLES)
5248	FIRE TUBE	TUBE DE FUMEE	RAUCHROHR
5249	FIRE TUBE BOILER	CHAUDIERE IGNITUBULAIRE, CHAUDIERE A TUBES DE FUMEE, CHAUDIERE TUBULAIRE	HEIZRÖHRENKESSEL, RAUCHRÖHRENKESSEL
5250	FIRE-REFINED COPPER	CUIVRE AFFINE AU FEU	KUPFER (THERMISCH VEREDELTES), KUPFER (THERMISCH VERGÜTETES)
5251	FIRE-WALL	CLOISON PARE-FEU, TABLIER DU MOTEUR	SPRITZWAND
5252	FIREBRICK	BRIQUE REFRACTAIRE	FEUERFESTERZIEGEL, FEUERZIEGEL
5253	FIRECLAY	ARGILE REFRACTAIRE	FEUERTON
5254	FIRECLAY BRICK	BRIQUE REFRACTAIRE ALUMINEUSE	SCHAMOTTESTEIN
5255	FIRECLAY MORTAR	CIMENT REFRACTAIRE	SCHAMOTTEMÖRTEL
5256	FIRELESS LOCOMOTIVE	LOCOMOTIVE SANS FOYER	FEUERLOSE LOKOMOTIVE
5257	FIREPROOF, REFRACTORY	REFRACTAIRE, IGNIFUGE	FEUERBESTÄNDIG, FEUERFEST, FEUERSICHER
5258	FIREWOOD, FUEL WOOD	BOIS DE CHAUFFAGE, BOIS A BRULER	BRENNHOLZ
5259	FIRING	CHAUFFAGE	HEIZEN, HEIZUNG, FEUERN
5260	FIRING ORDER	ORDRE D'ALLUMAGE	ZÜNDFOLGE
5261	FIRING TOOL	ATTIRAIL DE CHAUFFE	FEUERUNGSGERÄT
5262	FIRING UP, STOKING	CHARGEMENT, ALIMENTATION D'UN FOYER	BESCHICKUNG EINER FEUERUNG
5263	FIRING, STARTING A BOILER	ALLUMAGE DU FOYER D'UNE CHAUDIERE	ANHEIZEN EINES KESSELS, ANFEUERN EINES KESSELS
5264	FIRMER CHISEL	FERMOIR	STEMMEISEN, STEMMBEITEL
5265	FIRST FILTER	FILTRE PREPARATOIRE, PREMIER FILTRE	VORFILTER
5266	FIRST FORMED RIVET HEAD	PREMIERE TETE DE RIVET	SETZKOPF
5267	FIRST RUNNINGS	TETE DE DISTILLATION	VORLAUF (DER DESTILLATION)
5268	FIRST TAP, ENTERING TAP, TAPER TAP	TARAUD CONIQUE, TARAUD EBAUCHEUR	GEWINDEVORSCHNEIDER
5269	FISH BOLT	BOULON D'ECLISSE	LASCHENBOLZEN
5270	FISH PLATES	ECLISSES	VERBINDUNGSLASCHEN
5271	FISH TAIL	CAVITE EN V	V-LUNKER
5272	FISH-BELLIED	A VENTRE DE POISSON	FISCHBAUCHFÖRMIG
5273	FISSURE	GERCURE	BORSTE
5274	FISSURING	FENDILLEMENT	SPRUNG-BILDUNG
5275	FIT	AJUSTAGE, MONTAGE, EMMANCHEMENT	PASSUNG
5276	FIT (TO)	AJUSTER	EINPASSEN, ANPASSEN, JUSTIEREN
5277	FIT PIPES INTO EACH OTHER TO	EMBOITER, ABOUCHER DES TUYAUX, ASSEMBLER DES TUYAUX PAR EMBOITEMENT	ROHRE INEINANDERSTECKEN

5278	FITTER	AJUSTEUR, AJUSTEUR-MECANICIEN	MASCHINENSCHLOSSER, SCHLOSSER
5279	FITTER'S HAMMER	MARTEAU D'AJUSTEUR	MONTIERHAMMER
5280	FITTING	RACCORD, AJUSTAGE	ANSCHLUSS, EINPASSEN, JUSTIEREN
5281	FITTING PIPES INTO EACH OTHER	EMBOITEMENT, EMBOITAGE, EMMANCHEMENT DE TUBES	INEINANDERSTECKEN VON ROHREN
5282	FITTING SHOP	ATELIER D'AJUSTAGE	SCHLOSSERWERKSTATT, TEILMONTAGE-WERKSTATT
5283	FITTINGS	ACCESSOIRES, GARNITURES, ARMATURE	ZUBEHÖR, ARMATUREN, ARMATUR, AUSRÜSTUNG
5284	FIXED AT BOTH ENDS	ENCASTRE AUX DEUX EXTREMITES	BEIDERSEITIG EINGESPANNT
5285	FIXED AT ONE END	ENCASTRE PAR UNE EXTREMITE	EINSEITIG EINGESPANNT
5286	FIXED BEAM.	POUTRE ENCASTREE	EINGESPANNTER TRÄGER
5287	FIXED BLOCK	POULIE FIXE (D'UN PALAN)	FLASCHE (FESTE), ROLLE
5288	FIXED BLOCK FORMAT	FORMAT A BLOC FIXE	EINGABEFORMAT (FESTES)
5289	FIXED CARBON	CARBONE COMBINE	GEBUNDENER KOHLENSTOFF
5290	FIXED CIRCLE	CERCLE FIXE	KREIS (FESTER)
5291	FIXED CRANE	GRUE FIXE, STATIONNAIRE	KRAN (STANDFESTER) (ORTSFESTER)
5292	FIXED CYCLE	CYCLE FIXE	FESTER ZYKLUS
5293	FIXED GAUGE	CALIBRE FIXE	SCHABLONE (FIXE)
5294	FIXED OIL	HUILE GRASSE, HUILE FIXE	ÖL (FETTES)
5295	FIXED POINT	POINT FIXE, APPUI FIXE	FESTPUNKT, FIXPUNKT
5296	FIXED ROOF	TOIT FIXE	FESTDACH
5297	FIXED SEQUENTIAL FORMAT	FORMAT A SEQUENCE FIXE	EINGABEFORMAT IN FESTER WORTFOLGESCHREIBWEISE
5298	FIXED-BED MILLING MACHINE	FRAISEUSE A BANC FIXE	FESTBETTFRÄSMASCHINE
5299	FIXED, RIGID VICE JAW	MACHOIRE FIXE D'UN ETAU	SCHRAUBSTOCKBACKEN (FESTER)
5300	FIXING THE TEST BAR	ENCASTREMENT, FIXATION D'UNE EPROUVETTE	EINSPANNEN EINES PROBESTABES
5301	FIXTURE	MONTAGE	AUFSPANNVORRICHTUNG
5302	FLAG, FLAGSTONE	DALLE	FUSSBODENPLATTE
5303	FLAKE	FLOCON, LAMELLE	FLOCKE, LAMELLE
5304	FLAKE CRACK	TAPURE, CRIQUE DE TENSION, GERCURE	SPANNUNGSRISS
5305	FLAKE GRAPHITE	GRAPHITE LAMELLAIRE	FLOCKENGRAPHIT, SCHUPPENGRAPHIT
5306	FLAKING	ECAILLAGE	ABBLÄTTERUNG
5307	FLAME BLOWPIPE	CHALUMEAU	SCHWEISSBRENNER, SCHNEIDBRENNER
5308	FLAME BRIDGE	PONT DE CHAUFFE	FEUERBRÜCKE
5309	FLAME CHIPPING	NETTOYER AU CHALUMEAU	BRENNPUTZEN, FLAMMPUTZEN
5310	FLAME CLEANING	DECALAMINAGE AU CHALUMEAU	FLAMMENENTZUNDERUNG
5311	FLAME CONDITIONING TORCH	CHALUMEAU DEROUILLEUR	ENTROSTUNG SBRENNER
5312	FLAME CUTTER	CHALUMEAU COUPEUR	SCHNEIDBRENNER (AUTOGEN)
5313	FLAME DESCALING	DECALAMINAGE A LA FLAMME	ENTZUNDERUNG (THERMISCHE)
5314	FLAME FAILURE CONTROLLER	CONTROLEUR DE FLAMME	FLAMMENWÄCHTER
5315	FLAME HARDENING	TREMPE AU CHALUMEAU	AUTOGENE OBERFLÄCHENHÄRTUNG, BRENNHÄRTEN
5316	FLAME IGNITION	ALLUMAGE A FLAMME	FLAMMENZÜNDUNG
5317	FLAME PLATING	PLACAGE A LA FLAMME	FLAMMPLATTIERUNG
5318	FLAME SHAPING	COUPAGE A LA FLAMME	AUTOGENE FORMGEBUNG
5319	FLAME SOFTENING	RECUIT A LA FLAMME	THERMISCHE ERWEICHUNG
5320	FLAME SPECTROSCOPY	SPECTROSCOPIE DE LA FLAMME	FLAMMENSPEKTROSKOPIE
5321	FLAME-ARRESTOR	ARRETE-FLAMMES	FLAMMEN-LÖSCHER

5322	**FLAME-PROOF**	ANTIDEFLAGRANT	FLAMMENSICHER
5323	**FLANGE**	BRIDE, FLASQUE	FLANSCH, LAGERDECKEL
5324	**FLANGE (TO)**	EMBOUTIR	KÜMPELN
5325	**FLANGE ADAPTER.**	ADAPTATEUR DE BRIDE	FLANSCHANPASSSTÜCK
5326	**FLANGE BACK-FACE**	DOS DE LA FACE DE BRIDE	FLANSCHRÜCKFLÄCHE
5327	**FLANGE FACE**	FACE DE BRIDE	FLANSCHFLÄCHE
5328	**FLANGE HOLES**	TROUS DE BRIDE	FLANSCHBOHRUNGEN
5329	**FLANGE HUB**	MOYEU D'UNE BRIDE	FLANSCHNABE
5330	**FLANGE OF A TEE IRON**	AILE (D'UN FER A T)	FLANSCH EINES EISENS
5331	**FLANGE OF A WHEEL**	BOURRELET D'UNE ROUE	SPURKRANZ EINES RADES
5332	**FLANGE OF PULLEY**	REBORD, JOUE D'UNE POULIE	BORDSCHEIBE EINER RIEMENSCHEIBE
5333	**FLANGE ROLLING MACHINE**	MACHINE A MANDRINER LES BRIDES	FLANSCHENAUFWALZMASCHINE
5334	**FLANGE TURNING LATHE**	TOUR A BRIDES	FLANSCHENDREHBANK
5335	**FLANGE WELD**	SOUDURE SUR BORDS RELEVES	BÖRDELNAHT
5336	**FLANGED BELT PULLEY**	POULIE AVEC JOUES, POULIE AVEC REBORDS	RIEMENSCHEIBE MIT BORD SCHEIBE
5337	**FLANGED BRANCH PIECE**	TUBULURE A BRIDE, TUBULURE A COLLET DE RACCORDEMENT	FLANSCHSTUTZEN
5338	**FLANGED COUPLING, FACE PLATE COUPLING**	ACCOUPLEMENT PAR MANCHON A PLATEAU, ACCOUPLEMENT A PLATEAUX BOULONNES	SCHEIBENKUPPLUNG, FLANSCHENKUPPLUNG
5339	**FLANGED ENDS**	ORIFICES A BRIDES	FLANSCHENANSCHLUSS, GEFLANSCHTE ENDEN
5340	**FLANGED HEX (AGONAL) NUT**	ECROU HEXAGONAL (6-PANS) A EMBASE	SECHSKANTBUNDMUTTER
5341	**FLANGED NOZZLE**	TUBULURE A BRIDE	ROHRSTUTZEN
5342	**FLANGED NUT**	ECROU A EMBASE	BUNDMUTTER
5343	**FLANGED PIPE**	TUYAU A BRIDES	FLANSCHENROHR
5344	**FLANGED PIPE JOINT**	ASSEMBLAGE A BRIDES	FLANSCHENVERSCHRAUBUNG, FLANSCHENVERBINDUNG
5345	**FLANGING**	EMBOUTISSAGE	KÜMPELN
5346	**FLANGING MACHINE**	MACHINE A EMBOUTIR, EMBOUTISSEUSE, MACHINE A RETROUSSER, SERTISSEUSE	KÜMPELMASCHINE, BÖRDELMASCHINE, KREMPMASCHINE
5347	**FLANGING TEST**	ESSAI DE RABATTEMENT	BÖRDELPROBE
5348	**FLANK**	FLANC	FLANKE
5349	**FLANK CLEARANCE**	JEU ENTRE LES POINTS DE CONTACT DES DENTS	FLANKENSPIEL (ZAHNRÄDERN)
5350	**FLANK OF THREAD**	FLANC DU FILET	FLANKE EINES GEWINDES
5351	**FLANK SHOULDER OF TOOTH**	FLANC DE LA DENT	ZAHNFLANKE (UNTER DEM TEILKREIS)
5352	**FLANNEL**	FLANELLE	FLANELL
5353	**FLAP VALVE, CLACK VALVE**	CLAPET, SOUPAPE A CLAPET, SOUPAPE A CHARNIERE, DISTRIBUTEUR, OBTURATEUR A LEVEE ANGULAIRE	KLAPPE, KLAPPENVENTIL
5354	**FLAPPING**	RUPTURE DE LA COUCHE DE SCORIES	AUFBRECHEN DER SCHLACKENSCHICHT
5355	**FLARE**	TORCHERE	FACKEL
5356	**FLARE (TO)**	ELARGIR, S'ELARGIR, S'ETENDRE	AUFWEITEN, SICH ERWEITERN
5357	**FLASH**	BAVURE, ARETE (DE SOUDURE) DUE AU REFOULEMENT	GRAT, STAUCHGRAT, SCHWEISSRIPPE
5358	**FLASH (BUTT) WELDING**	SOUDAGE PAR ETINCELAGE	ABBRENNSCHWEISSEN

5359	**FLASH (TO), BURST FORTH INTO FLAME (TO)**	JETER DE LA FLAMME	AUFFLAMMEN
5360	**FLASH BAKER**	ETUVE A SECHAGE RAPIDE	SCHNELLTROCKNER, BLITZTROCKNER
5361	**FLASH COATING**	METALLISATION AU PISTOLET	METALLSPRITZEN
5362	**FLASH HEAT**	ECHAUFFEMENT RAPIDE	ANWÄRMHITZE (KURZE)
5363	**FLASH POINT**	POINT D'INFLAMMATION, POINT DE FLAMME	FLAMMPUNKT
5364	**FLASH TRIM (TO)**	EBARBER, EBAVURER	ENTGRATEN
5365	**FLASHBACK**	RENTREE DE FLAMME	FLAMMENRÜCKSCHLAG
5366	**FLASHER LIGHTS**	CLIGNOTANTS	BLINKLEUCHTE
5367	**FLASHING**	PLAGE D'OXYDE, INCRUSTATION DE CALAMINE	ZUNDERFLECK, ABSATZSTELLEN
5368	**FLASHING POINT**	POINT D'ECLAIR, POINT D'INFLAMMABILITE	FLAMMPUNKT
5369	**FLASK**	CHASSIS DE MOULAGE	FORMKASTEN
5370	**FLASK**	BALLON (CHIM.)	KOLBEN, GLASKOLBEN
5371	**FLASK ANNEALING**	RECUIT EN CAISSE	KASTENGLÜHEN
5372	**FLASK PIN (US)**	GOUPILLE DE CHASSIS	FORMKASTENSTIFT
5373	**FLAT ARCH**	ARC EN ANSE DE PANIER	KORBBOGEN
5374	**FLAT BAR**	BARRE PLATE	FLACHSTAB
5375	**FLAT BAR IRON, FLATS**	FER PLAT, PLATS, FER EN BANDES, BANDELETTE	FLACHEISEN, UNIVERSALEISEN
5376	**FLAT BULB IRON**	FER PLAT A BOUDIN	FLACHWULSTEISEN
5377	**FLAT CHISEL**	BURIN	FLACHMEISSEL
5378	**FLAT CURVE SWEEP**	COURBE APLATIE, COURBE A GRAND RAYON	FLACHE KURVE
5379	**FLAT DISC WHEEL**	ROUE A VOILE DROIT	SCHEIBENRAD (GERADES)
5380	**FLAT DRILL**	FORET A LANGUE D'ASPIC, LANGUE D'ASPIC, MECHE PLATE	SPITZBOHRER
5381	**FLAT FILE**	LIME PLATE	FLACHFEILE
5382	**FLAT FLANGE**	BRIDE PLATE	FLANSCH (GLATTER)
5383	**FLAT FOUR ENGINE**	MOTEUR A 4 CYLINDRES OPPOSES HORIZONTAUX	VIER-ZYLINDER-BOXER-MOTOR
5384	**FLAT GUIDE**	GLISSIERE PLATE	FLACHFÜHRUNG
5385	**FLAT HEAD SCREW**	VIS A TETE PLATE	SENKSCHRAUBE
5386	**FLAT HEADED BOLTS**	BOULONS A TETE PLATE	SCHEIBENBOLZEN
5387	**FLAT HEARTH TYPE MIXER**	MELANGEUR A SOLE PLATE	FLACHHERDMISCHER
5388	**FLAT KEY**	CLAVETTE PLATE, CLAVETTE A MEPLAT, CLAVETTE POSEE A PLAT	FLACHKEIL
5389	**FLAT OF THREAD**	TRONCATURE D'UN FILET	ABFLACHUNG DES GEWINDES
5390	**FLAT PILE**	LIME PLATE POINTUE	FLACHFEILE
5391	**FLAT ROLL**	CYLINDRE LISSE	FLACHWALZE
5392	**FLAT ROLLED STEEL**	ACIERS LAMINES PLATS	FLACHSTAHL
5393	**FLAT ROPE**	CABLE PLAT	FLACHSEIL, BANDSEIL
5394	**FLAT SCREW HEAD**	TETE PLATE D'UNE VIS	SCHRAUBENKOPF (FLACHER)
5395	**FLAT SHEET**	TOLE DRESSEE	BLECH (GERICHTETES)
5396	**FLAT SPIRAL SPRING**	RESSORT SPIRALE	SPIRALFEDER, SCHNECKENFEDER
5397	**FLAT TUBULAR WOVEN BELTING**	COURROIE EN TISSU TUBULAIRE	SCHLAUCHGEWEBERIEMEN
5398	**FLAT TWIN ENGINE**	MOTEUR A 2 CYLINDRES OPPOSES HORIZONTAUX	ZWEIZYLINDER-BOXER-MOTOR
5399	**FLAT WASHER**	RONDELLE PLATE	FLACHSCHEIBE
5400	**FLAT WEDGE**	COIN PLAT	FLACHKEIL

5401	FLAT WELD	CORDON DE SOUDURE PLAT	RAUPE (FLACHE)
5402	FLAT WIRE	FIL PLAT, FIL RECTANGULAITRE, FIL MEPLAT	FLACHDRAHT, DRAHT (RECHTECKIGER)
5403	FLAT WIRE PLIERS	PINCE PLATE	FLACHZANGE, DRAHTZANGE (FLACHE)
5404	FLAT-FACED PULLEY, PULLEY WITH STRAIGHT RIM FLAT FACE	POULIE DROITE, CYLINDRIQUE	RIEMENSCHEIBE GERADE
5405	FLATTEN (TO)	PLANER	AUSBEULEN
5406	FLATTENED	APLATI	ABGEFLACHT, ABGEPLATTET
5407	FLATTENED-STRAND WIRE ROPE	CABLE METALLIQUE A TORONS MEPLATS	DRAHTSEIL (FLACHLITZIGES)
5408	FLATTENING	LISSAGE, POLISSAGE	ABPLATTEN, ABFLACHEN, GLÄTTEN
5409	FLATTENING TEST	ESSAI D'APLATISSEMENT	STRECKPROBE, AUSBREITEPROBE
5410	FLATTER	CHASSE A PARER	FLACHHAMMER, GERADER SETZHAMMER
5411	FLAW	DEFAUT, AMORCE DE CRIQUE	ANRISS, FEHLER, DEFEKT
5412	FLAX	LIN	FLACHS
5413	FLAX TOW	ETOUPE DE LIN	FLACHSWERG
5414	FLESH SIDE OF LEATHER	COTE CHAIR DU CUIR	FLEISCHSEITE DES LEDERS
5415	FLEXIBILITY, PLIABILITY	FLEXIBILITE, SOUPLESSE	BIEGSAMKEIT, BIEGBARKEIT
5416	FLEXIBLE CONNECTION	FIXATION, LIAISON, JOINT SOUPLE	VERBINDUNG (NACHGIEBIGE)
5417	FLEXIBLE COUPLING	MANCHON FLEXIBLE, ACCOUPLEMENT FLEXIBLE	KUPPLUNG (BEWEGLICHE), KUPPLUNG (NACHGIEBIGE), KUPPLUNG (ELASTISCHE)
5418	FLEXIBLE JOINT PIPE, ARTICULATED PIPE, PIPE WITH BALL JOINT	TUYAU A JOINT SPHERIQUE	GELENKROHR
5419	FLEXIBLE METAL HOSE TUBING	TUYAU METALLIQUE FLEXIBLE	METALLSCHLAUCH
5420	FLEXIBLE SHAFT	ARBRE FLEXIBLE, TRANSMISSION FLEXIBLE	WELLE (BIEGSAME)
5421	FLEXILBLE PIPE	TUYAU METALLIQUE FLEXIBLE	METALLSCHLAUCH
5422	FLEXION	FLEXION	BIEGUNG
5423	FLEXURAL RIGIDITY	RIGIDITE A LA FLEXION	BIEGESTEIFIGKEIT
5424	FLEXURAL STRENGTH	RESISTANCE A LA FLEXION	BIEGEFESTIGKET
5425	FLEXURAL STRESS	TENSION DE FLEXION	BIEGESPANNUNG
5426	FLEXURE	FLECHE, DEFORMATION PAR FLEXION	BIEGUNG, DURCHBIEGUNG
5427	FLINT	FLINT, PIERRE A FUSIL, SILEX PYROMAQUE, PIERRE A BRIQUET	FEUERSTEIN, FLINT
5428	FLINT GLASS, LEAD GLASS	FLINT-GLASS	FLINTGLAS
5429	FLIP-FLOP	BASCULE BINAIRE	KIPPSCHALTUNG (BI-STABILE)
5430	FLOAT	FLOTTEUR	SCHWIMMER
5431	FLOAT CHAMBER	CUVE A NIVEAU CONSTANT	SCHWIMMERKAMMER
5432	FLOAT TRAP	PURGEUR D'EAU CONDENSEE A FLOTTEUR	SCHWIMMERKONDENSTOPF
5433	FLOAT VALVE	ROBINET A FLOTTEUR, SOUPAPE A FLOTTEUR	SCHWIMMERHAHN, SCHWIMMERVENTIL
5434	FLOATING	NUANCAGE, FUSEES	AUSSCHWIMMEN
5435	FLOATING AXE	ESSIEU-FLOTTANT	SCHWEBEACHSE
5436	FLOATING BODY	CORPS FLOTTANT	KÖRPER (SCHWIMMENDER)
5437	FLOATING DIE	MATRICE FLOTTANTE	SCHWEBEMANTELMATRIZE
5438	FLOATING PLUG	MANDRIN FLOTTANT	PENDELDORN, PENDELFUTTER
5439	FLOATING ROOF TANK	RESERVOIR A TOIT FLOTTANT	SCHWIMMDACHTANK
5440	FLOATING ZERO	ZERO FLOTTANT	NULLPUNKT (BELIEBIGER)
5441	FLOCCULATION	FLOCULATION	AUSFLOCKUNG

5442	FLOOD LUBRICATION, LUBRICATION BY THE CIRCULATING SYSTEM	GRAISSAGE SOUS PRESSION A CIRCULATION CONTINUE, GRAISSAGE A POMPE ET CIRCULATION D'HUILE	UMLAUFSCHMIERUNG, ZIRKULATIONSSCHMIERUNG, SPÜLSCHMIERUNG
5443	FLOODING	NUANCAGE, FUSEES	AUSSCHWIMMEN
5444	FLOOR COUNTERSHAFT	RENVOI FIXE AU SOL	FUSSBODENVORGELEGE
5445	FLOOR MOULDING	MOULAGE SUR LE SOL, MOULAGE EN FOSSE	FORMUNG AUF DEM BODEN
5446	FLOOR PAN ASSEMBLY	PLANCHER DE VOITURE	BODENBLECH, BODENBRETT
5447	FLOOR PLATE, FOOT PLATE	TOLE POUR REVETEMENT DE SOL	FUSSBODENBELAGBLECH, BELAGBLECH FÜR FUSSBÖDEN
5448	FLOOR STAND	COLONNE DE MANOEUVRE	FLURSÄULE
5449	FLOOR-PLATE	PLATELAGE	BODENPLATTE
5450	FLOORING	PLANCHER COUVERTURE DE PLANCHER, COUVERTURE DE SOL	FUSSBODEN (BELAG), FUSSBODENBELAG
5451	FLOORING/FLOOR-PLATE	PLATELAGE	BODENBELAG
5452	FLOTATION	FLOTTATION	FLOTATION, SCHWIMMAUFBEREITUNG
5453	FLOUR (EMERY)	POTEE D'EMERI	SCHMIRGEL (GESCHLÄMMTER)
5454	FLOW	ECOULEMENT	FLIESSEN
5455	FLOW GAUGE	CALIBRE A DEBIT	DURCHFLUSSMESSER
5456	FLOW LIMIT	LIMITE D'ECOULEMENT, LIMITE ELASTIQUE	FLIESSGRENZE
5457	FLOW LINE	LIGNE D'ECOULEMENT	FLIESSLINIE, FLIESSMARK
5458	FLOW METER	DEBITMETRE	DURCHFLUSSMESSER, MENGENMESSER
5459	FLOW RATE	VITESSE DE FLUAGE, TEMPS D'ECOULEMENT, DEBIT	FLIESSGESCHWINDIGKEIT, FLIESSZEIT, LEISTUNG, DURCHFLUSSGRÖSSE
5460	FLOW STRESS	EFFORT DE FLUAGE	SCHUBSPANNUNG
5461	FLOW STRUCTURE	STRUCTURE DUE A LA DEFORMATION PLASTIQUE	VERSCHIEBUNGSTRUKTUR, FLIESSTEXTUR
5462	FLOW VELOCITY	VITESSE D'ECOULEMENT	STRÖMUNGSGESCHWINDIGKEIT
5463	FLOWABILITY	FLUIDITE	FLIESSFÄHIGKEIT
5464	FLOWERS OF SULPHUR, SUBLIMED SULPHUR	FLEUR DE SOUFRE, SOUFRE SUBLIME	SCHWEFELBLUMEN, SCHWEFELBLÜTE
5465	FLOWING OF THE MATERIAL, PLASTIC YIELDING OF THE MATERIAL	ECOULEMENT DU MATERIAU	FLIESSEN DES MATERIALS
5466	FLOWING WATER	EAU COURANTE	WASSER (STRÖMENDES), WASSER (FLIESSENDES)
5467	FLUE	CARNEAU	RAUCHKANAL, RAUCHFANG
5468	FLUE BOILER	CHAUDIERE A FOYER INTERIEUR, CHAUDIERE A TUBE FOYER	FLAMMROHRKESSEL
5469	FLUE DUST	ESCARBILLES, POUSSIERE DE GUEULARD, POUSSIERES DE GAZ DE HAUT FOURNEAU	GICHTSTAUB
5470	FLUE GASES	GAZ DU FOYER, GAZ DE LA COMBUSTION	RAUCHGASE, VERBRENNUNGSGASE, FEUERGASE
5471	FLUID	TRES FLUIDE, TRES MOBILE	DÜNNFLÜSSIG, LEICHTFLÜSSIG
5472	FLUID DRIVE	TRANSMISSION HYDRAULIQUE	HYDRAULISCHE ÜBERTRAGUNG
5473	FLUID FRICTION	FROTTEMENT INTERIEUR DES LIQUIDES	REIBUNG (INNERE), FLÜSSIGKEITSREIBUNG, ZÄHIGKEITSREIBUNG
5474	FLUIDITY, LIQUIDITY	FLUIDITE	FLÜSSIGKEITSGRAD DÜNNFLÜSSIGKEIT, LEICHTFLÜSSIGKEIT
5475	FLUIDIZER	FONDANT	FLUSSMITTEL, SCHLACKENZUSCHLAG
5476	FLUORESCENCE	FLUORESCENCE	FLUORESZENZ

5477	**FLUORESCENT SCREEN**	ECRAN LUMINEUX, ECRAN FLUORESCENT	LEUCHTSCHIRM, FLUORESZENZSCHIRM
5478	**FLUORINE**	FLUOR	FLUOR
5479	**FLUORSPAR, CALCIUM FLUORIDE**	FLUORINE, SPATH FLUOR, FLUORURE DE CALCIUM, CHAUX FLUATEE, FLUSSPATH	FLUSSSPAT, FLUORKALZIUM, KALZIUMFLUORID
5480	**FLUSH HEADLIGHT**	PHARE ENCASTRE	SCHEINWERFER (EINGELASSENER)
5481	**FLUSH WELD**	SOUDURE BOUT A BOUT SANS SUREPAISSEUR	FLACHNAHT
5482	**FLUSH, FAIR**	AFFLEURE	BÜNDIG
5483	**FLUTE**	GOUJURE, GORGE	SPANNUT, AUSKEHLUNG
5484	**FLUTED**	CANNELE, STRIE	RIEFIG, GERIEFT, GERIFFELT
5485	**FLUTED REAMER**	ALESOIR A RAINURES, ALESOIR A CANNELURES	REIBAHLE (GENUTETE)
5486	**FLUTED ROLLER**	CYLINDRE CANNELE, STRIE	RIFFELWALZE
5487	**FLUTED SPECTRUM**	SPECTRE CANNELE	BANDSPEKTRUM
5488	**FLUTING**	RUPTURE PAR FLEXION	BIEGEBRUCH
5489	**FLUX**	FONDANT	FLUSSMITTEL, ZUSCHLAG
5490	**FLUX FOR SOLDERING**	FONDANT POUR SOUDER	LÖTAUFBRINGEMITTEL, AUFBRINGEMITTEL, LÖTFLUSSMITTEL
5491	**FLUX OIL**	HUILE DE FLUXAGE	STELLÖL
5492	**FLUXED ELECTRODE**	ELECTRODE ENROBE	ELEKTRODE (UMMANTELTE)
5493	**FLY ASH**	CENDRES VOLANTES	FLUGASCHE
5494	**FLYCUTTER**	OUTIL A TREPANER	SCHLAGFRÄSER
5495	**FLYING SHEARS**	CISAILLE VOLANTE, CISAILLE A PORTE A FAUX	SCHERE (FLIBGENDE)
5496	**FLYWHEEL**	VOLANT-MOTEUR	SCHWUNGRAD
5497	**FLYWHEEL IN HALVES**	VOLANT EN DEUX SEGMENTS	SCHWUNGRAD (ZWEITEILIGES)
5498	**FLYWHEEL PIT**	FOSSE DU VOLANT	SCHWUNGRADGRUBE
5499	**FOAM**	MOUSSE	SCHAUM
5500	**FOCAL LENGTH DISTANCE**	DISTANCE FOCALE, LONGUEUR FOCALE	BRENNWEITE
5501	**FOCAL PLANE**	PLAN FOCAL	BRENNEBENE
5502	**FOCUS**	FOYER (OPT. ET GEOM.)	BRENNPUNKT, FOKUS
5503	**FOCUS-FILM DISTANCE**	DISTANCE FOCALE	BRENNWEITE, BRENNPUNKTABSTAND
5504	**FOCUSED BEAM**	FAISCEAU FOCALISE, FAISCEAU CONCENTRE	GEBÜNDELTER STRAHL
5505	**FOCUSSING**	MISE AU POINT D UN INSTRUMENT D OPTIQUE	FOKUSSIEREN, EINSTELLEN, EINSTELLUNG EINES OPTISCHEN INSTRUMENTS
5506	**FOG LIGHTS**	PHARE DE BROUILLARD	NEBELLAMPE
5507	**FOGGING**	ASSOMBRISSEMENT	BLINDWERDEN
5508	**FOIL**	FEUILLE, FEUILLE DE METAL	FOLIE, METALFOLIE
5509	**FOIL OF METAL**	FEUILLE MINCE DE METAL	BLATTMETALL, FOLIE
5510	**FOLD**	REPLIURE DE LAMINAGE	ÜBERWALZUNGSFEHLER
5511	**FOLD (DOUBLED PART)**	PLI, REPLI (TRAVAIL DES TOLES)	FALZ (BLECHBEARBEITUNG)
5512	**FOLDING FOOT RULE**	METRE PLIANT	GELENKMASSSTAB, KLAPPMASSSTAB, FALTMASSSTAB, GLIEDERMASSSTAB, ZOLLSTOCK
5513	**FOLIUM OF DESCARTES**	FOLIUM DE DESCARTES	BLATT (DESCARTESSCHES), FOLIUM KARTESISCHES
5514	**FOLLOWER REST**	LUNETTE MOBILE, LUNETTE A SUIVRE	BRILLE (LAUFENDE)
5515	**FOLLOWING EDGE**	ARETE DE SORTIE	AUSTRITTKANTE

5516	FOLLOWING SIDE OF A BELT, SLACK SIDE; LOOSE SIDE, ROPE ETC.	BRIN CONDUIT, BRIN MENE, BRIN MOU, BRIN LACHE, BRIN SORTANT (D UNE COURROIE SANS FIN)	TRUMM (GEZOGENES), TRUMM (RÜCKLAUFENDES), TRUMM (LOSES, ABLAUFENDES) TRUMM
5517	FOOD PROCESSING INDUSTRY	INDUSTRIE DE L'ALIMENTATION	LEBENSMITTELINDUSTRIE
5518	FOOT	PIED ANGLAIS	FUSS (ENGLISCHER)
5519	FOOT BRAKE	FREIN A PIED, A PEDALE	FUSSBREMSE
5520	FOOT LATHE	TOUR A PEDALE	FUSSTRITTDREHBANK
5521	FOOT LEVER, PEDAL	LEVIER A PEDALE	FUSSHEBEL, TRITTHEBEL
5522	FOOT POUND (F.P., FT-LB.)	LIVRE-PIED	FUSSPFUND
5523	FOOT VALVE	CLAPET DE FOND, CLAPET DE PIED, CLAPET-CREPINE	FUSSVENTIL, BODENVENTIL, GEGENDRUCKVENTIL, FUSSVENTIL
5524	FOOTSTEP BEARING, PIVOT BEARING, STEP BEARING	PALIER VERITICAL, PALIER DE PIED; CRAPAUDINE	SPURLAGER, FUSSLAGER, STÜTZLAGER
5525	FORAGE AND SILAGE HARVESTERS	FAUCHEUSES A FOURRAGES	MÄHHÄCKSLER
5526	FORCE DIAGRAM	DIAGRAMME DES FORCES, DIAGRAMME DES EFFORTS	KRÄFTEPLAN
5527	FORCE FIT, PRESS FIT	EMMANCHEMENT A LA PRESSE, AJUSTAGE A LA PRESSE, MONTAGE A LA PRESSE	PRESSSITZ
5528	FORCE OF A SPRING	FORCE D'UN RESSORT	SPANNKRAFT EINER FEDER
5529	FORCE OF ATTRACTION, ATTRACTIVE FORCE	FORCE ATTRACTIVE	ANZIEHUNGSKRAFT
5530	FORCE OF REPULSION, REPULSIVE FORCE	FORCE REPULSIVE	ABSTOSSUNGSKRAFT
5531	FORCE PUMP, PRESSURE PUMP	POMPE FOULANTE	DRUCKPUMPE
5532	FORCE, EFFORT	FORCE, EFFORT	KRAFT
5533	FORCED DRAUGHT	TIRAGE FORCE	DRUCKZUG
5534	FORCED FLOW	COURANT FORCE	STRÖMUNG (AUFGEZWUNGENE)
5535	FORCED VIBRATION	OSCILLATIONS FORCEES	SCHWINGUNG (ERZWUNGENE)
5536	FORCEPS	PINCETTE	PINZETTE, FEDERZANGE
5537	FORCES ACTING IN OPPOSITE DIRECTIONS	FORCES OPPOSEES EN DIRECTION	KRÄFTE (ENTGEGENGESETZT GERICHTETE)
5538	FORCES ACTING IN THE SAME DIRECTION	FORCES DE MEME SENS	KRÄFTE (GLEICHGERICHTETE)
5539	FOREHAND WELDING	SOUDURE POUSSEE VERS LA GAUCHE	LINKSSCHWEISSUNG
5540	FOREIGN PATENT	BREVET ETRANGER	AUSLANDSPATENT
5541	FOREMAN	CONTREMAITRE	WERKFÜHRER, WERKMEISTER, MEISTER
5542	FORGE (TO)	FORGER	SCHMIEDEN
5543	FORGE COAL, SMITHY COAL	HOUILLE MARECHALE, CHARBON DE FORGE	SCHMIEDEKOHLE
5544	FORGE PIG	FONTE D'AFFINAGE	PUDDELROHEISEN
5545	FORGE TEST	ESSAI DE FORGEAGE	SCHMIEDEPROBE
5546	FORGE WELDING	SOUDAGE A LA FORGE	SCHMIEDESCHWEISSUNG
5547	FORGE, BLACK SMITH'S SHOP, SMITHY	FORGE, ATELIER DE FORGE	SCHMIEDE, SCHMIEDEWERKSTATT
5548	FORGE, FORGE FIRE, SMITH'S FIRE, BLACKSMITH'S HEARTH	FORGE, FORGE DE MARECHALE, FEU DE FORGE	SCHMIEDEFEUER, SCHMIEDEESSE, SCHMIEDEHERD
5549	FORGED CARBON STEEL FITTING	RACCORD ACIER FORGE	SCHMIEDESTAHLANSCHLUSS
5550	FORGED CRANK	MANIVELLE FORGEE, MANIVELLE COUDEE A LA FORGE, MANIVELLE VENUE DE FORGE	KURBEL (GESCHMIEDETE)
5551	FORGED EYEBAR	BARRE A OEILLETS	AUGENSTAB (GESCHMIEDETER)

5552	**FORGED IN THE SOLID**	FORGE D'UNE PIECE DANS LA MASSE	GESCHMIEDET (AUS DEM VOLLEN)
5553	**FORGED SCRAP IRON**	FER DE RIBLONS, FER DE MASSE, MITRAILLE	ABFALLEISEN
5554	**FORGED STEEL**	ACIER FORGE	SCHMIEDESTAHL, STAHL (GESCHMIEDETER)
5555	**FORGING**	FORGEAGE	SCHMIEDEN
5556	**FORGING (FORGED WORK)**	PIECE FORGEE, PIECE TRAVAILLEE A LA FORGE, PIECE DE FORGE	SCHMIEDESTÜCK
5557	**FORGING (MAKING FORGED WORK)**	FORGEAGE	SCHMIEDEN
5558	**FORGING BRASS**	LAITON A FORGER	SCHMIEDEMESSING
5559	**FORGING HAMMER**	MARTEAU DE FORGE, MARTEAU MECANIQUE	SCHMIEDEHAMMER (MECHANISCHER)
5560	**FORGING HEAT**	TEMPERATURE DE FORGEAGE	SCHMIEDETEMPERATUR
5561	**FORGING MACHINE**	MACHINE A FORGER, MACHINE A REFOULER	SCHMIEDEMASCHINE, STAUCHMASCHINE
5562	**FORGING PRESS**	PRESSE A FORGER	SCHMIEDEPRESSE
5563	**FORGING ROLLS**	LAMINOIR A FORGER	SCHMIEDEWALZWERK
5564	**FORGING STRAINS**	EFFORTS DUS AU FORGEAGE	SCHMIEDESPANNUNG
5565	**FORGING TEMPERATURE**	TEMPERATURE DE FORGEAGE	SCHMIEDETEMPERATUR
5566	**FORGINGS**	FERS FORGES	SCHMIEDESTÜCKE
5567	**FORK JOINT**	CHAPE, CARDAN	GABELGELENK
5568	**FORKED CONNECTING ROD END**	PIED DE LA BIELLE A FOURCHETTE, CHAPE-FOURCHETTE	SCHUBSTANGENKOPF (GEGABELTER), SCHUBSTANGENGABEL
5569	**FORKED LEVER**	LEVIER A FOURCHE	HEBEL (GEGABELTER)
5570	**FORM**	MOULE	FORM
5571	**FORM (TO), SHAPE (TO)**	FACONNER, PROFILER	FORMEN, FORM GEBEN
5572	**FORM FACTOR**	FACTEUR DE FORME	FORMFAKTOR
5573	**FORMALDEHYDE, FORMIC ALDEHYDE**	ALDEHYDE FORMIQUE, ALDEHYDE METHYLIQUE, FORMOL, FORMALDEHYDE, METHANAL, OXYDE DE METHYLENE	FORMALIN, FORMALDEHYD, FORMOL
5574	**FORMAT**	FORMAT	FORMAT
5575	**FORMATION OF MILDEW**	FORMATION DE LA MOISISSURE	SCHIMMELBILDUNG
5576	**FORMATION VOLTAGE**	TENSION DE FORMATION	FORMIERUNGSSPANNUNG
5577	**FORMED CUTTERS**	FRAISES-MERES	FORMFRÄSER
5578	**FORMIC ACID**	ACIDE FORMIQUE	AMEISENSÄURE, FORMYLSÄURE
5579	**FORMING**	CINTRAGE, FORMAGE, FORMATION	FORMGEBUNG, FORMIERUNG
5580	**FORMING DIE**	ESTAMPE D'EMBOUTISSAGE	PRÄGESTEMPEL
5581	**FORMING, SHAPING**	FACONNAGE	FORMEN, FORMGEBUNG
5582	**FORMULA**	FORMULE	FORMEL
5583	**FORMULA (CONSTITUTIONAL), FORMULA (RATIONAL)**	FORMULE DE CONSTITUTION	RATIONELLE FORMEL, KONSTITUTIONSFORMEL
5584	**FORWARD MOVEMENT**	MOUVEMENT D'AVANCE, AVANCEMENT	VORWÄRTSBEWEGUNG
5585	**FORWARD STROKE OF PISTON**	COURSE AVANT, COURSE ALLER, COURSE DIRECTE DU PISTON, AVANCE, MARCHE EN AVANT DU PISTON	KOLBENHINGANG
5586	**FOSSIL FUEL**	COMBUSTIBLE FOSSILE	BRENNSTOFF (FOSSILER)
5587	**FOUL (TO)**	GENER MUTUELLEMENT (SE)	HINDERN, (SICH BEI DER BEWEGUNG GEGENSEITIG) KOLLIDIEREN
5588	**FOUL ELECTROLYTE**	ELECTROLYTE IMPUR	ELEKTROLYT (VERBRAUCHTER)
5589	**FOUNDATION**	FONDATION, MASSIF DE FONDATION	UNTERBAU, FUNDAMENT, ERDUNG, FUNDIERUNG

5590	FOUNDATION BOLT, HOLDDOWN BOLT, TIE BOLT	BOULON DE FONDATION	ANKERBOLZEN, ANKERSCHRAUBE, ANKER, FUNDAMENTSCHRAUBE, FUNDAMENTBOLZEN
5591	FOUNDER	FONDEUR, OUVRIER FONDEUR	GIESSER
5592	FOUNDING	FONDERIE (ART), FONDERIE, COULEE	GIESSEREI, GIESSKUNST, GIESSEN
5593	FOUNDRY	FONDERIE	GUSSWERK, GIESSEREI
5594	FOUNDRY ALLOY	PRE-ALLIAGE	VORLEGIERUNG
5595	FOUNDRY COKE	COKE DE CUBILOT	GIESSEREIKOKS, KUPOLOFENKOKS
5596	FOUNDRY PIG (IRON)	FONTE DE MOULAGE, FONTE GRISE	GIESSEREIROHEISEN ROHEISEN (GRAUES)
5597	FOUNDRY SHOP	FONDERIES (ETABLISSEMENT)	GIESSEREI, GIESSHAUS
5598	FOUNDRYMEN'S NAILS	POINTES POUR FONDERIE	FORMERSTIFTE
5599	FOUR DOOR SEDAN	BERLINE 4 PORTES	VIERTÜRIGE LIMOUSINE
5600	FOUR STROKE CYCLE ENGINE, FOUR STROKE ENGINE	MOTEUR A QUATRE TEMPS	VIERTAKTMOTOR
5601	FOUR WHEEL DRIVE	PROPULSION A 4 ROUES MOTRICES	VIERRADANTRIEB
5602	FOUR-CUSPED HYPOCYCLOID	ASTROIDE	STERNKURVE, ASTROIDE, HYPOZYKLOIDE (VIERSPITZIGE)
5603	FOUR-CYCLE ENGINE	MOTEUR A QUATRE TEMPS	VIERTAKTMOTOR
5604	FOUR-JAW INDEPENDENT CHUCK	MANDRIN A QUATRE MORS INDEPENDANTS	VIERBACKENFUTTER
5605	FOUR-WAY COCK	ROBINET A QUATRE VOIES	VIERWEGHAHN
5606	FOUR-WAY VALVE	ROBINET A QUATRE VOIES	VIERWEGVENTIL
5607	FOURDRINIER WIRE	FIL DE BRONZE PHOSPHOREUX	FOURDRINIERDRAHT
5608	FOURTH POWER	QUATRIEME PUISSANCE	POTENZ (VIERTE)
5609	FOURTH ROOT	RACINE BIQUADRATIQUE	WURZEL (VIERTE), WURZEL (BIQUADRATISCHE)
5610	FOXEY TIMBER	BOIS A COEUR POURRI, BOIS POUILLEUX	HOLZ (ROTFAULES), HOLZ (KERNFAULES)
5611	FRACTION	FRACTION DE DISTILLATION	FRAKTION (EINER DESTILLATION)
5612	FRACTIONAL DISTILLATION	DISTILLATION FRACTIONNEE	DESTILLATION (STUFENWEISE), DESTILLATION (UNTERBROCHENE), DESTILLATION (FRAKTIONIERTE)
5613	FRACTIONAL NUMBER	FRACTION, NOMBRE FRACTIONNAIRE	BRUCH, ZAHL (GEBROCHENE)
5614	FRACTIONATING TOWER	TOUR DE FRACTIONNEMENT	FRAKTIONIERTURM
5615	FRACTIONISE (TO), FRACTIONATE (TO)	FRACTIONNER	DESTILLIEREN (STUFENWEISE), FRAKTIONIEREN
5616	FRACTOGRAPHY	THEORIE DES CASSURES	BRUCHTHEORIE, FRAKTOGRAPHIE
5617	FRACTURE	CASSURE, RUPTURE	BRUCH
5618	FRACTURE STRESS	EFFORT DE RUPTURE	BRUCHSPANNUNG
5619	FRACTURE TEST	ESSAI DE RUPTURE	BRUCHPROBE
5620	FRACTURED SURFACE, SURFACE OF FRACTURE	SECTION DE RUPTURE, CASSURE	BRUCH, BRUCHFLÄCHE
5621	FRAGILE	FRAGILE	ZERBRECHLICH
5622	FRAGILITY	FRAGILITE	ZERBRECHLICHKEIT
5623	FRAGMENTATION OF GRAINS	FRAGMENTATION DE GRAINS	KORNVERSTÜCKELUNG
5624	FRAME	CHARPENTE, CHASSIS, CADRE, CARCASSE	GERÜST, GERIPPE, FAHRGESTELL, POLGEHÄUSE, GESTELL, RAHMEN
5625	FRAME HORN	BRANCARD	BODENLÄNGSTRÄGER
5626	FRAME OF AN ENGINE, FRAME OF A MACHINE	BATI D'UNE MACHINE	MASCHINENRAHMEN
5627	FRAME RAIL	LONGERON	RAHMENLÄNGSTRÄGER
5628	FRAME SAW	SCIE A REFENDRE	ÖRTERSÄGE, SPANNSÄGE, TISCHLERSÄGE

5629	FRAME SAWING MACHINE, RECIPROCATING SAW	SCIE A CADRE, A CHASSIS, SCIERIE A MOUVEMENT ALTERNATIF, MACHINE ALTERNATIVE A SCIER	GATTERSÄGE
5630	FRAMEWORK	CHARPENTE, OSSATURE	BALKENWERK, RAHMENWERK
5631	FRAUNHOFER LINES	RAIES DE FRAUNHOFER	FRAUNHOFERSCHE LINIEN
5632	FREE CYANIDE	CYANURE LIBRE	ZYANID (FREIES)
5633	FREE FALL	CHUTE LIBRE	FREIFALL
5634	FREE FLOW	COURANT LIBRE	STRÖMUNG (FREIE)
5635	FREE FROM ACID, ACIDLESS	EXEMPT D'ACIDE	SÄUREFREI
5636	FREE FROM ASH	EXEMPT DE CENDRES	ASCHEFREI
5637	FREE FROM DUST (TO)	DEPOUSSIERER	ENTSTAUBEN
5638	FREE FROM SLAG, SLAGLESS	SANS SCORIES	SCHLACKENFREI
5639	FREE MACHINING STEEL	ACIER DE DECOLLETAGE RAPIDE	AUTOMATENSTAHL
5640	FREE MAGNETISM	MAGNETISME LIBRE	MAGNETISMUS (FREIER)
5641	FREE STANDING ACCESSWAY	ESCALIER DROIT	TREPPE (GERADE)
5642	FREE VENT	EVENT LIBRE	LUFTABZUG (FREIER)
5643	FREE VIBRATION	OSCILLATION LIBRE	SCHWINGUNG (FREIE)
5644	FREE WHEEL	ROUE LIBRE	FREILAUFRAD
5645	FREE-CUTTING BRASS	LAITON DE DECOLLETAGE	AUTOMATENMESSING
5646	FREE-STANDING ACCESSWAY	ESCALIER DROIT	TANKZUGANGSLEITER
5647	FREE, NATIVE, VIRGIN	NATIF, A L'ETAT NATIF, VIERGE (MIN.)	GEDIEGEN (MIN.)
5648	FREEHAND DRAWING	DESSIN A MAIN LEVEE	FREIHANDZEICHNUNG, HANDZEICHNUNG
5649	FREELY SUSPENDED	SUSPENDU LIBREMENT	FREI AUFGEHÄNGT
5650	FREELY, SIMPLY SUPPORTED	REPOSANT LIBREMENT	FREI AUFLIEGEND, FREI GELAGERT
5651	FREEWHEEL	ROUE LIBRE	FREILAUF
5652	FREEWHEELING MECHANISM	MECANISME DE ROUE LIBRE	FREILAUFMECHANISMUS
5653	FREEZING	SOLIDIFICATION	ERSTARRUNG
5654	FREEZING MIXTURE	MELANGE REFRIGERANT	KÄLTEMISCHUNG
5655	FREEZING OF FURNACE	CONGELATION DU FOUR	EINFRIEREN DES OFENS
5656	FREEZING POINT	POINT DE CONGELATION, POINT DE SOLIDIFICATION	GEFRIERPUNKT, ERSTARRUNGSPUNKT
5657	FREEZING PREVENTIVE	ANTI-GEL	FROSTSCHUTZMITTEL
5658	FREEZING RANGE	INTERVALLE DE SOLIDIFICATION, ZONE DE SOLIDIFICATION	ERSTARRUNGS-BEREICH, ERSTARRUNGS-INTERVALL
5659	FRENCH CHALK	STEATITE, TALC	SPECKSTEIN, TALKUM
5660	FRENCH CURVES	REGLE COURBE, PISTOLET	KURVENLINEAL, KURVENSCHIENE
5661	FREQUENCY	FREQUENCE	PERIODENZAHL, FREQUENZ
5662	FREQUENCY FUNCTION	FREQUENCE EMPIRIQUE	HÄUFIGKEITSFUNKTION
5663	FRESH WATER	EAU DOUCE	SÜSSWASSER
5664	FRESH-WATER LIMESTONE	CALCAIRE GROSSIER, CALCAIRE D'EAU DOUCE	GROBKALK
5665	FRET SAW	SCIE A DECOUPER, MACHINE A CHANTOURNER, SAUTEUSE	LAUBSÄGE
5666	FRETTING	CORROSION PAR FROTTEMENT	PASSFLÄCHENKORROSION, REIBUNGSOXYDATION
5667	FRETTING CORROSION	CORROSION DES FACES EN CONTACT, OXYDATION PAR FROTTEMENT	FASSFLÄCHENKORROSION, REIBUNGSOXYDATION
5668	FRICTION	FRICTION, FROTTEMENT	REIBUNG
5669	FRICTION BRAKE	FREIN A FRICTION	REIBUNGSBREMSE
5670	FRICTION COEFFICIENT	COEFFICIENT DE FROTTEMENT	REIBUNGSZAHL

5671	FRICTION CONE	CONE DE FRICTION	REIBUNGSKEGEL
5672	FRICTION COUPLING	EMBRAYAGE A FRICTION, ACCOUPLEMENT A FRICTION	REFBUNGSKUPPLUNG
5673	FRICTION DISC	PLATEAU DE FRICTION	REIBSCHEIBE, FRIKTIONSSCHEIBE
5674	FRICTION GEAR	ENGRENAGE A FRICTION	REIBGETRIEBE
5675	FRICTION GRIPPING PAWL	CLIQUET DE FROTTEMENT	KLEMMBACKE, KLEMMKLINKE, REIBUNGSKLINKE
5676	FRICTION GRIPPING PAWL MOTION	ENCLIQUETAGE A FROTTEMENT	REIBUNGSGESPERRE, REIBUNGSSCHALTWERK, KLEMMGESPERRE, KLEMMBACKENSCHALTGETRIEBE
5677	FRICTION OF MOTION, KINETIC FRICTION	FROTTEMENT PENDANT LE MOUVEMENT, FROTTEMENT EN MARCHE	REIBUNG DER BEWEGUNG
5678	FRICTION OF REPOSE, OF REST, STATIC FRICTION	FROTTEMENT AU DEPART, FROTTEMENT AU DEMARRAGE	REIBUNG (RUHENDE), REIBUNG DER RUHE, HAFTREIBUNG
5679	FRICTION ROLLER	GALET DE FRICTION	REIBROLLE
5680	FRICTION SAWING	SCIAGE PAR FRICTION	REIBUNGSSÄGEN
5681	FRICTION STRAP BRAKE, BAND BRAKE, RIBBON BRAKE	FREIN A BANDE, FREIN A RUBAN, FREIN A COLLIER, FREIN A ENROULEMENT	BANDBREMSE
5682	FRICTION WHEEL	ROUE DE FRICTION	REIBRAD
5683	FRICTIONAL FORCE, FORCE OF FRICTION	FORCE DU FROTTEMENT	REIBUNGSKRAFT
5684	FRICTIONAL GEARING	TRANSMISSION PAR FRICTION	REIBUNGSGETRIEBE
5685	FRICTIONAL LOSS	PERTE DUE AU FROTTEMENT	REIBUNGSVERLUST
5686	FRICTIONAL RESISTANCE	RESISTANCE DU FROTTEMENT	REIBUNGSWIDERSTAND
5687	FRICTIONAL SURFACE	SURFACE FROTTANTE	REIBUNGSFLÄCHE
5688	FRICTIONLESS	SANS FROTTEMENT	REIBUNGSLOS
5689	FRINGE LINES	LIGNES LIMITES	GRENZLINIEN
5690	FRITTING	FRITTAGE	SINTERUNG
5691	FRONT AXLE	ESSIEU AVANT	VORDERACHSE
5692	FRONT BODY PANEL	TABLIER	QUERWAND
5693	FRONT CROSS MEMBER	TRAVERSE AVANT	VORDERQUERTRÄGER
5694	FRONT ELEVATION, FRONT VIEW	VUE DE FACE	VORDERANSICHT
5695	FRONT END ASSEMBLY	AUVENT	WINDLAUF
5696	FRONT END GEOMETRY	GEOMETRIE DU TRAIN AVANT	VORDERACHSGEOMETRIE
5697	FRONT OPERATED LATHE	TOUR FRONTAL (TOUR EN L'AIR)	DREHMASCHINE MIT FRONTBEDIENUNG
5698	FRONT RAKE, TOP RAKE	ANGLE DE DEPOUILLE, ANGLE DE DEGAGEMENT DU COPEAU	BRUSTWINKEL
5699	FRONT WHEEL DRIVE	TRACTION AVANT	VORDERRADANTRIEB
5700	FROST SHAKE IN TIMBER	GELIVURE DU BOIS	EISKLUFT, FROSTRISS DES HOLZES
5701	FROSTED GROUND GLASS	VERRE DEPOLI	MATTGLAS
5702	FROSTING	GIVRAGE	EISBLUMENBILDUNG
5703	FROTHING	BULLAGE	SCHAUMBILDUNG
5704	FRUIT AND VEGETABLE GRADING AND WASHING MACHINERY	LAVEUSES ET TRIEURS DE RACINES ET DE FRUITS	SORTIER-UND WASCHMASCHINEN FÜR FRÜCHTE UND GEMÜSE
5705	FUEL	COMBUSTIBLE	BRENNMITTEL, BRENNSTOFF, HEIZSTOFF
5706	FUEL BUNKER	SOUTE A MAZOUT	ÖLBUNKER
5707	FUEL ECONOMY	ECONOMIE DE COMBUSTIBLE	BRENNSTOFFERSPARNIS
5708	FUEL FILTER	FILTRE A CARBURANT, FILTRE A COMBUSTIBLE	KRAFTSTOFFILTER
5709	FUEL GAS	GAZ DE CHAUFFAGE, GAZ COMBUSTIBLE	HEIZGAS, BRENNGAS

5710	FUEL GAUGE	INDICATEUR DE NIVEAU D'ESSENCE	KRAFTSTOFFSTANDMESSER
5711	FUEL INJECTION PUMP	POMPE D'INJECTION	EINSPRITZPUMPE
5712	FUEL INJECTION PUMP HOUSING	CARTER DE POMPE D'INJECTION	EINSPRITZPUMPEN-GEHÄUSE
5713	FUEL LIFT PUMP	POMPE D'ALIMENTATION	FÖRDERPUMPE
5714	FUEL OIL	MAZOUT, FUEL, HUILE LOURDE (POUR FORCE MOTRICE)	HEIZÖL, SCHWERÖL, TREIBÖL
5715	FUEL PUMP	POMPE D'ALIMENTATION	KRAFTSTOFFPUMPE
5716	FUEL TANK	RESERVOIR, RESERVOIR DE CARBURANT, RESERVOIR D'ESSENCE	KRAFTSTOFFBEHÄLTER
5717	FULCRUM	CENTRE DE ROTATION, POINT D'ARTICULATION	DREHPUNKT, GELENKPUNKT
5718	FULL ADMISSON TURBINE	TURBINE A ADMISSION TOTALE	VOLLTURBINE
5719	FULL ANNEALING	RECUIT A GROSGRAIN	HOCHGLÜHEN
5720	FULL FILLET JOINT	JOINT SOUDE A SIMPLE CLIN	EINZELKEHLNAHT
5721	FULL HARDENING	DURCISSEMENT A COEUR	DURCHHÄRTUNG
5722	FULL LOAD	PLEINE CHARGE	BELASTUNG (VOLLE)
5723	FULL SIZE (TO)	EN GRANDEUR NATURELLE, EN GRANDEUR D'EXECUTION	NATÜRLICHER GRÖSSE (IN), MASSSTAB NATÜRLICHEM (IN), ORIGINALGRÖSSE (IN)
5724	FULL SIZE DRAWING	DESSIN EN GRANDEUR NATURELLE, DESSIN NATURE	ZEICHNUNG IN NATÜRLICHER GRÖSSE
5725	FULL, UNBROKEN, SOLID LINE	LIGNE PLEINE	LINIE (AUSGEZOGENE)
5726	FULLER	ESTAMPE A DEGROSSIR, ESTAMPE A ETIRER	STRECKGESENK
5727	FULLER'S EARTH	TERRE A FOULON	WALKERDE, FULLERERDE, BLEICHERDE
5728	FULLERING	DEGORGEMENT	EINSCHNÜRUNG
5729	FULLERING TOOL	DEGORGEOIR	SETZHAMMER (RUNDER)
5730	FULLY MANUFACTURED ARTICLE, FINISHED PRODUCT, WORK, FULLY MANUFACTURED PRODUCT	PIECE FINIE, PRODUIT FINI, FABRIQUE	FERTIGERZEUGNIS, GANZFABRIKAT
5731	FULMINIC ACID	ACIDE FULMINIQUE	KNALLSÄURE
5732	FUMES	FUMEES	RAUCHGASE
5733	FUMING NORDHAUSEN SULPHURIC ACID, DISULPHURIC PYROSULPHURIC ACID	ACIDE SULFURIQUE FUMANT, ACIDE DE NORDHAUSEN, ACIDE DE SAXE	SCHWEFELSÄURE (RAUCHENDE), NORDHÄUSER VITRIOLÖL, OLEUM
5734	FUNCTION (MATH.)	FONCTION (MATH.)	FUNKTION (MATH.)
5735	FUNDAMENTAL RESEARCH	RECHERCHE FONDAMENTALE	GRUNDLAGENFORSCHUNG
5736	FUNDAMENTAL VIBRATION	VIBRATION PROPRE, VIBRATION FONDAMENTALE	GRUNDSCHWINGUNG, EIGENSCHWINGUNG, FUNDAMENTALSCHWINGUNG
5737	FUNICULAR LINK STRING POLYGON	POLYGONE FUNICULAIRE, POLYGONE ARTICULE	SEILECK, SEILZUG, SEILPOLYGON
5738	FUNNEL	ENTONNOIR	TRICHTER
5739	FURNACE	FOUR, FOURNEAU	OFEN
5740	FURNACE BRAZING	BRASAGE AU FOUR	OFENHARTLÖTEN
5741	FURNACE CHROME	CHROME A FOUR	CHROMERZMÖRTEL
5742	FURNACE COOLING	REFROIDISSEMENT DU FOUR	OFENKÜHLUNG
5743	FURNACE FLUE	CARNEAU DE FUMEE	FEUERKANAL, FEUERZUG, HEIZKANAL
5744	FURNACE FOR GASEOUS FUEL, GASEOUS FUEL FURNACE	FOYER A GAZ	GASFEUERUNG
5745	FURNACE FOR OIL FUEL, OIL FUEL FURNACE	FOYER A PETROLE, FOYER A HUILE LOURDE	ÖLFEUERUNG

5746	FURNACE FOR PULVERISED COAL, PULVERISED COAL FURNACE	FOYER A CHARBON PULVERISE	KOHLENSTAUBFEUERUNG
5747	FURNACE INSTALLATION	FOYER DE CHAUDIERE, FOUR DE CHAUFFERIE	FEUERUNGSANLAGE
5748	FURNACE LINING	GARNISSAGE DU FOUR	OFENFUTTER
5749	FURNACE SHELL	CHEMISE DU FOUR	OFENMANTEL
5750	FURTHER TREATMENT, WORKING, MACHINING	USINAGE ULTERIEUR	WEITERVERARBEITUNG
5751	FUSE	FUSIBLE	SICHERUNG
5752	FUSE BOX	BOITE A FUSIBLES	SICHERUNGSKASTEN
5753	FUSED ALUMINA	ELECTRO-CORINDON	ELEKTROKUNSTKORUND
5754	FUSEL OIL	HUILE DE POMMES DE TERRE	FUSELÖL
5755	FUSIBILITY	FUSIBILITE	SCHMELZBARKEIT
5756	FUSIBLE	FUSIBLE, LIQUEFIABLE	SCHMELZBAR
5757	FUSIBLE ALLOY	ALLIAGE FUSIBLE	LEGIERUNG (LEICHT SCHMELZBARE), SCHMELZLEGIERUNG
5758	FUSIBLE PLUG, SAFETY PLUG	BOUCHON, PLOMB FUSIBLE, FUSIBLE	SCHMELZPFROPFEN
5759	FUSING MELTING POINT	POINT DE FUSION	SCHMELZPUNKT
5760	FUSION	FUSION	SCHMELZEN
5761	FUSION RANGE	ZONE DE FUSION	SCHMELZBEREICH
5762	FUSION WELDING	SOUDAGE PAR FUSION	SCHMELZSCHWEISSEN
5763	FUSION ZONE	ZONE DE FUSION	SCHMELZZONE
5764	FUSION, MELTING	FUSION IGNEE, LIQUEFACTION	SCHMELZEN, SCHMELZUNG
5765	GABBRO	GABBRO	GABBRO
5766	GADGET	DISPOSITIF ACCESSOIRE	ÜBERFLÜSSIGE ZUTAT
5767	GADOLINIUM	GADOLINIUM	GADOLINIUM
5768	GAGE	INDICATEUR, JAUGE, ETALON, GABARIT	EICHMASS, LEHRE, MESSGERÄT
5769	GAGE BLOCKS	TAMPONS DE CONTROLE, ETALONS, CALIBRES, JAUGES, GABARITS	PARALLELENDMASS, LEHRE
5770	GAGE CALIBRATION	ETALONNAGE D'UNE JAUGE	PEILEICHUNG
5771	GAGE PRESSURE	PRESSION MANOMETRIQUE	DRUCK (MANOMETRISCHER), MESSERDRUCK
5772	GAGGER	CROCHET, TIRETTE, TIRANT D'ENLEVEMENT	SANDHAKEN, AUSHEBEBAND
5773	GAIN	GAIN	VERSTÄRKUNG
5774	GAIN IN SPACE	REDUCTION DES EMPLACEMENTS	PLATZERSPARNIS, RAUMERSPARNIS
5775	GALENA, LEAD GLANCE, LEAD SULPHIDE	PLOMB SULFURE, GALENE	BLEIGLANZ, GALENIT, SCHWEFELBLEI
5776	GALLERY, PLAFTORM	PLATEFORME, PASSERELLE	BÜHNE, BEDIENUNGSBÜHNE
5777	GALLIC ACID	ACIDE GALLIQUE	GALLUSSÄURE, TRIOXYBENZOESÄURE
5778	GALLING	USURE PAR FROTTEMENT, GRIPPAGE	FRESSEN, ANFRESSUNG
5779	GALLIUM	GALLIUM	GALLIUM
5780	GALLO TANNIC ACID, TANNIN	ACIDE TANNIQUE, ACIDE DIGALLIQUE, TANNIN	GERBSTOFF, GALLUSGERBSÄURE, TANNIN
5781	GALLON	GALLON (MESURE ANGLAISE)	GALLONE
5782	GALVANIC CELL	PILE GALVANIQUE, PILE HYDRO-ELECTRIQUE, ELEMENT GALVANIQUE, COUPLE ELECTRO-CHIMIQUE	ELEMENT (GALVANISCHES), ZELLE (GALVANISCHE)
5783	GALVANIC CORROSION	CORROSION GALVANIQUE	KORROSION (GALVANISCHE)
5784	GALVANISE (TO)	GALVANISER, ZINGUER	VERZINKEN
5785	GALVANISED SHEET IRON	TOLE GALVANISEE, TOLE ZINGUEE	BLECH (VERZINKTES), BLECH (GALVANISIERTES)

5786	GALVANISED WIRE	FIL GALVANISE, FIL ZINGUE	DRAHT (VERZINKTER), DRAHT (GALVANISIERTER)
5787	GALVANISING BY DIPPING, HOT COMMON ORDINARY GALVNISING	GALVANISATION, ZINGAGE	VERZINKUNG (HEISSE), FEUERVERZINKUNG
5788	GALVANIZED STRIPS	BANDES GALVANISEES	EISENBLECHE (VERZINKTE)
5789	GALVANIZING	GALVANISATION, ZINGAGE	GALVANISIEREN, VERZINKEN
5790	GALVANIZING EMBRITTLEMENT	FRAGILITE AU ZINGAGE	VERSPRÖDUNG BEIM VERZINKEN
5791	GALVANOMETER	GALVANOMETRE	GALVANOMETER
5792	GALVANOPLASTIC	GALVANOPLASTIQUE	GALVANOPLASTISCH
5793	GAMBOGE	GOMME-GUTTE	GUMMIGUTT
5794	GAMMA FUNCTION	FONCTION GAMMA	GAMMA-FUNKTION
5795	GAMMA IRON	FER GAMMA	GAMMA-EISEN
5796	GAMMA RAYS	RAYONS GAMMA	GAMMA-STRAHLEN
5797	GAMMA STRUCTURE	STRUCTURE GAMMA	GAMMA-STRUKTUR
5798	GAMMA-GRAPH	GAMMA-GRAPHIE	GAMMA-STRAHLENBILD
5799	GANG DRILL	PERCEUSE A TETES MULTIPLES	MEHRSPINDEL-BOHRMASCHINE
5800	GANG MILLING	FRAISAGE EN TRAIN	SATZFRÄSER
5801	GANG TOOLS	PORTE-OUTIL A PLUSIEURS OUTILS	MEISSELSATZ
5802	GANGUE	GANGUE, ROCHE-MERE	GANGART, GANGGESTEIN, GANGMINERAL, NEBENGESTEIN
5803	GANISTER	GANNISTER, BRIQUE DINAS	GANISTER, FLICKMASSE, KALKDINAS
5804	GANTRY CRANE, GOLIATH CRANE	PONT-ROULANT DE CHANTIER/DE GARE	BOCKKRAN, GERÜSTKRAN
5805	GAP SIZED GRADING	MELANGE DE GRAIN SANS GRAINS INTERMEDIAIRES	KORNGEMISCH OHNE MITTELKORN
5806	GAPING OF A JOINT	BAILLEMENT, BEANT D'UN JOINT	KLAFFEN EINER FUGE
5807	GARNET	GRENAT	GRANAT
5808	GAS	CORPS GAZEUX, GAZ, FLUIDE GAZEUX	GASFÖRMIGER KÖRPER, GAS
5809	GAS ANALYSIS	ANALYSE DES GAZ	GASANALYSE
5810	GAS BRAZING	BRASAGE AU CHALUMEAU	GASLÖTEN, FLAMMLÖTEN
5811	GAS BURNER	BRULEUR A GAZ, BEC A GAZ	GASBRENNER
5812	GAS CARBON	CHARBON DE CORNUE	RETORTENKOHLE, RETORTENGRAPHIT
5813	GAS COAL	HOUILLE A GAZ	GASKOHLE
5814	GAS COCK	ROBINET A GAZ	GASHAHN
5815	GAS CONSTANT	CONSTANTE D'UN GAZ	GASKÓNSTANTE
5816	GAS CURRENT	COURANT GAZEUX	GASSTROM
5817	GAS CUTTING	COUPAGE A L'AUTOGENE, COUPAGE AU CHALUMEAU, OXYCOUPAGE	BRENNSCHNEIDEN
5818	GAS CYANIDING	CEMENTATION A L'AZOTE	NITROZEMENTIERUNG
5819	GAS ENGINE	MOTEUR A ESSENCE	GASKRAFTMASCHINE, GASMOTOR
5820	GAS FLARE	TORCHERE	GASFACKEL
5821	GAS FLUE	CARNEAU A GAZ	GASABZUG
5822	GAS GOUGING	RAINURAGE A LA FLAMME	AUTOGENE RILLUNG
5823	GAS GROOVES	ONDULATIONS DUES AU GAZ	GASRILLEN
5824	GAS HEATED SOLDERING IRON	FER A SOUDER AU GAZ	GASLÖTKOLBEN
5825	GAS HOLE	SOUFFLURE	GASEINSCHLUSS, GASBLASE
5826	GAS LAMP	LAMPE A GAZ D'ECLAIRAGE	GASLAMPE, LEUCHTGASLAMPE
5827	GAS LIGHT	LUMIERE DU GAZ	GASLICHT
5828	GAS LIGHTING	ECLAIRAGE AU GAZ	GASBELEUCHTUNG
5829	GAS METER	COMPTEUR A GAZ	GASMESSER, GASUHR

5830	GAS NITRIDING	NITRURATION GAZEUSE	GASNITRIEREN
5831	GAS OIL	GAZOLE	GASÖL
5832	GAS PICKLING	DECAPAGE AU GAZ	GASBEIZUNG
5833	GAS PIPE	TUYAU A GAZ	GASROHR
5834	GAS PIPE LINE	CONDUITE DE GAZ, DISTRIBUTION DE GAZ	GASLEITUNG
5835	GAS PLIERS, GAS TONGS	PINCES A GAZ	GASROHRZANGE
5836	GAS POCKET	INCLUSION DE GAZ, SOUFFLURE	GASEINSCHLUSS, GASBLASE
5837	GAS PRESSURE	PRESSION DU GAZ	GASDRUCK
5838	GAS PRODUCER	GAZOGENE, GENERATEUR A GAZ	GASERZEUGER, GASGENERATOR
5839	GAS PURIFICATION	EPURATION DES GAZ	GASREINIGUNG
5840	GAS QUENCHING	TREMPE AU GAZ	GASHÄRTUNG
5841	GAS RECUPERATION WITHOUT COMBUSTION	CAPTAGE DES GAZ SANS COMBUSTION	GASABZUG OHNE VERBRENNUNG MIT WIEDERGEWINNUNG
5842	GAS TAKE	COLLECTEUR DE GAZ	GASFANG
5843	GAS THREAD	PAS POUR TUBES A GAZ	GASROHRGEWINDE
5844	GAS TURBINE	TURBINE A GAZ	GASTURBINE
5845	GAS WASHING BOTTLE	BARBOTEUR POUR LAVAGES, FLACON LAVEUR	GASWASCHFLASCHE
5846	GAS WATER-HEATER	CHAUFFE EAU A GAZ	DURCHLAUFERHITZER
5847	GAS WELDING	SOUDURE AUTOGENE	GASSCHWEISSEN
5848	GAS WORKS	USINE A GAZ	GASFABRIK, GASWERK, GASANSTALT
5849	GAS WORKS COKE	COKE DE GAZ, COKE D'USINE A GAZ	GASKOKS
5850	GAS-ELECTRIC GENERATOR	GENERATRICE A GAZ	GASDYNAMO
5851	GAS-FIRED FURNACE	FOUR CHAUFFE AU GAZ	OFEN (GASGEHEIZTER)
5852	GAS-SHIELD ARC WELDING	SOUDAGE A L'ARC SOUS PROTECTION GAZEUSE	SCHUTZGAS-LICHTBOGENSCHWEISSEN
5853	GAS-TIGHT	ETANCHE AUX GAZ	GASDICHT
5854	GASEOUS FUEL	COMBUSTIBLE GAZEUX	BRENNSTOFF (GASFÖRMIGER)
5855	GASEOUS INCLUSION	INCLUSION GAZEUSE	GASEINSCHLUSS
5856	GASEOUS MIXTURE	MELANGE GAZEUX	GASGEMISCH
5857	GASEOUS STATE	ETAT GAZEUX	AGGREGAT- ZUSTAND (GASFÖRMIGER)
5858	GASHING	AMORCAGE (DE COUPE)	VORFRÄSEN
5859	GASHOLDER, GASOMETER	GAZOMETRE	GASBEHÄLTER, GASGLOCKE, GASOMETER
5860	GASIFIABLE	GAZEIFIABLE	VERGASBAR
5861	GASIFICATION	GAZEIFICATION	VERGASUNG
5862	GASIFY (TO)	GAZEIFIER	VERGASEN
5863	GASKET	TRESSE, JOINT, JOINT D'ETANCHEITE	ZOPF FÜR PACKUNGEN, DICHTUNGSZOPF, PACKUNGSZOPF, DICHTUNG, DICHTFLANSCH
5864	GASOLINE, PETROL	ESSENCE	BENZIN
5865	GASSING	GAZEIFICATION	VERGASUNG
5866	GATE	ATTAQUE DE COULEE	AUSCHNITT
5867	GATE VALVE	VANNE, VANNE A PASSAGE DIRECT	ABSPERRSCHIEBER, DURCHGANGSVENTIL
5868	GATED PATTERN	MODELE A ENTONNOIRS	MODELL MIT EINGUSSTRICHTERN
5869	GATHERING STOCK	AUGMENTATION DE LA SECTION TRANSVERSALE	QUERSCHNITTSVERGRÖSSERUNG
5870	GATING SYSTEM	SYSTEME DE COULEE ET D'ALIMENTATION	ANSCHNITTSYSTEM

5871	GAUGE (METR), GAUGE (RAILWAY)	JAUGE, CALIBRE, INSTRUMENT DE VERIFICATION, VERIFICATEUR (METR), ECARTEMENT DE LA VOIE (CH. DE FER)	LEHRE, KALIBER (METR), SPURWEITE (EISENBAHN)
5872	GAUGE (TO)	ETALONNER, TARER, TIMBRER	EICHEN
5873	GAUGE BLOCKS	CALES-ETALONS	LEHRSTÜCKE
5874	GAUGE CALIBRATION	ETALONNAGE D'UNE JAUGE	PEILEICHUNG
5875	GAUGE COCK, GAUGE VALVE	ROBINET DE JAUGE	PROBIERHAHN, PROBIERVENTIL
5876	GAUGE FLOAT	FLOTTEUR (DE JAUGE)	SCHWIMMER
5877	GAUGE HEAD	TETE DE JAUGE	PEILKOPF
5878	GAUGE LENGTH	LONGUEUR DE REFERENCE	BEZUGSLÄNGE
5879	GAUGE MARK	REPERE	ENDMARKE
5880	GAUGE OF SHEET METAL	NUMERO DE JAUGE D'UNE TOLE	BLECHNUMMER
5881	GAUGE OF WIRE	NUMERO DE JAUGE D'UN FIL	DRAHTNUMMER
5882	GAUGE PRESSURE	PRESSION RELATIVE	ÜBERDRUCK
5883	GAUGE TAP	BOUCHON DE JAUGE	KALIBERSTOPFEN
5884	GAUGE-BLOCK BUILD-UP	MONTAGE DE CALES-ETALONS	LEHRSTÜCKENMONTAGE
5885	GAUGER'S PLATFORM	PLATEFORME DE JAUGEAGE	PEILBÜHNE
5886	GAUGING	ETALONNAGE, ETALONNEMENT, TARAGE, TIMBRE	EICHEN, EICHUNG
5887	GAUSSIAN DISTRIBUTION	DISTRIBUTION GAUSSIENNE	GAUSSSCHE VERTEILUNG
5888	GAUZE	GAZE, TOILE METALLIQUE	GAZE, DRAHTGEWEBE
5889	GEAR	ENGRENAGE, PIGNON	ZAHNRAD
5890	GEAR BOX	CARTER D'ENGRENAGE, BOITE DE VITESSE	RÄDERKASTEN, SCHALTGETRIEBE
5891	GEAR CUTTING MACHINE	MACHINE A TAILLER LES ENGRENAGES, TAILLEUSE D'ENGRENAGES	ZAHNRADSCHNEIDEMASCHINE
5892	GEAR DRIVE	COMMANDE PAR ENGRENAGE	ZAHNRADANTRIEB
5893	GEAR DRIVEN MACHINE	MACHINE A COMMANDE PAR ENGRENAGE	MASCHINE MIT ZAHNRADANTRIEB
5894	GEAR MILLING MACHINE	FRAISEUSE A TAILLER LES ENGRENAGES	RÄDERFRÄSMASCHINE
5895	GEAR PUMP	POMPE A ENGRENAGE	ZAHNRADPUMPE
5896	GEAR RATIO	RAPPORTS D'ENGRENAGES	ÜBERSETZUNGSVERHÄLTNIS
5897	GEAR RATIO OF A TRAIN OF WHEELS	RAPPORT D'ENGRENAGE	RÄDERÜBERSETZUNG (VERHÄLTNIS)
5898	GEAR TOOTH MILLING CUTTER, HOB	FRAISE A TAILLER LES ENGRENAGES	ZAHNRADFRÄSER
5899	GEAR TOOTH VERNIER	PIED A DENTURE	ZAHNMESS-SCHIEBLEHRE
5900	GEAR TOOTH VERNIER CALIPER	PIED A COULISSE A VERNIER POUR DENTS DE PIGNON	ZAHNRAD-NONIUSSCHUBLEHRE
5901	GEAR TRAIN, TRAIN OF GEARS OF GEARING OF WHEELS, WHEEL TRAIN	TRAIN D'ENGRENAGES, JEU D'ENGRENAGES	RÄDERGRUPPE, RÄDERGETRIEBE (ZUSAMMENGESETZTES)
5902	GEAR WHEEL TESTING MACHINE	BANC POUR VERIFIER LES ENGRENAGES	ZAHNRÄDERPRÜFMASCHINE
5903	GEAR WHEEL, TOOTHED WHEEL	ROUE DENTEE, ROUE D'ENGRENAGE	ZAHNRAD
5904	GEAR-WHEEL STEEL	ACIER POUR ENGRENAGES	ZAHNRADSTAHL
5905	GEARED BRACE, HAND DRILL	VILEBREQUIN A ENGRENAGE	RÄDERBOHRER, ECKENBOHRER, ECKBOHRWINDE
5906	GEARED FLYWHEEL	VOLANT DENTE, VOLANT ENGRENAGE	SCHWUNGRAD (GEZAHNTES), SCHWUNGRAD (VERZAHNTES)
5907	GEARING	HARNAIS D'ENGRENAGES, ENGRENAGES	RÄDERGETRIEBE, ZAHNRADGETRIEBE

5908	**GEARING DOWN**	DEMULTIPLICATION	UNTERSETZUNG
5909	**GEARING UP**	MULTIPLICATION	ÜBERSETZUNG
5910	**GEARING, MESHING**	ENGRENEMENT	EINGRIFF, INEINANDERGREIFEN, KÄMMEN VON ZAHNRÄDERN
5911	**GEIGER COUNTER**	COMPTEUR GEIGER	GEIGER-ZÄHLER
5912	**GEL**	GEL	GEL
5913	**GELATINE**	GELATINE PURE	GELATINE
5914	**GENERAL LAYOUT**	DISPOSITION GENERALE, PLAN DE DISPOSITION	DISPOSITIONSZEICHNUNG, GRUNDRISS
5915	**GENERAL MANAGER**	DIRECTEUR D'USINE	FABRIKDIREKTOR
5916	**GENERATE (TO), PRODUCE (TO) (STEAM, GAS)**	PRODUIRE (DE LA VAPEUR, DES GAZ)	ERZEUGEN, (DAMPF, GAS)
5917	**GENERATE AN ELECTRIC CURRENT (TO)**	PRODUIRE UN COURANT ELECTRIQUE	ELEKTRISCHEN STROM ERZEUGEN
5918	**GENERATING CONE**	CONE GENERATEUR, CONE COMPLEMENTAIRE	ERGÄNZUNGSKEGEL
5919	**GENERATING FUNCTION**	FONCTION GENERATRICE (DES MOMENTS)	ERZEUGENDE FUNKTION
5920	**GENERATING PLANT**	INSTALLATION GENERATRICE DE COURANT	STROMERZEUGUNGSANLAGE
5921	**GENERATING ROLLING MOVING CIRCLE**	CERCLE ROULANT, CERCLE MOBILE	ROLLENDER KREIS, ROLLKREIS, WÄLZKREIS
5922	**GENERATION OF POWER**	PRODUCTION D'ENERGIE	ENERGIEERZEUGUNG
5923	**GENERATION OF STEAM, STEAM RAISING**	GENERATION DE VAPEUR	DAMPFERZEUGUNG
5924	**GENERATOR, DYNAMOELECTRIC MACHINE**	DYNAMO, GENERATRICE DE COURANT	DYNAMOMASCHINE, DYNAMOELEKTRISCHE MASCHINE, GENERATOR (ELEKTRISCHER), STROMERZEUGER, LICHTMASCHINE
5925	**GENERATOR, GENERATING LINE**	GENERATRICE (GEOM.)	ERZEUGENDE, GENERATRIX
5926	**GEOMETRIC REPRESENTATION**	REPRESENTATION GEOMETRIQUE	DARSTELLUNG (GEOMETRISCHE)
5927	**GEOMETRICAL**	GEOMETRIQUE	GEOMETRISCH
5928	**GEOMETRICAL DRAWING**	DESSIN GEOMETRIQUE	ZEICHNUNG (GEOMETRISCHE)
5929	**GEOMETRICAL MEAN**	MOYENNE GEOMETRIQUE	MITTEL (GEOMETRISCHES)
5930	**GEOMETRICAL PROGRESSION**	PROGRESSION GEOMETRIQUE	REIHE (GEOMETRISCHE)
5931	**GEOMETRY**	GEOMETRIE	GEOMETRIE
5932	**GERMAN SILVER**	ARGENTAN, MAILLECHORT, ARGENTAL, ARGENT ALLEMAND, ARGENT DE CHINE	NEUSILBER, ARGENTAN, WEISSKUPFER, PACKFONG
5933	**GERMANIUM**	GERMANIUM	GERMANIUM
5934	**GERMINATION**	GERMINATION, FORMATION DE GERMES	KEIMBILDUNG
5935	**GERMINATIVE CONDITION**	CONDITION CRITIQUE DE DEFORMATION	SPANNUNGSZUSTAND (KRITISCHER)
5936	**GEYSER**	GEYSER (NATURE)	GEISER
5937	**GIB**	CONTRE-CLAVETTE	GEGENKEIL, KEILBEILAGE, BEILAGE
5938	**GIB AND COTTER**	CLAVETTE ET CONTRE-CLAVETTE	DOPPELKEIL
5939	**GIBHEADED KEY**	CLAVETTE A TALON, CLAVETTE A TETE, CLAVETTE A MENTONNET	NASENKEIL
5940	**GIBS**	LARDONS	LEISTEN
5941	**GILD (TO)**	DORER	VERGOLDEN
5942	**GILDING**	DORURE	VERGOLDEN, VERGOLDUNG
5943	**GILDING METAL**	ALLIAGE CUIVRE-ZINC	KUPFERZINKLEGIERUNG
5944	**GIMLET**	VRILLE, AVANT-CLOU	NAGELBOHRER, BORBOHRER

5945	GIN TRIPORD FORM, SHEAR LEGS, SHEAR POLES, SHEERS, SHEARS	CHEVRE A TROIS PIEDS, GIBET	HEBEBOCK, DREIBEIN
5946	GIRDER	POUTRE	TRÄGER
5947	GIRDER FLANGE	SEMELLE D'UNE POUTRE	TRÄGERFUSSPLATTE
5948	GIRTH SEAM	SOUDURE HORIZONTALE	RUNDNAHTSCHWEISSUNG
5949	GIVE (TO), YIELD (TO)	CEDER	NACHGEBEN
5950	GIVE OFF HEAT (TO)	DEPENSER DE LA CHALEUR	WÄRME ABGEBEN
5951	GIVEN QUANTITY	DONNEE	GRÖSSE (GEGEBENE)
5952	GLACIAL ACETIC ACID	ACIDE ACETIQUE CRISTALLISABLE	EISESSIG
5953	GLAND	PRESSE-ETOUPE, CHAPEAU DE LA BOITE A ETOUPE	STOPFBÜCHSE
5954	GLAND COCK	ROBINET A BOURRAGE	PACKHAHN, STOPFBÜCHSENHAHN
5955	GLAND FLANGE	BRIDE DE FOULOIR	STOPFBÜCHSENFLANSCH
5956	GLASS	VERRE	GLAS
5957	GLASS BRICK STONE	PIERRE DE VERRE	GLASBAUSTEIN
5958	GLASS CLOTH	TOILE VERREE	GLASLEINWAND
5959	GLASS HALF CELL	ELECTRODE DE VERRE	GLAS-HALBZELLE
5960	GLASS MEASURE, GRADUATED MEASURING CYLINDER	TUBE CYLINDRIQUE GRADUE	MESSZYLINDER
5961	GLASS PAPER	PAPIER DE VERRE	GLASPAPIER
5962	GLASS POWDER	VERRE PILE	GLASMEHL
5963	GLASS SHEET	PLAQUE DE VERRE	GLASTAFEL, GLASPLATTE, GLASSCHEIBE
5964	GLASS TILE	TUILE EN VERRE	GLASZIEGEL
5965	GLASS TUBE	TUBE EN VERRE	GLASRÖHRE
5966	GLASS WOOL	LAINE DE VERRE, VERRE FILE	GLASWOLLE
5967	GLAUBER'S SALT	SEL DE GLAUBER	GLAUBERSALZ
5968	GLAZE	GLACURE, COUVERTE	GLASUR
5969	GLAZE (TO)	GLACER, VITRER	GLASIEREN, VERGLASEN
5970	GLAZED BRICK	BRIQUE VERNISSEE, BRIQUE EMAILLEE	GLASURSTEIN
5971	GLAZIER'S PUTTY	MASTIC DES VITRIERS	GLASERKITT
5972	GLAZING	GLACURE, EMOUSSAGE, GLACAGE, VITRAGE	GLASUR, ABSCHLEIFEN, VERGLASEN, GLASIEREN
5973	GLIDE	GLISSEMENT	GLEIT
5974	GLIDE MODULUS, COEFFICIENT OF TRANSVERSE ELASTICITY OF RIGIDITY	COEFFICIENT D'ELASTICITE DE CISAILLEMENT, MODULE D'ELASTICITE TRANVERSALE	GLEITMASS, GLEITMODUL, SCHUBELASTIZITÄTSMODUL
5975	GLIDE PLANE	PLAN DE GLISSEMENT	GLEITEBENE
5976	GLISSETTE	GLISSETTE	GLEITKURVE
5977	GLOBE VALVE	ROBINET A SOUPAPE, ROBINET DROIT	ABSPERRVENTIL, KUGELVENTIL, DURCHGANGSVENTIL
5978	GLOBOID WORM	VIS GLOBIQUE	GLOBOIDSCHRAUBE, GLOBOIDSCHNECKE
5979	GLOBULAR POWDER	POUDRE GLOBULAIRE	PULVER (KUGELIGES)
5980	GLOVE COMPARTMENT	BOITE A GANTS	HANDSCHUHKASTEN
5981	GLOW	CHAUDE	GLÜHHITZE
5982	GLOW (TO)	PORTER AU ROUGE	GLÜHEN
5983	GLOWING	INCANDESCENT	GLÜHEND
5984	GLUE	COLLE	LEIM
5985	GLUE (TO)	COLLER	LEIMEN
5986	GLUED JOINT	JOINT COLLE	LEIMVERBAND
5987	GLUING	COLLAGE	LEIMEN

5988	GLYCERINE, GLYCEROL	GLYCERINE	GLYZERIN, ÖLSÜSS
5989	GNEISS	GNEISS	GNEIS
5990	GNOMONIC PROJECTION	PROJECTION GNOMONIQUE	PROJEKTION (GNOMONISCHE)
5991	GO-NO GO GAUGE	CALIBRE PASSE-PASSE PAS	GUT-UND-AUSSCHUSS-GRENZLEHRE
5992	GOGGLES	LUNETTES DE PROTECTION	SCHUTZBRILLEN
5993	GOLD	OR	GOLD
5994	GOLD BEATING	BATTEMENT D'OR	BLATTGOLDSCHLÄGEREI
5995	GOLD BRONZE	BRONZE D'OR	GOLDBRONZE
5996	GOLD COINAGE	OR DE MONNAIE	MÜNZGOLD
5997	GOLD CYANIDE	CYANURE D'OR	ZYANGOLD, GOLDZYANID
5998	GOLD DUST	POUSSIERE D'OR	GOLDSTAUB
5999	GOLD FILLED	DOUBLE	DOUBLE, DUBLEE
6000	GOLD LEAF	FEUILLE D'OR, OR BATTU	BLATTGOLD
6001	GOLD PAINT	LAQUE DE BRONZE	BRONZELACK
6002	GOLD PLATING	DORAGE	GOLDPLATTIERUNG
6003	GOLDSMITHING	ORFEVRERIE	GOLDSCHMIEDEKUNST
6004	GONIOMETER	GONIOMETRE	GONIOMETER, WINKELMESSER
6005	GOOD WELD	SOUDURE BIEN FAITE	SCHWEISSVERBINDUNG (GUTE)
6006	GOODNESS OF FIT TEST	TEST D'AJUSTEMENT	PASSUNGSPRÜFUNG
6007	GOODS LIFT, HOIST	MONTE-CHARGE	LASTENAUFZUG, WARENAUFZUG, AUFZUG
6008	GOOSE FLESH	PEAU D'ORANGE	APFELSINEN(SCHALEN-)EFFEKT
6009	GOOSENECK	COL DE CYGNE	S-BOGEN
6010	GOUGE	GOUGE DE MENUISIER	HOHLEISEN
6011	GOUGING	GOUGEAGE	FLÄMMHOBELN
6012	GOUGING MACHINE	MACHINE A DECOUPER	DEKUPIERMASCHINE
6013	GOVERNOR	REGULATEUR	DREHZAHLREGLER
6014	GOVERNOR ARM	BRAS/BRANCHE DE REGULATEUR	REGLERARM, KUGELARM, PENDELARM
6015	GOVERNOR BALL	BOULE/SPHERE DU REGULATEUR	REGLERKUGEL, SCHWUNGKUGEL, SCHWINGGEWICHT
6016	GOVERNOR HOUSING	CARTER DE REGULATEUR	REGLERGEHÄUSE
6017	GOVERNOR SLEEVE	MANCHON/DOUILLE DU REGULATEUR	REGLERMUFFE, REGLERHÜLSE
6018	GOVERNOR SPINDLE	TIGE VERTICALE, AXE DU REGULATEUR	REGLERSPINDEL, REGLERWELLE
6019	GOVERNOR SPRING	RESSORT DE REGULATEUR	REGLERFEDER
6020	GRACE PERIOD	FRANCHISE (DELAI AVANT)	KARENZZEIT
6021	GRADE	GRADE, DEGRE, NIVEAU	GRAD, STUFE
6022	GRADE OF STEEL	NUANCE/QUALITE DE L'ACIER	STAHLSORTE
6023	GRADING ANALYSIS	ANALYSE GRANULOMETRIQUE	KORNVERTEILUNGSBESTIMMUNG
6024	GRADUAL CHANGE OF TEMPERATURE/OF SPEED	VARIATION PROGRESSIVE DE TEMPERATURE/DE VITESSE	ÄNDERUNG (ALLMÄHLICHE DER TEMPERATUR), ÄNDERUNG DER GESCHWIDIGKEIT
6025	GRADUATE (TO)	GRADUER	GRADEINTEILUNG VERSEHEN (MIT), GRADUIEREN
6026	GRADUATED CIRCLE, LIMB	CERCLE GRADUE, CERCLE DIVISE, LIMBE	TEILKREIS, LIMBUS
6027	GRADUATED DIAL	CADRAN GRADUE	SKALENSCHEIBE
6028	GRADUATED STRAIGHT PIPETTE	PIPETTE DROITE GRADUEE	MESSPIPETTE
6029	GRAIN	GRAIN (MESURE ANGLAISE)	GRAIN
6030	GRAIN	CRISTAL, GRAIN	KORN, KRISTALL
6031	GRAIN (WITH THE), FIBRES (ALONG THE)	FIBRES (DANS LE SENS)	FASER (PARALLEL ZUR), FASERN (IN DER RICHTUNG)

6032	**GRAIN ASSORTMENT**	GRANULOMETRIE	KORNUNG
6033	**GRAIN BOUNDARIES**	LIMITE/JOINT DE GRAIN	KORNGRENZE
6034	**GRAIN ELEVATORS CONVEYORS**	ELEVATEURS DE GRAINS/AERO ENGRANGEURS	GETREIDEFÖRDERER UND HEBER
6035	**GRAIN GROWTH**	GROSSISSEMENT DU GRAIN, CROISSANCE DES GRAINS	KORNWASCHSTUM
6036	**GRAIN ORIENTED SHEET**	TOLE A GRAIN ORIENTE	BLECH (KORNORIENTIERTES)
6037	**GRAIN REFINEMENT**	AFFINEMENT DU GRAIN	KORNVERFEINERUNG
6038	**GRAIN SIDE, HAIR SIDE OF LEATHER**	COTE FLEUR DU CUIR, COTE POIL DU CUIR, FLEUR DU CUIR	HAARSEITE, NARBENSEITE DES LEDERS
6039	**GRAIN SIZE**	GROSSEUR DU GRAIN	KORNGRÖSSE
6040	**GRAIN SIZE DISTRIBUTION**	GRANULOMETRIE	KÖRNUNG, KORNGRÖSSENVERTEILUNG
6041	**GRAIN STRUCTURE**	STRUCTURE DU GRAIN	KORNSTRUKTUR
6042	**GRAIN TIN**	ETAIN EN GRENAILLES/EN GRAINS	ZINN IN KÖRNERN, GRANALIENZINN, GRANULIERTES ZINN
6043	**GRAINING**	GRANULATION	KÖRNUNG
6044	**GRAM MOLECULE**	MOLECULE-GRAMME	GRAMMOLEKÜL, GRAMMOL, MOL
6045	**GRAMME CALORIE**	PETITE CALORIE, CALORIE GRAMME-DEGRE	GRAMMKALORIE
6046	**GRAMME, GRAM**	GRAMME	GRAMM
6047	**GRANITE**	GRANIT	GRANIT
6048	**GRANITE SURFACE PLATEAU**	MARBRE	ABRICHTPLATTE
6049	**GRANT OF A PATENT**	DELIVRANCE D'UN BREVET, ACCORD DU BREVET	PATENTERTEILUNG, ERTEILUNG EINES PATENTS
6050	**GRANULAR**	GRANULAIRE	KÖRNIG
6051	**GRANULAR FRACTURE**	CASSURE GRENUE	BRUCH (KÖRNIGER), BRUCH (GRIESIGER)
6052	**GRANULAR STRUCTURE**	STRUCTURE GRANULAIRE	STRUKTUR (KÖRNIGE)
6053	**GRANULAR STRUCTURE, TEXTURE**	TEXTURE GRENUE	GEFÜGE (KÖRNIGES)
6054	**GRANULATED METAL**	METAL A GRENAILLES	GRANALIEN, KÖRNER
6055	**GRANULATED SLAG**	LAITIER GRANULE	SCHLACKE (GEKÖRNTE), SCHLACKE (GRANULIERTE)
6056	**GRANULATING PIT**	PUITS A GRANULATION DU LAITIER	GRANULIERGRUBE
6057	**GRANULATION**	GRANULATION	GRANULIERUNG, KÖRNUNG
6058	**GRAPE SUGAR, GLUCOSE, DEXTROSE**	SUCRE DE RAISIN, GLUCOSE	TRAUBENZUCKER, KARTOFFELZUCKER, STÄRKEZUCKER, GLUKOSE, GLYKOSE, DEXTROSE
6059	**GRAPHIC CARBON, GRAPHITIC CARBON**	CARBONE A L'ETAT GRAPHITOIDE	KOHLENSTOFF (GRAPHITISCHER)
6060	**GRAPHIC SOLUTION, GRAPHICAL CONSTRUCTION**	SOLUTION GRAPHIQUE, CALCUL GRAPHIQUE	LÖSUNG (ZEICHNERISCHE)
6061	**GRAPHIC STATICS**	GRAPHOSTATIQUE, STATIQUE GRAPHIQUE	STATIK (GRAPHISCHE)
6062	**GRAPHITE**	GRAPHITE	GRAPHIT
6063	**GRAPHITE OIL GREASE**	HUILE GRAPHITEE, GRAISSE GRAPHITEE, GRAISSE AMELIOREE, CAMBOUIS	GRAPHITÖL, GRAPHITSCHMIERE
6064	**GRAPHITE PLUMBAGO CRUCIBLE**	CREUSET EN GRAPHITE	GRAPHITTIEGEL
6065	**GRAPHITE PYROMETER**	PYROMETRE A GRAPHITE	GRAPHITPYROMETER
6066	**GRAPHITIZATION**	GRAPHITISATION	GRAPHITIERUNG, TEMPERKOHLEABSCHEIDUNG
6067	**GRAPHITIZER**	GRAPHITISANT	GRAPHITIERUNGSMITTEL
6068	**GRATE AREA, GRATE SURFACE**	SURFACE DE GRILLE	ROSTFLÄCHE
6069	**GRATE BARS**	BARRES POUR GRILLAGES	ROSTSTÄHLE

6070	**GRATE, FURNACE GRATE**	GRILLE	FEUERROST
6071	**GRATING**	GRILLAGE	GITTER
6072	**GRATUATED HYDROMETER**	AREOMETRE A POIDS CONSTANT	SKALENARÄOMETER
6073	**GRAVEL**	GRAVIER	KIES
6074	**GRAVEL FILTER**	FILTRE A GRAVIER	KIESFILTER
6075	**GRAVEL WIRE NETTING**	GRILLE A GRAVIER	KIESROST
6076	**GRAVELLY SOIL**	TERRAIN GRAVELEUX	KIESBODEN
6077	**GRAVIMETRIC ANALYSIS**	ANALYSE GRAVIMETRIQUE, ANALYSE PONDERALE	GEWICHTSANALYSE, BESTIMMUNG (GRAVIMETRISCHE)
6078	**GRAVITATION, GRAVITY**	GRAVITE, PESANTEUR	SCHWERE, SCHWERKRAFT, GRAVITATION
6079	**GRAVITY CASTING**	COULEE SOUS PRESSION PAR GRAVITE	SCHWERKRAFTDRUCKGUSS, STANDGUSS
6080	**GRAVITY DROP HAMMER**	MARTEAU-PILON	FREIFALLHAMMER
6081	**GRAVITY INCLINE, INCLINED PLANE, SELF-ACTING INCLINE D PLANE**	PLAN INCLINE, PLAN AUTOMOTEUR	BREMSBERG
6082	**GRAVITY ROLLER CONVEYOR**	TRANSPORTEUR PAR GRAVITE A ROULEAUX	ROLLENFÖRDERER, ROLLENTRANSPORTEUR, ROLLBAHN, ROLLGANG
6083	**GRAY BODY**	CORPS GRIS	GRAUSTRAHLER
6084	**GREASE (TO)**	GRAISSER (ENDUIRE DE GRAISSE)	EINFETTEN
6085	**GREASE BOX**	BOITE A GRAISSE	ACHSBÜCHSE, SCHMIERBÜCHSE
6086	**GREASE FITTING**	GRAISSEUR	SCHMIERNIPPEL
6087	**GREASING, GREASE LUBRICATION**	GRAISSAGE (ENDUCTION A LA GRAISSE), GRAISSAGE A LA GRAISSE CONSISTANCE, GRAISSAGE AU SUIF	EINFETTEN, FETTSCHMIERUNG, STARRSCHMIERUNG
6088	**GREEN**	AGGLOMERE	PRESSLING, PRESSKÖRPER
6089	**GREEN CARBONATE OF COPPER, MALACHITE**	MALACHITE	MALACHIT
6090	**GREEN COMPACT**	EBAUCHE DE COMPACT	GRÜNLING
6091	**GREEN CROP LOADERS**	RAMASSEUSES DE FOURRAGES	GRÜNFUTTERLADER
6092	**GREEN GOLD**	OR VERT	GOLD (GRÜNES)
6093	**GREEN VITRIOL, COPPERAS, FERROUS SULPHATE**	VITRIOL VERT, VITRIOL MARTIAL, SULFATE DE FER, COUPEROSE VERTE	EISENVITRIOL, VITRIOL (GRÜNES), FERROSULFAT, EISENOXYDULSUFAT, SCHWEFELSAURES EISENOXYDUL
6094	**GREY PIG IRON, GREY IRON, GREY PIG**	FONTE GRISE	ROHEISEN (GRAUES), GRAUGUSS
6095	**GREYWACKE, GRAUWACKE**	GRAUWACKE	GRAUWACKE
6096	**GRID**	GRILLE, GRILLAGE, RESEAU D'ELECTRIFICATION	GITTER, ROST, STROMVERSORGUNGSNETZ
6097	**GRID ACCUMULATOR**	ACCUMULATEUR A GRILLAGE	GITTERAKKUMULATOR
6098	**GRID BAR IRON, FIRE BAR IRON**	FER A BARREAUX DE GRILLE	ROSTSTABEISEN
6099	**GRID-LINE**	QUADRILLAGE	LINIENNETZ (QUADRATISCHES)
6100	**GRIDIRON VALVE**	TIROIR A GRILLE	GITTERSCHIEBER, SPALTSCHIEBER, ROSTSCHIEBER
6101	**GRILL-FLOORING**	CAILLEBOTIS, PLATELAGE	GITTERROST
6102	**GRILL-FLOORING TREADS**	MARCHES EN CAILLEBOTIS	GITTERROSTSTUFEN
6103	**GRIND (TO), CRUSH (TO)**	BROYER, PULVERISER, REDUIRE EN POUDRE	MAHLEN
6104	**GRIND IN (TO)**	RODER A L'EMERI	EINSCHLEIFEN, EINSCHMIRGELN
6105	**GRINDER**	AIGUISEUR, AFFUTEUR, REMOULEUR, RECTIFIEUR	SCHLEIFER
6106	**GRINDING**	MEULAGE, RODAGE, BROYAGE FIN (FRITTAGE), ABRASION, RECTIFICATION	SCHLEIFEN, MAHLEN, FEINVERKLEINERUNG, SCHLIFF

6107	**GRINDING AGENT, ABRASIVE MATERIAL, ABRADING AGENT**	MATIERE A AIGUISER, SUBSTANCE USANTE	SCHLEIFMITTEL, SCHÄRFMITTEL
6108	**GRINDING AND LAPPING COMPOUNDS**	PATE A SURFACER ET A RODER	OBERFLÄCHEN-U. EINSCHLEIFMASSE
6109	**GRINDING COOLANT**	REFRIGERANT DE MEULAGE	SCHLEIFKÜHLMITTEL
6110	**GRINDING FLOUR**	POUDRE ABRASIVE	SCHLEIFPULVER
6111	**GRINDING IN**	RODAGE A L'EMERI	EINSCHLEIFEN, EINSCHMIRGEIN
6112	**GRINDING LUBRICANT**	LUBRIFIANT DE RODAGE	SCHLEIFSCHMIERSTOFF
6113	**GRINDING MACHINE**	MACHINE A MEULER, MACHINE A DRESSER, MEULEUSE, MACHINE A RECTIFIER A LA MEULE	SCHLEIFMASCHINE (ZUM PUTZEN UND SCHLICHTEN), FEINSCHLEIFMASCHINE, PRÄZISIONSSCHLEIFMASCHINE
6114	**GRINDING PLATE**	MARBRE ·	SCHLEIFPLATTE
6115	**GRINDING WHEEL**	ROUE A MEULER, MEULE	SCHLEIFSCHEIBE
6116	**GRINDING WHEEL BALANCING DEVICES**	APPAREIL A EQUILIBRER LES MEULES	SCHLEIFSCHEIBEN-AUSGLEICHAPPARATE
6117	**GRINDING, CRUSHING**	BROYAGE, TRITURATION, PULVERISATION	MAHLEN
6118	**GRINDSTONE**	MEULE EN GRES, MEULE D'AFFUTAGE	SCHLEIFSTEIN, SCHLEIFSCHEIBE
6119	**GRINNING**	POUVOIR OPACIFIANT INSUFFISANT	UNGENÜGENDE DECKKRAFT
6120	**GRIP**	MORDACHES	FINSPANNBACKEN, PRESSBACKEN
6121	**GRIP DIE**	MANDRIN DE SERRAGE	SPANNBACKE
6122	**GRIPPING JAW**	MACHOIRE (D'ETAU)	SCHRAUBSTOCKBACKE
6123	**GRIT**	GRENAILLE FINE, ABRASIF	GRIT
6124	**GRIT BLASTING**	GRENAILLAGE (FIN)	SCHLEUDERSTRAHLEN, METALLSANDSTRAHLUNG
6125	**GROG**	ARGILE CALCINEE FONDUE, COULIS REFRACTAIRE	TON (GEBRANNTER), SCHAMOTTE (GEMAHLENE)
6126	**GROG FIRECLAY MORTAR**	COULIS REFRACTAIRE	SCHAMOTTEMÖRTEL
6127	**GROOVE**	RAINURE, GORGE, JOINT	NUT, FUGE, RILLE, RINNE
6128	**GROOVE (TO)**	RAINER, RAINURER	NUTEN, MIT EINER NUT VERSEHEN
6129	**GROOVE ANGLE**	ANGLE D'OUVERTURE DE LA RAINURE	FUGENÖFFNUNGSWINKEL
6130	**GROOVE FACE**	BORD A SOUDER	FUGENFLANKE
6131	**GROOVE OF SHEAVE PULLEY**	GORGE D'UNE POULIE	SEILSCHEIBENRILLE
6132	**GROOVE RADIUS**	RAYON D'ECARTEMENT ENTRE LES BORDS	FUGENRADIUS
6133	**GROOVE WELD**	SOUDURE DE BORD	FUGENNAHT
6134	**GROOVED DISC CAM**	CAME A RAINURE	NUTENSCHEIBE
6135	**GROOVED PULLEY**	POULIE A GORGE	SEILSCHEIBE
6136	**GROOVED-SERRATED ROLLER**	CYLINDRE CANNELE	WALZE (GERIFFELTE), PROFILWALZE
6137	**GROOVING**	RAINURAGE, RAINAGE	NUTEN
6138	**GROOVING MACHINE**	MACHINE A RAINER, MACHINE A CANNELER, RAINEUSE	RIFFELMASCHINE
6139	**GROOVING PLANE**	BOUVET FEMELLE	NUTHOBEL
6140	**GROOWED PULLEY**	POULIE A GORGE	RILLENSCHEIBE
6141	**GROSS WEIGHT**	POIDS BRUT	BRUTTOGEWICHT
6142	**GROUND AND POLISHED SURFACE**	SURFACE POLIE	SCHLIFFFLÄCHE
6143	**GROUND CLEARANCE**	GARDE AU SOL	BODENFREIHEIT
6144	**GROUND CORK**	ROGNURES DE LIEGE, POUDRE DE LIEGE	KORKSCHROT, KORKKLEIN, KORKMEHL
6145	**GROUND GLASS, FOCUSING SCREEN**	VERRE DEPOLI, GLACE DEPOLIE (PHOT.)	MATTSCHEIBE, VISIERSCHEIBE
6146	**GROUND IN STOPPER**	BOUCHON A L'EMERI	STÖPSEL (EINGESCHLIFFENER)

6147	**GROUND WATER**	EAUX SOUTERRAINES, NAPPE D'EAU SOUTERRAINE	GRUNDWASSER
6148	**GROUND WATER LEVEL**	NIVEAU DE L'EAU SOUTERRAINE	GRUNDWASSERSPIEGEL
6149	**GROUNDING OF A TANK, EARTHING OF A TANK**	MISE A LA TERRE	ERDUNG
6150	**GROUP DRIVE**	COMMANDE PAR GROUPES	GRUPPENANTRIEB
6151	**GROUP OF MACHINES**	GROUPE DE MACHINES	MASCHINENSATZ, MASCHINENAGGREGAT
6152	**GROUT**	LAIT DE CIMENT, COULIS	ZEMENTMILCH
6153	**GROUT (TO)**	COULER, BOURRER, SCELLER DE CIMENT	VERGIESSEN (MIT MÖRTEL, ZEMENT)
6154	**GROUTING**	COULAGE DE CIMENT	VERGIESSEN (MIT ZEMENT, MÖRTEL)
6155	**GROWTH**	GONFLEMENT, CROISSANCE	WACHSEN, WACHSTUM
6156	**GRUB SCREW**	VIS SANS TETE	SCHNITTSCHRAUBE, MADENSCHRAUBE, WURMSCHRAUBE, GEWINDESTIFT
6157	**GUARD CRADLE, NET, CATCH NET**	FILET PROTECTEUR	SCHUTZNETZ
6158	**GUDGEON PIN**	AXE DE PISTON	KOLBENBOLZEN
6159	**GUDGEON, PIN FOR FORKED END OF A ROD**	AXE D'ARTICULATION D'UNE FOURCHE	GABELZAPFEN
6160	**GUIDE**	GLISSIERE, GUIDE, PLAQUE DE GARDE, GUIDAGE	GLEITBAHN, FÜHRUNG, ABSTREIFPLATTE
6161	**GUIDE PULLEY**	GALET-GUIDE, GALET DE GUIDAGE, POULIE DE RENVOI, POULIE-GUIDE	LEITROLLE, FÜHRUNGSROLLE
6162	**GUIDE RAIL**	RAIL-GUIDE	GLEITSCHIENE, FÜHRUNGSSCHIENE
6163	**GUIDE RING**	BAGUE DE GUIDAGE	FÜHRUNGSRING
6164	**GUIDE ROD**	TIGE DE GUIDAGE, TRINGLE DE GUIDAGE, BARRE DE GUIDAGE	FÜHRUNGSSTANGE
6165	**GUIDE SCREW**	VIS-MERE	LEITSPINDEL
6166	**GUIDE SURFACE**	SURFACE DE GUIDAGE	FÜHRUNGSBAHN
6167	**GUIDE VANE, STATIONARY BLADE OF A TURBINE**	AUBE DIRECTRICE, AUBE FIXE D'UNE TURBINE	LEITSCHAUFEL EINER TURBINE
6168	**GUIDE WHEEL, GUIDE RING OF A TURBINE**	SATELLITE, COURONNE FIXE, COURONNE DIRECTRICE D'UNE TURBINE	LEITRAD, LEITRAD EINER TURBINE
6169	**GUIDES, SLIDE BARS, GUIDE BARS, MOTION BARS (FOR CROSSHEAD)**	GLISSIERE DE CROSSE, GUIDE DE LA CROSSE	GERADFÜHRUNG (DAMPFM.), GLEITBAHN (DAMPFM.), GLEITSCHIENE (DAMPFM.)
6170	**GUILLOTINE SHEARS**	CISAILLES A GUILLOTINE	RAHMENBLECHSCHERE
6171	**GUINIER-PRESTON ZONES**	ZONE DE GUINIER-PRESTON	GUINIER-PRESTON ZONE
6172	**GUM ARABIC, GUM ACACIA**	GOMME ARABIQUE	GUMMI (ARABISCHES)
6173	**GUM RESIN**	GOMME-RESINE	GUMMIHARZ, SCHLEIMHARZ
6174	**GUM TRAGACANTH, GUM DRAGON**	GOMME ADRAGANTE	TRAGANT, GUMMITRAGANT, TRAGANTGUMMI
6175	**GUN CARRIAGE**	AFFUT	LAFETTE
6176	**GUN COTTON, PYROXYLIN, CELLULOSE HEXANITRATE, NITROCELLULOSE**	FULMICOTON, COTONPOUDRE, NITROCELLULOSE, CELLULOSE NITREE, PYROXYLE, PYROXYLINE, XYLOIDINE	SCHIESSBAUMWOLLE, NITROZELLULOSE, PYROXYLIN, ZELLULOSENITRAT
6177	**GUN DRILL**	FORET A CANON	KANONENBOHRER
6178	**GUN METAL**	BRONZE A CANON	KANONENGUT, KANONENMETALL, GESCHÜTZBRONZE
6179	**GUN SPRAYING**	PEINTURE AU PISTOLET	FARBSPRITZVERFAHREN
6180	**GUN WELDING MACHINE**	PISTOLET DE SOUDAGE PAR POINTS	STOSSPUNKTER

6181	GUNPOWDER, BLACK POWDER	POUDRE NOIRE	SCHIESSPULVER
6182	GUSSET	GOUSSET	ECKBLECH, ZWICKEL
6183	GUT BAND	COURROIE EN BOYAUX	DARMSAITENRIEMEN
6184	GUTTA PERCHA	GOMME PLASTIQUE, GOMME DE SUMATRA, GOMME GETTANIA	GUTTAPERCHA
6185	GUTTER	CANAL DE TROP-PLEIN, RIGOLE, GOUTTIERE	ÜBERLAUFROHR, RINNE
6186	GUY CABLES	TRIRANTS	HALTESEILE, ANKERSEILE
6187	GUY, STAY ROPE	HAUBAN	ANKERDRAHT, VERANKERUNGSDRAHT, SPANNDRAHT
6188	GYPSUM CEMENT MOLD	MOULE EN PLATRE	GIPSFORM
6189	GYPSUM CEMENT PATTERN	MODELE EN PLATRE	GIPSMODELL
6190	GYPSUM, HYDROUS SULPHATE OF CALCIUM	SULFATE DE CALCIUM, GYPSE, PIERRE A PLATRE	GIPS, SCHWEFELSAURER KALK, KALZIUMSULFAT
6191	GYROSCOPE	GYROSCOPE	KREISEL
6192	GYROSCOPIC ACTION	ACTION GYROSCOPIQUE	KREISELWIRKUNG
6193	GYROSCOPIC MOTION	MOUVEMENT GYROSCOPIQUE	KREISELBEWEGUNG
6194	H-IRON	FER EN DOUBLE T, FER EN I	DOPPEL-T-EISEN, I-EISEN
6195	HACKLY FRACTURE	CASSURE HACHEE, CASSURE CROCHUE	BRUCH (ZACKIGER), BRUCH (HAKIGER)
6196	HACKSAW	SCIE A METAUX	BÜGELSÄGE
6197	HACKSAW BLADE	LAME POUR SCIE A METAUX	METALLSÄGEBLATT
6198	HACKSAWING MACHINE	MACHINE A SCIER ALTERNATIVE	BÜGELSÄGEMASCHINE
6199	HAFNIUM	HAFNIUM	HAFNIUM
6200	HAIR BELT	COURROIE EN POILS, COURROIE EN CRIN	HAARRIEMEN
6201	HAIR COMPASSES	COMPAS A CHEVEU, COMPAS DE PRECISION	HAARZIRKEL
6202	HAIR CRACK	FISSURE CAPILLAIRE/MICROCRIQUE	HAARRISS
6203	HALATION	FORMATION DE HALO	LICHTHOFBILDUNG
6204	HALF CELL	DEMI-CELLULE, HEMICELLULE	HALBZELLE
6205	HALF OF THE COUPLING	DEMI-MANCHON	KUPPLUNGSHÄLFTE, KUPPELSCHEIBE
6206	HALF SIDE MILLING CUTTERS	FRAISES A UNE TAILLE LATERALE	SCHEIBENFRÄSER MIT EINSEITIGER STIRNVERZAHNUNG
6207	HALF-CROSSED QUATERING BELT	COURROIE DEMI-CROISEE, COURROIE TORDUE	RIEMENTRIEB (HALBGESCHRÄNSKTER), HALBKREUZRIEMENTRIEB
6208	HALF-ROUND FILE	LIME DEMI-RONDE	HALBRUNDFEILE
6209	HALF-ROUND IRON	PROFILE MI-ROND	HALBRUNDPROFILEISEN
6210	HALF-ROUND TIMBER	BOIS MI-PLAT	HALBHOLZ
6211	HALF-VALUE THICKNESS	COUCHE DE DEMI-ATTENUATION	HALBWERTSCHICHT
6212	HALF-WAVE RECTIFICATION	RECTIFICATION DEMI-ONDE	HALBWELLENGLEICHRICHTEN
6213	HALO	HALO	HALO
6214	HALOGEN	HALOGENE	SALZBILDNER, HALOGEN
6215	HAMMER	MARTEAU	HAMMER
6216	HAMMER (TO)	MARTELER	HÄMMERN
6217	HAMMER SCALE, FORGE SCALE	BATTITURES, PAILLE DE FER, MARTELURES, SCORIES DE FORGE	GLÜHSPAN, ZUNDER, HAMMERSCHLAG, SCHMIEDESINTER
6218	HAMMER WELD	SOUDAGE A LA FORGE	HAMMERSCHWEISSEN
6219	HAMMER-HARDENING	ECROUISSAGE	HÄRTEN
6220	HAMMERED	MARTELE	GEHÄMMERT
6221	HAMMERED METALWORK	METAL MARTELLE	METALL (GEHÄMMERTES)
6222	HAMMERED STEEL	ACIER BATTU	HAMMERSTAHL

6223	HAMMERING	MARTELAGE	HÄMMERN
6224	HAMMERING SHARP CLOSING OF THE VALVE	CHOC DE LA SOUPAPE	VENTILSCHLAG, SCHLAGEN, ANPRALL DES VENTILS
6225	HAMMERMANN	FORGERON A MAIN	ZUSCHLÄGER
6226	HAND	AIGUILLE (DE MANOMETRE)	ZEIGER
6227	HAND BRAKE	FREIN A MAIN	HANDBREMSE
6228	HAND BURNISHING	POLISSAGE A LA MAIN	HANDGLANZSCHLEIFEN
6229	HAND CAPSTAN	CABESTAN A BRAS	GANGSPILL
6230	HAND CART	CHARRETTE A BRAS	HANDKARREN
6231	HAND CRANK	MANIVELLE A MAIN	HANDKURBEL, GRIFFKURBEL
6232	HAND CUTTING	COUPE A LA MAIN	HANDSCHNEIDEN
6233	HAND DRIVING	COMMANDE A LA MAIN	HANDANTRIEB, ANTRIEB VON HAND, HANDBETRIEB
6234	HAND FILE	LIME A MAIN	HANDFEILE
6235	HAND HAMMER	MARTEAU A MAIN	SCHMIEDEHAMMER, HANDHAMMER, BANKHAMMER, FAUSTHAMMER
6236	HAND HOLE	TROU A MAIN, TROU DE BRAS	HANDLOCH
6237	HAND LABOUR, MANUAL LABOUR	TRAVAIL MANUEL	HANDARBEIT
6238	HAND LADLE	POCHE DE COULEE A MAIN	HANDPFANNE
6239	HAND LEVER	LEVIER A POIGNEE	HANDHEBEL
6240	HAND MADE	TRAVAILLE A LA MAIN	HANDGEFERTIGT, HANDGEARBEITET
6241	HAND OPERATED, MOVED BY MANUAL POWER, ACTUATED BY HAND	A COMMANDE A LA MAIN	HAND ANGETRIEBEN, HANDBETRIEB (FÜR)
6242	HAND PART PROGRAMMING	PROGRAMMATION MANUELLE DE PIECE	HAND-TEILEPROGRAMM
6243	HAND PRINTING	ECRITURE PERPENDICULAIRE, ECRITURE DROITE	STEILSCHRIFT
6244	HAND PUMP	POMPE A MAIN, A BRAS	HANDPUMPE
6245	HAND RAILLING	FER MAIN-COURANTE	HANDLEISTENEISEN
6246	HAND RAMMER	FOULOIR A MAIN	HANDSTAMPFER
6247	HAND REAMER	ALESOIR A MAIN	HANBREIBAHLE
6248	HAND RIVETING	RIVETAGE A LA MAIN, RIVETAGE AU MARTEAU	HANDNIETUNG
6249	HAND SAW	SCIE A MAIN	HANDSÄGE
6250	HAND SHANK LADLE	POCHE A MAIN A MANCHE	STIELPFANNE
6251	HAND SHEARS	CISAILLES A MAIN	HANDSCHERE
6252	HAND SHIELD	MASQUE, ECRAN	HANDSCHIRM
6253	HAND TOOL	OUTIL A MAIN	HANDWERKZEUG
6254	HAND VICE	ETAU A MAIN	FEILKLOBEN, HANDKLOBEN
6255	HAND WHEEL	VOLANT DE MANOEUVRE	HANDRAD
6256	HAND WINCH	TREUIL A BRAS, TREUIL A MANIVELLE	HANDWINDE
6257	HAND WORKED TAP	TARAUD A MAIN	HANDGEWINDEBOHRER
6258	HAND-DRILL	CHIGNOLE	HANDBOHRMASCHINE
6259	HAND-RAIL	GARDE-CORPS	TREPPENGELÄNDER
6260	HAND-WARM PREHEATING	DEGOURDISSAGE (PRECHAUFFAGE AVANT SOUDAGE)	ANWÄRMEN, VORWÄRMEN
6261	HAND-WROUGHT	FORGE A LA MAIN	HANDGESCHMIEDET
6262	HANDLE	POIGNEE, MANCHE	HANDGRIFF, HANDHABE
6263	HANDLE OF A TOOL	MANCHE D'UN OUTIL	HEFT EINES WERKZEUGES, GRIFF EINES WERKZEUGS
6264	HANDLING HOLES	TROUS DE MANIPULATION	ANGREIFLÖCHER

6265	HANDRAIL IRON	FER A MAIN COURANTE	HANDLÄUFEREISEN, HANDLEISTENEISEN, GELÄNDEREISEN
6266	HANDRAIL POST	MONTANT DE GARDE-CORPS	GELÄNDERSTÜTZE
6267	HANDWHEEL	VOLANT A MAIN	HANDRAD
6268	HANDWHEEL RETAINING NUT	ECROU DE VOLANT	HANDRADMUTTER
6269	HANDY, MANAGEABLE	MANIABLE	HANDLICH
6270	HANGER WITH DOUBLE SUPPORT	PALIER PENDANT FERME	HÄNGELAGER (GESCHLOSSENES)
6271	HANGER WITH SINGLE SUPPORT	PALIER PENDANT OUVERT	HÄNGELAGER (OFFENES)
6272	HANSGIRG PROCESS	PROCEDE HANSGIRG	VERFAHREN (HANSGIRGSCHE)
6273	HARD BURNT STOCK BRICK	BRIQUE DURE	HARTBRANDSTEIN
6274	HARD DRAWN	TREMPE PRODUITE PAR ETIRAGE A FROID	HÄRTUNGSGRAD BEIM KALTZIEHEN
6275	HARD GREASE	GRAISSE CONSISTANTE	STARRFETT, FETT KONSISTENTES, STARRSCHMIERE
6276	HARD LEAD	PLOMB ANTIMONIEUX, PLOMB DURCI, METAL BLANC DUR	HARTBLEI, ANTIMONBLEI, WEISSMETALL (HARTES)
6277	HARD METAL	METAL DUR	HARTMETALL
6278	HARD SETTING	POSE DE PLAQUETTES DE COUPE METAL DUR	HARTMETALLBESTÜCKUNG
6279	HARD SOLDER, BRAZING METAL	BRASURE, SOUDURE FORTE (COMPOSITION FUSIBLE)	HARTLOT, SCHLAGLOT, STRENGLOT
6280	HARD STEEL	ACIER DUR	STAHL (HARTER), HARTSTAHL
6281	HARD WATER	EAU DURE, EAU CALCAIRE	WASSER (HARTES), WASSER (KALKHALTIGES)
6282	HARD WOOD	BOIS DUR, LOURD	HARTHOLZ, HARTES HOLZ
6283	HARD-SURFACING	RECHARGEMENT DUR	HARTMETALL-AUFTRAGSCHWEISSUNG
6284	HARD-SURFACING ALLOY	ALLIAGE DE RECHARGEMENT DUR	HARTMETALL-AUFTRAGSLEGIERUNG
6285	HARDEN A METAL (TO)	TREMPER UN METAL	METALL HÄRTEN (EIN)
6286	HARDEN IN THE AIR (TO)	DURCIR, FAIRE PRISE, SE SOLIDIFIER A L'AIR (CIMENT)	AN DER LUFT ERHÄRTEN (ZEMENT)
6287	HARDEN UNDER WATER (TO)	DURCIR, FAIRE PRISE, SE SOLIDIFIER DANS L'EAU (CIMENT)	IM WASSER ERHÄRTEN (ZEMENT)
6288	HARDENABILITY	APTITUDE A LA TREMPE, TREMPABILITE	HÄRTBARKEIT
6289	HARDENED STEEL	ACIER TREMPE	STAHL (GEHÄRTETER)
6290	HARDENER	DURCISSEUR (ALLIAGE)	HÄRTUNGSMITTEL
6291	HARDENING	DURCISSEMENT, TREMPE	HÄRTUNG, AUSHÄRTUNG
6292	HARDENING (CEMENT, MORTAR)	DURCISSEMENT, SOLIDIFICATION	ERHÄRTEN, ERHÄRTUNG (MÖRTEL, ZEMENT)
6293	HARDENING (METALS)	TREMPE DES METAUX	HÄRTEN VON METALLEN
6294	HARDENING BATH, MIXTURE	BAIN DE TREMPE	HÄRTUNGSBAD, HÄRTUNGSFLÜSSIGKEIT
6295	HARDENING OIL	HUILE DE TREMPE	HÄRTEÖL
6296	HARDENING QUALITY OF A METAL	FACULTE DE PRENDRE BIEN LA TREMPE	HÄRTBARKEIT EINES METALLES
6297	HARDNESS	DURETE	HÄRTE
6298	HARDNESS OF WATER	DURETE DE L'EAU, DEGRE HYDROTIMETRIQUE	HÄRTE DES WASSERS
6299	HARDNESS TEMPER OF STEEL	DEGRE DE TREMPE DE L'ACIER	HÄRTEGRAD DES STAHLES
6300	HARDNESS TEST, HARDNESS TESTING	ESSAI DE DURETE	HÄRTEPRÜFUNG, HÄRTEPROBE
6301	HARDWOOD	BOIS FEUILLU	HARTHOLZ
6302	HARMONIC MOTION	MOUVEMENT PENDULAIRE	SINUSSCHWINGUNG, SCHWINGUNGZ HARMONISCHE BEWEGUNG

6303	HARROWS : CHAIN, ZIG-ZAG, FLEXIBLE, SADDLE-BACK, ETC.	HERSES (A CHAINE, HERSES A ZIG-ZAG, HERSES FLEXIBLE)	EGGEN (GLEIDEREGGEN, ZIGZAGEGGEN)
6304	HATCH (TO)	HACHURER	SCHRAFFIEREN
6305	HATCHET	HACHETTE, HACHEREAU, HACHERON	BEIL
6306	HATCHING	HACHURE	SCHRAFFUR, SCHRAFFIERUNG
6307	HAULAGE HOOKS	CROCHETS DE TRACTION	ZUGHAKEN
6308	HAVING A HIGH MELTING POINT	A POINT DE FUSION ELEVE	HOCHSCHMELZEND
6309	HAWSER-LAID ROPE	AUSSIERE	SEIL (TROSSWEISE GESCHLAGENES)
6310	HAY FORKS : GRABS AND STACKERS	DECHARGEURS A GRIFFES (MAT. ET GRUE)	HEUGABELN UND GREIFER
6311	HAY MAKERS : COMBINE	RATELEUSES RAMASSEUSES	HEUMÄHDRESCHBINDER
6312	HAY RAKES	RATEAUX	HEURECHEN
6313	HAZE	VOILE	SCHLEIERBILDUNG
6314	HEAD	FOND, EXTREMITE, PRESSION (CHAUD.)	TANKBODEN, KOPF, DRUCK
6315	HEAD LOSS	PERTE DE CHARGE	DRUCKVERLUST
6316	HEAD OF A DRUM	FOND D'UN BALLON	KESSELTROMMELBODEN
6317	HEAD OF A HAMMER	CORPS DU MARTEAU	HAMMERKOPF, KOPF EINES HAMMERS
6318	HEAD OF KEY, KEY HEAD	TALON D'UNE CLAVETTE, TETE D'UNE CLAVETTE, NEZ D'UNE CLAVETTE, MENTONNET D'UNE CLAVETTE	KEILNASE
6319	HEAD RACE	CANAL D'ARRIVEE, BIEF D'AMONT	ZULEITUNGSKANAL, OBERWASSERKANAL
6320	HEAD REST	APPUIE-TETE	KOPFLEHNE
6321	HEAD SHAFT HANGER, HANG DOWN	PALIER PENDANT, PENDANT, PALIER A POTENCE	HÄNGELAGER, DECKENLAGER
6322	HEAD WATER	EAU D'AMONT	OBERWASSER
6323	HEAD, PRESSURE HEAD, STATIC HEAD	CHARGE D'EAU	DRUCKHÖHE (HÖHE DER FLÜSSIGKEITSSÄULE)
6324	HEADER	BOUTISSE	BINDER (BAUW.)
6325	HEADING TOOL	CLOUTIERE, CLOUERE, CLOUIERE, CLOUTERE, CLOUVIERE	NAGELEISEN
6326	HEADLAMP	PHARE, PROJECTEUR	SCHEINWERFER
6327	HEADLIGHT	PHARE	SCHEINWERFER
6328	HEADSTOCK	POUPEE DE TOUR, POUPEE FIXE	REITSTOCK, SPINDELSTOCK
6329	HEART CAM	CAME, EXCENTRIQUE EN COEUR	HERZSCHEIBE, HERZEXZENTER
6330	HEART-SHAPED THIMBLE	COSSE OVALE	HERZKAUSCHE
6331	HEARTH	CREUSET, SOLE	HERD
6332	HEARTH FURNACE	FOUR A SOIE	HERDOFEN
6333	HEAT	CHALEUR, PIQUEE, COULEE	HITZE, WÄRME, ABSTICH, SCHMELZE
6334	HEAT (TO), RUN HOT (TO)	ECHAUFFER, RECHAUFFER	WARMLAUFEN, HEISSLAUFEN, ERHITZEN, ERWÄRMEN
6335	HEAT ABSORPTION CAPACITY, ABSORBING CAPACITY FOR HEAT	CAPACITE CALORIFIQUE	WÄRMEAUFNAHMEFÄHIGKEIT, WÄRMEKAPAZITÄT
6336	HEAT ACCUMULATOR	ACCUMULATEUR THERMIQUE DE CHALEUR	WÄRMESPEICHER, WÄRMEAKKUMULATOR
6337	HEAT BALANCE	BILAN CALORIFIQUE, BILAN THERMIQUE	WÄRMEBILANZ
6338	HEAT BUILD UP	ECHAUFFEMENT INTERNE	WÄRMEENTWICKLUNG
6339	HEAT CAPACITY	CAPACITE CALORIFIQUE	WÄRMEKAPAZITÄT
6340	HEAT CONDUCTOR	CONDUCTEUR DE LA CHALEUR	WÄRMELEITER
6341	HEAT CONSTANT	CONSTANTE CALORIFIQUE	WÄRMEKONSTANTE

6342	HEAT CONTENT	CAPACITE CALORIQUE	WÄRMEINHALT
6343	HEAT DUE TO FRICTION	CHALEUR PRODUITE PAR FROTTEMENT	REIBUNGSWÄRME
6344	HEAT ENERGY	ENERGIE CALORIFIQUE, ENERGIE THERMIQUE	WÄRMEENERGIE
6345	HEAT ENGINE, THERMODYNAMIC ENGINE	MACHINE, MOTEUR THERMIQUE	WÄRMEKRAFTMASCHINE
6346	HEAT ENTROPY DIAGRAM	DIAGRAMME ENTROPIQUE	WÄRMEDIAGRAMM, TS-DIAGRAMM, ENTROPIEDIAGRAMM
6347	HEAT EQUIVALENT OF THE WORK DONE	EQUIVALENT CALORIFIQUE DU TRAVAIL	KALORISCHES ARBEITSÄQUIVALENT, WÄRMEWERT DER ARBEITSEINHEIT
6348	HEAT EVOLVED, ABSORBED IN REACTIONS	NOMBRE DE CALORIES DEGAGEES DANS UNE REACTION, CHALEUR DEGAGEE DANS UNE REACTION	WÄRMETÖNUNG
6349	HEAT EXCHANGER	ECHANGEUR DE CHALEUR, ECHANGEUR THERMIQUE	WÄRMEAUSTAUSCHER
6350	HEAT EXCHANGER TUBE	TUBE D'ECHANGEUR THERMIQUE	WÄRMETAUSCHERROHR
6351	HEAT IN CONTACT WITH AIR (TO)	CHAUFFER EN PRESENCE, CHAUFFER AU CONTACT DE L'AIR	UNTER LUFTZUTRITT ERHITZEN
6352	HEAT INSULATOR	SUBSTANCE CALORIFUGE, CALORIFUGE	WÄRMESCHUTZSTOFF, WÄRMESCHUTZMITTEL
6353	HEAT LOSS	PERTE DE CHALEUR, PERTE THERMIQUE	WÄRMEVERLUST
6354	HEAT OF ABSORPTION	CHALEUR D'ABSORPTION	ABSORPTIONSWÄRME
6355	HEAT OF CARBURIZATION	CHALEUR DE CARBURATION	KARBONISIERUNGSWÄRME
6356	HEAT OF COMBUSTION	CHALEUR DE COMBUSTION	VERBRENNUNGSWÄRME
6357	HEAT OF DECOMPOSITION	CHALEUR DE DECOMPOSITION	ZERSETZUNGSWÄRME
6358	HEAT OF DISSOCIATION	CHALEUR DE DISSOCIATION	DISSOZIATIONSWÄRME
6359	HEAT OF EVAPORATION	CHALEUR D'EVAPORATION	VERDAMPFUNGSWÄRME, VERDUNSTUNGSWÄRME
6360	HEAT OF FUSION	CHALEUR DE FUSION, TEMPERATURE DE FUSION	SCHMELZWÄRME
6361	HEAT OF OXIDATION	CHALEUR D'OXYDATION	OXYDATIONSWÄRME
6362	HEAT OF REDUCTION	CHALEUR DE REDUCTION	REDUKTIONSWÄRME
6363	HEAT OF SUBLIMATION	CHALEUR DE SUBLIMATION	SUBLIMATIONSWÄRME
6364	HEAT OF TRANSFORMATION	CHALEUR DE TRANSFORMATION	UMWANDLUNGSWÄRME
6365	HEAT OF TRANSITION	CHALEUR LATENTE	WÄRME (LATENTE)
6366	HEAT OF VAPORIZATION	CHALEUR DE VAPORISATION	VERDAMPFUNGSWÄRME, VERDUNSTUNGSWÄRME
6367	HEAT OUT OF CONTACT WITH AIR (TO)	CHAUFFER EN VASE CLOS	ERHITZEN (UNTER LUFTABSCHLUSS)
6368	HEAT PENETRATION	PENETRATION DE LA CHALEUR	WÄRMEEINDRINGTIEFE
6369	HEAT RAY	RAYON CALORIFIQUE	WÄRMESTRAHL
6370	HEAT SET FREE, HEAT EVOLVED	CHALEUR DEGAGEE	WÄRME (FREIGEWORDENE)
6371	HEAT TINTING	COLORATION THERMIQUE	WÄRMETÖNUNG
6372	HEAT TRANSFER	TRANSMISSION, TRANSFERT DE CHALEUR	WÄRMEÜBERGANG, WÄRMEÜBERTRAGUNG
6373	HEAT TRANSFER COEFFICIENT	COEFFICIENT DE CONDUCTIBILITE CALORIFIQUE	WÄRMEÜBERGANGSZAHL
6374	HEAT TRANSMISSION	TRANSMISSION DE CHALEUR, TRANSPORT DE CHALEUR	WÄRMEÜBERGANG, WÄRMEÜBERTRAGUNG
6375	HEAT TREATING	TRAITEMENT THERMIQUE	WÄRMEBEHANDLUNG, VERGÜTUNG
6376	HEAT TREATING PLANT	INSTALLATION DE REVENU	VERGÜTEANLAGE
6377	HEAT TREATMENT, THERMAL TREATMENT	TRAITEMENT THERMIQUE	WARMBEHANDLUNG, GLÜHBEHANDLUNG
6378	HEAT VALUE	POUVOIR CALORIFIQUE	HEIZWERT

6379	HEAT WAVE	ONDE CALORIFIQUE	WÄRMEWELLE
6380	HEAT-AFFECTED ZONE	ZONE INFLUENCEE PAR LA CHALEUR	WÄRMEEINFLUSSZONE
6381	HEAT-PRODUCING LOSSES	PERTES THERMOGENES	WÄRMEERZEUGENDE VERLUSTE
6382	HEAT-TIME	TEMPS D'ECOULEMENT DE COURANT, TEMPS DE SOUDAGE EFFECTIF	STROMZEIT
6383	HEAT-TREATABLE ALLOY	ALLIAGE DE TRAITEMENT	VERGÜTUNGSLEGIERUNG
6384	HEATED PISTON	PISTON CHAUFFE	KOLBEN (GEHEIZTER)
6385	HEATER	RECHAUFFEUR	ANWÄRMER, ERHITZER
6386	HEATER FILAMENT	FILAMENT	GLÜHDRAHT
6387	HEATER PLUG	BOUGIE DE PRECHAUFFAGE	GLÜHKERZE
6388	HEATER PLUG RESISTANCE	RESISTANCE DE BOUGIE DE PRECHAUFFAGE	GLÜHKERZENWIDERSTAND
6389	HEATER PLUG SWITCH	COMMUTATEUR DES BOUGIES DE PRECHAUFFAGE	GLÜHANLASSSCHALTER
6390	HEATING	CHAUFFAGE, CHAUFFE	ERHITZEN, ERWÄRMEN, ERWÄRMUNG
6391	HEATING BLOWPIPE	CHALUMEAU DE PRECHAUFFAGE	ANWÄRMBRENNER
6392	HEATING CALORIFIC THERMAL VALUE, FUEL VALUE	PUISSANCE CALORIFIQUE, POUVOIR CALORIFIQUE	HEIZWERT
6393	HEATING GATE	TROU DE SOUFFLAGE	BLASLOCH
6394	HEATING INSTALLATION	INSTALLATION DE CHAUFFAGE CENTRAL	HEIZUNGSANLAGE
6395	HEATING OF BUILDINGS	CHAUFFAGE DES BATIMENTS	HEIZUNG VON GEBÄUDEN
6396	HEATING PIPE	TUYAU DE CHAUFFAGE CENTRAL	HEIZUNGSROHR
6397	HEATING STEAM	VAPEUR DE CHAUFFAGE	HEIZDAMPF
6398	HEATING SURFACE, GENERATING SURFACE	SURFACE DE CHAUFFE	HEIZFLÄCHE
6399	HEATING TORCH	CHALUMEAU DE CHAUFFAGE	ANWÄRMBRENNER
6400	HEATING UP, FIRING OF BEARINGS	ECHAUFFEMENT DES COUSSINETS	WARMLAUFEN, HEISSLAUFEN DER LAGER
6401	HEAVY CURRENT	COURANT DE GRANDE INTENSITE, COURANT FORT	STARKSTROM
6402	HEAVY DUTY GEAR WHEEL	ENGRENAGE DE FORCE, ENGRENAGE DE GRANDE FATIGUE	KRAFTZAHNRAD
6403	HEAVY DUTY OIL	HUILE SPECIALE, HUILE H.D.	HEAV DUTY ÖL, H.D. ÖL
6404	HEAVY DUTY PLAIN MOLLING CUTTERS	FRAISES SIMPLES POUR TRAVAUX LOURDS	FRÄSER (EINFACHER) FÜR SCHWERE SCHNITTE
6405	HEAVY DUTY TRUCK (U.S.)	CAMION POIDS LOURD	SCHWERKRAFTLASTWAGEN
6406	HEAVY FORGING	PIECE DE GROSSE FORGE	SCHMIEDESTÜCK (SCHWERES)
6407	HEAVY METAL	METAL LOURD	SCHWERMETALL
6408	HEAVY OIL, THICK OIL	HUILE LOURDE, MAZOUT	SCHWERÖL, MASUT
6409	HEAVY PLATE	TOLE FORTE	GROBBLECH
6410	HEAVY PLATE MILL	LAMINOIR POUR TOLES FORTES	GROBWALZWERK
6411	HEAVY PLATE PRESS	PRESSE POUR TOLES FORTES	GROBBLECHPRESSE
6412	HEAVY RUNNING	MARCHE LOURDE	GANG (SCHWERER), GANG (HARTER), GANG (STOSSENDER)
6413	HEAVY SECTIONS	PROFILES LOURDS	PROFILE, SCHWERE
6414	HEAVY WATER	EAU LOURDE	SCHWERWASSER
6415	HECTOLITRE	HECTOLITRE	HEKTOLITER
6416	HECTOWATT	HECTOWATT	HEKTOWATT
6417	HECTOWATT/HOUR	HECTOWATT-HEURE	HECTOWATTSTUNDE

6418	HEDGE CLIPPERS AND CUTTERS	MACHINES POUR COUPER LES HAIES	HECKENSCHNEIDER
6419	HEFNER CANDIE	BOUGIE, ETALON HEFNER	HEFNERKERZE
6420	HEIGHT	HAUTEUR	HÖHE
6421	HEIGHT FALLEN THROUGH	HAUTEUR DE CHUTE	FALLHÖHE
6422	HEIGHT OF THE BAROMETER, BAROMETRIC HEIGHT	NIVEAU BAROMETRIQUE	BAROMETERSTAND
6423	HEIGHT UNDER CROSS-RAIL	HAUTEUR SOUS TRAVERSE	HÖHE UNTER QUERBALKEN
6424	HELIARC TORCH	TORCHE A HELIUM	HELIARC-BRENNER
6425	HELICAL GEAR WHEEL	ROUE HELICOIDALE, ENGRENAGE HELICOIDAL	SCHRÄGZAHNRAD, SPIRALZAHNSTIRNRAD, SCHRAUBENRAD
6426	HELICAL MOTION	MOUVEMENT HELICOIDAL	BEWEGUNG AUF EINER SCHRAUBENLINIE, SCHRAUBENBEWEGUNG
6427	HELICAL SPRING	RESSORT A BOUDIN, RESSORT EN HELICE	SCHRAUBENFEDER
6428	HELICOIDAL SURFACE	SURFACE HELICOIDALE, HELICOIDE	SCHRAUBENFLÄCHE
6429	HELIUM	HELIUM	HELIUM
6430	HELIX	HELICE	SCHRAUBENLINIE, SCHNECKE
6431	HELIX ANGLE	ANGLE HELICOIDAL	STEIGUNGSWINKEL
6432	HELMET SHIELD	CASQUE, MASQUE DE SOUDAGE	SCHUTZHAUBE
6433	HELVE	MANCHE D'UNE HACHE	HELM EINER AXT, HELM EINES BEILES
6434	HEMATITE	HEMATITE	HÄMATIT
6435	HEMATITE PIG IRON	FONTE HEMATITE	HÄMATITROHEISEN
6436	HEMIHEDRAL CRYSTAL	CRISTAL HEMIEDRIQUE	HEMIEDERKRISTALL
6437	HEMIMORPHIC CRYSTAL	CRISTAL HEMIMORPHE	KRISTALL (HEMIMORPHER)
6438	HEMISPHERE	DEMI-SPHERE, HEMISPHERE	HALBKUGEL
6439	HEMP	CHANVRE, FILASSE DE CHANVRE	HANF
6440	HEMP CORE	AME EN CHANVRE	HANFSEELE
6441	HEMP GASKET	TRESSE EN CHANVRE	HANFZOPF
6442	HEMP GASKET GREASED WITH TALLON	TRESSE DE CHANVRE SUIFFEE	HANFZOPF (GEFETTETER)
6443	HEMP ROPE	CABLE DE CHANVRE, CORDE DE CHANVRE	HANFSEIL
6444	HEMP TOW	ETOUPE DE CHANVRE	HANFWERG
6445	HEMPEN PACKING	BOURRAGE, GARNITURE EN CHANVRE	HANFPACKUNG, HANFDICHTUNG, HANFLIDERUNG
6446	HEMPSEED OIL	HUILE DE CHENEVIS	HANFÖL
6447	HERMETICALLY CLOSED, SEALED	HERMETIQUEMENT CLOS, FERME	LUFTDICHT, HERMETISCH VERSCHLOSSEN
6448	HERRINGBONE GEARS	ENGRENAGES A CHEVRONS	PFEILZAHNRÄDER
6449	HESSIAN, PACKING CANVAS	TOILE D'EMBALLAGE, TOILE DE JUTE	PACKLEINWAND
6450	HETEROCHROMATIC X-RADIATION	RAYONNEMENT HETEROGENE	HETEROGENE STRAHLUNG
6451	HETEROGENEOUS	HETEROGENE	HETEROGENER
6452	HETEROGENEOUS TEXTURE	TEXTURE HETEROGENE	GEFÜGE (UNGLEICHARTIGES), GEFÜGE (INHOMOGENES), GEFÜGE (HETEROGENES)
6453	HEXAGON	HEXAGONE	SECHSECK
6454	HEXAGON BAR	FER HEXAGONAL	SECHSKANTEISEN
6455	HEXAGON NUT	ECROU A SIX PANS	SECHSKANTMUTTER
6456	HEXAGON NUT ANGLE GAUGE	EQUERRE A SIX PANS	SECHSKANTWINKEL, SECHSKANTE
6457	HEXAGON-HEAD SCREW	VIS, BOULON A TETE A SIX PANS	SECHSKANTSCHRAUBE

6458	**HEXAGONAL CLOSE-PACKED STRUCTURE**	STRUCTURE HEXAGONALE COMPACTE	HEXAGONAL-DICHTGEPACKTE STRUKTUR
6459	**HEXAGONAL CRYSTAL**	CRISTAL HEXAGONAL	HEXAGONALER KRISTALL
6460	**HEXAGONAL PRISM**	PRISME HEXAGONAL	PRISMA (SECHSSEITIGES)
6461	**HEXAGONAL SCREW HEAD, BOLT HEAD**	TETE A SIX PANS D'UNE VIS	SECHSKANTSCHRAUBENKOPF
6462	**HEXAGONAL WIRE**	FIL HEXAGONAL	SECHSKANTDRAHT
6463	**HEXAHEDRON**	HEXAEDRE	SECHSFLACH, SECHSFLÄCHNER, HEXAEDER
6464	**HIGH AND LOW SPOTS**	IMPERFECTIONS DE SURFACE	UNEBENHEITEN
6465	**HIGH BREAST WHEEL**	ROUE DE POITRINE	WASSERRAD (RÜCKENSCHLÄCHTIGES), BRUSTRAD
6466	**HIGH EFFICIENCY ENGINE**	MACHINE A GRAND RENDEMENT, MACHINE A GRAND DEBIT	HOCHLEISTUNGSMASCHINE
6467	**HIGH FREQUENCY**	HAUTE FREQUENCE	HOCHFREQUENZ
6468	**HIGH FREQUENCY CURRENT**	COURANT A HAUTE FREQUENCE	HOCHFREQUENZSTROM
6469	**HIGH FREQUENCY FURNACE**	FOUR A HAUTE FREQUENCE	HOCHFREQUENZOFEN
6470	**HIGH HELIX DRILL**	FORET A HELICE SERREE	SPIRALBOHRER MIT GROSSEM DRALLWINKEL
6471	**HIGH HELIX PLAIN MILLING CUTTERS**	FRAISES SIMPLES A HELICE RAPIDE	SPIRALFRÄSER MIT GROSSEM DRALLWINKEL
6472	**HIGH LIGHTS**	POINTS BRILLANTS	GLANZPUNKTE, GLANZLICHTER
6473	**HIGH PRESSURE**	HAUTE PRESSION, FORTE PRESSION	DRUCK (HOHER), HOCHDRUCK
6474	**HIGH SPEED STEEL**	ACIER RAPIDE	SCHNELLDREHSTAHL
6475	**HIGH SPEED STEEL CUTTING FOOLS**	OUTILS EN ACIER RAPIDE	SCHNEIDWERKZEUGE AUS SCHNELLSTAHL
6476	**HIGH TEMPERATURE**	HAUTE TEMPERATURE	TEMPERATUR (HOHE)
6477	**HIGH VACUUM**	VIDE POUSSE	HOCHVAKUUM
6478	**HIGH VOLTAGE**	HAUTE TENSION	HOCHSPANNUNG
6479	**HIGH VOLTAGE PRESSURE TENSION**	HAUTE TENSION, HAUT VOLTAGE	HOCHSPANNUUG (ELEKT.)
6480	**HIGH-BRASS**	LAITON DE QUALITE	MESSING (HOCHWERTIGES)
6481	**HIGH-DENSITY ALLOY**	ALLIAGE A HAUTE DENSITE	LEGIERUNG GROSSER DICHTE
6482	**HIGH-DENSITY METAL**	METAL A HAUTE DENSITE	METALL GROSSER DICHTE
6483	**HIGH-GRADE**	A HAUTE TENEUR	HOCHPROZENTIG
6484	**HIGH-GRADE FUEL**	COMBUSTIBLE DE TRES BONNE QUALITE	BRENNSTOFF (HOCHWERTIGER)
6485	**HIGH-GRADE ORE**	MINERAI RICHE, MINERAI DE HAUTE TENEUR	ERZ (REICHES)
6486	**HIGH-GRADE STEEL**	ACIER DE QUALITE SUPERIEURE, ACIER SUPERIEUR	STAHL (HOCHWERTIGER)
6487	**HIGH-PRESSURE BOILER**	CHAUDIERE A HAUTE PRESSION	HOCHDRUCKKESSEL
6488	**HIGH-PRESSURE CYLINDER**	CYLINDRE A HAUTE PRESSION	HOCHDRUCKZYLINDER
6489	**HIGH-PRESSURE HOT WATER HEATING**	CHAUFFAGE PAR L'EAU CHAUDE A HAUTE PRESSION	HOCHDRUCKWASSERHEIZUNG, HEISSWASSERHEIZUNG
6490	**HIGH-PRESSURE PIPING**	CONDUITE POUR HAUTE PRESSION	HOCHDRUCKLEITUNG
6491	**HIGH-PRESSURE STEAM**	VAPEUR A HAUTE PRESSION	HOCHDRUCKDAMPF, HOCH GESPANNTER DAMPF
6492	**HIGH-PRESSURE STEAM ENGINE**	MACHINE A HAUTE PRESSION	HOCHDRUCKDAMPFMASCHINE
6493	**HIGH-PRESSURE STEAM HEATING**	CHAUFFAGE PAR LA VAPEUR A HAUTE PRESSION	HOCHDRUCKDAMPFHEIZUNG
6494	**HIGH-PRESSURE TURBINE**	TURBINE A HAUTE PRESSION	HOCHDRUCKTURBINE
6495	**HIGH-SPEED DRILLING MACHINE**	PERCEUSE A GRANDE VITESSE	SCHNELLBOHRMASCHINE
6496	**HIGH-SPEED ENGINE**	MACHINE A GRANDE VITESSE, MACHINE A MARCHE RAPIDE	SCHNELLAUFENDE MASCHINE, SCHNELLÄUFER

6497	**HIGH-SPEED STEEL**	ACIER RAPIDE	SCHNELLARBEITSSTAHL, SCHNELLDREHSTAHL
6498	**HIGH-TEMPERATURE OXIDATION**	OXYDATION A TEMPERATURE ELEVEE	OXYDATION BEI HOCHTEMPERATUR
6499	**HIGH-TEMPERATURE TESTING**	ESSAI A HAUTE TEMPERATURE	PRÜFUNG BEI HOCHTEMPERATUR
6500	**HIGH-TENSION CURRENT**	COURANT A HAUTE TENSION	HOCHSPANNUNGSSTROM, STROM (HOCHGESPANNTER)
6501	**HIGH-TENSION LINE**	LIGNE A HAUTE TENSION	HOCHSPANNUNGSLEITUNG
6502	**HIGH-TENSION MOTOR**	MOTEUR A HAUTE TENSION	HOCHSPANNUNGSMOTOR
6503	**HIGHLY INFLAMMABLE**	INFLAMMABLE, DANGEREUX	FEUERGEFÄHRLICH
6504	**HINDERED CONTRACTION**	RETRAIT CONTRARIE	SCHWINDUNG (BEHINDERTE)
6505	**HINGE**	CHARNIERE, BRAS D'ARTICULATION	GELENKBAND, SCHARNIER, FÜHRUNGSARM
6506	**HINGE BUCKLE**	PLI CHARNIERE	SCHARNIERHEBEL
6507	**HINGE PILLAR**	MONTANT DE PORTE	TÜRPFOSTEN
6508	**HINGE PIN**	CHARNIERE (MALE)	SCHARNIERSTIFT
6509	**HINGED BOLT**	BOULON A CHARNIERE, BOULON A BASCULE	KLAPPSCHRAUBE, SCHARNIERSCHRAUBE
6510	**HINGED LID COVER**	COUVERCLE A CHARNIERE	KLAPPDECKEL, AUFKLAPPBARER DECKEL
6511	**HINGED, TAIL VICE**	ETAU ORDINAIRE, ETAU A PIED, ETAU DU NORD	FLASCHENSCHRAUBSTOCK
6512	**HOB**	PLATEAU DE FOUR, FRAISE-MERE	OFENPLATTE, ABWALZFRÄSEN
6513	**HOBBING**	FINITION A LA FRAISE-MERE	ABWÄLZ-FRÄSEN
6514	**HOBRED CAVITY**	DECOUPAGE A L'EMPORTE-PIECE	LOCHEISENHOHLRÄUME
6515	**HOES, TRACTOR TRAILED AND MOUNTED.**	HOUES (A TRACTEUR TRAINEE OU MONTEE)	HAUEN (DURCH TRAKTOREN ANGETRIEBEN)
6516	**HOIST**	PALAN	HEBEZEUG
6517	**HOIST (TO), WIND (TO)**	MONTER, LEVER	AUFWINDEN, HOCHWINDEN
6518	**HOISTING CHAIN**	CHAINE-CABLE, CHAINE DE LEVAGE	LASTKETTE, HUBKETTE
6519	**HOISTING DEVICE, LIFTING MECHANISM, APPLIANCE**	APPAREIL DE LEVAGE, ENGIN	HEBEMASCHINE, HEBEZEUG
6520	**HOISTING ROPE**	CABLE DE LEVAGE, CABLE D'EXTRACTION	FÖRDERSEIL
6521	**HOLD**	ARRET DE SUSPENSION	HALT (WAHLWEISER)
6522	**HOLD DOWN**	GOUPILLE DE RETENUE	STELLSPINDEL
6523	**HOLD TIME**	TEMPS DE MAINTIEN DE L'EFFORT	NACHHALTEZEIT
6524	**HOLD UP (TO)**	TENIR LE COUP, TENIR (RIVETAGE), SOUTENIR LE TAS	VORHALTEN, GEGENHALTEN
6525	**HOLD-OPEN MECHANISM**	MECANISME DE RETENUE EN POSITION D'OUVERTURE	OFFENHALTUNGSVORRICHTUNG
6526	**HOLDING**	MAINTIEN	WARMHALTEN
6527	**HOLDING FURNACE**	FOUR DE MAINTIEN	WARMHALTEOFEN
6528	**HOLDING TIME**	MAINTIEN EN TEMPERATURE	HALTEZEIT
6529	**HOLDING-UP HAMMER**	CONTRE-BOUTEROLLE A MANCHE	VORHALTHAMMER
6530	**HOLE**	TROU, ORIFICE, LACUNE	LOCH, BOHRUNG, ÖFFNUNG
6531	**HOLE GAUGE**	TAMPON DE CONTROLE D'ALESAGE	BOHRUNGSLEHRE
6532	**HOLE-IN-THE-DIE**	CALIBRE DE MATRICE	LOCHRINGÖFFNUNG
6533	**HOLED CIRCULAR NUT**	ECROU A TROUS, ECROU A ENCOCHES	LOCHMUTTER
6534	**HOLIDAY DETECTOR**	BALAI ELECTRIQUE	LÜCKENSUCHGERÄT
6535	**HOLIDAY TEST**	ESSAI (OU CONTROLE) AU BALAI ELECTRIQUE	PRÜFUNG MIT DEM LÜCKENSUCHGERÄT
6536	**HOLIDAYS**	LACUNES (AFNOR), MANQUES (AFNOR)	BLANKE STELLE

6537	**HOLLOW**	CONGE DE RACCORDEMENT, ARRONDI	ABRUNDUNGSBOGEN, AUSRUNDUNG
6538	**HOLLOW BODY**	CORPS CREUX	HOHLKÖRPER
6539	**HOLLOW CASTING**	PIECE DE MOULAGE CREUSE	HOHLGUSS
6540	**HOLLOW CYLINDER**	CYLINDRE CREUX	HOHLZYLINDER
6541	**HOLLOW HALF-ROUND IRON**	FER DEMI-ROND CREUX	HALBRUNDEISEN (HOHLES)
6542	**HOLLOW KEY, SADDLE KEY**	CLAVETTE A FRICTION, CLAVETTE CREUSE, CLAVETTE EVIDEE	HOHLKEIL
6543	**HOLLOW OF WAVE, WAVE HOLLOW TROUGH**	CREUX, VIDE ENTRE LES ONDES	WELLENTAL
6544	**HOLLOW PLANE**	MOUCHETTE	LEISTENHOBEL
6545	**HOLLOW PUNCH**	DECOUPOIR, EMPORTE-PIECE CYLINDRIQUE	LOCHEISEN, LOCHSTAHL, AUSSCHLAGEISEN, AUSSCHLAGSTAHL, AUSSCHLAGPUNZE
6546	**HOLLOW SET, SMITH'S GOUGE**	GOUGE DE FORGERON	HALBRUNDER MEISSEL
6547	**HOLLOW SHAFT**	ARBRE CREUX	WELLE (HOHLE)
6548	**HOLLOW SPACE BETWEEN TEETH**	CREUX DE LA DENT, VIDE ENTRE DEUX DENTS	ZAHNLÜCKE
6549	**HOLLOW SPHERE**	SPHERE CREUSE	HOHLKUGELWALZE
6550	**HOLLOW SQUARE SECTION**	SECTION RECTANGULAIRE CREUSE, SECTION A NERVURES	QUERSCHNITT (HOHLRECHTECKIGER)
6551	**HOLMIUM**	HOLMIUM	HOLMIUM
6552	**HOLOHEDRAL CRYSTAL**	CRISTAL HOLOEDRE	HOLOEDERKRISTALL
6553	**HOLSTING, WINDING**	LEVAGE	AUFWINDEN, HOCHWINDEN
6554	**HOME SCRAP**	FERAILLES DE PRODUCTION PROPRE	HÜTTENSCHROTT
6555	**HOMOGENEOUS**	HOMOGENE	HOMOGEN, GLEICHMÄSSIG
6556	**HOMOGENEOUS BONDING**	UNION HOMOGENE	VERBINDUNG (HOMOGENE)
6557	**HOMOGENEOUS EQUATION**	EQUATION HOMOGENE	GLEICHUNG (HOMOGENE)
6558	**HOMOGENEOUS MATTER**	SUBSTANCE HOMOGENE	GRUNDSTOFF (HOMOGENE)
6559	**HOMOGENEOUS RADIATION**	RAYONNEMENT HOMOGENE	STRAHLUNG (HOMOGENE)
6560	**HOMOGENEOUS TEXTURE**	TEXTURE HOMOGENE	GEFÜGE (GLEICHARTIGES), GEFÜGE (HOMOGENES)
6561	**HOMOGENIZING**	HOMOGENEISATION	HOMOGENISIERUNG
6562	**HOMOLOGOUS SERIES**	SERIE HOMOLOGUE	REIHE (HOMOLOGE) (CHEM.)
6563	**HONEYCOMBED POROUS SPONGY BLOWN CASTING**	FONTE A SOUFFLURES	GUSS (BLASIGER)
6564	**HONING**	RECTIFICATION INTERIEURE, RODAGE AU LIQUIDE	HONEN, ZIEHSCHLEIFEN
6565	**HONING AND LAPPING MACHINE**	MACHINE A RODER	ZIEHSCHLEIF- U. LÄPPMASCHINE
6566	**HOOD**	CAPOT DE VOITURE	HAUBE
6567	**HOOF DUST, CRUSHED HOOFS**	SABOTS D'ANIMAUX RAPES	KLAUENMEHL, HUFMEHL
6568	**HOOK**	CROCHET	HAKEN
6569	**HOOK BOLT**	BOULON A CROCHET	HAKENSCHRAUBE
6570	**HOOK LINK CHAIN, WIDE LINK CHAIN, LADDER CHAIN**	CHAINE DE VAUCANSON	HAKENKETTE, VAUCANSONSCHE KETTE
6571	**HOOK WRENCH, HAND HOOK**	GRIFFE DE FORGERON	RICHTHAKEN, RICHTHORN
6572	**HOOKE'S JOINT COUPLING, UNIVERSAL COUPLING**	MANCHON ARTICULE UNIVERSEL	UNIVERSAL-GELENKKUPPLUNG
6573	**HOOKE'S JOINT, UNIVERSAL JOINT**	JOINT DE CARDAN, ARTICULATION A CARDAN, CARDAN, JOINT UNIVERSEL, CHARNIERE UNIVERSELLE	KREUZGELENK, KARDANGELENK, KARDANISCHES GELENK, UNIVERSALGELENK, HOOKESCHER SCHLÜSSEL
6574	**HOOKE'S LAW**	LOI DE HOOKE	HOOKESCHES GESETZ
6575	**HOOP IRON**	FEUILLARD DE FER, FER FEUILLARD	BANDEISEN
6576	**HOOP STRESS**	CONTRAINTE CIRCONFERENTIELLE	UMFANGSSPANNUNG

6577	HOPPER	TREMIE	FÜLLTRICHTER, SCHÜTTRUMPF
6578	HORIZONTAL	HORIZONTAL	LIEGEND, WAAGRECHT, HORIZONTAL
6579	HORIZONTAL BOILER	CHAUDIERE HORIZONTALE	KESSEL (LIEGENDER)
6580	HORIZONTAL BULLET DRUM	RESERVOIR BALLON	BOILER (WAAGERECHTER)
6581	HORIZONTAL DRILLING MACHINE	PERCEUSE HORIZONTALE	BOHRMASCHINE (LIEGENDE), HORIZONTALBOHRMASCHINE
6582	HORIZONTAL ENGINE	MACHINE HORIZONTALE	MASCHINE (LIEGENDE)
6583	HORIZONTAL FILLET WELD	SOUDURE EN ANGLE A PLAT	HORIZONTALKEHLNAHTSCHWEISSEN
6584	HORIZONTAL GRATE	GRILLE HORIZONTALE	PLANROST
6585	HORIZONTAL LINE	LIGNE HORIZONTALE	WAAGERECHTE, HORIZONTALE
6586	HORIZONTAL NECK JOURNAL BEARING	SUPPORT, APPUI INTERMEDIAIRE D'UN ARBRE	HALSLAGER EINER LIEGENDEN WELLE
6587	HORIZONTAL SHAFT	ARBRE HORIZONTAL	WELLE (LIEGENDE)
6588	HORIZONTAL SPINDLE GRINDER	MACHINE A MEULER A AXE HORIZONTAL	WAAGERECHTSPINDELSCHLEIFBOCK
6589	HORIZONTAL SPINDLE MILLING MACHINE	FRAISEUSE HORIZONTALE	LIEGENDE FRÄSMASCHINE, FRÄSMASCHINE (HORIZONTALE), FRÄSMASCHINE (WAAGERECHTE)
6590	HORIZONTAL THRUST	POUSSEE HORIZONTALE	HORIZONTALSCHUB
6591	HORIZONTAL WELDING	SOUDURE EN CORNICHE	SCHWEISSEN (WAAGERECHTES)
6592	HORN	AVERTISSEUR SONORE, KLAXON, BRAS DE L'ELECTRODE	HORN, HUPE, ELEKTRODENARM
6593	HORN BLOCK	PLAQUE DE GARDE	ACHSENGABEL
6594	HORN CENTRE	CENTRE A COMPAS	ZENTRUMSCHEIBE (FÜR ZIRKEL)
6595	HORN CLIPPINGS	CORNE RAPEE	HORNSPÄNE
6596	HORN OF PLATEN SPACING	DISTANCE ENTRE LES BRAS	ARMABSTAND
6597	HORN OF SADDLE SUPPORT	EXTREMITE DU BERCEAU	SATTELENDE
6598	HORNBLENDE	HORNBLENDE, AMPHIBOLE	HORNBLENDE
6599	HORNBLENDE ASBESTOS, AMPHIBOLE ASBESTOS	AMIANTE	HORNBLENDEASBEST
6600	HORNGATE	ATTAQUE EN CORNICHON	HORNEINGUSS, HORNZULAUF
6601	HORSE NAILS	CLOUS DE FER A CHEVAL	HUFNÄGEL
6602	HORSE POWER (H.P.)	CHEVAL-VAPEUR	PFERDESTÄRKE, PFERDEKRAFT
6603	HORSE-POWER/HOUR (H.P.-HR., H.P.-HR.)	CHEVAL-HEURE	PFERDESTÄRKE-STUNDE
6604	HORSEHAIR	CRIN DE CHEVAL	ROSSHAAR, PFERDEHAAR
6605	HORSESHOE IRON	FER CAVALIER	HUFSTABEISEN
6606	HORSESHOE MAGNET	AIMANT EN U, EN FER A CHEVAL	HUFEISENMAGNET
6607	HORSESHOE STEP GAUGE	JAUGE-MACHOIRE A UNE BRANCHE POUR COTES 'MINI' ET 'MAXI'	EINSEITIGE GRENZRACHENLEHRE MIT ZWEI MESSSTELLEN FÜR 'GUT' UND 'AUSSCHUSSSEITE'
6608	HORSESHOE, SNAP, CALLIPER GAUGE	CALIBRE EN FER A CHEVAL	OFFENE LEHRE, LOCHLEHRE, TASTERLEHRE, RACHENLEHRE, KLINKE
6609	HOSE	DURITE	VERBINDUNG
6610	HOSE CLAMP	COLLIER DE SERRAGE (POUR TUYAU)	SCHLAUCHSCHELLE
6611	HOSE CLIP	COLLIER A VIS POUR TUYAUX FLEXIBLES	SCHLAUCHSCHELLE, SCHLAUCHKLEMME, SCHLAUCHKLAMMER
6612	HOSE COUPLING	RACCORD POUR TUYAUX FLEXIBLES	SCHLAUCHKUPPLUNG
6613	HOSE DRAIN NOZZLE	TUBULURE POUR PURGE SOUPLE	ABLASSSCHLAUCH-STUTZEN
6614	HOSE PIPE	TUBE FLEXIBLE, BOYAU	SCHLAUCH
6615	HOT AIR ENGINE	MOTEUR A AIR CHAUD	HEISSLUFTMASCHINE, LUFTMOTOR
6616	HOT AIR HEATING	CHAUFFAGE A AIR CHAUD	LUFTHEIZUNG
6617	HOT BED	BANC REFROIDISSEUR	KÜHLBETT, WARMLAGER

6618	HOT BEND TEST	ESSAI DE PLIAGE A CHAUD	WARMBIEGEPROBE
6619	HOT BENDING	FLEXION A CHAUD, PLIAGE A CHAUD	WARMBIEGUNG
6620	HOT BLAST IRON	FONTE A L'AIR CHAUD	ROHEISEN (HEISS ERBLASENES)
6621	HOT CRACK	RUPTURE A CHAUD, CRIQUE A CHAUD	WARMBRUCH, WARMRISS
6622	HOT CUTTING	COUPAGE A CHAUD	WARMSCHNEIDEN
6623	HOT DRAW (TO)	ETIRER A CHAUD	WARMZIEHEN
6624	HOT DRAWING	ETIRAGE A CHAUD, TREFILAGE A CHAUD	WARMZIEHEN
6625	HOT EMBRITTLEMENT	FRAGILISATION A CHAUD	WARMVERSPRÖDUNG
6626	HOT FORGING	FORGEAGE A CHAUD	WARMSCHMIEDEN
6627	HOT FORMING	FACONNAGE A CHAUD, FORMAGE A CHAUD	WARMFORMUMG,
6628	HOT GALVANIZING	GALVANISATION PAR TREMPE	FEUERVERZINKUNG
6629	HOT HARDNESS	DURETE A CHAUD	HÄRTE BEI HOCHTEMPERATUR
6630	HOT IRON SAW	SCIE A CHAUD, SCIE POUR METAUX AU ROUGE	WARMSÄGE, HEISSSÄGE
6631	HOT METAL	METAL EN ETAT DE FUSION, METAL EN FUSION	METALL IM SCHMELZZUSTAND
6632	HOT PRESSING	ESTAMPAGE A CHAUD	WARMPRESSEN
6633	HOT QUENCHING	TREMPE ECHELONNEE	STUFENHÄRTUNG
6634	HOT RIVET (TO)	RIVER A CHAUD	WARMNIETEN
6635	HOT RIVETING	RIVURE A CHAUD	WARMNIETEN
6636	HOT ROLL (TO)	LAMINER A CHAUD	WARMWALZEN
6637	HOT ROLLED PLATE	TOLE LAMINEE A CHAUD	BLECH (WARMGEWALZTES)
6638	HOT ROLLING	LAMINAGE A CHAUD	WARMWALZEN
6639	HOT SERVICE THERMAL INSULATION	CALORIFUGEAGE	WÄRMEDÄMMUNG
6640	HOT SET	TRANCHE A CHAUD	WARMSCHROTMEISSEL
6641	HOT SHORTNESS	FRAGILITE A CHAUD	MESSBRÜCHIGKEIT, WARMBRÜCHIGKEIT
6642	HOT SPOT	POINT CHAUD	STELLE (HEISSE), STELLE (WARME)
6643	HOT STAMPING	ETAMPAGE A CHAUD	WARMGESENKDRÜCKEN
6644	HOT SURFACE	SURFACE TRES POREUSE	OBERFLÄCHE (SAUGENDE)
6645	HOT TEAR	CRIQUE DE RETRAIT	WARMRISS, SCHWINDUNGSRISS
6646	HOT TEST	ESSAI DE RESISTANCE 'A CHAUD	WARMVERSUCH
6647	HOT TRIMMING	EBARBAGE A CHAUD	HEISSFERTIGPUTZEN
6648	HOT WATER HEATING	CHAUFFAGE PAR L'EAU CHAUDE	WASSERHEIZUNG
6649	HOT WATER PIPE LINE	CONDUITE D'EAU CHAUDE	WARMWASSERLEITUNG
6650	HOT WORKING	TRAVAIL A CHAUD, USINAGE A CHAUD	WARMBEARBEITUNG
6651	HOT-BLAST MAIN	CONDUITE DE VENT CHAUD	HEISSWINDLEITUNG
6652	HOT-CATHODE TUBE	TUBE A CATHODE CHAUDE	GLÜHKATODENRÖHRE
6653	HOT-DIP GALVANIZING	GALVANISATION A CHAUD	FEUERTAUCHVERZINKEN
6654	HOT-FACE TEMPERATURE	TEMPERATURE SUPERFICIELLE MAXIMALE	MAXIMALE OBERFLÄCHENTEMPERATUR
6655	HOT-FINISHING	FINISSAGE A CHAUD	WARMVERGÜTUNG
6656	HOT-ROLLING	LAMINAGE A CHAUD	WARMWALZEN
6657	HOT-SAWING	SCIAGE A CHAUD	WARMSÄGEN
6658	HOT-SHORT	CASSANT A CHAUD	WARMBRÜCHIG
6659	HOTDIP COATING PROCESS	PROCEDE DE REVETEMENT EN BAIN CHAUD	SCHMELZTAUCHVERFAHREN
6660	HOUSING	LOGEMENT, BOITIER	LAGER, GEHÄUSE

6661	HUB	MOYEU	NABE
6662	HUB CAP	ENJOLIVEUR DE ROUE	RAD(NABEN)KAPPE
6663	HUB FLANGE	BRIDE A MOYEU	NABENFLANSCH
6664	HUMECTANT	HUMIDIFIANT	BEFEUCHTER
6665	HUMIC ACID	ACIDE HUMIQUE	HUMUSSÄURE
6666	HUMIDITY MOISTURE OF THE AIR, ATMOSPHERIC MOISTURE	ETAT HYGROMETRIQUE DE L'AIR, HUMIDITE DE L'AIR	LUFTFEUCHTIGKEIT
6667	HUMUS, VEGETABLE MOULD	HUMUS	HUMUS
6668	HUNDREDWEIGHT (CWT.) (100 LBS)	QUINTAL (50 KGS ENV.)	ZENTNER (50 KGS ENV.)
6669	HUNG ON KNIFE EDGE	SUSPENDU SUR UN COUTEAU	AUF EINER SCHNEIDE GELAGERT
6670	HUNTING	POMPAGE	PENDELN
6671	HUNTING OF GOVERNOR	OSCILLATIONS DU REGULATEUR	TANZEN DES REGLERS
6672	HYDRANT	BOUCHE D'ARROSAGE, BOUCHE D'EAU	SCHACHTHYDRANT, HYDRANT
6673	HYDRATE, HYDROXIDE	HYDRATE, HYDROXYDE	HYDRAT, HYDROXYD
6674	HYDRATED	HYDRATE	WASSERHALTIG
6675	HYDRATION WATER, WATER OF HYDRATION	EAU D'HYDRATATION	HYDRATWASSER
6676	HYDRAULIC	HYDRAULIQUE	HYDRAULISCH
6677	HYDRAULIC ACCUMULATOR	ACCUMULATEUR HYDRAULIQUE	DRUCKWASSERSPEICHER, AKKUMULATOR (HYDRAULISCHER)
6678	HYDRAULIC BRAKE	AMORTISSEUR A LIQUIDE, FREIN HYDRAULIQUE, FREIN HYDROPNEUMATIQUE	FLÜSSIGKEITSBREMSE
6679	HYDRAULIC CHUCKS	MANDRINS HYDRAULIQUES	FUTTER (HYDRAULISCHE)
6680	HYDRAULIC COMPRESSOR	COMPRESSEUR HYDRAULIQUE	VERDICHTER (HYDRAULISCHER)
6681	HYDRAULIC COUPLING	ACCOUPLEMENT HYDRAULIQUE, EMBRAYAGE HYDRAULIQUE	KUPPLUNG (HYDRAULISCHE)
6682	HYDRAULIC DRAWING PRESS	PRESSE HYDRAULIQUE A EMBOUTIR	ZIEHPRESSE (HYDRAULISCHE)
6683	HYDRAULIC DRIVE	COMMANDE HYDRAULIQUE	DRUCKWASSERANTRIEB, ANTRIEB (HYDRAULISCHER)
6684	HYDRAULIC EFFICIENCY	RENDEMENT HYDRAULIQUE	WIRKUNGSGRAD (HYDRAULISCHER)
6685	HYDRAULIC ENGINEERING	CONSTRUCTION HYDRAULIQUE	WASSERBAU
6686	HYDRAULIC FIELD	REMBLAI HYDRAULIQUE	SPÜLVERSATZ
6687	HYDRAULIC FORGING PRESS, HYDRAULIC HAMMER	PRESSE A FORGER HYDRAULIQUE, PRESSE HYDRAULIQUE A MARTEAU-PILON	SCHMIEDEPRESSE (HYDRAULISCHE), DRUCKWASSERSCHMIEDEPRESSE
6688	HYDRAULIC JACK	VERIN HYDRAULIQUE	HEBEBOCK (HYDRAULISCHER), DAUMENKRAFT
6689	HYDRAULIC LIME	CHAUX HYDRAULIQUE	WASSERKALK, KALK (HYDRAULISCHER)
6690	HYDRAULIC MORTAR	MORTIER HYDRAULIQUE	WASSERMÖRTEL, MÖRTEL (HYDRAULISCHER)
6691	HYDRAULIC MOTOR, WATER MOTOR	MOTEUR HYDRAULIQUE, MACHINE HYDRAULIQUE	WASSERKRAFTMASCHINE, WASSERMOTOR, MOTOR (HYDRAULISCHER)
6692	HYDRAULIC POWER PLANT	USINE HYDRAULIQUE, CENTRALE HYDRAULIQUE	WASSERKRAFTANLAGE
6693	HYDRAULIC PRESS	PRESSE HYDRAULIQUE	DRUCKWASSERPRESSE, PRESSE (HYDRAULISCHE)
6694	HYDRAULIC RAM	BELIER HYDRAULIQUE	WIDDER (HYDRAULISCHER), STOSSHEBER
6695	HYDRAULIC TEST, WATER PRESSURE TEST	EPREUVE HYDRAULIQUE	DRUCKWASSERPROBE, WASSERDRUCKPROBE, PROBE (HYDRAULISCHE)
6696	HYDRAULIC TURBINE, WATER TURBINE	TURBINE HYDRAULIQUE	WASSERTURBINE

6697	HYDRAULICALLY OPERATED VALVE	SOUPAPE A COMMANDE HYDRAULIQUE	VENTIL (HYDRAULISCH BETÄTIGTES)
6698	HYDRAULICS	HYDRAULIQUE	HYDRAULIK
6699	HYDRID PROCESS	PROCEDE A PARTIR DES PRODUITS RESULTANT DE LA DECOMPOSITION DE L'HYDRURE.	HYBRID-VERFAHREN
6700	HYDRO-ELECTRIC GENERATOR	GENERATRICE HYDROELECTRIQUE	WASSERDYNAMO
6701	HYDRO-ELECTRIC POWER STATION	USINE HYDROELECTRIQUE	ZENTRALE (WASSERELEKTRISCHE), ZENTRALE (HYDROELKTRISCHE)
6702	HYDRO-EXTRACT (TO)	ESSORER PAR FORCE CENTRIFUGE	AUSSCHLEUDERN, ZENTRIFUGIEREN
6703	HYDRO-EXTRACTOR, CONTRIFUGAL MACHINE	ESSOREUSE CENTRIFUGE	SCHLEUDER, TROCKENSCHLEUDER, ZENTRIFUGE
6704	HYDROBROMIC ACID	ACIDE BROMHYDRIQUE	BROMWASSERSTOFFSÄURE
6705	HYDROCARBON	CARBURE D'HYDROGENE, HYDROCARBURE	KOHLENWASSERSTOFF, HYDROKARBÜR
6706	HYDROCHLORIC ACID, MURIATIC ACID, HYDROGEN CHLORIDE, SPIRITS OF SALT	ACIDE CHLORHYDRIQUE, HYDROCHLORIQUE, MURIATIQUE, ESPRIT DE SEL	SALZSÄURE, CHLORWASSERSTOFFSÄURE
6707	HYDROCYANIC ACID, PRUSSIC ACID	ACIDE CYANHYDRIQUE, ACIDE PRUSSIQUE	ZYANWASSERSTOFF, BLAUSÄURE, ZYANWASSERSTOFFSÄURE
6708	HYDRODYNAMIC	HYDRODYNAMIQUE	HYDRODYNAMISCH
6709	HYDRODYNAMIC DAMPENER	FREIN HYDRAULIQUE	DÄMPFUNGVORRICHTUNG (HYDRODYNAMISCHE)
6710	HYDRODYNAMICS	HYDRODYNAMIQUE	HYDRODYNAMIK
6711	HYDROFLUORIC ACID	ACIDE FLUORHYDRIQUE	FLUORWASSERSTOFFSÄURE, FLUSSSPATSÄURE, FLUSSSÄURE
6712	HYDROGEN	HYDROGENE	WASSERSTOFF
6713	HYDROGEN EMBRITTLEMENT	FRAGILITE DE DECAPAGE	BEIZSPRÖDIGKEIT
6714	HYDROGEN FLAME	FLAMME D'HYDROGENE	WASSERSTOFFFLAMME
6715	HYDROGEN ION CONCENTRATION	CONCENTRATION DES IONS HYDROGENE	WASSERSTOFFIONENKONZENTRATION
6716	HYDROGEN PEROXIDE, HYDROGEN DIOXIDE	EAU OXYGENEE, BIOXYDE D'HYDROGENE	WASSERSTOFFSUPEROXYD
6717	HYDROGEN SULPHIDE, SULPHURETTED HYDROGEN	ACIDE SULFHYDRIQUE, HYDROGENE SULFURE, GAZ PUANT	SCHWEFELWASSERSTOFF, WASSERSTOFFSULFID
6718	HYDROGEN TYPE CORROSION	CORROSION AVEC DEGAGEMENT D'HYDROGENE	KORROSION UNTER WASSERSTOFFENTWICKLUNG
6719	HYDROLYSIS	HYDROLYSE	HYDROLYSE
6720	HYDROMECHANICS	MECANIQUE DES CORPS LIQUIDES	HYDROMECHANIK
6721	HYDROMETALLURGY	HYDROMETALLURGIE	HYDROMETALLURGIE
6722	HYDROMETER	AREOMETRE, PESE-LIQUEURS	SENKWAAGE, SCHWIMMWAAGE, ARÄOMETER, GRAVIMETER
6723	HYDROMETER WITH SCALE PANS	AREOMETRE A POIDS VARIABLE	GEWICHTSARÄOMETER
6724	HYDROPHILIC	HYDROPHILE	HYDROPHIL
6725	HYDROSTATIC BALANCE	BALANCE HYDROSTATIQUE	WAAGE (HYDROSTATISCHE)
6726	HYDROSTATIC FLUID PRESSURE	PRESSION HYDROSTATIQUE	FLÜSSIGKEITSDRUCK, DRUCK (HYDROSTATISCHER)
6727	HYDROSTATIC TESTING	ESSAI HYDROSTATIQUE	PROBE (HYDROTATISCHE)
6728	HYDROSTATICS	HYDROSTATIQUE	HYDROSTATIK
6729	HYGROMETER	HYGROMETRE	FEUCHTIGKEITSMESSER, HYGROMETER
6730	HYGROSCOPIC	HYGROMETRIQUE, AVIDE D'EAU	WASSERAUFSAUGEND, WASSERANZIEHEND, WASSERGIERIG, HYGROSKOPISCH
6731	HYGROSCOPIC MOISTURE	EAU HYGROMETRIQUE	WASSER (HYGROSKOPISCHES), FEUCHTIGKEIT (HYGROSKOPISCHE)

6732	HYGROSCOPICITY	HYGROMETRICITE	HYGROSKOPISCHE EIGENSCHAFT, WASSERGIER, HYGROSKOPIZITÄT
6733	HYPER-EUTECTOID	HYPEREUTECTOIDE	ÜBEREUTEKTOID
6734	HYPERBOLA	HYPERBOLE	HYPERBEL
6735	HYPERBOLIC	HYPERBOLIQUE	HYPERBOLISCH
6736	HYPERBOLIC CYLINDER	CYLINDRE HYPERBOLIQUE	ZYLINDER HYPERBOLISCHER
6737	HYPERBOLIC FUNCTION	FONCTION HYPERBOLIQUE	HYPERBELFUNKTION
6738	HYPERBOLIC SPIRAL	SPIRALE HYPERBOLIQUE	SPIRALE (HYPERBOLISCHE)
6739	HYPERBOLOID OF ONE SHEET	HYPERBOLOIDE A UNE NAPPE	HYPERBOLOID (EINSCHALIGES)
6740	HYPERBOLOID OF TWO SHEETS	HYPERBOLOIDE A DEUX NAPPES	HYPERBOLOID (ZWEISCHALIGES)
6741	HYPO-EUTECTOID	HYPOEUTECTOIDE	UNTEREUTEKTOID
6742	HYPOCHLOROUS ACID	ACIDE HYPOCHLOREUX	SÄURE (UNTERCHLORIGE)
6743	HYPOCYCLOID	HYPOCYCLOIDE	HYPOZYKLOIDE
6744	HYPOCYCLOIDAL GEAR	DENTS A FLANCS HYPOCYCLOIDAUX, ENGRENAGE HYPOCYCLOIDAL	HYPOZYKLOIDENVERZAHNUNG
6745	HYPOID GEARS	ENGRENAGES HYPOIDES	HYPOIDGETRIEBE
6746	HYPOTENUSE	HYPOTENUSE	HYPOTENUSE
6747	HYSTERESIS	HYSTERESIS	REIBUNG (MAGNETISCHE), HYSTERESIS, HYSTERESE
6748	I BEAM	POUTRELLE EN I	I-TRÄGER
6749	ICE CALORIMETER	CALORIMETRE A FUSION DE LA GLACE	EISKALORIMETER
6750	ICE MAKING MACHINE	MACHINE A GLACE	EISMASCHINE
6751	IDIOMORPHIC CRYSTAL	CRISTAL IDIOMORPHE	KRISTALL (IDIOMORFER)
6752	IDLE RUNNING	MARCHE A VIDE	LEERLAUF
6753	IDLE TIME	TEMPS MORT	LEERZEIT, LEERLAUFZET
6754	IDLE WHEEL, IDLER	ROUE INTERMEDIAIRE, ROUE DE TRANSPORT, POULIE FOLLE	ZWISCHENTRIEBRAD, ÜBERTRAGUNGSRAD, TRANSPORTRAD
6755	IDLING	RALENTI	LEERLAUF
6756	IDLING SHAFT	ARBRE DE RENVOI, ARBRE INTERMEDIAIRE	VORGELEGEWELLE
6757	IGNITE (TO)	ALLUMER, ENFLAMMER (S')	ANZÜNDEN, ENTZÜNDEN
6758	IGNITER	ALLUMEUR	ZÜNDVORRICHTUNG, ZÜNDER
6759	IGNITING POINT	POINT D'ALLUMAGE, D'INFLAMMATION	ZÜNDPUNKT
6760	IGNITION	INFLAMMATION, IGNITION, ALLUMAGE	ENTZÜNDUNG, ENTFLAMMUNG, ZÜNDUNG
6761	IGNITION COIL	BOBINE D'ALLUMAGE	ZÜNDSPULE
6762	IGNITION DISTRIBUTOR	ALLUMEUR, DISTRIBUTEUR D'ALLUMAGE	ZÜNDVERTEILER
6763	IGNITION LOSS	PERTES AU FEU	GLÜHVERLUST
6764	IGNITION POINT	POINT D'INFLAMMATION, POINT DE FLAMME	ENTZÜNDUNGSPUNKT, FLAMMPUNKT
6765	IGNITION RESIDUE	RESIDU DE CALCINATION, IMBRULES	GLÜHRÜCKSTAND
6766	IGNITION SWITCH	CONTRACTEUR D'ALLUMAGE	ZÜNDSCHALTER
6767	IGNITION TEMPERATURE	TEMPERATURE D'ALLUMAGE, D'INFLAMMATION	ENTZÜNDUNGSTEMPERATUR
6768	IGNITION TIMING	REGLAGE DE L'ALLUMAGE	ZÜNDZEITVERSTELLUNG
6769	IGNITION WARNING-LIGHT	TEMOIN D'ALLUMAGE	ZÜNDUNGPRÜFLAMPE
6770	IGNITION, FIRING	ALLUMAGE (DANS LES MOTEURS A EXPLOSION)	ZÜNDUNG
6771	ILLINIUM	ILLINIUM, PROMETHEUM	PROMETHIUM

6772	ILLUMINANT	APPAREIL D'ECLAIRAGE	BELEUCHTUNGSKÖRPER
6773	ILLUMINATED BODY	CORPS ECLAIRE	BELEUCHTETER KÖRPER
6774	ILLUMINATING DEVICE	DISPOSITIF D'ECLAIRAGE	BELEUCHTUNGSVORRICHTUNG
6775	ILLUMINATING GAS, COAL GAS	GAZ D'ECLAIRAGE, GAZ LUMIERE	LEUCHTGAS
6776	ILLUMINATING POWER	POUVOIR ECLAIRANT	LEUCHTKRAFT
6777	ILLUMINATION	INTENSITE D' ECLAIREMENT	BELEUCHTUNGSSTÄRKE, BELEUCHTUNG
6778	ILLUSTRATION	ILLUSTRATION, FIGURE	ABBILDUNG
6779	IMAGE	IMAGE	BILD (OPT.)
6780	IMAGE POINT	POINT D'IMAGE	BILDPUNKT
6781	IMAGINARY NUMBER	NOMBRE IMAGINAIRE	ZAHL (IMAGINÄRE)
6782	IMCOMPLETE FUSION	MANQUE DE FUSION	BINDEFEHLER
6783	IMPACT	CHOC	SCHLAG, STOSS
6784	IMPACT EXTRUSION	EXTRUSION PAR CHOC	SCHLAGFLIESSPRESSEN
6785	IMPACT STRENGTH	RESILIENCE, RESISTANCE AU CHOC	STOSSFESTIGKEIT, SCHLAGFESTIGKEIT
6786	IMPACT TEST	ESSAI DE RESISTANCE AU CHOC, ESSAI DE RUPTURE AU CHOC, ESSAI PAR CHOC	SCHLAGVERSUCH, STOSSVERSUCH
6787	IMPACT TESTING MACHINE	MOUTON-PENDULE DE CHARPY	CHARPY-PRÜFMASCHINE, PENDELHAMMER
6788	IMPACT VALVE	RESILIENCE	KERBSCHLAGZÄHIGKEIT
6789	IMPALPABLE POWDER, FINELY GROUND POWDER	POUDRE IMPALPABLE, SUBSTANCE FINEMENT BROYEE, SUBSTANCE EN POUDRE FINE	PULVER (FEINGEMAHLENES)
6790	IMPART (TO), GIVE A FINE EDGE (TO)	AFFILER, DONNER LE FIL A UN INSTRUMENT TRANCHANT	ABZIEHEN, NACHSCHÄRFEN
6791	IMPART TO, APPLY A PRESERVATIVE SKIN	APPLIQUER UNE COUCHE PROTECTRICE	SCHUTZSCHICHT AUFBRINGEN, SCHUTZSCHICHT (EINE) AUFTRAGEN
6792	IMPARTING APPLICATION OF A PRESERVATIVE SKIN	APPLICATION D'UNE COUCHE PROTECTRICE	AUFBRINGEN EINER SCHUTZSCHICHT, AUFTRAGEN EINER SCHUTZSCHICHT
6793	IMPARTING, GIVING A FINE EDGE	AFFILAGE	ABZIEHEN, NACHSCHÄRFEN
6794	IMPEDANCE, APPARENT RESISTANCE	IMPEDANCE, RESISTANCE APPARENTE	SCHEINWIDERSTAND, WIDERSTAND (SCHEINBARER), IMPEDANZ
6795	IMPELLER	ROTOR, TURBINE, TURBINE DE POMPE CENTRIFUGE	LAUFRAD, VERDICHTERRAD
6796	IMPERMEABILITY TO WATER	IMPERMEABILITE A L'EAU	WASSERUNDURCHLÄSSIGKEIT
6797	IMPERMEABILITY, IMPERVIOUSNESS	IMPERMEABILITE	DICHTHEIT, UNDURCHLÄSSIGKEIT
6798	IMPERMEABLE, IMPERVIOUS	IMPERMEABLE	UNDURCHLÄSSIG
6799	IMPINGEMENT ATTAK	CORROSION SOUS EROSION	AUFPRALLKORROSION
6800	IMPONDERABLE	IMPONDERABLE	UNWÄGBAR
6801	IMPOVERISHMENT	APPAUVRISSEMENT	VERARMUNG
6802	IMPREGNATE (TO)	IMPREGNER, IMBIBER	DURCHTRÄNKEN, IMPRÄGNIEREN
6803	IMPREGNATED PAPER GAS PIPE	TUYAU A GAZ EN PAPIER IMPREGNE DE BITUME	ASPHALTROHR, PAPIERROHR
6804	IMPREGNATING MEDIUM AGENT	AGENT D'IMPREGNATION	TRÄNKUNGSMITTEL, IMPRÄGNIERUNGSMITTEL
6805	IMPREGNATION	IMPREGNATION, IMBIBITION, INFILTRATION	TRÄNKEN, DURCHTRÄNKEN, IMPRÄGNIEREN, TRÄNKUNG
6806	IMPRESSION, CURVED DEPRESSION	EMPREINTE	EINDRUCK, EINPRÄGUNG
6807	IMPROPER FRACTION	EXPRESSION FRACTIONNAIRE	BRUCH (UNECHTER)
6808	IMPROVE THE STEEL BY HEAT TREATMENT (TO)	CORRIGER L'ACIER PAR UN REVENU	STAHL VERGÜTEN (DEN)
6809	IMPULSE	IMPULSION, MOUVEMENT MOTEUR	ANTRIEB EINER KRAFT, IMPULS

6810	**IMPULSE, ACTION TURBINE**	TURBINE A ACTION, TURBINE A IMPULSION, TURBINE A LIBRE DEVIATION	DRUCKTURBINE, AKTIONSTURBINE
6811	**IMPULSIVE FORCE**	FORCE IMPULSIVE	STOSSKRAFT, MOMENTANKRAFT
6812	**IMPULSIVE SUDDEN LOAD**	CHARGE PERIODIQUE	BELASTUNG (STOSSWEISE), BELASTUNG (PULSIEREND)
6813	**IMPURITIES, ADMIXTURES**	CORPS ETRANGERS, SUBSTANCES ETRANGERES, IMPURETES	BEIMENGUNGEN, UNREINIGKEITEN, VERUNREINIGUNGEN, FREMDSTOFFE
6814	**IMPURITY**	IMPURETE	UNREINHEIT
6815	**IN DIRECT RATIO**	EN RAISON DIRECTE	VERHÄLTNIS (IN GERADEM), VERHÄLTNIS (IN DIREKTEM)
6816	**IN INVERSE RATIO**	EN RAISON INVERSE	VERHÄLTNIS (IN UMGEKEHRTEM)
6817	**IN THE FORM OF VAPOUR**	SOUS FORME DE VAPEUR	DAMPFFÖRMIG
6818	**IN UNCOMBINED CONDITION**	NON COMBINE, A L'ETAT LIBRE	UNGEBUNDEN, ZUSTAND (IN UNGEBUNDENEM), ZUSTAND (IN FREIEM) (CHEM.)
6819	**IN WORKING ORDER**	EN ETAT DE SERVICE	BETRIEBSFÄHIG, BETRIEBSFÄHIGEM ZUSTAND (IN)
6820	**INACCESSIBILITY**	INACCESSIBILITE	UNZUGÄNGLICHKEIT
6821	**INACCESSIBLE**	INACCESSIBLE	UNZUGÄNGLICH
6822	**INCANDESCENCE**	INCANDESCENCE	GLÜHUNG
6823	**INCANDESCENCE, WHITE HEAT**	CHALEUR BLANCHE, BLANC EBLOUISSANT, BLANC INCANDESCENT	WEISSGLUT
6824	**INCANDESCENT**	CHAUFFE A BLANC, PORTE A INCANDESCENCE	WEISSGLÜHEND
6825	**INCANDESCENT ELECTRIC LAMP, GLOW LAMP**	LAMPE A INCANDESCENCE	GLÜHLAMPE
6826	**INCANDESCENT ELECTRIC LIGHT**	LUMIERE A INCANDESCENCE	GLÜHLICHT
6827	**INCANDESCENT GAS**	GAZ INCANDESCENT	GAS (GLÜHENDES)
6828	**INCANDESCENT GAS LAMP**	LAMPE A GAZ, LAMPE A INCANDESCENCE	GASGLÜHLICHTLAMPE
6829	**INCANDESCENT GAS LIGHT**	LUMIERE DU GAZ A INCANDESCENCE	GASGLÜHLICHT
6830	**INCH**	POUCE ANGLAIS	ZOLL (ENGLISCHER)
6831	**INCIDENCE**	INCIDENCE	EINSTRAHLUNG
6832	**INCIDENT RAY**	RAYON INCIDENT	STRAHL (AUFTREFFENDER), STRAHL (EINFALLENDER)
6833	**INCIPIENT CRACK**	AMORCE DE CRIQUE, AMORCE DE FISSURE	ANRISS
6834	**INCIPIENT FRACTURE**	FENTE SUPERFICIELLE	ANBRUCH, ANRISS
6835	**INCLINABLE, TILTING**	INCLINABLE, BASCULANT	KIPPBAR, UMKIPPBAR, UMBIEGBAR
6836	**INCLINATION, INCLINE, GRADIENT, SLOPE**	INCLINAISON, PENTE	NEIGUNG
6837	**INCLINED GRATE**	GRILLE INCLINEE	SCHRÄGROST. SCHÜTTROST
6838	**INCLINED PLANE**	PLAN INCLINE	EBENE (SCHIEFE)
6839	**INCLINED TOOTH CLUTCH**	EMBRAYAGE A DENTURE CONIQUE	KEGELKUPPLUNG
6840	**INCLUSION**	INCLUSION	EINSCHLUSS
6841	**INCLUSION STRINGERS**	INCLUSION ALLONGEE	LANGGESTRECKETER EINSCHLUSS
6842	**INCOMBUSTIBILITY**	INCOMBUSTIBILITE	UNVERBRENNBARKEIT, UNVERBRENNLICHKEIT
6843	**INCOMBUSTIBLE**	IMCOMBUSTIBLE	NICHT BRENNBAR, UNVERBRENNBAR, UNVERBRENNLICH
6844	**INCOMING SUPPLY**	ALIMENTATION ARRIVEE	ANKOMMENDER STROM
6845	**INCOMPLETE COMBUSTION**	COMBUSTION INCOMPLETE	VERBRENNUNG (UNVOLLKOMMENE)
6846	**INCOMPRESSIBILITY**	INCOMPRESSIBILITE	UNZUSAMMENDDRÜCKBARKEIT

6847	INCOMPRESSIBLE	INCOMPRESSIBLE	UNZUSAMMENDRÜCKBAR, NICHT ZUSAMMENDRÜCKBAR, UNKOMPRIMIERBAR
6848	INCREASE OF PRESSURE	AUGMENTATION DE LA PRESSION	DRUCKSTEIGERUNG, DRUCKANSTIEG, DRUCKZUNAHME
6849	INCREASE OF SPEED	ACCROISSEMENT DE VITESSE	GESCHWINDIGKEITZUNAHME
6850	INCREASE OF STRENGTH	AUGMENTATION DE LA RESISTANCE	VERFESTIGUNG
6851	INCREASE THE VELOCITY (TO)	AUGMENTER LA VITESSE, ACCELERER	GESCHWINDIGKEIT ERHÖHEN, GESCHWINDIGKEIT STEIGERN
6852	INCREASING SPEED	VITESSE CROISSANTE	GESCHWINDIGKEIT (ZUNEHMENDE)
6853	INCREMENTAL COLLAPSE	DEFORMATION PROGRESSIVE	VERFORMUNG (STUFENLOSE)
6854	INCREMENTAL DIMENSION	COTATION RELATIVE	KETTENMASSSYSTEM
6855	INCREMENTAL PRESSURE	AUGMENTATION DE PRESSION	STEIGDRUCK
6856	INCRUSTATION, FOULING	FORMATION D'INCRUSTATIONS, ENTARTRAGE	KRUSTENBILDUNG; KESSELSTEINBILDUNG
6857	INCRUSTED, FOULED	INCRUSTE, ENTARTRE	VERKRUSTET
6858	INDEFINITE INTEGRAL	INTEGRALE INDEFINIE	INTEGRAL (UNBESTIMMTES)
6859	INDENTATION	EMPREINTE DE DURETE	HÄRTE-EINDRUCK
6860	INDENTATION HARDNESS	DURETE VICKERS	VICKERSHÄRTE
6861	INDENTATION TEST	ESSAI DE DURETE PAR EMPREINTE DE BILLE	KUGELDRUCKHÄRTEPRÜFUNG
6862	INDENTED	DENTELE	GEZÄHNELT, AUSGEZACKT, ZACKIG
6863	INDEPENDENT MACHINE	MACHINE A COMMANDE INDEPENDANTE	MASCHINE MIT EINZELANTRIEB
6864	INDEX	INDICE (MATH.)	INDEX (MATH.)
6865	INDEX CRANK	MANIVELLE D'INDEXAGE	INDEXHANDKURBEL
6866	INDEX OF REFRACTION, REFRACTIVE INDEX	INDICE DE REFRACTION	BRECHUNGSZAHL, BRECHUNGSEXPONENT, BRECHUNGSINDEX
6867	INDEX PLATE	SECTEUR GRADUE	TEILPLATTE
6868	INDEX, POINTER, FINGER, INDICATOR	INDICE, AIGUILLE INDICATRICE	ZEIGER (EINES INSTRUMENTS)
6869	INDEXING HEADS	MECANISME D'INDEXAGE	INDEXSCHALTEINRICHTUNG
6870	INDEXING PLATES	PLATEAUX DIVISEURS	TEILSCHEIBEN
6871	INDIAN INK, CHINESE INK	ENCRE DE CHINE	TUSCHE, AUSZIEHTUSCHE
6872	INDICATED EFFICIENCY	RENDEMENT INDIQUE	WIRKUNGSGRAD (INDIZIERTER)
6873	INDICATED HORSE POWER (I.H.P.)	CHEVAL NOMINAL	PFERDESTÄRKE (INDIZIERTE), PSI
6874	INDICATED HORSEPOWER	PUISSANCE INDIQUEE (OU FISCALE)	PFERDESTÄRKE (ANGEGEBENE)
6875	INDICATED POWER	PUISSANCE INDIQUEE	LEISTUNG (INDIZIERTE)
6876	INDICATED WORK	TRAVAIL INDIQUE	ARBEIT (INDIZIERTE)
6877	INDICATING MICROMETER	MICROMETRE A CADRAN	SKALENSCHEIBENMIKROMETER
6878	INDICATOR	INDICATEUR DE PRESSION	INDIKATOR (DAMPFM.)
6879	INDICATOR	INDICATEUR, REACTIF INDICATEUR	INDIKATOR (CHEM.)
6880	INDICATOR CARD	PAPIER D'INDICATEUR	INDIKATORPAPIER
6881	INDICATOR DIAGRAM, DIAGRAM OF WORK	DIAGRAMME D'INDICATEUR, DIAGRAMME DE TRAVAIL	ARBEITSDIAGRAMM, INDIKATORDIAGRAMM, SPANNUNGSDIAGRAMM, DRUCKVOLUMENDIAGRAMM, PV-DIAGRAMM
6882	INDIFFERENT NEUTRAL EQUILIBRIUM	EQUILIBRE INDIFFERENT	GLEICHGEWICHT (INDIFFERENTES)
6883	INDIRECT ACTING	A ACTION INDIRECTE	INDIREKT WIRKEND
6884	INDIRECT EXTRUSION	FILAGE AVEC REMONTEE DE MATIERES	RÜCKWÄRTSFLIESSPRESSEN
6885	INDIUM	INDIUM	INDIUM
6886	INDIVIDUAL DRIVE	COMMANDE INDIVIDUELLE	EINZELANTRIEB
6887	INDOOR LIGHTING	ECLAIRAGE INTERIEUR	INNENBELEUCHTUNG
6888	INDUCE (TO)	INDUIRE	INDUZIEREN

6889	**INDUCED CURRENT**	COURANT D'INDUCTION	INDUKTIONSSTROM, STROM (INDUZIERTER)
6890	**INDUCED DRAUGHT**	TIRAGE INDUIT	SAUGZUG
6891	**INDUCTANCE**	INDUCTANCE	SELBSTINDUKTIONSWIDERSTAND, INDUKTANZ
6892	**INDUCTION**	INDUCTION	INDUKTION
6893	**INDUCTION BRAZING**	BRASAGE PAR INDUCTION	INDUKTIONSHARTLÖTEN
6894	**INDUCTION COIL**	BOBINE D'INDUCTION	SELBSTINDUKTIONSSPULE
6895	**INDUCTION FURNACE**	FOUR A INDUCTION	INDUKTIONSOFEN
6896	**INDUCTION HARDENING**	TREMPE PAR INDUCTION	INDUKTIONSHÄRTUNG
6897	**INDUCTION HEATING**	CHAUFFAGE PAR INDUCTION	INDUKTIONSHEIZUNG
6898	**INDUCTOR**	INDUCTEUR	ERREGERWICKLUNG
6899	**INDUSTRIAL FUMES**	FUMEES INDUSTRIELLES	INDUSTRIELLE RAUCHGASE
6900	**INDUSTRIAL FURNACE**	FOUR INDUSTRIEL	INDUSTRIEOFEN
6901	**INDUSTRIAL ORGANISATION, ADMINISTRATION**	ORGANISATION DES USINES	FABRIKORGANISATION
6902	**INDUSTRIAL WASTE WATER, MILL EFFLUENTS**	EAUX RESIDUAIRES	FABRIKABWÄSSER, INDUSTRIEABWÄSSER
6903	**INDUSTRY**	INDUSTRIE	INDUSTRIE
6904	**INELASTIC**	INELASTIQUE	UNELASTISCH
6905	**INERT GAS**	GAZ INERTE, GAZ DE PROTECTION	EDELGAS, INDIFFERENTES GAS, INAKTIVES GAS, SCHUTZGAS
6906	**INERT-GAS TUNGSTEN-ARC WELDING**	SOUDAGE T.I.G (A L'ARC DE TUNGSTENE SOUS GAZ INERTE)	WOLFRAM-INERTGAS-SCHWEISSEN
6907	**INERTIA**	INERTIE, FORCE, EFFORT D'INERTIE	TRÄGHEIT, TRÄGHEITSKRAFT, BEHARRUNGSVERMÖGEN
6908	**INFANT MORTALITY**	DEFAILLANCE PRECOCE	ZEIT DER ANFANGSAUSFÄLLE
6909	**INFINITE SERIES**	SERIE INFINIE	REIHE (UNENDLICHE)
6910	**INFINITELY GREAT**	INFINIMENT GRAND	UNENDLICH GROSS
6911	**INFINITELY SMALL**	INFINIMENT PETIT	UNENDLICH KLEIN
6912	**INFINITESIMAL CALCULUS**	CALCUL INFINITESIMAL	INFINITESIMALRECHNUNG
6913	**INFLAMMABILITY**	INFLAMMABILITE	ENTFLAMMBARKEIT, ENTZÜNDBARKEIT, ZÜNDFÄHIGKEIT
6914	**INFLAMMABLE**	INFLAMMABLE	ENTFLAMMBAR, ENTZÜNDBAR
6915	**INFLOW SUPPLY OF WATER**	ARRIVEE D'EAU, ENTREE D'EAU	ZUFLUSS, WASSERZUFLUSS
6916	**INFRINGEMENT OF A PATENT**	ATTEINTE PORTEE AUX DROITS DU BREVETE	PATENTVERLETZUNG, VERLETZUNG EINES PATENTES
6917	**INFUSIBILITY**	INFUSIBILITE	UNSCHMELZBARKEIT
6918	**INFUSIBLE**	INFUSIBLE	UNSCHMELZBAR
6919	**INGOT**	LINGOT	BLOCK, METALLBLOCK, ROHBLOCK
6920	**INGOT CAR**	CAR A LINGOTS (TRANSPORT DE LINGOT)	BLOCKWAGEN
6921	**INGOT IRON**	FER FONDU	FLUSSEISEN
6922	**INGOT LATHE**	TOUR A LINGOT	BLOCKDREHBANK
6923	**INGOT MOLD**	LINGOTIERE	BLOCKFORM, STAHLWERKSKOKILLE
6924	**INGOT STEEL**	ACIER FONDU	FLUSSSTAHL
6925	**INGOTISM**	SEGREGATION MAJEURE	BLOCKSEIGERUNG
6926	**INHIBIT (TO)**	EMPECHER, LIMITER, RETARDER	HEMMEN, HINDERN, VERZÖGERN
6927	**INHIBITOR**	INHIBITEUR	INHIBITOR, HEMMSTOFF
6928	**INITIAL CREEP**	FREINAGE INITIAL	ANFANGSKRIECHEN
6929	**INITIAL PRESSURE**	PRESSION INITIALE	ANFANGSDRUCK
6930	**INITIAL STATE**	ETAT INITIAL	ANFANGSZUSTAND

6931	INITIAL TEMPERATURE	TEMPERATURE INITIALE	ANFANGSTEMPERATUR, AUSGANGSTEMPERATUR
6932	INITIAL VALUE	VALEUR INITIALE	ANFANGSWERT
6933	INITIAL VELOCITY	VITESSE INITIALE	ANFANGSGESCHWINDIGKEIT
6934	INJECT (TO)	INJECTER, PROJETER DANS...	EINSPRITZEN
6935	INJECTION	INJECTION	EINSPRITZEN, EINSPRITZUNG
6936	INJECTION ORDER	ORDRE D'INJECTION	EINSPRITZFOLGE
6937	INJECTION VALVE	SOUPAPE D'INJECTION	EINSPRITZVENTIL
6938	INJECTOR	INJECTEUR, GIFFARD	EINSPRITZDÜSE, INJEKTOR
6939	INK ERASER	GOMME A ENCRE	TINTENGUMMI
6940	INK IN (TO)	PASSER A L'ENCRE, ENCRER	MIT TUSCHE AUSZIEHEN
6941	INKED-IN DRAWING	DESSIN PASSE A L'ENCRE	ZEICHNUNG (AUSGEZOGENE)
6942	INKING IN	PASSAGE A L'ENCRE	AUSZIEHEN MIT TUSCHE
6943	INLET	ENTREE, ADMISSION	EINGANG, ZUFLUSS, EINLASS
6944	INLET CONNECTING TERMINAL	BORNE D'ARRIVEE	EINGANGSKLEMME
6945	INLET MANIFOLD	COLLECTEUR D'ADMISSION	ANSAUG-ROHR, SAUGROHR
6946	INLET NOZZLE / FILLING NOZZLE	TUBULURE D'ARRIVEE, TUBULURE D'ENTREE	EINGANGS-ROHRSTUTZEN
6947	INLET OPENING	ORIFICE D'ADMISSION, ORIFICE D'ENTREE, ORIFICE D'ARRIVEE, ORIFICE D'ADDUCTION, ORIFICE D'INTRODUCTION	EINLASSÖFFNUNG, EINFLUSSÖFFNUNG
6948	INLET PIPE	TUBULURE D'ADMISSION	SAUGLEITUNG
6949	INLET VALVE	SOUPAPE D'ADMISSION	EINLASSVENTIL
6950	INLET VELOCITY, VELOCITY OF APPROACH	VITESSE D'ARRIVEE, VITESSE A L'ARRIVEE, VITESSE D'ENTREE	ZUSTRÖMGESCHWINDIGKEIT, ZUFLUSSGESCHWINDIGKEIT, EINTRITTGESCHWINDIGKEIT
6951	INNER CONE	DARD	INNENKONUS
6952	INNER FLAME CONE	DARD INTERIEUR	FLAMMENKEGEL (INNERER)
6953	INNER RACE OF BALL BEARING	BAGUE INTERIEURE D'UN ROULEMENT A BILLES	RING (INNERER) EINES KUGELLAGERS, INNENRING
6954	INNER SURFACE OF A PIPE	SURFACE INTERIEURE D'UN TUYAU	INNENWANDUNG EINES ROHRES
6955	INNER TUBE	CHAMBRE A AIR	LUFTSCHLAUCH
6956	INOCULATION	INOCULATION	IMPFUNG
6957	INORGANIC MATTER	MATIERE INORGANIQUE	ANORGANISCHER STOFF, UNORGANISCHER STOFF
6958	INORGANIC MINERAL CHEMISTRY	CHIMIE MINERALE OU INORGANIQUE	CHEMIE (ANORGANISCHE)
6959	INOXIDISABLE, UNOXIDISABLE	INOXYDABLE	NICHT OXYDIERBAR
6960	INPUT MEDIUM	SUPPORT D'INFORMATION D'ENTREE	EINGABEDATENTRÄGER
6961	INSCRIBABLE QUADRILATERAL	QUADRILATERE INSCRIPTIBLE	KREISVIERECK
6962	INSCRIBED ANGLE	ANGLE INSCRIT, ANGLE DANS LE SEGMENT	PERIPHERIEWINKEL
6963	INSCRIBED CIRCLE	CERCLE INSCRIT, CIRCONFERENCE INSCRITE	EINGESCHRIEBENER KREIS, EINLIEGENDER KREIS, INKREIS
6964	INSCRIBED POLYGON	POLYGONE INSCRIT	POLYGON (EINGESCHRIEBENES), SEHNENPOLYGON
6965	INSERT	INSERTION, PIECE RAPPORTEE, INSERT	EINLAGE, EINSATZSTÜCK
6966	INSERT DIE	MATRICE AMOVIBLE	EINSATZMATRIZE, EINSATZGELENK
6967	INSERT THE RIVETS (TO)	POSER LES RIVETS	NIETEN EINSETZEN (DIE), NIETEN EINZIEHEN (DIE)
6968	INSERTED BEAM	POUTRE ENCASTREE	EINBAUTRÄGER
6969	INSERTED NOZZLE	TUBULURE ENCASTREE	ROHRSTUTZEN (EINGELASSENER)
6970	INSERTED PIECE	PIECE RAPPORTEE	STÜCK (EINGESETZTES)

6971	**INSERTED TOOTH MILLING CUTTER**	FRAISE A DENTS RAPPORTEES, A LAMES RAPPORTEES	FRÄSER MIT EINGESETZTEN ZÄHNEN, MESSERKOPF
6972	**INSERTING THE RIVETS**	POSE DES RIVETS	EINSETZEN DER NIETEN, EINZIEHEN DER NIETEN
6973	**INSERTION RUBBER**	CAOUTCHOUC AVEC TOILE INTERCALEE	GUMMI MIT GEWEBEEINLAGE
6974	**INSIDE**	COTE INTERIEUR	INNENSEITE
6975	**INSIDE CALLIPERS**	COMPAS DE DIAMETRE INTERIEUR, MAITRE A DANSER	LOCHTASTER, INNENTASTER, LOCHZIRKEL, HOHLZIRKEL, TANZMEISTER
6976	**INSIDE CRANK, CENTRE CRANK**	COUDE D'UN ARBRE, MANIVELLE DU VILEBREQUIN	KURBEL (GEKRÖPFTE), MITTENKURBEL, KRÖPFUNG DER WELLE, KRUMMACHSE
6977	**INSIDE DIAMETER**	DIAMETRE INTERIEUR	INNENDURCHMESSER
6978	**INSIDE LENGTH OF CHAIN LINK**	PAS D'UNE CHAINE	KETTENTEILUNG, BAULÄNGE EINES KETTENGLIEDES, INNERE, GLIEDERLÄNGE
6979	**INSIDE MICROMETER CALIPERS**	CALIBRES MICROMETRIQUES D'INTERIEUR	INNENTASTER (MIKROMETRISCHER)
6980	**INSIDE SCREW GATE VALVE**	VANNE A VIS INTERIEURE	ABSPERRSCHIEBER MIT INNENSPINDEL, ABSPERRSCHIEBER MIT INNENLIEGENDEM SPINDELGEWINDE
6981	**INSIDE SCREW STEM**	TIGE A VIS INTERIEURE	SPINDEL MIT INNENLIEGENDEM GEWINDE
6982	**INSIDE WIDTH OF CHAIN LINK**	LARGEUR INTERIEURE D'UN MAILLON	INNERE LICHTE EINER KETTE, GLIEDERBREITE EINER KETTE
6983	**INSOLUBILITY**	INSOLUBILITE	UNLÖSLICHKEIT
6984	**INSOLUBLE**	INSOLUBLE	UNLÖSLICH
6985	**INSPECTION**	INSPECTION	INSPEKTION, REVISION
6986	**INSPECTION LINE**	LIGNE D'INSPECTION	PRÜFSTRASSE
6987	**INSPECTION OR EXAMINATION**	CONTROLE	PRÜFUNG
6988	**INSTALLATION**	INSTALLATION INTERIEURE	INSTALLATION
6989	**INSTANTANEOUS MOMENTARY VIRTUAL CENTRE**	CENTRE INSTANTANE DE ROTATION	MOMENTANZENTRUM
6990	**INSTANTANEOUS VALUE**	VALEUR INSTANTANEE	MOMENTANWERT
6991	**INSTRUCTION**	INSTRUCTION	ANWEISUNG, BEFEHL
6992	**INSTRUCTIONS FOR USE**	INSTRUCTION, MODE D'EMPLOI	GEBRAUCHSANWEISUNG
6993	**INSTRUMENT PANEL**	TABLEAU DE BORD	INSTRUMENTENBRETT
6994	**INSULATE (TO)**	ISOLER	ISOLIEREN
6995	**INSULATED TANK**	RESERVOIR CALORIFUGE	TANK (WÄRMEISOLIERTER)
6996	**INSULATED WIRE**	FIL ISOLE	DRAHT (ISOLIERTER)
6997	**INSULATING COMPOUND, NON-CONDUCTING COMPOSITION**	COMPOSE ISOLANT	ISOLIERMASSE, SCHUTZMASSE
6998	**INSULATING MATERIAL, INSULATION**	CALORIFUGE, MATIERE ISOLANTE, ISOLANT	ISOLIERSTOFF, ISOLIERMITTEL, ISOLIERMATERIAL
6999	**INSULATING OIL**	HUILE POUR ISOLEMENT	ISOLIERÖL
7000	**INSULATING PROPERTY**	POUVOIR ISOLANT	ISOLIERVERMÖGEN
7001	**INSULATING TAPE**	RUBAN ISOLANT	ISOLIERBAND
7002	**INSULATING TUBE**	TUBE ISOLANT	ISOLIERROHR
7003	**INSULATION**	ISOLATION THERMIQUE, ISOLEMENT, ISOLATION	ISOLIERUNG, WÄRMESCHUTZ, KÄLTESCHUTZ, ISOLIEREN
7004	**INSULATION FROM SOUND**	ISOLATION CONTRE LE BRUIT, LE SON	ISOLIERUNG GEGEN SCHALL
7005	**INSULATION RESISTANCE**	RESISTANCE D'ISOLEMENT	ISOLIERFESTIGKEIT
7006	**INSULATOR**	ISOLANT, ISOLATEUR	ISOLIERKÖRPER, ISOLATOR
7007	**INTAKE MANIFOLD**	COLLECTEUR D'ADMISSION	ANSAUGLEITUNG
7008	**INTAKE STROKE**	TEMPS D'ADMISSION	EINLASSHUB, SAUGHUB

7009	**INTEGRAL**	INTEGRALE	INTEGRAL
7010	**INTEGRAL BEAM**	POUTRE INCORPOREE	BALKEN (EINGEBAUTER)
7011	**INTEGRAL CALCULUS**	CALCUL INTEGRAL	INTEGRALRECHNUNG
7012	**INTEGRAL NUMBER**	NOMBRE ENTIER	ZAHL (GANZE), ZAHL (NATÜRLICHE)
7013	**INTEGRAPH**	INTEGRAPHE	INTEGRAPH
7014	**INTEGRATE (TO)**	INTEGRER	INTEGRIEREN
7015	**INTEGRATION**	INTEGRATION	INTEGRATION
7016	**INTEGRATOR**	INTEGRATEUR	INTEGRATOR
7017	**INTENSIFYING SCREEN**	ECRAN RENFORCATEUR	VERSTÄRKERFOLIE
7018	**INTENSITY**	INTENSITE	INTENSITÄT
7019	**INTENSITY OF RADIATION**	INTENSITE DE RADIATION, INTENSITE LUMINEUSE	STRAHLUNGSINTENSITÄT, AUSSTRAHLUNGSSTÄRKE
7020	**INTENSITY REGULATOR**	REGULATEUR D'INTENSITE	STROMSTÄRKEREGLER
7021	**INTENSITY STRENGTH OF CURRENT**	INTENSITE D'UN COURANT	STROMSTÄRKE
7022	**INTERCALATE (TO), INSERT (TO)**	INTERCALER, INSERER	EINSCHALTEN, EINSCHIEBEN
7023	**INTERCALATION, INSERTION**	INTERCALATION, INSERTION	EINSCHALTEN, EINSCHIEBEN
7024	**INTERCEPT RAYS (TO)**	INTERCEPTER DES RAYONS	STRAHLEN AUFFANGEN
7025	**INTERCHANGEABILITY**	INTERCHANGEABILITE	AUSTAUSCHBARKEIT, AUSWECHSELBARKEIT
7026	**INTERCHANGEABLE**	INTERCHANGEABLE	AUSWECHSELBAR, AUSTAUSCHBAR
7027	**INTERCHANGEABLE GEARS**	ROUES D'ASSORTIMENT, DE SERIE	SATZRÄDER, RÄDER (AUSTAUSCHBARE)
7028	**INTERCHANGEABLE MANUFACTURE**	INTERCHANGEABILITE DE FABRICATION	AUSTAUSCHBARKEIT
7029	**INTERCOMMUNICATING POROSITY**	POROSITE A COMMUNICATION INTERNE	VERBUNDPOROSITÄT
7030	**INTERCRYSTALLINE CORROSION**	CORROSION INTERCRISTALLINE, CORROSION INTERGRANULAIRE	KORROSION (INTERKRISTALLINE), KORNGRENZENBRUCH
7031	**INTERDENDRITIC ATTACK**	ATTAQUE INTERDENDRITIQUE	ANGRIFF (INTERDENDRITISCHER)
7032	**INTERFACE**	INTERFACE	ZWISCHENFLÄCHE
7033	**INTERFACIAL TENSION**	TENSION SUPERFICIELLE	OBERFLÄCHENSPANNUNG
7034	**INTERFACIAL ZONE**	ZONE DE SURFACE DE SEPARATION	TRENNUNGSFLÄCHENGEBIET
7035	**INTERFERENCE**	INTERFERENCE	INTERFERENZ
7036	**INTERFERENCE BAND**	BANDE D'INTERFERENCE	INTERFERENZSTREIFEN
7037	**INTERFERENCE COLOUR**	COULEUR D'INTERFERENCE	INTERFERENZFARBE
7038	**INTERFERENCE COMPARATOR**	COMPARATEUR INTERFERENTIEL	INTERFERENZKOMPARATOR
7039	**INTERFERENCE FIGURE**	FIGURE D'INTERFERENCE	INTERFERENZFIGUR
7040	**INTERFERENCE FRINGE**	FRANGE D'INTERFERENCE	INTERFERENZSTREIFEN
7041	**INTERGRANULAR CORROSION**	CORROSION INTERCRISTALLINE, CORROSION FISSURANTE	KORROSION (INTERKRISTALLINE)
7042	**INTERGRANULAR FRACTURE**	CASSURE INTERGRANULAIRE	KORNGRENZENBRUCH
7043	**INTERMEDIARY PRODUCT**	PRODUIT INTERMEDIAIRE	ZWISCHENERZEUGNIS, ZWISCHENPRODUKT
7044	**INTERMEDIATE BEARING, AUXILIARY BEARING**	PALIER AUXILIAIRE, PALIER SECONDAIRE	NEBENLAGER
7045	**INTERMEDIATE BELT GEARING, BELT COUNTERSHAFT**	RENVOI A COURROIE	RIEMENVORGELEGE
7046	**INTERMEDIATE COOLER**	REFROIDISSEUR INTERNE	KÜHLKÖRPER (EINGEGOSSENER)
7047	**INTERMEDIATE GIRDER**	RAIDISSEUR INTERMEDIAIRE	ZWISCHENTRÄGER
7048	**INTERMEDIATE LANDING**	PALIER INTERMEDIAIRE	ZWISCHENPODEST
7049	**INTERMEDIATE SHAFT, SUBSIDARY SHAFT**	ARBRE INTERMEDIAIRE, ARBRE TROISIEME MOTEUR	ZWISCHENWELLE, HILFSWELLE, ÜBERTRAGUNGSWELLE

7050	INTERMEDIATE TOOTHED GEARING, GEARED COUNTERSHAFT	ENGRENAGE INTERMEDIAIRE, RENVOI A ENGRENAGES, HARNAIS D'ENGRENAGES	RÄDERVORGELEGE, ZAHNRADVORGELEGE
7051	INTERMEDIATE VALUE	VALEUR INTERMEDIAIRE	ZWISCHENWERT
7052	INTERMETALLIC COMPOUND	COMPOSE INTERMETALLIQUE	INTERMETALLISCHE VERBINDUNG
7053	INTERMITTENT MOTION	MOUVEMENT INTERMITTENT, MOUVEMENT DISCONTINU, MOUVEMENT SACCADE	BEWEGUNG (AUSSETZENDE), BEWEGUNG (UNTERBROCHENE), BEWEGUNG (STOSSWEISE), BEWEGUNG (RUCKWEISE), BEWEGUNG (INTERMITTIERENDE)
7054	INTERMITTENT WELD	SOUDURE INTERMITTENTE	NAHT (UNTERBROCHENE)
7055	INTERMITTENT WELDING	SOUDURE DISCONTINUE	NAHT (UNTERBROCHENE)
7056	INTERMITTENT WORKING, INTERMITTENT RUNNING	FONCTIONNEMENT INTERMITTENT, SERVICE INTERMITTENT	BETRIEB (AUSSETZENDER), BETRIEB (INTERMITTIERENDER)
7057	INTERNAL ANGLE	ANGLE INTERIEUR, ANGLE INTERNE	INNENWINKEL
7058	INTERNAL ANNULAR GEARING	ENGRENAGE INTERIEUR	GETRIEBE MIT INNENVERZAHNUNG, INNENGETRIEBE
7059	INTERNAL BURR REMOVER FOR TUBES	FRAISE EBARBEUSE MALE, FRAISE D'INTERIEUR POUR TUBES	ROHRFRÄSER ZUM INNENFRÄSEN
7060	INTERNAL CHASER	PEIGNE POUR L'INTERIEUR, PEIGNE A TARAUDER	INNENSTRÄHLER
7061	INTERNAL COMBUSTION ENGINE	MOTEUR A COMBUSTION INTERNE, MOTEUR A EXPLOSIONS	VERBRENNUNGSMOTOR
7062	INTERNAL COMBUSTION ENGINE	MOTEUR A COMBUSTION INTERNE	VERBRENNUNGSMOTOR
7063	INTERNAL DISPLACEMENT	DEPLACEMENT INTERNE	VERSCHIEBUNG (INNERE)
7064	INTERNAL FORCE	FORCE INTERIEURE	KRAFT (INNERE)
7065	INTERNAL FRICTION	FRICTION INTERNE, FROTTEMENT INTERNE	REIBUNG (INNERE), DÄMPFUNG (INNERE)
7066	INTERNAL GAUGE	JAUGE D'INTERIEUR	LOCHLEHRE
7067	INTERNAL GEARS	ENGRENAGES INTERIEURS	INNENGETRIEBE
7068	INTERNAL GRINDER	MACHINE A RECTIFIER LES SURFACES INTERIEURES	HOHLSCHLEIFMASCHINE, INNENSCHLEIFMASCHINE
7069	INTERNAL GRINDING	RECTIFICATION INTERIEURE	INNENSCHLIFF
7070	INTERNAL HEATING	CHALEUR INTERNE	WÄRME (INNERE)
7071	INTERNAL INSIDE DIAMETER	DIAMETRE INTERIEUR	DURCHMESSER (INNERER)
7072	INTERNAL RESISTANCE	RESISTANCE INTERNE	WIDERSTAND (INNERER)
7073	INTERNAL SHRINKAGE	CAVITE, RETASSURE	INNENLUNKER
7074	INTERNAL STRESS	TENSION INTERNE	SPANNUNG (INNERE)
7075	INTERNAL STRESSES IN CASTINGS, STRESSES SET UP IN CASTINGS	TENSIONS INTERNES DANS LA FONTE	GUSSSPANNUNG
7076	INTERNAL TAPER GAUGE	JAUGE-TAMPON CONIQUE	KEGELLEHRDORN
7077	INTERNAL TEETH	DENTURE INTERIEURE	INNENVERZAHNUNG
7078	INTERNAL THREAD	FILET (AGE) INTERIEUR, TARAUDAGE (INTERIEUR)	INNENGEWINDE
7079	INTERNAL TOOTHED WHEEL, ANNULAR WHEEL	ROUE A DENTURE INTERIEURE	HOHLZAHNRAD
7080	INTERNAL WELDING	SOUDAGE INTERIEUR	INNENSCHWEISSEN
7081	INTERNAL WORK	TRAVAIL INTERIEUR	ARBEIT (INNERE)
7082	INTERNALLY RIBBED TUBE	TUBE A AILETTES INTERIEURES	RIPPENROHR MIT INNENRIPPEN
7083	INTERNATIONAL METRIC TAREAD	FILET METRIQUE INTERNATIONAL	GEWINDE (INTERNATIONALES METRISCHES)
7084	INTERNATIONAL STANDARD THREAD, MILLIMETRE PITCH THREAD, METRIC THREAD	PAS SYSTEME INTERNATIONAL A BASE METRIQUE (S.I.)	GEWINDE (METRISCHES, INTERNATIONALES), S-I-GEWINDE
7085	INTERPASS ANNEALING	RECUIT INTERMEDIAIRE ENTRE DEUX ETIRAGES	GLÜHEN ZWISCHEN ZWEI ZÜGEN

7086	INTERPASS TEMPERATURE	TEMPERATURE DE LA PASSE INTERMEDIAIRE	ZWISCHENLAGENTEMPERATUR
7087	INTERPLANAR DISTANCE	DISTANCE RETICULAIRE	NETZEBENEN-ABSTAND
7088	INTERPOLATE (TO)	INTERPOLER	EINSCHALTEN, INTERPOLIEREN
7089	INTERPOLATION	INTERPOLATION	EINSCHALTUNG, INTERPOLATION
7090	INTERRUPT THE ELECTRIC CURRENT (TO), BREAK (TO)	COUPER LE COURANT, INTERROMPRE LE CIRCUIT ELECTRIQUE	ELEKTRISCHEN STROM UNTERBRECHEN (DEN)
7091	INTERRUPTED AGING	VIEILLISSEMENT INTERROMPU	ALTERUNG (UNTERBROCHENE)
7092	INTERRUPTION OF SERVICE, STOPPAGE OF WORK	INTERRUPTION DE SERVICE, ARRET D'USINE, CHOMAGE	BETRIEBSUNTERBRECHUNG
7093	INTERRUPTION OF THE ELECTRIC CURRENT, BREAK	INTERRUPTION DE PASSAGE DU COURANT ELECTRIQUE	STROMUNTERBRECHUNG, UNTERBRECHUNG DES ELEKTRISCHEN STROMES
7094	INTERSECT (TO)	COUPER (SE) (GEOM.)	SCHNEIDEN (SICH) (GEOM.)
7095	INTERSECTING PLANE	PLAN D'INTERSECTION	SCHNITTEBENE
7096	INTERSECTING SHAFTS	ARBRES CONCOURANTS	SCHNEIDENDE WELLEN (SICH)
7097	INTERSECTION RADIOGRAPHIC CONTROL	RADIO DES NOEUDS DE SOUDURE	RADIOGRAMM EINER SCHWEISSNAHT
7098	INTERSTAND	CAGE MEDIANE	MITTEL-(WALZ-)GERÜST
7099	INTERURBAN BUS	AUTOCAR	ÜBERLANDBUS
7100	INTERVAL	ARRET DE SERVICE, REPOS	BETRIEBSPAUSE, ARBEITSPAUSE
7101	INTERVAL, SPACE OF TIME	INTERVALLE	ZWISCHENZEIT, ZEITSTRECKE, INTERVALL
7102	INTIMATE THOROUGH MIXING	MELANGE INTIME	INNIGES MISCHEN, DURCHMISCHEN
7103	INTRA-RED RAYS	RAYONS INFRA-ROUGES	STRAHLEN (INFRAROTE)
7104	INTRINSIC BRIGHTNESS	INTENSITE LUMINEUSE PAR UNITE DE SURFACE	LEUCHTDICHTE, FLÄCHENHELLIGKEIT
7105	INVAR	METAL INVAR, INVAR	INVAR
7106	INVARIANT	INVARIANT	INVARIANTE
7107	INVENTION	INVENTION	ERFINDUNG
7108	INVENTOR	INVENTEUR, AUTEUR D'UNE INVENTION	ERFINDER
7109	INVERSE CHILL	TREMPE INVERSE	HARTGUSS (UMGEKEHRTER)
7110	INVERSION	INVERSION (MATH.)	UMKEHRUNG, INVERSION
7111	INVERTED PLUG VALVE	ROBINET A BOISSEAU RENVERSE	HAHN (SELBSTDICHTENDER)
7112	INVERTED TROUGH IRON	FER POUR TABLIERS DE PONT, FER ZORES	BELAGEISEN, ZORESEISEN
7113	INVERTED WELDING	SOUDAGE AU PLAFOND	ÜBERKOPFSCHWEISSEN
7114	INVESTIGATION, EXAMINATION, RESEARCH WITH THE MICROSCOPE, MICROSCOPIC EXAMINATION	EXAMEN MICROSCOPIQUE	UNTERSUCHUNG (MIKROSKOPISCHE), BEOBACHTUNG
7115	INVESTMENT CASTING	MOULAGE A CIRE PERDUE	GIESSEN MIT VERLORENER GIESSFORM, INVESTMENTGUSS
7116	INVESTMENT PATTERN	MODELE A CIRE PERDUE	AUSSCHMELZMODELL
7117	INVOLUTE	DEVELOPPANTE	EVOLVENTE, INVOLUTE
7118	INVOLUTE OF THE CIRCLE	DEVELOPPANTE DU CERCLE	KREISEVOLVENTE
7119	INVOLUTE TEETH	ENGRENAGE A DEVELOPPANTE DE CERCLE	EVOLVENTENVERZAHNUNG
7120	IODINE	IODE	JOD
7121	ION	ION	ION
7122	ION CONCENTRATION	CONCENTRATION D'IONS	IONENKONZENTRATION
7123	ION MIGRATION	MIGRATION D'IONS	IONENWANDERUNG
7124	IONIZATION	IONISATION	IONISATION, IONISIERUNG

7125	IONIZATION CHAMBER	CHAMBRE D'IONISATION	IONISATIONSKAMMER
7126	IONIZATION CHAMBER CURRENT	COURANT DE LA CHAMBRE D'IONISATION	KAMMERSTROM (IONISATION)
7127	IONIZATION METHOD	METHODE D'IONISATION	IONISATIONSMETHODE
7128	IONIZED REGION	ZONE IONISEE	ZONE (IONISIERTE)
7129	IONOGEN	IONOGENE	IONOGEN
7130	IONOGEN SOLVENT	SOLVANT IONOGENE	LÖSUNGSMITTEL (IONISIERENDES)
7131	IRIDIUM	IRIDIUM	IRIDIUM
7132	IRIS DIAPHRAGM	DIAPHRAGME IRIS	IRISBLENDE
7133	IRON	FER	EISEN
7134	IRON ACETATE	ACETATE DE FER	ESSIGSAURES EISEN, EISENAZETAT
7135	IRON ALUM	ALUN DE FER	EISENALAUN
7136	IRON AND SECTION IRON SHEARING MACHINE	MACHINE A CISAILLER LES FERS ET PROFILES	EISEN- UND FORMSTAHLSCHEREN
7137	IRON ANGLES FROM REROLLED SCRAP	CORNIERES DE FER AU PAQUET	WINKELEISEN AUS PAKETEISEN
7138	IRON CARBIDE	CARBURE DE FER	EISENKARBID
7139	IRON CARBON DIAGRAM	DIAGRAMME FER-CARBURE	EISEN-KOHLENSTOFF-DIAGRAMM
7140	IRON CARBONYL	CARBONYLE DE FER	EISENKARBONYL
7141	IRON CEMENT	MASTIC DE FER, MASTIC POUR FONTE	ROSTKITT, EISENKITT
7142	IRON FOUNDRY	FONDERIE DE FER/DE FONTE	EISENGIESSEREI
7143	IRON ORE	MINERAI DE FER	EISENERZ
7144	IRON OXIDE	OXYDE DE FER	EISENOXYD
7145	IRON POWDER	POUDRE DE FER	EISENPULVER
7146	IRON PYRITES, IRON DISULPHIDE	PYRITE JAUNE, MARTIALE	SCHWEFELKIES, EISENKIES, PYRIT
7147	IRON ROPE	CABLE EN FILS DE FER	EISENDRAHTSEIL
7148	IRON SESQUIOXIDE, FERRIC OXIDE, COLCOTHAR	PEROXYDE, SESQUIOXYDE DE FER, OXYDE FERRIQUE, ROUGE DE PRUSSE, ROUGE ANGLAIS, ROUGE A POLIR, COLCOTHAR	ENGLISCHROT, POLIERROT, KOLKOTHAR, EISENSESQUIOXYD, EINSEMENNIGE
7149	IRON SILICATE	SILICATE DE FER	EISENSILIKAT
7150	IRON SLEEPER	TRAVERSE EN FER	EISENSCHWELLE
7151	IRON SPONGE	EPONGE DE FER	EISENSCHWAMM
7152	IRON STEEL STRUCTURE	CONSTRUCTION METALLIQUE, CHARPENTE METALLIQUE	EISENBAU, EISENKONSTRUKTION
7153	IRON SULFIDE	SULFURE FERREUX/FERRIQUE	EISENSULFID
7154	IRON TOOTH	DENT EN FONTE	EISENZAHN
7155	IRON WIRE	FIL DE FER	EISENDRAHT
7156	IRON-BASE ALLOY	ALLIAGE DE FER	EISENLEGIERUNG
7157	IRON, CAST IRON	FONTE	GUSSEISEN
7158	IRONING	EMBOUTISSAGE PROFOND	TIEFZIEHEN
7159	IRONWORK (FITTINGS, MOUNTINGS)	FERRURE	BESCHLAG, EISENBESCHLAG, BESCHLÄGE
7160	IRONWORKS	FORGE (USINE)	EISENHÜTTE, EISENWERK
7161	IRRADIATION	IRRADIATION	BESTRAHLUNG
7162	IRRATIONAL NUMBER	NOMBRE IRRATIONNEL/ INCOMMENSURABLE	ZAHL (IRRATIONALE)
7163	IRREGULAR	IRREGULIER	UNREGELMÄSSIG
7164	IRREGULAR POLYGON	POLYGONE IRREGULIER	POLYGON (UNREGELMÄSSIGES), POLYGON (IRREGULÄRES)
7165	IRREGULAR RUNNING	MARCHE IRREGULIERE, BOITEMENT	GANG (UNRUHIGER)
7166	IRREGULAR TRACE, LINE WITH SHARP TURNS	LIGNE BRISEE	LINIE (GEBROCHENE)

7167	**IRREGULARITY OF MOTION**	IRREGULARITE DU MOUVEMENT	UNGLEICHFÖRMIGKEIT DER BEWEGUNG
7168	**IRREVERSIBILITY**	IRREVERSIBILITE, NON-REVERSIBILITE	NICHTUMKEHRBARKEIT
7169	**IRREVERSIBLE**	IRREVERSIBLE, NON-REVERSIBLE	NICHT UMKEHRBAR
7170	**IRREVERSIBLE CHANGE OF STATE**	CHANGEMENT D'ETAT NON REVERSIBLE	ZUSTANDSÄNDERUNG (NICHTUMKEHRBARE)
7171	**IRREVERSIBLE ELECTROLYTIC PROCESS**	REACTION ELECTROLYTIQUE IRREVERSIBLE	REAKTION (IRREVERSIBEL ELEKTROLYTISCHE)
7172	**IRRIGATION PLANT**	MATERIEL D'IRRIGATION	BEWÄSSERUNGSANLAGEN
7173	**ISENTROPIC LINE**	LIGNE ISENTROPIQUE	ISENTROPE
7174	**ISINGLASS, FISH GLUE**	COLLE DE POISSON, ICHTYOCOLLE	HAUSENBLASE, FISCHLEIM
7175	**ISO ELECTRIC POINT**	POINT ISO-ELECTRIQUE	PUNKT (ISOELEKTRISCHER)
7176	**ISOBAR**	ISOBARE	ISOBARE
7177	**ISOBAR, ISOPIESTIC LINE**	COURBE ISOBARIQUE	LINIE GLEICHEN DRUCKES, ISOBARE
7178	**ISOCHRONISM**	ISOCHRONISME	ISOCHRONISMUS
7179	**ISOCHRONOUS**	ISOCHRONE	ISOCHRON
7180	**ISOCHRONOUS GOVERNOR**	REGULATEUR ISOCHRONE	REGLER (ISOCHRONER)
7181	**ISOCLINIC LINE**	LIGNE ISOCLINE	ISOKLINE
7182	**ISODYNAMIC LINE**	LIGNE ISODYNAME, LIGNE ISODYNAMIQUE	KURVE (ISODYNAME), KURVE (ISODYNAMISCHE)
7183	**ISOLATED, CONCENTRATED LOAD**	CHARGE ISOLEE	EINZELLAST
7184	**ISOLATION**	ISOLATION	ISOLIERUNG, ABTRENNUNG
7185	**ISOMERICAL**	ISOMERE	ISOMER
7186	**ISOMERISM**	ISOMERIE	ISOMERISMUS, ISOMERIE
7187	**ISOMETRIC LINE, ISOPLERE**	ISOPLERE, LIGNE COURBE DE VOLUME CONSTANT	LINIE GLEICHEN RAUMINHALTES, ISOPLERE
7188	**ISOMETRIC PROJECTION**	PROJECTION ISOMETRIQUE	PROJEKTION (ISOMETRISCHE)
7189	**ISOMORPHISM**	ISOMORPHIE	ISOMORPHIE, ISOMORPHISMUS
7190	**ISOMORPHOUS**	ISOMORPHE	ISOMORPH
7191	**ISOMORPHOUS MIXTURE**	ISOMORPHISME	HOMÖOMORPHIE, ISOMORPHISMUS
7192	**ISOSCELES TRIANGLE**	TRIANGLE ISOCELE	DREIECK (GLEICHSCHENKLIGES)
7193	**ISOTHERMAL CURVE**	LIGNE COURBE ISOTHERMIQUE	ISOTHERME
7194	**ISOTHERMAL HEAT TREATING**	TRAITEMENT ISOTHERME	BEHANDLUNG (ISOTHERMISCHE)
7195	**ISOTHERMAL QUENCHING**	TREMPE ISOTHERME	HÄRTUNG (ISOTHERMISCHE)
7196	**ISOTHERMAL TRANSFORMATION**	TRANSFORMATION ISOTHERME	UMWANDLUNG (ISOTHERMISCHE)
7197	**ISOTOPES**	ISOTOPES	ISOTOPEN
7198	**ISOTROPIC**	ISOTROPE	ISOTROP
7199	**IVORY**	IVOIRE	ELFENBEIN
7200	**IZOD IMPACT TEST**	ESSAI IZOD	IZOD-PROBE
7201	**JACK PLANE**	RIFLARD	SCHROPPHOBEL, SCHROBHOBEL, SCHRUPPHOBEL, SCHROTHOBEL, SCHURFHOBEL, SCHÜRFHOBEL
7202	**JACK SHAFT**	ARBRE INTERMEDIAIRE	ZWISCHENWELLE
7203	**JACKET**	FRETTE	SCHRUMPFRING
7204	**JACKET COOLING**	REFROIDISSEMENT PAR CHEMISE D'EAU	ABKÜHLUNG MIT WASSERMANTEL
7205	**JACKETED VALVE**	VANNE A ENVELOPPE CHAUFFANTE	HEIZMANTELSCHIEBER
7206	**JACKING BRACKET LUGS**	PATTES DE LEVAGE PAR VERINS	HEBESTUTZEN
7207	**JACKING SHOE**	SABOT POUR DISPOSITIF DE LEVAGE	SCHUH FÜR HEBEVORRICHTUNG
7208	**JACOBS COLLET CHUCK**	MANDRIN EXPANSIBLE JACOBS	JACOBS-SPREIZZANGENFUTTER

7209	JAG BOLT	BOULON DE SCELLEMENT A BARBELURES	STEINSCHRAUBE MIT AUFGEHAUENEN KANTEN
7210	JAM (TO)	COINCER (SE)	KLEMMEN (SICH), ECKEN, HÄNGENBLEIBEN
7211	JAMB	MONTANT LATERAL D'ENTREE	TÜRSTEIN
7212	JAMMING	COINCEMENT, PINCEMENT	ECKEN, KLEMMEN
7213	JAPAN WAX	CIRE DU JAPON	WACHS (JAPANISCHES)
7214	JAPANNING	REVETEMENT AVEC VERNIS FIN	BRANDLACKIEREN
7215	JARRING MACHINE	MACHINE A MOULER RAPIDE A SECOUSSES	RÜTTELFORMMASCHINE
7216	JAW OF WRENCH	MACHOIRE DE CLEF, PAND D'UNE CLEF	BACKEN EINES SCHRAUBENSCHLÜSSELS
7217	JENA GLASS	VERRE D'IENA	JENAER GLAS
7218	JET	GICLEUR	DÜSE
7219	JET CARRIER	PORTE-GICLEUR	DÜSENTRÄGER
7220	JET CONDENSER	CONDENSEUR PAR INJECTION	EINSPRITZKONDENSATOR
7221	JET OF FLUID	JET LIQUIDE	FLÜSSIGKEITSSTRAHL
7222	JET PIERCING	PERCAGE PAR JET DE FLAMME	STRAHLDÜSENBOHREN
7223	JET PUMP	INJECTEUR	STRAHLPUMPE
7224	JIB CRANE	GRUE A FLECHE	AUSLEGERKRAN
7225	JIB LENGTH	LONGUEUR DE LA FLECHE	AUSLERGERLÄNGE
7226	JIB OF A CRANE	FLECHE PRINCIPALE D'UNE GRUE	HAUPTAUSLEGER
7227	JIG	DISPOSITIF DE FIXATION, DISPOSITIF DE SERRAGE	EINSPANNVORRICHTUNG
7228	JIG BORING MACHINE	MACHINE A POINTER	LEHRENBOHRMASCHINE
7229	JOB	AFFAIRE	ARBEIT, GESCHÄFT
7230	JOBBING FOUNDRY	FONDERIE SUR MODELES	KUNDENGIESSEREI
7231	JOGGLE JOINT	RENVOI A LA CHASSE	ZAPFENFUGE
7232	JOGGLING MACHINES	MACHINES A COUDER	KRÖPFMASCHINEN
7233	JOINER	MENUISIER	SCHREINER, TISCHLER
7234	JOINER'S BENCH	ETABLI, BANC DE MENUISIER	HOBELBANK
7235	JOINER'S SHOP	MENUISERIE (ATELIER)	SCHREINEREI, TISCHLEREI (WERKSTÄTTE)
7236	JOINERY	MENUISERIE	TISCHLEREI, SCHREINEREI (HANDWERK)
7237	JOINT	JOINT SOUDE	SCHWEISSVERBINDUNG
7238	JOINT (MEC.)	JOINT (POINT DE JONCTION), ARTICULATION (MEC.)	STOSSSTELLE, STOSS, FUGE, GELENK (MECH.)
7239	JOINT DISTRIBUTION FUNCTION	FONCTION DE REPARTITION COMPOSEE	VERTEILUNG (MEHRDIMENSIONALE)
7240	JOINT GEOMETRY	GEOMETRIE DU JOINT	VERBINDUNGSGEOMETRIE
7241	JOINT OF BEARING	JOINT DE SEPARATION D'UN PALIER	LAGERFUGE, TEILFUGE, TRENNUNGSFUGE EINES LAGERS
7242	JOINT PIN, LINK PIN	TOURILLON D'ARTICULATION	GELENKZAPFEN, GELENKBOLZEN
7243	JOINT PIPES WITH FLANGES (TO), CONNECT PIPES IN LINE WITH FLANGES (TO)	BRIDER DES TUYAUX	ROHRE DURCH FLANSCHEN VERBINDEN
7244	JOINT UP PIPES (TO)	RACCORDER DES TUYAUX	ROHRE VERBINDEN
7245	JOINTER PLANE	VARLOPE	RAUHBANK, LANGHOBEL
7246	JOINTER, JOINTING MACHINE	MACHINE A FAIRE LES JOINTS	FÜGEMASCHINE
7247	JOINTING MATERIAL, PACKING MATERIAL	MATIERE POUR JOINT, JOINT, BOURRAGE	DICHTUNGSMITTEL, DICHTUNGSSTOFF, DICHTUNGSMATERIAL
7248	JOINTING PIPES, CONNECTING TOGETHER LENGTHS OF PIPING	RACCORDEMENT DE TUYAUX	VERBINDEN VON ROHREN

7249	JOLT AND JUMBLE TEST	ESSAI DE CHOC ET BALLOTTEMENT	STOSS-UND RÜTTELPRÜFUNG
7250	JOLT MOLDING MACHINE	MACHINE A MOULER A SECOUSSES	RÜTTELFORM-MASCHINE
7251	JOLT ROLL-OVER MOLDING MACHINE	MACHINE A MOULER A SECOUSSES AVEC PLAQUE REVERSIBLE	UMROLL-RÜTTELFORM MASCHINE
7252	JOLT-SQUEEZE MACHINE	MACHINE A SERRAGE PAR SECOUSSE ET PRESSION SANS DEMOULAGE	RÜTTELPRESS-FORMMASCHINE OHNE ENTFORMUNG
7253	JOLT-SQUEEZE STRIPPER MACHINE	MACHINE A MOULER A SECOUSSES ET A SERRAGE COMBINES AVEC DEMOULAGE	RÜTTELPRESSFORMMASCHINE
7254	JOMINY TEST	ESSAI JOMINY, ESSAI DE REBONDISSEMENT BRUSQUE	JOMINY-PROBE, STIRNABSCHRECKPROBE
7255	JOULE	JOULE, VOLT-COULOMB	JOULE
7256	JOURNAL	PORTEE, TOURILLON	DREHZAPFEN, ZAPFEN, TRAGZAPFEN
7257	JOURNAL BEARING	PALIER A CHARGE RADIALE/ TRANSVERSALE	TRAGLAGER, QUERDRUCKLAGER
7258	JOURNAL FRICTION	FROTTEMENT DES TOURILLONS	ZAPFENREIBUNG
7259	JOURNAL PRESSURE	PRESSION/REACTION SUR LES TOURILLONS	ZAPFENDRUCK
7260	JUMP SCAFFOLD	ECHAFAUDAGE VOLANT	SPRUNGGERÜST
7261	JUMP SEAT	STRAPONTIN	KLAPPSITZ
7262	JUMP TEST	ESSAI D'ECRASEMENT	STAUCHVERSUCH
7263	JUMP UP (TO), UPSET (TO)	ECRASER, REFOULER	ZUSAMMENSTAUCHEN
7264	JUMPER CABLE	CABLE VOLANT	SCHALTDRAHT
7265	JUMPING UP, UPSETTING	ECRASEMENT, REFOULEMENT, REFOULAGE	ZUSAMMENSTAUCHEN
7266	JUNK RING	COUVERCLE DU PISTON	KOLBENDECKEL
7267	JURASSIC LIMESTONE	CALCAIRE JURASSIQUE	JURAKALK
7268	JUTE	JUTE	JUTE
7269	K-FACTOR (THERMAL CONDUCTIVITY)	COEFFICIENT CALORIFIQUE (OU DE CONDUCTIVITE THERMIQUE)	WÄRMELEITZAHL
7270	KAHLBAUM IRON	FER KAHLBAUM	KAHLBAUM-EISEN
7271	KALDO PROCESS	PROCEDE FOUR KALDO	KALDOVERFAHREN
7272	KALE CUTTERS	COUPE-RAVES	BLATTKOHLSCHNEIDER
7273	KEELBLOCK	LINGOT-EPROUVETTE (EN FORME DE QUILLE DE NAVIRE)	KIELBLOCK, KIELKLOTZ
7274	KERF	SAIGNEE	SCHNITTFUGE
7275	KEROSENE OIL, PARAFFIN OIL, COMMON LAMP OIL, ILLUMINATING OIL	HUILE LAMPANTE DE PETROLE, HUILE D'ECLAIRAGE, PETROLE LAMPANT, KEROSENE	LEUCHTÖL, LEUCHTPETROLEUM
7276	KEY	CHEVILLE, CLAVETTE, CLAVETTE LONGITUDINALE	DÜBEL, KEIL, LÄNGSKEIL
7277	KEY (TO), WEDGE (TO)	CLAVETER, CALER	VERKEILEN
7278	KEY DRIFT	CHASSE-CLAVETTE	KEILTREIBER
7279	KEY ON TO (TO)	CLAVETER, CALER SUR...	AUFKEILEN
7280	KEY PLATE	CLAME D'ASSEMBLAGE	SPANNFINGER
7281	KEY WAY MILLING MACHINE	MACHINE A FRAISER LES RAINURES	LANGLOCHFRÄSMASCHINE, KEILNUTENFRÄSMASCHINE, NUTENFRÄSMASCHINE
7282	KEY WAY, KEY BED, KEY SCATING	RAINURE, CANNELURE, LOGEMENT D'UNE CLAVETTE, MORTAISE A CLAVETTE, RAINURE DE CLAVETAGE	KEILNUT, KEILRILLE
7283	KEY-SEAT GAUGE	JAUGE POUR RAINURE DE CLAVETTE	KEILNUTENSCHABLONE

7284	KEY, FIXED SPANNER	CLEF DE CALIBRE, CLEF A FOURCHE	SCHRAUBENSCHLÜSSEL (GEWÖHNLICHER)
7285	KEYED JOINT, COTTERED JOINT	ASSEMBLAGE PAR CLAVETTE	KEILVERBINDUNG
7286	KEYHEAD	TALON DE CLAVETTE	KEILNASE
7287	KEYHOLE	TROU DE COULEE	SCHLÜSSELOCH, STICHLOCH
7288	KEYING	CLAVETAGE	KEILVERBINDUNG
7289	KEYSEAT CLAMPS	BRIDES DE MORTAISAGE	KEILNUTENFLANSCHEN
7290	KEYSEAT RULES	REGLES DE MORTAISAGE	KEILNUTENSCHABLONE
7291	KEYWAY	RAINURE DE CLAVETTE	KEILNUT
7292	KIESELGUHR, DIATOM EARTH, DIATOMITE, MOUNTAIN MEAL, FOSSIL MEAL	KIESELGUHR, TRIPOLI SILICEUX, FARINE FOSSILE, SILICE FARINEUSE, TERRE POURRIE, TERRE D'INFUSOIRES	KIESELGUR, INFUSORIENERDE, DIATOMEENERDE
7293	KILLED SPIRITS, SOLUTION OF ZINC CHLORIDE	EAU A SOUDER, LIQUIDE DE DECAPAGE, ESPRIT DE SEL DENATURE	LÖTWASSER
7294	KILLED STEEL	ACIER CALME	BERUHIGTER STAHL
7295	KILN	FOUR DE CALCINATION	BRENNOFEN
7296	KILODYNE	KILODYNE	KILODYN
7297	KILOGRAMME	KILOGRAMME	KILOGRAMM, KILO
7298	KILOGRAMME CALORIE, MAJOR CALORIE	GRANDE CALORIE, CALORIE KILOGRAMME-DEGRE	KILOGRAMMLKALORIE
7299	KILOGRAMMETRE (KGM)	KILOGRAMMETRE	METERKILOGRAMM, KILOGRAMMETER, MKG
7300	KILOMETER	KILOMETRE	KILOMETER
7301	KILOVOLT	KILOVOLT	KILOVOLT
7302	KILOVOLT-AMPERE	KILOVOLT-AMPERE	KILOVOLTAMPERE
7303	KILOWATT	KILOWATT	KILOWATT
7304	KILOWATT/OUR (KW/HR), BOARD OF TRADE UNIT (B.T.U.), KELVIN	KILOWATT-HEURE	KILOWATTSTUNDE
7305	KINEMATICAL	CINEMATIQUE	KINEMATISCH
7306	KINEMATICS	CINEMATIQUE	GETRIEBELEHRE, KINEMATIK
7307	KINETIC	CINETIQUE	KINETISCH
7308	KINETIC ENERGY	ENERGIE CINETIQUE, ENERGIE ACTUELLE, PUISSANCE, FORCE VIVE	ENERGIE DER BEWEGUNG, ENERGIE (KINETISCHE), ENERGIE DYNAMISCHE, LEBENDIGE KRAFT, WUCHT
7309	KINETIC FRICTION	FROTTEMENT CINETIQUE	REIBUNGSGRENZE
7310	KINETICS	CINETIQUE	BEWEGUNGSLEHRE, KINETIK
7311	KING PIN	AXE (DE PIVOT), AXE DE PIVOTEMENT	ACHSSCHENKELBOLZEN
7312	KING PIN INCLINATION ANGLE	ANGLE D'INCLINATION DES PIVOTS	STURZ
7313	KINK	NOEUD, COQUE	KINKE, KNICK
7314	KINK BAND	BANDE DE PLIAGE	KNICKBAND
7315	KINKING	DEFORMATION EN GENOU	KNICKUNG
7316	KIP = KILOPOUND (US)	KIP = 1000 LIVRES = 453,59 KG	KIP = 1000 PFD = 453,59 KG
7317	KISH	ECUME DE GRAPHITE	GARSCHAUMGRAPHIT
7318	KLINGERIT	KLINGERITE	KLINGERIT
7319	KNEE-AND-COLUMN MILLING MACHINE	FRAISEUSE A CONSOLE ET COLONNE	KONSOL-UND SÄULENFRÄSMASCHINE
7320	KNEE, ELBOW	COUDE D'EQUERRE	KNIEROHR, KNIESTÜCK
7321	KNEE, SHOULDER	EPAULEMENT	KRÖPFUNG
7322	KNEETYPE MILLING MACHINE	MACHINE A FRAISER A CONSOLE	KONSOLFRÄSMASCHINE
7323	KNIFE	COUTEAU	MESSER

7324	**KNIFE (EDGED) FILE**	LIME-COUTEAU	MESSERFEILE
7325	**KNIFE-EDGE BEARING**	SUPPORT A COUTEAU, COUTEAU	SCHNEIDENLAGER
7326	**KNIFE-EDGE SUSPENSION**	SUSPENSION A COUTEAU	SCHNEIDENAUFHÄNGUNG
7327	**KNOCK-ON**	COLLISION (DE NEUTRONS AVEC DES ATOMES)	ANSTOSS
7328	**KNOCK-OUT**	DECOCHAGE	AUSSCHLAGEN, AUSLEEREN
7329	**KNOCKED-DOWN**	DEMONTE, PIECES DETACHEES (EN)	ZERLEGT
7330	**KNOCKING**	DETONATION, COGNEMENT	KNALL, KLOPFEN
7331	**KNOCKING OF THE PUMP**	TAPAGE D'UNE POMPE	SCHLAGEN DER PUMPE
7332	**KNOCKOUT BAR**	CHASSE-POINTE	DURCHSCHLAG
7333	**KNOCKOUT PIN (US)**	BARRE DE PIQUAGE	STOSSEISEN, BRECHSTANGE
7334	**KNOCKOUT PLATE (US)**	GRILLE DE DECOCHAGE	AUSSCHLAGROST, AUSLEERROST
7335	**KNOWN QUANTITY**	QUANTITE CONNUE	GRÖSSE (BEKANNTE)
7336	**KNUCKLE JOINT**	FERMETURE A GRENOUILLERE, ROTULE	KNIEHEBELVERCHLUSS, KNOCHENGELENK
7337	**KNURL DRIVE**	COMMAMDE PAR VIS MOLETEE	RÄNDELSCHRAUBENANTRIEB
7338	**KNURLING**	MOLETAGE	KORDELUNG, RÄNDELUNG
7339	**KRYPTON** ·	KRYPTON	KRYPTON
7340	**KYANISING**	IMPREGNATION DU BOIS AU SUBLIME CORROSIF PAR LE PROCEDE KYAN	KYANISIEREN DES HOLZES
7341	**L.O.I= LOSS ON IGNITION**	PERTE AU ROUGE, PERTE AU FEU	GLÜHVERLUST
7342	**L.P.G (LIQUEFIED PETROLEUM GAS)**	GAZ LIQUEFIE	FLÜSSIGGAS
7343	**LABILE**	LABILE	LABIL
7344	**LABORATORY**	LABORATOIRE	LABORATORIUM
7345	**LABORATORY SET**	JEU DE LABORATOIRE	LABORGERÄTESATZ
7346	**LABOUR COSTS, COST OF LABOUR**	FRAIS, PRIX DE MAIN-D'OEUVRE	ARBEITSKOSTEN
7347	**LABYRINTH PACKING**	GARNITURE, JOINT A LABYRINTHE, JOINT A CHICANE	LABYRINTHDICHTUNG
7348	**LACED BELT JOINT**	JONCTION DES COURROIES PAR LANIERES EN CUIR	RIEMENVERBINDUNG (GENÄHTE)
7349	**LACED LEATHER BELT FLEXIBLE COUPLING**	MANCHON A BANDE DE COURROIE	BANDKUPPKLUNG, RIEMENKUPPLUNG, LEDERRIEMENKUPPLUNG
7350	**LACED SEWED BELT**	COURROIE COUSUE	RIEMEN (GENÄHTER)
7351	**LACK OF FUSION**	COLLAGE	BINDEFEHLER
7352	**LACK OF UNIFORMITY IN THE MATERIAL**	INHOMOGENEITE D'UN MATERIEL	UNGLEICHFÖRMIGKEIT DES MATERIALS
7353	**LACQUER (TO), VARNISH (TO), COAT WITH LACQUER (TO)**	LAQUER, VERNIR	LACKIEREN
7354	**LACQUER, SPIRIT VARNISH**	LAQUE, VERNIS A L'ALCOOL	SPIRITUSLACK, LACK (ALKOHOLISCHER), WEINGEISTLAK, WEINGEISTFIRNIS, SPIRITUSFIRNIS, ALKOHOLFIRNIS
7355	**LACTIC ACID**	ACIDE LACTIQUE	MILCHSÄURE
7356	**LADLE**	POCHE DE COULEE	PFANNE, GIESSPFANNE
7357	**LADLE ADDITION**	ADDITION EN POCHE	PFANNENZUSATZ
7358	**LADLE ANALYSIS**	ANALYSE DE COULEE	PFANNENANALYSE
7359	**LADLE BRICK**	BRIQUE DE POCHE	PFANNENZIEGEL, PFANNENSTEIN
7360	**LADLE CHILL**	SOLIDIFICATION DE LA MASSE FONDUE DANS LA POCHE	ERSTARRUNG DER SCHMELZE IN DER PFANNE
7361	**LADLE LIP**	BEC DE COULEE DE LA POCHE	PFANNENAUSGUSS
7362	**LADLE MAN, LADLE POURER**	COULEUR	PFANNENFÜHRER
7363	**LADLE NOSE**	BEC DE LA POCHE DE COULEE	PFANNENAUSGUSS
7364	**LAEVO-TARTARIC ACID**	ACIDE TARTRIQUE LEVOGYRE	LINKSWEINSÄURE

7365	**LAEVOROTATORY, LAEVOGYRATE**	LEVOGYRE	LINKSDREHEND
7366	**LAG (TO)**	RETARDER	NACHEILEN
7367	**LAGGING**	REVETEMENT CALORIFUGE	WÄRMESCHUTZ
7368	**LAGGING OF PHASE**	DEPHASAGE, DECALAGE EN ARRIERE	PHASENNACHEILUNG
7369	**LAGGING OF PIPES, PIPE COVERING**	RECOUVREMENT DE TUYAUX, ENVELOPPE DE TUYAUX	ROHRVERKLEIDUNG, VERKLEIDEN, UMHÜLLEN VON ROHREN
7370	**LAKE COPPER**	CUIVRE NATIF, CUIVRE DU LAC SUPERIEUR	KUPFER (GEDIEGENES), OBERSEE KUPFER, KUPFER VON LAKE-SUPERIOR-ERZEN
7371	**LAMELLA**	LAMELLE	LAMELLE
7372	**LAMELLAR**	LAMELLAIRE	LAMELLAR
7373	**LAMELLAR STRUCTURE**	STRUCTURE LAMELLAIRE	BLÄTTRIGE STRUKTUR
7374	**LAMINAR TEXTURE**	TEXTURE LAMELLEUSE	GEFÜGE (BLÄTTRIGES)
7375	**LAMINATED**	LAMINE	LAMELLIERT, GESCHICHTET
7376	**LAMINATED BELT, BELT MADE UP OF BUILT-ON EDGE STRIPS**	COURROIE EN LANIERES DE CUIR TRAVAILLANT SUR CHAMP	HOCHKANTRIEMEN, LEDERHOCKANTRIEMEN
7377	**LAMINATED COMPOUND MAGNET, MAGNETIC BATTERY**	AIMANT A LAMES SUPERPOSEES	BLÄTTERMAGNET, LAMELLENMAGNET, MAGNETISCHES MAGAZIN
7378	**LAMINATED FRACTURE**	CASSURE LAMELLAIRE	BRUCH (BLÄTTRIGER)
7379	**LAMINATED METAL**	METAL STRATIFIE	MEHRSCHICHTMETALL
7380	**LAMINATED RECTANGULAR PLATE SPRING**	RESSORT RECTANGULAIRE A LAMES SUPERPOSEES	RECHTECKFEDER (GESCHICHTETE)
7381	**LAMINATED TRIANGULAR PLATE SPRING**	RESSORT TRIANGULAIRE A LAMES SUPERPOSEES	DREIECKFEDER (GESCHICHTETE)
7382	**LAMINATION**	STRATIFICATION, FEUILLETAGE (DEDOUBLEMENT DES BORDS) DE TOLE	LAGE, SCHICHTUNG, DOPPLUNG
7383	**LAMINATION CRACK**	FISSURE CRIQUE DE LAMINAGE	LAGENRISS, SCHICHTENRISS, WALZSPLITTER
7384	**LAMP**	LAMPE	LAMPE
7385	**LAMP, BULB**	AMPOULE	LAMPE, BIRNE
7386	**LAMPBLACK**	NOIR DE FUMEE, NOIR A NOIRCIR	KIENRUSS
7387	**LANCE CUTTING**	OXYCOUPAGE	BRENNSCHNEIDEN
7388	**LANDING**	PALIER	TREPPENABSATZ
7389	**LANG LAY**	CABLAGE, TORSION LANG	ALBERTSCHLAG
7390	**LANG LAY WIRE ROPE**	CABLE METALLIQUE A FILS PARALLELES, CABLE A CABLAGE LANG OU ALBERT	DRAHTSEIL NACH ALBERTSCHLAG, GLEICHSCHLAGRAHTSEIL
7391	**LANOLINE**	LANOLINE	LANOLIN
7392	**LANTERN RING**	LANTERNE D'UNE VANNE, LANTERNE DE PRESSE-ETOUPE	ZWISCHENRING, AUFSATZRING, LATERNE EINES SCHIEBERS
7393	**LANTERN WHEEL**	LANTERNE	LATERNE, DREHLING
7394	**LANTHANUM**	LANTHANE	LANTHAN
7395	**LAP**	REPLIURE DE LAMINAGE, REPLI	ÜBERWALZUNGSFEHLER
7396	**LAP JOINT**	ASSEMBLAGE PAR RECOUVREMENT	ÜBERLAPPUNG, ÜBERLAPPTER STOSS
7397	**LAP JOINT**	JOINT A RECOUVREMENT, ASSEMBLAGE PAR RECOUVREMENT	ÜBERLAPPUNGSVERBINDUNG, ÜBERLAPPTER STOSS
7398	**LAP JOINT FLANGE**	BRIDE TOURNANTE	ÜBERLAPPUNGSFLANSCH
7399	**LAP OF SLIDE VALVE**	RECOUVREMENT DU TIROIR	ÜBERDECKUNG DES SCHIEBERS, ÜBERLAPPUNG DES SCHIEBERS
7400	**LAP RIVETED JOINT**	RIVURE A RECOUVREMENT	ÜBERBLATTUNGSNIETUNG, ÜBERLAPPUNGSNIETUNG

7401	LAP SEAM WELDING	SOUDAGE CONTINU PAR RECOUVREMENT	ÜBERLAPPNAHTSCHWEISSEN
7402	LAP SPOT WELDING	SOUDAGE PAR POINTS A RECOUVREMENT	ÜBERLAPPUNGSPUNKTNAHT
7403	LAP WELD	SOUDAGE PAR RECOUVREMENT	NAHT (ÜBERLAPPTE)
7404	LAP WELD (TO)	SOUDER EN BISEAU	ÜBERLAPPT SCHWEISSEN
7405	LAP WELDED JOINT	SOUDURE PAR RECOUVREMENT, SOUDURE EN BISEAU	ÜBERLAPPTE SCHWEISSUNG, ÜBERLAPPUNGSSCHWEISSUNG, SCHWEISSVERBINDUNG (ÜBERLAPPTE)
7406	LAP WELDED TUBE	TUBE SOUDE PAR RECOUVREMENT, TUBE A RECOUVREMENT	ROHR (ÜBERLAPPT GESCHWEISSTES)
7407	LAPPING	REPRISES VISIBLES, RODAGE	ANSÄTZE, LÄPPEN
7408	LAPSE OF A PATENT	DECHEANCE D'UN BREVET	ERLÖSCHEN EINES PATENTES
7409	LARD OIL	HUILE DE GRAISSE, HUILE DE SAINDOUX	SPECKÖL, SCHMALZÖL, LARDÖL
7410	LARGE GAS ENGINE	MOTEUR A ESSENCE DE GRANDE PUISSANCE	GROSSGASMASCHINE
7411	LARGE RAKE ANGLE	GRANDE PENTE DE COUPE EFFECTIVE	WEITERSPANNWINKEL
7412	LARGE WATER-CAPACITY BOILER, BOILER WITH LARGE WATER SPACE	CHAUDIERE A GRAND VOLUME	GROSSWASSERRAUMKESSEL
7413	LAST RUNNINGS	QUEUE DE DISTILLATION	NACHLAUF DER DESTILLATION
7414	LATCH	LOQUET	FALLKLINKE, FALLE
7415	LATENT HEAT	CHALEUR LATENTE	WÄRME (LATENTE)
7416	LATENT HEAT OF VAPORISATION	CHALEUR LATENTE DE VAPORISATION	VERDAMPFUNGSWÄRME (LATENTE)
7417	LATENT MAGNETISM	MAGNETISME LATENT	MAGNETISMUS (GEBUNDENER)
7418	LATERAL BUCKLING	GAUCHISSEMENT LATERAL	VERWERFUNG (SEITLICHE), VERZIEHEN (SEITLICHES)
7419	LATERAL DEVIATION	DEVIATION LATERALE	ABLENKUNG (SEITLICHE)
7420	LATERAL DISPLACEMENT	DEPLACEMENT LATERAL, DEPLACEMENT PERPENDICULAIRE A L'AXE	QUERVERSCHIEBUNG
7421	LATERAL EXPANSION	DILATATION LATERALE	QUERDEHNUNG, QUERAUSBAUCHUNG
7422	LATERAL PRESSURE	PRESSION LATERALE	DRUCK (SEITLICHER), SEITENDRUCK
7423	LATERAL TRANSLATION	DEPORT LATERAL	VERSCHIEBUNG (SEITLICHE)
7424	LATERAL TRANSVERSE CONTRACTION	STRICTION	QUERKONTRAKTION
7425	LATEX LATICES	LATEX	LATEX
7426	LATH; BATTEN	LATTE	HOLZLATTE
7427	LATHE	TOUR	DREHBANK
7428	LATHE WITH LEADING SCREW	TOUR A COMMANDES PAR VIS-MERE	LEITSPINDELDREHBANK
7429	LATTICE	RESEAU CRISTALLIN, TREILLIS METALLIQUE	GITTER
7430	LATTICE CONSTANTS	CONSTANTES DU RESEAU	GITTERKONSTANTE
7431	LATTICE GIRDER	POUTRE EN TREILLIS	FACHWERKTRÄGER, GITTERTRÄGER
7432	LATTICE PLANE	PLAN RETICULAIRE	NETZEBENE
7433	LATTICE WORK	TREILLIS	FACHWERK
7434	LATTICE-DISTORTION THEORY	THEORIE DE LA DISTORSION DU RESEAU	GITTERVERZERRUNGSTHEORIE
7435	LAUE DIAGRAMME	DIAGRAMME DE LAUE	RÜCKAUFNAHME LAUEDIAGRAMM
7436	LAUNDER	CANAL DE COULEE	ABSTICHRINNE
7437	LAVA	LAVE	LAVA
7438	LAVE METHOD	METHODE DE LAVE	LAVE-VERFAHREN

7439	LAWN MOWERS AND CUTTERS	TONDEUSES A GAZON	RASENMÄHER
7440	LAY BARGES	BARGES DE POSE	VERLEGESCHIFF
7441	LAY PIPES (TO)	POSER DES TUYAUX	ROHRE VERLEGEN
7442	LAYER	COUCHE	SCHICHT, LAGE
7443	LAYER CORROSION	CORROSION PAR INFILTRATION STRATIFIEE	SCHICHTKORROSION
7444	LAYER LINE	STRATE	SCHICHTLINIE
7445	LAYOUT	COMPARAISON DE DIMENSIONS	VERGLEICH VON ABMESSUNGEN
7446	LEACHING	LESSIVAGE	AUSLAUGUNG
7447	LEAD	PLOMB	BLEI
7448	LEAD (TO)	AVANCER	VOREILEN
7449	LEAD ACETATE PAPER	PAPIER A L'ACETATE DE PLOMB	BLEIZUCKERPAPIER
7450	LEAD ACETATE, SUGAR OF LEAD	ACETATE DE PLOMB	BLEIESSIGSÄURE, BLEIAZETAT, BLEIZUCKER
7451	LEAD ALLOY	ALLIAGE A BASE DE PLOMB	BLEILEGIERUNG
7452	LEAD ASSY	FAISCEAU DE FILS	KABELBÜNDEL
7453	LEAD BASE ALLOY	ALLIAGE AU PLOMB	BLEILEGIERUNG
7454	LEAD BATH	BAIN DE PLOMB	BLEIBAD
7455	LEAD BATTERY	ACCUMULATEUR AU PLOMB	BLEIAKKUMULATOR
7456	LEAD BRONZE	BRONZE AU PLOMB	BLEIBRONZE
7457	LEAD CAST	REPRODUCTION EN PLOMB FONDU	BLEIABGUSS
7458	LEAD CHLORIDE	CHLORURE DE PLOMB	CHLORBLEI, BLEICHLORID
7459	LEAD COATED WIRE	FIL PLOMBE	DRAHT (VERBLEITER)
7460	LEAD COVERED SHEET IRON	TOLE PLOMBEE	BLECH (VERBLEITES)
7461	LEAD DIOXIDE	DIOXYDE DE PLOMB, OXYDE PUCE	BLEISUPEROXYD, BLEIDIOXYD, BLEIPEROXYD
7462	LEAD FOIL	FEUILLE DE PLOMB, PLOMB EN FEUILLES	BLEIFOLIE
7463	LEAD HAMMER	MASSETTE EN PLOMB	BLEIHAMMER
7464	LEAD JOINT FOR PIPING	JOINT A BAGUE DE PLOMB MATEE POUR TUYAUTERIES	BLEIDICHTUNG FÜR ROHRE
7465	LEAD ORE	MINERAI DE PLOMB	BLEIERZ
7466	LEAD PAINT	MATIERE COLORANTE A BASE DE PLOMB	BLEIFARBE
7467	LEAD PENCIL DRAWING	DESSIN AU CRAYON	BLEISTIFTZEICHNUNG
7468	LEAD PIPE	TUYAU EN PLOMB	BLEIROHR
7469	LEAD PIPE COATED WITH TIN	TUYAU EM PLOMB ETAME A L'INTERIEUR	MANTELROHR
7470	LEAD RING	BAGUE EN PLOMB	BLEIRING
7471	LEAD SCREW	VIS MERE	LEITSPINDEL
7472	LEAD SHOT	GRENAILLE DE PLOMB	SCHROT, BLEISCHROT
7473	LEAD SUPHATE	SULFATE DE PLOMB	BLEI (SCHWEFESAURES), BLEISULFAT
7474	LEAD THIOSULFATE	HYPOSULFITE DE PLOMB	BLEI (UNTERSCHWEFLIGSAURES), BLEITHIOSULFAT
7475	LEAD TIN SOLDER	SOUDURE CLAIRE	BLEI-ZINNLOT
7476	LEAD WELDING	SOUDAGE DE PLOMB	BLEISCHWEISSEN
7477	LEAD WIRE	FIL DE PLOMB	BLEIDRAHT
7478	LEAD, BLACK LEAD PENCIL	CRAYON	ZEICHENSTIFT, BLEISTIFT
7479	LEADED	PLOMBE	VERBLEIT
7480	LEADEN SEAL	PLOMB (SCEAU EN PLOMB)	PLOMBE
7481	LEADING DIMENSIONS	DONNEES PRINCIPALES, DIMENSIONS D'ENCOMBREMENT	HAUPTMASSE, HAUPTABMESSUNGEN

7482	LEADING HAND	CHEF D'EQUIPE	VORARBEITER
7483	LEADING OF PHASE	DEPHASAGE, DECALAGE EN AVANCE	PHASENVOREILUNG
7484	LEADWORKING	PLOMBERIE	BLEIVERARBEITUNG
7485	LEAF SPRING	RESSORT A LAMES	BLATTFEDER
7486	LEAK	FUITE AUX JOINTS	UNDICHTIGKEIT, UNDICHTE STELLE, LECK
7487	LEAK (TO)	FUIR	LECKEN, LECK SEIN, RINNEN, TROPFEN
7488	LEAK, SEEPAGE	FUITE, SUINTEMENT	LECK, VERSICKERUNG
7489	LEAKAGE	COULURE	LECKSTELLE
7490	LEAKAGE CURRENT	COURANT DE FUITE, COURANT DE DISPERSION	STREUSTROM, VERLUSTSTROM
7491	LEAKINESS	INETANCHEITE, ETANCHEITE (MANQUE D')	UNDICHTIGKEIT, UNDICHTHEIT
7492	LEAKING	FUITE, COULAGE (D'UN LIQUIDE)	LECKEN, TROPFEN, RINNEN
7493	LEAKY	PERD (QUI), FUIT (QUI)	UNDICHT
7494	LEATHER	CUIR	LEDER
7495	LEATHER BELT	COURROIE EN CUIR	LEDERRIEMEN
7496	LEATHER CHAIN BELT, LEATHER LINK BELT	COURROIE EN CUIR A MAILLONS	GLIEDERRIEMEN AUS LEDER, LEDERGLIEDERRIEMEN, KETTENBAND
7497	LEATHER CLIPPINGS	ROGNURES DE CUIR	LEDERABFÄLLE
7498	LEATHER DIAPHRAGM	DIAPHRAGME EN CUIR	LEDERMEMBRAN
7499	LEATHER HOSE	TUYAU EN CUIR	LEDERSCHLAUCH
7500	LEATHER LACE, BELT LACE	LANIERE POUR COURROIES	NÄHRIEMEN, BINDERIEMEN
7501	LEATHER LINK FLEXIBLE COUPLING	MANCHON A LANIERES DE CUIR	LEDERLASCHENKUPPLUNG
7502	LEATHER PACKING	GARNITURE EN CUIR	LEDERPACKUNG, LEDERDICHTUNG
7503	LEATHER STRAP	COURROIE EN CUIR	LEDERRIEMEN
7504	LEDEBURITE	LEDEBURITE	LEDEBURIT
7505	LEE	COTE SOUS LE VENT	LEESEITE
7506	LEFT HAND SCREW	VIS FILETEE A GAUCHE	SCHRAUBE (LINKSGÄNGIGE)
7507	LEFT-HAND OFFSET	DECALAGE A GAUCHE	LINKSVERSCHIEBUNG
7508	LEFT-HAND THREAD	PAS A GAUCHE	GEWINDE (LINKSGÄNGIGES), LINKSGEWINDE
7509	LEFT-HANDED HELICAL TEETH	DENTURE A PAS INCLINE A GAUCHE	VERZAHNUNG (LINKSSTEIGENDE)
7510	LEFT-HANDED SCREW	VIS A GAUCHE	SCHRAUBE (LINKSGÄNGIGE)
7511	LEFT-LAID ROPE	CABLE TORDU A GAUCHE	SEIL (LINKSGESCHLAGENES)
7512	LEG OF AN ANGLE	COTE D'UN ANGLE	SCHENKEL EINES WINKELS
7513	LEGEND	LEGENDE	LEGENDE
7514	LEMNISCATE	LEMNISCATE	LEMNISKATE, SCHLEIFENKURVE
7515	LENARD RAYS	RAYONS DE LENARD	LENARD-STRAHLEN
7516	LENGTH	LONGUEUR	LÄNGE
7517	LENGTH OF ARC	LONGUEUR D'ARC	BOGENLÄNGE
7518	LENGTH OF OVERHANG	LONGUEUR LIBRE	FREILÄNGE
7519	LENGTH OF PISTON STROKE, STROKE TRAVEL OF PISTON	COURSE DU PISTON, JEU DU PISTON	KOLBENHUB, KOLBENWEG, HUB, HUBLÄNGE, HUBHÖHE EINES KOLBENS
7520	LENGTH OF WELD	LONGUEUR DU CORDON	SCHWEISSNAHTLÄNGE
7521	LENGTHENING BAR	RALLONGE D'UN COMPAS	ZIRKELVERLÄNGERUNG, VERLÄNGERUNGSSTÜCK FÜR ZIRKEL
7522	LENS	LENTILLE	LINSE
7523	LEONARD EFFECT	EFFET LEONARD, SOUFFLURE	LEONARD-EFFEKT, GASBLASE
7524	LETTERING OF A DRAWING	INDICATIONS D'UN DESSIN	BESCHRIFTUNG EINER ZEICHNUNG

7525	LEVEL	NIVEAU	NIVEAU (HÖHENLAGE)
7526	LEVEL GAUGE	INDICATEUR DE NIVEAU	STANDPEILER
7527	LEVEL SURFACE, EQUIPOTENTIAL SURFACE	SURFACE DE NIVEAU, SURFACE EQUIPOTENTIELLE	SCHICHTFLÄCHE, EQUIPOTENTIALFLÄCHE, NIVEAUFLÄCHE
7528	LEVELLING ACTION	EFFET D'EGALISATION	VERLAUFEFFEKT
7529	LEVELLING SCREW, FOOT SCREW	VIS EGALISATRICE	FUSSSCHRAUBE, NIVELLIERSCHRAUBE
7530	LEVER	LEVIER	HEBEL
7531	LEVER ARM	BRAS DE LEVIER	HEBELARM
7532	LEVER BRAKE	FREIN A LEVIER	HEBELBREMSE
7533	LEVER JACK	CRIC A LEVIER	HEBELADE
7534	LEVER PRESS	PRESSE A LEVIER	HEBELPRESSE
7535	LEVER WEIGHT (SAFETY) VALVE	SOUPAPE DE SURETE A LEVIER ET CONTREPOIDS	VENTIL MIT HEBELGEWICHTSBELASTUNG
7536	LEVERAGE	RAPPORT DES BRAS DE LEVIER, LONGUEUR DU BRAS DE LEVIER	HEBELÜBERSETZUNG (VERHÄLTNIS)
7537	LEVIGATION	LEVIGATIONS, DEBOURBAGE	ABSCHLÄMMEN
7538	LIBERATION TANK	BAIN DE DECOMPOSITION DE L ELECTROLYTE	ELECTROLYTREINIGUNGSBAD
7539	LICENCE	LICENCE D'EXPLOITATION	LIZENZ (PATENTRECHTLICHE)
7540	LICENSEE	CONCESSIONNAIRE DE LICENCE	LIZENZINHABER
7541	LIFE	DUREE	LEBENSDAUER
7542	LIFT (TO), RAISE A LOAD (TO)	SOULEVER, ELEVER UN FARDEAU	HEBEN (EINE LAST)
7543	LIFT CHECK VALVE	CLAPET DE RETENUE A SOUPAPE	RÜCKSCHLAGKLAPPE, RÜCKSCHAGVENTIL
7544	LIFT OF VALVE, VALVE LIFT	SOULEVEMENT DE LA SOUPAPE, LEVEE DE LA SOUPAPE	VENTILHUB, VENTILAUSSCHLAG
7545	LIFT PUMP	POMPE ELEVATOIRE	HUBPUMPE
7546	LIFT ROPE	CABLE D'ASCENSEUR	AUFZUGSEIL
7547	LIFT VALVE, POPPET, MUSHROOM VALVE	SOUPAPE A CLAPET, SOUPAPE CHAMPIGNON	HUBVENTIL
7548	LIFTER	HAPPE	TIEGELZANGE
7549	LIFTING	DETREMPAGE	AUFZIEHEN
7550	LIFTING CYLINDER	CYLINDRE DE LEVAGE	HUBZYLINDER
7551	LIFTING HEIGHT OF A CRANE	HAUTEUR DE LEVAGE D'UNE GRUE	HUBHÖHE, FÖRDERHÖHE EINES KRANS
7552	LIFTING HOOK	CROCHET DE LEVAGE	LASTHAKEN
7553	LIFTING LEVER	LEVIER DE PURGE LIBRE	ANLÜFTHEBEL
7554	LIFTING MAGNET	ELECTRO-AIMANT PORTEUR	LASTMAGNET, HUBMAGNET, TRAGMAGNET
7555	LIFTING PISTON	PISTON DE LEVAGE	HUBKOLBEN, HEBEKOLBEN
7556	LIFTING SCREW	VIS DE RELEVAGE	ABDRÜCKSCHRAUBE
7557	LIFTING TONGS	PINCE (D'UN APPAREIL DE LEVAGE)	LASTZANGE
7558	LIFTING TRUNNION	TOURILLON CROCHET DE LEVAGE	HEBEHAKEN
7559	LIFTING, RAISING A LOAD	SOULEVEMENT D'UN FARDEAU, ASCENSION D'UNE CHARGE	HEBEN EINER LAST
7560	LIGHT	LUMIERE	LICHT
7561	LIGHT FILLET WELD	SOUDURE D'ANGLE CONCAVE, SOUDURE EN CONGE	HOHLKEHLNAHT
7562	LIGHT LOAD	CHARGE INCOMPLETE, CHARGE INFERIEURE A LA NORMALE	UNTERBELASTUNG, UNTERLAST
7563	LIGHT LORRY	CAMIONNETTE	KLEINLASTWAGEN
7564	LIGHT METAL	METAL LEGER, ALLIAGE LEGER	LEICHTMETALL

7565	LIGHT METAL TURNING LATHE	TOUR POUR METAUX LEGERS	LEICHTMETALLDREHBANK
7566	LIGHT PHENOMENON	PHENOMENE LUMINEUX	ICHTERSCHEINUNG, LICHTPHÄNOMEN
7567	LIGHT RUNNING	MARCHE DOUCE	GANG (WEICHER), GANG (LEICHTER), GANG (STOSSFREIER)
7568	LIGHT SOURCE	SOURCE LUMINEUSE	LICHTQUELLE
7569	LIGHT SPEED	VITESSE DE LA LUMIERE	LICHTGESCHWINDIGKEIT
7570	LIGHT THIN OIL	HUILE LEGERE	LEICHTÖL
7571	LIGHT WAVE	ONDE LUMINEUSE	LICHTWELLE
7572	LIGHT-TIGHT	IMPERMEABLE A LA LUMIERE	LICHTUNDURCHLÄSSIG, LICHTDICHT
7573	LIGHTING FLARE	FUSEE ECLAIRANTE	LEUCHTBOMBE
7574	LIGHTING, ILLUMINATION	ECLAIRAGE, ILLUMINATION	BELEUCHTUNG
7575	LIGHTNING ARRESTER	PARAFOUDRE	BLITZSCHUTZVORRICHTUNG
7576	LIGHTNING CONDUCTOR	PARATONNERRE	BLITZABLEITER
7577	LIGNIN, LIGNONE	LIGNINE, LIGNONE, LIGNOSE	LIGNIN
7578	LIGNITE	LIGNITE, LIGNEUX	LIGNIT
7579	LIGNITE WAX	CIRE DE LIGNITE, CIRE DE PARAFFINE	MONTANWACHS
7580	LIGROIN, LIGHT PETROLEUM	LIGROINE, BENZOLINE	LIGROIN
7581	LIKE, SIMILAR POLES	POLES SEMBLABLES	POLE (GLEICHNAMIGE)
7582	LIME	CHAUX	KALK
7583	LIME TITANIA TYPE COATING	REVETEMENT PAR COMPOSITION KB TI	KB-TI-MISCHTYP-ÜBERZUG
7584	LIME TYPE COATING	ENROBAGE BASIQUE	KALKBASISCHE UMHÜLLUNG
7585	LIME WATER	EAU DE CHAUX	KALKWASSER
7586	LIMESTONE	CALCAIRE	KALKSTEIN
7587	LIMIT	TOLERANCE LIMITE, LIMITE	GRENZMASS
7588	LIMIT OF ELASTICITY	LIMITE ELASTIQUE	ELASTIZITÄTSGRENZE
7589	LIMIT OF ERROR	LIMITE D'ERREURS	FEHLERGRENZE
7590	LIMIT OF PROPORTIONALITY	LIMITE D'ELASTICITE PROPORTIONNELLE	PROPORTIONALITÄTSGRENZE, GLEICHMASSGRENZE
7591	LIMIT OF SATURATION	LIMITE DE SATURATION	SÄTTIGUNGSGRENZE
7592	LIMIT OF TEMPERATURE	LIMITE DE TEMPERATURE	TEMPERATURGRENZE
7593	LIMIT PRESSURE	PRESSION LIMITE	GRENZDRUCK
7594	LIMIT VALUE	VALEUR LIMITE	GRENZWERT
7595	LIMITS	LIMITES DE TOLERANCE	TOLERANZGRENZE, GRENZEN
7596	LIMONITE, BOG IRON ORE, BROWN HAEMATITE	LIMONITE, HEMATITE BRUNE	BRAUNEISENSTEIN, BRAUNEISENERZ, RASENEISENERZ, SUMPFERZ, LIMONIT
7597	LINE	LIGNE, RAIE	LINIE
7598	LINE (TO)	REVETIR, DOUBLER	AUSFÜTTERN, AUSKLEIDEN
7599	LINE FOCUS	SOURCE LINEAIRE	STRICHFOKUS
7600	LINE INTEGRAL	INTEGRALE LINEAIRE	LINIENINTEGRAL
7601	LINE OF COLLIMATION	LIGNE DE COLLIMATION	KOLLIMATIONSLINIE, ABSEHLINIE, ZIELLINIE
7602	LINE OF CONTACT ACTION	LIGNE D'ENGRENEMENT	EINGRIFFSLINIE
7603	LINE OF FORCE	LIGNE DE FORCE	KRAFTLINIE
7604	LINE OF GAUGE MARK	LIGNE DE TRUSQUINAGE	PARALLELREISSLINIE
7605	LINE OF INTERSECTION	LIGNE D'INTERSECTION	SCHNITTLINIE
7606	LINE OF RAILS	VOIE FERREE, CHEMIN DE ROULEMENT	GLEIS
7607	LINE OF SHAFTING	LIGNE D'ARBRES, LIGNE DE TRANSMISSION	WELLENSTRANG
7608	LINE POLE	POTEAU DE LIGNE ELECTRIQUE	LEITUNGSMAST

7609	LINE SHAFT DRIVE	COMMANDE PAR ARBRE DE TRANSMISSION	WELLENANTRIEB, TRANSMISSIONSANTRIEB
7610	LINE SPECTRUM	SPECTRE DISCONTINU	LINIENSPEKTRUM
7611	LINEAR ACCELERATION	ACCELERATION LINEAIRE	BESCHLEUNIGUNG (LINEARE)
7612	LINEAR EXPANSION, EXTENSION	ALLONGEMENT LONGITUDINAL, DILATATION LINEAIRE	LANGSDEHNUNG, AUSDEHNUNG (LINEARE)
7613	LINEAR INTERPOLATION	INTERPOLATION LINEAIRE	INTERPOLATION (LINEAIRE)
7614	LINEAR PERSPECTIVE	PERSPECTIVE LINEAIRE	PERSPEKTIVE (MATHEMATISCHE), PERSPEKTIVE (LINEARE)
7615	LINEAR PITCH	PAS LINEAIRE	GERADLINIGE LÄNGSTEIGUNG
7616	LINEAR SIMPLE EQUATION	EQUATION LINEAIRE, EQUATION DU PREMIER DEGRE	GLEICHUNG ERSTEN GRADES, GLEICHUNG (LINEARE)
7617	LINEAR VELOCITY	VITESSE LINEAIRE	GESCHWINDIGKEIT (LINEARE)
7618	LINED LADLE	POCHE GARNIE, POCHE BRIQUETEE	GIESSPFANNE (GEMAUERTE)
7619	LINEN	TOILE	LEINEN, LEINWAND
7620	LINER	COLONNE PERDUE, GARNITURE	ROHR (VERLORENES), EINLEGESTREIFEN
7621	LINER, SHIM	FEUILLE DE CLINQUANT, CLINQUANT D'EPAISSEUR	FUGEINLAGE, ZWISCHENLAGE, BEILAGEBLECH, PASSBLECH
7622	LINES OF LUDERS	LIGNES DE LUDERS	LÜDERSFLIESSFIGUREN
7623	LINING	GARNISSAGE, FOURRURE, CALORIFUGEAGE, REVETEMENT, REVETEMENT INTERIEUR, CHEMISE, DOUBLAGE, DOUBLURE	FUTTER, WÄRMEISOLIERUNGSBEKLEIDUNG, AUSFÜTTERUNG, AUSKLEIDUNG, FÜTTERN, AUSFÜTTERN, AUSKLEIDEN
7624	LINING FOR BEARINGS	FOURRURE DUN PALIER, GARNITURE DE METAL ANTIFRICTION, REVETEMENT D'ANTIFRICTION	LAGERFUTTER
7625	LINK BLOCK, DIE BLOCK	COULISSEAU MOBILE, CURSEUR, VOYAGEUR	GELENKSTEIN, KULISSENSTEIN
7626	LINK MOTION	DISTRIBUTION PAR COULISSE	KULISSENSTEUERUNG
7627	LINK OF CHAIN	CHAINON, MAILLON, ANNEAU, MAILLE D'UNE CHAINE	KETTENGLIED, SCHAKE
7628	LINKAGE	SYSTEME ARTICULE	GELENKSYSTEM
7629	LINKED, FLEXIBLY CONNECTED	ARTICULE	GELENKIG VERBUNDEN
7630	LINOLEUM	LINOLEUM	LINOLEUM
7631	LINSEED OIL	HUILE DE LIN	LEINÖL
7632	LIP	BEC DE COULEE	PFANNENAUSGUSS
7633	LIPS	LEVRES DE COUPES	SCHNEIDLIPPEN
7634	LIQUATED SURFACE	SURFACE DE LIQUATION	OBERFLÄCHE (AUSGESEIGERTE)
7635	LIQUATION, ELIQUATION	LIQUATION, SEGREGATION	SAIGERUNG, SEIGERUNG
7636	LIQUEFACTION OF A GAS	LIQUEFACTION D'UN GAZ	VERFLÜSSIGUNG EINES GASES
7637	LIQUEFIED GAS	GAZ LIQUEFIE	GAS (VERFLÜSSIGTES)
7638	LIQUEFIED PETROLEUM GAS (L.P.G)	GAZ DE PETROLE LIQUEFIE	FLÜSSIGGAS
7639	LIQUEFLABLE	LIQUEFIABLE	VERFLÜSSIGBAR
7640	LIQUEFY A GAS (TO)	LIQUEFIER UN GAZ	GAS VERFLÜSSIGEN (EIN)
7641	LIQUID	LIQUIDE, CORPS LIQUIDE, FLUIDE, FLUIDE LIQUIDE	KÖRPER (FLÜSSIGER), FLÜSSIGKEIT
7642	LIQUID	LIQUIDE	TROPFBAR, FLÜSSIG
7643	LIQUID AIR	AIR LIQUIDE	LUFT (FLÜSSIGE)
7644	LIQUID AMMONIA	AMMONIAQUE	AMMONIAK
7645	LIQUID CARBON DIOXYDE	ACIDE CARBONIQUE LIQUIDE, GAZ CARBONIQUE LIQUEFIE	KOHLENSÄURE (FLÜSSIGE)
7646	LIQUID CARBURIZING	CEMENTATION EN BAIN	BADZEMENTIEREN
7647	LIQUID CHARGE	CHARGE LIQUIDE	EINSATZ (FLÜSSIGER)

7648	LIQUID CONTRACTION	CONTRACTION DE REFROIDISSEMENT	ABKÜHLUNGSSCHRUMPFUNG
7649	LIQUID DAMPING	AMORTISSEMENT PAR LIQUIDE	FLÜSSIGKEITSDÄMPFUNG
7650	LIQUID FUEL, OIL FUEL	COMBUSTIBLE LIQUIDE	BRENNSTOFF (FLÜSSIGER)
7651	LIQUID HARDENING	TREMPE EN BAIN DE SEL	SALZBADHÄRTUNG
7652	LIQUID HEAD	CHARGE HYDROSTATIQUE	DRUCK (HYDROSTATISCHER), BELASTUNG (HYDROSTATISCHE)
7653	LIQUID HONING	HONAGE AU JET DE VAPEUR	STRAHLHONVERFAHREN
7654	LIQUID LEVEL REGULATOR	REGULATEUR DE NIVEAU	NIVEAU REGLER
7655	LIQUID LEVEL-CONTROLLER	CONTROLEUR DE NIVEAU	NIVEAUSTANDSREGLER
7656	LIQUID PENETRANT EXAMINATION	EXAMEN PAR RESSUAGE	SEIGERUNGSPRÜFUNG
7657	LIQUID STATE	ETAT LIQUIDE	AGGREGAT-ZUSTAND (FLÜSSIGER)
7658	LIQUID TIGHT COMPARTMENTS	CAISSONS ETANCHES	SENKKASTEN (FLÜSSIGKEITSDICHTER)
7659	LIQUID VEIN TRANSFER	TRANSFERT PAR VEINE LIQUIDE	METALLÜBERGANG (FLÜSSIGER)
7660	LIQUIDS	LIQUIDES	FLÜSSIGKEITEN
7661	LIQUIDUS	LIQUIDUS, LIGNE DU LIQUIDUS	LIQUIDUSLINIE
7662	LIST OF PARTS	NOMENCLATURE DES PIECES	STÜCKLISTE
7663	LITHARGE, LEAD MONOXIDE	PROTOXYDE DE PLOMB, LITHARGE	BLEIGLÄTTE, BLEIOXYD
7664	LITHIUM	LITHIUM	LITHIUM
7665	LITHIUM CARBONATE	CARBONATE DE LITHIUM	LITHIUMKARBONAT
7666	LITHIUM CHLORIDE	CHLORURE DE LITHIUM	LITHIUMCHLORID
7667	LITHIUM FLUORIDE	FLUORURE DE LITHIUM	LITHLUMFLUORID
7668	LITMUS PAPER	PAPIER DE TOURNESOL	LACKMUSPAPIER
7669	LITRE	LITRE	LITER
7670	LIVE CONDUCTOR, CURRENT CARRYING CONDUCTOR	CONDUCTEUR PARCOURU, TRAVERSE PAR UN COURANT	STROMFÜHRENDER LEITER, LEITER (UNTER SPANNUNG STEHENDER)
7671	LIVE LOAD	CHARGE VIVE	LAST (BEWEGLICHE)
7672	LIVE LOAD, VARIABLE INTERMITTENT FLUCTUATING LOAD	CHARGE VARIABLE, CHARGE INTERMITTENTE	BELASTUNG (WECHSELNDE), BELASTUNG (UNSTETE), BELASTUNG (INTERMITTIERENDE)
7673	LIVE STEAM	VAPEUR FRAICHE	FRISCHDAMPF
7674	LIVER OF SULPHUR	FOIE DE SOUFRE	SCHWEFELLEBER
7675	LIVERING	EPAISSISSEMENT (PEINTURE)	VERDICKUNG
7676	LIXIVIATE (TO)	LESSIVER, LIXIVIER	AUSLAUGEN
7677	LIXIVIATION	LESSIVAGE, LIXIVIATION	AUSLAUGEN, AUSLAUGUNG
7678	LNG(LIQUEFIED NATURAL GAS) TANK	RESERVOIR DE GNL	TANK FÜR FLÜSSIGES ERDGAS
7679	LOAD (ING)	CHARGE, SOLLICITATIONS, EFFORT EXTERIEUR	BELASTUNG, BEANSPRUCHUNG, LAST, AUFLAST, LADUNG
7680	LOAD (TO)	CHARGER (RESISTANCE DES MATERIAUX)	BELASTEN
7681	LOAD BRAKE	FREIN DE RETENUE	LASTDRUCKBREMSE, SENKSPERRBREMSE
7682	LOAD CARRYING CAPACITY	CAPACITE DE CHARGE	TRAGFÄHIGKEIT
7683	LOAD DIAGRAM	PLAN DE CHARGE	BELASTUNGSDIAGRAMM
7684	LOAD DISTRIBUTION	REPARTITION DES CHARGES	LASTVERTEILUNG
7685	LOAD WITHOUT IMPACT	CHARGE SANS CHOCS	BELASTUNG (STOSSFREIE)
7686	LOADED ENGINE	MOTEUR SOUS TENSION	MOTOR (BELASTETER)
7687	LOADERS : GRASS AND GREEN CROP	CHARGEUR (OU CHARGEUSE) POUR FOURRAGE VERT	GRÜNFUTTERLADER
7688	LOADERS AND HOISTS (SACK, BAG AND SUGAR BEET)	CHARGEURS DE SACS ET DE BETTERAVES	HEBEMASCHINEN (SÄCKE, BÜNDEL UND ZUCKE PRÜBEN)
7689	LOADERS, MANURE	CHARGEURS DE FUMIER	DÜNGERLADER

7690	LOADING AGENT	MATIERE CHARGEANTE	BESCHWERUNGSMITTEL
7691	LOAM	TERRE GLAISE, GLAISE	LEHM
7692	LOAM BRICK	BRIQUE CRUE, BRIQUE SECHEE A L'AIR	LEHMSTEIN, LEHMPATZEN, LUFTZIEGEL
7693	LOAM MOLD	MOULE EN TERRE GLAISE	LEHMFORM
7694	LOCAL CELL	PILE LOCALE	LOKALELEMENT
7695	LOCAL CORROSION	CORROSION LOCALE	LOKALANGRIFF, LOKALKORROSION
7696	LOCAL HEATING	CHAUFFAGE PARTIEL	LOKALHEIZUNG
7697	LOCAL QUENCHING	TREMPE PARTIELLE	HÄRTEN (PARTIELLES)
7698	LOCK	ECLUSE	SCHLEUSE
7699	LOCK (TO)	VERROUILLER	VERRIEGELN, SPERREN
7700	LOCK A NUT (TO), A KEY	BLOQUER UN ECROU, BLOQUER UNE CLAVETTE	SICHERN (EINEN KEIL), SICHERN (EINE MUTTER)
7701	LOCK ANGLE	ANGLE DE BRAQUAGE	EINSCHLAGWINKEL
7702	LOCK NUT, CHECK NUT, JAM NUT	ECROU DE SURETE, CONTRE-ECROU, ECROU DE BLOCAGE	STELLMUTTER, KONTERMUTTER, GEGENMUTTER, SICHERHEITSMUTTER
7703	LOCK WASHER	RONDELLE FREIN, RONDELLE DE SERRAGE, RONDELLE DE BLOCAGE	UNTERLAGSCHEIBE (FERERNDE), SICHERUNGSSCHEIBE
7704	LOCKING A KEY	BLOCAGE, FIXATION D'UN COIN DE SERRAGE	KEILSICHERUNG
7705	LOCKING DEVICE	DISPOSITIF DE BLOCAGE, ENCLIQUETAGE	FESTSTELLVORRICHTUNG, SPERRE, ARRETIERUNG
7706	LOCKING LEVER	MANETTE DE COMMANDE	STELLHEBEL, FESTSTELLHEBEL
7707	LOCKING OF BOLTS AND NUTS	BLOCAGE, CONSOLIDATION, IMMOBILISATION DES ECROUS ET BOULONS	SICHERUNG VON SCHRAUBEN, SCHRAUBENSICHERUNG
7708	LOCKING SCREW	VIS DE BLOCAGE	SICHERUNGSSCHRAUBE
7709	LOCKING STOP GEAR	MECANISME D'ARRET	SPERRWERK, GESPERRE
7710	LOCKING, BOLTING	VERROUILLAGE	VERRIEGELUNG, VERIEGEIN
7711	LOCKING, CLAMPING	BLOCAGE, ASSUJETTISSEMENT	SICHERUNG GEGEN VERSCHIEBEN, LAGENSICHERUNG
7712	LOCOMOTION	LOCOMOTION	ORTSVERÄNDERUNG, FORTBEWEGUNG
7713	LOCOMOTIVE	LOCOMOTIVE	LOKOMOTIVE
7714	LOCOMOTIVE BOILER	CHAUDIERE TYPE LOCOMOTIVE	LOKOMOTIVKESSEL
7715	LOCUS	LIEU GEOMETRIQUE	ORT (GEOMETRISCHER)
7716	LOESS	LOESS	LÖSS
7717	LOG	PROFIL, COUPE, DIAGRAMME, GRAPHIQUE, RELEVE	DIAGRAMM
7718	LOGARITHM	LOGARITHME	LOGARITHMUS
7719	LOGARITHMIC SERIES	SERIE LOGARITHMIQUE	REIHE (LOGARITHMISCHE)
7720	LOGARITHMIC SPIRAL	SPIRALE LOGARITHMIQUE	SPIRALE (LOGARITHMISCHE)
7721	LOGARITHMIC TABLE, TABLE OF LOGARITHMS	TABLE DES LOGARITHMES	LOGARITHMENTAFEL
7722	LOGOCYCLIC CURVE, STROPHOID, FOLLATE	STROPHOIDE	STROPHOIDE
7723	LOGS, UNHEWN TIMBER	BOIS EN GRUME	GANZHOLZ
7724	LONG FLAME COAL	HOUILLE A LONGUE FLAMME	KOHLE (LANGFLAMMIGE)
7725	LONG LEVER	ANSPECT	HEBEBAUM
7726	LONG NIPPLE	LONGUE-VIS	LANGNIPPEL
7727	LONG RUN WORK	FABRICATION EN GRANDES SERIES	GROSSSERIENANFERTIGUNG
7728	LONG-DISTANCE NETWORK	RESEAU A GRANDE DISTANCE	FERNLEITUNGSNETZ
7729	LONG-LINE CURRENT EFFECT	EFFET DE LIGNE DE TRANSMISSION	LANGLEITUNGSEFFEKT

7730	LONG-LINK CHAIN	CHAINE A MAILLONS LONGS	KETTE (LANGGLIEDRIGE)
7731	LONG-STROKE ENGINE	MACHINE A GRANDE COURSE	MASCHINE (LANGHÜBIGE)
7732	LONG-THROW CRANK	MANIVELLE A GRANDE COURSE	KURBEL (LANGHÜBIGE)
7733	LONGITUDINAL AXIS	AXE LONGITUDINAL	LÄNGSACHSE
7734	LONGITUDINAL BAR	LONGRINE	LÄNGSRIEGEL
7735	LONGITUDINAL DUCTILITY	DUCTILITE LONGITUDINALE	LÄNGSDUKTILITÄT
7736	LONGITUDINAL PITCH OF RIVETS	DISTANCE DES RIVETS DE CENTRE A CENTRE EN LIGNE CONTINUE	LÄNGSNIETTEILUNG
7737	LONGITUDINAL SEAM WELDING	SOUDAGE LONGITUDINAL	LÄNGSNAHTSCHWEISSEN
7738	LONGITUDINAL SECTION	PROFIL EN LONG, COUPE LONGITUDINALE	LÄNGSSCHNITT, LÄNGENPROFIL
7739	LONGITUDINAL VIBRATION	OSCILLATION LONGITUDINALE	LÄNGESCHWINGUNG
7740	LONGITUDINALLY RUBBED TUBE, TUBE RIBBED IN ITS LONGITUDINAL SECTION	TUBE A NERVURES LONGITUDINALES, TUBE SERVE	RIPPENROHR MIT LÄNGSRIPPEN
7741	LOOP OF A CURVE	LACET, NOEUD D'UNE COURBE	SCHLEIFE EINER KURVE
7742	LOOP OF A ROPE	BOUCLE D'UNE CORDE	SEILSCHLINGE
7743	LOOSE COLLAR, SET COLLAR	BAGUE D'ARRET, BAGUE AMOVIBLE, RONDELLE GOUPILLEE	STELLRING
7744	LOOSE PATTERN	MODELE DEMONTABLE	MODELL (ZELEGBARES)
7745	LOOSE PIECE	PARTIE DEMONTABLE	TEIL (LOSER)
7746	LOOSE PIPE FLANGE	BRIDE FOLLE	FLANSCH (LOSER), ÜBERWURFFLANSCH
7747	LOOSE PULLEY	POULIE FOLLE	LOSSCHEIBE, LEERSCHEIBE, RIEMENSCHEIBE (LOSE)
7748	LOOSE PULLEY BUSH	DOUILLE POUR POULIE FOLLE	LEERLAUFBÜCHSE
7749	LOOSELY MOUNTED ON THE SHAFT	PLACE FOU (SUR L'ARBRE)	LOSE AUFGESETZT (AUF DIE WELLE)
7750	LOOSEN A KEY (TO)	DECLAVETER	LOSKEILEN, EINEN KEIL LÖSEN
7751	LOOSENING A KEY	DEMONTAGE D'UNE CLAVETTE, DECLAVETAGE	LÖSEN EINES KEILS, LÖSUNG EINES KEILS
7752	LORRY (GB)	CAMION	LASTWAGEN
7753	LORRY, TRUCK (U.S.A.)	CAMION A RIDELLES	LASTKRAFTWAGEN
7754	LOSS	PERTE	VERLUST
7755	LOSS BY FRICTION	PERTE DE FROTTEMENT	REIBUNGSVERLUST
7756	LOSS DUE TO LEAKAGE	PERTE PAR FUITES	UNDICHTIGKEITSVERLUST
7757	LOSS DUE TO SHOCK	PERTE PAR CHOCS	STOSSVERLUST
7758	LOSS OF ENERGY	PERTE D'ENERGIE	KRAFTVERLUST, ENERGIEVERLUST
7759	LOSS OF HEAD	PERTE DE CHARGE	WIDERSTANDSVERLUST, DRUCKVERLUST
7760	LOSS OF HEAT	PERTE DE CHALEUR, DEPERDITION DE CHALEUR	WÄRMEVERLUST
7761	LOSS OF IGNITION	PERTE AU FEU	ABBRANDVERLUST
7762	LOSS OF OUTPUT	PERTE DE PUISSANCE	LEISTUNGSVERLUST
7763	LOSS OF PRESSURE	PERTE DE PRESSION	DRUCKVERLUST
7764	LOSS OF WEIGHT	PERTE DE POIDS	GEWICHTSVERLUST
7765	LOSS OF WORK	PERTE DE TRAVAIL	ARBEITSVERLUST
7766	LOT	LOT	LOS
7767	LOW ANGLE BOUNDARY (OR SUB-BOUNDARY)	SOUS-JOINT A FAIBLE	
7768	LOW BRASS	LAITON DE FAIBLE ALLIAGE	MESSING (NIEDRIGLEGIERTES)
7769	LOW BREAST WHEEL	ROUE DE COTE (HYDRAULIQUE)	WASSERRAD (MITTELSCHÄCHTIGES)
7770	LOW CARBON STEEL	ACIER A FAIBLE TENEUR EN CARBONE, ACIER DOUX	STAHL (KOHLENSTOFFARMER), STAHL (WEICHER)

7771	LOW FREQUENCY	FAIBLE FREQUENCE	NIEDERFREQUENZ
7772	LOW FREQUENCY FURNACE	FOUR A INDUCTION A BASSE FREQUENCE	NIEDERFREQUENZOFEN
7773	LOW MELTING TEMPERATURE ALLOYS	ALLIAGES FUSIBLES	LEGIERUNGEN (NIEDRIG SCHMELZENDE)
7774	LOW PRESSURE	BASSE PRESSION	DRUCK (NIEDERER), NIEDERDRUCK
7775	LOW TEMPERATURE	BASSE TEMPERATURE	TEMPERATUR (NIEDRIGE), TEMPERATUR (TIEFE)
7776	LOW TEMPERATURE CARBONISATION	DISTILLATION DES COMBUSTIBLES A BASSE TEMPERATURE, CARBONISATION INCOMPLETE	SCHWELEN, SCHWELEREI
7777	LOW TENSION CURRENT	COURANT A BASSE TENSION	NIEDERSPANNUNGSSTROM, NIEDRIGGESPANNTER STROM
7778	LOW VOLTAGE PRESSURE TENSION	BASSE TENSION	NIEDERSPANNUNG (EL.)
7779	LOW-DENSITY METAL	METAL LEGER	LEICHTMETALL
7780	LOW-GRADE INFERIOR FUEL	MAUVAIS COMBUSTIBLE	BRENNSTOFF MINDERWERTIGER
7781	LOW-GRADE ORE	MINERAI PAUVRE, MINERAI DE BASSE TENEUR	ERZ (ARMES)
7782	LOW-PRESSURE BOILER	CHAUDIERE A BASSE PRESSION	NIEDERDRUCKKESSEL
7783	LOW-PRESSURE CYLINDER	CYLINDRE A BASSE PRESSION	NIEDERDRUCKZYLINDER
7784	LOW-PRESSURE HOT WATER HEATING	CHAUFFAGE PAR L'EAU CHAUDE A BASSE PRESSION	NIEDERDRUCKWASSERHEIZUNG, WARMWASSERHEIZUNG
7785	LOW-PRESSURE STEAM	VAPEUR A BASSE PRESSION	NIEDERDRUCKDAMPF, NIEDRIG GESPANNTER DAMPF, DAMPF VON NIEDRIGER SPANNUNG
7786	LOW-PRESSURE STEAM ENGINE	MACHINE A PRESSION ORDINAIRE	NIEDERDRUCKDAMPFMASCHINE
7787	LOW-PRESSURE STEAM HEATING	CHAUFFAGE PAR LA VAPEUR A BASSE PRESSION	NIEDERDRUCKDAMPFHEIZUNG
7788	LOW-PRESSURE TURBINE	TURBINE A BASSE PRESSION	NIEDERDRUCKTURBINE
7789	LOW-SPEED ENGINE	MACHINE A PETITE VITESSE, MACHINE A MARCHE LENTE	MASCHINE (LANGSAMLAUFENDE)
7790	LOW-TEMPERATURE JOINING	ASSEMBLAGE A BASSE TEMPERATURE	NIEDRIGTEMPERATURVERBINDUNG
7791	LOW-TENSION MOTOR	MOTEUR A BASSE TENSION	NIEDERSPANNUNGSMOTOR
7792	LOWENHERZ, DELISLE THREAD	PAS SYSTEME LOWENHERZ	LÖWENHERZGEWINDE
7793	LOWER A LOAD (TO)	ABAISSER UN FARDEAU	LAST SENKEN (EINE)
7794	LOWER PUNCH	POINCON INFERIEUR	UNTERSTEMPEL
7795	LOWER RAIL OR MID-BAR	SOUS-LISSE	MITTELSTANGE
7796	LOWER SAW GUIDE	GUIDE-RUBAN INFERIEUR	GATTERSTAB (UNTERER)
7797	LOWERING A LOAD	DESCENTE D'UN FARDEAU	SENKEN EINER LAST
7798	LOWERING OF THE WATER LEVEL	DENIVELLEMENT, DENIVELLATION	SENKUNG, ABSENKUNG DES WASSERSPIEGELS
7799	LOZENGE RIVETED JOINT	RIVURE EN LOSANGE	NIETUNG (VERJÜNGTE)
7800	LUBRICANT	LUBRIFIANT	SCHMIERMITTEL, SCHMIERSTOFF, SCHMIERMATERIAL
7801	LUBRICATE (TO)	LUBRIFIER, GRAISSER	SCHMIEREN
7802	LUBRICATED PLUG VALVE	ROBINET A BOISSEAU LUBRIFIE	HAHN MIT GEHÄUSESCHMIERUNG, HAHNVENTIL (GEÖLTES)
7803	LUBRICATING	LUBRIFICATION, GRAISSAGE	SCHMIERUNG
7804	LUBRICATING FAT GREASE	GRAISSE DE LUBRIFICATION	SCHMIERFETT
7805	LUBRICATING OIL	HUILE LUBRIFIANTE, HUILE DE GRAISSAGE	SCHMIERÖL
7806	LUBRICATING PROPERTY	POUVOIR LUBRIFIANT, QUALITE LUBRIFIANTE	SCHMIERWERT
7807	LUBRICATING WICK	MECHE DE GRAISSAGE	SCHMIERDOCHT

7808	**LUBRICATION**	LUBRIFICATION, GRAISSAGE	SCHMIEREN, SCHMIERUNG
7809	**LUBRICATION EFFECTED BY A MECHANICAL LUBRICATOR**	GRAISSAGE PAR OLEO-COMPRESSEUR	ZENTRALSCHMIERUNG
7810	**LUBRICATION GROOVE, OIL GROOVE, GREASE CHANNEL**	RIGOLE DE GRAISSAGE, PATTE D'ARAIGNEE	SCHMIERNUT
7811	**LUBRICATION SCREW**	BOULON GRAISSEUR	SCHMIERSCHRAUBE
7812	**LUBRICATOR, OIL FEED**	GRAISSEUR, GRAISSAGE, ORGANE DE LUBRIFICATION	SCHMIERVORRICHTUNG, SCHMIERGEFÄSS
7813	**LUBRICITY UNCTUOUSNESS OF A LUBRICANT**	ONCTUOSITE DU LUBRIFIANT	SCHLÜPFRIGKEIT EINES SCHMIERSTOFFES, LUBRIZITÄT
7814	**LUBRIFICATION**	GRAISSAGE	SCHMIERUNG
7815	**LUDER'S LINES**	LIGNES DE LUDERS, LIGNES D'HARTMANN	LINIEN (LÜDERSCHE)
7816	**LUG**	BERCEAU	SITZ (ANGEGOSSENER)
7817	**LUG, CAR (ON A CASTING)**	OREILLE, ATTACHE, BRIDE, PATTE	AUGE (ANGEGOSSENES)
7818	**LUMEN**	LUMEN	LUMEN
7819	**LUMINESCENCE**	LUMINESCENCE	LUMINESZENZ
7820	**LUMINOUS FLAME**	FLAMME ECLAIRANTE, FLAMME REDUCTRICE, FEU DE REDUCTION	FLAMME (LEUCHTENDE)
7821	**LUMINOUS FLUX**	FLUX LUMINEUX	LICHTSTROM
7822	**LUMINOUS INTENSITY, INTENSITY OF LIGHT**	INTENSITE LUMINEUSE	LICHSTÄRKE, LICHTINTENSITÄT
7823	**LUMINOUS POINT**	POINT LUMINEUX	PUNKT (LEUCHTENDER)
7824	**LUMINOUS RAY, RAY OF LIGHT, BEAM**	RAYON LUMINEUX	LICHTSTRAHL
7825	**LUMINOUS VIBRATION**	VIBRATION LUMINEUSE	LICHTSCHWINGUNG
7826	**LUMP COAL**	CHARBON GROS	STÜCKKOHLE, GROBKOHLE
7827	**LUMP FUEL**	COMBUSTIBLE EN GROS MORCEAUX	BRENNSTOFF (STÜCKIGER)
7828	**LUNAR CAUSTIC, SILVER NITRATE**	NITRATE D'ARGENT, AZOTATE D'ARGENT, PIERRE INFERNALE	HÖLLENSTEIN, SILBERNITRAT
7829	**LUNE**	CROISSANT	KREISSICHELSTÜCK, KREISBOGENSICHEL
7830	**LUSTRE**	ECLAT (MIN.)	GLANZ (MIN.)
7831	**LUTE (TO)**	LUTTER	ABDICHTEN
7832	**LUTE OF A PIPE, LOOP OF A PIPE**	LYRE (D'UN TUYAU)	ROHRBOGEN
7833	**LUTECIUM**	LUTECIUM	CASSIOPEIUM, LUTETIUM
7834	**LUX**	LUX	LUX
7835	**LYE**	LESSIVE	LAUGE
7836	**MACERATE (TO)**	MACERER	MAZERIEREN
7837	**MACERATION**	MACERATION	MAZERATION
7838	**MACHINABILITY**	USINABILITE	BEARBEITBARKEIT
7839	**MACHINE**	MACHINE DE TRAVAIL, MACHINE RECEPTRICE	MASCHINE, ARBEITSMASCHINE
7840	**MACHINE BOLT**	BOULON FAIT A LA MACHINE, BOULON MECANIQUE	MASCHINENSCHRAUBE
7841	**MACHINE CUT NAIL**	CLOU DECOUPE	NAGEL (GESCHNITTENER), SCHNITTNAGEL
7842	**MACHINE CUTTING**	DECOUPAGE A LA MACHINE	MASCHINABSCHNEIDEN
7843	**MACHINE DRAWING**	DESSIN DE MACHINE	MASCHINENZEICHNUNG
7844	**MACHINE FINISHING**	FINISSAGE A LA MACHINE	FERTIGBEARBEITUNG (MASCHINELLE)
7845	**MACHINE FOR MAKING BENDING TESTS**	MACHINE POUR ESSAIS DE FLEXION	BIEGEMASCHINE (FÜR VERSUCHE), BIEGEPRESSE (FÜR VERSUCHE)

7846	**MACHINE FOR MAKING TENSILE TESTS**	DYNAMOMETRE, BAN DE TRACTION, APPAREIL POUR LES RUPTURES A LA TRACTION	ZERREISSMASCHINE
7847	**MACHINE FORGING**	FORGEAGE A LA MACHINE	MASCHINENSCHMIEDEN
7848	**MACHINE FOUNDATION**	FONDATION D'UNE MACHINE	FUNDAMENT EINER MASCHINE, MASCHINENFUNDAMENT
7849	**MACHINE GRINDING**	MEULAGE A LA MACHINE	MASCHINENSCHLEIFEN
7850	**MACHINE MADE**	TRAVAILLE A LA MACHINE	MECHANISCH, MASCHINELL, MASCHINEN HERGESTELLT (MIT)
7851	**MACHINE MOLDING**	MOULAGE MECANIQUE	MASCHINENFORMUNG
7852	**MACHINE PART, ENGINE DETAIL**	ELEMENT, ORGANE, PIECE DE MACHINE, ORGANE, PIECE MECANIQUE	MASCHINENTEIL
7853	**MACHINE REAMER**	ALESOIR DE MACHINE	MASCHINENREIBAHLE
7854	**MACHINE RIVETING**	RIVETAGE MECANIQUE	MASCHINENNIETUNG
7855	**MACHINE SHOP**	ATELIER DE MACHINES, SALLE DES MACHINES, ATELIER DE MECANIQUE	MASCHINENHALLE, MASCHINENWERKSTATT
7856	**MACHINE SHOP**	ATELIER DE MECANIQUE	MASCHINENWERKSTATT
7857	**MACHINE STEEL**	ACIER A OUTILS	WERKZEUGSTAHL
7858	**MACHINE TAP**	TARAUD A LA MACHINE, TARAUDEUSE	MASCHINENGEWINDEBOHRER
7859	**MACHINE TOOL**	MACHINE-OUTIL	WERKZEUGMASCHINE
7860	**MACHINE WORK**	TRAVAIL MECANIQUE (TRAVAIL FAIT A LA MACHINE)	MASCHINENARBEIT
7861	**MACHINED**	USINE	BEARBEITET
7862	**MACHINERY**	MACHINERIE	MASCHINENEINRICHTUNG
7863	**MACHINERY CASTINGS**	FONTE MECANIQUE	MASCHINENGUSS
7864	**MACHINING**	USINAGE	BEARBEITUNG
7865	**MACHINING ALLOWANCE**	SUREPAISSEUR D'USINAGE	BEARBEITUNGSZUGABE
7866	**MACHINING STEP**	DEGRE D'USINAGE	BEARBEITUNGSSTUFE
7867	**MACHINING WIDTH**	LARGEUR D'USINAGE	ARBEITSBREITE
7868	**MACHINIST**	OUVRIER MECANICIEN, OUVRIER DE MACHINE -OUTIL	MASCHINENARBEITER
7869	**MACRO STRUCTURE**	STRUCTURE A GROS GRAINS, MACROSTRUCTURE	GROBGEFÜGE
7870	**MACROETCHING**	ATTAQUE MACROGRAPHIQUE	MAKROÄTZUNG
7871	**MACROGRAPHY**	MACROGRAPHIE	MAKROGRAPHIE
7872	**MACROSCOPIC**	MACROSCOPIQUE	MAKROSKOPISCH
7873	**MACROSCOPIC EXAMINATION**	OBSERVATION MACROSCOPIQUE A L'OEIL NU	BEOBACHTUNG (MAKROSKOPISCHE), UNTERSUCHUNG (MAKROSKOPISCHE)
7874	**MACROSCOPIC STRESS**	EFFORT MACROSCOPIQUE	BEANSPRUCHUNG (MAKROSKOPISCHE)
7875	**MACROSECTION**	SECTION MACROSCOPIQUE	QUERSCHNITT (MAKROSKOPISCHER)
7876	**MACROSEGREGATION**	MACROSEGREGATION	MAKROSEIGERUNG
7877	**MAGNAFLUX TESTING METHOD**	MAGNETOSCOPIE	MAGNETPULVER-PRÜFVERFAHREN
7878	**MAGNALIUM**	MAGNALIUM	MAGNALIUM
7879	**MAGNESIA, MAGNESIUM OXIDE**	MAGNESIE, OXYDE DE MAGNESIUM	MAGNESIA, MAGNESIUMOXYD
7880	**MAGNESITE**	GLOBERTITE, MAGNESITE	MAGNESIT, TALKSPAT, BITTERSPAT
7881	**MAGNESITE BRICK**	AGGLOMERE MAGNESIEN	MAGNESITSTEIN
7882	**MAGNESIUM**	MAGNESIUM	MAGNESIUM
7883	**MAGNESIUM BICARBONATE**	BICARBONATE DE MAGNESIUM	ZWEIFACHKOHLENSÄURE, DOPPELTKOHLENSÄURE MAGNESIA, MAGNESIUMBIKARBONAT

7884	**MAGNESIUM CARBONATE, CARBONATE OF MAGNESIA**	CARBONATE DE MAGNESIE	MAGNESIAKOHLENSÄURE, MAGNESIUMKARBONAT
7885	**MAGNESIUM CHLORIDE**	CHLORURE DE MAGNESIUM	MAGNESIUMCHLORID, CHLORMAGNESIUM
7886	**MAGNESIUM FLUORIDE**	FLUORURE DE MAGNESIUM	MAGNESIUMFLUORID, FLUORMAGNESIUM
7887	**MAGNESIUM HYDROXIDE**	HYDRATE DE MAGNESIE	MAGNESIUMHYDROXYD
7888	**MAGNESIUM SILICATE**	SILICATE DE MAGNESIUM	MAGNESIA-KIESELSÄURE, MAGNESIUMSILIKAT
7889	**MAGNESIUM SULPHATE, SULPHATE OF MAGNESIA, EPSOM SALT**	SULFATE DE MAGNESIUM, SEL ANGLAIS, SEL D'EPSOM, SEL DE SEDLITZ	BITTERSALZ, MAGNESIA-SCHWEFELSÄURE, MAGNESIUMSULFAT (ENGLISCHES), SEDLITZERSALZ, EPSOMER SALZ
7890	**MAGNESIUM-BASE ALLOY**	ALLIAGE A BASE DE MAGNESIUM	MAGNESIUMLEGIERUNG
7891	**MAGNET**	AIMANT	MAGNET
7892	**MAGNET STEEL**	ACIER POUR AIMANTS	MAGNETSTAHL
7893	**MAGNETIC**	MAGNETIQUE	MAGNETISCH
7894	**MAGNETIC ANALYSIS**	ANALYSE MAGNETIQUE	ANALYSE (MAGNETISCHE)
7895	**MAGNETIC ATTRACTION**	ATTRACTION MAGNETIQUE	ANZIEHUNG (MAGNETISCHE)
7896	**MAGNETIC BRAKE**	FREIN MAGNETIQUE	BREMSE (MAGNETISCHE)
7897	**MAGNETIC CHUCK BLOCKS**	CALES MAGNETIQUES DE FIXATION	MAGNETSPANNFUTTER
7898	**MAGNETIC CHUCKS**	MANDRINS MAGNETIQUES	MAGNETFUTTER
7899	**MAGNETIC DENSITY, FLUX DENSITY**	DENSITE MAGNETIQUE	DICHTE (MAGNETISCHE)
7900	**MAGNETIC ENERGY**	ENERGIE MAGNETIQUE	ENERGIE (MAGNETISCHE)
7901	**MAGNETIC FIELD**	CHAMP MAGNETIQUE	FELD (MAGNETISCHES)
7902	**MAGNETIC FIGURES**	SPECTRE MAGNETIQUE	FEILSPÄNKURVEN, KRAFTLINIENBILD
7903	**MAGNETIC FLUX**	FLUX MAGNETIQUE	MAGNETISCHE STRÖMUNG, MAGNETISCHER KRAFTFLUSS, KRAFTLINIENFLUSS, KRAFTLINIENSTROM
7904	**MAGNETIC FORCE**	FORCE MAGNETIQUE	KRAFT (MAGNETISCHE)
7905	**MAGNETIC GAUGE**	JAUGE MAGNETIQUE	MESSER (MAGNETISCHER)
7906	**MAGNETIC HYSTERESIS**	MAGNETISATION	MAGNETISIERUNG
7907	**MAGNETIC HYSTERESIS LOSS**	PERTES DE MAGNETISATION	MAGNETISIERUNGSVERLUSTE
7908	**MAGNETIC INDUCTION**	INDUCTION MAGNETIQUE	INDUKTION (MAGNETISCHE)
7909	**MAGNETIC NEEDLE**	AIGUILLE AIMANTEE	MAGNETNADEL
7910	**MAGNETIC PARTICLE INSPECTION (MAGNAFLUX)**	MAGNETOSCOPIE	MAGNETOSKOPIE
7911	**MAGNETIC PERMEABILITY**	PERMEABILITE MAGNETIQUE	DURCHLÄSSIGKEIT (MAGNETISCHE), PERMEABILITÄT (MAGNETISCHE)
7912	**MAGNETIC PLATES**	PLATEAUX MAGNETIQUES	MAGNET-PLANSCHEIBEN
7913	**MAGNETIC POLE**	POLE MAGNETIQUE	MAGNETPOL
7914	**MAGNETIC PYRITES, PYRRHOTITE**	PYRRHOTINE, PYRITE MAGNETIQUE	MAGNETKIES, PYRRHOTIN
7915	**MAGNETIC REPULSION**	REPULSION MAGNETIQUE	ABSTOSSUNG (MAGNETISCHE)
7916	**MAGNETIC SATURATION**	SATURATION MAGNETIQUE	SÄTTIGUNG (MAGNETISCHE)
7917	**MAGNETIC SEPARATION**	SEPARATION MAGNETIQUE	TRENNUNG (MAGNETISCHE)
7918	**MAGNETIC SUSCEPTIBILITY**	SUSCEPTIBILITE MAGNETIQUE	SUSZEPTIBILITÄT (MAGNETISCHE), MAGNETISIERUNGSKOEFFIZIENT
7919	**MAGNETIC TAPE**	BANDE MAGNETIQUE	MAGNETBAND
7920	**MAGNETIC TRANSFORMATION TEMPERATURE**	POINT DE CURIE	CURIEPUNKT
7921	**MAGNETIC TRIM HAMMER**	MARTEAU DE TAPISSIER AIMANTE	MAGNETHAMMER
7922	**MAGNETISABLE**	MAGNETISABLE	MAGNETISIERBAR
7923	**MAGNETISATION**	AIMANTATION	MAGNETISIEREN, MAGNETISIERUNG
7924	**MAGNETISE (TO)**	AIMANTER	MAGNETISIEREN

7925	**MAGNETISING FORCE**	FORCE MAGNETISANTE	KRAFT (MAGNETISIERENDE)
7926	**MAGNETISM**	MAGNETISME	MAGNETISMUS, MAGNETISM
7927	**MAGNETITE, FERROSOFERRIC OXIDE, MAGNETIC OXIDE OF IRON, LOADSTONE**	MAGNETITE, FER OXYDULE, MINERAI DE FER MAGNETIQUE	MAGNETEISENSTEIN, MAGNETEISENERZ, MAGNETIT, MAGNETISCH. EISENOXYD, EISENOXYDULOXYD, MAGNETSTEIN
7928	**MAGNETIZATION**	MAGNETISATION, AIMANTATION	MAGNETISIERUNG
7929	**MAGNETIZING FORCE**	FORCE MAGNETOMOTRICE SPECIFIQUE	MAGNETOMOTORISCHE KRAFT
7930	**MAGNETO-ELECTRIC MACHINE**	MACHINE MAGNETO-ELECTRIQUE, MAGNETO	MASCHINE (MAGNETELEKTRISCHE)
7931	**MAGNETOGRAPHIC INSPECTION**	EXAMEN MAGNETOGRAPHIQUE	PRÜFUNG (MAGNETOGRAPHISCHE)
7932	**MAGNETOMOTIVE FORCE (M M F)**	FORCE MAGNETOMOTRICE	KRAFT (MAGNETOMOTORISCHE)
7933	**MAGNETOSTRICTION**	MAGNETOSTRICTION	MAGNETOSTRIKTION
7934	**MAGNETTE CLUTCH**	EMBRAYAGE MAGNETIQUE	KUPPLUNG (MAGNETISCHE), KUPPUNG (ELEKTROMAGNETISCHE)
7935	**MAGNIFICATION**	GROSSISSEMENT, AGRANDISSEMENT	VERGRÖSSERUNG (OPT.)
7936	**MAGNIFYING GLASS**	LOUPE, VERRE GROSSISSANT	LUPE, VERGRÖSSERUNGSGLAS
7937	**MAGNIFYING POWER**	GROSSISSEMENT, POUVOIR GROSSISSANT	VERGRÖSSERUNGSKRAFT
7938	**MAGNIFYING POWER OF A LENS**	GROSSISSEMENT D'UNE LENTILLE	VERGRÖSSERUNG EINER LINSE
7939	**MAGNITUDE OF A FORCE**	GRANDEUR, INTENSITE D'UNE FORCE	GRÖSSE EINER KRAFT, INTENSITÄT EINER KRAFT
7940	**MAIN (ELECTRIC) POWER SUPPLY**	ALIMENTATION, RESEAU	HAUPTSPEISELEITUNG
7941	**MAIN BEARING**	COUSSINET DE PALIER, PALIERS DE VILEBREQUIN	HAUPTLAGER, GRUNDLAGER
7942	**MAIN JET**	GICLEUR PRINCIPAL	HAUPTDÜSE
7943	**MAIN SHAFT (GB)**	\ ARBRE PRIMAIRE OU PRINCIPAL	HAUPTWELLE,ABTRIEBSWELLE
7944	**MAIN SHAFTING, LINE SHAFTING**	ARBRE PRINCIPAL, TRANSMISSION PRINCIPALE, ARBRE DE TRANSMISSION, ARBRE D'ATTAQUE, ARBRE SECOND MOTEUR, ARBRE DEUXIEME MOTEUR	HAUPTWELLE, TRIEBWERKSWELLE, TRANSMISSIONSWELLE
7945	**MAINTENANCE COSTS, COST OF UPKEEP, OF MAINTENANCE**	FRAIS D'ENTRETIEN	INSTANDHALTUNGSKOSTEN, UNTERHALTUNGSKOSTEN
7946	**MAINTENANCE, UPKEEP**	ENTRETIEN	UNTERHALTUNG, INSTANDHALTUNG, WARTUNG
7947	**MAJOR AXIS OF ELLIPSE**	GRAND AXE D'UNE ELLIPSE	GROSSE ACHSE EINER ELLIPSE
7948	**MAKE (TO), MANUFACTURE (TO)**	FABRIQUER, MANUFACTURER	HERSTELLEN, ERZEUGEN, FABRIZIEREN
7949	**MAKE AND BREAK COMPONENT**	CONJONCTEUR-DISJONCTEUR	RÜCKSTROMSCHALTELEMENT
7950	**MAKE CONTACT (TO)**	REALISER, ETABLIR UN CONTACT	KONTAKT HERSTELLEN (EINEN)
7951	**MAKE FLUSH (TO)**	AFFLEURER	BÜNDIG MACHEN
7952	**MAKE MALLEABLE CASTINGS (TO)**	MALLEABILISER LA FONTE	TEMPERN
7953	**MAKE TIGHT (TO), CAULK (TO)**	RENDRE ETANCHE, ETANCHER	ABDICHTEN
7954	**MAKING TIGHT, CAULKING**	ETANCHEMENT	ABDICHTEN, DICHTEN
7955	**MALE EXTERNAL SCREW THREAD**	FILETAGE EXTERIEUR, FILETAGE MALE	AUSSENGEWINDE, BOLZENGEWINDE, VATERGEWINDE
7956	**MALE TAPER GAUGE**	JAUGE-TAMPON CONIQUE	KEGELLEHRDORN
7957	**MALE-FEMALE FACING**	EMBOITEMENT SIMPLE (M ET F)	VOR-UND RÜCKSPRUNG
7958	**MALLEABILITY, FORGEABILITY**	MALLEABILITE, FORGEABILITE	HÄMMERBARKEIT, SCHMIEDBARKEIT
7959	**MALLEABLE CAST IRON, MALLEABLE CASTING**	FONTE MALLEABLE	GUSS (SCHMIEDBARER), TEMPERGUSS, TEMPERSTAHLGUSS, WEICHGUSS
7960	**MALLEABLE, FORGEABLE**	MALLEABLE, FORGEABLE	HÄMMERBAR, SCHMIEDBAR

7961	MALLEABLEIZING	MALLEABILISATION	SCHMIEDBARMACHEN
7962	MALLET, WOODEN HAMMER	MAILLET EN BOIS	SCHLÄGEL, HOLZHAMMER
7963	MAN HOLE	TROU D'HOMME, OUVERTURE, JOINT AUTOCLAVE	MANNLOCH
7964	MANDREL	POUPEE (DE TOUR), MANDRIN DE MONTAGE, MANDRIN DE REPRISE	DORN, AUFNAHMEDORN
7965	MANE HAIR	POILS DE LA CRINIERE, CRINS	MÄHNENHAAR
7966	MANGANESE	MANGANESE	MANGAN
7967	MANGANESE BRONZE	BRONZE AU MANGANESE, LAITON AU MANGANESE	MANGANBRONZE
7968	MANGANESE STEEL	ACIER AU MANGANESE	MANGANSTAHL
7969	MANGANESE-KILLED STEEL	ACIER CALME AU MANGANESE	STAHL MIT MANGAN (BERUHIGTER)
7970	MANGANIC OXIDE	SESQUIOXYDE DE MANGANESE	MANGAN-SESQUIOXYD
7971	MANGANIFEROUS	MANGANESIFERE	MANGANHALTIG
7972	MANGANOUS CHLORIDE	CHLORURE MANGANEUX	MANGANCHLORÜR, MANGANOCHLORID, CHLORMANGAN
7973	MANGANOUS OXIDE	PROTOXYDE DE MANGANESE	MANGANOXYDUL, MANGANMONOXYD
7974	MANGANOUS SULPHATE	SULFATE MANGANEUX	MANGANOSULFAT
7975	MANGLE WHEEL	ROUE DENTEE INTERROMPUE	MANGELRAD, WENDERAD
7976	MANHOLE	TROU D'HOMME	EINSTEIGELOCH
7977	MANHOLE COVER	TAMPON DE TROU D'HOMME	MANNLOCHDECKEL
7978	MANHOLE NECK	COLLET DE TROU D'HOMME	MANNLOCHRING
7979	MANIFOLD	COLLECTEUR	SAMMELLEITUNG
7980	MANILLA HEMP	CHANVRE DE MANILLE, MANILLE, ABACA	MANILAHANF
7981	MANILLA ROPE	CORDE EN MANILLE	MANILAHANFSEIL
7982	MANIPULATOR	MACHINE DE MANIPULATION, CULBUTEUR DE LINGOTS	KANTVORRICHTUNG
7983	MANNESMANN TUBE	TUBE MANNESMANN	MANNESMANNROHR
7984	MANOMETRIC, PRESSURE EFFICIENCY	RENDEMENT MANOMETRIQUE	WIRKUNGSGRAD (MANOMETRISCHER)
7985	MANTISSA	MANTISSE	MANTISSE
7986	MANTLE	ENVELOPPE, GAINE	MANTEL
7987	MANUAL DATA INPUT	INTRODUCTION MANUELLE DES DONNEES	DATENEINGABE (MANUELLE)
7988	MANUAL WELD	SOUDURE DEPOSEE A LA MAIN	NAHT (HANDGESCHWEISSTE)
7989	MANUFACTURE	FABRICATION, CONSTRUCTION, PRODUIT MANUFACTURE	HERSTELLUNG, FABRIKATION, ERZEUGNIS, FABRIKAT
7990	MANUFACTURE OF BRIQUETTES	AGGLOMERATION, FABRICATION DES AGGLOMERES	BRIKETTIERUNG
7991	MANUFACTURE OF COKE	FABRICATION DU COKE, CARBONISATION DE LA HOUILLE	KOKEREI, KOKSBRENNEREI, KOKSFABRIKATION
7992	MANUFACTURER, MAKER	FABRICANT, INDUSTRIEL, CONSTRUCTEUR	HERSTELLER, ERZEUGER, FABRIKANT
7993	MANUFACTURER'S DATA REPORT	ETAT DESCRIPTIF DU CONSTRUCTEUR	HERSTELLERBESCHREIBUNG
7994	MANUFACTURING RANGE	GAMME DE FABRICATION, PROGRAMME DE FABRICATION	FERTIGUNGSUMFANG, FERTIGUNGSBEREICH, FERTIGUNGSPROGRAMM
7995	MANURE SPREADERS : FARMYARD	EPANDEURS DE FUMIER	DÜNGERAUSBREITEMASCHINEN
7996	MANUSCRIPT	MANUSCRIT	MANUSKRIPT
7997	MANWAY	TROU D'HOMME, T.H.	MANNLOCH
7998	MANWAY NECK	MANCHON DE T.H., COLLET DE T.H.	MANNLOCHRING
7999	MAPPING PEN	PLUME A DESSIN	ZEICHENFEDER
8000	MARAGING STEEL	ACIER AUTOTREMPANT	STAHL (NATURHARTER), SS-STAHL, MARTENSIT, STAHL (AUSHÄRTENDER)

8001	**MARBLE**	MARBRE	MARMOR
8002	**MARGINAL PLATE**	TOLE MARGINALE	RANDBLECH
8003	**MARINE BOILER**	CHAUDIERE MARINE	SCHIFFSKESSEL
8004	**MARINE CHAINS**	CHAINES DE MARINE	SCHIFFSKETTEN
8005	**MARINE ENGINE**	MOTEUR MARIN, MACHINE MARINE	SCHIFFSMASCHINE
8006	**MARINE GLUE**	GLU MARINE	MARINELEIM
8007	**MARINE PATTERN CONNECTING ROD END**	TETE DE BIELLE A CAGE OUVERTE	OFFENER SCHUBSTANGENKOPF, MARINEKOPF, SCHIFFSKOPF
8008	**MARK (TO)**	REPERER, MARQUER	ANZEICHNEN, MARKIEREN
8009	**MARK OF A SCALE**	TRAIT D'UNE ECHELLE	TEILSTRICH, GRADSTRICH
8010	**MARK OUT (TO), SET OUT (TO), LINE OUT (TO), LAY OUT (TO), SCRIBE (TO)**	TRACER, MARQUER	ANREISSEN, VORREISSEN
8011	**MARK WITH THE CENTRE PUNCH (TO)**	MARQUER UN REPERE, REPERER, AMORCER AU MOYEN D'UN COUP DE POINTEAU	ANKÖRNEN
8012	**MARK, GUILDING MARK**	REPERE	ZEICHEN, MERKZEICHEN
8013	**MARKER OUT, LINER OUT**	TRACEUR-MECANICIEN	ANREISSER, VORREISSER (ARBEITER)
8014	**MARKET LEAD, SOFT LEAD**	PLOMB MARCHAND, PLOMB DOUX	WEICHBLEI, HÜTTENBLEI, KAUFBLEI
8015	**MARKING**	REPERAGE, MARQUAGE	ANZEICHNEN, MARKIEREN
8016	**MARKING GAUGE**	TRUSQUIN A POINTE, CALIBRE DE TRACAGE	PARALLELMASS, STREICHMASS, REISSMASS, PARALLELREISSER, STREICHMODELL, REISSMODELL, ANDREISSSCHABLONE
8017	**MARKING OUT, SETTING OUT, LINING OUT, LAYING OUT, SCRIBING**	TRACAGE, TRACE	ANREISSEN, VORREISSEN
8018	**MARKING-OUT OF PLATES**	TRACAGE DES TOLES	BLECHANREISSEN
8019	**MARL**	MARNE	MERGEL
8020	**MARSEILLE SOAP**	SAVON DE MARSEILLE	MARSEILLER SEIFE
8021	**MARTEMPERING**	TREMPE EN BAIN CHAUD, TREMPE MARTENSITIQUE	WARMBADHÄRTUNG
8022	**MARTENSITE**	MARTENSITE	HARDENIT, MARTENSIT
8023	**MARTIN-SIEMENS STEEL, OPEN HEARTH STEEL**	ACIER SIEMENS- MARTIN	SIEMENS-MARTINSTAHL
8024	**MASH WELD**	SOUDURE A L'ECRASEMENT	ROLLENQUETSCHNAHT
8025	**MASONRY**	MACONNERIE, OUVRAGE DE MACONNERIE	MAUERWERK
8026	**MASS**	MASSE (MEC.)	MASSE
8027	**MASS EFFECT**	EFFET DE MASSE, SENSIBILITE A L'EPAISSEUR	MASSENEFFEKT
8028	**MASS PRODUCTION, BULK PRODUCTION**	CONSTRUCTION, PRODUCTION, FABRICATION, CONFECTION, TRAVAIL EN GRANDE SERIE	MASSENHERSTELLUNG, MASSENFABRIKATION
8029	**MASS SPECTROMETER**	SPECTROMETRE DE MASSE	MASSENSPEKTROMETER
8030	**MASS-ABSORPTION COEFFICIENT**	COEFFICIENT D'ABSORPTION MASSIQUE	MASSENABSORPTIONSKOEFFIZIENT
8031	**MASSICOT, LEAD MONOXIDE**	MASSICOT	MASSICOT, BLEIOXYD (GELBES), BLEIGELB, NEUGELB, KÖNIGSGELB
8032	**MASTER**	MODELE ETALON	MUSTERMODELL
8033	**MASTER BLOCK**	PORTE-ESTAMPES, PORTE-POINCONS	STEMPELHALTER
8034	**MASTER FORM**	PIECE DE REFERENCE	KOPIERMODELL, BEZUGSFORMSTÜCK
8035	**MASTER GAUGE**	JAUGE DE REFERENCE	PRÜFLEHRE
8036	**MASTER SET**	JEU ETALON	BEZUGSGERÄTESATZ
8037	**MASTER-CYLINDER**	MAITRE-CYLINDRE	HAUPTZYLINDER

8038	MASTIC	MASTIC, LUT	KITT
8039	MASTIC RESIN	MASTIC EN LARMES, MASTIC EN GRAINS	MASTIX
8040	MASURIUM	MASURIUM,TECHNETIUM	MASURIUM, TECHNETIUM
8041	MASUT	MAZOUT	MASUT
8042	MATCH (TO)	ASSORTIR, ACCOSTER	ZUSAMMENPASSEN, ANHAFTEN
8043	MATCH LINE	FACE DE REFERENCE	BEZUGSKANTE
8044	MATCH PLATE PATTERN	PLAQUE MODELE DOUBLE FACE	WENDEPLATTE, MODELLPLATTE (ZWEISEITIGE)
8045	MATCH PLATES (TO) (DURING ERECTION)	ACCOSTER DES TOLES (MONTAGE)	BLECHE ANEINANDERLEGEN
8046	MATCHING (OF PLATES)	ACCOSTAGE DE TOLES	BLECHANHAFTUNG
8047	MATCHING PLANE	BOUVET A FOURCHEMENT	SPUNDHOBEL
8048	MATE, HELPER	AIDE	HILFSARBEITER
8049	MATERIAL COSTS, COST OF MATERIALS	FRAIS DE MATIERE, FRAIS DE CONSOMMATION	MATERIALKOSTEN
8050	MATERIAL OF CONSTRUCTION ENGINEERING	MATERIEL DE CONSTRUCTION	BAUSTOFF, MATERIAL, BAUMATERIAL
8051	MATHEMATICAL INSTRUMENTS	INSTRUMENTS DE MATHEMATIQUES, BOITE DE COMPAS	REISSZEUG
8052	MATHEMATICAL TABLE	BAREME	RECHENTAFEL, RECHENTABELLE
8053	MATHEMATICS	MATHEMATIQUE	MATHEMATIK
8054	MATING FLANGE	CONTREBRIDE	GEGENFLANSCH
8055	MATRIX	MATRICE, MASSE PRINCIPALE	GRUNDMASSE, MATRIZE
8056	MATRIX BRASS	LAITON DE BASE	GRUNDMESSING
8057	MATRIX METAL	METAL DE BASE	GRUNDMETALL
8058	MATTE	MATTE	LECH, STEIN
8059	MATTE DIP	BAIN DE DECAPAGE MAT, BAIN DE BLANCHIMENT	MATTBRENNE
8060	MATTE SURFACE	SURFACE MATE	OBERFLÄCHE (GLANZLOSE)
8061	MATTER IN SOLUTION	MATIERES EN DISSOLUTION	LÖSUNGSSTOFFE
8062	MATTER IN SUSPENSION, SUSPENDED MATTER	MATIERES EN SUSPENSION	SCHWEBESTOFFE, SINKSTOFFE
8063	MAXIMUM AND MINIMUM THERMOMETER	THERMOMETRE A MAXIMA ET MINIMA	MAXIMUM-UND MINIMUM-THERMOMETER
8064	MAXIMUM DEFLECTION	FLECHE MAXIMUM	DURCHBIEGUNG (GRÖSSTE)
8065	MAXIMUM LOAD, PEAK LOAD	CHARGE MAXIMUM	SPITZENLAST, SPITZENBELASTUNG, HÖCHSTBELASTUNG
8066	MAXIMUM OUTPUT	PUISSANCE MAXIMUM	HÖCHSTLEISTUNG
8067	MAXIMUM SPEED	VITESSE MAXIMUM	HÖCHSTGESCHWINDIGKEIT
8068	MAXIMUM TEMPERATURE	TEMPERATURE MAXIMUM	HÖCHSTTEMPERATUR, MAXIMALTEMPERATUR
8069	MAXIMUM VALUE	VALEUR MAXIMUM	HÖCHSTWERT, GRÖSSTWERT, MAXIMALWERT
8070	MAXIMUM WEIGHT	POIDS MAXIMUM	HÖCHSTGEWICHT, MAXIMALGEWICHT
8071	MCQUAID-EHN TEST	TEST DE MCQUAID-EHN	MCQUAID-EHN-PRÜFUNG
8072	MEAN	MOYEN	MITTEL
8073	MEAN AVERAGE VALUE	VALEUR MOYENNE	WERT (MITTLERER), MITTELWERT, DURCHSCHNITTSWERT
8074	MEAN DEVIATION	ECART MOYEN	ABWEICHUNG (DURCHSCHNITTLICHE)
8075	MEAN PRESSURE	PRESSION MOYENNE	DRUCK (MITTLERER)
8076	MEAN SPECIFIC HEAT	CHALEUR SPECIFIQUE MOYENNE	WÄRME (MITTLERE SPEZIFISCHE)
8077	MEASURE	MESURE	MASS

8078	MEASURE (TO)	MESURER	MESSEN, AUSMESSEN, ABMESSEN
8079	MEASURE OF CAPACITY	MESURE DE CAPACITE	HOHLMASS
8080	MEASURE OF LENGTH, LINEAR MEASURE	MESURE LINEAIRE	LÄNGENMASS
8081	MEASURE OF SURFACE, SQUARE MEASURE	MESURE DE SURFACE	FLÄCHENMASS
8082	MEASURE OF VOLUME, CUBIC SOLID MEASURE	MESURE DE VOLUME	KÖRPERMASS
8083	MEASURE THE POWER BY A BRAKE TEST (TO)	MESURER LA PUISSANCE DES FREINS	LEISTUNG ABBREMSEN (DIE)
8084	MEASURE WITH THE PLANIMETER (TO)	PLANIMETRER	PLANIMETRIEREN
8085	MEASURE, MEASURING ROD, RULE	REGLE GRADUEE	MASSSTAB, MESSSTAB
8086	MEASUREMENT OF TEMPERATURE	MESURE DE LA TEMPERATURE	TEMPERATURMESSUNG
8087	MEASUREMENT WITH THE PLANIMETER	PLANIMETRAGE	PLANIMETRIERUNG
8088	MEASUREMENT, MEASURING	MESURAGE	MESSUNG, MESSEN
8089	MEASURING ERROR, ERROR IN MEASUREMENT	ERREUR DE MESURE	MESSFEHLER
8090	MEASURING INSTRUMENT APPLIANCE	INSTRUMENT DE MESURE	MESSGERÄT, MESSINSTRUMENT
8091	MEASURING POINT	POINT DE MESURE	MESSPUNKT
8092	MEASURING TAPE	METRE A RUBAN, RUBAN, ROULETTE D'ARPENTAGE	MESSBAND, BANDMASS, ROLLBANDMASS
8093	MECHANICAL	MECANIQUE	MECHANISCH
8094	MECHANICAL ADVANTAGE	GAIN DE TRAVAIL, BENEFICE DANS LE TRAVAIL	ARBEITSGEWINN, GEWINN AN ARBEIT
8095	MECHANICAL CHUCKS	MANDRINS MECANIQUES	FUTTER (MECHANISCHES)
8096	MECHANICAL CLEANING	NETTOYAGE MECANIQUE	REINIGUNG (MECHANISCHE)
8097	MECHANICAL COMPARATOR	COMPARATEUR MECANIQUE	VERGLEICHER (MECHANISCHER)
8098	MECHANICAL DRAUGHT	TIRAGE ARTIFICIEL, TIRAGE A SOUFFLERIE	ZUG (KÜNSTLICHER)
8099	MECHANICAL DRAWING	DESSIN INDUSTRIEL	ZEICHNUNG (TECHNISCHE)
8100	MECHANICAL EFFICIENCY	RENDEMENT MECANIQUE, RENDEMENT ORGANIQUE	WIRKUNGSGRAD (MECHANISCHER)
8101	MECHANICAL ENERGY	ENERGIE MECANIQUE	ARBEITSVERMÖGEN (MECHANISCHES), ENERGIE (MECHANISCHE)
8102	MECHANICAL ENGINEER	INGENIEUR-MECANICIEN	MASCHINENBAUINGENIEUR
8103	MECHANICAL ENGINEERING	CONSTRUCTION MECANIQUE, CONSTRUCTION, FABRICATION DES MACHINES	MASCHINENBAU
8104	MECHANICAL EQUIVALENT OF HEAT	EQUIVALENT MECANIQUE DE LA CHALEUR	WÄRMEÄQUIVALENT (MECHANISCHES)
8105	MECHANICAL FORCE FEED LUBRICATOR	GRAISSEUR MOLLERUP, APPAREIL DE GRAISSAGE SOUS PRESSION MECANIQUE	SCHMIERPRESSE
8106	MECHANICAL HANDLING OF MATERIALS IN THE WORKS, WORKS TRANSPORT	MANUTENTION MECANIQUE	FÖRDERUNG, BEFÖRDERUNG, TRANSPORT IM WERK
8107	MECHANICAL METALLURGY	METALLURGIE MECANIQUE	METALLURGIE (MECHANISCHE)
8108	MECHANICAL PLATING	PLACAGE MECANIQUE	PLATTIERUNG (MECHANISCHE)
8109	MECHANICAL PROPERTIES	CARACTERISTIQUES MECANIQUES	EIGENSCHAFTEN (MECHANISCHE)
8110	MECHANICAL STOKER	CHARGEUR, FOYER AUTOMATIQUE	SELSTTÄTIGE MECHANISCHE ROSTBESCHICKUNGSVORRICHTUNG
8111	MECHANICAL STOKING	CHAUFFAGE MECANIQUE	SELBSTTÄTIGE MECHANISCHE ROSTBESCHICKUNG
8112	MECHANICAL TESTING	ESSAIS MECANIQUES	PRÜFUNG (MECHANISCHE)

8113	MECHANICAL TWINS	MACLE DE DEFORMATION	VERZERRUNGSZWILLINGSKRISTALL
8114	MECHANICAL VIBRATION	VIBRATION MECANIQUE	SCHWINGUNG (MECHANISCHE)
8115	MECHANICAL WELDING	SOUDAGE MECANIQUE (AUTOMATIQUE)	MASCHINENSCHWEISSEN
8116	MECHANICAL WOOD PULP	PATE MECANIQUE DE BOIS	HOLZSTOFF, HOLZSCHLIFF
8117	MECHANICAL WORK	TRAVAIL MOTEUR, TRAVAIL MECANIQUE	ARBEIT (MECHANISCHE)
8118	MECHANICAL-OPTICAL COMPARATOR	COMPARATEUR OPTICO-MECANIQUE	VERGLEICHER (OPTISCH-MECHANISCHER)
8119	MECHANICALLY DRIVEN OPERATED, OPERATED BY MACHINERY, POWER DRIVEN, ACTUATED BY POWER	COMMANDE MECANIQUE, COMMANDE PAR MOTEUR	MECHANISCH, MASCHINELL ANGETRIEBEN, FÜR MASCHINENBETRIEB
8120	MECHANICALLY OPERATED VALVE	SOUPAPE COMMANDEE	VENTIL (GESTEUERTES)
8121	MECHANICS	MECANIQUE	MECHANIK
8122	MECHANISM, GEAR, MOTION	MECANISME, MOUVEMENT	GETRIEBE, TRIEBWEK, MECHANISMUS
8123	MEDIAL SECTION	DIVISION D'UNE LIGNE EN MOYENNE ET EXTREME RAISON	SCHNITT (GOLDENER)
8124	MEDIAN LINE	MEDIANE	MITTELLINIE, SCHWERLINIE, SEITENHALBIERENDE, MEDIANE, TRANSVERSALE
8125	MEDIUM OIL	HUILE MOYENNE	MITTELÖL
8126	MEDIUM WATER-CAPACITY BOILER, BOILER WITH MEDIUM WATER SPACE	CHAUDIERE A MOYEN VOLUME	MITTELWASSERRAUMKESSEL
8127	MEDIUM-PRESSURE BOILER	CHAUDIERE A MOYENNE PRESSION	MITTELDRUCKKESSEL
8128	MEDULLARY RAY	RAYON MEDULLAIRE	MARKSTRAHL
8129	MEGADYNE	MEGADYNE	MEGADYN
8130	MEGAVOLT	MEGAVOLT	MEGAVOLT
8131	MEGERG	MEGERG	MEGERG
8132	MEGOHM	MEGOHM	MEGOHM
8133	MELAPHYRE	MELAPHYRE	MELAPHYR
8134	MELT	FUSION, COULEE	SCHMELZE
8135	MELT (TO), FUSE TO)	FONDRE, LIQUEFIER, PASSER A L'ETAT LIQUIDE	SCHMELZEN
8136	MELTING	FUSION	SCHMELZEN
8137	MELTING BATH / MOLTEN POOL	BAIN DE FUSION	SCHMELZBAD
8138	MELTING FURNACE	FOUR DE FUSION	SCHMELZOFEN
8139	MELTING LEE	GLACE FONDANTE	EIS SCHMELZENDES
8140	MELTING LOSS	PERTES DE FUSION	SCHMELZVERLUSTE
8141	MELTING POINT	POINT DE FUSION	SCHMELZPUNKT
8142	MELTING POT	CREUSET	SCHMELZTOPF
8143	MELTING RANGE	INTERVALLE DE FUSION	SCHMELZBEREICH
8144	MELTING RATE	VITESSE DE FUSION	SCHMELZGESCHWINDIGKEIT
8145	MEMBER	MEMBRURE, ELEMENT	BAUGLIED, ELEMENT
8146	MEMORY	MEMOIRE	SPEICHER
8147	MENISCUS LENS, CONCAVO-CONVEX LENS	MENISQUE CONVERGENT	LINSE (KONKAVKONVEXE)
8148	MERCHANT IRON, COMMERCIAL IRON	FER MARCHAND	HANDELSEISEN
8149	MERCHANT SHAPES	PROFILES MARCHANDS	HANDELSPROFILE
8150	MERCURIAL BAROMETER	BAROMETRE A MERCURE	QUECKSILBERBAROMETER
8151	MERCURIAL PRESSURE GAUGE, MERCURY GAUGE	MANOMETRE A MERCURE	QUECKSILBERMANOMETER
8152	MERCURIAL THERMOMETER	THERMOMETRE A MERCURE	QUECKSILBERTHERMOMETER

8153	MERCURIC CHLORIDE, CORROSIVE SUBLIMATE	CHLORURE MERCURIQUE, BICHLORURE DE MERCURE, SUBLIME CORROSIF	QUECKSILBERCHLORID, MERKURICHLORID, SUBLIMAT
8154	MERCURIC FULMINATE	FULMINATE DE MERCURE	KNALLQUECKSILBER
8155	MERCURIC IODIDE	IODURE DE MERCURE	QUECKSILBERJODID, MERKURIJODID, ZWEIFACHJODQUECKSILBER, ROTES JODQUECKSILBER, JODZINNOBER, JODINROT
8156	MERCURIC NITRATE	NITRATE DE MERCURE, AZOTATE MERCURIQUE	QUECKSILBEROXYD SALPETERSAURES, QUECKSILBEROXYDNITRAT, MERKURINITRAT
8157	MERCURIC OXIDE, RED PRECIPITATE	OXYDE DE MERCURE, OXYDE MERCURIQUE	QUECKSILBEROXYD, MERKURIOXYD, ROTES PRÄZIPITAT, ROTOXID
8158	MERCURIC SULPHATE	SULFATE MERCURIQUE	QUECKSILBEROXYDSULFAT, MERKURISULFAT
8159	MERCUROUS CHLORIDE, CALOMEL	CHLORURE MERCUREUX, PROTOCHLORURE DE MERCURE, CALOMEL	QUECKSILBERCHLORÜR, MERKUROCHLORID, KALOMEL
8160	MERCUROUS NITRATE	NITRATE MERCUREUX, AZOTATE MERCUREUX	QUECKSILBEROXYDUL (SALPETERSAURES), MERKURONITRAT, QUECKSILBEROXYDULNITRAT
8161	MERCUROUS OXIDE	PROTOXYDE DE MERCURE, OXYDE MERCUREUX	QUECKSILBEROXYDUL, MERKUROOXYD
8162	MERCUROUS SULPHATE	SULFATE MERCUREUX	QUECKSILBEROXYDULSULFAT, MERKUROSULFAT
8163	MERCURY AIR PUMP	POMPE A MERCURE	QUECKSILBERLUFTPUMPE
8164	MERCURY ARC RECTIFIER	REDRESSEUR A VAPEUR DE MERCURE	QUECKSILBERDAMPF-GLEICHRICHTER
8165	MERCURY CELL	CELLULE AU MERCURE	QUECKSILBERZELLE
8166	MERCURY SWITCH	CONTACTEUR A MERCURE	QUECKSILBER-SCHALTER
8167	MERCURY VAPOUR LAMP	LAMPE A VAPEUR DE MERCURE	QUECKSILBERDAMPFLAMPE
8168	MERCURY, QUICKSILVER	MERCURE	QUECKSILBER
8169	MESH	MAILLE (D'UN CRIBLE)	SIEBMASCHE, MASCHE
8170	MESH OF A SIEVE MESH OF A STEEL FABRIC	MAILLE D'UN TAMIS, MAILLE D'UNE TOILE METALLIQUE	MASCHENGRÖSSE
8171	MESHED, IN MESH, IN GEAR	EN PRISE	IM EINGRIFF
8172	METACENTRE	METACENDRE	METAZENTRUM
8173	METAL	METAL	METALL
8174	METAL ARC	ARC METALLIQUE	METALL-LICHTBOGEN
8175	METAL ARC CUTTING	COUPAGE A L'ARC METALLIQUE	METALL-LICHTBOGENSCHNEIDEN
8176	METAL ARC WELDING	SOUDAGE A L'ARC ELECTRIQUE	METALL-LICHTBOGENSCHWEISSEN
8177	METAL CLAD CABLE	CABLE A GAINE METALLIQUE	KABEL MIT METALLMANTEL
8178	METAL CUTTING MACHINE	MACHINE-OUTIL A TRAVAILLER LES METAUX, MACHINE A METAUX	METALLBEARBEITUNGSMASCHINE
8179	METAL DISTRIBUTION RATIO	RAPPORT DE DISTRIBUTION DU METAL	NIEDERSCHLAGSVERTEILUNGSVERHÄLTNIS
8180	METAL ELECTRODE	ELECTRODE METALLIQUE	METALLELEKTRODE, SCHWEISSDRAHT
8181	METAL FILAMENT LAMP	LAMPE A FILAMENT METALLIQUE	METALLDRAHTLAMPE, METALLFADENLAMPE
8182	METAL FOG	BROUILLARD METALLIQUE	METALLNEBEL
8183	METAL FOR MEDALS, COINAGE BRONZE	BRONZE DES MONNAIES	MÜNZENBRONZE, MEDAILLENBRONZE
8184	METAL LEAF	FEUILLE METALLIQUE	METALLFOLIE
8185	METAL POWDER	METAL PULVERISE	METALLPULVER
8186	METAL SAW, HACK SAW	SCIE A MONTURE METALLIQUE, SCIE A METAUX	BÜGELSÄGE, BOGENSÄGE, METALLSÄGE

8187	**METAL SCREW**	VIS A METAUX	METALLSCHRAUBE
8188	**METAL SHOE**	SABOT MAGNETIQUE	MAGNETBREMSKLOTZ
8189	**METAL SPINNING, BURNISHING**	REPOUSSE, REPOUSSAGE, EMBOUTISSAGE	METALLDRÜCKEN, DRÜCKEN, METALLDRÜCKEREI
8190	**METAL SPRAYING**	METALLISATION	METALLSPRITZEN
8191	**METAL TURNING LATHE**	TOUR A METAUX	METALLDREHBANK
8192	**METAL WORKING**	TRAVAIL DES METAUX, USINAGE DES METAUX	METALLBEARBEITUNG
8193	**METAL-SLITTING SAN**	FRAISE-SCIE , SCIE A MORTAISER	SCHLITZSÄGE
8194	**METALLIC COATING, COAT OF METAL**	REVETEMENT METALLIQUE	METALLÜBERZUG, ÜBERZUG (METALLISCHER)
8195	**METALLIC LUSTRE**	ECLAT METALLIQUE	METALLGLANZ, GLANZ (METALLISCHER)
8196	**METALLIC PACKING**	GARNITURE METALLIQUE	PACKUNG (METALLISCHE), METALLPACKUNG, METALLIDERUNG
8197	**METALLIFEROUS**	METALLIFERE	METALLHALTIG
8198	**METALLIFEROUS MINE**	MINE METALLIQUE	ERZGRUBE, ERZBERGWERK
8199	**METALLISED PAPER**	PAPIER METALLISE	PAPIER (METALLISIERTES)
8200	**METALLIZED COATING**	REVETEMENT METALLISE	ÜBERZUG (METALLBEDAMPFTER)
8201	**METALLIZING**	METALLISATION	METALL-AUFSPRITZEN
8202	**METALLOGRAPHIC**	METALLOGRAPHIQUE	METALLOGRAPHISCH
8203	**METALLOGRAPHY**	METALLOGRAPHIE	METALLOGRAPHIE
8204	**METALLOGRAPIC ETCHING**	ATTAQUE METALLOGRAPHIQUE	METALLOGRAPHISCHE ÄTZUNG
8205	**METALLOID**	METALLOIDE	METALLOID
8206	**METALLURGICAL**	METALLURGIQUE	HÜTTENMÄNNISCH, METALLURGISCH
8207	**METALLURGICAL FURNACE**	FOUR METALLURGIQUE	OFEN (METALLURGISCHER)
8208	**METALLURGY**	METALLURGIE	HÜTTENWESEN, METALLURGIE
8209	**METALLURGY OF IRON**	METALLURGIE DU FER, SIDERURGIE	EISENHÜTTENWESEN
8210	**METAPHOSPHORIC ACID**	ACIDE METAPHOSPHORIQUE	METAPHOSPHORSÄURE
8211	**METASILICIC ACID**	ACIDE METASILICIQUE	METAKIESELSÄURE
8212	**METASTABLE**	METASTABLE	METASTABIL
8213	**METEORIC WATER**	EAU METEORIQUE	NIEDERSCHLAGWASSER, METEORISCHES WASSER
8214	**METER**	COMPTEUR	ZÄHLER (FLÜSSIGKEITS-)
8215	**METHANE, MARSH GAS, LIGHT CARBURETTED HYDROGEN**	METHANE, FORMENE, PROTOCARBURE D'HYDROGENE, HYDRURE METHYLIQUE, GAZ DES MARAIS	METHAN, GRUBENGAS, SUMPFAGS, LEICHTES KOHLENWASSERSTOFFGAS
8216	**METHOD**	METHODE	VERFAHREN
8217	**METHOD OF GEARING, MODE OF DRIVING**	MODE D'ATTAQUE	ANTRIEBART
8218	**METHOD OF MANUFACTURE, OF PRODUCTION**	PROCEDE DE FABRICATION	HERSTELLUNGSVERFAHREN, FABRIKATIONSVERFAHREN
8219	**METHOD OF MEASUREMENT**	METHODE DE MESURE	MESSVERFAHREN
8220	**METHOD OF WORKING, MODE OF OPERATION**	METHODE, MODE OPERATOIRE, METHODE DE TRAVAIL	ARBEITSWEISE, ARBEITSVERFAHREN, ARBEITSMETHODE
8221	**METHYL CHLORIDE**	CHLORURE DE METHYLE	METHYLCHLORID, CHLORMETHYL
8222	**METHYL ETHER**	ETHER OXYDE METHYLIQUE, OXYDE DE METHYLE	METHYLÄTHER, HOLZÄTHER, METHYLOXYD, METHYLENHYDRAT
8223	**METHYL ORANGE, HELIANTHINE, TROPAEOLINE D**	HELIANTHINE, METHYLORANGE	METHYLORANGE, HELIANTHIN
8224	**METHYLATED SPIRIT, DENATURED ALCOHOL**	ALCOOL DENATURE	ALKOHOL (VERGÄLLTER), WEINGEIST, DENATURIERTER SPIRITUS
8225	**METRE**	METRE (UNITE DE MESURE)	METER

8226	**METRE**	METRE, METRE DROIT, METRE RIGIDE	METERMASS, METERMASSSTAB
8227	**METRIC MEASURE**	MESURE METRIQUE	MASS (METRISCHES)
8228	**METRIC MICROMETER**	MICROMETRE METRIQUE	MIKROMETER (METRISCHER)
8229	**METRIC SYSTEM**	SYSTEME METRIQUE	SYSTEM (METRISCHES)
8230	**METRIC TAPER PLUG GAUGE**	JAUGE-TAMPON AU CONE METRIQUE	KEGELLEHRHÜLSE (METRISCHE)
8231	**MICA**	MICA	GLIMMER
8232	**MICA-SCHIST**	MICASCHISTE	GLIMMESCHIEFER
8233	**MICANITE**	MICANITE	MIKANIT
8234	**MICRO FISSURES**	MICROFISSURES	MIKRORISSE
8235	**MICRO SEGREGATION**	SEGREGATION MINEURE	MIKROSEIGERUNG
8236	**MICRO-SET ADJUSTABLE CENTRE**	CONTREPOINTE MICROMETRIQUE	KÖRNERSPITZE (MIKROMETRISCHE EINSTELLBARE)
8237	**MICROAMMETER**	MICROAMPEREMETRE	MIKROAMPEREMETER
8238	**MICROAMPERE**	MICROAMPERE	MIKROAMPERE
8239	**MICROBEAM**	FOYER FIN (TUBE DE RAYONS	
8240	**MICROBORE BORING BAR**	BARRE D'ALESAGE MICROMETRIQUE	MIKROBOHRWERKZEUG
8241	**MICROCOSMICSALT, SODIUM AMMONIUM PHOSPHATE**	SEL DE PHOSPHORE, PHOSPHATE DE SODIUM ET D'AMMONIUM, SEL MICROCOSMIQUE	NATRIUMAMMONIUMPHOSPHAT, PHOSPHORSALZ
8242	**MICROFARAD**	MICROFARAD	MIKROFARAD
8243	**MICROGRAPHIC**	MICROGRAPHIQUE	MIKROGRAPHISCH
8244	**MICROGRAPHY**	MICROGRAPHIE	MIKROAUFNAHME
8245	**MICROHARDNESS**	MICRODURETE	MIKROHÄRTE
8246	**MICROHARDNESS TESTING**	ESSAI DE MICRODURETE	MIKROHÄRTEPRÜFUNG
8247	**MICROHM**	MICROHM	MIKROHM
8248	**MICROINCH**	MICRO POUCE	MILLIONSTEL ZOLL
8249	**MICROMETER**	PALMER, MICROMETRE	BÜGELMESSSCHRAUBE, MILROMETER
8250	**MICROMETER CALIPER STAND**	PORTE-MICROMETRE	MIKROMETERHALTER
8251	**MICROMETER CALIPERS**	CALIBRES MICROMETRIQUES	LEHRE (MIKROMETRISCHE)
8252	**MICROMETER DEPTH GAUGE**	MICROMETRE DE PROFONDEUR	TIEFENMIKROMETER
8253	**MICROMETER DIAL**	ECHELLE CIRCULAIRE MICROMETRIQUE	MIKROMETERRUNDSKALA
8254	**MICROMETER EYEPIECE**	OCULAIRE-MICROMETRE	OKULARMIKROMETER
8255	**MICROMETER SCREW**	VIS MICROMETRIQUE	MIKROMETERSCHRAUBE
8256	**MICROMETRIC GAUGE**	JAUGE MICROMETRIQUE	MILLIMETERTASTER
8257	**MICRON**	MILLIEME DE MILLIMETRE, MICRON	MIKROMILLIMETER
8258	**MICROPHOTOGRAPH**	MICROGRAPHIE	MIKROPHOTOGRAMM
8259	**MICROSCOPE**	MICROSCOPE	MIKROSKOP
8260	**MICROSCOPIC EXAMINATION**	EXAMEN MICROSCOPIQUE	UNTERSUCHUNG (MIKROSKOPISCHE)
8261	**MICROSECTION**	COUPE MICROGRAPHIQUE	MIKROSCHLIFF
8262	**MICROSTRUCTURE**	MICROSTRUCTURE	FEINGEFÜGE, MIKROGEFÜGE
8263	**MICROTOME**	MICROTOME	MIKROTOM
8264	**MICROVOLT**	MICROVOLT	MIKROVOLT
8265	**MID POSITION**	POSITION INTERMEDIAIRE, DE MILIEU, MI-POSITION	MITTELLAGE, MITTELSTELLUNG
8266	**MIDDLE CUT FILE**	LIME A TAILLE MOYENNE	MITTELHIEBFEILE
8267	**MIDDLINGS**	FRAGMENTS MENUS	GRUBENKLEIN (MITTELGROSSES)
8268	**MIGRATION**	MIGRATION	WANDERUNG
8269	**MILD ABRASIVE**	ABRASIF DOUX	SCHLEIFMITTEL (MILDES)
8270	**MILD CARBON STEEL**	ACIER DOUX, ACIER A BAS CARBONE	STAHL (KOHLENSTOFFARMER), STAHL (WEICHER)

8271	MILDEW	MOISISSURE	SCHIMMEL
8272	MILE	MILLE (ANGLAIS)	MEILE (ENGLISCHE)
8273	MILK GLASS	VERRE OPAQUE	MILCHGLAS
8274	MILK OF LIME	LAIT DE CHAUX	KALKMILCH
8275	MILKING MACHINES	MACHINES A TRAIRE	MOLKMASCHINEN
8276	MILL (TO)	FRAISER	FRÄSEN
8277	MILL A SCREW HEAD (TO)	MOLETTER UNE TETE DE VIS	SCHRAUBENKOPF RÄNDELN (EINEN), SCHRAUBENKOPF KORDIEREN (EINEN)
8278	MILL FINISHING	FINISSAGE AU LAMINOIR	WALZFERTIGUNG
8279	MILL SCALE	PAILLE, BATITURE	WALZZUNDER
8280	MILLED NUT	ECROU MOLETE	RÄNDELMUTTER, MUTTER (GERÄNDELTE)
8281	MILLED SCREW HEAD	TETE MOLETEE D'UNE VIS	SCHRAUBENKOPF (GERÄNDELTER)
8282	MILLER INDICES	INDICES DE MILLER	INDIZES (MILLERSCHE)
8283	MILLI AMMETER	MILLIAMPERE/METRE	MILLIAMPEREMETER
8284	MILLIAMPERE	MILLIAMPERE	MILLIAMPERE
8285	MILLIGRAMME	MILLIGRAMME	MILLIGRAMM
8286	MILLIMETRE	MILLIMETRE	MILLIMETER
8287	MILLING	BROYAGE, FRAISAGE, MOLETAGE	MAHLEN, FRÄSEN, RÄNDELN
8288	MILLING A SCREW HEAD	MOLETAGE D'UNE TETE DE VIS	RÄNDELN EINES SCHRAUBENKOPFES, KORDIEREN EINES SCHRAUBENKOPFES
8289	MILLING CUTTER	FRAISE	FRÄSER, FRÄSE
8290	MILLING DOWELS, HOLDING TABLES	DES DE FRAISAGE, TABLES DE FIXATION	FRÄSSTIFTE, SPANNTISCHE
8291	MILLING MACHINE	FRAISEUSE, MACHINE A FRAISER	FRÄSMASCHINE
8292	MILLING MACHINIST	FRAISEUR	FRÄSER (ARBEITER)
8293	MILLING SHOP	ATELIER DE FRAISAGE	FRÄSEREI
8294	MILLIVOLT	MILLIVOLT	MILLIVOLT
8295	MILLS/COMBINE	MOULINS/COMBINE	MÜHLEN/(KOMBINIERT)
8296	MILLS/CRUSHING AND GRINDING	MOULINS/APLATISSEURS	QUETSCH-UND SCHROTMÜHLEN
8297	MILLS/HAMMER	MOULINS/BROYEUR	HAMMERMÜHLEN
8298	MILLSTONE	MEULE DE MOULIN	MÜHLSTEIN, MAHLSTEIN
8299	MILLSTONE GRIT	PIERRE MEULIERE, MEULIERE	MÜHLSTEIN (GEOL.)
8300	MINE	MINE	BERGWERK, GRUBE, ZECHE
8301	MINERAL	MINERAL	MINERAL
8302	MINERAL ACID	ACIDE MINERAL	MINERALSÄURE
8303	MINERAL DEPOSIT	GISEMENT MINERAL, GITE MINIER	LAGERSTÄTTE EINES MINERALS
8304	MINERAL OIL	HUILE MINERALE	MINERALÖL
8305	MINETTE	MINETTE	MINETTE
8306	MINIMUM TEMPERATURE	TEMPERATURE MINIMUM	TEMPERATUR (TIEFSTE), MINIMALTEMPERATUR
8307	MINIMUM VALUE	VALEUR MINIMUM	MINDESTWERT, KLEINSTWERT, MINIMALWERT
8308	MINIMUM VALUE OF UPPER YIELD POINT	LIMITE ELASTIQUE MINIMALE GARANTIE	MINDESTSTRECKGRENZE
8309	MINIMUM WAVELENGTH	LONGUEUR D'ONDE MINIMALE	GRENZWELLENLÄNGE
8310	MINIMUM WEIGHT	POIDS MINIMUM	MINDESTGEWICHT, MINIMALGEWICHT
8311	MINING	EXPLOITATION MINIERE, INDUSTRIE MINIERE	BERGBAU
8312	MINING ENGINEER	INGENIEUR DES MINES	BERGINGENIEUR
8313	MINOR AXIS OF ELLIPSE	PETIT AXE D'UNE ELLIPSE	ACHSE (KLEINE) EINER ELLIPSE
8314	MINOR DIAMETER	DIAMETRE INTERIEUR	KERNDURCHMESSER

8315	MINOR OF A DETERMINANT	DETERMINANT MINEUR	UNTERDETERMINANTE
8316	MINUEND	MINUENDE	MINUEND
8317	MINUS MESH	GROSSEUR DE GRAIN MINIMALE	MINUSKORNGRÖSSE
8318	MINUS-PRESSURE, VACUUM STEAM HEATING	CHAUFFAGE PAR LA VAPEUR A UNE PRESSION INFERIEURE A LA PRESSION ATMOSPHERIQUE	VAKUUMHEIZUNG
8319	MISCELLANEOUS FONCTION	FONCTION AUXILIAIRE	HILFSFUNKTION
8320	MISCH METAL	MISCHMETAL	MISCHMETALL
8321	MISRUN	MAL VENUE, REBUT	SCHLECHTAUSGELAUFEN
8322	MISTRIMMED FORGING	PIECE FORGEE A EBAVURAGE EXCESSIF	SCHMIEDESTÜCK (ÜBERENTGRATETES)
8323	MITRE GEARS	ENGRENAGES CONCOURANTS	KEGELRADPAAR MIT ÜBERSETZUNGS-VERHÄLTNIS 1:1
8324	MITRE JOINT	ASSEMBLAGE A ONGLET	GEHRFUGE, STOSS AUF GEHRUNG, GEHRSTOSS
8325	MITRE SQUARE	EQUERRE A ONGLET	GEHRMASS, ACHTELWINKELMASS, GEHRDREIECK
8326	MITRE, MITER	ONGLET	GEHRUNG
8327	MIX	MELANGE	GEMISCH, MISCHUNG
8328	MIX (TO)	MELANGER, AGITER	MISCHEN, RÜHREN
8329	MIXED CRYSTALS	CRISTAUX MIXTES	MISCHKRISTALLE
8330	MIXED FLOW TURBINE	TURBINE MIXTE, TURBINE AMERICAINE	TURBINE (GEMISCHTE)
8331	MIXER -MIXING DRUM	MELANGEUR	MISCHER, MISCHTROMMEL
8332	MIXERS, FOOD AND CONCRETE	MELANGEURS	NÄHRMITTEL- UND BETONMISCHMASCHINEN
8333	MIXING LADLE	POCHE MELANGEUSE	MISCHERPFANNE
8334	MIXING MACHINE	MELANGEUR, MALAXEUR	MISCHMASCHINE
8335	MIXING VALVE	ROBINET MELANGEUR	MISCHVENTIL
8336	MIXING, MIXTURE	MELANGE (ACTION), MELANGE (CHIMIQUE)	PULVERMISCHUNG, MISCHEN, MISCHUNG
8337	MIXTURE (AIR AND FUEL)	MELANGE CARBURE	KRAFTSTOFFLUFTGEMISCH
8338	MODEM, MODULATOR, DEMODULATOR	MODEM, MODULATEUR, DEMODULATEUR	MODEM, MODULATOR, DEMODULATOR
8339	MODULUS	MODULE	MODUL
8340	MODULUS COEFFICIENT OF ELASTICITY, ELASTIC MODULUS, YOUNG'S MODULUS, STRETCH MODULUS	COEFFICIENT DE RESISTANCE A L'ALLONGEMENT, MESURE DE L'ELASTICITE	ELASTIZITÄTSMODUL, ELASTIZITÄTSMASS
8341	MODULUS OF ELASTICITY	MODULE D'ELASTICITE	ELASTIZITÄTSMODUL, SCHWUNGSZAHL
8342	MODULUS OF RIGIDITY	MODULE DE CISAILLEMENT, MODULE DE GLISSEMENT	GLEITMODUL, SCHUBELASTIZITÄTSMODUL
8343	MODULUS OF RUPTURE	MODULE DE RUPTURE	BRUCHMODUL
8344	MOE DRAINERS, DITCH CUTTERS	SOUS-SOLEUSES, CHARRUES RIGOLEURSE	GRABENPFLÜGE, MASCHINEN FÜR MAULWURFS DRÄNUNG
8345	MOHR'S CLIP	PINCE DE MOHR	QUETSCHHAHN
8346	MOHS SCALE	ECHELLE DE DURETE DE MOHS	HÄRTESKALA (MOHSCHE)
8347	MOISTEN (TO)	HUMECTER, MOUILLER	ANFEUCHTEN
8348	MOISTENING, DAMPING	HUMIDIFICATION, MOUILLAGE	ANFEUCHTEN
8349	MOISTURE-PROOF	IMPERMEABLE A L'HUMIDITE, HYDROFUGE	WASSERABWEISEND
8350	MOLAR CONDUCTIVITY	CONDUCTANCE MOL(ECUL)AIRE	LEITFÄHIGKEITSMOLARE
8351	MOLAR RESISTIVITY	RESISTANCE MOL(ECUL)AIRE	WIDERSTANDSMOLARE
8352	MOLD (U.S) MOULD (GB)	MOULE	FORM

8353	MOLD CAVITY	CAVITE DU MOULE	FORMENHOHLRAUM
8354	MOLD JACKET (U.S), SLIP JACKET	JAQUETTE	GIESSRAHMEN, SCHUTZRAHMEN
8355	MOLD WASH	ENDUIT POUR LINGOTIERES	FORMSCHLICHTE
8356	MOLD WEIGHT	POIDS DE CHARGE	BELASTUNGSGEWICHT, BESCHWEREISEN
8357	MOLDERS'RULE	SONDE DE MOULEUR	TIEFENMASS
8358	MOLDING MACHINE	MACHINE A MOULER	FORMMASCHINE
8359	MOLECULAR ATTRACTION	ATTRACTION MOLECULAIRE	ANZIEHUNG (MOLEKULARE)
8360	MOLECULAR FORCE	FORCE MOLECULAIRE	MOLEKULARKRAFT
8361	MOLECULAR STRUCTURE	STRUCTURE MOLECULAIRE	MOLEKÜLSTRUKTUR
8362	MOLECULAR VOLUME	VOLUME MOLECULAIRE	MOLEKULARVOLUMEN
8363	MOLECULAR WEIGHT	POIDS MOLECULAIRE	MOLEKULARGEWICHT
8364	MOLECULE	MOLECULE	MOLEKEL, MOLEKÜL
8365	MOLTEN METAL	MATIERE EN FUSION, METAL EN FUSION	SCHMELZE, SCHMELZFUSS
8366	MOLYBDENITE, MOLYBDENITE GLANCE	MOLYBDENITE	MOLYBDÄNGLANZ
8367	MOLYBDENUM	MOLYBDENE	MOLYBDÄN
8368	MOLYBDENUM STEEL	ACIER AU MOLYBDENE	MOLYBDÄNSTAHL
8369	MOLYBDIC ACID	ACIDE MOLYBDIQUE	MOLYBDÄNSÄURE
8370	MOMENT OF A FORCE	MOMENT D'UNE FORCE	MOMENT EINER KRAFT
8371	MOMENT OF FRICTION	MOMENT DU FROTTEMENT	REIBUNGSMOMENT
8372	MOMENT OF INERTIA	MOMENT D'INERTIE	TRÄGHEITSMOMENT
8373	MOMENT OF MOMENTUM	MOMENT DE LA QUANTITE DE MOUVEMENT	SCHWUNGMOMENT
8374	MOMENT OF RESISTANCE	MOMENT RESISTANT, COUPLE RESISTANT, COUPLE REACTION	WIDERSTANDSMOMENT
8375	MOMENT OF STABILITY	MOMENT DE STABILITE	STABILITÄTSMOMENT
8376	MOMENTUM, QUANTITY OF MOTION	QUANTITE DE MOUVEMENT	BEWEGUNGSGRÖSSE
8377	MONATOMIC, MONOVALENT	MONOATOMIQUE, UNIVALENT	EINWERTIG, EINATOMIG
8378	MOND GAS	GAZ MOND	MONDGAS
8379	MONEL METAL	METAL MONEL, MONEL	MONELMETALL
8380	MONKEY COOLER	REFROIDISSEUR DU TROU DE LAITIER	SCHLACKENFORMKÜHLER, KÜHLKASTEN
8381	MONKEY WALL	PAROI DE LA BOITE DE REFROIDISSEMENT	KÜHLKASTENKOPFWAND
8382	MONOCHROMATIC EMISSIVITY	POUVOIR EMISSIF MONOCHROMATIQUE	EMISSIONSVERMÖGEN (MONOCHROMATISCHES)
8383	MONOCHROMATIC RADIATION	RAYONNEMENT MONOCHROMATIQUE	STRAHLUNG (MONOCHROMATISCHE)
8384	MONOCLINIC CRYSTAL	CRISTAL MONOCLINIQUE	KRISTALL (MONOKLINER)
8385	MONOLITHIC LINING	GARNITURE MONOLITHIQUE	OFENFUTTER (MONOLITHISCHES)
8386	MONOTECTOID REACTION	REACTION MONOTECTOIDE	MONOTEKTOIDE REAKTION
8387	MONOTRON	MONOTRON	MONOTRON
8388	MONOTROPIC	MONOTROPE	MONOTROP
8389	MONTEJUS	MONTE-JUS	SAFTHEBER, MONTEJUS
8390	MOORING	MOUILLAGE	ANKERPLATZ
8391	MORDANT	MORDANT	BEIZE
8392	MORSE TAPER PLUG GAUGE	JAUGE-TAMPON AU CONE MORSE	MORSEKEGELLEHRDORN
8393	MORSE TAPER RING GAUGE	CALIBRE DE CONICITE NORMAL	MORSEKEGELLEHRE
8394	MORTAR	MORTIER (A CONSTRUIRE)	MÖRTEL
8395	MORTAR	MORTIER (VASE EN PORCELAINE, EN AGATE)	REIBSCHALE
8396	MORTAR	MORTIER (VASE EN BRONZE)	MÖRSER
8397	MORTAR GUN CARRIAGE	AFFUT DE MORTIER	MÖRSERLAFETTE

8398	MORTISE CHISEL	BEDANE DE MENUISIER	LOCHBEITEL
8399	MORTISE GAUGE	TRUSQUIN D'ASSEMBLAGE	ZAPFENSTREICHMASS
8400	MORTISE WHEEL, COG WHEEL	ROUE A DENTS EN BOIS	KAMMZAHNRAD, HOLZEISENRAD
8401	MORTISE, MORTICE	MORTAISE	ZAPFENLOCH
8402	MORTISING MACHINE	MORTAISEUSE A BOIS	STEMMASCHINE
8403	MOSAIC STRUCTURE	STRUCTURE MOSAIQUE	MOSAIKSTRUKTUR
8404	MOSSY ZINC	GRENAILLE DE ZINC	ZINKGRIES
8405	MOTHER LIQUOR	EAU MERE, BITTERN	MUTTERLAUGE
8406	MOTION	MOUVEMENT	BEWEGUNG
8407	MOTION OF TRANSLATION	MOUVEMENT DE TRANSLATION	BEWEGUNG (FORTSCHREITENDE), TRANSLATIONSBEWEGUNG
8408	MOTIVE AGENT	AGENT MOTEUR	TREIBMITTEL, BETRIEBSSTOFF
8409	MOTIVE POWER	PUISSANCE MOTRICE, FORCE MOTRICE	TRIEBKRAFT, BETRIEBSKRAFT
8410	MOTOR BUS	AUTOBUS	STADTOMNIBUS
8411	MOTOR CAR, AUTOMOBILE, AUTOCAR	VOITURE AUTOMOBILE, AUTOMOBILE, AUTO	KRAFTWAGEN, AUTOMOBIL, AUTO
8412	MOTOR COACH	AUTOCAR	REISEOMNIBUS
8413	MOTOR GENERATOR	MOTEUR-GENERATEUR, GROUPE MOTEUR-GENERATEUR	MOTORGENERATOR
8414	MOTOR LORRY, MOTOR TRUCK, AUTOTRUCK	CAMION AUTOMOBILE, AUTOMOBILE INDUSTRIELLE	LASTKRAFTWAGEN, LASTAUTO
8415	MOTOR OIL	HUILE POUR MOTEUR	MOTORÖL
8416	MOTOR OPERATED GATE VALVE	VANNE MOTORISEE	MOTORSCHIEBER
8417	MOTOR OPERATED VALVE	ROBINET A MOTEUR	MOTORVENTIL
8418	MOTOR STARTER	DEMARREUR	ANLASSER (ELEKTR.)
8419	MOTOR TUNE-UP	MISE AU POINT	MOTOREINSTELLUNG
8420	MOTTLED IRON	FONTE TRUITEE	ROHEISEN (HALBIERTES), ROHEISEN (MELIERTES)
8421	MOTTLING	LIQUETURE	TÜPPELUNG
8422	MOULD	CALIBRE POUR PROFILS, LINGOTIERE	FORMKALIBER, GUSSFORM
8423	MOULD (FOUND.), MOULD	MOULE (FOND.), TERRE VEGETALE	GUSSFORM (GIESS.), DAMMERDE
8424	MOULD (TO)	MOULER	ABFORMEN
8425	MOULDED BRICK	BRIQUE PROFILEE	FORMSTEIN
8426	MOULDED CONCRETE	BETON MOULE	GUSSBETON
8427	MOULDER	MOULEUR, OUVRIER MOULEUR	FORMER
8428	MOULDING	MOULAGE (REPRODUCTION A L'AIDE D'UN MOULE)	FORMEN, ABFORMEN
8429	MOULDING MACHINE	MACHINE A MOULER, MACHINE A MOULURER	FORMMASCHINE, SANDFORMMASCHINE, KEHLMASCHINE
8430	MOULDING SAND	SABLE DE MOULAGE, SABLE DE FONDERIE	FORMSAND
8431	MOUNT A TRACING (TO)	MONTER UN CALQUE	PAUSE AUFZIEHEN (EINE)
8432	MOUNTED ON A POINTED PIVOT	REPOSANT, SUSPENDU SUR UN PIVOT	GELAGERT (AUF EINER SPITZE)
8433	MOUTH BLOW PIPE	CHALUMEAU A BOUCHE	BLASLÖTROHR, LÖTROHR MIT MUNDSTÜCK
8434	MOUTH OF A PLANE	LUMIERE D'UN RABOT	KEILLOCH EINES HOBELS
8435	MOVABLE BLOCK	POULIE MOBILE (D'UN PALAN)	FLASCHE (LOSE), ROLLE
8436	MOVABLE PLATEN	CHARIOT D'ECRASEMENT	STAUCHSCHLITTEN
8437	MOVABLE VICE JAW	MACHOIRE MOBILE D'UN ETAU	SCHRAUBSTOCKBACKEN (BEWEGLICHER)
8438	MOVEMENT, MOTION	MOUVEMENT	BEWEGUNG

8439	MOWERS : TRACTOR TRAILED, MOUNTED AND SEMI-MOUNTED	FAUCHEUSES A TRACTEUR-TRAINEES/MONTEES	MÄHMASCHINEN (DURCH TRAKTOREN ANGETRIEBEN) UND HALBSATTELMÄHMASCHINEN
8440	MUCILAGE	MUCILAGE, EXTRAIT MUCILAGINEUX	AUFQUELLUNG
8441	MUCK BAR	BARRE DE FER BRUT, FER EBAUCHE	ROHSCHIENE
8442	MUD HOLE	TROU DE VIDANGE	SCHLAMMLOCH
8443	MUD-FLAP	PARE-BOUE	SCHMUTZFÄNGER
8444	MUD, SLUDGE	BOUE, BOUES	SCHLAMM
8445	MUDDY	BOUEUX	SCHLAMMIG
8446	MUDDY WATER	EAU BOUEUSE	WASSER (SCHLAMMIGES)
8447	MUFFLE, CHAMBER FURNACE	FOUR A MOUFLE	MUFFELOFEN
8448	MUFFLER (U.S)	SILENCIEUX, POT D'ECHAPPEMENT	SCHALLDÄMPFER, AUSPUFFTOPF, AUSPUFFGERÄUSCHDÄMPFER
8449	MULLER	MALAXEUR	MISCHKOLLERGANG
8450	MULLING (U.S.)	FROTTAGE	KOLLERN
8451	MULLING MACHINE	FROTTEUR	MISCHKOLLERGANG
8452	MULTI HOLE TYPE NOZZLE	INJECTEUR A TROUS	MEHRLOCHDÜSE
8453	MULTI-CYLINDER ENGINE	MOTEUR POLYCYLINDRIQUE	MEHRZYLINDERMASCHINE
8454	MULTI-DISC BRAKE	FREIN A LAME, FREIN A DISQUE	LAMELLENBREMSE
8455	MULTI-PASS WELDING	SOUDAGE A PASSES MULTIPLES	MEHRLAGENSCHWEISSEN
8456	MULTI-PURPOSE MACHINE	MACHINE MULTIPLE	MEHRZWECKMACHINE
8457	MULTI-ROLLER PROFILING MACHINE	MACHINE A PROFILER A GALETS MULTIPLES	MEHRROLLENPROFILIERMASCHINE
8458	MULTI-SPINDLE DRILLING HEAD	TETE DE PERCAGE MULTIBROCHES	BOHRKÖPFE (MEHRSPINDLIGE)
8459	MULTI-SPINDLE HEAD	TETE MULTI-BROCHES	MEHRSPINDELBOHRKOPF
8460	MULTI-SPINDLE IN LINE DRILLING MACHINE	PERCEUSE MULTIBROCHE EN LIGNE	MEHRSPINDLIGE BOHRMASCHINE IN REIHENANORDNUNG
8461	MULTI-STAGE (TRANSFER) PRESSES	PRESSES TRANSFERT A POINCONS MULTIPLES	STUFENPRESSEN (MECHANISCHE)
8462	MULTIPHASE, POLYPHASE CURRENT	COURANT POLYPHASE	MEHRPHASENSTROM
8463	MULTIPLE BELT	COURROIE MULTIPLE, COURROIE A PLUSIEURS EPAISSEURS	MEHRFACHRIEMEN
8464	MULTIPLE ELECTRODE SPOT WELDING	SOUDAGE PAR POINTS MULTIPLES	VIELPUNKTSCHWEISSEN
8465	MULTIPLE INTEGRAL	INTEGRALE MULTIPLE	INTEGRAL (MEHRFACHES), INTEGRAL (VIELFACHES)
8466	MULTIPLE MOLD	MOULE MULTIPLE	MEHRFACHFORM
8467	MULTIPLE PORT PLUG	ROBINET DE DISTRIBUTEUR	MEHRWEGEVENTIL
8468	MULTIPLE PROJECTION WELDING	SOUDAGE PAR BOSSAGES MULTIPLES	MEHRFACHBUCKELSCHWEISSEN
8469	MULTIPLE SPINDLE AUTOMATIC LATHE	TOUR AUTOMATIQUE A BROCHES MULTIPLES	MEHRSPINDELAUTOMAT
8470	MULTIPLE SPINDLE DRILLING MACHINE	MACHINE A PERCER A BROCHES MULTIPLES	BOHRMASCHINE (MEHRSPINDLIGE)
8471	MULTIPLE SPINDLE MILLING MACHINE	FRAISEUSE MULTIPLE	FRÄSMASCHINE MIT MEHREREN SPINDELN
8472	MULTIPLE SYSTEM	SYSTEME MULTIPLE	MEHRFACHSYSTEM
8473	MULTIPLE SYSTEM OF ROPE DRIVING	TRANSMISSION PAR POULIE A GORGES MULTIPLES	SEILTRIEB (MEHRFACHER)
8474	MULTIPLE THREADED SCREW	VIS A PLUSIEURS FILETS	SCHRAUBE (MEHRGÄNGIGE)
8475	MULTIPLE VALVE (COAXIAL)	SOUPAPE ETAGEE	VENTIL (MEHRSTÖCKIGES), STUFENVENTIL, ETAGENVENTIL
8476	MULTIPLE VALVE (COPLANAR)	SOUPAPE MULTIPLE	VENTIL (MEHRFACHES), GRUPPENVENTIL

8477	**MULTIPLE-EXPANSION ENGINE**	MACHINE A MULTIPLE EXPANSION	MEHRFACHEXPANSIONSMASCHINE
8478	**MULTIPLE-THROW CRANK SHAFT**	VILEBREQUIN A PLUSIEURS COUDES	WELLE (MEHRKURBELIGE), KURBELWELLE (MEHRFACH GEKRÖPFTE)
8479	**MULTIPLICAND**	MULTIPLICANDE	MULTIPLIKAND
8480	**MULTIPLICATION**	MULTIPLICATION	VERVIELFACHEN, MULTIPLIKATION
8481	**MULTIPLICATION, REPRODUCTION**	REPRODUCTION	VERVIELFÄLTIGEN, REPRODUZIEREN
8482	**MULTIPLIER**	MULTIPLICATEUR	MULTIPLIKATOR
8483	**MULTIPLY (TO)**	MULTIPLIER	VERVIELFACHEN, MULTIPLIZIEREN
8484	**MULTIPLY (TO), REPRODUCE (TO)**	REPRODUIRE	VERVIELFÄLTIGEN, REPRODUZIEREN
8485	**MULTISPINDLE FINISHING LATHE**	TOUR DE REPRISE MULTIBROCHES	NACHDREHMASCHINE (MEHRSPINDLIGE)
8486	**MUNTZ METAL, YELLOW PATENT METAL**	METAL MUNTZ	MUNTZMETALL, SCHMIEDEMESSING
8487	**MUSCHELKALK**	CALCAIRE COQUILLIER, MUSCHELKALK	MUSCHELKALK
8488	**MUSHY STAGE**	ETAT PATEUX	TEIGZUSTAND
8489	**MUTUAL INDUCTION**	INDUCTION MUTUELLE	INDUKTION (GEGENSEITIGE), INDUKTION (WECHSELSEITIGE)
8490	**NAGELFLUH**	POUDINGUE	NAGELFLUH
8491	**NAIL**	CLOU	NAGEL
8492	**NAIL (TO)**	CLOUER	NAGELN
8493	**NAPHTHA**	NAPHTE	NAPHTHA
8494	**NAPHTHALENE**	NAPHTALINE, NAPHTALENE	NAPHTHALIN, STEINKOHLENTEERKAMPFER, NAPHTHYLWASSERSTOFF
8495	**NAPHTHOL**	NAPHTOL	NAPHTHOL (A-NAPHTHOL, SS-NAPHTHOL)
8496	**NARROW EDGE MILLING CUTTER**	FRAISE A DISQUE	SCHEIBENFRÄSER
8497	**NASCENT**	NAISSANT, LIBERANT (SE) (CHIM.)	FREIWERDEND (CHEM.)
8498	**NATIVE ALLOY**	ALLIAGE NATUREL	LEGIERUNG (NATÜRLICHE)
8499	**NATURAL ABRASIVE**	ABRASIF NATUREL	SCHLEIFMITTEL (NATÜRLICHES)
8500	**NATURAL AGING**	VIEILLISSEMENT NATUREL	ALTERUNG (NATÜRLICHE)
8501	**NATURAL DRAUGHT**	TIRAGE NATUREL	ZUG (NATÜRLICHER)
8502	**NATURAL EVAPORATION**	EVAPORATION A L'AIR LIBRE	VERDUNSTUNG
8503	**NATURAL FUEL**	COMBUSTIBLE NATUREL	BRENNSTOFF (NATÜRIICHER)
8504	**NATURAL GAS**	GAZ NATUREL	NATURGAS, ERDGAS
8505	**NATURAL LIGHTING**	ECLAIRAGE NATUREL	BELEUCHTUNG (NATÜRLICHE)
8506	**NATURAL MAGNET**	AIMANT NATUREL, PIERRE D'AIMANT	MAGNET (NATÜRLICHER)
8507	**NATURAL STONE**	PIERRE NATURELLE	STEIN (NATÜRLICHER), NATURSTEIN
8508	**NATURAL VENTILATION**	VENTILATION NATURELLE	LÜFTUNG (NATÜRLICHE)
8509	**NATURAL, HYPERBOLIC LOGARITHM**	LOGARITHME NATUREL, LOGARITHME HYPERBOLIQUE	LOGARITHMUS (NATÜRLICHER), LOGARITHMUS (HYPERBOLISCHER)
8510	**NAVAL ARCHITECTURE**	CONSTRUCTION NAVALE	SCHIFFBAU
8511	**NAVAL BRASS**	LAITON ALPHA-BETA, LAITON DE L'AMIRAUTE	SONDERMESSING (SEEWASSERFESTES)
8512	**NECK**	CONGE, COL	KEHLE, HALS
8513	**NECK JOURNAL**	TOURILLON INTERMEDIAIRE	HALSZAPFEN, WELLENHALS
8514	**NECK NOZZLE**	TUBULURE A COLLET	HALSSTUTZEN
8515	**NECKING**	EXECUTION D'UNE GORGE, STRICTION	STECHEN, EINSCHNÜRUNG
8516	**NECKING DOWN**	STRICTION	EINHALSUNG
8517	**NEEDLE**	PARTICULE ACICULAIRE	TEILCHEN (NADELFÖRMIGES)
8518	**NEEDLE BEARING**	ROULEMENT A AIGUILLES	NADELLAGER

8519	NEEDLE FILE	LIME AIGUILLE	NADELFEILE
8520	NEEDLE LUBRICATOR	GRAISSEUR A TIGE, GRAISSEUR A EPINGLETTE, GODET EN VERRE A TIGE	STIFTÖLER, NADELÖLER
8521	NEEDLE POINT	BRANCHE PORTE-AIGUILLE D'UN COMPAS	NADELEINSATZ FÜR ZIRKEL
8522	NEEDLE VALVE	ROBINET A POINTEAU (A AIGUILLE), POINTEAU	NADELVENTIL, SCHWIMMERNADELVENTIL
8523	NEEDLED STEEL	ACIER A STRUCTURE ACICULAIRE	STAHL MIT NADELIGER STRUKTUR
8524	NEGATIVE (PHOTOGRAPHIC)	EPREUVE NEGATIVE, CLICHE PHOTOGRAPHIQUE	NEGATIV (PHOTOGRAPHISCHES)
8525	NEGATIVE ACCELERATION, RETARDATION	ACCELERATION NEGATIVE	BESCHLEUNIGUNG (NEGATIVE), VERZÖGERUNG
8526	NEGATIVE ELECTRICITY	ELECTRICITE NEGATIVE	ELEKTRIZITÄT (NEGATIVE)
8527	NEGATIVE HARDENING	TREMPE NEGATIVE	NEGATIVE HÄRTUNG
8528	NEGATIVE HEAT OF FORMATION	CHALEUR DE DECOMPOSITION	TRENNUNGSWÄRME
8529	NEGATIVE MATRIX	MATRICE NEGATIVE	NEGATIVE
8530	NEGATIVE PLATE	PLAQUE NEGATIVE	PLATTE (NEGATIVE), MINUSPLATTE
8531	NEGATIVE POLE	POLE NEGATIF	POL (NEGATIVER)
8532	NEGATIVE SIGN	SIGNE NEGATIF	VORZEICHEN (NEGATIVES), MINUSZEICHEN
8533	NEODYMIUM	NEODYME	NEODYM
8534	NEODYMIUM	NEODYMIUM	NEODYM
8535	NEON	NEON	NEON
8536	NEON LAMP	LAMPE AU NEON	NEONLAMPE
8537	NEPHELOMETRY	ANALYSE NEPHELOMETRIQUE	TRÜBUNGSANALYSE
8538	NEST OF TUBES	FAISCEAU TUBULAIRE	ROHRBÜNDEL, RÖHRENBÜNDEL
8539	NETWORK STRUCTURE	STRUCTURE RETICULAIRE	NETZSTRUKTUR, NETZGEFÜGE
8540	NEUMANN BANDS	BANDES DE NEUMANN	LINIEN (NEUMANNSCHE)
8541	NEUTRAL AXIS LINE	AXE NEUTRE	NULLACHSE, ACHSE (NEUTRALE), NULLINIE
8542	NEUTRAL FIBRE	FIBRE NEUTRE	FASER (NEUTRALE)
8543	NEUTRAL FLAME	FLAMME NEUTRE, FLAMME NORMALE	NORMALFLAMME
8544	NEUTRAL LAYER OF FIBRES	COUCHE DES FIBRES INVARIABLES	FASERSCHICHT (NEUTRALE), NULLSCHICHT
8545	NEUTRAL LINE	LIGNE NEUTRE	NULLEITUNG
8546	NEUTRAL NORMAL SALT	SEL NEUTRE	SALZ (NEUTRALES)
8547	NEUTRAL POINT	ZONE NEUTRE	FLIESSSCHEIDE
8548	NEUTRALISATION	NEUTRALISATION	NEUTRALISATION, ABSTUMPFUNG
8549	NEUTRALISE (TO)	NEUTRALISER	NEUTRALISIEREN, ABSTUMPFEN
8550	NEW ZEALAND FLAX	LIN DE LA NOUVELLE-ZELANDE	NEUSEELANDHANF
8551	NIB	COMPACT TERMINE	FERTIGER PRESSLING
8552	NIBBLING	GRIGNOTAGE	DEKUPIEREN, KNABBERN
8553	NIBBLING MACHINE	GRIGNOTEUSE	AUSHAUMASCHINE
8554	NICK (TO), NOTCH (TO)	ENTAILLER, PRATIQUER UNE SAIGNEE DANS UNE PIECE, SAIGNER	KERBEN, EINKERBEN
8555	NICK-BREAK TEST	ESSAI DE FLEXION SUR EPROUVETTE ENTAILLEE	KERBBIEGEVERSUCH
8556	NICKEL	NICKEL	NICKEL
8557	NICKEL ALLOY	ALLIAGE A BASE DE NICKEL	NICKELLEGIERUNG
8558	NICKEL AMMONIUM SULPHATE	SULFATE DOUBLE DE NICKEL ET D'AMMONIAQUE	SCHWEFELSAURES NICKELOXYDULAMMONIAK, NICKELAMMONIUMSULFAT

8559	**NICKEL BRASS**	LAITON AU NICKEL	NICKELMESSING
8560	**NICKEL BRONZE**	BRONZE AU NICKEL	NICKELBRONZE
8561	**NICKEL CARBONYL**	CARBONYLE DE NICKEL	NICKELKOHLENOXYD, KOHLENOXYDNICKEL, NICKELKARBONYL
8562	**NICKEL CHLORIDE**	CHLORURE NICKELEUX	NICKELCHLORÜR, CHLORNICKEL
8563	**NICKEL MATTE**	MATTE DE NICKEL	NICKELSTEIN
8564	**NICKEL ORE**	MINERAI DE NICKEL	NICKELERZ
8565	**NICKEL OXIDE**	PROTOXYDE DE NICKEL	NICKELOXYDUL
8566	**NICKEL PLATE (TO)**	NICKELER	VERNICKELN
8567	**NICKEL PLATING**	NICKELAGE, PLAQUAGE AU NICKEL	VERNICKELUNG, NICKELPLATTIERUNG
8568	**NICKEL SHOT**	GRENAILLE DE NICKEL	NICKELSCHROT
8569	**NICKEL SILVER**	ARGENTAN, ALPACCA, MAILLECHORT	NEUSILBER, ALPAKA, ARGENTAN
8570	**NICKEL STEEL**	ACIER AU NICKEL	NICKELSTAHL
8571	**NICKEL SULPHATE**	SULFATE DE NICKEL	SCHWEFELSAURES NICKELOXYDUL, NICKELSULFAT, NICKELVITRIOL
8572	**NICKEL-ALUMINIUM BRONZE**	BRONZE DE NICKEL ET ALUMINIUM	NICKELALUMINIUMBRONZE
8573	**NICKEL-CLAD STEEL**	ACIER NICKELE	STAHL (NICKELPLATTIERTER)
8574	**NICKELIC OXIDE**	OXYDE DE NICKEL	NICKELOXYD
8575	**NIGHT SHIFT**	TRAVAIL DE NUIT, TACHE DE NUIT	NACHTSCHICHT
8576	**NIOBIUM, COLUMBIUM**	NIOBIUM	NIOBIUM
8577	**NIP ANGLE**	ANGLE D'ATTAQUE	GREIFWINKEL
8578	**NIPPLE**	MAMELON, RACCORD MALE	NIPPEL
8579	**NITRATE**	NITRATE, AZOTATE	SALPETERSAURES SALZ, NITRAT
8580	**NITRIC ACID, ACQUA FORTIS**	ACIDE AZOTIQUE, ACIDE NITRIQUE, EAU FORTE	SALPETERSÄURE, SCHEIDEWASSER, AQUA FORTIS, STICKSTOFFSÄURE
8581	**NITRIC OXIDE**	BIOXYDE D'AZOTE, OXYDE AZOTIQUE	STICKSTOFF MONOXYD, STICKOXYD, SALPETERGAS
8582	**NITRIDING**	TREMPE PAR NITRURATION	NITRIERHÄRTUNG
8583	**NITRIDING ATMOSPHERE**	ATMOSPHERE DE NITRURATION	NITRIERATMOSPHÄRE
8584	**NITRIDING FURNACE**	FOUR A NITRURATION	NITRIEROFEN
8585	**NITRIDING STEEL**	ACIER DE NITRURATION	NITRIERSTAHL
8586	**NITRITE**	NITRITE, AZOTITE	SALPETRIGSAURES SALZ, NITRIT
8587	**NITROBENZENE, ESSENCE OF MIRBANE**	NITROBENZINE, NITROBENZENE, ESSENCE DE MIRBANE	MIRBANÖL, NITROBENZOL
8588	**NITROGEN**	AZOTE, NITROGENE	STICKSTOFF
8589	**NITROGEN HARDENING**	NITRURATION	NITRIERUNG
8590	**NITROGEN PENTOXIDE**	ACIDE AZOTIQUE ANHYDRE, ANHYDRIDE AZOTIQUE	STICKSTOFFPENTOXYD, SALPETERSSÄUREANHYDRID
8591	**NITROGEN TETROXIDE, PEROXIDE**	AZOTYLE, HYPOAZOTIDE, ANHYDRIDE, HYPOAZOTIQUE, PEROXYDE D'AZOTE	STICKSTOFFPEROXYD, STICKSTOFFTETROXYD, STICKSTOFFDIOXYD, UNTERSALPETERSÄURE
8592	**NITROGEN TRIOXIDE**	ANHYDRIDE AZOTEUX	STICKSTOFFTRIOXYD, STICKSTOFFSESQUIOXYD, SALPETRIGSÄUREANHYDRID
8593	**NITROGLYCERINE**	NITROGLYCERINE, TRINITRINE	NITROGLYZERIN
8594	**NITROMETER**	NITROMETRE	NITROMETER
8595	**NITROSULPHURIC ACID**	MELANGE SULFONITRIQUE	MISCHSÄURE, NITRIERSÄURE
8596	**NITROUS ACID**	ACIDE AZOTEUX	SÄURE (SALPETRIGE)
8597	**NITROUS OXYDE**	PROTOXYDE D'AZOTE, OXYDE AZOTEUX	STICKSTOFFOXYDUL
8598	**NO-LOAD CONDITIONS, RUNNING LIGHT**	MARCHE A VIDE	LEERLAUF

8599	NO-LOAD LOSS	PERTE A VIDE	LEERLAUFVERLUST
8600	NO-LOAD RESISTANCE	RESISTANCE DANS LA MARCHE A VIDE	LEERLAUFWIDERSTAND
8601	NO-LOAD WORK	TRAVAIL A VIDE	LEERLAUFARBEIT
8602	NOBLE METAL	METAL NOBLE OU PRECIEUX	EDELMETALL
8603	NOBLE PRECIOUS METAL	METAL PRECIEUX	EDELMETALL, METALL EDLES
8604	NODE CIRCLE (FOR CALCULATION OF PRESSURE TANK FRAMES)	CERCLE DE POINT NEUTRE (CALCUL DES CHARPENTES DE RESISTANCE SOUS PRESSION)	TOTLAGEKREIS
8605	NODE OF OSCILLATION, VIBRATION NODE	NOEUD DE VIBRATION, NOEUD D'OSCILLATION	SCHWINGUNGSKNOTEN
8606	NODULAR	SPHEROIDAL, NODULAIRE	KNOLLIG, KNODIG, NODULAR, KUGELIG
8607	NODULAR CAST-IRON	FONTE NODULAIRE	KUGELGRAPHITGUSSEISEN
8608	NOISE	BRUIT	GERÄUSCH
8609	NOISELESS QUIET SILENT RUNNING	MARCHE SILENCIEUSE	GANG (GERÄUSCHLOSER)
8610	NOISY RUNNING	MARCHE BRUYANTE	GANG (GERÄUSCHVOLLER)
8611	NOMINAL DIAMETER	DIAMETRE NOMINAL	NENNDURCHMESSER, SOLLDURCHMESSER
8612	NOMINAL DIMENSION	DIMENSION NOMINALE	NENNABMESSUNG
8613	NOMINAL PRESSURE	PRESSION NOMINALE	NENNDRUCK (ND)
8614	NOMINAL PRESSURE RATING	TIMBRE D'UN RESERVOIR (A. PRESSION)	GENEHMIGUNGSDRUCK
8615	NOMINAL SIZE	DIMENSION NOMINALE, ORIFICE NOMINAL	NENNMASS, NENNWEITE (NW)
8616	NOMINAL VALUE	VALEUR NOMINALE	NENNWERT
8617	NOMOGRAM	ABAQUE	FLUCHTLINIENTAFEL
8618	NON CONSUMABLE ELECTRODE	ELECTRODE NON-CONSOMMABLE	ELEKTRODE (NICHTSCHMELZENDE)
8619	NON FERROUS ALLOY	ALLIAGE NON-FERREUX	NICHTEISENLEGIERUNG
8620	NON METALLIC INCLUSIONS	INCLUSIONS NON-METALLIQUES	EINSCHLÜSSE (NICHTMETALLISCHE)
8621	NON REACTIVE	INERTE	INERT
8622	NON REFRACTORY ALLOY	ALLIAGE NON REFRACTAIRE	LEGIERUNG (NICHT-FEUERFESTE)
8623	NON RISING STEM	TIGE FIXE	SPINDEL (NICHTSTEIGENDE)
8624	NON RISING STEM GATE VALVE	VANNE A TIGE FIXE	ABSPERRSCHIEBER MIT NICHTSTEIGENDER SPINDEL
8625	NON-ADJUSTABLE GAUGE	JAUGE (NON- VARIABLE)	LEHRE (UNVERSTELLBARE)
8626	NON-CAKING FREE BURNING COAL	HOUILLE NON COLLANTE	KOHLE (NICHTBACKENDE)
8627	NON-CONDENSING STEAM ENGINE	MACHINE SANS CONDENSATION, ECHAPPEMENT LIBRE	AUSPUFFDAMPFMASCHINE
8628	NON-CONDUCTIVE OF HEAT, HEAT INSULATING	CALORIFUGE	ISOLIEREND GEGEN WÄRME, WÄRMESCHUTZ-
8629	NON-CONSTRAINED MOTION, NON-DEFINITELY CONTROLLED MOTION, NON-POSITIVE, FORCE-CLOSED MOTION	MOUVEMENT A COMMANDE ELASTIQUE	BEWEGUNG (KRAFTSCHLÜSSIGE)
8630	NON-DESTRUCTIVE TESTING	ESSAI NON DESTRUCTIF	PRÜFUNG (ZERSTÖRUNGSFREIE)
8631	NON-FERROUS	NON-FERREUX	NICHTEISEN-
8632	NON-FERROUS METAL	METAL AUTRE QUE LE FER, PETIT METAL	NICHTEISENMATALL
8633	NON-FREEZE COATED PALLET	CLAPET ANTIGEL	FROSTSCHUTZVENTIL
8634	NON-LIABILITY TO FREEZE	INCONGELABILITE	FROSTBESTÄNDIGKEIT
8635	NON-LIABLE TO FREEZE	INCONGELABLE	KÄLTEBESTÄNDIG, FROSTBESTÄNDIG
8636	NON-LUMINOUS FLAME	FLAMME INCOLORE, FLAMME OXYDANTE, FEU D'OXYDATION	FLAMME (NICHTLEUCHTENDE)
8637	NON-MAGNETIC	AMAGNETIQUE	UNMAGNETISCH

8638	**NON-MAGNETIC STEEL**	ACIER NON MAGNETIQUE	STAHL (UNMAGNETISIERBARER)
8639	**NON-METAL**	NON-METALLIQUE	NICHT-METALLISCH
8640	**NON-METAL, METALLOID**	METALLOIDE	METALLOID
8641	**NON-METALLIC COATING**	REVETEMENT NON-METALLIQUE	ÜBERZUG (NICHT-METALLISCHER)
8642	**NON-POSITIVE DRIVE**	COMMANDE ELASTIQUE	ANTRIEB (KRAFTSCHLÜSSIGER)
8643	**NON-PRESSURE WELDING**	SOUDAGE PAR FUSION	SCHMELZSCHWEISSEN
8644	**NON-RUSTING STEEL**	ACIER INOXYDABLE	STAHL (NICHTROSTENDER)
8645	**NON-SLIP DIFFERENTIAL**	SYSTEME ANTI-DERAPANT	GLEITSCHUTZDIFFERENTIAL
8646	**NON-SPINNING ROPE**	CABLE ANTIGIRATOIRE	SEIL (DRALLFREIES)
8647	**NON-UNIFORM MOTION**	MOUVEMENT NON UNIFORME	BEWEGUNG (UNGLEICHFÖRMIGE)
8648	**NON-UNIFORMLY ACCELERATED MOTION**	MOUVEMENT NON UNIFORMEMENT ACCELERE	BEWEGUNG (UNGLEICHFÖRMIG BESCHLEUNIGTE)
8649	**NON-UNIFORMLY RETARDED MOTION**	MOUVEMENT NON UNIFORMEMENT RETARDE	BEWEGUNG (UNGLEICHFÖRMIG VERZÖGERTE)
8650	**NONMETALLIC MOULD**	MOULE NON-METALLIQUE	FORM (NICHT-METALLISCHE)
8651	**NOOSE, RUNNING KNOT**	NOEUD COULANT	SCHLAUFE, ZUGKNOTEN
8652	**NORMAL**	NORMALE (GEOM.)	WINKELRECHT, NORMAL (GEOM.)
8653	**NORMAL ACCELERATION**	ACCELERATION NORMALE	NORMALBESCHLEUNIGUNG
8654	**NORMAL CROSS SECTION**	SECTION NORMALE, SECTION DROITE	NORMALSCHNITT
8655	**NORMAL LOAD**	CHARGE NORMALE	GRUNDLAST, GRUNDBELASTUNG
8656	**NORMAL MAGNESIUM PHOSPHATE**	PHOSPHATE DE MAGNESIE	PHOSPHORSAURE MAGNESIA, MAGNESIUMPHOSPHAT
8657	**NORMAL PRESSURE**	PRESSION NORMALE	NORMALDRUCK
8658	**NORMAL RATED OUTPUT**	DEBIT NORMAL	NENNLEISTUNG, REGELLEISTUNG, NORMALLEISTUNG
8659	**NORMAL RATED VOLTAGE**	VOLTAGE NORMAL	NENNSPANNUNG, BETRIEBSSPANNUNG (ELEKTR.)
8660	**NORMAL SEGREGATION**	SEGREGATION MAJEURE	BLOCKSEIGERUNG
8661	**NORMAL STRESS**	CHARGE NORMALE	NORMALSPANNUNG
8662	**NORMAL TO A CURVE**	NORMALE D'UNE COURBE	NORMALE EINER KURVE
8663	**NORMAL WORKING SPEED**	VITESSE DE REGIME, VITESSE NORMALE DE FONCTIONNEMENT, REGIME DE VITESSE	BETRIEBSGESCHWINDIGKEIT
8664	**NORMALIZING**	RECUIT DE NORMALISATION	NORMALGLÜHEN
8665	**NORTH SEEKING POLE OF A MAGNET**	POLE NORD D'UN AIMANT	NORDPOL EINES MAGNETEN
8666	**NOSE (OF A TOOL)**	BEC (D'UN OUTIL)	STAHLSPITZE
8667	**NOSE, PROJECTION**	NEZ, TALON	NASE
8668	**NOTCH**	ENTAILLE	KERBE
8669	**NOTCH BRITTLENESS**	FRAGILITE D'ENTAILLE	KERBSPRÖDIGKEIT
8670	**NOTCH SENSITIVITY**	SENSIBILITE A L'EFFET D'ENTAILLE	KERBEMPFINDLICHKEIT
8671	**NOTCH TOUGHNESS**	TENACITE A L'ENTAILLE, RESILIENCE	KERBZÄHIGKEIT, KERBSCHLAGZÄHIGKEIT
8672	**NOTCH, NICK**	ENTAILLE, ECHANCRURE, ENCOCHE, CRAN	KERBE, EINKERBUNG, RAST
8673	**NOTCHED BAR IMPACT TEST**	ESSAI DE CHOC SUR BARREAU ENTAILLE, ESSAI DE RESILIENCE	KERBSCHLAGPROBE
8674	**NOTCHED INGOT**	LINGOT ENTAILLE	VORGEKERBTER BLOCK
8675	**NOTCHED TEST BAR**	EPROUVETTE AVEC ENTAILLE, BARREAU ENTAILLE	PROBESTAB MIT EINKERBUNG
8676	**NOTCHING**	ENCOCHAGE	KERBEN
8677	**NOTCHING MACHINE**	GRUGEOIR	AUSKLINKMASCHINE
8678	**NOWEL**	MOULE DE DESSOUS	UNTERKASTEN

8679	**NOZZLE**	TUBULURE, PIETEMENT, LANCE, AJUSTAGE CONIQUE	STUTZEN, STRAHLROHR, DÜSE
8680	**NOZZLE DISTANCE/HEIGHT**	HAUTEUR DE BUSE	DÜSENHÖHE
8681	**NOZZLE DRILLING MACHINE**	MACHINE A PERCER LES TUYERES	DÜSENBOHRMASCHINE
8682	**NOZZLE HOLDER**	PORTE-INJECTEUR	DÜSENHALTER
8683	**NOZZLE NECK**	TUBULURE A COLLET	FLANSCHROHRSTUTZEN
8684	**NOZZLE NEEDLE**	AIGUILLE D'INJECTEUR	DÜSENNADEL
8685	**NOZZLE OF A BLOWER**	TUYERE, BUSE	GEBLÄSEDÜSE, DÜSE EINES GEBLÄSES
8686	**NOZZLE SPRING**	RESSORT D'INJECTEUR	DÜSENFEDER
8687	**NUCLEATION**	GERMINATION	KERNBILDUNG, KEIMBILDUNG
8688	**NUCLEUS**	GERME (DE CRISTAL), CRISTALLIN	KERN, KRISTALLKERN, KEIM
8689	**NUMBER OF A LOGARITHM**	NOMBRE D'UN LOGARITHME	NUMERUS EINES LOGARITHMUS
8690	**NUMBER OF REVOLUTIONS PER MINUTE, R. P. M.**	NOMBRE DE TOURS, TOURS A LA MINUTE	DREHZAHL, UMDREHUNGSZAHL, UMLAUFZAHL, TOURENZAHL
8691	**NUMBER OF TEETH**	NOMBRE DE DENTS	ZÄHNEZAHL
8692	**NUMBER OF THREADS**	NOMBRE DE FILETS	GEWINDEGANGZAHL, GANGZAHL EINES GEWINDES
8693	**NUMERATOR**	NUMERATEUR	ZÄHLER, (MATH.)
8694	**NUMERIC SOLUTION**	SOLUTION PAR LE CALCUL	LÖSUNG (RECHNERISCHE), LÖSUNG (ZAHLENMÄSSIGE)
8695	**NUMERICAL CONTROL SYSTEM**	SYSTEME DE COMMANDE NUMERIQUE	STEUERUNGS-SYSTEM (NUMERISCHES)
8696	**NUMERICAL DATA**	DONNEES NUMERIQUES	ANZEIGE (DIGITALE), DATEN
8697	**NUMERICAL VALUE**	VALEUR NUMERIQUE	ZAHLENWERT
8698	**NUT**	ECROU	MUTTER, SCHRAUBENMUTTER
8699	**NUT AUTOMATIC LATHE**	TOUR A DECOLLETER LES ECROUS	MUTTERAUTOMAT
8700	**NUT FORGED IN THE PRESS**	ECROU EMBOUTI	MUTTER (GEDRÜCKTE), MUTTER (GEPRESSTE)
8701	**NUT LOCK, BOLT LOCK**	MECANISME D'ARRET DES ECROUS, FREIN D'ECROU	SCHRAUBENSICHERUNGSVORRICHTUNG
8702	**NUTATION**	NUTATION	NUTATION
8703	**O-RING**	JOINT TORIQUE	O-RING, TORUSRING
8704	**OAK-TANNED LEATHER**	CUIR TANNE A L'ECORCE DE CHENE	LOHGARES, ROTGARES LEDER
8705	**OBELISK**	OBELISQUE	OBELISK, SPITZSÄULE
8706	**OBJECT POINT**	POINT D'OBJET	OBJEKTPUNKT
8707	**OBJECTION AGAINST A PATENT**	ACTION EN NULLITE RELATIVE A UN BREVET	EINSPRUCH GEGEN EIN PATENT, PATENTEINSPRUCH
8708	**OBJECTIVE**	VERRE, SYSTEME OBJECTIF, OBJECTIF	OBJEKTIV
8709	**OBLIQUE CONE**	CONE OBLIQUE	KEGEL (SCHIEFER)
8710	**OBLIQUE COORDINATES**	COORDONNEES OBLIQUES	KOORDINATEN (SCHIEFWINKLIGE)
8711	**OBLIQUE CROSS SECTION**	SECTION OBLIQUE	SCHNITT (SCHIEFER)
8712	**OBLIQUE CYLINDER**	CYLINDRE OBLIQUE	ZYLINDER (SCHIEFER)
8713	**OBLIQUE PARALLELOPIPED**	PARALLELEPIPEDE OBLIQUE	PARALLELEPIPED (SCHIEFES)
8714	**OBLONG HOLE, SLOT HOLE (ROUNDING AT THE ENDS)**	MORTAISE A EXTREMITES ARRONDIES, TROU OBLONG	LANGLOCH
8715	**OBROUND**	OBLONG	ELLIPSOID
8716	**OBSIDIAN**	OBSIDIENNE	OBSIDIAN
8717	**OBTUSE ANGLE**	ANGLE OBTUS	WINKEL (STUMPFER)
8718	**OBTUSE-ANGLED TRIANGLE**	TRIANGLE OBTUSANGLE	DREIECK (STUMPFWINKLIGES)
8719	**OCCLUSION**	OCCLUSION (CHIM.)	OKKLUSION (CHEM.), VERSTOPFUNG

8720	**OCHRE**	OCRE, SANGUINE, BOL DIARMENIE, CRAIE ROUGE	OCKER
8721	**OCTAGON**	OCTOGONE, OCTANGLE	ACHTECK
8722	**OCTAHEDRON**	OCTAEDRE	ACHTFLACH, ACHTFLÄCHNER, OKTAEDER
8723	**OCTANE RATING**	INDICE D'OCTANE	OKTANZAHL
8724	**OCTANT**	OCTANT	ACHTELKREIS, OKTANT
8725	**ODD NUMBER**	NOMBRE IMPAIR	ZAHL (UNGERADE)
8726	**ODONTOGRAPH**	ODONTOGRAPHE	ZAHNFLANKENZIRKEL, ODONTOGRAPH
8727	**OF ONE DIMENSION**	DIMENSION (A UNE)	EINDIMENSIONAL
8728	**OF THREE DIMENSIONS**	DIMENSIONS (A TROIS)	DREIDIMENSIONAL
8729	**OF TWO DIMENSIONS**	DIMENSION (A DEUX)	ZWEIDIMENSIONAL
8730	**OFF CENTER**	DESAXE, DECALE, DECENTRE	EXZENTRISCH, VERSCHOBEN
8731	**OFF CENTERED**	DECALE, HORS D'AXE	SEITLICH VERSCHOBEN
8732	**OFF HEAT**	COULEE RATEE	FEHLSCHMELZE
8733	**OFF IRON**	FONTE DE TRANSITION	ÜBERGANGSROHEISEN
8734	**OFF TIME**	TEMPS DE REPOS, COUPURE DE COURANT	STROMPAUSE
8735	**OFF-LINE OPERATION**	TRAITEMENT INDIRECT	OFF-LINE-BETRIEB
8736	**OFF-LOADING PORT**	PORT DE DECHARGEMENT (DE NAVIRES MARCHANDS)	ENTLADEHAFEN, ENTLADESTATION
8737	**OFF-SET (JOINT)**	DEFAUT D'ALIGNEMENT OU DENIVELLATION (D'UN JOINT SOUDE)	VERSETZUNG
8738	**OFF-SHORE**	EN MER, AU LARGE	OFF-SHORE, SCHELF...., KÜSTENNAHE.....
8739	**OFF-SHORE QUALITY STEEL**	ACIER POUR PLATE-FORME DE FORAGE EN MER	STAHL FÜR KÜSTENNAHE ÖL BOHRUNG
8740	**OFFICIAL TEST**	EPREUVE OFFICIELLE	PRÜFUNG (AMTLICHE)
8741	**OGEE-ARCH**	ARC EN DOS D'ANE	ESELRÜCKENBOGEN
8742	**OHM**	OHM	OHM
8743	**OIL**	HUILE	ÖL
8744	**OIL (TO)**	GRAISSER A L'HUILE, HUILER	ÖLEN
8745	**OIL BATH**	BAIN D'HUILE	ÖLBAD
8746	**OIL BATH AIR CLEANER**	FILTRE A AIR A BAIN D'HUILE	ÖLBADLUFTFILTER
8747	**OIL BATH RETURN GAUGE**	JAUGE AVEC RENVOI A BAIN D'HUILE	GEGENSTROMMESSER IN ÖLBAD
8748	**OIL BUNKERS**	SOUTES A COMBUSTIBLE	ÖLBUNKER
8749	**OIL CAN**	BURETTE DE GRAISSAGE, BURETTE A HUILE	ÖLKANNE, SCHMIERKANNE
8750	**OIL CONTROL RING**	SEGMENT RACLEUR	ÖLABSTREIFRING
8751	**OIL COOLER**	REFROIDISSEUR D'HUILE	ÖLKÜHLER
8752	**OIL COOLING**	REFROIDISSEMENT A L'HUILE	ÖLKÜHLUNG
8753	**OIL CUP, GREASE CUP**	GODET GRAISSEUR, GODET HUILEUR, GODET DE GRAISSAGE	SCHMIERBÜCHSE, SCHMIERKELCH, SCHMIERNAPF, ÖLVASE
8754	**OIL CUSHION**	MATELAS D'HUILE	ÖLPOLSTER (DÄMPFUNG)
8755	**OIL DASHPOT**	FREIN HYDRAULIQUE, FREIN A HUILE	ÖLBREMSE, ÖLPUFFER
8756	**OIL ENGINE**	MOTEUR A CARBURANT	ÖLMASCHINE
8757	**OIL FILTER**	FILTRE A HUILE	ÖLFILTER
8758	**OIL GAS**	GAZ D'HUILE, GAZ RICHE	ÖLGAS, FETTGAS, REICHGAS
8759	**OIL GROOVE**	PATTE D'ARAIGNEE	ÖLNUT
8760	**OIL HARDENING**	TREMPE A L'HUILE, TREMPE DOUCE	ÖLHÄRTUNG

8761	OIL HOLE	ORIFICE DE GRAISSAGE, TROU GRAISSEUR	SCHMIERLOCH, SCHMIERBOHRUNG
8762	OIL HOLE DRILL	FORET A CANALISATION D'HUILE	SCHMIERBOHRER
8763	OIL IMPELLER	TURBINE DE RETOUR D'HUILE	ÖLFLÜGELRAD
8764	OIL LEVEL DIPSTICK	JAUGE D'HUILE	ÖLMESSSTAB
8765	OIL OF TURPENTINE	ESSENCE DE TEREBENTHINE	TERPENTINÖL
8766	OIL PAINT, OIL COLOUR	PEINTURE A L'HUILE	ÖLFARBE
8767	OIL PAN	CARTER D'HUILE (INFERIEUR)	KURBELGEHÄUSE-UNTERTEIL
8768	OIL PRESSURE INDICATOR, GAUGE	INDICATEUR DE PRESSION D'HUILE	ÖLDRUCKMESSER
8769	OIL PUMP	POMPE DE GRAISSAGE, POMPE HYDRAULIQUE, POMPE A HUILE	HYDRAULIKPUMPE, ÖLPUMPE, SCHMIERÖLPUMPE
8770	OIL PUMP STRAINER	CREPINE D'HUILE	ÖLPUMPENSIEB
8771	OIL PUTTY	MASTIC A L'HUILE	ÖLKITT
8772	OIL QUENCHING	TREMPE A L'HUILE	ÖLHÄRTUNG
8773	OIL RESERVOIR CHAMBER	RESERVOIR D'HUILE	ÖLSPEICHER
8774	OIL RING	BAGUE DE GRAISSAGE	ÖLRING, SCHMIERRING
8775	OIL SCRAPER-RING	SEGMENT RACLEUR	ÖLABSTREIFRING
8776	OIL SEAL	JOINT (DE RETENUE) D'HUILE	ÖLFANGRING
8777	OIL SEPARATION	EXTRACTION DE L'HUILE	ENTÖLUNG, ÖLABSCHEIDUNG
8778	OIL SEPARATOR	SEPARATEUR D'HUILE	ENTÖLER, ÖLABSCHEIDER
8779	OIL SEPARATOR FOR STEAM	DESHUILEUR DE VAPEUR	ABDAMPF-ENTÖLER, ÖLABSCHEIDER
8780	OIL SHALE	SCHISTE BITUMINEUX, NAPHTOSCHISTE	ÖLSCHIEFER
8781	OIL STONE, HONE	PIERRE A HUILE	ÖLSTEIN
8782	OIL SUMP	CARTER D'HUILE (INFERIEUR)	KURBELGEHÄUSE-UNTERTEIL, ÖLWANNE
8783	OIL SUPPLY	AMENEE D'HUILE	ÖLZUFÜHRUNG
8784	OIL SYRINGE	SERINGUE A HUILE	ÖLSPRITZE
8785	OIL TESTER, OIL TESTING MACHINE	MACHINE A ESSAYER LES HUILES	ÖLPRÜFMASCHINE
8786	OIL VARNISH, BOILED OIL	HUILE DE LIN A VERNIS, HUILE DE LIN BOUILLIE, VERNIS A L'HUILE	LEINÖLFIRNIS
8787	OIL VISCOSITY INDEX	INDICE DE VISCOSITE	ÖLVISKOSITÄTSINDEX
8788	OILED PAPER	PAPIER HUILE	ÖLPAPIER
8789	OILING POINT	POINT A LUBRIFIER	SCHMIERSTELLE
8790	OILING RING	BAGUE DE GRAISSAGE	SCHMIERRING
8791	OILING, OIL LUBRICATION	GRAISSAGE A L'HUILE, HUILAGE	ÖLSCHMIERUNG, ÖLEN
8792	OILPROOF	INATTAQUABLE A L'HUILE	ÖLFEST
8793	OLD PROCESS	PROCEDE OLD	OLD VERFAHREN (OXYGENE-LINZ-DONAWITZ)
8794	OLDHAM'S COUPLING	JOINT D'OLDHAM, JOINT A DOUBLE TOURNEVIS	KREUZSCHEIBENKUPPLUNG, OLDHAMKUPPLUNG
8795	OLEIC ACID	ACIDE OLEIQUE	ÖLSÄURE, OLEINSÄURE, STEARINÖL
8796	OLIVINE	OLIVINE, PERIDOT	OLIVIN
8797	OLSEN CUP TEST	ESSAI D'EMBOUTISSAGE ERICKSEN	ERICHSEN-TIEFZIEHVERSUCH, EINBEULVERSUCH
8798	ON-COST	FRAIS GENERAUX	UNKOSTEN (ALLGEMEINE)
8799	ON-LINE OPERATION	TRAITEMENT DIRECT	ON-LINE-BETRIEB
8800	ON-SITE ERECTION	MONTAGE SUR CHANTIER	BAUSTELLENMONTAGE
8801	OOLITIC LIMESTONE	CALCAIRE OOLITHIQUE, PISOLITHIQUE	OOLITHKALK, ROGENSTEIN
8802	OPACITY, OPAQUENESS	OPACITE	UNDURCHSICHTIGKEIT
8803	OPAQUE	OPAQUE	UNDURCHSICHTIG

8804	**OPEN BELT**	COURROIE DROITE, COURROIE OUVERTE	RIEMENTRIEB (OFFENER)
8805	**OPEN CAM**	EXCENTRIQUE PLAN	EXZENTER (OFFENES)
8806	**OPEN CHAIN HYDROCARBONS**	HYDROCARBURES DE LA SERIE GRASSE	KOHLENWASSERSTOFFE DER FETTREIHE
8807	**OPEN CHANNEL**	CONDUITE LIBRE, CANAL DECOUVERT	LEITUNGSKANAL
8808	**OPEN CIRCUIT VOLTAGE**	TENSION A VIDE	LEERLAUFSPANNUNG
8809	**OPEN CORNER JOINT**	SOUDURE D'ANGLE OUVERTE	ECKNAHTVERBINDUNG MIT LUFTSPALT
8810	**OPEN DIES**	MATRICE OUVERTE	GESENK (OFFENES)
8811	**OPEN DOUBLE-BEVEL BUTT WELD**	SOUDURE EN K AVEC ECARTEMENT	K-NAHT MIT LUFTSPALT
8812	**OPEN DOUBLE-J BUTT WELD**	SOUDURE DOUBLE OUVERTE	DOPPEL-J-NAHT MIT LUFTSPALT
8813	**OPEN FIRE**	FEU NU	OFFENESFEUER, FREIES FEUER
8814	**OPEN FRONT MECHANICAL PRESS**	PRESSE MECANIQUE A BATI COL DE CYGNE	EINSTÄNDERPRESSE (MECHANISCHE)
8815	**OPEN HEARTH FURNACE**	FOUR MARTIN	SIEMENS-MARTIN-OFEN
8816	**OPEN HEARTH STEEL**	ACIER MARTIN	SIEMENS-MARTIN-STAHL, MARTIN-STAHL
8817	**OPEN LOOP SYSTEM**	SYSTEME DE COMMANDE EN BOUCLE OUVERTE	STEUERKETTE
8818	**OPEN MOTOR**	MOTEUR OUVERT, MOTEUR NON PROTEGE	OFFENER MOTOR
8819	**OPEN PIG IRON**	FER A GROS GRAIN	GROBKORNEISEN, GROSSLUCKIGES EISEN
8820	**OPEN SAND CASTING**	MOULAGE A DECOUVERT	HERDGUSS
8821	**OPEN SIDE PLANING MACHINE**	MACHINE A RABOTER OUVERTE SUR LE COTE, MACHINE A RABOTER A UN SEUL MONTANT	TISCHHOBELMASCHINE (EINSEITIG OFFENE), EINSTÄNDERHOBELMASCHINE, EINPILASTERHOBELMASCHINE
8822	**OPEN SINGLE-BEVEL BUTT WELD**	SOUDURE EN DEMI-V AVEC ECARTEMENT	HALB-V-HAHT MIT LUFTSPALT
8823	**OPEN SINGLE-J-BUTT WELD**	SOUDURE EN J AVEC ECARTEMENT	J-NAHT MIT LUFTSPALT
8824	**OPEN SQUARE BUTT WELD**	SOUDURE EN I AVEC ECARTEMENT	I-NAHT MIT LUFTSPALT
8825	**OPEN STEEL**	ACIER SEMI-CALME	STAHL (UNVOLLSTÄNDIG DESOXYDIERTER)
8826	**OPEN THE CIRCUIT (TO)**	OUVRIR LE CIRCUIT ELECTRIQUE	STROMKREIS ÖFFNEN (DEN)
8827	**OPEN TUBE PRESSURE GAUGE**	MANOMETRE A AIR LIBRE	GEFÄSSMANOMETER
8828	**OPEN-FRONT HYDRAULIC PRESS**	PRESSE HYDRAULIQUE A COL DE CYGNE	EINSTÄNDERPRESSE (HYDRAULICHE)
8829	**OPEN-HEARTH FURNACE**	FOUR MARTIN	SIEMENS-MARTIN-OFEN
8830	**OPEN-HEARTH PROCESS**	PROCEDE SIEMENS-MARTIN	SIEMENS-MARTINS-VERFAHREN
8831	**OPENING**	EVIDEMENT	AUSSPARUNG
8832	**OPENING OF THE VALVE**	OUVERTURE DE LA SOUPAPE	VENTILERÖFFNUNG
8833	**OPERATING BRAKE**	LEVIER DE MANOEUVRE	STEUERHEBEL
8834	**OPERATING CONDITIONS**	CONDITIONS DE SERVICE	BETRIEBSVERHÄLTNISSE
8835	**OPERATING INSTRUCTIONS**	INSTRUCTIONS, REGLES DE SERVICE	BEDIENUNGSVORSCHRIFTEN
8836	**OPERATING TEMPERATURE**	TEMPERATURE DE SERVICE, TEMPERATURE DE TRAVAIL	BETRIEBSTEMPERATUR, ARBEITSTEMPERATUR
8837	**OPERATION**	EXPLOITATION, FONCTIONNEMENT, SERVICE	BETRIEB
8838	**OPERATION NOMBER**	NUMERO D'OPERATION EN COURS	NUMMER DES ARBEITSGANGES
8839	**OPERATOR'S STAND**	POSTE, STAND DE L'OUVRIER	ARBEITSPLATZ, ARBEITSSTAND
8840	**OPTICAL**	OPTIQUE	OPTISCH
8841	**OPTICAL AXIS**	AXE OPTIQUE	ACHSE (OPTISCHE)
8842	**OPTICAL COMPARATOR**	COMPARATEUR OPTIQUE	VERGLEICHER (OPTISCHER)

8843	OPTICAL FLAT	PLAN OPTIQUE	PLANFLÄCHE (OPTISCHE)
8844	OPTICAL INSTRUMENT	INSTRUMENT D'OPTIQUE	INSTRUMENT (OPTISCHES)
8845	OPTICAL PYROMETER	PYROMETRE OPTIQUE	PYROMETER (OPTISCHES)
8846	OPTICAL PYROMETRY	PYROMETRIE OPTIQUE	GLÜHFADENPYROMETRIE
8847	OPTICAL READERS	LECTEURS OPTIQUES	ABLESEGERÄTE (OPTISCHE)
8848	OPTICAL SYSTEM	SYSTEME OPTIQUE	SYSTEM (OPTISCHES)
8849	OPTICS	OPTIQUE	OPTIK
8850	OPTIONAL STOP	ARRET FACULTATIF	HALT (WAHLWEISER)
8851	ORANGE PEEL EFFECT	EFFET DE PEAU D'ORANGE	ORANGENSCHALENEFFEKT, APPELSINENSCHALENEFFEKT
8852	ORDER OF MAGNITUDE	ORDRE DE GRANDEUR	BEDEUTUNG, GRÖSSENORDNUNG
8853	ORDINARY ETHER, DIETHYL ETHER	ETHER ORDINAIRE, ETHER SULFURIQUE, OXYDE D'ETHYLE	ÄTHER, ÄTHYLÄTHER, SCHWEFELÄTHER
8854	ORDINARY LIME MORTAR	MORTIER AERIEN	KALKMÖRTEL, LUFTMÖRTEL
8855	ORDINARY LINK CHAIN, OPEN LINK CHAIN	CHAINE ORDINAIRE	GLIEDERKETTE, SCHAKENKETTE
8856	ORDINARY SULPHURIC ACID	ACIDE SULFURIQUE COMMERCIAL	SCHWEFELSÄURE (ROHE), HANDELSSCHWEFELSÄURE
8857	ORDINARY WORKING CONDITIONS	REGIME	BETRIEBSZUSTAND (NORMALER)
8858	ORDINATE	ORDONNEE	ORDINATE
8859	ORE	MINERAI	ERZ
8860	ORE BRIQUETTE	BRIQUETTE DE MINERAI, MINERAI BRIQUETTE	ERZBRIKETT, ERZZIEGEL, ERZPRESSSTEIN, BRIKETTIERTES ERZ
8861	ORE DRESSING	ENRICHISSEMENT DU MINERAI, TRAITEMENT MECANIQUE, PREPARATION DES MINERAIS	AUFBEREITUNG VON ERZEN, ERZANREICHERUNG
8862	ORE ROASTING	GRILLAGE DU MINERAI	ERZRÖSTEN
8863	ORES AND FLUXES	MINERAIS ET FONDANTS	ERZ UND SCHMELZMITTEL
8864	ORGANIC ACID	ACIDE ORGANIQUE	SÄURE (ORGANISCHE)
8865	ORGANIC CHEMISTRY	CHIMIE ORGANIQUE	CHEMIE (ORGANISCHE)
8866	ORGANIC MATTER	MATIERE ORGANIQUE	STOFF (ORGANISCHER)
8867	ORIENTATION	ORIENTATION	ORIENTIERUNG
8868	ORIGIN OF COORDINATES	ORIGINE DES COORDONNEES	ANFANGSPUNKT DER KOORDINATEN, NULLPUNKT DER KOORDINATEN, URSPRUNG DER KOORDINATEN, KOORDINATENANFANG
8869	ORNAMENTAL, ECCENTRIC TURNING	TOURNAGE DES SURFACES FIGUREES	PASSIGDREHEN, UNRUNDDREHEN
8870	ORTHOGRAPHIC PARALLEL PROJECTION	PROJECTION ORTHOGONALE	PARALLELPROJEKTION (RECHTWINKLIGE), PARALLELPROJEKTION ORTHOGONALE, PARALLELPROJEKTION ORTHOGRAPHISCHE
8871	ORTHOHEXAGONAL CRYSTAL AXES	AXES CRISTALLINS ORTHOHEXAGONAUX	KRISTALLACHSEN (ORTHOHEXAGONALE)
8872	ORTHOMORPHIC CONFORM PROJECTION REPRESENTATION	REPRESENTATION CONFORME	WINKELTREUE KONFORME ABBILDUNG
8873	ORTHOPHOSPHORIC ACID	ACIDE ORTHOPHOSPHORIQUE	PHOSPHORSÄURE, ORTHOPHOSPHORSÄURE, KNOCHENSÄURE
8874	ORTHORHOMBIC CRYSTALS	CRISTAUX ORTHORHOMBIQUES	KRISTALLE (ORTHORHOMBISCHE)
8875	ORTHOSILICIC ACID	ACIDE ORTHOSILICIQUE	ORTHOKIESELSÄURE
8876	ORTHOTUNGSTIC ACID	ACIDE TUNGSTIQUE	WOLFRAMSÄURE, SCHEELSÄURE, TUNGSTEINSÄURE
8877	OSCILLATE (TO), VIBRATE (TO)	OSCILLER, VIBRER	SCHWINGEN, OSZILLIEREN
8878	OSCILLATING CAM	CAME OSCILLANTE	SCHWINGNOCKEN
8879	OSCILLATING CRYSTAL METHOD	METHODE DE CRISTAL OSCILLANT	SCHWINGKRISTALLMETHODE

8880	OSCILLATING ENGINE	MACHINE A CYLINDRE OSCILLANT	DAMPFMASCHINE MIT SCHWINGENDEM ZYLINDER, OSZILLIERENDE DAMPFMASCHINE
8881	OSCILLATING MOTION, VIBRATING MOTION	MOUVEMENT OSCILLATOIRE	BEWEGUNG (SCHWINGENDE), SCHWINGBEWEGUNG
8882	OSCILLATING PISTON	PISTON OSCILLANT	KOLBEN (SCHWINGENDER), SCHWINGKOLBEN
8883	OSCILLATION DUE TO RESONANCE	VIBRATIONS DUES A LA RESONANCE	RESONANZCHWINGUNG
8884	OSCILLATION, VIBRATION	OSCILLATION, VIBRATION	SCHWINGUNG, OSZILLATION
8885	OSCILLOGRAPH	OSCILLOGRAPHE	OSZILLOGRAPH
8886	OSCILLOSCOPE	OSCILLOSCOPE	OSZILLOSKOP
8887	OSCULATING PLANE	PLAN OSCULATEUR	SCHMIEGUNGSEBENE
8888	OSMIRIDIUM	OSMIRIDIUM	OSMIRIDIUM
8889	OSMIUM	OSMIUM	OSMIUM
8890	OSMOSIS	OSMOSE	OSMOSE
8891	OSMOTIC PRESSURE	PRESSION OSMOTIQUE	DRUCK (OSMOTISCHER)
8892	OUT OF GEAR	DESENGRENE	AUSSER EINGRIFF
8893	OUT OF LINE WITH THE AXIAL CENTRE	DESAXE	AUSSERACHSIG, DEZENTRIERT
8894	OUT-OF-ROUNDNESS	FAUX-ROND	UNRUNDHEIT
8895	OUT-OF-SQUARE	HORS D'EQUERRE	NICHT RECHTWINKELIG
8896	OUT-PUT	DEBIT	LEISTUNG
8897	OUTDOOR LIGHTING	ECLAIRAGE EXTERIEUR	AUSSENBELEUCHTUNG
8898	OUTER RACE OF BALL BEARING	BAGUE EXTERIEURE D'UN ROULEMENT A BILLES	ÄUSSERER RING, AUSSENRING EINES KUGELLAGERS
8899	OUTFLOW DISCHARGE OF WATER	ECOULEMENT D'EAU	ABFLUSS, WASSERABFLUSS
8900	OUTLET	SORTIE	AUSGANG, ABFLUSS
8901	OUTLET NOZZLE	TUBULURE DE SORTIE	AUSLASSSTUTZEN
8902	OUTLET OPENING , DISCHARGE ORIFICE	ORIFICE DE DECHARGE, ORIFICE DE SORTIE, ORIFICE D'ECHAPPEMENT, ORIFICE D'ECOULEMENT	AUSLASSÖFFNUNG, AUSFLUSSÖFFNUNG, AUSFLUSSMÜNDUNG
8903	OUTLET VELOCITY, VELOCITY OF DISCHARGE	VITESSE DE SORTIE, VITESSE A LA SORTIE, VITESSE D'ECOULEMENT	ABSTRÖMGESCHWINDIGKEIT, ABFLUSSGESCHWINDIGKEIT, AUSFLUSSGESCHWINDIGKEIT, AUSTRITTGESCHWINDIGKEIT
8904	OUTLET WEIR	BARRAGE DE SORTIE	FLUCHTSCHLEUSE
8905	OUTLINE DRAWING	DESSIN AU TRAIT, DESSIN LINEAIRE, PLAN D'ENSEMBLE, PLAN DE MASSE	STRICHZEICHNUNG, LINEARZEICHNUNG
8906	OUTLINE OF A CAM	PROFIL, CONTOUR D'UNE CAME	BEGRENZUNGSKURVE EINES EXZENTERS
8907	OUTLINE, CONTOUR, BOUNDARY OF A FIGURE	CONTOUR	UMRISS, UMRISSLINIE, KONTUR
8908	OUTPUT	DEBIT	LEISTUNG
8909	OUTPUT GOVERNOR	REGULATEUR DE PUISSANCE, REGULATEUR DE DEBIT	LEISTUNGSREGLER
8910	OUTPUT SHAFT	ARBRE SECONDAIRE	ABTRIEBSWELLE
8911	OUTSIDE	COTE EXTERIEUR	AUSSENSEITE
8912	OUTSIDE CALLIPERS	COMPAS D'EPAISSEUR	AUSSENTASTER, DICKZIRKEL
8913	OUTSIDE DIAMETER	DIAMETRE EXTERIEUR	AUSSENDURCHMESSER
8914	OUTSIDE SCREW GATE VALVE	VANNE A VIS EXTERIEURE	ABSPERRSCHIEBER MIT AUSSENSPINDEL, APSPERRSCHIEBER MIT AUSSENLIEBGENDEM SPINDELGEWINDE

8915	OUTSIDE SCREW STEM	TIGE A VIS EXTERIEURE	SPINDEL MIT AUSSENLIEGENDEM GEWINDE
8916	OVAL BAR IRON, OVALS	FER OLIVE, OLIVES	OVALEISEN
8917	OVAL CROSS SECTION	SECTION OVALE	QUERSCHNITT (EIRUNDER), QUERSCHNITT OVALER
8918	OVAL ELLIPTICAL FLANGE	BRIDE OVALE	FLANSCH (OVALER)
8919	OVAL TURNING LATHE	TOUR A OVALES	OVALDREHBANK
8920	OVEN	FOUR	OFEN
8921	OVER DRIVE	MULTIPLICATEUR DE VITESSE	ÜBERSETZUNGSGETRIEBE
8922	OVER-ALL DIMENSIONS	DIMENSIONS HORS-TOUT	AUSSENMASSE
8923	OVERAGING	SURVIEILLISSEMENT	ÜBERALTERUNG
8924	OVERALL HEIGHT	HAUTEUR DE CONSTRUCTION, HAUTEUR TOTALE	BAUHÖHE
8925	OVERALL LENGTH	LONGUEUR TOTALE, LONGUEUR DE CONSTRUCTION	BAULÄNGE, GESAMTLÄNGE, FABRIKATIONSLÄNGE
8926	OVERALL WIDTH	LARGEUR DE CONSTRUCTION, LARGEUR TOTALE	BAUBREITE
8927	OVERALL, TOTAL EFFICIENCY	RENDEMENT TOTAL	GESAMTWIRKUNGSGRAD, RESULTIERENDER WIRKUNGSGRAD
8928	OVERBALANCE (TO)	SURPASSER, EMPORTER SUR...(L')	ÜBERWIEGEN
8929	OVERCOME A RESISTANCE (TO)	VAINCRE UNE RESISTANCE	WIDERSTAND ÜBERWINDEN (EINEN)
8930	OVERDRIVE	VITESSE SURMULTIPLIEE	SCHNELLGANG
8931	OVERFALL	DEVERSOIR, TROP PLEIN	ÜBERFALL
8932	OVERFLOW PIPE	TUYAU DE TROP-PLEIN, TROP-PLEIN	ÜBERFALLROHR, ÜBERLAUFROHR
8933	OVERFLOW VALVE	SOUPAPE DE TROP PLEIN	SCHLABBERVENTIL, ÜBERSTRÖMVENTIL
8934	OVERHANG	PORTE-A-FAUX	ÜBERHANG
8935	OVERHANGING CYLINDER	CYLINDRE EN PORTE-A-FAUX	ZYLINDER (FREITRAGENDER), ZYLINDER (FREIHÄNGENDER), ZYLINDER (SCHWEBENDER)
8936	OVERHANGING, OVERHUNG	PORTE-A-FAUX	FLIEGEND GELAGERT, FREISCHWEBEND, ÜBERHÄNGEND, ÜBERSTEHEND, FREIHÄNGEND
8937	OVERHAUL (TO)	REVOIR, REVISER	NACHSEHEN, ÜBERHOLEN
8938	OVERHEAD CABLE	CABLE AERIEN	LUFTKABEL
8939	OVERHEAD CAMSHAFT	ARBRE A CAMES EN TETE	NOCKENWELLE, (OBENLIEGENDE)
8940	OVERHEAD COUNTERSHAFT	RENVOI DE PLAFOND	DECKENVORGELEGE
8941	OVERHEAD POSITION	SOUDURE EN POSITION PLAFOND	ÜBERKOPFSCHWEISSUNG
8942	OVERHEAD TRAVELLER, TRAVELLING CRANE	PONT-ROULANT D'USINE	LAUFKRAN
8943	OVERHEAD VALVES	SOUPAPES EN TETE	VENTILE (HÄNGENDE)
8944	OVERHEAD WELD	SOUDURE AU PLAFOND	ÜBERKOPFSCHWEISSNAHT
8945	OVERHEATED STEEL	ACIER SURCHAUFFE	STAHL, (ÜBERHITZTER)
8946	OVERHEATING	SURCHAUFFE	UBERHITZUNG
8947	OVERHUNG OVERHANGING CRANK, OUTSIDE CRANK	MANIVELLE FRONTALE, MANIVELLE EN BOUT, MANIVELLE EN PORTE-A-FAUX	STIRNKURBEL, ENDKURBEL, FLIEGEND ANGEORDNETE KURBEL
8948	OVERLAP	DEPASSEMENT, RECOUVREMENT, SUREPAISSEUR, BAVURE	ÜBERLAGERUNG, ÜBERLAPPUNG, NAHTÜBERHÖHUNG
8949	OVERLAP (OF A WELD)	BAVURE (DE SOUDURE)	ÜBERLAPPUNG
8950	OVERLAP (TO), LAP (TO)	CHEVAUCHER	ÜBEREINANDERLIEGEN, ÜBEREINANDERGREIFEN, ÜBEREINANDERSTEHEN
8951	OVERLAPPING, OVERLAP	RECOUVREMENT, CHEVAUCHEMENT	ÜBEREINANDERGREIFEN, ÜBERLAPPEN, ÜBERLAPPUNG

8952	OVERLAY	SURCOUCHE	AUFTRAGSSCHWEISSZUSATZ, ÜBERLAGERUNG
8953	OVERLOAD	SURCHARGE	ÜBERLAST, ÜBERBELASTUNG, ÜBERLADUNG
8954	OVERLOAD (TO)	SURCHARGER	ÜBERBELASTEN
8955	OVERPRESSURE	PRESSION SUPERIEURE A LA PRESSION AUTORISEE, SURPRESSION	ÜBERNORMALDRUCK, ÜBERDRUCK
8956	OVERRIDER	BUTOIR DE PARE-CHOCS	STOSSSTANGENHORN
8957	OVERRUNNING CLUTCH	EMBRAYAGE A ROUE LIBRE	FREILAUFKUPPLUNG
8958	OVERSEER, OVERLOOKER	SURVEILLANT	FABRIKAUFSEHER, AUFSEHER
8959	OVERSHOOT	DEPASSEMENT	ÜBERFAHREN
8960	OVERSHOT WATER WHEEL	ROUE PAR-DESSUS, ROUE A AUGETS, NORIA	OBERSCHLÄCHTIGES WASSERRAD
8961	OVERSIZE	SURCROIT DE DIMENSION	ÜBERGRÖSSE, ÜBERMASS
8962	OVERSTRAIN A MATERIAL (TO)	SURCHARGER, FATIGUER UN MATERIEL	MATERIAL ÜBERANSTRENGEN (EIN)
8963	OVERSTRESSING	SURCHARGE	ÜBERLASTUNG, ÜBERBEANSPRUCHUNG
8964	OVERTIME	HEURES SUPPLEMENTAIRES	ÜBERSTUNDEN
8965	OVERTURNING	RENVERSEMENT (CALCUL SOUS SEISMES)	UMSTÜRZEN, UMKEHRUNG
8966	OVERTURNING MOMENT	MOMENT DE RENVERSEMENT	KIPPMOMENT
8967	OVERVOLTAGE	SURVOLTAGE	ÜBERSPANNUNG
8968	OXALATE	OXALATE	OXALSAURES SALZ, OXALAT
8969	OXALIC ACID	ACIDE OXALIQUE	OXALSÄURE, ÄTHANDISÄURE, KLEESÄURE, SAUERKLEESÄURE
8970	OXIDATION	OXYDATION	OXYDATION, OXYDIERUNG
8971	OXIDE	OXYDE	OXYD
8972	OXIDISABLE, SUSCEPTIBLE OF OXIDATION	OXYDABLE	OXYDIERBAR
8973	OXIDISE (TO), OXIDATE (TO), CONVERT INTO AN OXYDE (TO)	OXYDER, OXYDER (S')	OXYDIEREN
8974	OXIDISING AGENT	OXYDANT	OXYDATIONSMITTEL
8975	OXIDISING FLAME	DARD DE CHALUMEAU, FLAMME OXYDANTE	STICHFLAMME, FLAMME (OXIDIERENDE)
8976	OXY-ACETYLENE DESEAMING	ELIMINATION DES DEFAUTS DE SURFACE AU CHALUMEAU	AUTOGENES BRENNSCHNEIDEN, BRENNPUTZEN (FLÄMMEN)
8977	OXYACETYLENE CUTTING	OXYCOUPAGE	AZETYLEN-SAUERSTOFF BRENNSCHNEIDEVERFAHREN
8978	OXYACETYLENE PRESSURE WELDING	SOUDAGE AUTOGENE PAR PRESSION	AUTOGEN-PRESS-SCHWEISSEN
8979	OXYACETYLENE WELDING	SOUDURE OXY-ACETYLENIQUE, SOUDAGE AUTOGENE	SAUERSTOFF-AZETYLEN-, AUTOGENSCHWEISSEN
8980	OXYGEN	OXYGENE	SAUERSTOFF
8981	OXYGEN LANCE	LANCE A OXYGENE	SAUERSTOFFLANZE
8982	OXYGEN PROCESS	PROCEDE A L'OXYGENE	SAUERSTOFFVERFAHREN
8983	OXYGEN-FREE HIGH CONDUCTIVITY COPPER	CUIVRE EXEMPT D'OXYGENE A HAUTE CONDUCTIVITE	LEITKUPFER OHNE SAUERSTOFF
8984	OXYGENIZATION	OXYGENATION, OXYDATION	OXYDIEREN
8985	OXYHYDROGEN	GAZ OXHYDRIQUE	KNALLGAS
8986	OXYHYDROGEN BLOW PIPE	CHALUMEAU OXHYDRIQUE	KNALLGASGEBLÄSE, SAUERSTOFFGEBLÄSE
8987	OXYHYDROGEN LIGHT, LIMELIGHT	LUMIERE OXHYDRIQUE, LUMIERE DE DRUMMOND	KALKLICHT, DRUMMONDSCHES LICHT

8988	OXYHYDROGEN WELDING	SOUDURE OXHYDRIQUE	SAUERSTOFF-WASSERSTOFF-SCHWEISSUNG, SCHWEISSUNG (HYDROOXYGENE
8989	OZONE	OZONE	OZON
8990	PACK (TO), INSERT PACKING MATERIAL (TO)	GARNIR DE MATIERE DE JOINT	PACKEN, VERPACKEN, LIDERN
8991	PACK (TO), MAKE UP INTO A PACKAGE (TO)	EMBALLER	VERPACKEN
8992	PACK CARBURIZING	CEMENTATION A LA POUDRE	PULVERZEMENTIEREN
8993	PACK FILM	PELLICULE RIGIDE, PLATE, FILM RIGIDE	PLANFILM
8994	PACK ROLLING	LAMINAGE EN PAQUET	PAKETWALZEN
8995	PACKAGING	CONDITIONNEMENT	VERPACKUNG
8996	PACKED OILS	HUILES CONDITIONNEES	AUFBEREITETE ÖLE
8997	PACKED PLUG VALVE	ROBINET A BOISSEAU	STOPFENHAHN, PACKHAHN, HAHNVENTIL
8998	PACKING	GARNITURE DE JOINT, GARNITURE DE SOUPAGE	DICHTUNG, PACKUNG
8999	PACKING	EMBALLAGE, ENROBAGE	VERPACKUNG, UMHÜLLUNG
9000	PACKING COLLAR	BAGUE DE FOND	GRUNDRING
9001	PACKING GLAND	DOUILLE-FOULOIR (DE PRESSE-ETOUPE)	DICHTUNGSBÜCHSE, STOPFBÜCHSE
9002	PACKING GLAND FLANGE	BRIDE DE PRESSE-ETOUPE	STOPFBÜCHSENBRILLE
9003	PACKING LIST	ETAT DE COLISAGE	KOLLISPEZIFIKATION
9004	PACKING MATERIAL	GARNITURE, BOURRAGE	PACKMATERIAL
9005	PACKING NUT	ECROU DE PRESSE-ETOUPE	STOPFBÜCHSENMUTTER
9006	PACKING PIECE	PIECE D'AJUSTAGE	PASSSTÜCK
9007	PACKING RING	BAGUE DE GARNITURE, BAGUE D'ETANCHEITE	DICHTUNGSRING, LIDERUNGSRING, PACKUNGSRING
9008	PACKING SPACE	BOITE A BOURRAGE, BOITE A GARNITURES, BOISSEAU DE LA BOITE A ETOUPE	STOPFBÜCHSTOPF, STOPFBÜCHSGEHÄUSE, PACKUNGSRAUM
9009	PACKING, JOINT	GARNITURE, BOURRAGE	LIDERUNG, DICHTUNG, PACKUNG
9010	PACKLESS VALVE	ROBINET SANS PRESSE-ETOUPE	VENTIL (STOPFBÜCHSLOSES)
9011	PAD	SOLE DE HAUT-FOURNEAU, ASPERITE, TAMPON, ASSISE, GALETTE	HOCHOFENBODEN, RAUHIGKEIT, STOPFEN, TRAGBELAG
9012	PADTYPE NOZZLE	TUBULURE AUTORENFORCEE	ROHRSTUTZEN (SELBSTVERSTEIFTER)
9013	PAINT	PEINTURE	ANSTRICH
9014	PAINT (TO), COAT WITH PAINT (TO)	PEINTURER, ENDUIRE	ANSTREICHEN
9015	PAINT, COLOUR	COULEUR, PEINTURE	ANSTRICHFARBE
9016	PAINTED PIPE	TUBE PROTEGE PAR UNE PEINTURE	ROHR (ANGESTRICHENES), ROHR (GESTRICHENES)
9017	PAINTER'S TROLLEY	NACELLE DE PEINTRE	SCHWERARBEITSSITZ, ANSTRICHGONDEL
9018	PAINTING	PEINTURE, PEINTURAGE	ANSTRICH, ANSTREICHEN
9019	PAINTING BRUSH	BROSSE, PINCEAU	MALPINSEL
9020	PAINTING BY SPRAYING	PEINTURE AU PISTOLET	SPRITZLACKIEREN
9021	PAINTING DEFECTS	DEFAUTS DE PEINTURE	ANSTRICHMÄNGEL
9022	PAIR OF COMPASSES	COMPAS	ZIRKEL
9023	PALLADIUM	PALLADIUM	PALLADIUM
9024	PALLET	CLAPET DE SOUPAPE	VENTILKLAPPE
9025	PALLETIZED TRUCKS	CAMIONS TRANSPALETTES	PALETTEN-HUBLASTWAGEN
9026	PALLETIZING	PALETTISATION	PALETTIEREN
9027	PALM OIL	HUILE DE PALME	PALMÖL, PALMBUTTER, PALMFETT

9028	PALMITIC ACID	ACIDE PALMITIQUE	PALMITINSÄURE
9029	PAN	RECIPIENT PLAT, CUVE DE FOUR	SCHALE, HERDWANNE
9030	PAN HEAD RIVET	RIVET A TETE TRONCONIQUE	NIET MIT TRAPEZPROFILKOPF
9031	PANE, PEEN, PENE, PEAN, PLEND OF A HAMMER	PANNE DU MARTEAU	HAMMERFINNE, HAMMERPINNE, FINNE, PINNE EINES HAMMERS
9032	PANEL	PLAQUE, MAILLE, PLANCHE, PANNEAU	KUNSTSTOFFPLATTE, RAHMENFÜLLUNG, TAFEL
9033	PANTOGRAPH	PANTOGRAPHE	STORCHSCHNABEL, PANTOGRAPH
9034	PAPER PULLEY	POULIE EN CARTON	RIEMENSCHEIBE AUS PAPIER, PAPIERRIEMENSCHEIBE
9035	PAPER PULP	PATE A PAPIER	PAPIERSTOFF, PAPIERZEUG
9036	PAPIER MACHE	PAPIER MACHE, PAPIER POURRI	PAPIERMACHE
9037	PARA RUBBER	CAOUTCHOUC DE PARA	PARAKAUTSCHUK, PARAGUMMI
9038	PARABOLA	PARABOLE	PARABEL
9039	PARABOLIC CYLINDER	CYLINDRE PARABOLIQUE	ZYLINDER (PARABOLISCHER)
9040	PARABOLIC INTERPOLATION	INTERPOLATION PARABOLIQUE	PARABELINTERPOLATION
9041	PARABOLIC MIRROR	MIROIR PARABOLIQUE	PARABOLSPIEGEL
9042	PARABOLOID OF REVOLUTION	PARABOLOIDE DE REVOLUTION	UMDREHUNGSPARABOLOID, DREHUNGSPARABOLOID, ROTATIONSPARABOLOID
9043	PARAFFIN	PARAFFINE	PARAFFIN
9044	PARAFFIN OIL	HUILE LOURDE DE PARAFFINE, HUILE PARAFFINE	PARAFFINÖL
9045	PARAFFIN WAX	CIRE DE PARAFFINE	HARTPARAFFIN
9046	PARAFFIN, PETROLEUM ENGINE	MOTEUR A PETROLE	PETROLEUMMOTOR
9047	PARALLAX	PARALLAXE	PARALLAXE
9048	PARALLAX ERROR	ERREUR PARALLACTIQUE	PARALLAXENFEHLER
9049	PARALLEL	PARALLELE	PARALLELE
9050	PARALLEL (ADJ)	PARALLELE	PARALLEL
9051	PARALLEL CONNECTION, CONNECTION IN PARALLEL	MONTAGE, GROUPEMENT, COUPLAGE EN QUANTITE, MONTAGE EN SURFACE, MONTAGE EN PARALLELE	NEBENEINANDERSCHALTUNG, PARALLELSCHALTUNG
9052	PARALLEL DISC WITH SPRING	OPERCULE PARALLELE A SERRAGE PAR RESSORT	PARALLELSCHIEBER MIT FEDERVERSCHLUSS
9053	PARALLEL DISC WITH WEDGE	OPERCULE PARALLELE A SERRAGE MECANIQUE	PARALLELSCHIEBER MIT MECHANISCHEM VERSCHLUSS
9054	PARALLEL ENTRY	ENTREE EN PARALLELE	PARALLELANZEIGE
9055	PARALLEL FLOW	COURANT A FILETS PARALLELES	STRÖMUNG (GEORDNETE), STRÖMUNG (LAMINARE), PARALLELSTRÖMUNG, SCHICHTENSTRÖMUNG
9056	PARALLEL FORCES	FORCES PARALLELES	KRÄFT (PARALLELE)
9057	PARALLEL MOTION	PARALLELOGRAMME ARTICULE	PARALLELOGRAMMGETRIEBE
9058	PARALLEL PROJECTION	PROJECTION PARALLELE	PARALLELPROJEKTION
9059	PARALLEL RULER	REGLE A TRACER DES PARALLELES	PARALLELLINEAL
9060	PARALLEL SEAT GATE VALVE	VANNE A SIEGES PARALLELES	PARALLELSCHIEBER
9061	PARALLEL SEAT GATE VALVE WITH SPRING	VANNE A LIBRE DILATATION	PARALLELSCHIEBER (SELBSTDICHTENDER)
9062	PARALLEL SEAT GATE VALVE, WITH WEDGE	VANNE A BLOCAGE MECANIQUE	SCHIEBER MIT MECHANISCHER ABSPERRUNG
9063	PARALLEL SEATS	SIEGES PARALLELES	SITZE (PARALLELE)
9064	PARALLEL SHAFTS	ARBRES PARALLELES	WELLEN (PARALLELE)
9065	PARALLEL THREAD	FILETAGE CYLINDRIQUE	GEWINDE (ZYLINDRISCHES)
9066	PARALLEL VICE	ETAU A SERRAGE PARALLELE	PARALLELSCHRAUBSTOCK
9067	PARALLEL, BLUNT FILE	LIME LARGE, LIME PLATE MAIN	STUMPFFEILE

9068	PARALLELEPIPED, PARALLELOPIPED	PARALLELEPIPEDE	PARALLELEPIPED
9069	PARALLELISM	PARALLELISME	PARALLELITÄT
9070	PARALLELOGRAM	PARALLELOGRAMME	PARALLELOGRAMM
9071	PARALLELOGRAM OF FORCES	PARALLELOGRAMME DES FORCES	PARALLELOGRAMM DER KRÄFTE, KRÄFTEPARALLELOGRAMM
9072	PARALLELOGRAM OF VELOCITIES	PARALLELOGRAMME DES VITESSES	PARALLELOGRAMM DER GESCHWINDIGKEITEN, GESCHWINDIGKEITSPARALLELOGRAMM
9073	PARAMAGNETIC	PARAMAGNETIQUE	PARAMAGNETISCH
9074	PARAMAGNETISM	PARAMAGNETISME	PARAMAGNETISMUS
9075	PARAMETER	PARAMETRE, CARACTERISTIQUE	KENNWERT, PARAMETER, GITTERKONSTANTE
9076	PARCEL PLATING	GALVANISATION PARTIELLE	ÜBERZUG (TEILWEISER)
9077	PARCHMENT	PARCHEMIN	PERGAMENT
9078	PARCHMENT PAPER, ARTIFICIAL IMITATION PARCHMENT	PARCHEMIN VEGETAL, PAPIER-PARCHEMIN, PAPYRINE	PERGAMENTPAPIER, PERGAMENT (VEGETABILISCHES)
9079	PARENT METAL	METAL DE BASE	MUTTERWERKSTOFF, GRUNDMETALL
9080	PARING CHISEL	CISEAU POUR MENUISIER	STECHBEITEL
9081	PARITY CHECK	CONTROLE DE PARITE	PARITÄTSPRÜFUNG
9082	PARKING BRAKE	FREIN DE STATIONNEMENT, FREIN A MAIN	FESTSTELLBREMSE, HANDBREMSE, FESTSTELLBREMSE
9083	PART	PIECE	TEIL
9084	PART PROGRAMMING	PROGRAMMATION DE PIECE	TEILEPROGRAMM
9085	PARTIAL ADMISSION TURBINE	TURBINE A ADMISSION PARTIELLE	PARTIALTURBINE
9086	PARTIAL DIFFERENTIAL EQUATION	EQUATION DIFFERENTIELLE PARTIELLE	PARTIELLE DIFFERENTIALGLEICHUNG
9087	PARTIAL PRESSURE	PRESSION PARTIELLE	TEILDRUCK
9088	PARTICLE	PARTICULE	TEILCHEN
9089	PARTICLE SIZE	GROSSEUR DE LA PARTICULE	TEILCHENGRÖSSE
9090	PARTICLE SIZE DISTRIBUTION	COMPOSITION GRANULOMETRIQUE	KORNGRÖSSENVERTEILUNG
9091	PARTING	PRECIPITATION	SCHEIDUNG
9092	PARTING COMPOUND	LUBRIFIANT DE MOULE	TRENNMITTEL
9093	PARTING LINE	LIGNE DE JOINT, PLAN DE JOINT	TRENNLINIE
9094	PARTING SAND	SABLE ISOLANT	STREUSAND
9095	PARTS LIST	NOMENCLATURE	TEILLISTE
9096	PASS	OPERATION, PASSE, COUCHE	ARBEITSGANG, LAGE
9097	PASS SEQUENCE	SEQUENCE DES CALIBRES	KALIBERFOLGE, STICHFOLGE
9098	PASSAGE OF HEAT	PASSAGE DE LA CHALEUR	WÄRMEÜBERGANG
9099	PASSAGE OF LIGHT	PASSAGE DE LA LUMIERE	LICHTDURCHGANG
9100	PASSAGE OPENING	ORIFICE DE PASSAGE	DURCHLASSÖFFNUNG, DURCHFLUSSÖFFNUNG, DURCHGANGSÖFFNUNG
9101	PASSENGER LIFT	ASCENSEUR	PERSONENAUFZUG, AUFZUG
9102	PASSING LIGHT	FEU DE DEPASSEMENT	ÜBERHOLUNGSLEUCHTE
9103	PASSIVATING	BAIN DE PASSIVATION	PASSIVIERBAD
9104	PASSIVATING FILM	PELLICULE PASSIVANTE	PASSIVIERUNGSSCHICHT
9105	PASSIVATION	PASSIVATION	PASSIVIERUNG
9106	PASSIVATOR	PASSIVANT	PASSIVIERUNGSMITTEL
9107	PASSIVITY	PASSIVITE	PASSIVITÄT
9108	PASTE SOLDER	PATE A SOUDER	LÖTPASTE
9109	PATCH	RAPIECAGE, RAPIECEMENT (RESULTAT)	FLICKEN
9110	PATCH (TO)	RAPIECER	FLICKEN

9111	**PATCHING**	RAPIECAGE, RAPIECEMENT (ACTION)	FLICKEN
9112	**PATENT**	BREVET D'INVENTION	PATENT, ERFINDUNGSPATENT
9113	**PATENT (TO)**	BREVETER	PATENTIEREN, PATENTRECHTLICH SCHÜTZEN
9114	**PATENT AGENT**	AGENT DE BREVETS D'INVENTION	PATENTANWALT
9115	**PATENT FEES**	TAXES DE BREVET	PATENTGEBÜHREN
9116	**PATENT LAW**	BREVETS D'INVENTION (LOI SUR LES)	PATENTGESETZ
9117	**PATENT LEATHER**	CUIR VERNI	GLANZLEDER
9118	**PATENT LEVELING**	PLANAGE	STRECKRICHTEN
9119	**PATENT MEDICINE**	SPECIALITE PHARMACEUTIQUE	SPEZIALITÄT (PHARMAZEUTISCHE)
9120	**PATENT OFFICE**	BREVETS D'INVENTION (BUREAU DES)	PATENTAMT
9121	**PATENT PRIVILEGES, RIGHTS**	DROITS DU BREVETE	PATENTRECHTE
9122	**PATENTABLE INVENTION**	INVENTION BREVETABLE, SUSCEPTIBLE D'ETRE BREVETEE	ERFINDUNG (PATENTFÄHIGE), ERFINDUNG (SCHUTZFÄHIGE)
9123	**PATENTEE**	BREVETE	PATENTINHABER, INHABER EINES PATENTS
9124	**PATENTING**	PATENTAGE	PATENTIEREN
9125	**PATH OF FLOW**	TRAJET SUIVI PAR UN FLUIDE	STRÖMUNGSWEG
9126	**PATHOGENIC ORGANISM**	MICROBE PATHOGENE	KRANKHEITSKEIM, KRANKHEITSERREGER
9127	**PATINA, PATINA FINISCH**	PATINE	PATINA, EDELROST
9128	**PATTERN**	GABARIT, MODELE, DIAGRAMME DE RX	SCHABLONE, MODELL, MUSTER, RÖNTGENBILD ODER DIAGRAM
9129	**PATTERN FOR CASTING**	MODELE POUR FONTE	GUSSMODELL
9130	**PATTERN MAKER**	OUVRIER MODELEUR	MODELLSCHREINER, MODELLTISCHLER
9131	**PATTERN MAKING**	MENUISERIE DE MODELES (ART)	MODELLTISCHLEREI (HERSTELLUNG VON MODELLEN)
9132	**PATTERN SHOP**	MENUISERIE DE MODELES (ATELIER)	MODELLTISCHLEREI (WERKSTÄTTE)
9133	**PATTERN, MODEL**	MODELE	MODELL
9134	**PATTERNED SURFACE**	PEAU D'ORANGE (AFNOR)	APFELSINEN-(SCHALEN-)EFFEKT
9135	**PATTERNMAKING**	MODELAGE	MODELLHERSTELLUNG
9136	**PAVEMENT, PAVING**	PAVE, DALLAGE	PFLASTER, PFLASTERUNG
9137	**PAVING STONE**	PAVE (PIERRE)	PFLASTERSTEIN
9138	**PAWL AND RATCHET MOTION**	ENCLIQUETAGE A ROCHET	ZAHNGESPERRE, KLINKENGETRIEBE, SCHALTGETRIEBE, SCHALTMECHANISMUS, KLINKEN, SCHALTWERK
9139	**PAYLOAD**	CHARGE UTILE	LADEFÄHIGKEIT
9140	**PEA COAL**	GRAINS POUR GAZOGENES	PERLKOHLE, ERBSKOHLE
9141	**PEA CUTTER AND CORN SWATHER**	RECOLTEUSES DE POIS	ERBSENSCHNEIDER
9142	**PEAR-SHAPED FLASK**	BALLON A FOND PLAT	STEHKOLBEN
9143	**PEARL ASH**	PERLASSE	PERLASCHE, POTTASCHE (GEREINIGTE)
9144	**PEARLITE**	PERLITE	PERLIT
9145	**PEARLITIC STEEL**	ACIER PERLITIQUE	STAHL (PERLITISCHER)
9146	**PEARLY LUSTRE**	ECLAT NACRE, PERLE	PERLMUTTERGLANZ
9147	**PEAT**	TOURBE	TORF
9148	**PEAT BRICK**	BRIQUE EN TOURBE	TORFSTEIN
9149	**PEAT BRIQUETTE, BRIQUETTED PEAT**	AGGLOMERE DE TOURBE	TORFBRIKETT
9150	**PEAT SLAB**	PLAQUE DE TOURBE	TORFPLATTE

9151	**PEAT TURF**	MOTTE DE TOURBE	TORFSODE, TORFKUCHEN
9152	**PEBBLE**	CAILLOU	KIESELSTEIN
9153	**PEBBLE TRAP**	PIEGE A CAILLOUX	STEINFÄNGER
9154	**PEDESTAL BEARING**	PALIER A CHAISE SUR LE SOL	BOCKLAGER
9155	**PEEL**	CROUTE	SCHALE
9156	**PEELING**	ECROUTAGE, ECAILLAGE, DECOLLEMENT	SCHÄLEN, ABBLÄTTERN ABLÖSEN
9157	**PEENING**	MARTELAGE, PANNAGE	HÄMMERN, DRESSIEREN, GLÄTTEN
9158	**PEEP HOLE**	REGARD	SCHAULOCH
9159	**PEG**	GOUPILLE	HALTESTEIN, HALTESTIFT
9160	**PEN POINT**	BRANCHE TIRE-LIGNE D'UN COMPAS	ZIEHFEDEREINSATZ, TINTENEINSATZ, TUSCHEINSATZ FÜR ZIRKEL
9161	**PENCIL**	PLUME, STYLET, STYLET TRACEUR, TRACEUR, CRAYON (D'UN APPAREIL ENREGISTREUR)	SCHREIBSTIFT (EINER REGISTRIERVORRICHTUNG)
9162	**PENCIL BUNDLE OF RAYS OF LIGHT**	FAISCEAU LUMINEUX, PINCEAU LUMINEUX	LICHTSTRAHLENBÜNDEL
9163	**PENCIL ERASER**	GOMME A CRAYON	BLEISTIFTGUMMI
9164	**PENCIL OF RAYS**	FAISCEAU DE RAYONS	STRAHLENBÜNDEL
9165	**PENCIL POINT**	BRANCHE PORTE-CRAYON, PORTE-CRAYON D'UN COMPAS	BLEISTIFTEINSATZ FÜR ZIRKEL
9166	**PENDANT BRACKET, HANG DOWN BRACKET**	CHAISE PENDANTE A DEUX JAMBES, CHAISE EN U	HÄNGEBOCK
9167	**PENDULUM**	PENDULE	PENDEL
9168	**PENDULUM GOVERNOR**	REGULATEUR PENDULE	PENDELREGLER
9169	**PENETRATING OIL**	LIQUIDE PENETRANT	ÖL (ROSTLÖSENDES)
9170	**PENETRATION**	PENETRATION	DURCHDRINGUNG
9171	**PENETROMETER**	PENETROMETRE	PENETROMETER
9172	**PENTAGON**	PENTAGONE	FÜNFECK
9173	**PENTAGONAL PRISM**	PRISME PENTAGONAL	PRISMA (FÜNFSEITIGES)
9174	**PENTAGONAL PYRAMID**	PYRAMIDE PENTAGONALE	PYRAMIDE (FÜNFSEITIGE)
9175	**PENTAHEDRON**	PENTAEDRE	FÜNFFLACH, FÜNFFLÄCHNER, PENTAEDER
9176	**PENTANE (NORMAL)**	PENTANE NORMAL	PENTAN (NORMALES)
9177	**PENUMBRA**	PENOMBRE	HALBSCHATTEN
9178	**PEPPER BLISTERS**	POROSITES SUPERFICIELLES	OBERFLÄCHENPOREN
9179	**PERCENTAGE BY VOLUME**	VOLUME (POUR CENT DU)	RAUMTEIL, RAUMPROZENT, VOLUMPROZENT
9180	**PERCENTAGE BY WEIGHT**	POIDS (POUR CENT DU)	GEWICHTSTEIL, GEWICHTSPROZENT
9181	**PERCOLATING WATER**	EAU D'INFILTRATION	SICKERWASSER
9182	**PERCUSSION PRESS**	PRESSE MECANIQUE A FRAPPER	SCHLAGPRESSE (MECHANISCHE)
9183	**PERCUSSIVE WELDING**	SOUDURE PAR PERCUSSION	SCHLAGSCHWEISSEN
9184	**PERFECT COMBUSTION**	COMBUSTION PARFAITE	VOLLKOMMENE VERBRENNUNG
9185	**PERFECT GAS**	GAZ PARFAIT	GAS (VOLLKOMMENES)
9186	**PERFORATE (TO)**	PERFORER, PERCER DES TROUS	DURCHLOCHEN, PERFORIEREN
9187	**PERFORATED LINER**	COLONNE PERDUE A TROUS	LOCHLINER
9188	**PERFORATED PULLEY**	POULIE A JANTE PERFOREE	RIEMENSCHEIBE (DURCHLOCHTE), RIEMENSCHEIBE MIT DURCHLOCHTEM KRANZ
9189	**PERFORATED SHEET**	TOLE PERFOREE, TOLE AJOUREE, TOLE DECOUPEE A JOUR	BLECH (GELOCHTES), BLECH (PERFORIERTES)
9190	**PERIMETER**	PERIMETRE	UMFANG (GEOM.)
9191	**PERIODIC**	PERIODIQUE	PERIODISCH
9192	**PERIODIC DECIMAL FRACTION**	FRACTION DECIMALE PERIODIQUE	DEZIMALBRUCH (PERIODISCHER)

9193	**PERIODIC INSPECTION OF BOILERS**	SURVEILLANCE DES CHAUDIERES	DAMPFKESSELÜBERWACHUNG, KESSELKONTROLLE
9194	**PERIODIC MOTION**	MOUVEMENT PERIODIQUE	BEWEGUNG (PERIODISCHE)
9195	**PERIODICAL INTERMITTENT LUBRICATION**	GRAISSAGE INTERMITTENT	SCHMIERUNG (UNTERBROCHENE)
9196	**PERIODICITY**	PERIODICITE	PERIODIZITÄT
9197	**PERIPHERAL RESISTANCE**	RESISTANCE SUIVANT LA TANGENTE	UMFANGSWIDERSTAND
9198	**PERIPHERAL SPEED, CIRCUMFERENTIAL VELOCITY**	VITESSE PERIPHERIQUE, VITESSE CIRCONFERENTIELLE	UMFANGSGESCHWINDIGKEIT
9199	**PERITECTIC REACTION**	REACTION PERITECTIQUE	REAKTION (PERITEKTISCHE)
9200	**PERLITIC CAST IRON**	FONTE PERLITIQUE	PERLITGUSS
9201	**PERMALLOY**	PERMALLOY	PERMALLOY
9202	**PERMANENT DEFORMATION**	DEFORMATION PERMANENTE	VERFORMUNG (BLEIBENDE)
9203	**PERMANENT HARDNESS OF WATER**	DURETE PERMANENTE DE L'EAU	BLEIBENDE HÄRTE DES WASSERS, PERMANENTE HÄRTE DES WASSERS, MINERALSÄURE HÄRTE DES WASSERS
9204	**PERMANENT MAGNET**	AIMANT PERMANENT	DAUERMAGNET
9205	**PERMANENT MAGNETIC CHUCK**	PLATEAU A AIMANTS PERMANENTS	DAUERMAGNET-SPANNPLATTE
9206	**PERMANENT MAGNETISM**	MAGNETISME PERMANENT	MAGNETISMUS (PERMANENTER)
9207	**PERMANENT MOLD**	COQUILLE	KOKILLE, DAUERFORM
9208	**PERMANENT MOLD CASTING**	COULEE EN COQUILLE	KOKILLENGUSS
9209	**PERMANENT SERVICE**	SERVICE CONTINU	DAUERBETRIEB
9210	**PERMANENT SET**	DEFORMATION PERMANENTE	FORMÄNDERUNG (BLEIBENDE)
9211	**PERMEABILITY, PERVIOUSNESS**	PERMEABILITE	PERMEABILITÄT, DURCHLÄSSIGKEIT, DURCHDRINGLICHKEIT
9212	**PERMEABLE TO AIR**	PERMEABLE A L'AIR	LUFTDURCHLÄSSIG
9213	**PERMEABLE, PERVIOUS**	PERMEABLE	DURCHLÄSSIG
9214	**PERMEAMETER**	PERMEAMETRE	PERMEAMETER
9215	**PERMISSIBLE ALLOWABLE WORKING STRESS**	CHARGE ADMISE	BEANSPRUCHUNG (ZULÄSSIGE), SPANNUNG (ZULÄSSIGE)
9216	**PERMUTATION**	PERMUTATION	VERTAUSCHUNG, VERSETZUNG, PERMUTATION
9217	**PERPENDICULAR**	VERTICALE, PERPENDICULAIRE, APLOMB (D')	LOT, SENKRECHTE, LOTRECHTE, VERTIKALE
9218	**PERPENDICULAR SIDE OF RIGHT-ANGLED TRIANGLE**	COTE DE L'ANGLE DROIT	KATHETE
9219	**PERPENDICULARITY**	PERPENDICULARITE, APLOMB	RICHTUNG (SENKRECHTE), STELLUNG
9220	**PERSPECTIVE**	PERSPECTIF, PERSPECTIVE	PERSPEKTIVISCH, PERSPEKTIVE
9221	**PESTLE**	PILON DE MORTIER	MÖRSERKEULE, STÖSSEL, REIBER
9222	**PETROL ; GASOLINE (U.S.)**	CARBURANT, ESSENCE, ESSENCE MINERALE, PETROLE, GAZOLINE	KRAFTSTOFF, BENZIN, PETROLEUMBENZIN
9223	**PETROL ENGINE (GASOLINE ENGINE) (U.S.)**	MOTEUR A ESSENCE DE PETROLE	BENZINMOTOR
9224	**PETROL LAMP**	LAMPE A ESSENCE MINERALE	BENZINLAMPE
9225	**PETROL LOCOMOTIVE**	LOCOMOTIVE A ESSENCE, LOCOMOTION A BENZINE	BENZINLOKOMOTIVE
9226	**PETROLEUM**	PETROLE	PETROLEUM, ERDÖL
9227	**PETROLEUM LAMP, KEROSENE LAMP**	LAMPE A PETROLE	PETROLEUMLAMPE
9228	**PETROLEUM SPIRIT**	ETHER DE PETROLE, GAZOLINE	PETROLIEUMÄTHER, GASÄTHER, GASOLIN
9229	**PETROLEUM, ROCK OIL**	PETROLE, HUILE DE PIERRE	PETROLEUM, ERDÖL, STEINÖL
9230	**PEWTER**	ETAIN DUR, METAL ANGLAIS	HARTZINN, PEWTER
9231	**PH, PH-VALUE**	PH.	PH-WERT
9232	**PHASE**	PHASE	PHASE

9233	PHASE ANGLE	ANGLE DE DEPHASAGE, ANGLE DE DECALAGE	PHASENVERSCHIEBUNGSWINKEL
9234	PHASE DISPLACEMENT	DEPHASAGE, DECALAGE DES PHASES	PHASENVERSCHIEBUNG
9235	PHASE METER, INDICATOR, POWER FACTOR INDICATOR, METER	INDICATEUR DE PHASES	PHASENMESSER, PHASENINDIKATOR, LEISTUNGSFAKTORMESSER
9236	PHASE RULE	REGLE DES PHASES	PHASENGESETZ
9237	PHASE SHIFT	DECALAGE DE PHASE	PHASENVERSCHIEBUNG
9238	PHENOLPHTHALEIN	PHENOLPHTALEINE, PHTALEINE DU PHENOL	PHENOLPHTALEIN
9239	PHENOLPHTHALEIN PAPER	PAPIER A LA PHENOIPHTALEINE	PHENOLPHTALEINPAPIER
9240	PHENOMENON	PHENOMENE	VORGANG
9241	PHOENIX COLUMN SECTION, PILLAR IRON	FER EN QUART DE ROND	QUADRANTEISEN
9242	PHONAUTOGRAPH	PHONAUTOGRAPHE	VIBROGRAPH, PHONAUTOGRAPH
9243	PHONOLITE	OLIVINE, PERIDOT	PHONOLITH
9244	PHOSPHATE	PHOSPHATE	PHOSPHORSAURES SALZ, PHOSPHAT
9245	PHOSPHATE COATINGS	COUCHE DE PHOSPHATE	PHOSPHATÜBERZUG
9246	PHOSPHOR BRONZE	BRONZE PHOSPHOREUX	PHOSPHORBRONZE, ZINNBRONZE
9247	PHOSPHORESCENCE	PHOSPHORESCENCE	PHOSPHORESZENZ
9248	PHOSPHORESCENT	PHOSPHORESCENT	PHOSPHORESZIEREND
9249	PHOSPHORIC OXIDE	ANHYDRIDE PHOSPHORIQUE	PHOSPHORSÄUREANHYDRID, PHOSPHORPENTOXYD
9250	PHOSPHORIZED COPPER	CUIVRE PHOSPHOREUX	PHOSPHORKUPFER
9251	PHOSPHORS	SUBSTANCES PHOSPHORESCENTES	SUBSTANZEN (SELBSTLEUCHTENDE)
9252	PHOSPHORUS	PHOSPHORE	PHOSPHOR
9253	PHOSPHORUS COPPER	CUIVRE PHOSPHOREUX	PHOSPHORKUPFER
9254	PHOSPHORUS PENTACHLORIDE	PENTACHLORURE DE PHOSPHORE	PHOSPHORPENTACHLORID, PHOSPHORSUPERCHLORID
9255	PHOSPHORUS TRIBROMIDE	TRIBOMURE DE PHOSPHORE	PHOSPHORBROMÜR, PHOSPHORTRIBROMID
9256	PHOSPHORUS TRICHLORIDE	TRICHLORURE DE PHOSPHORE	PHOSPHORTRICHLORID, PHOSPHORCHLORÜR
9257	PHOTO PRINTING PAPER	PAPIER PHOTOCALQUE	LICHTPAUSPAPIER
9258	PHOTO-DIODE	PHOTO-DIODE	FOTODIODE
9259	PHOTO-ELECTRIC CELL	PILE PHOTO-ELECTRIQUE	ZELLE (PHOTOELEKTRISCHE)
9260	PHOTOCELL	CELLULE PHOTO-ELECTRIQUE	PHOTOZELLE (LICHTELEKTRISCHE)
9261	PHOTOCONDUCTIVE CELL	CELLULE PHOTO-CONDUCTRICE	PHOTOWIDERSTAND
9262	PHOTOGRAPH (TO)	PHOTOGRAPHIER	PHOTOGRAPHISCH AUFNEHMEN, PHOTOGRAPHIEREN
9263	PHOTOGRAPH, PHOTOGRAPHIC IMAGE PICTURE	PHOTOGRAPHIE, REPRODUCTION PHOTOGRAPHIQUE	LICHTBILD, PHOTOGRAPHIE
9264	PHOTOGRAPHIC CAMERA	CHAMBRE NOIRE	APPARAT (PHOTOGRAPHISCHER), KAMERA
9265	PHOTOGRAPHY	PHOTOGRAPHIE	PHOTOGRAPHIEREN, PHOTOGRAPHIE
9266	PHOTOMACROGRAPH	MACROPHOTOGRAPHIE	MAKROPHOTOGRAPHIE
9267	PHOTOMETER	PHOTOMETRE	HELLIGKEITSMESSER, LICHTSTÄRKEMESSER, PHOTOMETER
9268	PHOTOMETER BENCH	BANC PHOTOMETRIQUE	PHOTOMETERBANK, BANK (OPTISCHE)
9269	PHOTOMETRIC	PHOTOMETRIQUE	PHOTOMETRISCH
9270	PHOTOMETRIC UNIT	UNITE DE LUMIERE, ETALON PHOTOMETRIQUE	LICHTEINHEIT, EINHEIT PHOTOMETRISCHE
9271	PHOTOMETRY	PHOTOMETRIE	LICHTISTÄRKEMESSUNG, PHOTOMETRIE

9272	PHOTOMICROGRAPH	MICROPHOTOGRAMME	MIKROPHOTOGRAMM
9273	PHOTOMISSIVE CELL	CELLULE PHOTO-EMISSIVE	EMISSIONSPHOTOZELLE
9274	PHOTOMULTIPLIER	PHOTOMULTIPLICATEUR	PHOTOVERVIELFACHER
9275	PHOTON	PHOTON	PHOTON
9276	PHOTOTYPE, SUN COPY, SUN PRINT	PHOTOCALQUE, PHOTOCOPIE	LICHTPAUSE
9277	PHOTOVOLTAIC CELL	CELLULE PHOTOVOLTAIQUE	SPERRSCHICHTPHOTOZELLE
9278	PHYLLITE	PHYLLITE	URTONSCHIEFER, PHYLLIT
9279	PHYSICAL	PHYSIQUE	PHYSIKALISCH
9280	PHYSICAL CHANGE	CHANGEMENT D'ETAT PHYSIQUE, ALTERNATION PHYSIQUE	UMWANDLUNG (PHYSISCHE), VERÄNDERUNG (PHYSIKALISCHE)
9281	PHYSICAL CHEMISTRY	CHIMIE PHYSIQUE	CHEMIE (PHYSIKALISCHE)
9282	PHYSICAL METALLURGY	METALLURGIE PHYSIQUE	METALLURGIE (PHYSIKALISCHE)
9283	PHYSICAL PROPERTIES	PROPRIETES PHYSIQUES	EIGENSCHAFTEN (PHYSIKALISCHE)
9284	PHYSICAL TESTING	ESSAI PHYSIQUE	PRÜFUNG (PHYSIKALISCHE)
9285	PHYSICS	PHYSIQUE	PHYSIK
9286	PHYSIO- CHEMICAL ENGINEERING	CHIMIE INDUSTRIELLE OU TECHNOLOGIQUE	CHEMIE (TECHNISCHE)
9287	PIANO WIRE, MUSIC WIRE	CORDE A PIANO, FIL D'ACIER POUR CORDES A PIANO	KLAVIERSAITENDRAHT
9288	PICKLE, PICKLING	DECAPAGE A L'ACIDE, DECAPAGE CHIMIQUE	BEIZEN, DEKAPIEREN
9289	PICKLING DIP	BAIN DE DECAPAGE, BAIN DE MORDANCAGE	BEIZBAD, BEIZLÖSUNG
9290	PICKLING SOLUTION	SOLUTION ACIDE, BAIN DE DECAPAGE A L'ACIDE	BEIZLÖSUNG
9291	PICRIC ACID	ACIDE PICRIQUE, ACIDE CARBAZOTIQUE	PIKRINSÄURE, TRINITROPHENOL
9292	PIECE OF TUBE, SECTION OF TUBING	TRONCON DE TUYAU	ROHRSTÜCK
9293	PIECE OF WORK	PIECE D'USINAGE, PIECE A TRAVAILLER	WERKSTÜCK, ARBEITSSTÜCK
9294	PIECE WORK	TRAVAIL AUX PIECES, TRAVAIL A LA TACHE	STÜCKARBEIT, AKKORDARBEIT
9295	PIECE WORK WAGES	SALAIRE AUX PIECES, SALAIRE A LA TACHE	STÜCKLOHN, AKKORDLOHN
9296	PIERCING MACHINE	PERCEUSE	LOCHMASCHINE
9297	PIEZOELECTRICITY	PIEZO-ELECTRICITE	PIEZOELEKTRIZITÄT
9298	PIG	GUEUSE	ROHEISENMASSEL
9299	PIG IRON	FONTE BRUTE, FONTE CRUE	GUSSEISEN, ROHEISEN, ROHGUSS
9300	PIG SKIN	PEAU DE PORC	SCHWEINSLEDER
9301	PIG TALL HOOK	CROCHET A QUEUE DE COCHON	SAUSCHWANZHAKEN
9302	PIG, SOW	GUEUSE	MASSEL
9303	PIGMENT, COLOUR, DYE	COLORANT, MATIERE COLORANTE, COULEUR	FARBE, FARBSTOFF
9304	PILE	PILE	STAPEL
9305	PILE CAP	DALLE SUPPORT / SUR PIEUX	PFAHLKAPPE, SOHLE AUF PFÄHLEN
9306	PILE DRIVING	BATTAGE DE PIEUX	PFAHLEINTREIBEN, EINRAMMEN
9307	PILE WOOD (TO)	EMPILER LE BOIS	AUFSTAPELN (HOLZ)
9308	PILING FURMACE	FOUR A RECHAUFFER EN PAQUETS	PAKETWÄRMOFEN
9309	PILING WOOD	EMPILAGE DU BOIS	AUFSTAPELN DES HOLZES
9310	PILL PRESS	PRESSE A PASTILLES, PASTILLEUSE	TABLETTENMASCHINE
9311	PILLAR	PILIER	PFEILER
9312	PILLAR DRILLING MACHINE, PILLAR TYPE DRILLING MACHINE	PERCEUSE A COLONNE, FOREUSE A COLONNE	SÄULENBOHRMASCHINE

9313	**PILLOW, BODY OF BEARING**	CORPS DU PALIER	LAGERKÖRPER
9314	**PIMPLING**	GRANULATION SUPERFICIELLE	PICKELBILDUNG
9315	**PIN**	PIVOT, GOUPILLE, CHEVILLE, GOUJON AIGUILLE, AXE	ZAPFEN, BOLZEN, STIFT, WELLE
9316	**PIN DRILL, PLUG CENTRE BIT**	FORET A TETON CYLINDRIQUE	ZAPFENBOHRER
9317	**PIN HEAD BLISTER**	TETE D'EPINGLE	NADELSTAPEL
9318	**PIN JOINT**	ARTICULATION A TOURILLON	ZAPFENGELENK, BOLZENGELENK
9319	**PIN OF CENTRE CRANK**	BOUTON DE VILEBREQUIN, MANETON	KRUMMZAPFEN
9320	**PIN OF ROLLER CHAIN**	GOUJON D'ASSEMBLAGE, SOIE, PIVOT, D'UNE CHAINE A ROULEAUX	BOLZEN, INNENBOLZEN EINER ROLLENKETTE, KETTENBOLZEN
9321	**PIN RIVETING**	RIVETAGE A TIGE, RIVETAGE A RIVET SANS TETE	STIFTNIETUNG
9322	**PIN SPANNER, FORK SPANNER**	CLEF A GRIFFES EN BOUT	GABELSCHLÜSSEL, STIFTSCHLÜSSEL
9323	**PIN STAVE OF LANTERN WHEEL**	FUSEAU D'UN ENGRENAGE A LANTERNE	TRIEBSTOCK
9324	**PIN WHEEL**	PIGNON A FUSEAUX	ZAPFENZAHNRAD, TRIEBSTOCKRAD, STIFTRAD
9325	**PIN WHEEL AND PINION GEAR**	ENGRENAGE A POINT, ENGRENAGE A FUSEAUX, ENGRENAGE A LANTERNE	TRIEBSTOCKGETRIEBE, STOCKGETRIEBE
9326	**PIN-PUNCH**	CHASSE-GOUPILLE	SPLINTTREIBER
9327	**PINACOID**	PINACOIDE	PINAKOID
9328	**PINCERS**	TENAILLE	NAGELZANGE
9329	**PINCH EFFECT**	EFFET DE PINCEMENT	EINSCHNÜRUNGSEFFEKT, PINCHEFFEKT
9330	**PINCH PASS**	LAMINAGE A FAIBLE PRESSION	KALTNACHWALZUNG
9331	**PINE TREE CRYSTALS**	DENDRITE	DENDRIT
9332	**PINHOLE**	TROU D'AXE OU DE GOUPILLE	STIFTLOCH, SPLINTLOCH
9333	**PINHOLE POROSITY**	PIQURE	SPLINTLOCH
9334	**PINHOLING**	PIQURES	KRATERBILDUNG
9335	**PINION**	PIGNON	RITZEL, ZAHNRAD KLEINERE
9336	**PINION DRIVE**	LANCEUR	ANLASSER, RITZEL
9337	**PINT**	PINT	PINT
9338	**PINTLE TYPE NOZZLE**	INJECTEUR A TETON	ZAPFENDÜSE
9339	**PIPE**	TUBE, TUYAU, CANALISATION	ROHR, SCHLAUCH, LEITUNG, TREIBSTOFFSYSTEM
9340	**PIPE (IN CASTINGS)**	RETASSURE	LUNKER, LUNGER, SAUGTRICHTER
9341	**PIPE BENDING**	CINTRAGE D'UN TUBE	BIEGEN EINES ROHRES
9342	**PIPE CAST HORIZONTALLY**	TUBE COULE HORIZONTAL	ROHR (LIEGEND GEGOSSENES)
9343	**PIPE CAST UPRIGHT ON END**	TUBE COULE DEBOUT	ROHR (STEHEND GEGOSSENES)
9344	**PIPE CLAMP**	COLLIER DE FIXATION (POUR TUBE)	ROHRSCHELLE
9345	**PIPE CLIP**	COLLIER POUR TUBES, COLLIER POUR SCELLEMENT DANS LES MURS	ROHRSCHELLE
9346	**PIPE COIL**	SERPENTIN	ROHRSCHLANGE
9347	**PIPE CONNECTION**	JOINT DE TUBES	ROHRVERBINDUNG
9348	**PIPE CUTTER**	COUPE-TUBES	ROHRABSCHNEIDER
9349	**PIPE CUTTING MACHINE**	MACHINE A COUPER LES TUBES	ROHRABSCHNEIDEMASCHINE
9350	**PIPE FITTINGS**	RACCORD POUR TUBES	FORMSTÜCK FÜR ROHRLEITUNGEN, ROHRFORMSTÜCK, FASSONROHR
9351	**PIPE FLANGE**	BRIDE RONDELLE DE TUBE, BRIDE DE TUBE	FLANSCH EINES ROHRES, ROHRFLANSCH
9352	**PIPE FLARING MACHINE**	MACHINE A ELARGIR LES TUBES	ROHRWEITEMASCHINE

9353	PIPE FRACTURE, FRACTURE OF A PIPE	RUPTURE, BRIS D'UN TUBE	ROHRBRUCH
9354	PIPE HOOK	CROCHET POUR TUBES	ROHRHAKEN
9355	PIPE JOINT	JOINT POUR TUYAUTERIE	ROHRDICHTUNG
9356	PIPE LAYING	POSE DE TUBES	VERLEGEN VON ROHREN
9357	PIPE LINE, PIPING, TUBING, SERIES OF PIPES	TUYAUTAGE, TUYAUTERIE, TUYAU DE CONDUITE, CONDUITE	ROHRLEITUNG, ROHRSTRECKE, ROHRSTRANG, RÖHRENFAHRT, RÖHRENTOUR
9358	PIPE MILL	LAMINOIR A TUBES	RÖHRENWALZWERK
9359	PIPE STANDARDS	STANDARDS POUR TUYAUTERIES	ROHRNORMALIEN
9360	PIPE THREAD	PAS POUR TUBES	ROHRGEWINDE
9361	PIPE THREAD CUTTING MACHINE	MACHINE A TARAUDER ET FILETER LES TUBES	ROHRGEWINDE-SCHNEIDEMASCHINE
9362	PIPE TONGS	PINCES A TUBES	ROHRZANGE
9363	PIPE VICE	ETAU A TUBES	ROHRSCHRAUBSTOCK
9364	PIPE WELDED BY THE AUTOGENOUS PROCESS	TUBE SOUDE A L'AUTOGENE	ROHR (AUTOGEN GESCHWEISSTES)
9365	PIPE WITH BRANCH CONNECTIONS	TUBE A RACCORDS	ABZWEIGROHR
9366	PIPE WRENCH, TUBE WRENCH	CLEF A TUBES, CLEF CROCODILE, MACHOIRE DE CROCODILE	ROHRSCHLÜSSEL
9367	PIPECLAY	TERRE DE PIPE	PFEIFENTON, PFEIFENERDE
9368	PIPETTE	PIPETTE	PIPETTE
9369	PISTOL GRIP HAND HACKSAW	SCIE A METAUX A POIGNEE PISTOLET	PISTOLENGRIFFMETALLSÄGE
9370	PISTON	PISTON	KOLBEN (MASCHINENB.)
9371	PISTON ACCELERATION	ACCELERATION DU PISTON	KOLBENBESCHLEUNIGUNG
9372	PISTON BLOWER	VENTILATEUR A PISTON	KOLGENGEBLÄSE
9373	PISTON BODY HEAD	CORPS DU PISTON	KOLBENKÖRPER
9374	PISTON COMPRESSOR	COMPRESSEUR A PISTON	KOLBENVERDICHTER
9375	PISTON PACKING	GARNITURE DE PISTON	KOLBENDICHTUNG
9376	PISTON PIN	AXE DE PISTON	KOLBENBOLZEN
9377	PISTON PRESSURE	PRESSION SUR LE PISTON	KOLBENDRUCK
9378	PISTON PUMP	POMPE A PISTON	KOLBENPUMPE
9379	PISTON RING	BAGUE, SEGMENT DE PISTON	KOLBENRING
9380	PISTON RING WITH BUTT JOINT	BAGUE DE PISTON AVEC FENTE D'EQUERRE	KOLBENRING MIT SENKRECHTER STOSSFUGE
9381	PISTON RING WITH INCLINED JOINT	BAGUE DE PISTON A JOINT EN BISEAU	KOLBENRING MIT SCHRÄGER STOSSFUGE
9382	PISTON RING WITH LAP JOINT	BAGUE DE PISTON AVEC JOINT (FENTE) A RECOUVREMENT	KOLBENRING MIT TREPPENSTOSS, KOLBENRING MIT ÜBERLAPPTER STOSSFUGE
9383	PISTON ROD	TIGE DU PISTON	KOLBENSTANGE
9384	PISTON ROD END	EXTREMITE, QUEUE DE LA TIGE DU PISTON	KOLBENSTANGENENDE
9385	PISTON ROD GUIDE	GUIDE DE LA TIGE DU PISTON	KOLBENSTANGENFÜHRUNG
9386	PISTON ROD PACKING	GARNITURE, BOURRAGE DE TIGE DE PISTON	KOLBENSTANGENPACKUNG, STANGENDICHTUNG
9387	PISTON ROD WITH TAIL ROD	TIGE DE PISTON TRAVERSANTE, BILATERALE	KOLBENSTANGE (DURCHLAUFENDE), KOLBENSTANGE (DURCHGEHENDE)
9388	PISTON SPEED VELOCITY	VITESSE DU PISTON	KOLBENGESCHWINDIGKEIT
9389	PISTON STROKE, STROKE OF PISTON	COUP DE PISTON	KOLBENHUB, HUB EINES KOLBENS, KOLBENBEWEGUNG
9390	PISTON VALVE	TIROIR CYLINDRIQUE, TIROIR A PISTON, TIROIR EQUILIBRE, PISTON DISTRIBUTEUR	KOLBENSCHIEBER

9391	**PISTON WITH TAIL ROD**	PISTON A CONTRE-TIGE, PISTON A DOUBLE TIGE	KOLBEN (SCHWEBENDER, VON DER STANGE GETRAGENER), FREISCHWEBENDER KOLBEN
9392	**PISTON WITHOUT TAIL ROD**	PISTON A TIGE UNIQUE	KOLBEN (AUFLIEGENDER), KOLBEN (VON DER ZYLINDERWAND GETRAGENER), SELBSTTRAGENDER KOLBEN, SCHLEIFKOLBEN
9393	**PIT**	PIQURE	GRÜBCHEN
9394	**PIT CASTING**	COULEE EN FOSSE	GRUBENGUSS
9395	**PIT PLANING MACHINE**	MACHINE A RABOTER A FOSSE	GRUBENHOBELMASCHINE
9396	**PIT SAND**	SABLE DE CARRIERE	GRUBENSAND
9397	**PIT WATER**	EAU DE PUITS DE MINE	GRUBENWASSER, BERGWERKSWASSER
9398	**PIT, SHAFT OF A MINE**	PUITS (D'UNE MINE)	SCHACHT, GRUBENSCHACHT
9399	**PITCH**	PAS DE FILETAGE	GEWINDESTEIGUNG, TEILUNG
9400	**PITCH**	POIX	PECH, SCHIFFSPECH, SCHUSTERPECH
9401	**PITCH ANGLE**	ANGLE PRIMITIF	TEILKEGELWINKEL
9402	**PITCH CHAIN WHEEL, SPROCKET WHEEL, LARGE SPROCKET**	ROUE, PIGNON DE CHAINE	KETTENZAHNRAD
9403	**PITCH CIRCLE**	CERCLE PRIMITIF DE DENTURE	ZAHNTEILKREIS
9404	**PITCH CIRCLE, PITCH LINE, PRIMITIVE CIRCLE CIRCUMFERENCE, PITCH CIRCOMFERENCE**	CERCLE PRIMITIF, CIRCONFERENCE PRIMITIVE	TEILKREIS
9405	**PITCH COAL**	LIGNITE NOIR	PECHKOHLE
9406	**PITCH CONE**	CONE PRIMITIF	GRUNDKEGEL
9407	**PITCH DIAMETER**	DIAMETRE PRIMITIF, DIAMETRE A FLANC DE FILET	TEILKREISDURCHMESSER, FLANKENDURCHMESSER
9408	**PITCH OF GEAR, PITCH OF TEETH**	PAS DE L'ENGRENAGE	ZAHNTEILUNG
9409	**PITCH OF RIVETS**	DISTANCE DE CENTRE A CENTRE DES RIVETS	NIETTEILUNG
9410	**PITCH OF THREAD**	PAS DE L'HELICE, PAS DE VIS	GANGHÖHE EINER SCHRAUBE
9411	**PITCHBLENDE**	PECHBLENDE, PECHURANE	PECHBLENDE
9412	**PITCHED CHAIN**	CHAINE CALIBREE	KETTE (KALIBRIERTE), KETTE (ADJUSTIERTE)
9413	**PITH**	MOELLE DU BOIS	MARK DES HOLZES
9414	**PITMAN ARM**	LEVIER DE COMMANDE	LENKSTOCKHEBEL
9415	**PITOT TUBE**	TUBE DE PITOT	PITOTROHR
9416	**PITS**	PIQURES	ÄTZGRÜBCHEN
9417	**PITTING**	PIQURE DE LAMINAGE	WALZNARBE
9418	**PITTING CORROSION**	CORROSION LOCALISEE	LOCHFRASS
9419	**PIVOT**	PIVOT	ZAPFEN, ANGEL
9420	**PIVOT**	TOURILLON, PIVOT	ZAPFEN, WELLENZAPFEN
9421	**PIVOT, JOURNAL OF FOOTSTEP BEARING**	PIVOT (D'UN ARBRE VERTICAL), PIVOT, TOURILLON D'APPUI, TOURILLON DE BUTEE POUR CRAPAUDINE	SPURZAPFEN, STÜTZZAPFEN
9422	**PIVOTED**	PIVOTANT	DREHBAR, AUF ZAPFEN GELAGERT
9423	**PLACE OF ERECTION OF INSTALLATION**	LIEU D'ETABLISSEMENT, LIEU D'INSTALLATION, LIEU DE MONTAGE	AUFSTELLUNGSORT
9424	**PLAIN CYLINDRICAL JOURNAL BEARING**	PALIER A COUSSINETS LISSES, ROULEMENT LISSE	GLEITLAGER
9425	**PLAIN MANDREL**	MANDRIN DE MONTAGE LISSE	DORN (FESTER)
9426	**PLAIN SLIDE VALVE**	TIROIR PLAT, PLAN	FLACHSCHIEBER
9427	**PLAIN STEEL**	ACIER AU C.	STAHL

9428	PLAIN TURNING LATHE	TOUR A CYLINDRER, TOUR PARALLELE	PARALLELDREHBANK, EGALISIERBANK
9429	PLAIN VICE	ETAU SIMPLE	SCHRAUBSTOCK (EINFACHER)
9430	PLAN OF SITE, GENERAL PLAN	PLAN, DESSIN, VUE D'ENSEMBLE	LAGEPLAN, ÜBERSICHTSPLAN, SITUATIONSPLAN
9431	PLAN RING GAUGE	CALIBRE FEMELLE LISSE	LEHRRING (NORMALE)
9432	PLANE	RABOT	HOBEL
9433	PLANE	PLAN	EBENE (GEOM.)
9434	PLANE (TO)	RABOTER	HOBELN
9435	PLANE CURVE	COURBE PLANE	KURVE (EBENE)
9436	PLANE IRON, CUTTING IRON OF A PLANE	FER DU RABOT	HOBELEISEN, HOBELSTAHL, HOBELMESSER, HOBELSTICHEL
9437	PLANE MIRROR	MIROIR PLAN	SPIEGEL (EBENER)
9438	PLANE OF PROJECTION	PLAN DE PROJECTION	PROJEKTIONSEBENE
9439	PLANE OF REFRACTION	PLAN DE REFRACTION	BRECHUNGSEBENE
9440	PLANE OF SYMETRY	PLAN DE SYMETRIE	SYMMETRIEEBENE
9441	PLANE SURFACE, FLAT FACE	SURFACE PLANE	FLÄCHE (EBENE)
9442	PLANE-FACING AND SURFACING MACHINE	MACHINE A DRESSER ET A SURFACER	ABRICHT- U. FLÄCHENSCHLEIFMASCHINE
9443	PLANER	RABOTEUSE, RABOTEUR, OUVRIER RABOTEUR	HOBELMASCHINE, HOBLER
9444	PLANET WHEEL	PIGNON, ROUE SATELLITE, SATELLITE, ROUE PLANETAIRE	PLANETENRAD
9445	PLANETARY	PLANETAIRE	PLANETARISCH
9446	PLANETARY GEAR	ENGRENAGE EPICYCLIQUE, TRAIN PLANETAIRE	PLANETENGETRIEBE
9447	PLANETARY MOTION	MOUVEMENT PLANETAIRE, SATELLITE	PLANETENBEWEGUNG
9448	PLANIMETER	PLANIMETRE	PLANIMETER
9449	PLANIMETRY	PLANIMETRIE	PLANIMETRIE
9450	PLANING	RABOTAGE, RABOTEMENT	HOBELN
9451	PLANING MACHINE	MACHINE A RABOTER, PLANEUSE	HOBELMASCHINE
9452	PLANING MACHINE, PLANER	MACHINE A RABOTER, RABOTEUSE	HOBELMASCHINE
9453	PLANING-MILLING MACHINE	RABOTEUSE-FRAISEUSE	HOBEL-UND FRÄSMASCHINE
9454	PLANISHING	PLANAGE	PLANIEREN, GLÄTTEN
9455	PLANISHING HAMMER	MARTEAU A PLANER	SCHLICHTHAMMER, PLANIERHAMMER
9456	PLANISHING MILL	LAMINOIR A POLIR	GLÄTTWALZWERK
9457	PLANK, DEAL	PLANCHE FORTE, MADRIER	BOHLE
9458	PLANO-CONCAVE LENS	LENTILLE PLAN-CONCAVE	LINSE (PLANKONKAVE)
9459	PLANO-CONVEX LENS	LENTILLE PLAN-CONVEXE	LINSE (PLANKONVEXE)
9460	PLANO-MILLER, PLANER TYPE MILLING MACHINE, SLAB MILLER, SLABBING MACHINE	FRAISEUSE-RABOTEUSE	LANGFRÄSMASCHINE IN HOBELMASCHINENFORM
9461	PLANT	INSTALLATION INDUSTRIELLE OU USINE	WERKE, ANLAGEN
9462	PLANT ECONOMY, ECONOMY OF AN UNDERTAKING, ENTERPRISE	PROSPERITE ECONOMIQUE D'UNE ENTREPRISE, RENTABILITE	WIRTSCHAFTLICHKEIT EINES BETRIEBES
9463	PLASTER	ENDUIT (MACON.)	PUTZ, VERPUTZ
9464	PLASTER (TO)	ENDUIRE (MACON.)	VERPUTZEN (BAUW.)
9465	PLASTER MOLD	MOULE EN PLATRE	GIPSFORM
9466	PLASTER OF PARIS	PLATRE	GIPS (GEBRANNTER)
9467	PLASTIC BODY	CORPS PLASTIQUE	KÖRPER (BILDSAMER), KÖRPER (FORMBARER), KÖRPER (PLASTISCHER)

9468	PLASTIC COATED WIRE	FIL GAINE DE PLASTIQUE	DRAHT (KUNSTSTOFFUMKLEIDETER)
9469	PLASTIC DEFORMATION	DEFORMATION PLASTIQUE	VERFORMUNG (PLASTISCHE)
9470	PLASTIC FIRECLAY	ARGILE PLASTIQUE	FORMENTON
9471	PLASTIC FLOW	FLUAGE PLASTIQUE, DEFORMATION PLASTIQUE	FLIESSEN (PLASTISCHES), KRIECHEN
9472	PLASTIC INSULATED CABLE	CABLE ISOLE AU P.V.C.	KABEL (KUNSTSTOFFISOLIERTES)
9473	PLASTIC REFRACTORY `	REFRACTAIRE PLASTIQUE	PLASTMASSE (FEUERFESTE)
9474	PLASTICITY	PLASTICITE	BILDSAMKEIT, PLASTIZITÄT
9475	PLASTICIZER	PLASTIFIANT	WEICHMACHER
9476	PLASTICS	MATIERES PLASTIQUES	KUNSTSTOFFE
9477	PLATE	PLAQUE, TOLE FORTE, TOLE EPAISSE	PLATTE, GROBBLECH
9478	PLATE (FLANGING QUALITY), SOFT SHEETS	TOLE POUR EMBOUTISSAGE	BÖRDELBLECH
9479	PLATE (TO)	PLAQUER	BEPLATTEN, PLATTIEREN
9480	PLATE BENDING MACHINE	MACHINE A CINTRER LES TOLES	BLECHBIEGEMASCHINE
9481	PLATE BENDING PRESS	PRESSE PLIEUSE	BLECHABKANTPRESSE
9482	PLATE CAM, DISC CAM	CAME A DISQUE, CAME A PLATEAU	SCHEIBENEXZENTER
9483	PLATE EDGE	RIVE DE TOLE	BLECHRAND, BLECHKANTE
9484	PLATE EDGE PLANING MACHINE	MACHINE A CHANFREINER LES TOLES	BLECHKANTENHOBELMASCHINE
9485	PLATE FLATTENING MACHINE	MACHINE A PLANER LES TOLES	BLECHRICHTMASCHINE
9486	PLATE FRICTION CLUTCH, DISC FRICTION CLUTCH	EMBRAYAGE A LAMES, A DISQUES	PLANSCHEIBENKUPPLUNG, LAMELLENKUPPLUNG
9487	PLATE GAUGE	CALIBRE D'UNE TOLE	BLECHDICKE
9488	PLATE GIRDER	POUTRE COMPOSEE	VERBUNDTRÄGER
9489	PLATE GIRDER STIFFENER	RAIDISSEUR DE POUTRES EN TOLE, ENTRETOISE	VERSTEIFUNGSBLECH
9490	PLATE GIRDER, BUILT-UP GIRDER	POUTRE COMPOSEE	BLECHTRÄGER
9491	PLATE GLASS	VERRE A GLACES	SPIEGELGLAS
9492	PLATE LEAF OF A SPRING	LAME, FEUILLE D'UN RESSORT	FEDERBLATT
9493	PLATE LEFT-OVER	CHUTE (RESTE) DE TOLE	BLECHABFALL
9494	PLATE SHEARING MACHINE	CISAILLE A TOLES	BLECHSCHERMASCHINE, BLECHSCHERE
9495	PLATE SPRING	RESSORT A LAME	BLATTFEDER
9496	PLATE TRIMMING	EBARBAGE DE TOLE EN BISEAU	BLECHKANTENABSCHRÄGUNG
9497	PLATE-CAMS	CAMES-DISQUES	STEUERSCHEIBEN
9498	PLATELET	PLAQUETTE	PLÄTTCHEN
9499	PLATEN AREA	PLATINE DE FIXATION	AUFSPANNPLATTE
9500	PLATES LAY-OUT SEQUENCE	TOLES (ORDRE DE MISE EN PLACE DES)	BLECHEINLEGEFOLGE
9501	PLATES OVERLAPPING	RECOUVREMENT DE TOLES	PLATTENÜBERLAPPUNG
9502	PLATFORM / LANDING	PALIER	TREPPENABSATZ, BÜHNE
9503	PLATING	PLACAGE, REVETEMENT METALLIQUE PAR GALVANOPLASTIE	GALVANISCHER UEBERZUG, BEPLATTUNG, PLATTIEREN
9504	PLATING LAY-OUT DRAWING	PLAN DE LA TOLERIE	BLECHKONSTRUKTIONSZEICHNUNG
9505	PLATING RACK	CHARIOT DE SUSPENSION POUR PIECES A TREMPER	EINHÄNGEGESTELL
9506	PLATING SALTS	SELS POUR BAINS	BADZUSÄTZE
9507	PLATINIC CHLORIDE	CHLORURE PLATINIQUE	PLATINCHLORID, PLATINTETRACHLORID, CHLORPLATIN
9508	PLATINITE	ACIER PLATINITE, PLATINITE	PLATINID

9509	**PLATINOUS CHLORIDE**	CHLORURE PLATINEUX	PLATINCHLORÜR
9510	**PLATINUM**	PLATINE (METAL)	PLATIN
9511	**PLATINUM BLACK**	NOIR DE PLATINE	PLATINMOHR, PLATINSCHWARZ
9512	**PLATINUM CRUCIBLE**	CREUSET EN PLATINE	PLATINTIEGEL
9513	**PLATINUM WIRE**	FIL DE PLATINE	PLATINDRAHT
9514	**PLATINUM-BLACK**	MOUSSE DE PLATINE	PLATINMOHR
9515	**PLATINUM-IRIDIUM**	PLATINE IRIDIE	PLATIN-IRIDIUM
9516	**PLAY, CLEARANCE**	JEU (MOUVEMENT), JEU DE MONTAGE, JEU UTILE, CHASSE	SPIEL (RAUM)
9517	**PLENUM SYSTEM VENTILATION**	VENTILATION PAR PULSION	DRUCKLÜFTUNG, PULSIONSVENTILATION
9518	**PLIANT, PLIABLE, FLEXIBLE**	FLEXIBLE	BIEGSAM, BIEGBAR
9519	**PLOT (TO)**	TRACER GRAPHIQUEMENT	AUFZEICHNEN
9520	**PLOT A DIAGRAM (TO)**	TRACER, RELEVER UN DIAGRAMME	DIAGRAMM AUFZEICHNEN (EIN)
9521	**PLOTTING A DIAGRAM**	TRACE D'UN DIAGRAMME	AUFZEICHNEN EINES DIAGRAMMES
9522	**PLOUGH**	GUILLAUME	SIMSHOBEL, GESIMSHOBEL
9523	**PLOUGHS DISC (TRAILED OR MOUTED)**	CHARRUES A DISQUES (TRAINEES OU PORTEES)	SCHEIBENPFLÜGE UND ANBAUPFLÜGE
9524	**PLOUGHS MOULDBOARD, TRACTOR MOUNTED**	CHARRUES PORTEES	SCHARPFLÜGE UND SCHLEPPERPFLÜGE
9525	**PLOUGHS MOULDBOARD, TRACTOR TRAILED**	CHARRUES TRAINEES	PFLÜGE DURCH TRAKTOREN ANGETRIEBENE
9526	**PLOUGHS ONE-WAY, TRAILED OR MOUNTED**	CHARRUES, BRABANT (TRAINEES OU PORTEES)	KEHRPFLÜGE UND ANHANGEPFLÜGE
9527	**PLOUGHS RIDGING**	CHARRUES BILLONEUSES	HÄUFEPFLÜGE
9528	**PLOW STEEL**	ACIER A CHARRUE	PATENT-PFLUGSTAHL
9529	**PLUG**	BOUCHON, FICHE DE PRISE DE COURANT, OBTURATEUR, TAMPON	STOPFENSTECKER, STOPFEN, KÜKEN, VERSCHLUSSSTÜCK, ABSPERRSTÜCK, PFLOCK, PFROPFEN
9530	**PLUG COCK**	ROBINET A CLEF	KEGELHAHN
9531	**PLUG GAGE**	TAMPON	LEHRDORN
9532	**PLUG OF A COCK**	CLEF, NOIX DE ROBINET	KÜKEN, WIRBEL, REIBER, SCHLÜSSEL, KONUS EINES HAHNES, HAHNKÜKEN, HAHNKEGEL, HAHNWIRBEL, HAHNSCHLÜSSEL, HAHNREIBER
9533	**PLUG SQUARE**	CARRE (POUR CLEF)	VIERKANT
9534	**PLUG TAP, PARALLEL TAP**	TARAUD CYLINDRIQUE, TARAUD FINISSEUR	GEWINDENACHSCHNEIDER
9535	**PLUG VALVE**	ROBINET A BOISSEAU	HAHNVENTIL
9536	**PLUG WELD**	SOUDURE EN BOUCHON	LOCHNAHTSCHWEISSUNG
9537	**PLUG WELDING**	SOUDAGE DE TROUS DE RIVETS	NIETLOCHSCHWEISSEN
9538	**PLUMB BOB, BOB**	PLOMB (DU FIL A PLOMB)	SENKELGEWICHT, SENKELBIRNE
9539	**PLUMB LINE**	FIL A PLOMB, APLOMB, PLOMB	SENKEL, SENKLOT, LOT
9540	**PLUMB RULE**	NIVEAU A FIL	BLEILOT, BLEIWAAGE
9541	**PLUMBAGO**	PLOMBAGINE, GRAPHITE	GRAPHIT
9542	**PLUMBAGO, BLACK LEAD**	PLOMBAGINE, MINE DE PLOMB	GRAPHIT, PLUMBAGO
9543	**PLUMMER BLOCK, PILLOW BLOCK**	PALIER A PATIN	STEHLAGER
9544	**PLUNGER**	POINCON	STEMPEL
9545	**PLUNGER**	PISTON PLONGEUR, PLONGEUR	TAUCHKOLBEN, ROHRKOLBEN, MÖNCHSKOLBEN, PLUNSCHER, PLUNGER
9546	**PLUNGER PUMP**	POMPE A PISTON PLONGEUR	TAUCHKOLBENPUMPE, PLUNSCHEPUMPE, PLUNGERPUMPE
9547	**PLUNGER SPRING WITH RETAINER**	RESSORT DE PISTON AVEC CUVETTE	KOLBENFEDER MIT TELLER
9548	**PLUS MESH**	REFUS DE CRIBLAGE	SIEBRÜCKSTAND

9549	**PNEUMATIC**	PNEUMATIQUE	PNEUMATISCH
9550	**PNEUMATIC CHUCKS**	MANDRINS PNEUMATIQUES	FUTTER (PNEUMATISCHE)
9551	**PNEUMATIC DROP HAMMER**	MARTEAU PNEUMATIQUE	DRUCKLUFTHAMMER
9552	**PNEUMATIC FORGING HAMMER**	MARTEAU-PILON	SCHMIEDEHAMMER
9553	**PNEUMATIC HAMMER**	MARTEAU PNEUMATIQUE, MARTEAU A AIR COMPRIME	PRESSLUFTHAMMER, DRUCKLUFTHAMMER, LUFTHAMMER
9554	**PNEUMATIC KNOCKOUT**	EJECTEUR PNEUMATIQUE, DECOCHAGE, DEMOULAGE	AUSWERFER, AUSSCHLAGEN
9555	**PNEUMATIC RIVETING HAMMER**	MARTEAU A RIVER PNEUMATIQUE, MARTEAU MECANIQUE A AIR COMPRIME	SCHLAGNIETHAMMER, PRESSLUFTNIETHAMMER, DRUCKLUFTNIETHAMMER
9556	**PNEUMATIC TOOL**	OUTIL PNEUMATIQUE	PRESSLUFTWERKZEUG
9557	**PNEUMATICALLY OPERATED VALVE**	SOUPAPE A COMMANDE PNEUMATIQUE	PRESSLUFTVENTIL, VENTIL (PNEUMATISCH GESTEUERTES)
9558	**PNEUMATICS**	PNEUMATIQUE	MECHANIK DER LUFT, AEROMECHANIK
9559	**POINT**	POINT	PUNKT
9560	**POINT IN SPACE**	POINT DANS L'ESPACE	RAUMPUNKT
9561	**POINT OF APPLICATION**	POINT D'APPLICATION	ANGRIFFSPUNKT
9562	**POINT OF CONTACT**	POINT DE CONTACT, POINT DE TANGENCE	BERÜHRUNGSSTELLE, BERÜHRUNGSPUNKT
9563	**POINT OF FRACTURE**	POINT DE RUPTURE	BRUCHSTELLE
9564	**POINT OF INFLECTION**	POINT D'INFLEXION	WENDEPUNKT, INFLEXIONSPUNKT (GEOM.)
9565	**POINT OF INTERSECTION**	POINT D'INTERSECTION, POINT DE RENCONTRE, POINT DE CONCOURS	SCHNITTPUNKT, TREFFPUNKT
9566	**POINT OF REFERENCE**	POINT DE REPERE	BEZUGSPUNKT
9567	**POINT OF SUPPORT**	POINT D'APPUI	STÜTZPUNKT, UNTERSTÜTZUNGSPUNKT
9568	**POINT OF SUSPENSION**	POINT DE SUSPENSION	AUFHÄNGEPUNKT
9569	**POINT OF TOOTH**	TETE DE LA DENT	ZAHNKOPF, ZAHNKRONE
9570	**POINT SUSPENSION**	SUSPENSION A PIVOT	SPITZENAUFHÄNGUNG
9571	**POINT TOOL**	GRAIN D'ORGE	SPITZSTAHL
9572	**POINT-TO-POINT CONTROL**	COMMANDE POINT-A-POINT	PUNKTSTEUERUNG
9573	**POINTED SOLDERING IRON**	FER A SOUDER DROIT	SPITZLÖTKOLBEN
9574	**POINTS**	CONTACTS	KONTAKTSTÜCKE
9575	**POKE A FIRE (TO)**	RINGARDER	SCHÜREN
9576	**POKE WELDING**	SOUDAGE PAR POINTS AU PISTOLET, SOUDAGE A LA CAROTTE	PUNKTSCHWEISSEN MIT HANDBETÄTIGTER SCHWEISSENELEKTRODE
9577	**POLAR**	POLAIRE	POLARE
9578	**POLAR COORDINATES**	COORDONNEES POLAIRES	POLARKOORDINATEN
9579	**POLAR DISTANCE**	DISTANCE POLAIRE	POLABSTAND
9580	**POLAR RAY**	RAYON POLAIRE	SEILSTRAHL, POLSTRAHL
9581	**POLAR TRIANGLE**	TRIANGLE POLAIRE	POLARDREIECK
9582	**POLARISATION**	POLARISATION	POLARISATION
9583	**POLARISATION MICROSCOPE**	MICROSCOPE POLARISANT	POLARISATIONSMIKROSKOP
9584	**POLARISCOPE**	POLARISCOPE	POLARISATIONSAPPARAT, POLARISKOP
9585	**POLARISED LIGHT**	LUMIERE POLARISEE	LICHT (POLARISIERTES)
9586	**POLARISER**	PRISME POLARISATEUR, POLARISEUR	POLARISATOR, POLARISATIONSPRISMA
9587	**POLARITY**	POLARITE	POLARITÄT
9588	**POLARITY INDICATOR**	INDICATEUR DU SENS DE COURANT	STROMRICHTUNGSANZEIGER

9589	**POLARIZATION**	POLARISATION	POLARISATION
9590	**POLAROGRAPHY**	POLAROGRAPHIE	POLAROGRAPHIE
9591	**POLE**	POLE (ELECTR.)	POL (ELEKTRISCHER)
9592	**POLE**	POLE	POL (GEOM.)
9593	**POLE FACES**	MASSES POLAIRES	POLSCHUHEN
9594	**POLE FINDING PAPER**	PAPIER CHERCHEUR DE POLES, PAPIER POLE	POLSUCHPAPIER
9595	**POLE OF A MAGNET**	POLE D'UN AIMANT	MAGNETPOL, POL EINES MAGNETEN
9596	**POLE STRENGTH, STRENGTH OF POLE**	INTENSITE DE POLE, FORCE POLAIRE	POLSTÄRKE
9597	**POLING**	TRAVAIL A LA PERCHE	POLEN
9598	**POLISH (TO), BUFF (TO)**	POLIR	GLANZ SCHLEIFEN, BLANK SCHLEIFEN, POLIEREN
9599	**POLISHABLE**	POLI (QUI PREND BIEN LE), POLI (SUSCEPTIBLE D'UN BEAU), POLI (SUSCEPTIBLE DE PRENDRE UN)	POLIERBAR
9600	**POLISHING**	POLISSAGE	POLIEREN
9601	**POLISHING AGENT**	MATIERE A POLIR, MATIERE ABRASIVE, ABRASIF	SCHLEIFMITTEL, POLIERMITTEL
9602	**POLISHING CYLINDER DRUM**	TONNEAU A POLIR	SCHEUERTROMMEL, POLIERTROMMEL, POLIERFASS, POLIERTONNE
9603	**POLISHING MACHINE**	MACHINE A POLIR	POLIERMASCHINE
9604	**POLISHING WHEEL**	ROUE POLISSEUSE, MEULE, MEUBLE A POLIR	POLIERSCHEIBE
9605	**POLISHING, BUFFING**	POLISSAGE	SCHLEIFEN, GLANZSCHLEIFEN, BLANKSCHLEIFEN, POLIEREN
9606	**POLONIUM**	POLONIUM	POLONIUM
9607	**POLYCHROMATIC**	POLYCHROME	VIELFARBIG
9608	**POLYCRYSTALLINE METAL**	METAL POLYCRISTALLIN	POLYKRISTALLINES METALL
9609	**POLYGON**	POLYGONE	VIELECK, POLYGON
9610	**POLYGON OF FORCES**	POLYGONE DES FORCES	KRAFTECK, KRÄFTEPOLYGON
9611	**POLYGONAL TRACE**	LIGNE POLYGONALE	POLYGONZUG
9612	**POLYHEDRON**	POLYEDRE	VIELFLACH, POLYEDER
9613	**POLYMER, POLYMERIC**	POLYMERE	POLYMEREKÖRPER, POLYMER
9614	**POLYMERISATION**	POLYMERISATION	POLYMERISATION
9615	**POLYMERISE (TO)**	POLYMERISER (SE)	POLYMERISIEREN (SICH)
9616	**POLYMERISM**	POLYMERIE	POLYMERIE
9617	**POLYMORPHISM**	POLYMORPHISME	METALL (POLYKRISTALLINES)
9618	**POLYTROPIC CURVE**	COURBE POLYTROPIQUE	KURVE (POLYTROPISCHE), POLYTROPE
9619	**PONDERABLE**	PONDERABLE	WÄGBAR
9620	**PONTON MANHOLE**	TROU D'HOMME, TROU DE VISITE D'UN CAISSON	PONTONSMANNLOCH
9621	**PONTOON DECK ROOF**	CAISSON (TOIT A...)	SCHWIMMBRÜCKENDACH
9622	**PONY ROUGHING MILL**	CAGE DEGROSSISSEUSE	VORSTRECKGERÜST
9623	**POOR LIME**	CHAUX MAIGRE	KALK (MAGERER), MAGERKALK
9624	**POPPY SEED OIL**	HUILE D'OEILLETTE	MOHNÖL
9625	**PORCELAIN**	PORCELAINE	PORZELLAN
9626	**PORCELAIN CRUCIBLE**	CREUSET EN PORCELAINE	PORZELLANTIEGEL
9627	**PORE**	PORE	PORE
9628	**PORE FORMING MATERIAL**	PRODUIT FORMANT DES POROSITES, POROGENE	MATERIAL (PORENERZEUGENDES)
9629	**POROSITY**	POROSITE, PIQUAGE	POROSITÄT, GASPORE, PORIGKEIT
9630	**POROUS**	POREUX	PORIG, PORÖS

9631	PORPHYRITIC TUFF	TUF PORPHYRITIQUE	PORPHYRTUFF
9632	PORPHYRY	PORPHYRE	PORPHYR
9633	PORT	ORIFICE	ÖFFNUNG
9634	PORT SIZE, ORIFICE SIZE	ORIFICE DE PASSAGE	DURCHGANGSWEITE
9635	PORTABLE	MOBILE, ROULANT	FAHRBAR
9636	PORTABLE DRILLING MACHINE	MACHINE A PERCER, PERCEUSE PORTATIVE	BOHRMASCHINE (FAHRBARE) (TRAGBARE)
9637	PORTABLE FORGE	FORGE PORTATIVE, FORGE VOLANTE	FELDSCHMIEDE
9638	PORTABLE PUMP	POMPE PORTATIVE	PUMPE (TRAGBARE)
9639	PORTABLE RAILWAY	CHEMIN DE FER A VOIE ETROITE, VOIE PORTATIVE, VOIE DECAUVILLE, DECAUVILLE	FELDBAHN
9640	PORTABLE STEAM ENGINE, LOCOMOBILE	LOCOMOBILE	LOKOMOBILE
9641	PORTAL CRANE	GRUE A PORTIQUE, PORTIQUE	PORTALKRAN
9642	PORTAL MILLING MACHINE	FRAISEUSE A PORTIQUE	PORTALFRÄSWERKE
9643	PORTION OF SHAFT	TRONCON D'ARBRE	WELLENSTÜCK
9644	PORTLAND CEMENT	CIMENT DE PORTLAND	PORTLANDZEMENT
9645	POSITION OF CRANK	POSITION DE LA MANIVELLE	KURBELSTELLUNG
9646	POSITION OF DEAD CENTRE	POSITION AU POINT MORT	TOTPUNKTLAGE, TOTPUNKTSTELLUNG
9647	POSITION OF EQUILIBRIUM	POSITION D'EQUILIBRE	GLEICHGEWICHTSLAGE
9648	POSITION OF REST	POSITION DE REPOS	RUHELAGE
9649	POSITION SENSOR	CAPTEUR DE POSITION	WEGMESSGERÄT
9650	POSITIONING CONTROL	COMMANDE DE POSITIONNEMENT	POSITIONSSTEUERUNG
9651	POSITIONING TIME	TEMPS DE POSITIONNEMENT	POSITIONIERZEIT
9652	POSITIVE ACCELERATION	ACCELERATION POSITIVE	BESCHLEUNIGUNG (POSITIVE)
9653	POSITIVE DISPLACEMENT PUMP	POMPE VOLUMETRIQUE	VERDRÄNGERPUMPE
9654	POSITIVE DRIVE	COMMANDE POSITIVE, COMMANDE MECANIQUE	ANTRIEB (ZWANGSLAÜFIGER)
9655	POSITIVE ELECTRICITY	ELECTRICITE POSITIVE	ELEKTRIZITÄT (POSITIVE)
9656	POSITIVE HEAT OF FORMATION	CHALEUR DE FORMATION, CHALEUR DE COMBINAISON	BILDUNGSWÄRME
9657	POSITIVE ION	CATION	KATION
9658	POSITIVE PLATE	PLAQUE POSITIVE	PLATTE (POSITIVE), PLUSPLATTE
9659	POSITIVE POLE	POLE POSITIF	POL (POSITIVER)
9660	POSITIVE RAYS	LUMIERE ANODIQUE	ANODENSTRAHLUNG
9661	POSITIVE SIGN	SIGNE POSITIF	VORZEICHEN (POSITIVES), PLUSZEICHEN
9662	POSITRON	ELECTRON POSITIF	POSITRON
9663	POST HANGER, STANCHION BRACKET ~ PEDESTAL	PALIER-CONSOLE SUR COLONNE	SÄULENLAGER
9664	POST HOLE BORERS	VRILLES	BOHRMASCHINEN FÜR MASTEN
9665	POST PROCESSOR	POST-PROCESSEUR	POST-PROZESSOR
9666	POST WELD HEAT TREATMENT	TRAITEMENT DE DETENTE APRES SOUDAGE	WÄRMENACHBEHANDLUNG
9667	POST-WELDING HEAT TREATMENT	RECUIT DE DETENTE	WÄRMEBEHANDLUNG NACH DEM SCHWEISSEN
9668	POSTHEATING	POSTCHAUFFAGE	NACHWÄRMEN
9669	POT FURNAGE	FOUR A CREUSET	TIEGELSCHMELZOFEN
9670	POTASH HARDENING	CEMENTATION AU PRUSSIATE DE POTASSE, CEMENTATION AU PRUSSIATE JAUNE	EINBRENNHÄRTUNG

9671	POTASH LYE, CAUSTIC POTASH SOLUTION	SOLUTION DE POTASSE CAUSTIQUE, LESSIVE DE POTASEE	KALILAUGE, ÄTZKALILAUGE, ÄTZALKALISCHE LÖSUNG
9672	POTASSIUM	POTASSIUM	KALIUM
9673	POTASSIUM ACETATE	ACETATE DE POTASSE	ESSIGSAURES KALI, KALIUMAZETAT
9674	POTASSIUM ALUMINATE	ALUMINATE DE POTASSIUM	KALIUMALUMINAT
9675	POTASSIUM ANTIMONYL TARTRATE, TARTAR EMETIC	TARTRATE DE POTASSE ET D'ANTIMOINE, EMETIQUE, TARTRE STIBIE	BRECHWEINSTEIN, WEINSAURES ANTIMONOXYDKALI
9676	POTASSIUM BICARBONATE, BICARBONATE OF POTASH, POTASSIUM HYDROGEN CARBONATE	BICARBONATE DE POTASSE	SAURES KOHLENSAURES KALI, DOPPELT KOHLENSAURES KALI, KALIUMBIKARBONAT, KALIUMHYDROKARBONAT
9677	POTASSIUM BICHROMATE	BICHROMATE DE POTASSIUM	KALIUMBICHROMAT, KALIUMDICHROMAT, ROTES CHROMKALI, ROTES CHROMSAURES KALI, DOPPELTCHROMSAURES KALI, ROTES CHROMSALZ
9678	POTASSIUM BROMIDE	BROMURE DE POTASSIUM	BROMKALIUM, KALIUMBROMID
9679	POTASSIUM CARBONATE, POTASH	CARBONATE NEUTRE DE POTASSE, POTASSE	POTTASCHE, KOHLENSAURES KALI, KALIUMKARBONAT
9680	POTASSIUM CHLORATE	CHLORATE DE POTASSIUM	CHLORSAURES KALI, KALIUMCHLORAT
9681	POTASSIUM CHLORIDE	CHLORURE DE POTASSIUM	CHLORKALIUM, KALIUMCHLORID
9682	POTASSIUM CHROMATE	CHROMATE NEUTRE DE POTASSIUM	NEUTRALES KALIUMCHROMAT, GELBES CHROMKALI, GELBES CHROMSAURES KALI
9683	POTASSIUM CYANIDE	CYANURE DE POTASSIUM	ZYANKALI, BLAUSAURES KALI, KALIUMZYANID
9684	POTASSIUM FERRICYANIDE, RED PRUSSIATE OF POTASH	FERRICYANURE DE POTASSIUM, PRUSSIATE ROUGE	ROTES BLUTLAUGENSALZ, FERRIZYANKALIUM, KALIUMEISENZYANID, ROTES ZYANEISENKALIUM
9685	POTASSIUM FERROCYANIDE, YELLOW PRUSSIATE OF POTASH	FERROCYANURE DE POTASSIUM, CYANOFERRURE DE POTASSIUM, PRUSSIATE JAUNE	GELBES BLUTLAUGENSALZ, FERROZYANKALIUM, GELBES ZYANEISENKALIUM, KALIUMEISENZYANÜR
9686	POTASSIUM FLUORIDE	FLUORURE DE POTASSIUM	KALIUMFLUORID, FLUORKALIUM
9687	POTASSIUM HYDROXIDE, POTASSIUM HYDRATE, CAUSTIC POTASH	POTASSE CAUSTIQUE, HYDRATE DE POTASSIUM	ÄTZKALI, KALIUMHYDROXYD, KALIUMOXYDHYDRAT, KALIHYDRAT
9688	POTASSIUM HYPOCHLORITE SOLUTION	CHLORURE DE POTASSE (SOLUTION DE)	JAVELLESCHE LAUGE, EAU DE JAVEL, KALIUMHYPOCHLORITLÖSUNG
9689	POTASSIUM IODIDE	IODURE DE POTASSIUM	JODKALIUM, KALIUMJODID
9690	POTASSIUM IODIDE STARCH PAPER	PAPIER IODOAMIDONNE	JODKALIUMSTÄRKEPAPIER
9691	POTASSIUM NITRITE	AZOTITE, NITRITE DE POTASSE, POTASSE AZOTEUSE, NITREUSE	SALPETRIGSAURES KALI, KALIUMNITRIT
9692	POTASSIUM OXALATE	OXALATE NEUTRE DE POTASSIUM	OXALSAURES KALI, NEUTRALES KALIUMOXALAT
9693	POTASSIUM PERCARBONATE	PERCARBONATE DE POTASSE	KALIUMPERKARBONAT, ÜBERKOHLENSAURES KALIUM
9694	POTASSIUM PERCHLORATE	PERCHLORATE DE POTASSE	ÜBERCHLORSAURES KALI, KALIUMPERCHLORAT
9695	POTASSIUM PERMANGANATE, PERMANGANATE OF POTASH	PERMANGANATE DE POTASSE	ÜBERMANGANSAURES, HYPERMANGANSAURES KALI, KALIUMPERMANGANAT
9696	POTASSIUM PHOSPHATE	PHOSPHATE DE POTASSIUM	PHOSPHORSAURES KALI, KALIUMPHOSPHAT
9697	POTASSIUM POLYSULPHIDE	POLYSULFURE DE POTASSIUM	KALIUMPOLYSULFID, POLYSULFID DES KALIUMS
9698	POTASSIUM SILICATE	SILICATE DE POTASSE	KALIWASSERGLAS, KIESELSAURES KALIUM, KALIUMSILIKAT

9699	POTASSIUM SULPHATE	SULFATE DE POTASSE	SCHWEFELSAURES KALI, KALIUMSULFAT
9700	POTASSIUM SULPHIDE	SULFURE DE POTASSE	SCHWEFELKALIUM
9701	POTASSIUM SULPHOCYANATE	SULFOCYANURE DE POTASSIUM	RHODANKALIUM, KALIUMRHODANID, SCHWEFELZYANKALIUM, KALIUMSULFOZYANAT
9702	POTATO COVERER	BUTTEUSES POUR POMMES DE TERRE	MASCHINEN ZUM BEDECKEN DER KARTOFFELPFLANZEN
9703	POTATO ELEVATOR DIGGERS AND SHAKERS	ARRACHEUSES DE TUBERCULES	KARTOFFELRODER
9704	POTATO HARVESTERS	RECOLTEUSES DE POMMES DE TERRE	KARTOFFELERNTEMASCHINEN
9705	POTATO PLANTERS	PLANTEUSES DE POMMES DE TERRE	KARTOFFELLEGEMASCHINEN
9706	POTATO SORTERS	TRIEURS DE POMMES DE TERRE	KARTOFFELSORTIERMASCHINEN
9707	POTATO SPINNERS AND DIGGERS	ARRACHEUSES DE POMMES DE TERRE	KARTOFFELPFLÜGE
9708	POTENTIAL	POTENTIEL	POTENTIAL, SPANNUNG
9709	POTENTIAL DIFFERENCE, DIFFERENCE OF POTENTIAL	DIFFERENCE DE POTENTIEL, TENSION AUX ELECTRODES	SPANNUNGSUNTERSCHIED, POTENTIALDIFFERENZ, ELEKTRODENSPANNUNG
9710	POTENTIAL ENERGY	ENERGIE POTENTIELLE, ENERGIE LATENTE	ENERGIE (POTENTIELLE), ENERGIE (STATISCHE)
9711	POTENTIOMETER	POTENTIOMETRE	POTENTIOMETER
9712	POUND (TO)	PILER, ECRASER	ZERSTOSSEN
9713	POURED ASPHALT	ASPHALTE COULE	GUSSASPHALT
9714	POURED CONCRETE	BETON COULE DANS DES TROUS	SCHÜTTBETON
9715	POURING	COULEE	GUSS, GIESSEN
9716	POURING CUP	ENTONNOIR DE COULEE	GIESSTRICHTER
9717	POURING/TEEMING	SORTIE EN LINGOTIERE	KOKILLENAUSTRITT, ABGUSS
9718	POWDER COMPACTING PRESS	PRESSE A COMPRIMER LES POUDRES	METALLPULVERPRESSE
9719	POWDER CUTTING	DECOUPAGE A LA POUDRE	PULVERBRENNSCHNEIDEN
9720	POWDER METALLURGY	METALLURGIE DES POUDRES	PULVERMETALLURGIE
9721	POWDER METHOD	METHODE DES POUDRES	PULVERMETHODE
9722	POWDER WELDING	SOUDAGE SOUS POUDRE	PULVERSPRITZSCHWEISSEN
9723	POWDERED ASPHALT	ASPHALTE EN POUDRE	ASPHALTSTEINPULVER
9724	POWDERED BRICK	BRIQUE PILEE	ZIEGELMEHL, SCHAMOTTEMEHL
9725	POWDERED MAGNESIA	MAGNESIE EN POUDRE	MAGNESIA (GEMAHLENE)
9726	POWDERED PEAT	POUSSIER DE TOURBE, TOURBE MENUE	TORFMULL, TORFMEHL
9727	POWDERED PUMICE	POUDRE DE PONCE	BIMSSAND
9728	POWDERY, PULVEROUS, PULVERULENT	PULVERULENT	PULVERFÖRMIG, PULVERIG, IN PULVERFORM
9729	POWER	PUISSANCE (MATH.)	POTENZ (MATH.)
9730	POWER	PUISSANCE, EFFET, DEBIT	ARBEITSLEISTUNG, EFFEKT, KRAFT, LEISTUNG
9731	POWER BRAKES	FREINS ASSISTES, SERVO-FREINS	SERVOBREMSEN
9732	POWER DRILL	CHIGNOLE, PERCEUSE ELECTRIQUE	HANDBOHRMASCHINE (ELEKTRISCHE)
9733	POWER DRIVE	FORCE MOTRICE	KRAFTANTRIEB
9734	POWER DRIVING	COMMANDE MECANIQUE, COMMANDE PAR MOTEUR	MASCHINENANTRIEB, MASCHINENBETRIEB, MECHANISCHER ANTRIEB MASCHINELLER
9735	POWER FACTOR	FACTEUR DE PUISSANCE	LEISTUNGSFAKTOR
9736	POWER GAS	GAZ POUR ACTIONNER LES MOTEURS	KRAFTGAS

9737	**POWER GENERATION**	PRODUCTION D'ENERGIE	KRAFTTERZEUGUNG
9738	**POWER OF MEN, MAN POWER**	FORCE DE L'HOMME	MENSCHENKRAFT
9739	**POWER PLANT**	INSTALLATION POUR LA PRODUCTION DE FORCE MOTRICE	KRAFTANLAGE
9740	**POWER RIVETING**	RIVETAGE POUR CONSTRUCTIONS METALLIQUES, RIVURE DE FORCE, RIVURE D'ASSEMBLAGE DE FORCE	KRAFTNIETUNG
9741	**POWER SEAT**	SIEGE A REGLAGE AUTOMATIQUE	SITZ (SELBSTEINSTELLENDER)
9742	**POWER STATION**	USINE DE FORCE MOTRICE, STATION GENERATRICE, STATION CENTRALE	KRAFTWERK, KRAFTSTATION, KRAFTZENTRALE
9743	**POWER STEERING**	DIRECTION ASSISTEE, SERVO-DIRECTION	SERVOLENKUNG
9744	**POWER SUPPLY, SUPPLY OF POWER**	DISTRIBUTION DE FORCE MOTRICE	KRAFTVERSORGUNG
9745	**POWER TAKE-OFF**	PRISE DE (FORCE) *Puissance*	ZAPFWELLE
9746	**POWER TRANSMISSION**	TRANSMISSION DE PUISSANCE	KRAFTÜBERTRAGUNG
9747	**POWER, MACHINE SAW, SAWING MACHINE, SAW DRIVEN BY MACHINERY**	MACHINE A SCIER, SCIERIE MECANIQUE, SCIE MECANIQUE	GESTELLSÄGE, MASCHINENSÄGE
9748	**POZZOLANA, PUZZUOLANA**	POUZZOLANE, POZZOLANE	PUZZOLANERDE
9749	**POZZUOLANIC CEMENT**	CIMENT POUZZOLANIQUE	PUZZOLANZEMENT
9750	**PRASEODYMIUM**	PRASEODYME	PRASEODYM
9751	**PRE-FILTER**	PREFILTRE	VORFILTER
9752	**PRE-SET TOOLING**	USINAGE PREREGLE	BEARBEITUNG (VOREINGESTELLTE)
9753	**PRECAUTIONS**	PRECAUTIONS	VORSICHTSMASSREGEIN
9754	**PRECESSION**	PRECESSION	PRÄZESSION
9755	**PRECICION GRINDING**	AFFUTAGE DE PRECISION	MASSSCHLEIFEN
9756	**PRECIOUS METAL**	METAL PRECIEUX	EDELMETALL
9757	**PRECIPITABLE**	PRECIPITABLE	AUSFÄLLBAR
9758	**PRECIPITANT**	PRECIPITANT	FÄLLMITTEL
9759	**PRECIPITATE**	PRECIPITE	NIEDERSCHLAG, BODENSATZ, PRÄZIPITAT
9760	**PRECIPITATE (TO)**	PRECIPITER, SEPARER (SE), PRECIPITER (SE)	AUSFÄLLEN, SICH ABSCHEIDEN, SICH AUSSCHEIDEN
9761	**PRECIPITATION**	PRECIPITATION	FÄLLEN, AUSFÄLLEN, NIEDERSCHLAG, AUSSCHEIDUNG
9762	**PRECIPITATION HARDENING**	DURCISSEMENT PAR PRECIPITATION	AUSSCHEIDUNGSHÄRTUNG
9763	**PRECISION**	PRECISION	GENAUIGKEIT
9764	**PRECISION BALANCE**	BALANCE DE PRECISION	FEINWAAGE, PRÄZISIONSWAAGE
9765	**PRECISION CASTING**	MOULAGE DE PRECISION	PRÄZISIONSGUSS, WACHSAUSSCHMELZGUSS
9766	**PRECISION GAUGE**	JAUGE DE PRECISION	PRÄZISIONSLEHRE
9767	**PRECISION GRADUATED SCALE**	REGLE DIVISEE DE PRECISION	PRÄZISIONSMASSSTAB
9768	**PRECISION HEIGHT GAUGE**	CALIBRE DE HAUTEUR DE PRECISION	PRÄZISIONSHÖHENMESSER
9769	**PRECISION INDEXING HEAD**	DIVISEUR DE PRECISION	GENAUIGKEITSTEILGERÄT
9770	**PRECISION MACHINE**	MACHINE DE PRECISION	PRÄZISIONSMASCHINE
9771	**PREFERRED ORIENTATION**	ORIENTATION PREFERENTIELLE	ORIENTIERUNG (VORZUGSWEISE)
9772	**PREFORM**	PREFORME	VORFORMLING
9773	**PREFORMING**	PREFORMAGE	VORFORMUNG
9774	**PREHEAT (TO)**	CHAUFFER PREALABLEMENT	VORWÄRMEN
9775	**PREHEATER**	RECHAUFFEUR	VORWÄRMER
9776	**PREHEATING**	EBULLITION PREALABLE, PRECHAUFFAGE	VORWÄRMEN

9777	**PREHEATING FURNACE**	FOUR PRECHAUFFEUR	VORWÄRMOFEN
9778	**PREIGNITION**	ALLUMAGE PREMATURE	FRÜHZÜNDUNG
9779	**PRELIMINARY TEST**	ESSAI PRELIMINAIRE	VORVERSUCH
9780	**PRELIMINARY TREATMENT**	TRAITEMENT PREPARATOIRE, PRELIMINAIRE	VORBEHANDLUNG
9781	**PRELOAD PRESTRESSING DEVICE**	DISPOSITIF DE PRECONTRAINTE	VORSPANNVORRICHTUNG
9782	**PRELOAD, PRESTRESSING, PRESTRESS**	PRECHARGE, PRECONTRAINTE	VORBELASTUNG, VORSPANNUNG
9783	**PREPACKED GRAVEL LINER**	CREPINE PREGRAVILLONNEE	LINER (KIESBEDECKTER)
9784	**PREPARATION**	PREPARATION, PRODUIT	PRÄPARAT
9785	**PREPARATION OF A SOLUTION**	PREPARATION D'UNE SOLUTION	ANSETZEN EINER LÖSUNG
9786	**PREPARATORY FUNCTION**	FONCTION PREPARATOIRE	WEGBEDINGUNG
9787	**PREPARATORY WORK, PRELIMINARY OPERATIONS**	TRAVAUX PREPARATOIRES, TRAVAUX PRELIMINAIRES, OPERATIONS PREPARATOIRES, OPERATIONS PRELIMINAIRES	VORARBEITEN
9788	**PREPARE (TO) A SOLUTION**	PREPARER UNE SOLUTION	ANSETZEN (EINE LÖSUNG)
9789	**PRESELECTOR**	PRESELECTEUR	VORWÄHLER
9790	**PRESERVATION OF TIMBER**	CONSERVATION DU BOIS	KONSERVIERUNG DES HOLZES
9791	**PRESERVATIVE**	AGENT DE CONSERVATION	KONSERVIERUNGSMITTEL
9792	**PRESERVATIVE COATINGS**	REVETEMENTS DE PROTECTION	SCHUTZÜBERZÜGE
9793	**PRESERVATIVE SKIN, PROTECTIVE COATING**	COUCHE PROTECTRICE	SCHUTZSCHICHT, SCHUTZÜBERZUG, SCHUTZANSTRICH
9794	**PRESINTERING**	PREFRITTAGE	VORSINTERUNG, VORSINTERN
9795	**PRESS**	PRESSE	PRESSE
9796	**PRESS BLOWERS**	SOUFFLETTES	PRESSLUFT-AUSWURFVORRICHTUNGEN
9797	**PRESS BUTTON, PUSH BUTTON**	POUSSOIR, BOUTON-POUSSOIR, BOUTON D'INTERRUPTEUR, BOUTON ELECTRIQUE	DRUCKKNOPF, KONTAKTKNOPF
9798	**PRESS FORGING**	FORGEAGE A LA PRESSE	PRESSSCHMIEDEN
9799	**PRESS SINTERING (HOT PRESSING)**	FRITTAGE SOUS PRESSION	DRUCKSINTERN
9800	**PRESS WITH THE FORGING MACHINE (TO)**	FORGER A LA PRESSE	PRESSEN (MIT DER SCHMIEDEPRESSE)
9801	**PRESS-SPAHN, PRESS BOARD, FULLER BOARD, GLAZED BOARD, PRESS PAPER**	PRESSSPAHN	PRESSSPAN, GLANZPAPPE, GLANZDECKEL
9802	**PRESSED CORK SLAB**	PLAQUE DE LIEGE AGGLOMERE	KORKPLATTE (GEPRESSTE)
9803	**PRESSED FLANGE**	BRIDE A COLLET RABATTU, COLLERETTE	BÖRDELFLANSCH
9804	**PRESSED PEAT**	TOURBE COMPRIMEE	PRESSTORF, MASCHINENTORF
9805	**PRESSED STEEL PULLEY**	POULIE EN TOLE EMBOUTIE	RIEMENSCHEIBE AUS GEPRESSTEM BLECH
9806	**PRESSING**	COMPRESSION, PRESSAGE	PRESSEN
9807	**PRESSING WITH THE FORGING MACHINE**	FORGEAGE A LA PRESSE	PRESSEN (MIT DER SCHMIEDEPRESSE)
9808	**PRESSURE**	PRESSION, TENSION	DRUCK, SPANNUNG
9809	**PRESSURE ABOVE ATMOSPHERIC**	PRESSION MANOMETRIQUE	ÜBERDRUCK
9810	**PRESSURE ANGLE, ANGLE OF OBLIQUITY OF PRESSURE**	ANGLE DE PRESSION	EINGRIFFSWINKEL
9811	**PRESSURE BELOW ATMOSPHERIC**	DEPRESSION	UNTERDRUCK
9812	**PRESSURE CASTING**	MOULAGE SOUS PRESSION, COULEE SOUS PRESSION	DRUCKGUSS
9813	**PRESSURE CONTROLLER**	CONTROLEUR DE PRESSION	DRUCKREGLER
9814	**PRESSURE DIFFERENCE**	DIFFERENCE DE PRESSION	DRUCKUNTERSCHIED
9815	**PRESSURE DROP**	CHUTE DE PRESSION	DRUCKABFALL

9816	PRESSURE DUE TO FRICTION	PRESSION DUE AU FROTTEMENT	REIBUNGSDRUCK
9817	PRESSURE FAN	VENTILATEUR SOUFFLANT, VENTILATEUR REFOULANT	BLASENDER LÜFTER, VENTILATOR, DRUCKLÜFTER
9818	PRESSURE FEED LUBRICATION, FORCED LUBRICATION, LUBRICATION BY THE PRESSURE SYSTEM	GRAISSAGE SOUS PRESSION	DRUCKSCHMIERUNG, PRESSSCHMIERUNG,
9819	PRESSURE GAS PRODUCER	GAZOGENE A VENT SOUFFLE	DRUCKGASGENERATOR
9820	PRESSURE GAUGE	MANOMETRE, INDICATEUR DE PRESSION	MANOMETER, DRUCKMESSER
9821	PRESSURE GAUGE, MANOMETER	MANOMETRE	DRUCKMESSER, MANOMETER
9822	PRESSURE IN A PIPE LINE	PRESSION INTERIEURE DANS UNE CONDUITE	DRUCK (INNERER), LEITUNGSDRUCK
9823	PRESSURE LOAD	CHARGE DE FORGEAGE	DRUCKBELASTUNG
9824	PRESSURE LUBRICATION	GRAISSAGE SOUS PRESSION	DRUCKSCHMIERUNG
9825	PRESSURE OF FLOW	PRESSION DU COURANT	STRÖMUNGSDRUCK
9826	PRESSURE ON EDGES	PRESSION SUR LES ARETES	KANTENPRESSUNG
9827	PRESSURE PIPE	CONDUIT DE PRESSION	DRUCKROHRLEITUNG
9828	PRESSURE RATING, STAMPING OF A VALVE	TIMBRE D'UNE SOUPAPE, TIMBRAGE	DRUCK (ZULÄSSIGER)
9829	PRESSURE RELIEF CONTROLLER	DISPOSITIF DE CONTROLE DE PRESSION	DRUCKREGLER
9830	PRESSURE RING	BAGUE DE PRESSION	DRUCKRING
9831	PRESSURE SCREW	VIS DE PRESSION	DRUCKSCHRAUBE
9832	PRESSURE SETTING	PRESSION DE REGLAGE	REGELDRUCK
9833	PRESSURE STAGE	ETAGE DE PRESSION	DRUCKSTUFE
9834	PRESSURE STROKE	COURSE DE REFOULEMENT	DRUCKHUB
9835	PRESSURE SWITCH	MANOSTAT	DRUCKSCHALTER, MANOSTAT
9836	PRESSURE TEST	EPREUVE DE PRESSION	DRUCKPROBE
9837	PRESSURE THERMIT WELDING	SOUDAGE ALUMINOTHERMIQUE PAR PRESSION	PRESS SCHWEISSEN (ALUMINOTHERMISCHES), THERMITPRESSSCHWEISSEN
9838	PRESSURE TIGHT WELDING	SOUDURE RESISTANTE A LA PRESSION	NAHT (DRUCKFESTE)
9839	PRESSURE VESSEL	APPAREIL CHAUDRONNE SOUS PRESSION, RECIPIENT SOUS PRESSION	BEHÄLTER (UNTER DRUCK TIEFGEZOGENER), DRUCKGEFÄSS
9840	PRESSURE WATER PIPING	CONDUITE D'EAU SOUS PRESSION,. D'EAU FORCEE	DRUCKWASSERLEITUNG
9841	PRESSURE WATER, WATER UNDER PRESSURE	EAU SOUS PRESSION	DRUCKWASSER
9842	PRESSURE WELDING	SOUDAGE PAR PRESSION	PRESSSCHWEISSEN
9843	PRESSURE-VACUUM VALVE	SOUPAPE PRESSION-DEPRESSION	SAUG-DRUCKVENTIL
9844	PRICK PUNCH	POINCON DE TRACAGE	KÖRNER
9845	PRIMARY BATTERY, GALVANIC BATTERY	BATTERIE DE PILES	BATTERIE (GALVANISCHE)
9846	PRIMARY CELL	PILE PRIMAIRE	PRIMÄRELEMENT
9847	PRIMARY COLOUR	COULEUR SIMPLE	GRUNDFARBE, ERSTFARBE
9848	PRIMARY CONSTITUENT	ELEMENT PRO-EUTECTIQUE	VOR-EUTEKTISCHER BESTANDTEIL
9849	PRIMARY CRYSTALLIZATION	CRISTALLISATION PRIMAIRE	KRISTALLISATION (PRIMÄRE)
9850	PRIMARY CURRENT RATIO	TAUX DE COURANT PRIMAIRE	PRIMÄRSTROMVERHÄLTNIS
9851	PRIMARY GRAPHITE	CHARBON GRAPHITIQUE	GRAPHITKOHLE
9852	PRIMARY MATERIAL	PRODUIT PRIMAIRE	AUSGANGSSTOFF, GRUNDSTOFF
9853	PRIMARY METAL	METAL VIERGE OU NAISSANT	METALL (GEDIEGENES)
9854	PRIMARY STRENGTH	RESISTANCE PRIMITIVE	URSPRUNGSFESTIGKEIT

9855	**PRIME**	APPRET	GRUNDIERLACK
9856	**PRIME A FORCE PUMP (TO), FETCH A LIFT PUMP (TO)**	AMORCER UNE POMPE	PUMPE ANSAUGEN (EINE) ANHEBEN LASSEN
9857	**PRIME COAT**	COUCHE PRIMAIRE, COUCHE DE FOND	UNTERSCHICHT, GRUNDSCHICHT
9858	**PRIME COST, INITIAL COST; COST OF INSTALLATION**	FRAIS DE PREMIER ETABLISSEMENT	ANSCHAFFUNGSKOSTEN, ANLAGEKOSTEN
9859	**PRIME MOVER, ENGINE, MOTOR**	MACHINE MOTRICE, MOTEUR, TRACTEUR	MASCHINE, ANTRIEBMACHINE, KRAFTMACHINE, MOTOR, TRAKTOR
9860	**PRIME NUMBER**	NOMBRE PREMIER	GRUNDZAHL, PRIMZAHL
9861	**PRIMER BRASS**	ALLIAGE CUIVRE-ZING	KUPFERZINKLEGIERUNG
9862	**PRIMES**	PRODUITS DE PREMIERE QUALITE	PRODUKTE (ERSTKLASSIGE)
9863	**PRIMING**	PRIMAGE (ENTRAINEMENT DE L'EAU PAR LA VAPEUR)	MITREISSEN VON WASSER IM DAMPF
9864	**PRIMING COAT, FIRST COAT OF PAINT**	COUCHE DE BASE, PEINTURE SOUS-JACENTE, PREMIERE COUCHE DE PEINTURE	GRUNDANSTRICH, GRUNDIERUNG
9865	**PRIMITIVE TRANSLATION**	TRANSLATION PRIMITIVE	PRIMITIVE TRANSLATION
9866	**PRINCIPAL AXIS**	AXE PRINCIPAL	HAUPTACHSE
9867	**PRINCIPAL DYNAMO**	DYNAMO, GENERATRICE	HAUPTDYNAMO, HAUPTMASCHINE
9868	**PRINCIPAL SECTION**	SECTION PRINCIPALE	HAUPTSCHNITT
9869	**PRINCIPAL STRESS**	TENSION PRINCIPALE	HAUPTSPANNUNG
9870	**PRINT**	EPREUVE, COPIE (PHOT.)	ABZUG (PHOT.)
9871	**PRINTED CIRCUIT (PC)**	CIRCUIT IMPRIME (C.I.)	SCHALTUNG (GEDRÜCKTE)
9872	**PRINTER**	IMPRIMANTE	DRUCKER
9873	**PRINTING ROLL KNURLING MACHINE**	MACHINE A MOLETER LES CYLINDRES D'IMPRIMERIE	DRUCKWALZENRÄNDELMASCHINE
9874	**PRINTING, COPYING FRAME**	CHASSIS POUR BLEUS	LICHTPAUSAPPARAT
9875	**PRISM**	PRISME	PRISMA
9876	**PRISM WITH OBLIQUE CROSS SECTION**	PRISME A SECTION OBLIQUE	PRISMA (SCHIEF ABGESCHNITTENES)
9877	**PRISMATIC PLANE**	PLAN PRISMATIQUE	EBENE (PRISMATISCHE)
9878	**PRISMATIC SPECTRUM**	SPECTRE PRODUIT PAR PRISME	PRISMENSPEKTRUM
9879	**PRISMATICAL**	PRISMATIQUE	PRISMATISCH
9880	**PRISMOID**	PRISMATOIDE	PRISMATOID
9881	**PROBABILITY DISTRIBUTION**	FONCTION DE DISTRIBUTION	WAHRSCHEINLICHKEITSVERTEILUNG
9882	**PROBE**	PALPEUR	SONDE, MESSFÜHLER
9883	**PROCESS**	PROCESSUS, PROCEDE	VORGANG, VERFAHREN
9884	**PROCESS ANNEALLING**	RECUIT INTERMEDIAIRE D'USINAGE	ZWISCHENGLÜHUNG
9885	**PROCESS BOOK**	MANUEL DE FABRICATION	HERSTELLUNGSHANDBUCH
9886	**PROCESS METALLURGY**	METALLURGIE D'EXTRACTION	EXTRAKTIONSMETALLURGIE
9887	**PROCESS OF MANUFACTURE**	PROCEDE D'USINAGE	ARBEITSVORGANG, ARBEITSPROZESS
9888	**PROCESSING INDUSTRY**	INDUSTRIE DE TRANSFORMATION	VEREDLUNGSINDUSTRIE
9889	**PROCESSOR**	PROCESSEUR	PROZESSOR
9890	**PRODUCER GAS**	GAZ DE GAZOGENE	GENERATORGAS, MAGERES GAS, SCHWACHGAS
9891	**PRODUCT**	PRODUIT (MATH.)	PRODUKT (MATH.)
9892	**PRODUCT**	PRODUIT	PRODUKT, ERZEUGNIS
9893	**PRODUCT OF COMBUSTION**	PRODUIT DE COMBUSTION	VERBRENNUNGSERZEUGNIS, VERBRENNUNGSPRODUKT
9894	**PRODUCTION GENERATION OF AN ELECTRIC CURRENT**	PRODUCTION DE COURANT	STROMERZEUGUNG
9895	**PRODUCTION OF COLD**	PRODUCTION DU FROID	KÄLTEERZEUGUNG

9896	PRODUCTION OF HEAT	PRODUCTION DE CHALEUR	WÄRMEERZEUGUNG
9897	PRODUCTIVITY	PRODUCTIVITE	ERTRAGSFÄHIGKEIT, PRODUKTIVITÄT
9898	PROFILE	PROFIL, COUPE	SCHNITT, PROFIL, SEITENANSICHT
9899	PROFILE MILLING CUTTER	FRAISE DE FORME, FRAISE PROFILEE	FORMFRÄSER, PROFILFRÄSER, FASSONFRÄSER
9900	PROFILE MILLING MACHINE	FRAISEUSE-FACONNEUSE, FRAISEUSE A COPIER	FORMFRÄSMASCHINE, PROFILFRÄSMASCHINE, KOPIERFRÄSMASCHINE
9901	PROFILING MACHINE FOR WOOD	TOUPIE	HOLZFRÄSMASCHINE, KREISELFRÄSER
9902	PROGRAM	PROGRAMME	PROGRAMM
9903	PROGRAM STOP	ARRET PROGRAMME	PROGRAMMIERTER HALT
9904	PROGRAMMING	PROGRAMMATION	PROGRAMMIEREN
9905	PROGRESS	AVANCEMENT	ARBEITSFORTSCHRITT
9906	PROGRESSIVE AGING	VIEILLISSEMENT PROGRESSIF	PROGRESSIVE ALTERUNG
9907	PROGRESSIVE CYLINDRICAL PLUG GAUGE	TAMPON A ECHELONS DE TOLERANCE	GRENZLEHRDORN
9908	PROGRESSIVE INDUCTION SEAM WELDING	BRASAGE PAR INDUCTION PROGRESSIVE	PROGRESSIVINDUKTIONSSCHWEISSEN
9909	PROJECT (TO)	PROJETER	PROJIZIEREN
9910	PROJECTING	SAILLANT	VORSPRINGEND, VORSTEHEND
9911	PROJECTING PLATE	TOLE EN SAILLIE	VORSPRINGENDES BLECH
9912	PROJECTION	PROJECTION	PROJEKTION
9913	PROJECTION	SAILLIE	VORSPRUNG, ÜBERKRAGUNG, AUSLADUNG
9914	PROJECTION LENS	LENTILLE DIVERGENTE	PROJEKTIV
9915	PROJECTION WELD	JOINT DE SOUDURE PAR BOSSAGES	BUCKELSCHWEISSNAHT
9916	PROJECTION WELDING	SOUDAGE PAR BOSSAGES	BUCKELSCHWEISSEN
9917	PROJECTION, SET OFF	RESSAUT, SAILLIE	AUSKRAGUNG, VORKRAGUNG, AUSLADUNG
9918	PROLATE CYCLOID	CYCLOIDE RALLONGEE	ZYKLOIDE (GEDEHNTE), ZYKLOIDE (VERLÄNGERTE)
9919	PROLONGATION OF A PATENT	PROLONGATION DE DUREE D'UN BREVET	PATENTVERLÄGERUNG, VERLÄNGERUNG EINES PATENTES
9920	PRONY BRAKE	FREIN DYMAMOMETRIQUE DE PRONY	ZAUM (PRONYSCHER)
9921	PROOF	EPREUVE, ESSAI	VERSUCH
9922	PROOF STRENGTH	LIMITE ELASTIQUE CONVENTIONNELLE OU APPARENTE	PRÜFDEHNGRENZE
9923	PROOF STRESS	LIMITE D'ALLONGEMENT, LIMITE ELASTIQUE CONVENTIONNELLE	DEHNGRENZE
9924	PROOF TEST	ESSAI NON DESTRUCTIF	VERSUCH (ZERSTÖRUNGSFREIER)
9925	PROP, STAY	ETAI	STÜTZE, STEIFE
9926	PROPAGATION OF WAVES	PROPAGATION D'ONDES	FORTPFLANZUNG VON WELLEN
9927	PROPANE	PROPANE, HYDRURE DE PROPYLE	PROPAN
9928	PROPELLER BLADE	AILE D'UNE HELICE	SCHRAUBENFLÜGEL
9929	PROPELLER FAN, HELICAL, SCREW BLOWER	VENTILATEUR A HELICE	SCHRAUBENGEBLÄSE, SCHNECKENGEBLÄSE, AXIALGEBLÄSE
9930	PROPELLER SHAFT	ARBRE D'ENTRAINEMENT, ARBRE D'HELICE	TREIBWELLE, SCHRAUBENWELLE
9931	PROPELLER-MIXER	HELICO-MELANGEUR	SPIRALMISCHER
9932	PROPER FRACTION	FRACTION PROPRE, PURE	BRUCH (ECHTER)
9933	PROPORTION	PROPORTION (MATH.)	VERHÄLTNIS, VERHÄLTNISGLEICHUNG, PROPORTION

9934	PROPORTION OF MIXTURE	PROPORTION, DOSAGE DU MELANGE	MISCHUNGSVERHÄLTNIS
9935	PROPORTIONAL	PROPORTIONNEL	VERHÄLTNISGLEICH, PROPORTIONAL, VERHÄLTNISSMÄSSIG
9936	PROPORTIONAL LIMIT	LIMITE (D'ELASTICITE) PROPORTIONNELLE	PROPORTIONALITÄTSGRENZE
9937	PROPORTIONAL REDUCTION COMPASSES	COMPAS DE REDUCTION	REDUKTIONSZIRKEL, PROPORTIONALZIRKEL
9938	PROPORTIONALITY	PROPORTIONNALITE	PROPORTIONALITÄT
9939	PROPYLENE	PROPYLENE	PROPYLEN
9940	PROTACTINIUM	PROTACTINIUM	PROTAKTINIUM
9941	PROTECTED MOTOR	MOTEUR PROTEGE	GESCHÜTZTER MOTOR
9942	PROTECTING SLEEVE	MANCHON PROTECTEUR	SCHUTZMUFFE
9943	PROTECTION	PROTECTION	SCHUTZ
9944	PROTECTION CAP	COUVERCLE PROTECTEUR	SCHUTZKAPPE, SCHUTZDECKEL
9945	PROTECTION FROM RUST	PROTECTION CONTRE LA ROUILLE	ROSTSCHUTZ
9946	PROTECTION OF INDUSTRIAL PROPERTY	PROTECTION DE LA PROPRIETE INDUSTRIELLE	SCHUTZ DES GEWERBLICHEN EIGENTUMS
9947	PROTECTIVE ATMOSPHERE	ATMOSPHERE PROTECTRICE	SCHUTZATMOSPHÄRE
9948	PROTECTIVE CASING	CAGE, BOITE, ENVELOPPE PROTECTRICE	SCHUTZGEHÄUSE
9949	PROTECTIVE COVERING	REVETEMENT PROTECTEUR	SCHUTZHÜLLE
9950	PROTECTIVE FILM	COUCHE PROTECTRICE	DECKSCHICHT
9951	PROTECTIVE GRATING	TREILLIS DE PROTECTION, TREILLAGE PROTECTEUR	SCHUTZGITTER
9952	PROTECTIVE MATERIAL	MATIERE PROTECTRICE	SCHUTZSTOFF
9953	PROTECTIVE WALL	PAROI PROTECTRICE	SCHUTZWAND
9954	PROTON	PROTON	PROTON
9955	PROTOTYPE	PROTOTYPE	URMUSTER
9956	PROTRACTOR	RAPPORTEUR	GRADBOGEN, WINKELMESSER, TRANSPORTEUR
9957	PSEUDO-ASTATIC GOVERNOR	REGULATEUR PSEUDO-ASTATIQUE	REGLER (PSEUDOASTATISCHER)
9958	PSEUDOBINARY SYSTEM	SYSTEME QUASI-BINAIRE	SYSTEM (QUASIBINÄRES)
9959	PSEUDOCARBURIZING	CEMENTATION BRILLANTE	BLINDAUFKOHLEN
9960	PSYCHROMETER, WET AND DRY BULB HYGROMETER	PSYCHROMETRE	PSYCHROMETER
9961	PUDDLE	BAIN DE FUSION	SCHWEISSBAD
9962	PUDDLED IRON	FER PUDDLE	PUDDELEISEN
9963	PUDDLED STEEL	ACIER PUDDLE	PUDDELSTAHL
9964	PUDDLING	PUDDLER	PUDDELN
9965	PUDDLING FURNACE	FOUR DE PUDDLAGE	PUDDELOFEN
9966	PUG MILL	MALAXEUR	LEHMMÜHLE
9967	PULL AT RIGHT ANGLES TO THE FIBRES	TRACTION NORMALEMENT A LA DIRECTION DES FIBRES	ZUG SENKRECHT ZUR FASER
9968	PULL IN THE DIRECTION OF THE FIBRES	TRACTION DANS LA DIRECTION DES FIBRES	ZUG PARALLEL ZUR FASER
9969	PULL THE RIVET HOLES INTO LINE COINCIDENCE WITH EACH OTHER (TO)	FAIRE CORRESPONDRE LES TROUS DE RIVETS, ASSURER LA CONCORDANCE DES TROUS DE RIVETS	NIETLÖCHER AUFEINANDERPASSEN DIE
9970	PULL, TENSION	TRACTION	ZUG (MECH.)
9971	PULLEY (FOR LIFTING WEIGHTS)	POULIE (APPAREIL DE LEVAGE)	ROLLE (ALS HEBEZEUG)
9972	PULLEY BLOCK	MOUFLE	FLASCHENZUG
9973	PULLEY BLOCKS, LIFTING TACKLE, BLOCKS AND TACKLE, PULLEY TACKLE, LIFTING BLOCKS	PALAN, MOUFLES, POULIES MOUFLEES	FLASCHENZUG, ROLLENZUG, KLOBENZUG, ZUG

9974	**PULLEY OUT OF TRUTH**	POULIE MAL TOURNEE	UNRUNDE RIEMENSCHEIBE
9975	**PULLEY TURNING LATHE**	FOUR A POULIES	RIEMENSCHEIBENDREHBANK
9976	**PULLEY WITH CURVED ARMS**	POULIE A BRAS COURBES, POULIE A BRAS PARABOLIQUES	RIEMENSCHEIBE MIT GESCHWEIFTEN ARMEN
9977	**PULLEY WITH STRAIGHT ARMS**	POULIE A BRAS DROITS	RIEMENSCHEIBE MIT GERADEN ARMEN
9978	**PULSATION WELDING**	SOUDAGE PAR PULSATION	PULSATIONSSCHWEISSVERFAHREN
9979	**PULSE**	IMPULSION	IMPULS
9980	**PULSOMETER**	PULSOMETRE	DAMPFDRUCKPUMPE, PULSOMETER
9981	**PULVERISABLE**	PULVERISABLE	PULVERISIERBAR
9982	**PULVERISE (TO), POWDER (TO)**	PULVERISER	PULVERISIEREN
9983	**PULVERISED**	PULVERISE	GEPULVERT, PULVERISIERT
9984	**PULVERISED POWDERED COAL**	CHARBON PULVERISE, POUDRE DE CHARBON	KOHLENSTAUB
9985	**PULVERIZATION**	PULVERISATION	PULVERISIERUNG, FEINSTMAHLUNG
9986	**PUMICE**	PIERRE PONCE	NATURBIMSSTEIN
9987	**PUMICE (TO)**	PONCER	ABBIMSEN
9988	**PUMICE STONE**	PIERRE PONCE, PONCE	BIMSSTEIN, BIMS
9989	**PUMICEOUS TUFF**	TUF PONCEUX	BIMSSTEINTUFF
9990	**PUMICING**	PONCAGE	BIMSEN, ABBIMSEN
9991	**PUMP**	POMPE	PUMPE
9992	**PUMP CYLINDER**	CYLINDRE DE POMPE	PUMPZYLINDER
9993	**PUMP INJECTOR**	INJECTEUR DE POMPE	EINSPRITZROHR
9994	**PUMP OILER, OIL CAN WITH FORCE PUMP**	BURETTE A PISTON	VENTILÖLKANNE
9995	**PUMP PLUNGER**	PISTON DE POMPE	PUMPENELEMENT
9996	**PUMPING SETS**	POMPES	PUMPANLAGEN
9997	**PUNCH**	POINCON, PERCOIR POUR TOLE	STEMPEL, DURCHSCHLAG, LOCHSTEMPEL, MÖNCH, LOCHER
9998	**PUNCH (TO), STAMP OUT (TO)**	DECOUPER, POINCONNER	STANZEN, SCHNEIDEN, AUSSCHNEIDEN
9999	**PUNCH HOLES (TO)**	POINCONNER	LOCHEN, LÖCHER STANZEN
10000	**PUNCH PLIERS**	EMPORTE-PIECE, PINCE EMPORTE-PIECE	LOCHZANGE
10001	**PUNCH PRESS**	POINCONNEUSE, PERFOREUSE	LOCHPRESS, LOCHSTANZE
10002	**PUNCHED CARD**	CARTE PERFOREE	LOCHKARTE
10003	**PUNCHED PLATE**	TOLE A TROUS POINCONNES	BLECH MIT GESTANZTEN LÖCHERN
10004	**PUNCHED RIVET HOLE**	TROU DE RIVET POINCONNE	NIETLOCH (GESTANZTES), NIETLOCH (GELOCHTES)
10005	**PUNCHED TAPE**	BANDE PERFOREE	LOCHSTREIFEN
10006	**PUNCHING**	PERCAGE, POINCONNAGE	LOCHEN, STANZEN, AUSSTANZEN
10007	**PUNCHING AND SHEARING MACHINE**	MACHINE A POINCONNER ET A CISAILLER, POINCONNEUSE-CISAILLE	LOCHMASCHINE MIT SCHERE
10008	**PUNCHING HOLES**	POINCONNAGE	LOCHEN, STANZEN VON LÖCHERN
10009	**PUNCHING MACHIME**	MACHINE A POINCONNER, POINCONNEUSE, POINCONNEUSE-DECOUPEUSE	LOCHMASCHINE, LOCHSTANZE, STANZE, STANZE (AUSSCHNEIDEMASCHINE)
10010	**PUNCHING TEST, DRIFT TEST**	ESSAI DE POINCONNAGE, ESSAI A LA PERFORATION	LOCHVERSUCH, AUFDORNPROBE
10011	**PUNCHING UNIT**	UNITE DE POINCONNAGE	STANZEINHEIT
10012	**PUNCHING, STAMPING OUT**	DECOUPAGE, POINCONNAGE	STANZEN, SCHNEIDEN, AUSSCHNEIDEN
10013	**PUNCTURE OF INSULATION**	PERFORATION PAR DECHARGE DISRUPTIVE	DURCHSCHLAG (ELEKTR.)

10014	**PURCHASE CRAB, WINCH**	TREUIL A ENGRENAGE	RÄDERWINDE
10015	**PURCHASE INSPECTION GAUGE**	JAUGE DE RECEPTION	ABNAHMELEHRE
10016	**PURE RESEARCH**	RECHERCHE PURE	GRUNDLAGENFORSCHUNG
10017	**PURGING**	EFFET DE NETTOYAGE	REINIGUNGSWIRKUNG
10018	**PURIFICATION OF GAS**	EPURATION DES GAZ	GASREINIGUNG
10019	**PURIFICATION OF WATER**	PURIFICATION DE L'EAU	REINIGUNG DES WASSERS, WASSERREINIGUNG
10020	**PURIFIER**	PURGEUR, EPURATEUR	REINIGER
10021	**PUSH ALONG THE FIBRES**	COMPRESSION DANS LA DIRECTION DES FIBRES	DRUCK PARALLEL ZUR FASER
10022	**PUSH AT RIGHT ANGLES TO THE FIBRES**	COMPRESSION NORMALEMENT A LA DIRECTION DES FIBRES	DRUCK SENKRECHT ZUR FASER
10023	**PUSH BENCH**	BANC D'ETIRAGE	ZIEHBANK
10024	**PUSH HEATING FURNACE**	FOUR A SECOUSSES	STOSSOFEN
10025	**PUSH IN FIT**	EMMANCHEMENT, AJUSTAGE, MONTAGE LACHE	SCHIEBESITZ
10026	**PUSH ROD**	TIGE DU CULBUTEUR	STOSSTANGE
10027	**PUSH-WELDING**	SOUDAGE A LA CAROTTE	HAND-STOSSELEKTRODENSCHWEISSEN
10028	**PUSH, COMPRESSION**	COMPRESSION	DRUCK, ZUSAMMENDRÜCKEN
10029	**PUSHER**	POUSSOIR	STOSSVORRICHTUNG
10030	**PUSHER-TYPE FURNACE**	FOUR POUSSANT	STOSSOFEN, DURCHSTOSSOFEN
10031	**PUSHROD**	TIGE DE CULBUTEURS	STÖSSELSTANGE
10032	**PUT A SPRING IN TENSION (TO)**	TENDRE UN RESSORT, BANDER UN RESSORT	FEDER SPANNEN (EINE)
10033	**PUT TO THROW INTO GEAR (TO), ENGAGE (TO)**	EMBRAYER	EINSCHALTEN, EINRÜCKEN, EINGRIFF BRINGEN IN, EINKUPPELN
10034	**PUT TO THROW OUT OF GEAR (TO), DISENGAGE (TO)**	DEBRAYER, DESEMBRAYER, DESENGRENER	AUSSCHALTEN, AUSRÜCKEN, AUSKUPPELN, ENTKUPPELN, LOSKUPPELN
10035	**PUTREFACTION**	PUTREFACTION (ACTION), POURRITURE (ETAT)	FÄULNIS
10036	**PUTREFY (TO)**	PUTREFIER (SE), FERMENTER, POURRIR	FAULEN, IN FÄULNIS ÜBERGEHEN
10037	**PUTTY**	MASTIC	KITT, DICHTUNGSMASSE
10038	**PYRAMID**	PYRAMIDE	PYRAMIDE
10039	**PYRAMID HARDNESS**	DURETE VICKERS A LA PYRAMIDE	VICKERS-PYRAMIDHÄRTE
10040	**PYRAMIDAL PLANE SYSTEM**	SYSTEME PYRAMIDAL	PYRAMIDENSYSTEM
10041	**PYRIDINE**	PYRIDINE	PYRIDIN
10042	**PYRO-ELECTRIC**	PYROELECTRIQUE	PYROELEKTRISCH
10043	**PYRO-ELECTRICITY**	PYROELECTRICITE	KRISTALLELEKTRIZITÄT, PYROELEKTRIZITÄT
10044	**PYROCONDUCTIVITY**	CONDUCTIVITE THERMO-ELECTRIQUE	LEITFÄHIGKEIT (THERMO-ELEKTRISCHE)
10045	**PYROGALLOL, PYROGALLIC ACID, TRIHYDROXYBENZENE**	ACIDE PYROGALLIQUE, PYROGALLOL	PYROGALLUSSÄURE, PYROGALLOL
10046	**PYROLIGNEOUS ACID**	ACIDE PYROLIGNEUX	HOLZESSIG, HOLZSÄURE
10047	**PYROLUSITE, MANGANESE DIOXIDE**	BIOXYDE DE MANGANESE, PEROXYDE DE MANGANESE	BRAUSTEIN, PYROLUSIT, MANGANDIOXYD, MANGANSUPEROXYD, MANGANHYPEROXYD
10048	**PYROMETALLURGY**	PYROMETALLURGIE	PYROMETALLURGIE
10049	**PYROMETER**	PYROMETRE	HITZEMESSER, PYROMETER
10050	**PYROMETRIC CONE**	CONE PYROMETRIQUE	SEGERKEGEL, SCHMELZKEGEL, PYROMETERKEGEL
10051	**PYROMETRIC CONE EQUIVALENT**	EQUIVALENT DU CONE PYROMETRIQUE	PYROMETERKEGEL-FALLPUNKT

10052	PYROMETRY	PYROMETRIE	PYROMETRIE
10053	PYROPHORIC ALLOY	ALLIAGE PYROPHORIQUE	ZÜNDLEGIERUNG
10054	PYROPHOSPHORIC ACID	ACIDE PYROPHOSPHORIQUE	PYROPHOSPHORSÄURE, PARAPHOSPHORSÄURE
10055	QUADRANGULAR PRISM	PRISME QUADRANGULAIRE, QUADRILATERE	VIERSEITIGES PRISMA
10056	QUADRANT	QUADRANT	VIERTELKREIS, QUADRANT
10057	QUADRATIC EQUATION	EQUATION QUADRATIQUE, EQUATION DU SECOND DEGRE	GLEICHUNG ZWEITEN GRADES, GLEICHUNG (QUADRATISCHE)
10058	QUADRILATERAL, QUADRANGLE	QUADRANGLE, QUADRILATERE	VIERECK
10059	QUADRUPLE-EXPANSION ENGINE	MACHINE A QUADRUPLE EXPANSION	VIERFACHEXPANSIONSMASCHINE
10060	QUADRUPLE-SEAT VALVE	SOUPAPE A QUADRUPLE SIEGE	VENTIL (VIERSITZIGES)
10061	QUALIMETER	PENETROMETRE	QUALIMETER
10062	QUALITATIVE ANALYSIS	ANALYSE QUALITATIVE	ANALYSE (QUALITATIVE)
10063	QUALITY CONTROL	CONTROLE DE QUALITE	QUALITÄTSKONTROLLE, GÜTEPRÜFUNG
10064	QUANTITATIVE ANALYSIS	ANALYSE QUANTITATIVE, DOSAGE	ANALYSE (QUANTITATIVE), MENGENBESTIMMUNG
10065	QUANTITY	GRANDEUR, QUANTITE	GRÖSSE, (MATH.)
10066	QUANTITY AMOUNT OF HEAT	QUANTITE DE CHALEUR	WÄRMEMENGE
10067	QUANTITY OF ELECTRICITY	QUANTITE D'ELECTRICITE	ELEKTRIZITÄTSMENGE
10068	QUANTITY OF MOTION	QUANTITE DE MOUVEMENT	BEWEGUNGSGRÖSSE
10069	QUANTITY PASSING	DEBIT	DURCHFLUSSMENGE
10070	QUANTITY TO BE MEASURED	QUANTITE MESUREE	MESSGRÖSSE
10071	QUANTUM EFFICIENCY	RENDEMENT QUANTIQUE	QUANTENAUSBEUTE
10072	QUANTUM LIMIT	LONGUEUR D'ONDES CRITIQUE	GRENZWELLENLÄNGE
10073	QUARRY	CARRIERE	STEINBRUCH
10074	QUARTER BEND	COUDE ROND A ANGLE DROIT, COUDE ROND AU 1/4	NORMALKRÜMMER
10075	QUARTER SAWING OF TIMBER	SCIAGE HOLLANDAIS, SCIAGE PAR RAYONNEMENT, DEBIT AU COIN DU BOIS	SPIEGELSCHNITT, RADIALSCHNITT DES HOLZES
10076	QUARTER TIMBER	QUARTIER, BOIS EN QUARTIERS	VIERTELHOLZ, KREUZHOLZ
10077	QUARTZ	QUARTZ	QUARZ
10078	QUARTZ FIBRE	FILAMENT DE QUARTZ	QUARZFADEN
10079	QUARTZ GLASS	VERRE DE QUARTZ	QUARZGLAS
10080	QUARTZ MERCURY LAMP	LAMPE A VAPEUR DE MERCURE EN QUARTZ	QUARZQUECKSILBERLAMPE
10081	QUARTZ PORPHYRY	PORPHYRE QUARTZIFERE	QUARZPORPHYR
10082	QUARTZ SAND, SILICEOUS SAND	SABLE QUARTZEUX, SILICEUX	QUARZSAND
10083	QUARTZITE	QUARTZITE	QUARZIT
10084	QUATERNARY ALLOY	ALLIAGE QUATERNAIRE	LEGIERUNG (QUATERNÄRE)
10085	QUATERNARY EUTECTIC ALLOY	ALLIAGE QUATERNAIRE EUTECTIQUE	LEGIERUNG (QUATERNÄRE EUTEKTISCHE)
10086	QUATREFOIL	QUATRE-FEUILLES	VIERBLATT
10087	QUENCH AGING	VIEILLISSEMENT PAR REFROIDISSEMENT RAPIDE	ABSCHRECKALTERUNG
10088	QUENCH AND HOT BEND TEST, BENDING TEST AFTER QUENCHING	FLEXION EFFECTUE SUR DES ACIERS TRAITES AVEC DES TREMPES (ESSAI DE...)	ABSCHRECKBIEGEPROBE
10089	QUENCH HARDENING	DURCISSEMENT PAR TREMPE	ABSCHRECKHÄRTUNG
10090	QUENCH TANK	CUVE DE TREMPE	ABSCHRECKBEHÄLTER
10091	QUENCH THE STEEL (TO)	REFROIDIR BRUSQUEMENT L'ACIER	STAHL ABSCHRECKEN (DEN)

10092	**QUENCHED AND TEMPERED**	TREMPE ET REVENU	HÄRTEN UND ANLASSEN, VERGÜTUNG
10093	**QUENCHED AND TEMPERED STEEL**	ACIER CALME ET TREMPE, ACIER TREMPE ET REVENU	STAHL (BERUHIGTER UND GEHÄRTETER), STAHL (VERGÜTETER)
10094	**QUENCHING**	TREMPE	ABSCHRECKEN (FONTE)
10095	**QUENCHING AND TEMPERING**	TREMPE ET REVENU	VERGÜTUNG
10096	**QUENCHING BATH**	BAIN DE TREMPE	HÄRTEBAD
10097	**QUENCHING CRACK**	CRIQUE, FISSURE DE TREMPE	HÄRTERISS
10098	**QUENCHING MEDIUM**	MILIEU DE TREMPE	ABSCHRECKMITTEL
10099	**QUENCHING OIL**	HUILE DE TREMPE	HÄRTEÖL
10100	**QUENCHING THE STEEL**	REFROIDISSEMENT BRUSQUE DE L'ACIER	ABSCHRECKEN DES STAHLES
10101	**QUESTIONNAIRE**	QUESTIONNAIRE	FRAGEBOGEN
10102	**QUICK CURVE SWEEP, SHARP CURVE**	COURBE VIVE, COURBE RAIDE, COURBE A PETIT RAYON, COURBE A FAIBLE RAYON	KURVE (STEILE)
10103	**QUICK OPENING GATE VALVE**	VANNE A MANOEUVRE RAPIDE	SCHNELLÖFFNUNGS-SCHIEBER
10104	**QUICK OPERATING VALVE**	ROBINET A MANOEUVRE RAPIDE	SCHNELLSCHLUSSVENTIL
10105	**QUICK PITCH THREAD**	FILET A PAS ALLONGE, FILET A PAS RAPIDE	GEWINDE MIT STARKER STEIGUNG
10106	**QUICK RETURN CRANK MOTION**	MECANISME POUR RETOUR RAPIDE, RETOUR RAPIDE	SCHNELLRÜCKLAUFGETRIEBE, RASCHER RÜCKLAUF
10107	**QUICK SETTING CEMENT**	CIMENT A PRISE RAPIDE	ZEMENT (SCHNELL BINDENDER)
10108	**QUICK-CHANGE ADAPTOR**	ADAPTEUR POUR ECHANGE RAPIDE	SCHNELLWECHSELEINRICHTUNG
10109	**QUIESCENT POURING (DURVILLE PROCESS)**	COULEE TRANQUILLE (PROCEDE DURVILLE)	GIESSEN (WIRBELFREIES) (DURVILLE-VERFAHREN)
10110	**QUINHYDRONE HALF-CELL**	QUINHYDRONE (HEMI-CELLULE A LA)	CHINHYDRONELEKTRODE
10111	**QUINTAL (METRIC)**	QUINTAL METRIQUE	DOPPELZENTNER, METRISCHER ZENTNER, METERZENTNER
10112	**QUINTIC EQUATION**	EQUATION QUINTIQUE, EQUATION DU CINQUIEME DEGRE	GLEICHUNG FÜNFTEN GRADES
10113	**QUOTE (TO)**	CITER	ANFÜHREN, ZITIEREN
10114	**QUOTIENT**	QUOTIENT	QUOTIENT
10115	**RABBET PLANE, REBATE PLANE**	FEUILLERET	FALZHOBEL
10116	**RABBLE**	RINGARD	SCHÜREISEN
10117	**RACE (TO)**	EMBALLER, EMBALLER (S')	DURCHGEHEN (MASCHINE)
10118	**RACING**	EMBALLEMENT D'UN MOTEUR	DURCHGEHEN EINER MASCHINE
10119	**RACK**	CREMAILLERE	ZAHNSTANGE
10120	**RACK AND PINION**	CREMAILLERE ET PIGNON, ENGRENAGE PAR ROUE DENTEE ET CREMAILLERE	ZAHNSTANGENGETRIEBE
10121	**RACK AND PINION JACK**	CRIC, LEVE-ROUE	ZAHNSTANGENWINDE, WAGENWINDE
10122	**RACK COMPASSES**	COMPAS A CREMAILLERE	BOGENZIRKEL MIT GEZAHNTEM BOGEN
10123	**RACK INDEXING ATTACHMENT**	MECANISME D'INDEXAGE POUR CREMAILLERE	ZAHNSTANGENTEILVORRICHTUNG
10124	**RACK LINK OU CONTROL ROD**	TIGE DE CREMAILLERE	ZAHNSTANGE
10125	**RADIAL**	RADIAL	RADIAL
10126	**RADIAL (PLY) TYRE**	PNEU A CARCASSE RADIALE	RADIALREIFEN
10127	**RADIAL ACCELERATION**	ACCELERATION CENTRIFUGE	FLIEHKRAFTBESCHLEUNIGUNG
10128	**RADIAL ARM**	BRAS RADIAL	AUSLEGERARM
10129	**RADIAL ARM DRILLING MACHINE**	MACHINE A PERCER RADIALE	RADIALBOHRMASCHINE

10130	**RADIAL BALL BEARING**	ROULEMENT A BILLES A CHARGE RADIALE, ROULEMENT A POUSSEE LATERALE	KUGELQUERDRUCKLAGER, RADIALKUGELLAGER
10131	**RADIAL BEAM / RAFTER**	POUTRE RAYONNANTE	RADIALTRÄGER, TRÄGER (EINGESPANNTER)
10132	**RADIAL COMPONENT**	COMPOSANTE RADIALE	RADIALKOMPONENTE
10133	**RADIAL DISPLACEMENT**	DEPLACEMENT RADIAL	RADIALE VERSCHIEBUNG
10134	**RADIAL DRILL PRESS, RADIAL DRILLING MACHINE**	PERCEUSE RADIALE	RADIALBOHRMASCHINE
10135	**RADIAL ENGINE**	MOTEUR EN ETOILE	STERNMOTOR
10136	**RADIAL FLOW TURBINE**	TURBINE RADIALE	RADIALTURBINE
10137	**RADIAL PROJECTION**	PROJECTION CENTRALE	ZENTRALPROJEKTION
10138	**RADIAN**	RADIAN	RADIAN
10139	**RADIANT ENERGY**	ENERGIE DE RAYONNEMENT	ENERGIE (STRAHLENDE)
10140	**RADIANT HEAT**	CHALEUR RADIANTE, CHALEUR RAYONNANTE	STRAHLUNGSWÄRME, WÄRME (STRAHLENDE)
10141	**RADIATE (TO)**	RAYONNER	STRAHLEN
10142	**RADIATING SURFACE**	SURFACE RAYONNANTE	FLÄCHE (STRAHLENDE), EMMISSIONSFLÄCHE
10143	**RADIATION**	RADIATION, RAYONNEMENT	AUSSTRAHLUNG, RADIATION, STRAHLENEMISSION
10144	**RADIATION CONSTANT**	CONSTANTE DE RADIATION	STRAHLUNGSKONSTANTE
10145	**RADIATION LOSS**	PERTE DE CHALEUR PAR RAYONNEMENT	STRAHLUNGSWÄRMEVERLUST
10146	**RADIATION OF HEAT**	RAYONNEMENT DE LA CHALEUR	WÄRMESTRAHLUNG
10147	**RADIATION OF LIGHT, LUMINOUS RADIATION**	RAYONNEMENT IRRADIATION DE LA LUMIERE	LICHTSTRAHLUNG
10148	**RADIATOR**	RADIATEUR	KÜHLER
10149	**RADIATOR (FOR COOLING)**	RADIATEUR DE REFROIDISSEMENT	KÜHLER (FÜR MOTOREN)
10150	**RADIATOR (FOR HEATING)**	RADIATEUR DE CHAUFFAGE	HEIZKÖRPER, HEIZELEMENT, RADIATOR
10151	**RADIATOR CAP**	BOUCHON DE RADIATEUR	KÜHLERVERSCHRAUBUNG
10152	**RADIATOR COWL**	CALANDRE DE RADIATEUR	KÜHLERVERKLEIDUNG
10153	**RADIATOR VALVE**	ROBINET DE RADIATEUR	HEIZUNGS-REGULIERVENTIL
10154	**RADICAL**	RADICAL	RADIKAL (CHEM.)
10155	**RADIO METALLURGY**	RADIOMETALLOGRAPHIE	ROENTGENMETALLOGRAPHIE
10156	**RADIO SUPRESSOR**	ANTI-PARASITE	STÖRSCHUTZ
10157	**RADIOACTIVE**	RADIO-ACTIF	RADIOAKTIV
10158	**RADIOACTIVE SUBSTANCES**	SUBSTANCES RADIOACTIVES	SUBSTANZEN (RADIOAKTIVE)
10159	**RADIOACTIVE, RADIUM EMANATION, NITON**	RADON, EMANATION	EMANATION DES RADIUMS
10160	**RADIOACTIVITY**	RADIO-ACTIVITE	RADIOAKTIVITÄT
10161	**RADIOGRAPH**	RADIOGRAPHIE	ROENTGENBILD
10162	**RADIOGRAPHIC INSPECTION**	EXAMEN RADIOGRAPHIQUE	ROENTGENUNTERSUCHUNG
10163	**RADIOGRAPHY**	RADIOGRAPHIE	RÖNTGENPHOTOGRAPHIE, RADIOGRAPHIE
10164	**RADIOLOGY**	RADIOLOGIE	RADIOLOGIE, STRAHLENKUNDE
10165	**RADIUM**	RADIUM	RADIUM
10166	**RADIUS**	RAYON	HALBMESSER, RADIUS
10167	**RADIUS CUTTING**	CHANTOURNAGE	AUSKEHLEN
10168	**RADIUS OF CURVATURE**	RAYON DE COURBURE	KRÜMMUNGSHALBMESSER, BIEGEHALBMESSER, ABRUNDUNGSHALBMESSER, WÖLBHALBMESSER
10169	**RADIUS OF INERTIA, RADIUS OF GYRATION**	RAYON DE GIRATION, RAYON DE ROTATION	TRÄGHEITSHALBMESSER, DREHUNGSHALBMESSER

10170	**RADIUS VECTOR**	RAYON VECTEUR	FAHRSTRAHL, LEITSTRAHL, RADIUSVEKTOR
10171	**RAG**	PIQURE	WALZNARBE
10172	**RAG BOLT, LEWIS BOLT**	BOULON DE SCELLEMENT	STEINSCHRAUBE
10173	**RAG WHEEL**	DISQUE EN DRAP	TUCHSCHEIBE, SCHWABBELSCHEIBE
10174	**RAGGED ROLL**	CYLINDRE RUGUEUX	WALZE (RAUHE)
10175	**RAIL**	RAIL	SCHIENE, FAHRSCHIENE
10176	**RAIL AND FISH PLATE DRILLING MACHINE**	MACHINE A PERCER RAILS ET ECLISSES	SCHIENEN U. LASCHENBOHRMASCHINE
10177	**RAIL STEEL PRODUCTS**	PRODUITS EN ACIER A RAIL	SCHIENENSTAHL (UMGEWALZTER)
10178	**RAIN SPOTTING**	TACHES D'EAU (DE PLUIE)	REGENWASSERFLECKEN
10179	**RAIN WATER**	EAU PLUVIALE	REGENWASSER
10180	**RAISE TO A POWER (TO)**	ELEVER A UNE PUISSANCE	ERHEBEN (IN EINE POTENZ, POTENZIEREN
10181	**RAISED PATTERN PLATE**	TOLE LARMEE	TRÄNENBLECH
10182	**RAISED WATER LEVEL**	RETENUE D'EAU	STAU, ANSTAUUNG
10183	**RAISED-UP FLANGE**	BRIDE A FACE SURELEVEE	FLANSCH (ÜBERHÖHTER)
10184	**RAISING TEST**	ESSAI D'EMBOUTISSAGE	TREIBPROBE
10185	**RAISING THE SPEED**	AUGMENTATION DE VITESSE	GESCHWINDIGKEITSERHÖHUNG, STEIGERUNG DER GESCHWINDIGKEIT
10186	**RAKE ANGLE**	ANGLE DE COUPE	SPANWINKEL, BRUSTWINKEL
10187	**RAKER SET**	DENTURE AVOYEE	SÄGEVERZAHNUNG (GESCHRÄNKTE)
10188	**RAM**	COULISSEAU, PISTON DE VERIN	STÖSSEL, SCHLITTEN, ARBEITSZYLINDER
10189	**RAM (TO), TAMP (TO)**	DAMER	FESTSTAMPFEN
10190	**RAM-TYPE VERTICAL RILLING-MACHINE**	FRAISEUSE RADIALE A COULISSE	RÄUMMASCHINE (SENKRECHTE)
10191	**RAM'S HORN**	CROCHET DOUBLE, CROCHET A TETE DE BELIER	DOPPELHAKEN, WIDDERKOPF
10192	**RAMIE FIBRE, CHINA GRASS, RHEA**	RAMIE, CHINA-GRASS	RAMIE, RAMIEFASER
10193	**RAMMED CONCRETE**	BETON DAME	STAMPFBETON
10194	**RAMMING**	SERRE, SERRAGE	STAMPFEN
10195	**RAMMING, TAMPING**	DAMAGE	STAMPFEN, FESTSTAMPFEN
10196	**RAMSBOTTOM PISTON**	PISTON SUEDOIS, PISTON RAMSBOTTOM	SCHWEDISCHER KOLBEN, RAMSBOTTOMKOLBEN
10197	**RANCID OIL**	HUILE RANCE	ÖL (RANZIGES)
10198	**RANDOM ACCESS**	ACCES DIRECT	ZUGRIFF (DIREKTER)
10199	**RANDOMLY ORIENTAETED**	ORIENTE AU HASARD	REGELLOS ORIENTIERT
10200	**RANGE**	GAMME	BEREICH
10201	**RANGE OF A MEASURING INSTRUMENT**	ETENDUE DE L'ECHELLE D'UN INSTRUMENT	MESSBEREICH, ANZEIGEBEREICH EINES INSTRUMENTS
10202	**RANGE OF SOUND**	PORTEE DU SON	REICHWEITE DES SCHALLES
10203	**RANKINE DIAGRAM**	DIAGRAMME RANKINISE, TOTALISE	DIAGRAMM (RANKINISIERTES)
10204	**RAPE SEED OIL; COLZA OIL**	HUILE DE COLZA, HUILE DE NAVETTE	RÜBÖL
10205	**RAPID COMBUSTION**	COMBUSTION VIVE	VERBRENNUNG (LEBHAFTE)
10206	**RAPID STEEL DRILL**	MECHE A COUPE RAPIDE	SCHNELLBOHRER
10207	**RAPID TOOL STEEL**	ACIER RAPIDE A OUTILS	WERKZEUGSCHNELLSTAHL
10208	**RAPID TRAVERSE**	AVANCE RAPIDE	SCHNELLGANG
10209	**RAPIDLY ROTATING, REVOLVING SHAFT**	ARBRE A GRANDE VITESSE	WELLE (SCHNELLAUFENDE)
10210	**RARE EARTHS**	TERRES RARES	ERDEN (SELTENE)
10211	**RAREFACTION OF A GAS**	RAREFACTION D'UN GAZ	VERDÜNNUNG EINES GASES

10212	**RAREFACTION OF AIR**	RAREFACTION DE L'AIR	LUFTVERDÜNNUNG
10213	**RAREFIED AIR**	AIR RAREFIE	LUFT (VERDÜNNTE)
10214	**RAREFY A GAS (TO)**	RAREFIER UN GAZ	VERDÜNNEN (EIN GAS)
10215	**RASP**	RAPE	RASPEL
10216	**RAT TAIL FILE**	LIME QUEUE DE RAT	RATTENSCHWANZ
10217	**RATATING MOORING SYSTEM**	FLOTTEUR DE STOCKAGE ET D'ACCOSTAGE	MOORINGSYSTEM (DREHBARES)
10218	**RATCHET**	CLIQUET (MECANISME POUR TRANSFORMER UN MOUVEMENT ALTERNATIF EN MOUVEMENT CONTINU)	RATSCHE, RÄTSCHE, KNARRE
10219	**RATCHET BRACE**	CLIQUET POUR PERCER, RACCAGNAC	BOHRKNARRE, BOHRRATSCHE
10220	**RATCHET EFFECT**	EFFET DE CLIQUET	SPERRKLINKEEFFEKT
10221	**RATCHET GEAR COUPLING**	ACCOUPLEMENT, MANCHON A CLIQUETS	FREILAUFKUPPLUNG
10222	**RATCHET JACK**	VERIN A CLIQUET	SCHRAUBENWINDE MIT RATSCHE
10223	**RATCHET LEVER**	LEVIER A CLIQUET	SPERRADHEBEL, SCHALTHEBEL, KLINKENHEBEL
10224	**RATCHET SPANNER**	CLEF A CLIQUET	KNARRENSCHRAUBENSCHLÜSSEL
10225	**RATCHET WHEEL, DOG WHEEL**	ROUE A ROCHET, ROCHET, ROUE DENTEE A CLIQUET	KLINKENRAD, SCHALTRAD, SPERRAD
10226	**RATE**	TAUX, VITESSE	SATZ, VERHÄLTNIS, GESCHWINDIGKEIT
10227	**RATE OF COMBUSTION**	VITESSE DE COMBUSTION	BRENNGESCHWINDIGKEIT
10228	**RATE OF DELIVERY OF A PUMP**	DEBIT D'UNE POMPE	FÖRDERMENGE, LIEFERMENGE EINER PUMPE
10229	**RATE OF DEPOSITION**	VITESSE DE DEPOT	ABSCHEIDUNGSGESCHWINDGKEIT
10230	**RATE OF FLAME PROPAGATION**	VITESSE DE PROPAGATION DE LA FLAMME	ZÜNDGESCHWINDIGKEIT
10231	**RATE OF HEATING**	MONTEE EN TEMPERATURE	TEMPERATURANSTIEG
10232	**RATE OF OIL FLOW**	VITESSE D'ECOULEMENT D'HUILE	ÖLDURCHLÄSSIGKEITSMASS
10233	**RATED HORSE POWER**	PUISSANCE FISCALE, PUISSANCE NOMINALE	STEUERLEISTUNG
10234	**RATIO**	RAPPORT, QUOTIENT	VERHÄLTNIS
10235	**RATIONAL NUMBER**	NOMBRE RATIONNEL, COMMENSURABLE	ZAHL (RATIONALE)
10236	**RATTAN CANE**	ROTIN, JONC DES INDES	ROHR (SPANISCHES)
10237	**RAW HIDE**	CUIR BRUT, CUIR VERT	ROHHAUT
10238	**RAW HIDE PINION**	PIGNON EN CUIR VERT	ROHHAUTRITZEL
10239	**RAW MATERIAL**	MATIERE PREMIERE	ROHSTOFF, ROHMATERIAL
10240	**RAW UNTREATED WATER**	EAU ORDINAIRE	ROHWASSER
10241	**RAYING**	IRRADIATION	BESTRAHLUNG
10242	**RAYS**	RAYONS	STRAHLEN
10243	**RE-ENTRANT CORNER ANGLE**	ANGLE RENTRANT	ECKE (EINSPRINGENDE), WINKEL (EINSPRINGENDER)
10244	**RE-ERECT (TO)**	REMONTER	WIEDERZUSAMMENSETZEN
10245	**RE-ERECTION**	REMONTAGE	WIEDERZUSAMMENSETZUNG
10246	**RE-UTILISATION**	REUTILISATION, REEMPLOI	WIEDERVERWERTUNG, WIEDERVERWENDUNG
10247	**REACTANCE**	REACTANCE	BLINDWIDERSTAND, REAKTANZ
10248	**REACTION**	FORCE REACTIVE, REACTION	RÜCKWIRKUNG, GEGENKRAFT, REAKTION
10249	**REACTION LIMIT**	LIMITE DE REACTION	REAKTIONSGRENZE

10250	**REACTION, PRESSURE TURBINE**	TURBINE A REACTION, TURBINE A PRESSION INTERIEURE	ÜBERDRUCKTURBINE, ·REAKTIONSTURBINE
10251	**REACTOR (NUCL.)**	REACTEUR	REAKTOR (KERN)
10252	**READ (TO)**	LIRE	LESEN
10253	**READ-OUT**	VISUALISATION, AFFICHAGE	ANZEIGE
10254	**READING**	LECTURE	ABLESEN, ABLESUNG
10255	**READJUST (TO)**	RETOUCHER LE REGLAGE	NACHSTELLEN
10256	**READJUSTEMENT**	RETOUCHE DE REGLAGE, REGLAGE CONSECUTIF	NACHSTELLEN, NACHSTELLUNG
10257	**REAGENT**	REACTIF	REAGENS
10258	**REAL FOCUS**	FOYER REEL	BRENNPUNKT (WIRKLICHER), BRENNPUNKT (REELLER)
10259	**REAL IMAGE**	IMAGE REELLE	BILD (WIRKLICHES), BILD (REELLES)
10260	**REAL NUMBER**	NOMBRE REEL	ZAHL (IRRATIONALE)
10261	**REAL STRENGTH**	RESISTANCE EFFECTIVE	WIRKWIDERSTAND
10262	**REALGAR**	BISULFURE D'ARSENIC, REALGAR	ARSENDISULFID, ARSENSULFÜR, ROTES SCHWEFELARSEN, REALGAR, RAUSCHROT, ROTER ARSENIK
10263	**REAMER**	ALESOIR	REIBAHLE
10264	**REAMER HOLDER**	PORTE-ALESOIRS	REIBAHLENHALTER
10265	**REAR AXLE**	PONT ARRIERE (MOTEUR), ESSIEU ARRIERE (NON-MOTEUR)	HINTERACHSE
10266	**REAR VIEW MIRROR**	RETROVISEUR	RÜCKBLICKSPIEGEL
10267	**REAUMUR SCALE**	ECHELLE DE REAUMUR	REAUMURSKALA
10268	**REAUMUR THERMOMETER**	THERMOMETRE REAUMUR	THERMOMETER (ACHTZIGTEILIGES), REAUMURTHERMOMETER
10269	**REBATE, RABBET**	FEUILLURE	FALZ (BAUWESEN)
10270	**REBORE (TO)**	REALESER, RECTIFIER AVEC L'ALESOIR	NACHBOHREN
10271	**REBORING**	REALESAGE, RECTIFICATION A L'ALESOIR	NACHBOHREN
10272	**REBOUND**	REBONDISSEMENT	RÜCKSPRUNG, RÜCKPRALL
10273	**REBOUND (TO)**	REBONDIR	ZURÜCKPRALLEN, ZURÜCKSPRINGEN
10274	**RECALESCENCE**	DEGAGEMENT DE CHALEUR, RECALESCENCE	WÄRMEENTWICKLUNG, REKALESZENZ
10275	**RECARBURISING THE IRON**	RECARBURATION DU FER	RÜCKKOHLUNG DES EISENS
10276	**RECARBURIZER**	RECARBURANT	RÜCKKOHLUNGSMITTEL
10277	**RECEIVER**	RECIPIENT (D'UNE CORNUE), MATRAS	VORLAGE, RETORTENVORLAGE
10278	**RECEIVER SET**	POSTE RECEPTEUR	RADIOEMPFÄNGER
10279	**RECESS**	EVIDEMENT, NICHE	AUSSPARUNG, NISCHE
10280	**RECIPROCAL**	QUANTITE RECIPROQUE	WERT (REZIPROKER)
10281	**RECIPROCAL LATTICE**	RESEAU RECIPROQUE	GITTER (REZIPROKES)
10282	**RECIPROCAL MILLING**	FRAISAGE RECIPROQUE	PENDELFRÄSER
10283	**RECIPROCATING MOTION, TO-AND-FRO MOTION**	MOUVEMENT DE VA-ET-VIENT, MOUVEMENT ALTERNATIF, MOUVEMENT PENDULAIRE	BEWEGUNG (HIN-UND HERGEHENDE)
10284	**RECIPROCATING PUMP**	POMPE A PISTON A MOUVEMENT RECTILIGNE ET ALTERNATIF	PUMPE MIT GERADLINIG HIN UND HERGEHENDEM KOLBEN, KOLBENPUMPE
10285	**RECIPROCATING STEAM ENGINE**	MACHINE A PISTON	KOLBENDAMPFMASCHINE
10286	**RECOIL**	CHOC EN RETOUR, CHOC EN ARRIERE	RÜCKSCHLAG, RÜCKSTOSS
10287	**RECONDITIONING**	REMISE EN ETAT	WIEDERINSTANDSETZUNG
10288	**RECORD (TO), REGISTER (TO)**	ENREGISTRER	AUFZEICHNEN, REGISTRIEREN

10289	**RECORDER, RECORDING INSTRUMENT**	ENREGISTREUR, INSTRUMENT ENREGISTREUR	SCHREIBER, SELBSTAUFZEICHNENDES INSTRUMENT, REGISTRIERENDES INSTRUMENT, SELBSTSCHREIBER
10290	**RECORDING DRUM**	TAMBOUR, CYLINDRE TOURNANT D'UN ENREGISTREUR	TROMMEL, WALZE EINER REGISTRIERVORRICHTUNG, REGISTRIERTROMMEL, REGISTRIERWALZE
10291	**RECORDING DYNAMOMETER, DYNAMOGRAPH**	DYNAMOGRAPHE	DYNAMOMETER MIT SCHREIBVORRICHTUNG, DYNAMOGRAPH
10292	**RECORDING REGISTERING PRESSURE GAUGE**	MANOMETRE ENREGISTREUR	MANOMETER (REGISTRIERENDES)
10293	**RECORDING REGISTERING THERMOMETER, THERMOMETROGRAPH**	THERMOMETRE ENREGISTREUR, THERMOGRAPHE	THERMOMETER (AUFZEICHNENDES), THERMOMETER SCHREIBENDES, THERMOMETER REGISTRIERENDES, THERMOGRAPH
10294	**RECORDING, SELF REGISTERING APPARATUS**	APPAREIL ENREGISTREUR	SCHREIBWERK, SCHREIBZEUG, SCHREIBVORRICHTUNG, REGISTRIERVORRICHTUNG
10295	**RECOVER (TO)**	RECUPERER	WIEDERGEWINNEN, ZURÜCKGEWINNEN
10296	**RECOVERY**	RECUPERATION, RESTAURATION	RÜCKGEWINNUNG, WIEDERVERWERTUNG, WIEDERGEWINNUNG, ZURÜCKGEWINNUNG
10297	**RECRYSTALLISATION**	RECRISTALLISATION	REKRISTALLISATION
10298	**RECRYSTALLIZATION TEMPERATURE**	TEMPERATURE DE RECRISTALLISATION	REKRISTALLISATIONSTEMPERATUR
10299	**RECTANGLE**	RECTANGLE	RECHTECK
10300	**RECTANGULAR COORDINATE ELECTRO-DISCHARGE JIG-BORING MACHINE**	MACHINE A POINTER PAR ETINCELAGE EN COORDONNEES RECTANGULAIRES	FUNKENEROSIONS-LEHRENBOHR-MASCHINE (IN RECHTECKIGEN KOORDINATEN ARBEITEND)
10301	**RECTANGULAR COORDINATES**	COORDONNEES ORTHOGONALES, COORDONNEES RECTANGULAIRES	KOORDINATEN (RECHTWINKLIGE)
10302	**RECTANGULAR CROSS SECTION**	SECTION RECTANGULAIRE	QUERSCHNITT (RECHTECKIGER)
10303	**RECTANGULAR LOAD**	CHARGE RECTANGULAIRE	RECHTECKLAST
10304	**RECTANGULAR MESH**	MAILLE RECTANGULAIRE	RECHTECKMASCHE
10305	**RECTANGULAR PARALLELOPIPED**	PARALLELEPIPEDE RECTANGLE	PARALLELEPIPED (RECHTWINKLIGES), RECHTKANT
10306	**RECTANGULAR PLATE SPRING**	RESSORT RECTANGULAIRE A LAME PLATE	RECHTECKFEDER
10307	**RECTANGULAR PLATE SPRING WITH END TAPERED**	RESSORT A LAME RECTANGULAIRE A PROFIL PARABOLOIDE	RECHTECKFEDER (ZUGESCHÄRFTE)
10308	**RECTANGULAR PROTRACTOR**	RAPPORTEUR A FORME D'EQUERRE	WINKELTRANSPORTEUR
10309	**RECTANGULAR SQUARE SCREW HEAD BOLT HEAD**	TETE CARREE D'UNE VIS	VIERKANTSCHRAUBENKOPF
10310	**RECTANGULAR, RIGHT-ANGLED**	RECTANGLE	RECHTWINKLIG
10311	**RECTIFICATION**	RECTIFICATION (CHIM.)	REKTIFIKATION (CHEM.)
10312	**RECTIFICATION**	REDRESSEMENT, DEMODULATION	GLEICHRICHTUNG, DEMODULATION
10313	**RECTIFIED SPIRIT**	ALCOOL RECTIFIE	ALKOHOL (REKTIFIZIERTER)
10314	**RECTIFIER**	REDRESSEUR DE COURANT	GLEICHRICHTER
10315	**RECTIFIER ANODE**	PLAQUE DE REDRESSEUR	GLEICHRICHTERANODE
10316	**RECTIFIER CATHODE**	CATHODE DE REDRESSEUR	GLEICHRICHTERKATODE
10317	**RECTIFIER TUBE**	TUBE REDRESSEUR	GLEICHRICHTERRÖHRE
10318	**RECTIFY (TO)**	RECTIFIER (CHIM.)	REKTIFIZIEREN (CHEM.)
10319	**RECTILINEAR**	RECTILIGNE	GERADLINIG
10320	**RECTILINEAR MOTION**	MOUVEMENT RECTILIGNE	BEWEGUNG (GERADLINIGE)
10321	**RECUPERATIVE FURNACE**	FOUR A REGENERATION	REKUPERATIVOFEN
10322	**RECUPERATOR**	RECUPERATEUR	REKUPERATOR

10323	**RECUT FILES (TO)**	RETAILLER LES LIMES	FEILEN AUFHAUEN
10324	**RECUTTING FILES**	RETAILLAGE DES LIMES	FEILENAUFHAUEN
10325	**RED BRASS**	LAITON ROUGE	ROTMESSING
10326	**RED BRONZE**	BRONZE AU ZING	ROTGUSS
10327	**RED FUMES**	FUMEES ROUSSES	RAUCH (BRAUNER)
10328	**RED HARDNESS**	DURETE A CHAUD	WARMHÄRTE
10329	**RED HEAT**	CHAUDE, CHALEUR ROUGE, ROUGE	ROTGLUT
10330	**RED HEAT (OF A)**	ROUGE AU FEU, CHAUFFE AU ROUGE	ROTWARM, ROTGLÜHEND
10331	**RED HEAT TEST**	ESSAI DE RESISTANCE AU CHAUD ROUGE	ROTBRUCHPROBE
10332	**RED LEAD PUTTY**	MASTIC AU MINIUM, MASTIC ROUGE	MENNIGEKITT, MINIUMKITT
10333	**RED LEAD, MINIUM**	MINIUM, OXYDE SALIN DE PLOMB	BLEIMENNIGE, ROTES BLEIOXYD
10334	**RED PHOSPHORUS**	PHOSPHORE ROUGE	ROTER PHOSPHOR, AMORPHER PHOSPHOR
10335	**RED RUST**	ROUILLE	ROST (EISENOXYD)
10336	**RED SANDSTONE**	GRES ROUGE	SANDSTEIN (ROTER)
10337	**RED SHORT**	CASSANT A CHAUD	WARMBRÜCHIG
10338	**RED SHORT IRON, HOT SHORT IRON**	FER CASSANT A CHAUD, FER DE COULEUR, FER METIS	EISEN (ROTBRÜCHIGES)
10339	**RED SHORTNESS**	FRAGILITE A CHAUD	WARMBRÜCHIGKEIT
10340	**REDDLE, RED CHALK**	TERRE RUBRIQUE, RUBRIQUE	RÖTEL
10341	**REDISTILLED ZINC**	ZINC REFONDU	ZINK (UMGESCHMOLZENES)
10342	**REDUCE (TO)**	REDUIRE (CHIM.)	REDUZIEREN (CHEM.)
10343	**REDUCED SCALE**	ECHELLE REDUITE	MASSSTAB (IN VERJÜNGTEM), MASSSTÄBLICH VERKLEINERT
10344	**REDUCED TEMPERATURE**	TEMPERATURE REDUITE	TEMPERATUR (HERABGESETZTE)
10345	**REDUCING**	REDUCTEUR, REDUCTRICE	REDUZIEREND, REDUKTIONS..., REDUZIER...
10346	**REDUCING AGENT**	REDUCTEUR, AGENT REDUCTEUR	REDUKTIONSMITTEL
10347	**REDUCING ATMOSPHERE**	ATMOSPHERE REDUCTRICE	REDUKTIONSATMOSPHÄRE
10348	**REDUCING COUPLING**	REDUCTION FEMELLE-FEMELLE	REDUZIERMUFFE
10349	**REDUCING FLAME**	FLAMME REDUCTRICE	FLAMME (REDUZIERENDE)
10350	**REDUCING FLANGE**	BRIDE A REDUCTION	ÜBERGANGSFLANSCH, REDUKTIONSFLANSCH
10351	**REDUCING NIPPLE**	MANCHON REDUCTEUR	REDUZIERNIPPEL
10352	**REDUCING PIPE, REDUCER, REDUCING PIECE**	TUBE DE REDUCTION, RACCORD CONIQUE	ÜBERGANGSROHR, VERJÜNGUNGSROHR, REDUKTIONSROHR
10353	**REDUCING ROLLS**	TRAIN REDUCTEUR	REDUZIERWALZSTRASSE
10354	**REDUCING TEMPERATURE**	TEMPERATURE DE REDUCTION	REDUKTIONSTEMPERATUR
10355	**REDUCING VALVE, PRESSURE REGULATOR, PRESSURE REGULTATING VALVE**	SOUPAPE DE REDUCTION, REDUCTEUR DE PRESSION, DETENDEUR	DRUCKMINDERVENTIL, DRUCKREGLER, DRUCKMINDERER, DRUCKMINDERUNGSVENTIL, REDUZIERVENTIL
10356	**REDUCTION**	REDUCTION (CHIM.)	REDUKTION (CHEM.)
10357	**REDUCTION**	DEMULTIPLICATION	UNTERSETZUNG
10358	**REDUCTION CONE**	CONE DE REDUCTION	EINSATZFUTTER
10359	**REDUCTION FURNACE**	FOUR REDUCTEUR	REDUZIEROFEN
10360	**REDUCTION GEAR RATIO**	RAPPORT DE REDUCTION	UNTERSETZUNGSVERHÄLTNIS
10361	**REDUCTION OF AREA**	STRICTION	EINSCHNÜRUNG, KONTRAKTION
10362	**REDUCTION OF AREA OF CROSS SECTION**	DIMINUTION, REDUCTION DE LA SECTION TRANSVERSALE	QUERSCHNITTSVERMINDERUNG
10363	**REDUCTION OF SPEED, REDUCING THE SPEED**	REDUCTION DE VITESSE	GESCHWINDIGKEITSVERMINDERUNG, HERABSETZUNG DER GESCHWINDIGKEIT

10364	**REDUNDANCY**	REDONDANCE	REDUNDANZ
10365	**REED COMPARATOR**	COMPARATEUR A LAMES	LAMELLENVERGLEICHSMESSER
10366	**REFERENCE GAUGE**	RAPPORTEUR FIXE, JAUGE DE REFERENCE	PRÜFLEHRE
10367	**REFERENCE MARK**	INDICE DE RAPPEL	BEZUGSZEICHEN
10368	**REFERENCE POINT**	POINT DE REFERENCE	BEZUGSPUNKT
10369	**REFERENCE TEMPERATURE**	TEMPERATURE DE REPERE	BEZUGSTEMPERATUR
10370	**REFINE (TO)**	AFFINER	REINIGEN, LÄUTERN, RAFFINIEREN
10371	**REFINE A METAL (TO)**	AFFINER UN METAL	METALL VEREDELN (EIN)
10372	**REFINED**	PURIFIE, RAFFINE	RAFFINIERT
10373	**REFINED OIL**	HUILE EPUREE	ÖL (GELÄUTERTES), ÖL (RAFFINIERTES)
10374	**REFINED PRODUCT**	PRODUIT DE RAFFINAGE	RAFFINAT
10375	**REFINERY**	RAFFINERIE	RAFFINERIE
10376	**REFINERY FLARE**	TORCHERE DE RAFFINERIE	RAFFINERIEFACKEL
10377	**REFINING**	AFFINAGE, RAFFINAGE	AFFINIEREN, VEREDELUNG, REINIGEN, LÄUTERN, RAFFINIEREN
10378	**REFINING A METAL**	RAFFINAGE D'UN METAL	VEREDELUNG EINES METALLS
10379	**REFINING FURNACE**	FOUR D'AFFINAGE	FRISCHOFEN
10380	**REFINING PROCESS**	PROCEDE D'AFFINAGE	FRISCHVERFAHREN
10381	**REFINING REMPERATURE**	TEMPERATURE D'AFFINAGE	FRISCHUNGSTEMPERATUR
10382	**REFLECT (TO)**	REFLETER, REFLECHIR	ZURÜCKSTRAHLEN, ZURÜCKWERFEN, REFLEKTIEREN
10383	**REFLECTED LIGHT**	LUMIERE REFLECHIE	LICHT (ZURÜCKGEWORFENES), LICHT (REFLEKTIERTES)
10384	**REFLECTED RAY**	RAYON REFLECHI	STRAHL (ZUÜCKGEWORFENER), STRAHL (REFLEKTIERTER)
10385	**REFLECTING SURFACE**	SURFACE REFLECHISSANTE	FLÄCHE (ZURÜCKSTRAHLENDE), FLÄCHE (REFLEKTIERENDE)
10386	**REFLECTING TELESCOPE**	TELESCOPE	SPIEGELFERNROHR, TELESKOP
10387	**REFLECTION**	DIFFRACTION, REFLEXION	REFLEKTION, REFLEXION
10388	**REFLECTION GRATING**	RESEAU DE REFLEXION	REFLEXIONSGITTER
10389	**REFLECTION LOSS**	PERTE PAR REFLEXION	REFLEXIONSVERLUST
10390	**REFLECTION OF LIGHT**	REFLEXION DE LA LUMIERE	RÜCKSTRAHLUNG, REFLEXION DES LICHTES
10391	**REFLECTION RATIO**	RAPPORT DE REFLEXION	REFLEXIONSVERHÄLTNIS
10392	**REFLECTIVE, REFLECTING POWER**	POUVOIR REFLECHISSANT	RÜCKSTRAHLUNGSVERMÖGEN, REFLEXIONSVERMÖGEN
10393	**REFLECTIVITY**	POUVOIR REFLECHISSANT	REFLEXIONSVERMÖGEN
10394	**REFLECTOR**	REFLECTEUR	REFLEKTOR, LICHTSPIEGEL
10395	**REFLEX GAUGE**	INDICATEUR A REFRACTION	REFLEXIONS-WASSERSTANDSANZEIGER
10396	**REFLEX GLASS**	GLACE A REFRACTION	REFLEXIONSGLAS
10397	**REFRACTED RAY**	RAYON REFRACTE, BRISE	STRAHL (GEBROCHENER)
10398	**REFRACTING TELESCOPE, REFRACTOR**	LUNETTE D'APPROCHE	FERNROHR
10399	**REFRACTION**	REFRACTION	STRAHLENBRECHUNG
10400	**REFRACTION OF LIGHT**	REFRACTION DE LA LUMIERE	BRECHUNG DES LICHTES, LICHTBRECHUNG, REFRAKTION
10401	**REFRACTIVE MEDIUM**	MILIEU REFRACTIF, REFRINGENT	MEDIUM (STRAHLENBRECHENDES)
10402	**REFRACTIVE, REFRACTING**	REFRINGENT	LICHTBRECHEND
10403	**REFRACTIVITY**	POUVOIR REFRINGENT	LICHTBRECHUNGSVERMÖGEN, REFRAKTIONSVERMÖGEN
10404	**REFRACTOMETER**	REFRACTOMETRE	REFRAKTOMETER

10405	REFRACTORINESS	PROPRIETE REFRACTAIRE	FEUERFESTIGKEIT, FEUERBESTÄNDIGKEIT
10406	REFRACTORY MORTAR	COULIS, MORTIER REFRACTAIRE	MÖRTEL (FEUERFESTER)
10407	REFRIGERATING MACHINE	MACHINE FRIGORIFIQUE, MACHINE A FROID	KÄLTEMASCHINE
10408	REFRIGERATING PLANT	INSTALLATION DE REFRIGERATION	KÜHLANLAGE
10409	REFRIGERATOR	REFRIGERANT, REFRIGERATEUR	KÜHLER
10410	REGENERATIVE FURNACE	FOYER A RECUPERATION INTERMITTENTE, FOUR A REGENERATION	REGENERATIVFEUERUNG, REKUPERATIVOFEN
10411	REGENERATIVE HEAT	CHALEUR EMMAGASINEE	SPEICHERWÄRME
10412	REGENERATOR	REGENERATEUR, RECUPERATEUR, SYSTEME SIEMENS	REGENERATOR, WÄRMESPEICHER
10413	REGISTER, REGISTERING FACE, RING	BAGUE DE CENTRAGE	ZENTRIERRING, ZENTRIERLEISTE
10414	REGULAR LAY	CABLAGE ALTERNATIF, TORSION ALTERNATIVE	KREUZSCHLAG
10415	REGULAR LAY WIRE ROPE	CABLE METALLIQUE TORDU ALTERNATIF	KREUZGESCHLAGENES DRAHTSEIL, SPIRALSEIL
10416	REGULAR POLYGON	POLYGONE REGULIER	POLYGON (REGELMÄSSIGES), POLYGON (REGULÄRES)
10417	REGULARITY OF MOTION	REGULARITE DU MOUVEMENT	GLEICHFÖRMIGKEIT DER BEWEGUNG
10418	REGULATING DEVICE	DISPOSITIF DE REGLAGE	REGELVORRICHTUNG
10419	REGULATION OF PRESSURE	REGLAGE DE LA PRESSION	DRUCKREGLUNG
10420	REGULATOR	DETENDEUR	DRUCKMINDERER
10421	REGULATOR, GOVERNOR	REGULATEUR DE VITESSE, MODERATEUR	REGLER, REGULATOR
10422	REGULUS	REGULE	LAGERMETALL
10423	REGULUS OF ANTIMONY	REGULE D'ANTIMOINE	ANTIMONREGULUS
10424	REHEAT (TO)	RECHAUFFER	WIEDERERHITZEN
10425	REHEATING	RECHAUFFAGE, RECHAUFFEMENT	WIEDERERHITZEN, NACHWÄRMEN
10426	REHEATING FURNACE	FOUR A RECUIRE	NACHWÄRMOFEN
10427	REINFORCEMENT OF WELD	RENFORCEMENT DE LA SOUDURE	SCHWEISSNAHTÜBERHÖHUNG
10428	REINFORCEMENT WELD	SOUDAGE DE RENFORCEMENT	VERSTÄRKUNGSSCHWEISSNAHT
10429	REINFORCING	RENFORT	VERSTÄRKUNG
10430	REINFORCING PLATE	TOLE DE RENFORT, TOLE DOUBLANTE	VERSTÄRKUNGSBLECH
10431	REJECTION	REFUS DE RECEPTION, REBUT	ABNAHMEVERWEIGERUNG, ZÜRÜCKWEISUNG
10432	RELATIVE HUMIDITY	HUMIDITE RELATIVE	FEUCHTIGKEIT (RELATIVE)
10433	RELATIVE HUMIDITY OF THE AIR	DEGRE HYGROMETRIQUE DE L'AIR, HUMIDITE RELATIVE DE L'AIR	FEUCHTIGKEITSGRAD, RELATIVE FEUCHTIGKEIT DER LUFT
10434	RELATIVE MOTION	MOUVEMENT RELATIF	BEWEGUNG (GEGENSEITIGE), BEWEGUNG (RELATIVE)
10435	RELAXATION	RELAXATION, DETENTE	ENTSPANNUNG
10436	RELEASE A BRAKE (TO)	DESSERRER UN FREIN	LÖSEN (EINE BREMSE)
10437	RELEASING CAM	DOIGT DE DECLENCHEMENT	AUSLÖSEDAUMEN, AUSLÖSEFINGER
10438	RELEASING LEVER	LEVIER DE DECLENCHEMENT	AUSLÖSCHEBEL
10439	RELIABILITY, SAFETY OF SERVICE	SECURITE DE BON FONCTIONNEMENT, FIDELITE DE SERVICE, FIABILITE	BETRIEBSSICHERHEIT, ZUVERLÄSSIGKEIT
10440	RELIABLE, SAFE	FONCTIONNEMENT SUR (DE)	BETRIEBSSICHER
10441	RELIEF VALVE	SOUPAPE DE DECHARGE, CLAPET DE DECHARGE	DRUCKBEGRENZUNGSVENTIL, ÜBERDRUCKVENTIL
10442	RELIEVING ARCH	ARC DE SOUTENEMENT	STÜTZBOGEN
10443	RELIEVING LATHE	TOUR A DETALONNER	HINTERDREHBANK

10444	**RELUCTANCE, MAGNETIC RESISTANCE**	RESISTANCE MAGNETIQUE, RELUCTANCE	WIDERSTAND (MAGNETISCHER), RELUKTANZ
10445	**REMANENCE, REMANENT FLUX, REMANENT RESIDUAL MAGNETISM**	REMANENCE, MAGNETISME REMANENT, MAGNETISME RESIDUEL, AIMANTATION REMANENTE, AIMANTATION RESIDUELLE	RÜCKSTAND (MAGNETISCHER), MAGNETISMUS (REMANENTER), REMANENZ
10446	**REMELT (TO)**	REFONDRE	EINSCHMELZEN, UMSCHMELZEN
10447	**REMELTING**	REFONTE	EINSCHMELZEN, UMSCHMELZEN
10448	**REMOTE CONTROL**	COMMANDE A DISTANCE, TELECOMMANDE	FERNSTEUERUNG
10449	**REMOTE CONTROL GAUGING**	JAUGEAGE A DISTANCE, TELEJAUGEAGE	FERNEICHUNG, FERNMESSUNG
10450	**REMOVABLE CRANK**	MANIVELLE AMOVIBLE	EINSTECKKURBEL
10451	**REMOVAL AND REPLACE**	DEPOSE ET REMONTAGE	ABBAUEN UND ERSETZEN
10452	**REMOVAL EXTRACTION OF DUST**	EVACUATION DES POUSSIERES	ENTSTAUBEN, ENTSTAUBUNG
10453	**REMOVAL OF WASTE WATER**	EVACUATION DES EAUX RESIDUAIRES	ABWÄSSERBESEITIGUNG
10454	**REMOVAL, ELIMINATION OF THE IRON IN THE WATER**	ELIMINATION DU FER DE L'EAU	ENTEISENUNG DES WASSERS
10455	**REMOVE GASES (TO)**	EVACUER LES GAZ, DEGAZER	GASE ABFÜHREN
10456	**RENDER SOLUBLE (TO)**	SOLUBILISER	LÖSLICH MACHEN
10457	**RENEW A MACHINE PART (TO)**	RENOUVELER UN ORGANE DE MACHINE	MASCHINENTEIL ERNEUERN (EINEN)
10458	**RENEWAL OF A MACHINE PART**	RENOUVELLEMENT D'UN ORGANE DE MACHINE	ERNEUERUNG EINES MASCHINENTEILS
10459	**REPAIR**	REPARATION	AUSBESSERUNG, WIEDERINSTANDSETZUNG, REPARATUR
10460	**REPAIR (TO), MEND (TO)**	REPARER	AUSBESSERN, WIEDERINSTANDSETZEN, REPARIEREN
10461	**REPAIRING SHOP**	ATELIER D'ENTRETIEN, ATELIER DE REPARATION	REPARATURWERKSTÄTTE
10462	**REPETITION WORK**	CONSTRUCTION, FABRICATION, PRODUCTION, CONFECTION, TRAVAIL EN SERIE	REIHENANFERTIGUNG, SERIENBAU, SERIENFABRIKATION
10463	**REPLACE (TO), INERCHANGE (TO)**	REMPLACER, RECHANGER	AUSWECHSELN, AUSTAUSCHEN
10464	**REPLACEMENT OF A MACHINE PART**	RECHANGE D'UN ORGANE DE MACHINE	AUSWECHSLUNG EINES MASCHINENTEILS, AUSTAUSCH EINES MASCHINENTEILS
10465	**REPLICA**	EMPREINTE	ABDRUCK
10466	**REPRACTORIES**	MATERIAUX REFRACTAIRES	STOFFE (FEUERFESTE)
10467	**REPRODUCTION**	REPRODUCTION (RESULTAT)	VERVIELFÄLTIGUNG, REPRODUKTION, KOPIE
10468	**REPULSION**	REPULSION	ABSTOSSUNG
10469	**REQUIRED QUANTITY**	GRANDEUR CHERCHEE	GRÖSSE (GESUCHTE)
10470	**REROLLING**	RELAMINAGE	NACHWALZEN
10471	**RESERVOIR**	RESERVOIR	SAMMELBEHÄLTER
10472	**RESIDUAL ELASTICITY, ELASTIC AFTERWORKING**	ELASTICITE RESIDUELLE	NACHWIRKUNG (ELASTISCHE)
10473	**RESIDUAL INDUCTION**	REMANENCE	INDUKTION (REMANENTE)
10474	**RESIDUAL STRESS**	CONTRAINTE RESIDUELLE	EIGENSPANNUNG
10475	**RESIDUE**	RESIDU	RÜCKSTAND
10476	**RESIDUE OF COMBUSTION**	RESIDU DE LA COMBUSTION	VEBRENNUNGSRÜCKSTAND
10477	**RESILIENCE**	RESILENCE, RESISTANCE VIVE	FORMÄNDERUNG (ELASTISCHE), KERRSCHLAGZÄHIGKEIT
10478	**RESIN, ROSIN**	RESINE	HARZ

10479	**RESINIFEROUS**	RENFERMANT DE LA RESINE	HARZHALTIG
10480	**RESINIFICATION, GUMMING**	RESINIFICATION	VERHARZUNG, VERHARZEN
10481	**RESINIFY (TO), GUM (TO)**	RESINIFIER (SE)	VERHARZEN
10482	**RESINOUS CEMENT**	MASTIC RESINEUX, MASTIC A LA RESINE	HARZKITT
10483	**RESINOUS WOOD, TIMBER FROM CONIFEROUS TREES**	BOIS RESINEUX	NADELHOLZ
10484	**RESISTANCE / RESISTOR**	RESISTANCE, FORCE RESISTANCE	BESTÄNDIGKEIT, FESTIGKEIT, WIDERSTAND, WIDERSTANDSKRAFT
10485	**RESISTANCE AIR-FURNACE**	FOUR A ARC A RESISTANCE	LICHT BOGEN-WIDERSTANDSOFEN
10486	**RESISTANCE ALLOYS**	ALLIAGES POUR RESISTANCES ELECTRIQUES	WIDERTANDSLEGIERUNGEN
10487	**RESISTANCE BRAZING**	BRASAGE PAR RESISTANCE	WIDERSTANDSHARTLÖTEN
10488	**RESISTANCE BUTT WELDING**	SOUDAGE EN BOUT PAR RESISTANCE	WIDERSTANDSSTUMPFSCHWEISSUNG
10489	**RESISTANCE DIAGRAM**	DIAGRAMME DES RESISTANCES	WIDERSTANDSDIAGRAMM
10490	**RESISTANCE FURNACE**	FOUR ELECTRIQUE A RESISTANCE	WIDERSTANDSOFEN (ELEKTRISCHER)
10491	**RESISTANCE OF A CONDUCTOR**	RESISTANCE D'UN CONDUCTEUR	LEITUNGSWIDERSTAND (ELEKTR.)
10492	**RESISTANCE OF THE AIR, AIR RESISTANCE, WIND RESISTANCE**	RESISTANCE DE L'AIR	LUFTWIDERSTAND, LUFTREIBUNG
10493	**RESISTANCE SPOT WELDING**	SOUDAGE PAR POINTS PAR RESISTANCE	WIDERSTANDSPUNKTSCHWEISSEN
10494	**RESISTANCE STRAIN GAUGE**	JAUGE DE CONTRAINTE A RESISTANCE	WIDERSTANDSDEHNUNGSMESSTREIFEN
10495	**RESISTANCE THERMOMETER, PYROMETER**	PYROMETRE A RESISTANCE, PYROMETRE ELECTRIQUE	WIDERSTANDSTHERMOMETER, PLATINTHERMOMETER, WIDERSTANDSPYROMETER
10496	**RESISTANCE TO ACIDS**	INATTAQUABILITE AUX ACIDES	SÄUREFESTIGKEIT, SÄUREBESTÄNDIGKEIT
10497	**RESISTANCE TO MOTION**	RESISTANCE AU MOUVEMENT, RESISTANCE PASSIVE	SCHÄDLICHER WIDERSTAND, VERLUSTWIDERSTAND, BEWEGUNGSWIDERSTAND
10498	**RESISTANCE TO SLIPPING, RESISTANCE OF SLIDING FRICTION**	RESISTANCE AU GLISSEMENT	GLEITWIDERSTAND
10499	**RESISTANCE TO WEAR**	RESISTANCE A L'USURE	VERSCHLEISSFESTIGKEIT, VERSCHLEISSWIDERSTAND
10500	**RESISTANCE WELDING**	SOUDAGE PAR RESISTANCE	WIDERSTANDSSCHWEISSEN, WIDERSTANDSSCHWEISSUNG
10501	**RESISTANCE WIRE**	FIL POUR RESISTANCES ELECTRIQUES	WIDERSTANDSDRAHT
10502	**RESISTIBILITY**	RESISTANCE	WIDERSTANDSFÄHIGKEIT, WIDERSTANDSVERMÖGEN, BESTÄNDIGKEIT
10503	**RESISTIVITY, SPECIFIC RESISTANCE**	RESISTIVITE, RESISTANCE SPECIFIQUE ELECTRIQUE	LEITWIDERSTAND (SPEZIFISCHER)
10504	**RESOLUTION**	RESOLUTION	AUFLÖSUNG
10505	**RESOLUTION OF FORCES**	DECOMPOSITION DES FORCES	ZERLEGUNG VON KRÄFTEN
10506	**RESONANCE**	RESONANCE	MITSCHWINGEN, RESONANZ
10507	**RESQUARING**	CISAILLAGE	ZUSCHNEIDEN
10508	**RESTART (TO)**	REMETTRE EN MARCHE	GANG SETZEN (WIEDER IN)
10509	**RESTARTING**	REMISE EN MARCHE	WIEDERINGANGSETZEN
10510	**RESTRAINER**	INHIBITEUR	SPARBEIZE
10511	**RESTRIKING**	RECTIFICATION	NACHRICHTEN
10512	**RESULT OF TEST, TEST RESULT**	RESULTAT D'UNE EPREUVE	VERSUCHSERGEBNIS, PRÜFUNGSERGEBNIS, PRÜFUNGSRESULTAT

10513	**RESULTANT**	RESULTANTE	MITTELKRAFT, RESULTIERENDE, RESULTANTE
10514	**RETAINER NUT**	ECROU D'ARRET	SICHERUNGSSCHRAUBENMUTTER
10515	**RETAINING RING**	BAGUE DE RETENUE	SPRENGRING
10516	**RETAINING STRIP**	BAGUETTE DE RETENUE	HALTESTREIFEN
10517	**RETARDATION OF BOILING**	RETARD A L'EBULLITION	SIEDEVERZUG
10518	**RETARDED MOTION**	MOUVEMENT RETARDE	BEWEGUNG (VERZÖGERTE)
10519	**RETARDED VALVE CLOSING**	RETARD A LA FERMETURE DE LA SOUPAPE	VENTILSCHLUSS (VERSPÄTETER)
10520	**RETENTION OF HARDNESS**	DURETE APRES REVENU	ANLASSHÄRTE
10521	**RETENTION OF SHAPE**	RESISTANCE A LA DEFORMATION	FORMBESTÄNDIGKEIT
10522	**RETEST**	CONTRE-ESSAI	WIEDERHOLUNGSVERSUCH
10523	**RETORT**	CORNUE	RETORTE, DESTILLIERKOLBEN
10524	**RETORT FURNACE**	FOUR A CORNUE	RETORTENOFEN, KAMMEROFEN
10525	**RETORT WITH TUBULURE**	CORNUE TUBULEE	RETORTE (TUBULIERTE)
10526	**RETRACTING RELEASE, RETURN SPRING**	RESSORT DE RAPPEL	RÜCKZUGFEDER, RÜCKZIEHFEDER
10527	**RETROFIT**	MONTAGE ULTERIEUR	NACHAUSRÜSTUNG
10528	**RETURN CRANK, FLY CRANK**	CONTREMANIVELLE	GEGENKURBEL
10529	**RETURN IDLE NON-CUTTING STROKE**	COURSE A VIDE (D'UNE MACHINE-OUTIL)	LEERGANG (EINER WERKZEUGMASCHINE)
10530	**RETURN STROKE OF PISTON**	COURSE DE RETOUR, COURSE RETROGRADE DU PISTON, RETOUR, MARCHE EN ARRIERE DU PISTON	KOLBENRÜCKGANG, KOLBENRÜCKKEHR
10531	**REVEAL MOULDINGS**	ENJOLIVEURS	ZIERLEISTEN
10532	**REVERBERATORY FURNACE**	FOUR A REVERBERE	HERDGLÜHOFEN, FLAMMOFEN, REVERBERIEROFEN
10533	**REVERSAL TIME**	TEMPS D'INVERSION	UMSTELLZEIT
10534	**REVERSE (TO)**	RENVERSER LE SENS DE LA MARCHE	UMSTEUERN, UMKEHREN (DIE BEWEGUNG)
10535	**REVERSE-BEND TEST**	ESSAI DE FLEXION ALTERNEE	HIN-UND HERBIEGEVERSUCH
10536	**REVERSED FEEDBACK**	CONTRE-REACTION	GEGENKOPPLUNG
10537	**REVERSIBILITY**	REVERSIBILITE	UMKEHRBARKEIT
10538	**REVERSIBLE**	REVERSIBLE	UMKEHRBAR
10539	**REVERSIBLE CHANGE OF STATE**	CHANGEMENT D'ETAT REVERSIBLE	ZUSTANDSÄNDERUNG (UMKEHRBARE)
10540	**REVERSIBLE MOTOR**	MOTEUR REVERSIBLE	MOTOR (UMKEHRBARER)
10541	**REVERSING AN ENGINE, REVERSAL OF AN ENGINE**	RENVERSEMENT DE LA MARCHE D'UNE MACHINE	UMSTEUERUNG EINER MASCHINE
10542	**REVERSING COUNTERSHAFT**	DISPOSITIF DE CHANGEMENT DE SENS DE MARCHE POUR TRANSMISSION	UMKEHRVORGELEGE
10543	**REVERSING ENGINE**	MACHINE A VAPEUR REVERSIBLE	UMKEHRDAMPFMASCHINE, DAMPFMASCHINE MIT UMKEHRUNG, UMSTEUERUNG, REVERSIERDAMPFMASCHINE
10544	**REVERSING GEAR**	MECANISME DE CHANGEMENT DE MARCHE	WENDEGETRIEBE, KEHRGETRIEBE
10545	**REVERSING GEAR OPERATED WITH BEVEL WHEELS**	DISPOSITIF DE CHANGEMENT DE SENS DE MARCHE PAR ENGRENAGE CONIQUE	KEGELRÄDERWENDEGETRIEBE
10546	**REVERSING GEAR OPERATED WITH FRICTION DISCS, (CONES)**	DISPOSITIF DE CHANGEMENT DE SENS DE MARCHE PAR ROUE DE FRICTION	REIBRÄDERWENDEGETRIEBE
10547	**REVERSING GEAR OPERATED WITH SPUR WHEELS**	DISPOSITIF DE CHANGEMENT DE SENS DE MARCHE PAR ENGRENAGE DROIT	STIRNRÄDERWENDEGETRIEBE

10548	**REVERSING GEARS**	INVERSEUR DE MARCHE	UMSTEUERGETRIEBE
10549	**REVERSING LEVER**	LEVIER DE CHANGEMENT DE MARCHE	UMSTEUERHEBEL
10550	**REVERSING SHAFT, WEIGH SHAFT**	ARBRE DE CHANGEMENT DE MARCHE, ARBRE DE RELEVAGE, ARBRE DE DISTRIBUTION	STEUERWELLE, UMSTEUERWELLE
10551	**REVERSING SHEAVE**	POULIE DE RENVOI	UMLENKSCHEIBE
10552	**REVERSING THE MOTION**	INVERSION DE MOUVEMENT	UMKEHRUNG EINER BEWEGUNG, BEWEGUNGSUMKEHR
10553	**REVERSING VALVE**	CLAPET INVERSEUR	UMSCHALTVENTIL
10554	**REVERSION**	REVERSION	RÜCKBILDUNG
10555	**REVETED BELT JOINT**	JONCTION DES COURROIES PAR RIVET	RIEMENVERBINDUNG (GENIETETE)
10556	**REVOLUTION COUNTER, ENGINE COUNTER, REVOLUTION INDICATOR**	COMPTEUR DE TOURS, COMPTE-TOURS, TACHYMETRE	DREHZAHLMESSER, UMLAUFZÄHLER, TOURENZÄHLER
10557	**REVOLUTION, TURN**	TOUR, REVOLUTION	UMLAUF, UMDREHUNG, TOUR, UMLAUFBEWEGUNG
10558	**REVOLUTIONS PER MINUTE (R.P.M.)**	REGIME DE ROTATION, REGIME	DREHZAHL
10559	**REVOLVING GRATE**	GRILLE ROTATIVE, GRILLE TOURNANTE	DREHROST
10560	**REVOLVING HEAD CENTRE**	CONTREPOINTE TOURNANTE	REITSTOCKSPITZE (UMLAUFENDE)
10561	**REVOLVING ROTATING MASS**	MASSE EN ROTATION, MASSE TOURNANTE	SCHWUNGMASSE
10562	**RHEOLOGY**	RHEOLOGIE	RHEOLOGIE, FLIESSKUNDE
10563	**RHEOSTAT**	RHEOSTAT	WIDERSTANDSREGLER, REGELWIDERSTAND, REGULIERWIDERSTAND, RHEOSTAT
10564	**RHODIUM**	RHODIUM	RHODIUM
10565	**RHOMBOHEDRAL CRYSTAL**	CRISTAL RHOMBOEDRIQUE	KRISTALL (RHOMBOEDRISHER)
10566	**RHOMBOHEDRON**	RHOMBOEDRE	RAUTENFLÄCHNER, RHOMBOEDER
10567	**RHOMBOID**	RHOMBOIDE	RHOMBOID
10568	**RHOMBUS, LOZENGE, DIAMOND**	LOSANGE, RHOMBE	RAUTE, RHOMBUS
10569	**RIB**	NERVURE, RENFORT	RIPPE, VERSTEIFUNG
10570	**RIB OF A PIPE**	NERVURE D'UN TUBE	ROHRRIPPE, VERSTÄRKUNGSRIPPE EINES ROHRES
10571	**RIBBED GILLED TUBE**	TUBE A AILERONS, TUBE A AILETTES, TUBES A NERVURES	ROHR (GERIPPTES), RIPPENROHR
10572	**RIDDLE**	CRIBLE, TAMIS	SIEB
10573	**RIDGERS : 2, 3, AND 4 ROW**	PORTES-BILLONNEUSES	FURCHENZIEHER (2-, 3-, UND 4-REIHIG)
10574	**RIFFLED SHEET**	TOLE STRIEE	RIFFELBLECH
10575	**RIFFLER, BOW FILE**	RIFLOIR	RIFFELFEILE
10576	**RIFFLES**	STRIES	RIFFEL
10577	**RIGHT ANGLE**	ANGLE DROIT, EQUERRE (D')	WINKEL (RECHTER)
10578	**RIGHT CONE**	CONE DROIT, VERTICAL	KEGEL (GERADER)
10579	**RIGHT CYLINDER**	CYLINDRE DROIT	ZYLINDER (GERADER)
10580	**RIGHT HAND SCREW**	VIS FILETEE A DROITE	SCHRAUBE (RECHTSGÄNGIGE)
10581	**RIGHT PARALLELOPIPED**	PARALLELEPIPEDE DROIT	PARALLELEPIPED (GERADES), PARALLELEPIPED (NORMALES)
10582	**RIGHT-ANGLED TRIANGLE**	TRIANGLE RECTANGLE	DREIECK (RECHTWINKLIGES)
10583	**RIGHT-HAND OFFSET**	DECALAGE A DROITE	RECHTSVERSCHIEBUNG
10584	**RIGHT-HAND SCREW**	VIS A DROITE	SCHRAUBE (RECHTSGÄNGIGE)
10585	**RIGHT-HAND THREAD**	PAS A DROITE, FILET A DROITE	GEWINDE (RECHTSÄNGIGES), RECHTSGEWINDE

10586	RIGHT-HANDED HELICAL TEETH	DENTURE A PAS INCLINE A DROITE	VERZAHNUNG (RECHTSSTEIGENDE)
10587	RIGHT-LAID ROPE	CABLE TORDU A DROITE	SEIL (RECHTSGESCHLAGENES)
10588	RIGID BODY	CORPS RIGIDE	KÖRPER (STARRER)
10589	RIGID CONNECTION	FIXATION RIGIDE, LIAISON RIGIDE, JOINT RIGIDE, ASSEMBLAGE RIGIDE	VERBINDUNG (STARRE)
10590	RIGID DRILLS	PERCEUSES RIGIDES	STARRBOHRMASCHINEN
10591	RIGIDITY, STIFFNESS	RIGIDITE, RAIDEUR	STEIFIGKEIT
10592	RIM	JANTE	RADKRANZ
10593	RIM OF FLYWHEEL	JANTE DE VOLANT	SCHWUNGRADKRANZ
10594	RIM OF GEAR WHEEL	JANTE (COURONNE) D'UNE ROUE D'ENGRENAGE	KRANZ EINES ZAHNRADES, ZAHNKRANZ
10595	RIM OF PULLEY	JANTE D'UNE POULIE	KRANZ EINER RIEMENSCHEIBE, SCHEIBENKRANZ
10596	RIM OF WHEEL	COURONNE D'UNE ROUE	RADKRANZ
10597	RIM VENT	EVENT D'ETANCHEITE	RANDLUFTÖFFNUNG
10598	RIMMED STEEL	ACIER EFFERVESCENT, ACIER FRETTE	RANDSTAHL, SINTERSTAHL, STAHL (UNBERUHIGTER)
10599	RIMMING, OR RIMMED STEEL	ACIER EFFERVESCENT, ACIER PARTIELLEMENT DESOXYDE, ACIER MOUSSEUX	SINTERSTAHL
10600	RING	ANNEAU, BAGUE	RING
10601	RING LUBRICATION OILING	GRAISSAGE A BAGUE	RINGSCHMIERUNG, ABSTREIFSCHMIERUNG
10602	RING OILED BEARING	PALIER GRAISSEUR A BAGUE	RINGSCHMIERLAGER
10603	RING SEGMENT	SEGMENT D'ANNEAU CIRCULAIRE	RINGSTÜCK
10604	RING TYPE JOINT	JOINT ANNULAIRE	RINGDICHTUNG
10605	RING VALVE, ANNULAR VALVE	SOUPAPE ANNULAIRE	RINGVENTIL
10606	RING-SHAPED ANNULAR PIVOT	PIVOT ANNULAIRE	RINGSPURZAPFEN
10607	RINSE (TO)	RINCER, LAVER A GRANDE EAU	SPÜLEN
10608	RINSING	RINCAGE	SPÜLEN
10609	RIP SAW	SCIE DE LONG	SPALTSÄGE, KRANSÄGE, DIELENSÄGE, BRETTSÄGE, LÄNGENSÄGE
10610	RIPPING FENCE	GUIDE A REFENDRE	LANGSÄGEFÜHRUNG
10611	RISE	LEVEE, HAUSSE, HAUTEUR D'ASCENSION	ANSTEIG, STEIGHÖHE, (MECH.)
10612	RISE INCREASE IN TEMPERATURE, TEMPERATURE RISE	AUGMENTATION OU ELEVATION DE LA TEMPERATURE	TEMPERATURZUNAHME, TEMPERATURSTEIGERUNG, TEMPERATURERHÖHUNG, TEMPERATURANSTIEG
10613	RISE OF AN ARCH	FLECHE, SAGETTE, AFFAISSEMENT	BOGENHÖHE, PFEILHÖHE, STICHHÖHE, STICH
10614	RISER	COLONNE MONTANTE, EVANT, CONTRE-MARCHE	STEIGLEITUNG, STEIGTRICHTER, FUTTER
10615	RISER, RISING MAIN	TUYAU D'ASCENSION, TUYAU ELEVATOIRE, TUYAU VERTICAL, MONTANT	STEIGROHR
10616	RISING STEM	TIGE MONTANTE	SPINDEL (STEIGENDE)
10617	RISING STEM GATE VALVE	VANNE A TIGE MONTANTE	ABSPERRSCHIEBER MIT STEIGENDER SPINDEL
10618	RISING TEMPERATURE	TEMPERATURE CROISSANTE	TEMPERATUR (STEIGENDE), TEMPERATUR (ZUNEHMENDE)
10619	RIVER SAND	SABLE DE RIVIERE	FLUSSSAND
10620	RIVER WATER	EAU FLUVIALE, EAU DE RIVIERE	FLUSSWASSER
10621	RIVET	RIVET	NIET

10622	**RIVET AT EDGE OF PLATE**	RIVET AU BORD DE LA TOLE	RANDNIET
10623	**RIVET FORGE**	FOUR A CHAUFFER LES RIVETS	NIETWÄRMEOFEN
10624	**RIVET HEAD**	TETE DE RIVET	NIETKOPF
10625	**RIVET HOLE**	TROU DES TOLES A ASSEMBLER, TROU DE RIVET	NIETLOCH
10626	**RIVET SNAP, RIVETING SET**	BOUTEROLLE A MAIN, CHASSE-RIVET	SETZEISEN, SCHELLEISEN, DÖPPER
10627	**RIVET STEEL**	ACIER A RIVETS	NIETEISEN
10628	**RIVET TAIL, RIVET POINT, TAIL POINT CLOSING HEAD**	RIVURE	SCHLIESSKOPF
10629	**RIVET TO**	RIVETER	VERNIETEN
10630	**RIVET TONGS**	TENAILLE A RIVETS	NIETZANGE
10631	**RIVET WELDING**	SOUDAGE DES RIVETS	NIETSCHWEISSEN
10632	**RIVET WITH CYLINDRICAL HEAD**	RIVET A TETE CYLINDRIQUE	NIET MIT ZYLINDRISCHEM KOPF
10633	**RIVET WITH ELLIPSOIDAL HEAD**	RIVET A TETE GOUTTE DE SUIF	NIET MIT KORBBOGENKOPF
10634	**RIVET WITH FLAT HEAD**	RIVET A TETE PLATE	NIET MIT FLACHEM KOPF
10635	**RIVET WITH FLUSH HEAD, FLUSH COUNTERSUNK RIVET**	RIVET A TETE NOYEE, RIVET A TETE PERDUE, RIVET A TETE FRAISEE, RIVET A TETE AFFLEUREE	NIET MIT VERSENKTEM KOPF, SENKNIET, VERSENKTES NIET, NIET MIT BÜNDIGEM KOPF
10636	**RIVET WITH SEGMENTAL HEAD**	RIVET A TETE RONDE	NIET MIT HALBRUNDKOPF
10637	**RIVET WITH SNAP HEAD, CUP HEAD, BUTTON HEAD, SPHERICAL HEAD**	RIVET A TETE HEMISPHERIQUE	NIET MIT RUNDKOPF, NIET MIT SCHELLKOPF
10638	**RIVETED FLANGE**	BRIDE RIVETEE	AUFGENIETETER FLANSCH, NIETFLANSCH
10639	**RIVETED JOINT, RIVET JOINT**	ASSEMBLAGE PAR RIVETS, RIVURE, CLOUURE	NIETVERBINDUNG, NIETUNG
10640	**RIVETED PIPE**	TUYAU RIVE	GENIETETES ROHR
10641	**RIVETER**	RIVEUR, OUVRIER RIVEUR	NIETER
10642	**RIVETER, RIVETING MACHINE**	MACHINE A RIVETER, RIVEUSE	NIETMASCHINE
10643	**RIVETING**	RIVETAGE	NIETEN, VERNIETEN, NIETUNG
10644	**RIVETING HAMMER**	MARTEAU-RIVOIR, RIVOIR, MARTEAU A RIVER	NIETHAMMER
10645	**RIVETING MACHINE**	MACHINE A RIVER	NIETMASCHINE
10646	**RIVETS**	RIVETS	NIETE
10647	**ROAD OIL**	ESSENCE CRAQUEE	STRASSENÖL
10648	**ROAST ORES (TO)**	GRILLER LES MINERAIS	ERZE RÖSTEN
10649	**ROASTING FURNACE**	FOUR DE GRILLAGE	RÖSTOFEN
10650	**ROASTING ORES**	GRILLAGE DES MINERAIS	RÖSTEN VON ERZEN
10651	**ROCK**	ROCHE	GESTEIN
10652	**ROCK CRYSTAL**	CRISTAL DE ROCHE	BERGKRISTALL
10653	**ROCK SALT**	SEL GEMME	STEINSALZ
10654	**ROCKER ARM, ROCKER LEVER**	CULBUTEUR	KIPPHEBEL
10655	**ROCKING GRATE**	GRILLE A SECOUSSES, GRILLE A BARREAUX MOBILES, GRILLE OSCILLANTE	SCHÜTTELROST
10656	**ROCKING LEVER**	LEVIER OSCILLANT	SCHWINGHEBEL
10657	**ROCKING SHAFT**	ARBRE OSCILLANT	SCHWINGENDE WELLE
10658	**ROCKING SHEAR**	CISAILLE BASCULANTE	KIPPSCHERE
10659	**ROCKWELL HARDNESS TEST**	ESSAI DE DURETE ROCKWELL	ROCKWELL-HÄRTEPRÜFUNG
10660	**ROD**	BARRE, BARREAU, BAGUETTE, TIGE	STAB, STANGE
10661	**ROD GAUGE**	CALIBRE D'ALESAGE	STICHMASS, ZYLINDERMASS
10662	**ROD OF PLANIMETER**	BRANCHE DE LA POINTE TRACANTE DU PLANIMETRE	FAHRARM DES PLANIMETERS

10663	**ROD, BAR**	TRINGLE, TIGE, BARRE	STANGE
10664	**ROENTGENOGRAPHY**	RADIOGRAPHIE	ROENTGENBILD
10665	**ROLL (TO)**	LAMINER, ROULER	WALZEN, AUSWALZEN, ROLLEN
10666	**ROLL DOWN**	REDUIRE PAR LAMINAGE	AUSWALZEN
10667	**ROLL FILM**	PELLICULE EN BOBINE	FILM ROLLEN
10668	**ROLL FORGING**	FORGEAGE PAR LAMINAGE, PROFILAGE	WALZSCHMIEDEN PROFILWALZEN
10669	**ROLL LINE**	TRAIN DE LAMINOIR	WALZENSTRASSE
10670	**ROLL TRAIN**	TRAIN DE LAMINAGE	WALZENSTRASSE
10671	**ROLL, ROLLER, CYLINDER**	CYLINDRE, ROULEAU, TAMBOUR, GALET	WALZE, ROLLE, ZYLINDER
10672	**ROLLED FLANGE**	BRIDE MANDRINEE	FLANSCH (AUFGEWALZTER), WALZFLANSCH
10673	**ROLLED GOLD**	DOUBLE D'OR	GOLDDUBLEE, WALZGOLD
10674	**ROLLED PLATE**	TOLE LAMINEE	BLECH (GEWALZTES)
10675	**ROLLED SECTION**	FER LAMINE PROFILE	WALZPROFIL
10676	**ROLLED STEEL, ROLLED IRON, ROLLED BARS**	FER LAMINE	WALZEISEN
10677	**ROLLED TUBE**	TUBE LAMINE	ROHR (GEWALZTES)
10678	**ROLLED WIRE**	FIL LAMINE	WALZDRAHT, DRAHT (GEWALZTER)
10679	**ROLLER**	ROULEAU, GALET	ROLLE
10680	**ROLLER BEARING**	ROULEMENT A GALETS (A ROULEAUX), PALIER A GALETS (A ROULEAUX)	ROLLENLAGER, WALZENLAGER
10681	**ROLLER BEARING WITH BARREL-SHAPED ROLLERS**	ROULEMENT A ROTULE SUR ROULEAUX	ROLLENLAGER MIT TONNENFÖRMIGEN ROLLEN, TONNENLAGER
10682	**ROLLER BEARING WITH CYLINDRICAL ROLLERS**	ROULEMENT A GALETS CYLINDRIQUES	ROLLENLAGER MIT ZYLINDRISCHEN ROLLEN
10683	**ROLLER BEARING WITH TAPER ROLLERS**	ROULEMENT A GALETS CONIQUES	ROLLENLAGER MIT KEGELROLLEN
10684	**ROLLER CHAIN**	CHAINE A ROULEAUX	ROLLENKETTE
10685	**ROLLER LEVELER**	DRESSEUSE A GALETS	ROLLENRICHTMASCHINE
10686	**ROLLER LEVER, ROLLING LEVER**	LEVIER ROULANT	WÄLZHEBEL
10687	**ROLLER MILL**	BROYEUR A CYLINDRES	WALZENMÜHLE
10688	**ROLLER OF A CHAIN**	ROULEAU D'UNE CHAINE	BÜCHSE EINER ROLLENKETTE
10689	**ROLLER RACE WAY, ROLLER PATH**	CHEMIN DE ROULEMENT DES GALETS	ROLLENBAHN (EINES LAGERS)
10690	**ROLLER SADDLE SUPPORT**	BERCEAU A GALETS	ROLLENTROMMELSATTEL
10691	**ROLLER STEP BEARING**	BUTEE A ROULEAUX (A GALETS)	ROLLENSPURLAGER, WALZENSPURLAGER
10692	**ROLLER TAPPET**	POUSSOIR A GALET	ROLLENSTÖSSEL
10693	**ROLLER THREAD GAUGE**	JAUGE-MACHOIRE POUR FILETS	GRENZGEWINDERACHENLEHRE MIT MESSROLLEN
10694	**ROLLER TYPE SADDLE SUPPORT**	BERCEAU A GALETS	SCHLITTEN MIT ROLLEN
10695	**ROLLERS, FLAT, RING AND RIDGE**	ROULEAUX, GALETS (POUR MISE SUR CHAMP)	GLATTWALZEN, ACKERWALZEN, ROLLEN
10696	**ROLLING**	LAMINAGE, ROULEMENT	WALZEN, AUSWALZEN, ROLLEN
10697	**ROLLING CURVE, ROULETTE**	ROULETTE, COURBE ROULANTE	ROLLKURVE
10698	**ROLLING DIRECTION**	SENS DE LAMINAGE	WALZRICHTUNG
10699	**ROLLING FLANGE**	BRIDE A MANDRINER	WALZFLANSCH
10700	**ROLLING FRICTION**	FROTTEMENT DE ROULEMENT	REIBUNG (ROLLENDE)
10701	**ROLLING MILL**	LAMINOIR, TRAIN DE LAMINOIR	WALZENSTRASSE, WALZSTRECKE, WALZWERK
10702	**ROLLING MOTION**	MOUVEMENT DE ROULEMENT	BEWEGUNG (ROLLENDE), ROLLBEWEGUNG

10703	**ROLLING RESISTANCE**	RESISTANCE AU ROULEMENT	ROLLWIDERSTAND
10704	**ROLLING STAND**	CAGE DE CYLINDRES	WALZGERÜST
10705	**ROLLING SURFACE**	SURFACE DE ROULEMENT	WÄLZFLÄCHE
10706	**ROLLING, TRAVELLING DYNAMIC LOAD, MOVING WEIGHT**	CHARGE ROULANTE, MOBILE	VERKEHRSLAST, LAST (WANDERNDE), LAST (FAHRENDE), LAST (BEWEGLICHE)
10707	**ROMAN CEMENT**	CIMENT ROMAIN	ROMANZEMENT
10708	**ROOF**	TOIT	DACH
10709	**ROOF LIGHT**	PLAFONNIER	DECKENLEUCHTE
10710	**ROOF MANHOLE**	TROU D'HOMME DE TOIT	DACH-MANNLOCH
10711	**ROOF NOZZLE**	TUBULURE DE TOIT	DACHSTUTZEN
10712	**ROOFING FELT**	CARTON BITUME OU ASPHALTE	DACHPAPPE, TEERPAPPE
10713	**ROOM TEMPERATURE**	TEMPERATURE AMBIANTE, TEMPERATURE NORMALE, ORDINAIRE DE LA SALLE	ZIMMERWÄRME, RAUMTEMPERATUR
10714	**ROOT**	RACINE (MATH.)	WURZEL, (MATH.)
10715	**ROOT AND TURNIP CUTTERS AND CLEANERS**	COUPE-RACINES	REINIGUNGS- UND SCHNEIDEMASCHINEN FÜR WURZELN UND RÜBEN
10716	**ROOT ANGLE**	ANGLE DE PIED	ZAHNFUSSWINKEL
10717	**ROOT CIRCLE**	CERCLE DE PIED	FUSSKREIS
10718	**ROOT CONE**	CONE DE PIED	ZAHNFUSSKEGEL
10719	**ROOT DIAMETER**	DIAMETRE DE PIED	KERNDURCHMESSER
10720	**ROOT EDGE**	FLANC DE BASE, RACINE	WURZELFLANKE
10721	**ROOT FACE**	FLANC DE RACINE	STEGFLANKE
10722	**ROOT LIFTERS**	SOULEVEUSES DE RACINES	ERNTEMASCHINEN FÜR HACKFRÜCHTE
10723	**ROOT LINE CIRCLE**	CERCLE DE PIED, CERCLE DE FOND, CERCLE DE CREUX, CIRCONFERENCE D'EVIDEMENT	WURZELKREIS, FUSSKREIS
10724	**ROOT OF A TOOTH**	PIED DE DENT	ZAHNWURZEL
10725	**ROOT OF TOOTH**	PIED DE LA DENT	ZAHNFUSS
10726	**ROOT OF WELD**	RACINE DE LA SOUDURE	NAHTWURZEL
10727	**ROOT OPENING**	LARGEUR DE LA SOUDURE DE BASE	WURZELSPALT
10728	**ROOT RADIUS**	RAYON D'ECARTEMENT ENTRE LES BORDS	FUGENRADIUS
10729	**ROOT'S BLOWER**	VENTILATEUR SYSTEME ROOT	ROOTGEBLÄSE
10730	**ROPE**	CORDE, CABLE	SEIL
10731	**ROPE BLOCK**	MOUFLE A CORDE	TAUKLOBEN
10732	**ROPE BRAKE**	FREIN A CORDE	SEILBREMSE
10733	**ROPE DIAMETER**	DIAMETRE D'UN CABLE	SEILSTÄRKE, SEILDICKE, SEILDURCHMESSER
10734	**ROPE DRIVE OR GEARING**	TRANSMISSION PAR CABLES, TRANSMISSION PAR CORDES COURROIES, TRANSMISSION FUNICULAIRE	SEILTRIEB
10735	**ROPE DRIVE, DRIVING, DRIVE BY ROPES**	COMMANDE PAR CABLE	SEILANTRIEB
10736	**ROPE DRIVEN MACHINE**	MACHINE A COMMANDE PAR CABLE	MASCHINE FÜR SEILANTRIEB
10737	**ROPE DRUM**	TAMBOUR D'ENROULEMENT	SEILTROMMEL
10738	**ROPE FLYWHEEL**	VOLANT A GORGES	SEILSCHEIBENSCHWUNGRAD
10739	**ROPE LAYING, TWISTING**	COMMETTAGE, CABLAGE	SEILSCHLAG, VERSEILEN
10740	**ROPE LINE**	BRIN DE CABLE	SEILSTRANG

10741	**ROPE PULLEY BLOCK**	PALAN A CORDE	SEILFLASCHENZUG
10742	**ROPE WHEEL, ROPE SHEAVE, SHEAVE WHEEL (FOR ROPE)**	POULIE A CABLE, POULIE POUR CABLE DE TRANSMISSION	SEILSCHEIBE, SEILROLLE
10743	**ROPE WIRE**	FIL POUR CABLE	SEILDRAHT
10744	**ROPE YARN**	FIL DE CARET	SEILGARN, SEILFADEN
10745	**ROPINESS**	CORDAGE	VERDICKUNG (FASERIGE)
10746	**ROSE**	POMME D'ARROSOIR	BRAUSE
10747	**ROSE REAMER**	ALESOIR EN BOUT	VERSENKER
10748	**ROSETTE COPPER**	CUIVRE ROSETTE	ROSETTENKUPFER
10749	**ROSIN OIL, RESIN OIL**	HUILE DE RESINE, HUILE DE PIN	HARZÖL
10750	**ROSOLIC ACID**	ACIDE ROSOLIQUE	ROSOLSÄURE
10751	**ROTARY BLOWER FAN**	VENTILATEUR A PISTON ROTATIF	KAPSELGEBLÄSE, FLÜGELGEBLÄSE
10752	**ROTARY COOLER**	REFROIDISSEUR CIRCULAIRE	UMLAUFKÜHLER
10753	**ROTARY CULTIVATORS HOES AND TILLERS**	MOTOCULTEURS A FRAISES ROTATIVES	HACKFRÄSEN UND BODENFRÄSEN
10754	**ROTARY ENGINE**	MOTEUR A PISTONS ROTATIFS	UMLAUFMOTOR
10755	**ROTARY FILE**	LIME ROTATIVE	DREHFEILE
10756	**ROTARY PISTON**	PISTON ROTATIF	DREHKOLBEN
10757	**ROTARY PRESS**	PRESSE ROTATIVE	KARUSSELLPRESSE
10758	**ROTARY PUMP**	POMPE A ROTOR, POMPE ROTATIVE, POMPE A PALETTES ROTATIVES	ZENTRIFUGALPUMPE, DREHKOLBENPUMPE, ROTATIONSPUMPE, KAPSELPUMPE, WALZENPUMPE, WÜRGELPUMPE, KREISKOLBENPUMPE
10759	**ROTARY SHEARS**	CISAILLES ROTATIVES	SCHERE (ROTIERENDE)
10760	**ROTARY SLIDE VALVE**	DISTRIBUTEUR GLISSANT A ROBINET, ROBINET DISTRIBUTEUR TOURNANT, OBTURATEUR, DISTRIBUTEUR OSCILLANT, VALVE OSCILLANTE	DREHSCHIEBER
10761	**ROTARY STEAM ENGINE**	MACHINE ROTATIVE	ROTIERENDE DAMPFMASCHINE, ROTATIONSDAMPFMASCHINE, DAMPFMASCHINE MIT UMLAUFENDEM KOLBEN
10762	**ROTARY TABLE**	PLATEAU TOURNANT	DREHTISCH
10763	**ROTATE (TO), REVOLVE (TO)**	TOURNER	DREHEN (SICH), UMLAUFEN, ROTIEREN
10764	**ROTATING SHAFT**	ARBRE ROTATIF	UMLAUFENDE WELLE, ROTIERENDE WELLE, DREHENDE WELLE (SICH)
10765	**ROTATION; REVOLUTION**	ROTATION	UMDREHUNG, DREHUNG, ROTATION
10766	**ROTATORY MOTION, MOTION OF ROTATION**	MOUVEMENT ROTATOIRE, MOUVEMENT DE ROTATION	DREHBEWEGUNG, BEWEGUNG (DREHENDE)
10767	**ROTOR**	ROTOR	LÄUFER, ROTOR
10768	**ROTTEN TIMBER**	BOIS CARIE	HOLZ (MORSCHES)
10769	**ROTTENNESS**	DECOMPOSITION	FÄULNIS, ZERSETZUNG
10770	**ROTTING OF TIMBER, DECAY IN TIMBER**	CARIE, NECROSE, POURRITURE DU BOIS	FÄULNIS DES HOLZES
10771	**ROUGH BORE (TO), DRILL (TO)**	PERCER, FORER PREALABLEMENT	VORBOHREN
10772	**ROUGH CALCULATION**	CALCUL RAPIDE APPROCHE	ÜBERSCHLÄGIGE RECHNUNG, ÜBERSCHLAGSRECHNUNG
10773	**ROUGH CAST GEAR WHEEL**	ROUE D'ENGRENAGE BRUTE DE FONDERIE, ENGRENAGE BRUT DE FONTE, ROUE A DENTURE BRUTE	ZAHNRAD (ROH GEGOSSENES)
10774	**ROUGH CUT FILE, RUBBER FILE**	LIME A GROSSE TAILLE, LIME A TAILLE RUDE, LIME FORTE	GROBFEILE, ARMFEILE, STROHFEILE

10775	**ROUGH DOWN (TO)**	DEGROSSIR, EBAUCHER	SCHRUPPEN, ZURICHTEN, VORBEARBEITEN, GROBEN BEARBEITEN (AUS DEM), ROHEN BEARBEITEN (AUS DEM)
10776	**ROUGH FLANGE**	BRIDE BRUTE DE FONTE	FLANSCH (ROHER), FLANSCH (UNBEARBEITETER)
10777	**ROUGH GRINDING**	DEGROSSISSAGE A LA MEULE	VORSCHLEIFEN
10778	**ROUGH ROLLED, COGGED**	BRUT DE LAMINAGE	VORGEWALZT
10779	**ROUGH SURFACE**	SURFACE RUGUEUSE	FLÄCHE (RAUHE)
10780	**ROUGH TUBE**	TUBE RUGUEUX	ROHR (RAUHES)
10781	**ROUGH WEIGHT**	POIDS BRUT	ROHGEWICHT, BRUTTOGEWICHT
10782	**ROUGH-DRILLED HOLE**	AVANT-TROU	LOCH (VORGEBOHRTES)
10783	**ROUGHEN (TO)**	RENDRE RUGUEUX	AUFRAUHEN
10784	**ROUGHENED SURFACE**	SURFACE RUGUEUSE	FLÄCHE (GERAUHTE)
10785	**ROUGHING**	EBAUCHAGE	ROHBEARBEITUNG
10786	**ROUGHING DOWN**	DEGROSSISSAGE, EBAUCHAGE	BEARBEITUNG, AUS DEM GROBEN, BEARBEITUNG AUS DEM ROHEN, VORBEARBEITUNG, ZURICHTEN, SCHRUPPEN, SCHRUPPARBEIT
10787	**ROUGHING DOWN OR SHINGLING**	CAGE DEGROSSISEURS	VORWALZE
10788	**ROUGHING LATHE**	TOUR A EBAUCHER	SCHRUPPDREHBANK
10789	**ROUGHING MILL**	TRAIN EBAUCHEUR	VORSTRASSE
10790	**ROUGHING TAPER REAMER**	ALESOIR EBAUCHEUR CONIQUE	VORREIBAHLE (KONISCHE)
10791	**ROUGHING TOOL**	OUTIL A DEGROSSIR	SCHRUPPSTAHL
10792	**ROUGHNESS**	RUGOSITE, ASPERITE	RAUHIGKEIT
10793	**ROUND AN EDGE (TO)**	ARRONDIR UNE ARETE	KANTE RUNDEN (EINE), ABRUNDEN
10794	**ROUND BACKED ANGLE**	FER CORNIERE A ANGLE ARRONDI	WINKELEISEN (INNEN VOLL), ABGERUNDET
10795	**ROUND BAR IRON**	FER ROND	RUNDEISEN, RUNDSTAB
10796	**ROUND FILE**	LIME RONDE	RUNDFEILE
10797	**ROUND FLANGE**	BRIDE RONDE	FLANSCH (RUNDER)
10798	**ROUND FLASK**	BALLON ORDINAIRE	RUNDKOLBEN
10799	**ROUND HAND**	ECRITURE RONDE	RUNDSCHRIFT
10800	**ROUND PLANE**	RABOT ROND	KEHLHOBEL
10801	**ROUND ROPE**	CABLE ROND	RUNDSEIL
10802	**ROUND THREAD**	FILET ROND	GEWINDE (RUNDES), KORDELGEWINDE
10803	**ROUND TIMBER**	RONDIN, BOIS EN RONDINS	RUNDHOLZ
10804	**ROUND WIRE**	FIL ROND	RUNDDRAHT, DRAHT (RUNDER)
10805	**ROUND WIRE PLIERS**	PINCE RONDE	RUNDZANGE, DRAHTZANGE (RUNDE)
10806	**ROUND-HAND PEN**	PLUME DE RONDE	RUNDSCHRIFTFEDER
10807	**ROUND-THREADED SCREW**	VIS A FILET ROND	SCHRAUBE MIT RUNDGEWINDE, RUNDGÄNGIGE SCHRAUBE
10808	**ROUNDED EDGE**	ARETE ARRONDIE	KANTE (ABGERUNDETE)
10809	**ROUNDING OF THREAD**	ARRONDI DU FILET	ABRUNDUNG DES GEWINDES
10810	**ROUTE**	ITINERAIRE MARITIME, PARCOURS	SEEWEG, FAHRT
10811	**ROUTINE**	PROGRAMME D'ORDINATEUR	RECHNERPROGRAMM
10812	**ROW**	LIGNE	ZEILE
10813	**ROW OF RIVETS**	LIGNE DE RIVETS, RANG DE RIVETS	NIETREIHE
10814	**RPM**	REVOLUTIONS PAR MINUTE, TOUR MINUTE	DREHZAHL, TOURENZAHL, UMDREHUNGEN PRO MINUTE, U/MIN
10815	**RUB THE RUST OFF (TO)**	DEROUILLER	ENTROSTEN
10816	**RUBBER**	CAOUTCHOUC	GUMMI

10817	**RUBBER BELT**	COURROIE EN CAOUTCHOUC, COURROIE EN TISSU CAOUTCHOUTE	GUMMIRIEMEN
10818	**RUBBER CLOTH**	TISSU CAOUTCHOUTE, TOILE CAOUTCHOUTEE, ETOFFE CAOUTCHOUTEE	GUMMISTOFF
10819	**RUBBER DIAPHRAGM**	DIAPHRAGME EN CAOUTCHOUC	GUMMIMEMBRAN
10820	**RUBBER GASKET**	JOINT A BAGUE DE CAOUTCHOUC	GUMMIRINGDICHTUNG, DICHTUNG MIT GUMMIRING
10821	**RUBBER HOSE**	TUYAU (FLEXIBLE) DE CAOUTCHOUC	GUMMISCHLAUCH
10822	**RUBBER INSULATING TAPE**	RUBAN CAOUTCHOUTE	GUMMIISOLIER BAND
10823	**RUBBER RING**	BAGUE EN CAOUTCHOUC, RONDELLE EN CAOUTCHOUC	GUMMIRING
10824	**RUBBER STOPPER**	BOUCHON DE CAOUTCHOUC	GUMMIPFROPFEN, KAUTSCHUKSTOPFEN
10825	**RUBBER, INDIA-RUBBER, CAOUTCHOUC**	CAOUTCHOUC, GOMME ELASTIQUE	KAUTSCHUK, FEDERHARZ
10826	**RUBBING OFF THE RUST**	DEROUILLEMENT	ENTROSTEN
10827	**RUBBLE STONE**	MOELLON	BRUCHSTEIN
10828	**RUBBLE WORK**	MACONNERIE EN MOELLONS	BRUCHSTEINMAUERWERK
10829	**RUBIDIUM**	RUBIDIUM	RUBIDIUM
10830	**RUHMKORFF COIL**	BOBINE DE RUHMKORFF	RUHMKORFF-INDUKTOR
10831	**RULER**	REGLE DE DESSINATEUR	LINEAL
10832	**RUN**	COUCHE, PASSE	LAGE
10833	**RUN AT A HIGH SPEED (TO)**	MARCHER A GRANDE VITESSE	SCHNELLAUFEN
10834	**RUN AT A LOW SPEED (TO)**	MARCHER A PETITE VITESSE	LANGSAMLAUFEN
10835	**RUN IN AN ENGINE (TO)**	RODER UN MOTEUR	EINLAUFEN LASSEN (EINE MASCHINE)
10836	**RUN OUT**	BAVURE	GRAT
10837	**RUN OUT BEARING**	FONDRE (LE COUSSINET FOND)	AUSGELAUFENES LAGER, AUSGESCHMOLZENES
10838	**RUN OUT TABLE**	TRAIN DE ROULEAUX DE SORTIE	AUSLAUFROLLGANG
10839	**RUN PIPE**	COLLECTEUR	TRANSPORTROHR
10840	**RUN UNTRUE (TO)**	TOURNER FAUX-ROND	UNRUND LAUFEN
10841	**RUN-OF-MINE COAL, ROUGH COAL**	TOUT-VENANT, HOUILLE TOUT-VENANT	FÖRDERKOHLE
10842	**RUN-OFF PLATE**	TOLE TECHNOLOGIQUE	AUSLAUFBLECH
10843	**RUNDOWN TANK**	RESERVOIR DE RECETTE	EMPFANGSTANK
10844	**RUNNER**	GALET DE ROULEMENT, COULEE, CANAL DE COULEE	LAUFROLLE, GUSSRINNE
10845	**RUNNER GATE**	COULEE, ATTAQUE	ANGUSS
10846	**RUNNING BACKWARD, GOING ASTERN**	MARCHE ARRIERE	RÜCKWÄRTSGANG
10847	**RUNNING FIT**	EMMANCHEMENT, AJUSTAGE, MONTAGE TOURNANT	LAUFSITZ
10848	**RUNNING FORWARD, GOING AHEAD**	MARCHE AVANT	VORWÄRTSGANG
10849	**RUNNING METRE, METRE RUN**	METRE COURANT	METER (LAUFENDES)
10850	**RUNNING OF AN ENGINE**	MARCHE, FONCTIONNEMENT, ALLURE D'UNE MACHINE	GANG EINER MASCHINE
10851	**RUNNING PART (IN LIFTING TACKLE)**	BRIN COURANT, COURANT D'UN PALAN	LAUFENDES TRUMM EINES FLASCHENZUGES
10852	**RUNNING SCREW**	VIS DE TRANSLATION	BEWEGUNGSSCHRAUBE
10853	**RUNNING SIDE DRIVING FACE OF A BELT**	COTE INTERIEUR D'UNE COURROIE	LAUFSEITE EINES RIEMENS
10854	**RUNNING WHEEL**	ROUE PORTEUSE, GALET DE ROULEMENT	LAUFRAD

10855	RUNNING-IN	RODAGE	EINFAHRZEIT
10856	RUNS, RUNNING	COULURES	LÄUFER
10857	RUNWAY (FOR A RODF ROLLING LADDER)	CHEMIN DE ROULEMENT (POUR ECHELLE ROULANTE DE TOIT)	FAHRFLÄCHE-LAUFBAHN, ROLLBAHN (FÜR ROLLEITER)
10858	RUPTURE	FRACTURE, RUPTURE, ARRACHEMENT	BRUCH
10859	RUPTURE (TO), REND (TO)	DECHIRER (SE)	REISSEN, EINREISSEN, AUFREISSEN
10860	RUPTURE DISC	MEMBRANE D'ECLATEMENT, DISQUE DE RUPTURE	SICHERHEITSMEMBRANE, BRUCHSCHEIBE
10861	RUPTURE, FRACTURE	RUPTURE, BRIS	BRUCH, BRECHEN
10862	RUSSIAN OIL OF TURPENTINE	ESSENCE DE TEREBENTHINE RUSSE	KIENÖL
10863	RUST	ROUILLE	ROST
10864	RUST (TO)	ROUILLER (SE)	VERROSTEN
10865	RUST FORMATION, FORMATION OF RUST	FORMATION DE ROUILLE	ROSTEN, VERROSTEN, ROSTBILDUNG
10866	RUST INHIBITOR	PRODUIT ANTI-ROUILLE	ROSTSCHUTZMITTEL
10867	RUST JOINT	JOINT AU MASTIC DE FONTE	ROSTDICHTUNG
10868	RUST PREVENTING	PRESERVANT DE LA ROUILLE	ROSTVERHÜTEND
10869	RUST PREVENTING AGENT	PRESERVATIF CONTRE LA ROUILLE, ANTI-ROUILLE	ROSTSCHUTZMITTEL
10870	RUST PREVENTIVE (AGENT)	PRODUIT ANTI-ROUILLE	ROSTSCHUTZMITTEL
10871	RUST PROOFING	PROTECTION CONTRE LA ROUILLE	ROSTSCHUTZ
10872	RUST REMOVING DIP	BAIN DE DEROUILLAGE	ENTROSTUNGSBAD
10873	RUSTY	ROUILLE	ROSTIG, VERROSTET
10874	RUTHENIUM	RUTHENIUM	RUTHENIUM
10875	S.S. (STAINLESS STEEL) STRAP	FEUILLARD INOX	STAHLBAND (ROSTFREIES)
10876	SADDLE	PIECE D'APPUI	SATTEL
10877	SADDLE SUPPORT	BERCEAU	STÜTZGESTELL, SATTELHALTER
10878	SAFE EDGED FILE	LIME A COTES LISSES	ANSATZFEILE
10879	SAFE LOAD, WORKING LOAD	CHARGE PRATIQUE, CHARGE DE SECURITE	BETRIEBSBELASTUNG
10880	SAFE WORKING STRESS	EFFORT ADMISSIBLE	BEANSPRUCHUNG (ZULÄSSIGE)
10881	SAFETY APPLIANCE	DISPOSITIF PROTECTEUR, DISPOSITIF DE SURETE, DISPOSITIF DE SECURITE, APPAREIL DE PROTECTION	SCHUTZVORRICHTUNG, SICHERHEITSVORRICHTUNG
10882	SAFETY BELT	CEINTURE DE SECURITE	SICHERHEITSGURT
10883	SAFETY BRAKE	FREIN DE SECURITE	SICHERHEITSBREMSE
10884	SAFETY COUPLING	MANCHON DE SURETE	SICHERHEITSKUPPLUNG
10885	SAFETY FACTOR	COEFFICIENT DE SECURITE	SICHERHEITSFAKTOR
10886	SAFETY FUSE	COUPE-CIRCUIT DE SURETE, PLOMB FUSIBLE, FUSIBLE DES COUPE-CIRCUIT, PLOMB DE SURETE	SICHERUNG (ELEKTR.)
10887	SAFETY HEAD	DISQUE DE RUPTURE	BRECHSICHERUNG
10888	SAFETY HOOK	CROCHET DE SURETE	SICHERHEITSHAKEN
10889	SAFETY RULES	MESURES DE SECURITE	SICHERHEITSVORSCHRIFTEN
10890	SAFETY VALVE WITH HIGH LIFT	SOUPAPE DE SURETE A GRANDE LEVEE	HOCHHUBSICHERHEITSVENTIL
10891	SAFETY VALVE, RELIEF VALVE	SOUPAPE DE SURETE	SICHERHEITSVENTIL
10892	SAG	EXCENTRATION, FLECHISSEMENT, COULURE POCHE, DEGOULINAGE	DURCHHÄNGEN, NASENBILDUNG
10893	SAG (TO)	FLECHIR	DURCHHÄNGEN
10894	SAG, SAGGING	FLECHISSEMENT	DURCHHANG
10895	SAGGING (OR SAGS)	COULURES	VORHÄNGEBILDUNG

10896	**SALAMANDER**	LOUP	OFENSAU
10897	**SALARY**	APPOINTEMENTS, SALAIRE	GEHALT
10898	**SALES ENGINEER**	INGENIEUR CHARGE DES VENTES	VERKAUFSINGENIEUR
10899	**SALICYLIC ACID**	ACIDE SALICYLIQUE	SALIZYLSÄURE
10900	**SALIENT ANGLE**	ANGLE SAILLANT	VORSPRINGENDE ECKE
10901	**SALINOMETER**	PESE-SELS	SALZSPINDEL, SALZWAAGE, SOLWAAGE, SALINOMETER
10902	**SALOON**	CONDUITE INTERIEURE, BERLINE	INNENLENKER, INNENSTEUERLIMOUSINE
10903	**SALPETRE, NITRE, POTASSIUM NITRATE**	SALPETRE, NITRE, NITRATE, AZOTATE DE POTASSIUM	SALPETER, KALISALPETER, SALPETERSAURES KALI, KALIUMNITRAT
10904	**SALT**	SEL (CHIM.)	SALZ (CHEM.)
10905	**SALT BATH**	BAIN DE SEL	SALZBAD
10906	**SALT CONTENT**	TENEUR EN SEL	SALZGEHALT
10907	**SALT OF SORREL, SALT OF LEMON**	SEL D'OSEILLE	SAUERKLEESALZ, KLEESALZ
10908	**SALT SOLUTION**	SOLUTION SALINE	SALZLÖSUNG
10909	**SALT SPRAY TEST**	ESSAI AU BROUILLARD SALIN	SALZSPRÜHNEBELPRÜFUNG
10910	**SALT WATER**	EAU SALINE	WASSER (SALZHALTIGES), SALZWASSER
10911	**SALT-BATH BRAZING**	BRASAGE AU BAIN SALIN	SALZBADLÖTEN
10912	**SALT-BATH FURNACE**	FOUR A BAIN DE SEL	SALZBADOFEN
10913	**SALT-BATH HARDENING**	TREMPE AU BAIN DE SEL	SALZBADHÄRTUNG
10914	**SAMARIUM**	SAMARIUM	SAMARIUM
10915	**SAMPLE, SPECIMEN**	ECHANTILLON, SPECIMEN	PROBE, PROBESTÜCK, MUSTER
10916	**SAMPLING**	PRISE D'UN ECHANTILLON, ECHANTILLONNAGE	ENTNAHME EINER PROBE, PROBENEHMEN, AUSWAHL, PROBEENTNAHME
10917	**SAND**	SABLE	SAND
10918	**SAND BATH**	BAIN DE SABLE	SANDBAD
10919	**SAND BLAST (TO)**	SABLER	SANDSTRAHL BEARBEITEN (MIT), ABSANDEN
10920	**SAND BLAST APPARATUS**	MACHINE AU JET DE SABLE, SABLEUSE	SANDSTRAHLGEBLÄSE
10921	**SAND BLASTING**	SABLAGE	BEARBEITEN MIT SANDSTRAHL, ABSANDEN, SANDTRAHLUNG
10922	**SAND CASTING**	COULEE EN SABLE	SANDGUSS
10923	**SAND FILTER**	FILTRE A SABLE	SANDFILTER
10924	**SAND MOLD**	MOULE EN SABLE	SANDFORM
10925	**SAND PAPER**	PAPIER SABLE	SANDPAPIER
10926	**SAND PAPERING MACHINE, GLASS PAPERING MACHINE**	MACHINE A PONCER AU PAPIER DE VERRE	SANDPAPIERMASCHINE
10927	**SANDARACH**	SANDARAQUE, RESINE DE VERNIS	SANDARAK
10928	**SANDSTONE**	GRES (GEOL.)	SANDSTEIN
10929	**SANDWICH ROLLING**	LAMINAGE EN SANDWICH, LAMINAGE STRATIFIE	SANDWICHWALZEN
10930	**SANDY SOIL**	TERRAIN SABLEUX	SANDBODEN
10931	**SANITRY FACILITIES**	INSTALLATIONS SANITAIRES, EQUIPEMENT SANITAIRE	EINRICHTUNG (SANITÄRE)
10932	**SAPONIFIABLE**	SAPONIFIABLE	VERSEIFBAR
10933	**SAPONIFICATION**	SAPONIFICATION	SEIFENBILDUNG, VERSEIFUNG
10934	**SAPONIFY (TO)**	SAPONIFIER	VERSEIFEN
10935	**SAPONITE**	SAPONITE, PIERRE DE SAVON	SEIFENSTEIN, SAPONIT
10936	**SAPWOOD, ALBURNUM**	AUBIER	SPLINT, SPLINTHOLZ
10937	**SASH BAR IRON**	FER (MOULURE) A VITRAGES	SPROSSENEISEN

10938	SATIN FINISH	FINI SATINE	MATTGESCHLIFFEN
10939	SATURATE (TO)	SATURER	SÄTTIGEN
10940	SATURATED HYDROCARBONS	HYDROCARBURES SATURES, HYDROCARBURES PARAFFINIQUES	KOHLENWASSERSTOFFE (GESÄTTIGTE)
10941	SATURATED SOLUTION	SOLUTION SATUREE	LÖSUNG (GESÄTTIGTE)
10942	SATURATED STEAM	VAPEUR SATUREE, SATURANTE	DAMPF (GESÄTTIGTER), SATTDAMPF
10943	SATURATION	SATURATION	SÄTTIGUNG
10944	SATURATION PRESSURE	PRESSION DE SATURATION	SÄTTINGUNGSDRUCK
10945	SATURATION TEMPERATURE	TEMPERATURE DE SATURATION	SÄTTIGUNGSTEMPERATUR
10946	SAUCER	GODET (DU DESSINATEUR)	TUSCHNAPF, TUSCHSCHALE
10947	SAVING OF LABOUR	ECONOMIE DANS LE TRAVAIL, ECONOMIE DE MAIN-D'OEUVRE	ARBEITSERSPARNIS
10948	SAVING OF POWER	ECONOMIE DE FORCE MOTRICE	KRAFTERSPARNIS
10949	SAVING OF TIME, ECONOMY OF TIME	ECONOMIE DE TEMPS, GAIN DE TEMPS	ZEITERSPARNIS
10950	SAW	SCIE	SÄGE
10951	SAW (TO)	SCIER	SÄGEN
10952	SAW BLADE	LAME DE SCIE	SÄGEBLATT
10953	SAW DUST	SCIURE DE BOIS , BRAN DE SCIE	SÄGEMEHL, SÄGESPÄNE
10954	SAW FILE	LIME A SCIES	SÄGEFEILE
10955	SAW FILER'S CLAMP, VICE	ETAU D'AFFUTAGE POUR SCIES	SÄGEFEILKLUPPE
10956	SAW MILL	SCIERIE (USINE)	SÄGEWERK
10957	SAW SET	TOURNE-A-GAUCHE POUR AVOYER, TOURNE-A-GAUCHE POUR DONNER DU PAS/DE VOIE A UNE SCIE	SCHRÄNKEISEN
10958	SAW SETTING	MISE EN VOIE DES DENTS D'UNE SCIE, AVOYAGE D'UNE SCIE	SCHRÄNKEN, AUSSETZEN DER SÄGEZÄHNE
10959	SAW TOOTH	DENT DE SCIE	SÄGEZAHN
10960	SAW TOOTH CLUTCH	EMBRAYAGE A DENTS DE SCIE	SÄGEZAHNKUPPLUNG
10961	SAWING	SCIAGE	SÄGEN
10962	SAWN TIMBER	BOIS DEBITE	SCHNITTHOLZ
10963	SAWS AND SAW BENCHES	SCIES A BUCHES	SÄGEN UND SÄGEBÄNKE
10964	SCAB	ECAILLE	SCHUPPE
10965	SCAFFOLD	ECHAFAUDAGE	GERÜST, ARBEITSGERÜST, BAUGERÜST
10966	SCALAR PRODUCT	PRODUIT SCALAIRE	PRODUKT (SKALARES)
10967	SCALAR QUANTITY	SCALAIRE	SKALAR
10968	SCALE	ECHELLE, ECAILLE	MASSSTAB, SCHUPPE
10969	SCALE	DEPOTS CALCAIRES, INCRUSTATIONS DES CHAUDIERES, CALCAIRE, TARTRE DES CHAUDIERES	KESSELSTEIN
10970	SCALE (ON A LARGER)	REPRODUCTION (EN GRAND)	MASSSTAB (IN VERGRÖSSERTEM), MASSSTÄBLICH VERGRÖSSERT
10971	SCALE (TO), REMOVE THE SCALE (TO)	DESINCRUSTER, DETARTRER	KESSELSTEIN ENTFERNEN
10972	SCALE BEAM, BEAM OF A BALANCE	FLEAU DE BALANCE	WAAGEBALKEN
10973	SCALE OF A DRAWING	ECHELLE D'UN PLAN	MASSSTAB EINER ZEICHNUNG, ZEICHNUNGSMASSSTAB
10974	SCALE OF HARDNESS	ECHELLE, GAMME DE DURETE	HÄRTESKALA
10975	SCALE OFF (TO), PEEL OFF (TO), FLAKE (TO)	ECAILLER (S')	ABBLÄTTERN
10976	SCALE PAN	PLATEAU D'UNE BALANCE	WAAGSCHALE, GEWICHTSSCHALE
10977	SCALE PIT	FOSSE A BATTITURES	SINTERGRUBE

10978	SCALE PREVENTIVE, ANTI-INCRUSTATOR	DESINCRUSTANT PREVENTIF, SUBSTANCE ANTI-INCRUSTANTE, DETARTRANT	KESSELSTEINVERHÜTUNGSMITTEL
10979	SCALE REMOVING	DECALAMINAGE	ENTZUNDERUNG
10980	SCALE RESISTANCE	INOXYDABILITE	ZUNDERBESTÄNDIGKEIT
10981	SCALE, DIVISION, GRADUATION	DIVISION, GRADUATION, ECHELLE	TEILUNG, GRADEINTEILUNG, SKALA
10982	SCALE, RATIO OF DIMENSIONS	ECHELLE (MOYEN DE COMPARAISON)	MASSSTAB, GRÖSSENVERHÄLTNIS
10983	SCALED PROPORTIONAL DRAWING, DRAWING TO SCALE	DESSIN A L'ECHELLE	ZEICHNUNG (MASSSTÄBLICHE)
10984	SCALENE TRIANGLE	TRIANGLE SCALENE	DREIECK (UNGLEICHSEITIGES)
10985	SCALING	ECAILLAGE	ABBLÄTTERUNG
10986	SCALING HAMMER	MARTEAU A EBARBER LA FONTE, MARTEAU A PIQUER LES CHAUDIERES	GUSSPUTZHAMMER, KESSELSTEINHAMMER
10987	SCALING, REMOVAL OF SCALE	DESINCRUSTATION, DETARTRAGE	KESSELSTEINENTFERNUNG
10988	SCALPING	ECROUTAGE	SCHÄLUNG
10989	SCALY FLAKE GRAPHITE	GRAPHITE EN PAILLETTES	SCHUPPENGRAPHIT
10990	SCAN	BALAYER	ABTASTEN
10991	SCANDIUM	SCANDIUM	SKANDIUM
10992	SCANNING	BALAYAGE	ZONENABTASTEN
10993	SCARF	DECRIQUER (AU CHALUMEAU)	FLÄMMEN
10994	SCATTERING	DIFFUSION	STREUUNG
10995	SCAVENGER	ELIMINATEUR D'IMPURETES	REINIGUNGSMITTEL
10996	SCHEELE'S GREEN	VERT DE SCHEELE, ARSENITE DE CUIVRE	GRÜN (SCHEELESCHES), SCHWEDISCHGRÜN, MINERALGRÜN
10997	SCHEELITE	SCHEELITE, SCHEELIN CALCAIRE	SCHEELIT
10998	SCHELLAC VARNISH	VERNIS A LA GOMME-LAQUE	SCHELLACKFIRNIS
10999	SCHIST	SCHISTE	SCHIEFER (GEOL.)
11000	SCHWEINFURTH GREEN	VERT DE SCHWEINFURT	SCHWEINFURTERGRÜN
11001	SCIENTIFIC MANAGEMENT	DIRECTION/ ORGANISATION SCIENTIFIQUE DES ATELIERS	BETRIEBSFÜHRUNG (WISSENCHAFTLICHE)
11002	SCLEROMETER	SCLEROMETRE	RITZHÄRTEPRÜFER, SKLEROMETER
11003	SCLEROSCOPE HARDNESS	DURETE SHORE	KUGELFALLHÄRTE, SHOREHÄRTE
11004	SCORIFICATION	SCORIFICATION	VERSCHLACKUNG
11005	SCOUR (TO), PICKLE A METAL (TO)	DECAPER, DEROCHER, DEGRAISSER UN METAL	ABBEIZEN (EIN METALL), BLANKBEIZEN, ENTZUNDERN, DEKAPIEREN
11006	SCOURING, PICKLING A METAL	DECAPAGE, DEROCHAGE, DEGRAISSAGE D'UN METAL	BEIZEN, ABBEIZEN, BLANKBEIZEN, ENTZUNDERN, DEKAPIEREN EINES METALLES
11007	SCRAP, SCRAP IRON	RIBLONS DE FER, ROGNURES DE FER, FERRAILLE, MITRAILLE	ALTEISEN, SCHROTT
11008	SCRAPE (TO)	GRATTER, RACLER, CURER	SCHABEN
11009	SCRAPE, SCRAPER	GRATTOIR, RACLOIR, RACLETTE, ROGNOIR, CURETTE	SCHABER
11010	SCRAPER	TRANSPORTEUR A RACLETTES	KRATZER, SCHLEPPER, SCHLEPPKETTE, SEILFÖRDERER, SCHLEPPSEIL
11011	SCRAPING	GRATTAGE, RACLAGE, CURAGE	SCHABEN, KRATZEN
11012	SCRAPING MACHINE	MACHINE A GRATTER	SCHABEMASCHINE
11013	SCRATCH	GRIFFE	RITZ
11014	SCRATCH BRUSH	BROSSE METALLIQUE	KRATZBÜRSTE, PUTZBÜRSTE
11015	SCRATCH HARDNESS	RESISTANCE A L'ABRASION	KRATZFESTIGKEIT
11016	SCRATCH-BRUSH FINISH	FINISSAGE A LA BROSSE METALLIQUE	METALLBÜRSTEN-FERTIGBEARBEITUNG
11017	SCRATCHING TEST	ESSAI SCLEROMETRIQUE	RITZVERSUCH

11018	SCREEN	ECRAN, TAMIS	SCHIRM, SIEB
11019	SCREEN (OPT.)	ECRAN (OPT.)	SCHIRM (OPT.)
11020	SCREEN ANALYSIS	ANALYSE GRANULOMETRIQUE	SIEBANALYSE
11021	SCREW	VIS, BOULON A VIS	SCHRAUBE
11022	SCREW (TO)	VISSER, SERRER A VIS	VERSCHRAUBEN, ANSCHRAUBEN, FESTSCHRAUBEN, ZUSAMMENSCHRAUBEN
11023	SCREW AUGER, TWISTED AUGER	MECHE STYRIENNE, MECHE FACON SUISSE	SCHNECKENBOHRER
11024	SCREW AXIS	AXE DE SYMETRIE	SYMMETRIEACHSE
11025	SCREW CAP, CAP WITH A FEMALE THREAD	BOUCHON FILETE, BOUCHON FEMELLE, BOUCHON TARAUDE	VERSCHLUSSKAPPE, SCHRAUBVERSHLUSS, SCHRAUBKAPPE
11026	SCREW COUPLING	MANCHON A VIS	SCHRAUBENKUPPLUNG, GEWINDEKUPPLUNG
11027	SCREW CURRENT METER	MOULINET HYDRAULIQUE (POUR LA MESURE DE LA VITESSE D'ECOULEMENT D'UN COURANT D'EAU)	WASSERMESSFLÜGEL, WASSERMESSSCHRAUBE, HYDROMETRISCHER FLÜGEL
11028	SCREW CUTTING	FILETAGE (OPERATION)	SCHRAUBENSCHNEIDEN
11029	SCREW CUTTING LATHE	TOUR A FILETER	GEWINDEDREHBANK
11030	SCREW CUTTING, SCREW THREADING, CUTTING SCREW THREADS	CREUSAGE, CREUSEMENT D'UN FILET	SCHNEIDEN EINES GEWINDES, GEWINDESCHNEIDEN
11031	SCREW DISLOCATION	DISLOCATION-VIS (EN HELICE)	SCHRAUBENVERSETZUNG
11032	SCREW DOLLY	TAS AVEC CONTRE-BOUTEROLLE, TURC	NIETWINDE
11033	SCREW DOWN COCK	ROBINET A VIS DE PRESSION	NIEDERSCHRAUBHAHN
11034	SCREW GAUGE	PALMER	SCHRAUBLEHRE
11035	SCREW JACK	VERIN, CRIC MECANIQUE, CRIC A VIS	WINDE, WAGENHEBER, SCHRAUBENHEBER, SCHRAUBENWINDE
11036	SCREW MILLING MACHINE, THREAD MILLER	MACHINE A FRAISER LES VIS	GEWINDEFRÄSMASCHINE
11037	SCREW ON (TO)	VISSER SUR, MONTER A VIS	AUFSCHRAUBEN
11038	SCREW PITCH GAUGE	JAUGE DE FILETAGE	GEWINDEKONTROLLEHRE
11039	SCREW PLATE, DIE PLATE	FILIERE SIMPLE, FILIERE A TRUELLE	SCHNEIDEISEN, GEWINDEEISEN, SCHNEIDKLINGE, SCHRAUBENBLECH
11040	SCREW PRESS	PRESSE A VIS, BALANCIER A VIS	SPINDELPRESSE, SCHRAUBENPRESSE
11041	SCREW PROPELLER, PROPELLER	PROPULSEUR, HELICE PROPULSIVE	ANTRIEBSCHRAUBE, TREIBSCHRAUBE, PROPELLER
11042	SCREW PUMP	POMPE HELICOIDALE	SCHRAUBENPUMPE
11043	SCREW SPIKE	TIRE-FOND	SCHWELLENSCHRAUBE, SCHIENENSCHRAUBE
11044	SCREW TAP	TARAUD	GEWINDEBOHRER
11045	SCREW THREAD	FILET, PAS DE VIS, FILETAGE	GEWINDE, SCHRAUBENGEWINDE
11046	SCREW THREAD (TO), CUT SCREW THREADS (TO)	CREUSER UN FILER	GEWINDE SCHNEIDEN
11047	SCREW THREAD GAUGE	CALIBRE DE FILETAGE	GEWINDELEHRE
11048	SCREW THREAD MICROMETER COMPARATOR	COMPARATEUR MICROMETRIQUE POUR FILETAGES	MIKROMETRISCHER SCHRAUBENGEWINDE-KOMPARATOR
11049	SCREW THREAD PLUG GAUGE	TAMPON D'ALESAGE, TAMPON FILETE MALE	GEWINDELEHRDORN
11050	SCREW THREAD RING GAUGE	BAGUE TARAUDEE DU CALIBRE DE FILETAGE	GEWINDELEHRMUTTER
11051	SCREW WITH CYLINDRICAL HEAD, CHEESE HEAD SCREW	VIS A TETE RONDE	RUNDKOPFSCHRAUBE, SCHRAUBE MIT ZYLINDERKOPF

11052	SCREW WRENCH, MONKEY WRENCH, ADJUSTABLE SPANNER	CLEF AJUSTABLE, CLEF ANGLAISE	SCHRAUBENSCHLÜSSEL (VERSTELLBARER), SCHRAUBENSCHLÜSSEL (ENGLISCHER), UNIVERSALSCHRAUBENSCHLÜSSEL
11053	SCREWED ENDS (FEMALE)	ORIFICES TARAUDES (FEMELLES)	MUFFENANSCHLUSS
11054	SCREWED ENDS (MALE)	ORIFICES FILETES (MALES)	GEWINDEWEITEN
11055	SCREWED FLANGE	BRIDE A VISSER	GEWINDEFLANSCH
11056	SCREWED HOLE	TROU FILETE, ECROU	GEWINDELOCH
11057	SCREWED HOOK	CROCHET A VIS, PITON-VIS	SCHRAUBENHAKEN
11058	SCREWED JOINT, THREADED CONNECTION	RACCORD VISSE	SCHRAUBVERBINDUNG, VERSCHRAUBUNG
11059	SCREWED PIPE	TUBE TARAUDE, TUBE FILETE	GEWINDEROHR
11060	SCREWED PLUG, PLUG WITH A MALE THREAD	BOUCHON VISSE, BOUCHON A VIS, BOUCHON FILETE, BOUCHON MALE	VERSCHLUSSPFROPFEN, VERSCHLUSSSCHRAUBE ,GEWINDEPFROPFEN
11061	SCREWED SOCKET	MANCHON TARAUDE, MANCHON FEMELLE, ECROU A RACCORD, RACCORD FILETE	SCHRAUBMUFFE, GEWINDEMUFFE
11062	SCREWED SPINDLE	TIGE FILETEE	SPINDEL, SCHRAUBENSPINDEL, GEWINDESPINDEL
11063	SCREWED-ON FLANGE	BRIDE VISSEE	FLANSCH (AUFGESCHRAUBTER)
11064	SCREWED, SCREWING FLANGE	BRIDE TARAUDEE	GEWINDEFLANSCH
11065	SCREWING	VISSAGE	VERSCHRAUBEN, ANSCHRAUBEN, FESTSCHRAUBEN, ZUSAMMENSCHRAUBEN
11066	SCREWING MACHINE, SCREW CUTTING MACHINE	MACHINE A VISSER, MACHINE A FILETER ET TARAUDER, FILETEUSE, TARAUDEUSE	SCHRAUBENSCHNEIDMASCHINE, GEWINDESCHNEIDMASCHINE
11067	SCREWING ON	MONTAGE A VIS	AUFSCHRAUBEN
11068	SCRIBER, SCRIBING IRON	POINTE A TRACER, TRACERET, TRACOIR	VORREISSER, REISSNADEL, ANREISSNADEL, ANREISSSPITZE
11069	SCRIBING BLOCK, SURFACING GAUGE	TRUSQUIN A MARBRE	STREICHMASS (STEHENDES)
11070	SEA LEVEL	NIVEAU DE LA MER	MEERESSPIEGEL, MEERESHÖHE, SEEHÖHE
11071	SEA SAND, MARINE SAND	SABLE DE MER	MEERESSAND, SEESAND, DÜNENSAND
11072	SEA WATER	EAU DE MER	MEERWASSER, SEEWASSER
11073	SEAL PAN	CUVETTE	PLATTENABDICHTUNGSSCHALE
11074	SEAL PAN (OF TRAYS)	CUVETTE DE PLATEAUX	BECKEN DES ABSTELLTISCHES
11075	SEAL PLATE	PLAQUETTE	DICHTPLÄTTCHEN
11076	SEAL WELD	SOUDAGE ETANCHE	DICHTSCHWEISSEN
11077	SEAL WITH LEAD (TO)	PLOMBER (MARQUER D'UN SCEAU EN PLOMB)	PLOMBIEREN, MIT EINER PLOMBE VERSEHEN
11078	SEALANT	BOURRAGE PLASTIQUE ETANCHE, JOINT PLASTIQUE	KUNSTSTOFF-DICHTPACKUNG
11079	SEALED BEAM	BLOC OPTIQUE, MONOBLOC	SEALED-BEAM SCHEINMERFER
11080	SEALINE	CANALISATION EN MER OUVERTE, CONDUITE EN MER	SEEROHRLEITUNG
11081	SEALING	ETANCHEITE	ABDICHTUNG
11082	SEALING LIQUID	LIQUIDE OBTURATEUR	SPERRFLÜSSIGKEIT
11083	SEAM	SOUDURE	FUGE, NAHT
11084	SEAM OF RIVETED JOINT	LIGNE DE JONCTION DES TOLES	NIETNAHT
11085	SEAM OF TUBE	SOUDURE D'UN TUBE	NAHT EINES ROHRES, ROHRNAHT
11086	SEAM WELDING	SOUDURE CONTINE, SOUDAGE EN LIGNE CONTINUE A LA MOLETTE	NAHTSCHWEISSUNG, ROLLEN(NAHT)SCHWEISSEN
11087	SEAM-FOLDING MACHINE	PLIEUSE	FALZBIEGEMASCHINE
11088	SEAMING MACHINE	SERTISSEUSE	FALZMASCHINE

11089	SEAMLESS PIPE, SEAMLESS TUBES, SEAMLESS WELDLESS TUBE	TUBE SANS SOUDURE	ROHR (NAHTLOSES)
11090	SEARCH UNIT	PALPEUR	FÜHLER
11091	SEARCHLIGHT	PROJECTEUR DE LUMIERE	SCHEINWERFER
11092	SEASON CRACK	CRIQUE DE VIEILLISSEMENT, CRAQUELURE SAISONNIERE	ALTERSRISS
11093	SEASON CRACKING	CORROSION INTERCRISTALLINE DES LAITONS 70/30	SPANNUNGSRISSKORROSION
11094	SEASON TIMBER (TO)	SECHER LES BOIS A L'AIR LIBRE	HOLZ AUSTROCKNEN, HOLZ AN DER LUFT TROCKNEN
11095	SEASONING (TIMBER)	DESSECHAGE/DESSICCATION DES BOIS A L'AIR LIBRE	TROCKNEN DES HOLZES AN DER LUFT, AUSTROCKNEN DES HOLZES
11096	SEAT BELT	CEINTURE DE SECURITE	SITZGURT
11097	SEAT RING	SIEGE DE SOUPAPE	VENTILSITZ
11098	SEAT, SEATING	SIEGE	SITZ, SITZFLÄCHE
11099	SECANT	SECANTE	SEKANTE
11100	SECOND CUT FILE	LIME A TAILLE DEMI-DOUCE	HALBSCHLICHTFEILE
11101	SECOND FILTER	SECOND FILTRE	FEINFILTER
11102	SECOND TAP	TARAUD DEMI-CONIQUE, TARAUD INTERMEDIAIRE	GEWINDEMITTELSCHNEIDER
11103	SECONDARY COLOUR	COULEUR COMPOSEE	ZWEITFARBE, MISCHFARBE
11104	SECONDARY CREEP	FLUAGE SECONDAIRE	SEKUNDÄRES KRIECHEN, ZWEITES KRIECHSTADIUM
11105	SECONDARY HARDENING	TREMPE SECONDAIRE	SEKUNDÄRE HÄRTUNG
11106	SECONDARY INCLUSION	INCLUSION SECONDAIRE	EINSCHLUSS (SEKUNDÄRER)
11107	SECONDARY METAL	METAL DE RECUPERATION	UMSCHMELZMETALL
11108	SECONDARY SPECTRUM	SPECTRE SECONDAIRE	NEBENSPEKTRUM
11109	SECONDARY STRESS	CONTRAINTE SECONDAIRE	NEBENSPANNUNG
11110	SECONDARY WELDING CURRENT	COURANT SECONDAIRE DE SOUDAGE	SEKUNDÄRSCHWEISSSTROM
11111	SECTION	SECTION, COUPE TRANSVERSALE	QUERSCHNITT
11112	SECTION (ON LINE AB), SECTIONAL VIEW	COUPE D'UN DESSIN	SCHNITT, DURCHSCHNITT NACH A-B
11113	SECTION FOR MICROSCOPIC RESEARCH	OBJET A EXAMINER AU MICROSCOPE SOUS FORME DE COUPE MINCE	SCHLIFF FÜR MIKROSKOPIE, DÜNNSCHLIFF
11114	SECTION IRON(S) OR SHAPE(S) STRUCTURAL(S)	PROFILE(S)	FORMSTAHL
11115	SECTION MODULUS	MODULE D'INERTIE	WIDERSTANDSMOMENT
11116	SECTION OF A BEAM	SECTION (DROITE OU OBLIQUE) D'UNE POUTRE	SENKRECHTER ODER SCHRÄGER ABSCHNITT EINES BALKENS
11117	SECTION OF A PRESSURE VESSEL	ELEMENT D'UN RECIPIENT A PRESSION	DRUCKLUFTBEHÄLTERGLIED
11118	SECTION OF A STRAIGHT LINE	SEGMENT D'UNE DROITE	STRECKE (GEOM.)
11119	SECTION, SECTIONAL IRON	FER PROFILE FACONNE, PROFILE	FORMEISEN, PROFILEISEN, FASSONEISEN
11120	SECTIONAL AREA	SURFACE, AIRE DE LA SECTION	QUERSCHNITTFLÄCHE
11121	SECTIONAL AREA OF VALVE PASSAGE	SECTION DE PASSAGE D'UNE SOUPAPE, AIRE DE L'OUVERTURE MASQUEE PAR LA SOUPAPE	SPALTQUERSCHNITT EINES VENTILS
11122	SECTIONAL DRAWING	DESSIN EN COUPE, COUPE, SECTION	SCHNITTZEICHNUNG
11123	SECTIONAL GROOVE	CALIBRE DE PROFILAGE	PROFILKALIBER
11124	SECTIONAL PLAN, PLAN	PLAN, PROJECTION HORIZONTALE	GRUNDRISS, HORIZONTALPROJEKTION
11125	SECTIONAL STEEL	ACIER PROFILE	PROFILSTAHL

11126	**SECTOR OF CIRCLE**	SECTEUR D'UN CERCLE, SECTEUR CIRCULAIRE	KREISAUSSCHNITT, KREISSEKTOR
11127	**SECTOR OF SPHERE, SPHERICAL SECTOR**	SECTEUR SPHERIQUE	KUGELAUSSCHNITT, KUGELSEKTOR
11128	**SEDAN**	BERLINE	INNENLENKER
11129	**SEED BARROWS AND BROADCASTERS**	SEMOIRS A LA VOLEE	KARREN UND BREITSÄMASCHINEN
11130	**SEED CHARGE**	AGENT PRECIPITANT	FÄLLMITTEL
11131	**SEED LAC**	GOMME-LAQUE EN GRAINS	KÖRNERLACK, LACK IN KÖRNERN
11132	**SEEDERS AND SEEDER UNITS**	SEMOIRS POUR PORTES OUTILS	SÄMASCHINEN
11133	**SEEDING TRANSPLANTERS AND SISAL PLANTERS**	REPIQUEUSES	PIKIERMASCHINEN UND SISALPFLANZMASCHINEN
11134	**SEGER CONE**	CONE PYROMETRIQUE (DE SEGER), MONTRE DE SEGER, MONTRE FUSIBLE	SEGERKEGEL, SCHMELZKEGEL, BRENNKEGEL
11135	**SEGMENT**	SEGMENT	AUSSCHNITT
11136	**SEGMENT OF CIRCLE**	SEGMENT D'UN CERCLE	KREISABSCHNITT, KREISSEGMENT
11137	**SEGMENT OF SPHERE**	CALOTTE SPHERIQUE, SEGMENT SPHERIQUE A UNE BASE	KUGELHAUBE, KUGELSCHALE, KALOTTE
11138	**SEGMENTAL FLYWHEEL**	VOLANT EN PLUSIEURS SEGMENTS	SCHWUNGRAD (MEHRSTELLIGES)
11139	**SEGREGATION**	TRI, SEGREGATION	AUSSONDERUNG
11140	**SEIZE (TO)**	GRIPPER	FRESSEN, EINFRESSEN (SICH)
11141	**SEIZING**	GRIPPAGE, GRIPPEMENT	FRESSEN, EINFRESSEN, FESTFRESSEN
11142	**SELECTIVE ANNEALING**	RECUIT SELECTIF	AUSGLÜHEN (SELEKTIVES)
11143	**SELECTIVE CARBURIZING**	CEMENTATION SELECTIVE	AUFKOHLUNG (SELEKTIVE)
11144	**SELECTIVE FLOTATION**	FLOTTATION SELECTIVE	SCHWIMMAUFBEREITUNG (SELEKTIVE)
11145	**SELECTIVE HARDENING**	TREMPE PARTIELLE	TEILHÄRTUNG
11146	**SELENIUM**	SELENIUM	SELEN
11147	**SELF- LUMINOUS BODY**	CORPS LUMINEUX	KÖRPER (SELBSTLEUCHTENDER)
11148	**SELF-ACTING, AUTOMATIC**	AUTOMATIQUE	SELBSTTÄTIG, AUTOMATISCH
11149	**SELF-ACTUATING CLUTCH**	ACCOUPLEMENT AUTOMATIQUE	KUPPLUNG (SELBSTEINRÜCKENDE)
11150	**SELF-ADJUSTING BEARING**	PALIER AUTO-REGULATEUR	LAGER (SICH SELBSTEINSTELLENDES
11151	**SELF-CONTAINED MACHINE**	MACHINE ISOLEE	MASCHINE (FREISTEHENDE)
11152	**SELF-DRIVING CHUCKS**	MANDRINS AUTO-ENTRAINEURS	FUTTER (SELBSTTÄTIGE)
11153	**SELF-EXCITATION**	AUTO-EXCITATION	SELBSTERREGUNG, EIGENERREGUNG
11154	**SELF-HARDENING STEEL**	ACIER AUTO-TREMPANT	SELBSTHÄRTESTAHL
11155	**SELF-HEATING SOLDERING IRON**	FER A SOUDER A CHAUFFAGE AUTOMATIQUE	SELBSTWÄRMENDER LÖTKOLBEN, LÖTKOLBEN MIT SELBSTBEHEIZUNG
11156	**SELF-INDUCTION**	AUTO-INDUCTION, SELF-INDUCTION, INDUCTION PROPRE	SELBSTINDUKTION
11157	**SELF-LOCKING**	BLOCAGE AUTOMATIQUE (A)	SELBSTHEMMEND, SELBSTSPERREND
11158	**SELF-LUBRICATING BEARING**	COUSSINET AUTO-GRAISSEUR	LAGER (SELBSTSCHMIERENDES)
11159	**SELF-RELEASING TAPER**	CONE AUTO-DEMONTABLE	KEGEL (SELBSTLÖSENDER)
11160	**SELF-SUPPORTING CONICAL ROOF TANK**	RESERVOIR A TOIT CONIQUE AUTOPORTANT	TANK MIT SELBSTTRAGENDEM KONUSDACH
11161	**SELF-SUPPORTING FRAME**	CHARPENTE AUTOPORTANTE	GERIPPE (SELBSTTRAGENDES), KONSTRUKTION (SELBSTTRAGENDE), GERÜST (SELBSTTRAGENDES)
11162	**SELF-TAPPING SCREW**	VIS AUTOTARAUDEUSE	SCHRAUBE (SELBSTSCHNEIDENDE)
11163	**SELF-TAPPING WASHER**	RONDELLE AUTOTARAUDEUSE	SCHEIBE (SELBSTSCHNEIDENDE)
11164	**SELLERS COUPLING**	MANCHON D'ACCOUPLEMENT SELLERS	KLEMMKUPPLUNG, SELLERSKUPPLUNG
11165	**SEMI-ANTHRACITE, STEAM COAL**	HOUILLE ANTHRACITEUSE, HOUILLE MAIGRE A COURTE FLAMME	KOHLE (ANTHRAZITISCHE)

11166	SEMI-AUTOMATIC ARC WELD	SOUDAGE SEMI-AUTOMATIQUE A L'ARC	LICHTBOGEN-SCHWEISSEN (TEILAUTOMATISCHES)
11167	SEMI-AXIS	DEMI-AXE	HALBACHSE (GEOM.)
11168	SEMI-BITUMINOUS COAL, LEAN MEAGER COAL	HOUILLE MAIGRE	KOHLE (MAGERE)
11169	SEMI-CIRCULAR (ADJ)	DEMI-CIRCULAIRE	HALBKREISFÖRMIG
11170	SEMI-CIRCULAR, HALF-ROUND CROSS SECTION	SECTION MI-RONDE, SEMI-CIRCULAIRE	QUERSCHNITT (HALBRUNDER)
11171	SEMI-DIESEL (OIL INJECTION ENGINE, MIXED-CYCLE ENGINE)	MOTEUR SEMI-DIESEL, SEMI-DIESEL	HALBDIESELMOTOR
11172	SEMI-JIG BORING MACHINE	SEMI-POINTEUSE	HALBLEHRENBOHRMASCHINE
11173	SEMI-KILLED STEEL	ACIER SEMI-CALME	STAHL (HALBBERUHIGTER)
11174	SEMI-MANUFACTURED GOODS	PRODUIT SEMI-OUVRE	HALBZEUG, HALBFABRIKAT
11175	SEMI-PORTABLE STEAM ENGINE, STATIONARY LOCOMOBILE	MACHINE A VAPEUR DEMI-FIXE	LOKOMOBILE (FESTSTEHENDE), LOKOMOBILE (ORTSFESTE), LOKOMOBILE (STATIONÄRE)
11176	SEMI-ROTARY WING PUMP	POMPE A PISTON OSCILLANT	PUMPE MIT SCHWINGENDEM ODER OSZILLIERENDEM KOLBEN, FLÜGELPUMPE
11177	SEMI-STEEL	FONTE ACIEREE	GUSSEISEN MIT STAHLZUSATZ
11178	SEMI-TRAILER	SEMI-REMORQUE	SATTELSCHLEPPER
11179	SEMICIRCLE	DEMI-CERCLE	HALBKREIS
11180	SEMICIRCULAR PROTRACTOR	RAPPORTEUR DEMI-CERCLE	HALBKREISTRANSPORTEUR
11181	SENDZIMIR MILL	LAMINOIR SENDZIMIR	SENDZIMIRWALZWERK
11182	SENSE OF A FORCE	SENS D'UNE FORCE	KRAFTSINN, SINN EINER KRAFT
11183	SENSE OF MOTION	SENS DU MOUVEMENT, SENS DE LA MARCHE	BEWEGUNGSSINN
11184	SENSIBLE HEAT	CHALEUR SENSIBLE	WÄRME (FÜHLBARE)
11185	SENSING AND SORTING MACHINES	TRIEUSES ELECTRONIQUES	ABTAST-UND-SORTIERMASCHINEN
11186	SENSITIVE DRILLING MACHINE	MACHINE A PERCER SENSITIVE	GEFÜHLSBOHRMASCHINE
11187	SENSITIVE PLATE	PLAQUE SENSIBLE, PLAQUE PHOTOGRAPHIQUE	PLATTE (LICHTEMPFINDLICHE), PLATTE (PHOTOGRAPHISCHE)
11188	SENSITIVENESS OF AN INSTRUMENT	SENSIBILITE D'UN INSTRUMENT	EMPFINDLICHKEIT EINES INSTRUMENTS
11189	SEPARATE (TO)	SEPARER (CHIM.)	ABSCHEIDEN (CHEM.)
11190	SEPARATE EXCITATION	EXCITATION SEPAREE, INDEPENDANTE	FREMDERREGUNG, SONDERERREGUNG, ÄUSSERE ERREGUNG
11191	SEPARATION	SEPARATION (CHIM.)	ABSCHEIDUNG, ABSCHEIDEN
11192	SEPARATION OF A MIXTURE	SEPARATION D'UN MELANGE	ENTMISCHUNG
11193	SEQUENCE NUMBER	NUMERO DE SEQUENCE	SATZNUMMER
11194	SEQUENTIAL	SEQUENTIEL	SEQUENTIELL
11195	SEQUESTRATION	CHELATION	CHELATBILDUNG
11196	SERIAL	SERIE (EN)	SERIEN-
11197	SERIAL NUMBER	NUMERO D'ORDRE DE LA FABRICATION	FABRIKNUMMER
11198	SERIAL PRODUCTION	PRODUCTION EN SERIE	SERIENPRODUKTION
11199	SERIES CONNECTION, CONNECTION IN SERIES	MONTAGE, GROUPEMENT, COUPLAGE EN SERIE, COUPLAGE EN TENSION	REIHENSCHALTUNG, SERIENSCHALTUNG, HINTEREINANDERSCHALTUNG
11200	SERIES SPOT WELD	SOUDAGE PAR POINTS EN SERIE PAR RESISTANCE	SERIENPUNKTNAHT
11201	SERIES WOUND GENERATOR DYNAMO	DYNAMO EN SERIE	HAUPTSCHLUSSDYNAMO, REIHENSCHLUSSDYNAMO, SERIENDYNAMO, DYNAMO (IN SERIE GESCHALTETER
11202	SERIES WOUND MOTOR	MOTEUR-SERIE	HAUPTSCHLUSSMOTOR, REIHENSCHLUSSMOTOR, SERIENMOTOR

11203	**SERIES, PROGRESSION**	SERIE, SUITE, PROGRESSION	REIHE
11204	**SERPENTINE**	SERPENTINE, OPHITE	FELS (SERPENTIN)
11205	**SERPENTINE ASBESTOS, CHRYSOTILE**	CHRYSOTILE	SERPENTINASBEST
11206	**SERRATED ROLLER**	CYLINDRE CANNELE, STRIE	RIFFELWALZE
11207	**SERUM ALBUMIN**	ALBUMINE DU SERUM	BLUTALBUMIN
11208	**SERVICE CONDITIONS**	CONDITIONS D'EMPLOI	BETRIEBSBEDINGUNGEN
11209	**SERVICE INSTRUCTIONS**	INSTRUCTIONS RELATIVES AU TRAVAIL, REGLEMENTS D'USINE	BETRIEBSVORSCHRIFTEN
11210	**SERVO-MECHANISM**	SERVO-MECANISME	SERVO-STEUERUNG
11211	**SERVOMOTOR**	SERVO-MOTEUR	HILFSMOTOR, SERVOMOTOR
11212	**SESAME OIL**	HUILE DE SESAME	SESAMÖL
11213	**SET**	JEU, SERIE, ASSORTIMENT (DE PIECES), GROUPE (DE MACHINES ELECTRIQUES)	SATZ, GARNITUR
11214	**SET (TO)**	FAIRE PRISE	ABBINDEN, ANZIEHEN (ZEMENT)
11215	**SET A BOILER IN MASONRY (TO)**	ENTOURER UNE CHAUDIERE DE MACONNERIE	KESSEL EINMAUERN (EINEN)
11216	**SET A VALVE (TO)**	TARER UNE SOUPAPE	TARIEREN
11217	**SET CHISEL, ROD CHISEL**	TRANCHE	SCHROTMEISSEL, SETZMEISSEL
11218	**SET HAMMER**	CHASSE (OUTIL DE FORGERON)	SETZHAMMER
11219	**SET OF WEIGHTS**	ASSORTIMENT, SERIE, JEU DE POIDS	GEWICHTSATZ
11220	**SET PIN**	GOUPILLE DE BLOCAGE, GOUPILLE DE CLAVETAGE, GOUPILLE DE RETENUE, PRISONNIER	HALTESTIFT, STIFTSCHRAUBE
11221	**SET POINT**	VALEUR DE REGLAGE	SOLLWERT
11222	**SET SCREW**	VIS POINTEAU POUR ARRET DE BAGUES	STELLSCHRAUBE, FESTSTELLSCHRAUBE
11223	**SET SQUARE**	EQUERRE A DESSIN	DREIECK, WINKEL
11224	**SET THE TEETH OF A SAW (TO)**	DONNER LA VOIE A UNE SCIE, AVOYER UNE SCIE	SÄGEZÄHNE SCHRÄNKEN, SÄGEZÄHNE AUSSETZEN (DIE)
11225	**SETTING GAUGE**	JAUGE D'AJUSTAGE	EINSTELLLEHRE
11226	**SETTING OF CEMENT**	PRISE DU CIMENT	ABBINDEN DES ZEMENTS
11227	**SETTLE (TO)**	DEPOSER (SE)	ABSETZEN (SICH)
11228	**SETTLING**	SEDIMENTATION, DECANTATION	ABSETZEN, ABLAGERUNG
11229	**SETTLING TANK**	RESERVOIR DE DECANTATION, CUVE DE DECANTATION	KLÄRGEFÄSS, KLÄRBOTTICH
11230	**SEWAGE WATER**	EAUX D'EGOUTS	ABWÄSSER VON STÄDTEN
11231	**SEXE OF AN AXE**	OEIL D'UNE HACHE	HELMLOCH EINER AXT
11232	**SHACKLE**	MANILLE D'ASSEMBLAGE	SCHÄKEL
11233	**SHACKLES**	JUMELLES	FEDERGEHÄNGE
11234	**SHADE (TO)**	OMBRER	SCHATTIEREN
11235	**SHADED DRAWING**	DESSIN OMBRE	ZEICHNUNG (SCHATTIERTE)
11236	**SHADING**	OMBRE	SCHATTIERUNG
11237	**SHADING**	ACTION D'OMBRER	SCHATTIEREN
11238	**SHADOWGRAPH**	PROJECTEUR D'OMBRE	SCHATTENAUFNAHMEAPPARAT
11239	**SHAFT**	ARBRE (MEC.)	WELLE (MASCHB.), ACHSE
11240	**SHAFT COUPLING**	ACCOUPLEMENT D'ARBRES	WELLENKUPPLUNG, KUPPLUNG
11241	**SHAFT FURNACE**	FOUR A CUVE, FOUR A MANCHE	SCHACHTOFEN
11242	**SHAFT GOVERNOR**	REGULATEUR PLAN, REGULATEUR SUR L'ARBRE	ACHSENREGLER, FLACHREGLER
11243	**SHAFT OF COLUMN**	FUT D'UNE COLONNE	STAMM EINER SÄULE, SÄULENSTAMM

11244	SHAFT OR HANDLE OF A HAMMER	MANCHE D'UN MARTEAU	HAMMERSTIEL, STIEL EINES HAMMERS
11245	SHAFT STRAIGHTENER	MACHINE A DRESSER LES ARBRES	WELLENRICHTMASCHINE
11246	SHAFT TURNING LATHE	TOUR A ARBRES	WELLENDREHBANK
11247	SHAFT WITH OVERHANGING ENDS	ARBRE EN PORTE-A-FAUX DES DEUX COTES	WELLE MIT ÜBERSTEHENDEN ENDEN
11248	SHAFTING	TRANSMISSION	WELLENLEITUNG, TRANSMISSION
11249	SHAFTS AT RIGHT ANGLES (OR INCLINED) AND NOT INTERSECTING	ARBRES (PERPENDICULAIRES, OBLIQUES) SITUES DANS DES PLANS DIFFERENTS	WELLEN (SICH KREUZENDE, GESCHRÄNKTE)
11250	SHAKE IN TIMBER	GERCE, GERCURE DU BOIS, CRIQUE DANS LE BOIS	RISS IM HOLZ
11251	SHAKE-OUT	DEMOULAGE	AUSSCHLAGEN
11252	SHAKING SCREEN	CRIBLE OSCILLANT	SCHÜTTELSIEB
11253	SHAKING TRAY, PAN (SHAKING)	TRANSPORTEUR, GOUTIERE A SECOUSSES	SCHÜTTELRINNE, WIPPE, SCHWINGE
11254	SHAKING-DOWN	AGITATION DU BAIN	RÜHREN
11255	SHALE OIL TAR	GOUDRON DE SCHISTE	SCHIEFERTEER
11256	SHALE OIL, SCHIST OIL	HUILE DE SCHISTE	SCHIEFERÖL
11257	SHANK	TIGE, QUEUE	SCHAFT
11258	SHANK LADLE	POCHE DE COULEE A ANSE	GABELGIESSPFANNE
11259	SHANK MILLING CUTTER	FRAISE A QUEUE	SCHAFTFRÄSER, FINGERFRÄSER
11260	SHANK OF BOLT SCREW	TIGE D'UNE VIS, TIGE D'UN BOULON	SCHAFT, BOLZEN EINER SCHRAUBE, SCHRAUBENSCHAFT, SCHRAUBENBOLZEN
11261	SHANK OF CONNECTING ROD	CORPS DE LA BIELLE	SCHAFT DER SCHUBSTANGE
11262	SHANK OF KEY	CORPS D'UNE CLAVETTE, TIGE D'UNE CLAVETTE	KEILBOLZEN
11263	SHANK OF RIVET	CORPS DU RIVET, TIGE DU RIVET, FUT DU RIVET	NIETSCHAFT, NIETBOLZEN
11264	SHAPE	FORME	FORM
11265	SHAPED PART	PIECE FACONNEE	FORMSTÜCK
11266	SHAPED PLATES	TOLES DECOUPEES	BLECH (ZUGESCHNITTENES)
11267	SHAPERS	ETAUX-LIMEUR	WAAGERECHTSTOSSMASCHINEN
11268	SHAPING	FACONNAGE	FORMGEBUNG
11269	SHAPING MACHINE	ETAU-LIMEUR	FEILMASCHINE, SHAPINGMASCHINE, WAAGRECHTSTOSSMASCHINE
11270	SHARK SKIN, SHAGREEN	PEAU DE CHIEN	FISCHHAUT
11271	SHARP BACKED ANGLE	FER CORNIERE A ANGLE VIF	WINKELEISEN (INNEN SCHARF)
11272	SHARP EDGE	ARETE VIVE	KANTE (SCHARFE)
11273	SHARPEN (TO), WHET (TO), GRIND (TO)	AIGUISER, MEULER, EMOUDRE, AFFUTER, RECTIFIER, REPASSER	SCHÄRFEN, SCHARFSCHLEIFEN
11274	SHARPENING MACHINE	AFFUTEUSE	SCHÄRFMASCHINE
11275	SHARPENING, WHETTING, GRINDING	AIGUISAGE, MEULAGE, EMOULAGE, AFFUTAGE, RECTIFICATION	SCHLEIFEN, SCHÄRFEN, SCHARFSCHLEIFEN
11276	SHARPNESS OF IMAGE	NETTETE D'UNE IMAGE	BILDSCHÄRFE
11277	SHAVINGS, CHIPS, CHIPPINGS	ALESURES	BOHRSPÄNE
11278	SHAVINGS, CHIPS, METAL SHAVINGS	COPEAUX	SPÄNE
11279	SHEAR	CISAILLE, CISAILLEMENT	SCHERE, SCHUB
11280	SHEAR (TO)	CISAILLER (COUPER AVEC DES CISAILLES)	ABSCHEREN, MIT DER SCHERE ABSCHNEIDEN
11281	SHEAR ANGLE	ANGLE DE CISAILLEMENT	SCHERWINKEL
11282	SHEAR BLADE	LAME D'UNE CISAILLE	SCHERBLATT, SCHERMESSER, SCHERKLINGE, SCHERBACKEN, DRUCKBACKEN EINER SCHERE
11283	SHEAR PIN	GOUPILLE DE CISAILLEMENT	ABSCHERSTIFT

11284	SHEAR PLANE	PLAN DE CISAILLEMENT	SCHERFLÄCHE
11285	SHEAR STRENGTH	RESISTANCE AU CISAILLEMENT	SCHERFESTIGKEIT
11286	SHEAR STRESS	TENSION DE CISAILLEMENT	SCHUBVERFORMUNG
11287	SHEAR TEST	ESSAI DE CISAILLEMENT	SCHERVERSUCH
11288	SHEAR, SHEARING, SLIDE	CISAILLEMENT, GLISSEMENT TRANSVERSAL	GLEITUNG, SCHIEBUNG, SCHUB (FESTIGKEITSL.)
11289	SHEARINESS	EMBUS	EINSAUGUNG
11290	SHEARING	CISAILLEMENT, CISAILLAGE	SCHERUNG, ABSCHEREN, SCHNEIDEN
11291	SHEARING FORCE	FORCE DE CISAILLEMENT	SCHERKRAFT, SCHUBKRAFT, QUERKRAFT, TRANSVERSALKRAFT
11292	SHEARING MACHINE	CISAILLE MECANIQUE, MACHINE A CISAILLER, CISAILLEUSE	MASCHINENSCHERE
11293	SHEARING OF THE RIVET	RUPTURE DU RIVET PAR CISAILLEMENT	ABSCHEREN DES NIETES
11294	SHEARING PIN	GOUPILLE DE CISAILLEMENT	ABSCHERBOLZEN
11295	SHEARING STRAIN	EFFORT DE CISAILLEMENT, EFFORT TRANCHANT, TRAVAIL AU CISAILLEMENT	BEANSPRUCHUNG AUF SCHUB, ABSCHERUNG, SCHUBBEANSPRUCHUNG, SCHERBEANSPRUCHUNG
11296	SHEARING STRESS	EFFORT DE CISAILLEMENT PAR UNITE DE SECTION, CONTRAINTE DE CISAILLEMENT	SCHERSPANNUNG, SCHUBSPANNUNG
11297	SHEARS	CISAILLE, CISAILLES	SCHERE
11298	SHEAT	TOLE	BLECH
11299	SHEATHED ELECTRODE	ELECTRODE BLINDEE	ELEKTRODE (BLECHUMHÜLLTE)
11300	SHEATHED WIRE	FIL GAINE	DRAHT (UMHÜLLTER)
11301	SHEAVE, PULLEY OF A TACKLE	POULIE, REA D'UN MOUFLE	ROLLE EINES FLASCHENZUGS
11302	SHED (FOR STORAGE)	HANGAR, PARC COUVERT	LAGERSCHUPPEN, SCHUPPEN
11303	SHEEL	ENVELOPPE, CHEMISE, COQUILLE	MANTEL, HÜLSE, SCHALE, KOKILLE
11304	SHEET	TOLE FINE	FEINBLECH, STURZBLECH
11305	SHEET (METAL)	TOLE	BLECH
11306	SHEET BAR	LARGET, PLATINE	PLATINE, VORBLECH
11307	SHEET BRASS	LAITON EN FEUILLES	MESSINGBLECH
11308	SHEET COPPER	CUIVRE EN FEUILLES	KUPFERBLECH
11309	SHEET IRON	TOLE DE FER	EISENBLECH
11310	SHEET IRON COVER	CAPOT EN TOLE	BLECHMANTEL
11311	SHEET IRON METAL GAUGE, PLATE GAUGE	JAUGE POUR TOLES	BLECHLEHRE
11312	SHEET IRON PIPE	TUYAU EN TOLE	BLECHROHR
11313	SHEET IRON PLATE	FEUILLE DE TOLE	BLECHTAFEL
11314	SHEET LEAD	PLOMB EN FEUILLES	BLEIBLECH, WALZBLEI
11315	SHEET METAL	TOLE, FEUILLE METALLIQUE, PLAQUE METALLIQUE	BLECH
11316	SHEET METAL LEVELLING MACHINE	MACHINE A PLANER ET DRESSER LES TOLES	BLECHRICHTMASCHINE
11317	SHEET METAL PERFORATING MACHINES	MACHINES A PERFORER LES TOLES	BLECHPERFORIERMASCHINEN
11318	SHEET METAL PLANING MACHINE	MACHINES A PLANER LES TOLES	BLECHPLANIERMASCHINEN
11319	SHEET METAL SCORING MACHINE	MACHINE A ENTAILLER LES TOLES	BLECHRITZMASCHINE
11320	SHEET METAL WORKING MACHINES	MACHINES A TRAVAILLER LA TOLE	BLECHBEARBEITUNGSMASCHINEN
11321	SHEET MILL / STRIP MILL	TRAIN A BANDES	BLECHSTREIFENWALZWERK
11322	SHEET OF FELT	PLAQUE EN FEUTRE	FILZPLATTE
11323	SHEET OR BAND METAL SHEARING MACHINE	MACHINE A CISAILLER LES METAUX EN FEUILLES OU EN BANDES	BLECH- ODER BANDMETALLSCHERMASCHINE

11324	SHEET ROLLING MILL	TRAIN (DE LAMINOIRS) A TOLES FINES	BLECHWALZSTRASSE
11325	SHEET STEEL	TOLE D'ACIER	STAHLBLECH
11326	SHEET TIN	ETAIN EN FEUILLES	ZINNBLECH
11327	SHEET ZINC	FEUILLE DE ZINC, ZINC EN FEUILLES	ZINKBLECH
11328	SHEETING	EXFOLIATION	ABBLÄTTERUNG
11329	SHEETING RUBBER, SHEETING NEOPRENE	FEUILLE DE CAOUTCHOUC (NEOPRENE)	GUMMI- (ODER NEOPREN-) ISOLIERUNG
11330	SHELL	ROBE	WANDUNG
11331	SHELL EXTENSION	REHAUSSE ....DE TOIT FLOTTANT	AUFSATZ FÜR SCHWIMMDACH
11332	SHELL MANHOLE	TROU D'HOMME DE ROBE	INNERWANDUNGSMANNLOCH
11333	SHELL MOLDING	MOULAGE EN COQUILLE	MASKENFORMEN
11334	SHELL OF A COCK	BOISSEAU D'UN ROBINET	GEHÄUSE EINES HAHNES, HAHNGEHÄUSE
11335	SHELL PLATE	TOLE DE ROBE	MANTELBLECH
11336	SHELL REAMER	ALESOIR/FRAISE CREUX	AUFSTECKREIBAHLE
11337	SHELL-END MILLS	FRAISES EN BOUT ALESEES	WALZENSTIRNFRÄSER
11338	SHELLAC	GOMME-LAQUE, LAQUE EN ECAILLES, LAQUE EN PLAQUES, LAQUE EN FEUILLES, LAQUE PLATE	SCHELLACK
11339	SHELLING	ECAILLAGE	ABBLÄTTERUNG
11340	SHERARDISE (TO)	SHERARDISER	SHERARDISIEREN
11341	SHERARDISING	SHERARDISATION	SHERARDISIEREN
11342	SHIELD	ECRAN DE PROTECTION, BOUCLIER	SCHUTZSCHILD, SCHUTZVORRICHTUNG
11343	SHIELD PLATE	TOLE DE RENFORT OU DOUBLANTE, TOLE DE BLINDAGE	VERSTÄRKUNGSBLECH, PANZERBLECH, SCHUTZBLECH
11344	SHIELDED ELECTRODE	ELECTRODE ENROBEE	ELEKTRODE (DICKUMHÜLLTE)
11345	SHIELDED METAL ARC WELDING	SOUDAGE A L'ARC METALLIQUE SOUS GAZ PROTECTEUR	METALL-LICHTBOGEN-SCHWEISSEN MIT UMHÜLLTER ELEKTRODE
11346	SHIELDING	ECRAN, PROTECTION	SCHUTZVORRICHTUNG, SCHIRM
11347	SHIFT	JOURNEE DE TRAVAIL	SCHICHT (ARBEITSZEIT)
11348	SHIFT SQUAD	EQUIPE D'OUVRIERS	SCHICHT (ARBEITERGRUPPE)
11349	SHIFT THE BELT (TO)	DEPLACER LA COURROIE	RIEMEN SCHALTEN (DEN), RIEMEN VERSCHIEBEN (DEN)
11350	SHIFT WORK	TRAVAIL EN (DEUX OU TROIS) EQUIPES	SCHICHTARBEIT
11351	SHIM	RONDELLE D'EPAISSEUR	UNTERLEGBLECH
11352	SHIM-PLATE	CALE D'EPAISSEUR	EINLAGE, ZWISCHENLAGESCHEIBE, BEILEGESCHEIBE
11353	SHIMMY	SHIMMY	FLATTERN
11354	SHINGLE	GALETS (GEOL.)	GERÖLLE, GESCHIEBE
11355	SHINGLING	COMPRESSION, SERRAGE, TASSEMENT	VERDICHTUNG
11356	SHIP PLATE	TOLE (POUR CONSTRUCTION) NAVALE	SCHIFFSBLECH
11357	SHIP THE BELT (TO)	MONTER, INSTALLER LA COURROIE, METTRE EN PLACE LA COURROIE	RIEMEN AUFLEGEN (DEN)
11358	SHOCK ABSORBER	AMORTISSEUR DE CHOCS	DÄMPFER, STOSSDÄMPFER
11359	SHOCK, IMPACT, PERCUSSION	CHOC, PERCUSSION, A-COUP, SECOUSSE, HEURT	STOSS (MECH.)
11360	SHOE	SABOT	SCHUH
11361	SHOP DRAWING	PLAN D'EXECUTION (ATELIER)	WERKSTATTZEICHNUNG

11362	SHOP WELDING	SOUDAGE EN ATELIER	WERKSTATTSCHWEISSEN
11363	SHOP, WORKING DRAWING	EPURE D'UNE PIECE MECANIQUE	WERKZEICHNUNG, ARBEITSZEICHNUNG
11364	SHOP, WORKSHOP	ATELIER	WERKSTATT
11365	SHORE	CHANDELLE (ETAI VERTICAL)	STÜTZE, TRAGSTANGE
11366	SHORE'S SCLEROSCOPE	SCLEROSCOPE, REBONDIMETRE, APPAREIL SHORE	HÄRTEPRÜFER, SKLEROSKOP NACH SHORE
11367	SHORE'S SCLEROSCOPE TEST	ESSAI PAR REBONDISSEMENT DES BILLES (APPAREIL SHORE)	KUGELFALLPROBE NACH SHORE
11368	SHORT BRITTLE IRON	FER CASSANT	EISEN (SPRÖDES), EISEN (BRÜCHIGES)
11369	SHORT CIRCUIT	COURT-CIRCUIT	KURZSCHLUSS
11370	SHORT CIRCUIT TRANSFER	TRANSFERT PAR COURT-CIRCUIT	KURZSCHLUSSÜBERTRAGUNG
11371	SHORT FLAME COAL	HOUILLE A COURTE FLAMME	KOHLE (KURZFLAMMIGE)
11372	SHORT-CIRCUIT	COURT-CIRCUIT	KURZSCHLUSS
11373	SHORT-LINK CHAIN, CLOSE-LINK CHAIN	CHAINE A MAILLONS COURTS, CHAINE A MAILLONS SERRES	KETTE (KURZGLIEDRIGE)
11374	SHORT-STROKE ENGINE	MACHINE A FAIBLE COURSE	MASCHINE (KURZHÜBIGE)
11375	SHORT-THROW CRANK	MANIVELLE A FAIBLE COURSE	KURBEL (KURZHÜBIGE)
11376	SHORTENING	RACCOURCISSEMENT	VERKÜRZUNG VERKÜRZEN
11377	SHORTENING A ROPE, A BELT	RACCOURCISSEMENT D'UN CABLE, RACCOURCISSEMENT D'UNE COURROIE	KÜRZUNG EINES SEILES, KÜRZUNG EINES RIEMENS
11378	SHORTNESS	FRAGILITE	BRÜCHIGKEIT
11379	SHOT	MITRAILLE, GRENAILLE	SCHROT, GRANULAT
11380	SHOT BLASTING	GRENAILLAGE	STAHLSANDBLASEN, SCHROTSTRAHLPUTZEN
11381	SHOT PEENING	DECAPAGE PAR GRENAILLAGE	KUGELSTRAHLEN
11382	SHOTBLASTING	GRENAILLAGE (GROS)	GRANALIEN BLASEN
11383	SHOULDER	EPAULEMENT	ABSATZ, SCHULTER, STUFE
11384	SHOULDER OF A SHAFT	EPAULEMENT D'UN ARBRE, COLLET D'UN ARBRE, BUTEE D'UN ARBRE	ANLAUF EINER WELLE, BRUST EINER WELLE, SCHULTER EINER WELLE
11385	SHOVEL	PELLE	SCHAUFEL, SCHIPPE
11386	SHOVEL (TO)	PELLETER	SCHAUFELN
11387	SHOWER VALVE	ROBINET DE DOUCHE	DUSCHVENTIL, BRAUSEVENTIL
11388	SHRINK (TO)	SE RETRECIR, SE CONTRACTER	EINSCHRUMPFEN, SCHWINDEN
11389	SHRINK FIT	EMMANCHEMENT, MONTAGE A LA PRESSE A CHAUD	SCHRUMPFSITZ, WARMSITZ
11390	SHRINK ON (TO)	POSER, EMMANCHER, FRETTER A CHAUD	WARM AUFZIEHEN, AUFSCHRUMPFEN
11391	SHRINK RING	FRETTE POSEE A CHAUD, COLLIER POSE A CHAUD, ANNEAU MIS A CHAUD, BAGUE DE SERRAGE	SCHRUMPFRING, SCHWINDRING
11392	SHRINKAGE	CONTRACTION, RETRAIT	SCHWINDUNG, SCHRUMPFUNG, SCHWINDEN
11393	SHRINKAGE CAVITY	RETASSURE	LUNKER
11394	SHRINKAGE CONTRACTION OF A METAL (IN COOLING DOWN)	RETASSEMENT	SCHWINDEN EINES METALLS BEIM ERSTARREN
11395	SHRINKAGE CRACK	TAPURE DE RETRAIT	SCHWINDUNGSRISS
11396	SHRINKAGE OF THE WOOD	RETRAIT DU BOIS	SCHWINDEN DES HOLZES
11397	SHRINKAGE, SHRINKING	RETRECISSEMENT, RETRACTION, CONTRACTION (PENDANT LE REFROIDISSEMENT)	SCHRUMPFEN, EINSCHRUMPFEN, SCHWINDEN
11398	SHRINKING ON	FRETTAGE, EMMANCHEMENT A CHAUD	AUFSCHRUMPFEN, WARMAUFZIEHEN

11399	**SHRIVELLING**	RIDAGE	KRÄUSELN
11400	**SHROUD**	JOUE D'UNE ROUE D'ENGRENAGE, EPAULEMENT D'UNE ROUE D'ENGRENAGE	SEITENSCHEIBE EINES ZAHNRADES, BORDSCHEIBE EINES ZAHNRADES
11401	**SHROUD LAID ROPE**	GRELIN COMPOSE DE QUATRE AUSSIERES	KUGELWEISE GESCHLAGENES SEIL
11402	**SHROUDED GEAR WHEEL**	ROUE D'ENGRENAGE EPAULEE, ROUE D'ENGRENAGE AVEC JOUES	VERSTEIFTES ZAHNRAD, ZAHNRAD MIT BORDSCHEIBEN
11403	**SHUNT**	DERIVATION, SHUNT	NEBENSCHLUSS
11404	**SHUNT WOUND GENERATOR DYNAMO**	DYNAMO-SHUNT, DYNAMO EN DERIVATION	NEBENSCHLUSSDYNAMO
11405	**SHUNT WOUND MOTOR**	MOTEUR-SHUNT, MOTEUR EN DERIVATION	NEBENSCHLUSSMOTOR
11406	**SHUT DOWN A MACHINE (TO)**	METTRE UNE MACHINE HORS SERVICE	AUSSER BETRIEB SETZEN (EINE MASCHINE), STILL SETZEN
11407	**SHUT OFF A PIPE (TO)**	FERMER L'INTRODUCTION, OBTURER L'ADMISSION DANS UNE TUYAUTERIE	AUSSCHALTEN (EINE ROHRLEITUNG), ABSCHALTEN EINE ROHRLEITUNG, ABSPERREN EINE ROHRLEITUNG
11408	**SHUT OFF COCK / VALVE**	ROBINET D'ARRET	ABSTELLHAHN, ABSPERRHAHN
11409	**SHUT-DOWNS**	ARRETS DE TRAVAIL	STILLEGUNG
11410	**SHUTTING DOWN A MACHINE**	MISE HORS SERVICE D'UNE MACHINE	AUSSERBETRIEBSETZUNG EINER MASCHINE, STILLSETZEN EINER MASCHINE
11411	**SIDE DELIVERY RAKES**	RATEAUX A DEVERSEMENT LATERAL	SCHWADENMÄHER
11412	**SIDE MILLING CUTTER**	FRAISE DISQUE	SCHEIBENFRÄSER
11413	**SIDE OF A BELT, OF A CHAIN**	BRIN D'UNE COURROIE, BRIN D'UNE CHAINE SANS FIN	TRUMM EINES RIEMENS, TRUMM EINES KETTENTRIEBS
11414	**SIDE OF POLYGON**	COTE D'UN POLYGONE	SEITE EINES POLYGONS, POLYGONSEITE
11415	**SIDE PLANING MACHINE**	MACHINE A RABOTER LATERALE	SEITENHOBELMASCHINE
11416	**SIDE PLATE OF A ROLLER CHAIN**	FLASQUE D'UNE CHAINE A ROULEAUX, JOUE D'UNE CHAINE A ROULEAUX	SEITENLASCHE, LASCHE EINER ROLLENKETTE
11417	**SIDE RABBET PLANE**	GUILLAUME DE COTE	WANDHOBEL, WANGENHOBEL
11418	**SIDE RAKE ANGLE**	ANGLE DE PENTE LATERALE	SPANWINKEL
11419	**SIDE STRAIN**	ONDULATIONS	WELLIGKEIT
11420	**SIDE VIEW, LONGITUDINAL ELEVATION**	VUE DE COTE, PROFIL (A DROITE, A GAUCHE)	SEITENANSICHT
11421	**SIDERITE, SPATHIC IRON ORE, CHALYBITE**	SIDEROSE, FER SPATHIQUE	SPATEISENSTEIN, EISENSPAT, SIDERIT
11422	**SIEVE, RIDDLE, SCREEN**	CRIBLE, TAMIS, SAS	SIEB
11423	**SIFT (TO), SCREEN (TO)**	CRIBLER, TAMISER, SASSER	SIEBEN, ABSIEBEN, DURCHSIEBEN
11424	**SIFTING, SCREENING**	CRIBLAGE, TAMISAGE	SIEBEN
11425	**SIGHT FEED LUBRICATOR**	GRAISSEUR A DEBIT VISIBLE	SCHAUÖLER
11426	**SIGHT FLOW INDICATOR, SIGHT GLASS**	CONTROLEUR DE CIRCULATION	DURCHFLUSSANZEIGER
11427	**SIGHT HOLE**	REGARD	SCHAUÖFFNUNG, SCHAULOCH, SCHAUGLAS
11428	**SIGMA WELDING**	SOUDAGE A L'ARC METALLIQUE SOUS PROTECTION DE GAZ INERTE	SIGMA-SCHWEISSEN
11429	**SIGN**	SIGNE	VORZEICHEN
11430	**SILAGE MAKING MACHINERY**	ENSILEUSES	GÄRFUTTERZUBEREITUNGSMASCHINEN
11431	**SILENCER, MUFFLER**	POT D'ECHAPPEMENT, SILENCIEUX	AUSPUFFTOPF, SCHALLDÄMPFER
11432	**SILENT CHAIN**	CHAINE SILENCIEUSE	ZAHNKETTE, KETTE (GERÄUSCHLOSE)

11433	SILENT RATCHET MOTION	ENCLIQUETAGE SILENCIEUX	STUMMES GESPERRE
11434	SILICA	SILICE	SILIZIUMDIOXYD, KIESELERDE
11435	SILICA BRICK	BRIQUE SILICEUSE	SILIKASTEIN, DINASSTEIN
11436	SILICA, SILICON DIOXIDE, SILICIC ACID	SILICE, ANHYDRIDE, ACIDE SILICIQUE	KIESELSÄURE, KIESELSÄUREANHYDRID, KIESELERDE, SILIZIUMDIOXYD
11437	SILICATE	SILICATE	KIESELSAURES SALZ, SILIKAT
11438	SILICEOUS ELECTRIC CALAMINE, HEMIMORPHITE	ZINC SILICATE	KIESELGALMEI, KIESELZINKERZ
11439	SILICEOUS MARL	MARNE SILICEUSE	MERGEL (KIESELIGER)
11440	SILICO-SPIEGEL	SILICO-SPIEGEL	SILIZIUMMANGANEISEN, SILIKOSPIEGEL
11441	SILICOFLUORIC ACID	ACIDE FLUOSILICIQUE	KIESELFLUORWASSERSTOFFSÄURE, KIESELFLUSSSÄURE
11442	SILICON	SILICIUM	SILIZIUM
11443	SILICON BRASS	LAITON SILICEUX	SILIZIUMMESSING
11444	SILICON BRONZE	BRONZE SILICEUX	SILIZIUMBRONZE
11445	SILICON CARBIDE	CARBURE DE SILICIUM	SILIZIUMKARBID
11446	SILICON COPPER	CUIVRE SILICEUX	SILIZIUMKUPFER
11447	SILICON STEEL	ACIER AU SILICIUM	SILIZIUMSTAHL
11448	SILICON STEEL SHEET	TOLE ELECTRIQUE	ELEKTROBLECH
11449	SILICON STEEL, SILICEOUS STEEL	ACIER AU SILICIUM	SILIZIUMSTAHL
11450	SILICON-MANGANESE STEEL	ACIER MANGANO-SILICEUX	SILIZIUM-MANGANSTAHL
11451	SILIECOUS LIMESTONE	CALCAIRE SILICEUX	KIESELKALK
11452	SILK	SOIE	SEIDE
11453	SILKING	NUANCAGE	STREIFIGKEIT
11454	SILKY FRACTURE	CASSURE SOYEUSE	BRUCH (SEIDIGER)
11455	SILKY LUSTRE	ECLAT SOYEUX	SEIDENGLANZ, ATLASGLANZ
11456	SILOCONE	SILICONE	SILIKON
11457	SILTING UP OF A PIPE	ENVASEMENT D'UNE CONDUITE	VERSCHLAMMEN EINER ROHRLEITUNG
11458	SILVER	ARGENT	SILBER
11459	SILVER (TO)	ARGENTER	VERSILBERN
11460	SILVER ALLOY BRAZING	BRASURE A L'ALLIAGE D'ARGENT	SILBERHARTLÖTEN
11461	SILVER CHLORIDE	ARGENT CHLORURE, CHLORURE D'ARGENT	CHLORSILBER, SILBERCHLORID
11462	SILVER CYANIDE	CYANURE D'ARGENT	ZYANSILBER, SILBERZYANID
11463	SILVER FULMINATE	FULMINATE D'ARGENT	KNALLSILBER
11464	SILVER IODIDE	IODURE D'ARGENT	JODSILBER, SILBERJODID
11465	SILVER PLATING	ARGENTURE	VERSILBERUNG
11466	SILVER SOLDER	SOUDURE A L'ARGENT, PAILLON D'ARGENT	SILBERSCHLAGLOT
11467	SILVER SULPHATE	SULFATE D'ARGENT	SCHWEFELSAURES SILBER, SILBERSULFAT
11468	SILVER-BEARING COPPER	CUIVRE ARGENTIFERE	KUPFER (SILBERHALTIGES)
11469	SILVERING	ARGENTURE	VERSILBERN, VERSILBERUNG
11470	SIMILARITY	SIMILITUDE	ÄHNLICHKEIT (GEOM.)
11471	SIMPLE INTEGRAL	INTEGRALE SINGULIERE	INTEGRAL (SINGULÄRES)
11472	SIMPLE PISTON ROD	TIGE DE PISTON UNILATERALE	KOLBENSTANGE (EINSEITIGE)
11473	SIMPLE STEEL	ACIER AU CARBONE	KOHLENSTOFFSTAHL
11474	SIMPLY FREELY SUPPORTED AT BOTH ENDS	SOUTENU, REPOSANT LIBREMENT SUR DEUX APPUIS	GESTÜTZT (BEIDERSEITIG), FREI AUFLIEGEND
11475	SINE	SINUS	SINUS

11476	**SINE CURVE**	COURBE SINUEUSE	SINUSKURVE
11477	**SINGLE BELT**	COURROIE SIMPLE	EINFACHRIEMEN
11478	**SINGLE BUTT STRAP RIVETED JOINT, BUTT RIVETED JOINT WITH ONE WELD**	RIVURE A UNE SEULE BANDE DE RECOUVREMENT, RIVURE A PLAT-JOINT, RIVURE A COUVRE-JOINT SIMPLE	EINSEITIGE LASCHENNIETUNG
11479	**SINGLE COLLAR THRUST BEARING**	PALIER DE BUTEE A UNE SEULE EMBASE	EINRINGDRUCKLAGER
11480	**SINGLE CRYSTAL**	MONOCRISTAL	EINKRISTALL
11481	**SINGLE CUT FILE, FLOAT CUT FILE**	ECOUENNE, ECOUANE, ECOINE	FEILE (EINHIEBIGE)
11482	**SINGLE CUT, FLOAT CUT OF A FILE**	TAILLE SIMPLE D'UNE LIME	EINFACHER HIEB EINER FEILE
11483	**SINGLE FORCE**	FORCE UNIQUE	EINZELKRAFT
11484	**SINGLE PART**	PIECE DETACHEE	EINZELTEIL
11485	**SINGLE PASS WELDING**	SOUDURE MONOPASSE	SCHWEISSNAHT (EINLAGIGE)
11486	**SINGLE PURPOSE MACHINE**	MACHINE POUR TRAVAUX SPECIAUX	MASCHINE FÜR SONDERZWECKE, SPEZIALMASCHINE
11487	**SINGLE RIVETED JOINT**	RIVURE SIMPLE, RIVURE A SIMPLE CLOUURE, RIVURE A UN RANG, CLOUURE SIMPLE	NIETUNG (EINREIHIGE), NIETUNG (EINFACHE)
11488	**SINGLE SEATED VALVE**	ROBINET A SOUPAPE SIMPLE	EINSITZVENTIL
11489	**SINGLE SHEAR RIVETED JOINT**	RIVURE A UNE COUPE	NIETUNG (EINSCHNITTIGE)
11490	**SINGLE SHEAR STEEL**	ACIER, FER CORROYE, FER FORT	GÄRBSTAHL
11491	**SINGLE VALVE**	SOUPAPE UNIQUE	VENTIL (EINFACHES)
11492	**SINGLE WEDGE DISC**	OPERCULE SIMPLE	KEIL (EINTEILIGER)
11493	**SINGLE-ACTING**	EFFET SIMPLE (A)	EINFACHWIRKEND
11494	**SINGLE-ACTING PISTON**	PISTON A SIMPLE EFFET	KOLBEN EINFACHWIRKENDER
11495	**SINGLE-ACTING STEAM ENGINE**	MACHINE A SIMPLE EFFET	DAMPFMASCHINE EINFACHWIRKENDE
11496	**SINGLE-ACTION TOOL**	OUTIL A SIMPLE EFFET	WERKZEUG (EINFACHWIRKENDES)
11497	**SINGLE-BEAT VALVE**	SOUPAPE A SIMPLE SIEGE	EINSITZIGES VENTIL, EINSITZVENTIL
11498	**SINGLE-CYLINDER ENGINE**	MACHINE MONOCYLINDRIQUE	EINZYLINDERMASCHINE
11499	**SINGLE-ENDED SPANNER**	CLEF A FOURCHE SIMPLE, CLEF DE CALIBRE SIMPLE	SCHRAUBENSCHLÜSSEL (EINFACHER), SCHRAUBENSCHLÜSSEL (EINMÄULIGER)
11500	**SINGLE-EXPANSION ENGINE**	MACHINE A SIMPLE EXPANSION	EINFACHEXPANSIONSMASCHINE
11501	**SINGLE-PHASE, MONOPHASE, ONE-PHASE, UNIPHASE CURRENT**	COURANT MONOPHASE	EINPHASENSTROM
11502	**SINGLE-THREADED SCREW**	VIS A UN FILET	SCHRAUBE (EINGÄNGIGE)
11503	**SINGLE-THROW CRANK SHAFT**	ARBRE COUDE SIMPLE	WELLE (EINKURBELIGE); EINFACH GEKRÖPFTE KURBELWELLE
11504	**SINGLE-WELDED LAP JOINT**	JOINT A RECOUVREMENT SOUDE D'UN SEUL COTE	EINSEITIG GESCHWEISSTE ÜBERLAPPUNGSVERBINDUNG
11505	**SINK HOLE**	RETASSURE	LUNKER
11506	**SINKHEAD**	MASSELOTTE	GIESSKOPF
11507	**SINKING**	PENETRATION	EINSINKEN
11508	**SINKING (ROLLING) MILL**	LAMINOIR REDUCTEUR	REDUZIERWALZWERK
11509	**SINTER (TO)**	FRITTER, S'AGGLOMERER	ZUSAMMENSINTERN
11510	**SINTERED CARBIDE ALLOY**	ALLIAGE DUR AUX CARBURES FRITTES	HARTMETALLLEGIERUNG
11511	**SINTERING**	FRITTAGE	SINTERUNG, SINTERN
11512	**SINTERING FURNACE**	FOUR A FRITTER	SINTEROFEN
11513	**SINUSOID**	SINUOIDE	SINUSOIDE
11514	**SIPHON**	SIPHON (DE MANOMETRE)	WASSERSACKROHR
11515	**SIPHON BAROMETER**	BAROMETRE A SIPHON	HEBERBAROMETER

11516	SIPHON LUBRICATOR, SIPHON OIL CUP	GRAISSEUR A MECHE	DOCHTÖLER
11517	SIPHON, SYPHON	SIPHON, SYPHON	HEBER, SAUGHEBER, SIPHON
11518	SIREN, HOOTER	TROMPE, SIRENE D'ALARME	SIRENE
11519	SITE HUT	BARAQUE DE CHANTIER	BAUSTELLENBUDE
11520	SITE SUPERINTENDENT	CHEF DE CHANTIER	BAUSTELLENLEITER
11521	SIX LIGHT SALOON	LIMOUSINE	LIMOUSINE
11522	SIZE OF GRAIN	GROSSEUR DU GRAIN	KORNGRÖSSE
11523	SIZE OF MESH	ECARTEMENT DES MAILLES	MASCHENWEITE
11524	SIZING	CALIBRAGE, FINISSAGE DIMENSIONNEL, TRIAGE	KALIBRIERUNG, MASSABFERTIGUNG, SORTIERUNG
11525	SIZING PRESS	PRESSE A CALIBRER	KALIBRIERPRESSE
11526	SKELETON PATTERN	MODELE SQUELETTE	SKELETT-MODELL
11527	SKELP	BANDE A TUBES	RÖHRENSTREIFEN
11528	SKETCH	CROQUIS	SKIZZE, ENTWURFZEICHNUNG
11529	SKETCH (TO)	CROQUER, DESSINER A GRANDS TRAITS	SKIZZIEREN
11530	SKETCH PAPER, SQUARED PAPER, CROSS SECTION PAPER	PAPIER QUADRILLE	PAPIER (KARIERTES), GITTERPAPIER, NETZPAPIER
11531	SKETCH PLATE	TOLE MARGINALE	RANDBLECH
11532	SKETCH, HAND SKETCH	CROQUIS	SKIZZE
11533	SKETCHING	ACTION DE CROQUER	SKIZZIEREN
11534	SKEW BEVEL WHEEL	ENGRENAGE HYPERBOLOIDE, ENGRENAGE A DENTURE SPIRALE	HYPERBELRAD, HYPERBOLOIDRAD
11535	SKEW, GAUCHE SURFACE	SURFACE GAUCHE	WINDSCHIEFE FLÄCHE
11536	SKIMMER (SPOON)	ECUMOIRE	SCHAUMLÖFFEL
11537	SKIMP	MANQUE	MÄNGEL
11538	SKIN	CROUTE DE LA FONTE	GUSSHAUT
11539	SKIN GLUE	COLLE DES PEAUX	LEDERLEIM, HAUT-LEIM
11540	SKIN PACKAGING	EMBALLAGE MOULANT	SKINPACKUNG, HAUTPACKUNG
11541	SKIN PASS	PASSE DE DECALAMINAGE (LAMINAGE), PASSE DE FINISSAGE	KALTSTICH, POLIERSTICH
11542	SKIN PASS	PASSE DE FINISSAGE	POLIERSTICH
11543	SKIN ROLL	PASSE DE DRESSAGE, PASSE DE DEGAUCHISSAGE	RICHTSTICH
11544	SKIN, SKIN DUE TO ROLLING	PELLICULE, COUCHE SUPERFICIELLE, CROUTE DE LAMINAGE	FILM, RANDSCHICHT, WALZHAUT
11545	SKINNING	FORMATION DE PEAUX	HAUTBILDUNG
11546	SKIP	INSTRUCTION DE SAUT, INTERRUPTION	ÜBERSPRINGUNSBEFEHL, UNSTETIGKEIT
11547	SKIRT	JUPE	EINFASSUNG
11548	SKIRT OF A COVER-PLATE	MANCHETTE D'UN TAMPON	DECKELMANSCHETTE, MUFFE
11549	SKIRT OF A TANK	JUPE (D'UN RESERVOIR)	SCHÜRZE
11550	SKULL	LOUP	BÄR
11551	SLAB	DALLE, BRAME	PLATTE, BRAMME
11552	SLAB / SLABBING MILL	BRAME (TRAIN A BRAMES)	BRAMME (BRAMMENSTRASSE)
11553	SLAB BENDING	BRAMES (CINTRAGE DES)	BRAMMENBIEGEN
11554	SLAB SHEARS	CISAILLES A BRAMES	BRAMMENSCHERE
11555	SLABBING MILL	LAMINOIR A BRAMES	BRAMMENWALZWERK
11556	SLABBING MILL(S)	TRAIN(S) A BRAMES	BRAMMEN-STRASSE
11557	SLACK COAL , FINE SMALL COAL	HOUILLE MENUE, MENUS DE HOUILLLE	FEINKOHLE, GRUSKOHLE, KOHLENGRUS

11558	SLACKEN A BOLT SCREW (TO)	DESSERRER UN ECROU, DESSERRER UNE VIS	SCHRAUBE LOCKERN (EINE)
11559	SLACKEN BACK (TO), WORK LOOSE OUT (TO)	DESSERRER (SE)	LOCKER WERDEN, LOCKERN (SICH)
11560	SLACKING BACK, WORKING LOOSE	DESSERRAGE, DEBLOCAGE	LOCKERUNG, LOCKERWERDEN
11561	SLAG	LAITIER, SCORIE	SCHLACKE
11562	SLAG BEARING	CONTENANT DES SCORIES	SCHLACKENHALTIG
11563	SLAG BRICK	MOELLON EN LAITIER	SCHLACKENSTEIN
11564	SLAG CEMENT	CIMENT DE LAITIER, CIMENT DE MACHEFER	SCHLACKENZEMENT
11565	SLAG CONCRETE	BETON DE LAITIER, BETON DE MACHEFER	SCHLACKENBETON
11566	SLAG WOOL, MINERAL WOOL, SILICATE COTTON	LAINE MINERALE	SCHLACKENWOLLE
11567	SLAG, CINDER INCLUSION	SCORIE EMPRISONNEE	SCHLACKE (EINGESCHLOSSENE), SCHLACKE (EINGEWALZTE)
11568	SLAKE THE LIME (TO)	ETEINDRE LA CHAUX	KALK LÖSCHEN
11569	SLAKING THE LIME	EXTINCTION DE LA CHAUX, HYDRATATION DE LA CHAUX	LÖSCHEN DES KALKES
11570	SLAT CONVEYOR	TRANSPORTEUR A TABLIER METALLIQUE, TRANSPORTEUR A PALETTES	PLATTENFÖRDERER
11571	SLATE	ARDOISE	SCHIEFER (BAUW.)
11572	SLEDGE HAMMER	MASSE, COGNEE	SCHMIEDEHAMMER
11573	SLEDGE, SLEDGE HAMMER	MARTEAU A FRAPPER DEVANT, FRAPPE-DEVANT	ZUSCHLAGHAMMER, VORSCHLAGHAMMER
11574	SLEEPER	TRAVERSE DE VOIE	SCHWELLE, EISENBAHNSCHWELLE
11575	SLEEPER SCREWS	TIREFONDS	SCHIENENSCHRAUBEN
11576	SLEEVE	FOURREAU, MANCHETTE, DOUILLE	HÜLSE, MANSCHETTE
11577	SLEEVE COUPLING, FRICTION CLIP COUPLING	MANCHON A FRETTES	HÜLSENKUPPLUNG
11578	SLEEVE GOVERNOR	REGULATEUR A MANCHON, REGULATEUR A DOUILLE	MUFFENREGLER
11579	SLEEVE VALVE	SOUPAPE A CLOCHE INTERIEURE, SOUPAPE A MANCHON	ROHRVENTIL
11580	SLENDERNESS	ELANCEMENT	SCHLANKHEIT
11581	SLENDERNESS OF A BEAM	ELANCEMENT D'UNE POUTRE	SCHLANKHEIT (EINES TRÄGERS)
11582	SLENDERNESS RATIO	RAPPORT D'ELANCEMENT	SCHLANKHEITSGRAD
11583	SLEWING CRANE	GRUE PIVOTANTE	DREHKRAN, SCHWENKKRAN
11584	SLICING LATHE	TOUR A DECOLLETER	ABSTECHDREHMASCHINE
11585	SLIDE	SURFACE DE GLISSEMENT, COULISSEAU, GLISSOIRE	GLEITFLÄCHE, GLEITSTÜCK
11586	SLIDE (TO)	GLISSER, COULISSER	GLEITEN
11587	SLIDE BLOCK, GUIDE BLOCK, SLIPPER BLOCK, SLIPPER OF CROSSHEAD	PATIN, GLISSOIR, BLOC DE LA CROSSE	SCHUH, GLEITSCHUH, FÜHRUNGSSCHUH, GLEITKLOTZ DES KREUZKOPFES
11588	SLIDE FACE	GLISSIERE	GLEITBAHN, GLEITFLÄCHE
11589	SLIDE OF A MICROSCOPE	LAME DE VERRE	OBJEKTTRÄGER, GLAS
11590	SLIDE REST LATHE	TOUR A CHARIOTER, TOUR A CHARIOT	SUPPORTDREHBANK
11591	SLIDE RULE	REGLE A CALCUL	RECHENSCHIEBER
11592	SLIDE VALVE	DISTRIBUTEUR, OBTURATEUR GLISSANT, APPAREIL DE DISTRIBUTION A GLISSEMENT, TIROIR, ROBINET	SCHIEBER
11593	SLIDE VALVE DIAGRAM	DIAGRAMME DE DISTRIBUTION	SCHIEBERDIAGRAMM

11594	**SLIDE VALVE GEAR**	DISTRIBUTION PAR TIROIR	SCHIEBERSTEUERUNG
11595	**SLIDING**	GLISSEMENT	GLEITEN
11596	**SLIDING BEARING**	PALIER COULISSANT	GLEITLAGER, SCHLITTENLAGER
11597	**SLIDING CALIPER**	PIED A COULISSE	SCHUBLEHRE
11598	**SLIDING CHANGE GEAR**	ENGRENAGE BALADEUR	SCHIEBERÄDERGETRIEBE
11599	**SLIDING CLUTCH SLEEVE**	MANCHON D'EMBRAYAGE, MANCHON MOBILE	AUSRÜCKMUFFE, VERSCHIEBBARE KUPPLUNGSMUFFE
11600	**SLIDING FIT**	EMMANCHEMENT, AJUSTAGE, MONTAGE GLISSANT, EMMANCHEMENT GRAS, EMMANCHEMENT COULISSANT	GLEITSITZ
11601	**SLIDING FRICTION**	FROTTEMENT DE GLISSEMENT	REIBUNG (GLEITENDE)
11602	**SLIDING GEAR TRAIN**	TRAIN BALADEUR	ZAHNRADSATZ (VERSCHIEBBARER)
11603	**SLIDING LATHE**	FOUR A CHARIOTER	ZUGSPINDELDREHMASCHINE
11604	**SLIDING LID COVER**	COUVERCLE A GLISSIERE	SCHIEBEDECKEL
11605	**SLIDING MOTION**	MOUVEMENT DE GLISSEMENT	BEWEGUNG (GLEITENDE), GLEITBEWEGUNG
11606	**SLIDING PLATE SUPPORT**	BERCEAU A PATINS	UNTERGESTELL MIT GLEITSCHUH, SCHLITTEN MIT GLEITPLATTEN
11607	**SLIDING SCALE**	ECHELLE MOBILE	LOHNSKALA (GLEITENDE)
11608	**SLIDING SURFACE**	SURFACE DE GLISSEMENT	GLEITFLÄCHE
11609	**SLIME**	BOUE	SCHLAMM
11610	**SLING**	ELINGUE	SCHLINGE
11611	**SLIP (PING)**	GLISSEMENT	GLEITUNG
11612	**SLIP (TO)**	PATINER, GLISSER	GLEITEN, SCHLÜPFEN
11613	**SLIP BANDS**	BANDES DE GLISSEMENT	GLEITLINIEN, STREIFEN
11614	**SLIP CLUTCH**	EMBRAYAGE A FRICTION	REIBUNGSKUPPLUNG
11615	**SLIP CRACK**	CRIQUE DE CLIVAGE	SCHIEFERBRUCH
11616	**SLIP FORM (CONCRETE)**	COFFRAGE GLISSANT (POUR BETON)	GLEITVERSCHALUNG
11617	**SLIP JOINT**	JOINT COULISSANT	AUSDEHNUNGSKUPPLUNG
11618	**SLIP RING**	BAGUE COLLECTRICE, COLLECTEUR	SCHLEIFRING, KOLLEKTOR
11619	**SLIPPING CLUTCH**	EMBRAYAGE PROGRESSIF A FRICTION	RUTSCHKUPPLUNG
11620	**SLIPPING OF THE BELT**	GLISSEMENT DE LA COURROIE, PATINAGE DE LA COURROIE	GLEITEN DES RIEMENS, RUTSCHEN DES RIEMENS, SCHLÜPFEN DES RIEMENS, GLEITSCHLUPF
11621	**SLIPPING, SKIDDING OF WHEELS**	PATINAGE DES ROUES	SCHLÜPFEN DER RÄDER
11622	**SLITTING**	FENDAGE	SPALTEN
11623	**SLITTING FILE, FEATHER EDGE FILE**	LIME A LOSANGE	EINSTREICHFEILE, SCHRAUBENKOPFFEILE, SCHWERTFEILE
11624	**SLIVER**	CROUTE	SCHALE
11625	**SLIVER ORE**	MINERAI D'ARGENT	SILBERERZ
11626	**SLIVERS**	ARRACHEMENTS	WALZSPLITTER
11627	**SLOPE**	TALUS, PENTE, INCLINAISON	BÖSCHUNG, ABHANG, NEIGUNG
11628	**SLOPING SCREEN**	CRIBLE POUR LE SABLE, POUR LA TERRE	DURCHWURFSIEB, SANDSIEB, ERDSIEB, WURFGITTER
11629	**SLOT**	FENTE	SPALT
11630	**SLOT (TO)**	MORTAISER	STOSSEN, NUTEN STOSSEN
11631	**SLOT AND CRANK**	MANIVELLE A COULISSE, COULISSE ET MANIVELLE	KURBELSCHLEIFE
11632	**SLOT DRILLS**	FRAISES A RAINURES, FRAISES A COUTEAUX	NUTENFRÄSER
11633	**SLOT LINK, REVERSING LINK**	COULISSE	SCHLEIFE, SCHWINGE, KULISSE
11634	**SLOT MILLING CUTTER**	FRAISE POUR RAINURES	NUTENFRÄSER

11635	SLOT WELD	SOUDURE A ENTAILLE	SCHLITZNAHT
11636	SLOT WELDING	SOUDAGE A ENTAILLE	SCHLITZSCHWEISSEN
11637	SLOTTED HEAD SCREW	VIS A TETE FENDUE	SCHLITZSCHRAUBE
11638	SLOTTED LEVER	LEVIER A COULISSE	SCHLITZHEBEL, KULISSENHEBEL
11639	SLOTTED LINER	COLONNE PERDUE A FENTES	LINER (GESCHLITZER)
11640	SLOTTERS	MORTAISEUSES	SENKRECHTSTOSSMASCHINEN
11641	SLOTTING	MORTAISAGE	STOSSEN, NUTENSTOSSEN
11642	SLOTTING ATTACHEMENT	APPAREIL A MORTAISER	STOSSAPPARATE
11643	SLOTTING MACHINE	MACHINE A MORTAISER, MORTAISEUSE	STOSSMASCHINE, VERTIKALHOBELMASCHINE, NUTSTOSSMASCHINE
11644	SLOW COMBUSTION	COMBUSTION LENTE	VERBRENNUNG (LANGSAME)
11645	SLOW DOWN AN ENGINE (TO)	RALENTIR LA MARCHE D'UNE MACHINE	VERLANGSAMEN (DEN GANG EINER MASCHINE)
11646	SLOW IDLING JET	GICLEUR DE RALENTI	LEERLAUFDÜSE
11647	SLOW PITCH THREAD	FILET A PETIT PAS	GEWINDE MIT SCHWACHER STEIGUNG
11648	SLOW SETTING CEMENT	CIMENT A PRISE LENTE	ZEMENT (LANGSAM BINDENDER)
11649	SLOWING DOWN AN ENGINE	RALENTISSEMENT DE LA MARCHE	VERLANGSAMEN DES GANGES
11650	SLOWLY ROTATING, REVOLVING SHAFT	ARBRE A PETITE VITESSE	WELLE (LANGSAMLAUFENDE)
11651	SLUG	EBAUCHE	ROHLING
11652	SLUICE	VANNE	SCHÜTZ, SCHÜTZE
11653	SLUICE VALVE, GATE VALVE	ROBINET-VANNE	ABSPERRSCHIEBER
11654	SLURRY	BOUE	SCHLAMM
11655	SLUSH	BOUE	SCHLAMM
11656	SLUSH CASTING PROCESS	PROCEDE DE MOULAGE INVERSE	STÜRZGUSSVERFAHREN
11657	SMALL COKE	GRESILLON DE COKE	KOKSKLEIN
11658	SMALL GAS ENGINE	MOTEUR A GAZ DE FAIBLE PUISSANCE	KLEINGASMASCHINE
11659	SMALL HAND SAW	SCIE EGOINE	FUCHSSCHWANZ
11660	SMALL IRONWARE	QUINCAILLERIE	KLEINEISENWAREN
11661	SMALL MOTOR, FRATIONAL HORSEPOWER MOTOR	MOTEUR DE FAIBLE PUISSANCE	KLEINMOTOR
11662	SMALL SPROCKET	NOIX POUR CHAINES, NOIX D'ENTRAINEMENT POUR CHAINES-CABLES	KETTENNUSS
11663	SMALL WATER-CAPACITY BOILER, BOILER WITH SMALL WATER SPACE	CHAUDIERE A FAIBLE VOLUME	KLEINWASSERRAUMKESSEL
11664	SMELT (TO), MELT DOWN FROM THE DRE (TO)	TRAITER LES MINERAIS	VERHÜTTEN
11665	SMELTER	FOUR A FONDRE	SCHMELZOFEN
11666	SMELTING	FUSION, TRAITEMENT DES MINERAIS	SCHMELZEN, VERHÜTTUNG
11667	SMELTING FURNACE	FOUR, FOURNEAU DE FUSION	SCHMELZOFEN
11668	SMELTING WORKS	USINE METALLURGIQUE	HÜTTENWERK
11669	SMITH FORGING	FORGEAGE A LA MAIN, PIECE FORGEE A LA MAIN	HANDSCHMIEDEN
11670	SMITH HAMMER FORGING	PIECE FORGEE AU MARTEAU	FREIFORMSCHMIEDESTÜCK
11671	SMITH, BLACKSMITH	FORGERON	SCHMIED, GROBSCHMIED
11672	SMOKE	FUMEE	RAUCH
11673	SMOKE CONSUMING	FUMIVORE	RAUCHVERZEHREND
11674	SMOKE CONSUMING FURNACE	FOYER FUMIVORE	FEUERUNG FÜR RAUCHFREIE VERBRENNUNG

11675	**SMOKE FORMATION**	FORMATION DE LA FUMEE	RAUCHBILDUNG, RAUCHENTWICKLUNG
11676	**SMOKE NUISANCE**	NUISANCE PAR LA FUMEE	RAUCHBELÄSTIGUNG
11677	**SMOKE PREVENTION**	SUPPRESSION DE LA FUMEE	RAUCHVERHÜTUNG
11678	**SMOKY FLAME**	FLAMME FULIGINEUSE	FLAMME (RUSSENDE)
11679	**SMOOTH (TO)**	LISSER	GLÄTTEN
11680	**SMOOTH CUT FILE**	LIME A TAILLE DOUCE, LIME FINE	FEINHIEBFEILE, SCHLICHTFEILE
11681	**SMOOTH EVEN RUNNING**	MARCHE REGULIERE	GANG (RUHIGER)
11682	**SMOOTH FINISH, FINISHING**	ACHEVEMENT, PARACHEVEMENT, FINISSAGE, FINITION, RETOUCHE	NACHARBEITEN, NACHBEARBEITUNG, FERTIGBEARBEITUNG, SCHLICHTEN
11683	**SMOOTH FRACTURE**	CASSURE NETTE	BRUCH (GLATTER)
11684	**SMOOTH PLAIN TUBE**	TUBE LISSE	ROHR (GLATTES)
11685	**SMOOTH STARTING**	DEMARRAGE DOUX, DEMARRAGE SANS A-COUPS	ANLAUF (STOSSFREIER)
11686	**SMOOTHING**	LISSAGE	GLÄTTEN
11687	**SMOOTHING PLANE**	RABOT A REPASSER	SCHLICHTHOBEL
11688	**SMOULDER (TO)**	BRULER SANS FLAMME, COUVER	SCHWELEN
11689	**SNAGGING**	EBARBAGE, MEULAGE	PUTZEN
11690	**SNAP ACTION SWITH**	INTERRUPTEUR A ACTION INSTANTANEE	SPRINGSCHALTER, SCHNAPPSCHALTER
11691	**SNAP HOOK**	PORTE-MOUSQUETON, MOUSQUETON	KARABINERHAKEN
11692	**SNAP RING**	JONC D'ARRET	SPRENGRING
11693	**SNAP-GAUGES**	CALIBRES-MACHOIRES	RACHENLEHRE
11694	**SNIFTING VALVE**	RENIFLARD	SCHNARCHVENTIL, SCHNÜFFELVENTIL, SCHNÜFFLER, LUFTANSAUGEVENTIL
11695	**SNOW LOAD**	CHARGE DE NEIGE	SCHNEELAST
11696	**SNOW-CHAINS**	CHAINES DE NEIGE	SCHNEEKETTEN
11697	**SOAK CLEANING**	NETTOYAGE PAR IMMERSION	EINTAUCHREINIGUNG
11698	**SOAKING FURNACE**	FOUR D'EGALISATION	AUSGLEICHOFEN
11699	**SOAKING PIT**	PIT A TREMPER	DURCHWEICHUNGSGRUBE
11700	**SOAKING TEMPERATURE**	TEMPERATURE DE RECUIT	DURCHWÄRMETEMPERATUR
11701	**SOAP**	SAVON	SEIFE
11702	**SOAP SUDS**	EAU DE SAVON, EAU SAVONNEUSE	SEIFENWASSER, SEIFENLAUGE, SEIFENLÖSUNG
11703	**SOAPSTONE, STEATITE**	STEATITE, CRAIE DE BRIANCON	SPECKSTEIN, STEATIT
11704	**SOCKET CHISEL**	CISEAU A DOUILLE	ROHRMEISSEL
11705	**SOCKET OF A PIPE**	MANCHON DE RACCORD	MUFFE, ROHRMUFFE
11706	**SOCKET OF A ROPE**	AGRAFE DE JONCTION POUR CABLES	SEILSCHLOSS
11707	**SOCKET PIPE**	TUYAU A EMBOITEMENT	MUFFENROHR
11708	**SOCKET WELDING FITTING**	RACCORD A EMBOITEMENT POUR SOUDURE	EINFÜGESCHWEISSFITTING
11709	**SOCKET WRENCH**	CLE A DOUILLE	STECKSCHLÜSSEL
11710	**SOCKETED END OF PIPE**	BOUT FEMELLE D'UN TUYAU	MUFFENENDE EINES ROHRES
11711	**SODA CRYSTALS, CRYSTALLISED SODIUM CARBONATE**	CRISTAUX DE SOUDE	KRISTALLISIERTE SODA, KRISTALLSODA, GEWÄSSERTES NATRIUMKARBONAT
11712	**SODA LYE**	LESSIVE DE SOUDE	SODALAUGE
11713	**SODA, SODIUM CARBONATE**	SOUDE, CARBONATE DE SODIUM	SODA, NATRIUMKARBONAT
11714	**SODIUM**	SODIUM	NATRIUM
11715	**SODIUM ACETATE**	ACETATE DE SODIUM, DE SOUDE	NATRIUM (ESSIGSAURES), NATRIUMAZETAT

11716	**SODIUM ALUMINATE**	ALUMINATE DE SODIUM	TONERDENATRON, NATRIUMALUMINAT
11717	**SODIUM BISULPHATE**	SULFATE ACIDE DE SOUDE	NATRIUM (SAURES), NATRIUM (SCHWEFELSAURES), NATRIUMBISULFAT
11718	**SODIUM BISULPHITE**	SULFITE ACIDE, BISULFITE DE SODIUM	NATRON (SCHWEFLIGSAURES), NATRON (DOPPELTSCHWEFLIGSAURES), NATRIUMBISULFIT
11719	**SODIUM CITRATE**	CITRATE DE SODIUM	NATRON (ZITRONENSAURES), NATRIUMZITRAT
11720	**SODIUM CYANIDE**	CYANURE DE SODIUM	ZYANNATRIUM, NATRIUMZYANID
11721	**SODIUM FLUORIDE**	FLUORURE DE SODIUM	FLUORNATRIUM
11722	**SODIUM HYDROXIDE, CAUSTIC SODA**	SOUDE CAUSTIQUE, HYDRATE DE SODIUM	ÄTZNATRON, NATRIUMHYDROXYD, NATRIUMOXYDHYDRAT, NATRONHYDRAT, KAUSTISCHE SODA
11723	**SODIUM LINE**	RAIE DU SODIUM	NATRIUMLINIE
11724	**SODIUM MONOXIDE**	OXYDE DE SODIUM	NATRON, NATRIUMOXYD
11725	**SODIUM NITRITE**	AZOTITE, NITRITE DE SOUDE, SOUDE AZOTEUSE, SOUDE NITREUSE	NATRIUM (SALPETRIGSAURES), NATRIUMNITRIT
11726	**SODIUM OXALATE**	OXALATE DE SODIUM	NATRIUM (OXALSAURES), NATRIUM, NATRIUMOXALAT
11727	**SODIUM PEROXIDE**	BIOXYDE DE SODIUM	NATRIUMSUPEROXYD, NATRIUMPEROXYD
11728	**SODIUM PHOSPHATE**	PHOSPHATE DE SODIUM	NATRON (PHOSPHORSAURES), NATRIUMPHOSPHAT
11729	**SODIUM POTASSIUM TARTRATE, SEIGNETTE'S SALT, ROCHELLE SALT**	TARTRATE DE POTASSE ET DE SOUDE, SEL DE SEIGNETTE	KALINATRON (WEINSAURES), ROCHELLESALZ, SCHWANENSALZ, SEIGNETTESALZ, NATRONWEINSTEIN, KALIUMNATRIUMTARTRAT
11730	**SODIUM PYROPHOSPHATE**	PYROPHOSPHATE DE SOUDE	NATRIUMPYROPHOSPHAT, PYROPHOSPHORSAURES NATRON
11731	**SODIUM SILICATE**	SILICATE DE SOUDE	NATRONWASSERGLAS, KIESELSAURES NATRIUM, NATRIUMSILIKAT
11732	**SODIUM STANNATE**	STANNATE DE SODIUM	NATRIUM (ZINNSAURES), NATRIUMSTANNAT, SODASTANNAT, ZINNOXYDNATRON, ZINNSODA, PRÄPARIERSALZ
11733	**SODIUM SULPHATE**	SULFATE NEUTRE DE SODIUM	NATRON (SCHWEFELSAURES), NATRIUMSULFAT
11734	**SODIUM SULPHITE**	SULFITE DE SOUDE	NATRON (SCHWEFLIGSAURES), NATRIUMSULFIT
11735	**SODIUM TANNATE**	TANNATE DE SODIUM	NATRIUM (GERBSAURES), NATRIUMTANNAT
11736	**SODIUM THIOSULPHATE HYPOSULPHITE**	HYPOSULFITE DE SODIUM	NATRON (UNTERSCHWEFLIGSAURES), NATRIUMTHIOSULFAT, NATRIUMHYPOSULFIT
11737	**SODIUM TUNGSTATE**	TUNGSTATE DE SOUDE	NATRIUM (WOLFRAMSAURES), NATRIUMWOLFRAMAT
11738	**SOFT FLOWING METAL**	METAL MOU, TENDRE	WEICHMETALL
11739	**SOFT PACKING**	GARNITURE SOUPLE	PACKUNG (WEICHE)
11740	**SOFT SOAP**	SAVON MOU, SAVON VERT, SAVON NOIR, SAVON DE POTASSE	SCHMIERSEIFE
11741	**SOFT SOLDER (TO)**	SOUDER A L'ETAIN	WEICH LÖTEN
11742	**SOFT SOLDER, PLUMBER'S SOLDER**	SOUDURE A L'ETAIN	WEICHLOT, SCHNELLOT, WEISSLOT, ZINNLOT, LÖTZINN
11743	**SOFT SOLDERING**	SOUDURE A L'ETAIN	WEICHLÖTEN
11744	**SOFT STEEL**	ACIER DOUX	STAHL (WEICHER), FLUSSEISEN, FLUSSSTAHL
11745	**SOFT WATER**	EAU PURE, EAU PEU CHARGEE	WEICHES WASSER

11746	**SOFT WOOD**	BOIS LEGER, BOIS TENDRE, BOIS BLANC	WEICHHOLZ, WEICHES HOLZ
11747	**SOFT-FACED HAMMER**	MARTEAU A TETE PLASTIQUE	KUNSTSTOFFBAHNHAMMER
11748	**SOFTEN THE WATER (TO)**	EPURER, PURIFIER LES EAUX D'ALIMENTATION PAR VOIE CHIMIQUE	WASSER ENTHÄRTEN
11749	**SOFTENING**	ADOUCISSEMENT	ENTHÄRTUNG
11750	**SOFTENING POINT**	POINT D'AFFAISSEMENT	ERWEICHUNGSPUNKT
11751	**SOIL BEARING CAPACITY**	PORTANCE DU SOL	TRAGFÄHIGKEIT DES BODENS
11752	**SOIL SURVEY**	ETUDE DE TERRAIN, ETUDE DE SOL	BODENUNTERSUCHUNG, TRAGFÄHIGKEIT DES BODENS
11753	**SOLAR OIL**	HUILE SOLAIRE	SOLARÖL, BRAUNKOHLENBENZIN
11754	**SOLAR SPECTRUM**	SPECTRE SOLAIRE	SONNENSPEKTRUM
11755	**SOLDER**	ALLIAGE FUSIBLE POUR SOUDER OU BRASER, SOUDURE (COMPOSITION FUSIBLE)	LOT, LÖTMITTEL
11756	**SOLDER (TO)**	SOUDER	LÖTEN
11757	**SOLDER EMBRITTLEMENT**	FRAGILITE PAR PENETRATION DE SOUDURE	VERSPRÖDUNG DURCH EINDRINGEN VON LOT
11758	**SOLDERED BRAZED FLANGE**	BRIDE BRASEE, RONDELLE BRASEE	FLANSCH (AUFGELÖTETER)
11759	**SOLDERED BRAZED JOINT**	ASSEMBLAGE PAR SOUDURE	LÖTVERBINDUNG
11760	**SOLDERED BRAZED PIPE**	TUBE BRASE	ROHR (GELÖTETES)
11761	**SOLDERED SEAM**	SOUDURE (TRAVAIL FAIT EN SOUDANT AVEC INTERPOSITION D'UN ALLIAGE)	LÖTSTELLE
11762	**SOLDERING**	SOUDAGE, SOUDURE, BRASAGE	LÖTEN, LÖTUNG
11763	**SOLDERING FURNACE**	FOUR A SOUDER	LÖTOFEN
11764	**SOLDERING IRON**	FER A SOUDER	LÖTKOLBEN
11765	**SOLDERING IRON BIT, COPPER BIT**	FER A SOUDER	LÖTKOLBEN, LÖTEISEN
11766	**SOLDERING LAMP**	LAMPE A SOUDER	LÖTLAMPE
11767	**SOLDERING TONGS**	PINCE, TENAILLE A SOUDER	LÖTZANGE
11768	**SOLDERING WITH THE BLOW PIPE**	SOUDURE AU CHALUMEAU	FLAMMENLÖTUNG
11769	**SOLDERING WITH THE SOLDERING BIT**	SOUDURE AU FER A SOUDER	KOLBENLÖTUNG
11770	**SOLE LEATHER**	CUIR FORT	SOHLLEDER
11771	**SOLE OF A PLANE**	SEMELLE D'UN RABOT	SOHLE EINES HOBELS
11772	**SOLE PLATE**	SEMELLE DE PALIER	SOHLPLATTE (FÜR LAGER)
11773	**SOLENOID**	SOLENOIDE	SOLENOID
11774	**SOLENOID VALVE**	SOUPAPE ELECTROMAGNETIQUE, SOLENOIDE, ROBINET ELECTROMAGNETIQUE, VALVE ELECTROMAGNETIQUE	MAGNETVENTIL
11775	**SOLID**	SOLIDE, CORPS SOLIDE	FESTKÖRPER, KÖRPER (FESTER)
11776	**SOLID ANGLE**	ANGLE SOLIDE	WINKEL (KÖRPERLICHER)
11777	**SOLID BEARING**	PALIER FERME	LAGER (EINTEILIGES)
11778	**SOLID BORING BAR**	BARRE D'ALESAGE MONOPIECE	BOHRSTANGE (EINTEILIGE)
11779	**SOLID BOSS**	MOYEU D'UNE SEULE PIECE	NABE (UNGETEILTE)
11780	**SOLID BRICK**	BRIQUE PLEINE	VOLLSTEIN, VOLLZIEGEL
11781	**SOLID DIE**	FILIERE RONDE MONOBLOC	GEWINDESCHNEIDKOPF (RUNDER)
11782	**SOLID DRAWN TUBE**	TUBE ETIRE	ROHR (GEZOGENES)
11783	**SOLID END MILLS**	FRAISES EN BOUT PLEINES	SCHAFTFRÄSER
11784	**SOLID FLYWHEEL**	VOLANT D'UNE SEULE PIECE	SCHWUNGRAD (EINTEILIGES)
11785	**SOLID FUEL**	COMBUSTIBLE SOLIDE	BRENNSTOFF (FESTER)
11786	**SOLID HALF-ROUND IRON**	FER DEMI-ROND PLEIN	HALBRUNDEISEN (VOLLES)

11787	SOLID JET	JET COMPACT	STRAHL (ZUSAMMENHALTENDER)
11788	SOLID JOURNAL BEARING, DEAD EYE BEARING	PALIER EN UNE SEULE PIECE	LAGER (EINTEILIGES), LAGER (GESCHLOSSENES), AUGENLAGER, FROSCHLAGER
11789	SOLID MANDREL	MANDRIN LISSE	DORN (FESTER)
11790	SOLID OF REVOLUTION	CORPS DE REVOLUTION	DREHUNGSKÖRPER, UMDREHUNGSKÖRPER, ROTATIONSKÖRPER
11791	SOLID PISTON	PISTON PLEIN	SCHEIBENKOLBEN, EINSCHEIBENKOLBEN, TELLERKOLBEN
11792	SOLID PULLEY	POULIE EN UNE PIECE	RIEMENSCHEIBE (UNGETEILTE) (GANZE)
11793	SOLID SHAFT	ARBRE PLEIN	WELLE (VOLLE), WELLE (MASSIVE), RIEMENSCHEIBE
11794	SOLID SOLUTION	SOLUTION SOLIDE	MISCHKRISTALL, LÖSUNG (FESTE)
11795	SOLID STATE	ETAT SOLIDE	ZUSTAND (FESTER), FESTER AGGREGAT-ZUSTAND
11796	SOLID WEDGE DISC	OPERCULE MONOBLOC (COIN)	KEIL (MASSIVER)
11797	SOLIDIFICATION	SOLIDIFICATION	ERSTARRUNG, ERSTARREN
11798	SOLIDIFICATION FRONT	FRONT DE SOLIDIFICATION, FRONT DE CRISTALLISATION	ERSTARRUNGSFRONT, KRISTALLISATIONSFRONT
11799	SOLIDIFICATION OF A LIQUID	SOLIDIFICATION D'UN LIQUIDE	ERSTARRUNG EINER FLÜSSIGKEIT
11800	SOLIDIFICATION POINT	POINT DE SOLIDIFICATION	ERSTARRUNGSPUNKT
11801	SOLIDIFICATION RANGE	INTERVALLE DE SOLIDIFICATION	ERSTARRUNGSBEREICH
11802	SOLIDIFICATION SHRINKAGE	CONTRACTION DE SOLIDIFICATION	ERSTARRUNGSSCHWINDUNG
11803	SOLIDIFY (TO)	SOLIDIFIER (SE) PAR REFROIDISSEMENT	ERSTARREN
11804	SOLIDIFYING POINT	POINT DE SOLIDIFICATION, POINT DE PRISE	ERSTARRUNGSPUNKT, STOCKPUNKT
11805	SOLIDUS	SOLIDUS	SOLIDUS, SOLIDUSLINIE
11806	SOLUBILITY	SOLUBILITE	LÖSLICHKEIT
11807	SOLUBLE	SOLUBLE	LÖSLICH
11808	SOLUBLE GLASS, WATER GLASS	VERRE SOLUBLE, VERRE LIQUIDE	WASSERGLAS
11809	SOLUBLE IN WATER	SOLUBLE DANS L'EAU	WASSERLÖSLICH
11810	SOLUBLE OIL	HUILE SOLUBLE	ÖL (WASSERLÖSLICHES)
11811	SOLUTION	SOLUTION	LÖSUNG, AUFLÖSUNG
11812	SOLUTION HEAT TREATMENT	RECUIT DE MISE EN SOLUTION	LÖSUNGSGLÜHEN
11813	SOLVE (TO)	RESOUDRE, DISSOUDRE	AUFLÖSEN (MATH.)
11814	SOLVENT	DISSOLVANT, SOLVANT	LÖSUNGSMITTEL
11815	SONIC TESTING	ESSAI ACOUSTIQUE	PRÜFUNG (AKUSTISCHE)
11816	SOOT	SUIE	RUSS
11817	SORBITE	SORBITE	SORBIT
11818	SOUND VIBRATION	VIBRATION SONORE	SCHALLSCHWINGUNG
11819	SOUND WAVE	ONDE SONORE	SCHALLWELLE
11820	SOUND WOOD	BOIS SAIN	HOLZ (GESUNDES)
11821	SOUNDNESS OF A WELD	HOMOGENEITE D'UNE SOUDURE	FEHLERFREIHEIT DER SCHWEISSNAHT
11822	SOUR CRUDE	PETROLE BRUT SULFUREUX	ERDÖL (SCHWEFELHALTIGES)
11823	SOURCE OF ELECTRIC CURRENT	SOURCE D'ELECTRICITE	STROMQUELLE
11824	SOURCE OF ENERGY	SOURCE D'ENERGIE	ENERGIEQUELLE
11825	SOURCE OF ERROR	CAUSE D'ERREUR, SOURCE D'ERREURS	FEHLERQUELLE
11826	SOURCE OF LIGHT, LUMINOUS SOURCE	SOURCE LUMINEUSE	LICHTQUELLE
11827	SOUTH SEEKING POLE OF A MAGNET	POLE SUD D'UN AIMANT	SÜDPOL EINES MAGNETEN

11828	SPACE LATTICE	RESEAU SPATIAL	RAUMGITTER
11829	SPACE OCCUPIED, REQUIRED	ENCOMBREMENT	PLATZBEDARF, RAUMBEDARF, RAUMBEANSPRUCHUNG
11830	SPALLING	ECAILLAGE	ABBLÄTTERUNG
11831	SPAN	PORTEE (DISTANCE ENTRE DEUX POINTS D'APPUI), TROUEE	TRAGWEITE, SPANNWEITE, STÜTZWEITE
11832	SPANNER GAP, WIDTH OF JAWS, GAP OF SPANNER	OUVERTURE DE CLEF	SCHRAUBENSCHLÜSSELWEITE, SCHLÜSSELWEITE
11833	SPANNER, SCREW KEY	CLEF A ECROUS	SCHRAUBENSCHLÜSSEL, MUTTERSCHLÜSSEL
11834	SPARE PART, REPAIR PART, REPLACEMENT PART	PIECE DE RECHANGE, PIECE INTERCHANGEABLE	ERSATZTEIL
11835	SPARE WHEEL	ROUE DE SECOURS	RESERVERAD, ERSATZRAD
11836	SPARK	ETINCELLE	FUNKE
11837	SPARK ADVANCE	CORRECTEUR D'AVANCE	ZÜNDVERSTELLUNG
11838	SPARK PLUG	BOUGIE D'ALLUMAGE	ZÜNDKERZE
11839	SPARK PLUG BARREL OR BODY	CULOT DE BOUGIE	ZÜNDKERZENGEHÄUSE
11840	SPARK TEST	ESSAI A L'ETINCELLE	FUNKENPROBE
11841	SPARKING PLUG	BOUGIE D'ALLUMAGE	ZÜNDKERZE
11842	SPATTER	CRACHEMENT, ECLABOUSSURE, PERLE DE SOUDURE	SPRITZEN, SCHWEISSPERLE
11843	SPATTER LOSS	PERTES PAR CRACHEMENT	SPRITZVERLUSTE
11844	SPATULA	SPATULE	SPACHTEL
11845	SPECIAL BRANCH	DOMAINE, BRANCHE, SPECIALITE	FACHGEBIET, SONDERGEBIET, SPEZIALGEBIET
11846	SPECIAL PLATE, SHAPED PLATE, PROFILED SHEET IRON	TOLE FACONNEE, TOLE PROFILEE	FORMBLECH
11847	SPECIAL SECTION WIRE	FIL PROFILE	FORMDRAHT, FASSONDRAHT, PROFILDRAHT
11848	SPECIAL STEEL	ACIER SPECIAL	SONDERSTAHL
11849	SPECIALITY, SPECIAL ARTICLE OF MANUFACTURE	PRODUIT SPECIAL	SONDERERZEUGNIS, SPEZIALARTIKEL
11850	SPECIFIC GRAVITY	POIDS SPECIFIQUE, DENSITE	GEWICHT (SPEZIFISCHES), EINHEITSGEWICHT
11851	SPECIFIC GRAVITY FLASK, PYKNOMETER	PYCNOMETRE	PYKNOMETER
11852	SPECIFIC HEAT	CHALEUR SPECIFIQUE	WÄRME (SPEZIFISCHE)
11853	SPECIFIC PRESSURE	PRESSION EN KILOGRAMMES PAR UNITE DE SURFACE	DRUCK (SPEZIFISCHER)
11854	SPECIFIC VOLUME	VOLUME SPECIFIQUE	VOLUMEN (SPEZIFISCHES)
11855	SPECIFICATION	CAHIER DES CHARGES	LASTENHEFT
11856	SPECIFICATION OF A PATENT	DESCRIPTION, MEMOIRE DESCRIPTIF DU BREVET	PATENTSCHRIFT, PATENTBESCHREIBUNG
11857	SPECIFIED SIZE	DIMENSIONS NOMINALES	NENNMASS
11858	SPECIMEN	EPROUVETTE DE METAL	PROBE
11859	SPECTRAL COLOURS	COULEURS SPECTRALES	SPEKTRALFARBEN
11860	SPECTROGRAPH	SPECTROGRAPHE	SPEKTROGRAPH
11861	SPECTROGRAPHE ANALYSIS	ANALYSE SPECTRALE	SPEKTRALANALYSE
11862	SPECTROSCOPE, SPECTROMETER, SPECTROGRAPH	SPECTROSCOPE, SPECTROMETRE	SPEKTROSKOP, SPEKTROMETER, SPEKTROGRAPH, SPEKTRALAPPARAT
11863	SPECTRUM	SPECTRE	SPEKTRUM, WELLENBAND
11864	SPECTRUM ANALYSIS	ANALYSE SPECTRALE	SPEKTRALANALYSE
11865	SPECULAR IRON, HEMATITE, HAEMATITE	OLIGISTE, FER SPECULAIRE, HEMATITE ROUGE	EISENGLANZ, ROTEISENERZ, ROTEISENSTEIN, HÄMATIT
11866	SPECULAR SURFACE	SURFACE MIROITANTE	FLÄCHE (SPIEGELNDE)

11867	**SPECULUM METAL**	BRONZE POUR MIROIRS DE TELESCOPES	SPIEGELMETALL, SPIEGELBRONZE
11868	**SPEED**	VITESSE	GESCHWINDIGKEIT
11869	**SPEED INDICATOR, TACHOMETER, TACHYMETER**	TACHYMETRE	GESCHWINDIGKEITSMESSER, TACHOMETER
11870	**SPEED OF IGNITION**	VITESSE D'INFLAMMATION	ZÜNDGESCHWINDIGKEIT, ENTZÜNDUNGSGESCHWINDIGKEIT
11871	**SPEED PULLEY, BELT SPEED CONE, CONE PULLEY, STEPPED PULLEY, STEP CONE, CONE**	POULIE ETAGEE, CONE A ETAGES, POIRE DE VITESSE, CONE D'ATTAQUE, CONE DE TRANSMISSION	STUFENSCHEIBE
11872	**SPEED REDUCER**	REDUCTEUR DE VITESSE	GESCHWINDIGKEITSREGLER
11873	**SPEED REDUCTION GEAR**	REDUCTEUR DE VITESSE, ENGRENAGE DEMULTIPLICATEUR	REDUKTIONSGETRIEBE
11874	**SPEED REGULATOR**	REGULATEUR DE VITESSE	GESCHWINDIGKEITSREGLER, DREHZAHLREGLER
11875	**SPEED, VELOCITY**	VITESSE	GESCHWINDIGKEIT
11876	**SPEED, VELOCITY OF ROTATION**	VITESSE DE ROTATION	UMDREHUNGSGESCHWINDIGKEIT, UMLAUFGESCHWINDIGKEIT, DREHGESCHWINDIGKEIT
11877	**SPEEDOMETER**	COMPTEUR DE VITESSE	GESCHWINDIGKEITSMESSER
11878	**SPEISS**	MATTE	ROHSTEIN, LECH
11879	**SPELTER**	ZINC (COMMERCIAL)	HANDELSZINK
11880	**SPELTER SOLDER**	SOUDURE JAUNE	MESSINGLOT
11881	**SPERM OIL**	HUILE DE BLANC DE BALEINE	WALRATÖL
11882	**SPERMACETI**	SPERMACETI, BLANC DE BALEINE	WALRAT
11883	**SPHERE**	SPHERE, GLOBE	KUGEL (GEOM.)
11884	**SPHERICAL**	SPHERIQUE	KUGELFÖRMIG, KUGELIG, SPHÄRISCH
11885	**SPHERICAL ABERRATION**	ABERRATION DE SPHERICITE	ABWEICHUNG (SPHÄRISCHE), ABERRATION (SPHÄRISCHE)
11886	**SPHERICAL ANGLE**	FUSEAU SPHERIQUE	KUGELZWEIECK, KUGELWINKEL
11887	**SPHERICAL CALOTTE**	CALOTTE SPHERIQUE	KUGELSCHALE
11888	**SPHERICAL MIRROR**	MIROIR SPHERIQUE	KUGELSPIEGEL, SPIEGEL (SPHÄRISCHER)
11889	**SPHERICAL MOTION**	MOUVEMENT SPHERIQUE	KUGELBEWEGUNG, BEWEGUNG (SPHÄRISCHE)
11890	**SPHERICAL THRUST BEARING**	CRAPAUDINE A BILLES	KUGELSPURLAGER
11891	**SPHERICAL TRIANGLE**	TRIANGLE SPHERIQUE	KUGELDREIECK, DREIECK (SPHÄRISCHES)
11892	**SPHERICAL WEDGE**	ONGLET SPHERIQUE	KUGELKEIL
11893	**SPHERICAL, JOURNAL, PIVOT**	TOURILLON, PIVOT SPHERIQUE	KUGELZAPFEN
11894	**SPHEROIDAL CAST IRON**	FONTE A GRAPHITE SPHEROIDAL, FONTE NODULAIRE, FONTE DUCTILE	GUSSEISEN MIT KUGELGRAPHIT
11895	**SPHEROIDAL GRAPHITE IRON**	FONTE A GRAPHITE SPHEROIDAL	GUSSEISEN MIT KUGELGRAPHIT
11896	**SPHEROIDIZING**	GLOBULISATION, SPHEROIDISATION	KUGELBILDUNG
11897	**SPHEROMETER**	SPHEROMETRE	KRÜMMUNGSMESSER, SPHÄROMETER
11898	**SPIEGEL IRON**	SPIEGEL	SPIEGELEISEN
11899	**SPIEGELEISEN, SPIEGEL**	SPIEGEL, SPIEGELEISEN, FONTE SPECULAIRE	SPIEGELEISEN
11900	**SPIGOT AND SOCKET JOINT**	JOINT A EMBOITEMENT	MUFFENVERBINDUNG
11901	**SPIGOTED END OF PIPE**	BOUT MALE D'UN TUYAU	MANTELENDE EINES ROHRES, SCHWANZENDE EINES ROHRES, SPITZENDE EINES ROHRES, ZOPFENDE EINES ROHRES
11902	**SPIKES**	CROCHETS	HAKENNÄGEL
11903	**SPILLS**	SOUFFLURES	BLASEN

11904	SPINDLE	MANDRIN, AXE, BROCHE	DORN, SPINDEL, ASCHE (KLEINE)
11905	SPINDLE BEARING	CRAPAUDINE, PALIER D'UN TOURILLON	ZAPFENLAGER
11906	SPINDLE BORE	PASSAGE DE BROCHE	SPINDEL-DURCHLASS
11907	SPINDLE NUT, DISC BUSHING	ECROU D'OPERCULE	SPINDELVERSCHRAUBUNG
11908	SPINDLE OIL	HUILE MINERALE POUR BROCHES	SPINDELÖL
11909	SPINDLE SPEEDS	VITESSE DE BROCHE	SPINDELDREHZAHLEN
11910	SPINDLE SUPPORT	SUPPORT DE FUSEE	SPINDELHALTER
11911	SPINNING AND PLANISHING LATHE	TOUR A REPOUSSER ET A LISSER	DRUCK-UND PLANIERMASCHINE
11912	SPIRAL	SPIRALE	SPIRALE
11913	SPIRAL CAM	CAME A PLATEAU COURBE	SCHEIBE (GEWUNDENE), SCHEIBE (KRUMME)
11914	SPIRAL CHUTE	TOBOGGAN	WENDELRUTSCHE, SPIRALRUTSCHE, TOBOGGAN
11915	SPIRAL FLUTED REAMER	ALESOIR A RAINURES HELICOIDALES, A RAINURES TORSES	SPIRALGENUTETE REIBAHLE
11916	SPIRAL GEARING	COUPLE DE ROUES HELICOIDALES	SCHRAUBENRADGETRIEBE
11917	SPIRAL MILLING CUTTER	FRAISE A DENTURE HELICOIDALE	FRÄSER MIT SCHRAUBENFÖRMIGEN SCHNEIDEN, SPIRALFRÄSER
11918	SPIRAL RIVETED	RIVETE EN SPIRALE	SPIRALFÖRMIG GENIETET
11919	SPIRAL STAIRWAY	ESCALIER HELICOIDAL	WENDELTREPPE
11920	SPIRAL WELDED TUBE	TUBE SOUDE EN SPIRALE, EN HELICE	ROHR (SPIRALGESCHWEISSTES)
11921	SPIRAL WELDING	SOUDAGE EN SPIRALE	SCHWEISSEN (SPIRALFÖRMIGES)
11922	SPIRIT ALCOHOL THERMOMETER	THERMOMETRE A ALCOOL	WEINGEISTTHERMOMETER, ALKOHOLTHERMOMETER
11923	SPIRIT LAMP	LAMPE A ALCOOL	SPIRITUSLAMPE, SPIRITUSBRENNER
11924	SPLASH CONE	CONE DE DEFLECTION (DANS RESERVOIR)	UMLENKKONUS, ABLENKKEGEL
11925	SPLASH LUBRICATION	GRAISSAGE PAR BARBOTAGE	TAUCHBADSCHMIERUNG, TAUCHSCHMIERUNG
11926	SPLASH PANEL	DOUBLURE	ABLENKPLATTE
11927	SPLASH-APRON	GARDE-BOUE	KÜHLERSPRITZBLECH
11928	SPLASHED OIL, WASTE OIL	HUILE PROJETEE, HUILE VERSEE	ÖL (VERSPRITZTES)
11929	SPLASHINGS	ECLABOUSSURES DE METAL	METALLTROPFEN
11930	SPLICE	EPISSURE	SPLEISS
11931	SPLICE A ROPE (TO)	EPISSURER	SPLEISSEN (EIN SEIL), SPLISSEN (EIN SEIL)
11932	SPLICE MEMBER	ELEMENTS A COUVRE-JOINTS	STOSSLASCHENTEIL, LASCHENVERBINDUNGEN
11933	SPLICE OF A ROPE	EPISSURE (RESULTAT)	SPLISS, SEILSPLISS
11934	SPLICING A ROPE	EPISSURE (ACTION)	SPLEISSEN, SPLISSEN, SEILVERSPLEISSUNG
11935	SPLINED SHAFT	ARBRE CANNELE	KEILWELLE
11936	SPLINTERY FRACTURE	CASSURE CEROIDE, ECAILLEUSE, ESQUILLEUSE	BRUCH (SPLITTRIGER)
11937	SPLIT	FENTE	SPALT
11938	SPLIT BOSS	MOYEU EN DEUX PIECES	NABE (GETEILTE), NABE (GESCHLITZTE), KLEMMNABE
11939	SPLIT COLLAR	BAGUE D'ARRET EN DEUX PIECES	STELLRING (ZWEITEILIGER), KLEMMRING
11940	SPLIT COTTER PIN	GOUPILLE FENDUE	SPLINT
11941	SPLIT FLYWHEEL	VOLANT FENDU	SCHWUNGRAD (GESPRENGTES)

11942	SPLIT NUT	ECROU FENDU, ECROU A FENTE	MUTTER (GESCHLITZTE), MUTTER (GETEILTE), MUTTER (AUFGESCHNITTENE), SCHLITZMUTTER
11943	SPLIT PIN	GOUPILLE FENDUE	SPLINT
11944	SPLIT PISTON RING	BAGUE DE PISTON FENDUE	KOLBENRING (GESPALTENER), KOLBENRING (GESCHLITZTER)
11945	SPLIT PULLEY	POULIE EN DEUX PIECES, POULIE FENDUE	RIEMENSCHEIBE (GETEILTE)
11946	SPLIT UP (TO)	DEDOUBLER, SCINDER	SPALTEN (CHEM.)
11947	SPLIT-BEARINGS	DEMI-COUSSINETS	SCHALENLAGER (GETEILTES)
11948	SPLITTING OF THE WOOD	ECLATEMENT DU BOIS	REISSEN DES HOLZES
11949	SPLITTING TEST	ESSAI D'ECRASEMENT, ESSAI DE CISAILLEMENT	SCHERVERSUCH, STAUCHVERSUCH
11950	SPLITTING UP	DEDOUBLEMENT	SPALTEN (CHEM.)
11951	SPOKE CENTRE WHEEL, SPIDER WHEEL	ROUE A RAYONS	SPEICHENRAD
11952	SPONGE CLOTH, WIPER	CHIFFON A NETTOYER, CHIFFON D'ESSUYAGE	PUTZTUCH, PUTZLAPPEN
11953	SPONGE IRON	PAILLE DE FER	SCHWAMMEISEN
11954	SPONGY LEAD	PLOMB SPONGIEUX	BLEISCHWAMM
11955	SPONGY PLATINUM, PLATINUM SPONGE	MOUSSE DE PLATINE, EPONGE DE PLATINE	PLATINSCHWAMM
11956	SPONTANEOUS COMBUSTION, IGNITION	COMBUSTION, INFLAMMATION SPONTANEE, AUTO-INFLAMMATION	SELBSTENTZÜNDUNG
11957	SPOON BIT, SHELL AUGER	MECHE, FORET A CUILLER	LÖFFELBOHRER
11958	SPOT TEST	ESSAI A LA GOUTTE	TROPFPROBE
11959	SPOT WELDING	SOUDURE PAR POINTS	PUNKTSCHWEISSUNG, PUNKTSCHWEISSUNG
11960	SPOT-FACING	FINITION PARTIELLE, LAMAGE	TEILFERTIGUNG, STIRNSENKUNG
11961	SPOTTING OUT	FORMATION DES TACHES	AUSSCHLAGBILDUNG
11962	SPRAY GUN	PISTOLET A PROJETER, PISTOLET DE PULVERISATION	SPRITZPISTOLE
11963	SPRAY NOZZLE	TUYERE DE PULVERISATION	STREUDÜSE
11964	SPRAY NOZZLE	TUYERE DE PULVERISATION	SPRÜHDÜSE
11965	SPRAY PAINTING	PEINTURE PAR PULVERISATION AU PISTOLET	SPRITZLACKIERUNG
11966	SPRAY QUENCHING	TREMPE PAR PULVERISATION	SPRÜHHÄRTUNG
11967	SPRAYER	PULVERISATEUR	SPRITZDUSE ZERSTÄUBER, SPRITZER
11968	SPRAYING	PROJECTION, PULVERISATION	SPRITZEN
11969	SPRAYING AND DUSTING MACHINES (HAND AND KNAPSACK)	POUDREUSES PORTATIVES	RÜCKENSTÄUBER UND RÜCKENSPRITZGRERÄTE
11970	SPRAYING AND DUSTING MACHINES FRUIT AND HOPS	POUDREUSES SOUFREUSES	ZERSTÄUBUNGS-UND SPRITZMASCHINEN (FRÜCHTE UND HOPFEN)
11971	SPRAYING AND DUSTING MACHINES GRASS AND CROPS	POUDREUSES PULVERISATEURS	ZERSTÄUBUNGS-UND SPRITZMASCHINEN (GRAS UND GETREIDE)
11972	SPRAYING PROCESS	PROCEDE PAR ASPERSION	SPRITZVERFAHREN
11973	SPREADER	ECARTEUR	ABSTANDHALTER
11974	SPRING	RESSORT, RESSORT ANTAGONISTE	FEDER, GEGENFEDER
11975	SPRING AND WIRE TESTING MACHINE	MACHINE A ESSAYER LES RESSORTS ET LES FILS	FEDER-UND DRAHT-PRÜFMASCHINE
11976	SPRING BALANCE	PESON A RESSORT, BALANCE A RESSORT	FEDERWAAGE
11977	SPRING BLADE	LAME DE RESSORT	FEDERBLATT
11978	SPRING BRASS	LAITON DE RESSORT	FEDERMESSING

11979	SPRING COLLET CHUCK	DOUILLE A RESSORT	ZANGENSPANNFUTTER
11980	SPRING DASHPOT	BOITE A RESSORT	FEDERPUFFER
11981	SPRING FACTOR	FACTEUR DE VIBRATION	SCHWINGUNGSFAKTOR
11982	SPRING GREASE BOX	COMPRESSEUR A GRAISSE	FEDERSCHMIERBÜCHSE
11983	SPRING IN TENSION, LOADED SPRING	RESSORT CHARGE, TENDU, BANDE	FEDER (GESPANNTE)
11984	SPRING LOADED GOVERNOR	REGULATEUR A RESSORT	FEDERKRAFTREGLER
11985	SPRING LOADED LEVER (SAFETY) VALVE, SPRING BALANCE VALVE	SOUPAPE DE SURETE A LEVIER ET A RESSORT	VENTIL MIT HEBELFEDERBELASTUNG
11986	SPRING LOADED SAFETY VALVE	SOUPAPE DE SURETE A RESSORT	SICHERHEITSVENTIL (FEDERBELASTETES)
11987	SPRING PAWL CATCH	CLIQUET A RESSORT	FEDERKLINKE
11988	SPRING PLATE	COUPELLE DE RESSORT	FEDERPLATTE, FEDERTELLER
11989	SPRING PRESSURE GAUGE	MANOMETRE METALLIQUE, MANOMETRE A RESSORT	FEDERMANOMETER
11990	SPRING RING	CERCLE ELASTIQUE, CIRCLIP	SPANNRING, FEDERRING, SELBSTSPANNENDER KOLBENRING, SPREIZRING, SELBSTSPANNER
11991	SPRING SHACKLE	JUMELLE DE RESSORT	FEDERLASCHE
11992	SPRING STEEL, STEEL FOR SPRINGS	ACIER A RESSORT	FEDERSTAHL
11993	SPRING SUBJECTED TO BENDING	RESSORT DE FLEXION	BIEGEFEDER, BIEGUNGSFEDER
11994	SPRING SUBJECTED TO TORSION	RESSORT DE TORSION	DREHUNGSFEDER, TORSIONSFEDER
11995	SPRING WASHER	RONDELLE GROWER	FEDERRING
11996	SPRING WATER	EAU DE SOURCE	QUELLWASSER
11997	SPRING WIRE	FIL POUR RESSORT	FEDERDRAHT
11998	SPRING-LOADED SAFETY VALVE	SOUPAPE DE SURETE A RESSORT	SICHERHEITSVENTIL MIT FEDERBELASTUNG, FEDERSICHERHEITSVENTIL
11999	SPRINGING OFF OF RIVET HEADS	RUPTURE DES TETES DE RIVET	ABSPRINGEN DER NIETKÖPFE, ABREISSEN DER NIETKÖPFE
12000	SPRINKLE (TO)	ARROSER	EINSPRENGEN, BESPRENGEN, BENETZEN
12001	SPRINKLING	ARROSAGE	EINSPRENGEN, BESPRENGEN, BENETZEN
12002	SPROCKET CHAIN, FLAT LINK CHAIN, PLATE LINK CHAIN, GALLE'S CHAIN	CHAINE GALLE	GELENKKETTE, LASCHENKETTE, GALLESCHE KETTE
12003	SPRUE	ENTONNOIR, DESCENTE DE COULEE	EINGUSSTRICHTER
12004	SPRUE, GATE, GEAT, GIT	JET DE COULEE	ANGUSS, GUSSZAPFEN
12005	SPUN YARN	BITORD	SCHIEMANNSGARN
12006	SPUR AND HELICAL GEAR CUTTING MACHINES	MACHINES A TAILLER LES ENGRENAGES DROITS ET HELICOIDAUX	VERZAHNMASCHINEN FÜR STIRN- U. SCHRÄGVERZAHNUNGEN
12007	SPUR FRICTION GEAR WHEEL	ROUE DE FRICTION CYLINDRIQUE	REIBRAD (ZYLINDRISCHES)
12008	SPUR GEARING	ENGRENAGE DROIT, ENGRENAGE CYLINDRIQUE	STIRNRÄDERGETRIEBE
12009	SPUR WHEEL	ROUE D'ENGRENAGE DROITE, ROUE D'ENGRENAGE CYLINDRIQUE	STIRNRAD
12010	SPUR WHEEL DRIVE	COMMANDE PAR ENGRENAGE DROIT	STIRNRADANTRIEB
12011	SPUR WHEEL PULLEY BLOCK	PALAN A ENGRENAGE DROIT	STIRNRADFLASCHENZUG
12012	SQUAB	DOSSIER	SITZRÜCKENLEHNE
12013	SQUARE	CARRE (DEUXIEME PUISSANCE), CARRE (GEOM.)	ZWEITE POTENZ, QUADRAT (GEOM.)
12014	SQUARE	EQUERRE	WINKELMASS, WINKELHAKEN, WINKEL
12015	SQUARE BAR IRON	FER CARRE	QUADRATEISEN, VIERKANTEISEN

12016	SQUARE CENTIMETRE	CENTIMETRE CARRE	QUADRATZENTIMETER
12017	SQUARE CROSS SECTION	SECTION CARREE	QUERSCHNITT (QUADRATISCHER)
12018	SQUARE DECIMETRE	DECIMETRE CARRE	QUADRATDEZIMETER
12019	SQUARE FILE	LIME CARREE, CARREE, CARREAU, CARRELET, LIME QUATRE-QUARTS, LIME 4/4	VIERKANTFEILE
12020	SQUARE FOOT	PIED CARRE ANGLAIS	QUADRATFUSS (ENGLISCHER)
12021	SQUARE INCH	POUCE CARRE ANGLAIS	QUADRATZOLL (ENGLISCHER)
12022	SQUARE METRE	METRE CARRE	QUADRATMETER
12023	SQUARE MILLIMETRE	MILLIMETRE CARRE	QUADRATMILLIMETER
12024	SQUARE NUT	ECROU CARRE	VIERKANTMUTTER
12025	SQUARE PIECE	CARRE	VIERKANT
12026	SQUARE PYRAMID	PYRAMIDE QUADRANGULAIRE, QUADRILATERE	PYRAMIDE (VIERSEITIGE)
12027	SQUARE ROOT	RACINE CARREE	QUADRATWURZEL, WURZEL (ZWEITE)
12028	SQUARE ROPE	CABLE A SECTION CARREE	VIERKANTSEIL, QUADRATSEIL
12029	SQUARE SHAFT	ARBRE CARRE	VIERKANTWELLE
12030	SQUARE THREAD	FILET CARRE	GEWINDE (FLACHES), GEWINDE (FLACHGÄNGIGES), FLACHGEWINDE, RECHTECKGEWINDE
12031	SQUARE YARD	YARD CARRE	QUADRATYARD (ENGLISCHES)
12032	SQUARE-GROOVE WELD	SOUDURE SUR BORDS DROITS	I-STUMPFNAHT
12033	SQUARE-HEADED SCREW	VIS A TETE CARREE, BOULON A TETE CARREE	VIERKANTSCHRAUBE
12034	SQUARE-THREADED SCREW	VIS A FILET CARRE	SCHRAUBE MIT FLACHGEWINDE, FLACHGÄNGIGE SCHRAUBE
12035	SQUARED TIMBER	BOIS EQUARRI	KANTHOLZ
12036	SQUARING	QUADRATURE	QUADRATUR
12037	SQUARING SHEAR	CISAILLE A EBOUTER	KOPFSCHERE
12038	SQUEEZE INTERVAL	RETARD DE SOUDAGE	SCHWEISSVERZÖGERUNGSZEIT
12039	SQUEEZE OUT (TO), PRESS OUT (TO)	EXPRIMER, EXTRAIRE	AUSPRESSEN
12040	SQUIRREL CAGE OF A ROLLER BEARING	CAGE D'UN ROULEMENT A GALETS	KÄFIG EINES ROLLENLAGERS, ROLLENKÄFIG
12041	STABILITY	STABILITE	STANDFESTIGKEIT, STANDSICHERHEIT, STABILITÄT
12042	STABILIZER SHAFT	STABILISATEUR	STABILISATOR
12043	STABILIZING	STABILISATION	STABILISIERUNG
12044	STABILIZING ANNEAL	RECUIT DE DETENTE	GLÜHEN (SPANNUNGSFREIES)
12045	STABLE EQUILIBRIUM	EQUILIBRE STABLE	GLEICHGEWICHT (STABILES)
12046	STACK	CUVE	SCHACHT
12047	STAGE OF A MISCROSCOPE	PORTE-OBJET	OBJEKTTISCH
12048	STAGGERED	QUINCONE (EN), ALTERNE	GEGENEINANDER VERSETZT
12049	STAGGERED INTERMITTENT FILLET WELD	SOUDURE D'ANGLE DISCONTINUE A RANGEES ALTERNEES	KEHLNAHT (UNTERBROCHENE VERSETZTE)
12050	STAGNANT WATER	EAU STAGNANTE, EAU DORMANTE	WASSER, (TOTES) WASSER (STEGEBDES), WASSER (STAGNIERENDES), TOTWASSER
12051	STAINLESS STEEL	ACIER INOXYDABLE	EISEN (NICHT-ROSTENDES), STAHL (ROSTFREIER), NIROSTAHL
12052	STAIR-HORSE	LIMON	TREPPENWANGE
12053	STAIRWAY	ESCALIER	TREPPE
12054	STALAGMOMETRY	STALAGMOMETRIE	STALAGMOMETRIE
12055	STAMP	PILON DE BOCARD	STEMPEL, POCHSTEMPEL

12056	**STAMP (TO)**	ESTAMPER, ETAMPER, MATRICER	STANZEN, PRÄGEN
12057	**STAMP MILL**	BOCARD, MOULIN A BOCARDS	POCHWERK, STAMPFWERK
12058	**STAMPED LINK CHAIN**	CHAINE DECOUPEE	BANDKETTE
12059	**STAMPING**	ESTAMPAGE, ETAMPAGE, MATRICAGE	GESENKSCHMIEDEN, STANZEN, PRÄGEN
12060	**STAMPING MACHINE, PRESS**	ESTAMPEUSE, ETAMPEUSE	STANZE, PRÄGWERK, PRÄGEPRESSE
12061	**STAMPING TEST**	ESSAI D'ESTAMPAGE	STANZPROBE, PRÄGEPROBE
12062	**STANCHION, POST, UPRIGHT**	MONTANT, POTEAU	STÄNDER, PFOSTEN
12063	**STANCHION, UPRIGHT, POST**	MONTANT	TREPPENPFOSTEN
12064	**STAND FOR PEDESTAL BEARING**	CHAISE SUR LE SOL	STEHBOCK, LAGERSTUHL, LAGERBOCK
12065	**STAND OF ROLLS**	CAGE DE LAMINOIR	WALZGERÜST
12066	**STAND PIPE**	TUYAU DE PRISE D'EAU	STANDROHR, HYDRANTENSTANDROHR
12067	**STAND PIPE, PIEZOMETER**	PIEZOMETRE	STANDROHR, PIEZOMETER
12068	**STAND-BY ....**	DE RESERVE	RESERVE-, HILFS-
12069	**STAND-BY COMPRESSOR**	COMPRESSEUR DE RESERVE	HILFSKOMPRESSOR, HILFSVERDICHTER
12070	**STAND-BY ENGINE**	MACHINE DE RESERVE, MOTEUR DE RESERVE	AUSHILFSMASCHINE, RESERVEMASCHINE
12071	**STANDARD**	NORME, DIMENSION NORMALE, STANDARD	NORM, NORMALE
12072	**STANDARD CUBIC FOOT**	PIED CUBE NORMALISE	KUBIKFUSS (GENORMTER)
12073	**STANDARD DESIGN**	MODELE STANDARD, CONCEPTION STANDARD	STANDARDMODELL
12074	**STANDARD DEVIATION**	ECART TYPE	STANDARDABWEICHUNG
12075	**STANDARD FILLET WELD**	SOUDURE D'ANGLE A CORDON PLAT	FLACHKEHLNAHT
12076	**STANDARD GAUGE**	ETALON, CALIBRE NORMAL	EICHMASS, NORMALLEHRE
12077	**STANDARD MEASURE**	MESURE ETALON, MESURE DE CONTROLE, ETALON	NORMALMASS
12078	**STANDARD OF AN ENGINE, A MACHINE**	MONTANT D'UNE MACHINE	MASCHINENSTÄNDER
12079	**STANDARD OF REFERENCE**	GRANDEUR DE COMPARAISON	BEZUGSGRÖSSE
12080	**STANDARD PRESSURE GAUGE**	MANOMETRE ETALON	KONTROLLMANOMETER
12081	**STANDARD ROD**	METRE ETALON	NORMALMASSSTAB
12082	**STANDARD SAMPLE**	ETALON NORMALISE	NORMPROBE
12083	**STANDARD SECTION**	PROFIL NORMAL	NORMALPROFIL
12084	**STANDARD SOLUTION**	LIQUEUR TITREE, LIQUEUR NORMALE	NORMALLÖSUNG, MASSFLÜSSIGKEIT, TITRIERTE LÖSUNG, TITERFLÜSSIGKEIT
12085	**STANDARD THERMOMETER**	THERMOMETRE ETALON	NORMALTHERMOMETER
12086	**STANDARD TYPE**	TYPE STANDARD	NORMALMODELL
12087	**STANDARD TYPE PLUG GAUGE**	JAUGE TAMPON NORMALE	NORMAL LEHRDORN
12088	**STANDARD TYPE SOLID SNAP GAUGE**	JAUGE MACHOIRE NORMALE	NORMALRACHENLEHRE
12089	**STANDARD WIRE GAUGE (S. W. G.), BIRMINGHAM WIRE GAUGE (B. W. G.)**	JAUGE ANGLAISE, JAUGE DE BIRMINGHAM POUR FILS METALLIQUES	DRAHTLEHRE (ENGLISCHE) (B. W. G.)
12090	**STANDARDISE (TO)**	UNIFIER, NORMALISER, STANDARDISER	NORMEN, NORMALISIEREN
12091	**STANDARDIZATION**	UNIFICATION, NORMALISATION, STANDARDISATION	NORMUNG, NORMALISIERUNG
12092	**STANDING PART (IN LIFTING TACKLE)**	BRIN DORMANT, DORMANT (DU CABLE D'UN PALAN)	TRUMM (FESTES), TRUMM (STEHENDES), TRUMM (EINES FLASCHENZUGS)
12093	**STANDPIPE**	COLONNE MONTANTE	STANDROHR
12094	**STANDSTILL OF AN ENGINE**	ARRET D'UNE MACHINE (ETAT)	STILLSTAND EINER MASCHINE

12095	**STANNATE**	STANNATE	ZINNSÄURESALZ, STANNAT
12096	**STANNIC CHLORIDE**	CHLORURE STANNIQUE, TETRACHLORURE, BICHLORURE D'ETAIN	ZWEIFACH-CHLORZINN, ZINNTETRACHLORID, STANNICHLORID
12097	**STANNIC HYDROXIDE**	ACIDE STANNIQUE	ZINNSÄURE, ZINNOXYDHYDRAT, ZINNHYDROXYD
12098	**STANNIC OXIDE**	OXYDE STANNIQUE, BIOXYDE D'ETAIN, POTEE D'ETAIN	ZINNASCHE, ZINNDIOXYD, ZINNSÄUREANHYDRID
12099	**STANNIC SULPHIDE**	SULFURE STANNIQUE, BISULFURE D'ETAIN, OR MUSIF	ZINNDTSULFID, MUSIVGOLD
12100	**STANNOUS CHLORIDE**	CHLORURE STANNEUX, PROTOCHLORURE D'ETAIN, SEL D'ETAIN	ZINNSALZ, ZINNCHLORÜR, STANNOCHLORID, EINFACH-CHLORZINN, ZINNDICHLORID
12101	**STANNOUS HYDROXIDE**	HYDRATE STANNEUX	METAZINNSÄURE
12102	**STANNOUS OXIDE**	OXYDE STANNEUX	ZINNOXYDUL, STANNOOXYD
12103	**STANNOUS SULPHIDE**	SULFURE STANNEUX	ZINNMONOSULFID, ZINNSULFÜR
12104	**STAPLE / FINGER BAR**	CAVALIER DE FOND	DRAHTKLAMMER
12105	**STAPLE, DOG**	CRAMPON, CLAMEAU	KLAMMER, KLAMMERHAKEN, KRAMPE
12106	**STAPLE, WIRE STAPLE**	CAVALIER	DRAHTÖSE, KRAMPE
12107	**STAR POLYGON**	POLYGONE ETOILE	STERNPOLYGON
12108	**STAR SHAKE IN TIMBER**	CADRANURE, CADRAN DU BOIS	STRAHLENRISS, SPIEGELKLUFT DES HOLZES
12109	**STAR WHEEL**	ROUE ETOILEE	STERNRAD
12110	**STAR-SHAPED CROSS SECTION**	SECTION ETOILEE	QUERSCHNITT (STERNFÖRMIGER)
12111	**STARCH**	AMIDON, FECULE	STÄRKE
12112	**STARCH PASTE**	PATE, COLLE D'AMIDON, EMPOIS	KLEISTER, STÄRKEKLEISTER
12113	**START (TO)**	DEMARRER	ANLAUFEN (MASCHINE)
12114	**START AN ENGINE (TO), SET A MACHINE IN MOTION (TO)**	METTRE EN MARCHE, METTRE EN TRAIN, METTRE EN ROUTE UNE MACHINE, LANCER UN MOTEUR	ANLASSEN (EINE MASCHINE), ANLAUFEN LASSEN (EINE MASCHINE), IN GANG SETZEN (EINE MASCHINE)
12115	**START/STOP BUTTON**	BOUTON MARCHE/ARRET	START/STOP DRUCKKNOPF
12116	**STARTER**	DEMARREUR	ANLASSER
12117	**STARTERSWITCH**	COMMANDE DE DEMARREUR	ANLASSSCHALTER
12118	**STARTING**	MISE EN MARCHE, MISE EN SERVICE, DEMARRAGE	INBETRIEBSETZUNG, ANLAUF
12119	**STARTING GEAR**	APPAREIL DE MISE EN MARCHE	ANLASSVORRICHTUNG
12120	**STARTING LEVER**	LEVIER DE DEMARRAGE	ANLASSHEBEL
12121	**STARTING MOTOR**	MOTEUR DE DEMARRAGE	ANLASSMOTOR
12122	**STARTING OF AN ENGINE, OF A MACHINE**	DEMARRAGE D'UN MOTEUR	ANLAUFEN, ANLAUF EINER MASCHINE
12123	**STARTING PERIOD**	PERIODE DE DEMARRAGE OU DE MISE EN MARCHE D'UNE USINE	ANLAUFPERIODE
12124	**STARTING POINT OF A MOTION**	POINT DE DEPART D'UN MOUVEMENT	AUSGANGSPUNKT EINER BEWEGUNG
12125	**STARTING POINTS**	HYPOTHESES DE BASE	ANHALTSPUNKTE
12126	**STARTING POSITION**	POSITION DE DEPART	ANFANGSLAGE, AUSGANGSLAGE, ANFANGSSTELLUNG
12127	**STARTING PUNCH**	CHASSE-CLOU	NAGELTREIBER
12128	**STARTING RESISTANCE**	RESISTANCE AU DEMARRAGE	ANLASSWIDERSTAND, ANLAUFWIDERSTAND
12129	**STARTING TORQUE**	COUPLE DE DEMARRAGE	ANZUGMOMENT
12130	**STARTING UP AN ENGINE, A MACHINE**	MISE EN MARCHE, MISE EN TRAIN, MISE EN ROUTE D'UNE MACHINE, LANCEMENT D'UN MOTEUR	ANLASSEN, INGANGSETZEN EINER MASCHINE

12131	**STARTING VALVE**	VALVE DE DEMARRAGE	ANLASSVENTIL
12132	**STATE OF EQUILIBRIUM**	ETAT D'EQUILIBRE	GLEICHGEWICHTSZUSTAND
12133	**STATIC**	STATIQUE	STATISCH
12134	**STATIC FRICTION**	FROTTEMENT STATIQUE	HAFTREIBUNG
12135	**STATIC GOVERNOR**	REGULATEUR STATIQUE	REGLER (STATISCHER)
12136	**STATIC LIQUID HEAD**	PRESSION STATIQUE	DRUCK (STATISCHER)
12137	**STATIC PRESSURE**	PRESSION STATIQUE	DRUCK (RUHENDER), DRUCK (STATISCHER)
12138	**STATICAL LOAD**	CHARGE STATIQUE	BELASTUNG (STATISCHE)
12139	**STATICAL MOMENT**	MOMENT STATIQUE	MOMENT (STATISCHES)
12140	**STATICS**	STATIQUE	GLEICHGEWICHTSLIEHRE, STATIK
12141	**STATION WAGON**	COMMERCIALE, BREAK	KOMBINATIONS-KRAFTWAGEN, KOMBIWAGEN
12142	**STATIONARY HEART FURNACE**	FOUR STATIONNAIRE FIXE	OFEN (FESTSTEGENDER)
12143	**STATIONARY LAND BOILER**	CHAUDIERE INDUSTRIELLE	LANDDAMPFKESSEL
12144	**STATIONARY STEAM ENGINE, LAND ENGINE**	MACHINE A VAPEUR FIXE	DAMPFMASCHINE (ORTSFESTE), DAMPFMASCHINE (FESTSTEHENDE), DAMPFMASCHINE (STATIONÄRE)
12145	**STATIONARY WAVE**	ONDE STATIONNAIRE	WELLE (STEHENDE) (PHYS.)
12146	**STATOR**	STATOR	STÄNDER, STATOR (ELEKTR.)
12147	**STATUARY BRONZE, ART BRONZE**	BRONZE D'ART, BRONZE STATUAIRE	KUNSTBRONZE, STATUENBRONZE
12148	**STAUFFER LUBRICATOR**	GRAISSEUR STAUFFER, STAUFFER	STAUFFERBÜCHSE
12149	**STAVE**	DOUVE	DAUBE
12150	**STAY (TO), PROP (TO)**	ETAYER, SOUTENIR, ENTRETOISER, ETANCONNER	AUSSTEIFEN, ABSTEIFEN
12151	**STAY BOLT**	ENTRETOISE	STEHBOLZEN, DISTANZBOLZEN
12152	**STAYING WITH RIBS**	RENFORCEMENT PAR DES NERVURES	RIPPENVERSTEIFUNG
12153	**STAYING, PROPPING**	ETAIEMENT, ETAYAGE, ETAYEMENT, SOUTENEMENT, ENTRETOISEMENT	ABSTEIFEN, ABSTEIFUNG
12154	**STEADY REST**	LUNETTE FIXE	BRILLE (FESTSTEHENDE)
12155	**STEAM**	VAPEUR	DAMPF
12156	**STEAM ACCUMULATOR**	ACCUMULATEUR DE VAPEUR	DAMPFSPEICHER
12157	**STEAM BATH**	BAIN DE VAPEUR	DAMPFBAD
12158	**STEAM BOILER, ENGINE BOILER**	CHAUDIERE A VAPEUR, GENERATEUR DE VAPEUR	DAMPFKESSEL
12159	**STEAM BRAKE**	FREIN A VAPEUR	DAMPFBREMSE
12160	**STEAM CALORIMETER**	CALORIMETRE A CONDENSATION	DAMPFKALORIMETER
12161	**STEAM COIL, HEATING COIL**	SERPENTIN CHAUFFE PAR LA VAPEUR	HEIZSCHLANGE, DAMPFSCHLANGE
12162	**STEAM CYLINDER**	CYLINDRE A VAPEUR	ZYLINDER EINER DAMPFMASCHINE, DAMPFMASCHINEN-ZYLINDER
12163	**STEAM DOME**	DOME DE PRISE DE VAPEUR	DAMPFDOM, DAMPFSAMMLER
12164	**STEAM ENGINE**	MACHINE A VAPEUR, MOTEUR A VAPEUR	DAMPFKRAFTMASCHINE
12165	**STEAM ENGINE WORKED NON-EXPANSIVELY WITHOUT CUT OFF**	MACHINE A PLEINE PRESSION	VOLLDRUCKDAMPFMASCHINE
12166	**STEAM HAMMER**	MARTEAU-PILON, MARTEAU A VAPEUR	DAMPFHAMMER
12167	**STEAM HEATING**	CHAUFFAGE PAR LA VAPEUR	DAMPFHEIZUNG
12168	**STEAM JACKET**	CHEMISE, ENVELOPPE DE VAPEUR	DAMPFMANTEL
12169	**STEAM JENNY**	GENERATEUR A VAPEUR	DAMPFERZEUGER
12170	**STEAM JET**	JET DE VAPEUR	DAMPFSTRAHL

12171	STEAM JET AIR PUMP	INJECTEUR DE VAPEUR	DAMPFSTRAHLLUFTPUMPE
12172	STEAM JET BLOWER	EJECTEUR, SOUFFLERIE DE VAPEUR	DAMPFSTRAHLGEBLÄSE
12173	STEAM LOCOMOTIVE	LOCOMOTIVE A VAPEUR	DAMPFLOKOMOTIVE
12174	STEAM METER	COMPTEUR DE VAPEUR	DAMPFVERBRAUCHMESSER
12175	STEAM NOZZLE	TUYERE A VAPEUR	DAMPFDÜSE
12176	STEAM PASSAGE, PORT, STEAMWAY	LUMIERE (D'UN CYLINDRE A VAPEUR)	DAMPFWEG, DAMPFKANAL
12177	STEAM PIPE	TUYAU A VAPEUR, TUBE DE VAPEUR	DAMPFLEITUNGSROHR
12178	STEAM PIPING, PIPE LINE	CONDUITE, TUYAUTERIE, CANALISATION DE VAPEUR	DAMPFLEITUNG
12179	STEAM PISTON	PISTON A VAPEUR	DAMPFKOLBEN
12180	STEAM POWER PLANT	INSTALLATION DE MACHINE A VAPEUR	DAMPFKRAFTANLAGE, DAMPFMASCHINENANLAGE
12181	STEAM PRESSURE	PRESSION DE LA VAPEUR	DAMPFSPANNUNG, DAMPFDRUCK
12182	STEAM PRESSURE TEST	ESSAI DE LA CHAUDIERE, EPREUVE A CHAUD DE LA CHAUDIERE	DAMPFDRUCKPROBE
12183	STEAM PUMP	POMPE A VAPEUR	DAMPFPUMPE
12184	STEAM SEPARATOR, STEAM DRYER	SEPARATEUR DE VAPEUR	DAMPFABSCHEIDER, DAMPFTROCKENER
12185	STEAM SPACE OF A BOILER	CHAMBRE DE VAPEUR D'UNE CHAUDIERE	DAMPFRAUM EINES KESSELS
12186	STEAM STOP VALVE	SOUPAPE D'ARRET DE VAPEUR	DAMPFABSPERRVENTIL
12187	STEAM SUPPLY VALVE	SOUPAPE DE PRISE DE VAPEUR, ROBINET DE PRISE DE VAPEUR	DAMPFENTNAHMEVENTIL
12188	STEAM TRAP	SEPARATEUR D'EAU, PURGEUR D'EAU CONDENSEE	WASSERABSCHEIDER, KONDENSWASSERABSCHEIDER, KONDENSTOPF, ENTWÄSSERUNGSTOPF, WASSERSAMMLER, KONDENSATIONSWASSERABLEITER
12189	STEAM TURBINE	TURBINE A VAPEUR	DAMPFTURBINE
12190	STEAM WHISTLE	SIFFLET A VAPEUR	DAMPFPFEIFE
12191	STEAM-ELECTRIC GENERATOR	GENERATRICE A VAPEUR	DAMPFDYNAMO
12192	STEAM, AQUEOUS VAPOUR, WATER VAPOUR	VAPEUR D'EAU	WASSERDAMPF
12193	STEAMTIGHT	ETANCHE (A LA VAPEUR)	DAMPFDICHT
12194	STEARIC ACID	ACIDE STEARIQUE	STEARINSÄURE
12195	STEARINE, STEARIN	STEARINE	STEARIN
12196	STEEL	ACIER	STAHL
12197	STEEL BAND TAPE	BANDE D'ACIER	STAHLBAND
12198	STEEL BARS - STEEL RODS FOR CONCRETE REINFORCEMENT	FERS A BETON	BETONEISEN
12199	STEEL BEAM, IRON GIRDER	POUTRE EN FER	TRÄGER (EISERNER), EISENTRÄGER
12200	STEEL BELT	COURROIE EN ACIER	STAHLBANDRIEMEN
12201	STEEL CABLE	CABLE EN ACIER, CABLE EN GRELIN	STAHLKABEL
12202	STEEL CASTING	ACIER MOULE, ACIER COULE	STAHLFORMGUSS
12203	STEEL CONCRETE, REINFORCED, ARMOURED CONCRETE, FERROCONCRETE	BETON, CIMENT ARME	EISENBETON, BETON (BEWEHRTER), BETON (ARMIERTER)
12204	STEEL CYLINDER	TUBE D'ACIER (POUR GAZ LIQUEFIES)	STAHLFLASCHE
12205	STEEL FACE (TO), STEEL (TO)	ACIERER	STÄHLEN, VERSTÄHLEN, ANSTÄHLEN
12206	STEEL FACING, STEELING	ACIERAGE	STÄHLEN, VERSTÄHLEN, ANSTÄHLEN
12207	STEEL FOUNDRY	FONDERIE D'ACIER	STAHLGIESSEREI

12208	**STEEL FRAME**	CHARPENTE METALLIQUE	EISENKONSTRUKTION, STAHLKONSTRUKTION, STAHLGERIPPE
12209	**STEEL FREE FROM BLISTERS**	ACIER SANS SOUFFLURES	STAHL (BLASENFREIER)
12210	**STEEL INGOT**	LINGOT D'ACIER	STAHLBARREN, STAHLBLOCK, ROHSTAHLBLOCK
12211	**STEEL MILL**	ACIERIE	STAHLWERKE
12212	**STEEL PIPE**	TUBE D'ACIER	STAHLROHR
12213	**STEEL PLATE**	TOLE D'ACIER	STAHLBLECH
12214	**STEEL PROCESSING**	ELABORATION DE L'ACIER	STAHLERZEUGUNG
12215	**STEEL ROPE**	CABLE EN ACIER, CABLE EN FILIN	STAHLDRAHTSEIL
12216	**STEEL SHEET**	FEUILLARD (TOLE MINCE)	BANDSTAHL
12217	**STEEL STRANDS / OR WIRE ROPES**	TORONS EN ACIER POUR CABLES	STAHLLITZEN
12218	**STEEL STRUCTURE**	CONSTRUCTION EN ACIER / METALLIQUE	STAHLBAU, STAHLGERÜST, STAHLKONSTRUKTION
12219	**STEEL TUBE**	TUBE EN ACIER	STAHLROHR
12220	**STEEL TYRES FOR LOCOMOTIVES**	BANDAGES DE ROUE	EISENBAHNRADREIFEN
12221	**STEEL WIRE**	FIL D'ACIER	STAHLDRAHT
12222	**STEEL WORKS**	ACIERIE	STAHLWERK
12223	**STEEL, CARBON STEEL**	ACIER	STAHL
12224	**STEELY IRON**	FER ACIEREUX, ACIER EXTRA-DOUX	EISEN (STAHLARTIGES)
12225	**STEELYARD, ROMAN BALANCE**	BASCULE ROMAINE, BASCULE ROMAINE (A CURSEUR)	LAUFGEWICHTSWAAGE, SCHNELLWAAGE, STELLIT (RÖMISCHE)
12226	**STEERING**	CONDUITE, MANOEUVRE	STEUERUNG
12227	**STEERING ARM**	LEVIER D'ATTAQUE DE DIRECTION	LENKSTOCKHEBEL
12228	**STEERING COLUMN**	COLONNE DE DIRECTION	LENKSAÜLE
12229	**STEERING GEAR**	BOITIER DE DIRECTION	LENKGETRIEBE
12230	**STEERING KNUCKLE**	SUPPORT DE FUSEE	ACHSSCHENKEL
12231	**STEERING KNUCKLE TIE ROD**	BARRE D'ACCOUPLEMENT	SPURSTANGE
12232	**STEERING LINKAGE**	TIMONERIE, TRINGLERIE DE DIRECTION	LENKGESTÄNGE
12233	**STEERING WHEEL**	VOLANT DE DIRECTION	STEUERRAD, LENKRAD
12234	**STELLITE**	STELLITE	STELLIT
12235	**STEM THERMOMETER**	THERMOMETRE GRADUE SUR TIGE	STABTHERMOMETER, STOCKTHERMOMETER
12236	**STEM, ROD, SPINDLE OF SLIDE VALVE**	QUEUE DE ROBINET, TIGE DE SOUPAPE	SPINDEL, STANGE EINES SCHIEBERS, STANGE EINES VENTILS
12237	**STEM, SPINDLE**	TIGE	SPINDEL, SCHAFT
12238	**STENCIL**	POCHOIR	SCHABLONE
12239	**STEP**	MARCHE, DEGRE, STADE, PAS, GRADIN	STUFE, SCHRITT, ABSATZ
12240	**STEP ANNEALING**	RECUIT ECHELONNE	STUFENGLÜHEN
12241	**STEP BEARING**	CRAPAUDINE	SPURLAGER
12242	**STEP HARDENING**	TREMPE ETAGEE	STUFENHÄRTUNG
12243	**STEP QUENCHING**	TREMPE ETAGEE MARTENSITIQUE	MARTENSITHÄRTUNG (GESTAFFELTE)
12244	**STEP-DOWN TRANSFORMER**	TRANSFORMATEUR DEVOLTEUR, TRANSFORMATEUR REDUCTEUR DE POTENTIEL	ABSPANNTRANSFORMATOR, ABWÄRTSTRANSFORMATOR
12245	**STEP-UP TRANSFORMER**	TRANSFORMATEUR SURVOLTEUR, TRANSFORMATEUR AMPLIFICATEUR DE POTENTIEL	AUFWÄRTSTRANSFORMATOR
12246	**STEPPED**	A GRADINS, EN GRADINS, ECHELONNE	STUFENFÖRMIG, ABGESTUFT, TREPPENFÖRMIG, STUFEN...
12247	**STEPPED GEARING**	ENGRENAGE A DENTURE CROISEE, ENGRENAGE A DENTS RECROISEES	STUFENRAD

12248	STEPPED GRATE	GRILLE A GRADINS, GRILLE A ETAGES	TREPPENROST, STUFENROST
12249	STEPPED QUENCHING	TREMPE INTERROMPUE	HÄRTEN (GEBROCHENES)
12250	STEREOMETRICAL	STEREOMETRIQUE	STEREOMETRISCH
12251	STEREOMETRY	STEREOMETRIE	STEREOMETRIE
12252	STERILISING THE WATER	STERILISATION DE L'EAU	ENTKEIMUNG DES WASSERS
12253	STERLING SILVER	ARGENT A 92/5%	SILBER (92/5% FEIN)
12254	STICK LAC	GOMME-LAQUE EN BRANCHES	STOCKLACK, STANGENLACK, GUMMILACK, LACK IN STANGENFORM
12255	STICK SULPHUR, ROLL SULPHUR	SOUFRE EN CANON	SCHWEFELSTANGEN, STANGENSCHWEFEL
12256	STIFFENER	ELEMENT RAIDISSEUR	VERSTEIFUNGSELEMENT
12257	STIFFENING PLATE	PLAQUE DE RENFORT, TOLE DE RENFORT	VERSTEIFUNGSPLATTE
12258	STIFFNESS OF A ROPE, OF A BELT	RAIDEUR D'UNE CORDE, RAIDEUR D'UNE COURROIE	STEIFIGKEIT EINES SEILES, STEIFHEIT EINES RIEMENS
12259	STILL, DISTILLING APPARATUS	APPAREIL DISTILLATOIRE, ALAMBIC	DESTILLIERVORRICHTUNG
12260	STIRRING ROD	AGITATEUR, BAGUETTE EN VERRE	RÜHRSTAB
12261	STIRRUP	ETRIER, DEMI-CERCLE	BÜGEL
12262	STOCK OF A PLANE	FUT DE RABOT	HOBELKASTEN
12263	STOCK OF MATERIAL	STOCK, EXISTENCE	LAGERVORRAT, LAGERBESTAND, LAGER
12264	STOCK SHEARS, BLOCK SHEARS	CISAILLES D'ETABLI	STOCKSCHERE
12265	STOCK YARD	CHANTIER, DEPOT, PARC NON COUVERT	LAGERPLATZ
12266	STOCKS AND DIES	FILIERE A COUSSINETS	KLUPPE, SCHNEIDKLUPPE, GEWINDESCHNEID-KLUPPE
12267	STOICHIOMETRIC	STOECHIOMETRIQUE	STÖCHIOMETRISCH
12268	STOICHIOMETRY	STOECHIOMETRIE	STÖCHIOMETRIE
12269	STOKE (TO), FIRE UP (TO)	CHARGER UN FOYER, ALIMENTER UN FOYER	BESCHICKEN (EINE FEUERUNG)
12270	STOKE-HALL	CHAUFFERIE	FEUERRAUM, HEIZRAUM
12271	STOKER, FIRE STOKER	CHAUFFEUR	HEIZER
12272	STOKER, FIREMAN	CHAUFFEUR	HEIZER, KESSELWÄRTER
12273	STONE BREAKER	CONCASSEUR A MACHOIRES	STEINBRECHER, STEINBRECHMASCHINE
12274	STONE CEMENT	MASTIC A PIERRE	STEINKITT
12275	STONE SHIELD	PARE-PIERRE	STEINSCHLAGGITTER
12276	STONEWARE	GRES CERAME	STEINZEUG
12277	STONEWARE PIPE	TUBE EN GRES	STEINZEUGROHR
12278	STOP	BUTEE, BUTOIR, DISPOSITIF D'ARRET, ARRET, TAQUET, TOC LIMITANT LA COURSE	ANSCHLAG, ANSCHLAGVORRICHTUNG, HUBBEGRENZER
12279	STOP A LEAK (TO)	AVEUGLER UNE FUITE	VERSTOPFEN (EINE UNDICHTE STELLE)
12280	STOP AN ENGINE (TO)	ARRETER UNE MACHINE	ABSTELLEN (EINE MASCHINE), ANHALTEN (EINE MASCHINE)
12281	STOP COCK	ROBINET D'ARRET	ABSPERRHAHN, DURCHGANGSHAHN
12282	STOP VALVE	SOUPAPE D'ARRET	ABSPERRVENTIL
12283	STOP WATCH	CHRONOMETRE A STOP, COMPTE-SECONDES	STOPPUHR
12284	STOPPER	BOUCHON DE COULEE	ABSTICHVERSCHLUSS-STOPFEN
12285	STOPPER, STOPPLE	BOUCHON	STOPFEN, STÖPSEL
12286	STOPPING AN ENGINE	ARRET D'UNE MACHINE	ABSTELLEN EINER MASCHINE, ANHALTEN EINER MASCHINE

12287	**STOPPING OFF**	ISOLATION	ABDECKUNG
12288	**STORAGE**	EMMAGASINAGE, MAGASINAGE	EINLAGERUNG
12289	**STORAGE MEDIUM**	SUPPORT D'INFORMATION	DATENTRÄGER
12290	**STORAGE OF ENERGY**	ACCUMULATION/EMMAGASINAGE D'ENERGIE	SPEICHERUNG, AUFSPEICHERUNG VON ENERGIE
12291	**STORAGE OF HEAT**	ACCUMULATION/EMMAGASINAGE DE CHALEUR	WÄRMEAUFSPEICHERUNG, AUFSPEICHREUNG VON VÄRME
12292	**STORAGE TANKS**	RESERVOIRS DE STOCKAGE	LAGER, SAMMELBEHÄLTER
12293	**STORE**	DEPOT (ENDROIT)	LAGERRAUM
12294	**STORE (TO)**	EMMAGANISER, ENTREPOSER, STATIONNER, STOCKER	EINLAGERN, AUFSPEICHERN
12295	**STORE ENERGY (TO)**	EMMAGASINER DE L'ENERGIE	ENERGIE AUFSPEICHERN
12296	**STORE HEAT (TO)**	ACCUMULER LA CHALEUR, EMMAGASINER LA CHALEUR	WÄRME AUFSPEICHERN
12297	**STOREHOUSE, WAREHOUSE, MAGAZINE, DEPOT**	MAGASIN	LAGER, LAGERHAUS, SPEICHER, MAGAZIN
12298	**STOREKEEPER**	MAGASINIER, MANUTENTIONNAIRE	LAGERVERWALTER
12299	**STORING**	EMMAGASINAGE	LAGERUNG, EINLAGERN, AUFSPEICHERN
12300	**STOVE**	FOUR, FOURNEAU	OFEN
12301	**STRADDLE MILLING CUTTER**	FRAISAGE EN DUPLEX	SCHEIBENFRÄSERPAAR
12302	**STRAIGHT ARM OF PULLEY**	BRAS DROIT D'UNE POULIE	SCHEIBENARM (GERADER)
12303	**STRAIGHT EDGE**	REGLE DE JAUGE	RICHTSCHIENE, RICHTLINEAL
12304	**STRAIGHT FLUTED DRILL**	FORET A COUTURE DROITE	BOHRER (GERADEGENUTETER)
12305	**STRAIGHT FLUTED REAMER**	ALESOIR A RAINURES DROITES	REIBAHLE (GERADEGENUTETE)
12306	**STRAIGHT LEVER**	LEVIER DROIT	HEBEL (GERADARMIGER)
12307	**STRAIGHT MILLING CUTTER**	FRAISE A DENTURE DROITE	FRÄSER MIT GERADEN SCHNEIDEN
12308	**STRAIGHT PANE SLEDGE HAMMER**	MARTEAU A DEVANT, PANNE EN LONG	KREUZSCHLAG, KREUZHAMMER, LÄNGSFINNE
12309	**STRAIGHT PIPE**	TUBE DROIT	ROHR (GERADES)
12310	**STRAIGHT RIGHT LINE**	LIGNE DROITE, DROITE	GERADE, GERADE LINIE
12311	**STRAIGHT SET**	DENTURE DROITE	GERADVERZAHNUNG
12312	**STRAIGHT SPRING SUBJECTED TO BENDING**	RESSORT DE FLEXION DROIT	BIEGEFEDER (GERADE)
12313	**STRAIGHT TOOTH CLUTCK**	EMBRAYAGE A DENTURE DROITE	GERADVERZAHNUNGSKUPPLUNG
12314	**STRAIGHT-CUT CONTROL**	COMMANDE PARAXIALE	STRECKENSTEUERUNG
12315	**STRAIGHT-LINE MOTION**	GUIDE DU MOUVEMENT RECTILIGNE	GERADFÜHRUNG
12316	**STRAIGHT-SIDED MECHANICAL PRESS**	PRESSE MECANIQUE A MONTANT	PRESSE (GERADSEITIGE MECHANISCHE)
12317	**STRAIGHT-WAY VALVE**	SOUPAPE DROITE	DURCHGANGSVENTIL
12318	**STRAIGHTEN (TO)**	REDRESSER, DEGAUCHIR	RICHTEN, GERADERICHTEN, AUSRICHTEN
12319	**STRAIGHTENING**	REDRESSAGE, DEGAUCHISSAGE, DRESSAGE, DEGAUCHISSEMENT	DRESSIEREN, RICHTEN, GERADERICHTEN, AUSRICHTEN
12320	**STRAIGHTENING MACHINE**	MACHINE A REDRESSER, BANC A RECTIFIER	RICHTMASCHINE
12321	**STRAIGHTENING PLATE**	MARBRE	RICHTPLATTE
12322	**STRAIN**	DEFORMATION, CONTRAINTE	VERFORMUNG, SPANNUNG
12323	**STRAIN A BODY (TO), SUBJECT A BODY TO STRAINING FORCES (TO)**	SOUMETTRE UN CORPS A UNE CHARGE, EXERCER UN EFFORT SUR UN CORPS	BEANSPRUCHEN (EINEN KÖRPER)
12324	**STRAIN AGING**	VIEILLISSEMENT PAR LES EFFORTS	STAUCHALTERUNG, RECKALTERUNG
12325	**STRAIN GAUGE**	EXTENSOMETRE	DEHNUNGSMESSER
12326	**STRAIN HARDENING**	ECROUISSAGE (DEFORMATION)	KALTHÄRTUNG

12327	STRAIN-AGEING	VIEILLISSEMENT PAR DEFORMATION	VERFORMUNGSALTERUNG
12328	STRAIN-DEFORMATION	VIEILLISSEMENT PAR DEFORMATION	VERFORMUNGSALTERUNG
12329	STRAIN-GAGE	JAUGE DE CONTRAINTE (A FIL RESISTANT)	DEHNUNGSMESSSTREIFE
12330	STRAIN-RATE	VITESSE DE DEFORMATION	VERFORMUNGGESCHWINDIGKEIT
12331	STRAIN, INTENSITY OF STRAIN, UNIT STRAIN	EFFORT, ACTION D'UNE FORCE EXTERIEURE, SOLLICITATION, CHARGE, TRAVAIL	BEANSPRUCHUNG, INANSPRUCHNAHME
12332	STRAINER	CREPINE, SUCETTE	SEIHER, SAUGKORB, SAUGKOPF, SAUGSIEB
12333	STRAINER, SEDIMENT SEPARATOR	FILTRE	SCHMUTZFÄNGER, FILTER
12334	STRAINING, TIGHTENING, JOCKEY, BINDER, TENSION PULLEY	GALET, ROULEAU, TENDEUR, ROULEAU, POULIE DE TENSION, TENDEUR A GALET	SPANNROLLE
12335	STRAITS TIN, MALACCA TIN	ETAIN DE MALACCA	MALAKKAZINN
12336	STRAND	TORON	LITZE
12337	STRAND (TO)	TORONNER	VERLITZEN
12338	STRAND OF A CABLE-LAID ROPE	CORDON D'UN CABLE	STRANG EINES TAUES, SCHENKEL EINES TAUES
12339	STRAND OF A ROPE	TORON D'UNE CORDE	LITZE EINES SEILES, SEILLITZE
12340	STRANDING	TORONNAGE	VERLITZEN
12341	STRANDING MACHINE	TORONNEUSE	DRAHTSEILHERSTELLUNGSMASCHINE
12342	STRANGLER	DISPOSITIF DE DEPART A FROID	STARTERKLAPPE
12343	STRAP	FEUILLARD	STAHLBAND
12344	STRAP AND SCREW BRAKE	FREIN A VIS	SCHRAUBENBREMSE, SPINDELBREMSE
12345	STRAP CONNECTING ROD END	CHAPE DE LA TETE DE BIELLE	KAPPENKOPF EINER SCHUBSTANGE
12346	STRAP FORK, BELT FORK	FOURCHE DE DEBRAYAGE, FOURCHE GUIDE-COURROIE, FOURCHETTE	RIEMENGABEL
12347	STRAP TYPE HINGE	CHARNIERE TYPE PENTURE	BANDSCHARNIER
12348	STRAW FIBRE	FIBRE DE PAILLE	STROHFASER
12349	STRAY CURRENT	COURANT VAGABOND	STREUSTROM
12350	STRAY CURRENT CORROSION	CORROSION PAR COURANT VAGABOND	IRRSTROMKORROSION
12351	STREAM LINE	FILET FLUIDE, LIGNE AERODYNAMIQUE	STROMLINIE
12352	STRENGTH	RESISTANCE, FORCE	FESTIGKEIT, KRAFT
12353	STRENGTH IN FATIGUE, ENDURANCE AGAINST REPEATED STRESSES	RESISTANCE AUX CHOCS REPETES	SCHWINGUNGSFESTIGKEIT
12354	STRENGTH OF A SOLUTION	TITRE D'UNE SOLUTION	TITER EINER LÖSUNG
12355	STRENGTH OF MATERIALS	RESISTANCE DES MATERIAUX	MATERIALFESTIGKEIT, FESTIGKEIT DES MATERIALS
12356	STRENGTH TEST	EPREUVE DE RESISTANCE	FESTIGKEITSPRÜFUNG
12357	STRENGTH WELD	SOUDURE PORTANTE	NAHT (TRAGENDE)
12358	STRENGTHEN (TO), STIFFENT (TO)	RENFORCER	VERSTEIFEN
12359	STRENGTHENING, STIFFENING	RENFORCEMENT, RENFORT	VERSTEIFEN, VERSTEIFUNG
12360	STRESS CONCENTRATION	CONCENTRATION DE LA CONTRAINTE	SPANNUNGSKONZENTRATION
12361	STRESS CORROSION	CORROSION SOUS TENSION	SPANNUNGSKORROSION
12362	STRESS RELIEF ANNEALING	RECUIT DE DETENTE	SPANNUNGSFREIGLÜHEN
12363	STRESS RELIEVING	RECUIT DE RELAXATION	ENTSPANNUNGSGLÜHEN
12364	STRESS-CORROSION CRACKING	FISSURATION PAR CORROSION SOUS TENSION	KORROSIONSRISSBILDUNG DURCH LATENTE SPANNUNGEN
12365	STRESS-STRAIN CURVE	COURBE EFFORT-DEFORMATION	DEHNUNGSKURVE

12366	STRESS-STRAIN DIAGRAM	COURBE CHARGE-ALLONGEMENT, GRAPHIQUE DE LA RESISTANCE MECANIQUE	SPANNUNGS-DEHNUNGSKURVE, SPANNUNGSDIAGRAMM
12367	STRESS, INTENSITY OF STRESS, UNIT STRESS	CHARGE PAR UNITE DE SECTION, TENSION INTERIEURE, EFFORT INTERIEUR MOLECULAIRE, CONTRAINTE	SPANNUNG, BEANSPRUCHUNG
12368	STRESSED SURFACE	SURFACE CHARGEE	BELASTUNGSFLÄCHE, SPANNUNGSFLÄCHE
12369	STRETCHER	PANNERESSE	LÄUFER (BAUW.)
12370	STRETCHER STRAIN	TENSION D'ETIRAGE	RECKSPANNUNG
12371	STRETCHER STRAIN MARKINGS	LIGNES DE LUDERS	FLIESSFIGUREN, LUDERSLINIEN
12372	STRETCHER/LEVELLER	MACHINE A DRESSER, PLANEUSE	SPANNMASCHINE
12373	STRETCHING FORCE	EFFORT DE TENSION	SPANNKRAFT
12374	STRETCHING OF A BELT, OF A ROPE	ALLONGEMENT D'UNE COURROIE, ALLONGEMENT D'UN CABLE	LÄNGUNG EINES RIEMENS, EINES SEILES
12375	STRIKE	COUP	SCHLAG, AUFSCHLAG
12376	STRING BEAD	CORDON SOUDE RECTILIGNE	STRICHRAUPE
12377	STRINGER	LIMON	TREPPENWANGE
12378	STRIP	FEUILLARD	STREIFEN
12379	STRIP A SCREW THREAD (TO)	ABIMER UN FILET	ÜBERDREHEN (EIN GEWINDE)
12380	STRIP CHART RECORDER	ENREGISTREUR A DEROULEMENT CONTINU	BANDSCHREIBER
12381	STRIPPING AGENTS	SOLVANTS DE DECAPAGE	BEIZMITTEL
12382	STRIPPING BATH	BAIN DE DECAPAGE	BEIZBAD
12383	STRIPPING CRANE	PONT ROULANT DEMOULEUR	STRIPPERKRAN
12384	STRIPPING PLATE	PEIGNE	ABSTREIFPLATTE
12385	STROBOSCOPE	STROBOSCOPE	STROBOSKOP
12386	STROBOSCOPIC METHOD	METHODE STROBOSCOPIQUE	STROBOSKOPISCHE METHODE
12387	STROKE	TEMPS, COURSE	ARBEITSTAKT, HUB
12388	STRONG BACK	U DE MONTAGE	U-MONTAGETEIL
12389	STRONTIUM	STRONTIUM	STRONTIUM
12390	STRONTIUM FLUORIDE	FLUORURE DE STRONTIUM	STRONTIUMFLUORID
12391	STRUCTURAL COMPONENT	ELEMENT STRUCTURAL	STRUKTURELEMENT
12393	STRUCTURAL STEEL	ACIER DE CONSTRUCTION	BAUSTAHL, KONSTRUKTIONSSTAHL
12394	STRUCTURE	STRUCTURE	STRUKTUR, GEFÜGE
12395	STRUCTURE, TEXTURE	CONSTRUCTION (EDIFICE), STRUCTURE, TEXTURE, GRAIN	BAUWERK, BAU, KONSTRUKTION, GEFÜGE, STRUKTUR
12396	STRUT, BRACE	CONTREFICHE, JAMBE DE FORCE	STREBE
12397	STUD	GOUJON	STIFTSCHRAUBE, BOLZENZAPFEN, STIFT, DÜBEL
12398	STUD BOLT	BOULON A TIGE ENTIEREMENT FILETEE	GEWINDEBOLZEN
12399	STUD BOLT, STUD, DOUBLE-ENDED BOLT	GOUJON, BOULON PRISONNIER, PRISONNIER	STIFTSCHRAUBE
12400	STUD GUN	PISTOLETS A GOUJONS	DRUCKLUFTNIETHAMMER
12401	STUD LATHE	TOUR A BOULONS	BOLZENDREHBANK
12402	STUD OF CHAIN LINK	ETANCON, ETAI, ENTRETOISE D'UN CHAINON	STEG EINER KETTE, KETTENSTEG
12403	STUDDED CHAIN, STAYED LINK CHAIN	CHAINE A ETANCONS, CHAINE A MAILLES ETANCONNEES, CHAINE-CABLE A ETAIS	STEGKETTE
12404	STUDDED TYRE	PNEU A CRAMPONS, PNEU A CLOUS	GLEITSCHUTZREIFEN
12405	STUFFING BOX	PRESSE-ETOUPE, BOITE A ETOUPE	STOPFBÜCHSE, STOPFBÜCHSRAUM

12406	STUFFING BOX PACKING	GARNITURE DE PRESSE-ETOUPE, GARNITURE DE BOITE A BOURRAGE	STOPFBÜCHSENPACKUNG
12407	STUMPF UNIFLOW ENGINE	MACHINE A EQUICOURANT, MACHINE STUMPF	GLEICHSTROMDAMPFMASCHINE
12408	SUBBOUNDARY	SOUS-JOINT	UNTERVERBINOUNG
12409	SUBDIVISION	SUBDIVISION	UNTERTEILUNG
12410	SUBDIVISION OF LABOUR	DIVISION DU TRAVAIL	ARBEITSTEILUNG
12411	SUBLIMATE (TO)	SUBLIMER	SUBLIMIEREN
12412	SUBLIMATION	SUBLIMATION	SUBLIMATION
12413	SUBMERGE (TO)	SUBMERGER	UNTERTAUCHEN
12414	SUBMERGED-ARC WELDING	SOUDAGE A L'ARC SUBMERGE	UNTERPULVERSCHWEISSEN
12415	SUBROUTINE	SOUS-PROGRAMME	UNTERPROGRAMM
12416	SUBSCALING	SOUS-COUCHE (D'OXYDE)	UNTERLAGE, UNTEROXYDSCHICHT
12417	SUBSEQUENT TREATMENT, AFTER-TREATMENT	TRAITEMENT ULTERIEUR	NACHBEHANDLUNG
12418	SUBSIDENCE	AFFAISSEMENT	EINSINKEN
12419	SUBSTANCE, MATTER	MATIERE, SUBSTANCE	STOFF, SUBSTANZ, MATERIE
12420	SUBSTITUTE	SUCCEDANE	ERSATZ, ERSATZSTOFF
12421	SUBSTITUTION	SUBSTITUTION (MATH.)	SUBSTITUTION (MATH.)
12422	SUBSTRATE	METAL DE BASE	SUBSTRAT
12423	SUBSTRUCTURE	SOUS-STRUCTURE	SUBSTRUKTUR
12424	SUBTRACT	SOUSTRAIRE, RETRANCHER	ABZIEHEN, SUBTRAHIEREN
12425	SUBTRACTION	SOUSTRACTION	ABZIEHEN, SUBTRAKTION
12426	SUBTRAHEND	NOMBRE A SOUSTRAIRE	SUBTRAHEND
12427	SUCK (TO)	ASPIRER	ANSAUGEN
12428	SUCTION	ASPIRATION	SAUGEN, ANSAUGEN, ANSAUGUNG
12429	SUCTION FAN	VENTILATEUR ASPIRANT	SAUGENDER LÜFTER, VENTILATOR, SAUGLÜFTER
12430	SUCTION GAS PRODUCER	GAZOGENE A ASPIRATION	SAUGGASGENERATOR
12431	SUCTION GOVERNOR	REGULATEUR A DEPRESSION	UNTERDRUCKREGLER
12432	SUCTION HEAD, SUCTION LIFT OF A PUMP	HAUTEUR D'ASPIRATION D'UNE POMPE	SAUGHÖHE EINER PUMPE
12433	SUCTION PIPE	TUYAU D'ASPIRATION	SAUGROHR, SAUGLEITUNG
12434	SUCTION PUMP	POMPE ASPIRANTE	SAUGPUMPE
12435	SUCTION STROKE	COURSE D'ASPIRATION	SAUGHUB
12436	SUCTION VALVE	SOUPAPE, CLAPET D'ASPIRATION	SAUGVENTIL, SAUGKLAPPE
12437	SUDDEN CHANGE OF TEMPERATURE	VARIATION BRUSQUE DE TEMPERATURE	ÄNDERUNG (PLÖTZLICHE) DER TEMPERATUR
12438	SUGAR BEET DOWN THE ROW THINNER	DEMARIEUSE POUR BETTERAVES A SUCRE	ZUCKERRÜBENLICHTMASCHINEN
12439	SUGAR BEET HARVESTERS	RECOLTEUSES DE BETTERAVES	ZUCKERRÜBENVOLLERNTEGERÄTE
12440	SUGAR BEET PLOUGHS AND LIFTERS	SOULEVEUSES DE BETTERAVES	ZUCKERRÜBENHEBER UND PFLÜGE
12441	SUGAR BEET TOPPERS	MACHINES A ETETER LES BETTERAVES	ZUCKERRÜBENKÖPFER
12442	SULFIDE EMBRITTLEMENT	FATIGUE PAR L'HYDROGENE SULFURE	ERMÜDUNG DURCH SCHWEFELWASSERSTOFF
12443	SULL	COUCHE D'OXYDE FERREUX	EISENOXYDSCHICHT
12444	SULPHATE	SULFATE	SULFAT
12445	SULPHATE SODA, LEBLANC SODA	SOUDE LEBLANC	LEBLANC-SODA, SULFATSODA
12446	SULPHIDE	SULFURE	SULFID
12447	SULPHIDE OF IRON	SULFURE DE FER	EISENSULFID
12448	SULPHITE	SULFITE	SULFIT

12449	**SULPHUR**	SOUFRE	SCHWEFEL
12450	**SULPHUR DIOXIDE, SULPHUROUS ANHYDRIDE**	ACIDE, ANHYDRIDE, OXYDE SULFUREUX	SÄURE (SCHWEFLIGE), SCHWEFLIGSÄUREANHYDRID, SCHWEFELDIOXYD
12451	**SULPHUR MONOCHLORIDE**	CHLORURE DE SOUFRE	SCHWEFELCHLORÜR, SCHWEFELMONOCHLORID, CHLORSCHWEFEL
12452	**SULPHUR TRIOXIDE, SULPHURIC ANHYDRIDE**	ANHYDRIDE SULFURIQUE	SCHWEFELSÄUREANHYDRID, WASSERFREIE SCHWEFELSÄURE, SCHWEFELTRIOXYD
12453	**SULPHURIC ACID, OIL OF VITRIOL, HYDROGEN SULPHATE**	ACIDE SULFURIQUE, ACIDE VITRIOLIQUE, VITRIOL, HUILE DE VITRIOL	SCHWEFELSÄURE
12454	**SUM**	SOMME	SUMME
12455	**SUMMIT OF THREAD**	SOMMET, DESSUS D'UN FILET	KOPF, SPITZE EINES GEWINDES, GEWINDEKOPF, GEWINDESPITZE
12456	**SUN AND PLANET WHEELS MOTION**	MOUVEMENT, TRAIN PLANETAIRE	PLANETENGETRIEBE, UMLAUFGETRIEBE
12457	**SUN WHEEL**	ROUE CENTRALE D'UN TRAIN PLANETAIRE	SONNENRAD
12458	**SUNK KEY**	CLAVETTE NOYEE	KEIL (VERSENKTER), NUTKEIL
12459	**SUPER ALLOY**	ALLIAGE HAUTE-TEMPERATURE	SUPERLEGIERUNG
12460	**SUPER HEATING**	SURCHAUFFAGE	ÜBERHITZUNG
12461	**SUPER SATURATION**	SURSATURATION	ÜBERSÄTTIGUNG
12462	**SUPER STEEL**	ACIER FIN	EDELSTAHL
12463	**SUPERCOOL (TO)**	REFROIDIR AU-DESSOUS DE LA TEMPERATURE DE CONDENSATION, REFROIDIR AU DESSOUS DU POINT DE CONGELATION	UNTERKÜHLEN
12464	**SUPERCOOLED LIQUID**	LIQUIDE SURFONDU	UNTERKÜHLTE FLÜSSIGKEIT, ÜBERSCHMOLZENE FLÜSSKIGKEIT
12465	**SUPERCOOLING, SUPERFUSION**	SURFUSION, SURREFROIDISSEMENT	UNTERKÜHLEN, ÜBERSCHMELZEN
12466	**SUPERFICIAL EXPANSION, EXTENSION**	DILATATION SUPERFICIELLE	FLÄCHENAUSDEHNUNG
12467	**SUPERFINE CUT FILE, DEAD SMOOTH CUT FILE**	LIME A TAILLE TRES DOUCE	FEINSCHLICHTFEILE
12468	**SUPERHEAT (TO)**	SURCHAUFFER	ÜBERHITZEN
12469	**SUPERHEATED STEAM**	VAPEUR SURCHAUFFEE	DAMPF (ÜBERHITZTER)
12470	**SUPERHEATED WATER**	EAU SURCHAUFFEE	WASSER (ÜBERHITZTES)
12471	**SUPERHEATED, GASEOUS STEAM**	VAPEUR SURCHAUFFEE	DAMPF (ÜBERHITZTER), HEISSDAMPF
12472	**SUPERHEATER**	SURCHAUFFEUR	ÜBERHITZER, DAMPFÜBERHITZER
12473	**SUPERHEATING**	SURCHAUFFAGE, SURCHAUFFE	ÜBERHITZEN
12474	**SUPERINTEND (TO), SUPERVISE (TO)**	SURVEILLER	BEAUFSICHTIGEN, ÜBERWACHEN
12475	**SUPERLATTICE**	SURSTRUCTURE, SUPER-RESEAU	ÜBERSTRUKTUR
12476	**SUPERPOSITION OF VIBRATIONS**	SUPERPOSITION DE VIBRATIONS	ÜBERLAGERUNG VON SCHWINGUNGEN
12477	**SUPERSATURATE (TO)**	SURSATURER	ÜBERSÄTTIGEN
12478	**SUPERSATURATED SOLUTION**	SOLUTION SURSATUREE	LÖSUNG (ÜBERSÄTTIGTE)
12479	**SUPERSATURATION**	SURSATURATION	ÜBERSÄTTIGUNG
12480	**SUPERSONIC INSPECTION**	ESSAI AUX ULTRA-SONS	ULTRASCHALLPRÜFUNG
12481	**SUPERVISION, SUPERINTENDENCE**	SURVEILLANCE	BEAUFSICHTIGUNG, ÜBERWACHUNG, AUFSICHT
12482	**SUPPLEMENT OF AN ANGLE**	SUPPLEMENT D'UN ANGLE	SUPPLEMENT EINES WINKELS
12483	**SUPPLEMENTARY ANGLE**	ANGLE SUPPLEMENTAIRE	SUPPLEMENTWINKEL
12484	**SUPPLY CABLES**	CABLES D'ALIMENTATION	SPEISELEITUNG

12485	SUPPLY HEAT (TO)	ENVOYER DE LA CHALEUR, FOURNIR DE NOUVELLES CALORIES	WÄRME ZUFÜHREN
12486	SUPPLY OF AIR, AIR SUPPLY	ADMISSION D'AIR, INTRODUCTION D'AIR	LUFTZUFUHR
12487	SUPPLY OF ELECTRICAL ENERGY, ELECTRIC SUPPLY	DISTRIBUTION PUBLIQUE D'ENERGIE ELECTRIQUE	ELEKTRIZITÄTSVERSORGUNG, STROMVERSORGUNG, STROMLIEFERUNG
12488	SUPPLY OF HEAT	FOURNITURE DE CHALEUR	WÄRMEZUFUHR
12489	SUPPLY PIPE	TUYAU ADDUCTEUR, TUYAU D'AMENEE D'ARRIVEE, CONDUITE D'ARRIVEE DU COMBUSTIBLE	ZUFLUSSROHR, ZULEITUNGSROHR, ZULAUFLEITUNG
12490	SUPPORT	SUPPORT, APPUI	STÜTZE, STÄNDER, GESTELL, TRÄGER, HALTER, BOCK, UNTERLAGE, UNTERSATZ
12491	SUPPORT (TO)	APPUYER, SUPPORTER, MONTER	LAGERN, STÜTZEN
12492	SUPPORT RING	COURONNE SUPPORT DE PLATEAU	STÜTZRING
12493	SUPPORTING	SOUTENENENT, MONTAGE	LAGERN, LAGERUNG, STÜTZEN
12494	SUPPORTING DATA	DONNEES JUSTIFICATIVES	UNTERLAGEN
12495	SUPPORTING PLATE	TOLE-SUPPORT	STÜTZBLECH
12496	SURFACE	SURFACE	FLÄCHE
12497	SURFACE (TO)	SURFACER	PLANDREHEN
12498	SURFACE ANALYZER	RUGOSIMETRE ENREGISTREUR	OBERFLÄCHENPRÜFGERÄT
12499	SURFACE CONDENSER	CONDENSEUR PAR SURFACE	OBERFLÄCHENKONDENSATOR
12500	SURFACE CONDITION	ETAT DE SURFACE	OBERFLÄCHENBESCHAFFENHEIT
12501	SURFACE CRACK	FISSURE SUPERFICIELLE, GERCURE	HAUTRISS, OBERFLÄCHENRISS
12502	SURFACE DEFECT	DEFAUT DE SURFACE	OBERFLÄCHENFEHLER
12503	SURFACE DEVIATION	IRREGULARITE DE SURFACE	OBERFLÄCHENUNEBENHEIT
12504	SURFACE FINISH	ETAT DE SURFACE	OBERFLÄCHENBESCHAFFENHEIT
12505	SURFACE GRINDER	MACHINE A RECTIFIER LES SURFACES PLANES	FLÄCHENSCHLEIFMASCHINE, PLANSCHLEIFMASCHINE
12506	SURFACE GRINDING	RECTIFICATION DE SURFACE	PLANSCHLEIFEN, FLÄCHENSCHLEIFEN
12507	SURFACE GRINDING MACHINE	RECTIFIEUSE POUR SURFACE PLANES	FLÄCHENSCHLEIFMASCHINE
12508	SURFACE HARDENING	TREMPE SUPERFICIELLE	OBERFLÄCHENHÄRTUNG
12509	SURFACE INDICATOR	RUGOSIMETRE	OBERFLÄCHENPRÜFGERÄT
12510	SURFACE OF CONTACT	SURFACE DE CONTACT	BERÜHRUNGSFLÄCHE
12511	SURFACE OF CUT	COUPE, SURFACE DE LA COUPE, AIRE DE LA SECTION TRANSVERSALE	SCHNITTFLÄCHE
12512	SURFACE OF IMPRESSION	SURFACE DE L'EMPREINTE	EINDRUCKFLÄCHE
12513	SURFACE OF REVOLUTION	SURFACE DE REVOLUTION	DREHUNGSFLÄCHE, UMDREHUNGSFLÄCHE, ROTATIONSFLÄCHE
12514	SURFACE OF THE TIP OF THE TOOTH	SOMMET DE LA DENT	ZAHNSCHEITEL, ZAHNRÜCKEN
12515	SURFACE PLANING MACHINE	MACHINE A DEGAUCHIR, DEGAUCHISSEUSE	ABRICHTHOBELMASCHINE
12516	SURFACE PLATE, MARKING OFF TABLE	MARBRE A DRESSER, TABLE DRESSEE	RICHTPLATTE, ANREISSPLATTE
12517	SURFACE PRESSURE	PRESSION SUPERFICIELLE	OBERFLÄCHENDRUCK
12518	SURFACE ROUGHNESS	RUGOSITE DE SURFACE	OBERFLÄCHENRAUHIGKEIT
12519	SURFACE TENSION	TENSION SUPERFICIELLE	OBERFLÄCHENSPANNUNG
12520	SURFACE WATER	EAU SUPERFICIELLE, EAU DE SURFACE, EAU DU JOUR	OBERFLÄCHENWASSER, TAGWASSER
12521	SURFACE, AREA	SURFACE	FLÄCHE, OBERFLÄCHE

12522	**SURFACING**	SURFACAGE, CHARIOTAGE TRANSVERSAL	PLANDREHEN
12523	**SURFACTANT**	AGENT DE TRAITEMENT DE SURFACE	OBERFLÄCHENBEHANDLUNGSMITTEL
12524	**SUSPENDED DECK**	TOIT SUSPENDU	HÄNGEDECKE
12525	**SUSPENDED DECK TANK**	RESERVOIR A TOIT SUSPENDU	HÄNGEDACHTANK
12526	**SUSPENSION**	SUSPENSION	AUFHÄNGUNG
12527	**SUSPENSION HOOK**	CROCHET DE SUSPENSION	AUFHÄNGEHAKEN
12528	**SUSPENSION-BRIDGE**	PONT SUSPENDU	HÄNGEBRÜCKE
12529	**SWABBING**	ENDUCTION	ANSTREICHEN
12530	**SWAGE (TO)**	ESTAMPER, ETAMPER A CHAUD	SCHMIEDEN (IM GESENK)
12531	**SWAGE BLOCK**	ETAMPE	GESENKPLATTE, LOCHPLATTE, GESENKKLOTZ, GESENKSTOCK
12532	**SWAGING**	ESTAMPAGE, MATRICAGE	GESENKSCHMIEDEN
12533	**SWAGING MACHINE**	MACHINE A RETREINDRE	REDUZIERMASCHINE
12534	**SWAGING, DROP FORGING, DIE FORGING**	ESTAMPAGE, ETAMPAGE A CHAUD	GESENKSCHMIEDEN
12535	**SWAN NECK, S-PIPE**	COL DE CYGNE	ROHR (GEKRÖPFTES), S-ROHR, S-STÜCK
12536	**SWARF**	COPEAU	SPAN, FEILSPÄNE, ABFALL
12537	**SWATH TURNERS**	VIRE-ANDAINS	SCHWADENWENDER
12538	**SWAY**	DEPLACEMENT LATERAL, ROULIS	SEITENNEIGUNG
12539	**SWAY BAR**	BARRE STABILISATRICE	DREHSTABSTABILISATOR
12540	**SWEATING**	SUINTEMENT, EXSUDATION	AUSSCHWITZUNG
12541	**SWEEPS**	RAMASSE FOIN	HEURAFFER
12542	**SWELL (TO)**	GONFLER, FOISONNER	AUFQUELLEN
12543	**SWELLING**	GONFLEMENT (PAR LES GAZ DE FISSION)	SCHWELLUNG, AUFWACHSEN
12544	**SWELLING OF WOOD**	GONFLEMENT DU BOIS	QUELLEN DES HOLZES
12545	**SWING CHECK VALVE**	CLAPET DE RETENUE A BATTANT	RÜCKFLUSSVENTIL MIT KLAPPE, RÜCKSCHLAGKLAPPE
12546	**SWING JOINT, SWIVEL JOINT**	GENOUILLERE	KNIESTÜCK
12547	**SWING OVER BED**	DIAMETRE ADMIS AU-DESSUS DU BANC	ZUGELASSENER DREHDURCHMESSER ÜBER BETT
12548	**SWING VIBRATION OSCILLATION OF A PENDULUM**	OSCILLATION D'UN PENDULE	PENDELSCHWINGUNG
12549	**SWINGING ARM**	BRAS OSCILLANT	SCHWENKARM
12550	**SWINGING ARM DRILL**	PERCEUSE A BRAS ARTICULE	GELENKSPINDELBOHRMASCHINE
12551	**SWIRL CHAMBER**	CHAMBRE DE TURBULENCE	WIRBELKAMMER
12552	**SWISS THREAD, THURY THREAD**	PAS SYSTEME THURY	NORMALGEWINDE (SCHWEIZER)
12553	**SWITCH**	INTERRUPTEUR, INTERRUPTEUR-CONJONCTEUR, INTERRUPTEUR-DISJONCTEUR, CONTACTEUR ELECTRIQUE	SCHALTER, EINSCHALTER, AUSSCHALTER (ELEKT.)
12554	**SWITCH OFF (TO), SWITCH OUT (TO), CUT OUT (TO)**	METTRE HORS CIRCUIT	AUSSCHALTEN (ELEKTR.), ABSCHALTEN (ELEKTR.)
12555	**SWITCH ON (TO), SWITCH IN (TO), CUT IN (TO)**	METTRE EN CIRCUIT	EINSCHALTEN (ELEKTR.)
12556	**SWITCH POINT**	CHANGEMENT DE VOIE, AIGUILLAGE	WEICHE
12557	**SWITCHBOARD**	TABLEAU DISTRIBUTEUR, STANDARD TELEPHONIQUE, TABLEAU GENERAL	SCHALTTAFEL, SCHALTBRETT
12558	**SWITCHGEAR**	APPAREILS DE DISTRIBUTION (ELECTR.)	SCHALTANLAGE

12559	SWITCHING OFF, SWITCHING OUT, CUTTING OUT	MISE HORS CIRCUIT	AUSSCHALTEN, ABSCHALTEN (ELEKTR.)
12560	SWITCHING ON, SWITCHING IN, CUTTING IN	MISE EN CIRCUIT	EINSCHALTEN (ELEKTR.)
12561	SWIVEL	EMERILLON	WIRBEL (EINES HAKENS, EINER KETTE)
12562	SWIVEL BASE VISE	ETAU ORIENTABLE, ETAU PIVOTANT	SCHRAUBSTOCK (SCHWENKBARER)
12563	SWIVEL BEARING, SELLERS BEARING	PALIER, COUSSINET A ROTULE	KUGELSCHALENLAGER, SELLERSLAGER
12564	SWIVEL HOOK	CROCHET PIVOTANT	WIRBELHAKEN
12565	SYENITE	SYENITE	SYENIT
12566	SYMMETRICAL	SYMETRIQUE	SYMMETRISCH, SPIEGELGLEICH
12567	SYMMETRY	SYMETRIE	SYMMETRIE
12568	SYNCHRO	SYNCHRO	DREHMELDER
12569	SYNCHRONISM	SYNCHRONISME	SYNCHRONISMUS
12570	SYNCHRONOUS	SYNCHRONE	SYNCHRON
12571	SYNCHRONOUS MOTOR	MOTEUR SYNCHRONE	SYNCHRONMOTOR
12572	SYNTHESIS	SYNTHESE	SYNTHESE
12573	SYNTHETIC	SYNTHETIQUE	SYNTHETISCH
12574	SYNTONY	SYNTONIE	ABSTIMMUNG, SYNTONIE
12575	SYSTEM OF COORDINATES	SYSTEME DE COORDONNEES	ACHSENKREUZ, KOORDINATENSYSTEM
12576	SYSTEM OF LENSES	SYSTEME DE LENTILLES	LINSENSYSTEM
12577	SYSTEM OF LEVERS	SYSTEME ENSEMBLE DE LEVIERS	HEBELWERK, GESTÄNGE
12578	SYSTEM OF MEASURES	SYSTEME DE MESURE	MASSSYSTEM
12579	SYSTEM OF PIPES	RESEAU DE TUYAUX	ROHRNETZ
12580	SYSTEM OF WEIGHTS	SYSTEME DE POIDS	GEWICHTSSYSTEM
12581	T.SQUARE, TEE-SQUARE	TE DE DESSIN, EQUERRE DOUBLE	REISSSCHIENE, KREUZWINKEL, DOPPELTER ANSCHLAGWINKEL
12582	T-SLOT CUTTERS	FRAISES POUR RAINURES EN T	T-SPANN-NUTEN FRÄSER
12583	TAB SEQUENTIAL FORMAT	FORMAT A TABULATION	EINGABEFORMAT IN TABULATOR-SCHREIBWEISE
12584	TABLE	TABLE, TABLEAU (SERIE DE NOMBRES)	TAFEL, TABELLE
12585	TABLE CROSS-TRAVEL	COURSE TRANSVERSALE DE LA TABLE	TISCH-QUERBEWEGUNG
12586	TACHOGRAPH	TACHYMETRE ENREGISTREUR	GESCHWINDIGKEITSSCHREIBER, TACHOGRAPH
12587	TACHOMETER	TACHYMETRE, COMPTE-TOURS	TACHOMETER
12588	TACK	CLOU SEMENCE, SEMENCE	ZWECKE (NAGEL)
12589	TACK WELD	SOUDAGE PROVISOIRE PAR POINTS DE POINTAGE	HEFTSCHWEISSEN
12590	TACK-WELDING	SOUDURE PROVISOIRE PAR POINTS	HEFTSCHWEISSEN
12591	TACKINESS	POISSAGE	KLEBRIGKEIT
12592	TACKING RIVET	BROCHE POUR ASSEMBLAGE PROVISOIRE DES TOLES, BOULON D'ASSEMBLAGE PROVISOIRE	HEFTNIET
12593	TAGGER	TOLE TRES MINCE	FEINBLECH (SEHR DÜNNES)
12594	TAIL HAIR	CRIN, POILS DE QUEUE DE CHEVAL	SCHWEIFHAAR
12595	TAIL LIGHT	FEUX ARRIERE	SCHLUSSLATERNE
12596	TAIL RACE	CANAL DE DECHARGE, CANAL DE FUITE, BIEF D'AVAL	ABLAUFKANAL, UNTERWASSERKANAL
12597	TAIL WATER	EAU D'AVAL	UNTERWASSER
12598	TAILINGS	RESIDUS	BERGE, RÜCKSTÄNDE

12599	**TAILSTOCK**	CONTRE-POINTE, POUPEE MOBILE	REITSTOCK
12600	**TAKE TO PIECES (TO), DISMANTLE (TO)**	DEMONTER, DESASSEMBLER	ZERLEGEN, AUSEINANDERNEHMEN, DEMONTIEREN
12601	**TAKE UP A FORCE (TO) (PULL, PUSH)**	RECEVOIR UNE FORCE	AUFNEHMEN (EINE KRAFT)
12602	**TAKE UP THE SLACK OF A BELT (TO)**	RETENDRE UNE COURROIE	NACHSPANNEN (EINEN RIEMEN)
12603	**TAKE UP THE WEAR OF A BEARING (TO), OF A COUPLING**	COMPENSER LE JEU D'UN COUSSINET, RATTRAPER LE JEU D'EMBRAYAGE, RACHETER L'USURE	NACHSTELLEN EIN LAGER, EINE KUPPLUNG
12604	**TAKING TO PIECES, DISMANTLING**	DEMONTAGE, DESASSEMBLAGE	ZERLEGEN, ZERLEGUNG, AUSEINANDERNEHMEN, DEMONTAGE
12605	**TAKING UP THE SLACK OF A BELT**	ACTION DE RETENDRE UNE COURROIE	NACHSPANNEN EINES RIEMENS
12606	**TAKING UP THE WEAR OF A BEARING, OF A COUPLING**	RATTRAPAGE DU JEU, RECALAGE (D'UN COUSSINET)	NACHSTELLEN EINES LAGERS, NACHSTELLEN EINER KUPPLUNG
12607	**TALC POWDER**	TALC EN POUDRE	TALKPULVER, TALKUM
12608	**TALC, FRENCH CHALK**	TALC	TALK
12609	**TALLOW, SUET**	SUIF	TALG, UNSCHLITT
12610	**TAMPED ASPHALT**	ASPHALTE COMPRIME	STAMPFASPHALT
12611	**TAMPER**	MASSE, DEMOISELLE, DAME	ERDRAMMER
12612	**TAN CAKE**	TAN COMPRIME, TAN EN MOTTES	LOHKÄSE, LOHKUCHEN
12613	**TANDEM STEAM ENGINE**	MACHINE TANDEM	TANDEMDAMPFMASCHINE
12614	**TANG, SHANK OF A TOOL**	SOIE, QUEUE D'UN OUTIL	ANGEL EINES WERKZEUGS
12615	**TANGENT (OF AN ANGLE)**	TANGENTE	TANGENTE, TANGENS
12616	**TANGENT (TO A CURVE)**	TANGENTE (GEOM.)	TANGENTE, BERÜHRUNGSLINIE
12617	**TANGENT AT THE VERTEX**	TANGENTE AU SOMMET	SCHEITELTANGENTE
12618	**TANGENT KEY**	CLAVETTE TANGENTIELLE	TANGENTIALKEIL
12619	**TANGENT SCREW**	VIS TANGENTE	TANGENTENSCHRAUBE
12620	**TANGENTIAL ACCELERATION**	ACCELERATION TANGENTIELLE	TANGENTIALBESCHLEUNIGUNG
12621	**TANGENTIAL FLOW TURBINE**	TURBINE TANGENTIELLE	TANGENTIALTURBINE
12622	**TANGENTIAL FORCE**	FORCE TANGENTIELLE	UMFANGSKRAFT, TANGENTIALKRAFT
12623	**TANGENTIAL SPEED**	VITESSE TANGENTIELLE	GESCHWINDIGKEIT (TANGENTIALE)
12624	**TANGENTIAL STRESS**	EFFORT TANGENTIEL	TANGENTIALSPANNUNG
12625	**TANK**	RESERVOIR	BEHÄLTER, TANK
12626	**TANK CAR**	WAGON-RESERVOIR, WAGON-CITERNE	KESSELWAGEN, ZISTERNENWAGEN
12627	**TANK FARM**	PARC DE STOCKAGE D'HYDROCARBURE	TANKLAGER
12628	**TANK PAD**	ASSISE DE RESERVOIR	TANKUNTERBAU
12629	**TANK PIT**	CUVETTE DE RETENTION D'UN RESERVOIR	TANKGRUBE
12630	**TANK SHELL**	ROBE DU RESERVOIR	BEHÄLTERMANTEL
12631	**TANK WITH BOTTOM CONE-DOWN**	RESERVOIR A FOND CONCAVE	BEHÄLTER, TANK MIT KONKAVEM BODEN
12632	**TANK WITH BOTTOM CONE-UP**	RESERVOIR A FOND CONVEXE	BEHÄLTER, TANK MIT KONVEXEM BODEN
12633	**TANNATE**	TANNATE	SALZ (GERBSAURES), TANNAT
12634	**TANTALUM**	TANTALE	TANTAL
12635	**TANYARD REFUSE**	TAN EPUISE	LOHE (VERBRAUCHTE)
12636	**TAP (TO), CUT THREADS IN HOLES (TO)**	TARAUDER	BOHREN (GEWINDE), SCHNEIDEN (INNENGEWINDE)
12637	**TAP HOLE**	TROU DE COULEE	ABSTICHLOCH
12638	**TAP WRENCH**	TOURNE-A-GAUCHE	WINDEISEN, WENDEEISEN

12639	TAP, PLUG	COUVERCLE, ROBINET	STÖPSEL, HAHN
12640	TAP, SCREW TAP	TARAUD	GEWINDEBOHRER
12641	TAPE	RUBAN, BANDE	BAND
12642	TAPE BLOCK VALVE	VANNE D'ARRET	ABSPERRSCHIEBER
12643	TAPE FEED	ENTREE DES DONNEES PAR BANDE	LOCHSTREIFENEINGABE
12644	TAPE RECORDER	MAGNETOPHONE	TONBANDGERÄT
12645	TAPE-HOLE / TAPPING HOLE	SORTIE DE COULEE (TROU)	STICHLOCH, ABSTICHLOCH
12646	TAPER	CONICITE, DELARDAGE	KONIZITÄT, ABKANTEN
12647	TAPER CONNECTING PIECE	RACCORD CONIQUE	MUFFE (KONISCHE)
12648	TAPER FILE	LIME POINTUE	SPITZFEILE
12649	TAPER FIT	EMMANCHEMENT A CONE, EMMANCHEMENT CONIQUE	KEGELSITZ
12650	TAPER HAND REAMER	ALESOIR A MAIN CONIQUE	KEGELHANDREIBAHLE
12651	TAPER MANDREL	MANDRIN CONIQUE	KONUSDORN
12652	TAPER MICROMETER	MICROMETRE DE CONICITE, MICRO-PALPEUR DE CONICITE	KONIZITÄTSMIKROMETER
12653	TAPER OF KEY	SERRAGE DE CLAVETTE	KEILANSTELLUNG
12654	TAPER OF WEDGE	PENTE, CONICITE D'UNE CLAVETTE	ANZUG, STEIGUNG, NEIGUNG EINES KEILS
12655	TAPER PIN	GOUPILLE CONIQUE	KEGELSTIFT, STIFT (KONISCHER)
12656	TAPER PLUG GAUGE	JAUGE-TAMPON CONIQUE, CALIBRE MALE CONIQUE	KEGELLEHRDORN
12657	TAPER PLUG GAUGE WITH DRIVING PILOT	JAUGE-TAMPON CONIQUE AVEC TETON	KEGELLEHRDORN MIT MITNEHMER-LAPPEN
12658	TAPER REAMER	ALESOIR CONIQUE	KEGELREIBAHLE, REIBAHLE (KONISCHE), REIBAHLE (KEGELIGE)
12659	TAPER RING GAUGES	CALIBRES FEMELLES CONIQUES	KEGELLEHRHÜLSEN
12660	TAPER ROLLER	GALET CONIQUE, ROULEAU CONIQUE	KEGELROLLE, KEGELWALZE
12661	TAPER ROPE	CABLE DIMINUE, CABLE A SECTION DECROISSANTE, CABLE CONIQUE	SEIL (VERJÜNGTES)
12662	TAPER SEAT GATE VALVE	VANNE A SIEGES OBLIQUES	KEILSCHIEBER
12663	TAPER SEATS	SIEGES OBLIQUES	SITZE (SCHRÄGE)
12664	TAPER SHANK	QUEUE CONIQUE	KEGELSCHAFT
12665	TAPER SUNK KEY	CLAVETTE CONIQUE	TREIBKEIL
12666	TAPER THREAD	FILETAGE CONIQUE	GEWINDE (KONISCHES)
12667	TAPER-SHANK MANDREL	MANDRIN A QUEUE CONIQUE	KEGELSCHAFTSPANNDORN
12668	TAPER, CONICAL TURNING	CHARIOTAGE, TOURNAGE CONIQUE, TOURNAGE DES SURFACES CONIQUES	KEGELIGDREHEN, KONISCHDREHEN
12669	TAPERED CONE	CONE LISSE, POULIE-CONE	RIEMENKEGEL, RIEMENKONUS, KEGELTROMMEL, KEGELSCHEIBE, RIEMENKONOID
12670	TAPERED ROD	BARRE A SECTION DECROISSANTE	VERJÜNGTER STAB
12671	TAPERING OF A PLATE	DELARDAGE D'UNE TOLE	ABKANTEN
12672	TAPPET	POUSSOIR, MENTONNET	STÖSSEL, HEBLING
12673	TAPPET CLEARANCE	JEU DES SOUPAPES	STÖSSELSPIEL
12674	TAPPET ROLLER	GALET DE POUSSOIR	STÖSSEL
12675	TAPPING	PERCAGE, DEBOUCHAGE	ABSTECHE, KUPOLOFENABSTICH
12676	TAPPING ATTACHMENT	APPAREIL A TARAUDER	GEWINDESCHNEIDVORRICHTUNG
12677	TAPPING MACHINE	TARAUDEUSE	INNENGEWINDESCHNEIDMASCHINE
12678	TAPPING, CUTTING THREADS IN HOLES	TARAUDAGE	BOHREN, EINES GEWINDES, GEWINDEBOHREN, SCHNEIDEN EINES INNENGEWINDES
12679	TAR	GOUDRON	TEER

12680	**TAR (TO)**	GOUDRONNER	TEEREN
12681	**TAR OIL**	HUILE DE GOUDRON	TEERÖL
12682	**TARGET**	ANTICATHODE, CIBLE, OBJECTIF	ANTIKATHODE, FANGELEKTRODE, ZIEL
12683	**TARGET AREA**	SURFACE DE LA CIBLE (OFFERTE AUX RAYONNEMENTS)	AUFPRALLAFLÄCHE
12684	**TARGET TUBE**	TUBE A RAYONS-X	ROENTGENRÖHRE
12685	**TARNISH (TO)**	TERNIR (SE), OXYDER (S')	ANLAUFEN, MATTIEREN, GLANZ (VERLIEREN DEN)
12686	**TARPAULIN COVERED LORRY**	CAMION BACHE	LASTKRAFTWAGEN MIT PLANE UND SPRIEGEL
12687	**TARRED PIPE**	TUBE PROTEGE PAR GOUDRONNAGE	ROHR (GETEERTES)
12688	**TARRED ROPE**	CORDE NOIRE, CORDE GOUDRONNEE	SEIL (GETEERTES)
12689	**TARRING**	GOUDRONNAGE	TEEREN
12690	**TARTARIC ACID, DIOXY-SUCINIC ACID**	ACIDE TARTARIQUE	WEINSÄURE, WEINSTEINSÄURE, DIOXYBERNSTEINSÄURE
12691	**TARTRATE**	TARTRATE	WEINSAURES SALZ, TARTRAT
12692	**TAXE A READING (TO)**	LIRE	ABLESEN
12693	**TEA-POT LADLE**	POCHE THEIERE	SIPHONPFANNE
12694	**TECHNICAL**	TECHNIQUE	TECHNISCH
12695	**TECHNICS**	TECHNIQUE	TECHNIK
12696	**TEDDERS**	FANEUSES	GABELHEUWENDER
12697	**TEE**	TE	TEESTÜCK
12698	**TEE IRON, TEE BAR, T-IRON**	FER EN T	T-EISEN
12699	**TEE PIPE, T-PIECE**	TE A TROIS DIRECTIONS	T-STÜCK, DREIWEGESTÜCK
12700	**TEE-BAR**	PROFILE EN T	T-TRÄGER
12701	**TEE-HEADED BOLT**	BOULON A TETE DE MARTEAU	HAMMERKOPFSCHRAUBE
12702	**TEE-JOINT**	ASSEMBLAGE EN T	FLANKENKEHLNAHT, T-MUFFE
12703	**TEEMING**	COULEE, MOULAGE SORTIE EN LINGOTIERE	GIESSEN, ABGUSS
12704	**TEETH**	DENTURE	VERZAHNUNG
12705	**TELEDYNAMIC CABLE, FLY ROPE**	CABLE TELEDYNAMIQUE	FERNTRIEBSEIL
12706	**TELEDYNAMIC TRANSMISSION GEAR**	TRANSMISSION PAR CABLE TELEDYNAMIQUE	FERNTRIEB, SEILFERNTRIEB
12707	**TELEGRAPH**	TELEGRAPHE	TELEGRAPH
12708	**TELEGRAPH WIRE**	FIL TELEGRAPHIQUE	TELEGRAPHENDRAHT
12709	**TELEGRAPHY**	TELEGRAPHIE	TELEGRAPHIE
12710	**TELEPHONE**	TELEPHONE	FERNSPRECHER, TELEPHON
12711	**TELEPHONE WIRE**	FIL TELEPHONIQUE	FERNSPRECHERDRAHT, TELEPHONDRAHT
12712	**TELEPHONY**	TELEPHONIE	FERNSPRECHWESEN, TELEPHONIE
12713	**TELESCOPE GAUGES**	CALIBRES TELESCOPIQUES	TELESKOPLEHRE
12714	**TELETHERMOMETER**	PYROMETRE A DISTANCE	FERNTHERMOMETER
12715	**TELL-TALE HOLE**	TROU TEMOIN	BEOBACHTUNGSÖFFNUNG, ANZEIGEEINRICHTUNG
12716	**TELLURIUM**	TELLURE	TELLUR
12717	**TEMPER (TO)**	FAIRE REVENIR, ADOUCIR	ANLASSEN, TEMPERN
12718	**TEMPER (TO) (MORTAR, CEMENT)**	DELAYER, GACHER, MALAXER (DU MORTIER, DU CIMENT)	ANMACHEN (MÖRTEL, ZEMENT), ANRÜHREN (MÖRTEL, ZEMENT)
12719	**TEMPER (TO), LET DOWN THE STEEL (TO)**	FAIRE REVENIR, DETREMPER L'ACIER	ANLASSEN (DEN STAHL), NACHLASSEN (DEN STAHL)
12720	**TEMPER ROLLING**	DRESSAGE	DRESSIEREN
12721	**TEMPERATURE**	TEMPERATURE	WÄRMEGRAD, TEMPERATUR

12722	TEMPERATURE CONTROL DEVICE, THERMOSTAT	THERMOSTAT	THERMOSTAT
12723	TEMPERATURE CURVES	COURBES DE TEMPERATURE	TEMPERATURKURVEN
12724	TEMPERATURE DROP	CHUTE DE TEMPERATURE	WÄRMEGEFÄLLE, TEMPERATURGEFÄLLE
12725	TEMPERATURE GRAPHS	RELEVES DE TEMPERATURE	TEMPERATURDIAGRAMME
12726	TEMPERATURE INDICATOR	INDICATEUR DE TEMPERATURE, THERMOMETRE	THERMOMETER
12727	TEMPERATURE OF IGNITION, IGNITION TEMPERATURE	TEMPERATURE D'INFLAMMATION	ENTZÜNDUNGSTEMPERATUR, ZÜNDTEMPERATUR
12728	TEMPERATURE OF THE AIR	TEMPERATURE DE L'AIR	LUFTTEMPERATUR
12729	TEMPERATURE OF THE AMBIENT AIR	TEMPERATURE AMBIANTE	TEMPERATUR (DER UMGEBENDEN LUFT), UMGEBUNGSTEMPERATUR
12730	TEMPERATURE RANGE	GAMME DE TEMPERATURE	TEMPERATURBEREICH
12731	TEMPERING	REVENU	ANLASS
12732	TEMPERING COLOUR	COULEUR, TEINTE DE REVENU, TEINTE DE RECUIT	ANLAUFFARBE, ANLASSFARBE
12733	TEMPERING, LETTING DOWN OF THE STEEL	REVENU, DETREMPE DE L'ACIER	ANLASSEN, NACHLASSEN DES STAHLES
12734	TEMPLATE, TEMPLET	GABARIT, PROFIL, PATRON, TROUSSEAU	SCHABLONE
12735	TEMPORARY HARDNESS OF WATER	DURETE TEMPORAIRE DE L'EAU	HÄRTE DES WASSERS (SCHNINDENDE), HÄRTE DES WASSERS (VORÜBERGEHENDE), HÄRTE DES WASSERS (TEMPORÄRE)
12736	TEMPORARY STORAGE	MEMOIRE TEMPORAIRE	PUFFERSPEICHER
12737	TENACITY	TENACITE	ZÄHIGKEIT
12738	TENACITY OF NOTCHED BAR	TENACITE DES BARRFAIIX ENTAILLES	KERBZÄHIGKEIT
12739	TENDONS	CABLES	VORSPANNUNGSKABEL
12740	TENON	TENON	ZAPFEN (TISCHLEREI)
12741	TENON SAW, PANEL SAW	SCIE A ARASER, SCIE A TENONS, SCIE A CHEVILLES	ZAPFENSÄGE
12742	TENONING MACHINE	MACHINE A FAIRE LES TENONS	ZAPFENSCHNEIDMASCHINE
12743	TENSILE	MECANIQUE (ESSAI)	ZUG
12744	TENSILE DEFORMATION	DEFORMATION PAR TRACTION	ZUGVERFORMUNG
12745	TENSILE FORCE	FORCE DE TRACTION	ZUGKRAFT (FESTIGKEITSL.)
12746	TENSILE PROPERTIES	CARACTERISTIQUES MECANIQUES	FESTIGKEITSEIGENSCHAFTEN
12747	TENSILE STRAIN	EFFORT DE TRACTION, TRAVAIL A LA TRACTION	ZUGSPANNUNG, BEANSPRUCHUNG AUF ZUG, ZUGBEANSPRUCHUNG
12748	TENSILE STRENGHT TEST	ESSAI DE TRACTION	ZUGVERSUCH
12749	TENSILE STRENGTH	RESISTANCE A LA TRACTION, CHARGE DE RUPTURE	ZUGFESTIGKEIT, BRUCHLAST
12750	TENSILE STRESS	CONTRAINTE DE TRACTION (FORCE), EFFORT DE TRACTION PAR UNITE DE SECTION	ZUGSPANNUNG
12751	TENSILE TEST	ESSAI A LA TRACTION	ZUGVERSUCH, ZERREISSVERSUCH
12752	TENSILE TESTING MACHINE	MACHINE D'ESSAI DE TRACTION	ZERREISS(PRÜF)MASCHINE
12753	TENSION	TENSION	SPANNUNG
12754	TENSION BAR ROD	BARRE TRAVAILLANT A LA TENSION	ZUGSTAB, STAB (GEZOGENER)
12755	TENSION BOLT	BOUTON DE TENSION	ZUGBOLZEN
12756	TENSION CARRIAGE	CHASSIS-TENDEUR, RAIL-TENDEUR	SPANNSCHLITTEN, SPANNWAGEN
12757	TENSION IMPACT	ENERGIE DU CHOC	SCHLAGSTÄRKE
12758	TENSION SCREW	VIS DE TENSION	ZUGSCHRAUBE

12759	**TENSION SPRING, SPRING FOR TENSION**	RESSORT DE TRACTION	ZUGFEDER
12760	**TERBIUM**	TERBIUM	TERBIUM
12761	**TERM OF PATENT, DURATION OF A PATENT**	DUREE DU BREVET	DAUER EINES PATENTRECHTES
12762	**TERMINAL**	BORNE, POUPEE (ELECTR.), BORNE DE RACCORDEMENT	KLEMME (ELEKTR.)
12763	**TERMINAL POST**	BORNE	ANSCHLUSSPOL
12764	**TERMINAL VOLTAGE**	DIFFERENCE DE POTENTIEL AUX BORNES, TENSION, VOLTAGE AUX BORNES	KLEMMENSPANNUNG
12765	**TERMINALS**	COSSES	KLEMMEN
12766	**TERNARY SYSTEM**	SYSTEME TERNAIRE	DREISTOFFSYSTEM
12767	**TERRACOTTA**	TERRE CUITE	TERRAKOTTA
12768	**TERRAZZO MOSAIC**	TERRAZZO	TERRAZZO
12769	**TESSELATED PLATE**	TOLE GAUFREE	WAFFELBLECH, WARZENBLECH
12770	**TEST**	ESSAI, EPREUVE	PROBE, PRÜFUNG, VERSUCH
12771	**TEST BAR LATHE**	TOUR A EPROUVETTES	PROBESTABDREHBANK
12772	**TEST BAR, SPECIMEN BAR**	EPROUVETTE, BARREAU-EPROUVETTE, BARRE SOUMISE AUX ESSAIS	PROBESTAB
12773	**TEST BENCH, TEST BED**	POSTE D'ESSAI, BANC DE CONTROLE, BANC D'ESSAI, TABLE D'EXAMEN	PRÜFSTAND, VERSUCHSTAND, PROBIERSTAND
12774	**TEST BY REPEATED FLEXURE**	ESSAI DE FLEXION ALTERNE, ESSAI DES FILS AU PLIAGE	HIN- UND HERBIEGEPROBE
12775	**TEST CERTIFICATE**	CERTIFICAT D'EPREUVE	PRÜFUNGSZEUGNIS
12776	**TEST COCK, TRY COCK**	ROBINET DE JAUGE	PRÜFHAHN, PROBIERHAHN
12777	**TEST COUPON**	LINGOT EPROUVETTE	PROBEBLOCK
12778	**TEST DRILLING**	FORAGE D'ESSAI, SONDAGE	VERSUCHSBOHRUNG
12779	**TEST GAUGE**	JAUGE DE CONTROLE	PRÜFLEHRE
12780	**TEST GLASS, TEST TUBE**	EPROUVETTE	REAGENZGLAS, PRÜFGLAS, PROBIERGLAS
12781	**TEST LOAD, PROOF LOAD**	CHARGE D'ESSAI	PROBEBELASTUNG
12782	**TEST OF RAW MATERIALS**	ESSAI DES MATERIAUX	WERKSTOFFPRÜFUNG, MATERIALPRÜFUNG, PRÜFUNG DES MATERIALS
12783	**TEST PAPER**	PAPIER REACTIF	REAGENZPAPIER
12784	**TEST PIECE, TEST SAMPLE**	EPROUVETTE	PROBESTÜCK, PRÜFSTÜCK
12785	**TEST PLUG**	TAMPON, FICHE D'ESSAI	PRÜFSTÖPSEL
12786	**TEST PRESSURE**	PRESSION D'EPREUVE, PRESSION D'ESSAI	PRÜFDRUCK, PROBEDRUCK
12787	**TEST PUMP**	POMPE D'EPREUVE	PRÜFPUMPE
12788	**TEST SAMPLE**	EPROUVETTE	PROBE(STÜCK)
12789	**TEST SPECIMEN, TEST PIECE**	EPROUVETTE (MORCEAU D'ESSAI)	PROBESTÜCK, VERSUCHSSTÜCK
12790	**TEST TANK**	RESERVOIR DE JAUGEAGE	PRÜFTANK
12791	**TEST-TUBE**	EPROUVETTE (CHIM)	REAGENZGLAS
12792	**TEST, EXPERIMENT**	ESSAI, EPREUVE, EXPERIENCE, EXAMEN	VERSUCH, EXPERIMENT
12793	**TESTING MACHINE**	MACHINE A ESSAYER, APPAREIL D'ESSAI	FESTIGKEITS-PRÜFMASCHINE
12794	**TESTING TRACK**	PISTE D'ESSAI	PROBEFAHRBAHN
12795	**TETRAGONAL**	QUADRATIQUE	TETRAGONAL
12796	**TETRAHEDRITE, FAHLERZ, GREY COPPER ORE**	CUIVRE GRIS, PANABASE	FAHLERZ

12797	**TETRAHEDRON**	TETRAEDRE	VIERFLACH, VIERFLÄCHNER, TETRAEDER
12798	**TETRATOMIC, TETRAVALENT**	TETRATOMIQUE, QUADRIVALENT	VIERWERTIG, VIERATOMIG
12799	**TEXTILE FIBRE**	FIBRE TEXTILE	SPINNFASER, GESPINSTFASER, TEXTILFASER
12800	**TEXTURE**	TEXTURE, STRUCTURE	GEFÜGE, STRUKTUR
12801	**THALLIUM**	THALLIUM	THALLIUM
12802	**THAT CAN BE WORKED UP INTO...**	MANUFACTURABLE	VERARBEITBAR
12803	**THERMAL CAPACITY**	CAPACITE THERMIQUE	WÄRMEKAPAZITÄT
12804	**THERMAL CONDUCTIVITY**	CONDUCTIVITE THERMIQUE	WÄRMELEITFÄHIGKEIT
12805	**THERMAL CONTRACTION**	RETRAIT THERMIQUE	WÄRMESCHWINDUNG
12806	**THERMAL EFFICIENCY**	RENDEMENT THERMIQUE	WÄRMEWIRKUNGSGRAD, THERMISCHER WIRKUNGSGRAD
12807	**THERMAL ELECTROMOTIVE FORCE**	FORCE ELECTROMOTRICE THERMIQUE	KRAFT (THERMOELEKTROMOTORISCHE)
12808	**THERMAL EXPANSION**	DILATATION THERMIQUE	WÄRMEAUSDEHNUNG
12809	**THERMAL EXPANSION COEFFICIENT**	COEFFICIENT DE DILATATION THERMIQUE	WÄRMEAUSDEHNUNGSKOEFFIZIENT
12810	**THERMAL INSOLATION**	ISOLATION THERMIQUE	ISOLIERUNG (THERMISCHE)
12811	**THERMAL RADIATION**	RAYONNEMENT DE LA CHALEUR	WÄRMESTRAHLUNG
12812	**THERMAL SHOCK**	CHOC THERMIQUE	TEMPERATURSCHOCK, WÄRMESTOSS
12813	**THERMAL SHOCK RESISTANCE**	RESISTANCE AUX VARIATIONS DE TEMPERATURE OU AU CHOC THERMIQUE	TEMPERATURWECHSELBESTÄNDIGKEIT
12814	**THERMAL STRESS**	TENSION THERMIQUE	WÄRMESPANNUNG
12815	**THERMAL TREATMENT**	TRAITEMENT THERMIQUE	WÄRMEBEHANDLUNG
12816	**THERMAL UNIT, UNIT OF HEAT**	UNITE DE CHALEUR, UNITE THERMIQUE	WÄRMEEINHEIT
12817	**THERMIT**	THERMIT	THERMIT
12818	**THERMIT ALUMINOTHERMIC WELDING**	SOUDURE AVEC APPORT DE FER-THERMIT, SOUDURE ALUMINOTHERMIQUE	THERMITSCHWEISSUNG, ALUMINOTHERMISCHE SCHWEISSUNG
12819	**THERMIT WELDING**	SOUDAGE PAR ALUMINOTHERMIE	THERMITSCHWEISSUNG
12820	**THERMO ELECTRIC PILE**	PILE THERMOELECTRIQUE, PILE THERMIQUE	THERMOSÄULE
12821	**THERMO- ELECTRIC COUPLE**	COUPLE THERMOELECTRIQUE, ELEMENT DE PILE THERMOELECTRIQUE	THERMOELEMENT
12822	**THERMO-CHEMISTRY**	THERMOCHIMIE	THERMOCHEMIE
12823	**THERMO-ELECTRIC**	THERMO-ELECTRIQUE	THERMOELEKTRISCH
12824	**THERMO-ELECTRIC CURRENT**	COURANT THERMOELECTRIQUE	THERMOSTROM, THERMOELEKTRISCHER STROM
12825	**THERMO-ELECTRIC PYROMETER**	PYROMETRE THERMOELECTRIQUE	PYROMETER (THERMOELEKTRISCHES)
12826	**THERMO-ELECTRICITY**	THERMO-ELECTRICITE	THERMOELEKTRIZITÄT
12827	**THERMOCOUPLE**	COUPLE THERMO-ELECTRIQUE, THERMOCOUPLE	THERMOELEMENT
12828	**THERMODYNAMIC (ADJ)**	THERMODYNAMIQUE	THERMODYNAMISCH
12829	**THERMODYNAMICS**	THERMODYNAMIQUE	MECHANISCHE WÄRMELEHRE, THERMODYNAMIK
12830	**THERMOELECTRIC EFFECT**	EFFET THERMOELECTRIQUE	SEEBECK-EFFEKT
12831	**THERMOELECTRIC POWER**	FORCE THERMOELECTRIQUE	KRAFT (THERMOELEKTRISCHE)
12832	**THERMOGRAPH**	THERMOGRAPHE	WÄRMESCHREIBER
12833	**THERMOMETER**	THERMOMETRE	THERMOMETER, WÄRMEMESSER
12834	**THERMOMETER F. WATER**	INDICATEUR DE TEMPERATURE	WASSERTHERMOMETER

12835	THERMOMETRIC SCALE	ECHELLE THERMOMETRIQUE	THERMOMETEREINTEILUNG, THERMOMETERSKALA, TEMPERATURSKALA
12836	THERMOPILE	PILE THERMO-ELECTRIQUE	THERMOSÄULE
12837	THERMOSCOPE	THERMOSCOPE	THERMOSKOP
12838	THERMOSTAT	THERMOSTAT	KÜHLLUFTREGLER, THERMOSTAT
12839	THERMOSTATIC TRAP	PURGEUR D'EAU CONDENSEE A DILATATION	KONDENSTOPF (THERMOSTATISCHER)
12840	THERMOSTATIC VALVE	REGULATEUR DE TEMPERATURE	TEMPERATUR-REGELVENTIL, TEMPERATURREGLER
12841	THICK-WALLED	PAROI EPAISSE (A)	DICKWANDIG, STARKWANDIG
12842	THICKENER	SUBSTANCE EPAISSISSANTE	VERDICKER
12843	THICKNESS	EPAISSEUR	DICKE, STÄRKE
12844	THICKNESS OF A PIPE	EPAISSEUR DE LA PAROI (DU METAL) D'UN TUBE	WANDDICKE EINES ROHRES, WANDSTÄRKE EINES ROHRES
12845	THICKNESS OF SHEET METAL	EPAISSEUR D'UNE TOLE	BLECHSTÄRKE, BLECHDICKE
12846	THICKNESS OF TOOTH	EPAISSEUR DE LA DENT	ZAHNHÖHE, ZAHNDICKE, ZAHNSTÄRKE
12847	THICKNESSING MACHINE	MACHINE A RABOTER TIRANT LES BOIS D'EPAISSEUR	DICKENHOBELMASCHINE
12848	THIMBLE	COSSE	KAUSCHE, SEILKAUSCHE
12849	THIN DRAWN WIRE	FIL METALLIQUE FIN	DRAHT (FEINER), DRAHT (FEINGEZOGNER), FEINDRAHT
12850	THIN PLATE	TOLE FINE	FEINBLECH
12851	THIN-WALLED	PAROI MINCE (A)	DÜNNWANDIG
12852	THIOSULPHATE	HYPOSULFITE, THIOSULFATE	SALZ (UNTERSCHWEFLIGSAURES), THIOSULFAT, HYPOSULFIT
12853	THIOSULPHURIC, HYPOSULPHOROUS ACID	ACIDE HYPOSULFUREUX, ACIDE THIOSULFURIQUE	UNTERSCHWEFLIGE SÄURE, THIOSCHWEFELSÄURE
12854	THIXOTROPY	THIXOTROPIE	THIXOTROPIE
12855	THOMAS BASIC STEEL	ACIER THOMAS	THOMASSTAHL
12856	THORIUM	THORIUM	THORIUM
12857	THREAD	FILETAGE, PAS DE VIS	GEWINDE
12858	THREAD CLEARANCE, CLEARANCE PLAY AT THE APEX OF MALE THREAD	JEU A FOND DE FILET	SPITZENSPIEL IM GEWINDE
12859	THREAD CUTTING LATHE	TOUR A FILETER	GEWINDEDREHBANK
12860	THREAD GAUGES	JAUGES POUR FILETAGES	GEWINDE LEHREN
12861	THREAD PLUG GAUGE, SINGLE ENDED TYPE	JAUGE NORMALE - TAMPON FILETE	NORMALGEWINDELEHRDORN
12862	THREAD PLUG GAUGES	CALIBRES MALES POUR FILETAGES	GEWINDELEHRDORN
12863	THREAD RING GAUGE	JAUGE NORMALE BAGUE FILETEE	NORMALGEWINDELEHRRING
12864	THREAD ROOT	FOND DE FILET	GEWINDEKERN
12865	THREAD TOOL GAUGE	GABARIT POUR OUTILS A FILETER	GEWINDESTAHLLEHRE
12866	THREADED FLANGE	BRIDE TARAUDEE	GEWINDEFLANSCH
12867	THREADED MANDREL	MANDRIN FILETE	GEWINDEDORN
12868	THREADING	TARAUDAGE, FILETAGE	GEWINDESCHNEIDEN, INNENGEWINDE
12869	THREADING MACHINE	MACHINE A FILETER	AUSSENGEWINDESCHNEIDMASCHINE
12870	THREE SQUARE FILE	LIME TIERS-POINT	FEILE (DREIKANTIGE)
12871	THREE-AXLE LORRY	VEHICULE A TROIS ESSIEUX	LASTKRAFTWAGEN (DREIACHSIG)
12872	THREE-CENTRE, THREE-ARC CURVE, COMPOUND CURVE	ANSE DE PANIER	KORBBOGENLINIE
12873	THREE-CYLINDER ENGINE	MACHINE A TROIS CYLINDRES	DREIZYLINDERMASCHINE
12874	THREE-JAW UNIVERSAL CHUCK	MANDRIN UNIVERSEL A TROIS MORS	DREIBACKENFUTTER

12875	THREE-PHASE GENERATOR, ALTENATOR	GENERATRICE A COURANT TRIPHASE	DREIPHASENDYNAMO, DREHSTROMDYNAMO, DREHSTROMMASCHINE
12876	THREE-PHASE MOTOR	MOTEUR TRIPHASE, MOTEUR A CHAMP TOURNANT	DREHSTROMMOTOR, DREIPHASENMOTOR
12877	THREE-PHASE, TRIPHASE CURRENT	COURANT TRIPHASE	DREIPHASENSTROM, DREHSTROM
12878	THREE-PLY BELT	COURROIE TRIPLE	RIEMEN (DREIFACHER)
12879	THREE-THREADED, TREBLE-THREADED SCREW	VIS A TROIS FILETS	SCHRAUBE (DREIGÄNGIE)
12880	THREE-THROW CRANK SHAFT	VILEBREQUIN A TROIS COUDES	WELLE (DREIKURBELIGE), KURBELWELLE (DREIFACH GEKRÖPFTE)
12881	THREE-WAY COCK	ROBINET A TROIS VOIES	DREIWEGHAHN
12882	THREE-WAY VALVE	SOUPAPE A TROIS VOIES, ROBINET A TROIS VOIES	DREIWEGEVENTIL, WECHSELVENTIL, UMSTELLVENTIL
12883	THREE-WIRE METHOD	METHODE DITE A TROIS FILS	DREIFADENMETHODE
12884	THRESHING MACHINES	BATTEUSES	DRESCHMASCHINEN
12885	THROAT DEPTH	EPAISSEUR DE LA SOUDURE	ELEKTRODENARMAUSLADUNG
12886	THROAT OF FILLET WELD	EPAISSEUR D'UNE SOUDURE EN ANGLE	KEHLNAHTDICKE
12887	THROAT OPENING	LARGEUR D'UNE SOUDURE	FENSTERÖFFNUNG
12888	THROTTLE	VOLET DE GAZ, PAPILLON	DROSSELKLAPPE
12889	THROTTLE (TO)	ETRANGLER	DROSSELN
12890	THROTTLE VALVE, BUTTERFLY VALVE	PAPILLON DE REGLAGE, REGULATEUR, VALVE A PAPILLON	DROSSELKLAPPE, REGELVENTIL
12891	THROTTLING	ETRANGLEMENT	DROSSELUNG, DROSSELN
12892	THROUGH HARDENING	DURCISSEMENT A COEUR	DURCHHÄRTUNG
12893	THROUGH HOLE	TROU TRAVERSANT LA PIECE	LOCH (DURCHGEHENDES)
12894	THROW OF AN ECCENTRIC	COURSE DE L'EXCENTRIQUE, COURSE DE LA CAME	EXZENTERHUB
12895	THROW OFF THE BELT (TO)	DESCENDRE LA COURROIE	ABWERFEN (DEN RIEMEN)
12896	THROWAWAY INSERTS	PLAQUETTES UNISERVICE	WEGWERFPLATTEN
12897	THROWING INTO GEAR, ENGAGEMENT	EMBRAYAGE	EINSCHALTEN, EINSCHALTUNG, EINRÜCKEN, EINRÜCKUNG, EINKUPPELN, EINKUPPLUNG
12898	THROWING OUT OF GEAR, DISENGAGEMENT	DEBRAYAGE, DESEMBRAYAGE, DESENGRENAGE	AUSSCHALTEN, AUSSCHALTUNG, AUSRÜCKEN, AUSRÜCKUNG, AUSKUPPELN, AUSKUPPLUNG, ENTKUPPELN, ENTKUPPLUNG
12899	THROWING POWER	ACTION EN PROFONDEUR	TIEFENSTREUUNG
12900	THRUST BALL BEARING	PALIER DE BUTEE A BILLES	AXIALKUGELLAGER
12901	THRUST BALL BEARING, BALL THRUST BEARING	BUTEE A BILLES, ROULEMENT A BILLES A CHARGE AXIALE	KUGELSTÜTZLAGER, KUGELSPURLAGER, STÜTZKUGELLAGER, KUGELDRUCKLAGER
12902	THRUST BEARING	PALIER A CHARGE AXIALE, PALIER DE BUTEE, BUTEE	LÄNGSDRUCKLAGER, DRUCKLAGER, AXIALLAGER
12903	THRUST BLOCK, MULTIPLE COLLAR THRUST BEARING	PALIER DE BUTEE A CANNELURES, COUSSINET A CANNELURES	KAMMLAGER, SCHEIBENLAGER, VIELRINNDRUCKLAGER
12904	THRUST COLLAR	COLLET, EMBASE D'UN PIVOT CANNELE, BAGUE DE BUTEE, BAGUE D'ARRET	KAMM, RING, DRUCKRING, SPURRING, SPURKRANZ EINES DRUCKLAGERS, STELLRING
12905	THRUST DISC	GRAIN, CULOT D'UNE CRAPAUDINE	SPURPLATTE, SPURPFANNE
12906	THRUST JOURNAL	TOURILLON A CANNELURES, TOURILLON DE BUTEE CANNELE	KAMMZAPFEN, SCHEIBENZAPFEN
12907	THRUST OF THE GROUND	POUSSEE DES TERRES	ERDDRUCK
12908	THRUST ON CROWN OF AN ARCH	POUSSEE AU SOMMET	SCHEITELDRUCK
12909	THRUST SCREW	VIS DE BUTEE	GEGENDRUCKSCHRAUBE
12910	THULIUM	THULIUM	THULIUM

12911	THUMB NUT, FLY NUT, WING NUT	ECROU A OREILLES, ECROU PAPILLON, VIS VIOLON	FLÜGELMUTTER
12912	THUMB SCREW, WING SCREW	VIS AILEE, VIS A OREILLES	FLÜGELSCHRAUBE
12913	THYRISTOR	THYRISTOR	THYRISTOR
12914	TIE (TO), ANCHOR (TO)	ACCROCHER, ANCRER	VERANKERN
12915	TIE ROD	BARRE D'ACCOUPLEMENT	SPURSTANGE
12916	TIE-BAR	BARRE D'ACCOUPLEMENT	SPURSTANGE
12917	TIE, TIE ROD	TIRANT	ZUGSTANGE, ZUGANKER
12918	TIEING, ANCHORING	ACCROCHAGE, ANCRAGE	VERANKERUNG, VERANKERN
12919	TIES OF BRACES	HAUBANS DE SPHERE	ANKERZUGSTANGE, ANKERZUGSEILE
12920	TIGHT	ETANCHE	DICHT, DICHTHALTEND, DICHTSCHLIESSEND
12921	TIGHT JOINT	JOINT ETANCHE	FUGE (DICHTE)
12922	TIGHT JOINT RIVETING, RIVETING FOR TIGHT JOINTS	RIVET D'ETANCHEITE	HEFTNIETUNG, VERSCHLUSSNIETUNG, DICHTUNGSNIETUNG
12923	TIGHT ROPE	CORDE TENDUE	SEIL (GESPANNTES)
12924	TIGHTEN A KEY (TO)	SERRER UN COIN	ANZIEHEN (EINEN KEIL)
12925	TIGHTEN A SCREW NUT (TO)	SERRER UN ECROU, SERRER UNE VIS	ANZIEHEN (EINE SCHRAUBE)
12926	TIGHTEN THE STUFFING BOX (TO), SCREW DOWN THE GLAND (TO)	SERRER LA BOITE A ETOUPE	ANZIEHEN (DIE STOPFBUCHSE)
12927	TIGHTEN UP (TO), PULL UP A SCREW (TO)	SERRER A BLOC UN ECROU, SERRER A FOND UN ECROU	FEST ANZIEHEN (EINE SCHRAUBE), SCHARF ANZIEHEN (EINE SCHRAUBE)
12928	TIGHTENING A KEY	SERRAGE D'UN COIN	ANZUG EINES KEILES, ANZIEHEN EINES KEILES
12929	TIGHTENING BY KEYS	SERRAGE PAR COIN	KEILVERSPANNUNG
12930	TIGHTENING BY SCREW BOLTS	SERRAGE PAR ECROUS	SCHRAUBENVERSPANNUNG
12931	TIGHTENING TORQUE	COUPLE DE SERRAGE	ANZUGSDREHMOMENT
12932	TIGHTNESS	ETANCHEITE	DICHTIGKEIT, DICHTHALTEN
12933	TILE, PAVING TILE, FLOOR TILE, WALL TILE	CARREAU (CONSTR.)	FLIESE
12934	TILE, ROOFING TILE	TUILE	ZIEGEL, DACHZIEGEL
12935	TILTING LADLE	POCHE DE COULEE A RENVERSEMENT	KIPPFANNE
12936	TILTING MOMENT	MOMENT BASCULANT	KIPPMOMENT
12937	TIMBER	BOIS DE CONSTRUCTION, BOIS DE CHARPENTE	WERKHOLZ, NUTZHOLZ, BAUHOLZ
12938	TIMBER CUT LONGITUDINALLY THROUGH THE MEDULLARY RAYS	BOIS DE MAILLE	SPIEGELHOLZ
12939	TIMBER CUT WITH THE GRAIN	BOIS DE FIL	LANGHOLZ, LÄNGENHOLZ
12940	TIMBER FREE FROM KNOTS	BOIS SANS NOEUDS	HOLZ (ASTFREIES)
12941	TIMBER SLEEPER	TRAVERSE EN BOIS	HOLZSCHWELLE
12942	TIMBER WORK	CHARPENTE, CONSTRUCTION EN BOIS	HOLZBAU, HOLZKONSTRUKTION
12943	TIME	TEMPS	ZEIT
12944	TIME CONSTANT	CONSTANTE DE TEMPS	ZEITKONSTANTE
12945	TIME OF SWING	DUREE D'UNE OSCILLATION	SCHWINGUNGSDAUER
12946	TIME RECORDER	COMPTE-SECONDES, COMPTEUR A POINTAGES, HORLOGE POINCONNEUSE	STECHUHR
12947	TIMING CASE	CARTER DE DISTRIBUTION	STEUERGEHÄUSE
12948	TIMING CHAIN	CHAINE DE DISTRIBUTION	STEUERKETTE
12949	TIMING GEAR	PIGNON DE DISTRIBUTION	VENTILSTEUERUNGS-ZAHNRAD
12950	TIMING GEAR CHAIN	CHAINE DE DISTRIBUTION	STEUERKETTE

12951	TIMING MARKS	REPERES DE CALAGE	EINSTELLUNGSANGELPUNKTE
12952	TIMING SETTING	CALAGE DE LA DISTRIBUTION	STEUERUNGSEINSTELLUNG
12953	TIN	ETAIN	ZINN
12954	TIN (TO)	ETAMER	VERZINNEN
12955	TIN FOIL	FEUILLE D'ETAIN, PAILLON D'ETAIN	BLATTZINN, ZINNFOLIE, STANNIOL
12956	TIN ORE	MINERAI D'ETAIN	ZINNERZ
12957	TIN PLATE	FER-BLANC, TOLE ETAMEE	WEISSBLECH
12958	TIN SOLDER	SOUDAGE TENDRE	WEICHLÖTEN
12959	TIN STONE, CASSITERITE	CASSITERITE	ZINNSTEIN, KASSITERIT
12960	TIN TUBE	TUYAU EN ETAIN	ZINNROHR
12961	TIN, CAN	BIDON (POUR HUILE, ESSENCE, ECT.)	KANISTER, BLECHKANNE
12962	TINNED WIRE	FIL ETAME, FIL GALVANISE	DRAHT (VERZINNTER)
12963	TINNING	ETAMAGE	VERZINNUNG, VERZINNEN
12964	TINSMITH, WHITESMITH	FERBLANTIER	KLEMPNER, SPENGLER
12965	TINTED DRAWING, WASH DRAWING, DRAWING WITH COLOURS WASHED IN	DESSIN LAVE, LAVIS	ZEICHNUNG (GETUSCHTE), TUSCHZEICHNUNG
12966	TIP OF A NOZZLE	BUSE, EMBOUCHURE	DÜSE, MUNDSTÜCK
12967	TIPPING LORRY	VEHICULE A BENNE BASCULANTE	KIPPER
12968	TIRATION	TITRAGE	TITRIEREN, TITRIERUNG, TITRATION
12969	TISSUE PAPER	PAPIER PELURE, PAPIER JOSEPH, PAPIER DE SOIE	SEIDENPAPIER
12970	TITANIUM	TITANE	TITAN
12971	TITANIUM DIOXIDE	ANHYDRIDE TITANIQUE	TITANDIOXYD, TITANSÄUREANHYDRID
12972	TITANIUM STEEL	ACIER AU TITANE	TITANSTAHL
12973	TITRATE (TO)	TITRER	TITRIEREN
12974	TITRATION	TITRAGE	TITRIERUNG
12975	TOE-IN	PINCEMENT	VORSPUR
12976	TOE-OUT	OUVERTURE	VORSPUR (NEGATIVE)
12977	TOE-PLATE	ARRETE-PIEDS	FUSSLEISTE
12978	TOGGLE BRAKE	FREIN A GENOUILLERE	KNIEHEBELBREMSE
12979	TOGGLE JOINT, MECHANISM	LEVIER A GENOUILLERE, GENOUILLERE, LEVIER ARTICULE	KNIEHEBEL, KNIEHEBELVERBINDUNG
12980	TOGGLE LEVER PRESS	PRESSE A GENOUILLERES	KNIEHEBELPRESSE
12981	TOGGLES	SERRE-BARRES	STANGEN-SPANNZANGEN
12982	TOLERANCE	TOLERANCE	TOLERANZ
12983	TOLUENE, TOLUOL	TOLUENE, METHYLBENZENE, TOLUOL, HYDRURE DE BENZYLE, HYDRURE DE CRESYLE,	TOLUOL, METHYLBENZOL, BENZYLWASSERSTOFF
12984	TOMBAC, RED BRASS	TOMBAC	ROTGUSS, ROTMESSING, ROTMETALL, TOMBAK, MASCHINENBRONZE
12985	TOMMY	BROCHE, LEVIER DE VIS	SCHRAUBENHEBEL
12986	TON (METRIC)	TONNE (1000 KG)	TONNE (1000 KG)
12987	TONGS, PLIERS, PLYERS, PINCERS, NIPPERS	TENAILLE (S), PINCES (S)	ZANGE
12988	TONGUE (TO)	RAINER ET LANGUETTER, BOUVETER	SPUNDEN, NUT UND FEDER VERSEHEN (MIT)
12989	TONGUE AND GROOVE	LANGUETTE ET RAINURE	FEDER UND NUT, SPUND
12990	TONGUE-GROOVE FACING	EMBOITEMENT DOUBLE (M ET F)	FEDER UND NUT
12991	TONGUING	ASSEMBLAGE PAR RAINURE ET LANGUETTE, BOUVETAGE	SPUNDEN

12992	TONGUING AND GROOVING MACHINE	MACHINE A BOUVETER, MACHINE A FAIRE LES RAINURES ET LANGUETTES	SPUNDMASCHINE
12993	TONGUING PLANE	BOUVET MALE	FEDERHOBEL
12994	TOOL	OUTIL	WERKZEUG
12995	TOOL ANGLE	ANGLE DE L'OUTIL, ANGLE TAILLANT, TRANCHANT, ANGLE DE COIN, ANGLE D'AFFUTAGE	ZUSCHÄRFUNGSWINKEL, MEISSELWINKEL
12996	TOOL BAG	SAC A OUTILS	WERKZEUGTASCHE
12997	TOOL BOX	BOITE A OUTILS	WERKZEUGKOFFER
12998	TOOL CABINET	ARMOIRE A OUTILS	WERKZEUGSCHRANK
12999	TOOL CHANGE TIME	TEMPS DE CHANGEMENT D'OUTIL	WERKZEUGWECHSELZEIT
13000	TOOL CHANGER	CHANGEUR D'OUTILS	WERKZEUGWECHSELVORRICHTUNG
13001	TOOL FUNCTION	FONCTION OUTIL	WERKZEUGBEFEHL
13002	TOOL GRINDER, TOOL GRINDING MACHINE	MACHINE A AFFUTER LES OUTILS, AFFUTEUSE	WERKZEUGSCHLEIFMASCHINE, SCHÄRFMASCHINE
13003	TOOL HEAD	PORTE-OUTIL	STÖSSELKOPFSTAHLHALTER
13004	TOOL HOLDER TURRETS	TOURELLES PORTE-OUTILS	STAHLHALTER-REVOLVERKÖPFE
13005	TOOL HOLDERS	PORTE-OUTILS	STAHLHALTER
13006	TOOL MAKING	FABRICATION D'OUTILS A TAILLANTS, TAILLANDERIE	WERKZEUGFABRIKATION
13007	TOOL OFFSET	CORRECTION DE POSITION D'OUTIL	WERKZEUGVERSATZ
13008	TOOL SMITH	OUVRIER TREMPEUR, TREMPEUR D'OUTILLAGE	ZEUGSCHMIED, WERKZEUGSCHLOSSER
13009	TOOL STEEL	ACIER A OUTILS	WERKZEUGSTAHL
13010	TOOL-POST	MONTANT PORTE-OUTIL	STAHLHALTER
13011	TOOL, CUTTER	OUTIL COUPANT (D'UNE MACHINE OUTIL)	STAHL, SCHNEIDSTAHL, STICHEL
13012	TOOLBARS	PORTE OUTILS	ACKERSCHIENEN
13013	TOOLING CLAMPS	BRIDES D'USINAGE	BEARBEITUNGSFLANSCHE
13014	TOOLMAKERS'S HAMMER	MARTEAU D'OUTILLEUR	WERKZEUGBAUHAMMER
13015	TOOLS	OUTILLAGE	HANDWERKSZEUG, WERKZEUGAUSRÜSTUNG
13016	TOOTH	DENT	ZAHN
13017	TOOTH DEPTH	HAUTEUR DE DENT	ZAHNHÖHE
13018	TOOTH FLANK	FLANC DE DENT	ZAHNFLANKE
13019	TOOTH PROFILE, OUTLINE OF TOOTH	PROFIL D'UNE DENT	ZAHNFORM, ZAHNPROFIL
13020	TOOTH SHAPE	PROFIL DE DENTURE	ZAHNPROFIL
13021	TOOTHED GEARING, WHEEL GEARING	ENGRENAGE	RÄDERGETRIEBE, ZAHNRADGETRIEBE, ZAHNTRIEB
13022	TOOTHED RACK	CREMAILLERE	ZAHNSTANGE
13023	TOOTHED RIM	COURONNE DENTEE	ZAHNKRANZ
13024	TOOTHED SEGMENT, TOOTHED SECTOR	SECTEUR DENTE	ZAHNBOGEN, ZAHNSEKTOR, VERZAHNTER SEKTOR, GEZAHNTER SEKTOR
13025	TOOTHED WHEEL DRIVE, GEAR WHEEL DRIVE	COMMANDE PAR ENGRENAGES, PAR ROUE DENTEE	RÄDERANTRIEB, ZAHNRADANTRIEB
13026	TOOTHING	DENTURE	VERZAHNUNG
13027	TOOTHING PLANE	RABOT DENTE A DENTS	ZAHNHOBEL
13028	TOP ANGLE , CURB ANGLE	CORNIERE DE TETE(D'UN RESERVOIR)	OBERWINKELEISEN
13029	TOP BOLTING BAR	PLAT DE MAINTIEN	BEFESTIGUNSFLACHEISEN
13030	TOP BRASS	DEMI-COUSSINET SUPERIEUR, COQUILLE SUPERIEURE, CONTRE-COUSSINET	LAGERSCHALE (OBERE), OBERSCHALE
13031	TOP CASTING	COULEE DIRECTE	FALLENDGIESSEN

13032	TOP DEAD CENTER	POINT MORT HAUT (P.M.H)	TOTPUNKT (OBERER)
13033	TOP IRON, BACK IRON, NON-CUTTING IRON OF A PLANE	CONTRE-FER	DECKPLATTE, DECKEL, KAPPE EINES HOBELEISENS
13034	TOP PISTON-RING	SEGMENT DE FEU	KOMPRESSIONSRING (OBERER)
13035	TOP RAIL	LISSE	HANDLEISTE, HANDLAUF
13036	TOP SWAGE	ETAMPE, MATRICE DE DESSUS	OBERGESENK
13037	TOP VIEW, PLAN	VUE DE HAUT EN BAS	AUFSICHTSBILD, DRAUFSICHTSBILD
13038	TOPAZ	TOPAZE	TOPAS
13039	TOPOCHEMISTRY	TOPOCHIMIE	TOPOCHEMIE
13040	TOPPED CRUDE	PETROLE BRUT TOPPE	TOPPRÜCKSTAND
13041	TOPPING FILE	LIME OLIVE	FEILE MIT ZWEI RUNDEN KANTEN
13042	TORCH	TORCHE, CHALUMEAU	BRENNER, FACKEL
13043	TORE, TORUS, ANCHOR RING	TORE	RINGKÖRPER, (GEOM.)
13044	TORICONICAL HEAD	FOND TORICONIQUE	KEGELSENKKOPF
13045	TORISPHERICAL HEAD	FOND EN ANSE DE PANIER	KORBBOGENBODEN, BODEN (FLACHGEWÖLBTER)
13046	TORISPHERICAL ROOF	TOIT SURBAISSE	DACH (FLACHEWÖLBTES)
13047	TORISPHERICAL ROOF TANK	RESERVOIR A TOIT SURBAISSE	TIEFBEHÄLTER
13048	TORQUE	COUPLE MOTEUR, COUPLE	DREHMOMENT
13049	TORQUE WRENCH	CLE, BRAS DYNAMOMETRIQUE	DREHMOMENTSCHRAUBENSCHLÜSSEL
13050	TORQUEMETER	COUPLE METRE	DREHMOMENTMESSER
13051	TORSION	TORSION	TORSION
13052	TORSION BALANCE	BALANCE DE TORSION	VERDREHUNGSWAAGE, TORSIONSWAAGE
13053	TORSION BAR	BARRE DE TORSION	DREHSTAB
13054	TORSION, TWISTING, TWIST	TORSION	DREHUNG, VERDREHUNG, VERDRILLUNG, TORSION
13055	TORSIONAL FORCE	FORCE DE TORSION	TORSIONSKRAFT
13056	TORSIONAL STRAIN	EFFORT DE TORSION, TRAVAIL A LA TORSION	TORSION, DREHBEANSPRUCHUNG, VERDREHUNGSBEANSPRUCHUNG, TORSONSBEANSPRUCHUNG
13057	TORSIONAL STRENGTH	RESISTANCE A LA TORSION	VERDREHUNGSFESTIGKEIT, DREHUNGSFESTIGKEIT, TORSIONSFESTIGKEIT
13058	TORSIONAL STRESS	EFFORT DE TORSION PAR UNITE DE SECTION	DREHSPANNUNG, TORSIONSSPANNUNG
13059	TORSIONAL TEST	ESSAI A LA TORSION	VERDREHUNGSVERSUCH
13060	TOTAL INTERNAL REFLEXION	REFLEXION TOTALE	REFLEXION (TOTALE)
13061	TOTAL LIFT OF A PUMP	HAUTEUR D'ELEVATION D'UNE POMPE	FÖRDERHÖHE EINER PUMPE
13062	TOTAL LOAD	CHARGE TOTALE	GESAMTBELASTUNG
13063	TOTAL LOST MOTION	JEU TOTAL PERDU	GESAMTTOTGANG
13064	TOTAL OUTPUT	DEBIT TOTAL	GESAMTLEISTUNG
13065	TOTAL PRESSURE	PRESSION TOTALE	GESAMTDRUCK
13066	TOUCH (TO)	TOUCHER (GEOM.)	BERÜHREN (GEOM.)
13067	TOUGH	TENACE	ZÄH
13068	TOUGH IRON	FER TENACE	ZÄHES EISEN
13069	TOUGH PITCH COPPER	CUIVRE A OXYDE CUIVREUX	KUPFER (CUPRIOXYD-ENTHALTENDES)
13070	TOUGH PITCH COPPER, REFINED COPPER	CUIVRE AFFINE	KUPFER (HAMMERGARE), KUPFER (RAFFINIERTES) RAFFINADEKUPFER
13071	TOUGHNESS	TENACITE, RESILIENCE	RESILIENZ, ZÄHIGKEIT
13072	TOURMALINE	TOURMALINE	TURMALIN
13073	TOW, OAKUM	ETOUPE	WERG
13074	TOWN REFUSE	ORDURES	MÜLL

13075	TOWN WATER	EAU DE DISTRIBUTION	LEITUNGSWASSER
13076	TRACE	TRACE	SPUR
13077	TRACE (TO)	CALQUER	PAUSEN, DURCHPAUSEN
13078	TRACER	CALQUEUR	PAUSER
13079	TRACER OF PLANIMETER	POINTE TRACANTE DU PLANIMETRE	FAHRSTIFT DES PLANIMETERS
13080	TRACER-CONTROLLED MILLING MACHINES	FRAISEUSE A COMMANDE PAR DISPOSITIF DE COPIAGE	FRÄSMASCHINE (FÜHLERGESTEUERTE)
13081	TRACHYTE	TRACHYTE	TRACHYT
13082	TRACING	CALQUAGE, DECALQUE, CALQUE	PAUSEN, DURCHPAUSEN, PAUSE
13083	TRACING CLOTH	TOILE A CALQUER	PAUSLEINWAND, ZEICHENLEINWAND
13084	TRACING PAPER	PAPIER CALQUE, PAPIER TRANSPARENT	PAUSPAPIER
13085	TRACK, TRACKING WIDTH	VOIE, PISTE, GORGE DE ROULEMENT	SPUR, SPURWEITE, LAUFRILLE
13086	TRACKING CHECK	VERIFICATION DU PARALLELISME	RADAUSFLUCHTUNGSKONTROLLE
13087	TRACKING CONTROL	GUIDE DE CENTRAGE	ZENTRIERFÜHRUNG
13088	TRACTION	TRACTION DES VEHICULES	ZUGFÖRDERUNG, TRAKTION
13089	TRACTIVE EFFORT, FORCE	EFFORT DE TRACTION	ZUGKRAFT (EINES TRAKTORS)
13090	TRACTIVE, PORTATIVE FORCE OF A MAGNET, TRACTIVE LIFTING POWER OF A MAGNET	PUISSANCE DE LEVAGE D'UN AIMANT	TRAGKRAFT, ZUGKRAFT EINES MAGNETEN
13091	TRACTOR	TRACTEUR	ZUGMASCHINE
13092	TRACTOR AND TRAILER UNIT	VEHICULE ARTICULE	SATTELZUG
13093	TRACTOR HALF TRACK	TRACTEUR AGRICOLE DEMI TRAC	HALBRAUPENSCHLEPPER
13094	TRACTOR TRACKLAYING	TRACTEUR AGRICOLE A CHENILLES	RAUPENSCHLEPPER
13095	TRACTOR WHEELED	TRACTEUR AGRICOLE A ROUES	ACKERRADSCHLEPPER
13096	TRACTOR, HORTICULTURAL	TRACTEUR MARAICHERS	TRAKTOREN, GÄRTNEREI-
13097	TRACTOR, TRACTION ENGINE	TRACTEUR	ZUGWAGEN, ZUGMASCHINE, SCHLEPPER, TRAKTOR
13098	TRACTORY, TRACTRIX	TRACTRICE, TRACTOIRE	SCHLEPPKURVE, TRAKTRIX
13099	TRADE BALANCE	BALANCE DU COMMERCE EXTERIEUR	AUSSENHANDELSBILANZ
13100	TRADE MARK	MARQUE DE FABRIQUE, MARQUE DE COMMERCE	FABRIKMARKE, WARENZEICHEN, SCHUTZMARKE
13101	TRADE NAME	NOM COMMERCIAL	HANDELSBEZEICHNUNG, HANDELSÜBLICHE BEZEICHNUNG
13102	TRAILERS	REMORQUES	ANHÄNGER
13103	TRAIN OIL, WHALE OIL, BLUBBER OIL	HUILE DE POISSON, HUILE DE BALEINE	TRAN, FISCHTRAN, FISCHÖL
13104	TRAJECTORY	TRAJECTOIRE	BAHN, TRAJEKTORIE
13105	TRAMMELS	COMPAS A VERGE	STANGENZIRKEL
13106	TRAMP IRON	FER DANS LE BROYEUR	EISEN IM BRECHGUT
13107	TRANSCENDENTAL EQUATION	EQUATION TRANSCENDANTE	GLEICHUNG (TRANSZENDENTE)
13108	TRANSDUCER	TRANSDUCTEUR, CAPTEUR	TRANSDUCER, TRANSDUKTOR
13109	TRANSFER	DECALCOMANIE	ABZIEHBILD
13110	TRANSFER LADLE	POCHE-TONNEAU	TRANSPORTPFANNE
13111	TRANSFERENCE, TRANSMISSION OF HEAT	TRANSMISSION DE LA CHALEUR	WÄRMEÜBERTRAGUNG
13112	TRANSFORM (TO), CONVERT ERNERGY (TO)	TRANSFORMER L'ENERGIE	ENERGIE UMWANDELN
13113	TRANSFORMATION RANGE	DOMAINE DE TRANSFORMATION	UMWANDLUNGSBEREICH
13114	TRANSFORMATION TEMPERATURE	TEMPERATURE DE TRANSFORMATION	UMWANDLUNGSTEMPERATUR

13115	**TRANSFORMER**	TRANSFORMATEUR, TRANSFORMATEUR DE POTENTIEL	UMWANDLER, UMSPANNER, TRANSFORMATOR
13116	**TRANSFORMER PLATE**	TOLE POUR TRANSFORMATEURS	TRANSFORMATORBLECH
13117	**TRANSGRANULAR FRACTURE**	CASSURE TRANSCRISTALLINE	BRUCH (TRANSKRISTALLINER)
13118	**TRANSIENT CREEP**	FLUAGE TRANSITOIRE	VORLÄUFIGES FLIESSEN
13119	**TRANSISTOR**	TRANSISTOR	TRANSISTOR
13120	**TRANSITION LATTICE**	RESEAU DE TRANSLATION	TRANSLATIONSGITTER
13121	**TRANSITION PART**	ELEMENT OU PARTIE TRONCONIQUE, REDUCTION	REDUZIERANSCHLUSS
13122	**TRANSITION POINT**	POINT DE TRANSITION	ÜBERGANGSPUNKT
13123	**TRANSLATION**	DEPLACEMENT, TRANSLATION, TRADUCTION	PARALLELVERSCHIEBUNG, UMSETZUNG, ÜBERSETZUNG, TRANSLATION
13124	**TRANSLUCENCE**	TRANSLUCIDITE	TRANSPARENZ
13125	**TRANSLUCENT**	TRANSLUCIDE	DURCHSCHEINEND, TRANSPARENT
13126	**TRANSMISSION COMPONENTS**	ORGANES DE TRANSMISSION	ANTRIEBSELEMENTE
13127	**TRANSMISSION DYNAMOMETER**	DYNAMOMETRE DE TRANSMISSION	TRANSMISSIONSDYNAMOMETER
13128	**TRANSMISSION GEAR WHEEL**	ENGRENAGE DE TRANSMISSION	ARBEITSRAD, TRIEBWERKSRAD
13129	**TRANSMISSION OF ELECTRICAL ENERGY**	TRANSMISSION D'ENERGIE ELECTRIQUE	KRAFTÜBERTRAGUNG (ELEKTRISCHE)
13130	**TRANSMISSION OF MOTION**	TRANSMISSION D'UN MOUVEMENT	ÜBERTRAGUNG EINER BEWEGUNG, BEWEGUNGSÜBERTRAGUNG
13131	**TRANSMISSION OF POWER BY GEARING**	TRANSMISSION PAR ENGRENAGE	RÄDERÜBERSETZUNG, KRAFTÜBERTRAGUNG DURCH RÄDER
13132	**TRANSMISSION OF POWER BY LEVERS**	TRANSMISSION PAR LEVIER	HEBELÜBERSETZUNG, KRAFTÜBERTRAGUNG DURCH HEBEL
13133	**TRANSMISSION OF POWER ELECTRICALLY OVER LONG DISTANCES,**	TRANSMISSION D'ELECTRICITE, TRANSPORT D'ENERGIE ELECTRIQUE A GRANDE DISTANCE	FERNKRAFTÜBERTRAGUNG (ELEKTRISCHE), FERNÜBERTRAGUNG VON ELEKTRISCHER ENERGIE, ÜBERTRAGUNG VON ELEKTRISCHER ENERGIE
13134	**TRANSMISSION OF POWER, POWER TRANSMISSION**	TRANSPORT DE L'ENERGIE, TRANSMISSION DE PUISSANCE	KRAFTÜBERTRAGUNG, ENERGIETRANSPORT
13135	**TRANSMISSION, CONVEYANCE OF ENERGY OVER LONG DISTANCES**	TRANSPORT D'ENERGIE A GRANDE DISTANCE, DISTRIBUTION A DISTANCE DE L'ENERGIE	FERNLEITUNG VON ENERGIE
13136	**TRANSMUTATION OF ELEMENTS**	TRANSMUTATION (CHIM.)	UMWANDLUNG, ELEMENTUMWANDLUNG
13137	**TRANSPARENCY**	TRANSPARENCE	DURCHSICHTIGKEIT
13138	**TRANSPARENT**	TRANSPARENT	DURCHSICHTIG
13139	**TRANSPARENT MEDIUM**	MILIEU TRANSPARENT	MEDIUM (DURCHSICHTIGES)
13140	**TRANSPORTATION, TRANSPORT**	TRANSPORT	BEFÖRDERUNG, VERSAND, TRANSPORT
13141	**TRANSPORTER**	PONT TRANSBORDEUR, TRANSBORDEUR	BRÜCKENKRAN, VERLADEBRÜCKE
13142	**TRANSVERSE AXIS**	AXE TRANSVERSAL	QUERACHSE
13143	**TRANSVERSE AXIS OF HYPERBOLA**	AXE FOCAL, AXE TRANSVERSE D'UNE HYPERBOLE	HAUPTACHSE DER HYPERBEL
13144	**TRANSVERSE ELASTICITY**	ELASTICITE DE CISAILLEMENT, ELASTICITE TRANSVERSALE	SCHUBELASTIZITÄT
13145	**TRANSVERSE PITCH OF RIVETS**	RIVETS (DISTANCE DES) DE CENTRE A CENTRE D'UNE LIGNE A L'AUTRE	QUERNIETTEILUNG
13146	**TRANSVERSE VIBRATION**	VIBRATION, OSCILLATION TRANSVERSALE	QUERSCHWINGUNG
13147	**TRANSVERSELY RIBBED TUBE, TUBE RIBBED IN ITS TRANSVERSE SECTION**	TUBE A NERVURES TRANSVERSALES	RIPPENROHR MIT QUERRIPPEN
13148	**TRAPEZIUM**	TRAPEZE	TRAPEZ

13149	**TRAPEZOID**	TRAPEZOIDE	TRAPEZOID
13150	**TRAPEZOIDAL LOAD**	CHARGE TRAPEZOIDALE	TRAPEZLAST
13151	**TRAPEZOIDAL PLATE SPRING**	RESSORT TRAPEZOIDAL	TRAPEZFEDER
13152	**TRASS**	TRASS	TRASS
13153	**TRAVEL INDICATOR**	INDICATEUR D'OUVERTURE	ANZEIGEVORRICHTUNG, ZEIGER
13154	**TRAVEL OF THE VALVE**	JEU D'UNE SOUPAPE	VENTILSPIEL
13155	**TRAVELLING CRAB**	CHARIOT D'UN PONT-ROULANT, CHARIOT PORTE-PALAN	LAUFKATZE
13156	**TRAVELLING CRANE**	PONT ROULANT	LAUFKRAN
13157	**TRAVELLING CRANE BEAM**	POUTRE DE PONT ROULANT	LAUFKRANTRÄGER
13158	**TRAVELLING MOTION**	TRANSLATION	FAHRBEWEGUNG
13159	**TRAVELLING NUT**	ECROU MOBILE	WANDERMUTTER
13160	**TRAVELLING POISE**	CURSEUR D'UNE BALANCE	LAUFGEWICHT, LÄUFER EINER WAAGE
13161	**TRAVELLING PORTABLE CRANE**	GRUE MOBILE, GRUE ROULANTE	FAHRBARER KRAN
13162	**TRAVERSE (TO)**	CYLINDRER, CHARIOTER UNE SURFACE CYLINDRIQUE	ABDREHEN, LANGDREHEN, PARALLELDREHEN, EGALISIEREN
13163	**TRAVERSE HANDWHEEL**	VOLANT DE CHARIOTAGE	SCHLITTENKREUZBEWEGUNGS-HANDRAD
13164	**TRAVERSING**	CYLINDRAGE, CHARIOTAGE D'UNE SURFACE CYLINDRIQUE, CHARIOTAGE LONGITUDINAL	ABDREHEN, LANGDREHEN, PARALLELDREHEN, EGALISIEREN
13165	**TRAVERSING SCREW JACK**	VERIN A CHARIOT	SCHLITTENWINDE
13166	**TRAVERTINE**	TRAVERTIN	TRAVERTIN
13167	**TRAY**	PLATEAU	TABLETT
13168	**TRAY MANWAY**	TRAPPE DE VISITE DE PLATEAU	TABLETTENMANNLOCH
13169	**TRAY SPACING**	ECARTEMENT DES PLATEAUX	TABLETTABSTAND
13170	**TRAY SUPPORT RING**	COURONNE-SUPPORT DE PLATEAUX	TABLETTRINGHALTER
13171	**TRAYS TOWER**	TOUR A PLATEAUX	FRAKTIONIERTURM
13172	**TREAD**	MARCHE, CHAPE, BANDE DE ROULEMENT	STUFE, REIFENLAUFFLÄCHE
13173	**TREAD OF A WHEEL**	SURFACE DE ROULEMENT D'UNE ROUE	LAUFFLÄCHE EINES RADES
13174	**TREADLE, FOOT BOARD**	PEDALE	TRITT, TRETSCHEMEL
13175	**TREATMENT FOR MAKING MALLEABLE IRON CASTINGS**	MALLEABILISATION DE LA FONTE	GLÜHFRISCHEN, TEMPERN
13176	**TREBLE HELICAL SPUR WHEEL**	ENGRENAGE A DOUBLES CHEVRONS	DOPPELWINKELZAHNRAD, DOPPELPFEILZAHNRAD
13177	**TREBLE RIVETED JOINT**	RIVURE TRIPLE A TRIPLE CLOUURE, RIVURE A TROIS RANGS	NIETUNG (DREIREIHIGE)
13178	**TREBLE SHEAR RIVETED JOINT**	RIVURE A TROIS COUPES	NIETUNG (DREISCHNITTIGE)
13179	**TREENAIL**	CHEVILLE EN BOIS	HOLZNAGEL
13180	**TRELLIS WORK**	TREILLAGE	FLECHTWERK
13181	**TREND**	ALLURE (COURBE)	LAUF, VERLAUF
13182	**TREPANNING MACHINE**	MACHINE A FORER	HOHLBOHRMASCHINE
13183	**TREPEZOIDAL CROSS SECTION**	SECTION TRAPEZOIDALE	QUERSCHNITT (TRAPEZF:ORMIGER)
13184	**TRIAL ERECTION**	MONTAGE A BLANC	VERSUCHSMONTAGE
13185	**TRIANGLE**	TRIANGLE	DREIECK (GEOM.)
13186	**TRIANGLE OF FORCES**	TRIANGLE DE FORCES	KRÄFTEDREIECK
13187	**TRIANGULAR COMPASSES**	COMPAS A TROIS BRANCHES	ZIRKEL (DREISPITZIGER), DREISCHENKELZIRKEL
13188	**TRIANGULAR CROSS SECTION**	SECTION TRIANGULAIRE	QUERSCHNITT (DREIECKIGER)
13189	**TRIANGULAR FILE, THREE-SQUARE, THREE-CORNERED FILE**	LIME TIERS-POINT, LIME TROIS-QUARTS	DREIKANTFEILE

13190	**TRIANGULAR LOAD**	CHARGE TRIANGULAIRE	DREIECKLAST
13191	**TRIANGULAR PLATE SPRING**	RESSORT TRIANGULAIRE	DREIECKFEDER
13192	**TRIANGULAR PRISM**	PRISME TRIANGULAIRE	PRISMA (DREISEITIGES)
13193	**TRIANGULAR PYRAMID**	PYRAMIDE TRIANGULAIRE	PYRAMIDE (DREISEITIGE)
13194	**TRIANGULAR ROPE**	CABLE A SECTION TRIANGULAIRE	DREIKANTSEIL
13195	**TRIANGULAR STRAND WIRE ROPE**	CABLE A TORONS TRIANGULAIRES	DREIKANTLITZENSEIL
13196	**TRIANGULAR WIRE**	FIL TRIANGULAIRE	DRAHT (DREIKANTIGER)
13197	**TRIATOMIC, TRIVALENT**	TRIATOMIQUE, TRIVALENT	DREIWERTIG, DREIATOMIG
13198	**TRIBASIC**	TRIBASIQUE	DREIBASISCH
13199	**TRICLINIC CRYSTAL SYSTEM**	SYSTEME TRICLINIQUE	KRISTALLSYSTEM (TRIKLINE)
13200	**TRIGONOMETRICAL**	TRIGONOMETRIQUE	TRIGONOMETRISCH
13201	**TRIGONOMETRY**	TRIGONOMETRIE	TRIGONOMETRIE
13202	**TRIHEDRON**	TRIEDRE	DREIKANT, DREIFLACH, TRIEDER
13203	**TRILINEAR COORDINATES**	COORDONNEES DANS L'ESPACE	RAUMKOORDINATEN
13204	**TRIM (TO)**	PARER, ROGNER	BESCHNEIDEN
13205	**TRIM A PLATE (TO)**	EBARBER UNE TOLE	BLECH (EIN) ENTGRATEN, ABGRATEN (EIN BLECH)
13206	**TRIM OFF THE BURR (TO)**	ENLEVER LES BAVURES, EBARBER	ABGRATEN
13207	**TRIMANGANIC, TRIMANGANESE TETROXIDE**	OXYDE SALIN DE MANGANESE	MANGANOXYDULOXYD, ROTES MANGANOXYD
13208	**TRIMETRIC PROJECTION**	PROJECTION TRIMETRIQUE	PROJEKTION (TRIMETRISCHE)
13209	**TRIMMING**	ACTION DE PARER, ROGNAGE, EBARBAGE	BESCHNEIDEN, ABGRATEN
13210	**TRIMMING OFF THE BURR**	ENLEVEMENT DES BAVURES, EBARBAGE, EBARBEMENT, EBAVURAGE	ABGRATEN
13211	**TRIMMING PRESS**	PRESSE A EBARBER	ABGRATPRESSE
13212	**TRINIDAD ASPHALT**	ASPHALTE DE TRINIDAD	TRINIDADASPHALT
13213	**TRINOMIAL EQUATION**	EQUATION TRINOME	GLEICHUNG (DREIGLIEDRIGE), GLEICHUNG (TRINOMISCHE)
13214	**TRIP DOGS**	CRABOTS ENCLENCHEURS/DECLENCHEURS	ANSCHLAGBOLZEN
13215	**TRIPLE INTEGRAL**	INTEGRALE TRIPLE	INTEGRAL (DREIFACHES)
13216	**TRIPLE POINT**	POINT TRIPLE	TRIPELPUNKT
13217	**TRIPLE-EXPANSION ENGINE**	MACHINE A TRIPLE EXPANSION	DREIFACHEXPANSIONSMASCHINE
13218	**TRIPOD JACK**	VERIN A TREPIED	DREIFUSSWINDE
13219	**TRIPOD STAND, STAND**	TREPIED	STÄNDER, STATIV
13220	**TRIPOLI**	TRIPOLI	TRIPEL, POLIERSCHIEFER
13221	**TRIPPING, JUMPING CAM**	DISQUE A CAMES	DAUMENSCHEIBE, KNAGGENSCHEIBE, NOCKENSCHEIBE
13222	**TRITURATE (TO) RUB TO POWDER (TO)**	TRITURER	ZERREIBEN
13223	**TROCHOID**	TROCHOIDE	TROCHOIDE
13224	**TROLLEY BEAMS**	PORTIQUE DE ROULEMENT	PORTALKRAN
13225	**TROLLEY BUS**	TROLLEYBUS	OBUS
13226	**TROMMEL, REVOLVING SCREEN**	TROMMEL	TROMMELSIEB
13227	**TROOSTITE**	TROOSTITE	TROOSTIT
13228	**TROUGH, TUB, VAT**	BAC, BACHE, AUGE, CUVE	TROG, BOTTICH
13229	**TRUCK CAB**	CABINE DE CAMION	FÜHRERHAUS
13230	**TRUCK, TRANSVEYOR, TROLLEY**	CHARIOT (POUR MANUTENTION D'ATELIER ET MAGASIN)	FÖRDERKARREN, TRANSPORTKARREN
13231	**TRUCKS TROLLEYS**	CHARIOTS A BRAS	HANDKARREN, ROLLBÖCKE
13232	**TRUE UP (TO)**	DRESSER (RENDRE PLAN)	ABRICHTEN

13233	**TRUEING DEVICES**	APPAREILS A DIAMANTER	ABRICHTAPPARATE
13234	**TRUING UP**	DRESSAGE (ACTION DE RENDRE PLAN)	ABRICHTEN, RICHTEN
13235	**TRUNCATED CONE**	TRONC DE CONE, CONE TRONQUE	KEGEL (ABGESTUMPFTER), KEGELSTUMPF
13236	**TRUNCATED CONICAL SPRING**	RESSORT TRONCONIQUE	KEGELSTUMPFFEDER
13237	**TRUNCATED PARALLELIPIPED**	TRONC DE PARALLELEPIPEDE	PARALLELEPIPED (ABGESTUMPFTES)
13238	**TRUNCATED PYRAMID**	TRONC DE PYRAMIDE	PYRAMIDE (ABGESTUMPFTE), PYRAMIDEN$TUMPF
13239	**TRUNK (U.S.)**	COFFRE, MALLE	KOFFERRAUM
13240	**TRUNK LID**	COUVERCLE DU COFFRE	KOFFERRAUMDECKEL
13241	**TRUNK PISTON**	PISTON SANS TIGE	TRUNKKOLBEN
13242	**TRUNNION**	TOURILLON SERVANT D'AXE DE ROTATION, ARTICULATION	DREHZAPFEN, SCHWENKZAPFEN
13243	**TRUSSES**	FERMES, ARMATURE, TRAVERSE	TRAGBALKEN
13244	**TRUSSES FRAME**	CHARPENTE A FERMES	BUNDGESPÄRRE
13245	**TRY SQUARE**	EQUERRE A EPAULEMENT A CHAPEAU	ANSCHLAGWINKEL
13246	**TUBE (TYRE)**	CHAMBRE A AIR	REIFENSCHLAUCH
13247	**TUBE CLEANER, SCRAPER**	BROSSE RACLETTE POUR TUBES	ROHRREINIGER
13248	**TUBE CUTTER, PIPE CUTTER**	COUPE-TUBE	ROHRABSCHNEIDER
13249	**TUBE DRAWING**	ETIRAGE DES TUBES	ZIEHEN VON ROHREN
13250	**TUBE EXPANDER**	MANDRIN A ELARGIR LES TUBES	ROHRAUFWEITEDORN
13251	**TUBE IGNITION**	ALLUMAGE PAR INCANDESCENCE	GLÜHROHRZÜNDUNG
13252	**TUBE MILL**	TUBE BROYEUR	ROHRMÜHLE
13253	**TUBE OF FORCE**	TUBE DE FORCE	KRAFTRÖHRE
13254	**TUBE PLATE**	PLAQUE TUBULAIRE	ROHRPLATTE
13255	**TUBE STEM OF A THERMOMETER**	TUBE CAPILLAIRE, TIGE DU THERMOMETRE	THERMOMETERRÖHRE
13256	**TUBE VICE, PIPE VICE**	ETAU A TUBES, ETAU DE TUYAUTEUR	ROHRSCHRAUBSTOCK
13257	**TUBE, PIPE**	TUYAU, TUBE	ROHR, RÖHRE
13258	**TUBULAR BRACING**	RENFORCEMENT EN TUBES, ENTRETOISAGE EN TUBES	VERSTEIFUNGSROHR
13259	**TUBULAR SPIRIT LEVEL**	NIVEAU A BULLE D'AIR	LIBELLE, RÖHRENLIBELLE, WASSERWAAGE, NIVEAU
13260	**TUFA, TUFF**	TUF	TUFF
13261	**TUMBLING BARREL**	TAMBOUR DE NETTOTAGE, TONNEAU DE NETTOYAGE, TONNEAU DE FINISSAGE	REINIGUNGSTROMMEL, PUTZTROMMEL
13262	**TUNGSTEN**	TUNGSTENE	WOLFRAM
13263	**TUNGSTEN BRONZE**	BRONZE AU TUNGSTENE	WOLFRAMBRONZE
13264	**TUNGSTEN FILAMENT LAMP**	LAMPE A FILAMENT DE TUNGSTENE	WOLFRAMDRAHTLAMPE
13265	**TUNGSTEN STEEL**	ACIER AU TUNGSTENE	WOLFRAMSTAHL
13266	**TUNGSTEN TRIOXIDE**	ANHYDRIDE TUNGSTIQUE	WOLFRAMSÄUREANHYDRID, WOLFRAMTRIOXYD
13267	**TURBINE**	TURBINE	TURBINE
13268	**TURBINE PLATES**	AUBES DE TURBINES	TURBINENSCHAUFELN
13269	**TURBINE PUMP**	POMPE CENTRIFUGE	ZENTRIFUGALPUMPE, TURBOPUMPE
13270	**TURBO-BLOWER**	TURBO-VENTILATEUR	KREISELGEBLÄSE, TURBOGEBLÄSE
13271	**TURBO-GENERATOR**	TURBO-GENERATEUR	TURBODYNAMO, TURBOGENERATOR
13272	**TURBOCHARGER**	TURBO-COMPRESSEUR	TURBOKOMPRESSOR
13273	**TURBULENCE**	TURBULENCE	TURBULENZ

13274	TURBULENT FLOW	COURANT TURBULENT	STRÖMUNG (TURBULENTE), STRÖMUNG (UNGEORDNETE), STRÖMUNG (WIRBELIGE)
13275	TURMERIC PAPER	PAPIER DE CURCUMA	KURKUMAPAPIER
13276	TURN (TO) (ON THE LATHE)	TOURNER, TRAVAILLER AU TOUR	DREHEN (AUF DER DREHBANK)
13277	TURN CAMBERED SURFACES (TO)	TOURNER DES SURFACES BOMBEES	BALLIGDREHEN
13278	TURN OF A HELIX	TOUR D'UNE HELICE	WINDUNG EINER SCHRAUBENLINIE
13279	TURN OF A THREAD	REVOLUTION DU FILET	GANG EINES GEWINDES, GEWINDEGANG
13280	TURN ORNAMENTAL (TO), ECCENTRIC WORK	TOURNER DES SURFACES FIGUREES	PASSIG DREHEN, UNRUND DREHEN
13281	TURN SPHERICAL SURFACES (TO)	TOURNER DES SURFACES SPHERIQUES, TOURNER DES CORPS RONDS	KUGELIG DREHEN
13282	TURN TAPER (TO),CONICAL	TOURNER DES SURFACES OBLIQUES/ RAMPANTES	DREHEN (KEGELIG), DREHEN (KONISCH)
13283	TURN UP WITH A SHOVEL (TO)	REMUER A LA PELLE	UMSCHAUFELN
13284	TURN-KEY JOB	CLE EN MAIN (AFFAIRE)	SCHLÜSSELFERTIG
13285	TURNBUCKLE	TENDEUR A VIS	SPANNSCHLOSS, SCHRAUBENSCHLOSS, SPANNVORRICHTUNG, NACHSPANNVORRICHTUNG
13286	TURNER	TOURNEUR	DREHER
13287	TURNING (ON THE LATHE)	TOURNAGE, TRAVAIL AU TOUR	DREHEN (AUF DER DREHBANK)
13288	TURNING CAMBERED SURFACES	TOURNAGE DES SURFACES BOMBEES	BALLIGDREHEN
13289	TURNING LATHE	TOUR (MACHINE-OUTIL)	DREHBANK, WELLENDREHBANK
13290	TURNING SHOP, TURNERY	ATELIER DE TOURNAGE	DREHEREI
13291	TURNING SPHERICAL SURFACES	TOURNAGE DES SURFACES SPHERIQUES DES CORPS RONDS	KUGELDREHEN
13292	TURNING TOOL, LATHE TOOL	OUTIL DE TOUR	DREHSTAHL, DREHSTICHEL, STAHL
13293	TURNINGS	TOURNURES	DREHSPÄNE
13294	TURNSCREW, SCREW DRIVER	TOURNEVIS	SCHRAUBENZIEHER
13295	TURNTABLE	PLAQUE TOURNANTE	DREHSCHEIBE
13296	TURPENTINE VARNISH	VERNIS A L'ESSENCE	TERPENTINÖLLACK
13297	TURRET LATHE	TOUR REVOLVER	REVOLVERDREHMASCHINE
13298	TURRET SINGLE SPINDLE AUTOMATIC LATHE	TOUR MONOBROCHE AUTOMATIQUE A TOURELLE	EINSPINDEL-AUTOMAT MIT REVOLVERKOPF
13299	TUYERE	TUYERE D'AMENEE DE L'AIR SOUS PRESSION	WINDFORM, BLASFORM
13300	TWIN	MACLE	ZWILLING
13301	TWIN ENGINE	MOTEUR A CYLINDRES JUMELES	ZWILLINGSMASCHINE
13302	TWINNING	MACLAGE	ZWILLINGSBILDUNG
13303	TWIST (TO)	COMMETTRE LES TORONS, CABLER	VERSEILEN
13304	TWIST DRILL	FORET HELICOIDAL, FORET HELICOIDE, FORET A HELICE, FORET AMERICAIN	SPIRALBOHRER
13305	TWIST OF A ROPE	TORS, TORSION D'UNE CORDE	DRALL EINES SEILES
13306	TWIST TEST	ESSAI DE TORSION	VERDREHUNGSPROBE
13307	TWISTED CURVE, CURVE IN SPACE	COURBE DANS L'ESPACE, COURBE GAUCHE, COURBE A DOUBLE COURBURE	RAUMKURVE, RÄUMLICHE KURVE, GEWUNDENE KURVE, KURVE DOPPELTER KRÜMMUNG
13308	TWISTING MOMENT, TURNING MOMENT, TORQUE	MOMENT TORDANT, MOMENT DU COUPLE	DREHMOMENT
13309	TWISTING OFF A SCREW THREAD	RUPTURE BRUSQUE DU CORPS D'UNE VIS	ABWÜRGEN (EINER SCHRAUBE)
13310	TWISTING TEST	ESSAI DE FLEXION TOURNANTE	VERWINDEPROBE

13311	**TWO PACK PAINT SYSTEM**	SYSTEME DE PEINTURE A DEUX COMPOSANTS	ZWEIKOMPONENTENLACK
13312	**TWO STROKE CYCLE ENGINE**	MOTEUR A DEUX TEMPS	ZWEITAKTMOTOR
13313	**TWO-CYLINDER ENGINE**	MACHINE A DEUX CYLINDRES	ZWEIZYLINDERDAMPFMASCHINE
13314	**TWO-PHASE, BIPHASE CURRENT**	COURANT DIPHASE	ZWEIPHASENSTROM
13315	**TWO-THREAD, DOUBLE THREADED SCREW**	VIS A DEUX FILETS	SCHRAUBE (ZWEIGÄNGIGE), SCHRAUBE (DOPPELGÄNGIGE)
13316	**TWO-THROW CRANK SHAFT**	ARBRE VILEBREQUIN A DEUX COUDES	ZWEIKURBELIGE WELLE, DOPPELT GEKRÖPFTE KURBELWELLE
13317	**TWO-WAY COCK**	ROBINET A DEUX VOIES	ZWEIWEGEHAHN
13318	**TWO-WAY VALVE**	SOUPAPE A DEUX VOIES	ZWEIWEGEVENTIL
13319	**TYPE METAL**	METAL PROPRE A LA FABRICATION DES CARACTERES D'IMPRIMERIE	SCHRIFTMETALL
13320	**TYPE, MODEL**	TYPE	BAUART, AUSFÜHRUNGSFORM, TYP, KONSTRUKTION, MODELL
13321	**TYRE TURNING LATHE**	TOUR A BANDAGES	RADREIFENDREHBANK
13322	**TYRE, TIRE**	BANDAGE DE ROUE, PNEU	RADREIFEN
13323	**U-BEND**	COUDE EN U, COUDE A 180	DOPPELKRÜMMER, U-ROHR, RÜCKBOGEN
13324	**U-BOLT**	COLLIER DE TUBE, ETRIER	ROHRSCHELLE, BÜGELSCHRAUBE
13325	**U-TUBE**	TUBE COUDE A DEUX BRANCHES COMMUNIQUANT ENTRE ELLES	RÖHREN (KOMMUNIZIERENDE)
13326	**ULLAGE**	CREUX DE LA SPHERE	LEERRAUM
13327	**ULTIMATE BREAKING STRENGTH**	RESISTANCE A LA RUPTURE, RESISTANCE, TENACITE EXTREME	BRUCHFESTIGKEIT
13328	**ULTIMATE COMPRESSIVE CRUSHING STRENGTH**	RESISTANCE A LA COMPRESSION, RESISTANCE A L'ECRASEMENT, RESISTANCE AU RACCOURCISSEMENT	DRUCKFESTIGKEIT
13329	**ULTIMATE ELONGATION, ELONGATION UP TO THE BREAKING STRAIN**	ALLONGEMENT PROPORTIONNEL A LA LIMITE D'ELASTICITE	BRUCHDEHNUNG
13330	**ULTIMATE SHEARING STRENGTH**	RESISTANCE AU CISAILLEMENT, RESISTANCE A LA RUPTURE TRANSVERSALE, RESISTANCE AU GLISSEMENT TRANSVERSAL	SCHUBFESTIGKEIT, SCHERFESTIGKEIT
13331	**ULTIMATE TENSILE STRENGTH**	RESISTANCE A LA TRACTION, RESISTANCE A L'ALLONGEMENT, RESISTANCE A L'EXTENSION, CHARGE DE RUPTURE	ZUGFESTIGKEIT, BRUCHLAST, BRUCHBELASTUNG, BRUCHFESTIGKEIT
13332	**ULTRA-RED RAYS**	RAYONS ULTRA-ROUGES	STRAHLEN (ULTRAROTE)
13333	**ULTRA-VIOLET RAYS**	RAYONS ULTRA-VIOLETS	STRAHLEN (ULTRAVIOLETTE)
13334	**ULTRAMARINE**	OUTREMER	ULTRAMARIN, LASURBLAU, AZURBLAU
13335	**ULTRAMICROSCOPE**	ULTRAMICROSCOPE	ULTRAMIKROSKOP
13336	**ULTRAMICROSCOPIC TEST**	EXAMEN A L'ULTRAMICROSCOPE	UNTERSUCHUNG (ULTRAMIKROSKOPISCHE), BEOBACHTUNG (ULTRAMIKROSKOPISCHE)
13337	**ULTRASONIC EXAMINATION, ULTRASONIC INSPECTION**	CONTROLE AUX ULTRA-SONS	ULTRASCHALLUNTERSUCHUNG, ULTRASCHALLPRÜFUNG
13338	**ULTRASONIC WELDING**	SOUDAGE PAR ULTRASONS	ULTRASCHALLSCHWEISSEN
13339	**UMBRA**	OMBRE PORTEE	SCHLAGSCHATTEN
13340	**UNANNEALED, BRIGHT, HARD DRAWN WIRE**	FIL CLAIR	DRAHT (UNGEGLÜHTER), DRAHT (HARTGEZOGENER), DRAHT (BLANKER)
13341	**UNBALANCE, WANT, LACK OF BALANCE**	BALOURD, DESEQUILIBRE	UNSYMMETRIE, UNGLEICHGEWICHT, UNBALANZ
13342	**UNBREAKABILITY**	FRAGILITE (NON-)	UNZERBRECHLICHKEIT

13343	UNBREAKABLE	INCASSABLE	UNZERBRECHLICH
13344	UNBURNT	IMBRULE	UNVERBRANNT
13345	UNCOMBINED CARBON, TEMPER CARBON	CARBONE NON COMBINE AU FER	TEMPERKOHLE
13346	UNDAMPED OSCILLATION	VIBRATION NON AMORTIE	SCHWINGUNG (UNGEDÄMPFTE)
13347	UNDER COOLING	REFROIDISSEMENT (SOUS-)	UNTERKÜHLUNG
13348	UNDER FILLING	MANQUE DE MATIERE	STOFFMANGEL
13349	UNDER SIZE	DIMENSION INFERIEURE AUX PRESCRIPTIONS	UNTERMASS
13350	UNDERBEAD-CRACK	FISSURE SOUS CORDON	UNTERNAHTRISS
13351	UNDERCUT	COUPE INFERIEURE A LA COTE, CANIVEAU (SOUDURE)	UNTERSCHNITT, EINBRANDKERBEN
13352	UNDERFILM CORROSION	CORROSION SOUS-JACENTE	UNTERROSTUNG
13353	UNDERGROUND CABLE; UNDERGROUND CONDUIT	LIGNE SOUTERRAINE, CANALISATION ELECTRIQUE	LEITUNG (UNTERIRDISCHE)
13354	UNDERGROUND PIPING	CONDUITE SOUTERRAINE, CANALISATION	ROHRLEITUNG (UNTERIRDISCHE)
13355	UNDERSHOT WATER WHEEL	ROUE PAR-DESSOUS	WASSERRAD (UNTERSCHLÄCHTIGES)
13356	UNEQUAL SIDED ANGLE IRON	CORNIERE A AILES INEGALES	WINKELEISEN (UNGLEICHSCHENKLIGES)
13357	UNEVEN FRACTURE	CASSURE INEGALE	BRUCH (UNEBENER)
13358	UNGULA	ONGLET CYLINDRIQUE	ZYLINDERHUF
13359	UNHARDENED	DURCI (NON-), TREMPE (NON-)	UNGEHÄRTET
13360	UNIFLOW CURRENT	COURANTS DE MEME SENS	GLEICHSTROM (VON FLÜSSIGKEITEN)
13361	UNIFORM ACCELERATION	ACCELERATION UNIFORME	BESCHLEUNIGUNG (GLEICHBLEIBENDE)
13362	UNIFORM MOTION	MOUVEMENT UNIFORME	BEWEGUNG (GLEICHFÖRMIGE)
13363	UNIFORM VELOCITY	VITESSE UNIFORME CONSTANTE	GESCHWINDIGKEIT (UNVERÄNDERLICHE), GESCHWINDIGKEIT (KONSTANTE)
13364	UNIFORMITY OF THE MATERIAL	HOMOGENEITE D'UN MATERIEL	GLEICHFÖRMIGKEIT DES MATERIALS
13365	UNIFORMLY ACCELERATED MOTION	MOUVEMENT UNIFORMEMENT ACCELERE	BEWEGUNG (GLEICHMÄSSIG BESCHLEUNIGTE)
13366	UNIFORMLY DISTRUBUTED LOAD	CHARGE UNIFORMEMENT REPARTIE	LAST (GLEICHMÄSSIG VERTEILTE)
13367	UNIFORMLY RETARDED MOTION	MOUVEMENT UNIFORMEMENT RETARDE	BEWEGUNG (GLEICHMÄSSIG VERZÖGERTE)
13368	UNION	UNION, RACCORD	VERSCHRAUBUNG
13369	UNION, PIPE UNION	RACCORD TROIS PIECES, RACCORD-UNION, ECROU DE RAPPEL	ROHRVERSCHRAUBUNG
13370	UNIT ELONGATION	ALLONGEMENT RELATIF	DEHNUNG (RELATIVE)
13371	UNIT LOAD	CHARGE PAR UNITE DE SURFACE	FLÄCHENEINHEITSLAST
13372	UNIT OF AREA	UNITE DE SURFACE	FLÄCHENEINHEIT
13373	UNIT OF LENGTH	UNITE DE LONGUEUR	LÄNGENEINHEIT
13374	UNIT OF MASS	UNITE DE MASSE	MASSENEINHEIT
13375	UNIT OF MEASURE	UNITE DE MESURE	MASSEINHEIT
13376	UNIT OF TIME	UNITE DE TEMPS	ZEITEINHEIT
13377	UNIT OF VOLUME SPACE	UNITE DE VOLUME, UNITE DE CAPACITE	RAUMEINHEIT
13378	UNIT OF WEIGHT	UNITE DE POIDS	GEWICHTSEINHEIT
13379	UNIT OF WORK	UNITE DE TRAVAIL D'ENERGIE	ARBEITSEINHEIT, ENERGIEEINHEIT
13380	UNIT VECTOR	VECTEUR UNITAIRE	EINHEITSVEKTOR
13381	UNITIZED BODY	CARROSSERIE PORTANTE	KAROSSERIE (SELBSTTRAGENDE)
13382	UNIVERSAL CHUCK	MANDRIN UNIVERSEL	UNIVERSALFUTTER

13383	UNIVERSAL CYLINDRICAL GRINDING MACHINE	MACHINE UNIVERSELLE A RECTIFIER LES SURFACES DE REVOLUTION	UNIVERSAL-RUNDSCHLEIFMASCHINE
13384	UNIVERSAL GRINDING MACHINE	AFFUTEUSE UNIVERSELLE	UNIVERSALSCHLEIFMASCHINE
13385	UNIVERSAL JOINT	JOINT A CARDAN	KARDANGELENK
13386	UNIVERSAL MILL	LAMINOIR UNIVERSEL	UNIVERSALSTAHLWALZWERK
13387	UNIVERSAL MILLING MACHINE	FRAISEUSE UNIVERSELLE	UNIVERSALFRÄSMASCHINE
13388	UNIVERSAL TESTING MACHINE	MACHINE D'ESSAIS UNIVERSELLE	UNIVERSALPRÜFMASCHINE
13389	UNIVERSAL VISE	ETAU UNIVERSEL	UNIVERSALSCHRAUBSTOCK
13390	UNIVERSAL WELDING HEAD	TETE DE SOUDAGE UNIVERSELLE	UNIVERSALSCHWEISSKOPF
13391	UNKNOWN QUANTITY	QUANTITE INCONNUE, INCONNUE	GRÖSSE (UNBEKANNTE), UNBEKANNTE
13392	UNLIKE OPPOSITE POLES	POLES DISSEMBLABLES, POLES DE NOM CONTRAIRE	POLE (UNGLEICHNAMIGE)
13393	UNSATURATED HYDROCARBONS	HYDROCARBURES NONSATURES (HYDROCARBURES ETHYLENIQUES, HYDROCARBURES ACETYLENIQUES)	KOHLENWASSERSTOFFE (UNGESÄTTIGTE)
13394	UNSATURATED SOLUTION	SOLUTION NON SATUREE	LÖSUNG (UNGESÄTTIGTE)
13395	UNSCREW (TO)	DEVISSER, DEMONTER UN BOULON	ABSCHRAUBEN, LOSSCHRAUBEN, AUFSCHRAUBEN
13396	UNSCREWING	DEVISSAGE	ABSCHRAUBEN, LOSSCHRAUBEN, AUFSCHRAUBEN
13397	UNSHIELDED METALARC WELDING	SOUDAGE A L'ARC METALLIQUE SANS GAZ PROTECTEUR	METALL-LICHTBOGENSCHWEISSEN OHNE SCHUTZGAS
13398	UNSKILLED LABOURER	OUVRIER NON SPECIALISE, MANOEUVRE DE SERVICE	ARBEITER (UNGELERNTER)
13399	UNSOLDER (TO)	DESSOUDER	LOSLÖTEN
13400	UNSOLDERING	DESSOUDAGE	LOSLÖTEN
13401	UNSTABLE EQUILIBRIUM	EQUILIBRE INSTABLE	GLEICHGEWICHT (LABILES)
13402	UNWIELDY, UNMANAGEABLE	MANIABLE (PEU)	UNHANDLICH
13403	UNWIND	RATTRAPAGE DE JEU	NACHSTELLEN
13404	UNWIND (TO), UNCOIL (TO)	DEROULER	ABWICKELN, ABWINDEN
13405	UNWINDING, UNCOILING	DEROULEMENT	ABWICKELN, ABWINDEN
13406	UP AND DOWN STOKE, FORWARD AND RETURN STROKE OF PISTON	COURSE COMPLETE (ALLER ET RETOUR) DU PISTON	KOLBENSPIEL
13407	UP STREAM	AMONT (EN), AMONT (D')	STROMAUFWÄRTS, GEGEN DEN STROM
13408	UP-MILLING	FRAISAGE EN REMONTANT	GLEICHLÄUFIGES FRÄSEN
13409	UP-STROKE, ASCENT OF PISTON	MONTEE DU PISTON, COURSE ASCENDANTE, COURSE MONTANTE DU PISTON	KOLBENAUFGANG
13410	UP-TIME	TEMPS ACTIF	ARBEITSZEIT
13411	UPPER DEAD CENTER (U.D.C.)	POINT MORT HAUT (P.M.H.)	TOTPUNKT (OBERER)
13412	UPRIGHT DRILLING MACHINE	PERCEUSE SUR COLONNE STANDARD	STÄNDERBOHRMASCHINE
13413	UPSET BUTT WELD	SOUDURE BOUT A BOUT PAR RESISTANCE	WIDERSTANDSSTUMPFSCHWEISSEN
13414	UPSET FORCING	REFOULEMENT	STAUCHEN
13415	UPSETTING MACHINE	MACHINE A REFOULER	STAUCHMASCHINE
13416	UPSETTING TEST	ESSAI D'ECRASEMENT, ESSAI DE COMPRESSION	STAUCHVERSUCH
13417	UPSETTING WITH ELECTRIC RESISTANCE HEATING	REFOULEMENT AVEC CHAUFFAGE PAR EFFET JOULE	ELEKTROSTAUCHVERFAHREN
13418	UPTAKE	CARNEAU	FUCHS (EINER FEUERUNGSANLAGE)

13419	UPWARD VERTICAL POSITION	SOUDURE EN POSITON VERTICALE MONTANTE	SENKRECHTE SCHWEISSUNG, AUFWÄRTSSCHWEISSUNG
13420	URANIUM	URANIUM	URAN
13421	URBAN BUS	AUTOBUS	STADTBUS
13422	USEFUL CROSS SECTION	SECTION UTILE	QUERSCHNITT (NUTZBARER)
13423	USEFUL DIAMETER	DIAMETRE UTILE	NUTZDURCHMESSER
13424	USEFUL LENGTH	LONGUEUR UTILE	LÄNGE (NUTZBARE), NUTZLÄNGE
13425	USEFUL LOAD	CHARGE UTILE	NUTZLAST
13426	USEFUL RESISTANCE	RESISTANCE UTILE	NUTZWIDERSTAND
13427	USEFUL ROLLING WIDTH	LARGEUR UTILE DE LAMINAGE	WALZNUTZBREITE
13428	USEFUL WORK	TRAVAIL UTILE	ARBEIT (NUTZBARE), NUTZARBEIT
13429	UTILISATION	UTILISATION	AUSNUTZUNG, NUTZBARMACHUNG, VERWERTUNG
13430	UTILISE (TO)	UTILISER	AUSNÜTZEN, NUTZBAR MACHEN, VERWERTEN
13431	V-BELT	COURROIE TRAPEZOIDALE	KEILRIEMEN
13432	V-SHAPED GOUGE	GOUGE TRIANGULAIRE	GEISSFUSS (STEMMEISEN)
13433	VACANCY	LACUNE	LÜCKE, LEERSTELLE
13434	VACUUM	VIDE, DEPRESSION ATMOSPHERIQUE, PRESSION	LUFTLEERER, RAUM, VAKUUM, UNTERDRUCK
13435	VACUUM BOX	BOITE A VIDE	VAKUUMKASTEN
13436	VACUUM BRAKE	FREIN A VIDE, FREIN A DEPRESSION	LUFTSAUGEBREMSE, VAKUUMBREMSE, UNTERDRUCKBRENSE
13437	VACUUM BREAKER	SOUPAPE CASSE-VIDE	RÜCKSCHLAGVENTIL
13438	VACUUM CAN (G.B.), VACUUM BOX (U.S.A.)	BOITE A VIDE	VAKUUMGEHÄUSE
13439	VACUUM CASTING	MOULAGE SOUS VIDE	VAKUUMFORMEN
13440	VACUUM CHAMBER	CHAMBRE A VIDE, ENCEINTE A VIDE	VAKUUMKAMMER
13441	VACUUM ENVIRONNMENT	ATMOSPHERE RAREFIEE	ATMOSPHÄRE (VERDÜNNTE)
13442	VACUUM GAUGE	INDICATEUR DU VIDE, VACUOMETRE	LUFTLEEREMESSER, UNTERDRUCKMESSER, VAKUUMMETER
13443	VACUUM HEAT TREATMENT	TRAITEMENT THERMIQUE SOUS VIDE	VAKUUM WÄRMEBEHANDLUNG
13444	VACUUM MELTING	FUSION SOUS VIDE	VAKUUMSCHMELZEN
13445	VACUUM METALLIZING	METALLISATION SOUS VIDE	VAKUMMETALLISIERUNG
13446	VACUUM METALLURGY	METALLURGIE SOUS VIDE	VAKUUMMETALLURGIE
13447	VACUUM REFINING	RAFFINAGE SOUS VIDE	VAKUUMRAFFINIERUNG
13448	VACUUM SINTERING	FRITTAGE SOUS VIDE	VAKUUMSINTERUNG
13449	VACUUM SPACE	CHAMBRE A VIDE	VAKUUMKAMMER
13450	VACUUM SYSTEM VENTILATION	VENTILATION PAR APPEL	SAUGLÜFTUNG, ASPIRATIONSVENTILATION
13451	VACUUM TANK	BASSIN A VIDE	VAKUUMKESSEL
13452	VALENCE, VALENCY	VALENCE	WERTIGKEIT, VALENZ
13453	VALVE	VALVE, LAMPE, SOUPAPE, CLAPET, OBTURATEUR, DISTRIBUTEUR	VENTIL, RÖHRE, MEMBRANVENTIL
13454	VALVE BODY	CORPS D'UNE SOUPAPE	VENTILKÖRPER
13455	VALVE COCK	ROBINET-VALVE A SOUPAPE	VENTILHAHN
13456	VALVE CONTROL	EPURE DE DISTRIBUTION	VENTILSTEUERUNG
13457	VALVE DIAGRAM	DIAGRAMME D'OUVERTURE ET DE FERMETURE D'UNE SOUPAPE	VENTILDIAGRAMM
13458	VALVE EFFECT	EFFET DE SOUPAPE	VENTILWIRKUNG
13459	VALVE FLAP	CLAPET D'UNE SOUPAPE	VENTILKLAPPE

13460	**VALVE GEAR**	DISTRIBUTION PAR SOUPAPE, APPAREIL DE DISTRIBUTION DE LA VAPEUR	VENTILSTEUERUNG, STEUERUNG (DAMPFM.)
13461	**VALVE GUARD**	ARRET, BUTEE D'UNE SOUPAPE	HUBBEGRENZER EINES VENTILS, VENTILFÄNGER
13462	**VALVE GUIDE**	GUIDE DE SOUPAPE	VENTILFÜHRUNG
13463	**VALVE HEAD**	TETE D'UNE SOUPAPE	TELLER (EINES VENTILS), VENTILTELLER
13464	**VALVE HOOD / CAP**	BOUCHON DE VALVE	VENTILVERSCHRAUBUNG
13465	**VALVE LEVER**	LEVIER DE SOUPAPE	VENTILHEBEL
13466	**VALVE LIFTER**	POUSSOIR DE SOUPAPE	VENTILSTÖSSEL
13467	**VALVE LUBRICATOR**	GRAISSEUR A SOUPAPE	VENTILÖLER
13468	**VALVE SEAT**	SIEGE DE SOUPAPE	VENTILSITZ
13469	**VALVE SETTING / RATING**	TARAGE DE SOUPAPE	VENTILEINSTELLUNGTARIEREN
13470	**VALVE SPRING**	RESSORT DE SOUPAPE	VENTILFEDER
13471	**VALVE TIMING**	DISTRIBUTION	STEUERUNG
13472	**VALVE TUBE**	SOUPAPE	VENTILRÖHRE
13473	**VALVE-CLEARANCE ADJUSTER**	CALIBRE DE REGLAGE DE SOUPAPE	VENTILLEHRE
13474	**VAN**	FOURGON	LIEFERWAGEN
13475	**VANADIUM**	VANADIUM	VANADIUM, VANADIN
13476	**VANADIUM STEEL**	ACIER AU VANADIUM	VANADIUMSTAHL
13477	**VANE**	AILETTES DE REFROIDISSEMENT	KÜHLERRIPPE
13478	**VANE PUMP**	POMPE A PALETTES	FLÜGELPUMPE
13479	**VANE WHEEL, BLADE WHEEL**	ROUE A PALETTES A AILETTES, ROUE A AUBES	FLÜGELRAD
13480	**VANE, BLADE, BUCKET OF A TURBINE, TURBINE BUCKET**	AUBE D'UNE TURBINE	SCHAUFEL EINER TURBINE
13481	**VAPOR LOCK**	BOUCHON DE VAPEUR	DAMPFBLASENBILDUNG
13482	**VAPOR-PLATTING**	RECOUVREMENT PAR DECOMPOSITION D'UN GAZ	DAMPFMETALLISIEREN
13483	**VAPORIZATION**	VAPORISATION	VERDAMPFEN
13484	**VAPOUR**	VAPEUR	DAMPF
13485	**VAPOUR ABSORPTION PLANT**	INSTALLATION D'EVACUATION, ELIMINATION DES BUEES	ENTNEBELUNGSANLAGE, ENTDUNSTUNGSANLAGE
13486	**VAPOUR DEGREASING**	DEGRAISSAGE A LA VAPEUR	ENTFETTEN IM TRIDAMPF
13487	**VAPOUR DEPOSITION**	METALLISATION SOUS VIDE	BEDAMPFEN (METALLDAMPF)
13488	**VARIABLE**	VARIABLE	VERÄNDERLICHE, VARIABLE
13489	**VARIABLE BLOCK FORMAT**	FORMAT A BLOC VARIABLE	EINGABEFORMAT IN VARIABLER SATZSCHREIBWEISE
13490	**VARIABLE MOTION**	MOUVEMENT VARIE	BEWEGUNG (VARIABLE), BEWEGUNG (VERÄNDERLICHE)
13491	**VARIABLE PRESSURE**	PRESSION VARIABLE	DRUCK (VERÄNDERLICHER)
13492	**VARIABLE SPEED DRIVE ASSEMBLY**	ENSEMBLE DE COMMANDE A VITESSE VARIABLE	REGELANTRIEBSTEUERUNG
13493	**VARIABLE SPEED UNIT**	VARIATEUR DE VITESSE	DREHZAHLREGLER
13494	**VARIABLE VELOCITY**	VITESSE VARIEE, VARIABLE	GESCHWINDIGKEIT (VERÄNDERLICHE)
13495	**VARIATION**	VARIATION	SCHWANKUNG
13496	**VARIATION (MATH)**	VARIATION (MATH.)	VARIATION (MATH.)
13497	**VARIATION FLUCTUATION OF PRESSURE**	VARIATIONS DE PRESSION	DRUCKSCHWANKUNG
13498	**VARIATION OF TEMPERATURE**	VARIATION DE TEMPERATURE	WÄRMESCHWANKUNG, TEMPERATURSCHWANKUNG
13499	**VARIATIONS OF SPEED, FLUCTUATIONS IN SPEED**	VARIATIONS DE VITESSE, FLUCTUATIONS DE VITESSE	GESCHWINDIGKEITSSCHWANKUNGEN

	English	French	German
13500	**VARIATIONS OF WATER LEVEL**	VARIATIONS DE NIVEAU D'EAU	VERÄNDERUNG DES WASSERSPIEGELS, SCHWANKUNGEN DES WASSERSTANDES, NIVEAUSCHWANKUNGEN
13501	**VARIEGATED COPPER ORE, PEACOCK COPPER**	CUIVRE PANACHE	BUNTKUPFERERZ
13502	**VARIEGATED SANDSTONE**	GRES BIGARRE	BUNTSANDSTEIN
13503	**VARIETY OF TIMBER**	ESPECE DE BOIS, ESSENCE DE BOIS	HOLZART
13504	**VARNISH**	VERNIS	FIRNIS
13505	**VARNISH COATING**	VERNISSAGE	DECKLACK, ÜBERZUGSLACK
13506	**VARNISHED PAPER**	PAPIER GOMME/LAQUE	SCHELLACKPAPIER, PAPIER (GEFIRNISSTES), PAPIER (LACKIERTES)
13507	**VARYING ACCELERATION**	ACCELERATION NON UNIFORME	BESCHLEUNIGUNG (UNGLEICHFÖRMIGE)
13508	**VASELINE, PETROLEUM JELLY**	VASELINE, GRAISSE MINERALE, COSMOLINE, PETREOLINE, PETROLEINE, PIMELEINE	MINERALFETT, VASELIN
13509	**VECTOR PRODUCT**	PRODUIT VECTORIEL	VEKTORPRODUKT, PRODUKT (VEKTORISCHES)
13510	**VECTOR QUANTITY**	VECTEUR	VEKTOR
13511	**VEE-BELT, V-BELT**	COURROIE DE SECTION TRIANGULAIRE, COURROIE TRAPEZOIDALE	KEILRIEMEN
13512	**VEGETABLE OIL**	HUILE VEGETALE	PFLANZENÖL, ÖL (VEGETABILISCHES)
13513	**VELOCITY CURVE**	COURBE DE VITESSE	GESCHWINDIGKEITSIKURVE
13514	**VELOCITY DIAGRAM**	DIAGRAMME DES VITESSES	GESCHWINDIGKEITSDIAGRAMM
13515	**VELOCITY HEAD**	CHARGE DE LA VITESSE	STAUHÖHE, GESCHWINDIGKEITSHÖHE
13516	**VELOCITY OF FLOW**	VITESSE DU COURANT	STRÖMUNGSGESCHWINDIGKEIT
13517	**VELOCITY OF LIGHT**	VITESSE DE LA LUMIERE	LICHTGESCHWINDIGKEIT
13518	**VELOCITY OF PASSAGE**	VITESSE DE PASSAGE	DURCHFLUSSGESCHWINDIGKEIT
13519	**VELOCITY OF PROPAGATION**	VITESSE DE PROPAGATION	FORTPFLANZUNGSGESCHWINDIGKEIT
13520	**VELOCITY OF SOUND WAVES**	VITESSE DU SON	SCHALLGESCHWINDIGKEIT
13521	**VELOCITY OF TRANSFORMATION**	VITESSE DE TRANSFORMATION	TRANSFORMATIONSGESCHWINDIGKEIT
13522	**VELOCITY RATIO**	RAPPORT DES VITESSES	ÜBERSETZUNGSVERHÄLTNIS
13523	**VELOCITY STAGE**	ETAGE DE VITESSE	GESCHWINDIGKEITSSTUFE
13524	**VENEER**	FEUILLET, FEUILLE DE PLACAGE, PLACAGE	FURNIERHOLZ
13525	**VENEER CUTTING MACHINE**	MACHINE A FAIRE LES FEUILLES DE PLACAGE	SPANHOBELMASCHINE, FURNIERHOBELMASCHINE
13526	**VENEER SAW**	SCIE A PLACAGE	KLOBSÄGE, FURNIERSÄGE
13527	**VENICE TURPENTINE**	TEREBENTHINE DE VENISE	TERPENTIN (VENEZIANER), LÄRCHENTERPENTIN
13528	**VENT**	EVENT, TROU D'AIR	ÖFFNUNG, ENTLÜFTER, LÜFTUNGSÖFFNUNG
13529	**VENTILATED MOTOR**	MOTEUR BLINDE VENTILE	MOTOR (VENTILIERTER)
13530	**VENTILATION**	VENTILATION, AERAGE, AERATION	LÜFTUNG, VENTILATION
13531	**VENTILATION WINDOW, VENTILATOR WINDOW**	DEFLECTEUR	AUSSTELLFENSTER, SCHWENKFENSTER
13532	**VENTURI METER, TUBE**	COMPTEUR VENTURI, COMPTEUR D'EAU VENTURI, TUBE DE VENTURI	VENTURIMESSER, VENTURIROHR, WASSERMESSER (VENTURISCHER)
13533	**VERDIGRIS**	VERT-DE-GRIS, VERDET, ACETATE DE CUIVRE	GRÜNSPAN, KUPFERAZETAT
13534	**VERIFICATION**	VERIFICATION	PRÜFUNG

13535	VERMILLON, MERCURIC SULPHIDE	SULFURE ROUGE DE MERCURE, VERMILLON	ZINNOBER, ZINNOBERROT, VERMILLON
13536	VERNIER	VERNIER	NONIUS, VERNIER
13537	VERNIER CALLIPERS, CALLIPER RULE	PIED, COMPAS A COULISSE	SCHUBLEHRE, SCHIEBLEHRE
13538	VERNIER DEPTH GAUGE	CALIBRE A COULISSE DE PROFONDEUR	TIEFENLEHRE
13539	VERNIER DEPTH GAUGE WITH FINE MICROMETER ADJUSTMENT	CALIBRE A COULISSE DE PROFONDEUR A FIXATION MICROMETRIQUE	TIEFENLEHRE MIT MIKROMETERSCHRAUBE
13540	VERNIER MICROMETER	MICROMETRE A VERNIER	MIKROMETERSCHRAUBE
13541	VERNIER SCALE	ECHELLE VERNIER	SCHIEBELEHRE
13542	VERSUS	FONCTION DE (EN)	ABHÄNGIGKEIT VON (IN)
13543	VERTEX CORNER OF POLYGON	SOMMET D'UN POLYGONE	ECKE EINES POLYGONS
13544	VERTICAL BOILER	CHAUDIERE VERTICALE	KESSEL (STEHENDER)
13545	VERTICAL BORING AND TURNING MILL	TOUR VERTICAL	KARUSSELLDREHMASCHINE
13546	VERTICAL ENGINE	MACHINE VERTICALE	MASCHINE (STEHENDE)
13547	VERTICAL NECK JOURNAL BEARING	PALIER INTERMEDIAIRE, COLLIER POUR ARBRES VERTICAUX	HALSLAGER EINER STEHENDEN WELLE
13548	VERTICAL SPINDLE MILLING MACHINE	FRAISEUSE VERTICALE	FRÄSMASCHINE (STEHENDE), FRÄSMASCHINE VERTIKALE, SENKRECHTFRÄSMASCHINE
13549	VERTICAL TRAVEL	COURSE VERTICALE	BEWEGUNG (SENKRECHTE)
13550	VERTICAL UPRIGHT SHAFT	ARBRE VERTICAL	WELLE (STEHENDE), WELLE (SENKRECHTE), KÖNIGSWELLE
13551	VERTICAL, UPRIGHT	VERTICAL	STEHEND, AUFRECHT, VERTIKAL
13552	VERTICAL, UPRIGHT DRILLING MACHINE	MACHINE A PERCER VERTICALE	BOHRMASCHINE (STEHENDE), VERTIKALBOHRMASCHINE
13553	VESSELS CONNECTED BY U-TUBE	VASES COMMUNICANTS	GEFÄSSE (KOMMUNIZIERENDE)
13554	VIBRATION	VIBRATION	SCHWINGUNG
13555	VIBRATION DAMPER	AMORTISSEUR DE VIBRATIONS	SCHWINGUNGSDÄMPFER
13556	VIBRATION DUE TO BENDING STRESS	VIBRATION DUE A DES EFFORTS DE FLEXION	BIEGUNGSSCHWINGUNG
13557	VIBRATION DUE TO TORSIONAL STRESS, TORSIONAL VIBRATION	VIBRATION DUE A DES EFFORTS DE TORSION	DREHSCHWINGUNG, VERDREHUNGSSCHWINGUNG, TORSIONSSCHWINGUNG
13558	VIBRATION OF A SPRING	OSCILLATION D'UN RESSORT	SCHWINGUNG EINER FEDER
13559	VIBRATIONS	VIBRATIONS, TREPIDATIONS, EBRANLEMENTS	ERSCHÜTTERUNGEN
13560	VIBRATIONS OF ETHER	VIBRATIONS DE L'ETHER	ÄTHERSCHWINGUNGEN
13561	VICE	ETAU	SCHRAUBSTOCK
13562	VICE CLAWS, CLAMPS	MORDACHE A CHARNIERE	SPANNKLUPPE
13563	VICE CLAWS, VICE CLAMPS	MORDACHE D'UN ETAU	BACKENFUTTER EINES SCHRAUBSTOCKS
13564	VICE JAW, CHECK	MACHOIRE, MORS, MORD D'UN ETAU	BACKEN EINES SCHRAUBSTOCKS
13565	VICE JAWS FOR WHEEL-WRIGHT, TYRE SMITH'S VICE	ETAU A CHANFREIN, TENAILLE A CHANFREIN	REIFKLOBEN
13566	VICKERS HARDNESS	DURETE VICKERS	VICKERSHÄRTE
13567	VIEW	VUE	ANSICHT
13568	VIEWING SCREEN	ECRAN DE PROJECTION	LEUCHTSCHIRM
13569	VIOLLE STANDARD	ETALON VIOLLE, VIOLLE	PLATINEINHEIT (DER LICHTSTÄRKE)
13570	VIRGIN METAL	METAL VIERGE	METALL (GEDIEGENES)
13571	VIRTUAL FOCUS	FOYER VIRTUEL IMAGINAIRE	BRENNPUNKT (SCHEINBARER), BRENNPUNKT (VIRTUELLER)

13572	VIRTUAL IMAGE	IMAGE VIRTUELLE	SCHEINBILD, BILD (VIRTUELLES)
13573	VIRTUAL VELOCITY	VITESSE VIRTUELLE	GESCHWINDIGKEIT (VIRTUELLE)
13574	VISCO SI METER	VISCOSIMETRE	ZÄHIGKEITSMESSER, VISKOSIMETER
13575	VISCOSE	VISCOSE	VISKOSE
13576	VISCOSITY	VISCOSITE	ZÄHFLÜSSIGKEIT, DICKFLÜSSIGKEIT, VISKOSITÄT, ZÄHIGKEIT
13577	VISCOUS	VISQUEUX	DICKFLÜSSIG, SCHWERFLÜSSIG, STRENGFLÜSSIG, ZÄHFLÜSSIG, KLEBRIG, VISKOS
13578	VISCOUS FLOW	FLUAGE VISQUEUX	FLIESSEN, LAMINÄRE STRÖMUNG
13579	VISE	ETAU	SCHRAUBSTOCK
13580	VISIBLE SIGNAL	SIGNAL OPTIQUE	ZEICHEN (SICHTBARES), SICHTSIGNAL, SIGNAL (OPTISCHES)
13581	VISUAL ANGLE	ANGLE OPTIQUE, ANGLE VISUEL	GESICHTSWINKEL
13582	VITREOUS	VITREUX	GLASIG
13583	VITREOUS GLASSY LUSTRE	ECLAT VITREUX	GLASGLANZ
13584	VITRIFICATION	VITRIFICATION	SINTERUNG, VERGLASUNG
13585	VOID	LACUNE	FEHLSTELLE
13586	VOLATILE CONSTITUENT	PARTIE CONSTITUANTE VOLATILE	BESTANDTEIL (FLÜCHTIGER)
13587	VOLATILISATION	VOLATILISATION	VERFLÜCHTIGUNG
13588	VOLATILISE (TO)	VOLATILISER	VERFLÜCHTIGEN
13589	VOLATILITY	VOLATILITE	FLÜCHTIGKEIT
13590	VOLCANIC TUFA	TUF VOLCANIQUE	TUFF (VULKANISCHER)
13591	VOLT	VOLT	VOLT
13592	VOLT-AMPERE	VOLT-AMPERE	VOLT-AMPERE
13593	VOLTAGE	TENSION DU COURANT, NOMBRE DE VOLTS, VOLTAGE	SPANNUNG, VOLTZAHL
13594	VOLTAGE POTENTIAL DROP, FALL DROP OF POTENTIAL	PERTE, CHUTE DE TENSION, CHUTE DE POTENTIEL	SPANNUNGSABFALL, SPANNUNGSGEFÄLLE, SPANNUNGSVERLUST, POTENTIALGEFÄLLE
13595	VOLTAGE REGULATOR, POTENTIAL REGULATOR	REGULATEUR DE TENSION, D'INTENSITE, REGULATEUR DE COURANT	SPANNUNGSREGLER, SPANNUNGSREGULATOR
13596	VOLTAIC PILE	PILE VOLTAIQUE, DE VOLTA	SÄULE (VOLTASCHE)
13597	VOLTMETER	VOLTMETRE	SPANNUNGSMESSER, SPANNUNGSZEIGER, VOLTMETER
13598	VOLUME	VOLUME, CUBAGE	RAUMINHALT, VOLUMEN
13599	VOLUME OF STROKE	VOLUME ENGENDRE PAR LE DEPLACEMENT DU PISTON	HUBVOLUMEN
13600	VOLUME REGULATOR	REGULATEUR DE DEBIT	MENGENREGLER
13601	VOLUMETRIC ANALYSIS	ANALYSE VOLUMETRIQUE	MASSANALYSE (VOLUMETRISCHE)
13602	VOLUMETRIC EFFICIENCY	RENDEMENT VOLUMETRIQUE	WIRKUNGSGRAD (RÄUMLICHER), WIRKUNGSGRAD (VOLUMETRISCHER)
13603	VOLUMETRIC EXPANSION, EXTENSION	DILATATION CUBIQUE	AUSDEHNUNG (RÄUMLICHE), RAUMAUSDEHNUNG
13604	VOLUMMETER	DEBITMETRE	MENGENMESSER
13605	VOLUTE SPRING, CONICAL HELICAL SPRING OF RECTANGULAR CROSS SECTION	RESSORT CONIQUE A SECTION RECTANGULAIRE, RESSORT CONIQUE A LAME PLATE	KEGELFEDER MIT RECHTECKIGEM QUERSCHNITT
13606	VORTEX, WHIRL, EDDY	TOURBILLON, REMOUS	WIRBEL, (PHYS.)
13607	VULCANISATION	VULCANISATION	VULKANISIEREN
13608	VULCANISE (TO)	VULCANISER	VULKANISIEREN
13609	VULCANISED FIBRE	FIBRE VULCANISEE, FIBRE	FIBER, VULKANFIBER
13610	VULCANISED RUBBER	CAOUTCHOUC VULCANISE	KAUTSCHUK (VULKANISIERTER), WEICHKAUTSCHUK, WEICHGUMMI

13611	**VULGAR FRACTION**	FRACTION ORDINAIRE	BRUCH (GEMEINER)
13612	**WADDING**	OUATE	WATTE
13613	**WAGES**	PAIE, PAIEMENT, SALAIRE	LOHN, ARBEITSLOHN
13614	**WAGON DRILL**	SONDEUSE SUR CAMION	BOHRWAGEN
13615	**WAIVER**	RENONCIATION, DEROGATION	VERZICHT, ABWEICHUNG
13616	**WALKWAY , GANGAY**	PASSERELLE	BEDIENUNGSGANG, LAUFSTEG
13617	**WALL BOX**	NICHE	MAUERKASTEN
13618	**WALL BRACKET**	CHAISE D'APPLIQUE, CHAISE-CONSOLE, CONSOLE	WANDARM, WANDLAGERSTUHL, WANDKONSOLE
13619	**WALL COUNTERSHAFT**	RENVOI MURAL, RENVOI FIXE CONTRE UN MUR	WANDVORGELEGE
13620	**WALL ENGINE, WALL DRILLING MACHINE**	MACHINE A VAPEUR MURALE	WANDDAMPFMASCHINE, WANDBOHRMASCHINE
13621	**WALL PLATE**	CONTREPLAQUE, PLAQUE D'ANCRAGE	ANKERPLATTE, WANDPLATTE, MAUERPLATTE, WANDBETT, ANKERROSETTE
13622	**WALL PLUG AND SOCKET**	CONTACT A FICHE	STECKKONTAKT, STÖPSELKONTAKT
13623	**WALL SLEWING CRANE**	GRUE A POTENCE	WANDDREHKRAN
13624	**WALLS OF A HOLE**	FACE D'UN TROU	LOCHWAND, LOCHLEIBUNG
13625	**WARDING FILE**	LIME D'ENTREE	SCHLÜSSELFEILE
13626	**WARNING ORDER**	ORDRE DE FONCTIONNEMENT	BETRIEBFOLGE
13627	**WARP**	CHAINE D'UN TISSU	KETTE EINES GEWEBES
13628	**WARP (TO), WIND (TO)**	GAUCHIR, BOMBER, DEJETER (SE), DEVERSER (SE), VOILER (SE)	WERFEN (SICH), VERZIEHEN (SICH)
13629	**WARPING, WINDING**	GAUCHISSEMENT, DEFORMATION, DEFORMATION PAR ENROULEMENT	UMWICKLUNG, WERFEN, VERZIEHEN, KRUMMZIEHEN
13630	**WASH STAND TAP**	ROBINET DE LAVABO	STANDVENTIL
13631	**WASHER**	RONDELLE	UNTERLAGSCHEIBE
13632	**WASHING**	DELAVAGE	WASCHEN
13633	**WASHING BOTTLE FLASK**	PISSETTE	SPRITZFLASCHE
13634	**WASTE GAS**	GAZ PERDU	ABGAS
13635	**WASTE HEAT**	CHALEUR PERDUE	ABWÄRME
13636	**WASTE OIL**	HUILE D'EGOUTTAGE, HUILE EN EXCES	ÖL (ABLAUFENDES), TROPFÖL
13637	**WASTE PRODUCTS**	DECHETS	ABFÄLLE, ABFALLSTOFFE
13638	**WASTE WATER PURIFICATION**	EPURATION, PURIFICATION DES EAUX RESIDUAIRES	ABWÄSSERREINIGUNG
13639	**WATCHMAKER'S LATHE**	FOUR D'HORLOGER	DREHSTUHL
13640	**WATER**	EAU	WASSER
13641	**WATER BATH**	BAIN-MARIE	WASSERBAD (CHEM.)
13642	**WATER BEARING SOIL**	SOL AQUIFERE	BODEN (WASSERFÜHRENDER)
13643	**WATER CALORIMETER**	CALORIMETRE A EAU	WASSERKALORIMETER
13644	**WATER COCK**	ROBINET A EAU	WASSERHAHN
13645	**WATER CONTENT, PERCENTAGE OF WATER**	PROPORTION D'EAU, TENEUR EN EAU	WASSERGEHALT
13646	**WATER COOLING**	REFROIDISSEMENT A L'EAU	WASSERKÜHLUNG
13647	**WATER DESCALING**	DECALAMINAGE A L'EAU	WASSERENTZUNDERUNG
13648	**WATER DISPLACING FILM**	FILM ANTI-MOUILLANT	HYDROPHOBIERUNGSFILM
13649	**WATER FOR INDUSTRIAL PURPOSES**	EAU INDUSTRIELLE	NUTZWASSER, BRAUCHWASSER, BETRIEBSWASSER, WIRTSCHAFTSWASSER
13650	**WATER FREE FROM BACTERIA**	EAU EXEMPTE DE MICROBES, EAU STERILISEE	WASSER (KEIMFREIES)
13651	**WATER GAS**	GAZ A L'EAU	WASSERGAS

13652	**WATER GAUGE, GAUGE GLASS**	INDICATEUR DE NIVEAU, NIVEAU D'EAU	WASSERSTANDSANZEIGER, WASSERSTANDSGLAS, WASSERSTAND
13653	**WATER HAMMER**	COUP DE BELIER	WASSERSCHLAG, WIDDERSTOSS
13654	**WATER HARDENING**	TREMPE A L'EAU	WASSERHÄRTUNG
13655	**WATER JACKET CHAMBER**	CHEMISE D'EAU	WASSERMANTEL, KÜHLMANTEL
13656	**WATER JET**	JET D'EAU	WASSERSTRAHL
13657	**WATER JET AIR PUMP**	INJECTEUR HYDRAULIQUE	WASSERSTRAHLLUFTPUMPE
13658	**WATER JET BLOWER**	EJECTEUR HYDRAULIQUE	WASSERSTRAHLGEBLÄSE
13659	**WATER LEVEL**	NIVEAU D'EAU (HAUTEUR D'UN LIQUIDE)	WASSERSPIEGEL, WASSERSTAND
13660	**WATER LEVEL MARK**	MARQUE REPERE DE NIVEAU D'EAU	WASSERSTANDMARKE
13661	**WATER METER**	COMPTEUR A EAU	WASSERMESSER, WASSERUHR
13662	**WATER OF CONDENSATION**	EAU DE CONDENSATION	NIEDERSCHLAGWASSER, KONDENSWASSER, KONDENSAT
13663	**WATER OF CRYSTALLISATION**	EAU DE CRISTALLISATION	KRISTALLWASSER
13664	**WATER PAINT**	PEINTURE A L'EAU	WASSERFARBE
13665	**WATER PIPE**	TUYAU DE CONDUITE, TUYAU DE DISTRIBUTION D'EAU	WASSERLEITUNGSROHR
13666	**WATER PIPING MAIN**	CONDUITE, CANALISATION, DISTRIBUTION D'EAU	WASSERLEITUNG
13667	**WATER PISTON, HYDRAULIC PISTON**	PISTON A EAU, PISTON DE MOTEUR HYDRAULIQUE	WASSERKOLBEN
13668	**WATER POWER, HYDRAULIC POWER**	FORCE HYDRAULIQUE	WASSERKRAFT
13669	**WATER PRESSURE**	PRESSION D'EAU, PRESSION HYDRAULIQUE	WASSERDRUCK
13670	**WATER PRESSURE ENGINE**	MACHINE A COLONNE D'EAU	WASSERSÄULENMASCHINE
13671	**WATER PROOF**	INATTAQUABLE A L'EAU, IMPERMEABLE	WASSERBESTÄNDIG, WASSERFEST
13672	**WATER QUENCHING**	TREMPE A L'EAU	WASSERHÄRTUNG
13673	**WATER QUENCHING FOLLOWED BY TEMPERING**	TREMPE A L'EAU SUIVIE DE REVENU	WASSERVERGÜTEN
13674	**WATER REPELLENT PRODUCT**	PRODUIT HYDROFUGE	MITTEL (WASSERABWEISENDES)
13675	**WATER SEAL**	JOINT HYDRAULIQUE	WASSERVERSCHLUSS
13676	**WATER SOFTENING**	EPURATION CHIMIQUE DES EAUX D'ALIMENTATION	ENTHÄRTEN DES WASSERS, WASSERENTHÄRTUNG
13677	**WATER SPACE OF A BOILER**	CHAMBRE A EAU D'UNE CHAUDIERE	WASSERRAUM EINES KESSELS
13678	**WATER SUPPLY**	ALIMENTATION EN EAU, DISTRIBUTION D'EAU	WASSERVERSORGUNG
13679	**WATER TANK**	RESERVOIR A EAU	WASSERBEHÄLTER
13680	**WATER TOWER**	CHATEAU D'EAU	WASSERTURM
13681	**WATER TUBE**	TUBE D'EAU	WASSERROHR (DAMPFKESSEL)
13682	**WATER TUBE BOILER**	CHAUDIERE A TUBES D'EAU, CHAUDIERE A VAPORISATION RAPIDE, CHAUDIERE MULTITUBULAIRE, CHAUDIERE INEXPLOSIBLE	WASSERRÖHRENKESSEL
13683	**WATER-COOLED**	REFROIDI PAR L'EAU, REFROIDISSEMENT D'EAU (A)	WASSERGEKÜHLT
13684	**WATER-COOLED SURFACE CONDENSER**	CONDENSEUR PAR SURFACE AU MOYEN DE L'EAU	OBERFLÄCHENKONDENSATOR MIT WASSERKÜHLUNG
13685	**WATER-REPELLENT**	HYDROFUGE	WASSERABWEISEND
13686	**WATER-TIGHT, STANCH, STAUNCH**	ETANCHE A L'EAU	WASSERDICHT
13687	**WATERING CAN POT**	ARROSOIR	GIESSKANNE
13688	**WATERPROOF**	ETANCHE A L'EAU	WASSERDICHT

13689	**WATERPROOF (TO)**	IMPERMEABILISER	WASSERDICHT MACHEN
13690	**WATERPROOF CLOTH**	TISSU IMPERMEABLE	GEWEBE (WASSERDICHTES)
13691	**WATERPROOFING**	IMPERMEABILISATION	WASSERDICHTMACHEN
13692	**WATERWHEEL**	ROUE, RECEPTEUR HYDRAULIQUE	WASSERRAD
13693	**WATERWORKS**	INSTALLATION DE DISTRIBUTION D'EAU	WASSERWERK, WASSERVERSORGUNGSANLAGE
13694	**WATT**	WATT	WATT
13695	**WATT-HOUR (W/HR)**	WATT-HEURE	WATTSTUNDE
13696	**WATT-MINUTE**	WATT-MINUTE	WATTMINUTE
13697	**WATT-SECOND**	WATT-SECONDE	WATTSEKUNDE
13698	**WATTMETER**	WATTMETRE	WATTMETER
13699	**WAVE**	ONDE	WELLE (PHYS.)
13700	**WAVE LENGTH**	LONGUEUR D'ONDE	WELLENLÄNGE
13701	**WAVINESS**	ONDULATION	WELLIGKEIT
13702	**WAX**	CIRE	WACHS
13703	**WAX PAATERN**	MODELE EN CIRE	WACHSMODELL
13704	**WAXED THREAD**	FIL POISSE	PECHGARN, PECHDRAHT
13705	**WEAK CURRENT**	COURANT DE FAIBLE INTENSITE	SCHWACHSTROM
13706	**WEAK SOLUTION**	SOLUTION FAIBLE	LÖSUNG (SCHWACHE)
13707	**WEAKENING OF CROSS SECTION**	AFFAIBLISSEMENT DE LA SECTION	SCHWÄCHUNG, VERSCHWÄCHUNG DES QUERSCHNITTS
13708	**WEAKENING OF THE MATERIAL**	AFFAIBLISSEMENT DU MATERIEL	SCHWÄCHUNG DES MATERIALS
13709	**WEAR**	USURE	VERSCHLEISS, ABNÜTZUNG
13710	**WEAR AND TEAR**	USURE NORMALE, CONSOMMATION	ABNÜTZUNG (NATÜRLICHE), VERSCHLEISS
13711	**WEAR BLOCKS**	CALES D'USURE	VERSCHLEISSPLATTEN
13712	**WEAR BY RUBBING**	USURE PAR FROTTEMENT	ABRIEB
13713	**WEAR DOWN (TO), OFF, OUT**	USER (S')	ABNÜTZEN (SICH), VERSCHLEISSEN
13714	**WEAR FEST**	ESSAI D'USURE	VERSCHLEISSPRÜFUNG
13715	**WEAR HARDENING**	ECROUISSAGE	KALTHÄRTUNG
13716	**WEAR PLATE**	TOLE D'USURE	VERSCHLEISSBLECH
13717	**WEAR RESISTANCE**	RESISTANCE A L'USURE	VERSCHLEISSFESTIGKEIT
13718	**WEAR RESISTING STEEL**	ACIER RESISTANT A L'USURE	STAHL (VERSCHLEISSFESTER)
13719	**WEARING OVAL OF A BEARING**	OVALISATION D'UN COUSSINET	UNRUNDWERDEN EINES LAGERS
13720	**WEATHER SHIELD**	JOINT SECONDAIRE D'ETANCHEITE	SPRITZWASSERSCHUTZ
13721	**WEATHER SIDE**	COTE DU VENT	LUVSEITE
13722	**WEATHER STRIP**	CAOUTCHOUC D'ETANCHEITE	DICHTUNGSSTREIFEN
13723	**WEATHER-PROOF**	INALTERABLE, INATTAQUABLE AUX INTEMPERIES	WETTERBESTÄNDIG, WETTERFEST
13724	**WEAVE BEAD**	PASSE LARGE	PENDELRAUPE
13725	**WEAVING MOTION**	MOUVEMENT ALTERNATIF, VA-ET-VIENT.	BEWEGUNG (HIN-UND HERGEHENDE)
13726	**WEB**	AME	STEG, KERN
13727	**WEB CRIPPLING OF A PLATE GIRDER**	DEFORMATION PERMANENTE DE L'AME D'UNE POUTRE	DURCHBIEGUNG (BLEIBENDE) EINES TRÄGERSTEGS
13728	**WEB LEG OF AN ANGLE IRON**	AILE, BRANCHE D'UN FER CORNIERE	SCHENKEL EINES WINKELEISENS
13729	**WEB OF A BEAM**	AME D'UNE POUTRELLE	STEG
13730	**WEB, CENTRE WEB, STEM**	AME D'UN FER PROFILE	STEG EINES FORMEISENS
13731	**WEDGE**	COIN, CALE, CALE D'EPAISSEUR, EPAISSEUR	KEIL, UNTERLAGKEIL, ZWISCHENKEIL
13732	**WEDGE FRICTION WHEEL**	ROUE DE FRICTION (A JANTE EN FORME DE COIN)	KEILRAD, RILLENRAD

13733	WEDGE OF A PLANE	COIN DU RABOT	HOBELKEIL
13734	WEDGE PRESS	PRESSE A COIN	KEILPRESSE
13735	WEDGE SHAPE	SECTION CONIQUE	QUERSCHNITT (KEILFÖRMIGER)
13736	WEDGE SURFACE FRICTION GEAR	TRANSMISSION PAR ROUES DE FRICTION, ENGRENAGE A COIN	KEILRÄDERGETRIEBE
13737	WEDGE-SHAPED	CUNEIFORME	KEILFÖRMIG
13738	WEDGE-SHAPED BRICK	BRIQUE CIRCULAIRE ARRONDIE CINTREE	KEILSTEIN
13739	WEDGE, KEY, COTTER	COIN, CALE, CLAVETTE	KEIL
13740	WEDGING	CLAVETAGE, CALAGE	VERKEILUNG, VERKEILEN
13741	WEED THISTLE AND BRACKEN CUTTERS	COUPE CHARDONS ET FOUGERES	UNKRAUTDISTEL-UND FARNKRAUTSCHNEIDER
13742	WEEDERS AND TILLERS	SARCLEUSES	JÄTMASCHINEN UND ACKERFRÄSEN
13743	WEFT	TRAME	SCHUSS EINES GEWEBES, EINSCHLAG EINES GEWEBES
13744	WEIGH (TO)	PESER	ABWIEGEN, WÄGEN
13745	WEIGHBRIDGE, WEIGH-BRIDGE FOR RAILWAY WAGONS, WAGON WEIGHING MACHINE	PONT A BASCULE	GLEISWAAGE, BRÜCKENWAAGE
13746	WEIGHING	PESAGE, PESEE	WÄGUNG, WÄGEN, WIEGEN
13747	WEIGHT	POIDS	GEWICHT
13748	WEIGHT PER METER	POIDS AU METRE	METERGEWICHT
13749	WEIGHT-LOADED SAFETY VALVE	SOUPAPE DE SURETE A CONTRE-POIDS	GEWICHTSSICHERHEITSVENTIL, SICHERHEITSVENTIL MIT GEWICHTSBELASTUNG
13750	WEIGHTED JOCKEY PULLEY (LENIX TYPE)	ENROULEUR DE COURROIE LENIX LENEVEU	SPANNROLLE NACH LENIX
13751	WEIGHTED LEVER	LEVIER A CONTREPOIDS	HEBEL (BELASTETER), GEWICHTSHEBEL
13752	WEIGHTED PENDULUM GOVERNOR, LOADED WATT GOVERNOR, PORTER GOVERNOR, CENTRE WEIGHTED GOVERNOR	REGULATEUR A MASSE CENTRALE, REGULATEUR PORTER	GEWICHTSREGLER
13753	WEIR	DEVERSOIR, BARRAGE	STAUANLAGE, WEHR
13754	WEIR CREST	CRETE D'UN DEVERSOIR	WEHRKANTE, ÜBERFALLKANTE, WEHRKRONE
13755	WELD (TO)	SOUDER	SCHWEISSEN, ZUSAMMENSCHWEISSEN
13756	WELD BEAD	CORDON DE SOUDURE	SCHWEISSRAUPE
13757	WELD BUILT-UP	SOUDURE EN PASSES SUPERPOSEES	MEHRLAGENSCHWEISSUNG
13758	WELD BY THE AUTOGENOUS PROCESS (TO)	SOUDER A L'AUTOGENE	AUTOGENSCHWEISSEN
13759	WELD DECAY	CORROSION DE SOUDURE	KORROSION DER SCHWEISSNAHT
13760	WELD DEFECTS	DEFAUTS DE LA SOUDURE	NAHTFEHLER, SCHWEISSFEHLER
13761	WELD DEPOSIT METAL	METAL DEPOSE	SCHWEISSGUT
13762	WELD DEPOSIT OVERLAY	RECHARGEMENT D'UNE SOUDURE	AUFTRAGSCHWEISSUNG
13763	WELD METAL	METAL DE SOUDURE	SCHWEISSGUT
13764	WELD NUGGET	LENTILLE DE SOUDURE	SCHWEISSLINSE
13765	WELD STEEL	ACIER SOUDE	SCHWEISSSTAHL
13766	WELD-IRON	FER SOUDE	SCHWEISSEISEN
13767	WELD, WELDED JOINT	SOUDAGE, SOUDURE, JOINT SOUDE	SCHWEISSUNG, SCHWEISSVERBINDUNG
13768	WELDABILITY	SOUDABILITE	SCHWEISSBARKEIT
13769	WELDABLE	SOUDABLE	SCHWEISSBAR
13770	WELDED CHAIN	CHAINE A MAILLES SOUDEES	KETTE (GESCHWEISSTE)

13771	WELDED COLLAR	BAGUE SOUDEE	AUFGESCHWEISSTER BUND, VORSCHWEISSBUND
13772	WELDED CONNECTION	LIAISON SOUDEE	SCHWEISSVERBINDUNG
13773	WELDED FLANGE	BRIDE, RONDELLE SOUDEE	FLANSCH (AUFGESCHWEISSTER), VORSCHWEISSFLANSCH
13774	WELDED JOINT / SEAM	JOINT SOUDE	SCHWEISSVERBINDUNG
13775	WELDED PIPE	TUBE SOUDE	ROHR (GESCHWEISSTES)
13776	WELDED SEAM	JOINT SOUDE, SOUDURE	SCHWEISSNAHT
13777	WELDED TUBE	TUBE SOUDE	ROHR (GESCHWEISSTES)
13778	WELDER	SOUDEUR	SCHWEISSER
13779	WELDING	SOUDURE, SOUDAGE	SCHWEISSEN, SCHWEISSUNG
13780	WELDING CURRENT	INTENSITE DU COURANT DE SOUDAGE	SCHWEISSSTROMSTÄRKE
13781	WELDING FITTING	RACCORD A SOUDER	SCHWEISSFITTING
13782	WELDING GUN POSITION	POSITION DU PISTOLET DE SOUDAGE	STELLUNG (LAGE) DER SCHWEISSPISTOLE
13783	WELDING HEAD	TETE DE SOUDAGE	SCHWEISSKOPF
13784	WELDING HEAT	BLANC SOUDANT, CHAUDE SUANTE, CHALEUR SOUDANTE	SCHWEISSHITZE, SCHWEISSWÄRME
13785	WELDING JIG	GABARIT DE SOUDAGE	SCHWEISSVORRICHTUNG
13786	WELDING MACHINE	MACHINE A SOUDER	SCHWEISSMASCHINE
13787	WELDING PARAMETERS	PARAMETRES DE SOUDAGE	SCHWEISSPARAMETER
13788	WELDING PERIOD	DUREE DU CYCLE DE SOUDAGE	SCHWEISSSPIELZEIT
13789	WELDING ROD	FIL A SOUDER, ELECTRODE	SCHWEISSDRAHT, SCHWEISSELEKTRODE
13790	WELDING SEQUENCE	SUCCESSION DES OPERATIONS DE SOUDAGE	SCHWEISSFOLGE
13791	WELDING SYMBOLS	SYMBOLES DE SOUDAGE	SCHWEISSNAHTSINNBILDER
13792	WELDING TEST	ESSAI DE SOUDABILITE	SCHWEISSVERSUCH
13793	WELDING WIRE	FIL (BAGUETTE) A SOUDER	SCHWEISSDRAHT, SCHWEISSSTAB
13794	WELDING-NECK FLANGE, SLIP-ON WELDING FLANGE	BRIDE A SOUDER (EN BOUT, A L'INTERIEUR)	VORSCHWEISSFLANSCH, EINSCHWEISSFLANSCH
13795	WELDMENT	CONSTRUCTION SOUDEE	SCHWEISSKONSTRUKTION
13796	WELL WATER	EAU DE PUITS	BRUNNENWASSER
13797	WELL-SINKER'S CEMENT	MASTIC DES FONTAINIERS	BRUNNENMACHERKITT
13798	WESTON DIFFERENTIAL PULLEY BLOCK	PALAN DIFFERENTIEL, PALAN WESTON	DIFFERENTIALFLASCHENZUG
13799	WET AIR PUMP	POMPE A AIR HUMIDE	LUFTPUMPE (NASSE)
13800	WET ANALYSIS	ANALYSE PAR VOIE HUMIDE	ANALYSE AUF NASSEM WEGE
13801	WET COMPRESSOR	COMPRESSEUR HUMIDE	VERDICHTER (NASSER), WASSERSÄULENVERDICHTER
13802	WET CORROSION	CORROSION HUMIDE	FEUCHTIGKEITSKORROSION
13803	WET DRAWING	PASSE HUMIDE	NASSZUG
13804	WET LINER	CHEMISE HUMIDE	LAUFBÜCHSE (NASSE)
13805	WET STEAM	VAPEUR HUMIDE	NASSDAMPF
13806	WETNESS OF STEAM	HUMIDITE DE LA VAPEUR	DAMPFNÄSSE
13807	WETTING	HUMECTATION	FEUCHTEN
13808	WETTING AGENT	AGENT MOUILLANT	BENETZUNGSMITTEL
13809	WHEATSTONE'S BRIDGE	PONT DE WHEATSTONE	BRÜCKENVIERECK, BRÜCKE WHEATSTONESCHE
13810	WHEEL	ROUE, VOLANT	HANDRAD, RAD
13811	WHEEL ARM	RAYONS D'UNE ROUE	RADARM, RADSPEICHE
13812	WHEEL BARROW	BROUETTE	SCHUBKARREN

13813	WHEEL BASE	ECARTEMENT DES ESSIEUX	ACHSSTAND, RADSTAND
13814	WHEEL CENTRE, SPIDER	BRASSURE (D'UNE ROUE)	RADSTERN
13815	WHEEL GUARD COVER	COUVRE-ROUES, CARTER D'ENGRENAGES	RADVERDECK
13816	WHEEL LATHE	TOUR POUR ROUES	RÄDERDREHBANK
13817	WHEEL OF PLANIMETER	TAMBOUR, ROULETTE DU PLANIMETRE	MESSROLLE DES PLANIMETERS
13818	WHEEL RIM	JANTE DE ROUE	RADFELGE
13819	WHEEL TOOTH	DENT D'ENGRENAGE	RADZAHN
13820	WHEEL VANE, MOVING BLADE OF A TURBINE	AUBE RECEPTRICE, AUBE MOBILE D'UNE TURBINE	LAUFRADSCHAUFEL EINER TURBINE
13821	WHEEL WORK	ROUAGE	RÄDERWERK
13822	WHEEL-RIM PROFILING MACHINE	MACHINE A PROFILER LES JANTES DE ROUES	RADKRANZPROFILIERMASCHINE
13823	WHEEL, ROTOR OF A TURBINE	COURONNE MOBILE, ROUE D'UNE TURBINE	LAUFRAD EINER TURBINE, TURBINENRAD
13824	WHETSTONE	PIERRE A AFFUTER, PIERRE A AFFILER, PIERRE A AIGUISER	SCHLEIFSTEIN, ABZIEHSTEIN
13825	WHIPPING OF A BELT	FOUETTEMENT D'UNE COURROIE, FLOTTEMENT D'UNE COURROIE, BALLANT D'UNE COURROIE	FLATTERN DES RIEMENS, PEITSCHEN DES RIEMENS, SCHLAGEN DES RIEMENS
13826	WHIRLED THERMOMETER, SING THERMOMETER	THERMOMETRE-FRONDE	SCHLEUDERTHERMOMETER
13827	WHIRLING	TOURBILLONNEMENT	WIRBELUNG
13828	WHISKER	FIL (CROISSANCE EN FILAMENT), FIL MONOCRISTALLIN	WHISKER-WACHSTUM, FADENKRISTALL
13829	WHISTLE	SIFFLET	PFEIFE
13830	WHITE CAST IRON, WHITE IRON, WHITE PIG	FONTE BLANCHE	ROHEISEN (WEISSES)
13831	WHITE GOLD	OR BLANC	WEISSGOLD
13832	WHITE HEAT	INCANDESCENCE	WEISSGLUT
13833	WHITE IRONWOOD	BOIS DE FER-BLANC, BOIS DE SABLE, SIDEROXYLE	EISENHOLZ
13834	WHITE LEAD, BASIC CARBONATE OF LEAD	CERUSE, BLANC DE PLOMB, BLANC D'ARGENT, BLANC DE CERUSE	BLEIWEISS, BLEI (BASISCH), BLEI (KOHLENSAURES)
13835	WHITE MEAL TURNINGS	METAL BLANC DEFLOCULE	FLOCKEN VON WEISSMETALL
13836	WHITE METAL	METAL, ALLIAGE BLANC, METAL ANTI-FRICTION	WEISSMETALL
13837	WHITE METAL LINING	GARNITURE ANTI-FRICTION	WEISSMETALLAUSGUSS
13838	WHITE OILS	HUILES BLANCHES	WEISSES MINERALÖL
13839	WHITE UNTARRED ROPE	CORDE NON GOUDRONNEE	SEIL (UNGESTEURTES)
13840	WHITE-WALLED TYRES	PNEUS A FLANCS BLANCS	WEISSWANDREIFEN
13841	WHITING	BLANC DE MEUDON, BLANC DE TROYES, BLANC D'ESPAGNE	SCHLÄMMKREIDE
13842	WHITING	CHARGE NEUTRE, CRAIE, KAOLIN (POUR LES PNEUS)	KREIDE
13843	WHITWORTH MEASURING MACHINE	MACHINE A MESURER WHITWORTH	MESSMASCHINE VON WHITWORTH
13844	WHITWORTH THREAD, ENGLISH STANDARD THREAD, COMMON THREAD	PAS WHITWORTH, PAS ANGLAIS	GEWINDE (WHITWORTHSCHES)
13845	WHOLE DEPTH	HAUTEUR DE DENT	ZAHNHÖHE, ZAHNTIEFE
13846	WICK	MECHE (D'UNE LAMPE, D'UN GRAISSEUR)	DOCHT
13847	WICK SIPHON LUBRICATION	GRAISSAGE PAR MECHE	DOCHTSCHMIERUNG
13848	WIDE EDGE MILLING CUTTER, HOB	FRAISE CYLINDRIQUE, FRAISE RECTILIGNE	MANTELFRÄSER, WALZENFRÄSER, ZYLINDERFRÄSER

13849	WIDE FLANGED BEAM	POUTRELLE A LARGES AILES	BREITFLANSCHTRÄGER
13850	WIDTH	LARGEUR	BREITE
13851	WINCH, CRAB	TREUIL	WINDE, BOCKWINDE
13852	WINCHES	TREUILS	ACKERWINDEN
13853	WIND (TO), COIL (TO)	ENROULER	AUFWICKELN, AUFWINDEN
13854	WIND CORD	BOURRELET D'ETANCHEITE	FENSTERDICHTUNGSCHNUR
13855	WIND ENGINE, WIND MILL	MOTEUR A VENT, MOTEUR EOLIEN PNEUMATIQUE, MOULIN A VENT	WINDKRAFTMASCHINE, WINDMOTOR
13856	WIND GIRDER	POUTRE RAIDISSEUSE, POUTRE AU VENT	VERSTEIFUNGSBALKEN
13857	WIND LOAD	CHARGE DUE AU VENT	WINDBELASTUNG
13858	WIND PRESSURE	POUSSEE, PRESSION DU VENT	WINDDRUCK
13859	WIND TURBINE, AIR TURBINE	TURBINE A AIR	WINDTURBINE
13860	WIND VANE	GIROUETTE	WINDFAHNE
13861	WIND WHEEL	ROUE ATMOSPHERIQUE	WINDRAD
13862	WINDING, COILING	ENROULEMENT, BOBINAGE	WICKLUNG, AUFWICKELN, AUFWINDEN
13863	WINDLASS	TREUIL SIMPLE	HASPEL
13864	WINDOW	FENETRE	FENSTER
13865	WINDOW FRAME	CHASSIS DE FENETRE	FENSTERRAHMEN
13866	WINDOW GLASS	VERRE A VITRES	FENSTERGLAS
13867	WINDOW REGULATOR	LEVE-GLACE	FENSTERKURBEL
13868	WINDOW RIM CHANNEL	GLISSIERE (DE FENETRE)	FENSTERFÜHRUNG
13869	WINDROWER	ABATTEURS POUR MOISSONNAGE	MASCHINEN ZUM AUFZIEHEN DES HEUES IN SCHWADEN
13870	WINDSHIELD-WASHER	LAVE-GLACE	SCHEIBENWASCHER
13871	WINDSHIELD-WIPER	ESSUIE-GLACE	SCHEIBENWISCHER
13872	WINDSHIELD, WINDSCREEN	PARE-BRISE	WINDSCHUTZSCHEIBE
13873	WINDUP	JEU	GANG (TOTER)
13874	WING	AILE	KOTFLÜGEL
13875	WING PLATFORM	PLATE-FORME EN PORTE-A-FAUX	PLATTFORM (AUSKRAGENDE)
13876	WING, QUADRANT COMPASSES	COMPAS A QUART DE CERCLE	BOGENZIRKEL
13877	WINNING EXTRACTION OF COAL	EXTRACTION DE LA HOUILLE	GEWINNUNG DER KOHLE
13878	WIPER	CAME	HEBEDAUMEN, DÄUMLING
13879	WIPER RING	BAGUE LARMIER	ÖL-ABSTREIFRING
13880	WIRE	FIL METALLIQUE	DRAHT
13881	WIRE (TO)	POSER UN FIL CONDUCTEUR	VERLEGEN (EINEN LEITUNGSDRAHT)
13882	WIRE BAR	FIL EN VERGE POUR TREFILERIE	DRAHTBARREN, STANGENDRAHT
13883	WIRE BENDING MACHINE	MACHINE A PLIER LES FILS	DRAHTBIEGEMASCHINE
13884	WIRE BRUSH	BROSSE EN FILS METALLIQUES, BROSSE METALLIQUE	DRAHTBÜRSTE
13885	WIRE CHAIN MAKING-MACHINE	MACHINE A FAIRE LES CHAINES EN FIL DE FER	DRAHTKETTENMASCHINE
13886	WIRE CUTTER	COUPE-FIL	DRAHTSCHNEIDER
13887	WIRE DRAW DIE	FILIERE A TREFILER	DRAHTZIEHSTEIN
13888	WIRE DRAWING	ETIRAGE EN FIL, TREFILERIE	ZIEHEN VON DRAHT, DRAHTZIEHEN
13889	WIRE DRAWING MACHINE	MACHINE A TREFILER	DRAHTZIEHMASCHINE
13890	WIRE EDGE	MORFIL	SCHLEIFGRAT
13891	WIRE FACK MACHINE	FRAPPEUR DE POINTES	DRAHTSTIFTSCHLAGMASCHINE
13892	WIRE GAUGE	JAUGE POUR FILS METALLIQUES, JAUGE DE TREFILERIE	DRAHTLEHRE
13893	WIRE GAUZE	TOILE METALLIQUE, GAZE METALLIQUE	DRAHTGEWEBE, DRAHTGAZE

13894	WIRE GLASS, FERROGLAS	VERRE ARME, VERRE A FIL DE FER NOYE	DRAHTGLAS, GLAS (NICHT SPLITTERNDES)
13895	WIRE NAIL, FRENCH NAIL	POINTE DE PARIS	DRAHTSTIFT, PARISER STIFT, DRAHTNAGEL
13896	WIRE NETTING	TREILLIS METALLIQUE	DRAHTNETZ, DRAHTGEFLECHT
13897	WIRE NETTING MACHINE	MACHINE A FAIRE LES GRILLAGES	DRAHTFLECHTMASCHINEN
13898	WIRE REEL	COURONNE DE FIL, DEVIDOIR	DRAHTHASPEL, DRAHTABSPULER
13899	WIRE ROPE	CABLE METALLIQUE	DRAHTSEIL, KABEL
13900	WIRE SPEED	VITESSE DU FIL	DRAHTABSPULGESCHWINDIGKEIT
13901	WIRE SPRING COILING MACHINE	MACHINE A ENROULER LES RESSORTS A BOUDIN	DRAHTFEDERNWINDEMASCHINE
13902	WIRE STRAND	TORON METALLIQUE	DRAHTLITZE
13903	WIRE WINDING, UNWINDING MACHINE	ENROULEUSE, DEROULEUSE DE FIL	DRAHTAUF-U. ABWICKELMASCHINE
13904	WIRE WORKING MACHINE	MACHINE A TRAVAILLER LES FILS METALLIQUES	DRAHTVERARBEITUNGSMASCHINE
13905	WIRE WORKS	TREFILERIE (FABRIQUE)	DRAHTZIEHEREI
13906	WIRE-DRAWING	TREFILAGE	DRAHTZIEHEN
13907	WIRELESS TELEGRAPHY, RADIOTELEGRAPHY	TELEGRAPHIE SANS FIL, RADIOTELEGRAPHIE	TELEGRAPHIE (DRAHTLOSE)
13908	WIRELESS TELEPHONY, RADIOTELEPHONY	TELEPHONIE SANS FIL, RADIOTELEPHONIE	TELEPHONIE (DRAHTLOSE)
13909	WIRING	POSE D'UN FIL CONDUCTEUR	VERLEGUNG EINES LEITUNGSDRAHTES
13910	WITHDRAW HEAT FROM.. (TO)	SOUSTRAIRE DE LA CHALEUR	WÄRME ENTZIEHEN
13911	WITHERITE	WITHERITE, CARBONATE NATUREL DE BARYTE	BARYT (KOHLENSAURER), WITHERIT
13912	WOLFRAM	WOLFRAM	WOLFRAMIT
13913	WOOD	BOIS	HOLZ
13914	WOOD CHARCOAL	CHARBON DE BOIS	HOLZKOHLE
13915	WOOD LAGGING	REVETEMENT EN BOIS	HOLZVERSCHALUNG, HOLZVERKLEIDUNG
13916	WOOD MEAL	POUDRE DE BOIS	HOLZMEHL
13917	WOOD SAW	SCIE A BOIS	HOLZSÄGE
13918	WOOD SCREW	VIS METALLIQUE A BOIS	HOLZSCHRAUBE
13919	WOOD SHAVINGS	COPEAUX (DE BOIS)	HOBELSPÄNE
13920	WOOD SPIRIT, METHYL ALCOHOL	ALCOOL METHYLIQUE, HYDRATE DE METHYLE, METHYLENE, ESPRIT DE BOIS	HOLZGEIST, HOLZALKOHOL, HOLZSPIRITUS, HOLZNAPHTA, METHYLALKOHOL, KARBINOL, METHANOL
13921	WOOD TAR	GOUDRON DE BOIS, GOUDRON VEGETAL	HOLZTEER
13922	WOOD TURNING LATHE	TOUR A BOIS	HOLZDREHBANK
13923	WOOD WOOL	LAINE DE BOIS, COPEAUX DE FIBRES DE BOIS	HOLZWOLLE
13924	WOOD WORKING	TRAVAIL DU BOIS	HOLZBEARBEITUNG
13925	WOOD WORKING MACHINE	MACHINE-OUTIL A TRAVAILLER LE BOIS, MACHINE A BOIS	HOLZBEARBEITUNGSMASCHINE
13926	WOODEN HAMMER	MAILLET	HOLZHAMMER
13927	WOODEN LINING	CHEMISE, FOURRURE EN BOIS	HOLZFUTTER
13928	WOODEN PIPE	TUYAU EN BOIS	HOLZROHR, HOLZRÖHRE
13929	WOODRUFF KEY	CLAVETTE DEMI-RONDE, CLAVETTE WOODRUFF, CLAVETTE-DISQUE	SEGMENTKEIL, WOODRUFFKEIL , SCHEIBENFEDER
13930	WOODRUFF KEYSEAT CUTTERS	FRAISES A RAINURE WOODRUFF	FRÄSER FÜR WOODRUFFKEILE
13931	WOODWORKING MACHINERY	MACHINES A BOIS	HOLZBEARBEITUNGSMASCHINEN
13932	WOOL	LAINE (DE MOUTON)	SCHAFWOLLE

13933	WOOL FAT, YOLK, SUINT	SUINT	WOLLFETT
13934	WORD	MOT	DATENWORT
13935	WORD ADDRESS FORMAT	FORMAT A ADRESSES	EINGABEFORMAT IN ADRESSSCHREIBWEISE
13936	WORK (MOMENTUM)	TRAVAIL	ARBEIT
13937	WORK (TO), MACHINE (TO)	TRAVAILLER, TRAITER, USINER, OUVRAGER	BEARBEITEN
13938	WORK AN INVENTION ON A COMMERCIAL SCALE (TO)	EXPLOITER, METTRE EN EXPLOITATION UNE INVENTION	AUSNUTZEN (EINE ERFINDUNG)
13939	WORK DISTRIBUTION	REPARTITION DU TRAVAIL	ARBEITSVERTEILUNG
13940	WORK DONE, OUTPUT	TRAVAIL EFFECTUE	ARBEIT (GELEISTETE), ARBEIT (ABGEGEBENE), LEISTUNG (ABGEGEBENE)
13941	WORK OF COMPRESSION	TRAVAIL DE COMPRESSION	VERDICHTUNGSARBEIT
13942	WORK OF FRICTION, WORK LOST EXPANDED IN FRICTION	TRAVAIL DU FROTTEMENT	REIBUNGSARBEIT
13943	WORK ROOM	LIEU DE TRAVAIL, ATELIER	ARBEITSRAUM
13944	WORK TO OVERCOME RESISTANCE, WORK AGAINST RESISTANCE	TRAVAIL RESISTANT	WIDERSTANDSARBEIT
13945	WORK-HOLDING JAW	MACHOIRE DE BLOCAGE	FESTSPANNEINRICHTUNG
13946	WORKABILITY, EASE OF WORKING	USINABILITE, FACILITE DU TRAVAIL	BEARBEITBARKEIT
13947	WORKABLE	FACILE A TRAVAILLER	BEARBEITBAR, BEARBEITUNGSFÄHIG
13948	WORKING AN INVENTION ON A COMMERCIAL SCALE	EXPLOITATION D'UNE INVENTION	AUSNUTZUNG EINER ERFINDUNG
13949	WORKING CONDITIONS, CONDITIONS OF SERVICE	CONDITIONS DE SERVICE	BETRIEBSVERHÄLTNISSE
13950	WORKING COST, EXPENSES	FRAIS D'EXPLOITATON	BETRIEBSKOSTEN
13951	WORKING DEPTH	HAUTEUR D'ACTION	EINGRIFFSTIEFE
13952	WORKING DEPTH IN TOOTHED GEARING	PROFONDEUR D'ENGRENEMENT	EINGRIFFSTIEFE
13953	WORKING ELEMENT	ORGANE TRAVAILLANT	BESTANDTEIL (ARBEITENDER)
13954	WORKING ENGINE CYLINDER	CYLINDRE MOTEUR, CYLINDRE DE TRAVAIL	ZYLINDER, ARBEITSZYLINDER
13955	WORKING GAUGE	JAUGE DE FABRICATION	ARBEITSLEHRE
13956	WORKING LOAD	CHARGE ADMISSIBLE	BELASTUNG (ZULÄSSIGE)
13957	WORKING PRESSURE	PRESSION DE REGIME, PRESSION DE SERVICE, TENSION DE SERVICE	BETRIEBSDRUCK
13958	WORKING STRESS	TAUX DE TRAVAIL	BELASTBARKEIT IM GEBRAUCH
13959	WORKING, MACHINING	TRAVAIL, TRAITEMENT, USINAGE	BEARBEITEN, BEARBEITUNG
13960	WORKMAN, WORKER, HAND	OUVRIER, TRAVAILLEUR	ARBEITER
13961	WORKMANSHIP	MAIN-D'OEUVRE	AUSFÜHRUNG, ARBEIT
13962	WORKPEOPLE	PERSONNEL OUVRIER	ARBEITERSCHAFT
13963	WORKS MANAGEMENT	DIRECTION DES USINES, DIRECTION DES ATELIERS, DIRECTION DU TRAVAIL	BETRIEBSFÜHRUNG
13964	WORKS MANAGER	CHEF DE FABRICATION, CHEF DE SERVICE, CHEF D'ATELIER	BETRIEBSFÜHRER, BETRIEBSLEITER
13965	WORKS RAILWAY	VOIE FERREE D'ATELIER	FABRIKBAHN
13966	WORKS SIDING	EMBRANCHEMENT INDUSTRIEL	FABRIKANSCHLUSSGLEIS, ANSCHLUSSGLEIS, INDUSTRIEGLEIS
13967	WORKSHOP GAUGE	JAUGE DE FABRICATION	ARBEITSLEHRE
13968	WORKSHOP, SHOP	ATELIER	WERKSTATT, WERKSTÄTTE
13969	WORM	VIS SANS FIN	SCHNECKE

13970	**WORM CONVEYOR**	TRANSPORTEUR A HELICE, VIS TRANSPORTEUSE, VIS D'ARCHIMEDE	FÖRDERSCHNECKE, TRANSPORTSCHNECKE, SCHNECKE
13971	**WORM CUTTER**	FRAISE MERE	SCHNECKENFRÄSER
13972	**WORM GEAR MILLING CUTTER, WORM HOB, MILLING CUTTER HOB FOR WORM GEAR**	FRAISE POUR TAILLER DES ENGRENAGES HELICOIDAUX	SCHNECKENFRÄSER
13973	**WORM GEAR PULLEY BLOCK**	PALAN A VIS TANGENTE, PALAN A VIS SANS FIN	SCHRAUBENFLASCHENZUG, SCHNECKENFLASCHENZUG
13974	**WORM GEARING**	ENGRENAGE, MECANISME A VIS SANS FIN, COUPLE DE ROUE ET VIS TANGENTE	SCHNECKENRADGETRIEBE, SCHRAUBGETRIEBE, WURMGETRIEBE
13975	**WORM WHEEL**	ROUE A VIS SANS FIN	SCHNECKENRAD, MUTTERRAD
13976	**WORM-EATEN**	VERMOULU	WURMSTICHIG
13977	**WORM-EATEN TIMBER**	BOIS VERMOULU	HOLZ (WURMSTICHIGES)
13978	**WORM, COOLING COIL**	SERPENTIN DE REFROIDISSEMENT	KÜHLSCHLANGE
13979	**WORM, ENDLESS SCREW, PERPETUAL SCREW**	VIS SANS FIN	SCHNECKE, WURM, SCHRAUBE OHNE ENDE, ZYLINDERSCHRAUBE
13980	**WORN OUT BEARING**	COUSSINET OVALISE PAR L'USURE, COUSSINET USE	LAGER (UNRUNDGEWORDENES), LAGER (AUSGELAUFENES)
13981	**WORN OUT, WORN DOWN SCREW THREAD**	FILET FOIRE, FILET JARETE	GEWINDE (AUSGELEIERTES)
13982	**WORN OUT, WORN DOWN, WORN OFF, WORN AWAY**	USAGE	ABGENUTZT
13983	**WOVEN BELT**	COURROIE TISSEE	RIEMEN (GEWEBTER)
13984	**WRAPPING PAPER**	PAPIER D'EMBALLAGE	EINWICKELPAPIER, PACKPAPIER
13985	**WRINKLING**	FRISAGE	FALTENBILDUNG
13986	**WRIST PIN**	GOUPILLE	ANLENKBOLZEN
13987	**WROUGHT BRASS**	ARCOT	ROHMESSING
13988	**WROUGHT IRON, MALLEABLE IRON**	FER DOUX, FER FORGE	SCHMIEDEEISEN
13989	**WROUGHT IRON, MALLEABLE IRON PIPE**	TUYAU EN FER	ROHR (SCHMIEDEEISERNES)
13990	**WROUGHT NAIL**	CLOU FORGE	NAGEL (GESCHMIEDETER)
13991	**WULFENITE, LEAD MOLYBDATE**	MELINOSE, WULFENITE	MOLYBDÄNBLEISPAT, GELBBLEIERZ, WULFENIT
13992	**WULFF NET**	ABAQUE, RESEAU DE WULFF (STEREOGRAPHIQUE)	ABAKUS (WULFFSCHER)
13993	**X-RAY ANALYSIS**	EXAMEN RADIOGRAPHIQUE	RÖNTGENANALYSE
13994	**X-RAY APPARATUS**	APPAREIL RADIOGRAPHIQUE	RÖNTGENAPPARAT
13995	**X-RAY BEAM**	FAISCEAU DE RAYONS X	RÖNTGENSTRAHLENBÜNDEL
13996	**X-RAY CRYSTALLOGRAPHY**	RADIO-CRISTALLOGRAPHIE	RÖNTGENKRISTALLOGRAPHIE
13997	**X-RAY DIFFRACTION PATTERN**	DIAGRAMME DE DIFFRACTION DES RAYONS X.	RÖNTGENBEUGUNGSDIAGRAMM
13998	**X-RAY EXAMINATION**	CONTROLE RADIOGRAPHIQUE	RÖNTGENUNTERSUCHUNG
13999	**X-RAY FLUORESCENT ANALYSIS**	ANALYSE FLUORESCENTE	RÖNTGEN-FLUORESZENZ-ANALYSE
14000	**X-RAY SPECTROGRAPH**	SPECTROMETRE A RAYONS X	RÖNTGENSPEKTROMETER
14001	**X-RAY SPECTRUM**	SPECTRE DES RAYONS X	RÖNTGENSPEKTRUM
14002	**X-RAY TUBE**	TUBE A RAYONS X	RÖNTGENRÖHRE
14003	**X-RAYS**	RAYONS X	RÖNTGENSTRAHLEN
14004	**XENON**	XENON	XENON
14005	**XEROGRAPHY**	XEROGRAPHIE	XEROGRAPHIE
14006	**XYLENE**	XYLENE, XYLOL	XYLOL
14007	**YARD**	YARD	YARD
14008	**YELLOW BRASS**	LAITON	MESSING

14009	YELLOW PHOSPHORUS	PHOSPHORE BLANC	PHOSPHOR (GELBWEISSER), PHOSPHOR (KRISTALLINISCHER)
14010	YIELD	RENDEMENT	ARBEITSLEISTUNG
14011	YIELD POINT, BREAKING DOWN POINT	LIMITE D'ECOULEMENT, LIMITE TARDIVE, LIMITE ELASTIQUE	FLIESSGRENZE, STRECKGRENZE
14012	YIELD STRENGTH	LIMITE D'ECOULEMENT, LIMITE ELASTIQUE	STRECKGRENZE, FLIESSGRENZE, STRECKFESTIGKEIT
14013	YIELD STRESS	LIMITE ELASTIQUE	STRECKGRENZE, ELASTIZITÄTSGRENZE
14014	YOKE	ARCADE	BÜGEL
14015	YOKE BONNET	CHAPEAU A ARCADE	BÜGELAUFSATZ
14016	YOKE BUSHING, YOKE SLEEVE	DOUILLE D'ARCADE	AUFSATZBUCHSE
14017	YOKE OF A MAGNET	CULASSE D'UN AIMANT	JOCH EINES MAGNETEN
14018	YOKE RETAINING NUT	ECROU DE DOUILLE D'ARCADE	BÜGELVERSCHRAUBUNG
14019	YTTERBIUM	YTTERBIUM	YTTERBIUM
14020	YTTRIUM	YTTRIUM	YTTRIUM
14021	ZAMAK	ZAMAK	FEINZINKLEGIERUNG
14022	ZAPON ENAMEL	VERNIS ZAPON	ZAPONLACK
14023	ZED BAR, Z-BAR	FER EN Z	Z-EISEN
14024	ZERO DEVIATION	DEVIATION NULLE	NULLPUNKTABWEICHUNG
14025	ZERO MARK LINE	TRAIT ZERO	NULLMARKE, NULLSTRICH
14026	ZERO OFFSET	DECALAGE DU POINT D'ORIGINE	NULLPUNKTVERSATZ
14027	ZERO POINT	ZERO (D'UNE ECHELLE), POINT ZERO	NULLPUNKT
14028	ZERO POSITION	POSITION DE ZERO	NULLSTELLUNG
14029	ZERO PRESSURE LINE	LIGNE DE NULLE PRESSION (DU DIAGRAMME D'INDICATEUR)	NULLINIE (INDIKATOR-DIAGRAMM)
14030	ZERO RESET	REMISE A ZERO	NULLRÜCKSTELLUNG
14031	ZERO SHIFT	DECALAGE DU POINT D'ORIGINE	NULLPUNKTVERSATZ
14032	ZERO SYNCHRONIZATION	REMISE A ZERO	NULLRÜCKSTELLUNG
14033	ZERO-SUPPRESSION	SUPPRESSION DES ZEROS	NULLUNTERDRÜCKUNG
14034	ZIGZAG RIVETED JOINT	RIVURE EN QUINCONCE	ZICKZACKNIETUNG, NIETUNG (VERSETZTE)
14035	ZINC	ZINC	ZINK
14036	ZINC ACETATE	ACETATE DE ZINC	ZINK (ESSIGSAURES), ZINKAZETAT
14037	ZINC CARBONATE	CARBONATE DE ZINC	ZINK (KOHLENSAURES), ZINKKARBONAT
14038	ZINC CHLORIDE	CHLORURE DE ZINC	ZINK (SALZSAURES), ZINKCHLORID, CHLORZINK
14039	ZINC CHROMATE	CHROMATE DE ZINC, JAUNE DE ZINC	ZINK (CHROMSAURES), ZINKCHROMAT, ZINKCHROMGELB
14040	ZINC CUTTINGS	ROGNURES DE ZINC	ZINKSPÄNE
14041	ZINC CYANIDE	CYANURE DE ZINC	ZYANZINK
14042	ZINC ORE	MINERAI DE ZINC	ZINKERZ
14043	ZINC OXIDE, ZINC WHITE	OXYDE DE ZINC, BLANC DE ZINC, BLANC DE NEIGE	ZINKOXYD, ZINKWEISS
14044	ZINC PLATING	ZINGAGE, GALVANISATION	VERZINKUNG
14045	ZINC SPRAYING	METALLISATION AU ZINC	SPRITZVERZINKEN
14046	ZINC SULPHATE, WHITE VITRIOL	SULFATE DE ZINC, VITRIOL BLANC, COUPEROSE BLANCHE	ZINKVITRIOL, VITRIOL (WEISSER), ZINKSCHWEFEL (SAURER), ZINKSULFAT
14047	ZINC TUBE	TUYAU EN ZINC	ZINKROHR
14048	ZINC WIRE	FIL DE ZINC	ZINKDRAHT
14049	ZIRCONIUM	ZIRCONIUM	ZIRKONIUM
14050	ZONE MELTING	FUSION PAR ZONE	ZONENSCHMELZEN

14051   **ZONE OF SPHERE**   ZONE SPHERIQUE, SEGMENT   KUGELSCHICHT, KUGELZONE
                                     SPHERIQUE A DEUX BASES

INDEX FRANÇAIS

A CONTREFIL 138
A-COUP 11359
A-H 481
ABACA 7980
ABAISSEMENT DE CONCENTRATION 3767
ABAISSEMENT DE LA TEMPERATURE 5022
ABAISSER UN FARDEAU 7793
ABAISSER UNE PERPENDICULAIRE 4216
ABAISSER UNE VERTICALE 4216
ABAQUE 13992
ABAQUE 8617
ABAQUE 2273
ABAQUE 3834
ABATTEURS POUR MOISSONNAGE 13869
ABATTRE UN ANGLE 2230
ABERRATION 4077
ABERRATION 5
ABERRATION CHROMATIQUE 2369
ABERRATION DE REFRANGIBILITE 2369
ABERRATION DE SPHERICITE 11885
ABIMER UN FILET 12379
ABOUCHER DES TUYAUX 5277
ABRASIF 6123
ABRASIF 17
ABRASIF 9601
ABRASIF DOUX 8269
ABRASIF NATUREL 8499
ABRASION 6106
ABRASION 13
ABRI DE LA POUSSIERE (A L') 4393
ABSCISSE 27
ABSORBABLE 38
ABSORBANT 39
ABSORBER 37
ABSORBEUR 40
ABSORPTIOMETRE 43
ABSORPTION 44
ABSORPTION D'EAU 52
ABSORPTION DE CHALEUR 51
ABSORPTIVITE 56
ACCELERATEUR 66
ACCELERATION 62
ACCELERATION ANGULAIRE 537
ACCELERATION AUTOMATIQUE 836
ACCELERATION CENTRIFUGE 10127
ACCELERATION CENTRIPETE 2189
ACCELERATION DE LA PESANTEUR 65
ACCELERATION DE LA PESANTEUR 64
ACCELERATION DE VITESSE D'UN CORPS TOMBANT 63
ACCELERATION DU PISTON 9371
ACCELERATION LINEAIRE 7611

ACCELERATION NEGATIVE 8525
ACCELERATION NON UNIFORME 13507
ACCELERATION NORMALE 8653
ACCELERATION POSITIVE 9652
ACCELERATION TANGENTIELLE 12620
ACCELERATION UNIFORME 13361
ACCELERER 6851
ACCES DIRECT 10198
ACCESSIBILITE 72
ACCESSIBLE 73
ACCESSOIRE 812
ACCESSOIRES 74
ACCESSOIRES 5283
ACCESSOIRES DE CHAUDIERES 1411
ACCIDENT DU TRAVAIL 75
ACCORD DU BREVET 6049
ACCOSTAGE DE TOLES 8046
ACCOSTER 8042
ACCOSTER DES TOLES (MONTAGE) 8045
ACCOUPLEMENT 3254
ACCOUPLEMENT 10221
ACCOUPLEMENT A DEBRAYAGE 4033
ACCOUPLEMENT A DENTS 2461
ACCOUPLEMENT A FRICTION 5672
ACCOUPLEMENT A GRIFFES 2461
ACCOUPLEMENT A MANCHON 1510
ACCOUPLEMENT A PLATEAUX BOULONNES 5338
ACCOUPLEMENT AUTOMATIQUE 11149
ACCOUPLEMENT D'ARBRES 11240
ACCOUPLEMENT DEBRAYAGE 4032
ACCOUPLEMENT ELASTIQUE 4476
ACCOUPLEMENT FIXE 5037
ACCOUPLEMENT FLEXIBLE 5417
ACCOUPLEMENT HYDRAULIQUE 6681
ACCOUPLEMENT PAR CONE 2928
ACCOUPLEMENT PAR MANCHON A PLATEAU 5338
ACCOUPLEMENT RIGIDE 5037
ACCOUPLER 3253
ACCROCHAGE 12918
ACCROCHER 12914
ACCROISSEMENT DE LA FRAGILITE 4685
ACCROISSEMENT DE VITESSE 6849
ACCUMULATEUR 76
ACCUMULATEUR A GRILLAGE 6097
ACCUMULATEUR AU PLOMB 7455
ACCUMULATEUR DE VAPEUR 12156
ACCUMULATEUR ELECTRIQUE 4534
ACCUMULATEUR HYDRAULIQUE 6677
ACCUMULATEUR THERMIQUE DE CHALEUR 6336
ACCUMULATION 1705
ACCUMULATION/EMMAGASINAGE D'ENERGIE 12290

ACCUMULATION/EMMAGASINAGE DE CHALEUR **12291**
ACCUMULER LA CHALEUR **12296**
ACETATE **91**
ACETATE **85**
ACETATE D'ALUMINIUM **417**
ACETATE D'AMMONIUM **458**
ACETATE D'AMYLE **487**
ACETATE D'ETHYLE **88**
ACETATE DE BARYUM **993**
ACETATE DE CELLULOSE **2132**
ACETATE DE CUIVRE **13533**
ACETATE DE FER **7134**
ACETATE DE PLOMB **7450**
ACETATE DE POTASSE **9673**
ACETATE DE SODIUM **11715**
ACETATE DE SOUDE **11715**
ACETATE DE ZINC **14036**
ACETATE FERREUX **5120**
ACETATE FERRIQUE **5099**
ACETONE **90**
ACETYLE DE CELLULOSE **91**
ACETYLENE **93**
ACETYLURE DE CALCIUM **1848**
ACHEVEMENT **11682**
ACHROMATIQUE **98**
ACHROMATISATION **99**
ACHROMATISER **100**
ACHROMATISME **101**
ACICULAIRE **102**
ACIDE **727**
ACIDE **129**
ACIDE **12450**
ACIDE ACETIQUE **86**
ACIDE ACETIQUE CRISTALLISABLE **5952**
ACIDE ALDEHYDIQUE **84**
ACIDE ARSENIQUE **722**
ACIDE AZOTEUX **8596**
ACIDE AZOTIQUE **8580**
ACIDE AZOTIQUE ANHYDRE **8590**
ACIDE BENZOIQUE **1202**
ACIDE BENZOSULFONIQUE **1200**
ACIDE BORIQUE **1467**
ACIDE BROMHYDRIQUE **6704**
ACIDE BROMIQUE **1647**
ACIDE BUTYRIQUE **1817**
ACIDE CARBAZOTIQUE **9291**
ACIDE CARBOLIQUE **1938**
ACIDE CARBONIQUE **1949**
ACIDE CARBONIQUE LIQUIDE **7645**
ACIDE CHLORHYDRIQUE **6706**
ACIDE CHROMIQUE **2379**

ACIDE CITRIQUE **2441**
ACIDE CYANHYDRIQUE **6707**
ACIDE CYANIQUE **3545**
ACIDE CYANURIQUE **3552**
ACIDE DE NORDHAUSEN **5733**
ACIDE DE SAXE **5733**
ACIDE DES CHAMBRES **2228**
ACIDE DIGALLIQUE **5780**
ACIDE FLUORHYDRIQUE **6711**
ACIDE FLUOSILICIQUE **11441**
ACIDE FORMIQUE **5578**
ACIDE FULMINIQUE **5731**
ACIDE GALLIQUE **5777**
ACIDE GRAS **5055**
ACIDE HUMIQUE **6665**
ACIDE HYDROCHLORIQUE **6706**
ACIDE HYPOCHLOREUX **6742**
ACIDE HYPOSULFUREUX **12853**
ACIDE LACTIQUE **7355**
ACIDE METAPHOSPHORIQUE **8210**
ACIDE METASILICIQUE **8211**
ACIDE MINERAL **8302**
ACIDE MOLYBDIQUE **8369**
ACIDE MURIATIQUE **6706**
ACIDE NITRIQUE **8580**
ACIDE OLEIQUE **8795**
ACIDE ORGANIQUE **8864**
ACIDE ORTHOPHOSPHORIQUE **8873**
ACIDE ORTHOSILICIQUE **8875**
ACIDE OXALIQUE **8969**
ACIDE PALMITIQUE **9028**
ACIDE PHENIQUE **1938**
ACIDE PHENYLSULFUREUX **1200**
ACIDE PICRIQUE **9291**
ACIDE PRUSSIQUE **6707**
ACIDE PYROGALLIQUE **10045**
ACIDE PYROLIGNEUX **10046**
ACIDE PYROPHOSPHORIQUE **10054**
ACIDE ROSOLIQUE **10750**
ACIDE SALICYLIQUE **10899**
ACIDE SILICIQUE **11436**
ACIDE STANNIQUE **12097**
ACIDE STEARIQUE **12194**
ACIDE SULFHYDRIQUE **6717**
ACIDE SULFUREUX **104**
ACIDE SULFURIQUE **12453**
ACIDE SULFURIQUE COMMERCIAL **8856**
ACIDE SULFURIQUE CONCENTRE **2884**
ACIDE SULFURIQUE FUMANT **5733**
ACIDE TANNIQUE **5780**
ACIDE TARTARIQUE **12690**

ACIDE TARTRIQUE DEXTROGYRE 3826
ACIDE TARTRIQUE LEVOGYRE 7364
ACIDE THIOSULFURIQUE 12853
ACIDE TUNGSTIQUE 8876
ACIDE VITRIOLIQUE 12453
ACIDES 132
ACIDIFIANT 130
ACIDIMETRE 89
ACIDIMETRIE 131
ACIDULER 133
ACIER 11490
ACIER 12223
ACIER 12196
ACIER 4156
ACIER A BAS CARBONE 8270
ACIER A CHARRUE 9528
ACIER A FAIBLE TENEUR EN CARBONE 7770
ACIER A GROS GRAIN 2577
ACIER A OUTILS 13009
ACIER A OUTILS 7857
ACIER A RESSORT 11992
ACIER A RIVETS 10627
ACIER A STRUCTURE ACICULAIRE 8523
ACIER ALLIE 368
ACIER ANORMAL 8
ACIER AU C. 9427
ACIER AU CARBONE 11473
ACIER AU CARBONE 1954
ACIER AU CHROME-TUNGSTENE 2378
ACIER AU COBALT 2612
ACIER AU COBALT-CHROME 2613
ACIER AU CUIVRE 3130
ACIER AU FOUR ELECTRIQUE 4532
ACIER AU MANGANESE 7968
ACIER AU MOLYBDENE 8368
ACIER AU NICKEL 8570
ACIER AU NICKEL-CHROME 2386
ACIER AU SILICIUM 11447
ACIER AU SILICIUM 11449
ACIER AU TITANE 12972
ACIER AU TUNGSTENE 13265
ACIER AU VANADIUM 13476
ACIER AUSTENITIQUE 826
ACIER AUTO-TREMPANT 298
ACIER AUTO-TREMPANT 11154
ACIER AUTOTREMPANT 8000
ACIER BATTU 6222
ACIER BESSEMER 107
ACIER BESSEMER 1215
ACIER BRULE 1763
ACIER BRUT 3414

ACIER CALME 7294
ACIER CALME A L'ALUMINIUM 442
ACIER CALME AU MANGANESE 7969
ACIER CALME ET TREMPE 10093
ACIER CEMENTE 2138
ACIER CHROME 2389
ACIER CHROME-NICKEL 2376
ACIER COMPRIME 2849
ACIER COULE 12202
ACIER DE CARBURATION 2138
ACIER DE CEMENTATION 2138
ACIER DE CONSTRUCTION 12392
ACIER DE DECOLLETAGE RAPIDE 5639
ACIER DE NITRURATION 8585
ACIER DE QUALITE SUPERIEURE 6486
ACIER DEMI-DOUX 1959
ACIER DOUX 1958
ACIER DOUX 7770
ACIER DOUX 11744
ACIER DOUX 8270
ACIER DUR 6280
ACIER DUR A HAUTE TENEUR EN CARBONE 1957
ACIER ECROUI 2697
ACIER EFFERVESCENT 10599
ACIER EFFERVESCENT 10598
ACIER EFFERVESCENT 4462
ACIER ELABORE AU FOUR ELECTRIQUE 4511
ACIER ELECTRIQUE 4511
ACIER ELECTRIQUE ACIDE 113
ACIER EN BARRES 984
ACIER ETIRE 4251
ACIER EXTRA-DOUX 12224
ACIER FAIBLEMENT ALLIE 382
ACIER FIN 12462
ACIER FONDU 6924
ACIER FONDU AU CREUSET 2057
ACIER FORGE 5554
ACIER FRETTE 10598
ACIER INOXYDABLE 12051
ACIER INOXYDABLE 8644
ACIER MANGANO-SILICEUX 11450
ACIER MARTIN 1048
ACIER MARTIN 8816
ACIER MARTIN PAR LE PROCEDE ACIDE 116
ACIER MARTIN PAR LE PROCEDE BASIQUE 1047
ACIER MOULE 12202
ACIER MOULE 2055
ACIER MOUSSEUX 10599
ACIER NICKELE 8573
ACIER NON CALME 4462
ACIER NON MAGNETIQUE 8638

ACIER PARTIELLEMENT DESOXYDE **10599**
ACIER PERLITIQUE **9145**
ACIER PLAQUE **2446**
ACIER PLATINITE **9508**
ACIER POUR AIMANTS **7892**
ACIER POUR ENGRENAGES **5904**
ACIER POUR PLATE-FORME DE FORAGE EN MER **8739**
ACIER PROFILE **11125**
ACIER PUDDLE **9963**
ACIER RAPIDE **6474**
ACIER RAPIDE **6497**
ACIER RAPIDE A OUTILS **10207**
ACIER RESISTANT A L'USURE **13718**
ACIER SANS SOUFFLURES **12209**
ACIER SEMI-CALME **8825**
ACIER SEMI-CALME **11173**
ACIER SIEMENS- MARTIN **8023**
ACIER SOUDE **13765**
ACIER SPECIAL **11848**
ACIER SUPERIEUR **6486**
ACIER SUR SOLE BASIQUE **1048**
ACIER SURCHAUFFE **8945**
ACIER THOMAS **1044**
ACIER THOMAS **12855**
ACIER TREMPE **6289**
ACIER TREMPE ET REVENU **10093**
ACIERAGE **12206**
ACIERAGE **135**
ACIERAGE **2142**
ACIERATION **2142**
ACIERER **12205**
ACIERER (MET.) **2134**
ACIERIE **12222**
ACIERIE **12211**
ACIERS DE TARIERES POUR FORAGES **4270**
ACIERS LAMINES PLATS **5392**
ACIERS RESISTANT A LA CORROSION ET A LA CHALEUR
  3184
ACOUSTIQUE **137**
ACOUSTIQUE **136**
ACTINIQUE **139**
ACTINIUM **141**
ACTINOMETRE **142**
ACTION D'OMBRER **11237**
ACTION D'UNE FORCE **143**
ACTION D'UNE FORCE EXTERIEURE **12331**
ACTION DE CROQUER **11533**
ACTION DE PARER **13209**
ACTION DE RETENDRE UNE COURROIE **12605**
ACTION DIRECTE (A) **3980**
ACTION DU FREIN **1550**

ACTION EN NULLITE RELATIVE A UN BREVET **8707**
ACTION EN PROFONDEUR **12899**
ACTION GYROSCOPIQUE **6192**
ACTION INDIRECTE (A) **6883**
ACTIONNER **4289**
ACTIONS ATMOSPHERIQUES **792**
ACTIVATEUR **146**
ACTIVATION **144**
ACUTANGLE **154**
ADAPTATEUR DE BRIDE **5325**
ADAPTEUR **160**
ADAPTEUR POUR ECHANGE RAPIDE **10108**
ADDITIF **168**
ADDITION **394**
ADDITION **169**
ADDITION DE CARBONE **145**
ADDITION EN POCHE **7357**
ADDITIONNER (MATH.) **162**
ADDITIONNER DE L'EAU **163**
ADENT **3761**
ADHERENCE **174**
ADHERENCE **173**
ADHERENT **175**
ADHESIF **175**
ADHESION **174**
ADIABATIQUE **178**
ADIABATIQUE **179**
ADJACENT **181**
ADJUVANT **169**
ADJUVANT DE FILTRATION **5174**
ADMISSION **6943**
ADMISSION D'AIR **12486**
ADOUCIR **12717**
ADOUCISSEMENT **11749**
ADRESSE **171**
ADSORBER **202**
ADSORPTION **203**
ADULTERATION **205**
ADULTERER **204**
AERAGE **13530**
AERATEUR **211**
AERATION **13530**
AERATION DE L'EAU **210**
AERER UN SABLE **208**
AERO-MOTEUR **2843**
AERODYNAMIQUE **215**
AERODYNAMIQUE **214**
AEROSTATIQUE **218**
AEROSTATIQUE **217**
AFFAIBLISSEMENT DE LA SECTION **13707**
AFFAIBLISSEMENT DU MATERIEL **13708**

AFFAIRE **7229**
AFFAISSEMENT **10613**
AFFAISSEMENT **12418**
AFFICHAGE **10253**
AFFILAGE **6793**
AFFILER **6790**
AFFINAGE **10377**
AFFINAGE AU FEU **5245**
AFFINAGE AU VENT **300**
AFFINAGE ELECTROLYTIQUE **4558**
AFFINEMENT DU GRAIN **6037**
AFFINER **10370**
AFFINER UN METAL **10371**
AFFINITE CHIMIQUE **2292**
AFFLEURE **5482**
AFFLEURER **7951**
AFFOLEMENT DE LA SOUPAPE **2277**
AFFOUILLEMENT **4889**
AFFUT **6175**
AFFUT DE CAMPAGNE **5142**
AFFUT DE MORTIER **8397**
AFFUTAGE **11275**
AFFUTAGE DE PRECISION **9755**
AFFUTER **11273**
AFFUTEUR **6105**
AFFUTEUSE **13002**
AFFUTEUSE **11274**
AFFUTEUSE UNIVERSELLE **13384**
AGATE **222**
AGENT ANTI-PIQURE **616**
AGENT ANTIPUTREFIANT **629**
AGENT D'IMPREGNATION **6804**
AGENT DE BREVETS D'INVENTION **9114**
AGENT DE BRILLANTAGE **1615**
AGENT DE CEMENTATION **2139**
AGENT DE CHELATION **2289**
AGENT DE CONSERVATION **9791**
AGENT DE DISPERSION **4050**
AGENT DE REFRIGERATION **3086**
AGENT DE TRAITEMENT DE SURFACE **12523**
AGENT FRIGORIFIQUE **3086**
AGENT MOTEUR **8408**
AGENT MOUILLANT **13808**
AGENT PRECIPITANT **11130**
AGENT REDUCTEUR **10346**
AGENTS ATMOSPHERIQUES **793**
AGGLOMERANT **1253**
AGGLOMERATION **7990**
AGGLOMERE **6088**
AGGLOMERE **2791**
AGGLOMERE **1623**

AGGLOMERE DE HOUILLE **2560**
AGGLOMERE DE LIGNITE **1666**
AGGLOMERE DE TOURBE **9149**
AGGLOMERE MAGNESIEN **7881**
AGGLUTINANT **1445**
AGITATEUR **12260**
AGITATEUR **236**
AGITATEUR MECANIQUE **238**
AGITATION DU BAIN **11254**
AGITER **8328**
AGRAFE DE JONCTION POUR CABLES **11706**
AGRAFE POUR COURROIES **1146**
AGRANDISSEMENT **7935**
AGRANDISSEMENT D'UN DESSIN **4769**
AGREGAT **227**
AGREGATION **229**
AIDE **8048**
AIGREUR DU FER **2704**
AIGUILLAGE **12556**
AIGUILLE (DE MANOMETRE) **6226**
AIGUILLE AIMANTEE **7909**
AIGUILLE D'INJECTEUR **8684**
AIGUILLE INDICATRICE **6868**
AIGUILLES POUR ROULEMENTS **1108**
AIGUISAGE **11275**
AIGUISER **11273**
AIGUISEUR **6105**
AILE **13728**
AILE **13874**
AILE **5095**
AILE (D'UN FER A T) **5330**
AILE D'UNE HELICE **9928**
AILERON **5183**
AILETTE **5183**
AILETTE DE LA SOUPAPE **5060**
AILETTES DE REFROIDISSEMENT **13477**
AIMANT **7891**
AIMANT A LAMES SUPERPOSEES **7377**
AIMANT ARTIFICIEL **735**
AIMANT EN FORME DE BARREAU **985**
AIMANT EN U **6606**
AIMANT NATUREL **8506**
AIMANT PERMANENT **9204**
AIMANTATION **7928**
AIMANTATION **7923**
AIMANTATION REMANENTE **10445**
AIMANTATION RESIDUELLE **10445**
AIMANTER **7924**
AIMANTER (S') **1126**
AIR AMBIANT **450**
AIR ATMOSPHERIQUE **240**

AIR COMBURANT 277
AIR COMPRIME 2839
AIR DE LA SOUFFLERIE 243
AIR HUMIDE 3599
AIR LIQUIDE 7643
AIR PARASITE 4771
AIR RAREFIE 10213
AIR SEC 4339
AIRE 697
AIRE DE L'OUVERTURE MASQUEE PAR LA SOUPAPE 11121
AIRE DE LA SECTION 11120
AIRE DE LA SECTION TRANSVERSALE 12511
AIRE DU CERCLE 695
AJOUTER 162
AJUSTABLE 183
AJUSTAGE 10847
AJUSTAGE 11600
AJUSTAGE 10025
AJUSTAGE 5275
AJUSTAGE 5280
AJUSTAGE 4307
AJUSTAGE A LA PRESSE 5527
AJUSTAGE CONIQUE 8679
AJUSTER 5276
AJUSTEUR 5278
AJUSTEUR-MECANICIEN 5278
AJUTAGE CONVERGENT 3066
AJUTAGE CONVERGENT 2769
AJUTAGE CYLINDRIQUE 3576
AJUTAGE D'ECOULEMENT 196
AJUTAGE DIVERGENT 3738
AJUTAGE DIVERGENT 4096
ALAMBIC 12259
ALBATRE 304
ALBEDO 308
ALBUMINE 310
ALBUMINE DU SERUM 11207
ALCALI 328
ALCALI VOLATIL 660
ALCALIMETRE 331
ALCALIMETRIE 333
ALCALIMETRIQUE 332
ALCALINISER 330
ALCALINITE 340
ALCALOIDE 341
ALCOOL ABSOLU 28
ALCOOL AMYLIQUE 488
ALCOOL ANHYDRE 28
ALCOOL AQUEUX 659
ALCOOL CARBURE 1983
ALCOOL DENATURE 8224

ALCOOL ETHYLIQUE 313
ALCOOL METHYLIQUE 13920
ALCOOL ORDINAIRE 313
ALCOOL RECTIFIE 10313
ALCOOL VINIQUE 313
ALCOOLIQUE 314
ALCOOMETRE 316
ALDEHYDE 317
ALDEHYDE ACETIQUE 84
ALDEHYDE BENZYLIQUE 1198
ALDEHYDE ETHYLIQUE 84
ALDEHYDE FORMIQUE 5573
ALDEHYDE METHYLIQUE 5573
ALESAGE 3855
ALESAGE 1461
ALESAGE 1468
ALESAGE AU TOUR 1473
ALESAGE D'UNE POULIE 4978
ALESAGE D'UNE ROUE 4978
ALESAGE DE CYLINDRE 1461
ALESAGE DES TROUS DE RIVET 1638
ALESAGE DU CYLINDRE 3565
ALESER 1462
ALESER AU TOUR 1466
ALESER LES TROUS DE RIVET 1635
ALESEUSE 4279
ALESEUSE 1471
ALESEUSE DE PRECISION 5196
ALESOIR 10263
ALESOIR 1636
ALESOIR A CANNELURES 5485
ALESOIR A FINES RAINURES 5197
ALESOIR A LAMES MOBILES 188
ALESOIR A MAIN 6247
ALESOIR A MAIN CONIQUE 12650
ALESOIR A MAIN EXPANSIBLE 4917
ALESOIR A MAIN REGLABLE 184
ALESOIR A MISE DE CARBURE 1935
ALESOIR A RAINURES 5485
ALESOIR A RAINURES DROITES 12305
ALESOIR A RAINURES HELICOIDALES 11915
ALESOIR A RAINURES TORSES 11915
ALESOIR CONIQUE 12658
ALESOIR DE MACHINE 7853
ALESOIR EBAUCHEUR CONIQUE 10790
ALESOIR EN BOUT 10747
ALESOIR EXPANSIBLE 4923
ALESOIR EXTENSIBLE 188
ALESOIR FINISSEUR CONIQUE 5226
ALESOIR/FRAISE CREUX 11336
ALESURES 11277

ALFAMETRE 319
ALGEBRE 320
ALGEBRIQUE 321
ALIGNEMENT 325
ALIGNEMENT 324
ALIGNER 323
ALIMENTATEUR AUTOMATIQUE 5072
ALIMENTATION 5063
ALIMENTATION 5082
ALIMENTATION 2263
ALIMENTATION 7940
ALIMENTATION ARRIVEE 6844
ALIMENTATION D'UN FOYER 5262
ALIMENTATION DOUBLE 4138
ALIMENTATION EN EAU 13678
ALIMENTER 5064
ALIMENTER UN FOYER 12269
ALIZARINE 327
ALLIAGE 360
ALLIAGE 619
ALLIAGE A BASE D'ALUMINIUM 439
ALLIAGE A BASE D'ALUMINIUM 418
ALLIAGE A BASE DE CUIVRE 3111
ALLIAGE A BASE DE MAGNESIUM 7890
ALLIAGE A BASE DE NICKEL 8557
ALLIAGE A BASE DE PLOMB 7451
ALLIAGE A HAUTE DENSITE 6481
ALLIAGE ANTI-FRICTION 385
ALLIAGE AU BORE 1478
ALLIAGE AU PLOMB 7453
ALLIAGE BINAIRE 4378
ALLIAGE BINAIRE 1247
ALLIAGE BLANC 13836
ALLIAGE CUIVRE-ZINC 5943
ALLIAGE CUIVRE-ZING 9861
ALLIAGE D'ALUMINIUM FORGEABLE 433
ALLIAGE D'ALUMINIUM POUR MOULAGE 435
ALLIAGE D'ALUMINIUM-BERYLLIUM 440
ALLIAGE D'ALUMUNIUM FORGEABLE 437
ALLIAGE D'APPORT 5159
ALLIAGE D'AUER 817
ALLIAGE DE BRASAGE 386
ALLIAGE DE FER 7156
ALLIAGE DE RECHARGEMENT DUR 6284
ALLIAGE DE TRAITEMENT 6383
ALLIAGE DEGAZEUR 3709
ALLIAGE DUPLEX 4378
ALLIAGE DUR AUX CARBURES FRITTES 11510
ALLIAGE FUSIBLE 5757
ALLIAGE FUSIBLE 389
ALLIAGE FUSIBLE POUR SOUDER OU BRASER 11755

ALLIAGE HAUTE-TEMPERATURE 12459
ALLIAGE INOXYDABLE 387
ALLIAGE LEGER 7564
ALLIAGE MAGNETIQUE 392
ALLIAGE MAGNETIQUE 'ALNICO' 398
ALLIAGE NATUREL 8498
ALLIAGE NON REFRACTAIRE 8622
ALLIAGE NON-FERREUX 8619
ALLIAGE OSTEOPLASTIQUE 498
ALLIAGE POUR COULEE SOUS PRESSION 388
ALLIAGE POUR COUSSINET 1106
ALLIAGE POUR COUSSINETS 385
ALLIAGE PYROPHORIQUE 10053
ALLIAGE QUATERNAIRE 10084
ALLIAGE QUATERNAIRE EUTECTIQUE 10085
ALLIAGE REFRACTAIRE 393
ALLIAGE RESISTANT A L'USURE 383
ALLIAGE RESISTANT A LA CHALEUR 391
ALLIAGE RESISTANT A LA CHALEUR ET A LA CORROSION 390
ALLIAGE RESISTANT A LA CORROSION 387
ALLIAGE RESISTANT AUX ACIDES 384
ALLIAGES FUSIBLES 7773
ALLIAGES POUR RESISTANCES ELECTRIQUES 10486
ALLIAGES RESISTANT AUX ACIDES 125
ALLIAGES RESISTANTS A L'ABRASION ET A LA CORROSION 396
ALLIER 361
ALLOMERIE 350
ALLOMERIQUE 349
ALLOMORPHE 352
ALLOMORPHIE 351
ALLONGEMENT 4679
ALLONGEMENT 4680
ALLONGEMENT D'UN CABLE 12374
ALLONGEMENT D'UNE COURROIE 12374
ALLONGEMENT DES CRISTAUX 3426
ALLONGEMENT LONGITUDINAL 7612
ALLONGEMENT PROPORTIONNEL A LA LIMITE D'ELASTICITE 13329
ALLONGEMENT RELATIF 13370
ALLONGER (S') 1125
ALLOTRIOMORPHE 354
ALLOTROPIE 357
ALLOTROPIQUE 356
ALLUMAGE 6760
ALLUMAGE (DANS LES MOTEURS A EXPLOSION) 6770
ALLUMAGE A FLAMME 5316
ALLUMAGE AUTOMATIQUE 844
ALLUMAGE DU FOYER D'UNE CHAUDIERE 5263
ALLUMAGE ELECTRIQUE 4514

ALLUMAGE ELECTRONIQUE **4633**
ALLUMAGE PAR INCANDESCENCE **13251**
ALLUMAGE PREMATURE **9778**
ALLUME-CIGARETTE **2396**
ALLUMER **6757**
ALLUMER LE FOYER D'UNE CHAUDIERE **5236**
ALLUMEUR **6758**
ALLUMEUR **6762**
ALLURE (COURBE) **13181**
ALLURE D'UNE MACHINE **10850**
ALPACCA **8569**
ALPHABET A FRAPPER **1560**
ALTERATION CHIMIQUE **2296**
ALTERATION DE LA TEINTE **4026**
ALTERNATEUR **411**
ALTERNATEUR **407**
ALTERNATION PHYSIQUE **9280**
ALTERNATIVE **409**
ALTERNE **12048**
ALTERNO-MOTEUR **408**
ALUMINATE **414**
ALUMINATE DE BARYTE **994**
ALUMINATE DE POTASSIUM **9674**
ALUMINATE DE SODIUM **11716**
ALUMINATION **429**
ALUMINE **438**
ALUMINE **412**
ALUMINE **413**
ALUMINE HYDRATEE **423**
ALUMINIFERE **415**
ALUMINIUM **416**
ALUMINIUM **444**
ALUMINIUM **432**
ALUMINIUN EN LINGOT **424**
ALUMINO **430**
ALUMINOTHERMIE **431**
ALUN DE CHROME **2370**
ALUN DE FER **7135**
ALUN ORDINAIRE **2787**
ALUN POTASSIQUE **2787**
AMAGNETIQUE **8637**
AMALGAMATION **447**
AMALGAME **445**
AMALGAME **397**
AMALGAMER **446**
AMAS DE GUINIER-PRESTON **2542**
AMBRE JAUNE **449**
AME **3142**
AME **13726**
AME D'UN CABLE **3167**
AME D'UN FER PROFILE **13730**

AME D'UNE POUTRELLE **13729**
AME EN CHANVRE **6440**
AMENDE **5195**
AMENEE D'HUILE **8783**
AMIANTE **6599**
AMIANTE **1898**
AMIANTE **746**
AMIANTE **755**
AMIANTE TRESSE EN CORDELETTE **751**
AMIDON **12111**
AMMONIAC **454**
AMMONIAC **660**
AMMONIAQUE **453**
AMMONIAQUE **7644**
AMONT (D') **13407**
AMONT (EN) **13407**
AMORCAGE (DE COUPE) **5858**
AMORCE **1456**
AMORCE DE CRIQUE **6833**
AMORCE DE CRIQUE **5411**
AMORCE DE FISSURE **6833**
AMORCER AU MOYEN D'UN COUP DE POINTEAU **8011**
AMORCER UNE POMPE **9856**
AMORPHE **475**
AMORTIR UN CHOC **3640**
AMORTISSEMENT **3604**
AMORTISSEMENT DE VIBRATIONS **3606**
AMORTISSEMENT PAR LIQUIDE **7649**
AMORTISSEMENT PNEUMATIQUE **255**
AMORTISSEUR **3601**
AMORTISSEUR (ELECTR.) **3602**
AMORTISSEUR A LIQUIDE **6678**
AMORTISSEUR DE CHOCS **11358**
AMORTISSEUR DE VIBRATIONS **13555**
AMORTISSEUR PNEUMATIQUE **256**
AMOVIBLE **3808**
AMPERAGE **478**
AMPERE **479**
AMPERE-HEURE **481**
AMPERE-MINUTE **482**
AMPERE-SECONDE **483**
AMPERE-TOUR **480**
AMPEREMETRE **452**
AMPHIBOLE **6598**
AMPHOTERE **484**
AMPLIFICATEUR **485**
AMPLITUDE **486**
AMPOULE **7385**
AMPOULE DU THERMOMETRE **1717**
ANAEROBIQUE **489**
ANALOGIQUE **490**

ANALYSE 493
ANALYSE (DETERMINATION) DE STRUCTURE CRISTALLINE 3425
ANALYSE AU CHALUMEAU 1356
ANALYSE CALORIMETRIQUE 1880
ANALYSE CHIMIQUE 2293
ANALYSE DE CONTROLE 2280
ANALYSE DE COPEAUX 2345
ANALYSE DE COULEE 7358
ANALYSE DES GAZ 5809
ANALYSE ELECTROLYTIQUE 4594
ANALYSE ELEMENTAIRE 4655
ANALYSE FLUORESCENTE 13999
ANALYSE GRANULOMETRIQUE 6023
ANALYSE GRANULOMETRIQUE 11020
ANALYSE GRAVIMETRIQUE 6077
ANALYSE MAGNETIQUE 7894
ANALYSE NEPHELOMETRIQUE 8537
ANALYSE PAR VOIE HUMIDE 13800
ANALYSE PAR VOIE SECHE 4341
ANALYSE PONDERALE 6077
ANALYSE QUALITATIVE 10062
ANALYSE QUANTITATIVE 10064
ANALYSE SPECTRALE 11861
ANALYSE SPECTRALE 11864
ANALYSE THERMIQUE SELECTIVE 3919
ANALYSE VOLUMETRIQUE 13601
ANALYSER 491
ANALYSEUR 492
ANALYTIQUE 494
ANASTIGMAT 497
ANCRAGE 499
ANCRAGE 12918
ANCRE (CONSTR.) 500
ANCRER 12914
ANEMOMETRE 503
ANEROIDE 504
ANGLE (GEOM.) 505
ANGLE ADJACENT 180
ANGLE AIGU 155
ANGLE AU CENTRE 507
ANGLE COMPLEMENTAIRE 2811
ANGLE D'AFFUTAGE 12995
ANGLE D'ATTAQUE 8577
ANGLE D'ATTAQUE 3524
ANGLE D'ATTAQUE 1277
ANGLE D'AVANCE 515
ANGLE D'ENROULEMENT 517
ANGLE D'INCIDENCE 520
ANGLE D'INCIDENCE (D'UN OUTIL) 2484
ANGLE D'INCLINAISON 521

ANGLE D'INCLINATION DES PIVOTS 7312
ANGLE D'OUVERTURE DE LA RAINURE 6129
ANGLE DANS LE SEGMENT 6962
ANGLE DE BRAQUAGE 7701
ANGLE DE CALAGE 529
ANGLE DE CARROSSAGE 1889
ANGLE DE CISAILLEMENT 11281
ANGLE DE COIN 12995
ANGLE DE CONTACT 517
ANGLE DE COUPE 10186
ANGLE DE COUPE 3524
ANGLE DE DECALAGE 9233
ANGLE DE DECALAGE EN ARRIERE 522
ANGLE DE DECALAGE EN AVANCE 515
ANGLE DE DEGAGEMENT DU COPEAU 5698
ANGLE DE DEPHASAGE 9233
ANGLE DE DEPHASAGE EN ARRIERE 523
ANGLE DE DEPHASAGE EN AVANCE 524
ANGLE DE DEPOUILLE 5698
ANGLE DE DETALONNAGE DU CORPS 1388
ANGLE DE DEVIATION 518
ANGLE DE FLEXION 519
ANGLE DE FROTTEMENT 527
ANGLE DE L'INCLINAISON DU FILET 531
ANGLE DE L'OUTIL 12995
ANGLE DE PENTE LATERALE 11418
ANGLE DE PIED 10716
ANGLE DE PRESSION 9810
ANGLE DE REFLEXION 525
ANGLE DE REFRACTION 526
ANGLE DE RETARD 522
ANGLE DE RETRAIT 4197
ANGLE DE ROTATION 528
ANGLE DE SAILLIE 165
ANGLE DE TETE 4990
ANGLE DE TORSION 532
ANGLE DIEDRE 3944
ANGLE DROIT 10577
ANGLE DU CHANFREIN 1223
ANGLE DU FILET 530
ANGLE EXTERNE 4950
ANGLE HELICOIDAL 6431
ANGLE INSCRIT 6962
ANGLE INTERIEUR 7057
ANGLE INTERNE 7057
ANGLE LIMITE 3345
ANGLE OBTUS 8717
ANGLE OPTIQUE 13581
ANGLE POINTU 155
ANGLE PRIMITIF 9401
ANGLE RENTRANT 10243

ANGLE SAILLANT **10900**
ANGLE SOLIDE **11776**
ANGLE SUPPLEMENTAIRE **12483**
ANGLE TAILLANT **12995**
ANGLE VIF **155**
ANGLE VISUEL **13581**
ANGLOIR **1228**
ANGSTROM **536**
ANHYDRE **550**
ANHYDRIDE **8591**
ANHYDRIDE **11436**
ANHYDRIDE **548**
ANHYDRIDE ACETIQUE **87**
ANHYDRIDE ARSENIEUX **727**
ANHYDRIDE ARSENIQUE **723**
ANHYDRIDE AZOTEUX **8592**
ANHYDRIDE AZOTIQUE **8590**
ANHYDRIDE CARBONIQUE **1948**
ANHYDRIDE CHROMIQUE **2383**
ANHYDRIDE HYPOCHLOREUX **2355**
ANHYDRIDE PHOSPHORIQUE **9249**
ANHYDRIDE SULFUREUX **12450**
ANHYDRIDE SULFURIQUE **12452**
ANHYDRIDE TITANIQUE **12971**
ANHYDRIDE TUNGSTIQUE **13266**
ANHYDRIT **549**
ANILINE **552**
ANILINE DU COMMERCE **553**
ANION **556**
ANISOTROPE **558**
ANISOTROPIE **559**
ANNEAU **589**
ANNEAU **4234**
ANNEAU **7627**
ANNEAU **10600**
ANNEAU CYLINDRIQUE **3588**
ANNEAU D'AMIANTE **753**
ANNEAU DE ROULEMENT **964**
ANNEAU MIS A CHAUD **11391**
ANNEAUX A TOURILLONNER **3312**
ANNEXE D'UN BATIMENT **585**
ANNULAIRE **588**
ANODE **590**
ANODE **602**
ANODE D'EXCITATION **4886**
ANODE DE CUIVRE **3112**
ANODE INSOLUBLE **598**
ANODIQUE **603**
ANOLYTE **606**
ANSE DE PANIER **12872**
ANSPECT **7725**

ANTHRACENE **607**
ANTHRACITE **609**
ANTI CATHODE **611**
ANTI-ACIDE RESISTANT AUX ACIDES **124**
ANTI-GEL **5657**
ANTI-PARASITE **10156**
ANTI-ROUILLE **10869**
ANTI-TARTRE **4041**
ANTIBELIER **246**
ANTICATHODE **12682**
ANTIDEFLAGRANT **5322**
ANTIFRICTION **889**
ANTIFRICTION **619**
ANTIGEL **612**
ANTILOGARITHME **620**
ANTIMAGNETIQUE **614**
ANTIMOINE **623**
ANTIPUTRIDE **628**
ANTIPUTRIDE **629**
ANTISEPTIQUE **628**
ANTISEPTIQUE **629**
APERIODICITE **637**
APERIODIQUE **634**
APEX **638**
APLANAT **639**
APLATI **5406**
APLOMB **9219**
APLOMB **9539**
APLOMB (D') **9217**
APOCHROMATIQUE **640**
APPAREIL **641**
APPAREIL A BILLER **967**
APPAREIL A EQUILIBRER LES MEULES **6116**
APPAREIL A MORTAISER **11642**
APPAREIL A TARAUDER **12676**
APPAREIL CHAUDRONNE **4982**
APPAREIL CHAUDRONNE SOUS PRESSION **9839**
APPAREIL D'ALARME **305**
APPAREIL D'ALIMENTATION **5083**
APPAREIL D'ECLAIRAGE **6772**
APPAREIL D'EQUILIBRAGE **937**
APPAREIL D'ESSAI **12793**
APPAREIL D'ESSAI DE DUCTILITE **3462**
APPAREIL DE CONSTRUCTON **1444**
APPAREIL DE DISTILLATION **12259**
APPAREIL DE DISTRIBUTION A GLISSEMENT **11592**
APPAREIL DE DISTRIBUTION DE LA VAPEUR **13460**
APPAREIL DE GRAISSAGE SOUS PRESSION MECANIQUE **8105**
APPAREIL DE LEVAGE **6519**
APPAREIL DE MANUTENTION **3080**

APPAREIL DE MISE EN MARCHE **12119**

APPAREIL DE PROTECTION **10881**

APPAREIL DISTILLATOIRE **12259**

APPAREIL DIVISEUR **4105**

APPAREIL ELECTRO MAGNETIQUE POUR ESSAI DE DURETE **4614**

APPAREIL ENREGISTREUR **10294**

APPAREIL EVAPORATEUR **4869**

APPAREIL POUR LES RUPTURES A LA TRACTION **7846**

APPAREIL RADIOGRAPHIQUE **13994**

APPAREIL SECHEUR **4356**

APPAREIL SHORE **11366**

APPAREIL SOUFFLANT **1361**

APPAREILS A DIAMANTER **13233**

APPAREILS ACCESSOIRES **1411**

APPAREILS DE DEPOUSSIERAGE **4391**

APPAREILS DE DISTRIBUTION (ELECTR.) **12558**

APPAUVRISSEMENT **6801**

APPLICATION D'UN REVETEMENT PAR TREMPAGE **3970**

APPLICATION D'UNE COUCHE PROTECTRICE **6792**

APPLICATION D'UNE FORCE **647**

APPLICATION EN PASSES CROISEES **3366**

APPLICATION INDUSTRIELLE D'UNE INVENTION **2783**

APPLIQUER UN FREIN **650**

APPLIQUER UNE COUCHE PROTECTRICE **6791**

APPOINTEMENTS **10897**

APPONTEMENT **1603**

APPRET **9855**

APPROXIMATION **652**

APPUI **12490**

APPUI **1118**

APPUI CALE **57**

APPUI FIXE **5295**

APPUI INTERMEDIAIRE D'UN ARBRE **6586**

APPUI MEDIAN **2152**

APPUIE-TETE **6320**

APPUYER **12491**

APTITUDE A LA TREMPE **6288**

APTITUDE AU DEBOURRAGE **2718**

ARBRE (MEC.) **11239**

ARBRE A CAME S **1888**

ARBRE A CAMES **1896**

ARBRE A CAMES EN TETE **8939**

ARBRE A GRANDE VITESSE **10209**

ARBRE A PETITE VITESSE **11650**

ARBRE A VILEBREQUIN **3309**

ARBRE CANNELE **11935**

ARBRE CARRE **12029**

ARBRE CONDUIT **4299**

ARBRE COUDE **3309**

ARBRE COUDE SIMPLE **11503**

ARBRE CREUX **6547**

ARBRE D'ATTAQUE **7944**

ARBRE D'ENTRAINEMENT **9930**

ARBRE D'HELICE **9930**

ARBRE DE CHANGEMENT DE MARCHE **10550**

ARBRE DE COMMANDE **4315**

ARBRE DE COUCHE **4758**

ARBRE DE DISTRIBUTION **10550**

ARBRE DE RELEVAGE **10550**

ARBRE DE RENVOI **6756**

ARBRE DE RENVOI **3233**

ARBRE DE RENVOI **3240**

ARBRE DE TRANSMISSION **4296**

ARBRE DE TRANSMISSION **7944**

ARBRE DEUXIEME MOTEUR **7944**

ARBRE EN ACIER COMPRIME **2850**

ARBRE EN PORTE-A-FAUX DES DEUX COTES **11247**

ARBRE FLEXIBLE **5420**

ARBRE HORIZONTAL **6587**

ARBRE INTERMEDIAIRE **6756**

ARBRE INTERMEDIAIRE **7202**

ARBRE INTERMEDIAIRE **7049**

ARBRE MENE **4299**

ARBRE MOTEUR **4315**

ARBRE OSCILLANT **10657**

ARBRE PLEIN **11793**

ARBRE PORTE-CAMES **1888**

ARBRE PORTE-FRAISE **3520**

ARBRE PREMIER MOTEUR **4758**

ARBRE PRIMAIRE OU PRINCIPAL **7943**

ARBRE PRINCIPAL **7944**

ARBRE RECEPTEUR **4299**

ARBRE ROTATIF **10764**

ARBRE SECOND MOTEUR **7944**

ARBRE SECONDAIRE **8910**

ARBRE SUCCIN **449**

ARBRE TROISIEME MOTEUR **7049**

ARBRE VERTICAL **13550**

ARBRE VILEBREQUIN A DEUX COUDES **13316**

ARBRE-MANIVELLE **3309**

ARBRES (PERPENDICULAIRES, OBLIQUES) SITUES DANS DES PLANS DIFFERENTS **11249**

ARBRES CONCOURANTS **7096**

ARBRES PARALLELES **9064**

ARC AVEC ELECTRODE DE CARBONE **1941**

ARC CIRCULAIRE **674**

ARC D'ENROULEMENT **675**

ARC D'UNE COURBE **664**

ARC DE CERCLE **674**

ARC DE CONTACT **2997**

ARC DE SOUTENEMENT **10442**

ARC EN ANSE DE PANIER **5373**
ARC EN DOS D'ANE **8741**
ARC METALLIQUE **8174**
ARC NON PROTEGE AVEC ELECTRODE AU CHARBON **1944**
ARC PROTEGE AVEC ELECTRODE AU CHARBON **1945**
ARC VOLTAIQUE **4494**
ARC-BOUTANT **688**
ARCADE **14014**
ARCHET **1504**
ARCHITECTE **691**
ARCHITECTURE **2442**
ARCOT **13987**
ARDOISE **11571**
AREOMETRE **6722**
AREOMETRE A POIDS CONSTANT **6072**
AREOMETRE A POIDS VARIABLE **6723**
ARETE **4441**
ARETE (DE SOUDURE) DUE AU REFOULEMENT **5357**
ARETE ARRONDIE **10808**
ARETE COUPANTE **3528**
ARETE D'UN TROU **4447**
ARETE DE REBROUSSEMENT **4448**
ARETE DE SORTIE **5515**
ARETE TERMINALE **3625**
ARETE TRANCHANTE EMOUSSEE **4365**
ARETE VIVE **11272**
ARGENT **11458**
ARGENT A 92/5% **12253**
ARGENT ALLEMAND **5932**
ARGENT BROMURE **1645**
ARGENT CHLORURE **11461**
ARGENT DE CHINE **5932**
ARGENT ELECTROLYTIQUE **4608**
ARGENT FIN **5202**
ARGENT VERT **1645**
ARGENTAL **5932**
ARGENTAN **5932**
ARGENTAN **8569**
ARGENTER **11459**
ARGENTURE **11469**
ARGENTURE **11465**
ARGILE **2465**
ARGILE CALCINEE FONDUE **6125**
ARGILE CUITE **926**
ARGILE PLASTIQUE **9470**
ARGILE REFRACTAIRE **5253**
ARGILLITE **700**
ARGON **701**
ARITHMETIQUE **702**
ARMATURE **5283**
ARMATURE **707**

ARMATURE **13243**
ARMATURE D'UN AIMANT **710**
ARMATURE D'UN CABLE **715**
ARMATURE DE NOYAU **3143**
ARMOIRE A OUTILS **12998**
ARMURE **710**
ARRACHEMENT **10858**
ARRACHEMENTS **11626**
ARRACHEUSES DE POMMES DE TERRE **9707**
ARRACHEUSES DE TUBERCULES **9703**
ARRET **12278**
ARRET D'AXE DE PISTON **2405**
ARRET D'UNE MACHINE **12286**
ARRET D'UNE MACHINE (ETAT) **12094**
ARRET D'UNE SOUPAPE **13461**
ARRET D'USINE **7092**
ARRET DE SERVICE **7100**
ARRET DE SUSPENSION **6521**
ARRET FACULTATIF **8850**
ARRET PROGRAMME **9903**
ARRET TEMPORISE **4396**
ARRETE-FLAMMES **5321**
ARRETE-PIEDS **12977**
ARRETER UNE MACHINE **12280**
ARRETS DE TRAVAIL **11409**
ARRIVEE D'EAU **6915**
ARRONDI **6537**
ARRONDI DU FILET **10809**
ARRONDIR UNE ARETE **10793**
ARROSAGE **12001**
ARROSER **12000**
ARROSOIR **13687**
ARSENIATE **720**
ARSENIC **721**
ARSENIQUE BLANC **727**
ARSENITE **729**
ARSENITE DE CUIVRE **10996**
ARSENIURE **726**
ARSENIURE D'HYDROGENE **730**
ARTERE **5081**
ARTICULATION **13242**
ARTICULATION (MEC.) **7238**
ARTICULATION A CARDAN **6573**
ARTICULATION A TOURILLON **9318**
ARTICULE **7629**
ASBESTE **755**
ASCENSEUR **9101**
ASCENSEUR CONTINU **3020**
ASCENSION **1925**
ASCENSION D'UNE CHARGE **7559**
ASEPTISANT **628**

ASPERITE **9011**
ASPERITE **10792**
ASPHALTAGE **767**
ASPHALTE **768**
ASPHALTE COMPRIME **12610**
ASPHALTE COULE **9713**
ASPHALTE DE TRINIDAD **13212**
ASPHALTE EN POUDRE **9723**
ASPHALTER **763**
ASPIRATEUR **769**
ASPIRATEUR DE POUSSIERES **4390**
ASPIRATION **12428**
ASPIRATION PAR CAPILLARITE **1922**
ASPIRER **12427**
ASSECHEMENT **4210**
ASSEMBLAGE **3254**
ASSEMBLAGE **1778**
ASSEMBLAGE A BASSE TEMPERATURE **7790**
ASSEMBLAGE A BRIDES **5344**
ASSEMBLAGE A COUVRE-JOINTS **1782**
ASSEMBLAGE A ONGLET **8324**
ASSEMBLAGE A QUEUE D'ARONDE **4178**
ASSEMBLAGE EN ANGLE **2519**
ASSEMBLAGE EN T **12702**
ASSEMBLAGE PAR CLAVETTE **7285**
ASSEMBLAGE PAR RAINURE ET LANGUETTE **12991**
ASSEMBLAGE PAR RECOUVREMENT **7396**
ASSEMBLAGE PAR RECOUVREMENT **7397**
ASSEMBLAGE PAR RIVETS **10639**
ASSEMBLAGE PAR SOUDURE **11759**
ASSEMBLAGE PAR SOUDURE DE PIECES MOULEES **2060**
ASSEMBLAGE PAR VIS **1442**
ASSEMBLAGE RIGIDE **10589**
ASSEMBLER DES TUYAUX PAR EMBOITEMENT **5277**
ASSISE **9011**
ASSISE **2896**
ASSISE DE BRIQUES **3259**
ASSISE DE RESERVOIR **12628**
ASSOMBRISSEMENT **5507**
ASSORTIMENT **11219**
ASSORTIMENT (DE PIECES) **11213**
ASSORTIR **8042**
ASSUJETTISSEMENT **7711**
ASSURER LA CONCORDANCE DES TROUS DE RIVETS **9969**
ASTATIQUE **774**
ASTERISME **776**
ASTROIDE **5602**
ASYMETRIE **779**
ASYMETRIQUE **778**
ASYMPTOTE **780**
ASYMPTOTE **781**

ASYMPTOTIQUE **781**
ASYNCHRONE **784**
ASYNCHRONISME **782**
ATELIER **13968**
ATELIER **13943**
ATELIER **11364**
ATELIER **4764**
ATELIER **773**
ATELIER D'AJUSTAGE **5282**
ATELIER D'EBARBAGE **5130**
ATELIER D'ECROUTAGE **5130**
ATELIER D'ENTRETIEN **10461**
ATELIER DE CARROSSERIE **1398**
ATELIER DE CHAUDRONNERIE EN FER **1420**
ATELIER DE DESABLAGE **5130**
ATELIER DE FORGE **5547**
ATELIER DE FRAISAGE **8293**
ATELIER DE MACHINES **7855**
ATELIER DE MECANIQUE **7855**
ATELIER DE MECANIQUE **7856**
ATELIER DE MONTAGE **4811**
ATELIER DE REPARATION **10461**
ATELIER DE TOLERIE **1398**
ATELIER DE TOURNAGE **13290**
ATHERMANE **786**
ATHERMANEITE **785**
ATMOSPHERE **787**
ATMOSPHERE (UNITE DE PRESSION) **788**
ATMOSPHERE A USAGE SPECIAL **791**
ATMOSPHERE ARTIFICIELLE **789**
ATMOSPHERE DE CARBONITRURATION **1973**
ATMOSPHERE DE NITRURATION **8583**
ATMOSPHERE PROTECTRICE **9947**
ATMOSPHERE PROTECTRICE **790**
ATMOSPHERE RAREFIEE **13441**
ATMOSPHERE REDUCTRICE **10347**
ATOME **800**
ATOMISATION **811**
ATTACHE **7817**
ATTACHE **1522**
ATTACHE **2502**
ATTACHE DES COURROIES **1151**
ATTAQUE **813**
ATTAQUE **4297**
ATTAQUE **10845**
ATTAQUE (COR.) **4846**
ATTAQUE A L ACIDE **4846**
ATTAQUE DE COULEE **5866**
ATTAQUE EN CORNICHON **6600**
ATTAQUE INTERDENDRITIQUE **7031**
ATTAQUE MACROGRAPHIQUE **7870**

ATTAQUE METALLOGRAPHIQUE **8204**
ATTAQUE PROFONDE **3685**
ATTAQUER **3182**
ATTAQUER A L'ACIDE **4839**
ATTEINTE PORTEE AUX DROITS DU BREVETE **6916**
ATTENUATION **3604**
ATTIRAIL DE CHAUFFE **5261**
ATTRACTION **815**
ATTRACTION MAGNETIQUE **7895**
ATTRACTION MOLECULAIRE **8359**
AU LARGE **8738**
AUBE **1677**
AUBE D'UNE TURBINE **13480**
AUBE DIRECTRICE **6167**
AUBE FIXE D'UNE TURBINE **6167**
AUBE MOBILE D'UNE TURBINE **13820**
AUBE RECEPTRICE **13820**
AUBES DE TURBINES **13268**
AUBIER **10936**
AUGE **13228**
AUGITE **819**
AUGITE-SYENITE **820**
AUGMENTATION DE LA PRESSION **6848**
AUGMENTATION DE LA RESISTANCE **6850**
AUGMENTATION DE LA SECTION TRANSVERSALE **5869**
AUGMENTATION DE PRESSION **6855**
AUGMENTATION DE VITESSE **10185**
AUGMENTATION DE VOLUME **4912**
AUGMENTATION OU ELEVATION DE LA TEMPERATURE **10612**
AUGMENTER LA VITESSE **6851**
AULNE **318**
AUNE **318**
AURIFERE **822**
AUSSIERE **6309**
AUSTENITE **825**
AUSTENITISATION **827**
AUTEUR D'UNE INVENTION **7108**
AUTO **8411**
AUTO-EXCITATION **11153**
AUTO-INDUCTION **11156**
AUTO-INFLAMMATION **11956**
AUTOBUS **13421**
AUTOBUS **8410**
AUTOCAR **7099**
AUTOCAR **8412**
AUTOCLAVE **829**
AUTOMATICITE **837**
AUTOMATIQUE **11148**
AUTOMOBILE **8411**
AUTOMOBILE INDUSTRIELLE **8414**

AUTONOMIE **3416**
AUTOSOUDURE **834**
AUVENT **3276**
AUVENT **5695**
AVAL (EN) **4184**
AVAL (D') **4184**
AVANCE **5585**
AVANCE **206**
AVANCE **5063**
AVANCE D'UN OUTIL **5068**
AVANCE PAR DENT **5069**
AVANCE RAPIDE **10208**
AVANCEMENT **5584**
AVANCEMENT **9905**
AVANCER **7448**
AVANT-CLOU **5944**
AVANT-TROU **10782**
AVERTISSEUR SONORE **6592**
AVEUGLER UNE FUITE **12279**
AVIDE D'EAU **6730**
AVOYAGE D'UNE SCIE **10958**
AVOYER UNE SCIE **11224**
AXE **9315**
AXE **2151**
AXE **11904**
AXE **864**
AXE (DE PIVOT) **7311**
AXE D'ARTICULATION D'UNE FOURCHE **6159**
AXE D'OSCILLATION **867**
AXE D'UN CRISTAL **865**
AXE D'UN TUYAU **866**
AXE DE FIBRE **5132**
AXE DE PISTON **9376**
AXE DE PISTON **6158**
AXE DE PIVOTEMENT **7311**
AXE DE ROTATION **868**
AXE DE SOUDURE **870**
AXE DE SYMETRIE **11024**
AXE DE SYMETRIE **869**
AXE DE SYMETRIE DES CRISTAUX **863**
AXE DES ABCISSES **872**
AXE DES ORDONNEES **873**
AXE DES X **872**
AXE DES Y **873**
AXE DU CORDON DE SOUDURE **871**
AXE DU REGULATEUR **6018**
AXE FOCAL **13143**
AXE GEOMETRIQUE **874**
AXE LONGITUDINAL **7733**
AXE NEUTRE **8541**
AXE NON TRANSVERSE D'UNE HYPERBOLE **2953**

AXE OPTIQUE **8841**
AXE PRINCIPAL **9866**
AXE TRANSVERSAL **13142**
AXE TRANSVERSE D'UNE HYPERBOLE **13143**
AXES CRISTALLINS ORTHOHEXAGONAUX **8871**
AXES D'EQUERRE **858**
AZOTATE **1271**
AZOTATE **8579**
AZOTATE **4850**
AZOTATE D'ARGENT **7828**
AZOTATE DE BARYUM **1001**
AZOTATE DE BISMUTHYLE **1273**
AZOTATE DE COBALT **2610**
AZOTATE DE POTASSIUM **10903**
AZOTATE DE SODIUM **2330**
AZOTATE MERCUREUX **8160**
AZOTATE MERCURIQUE **8156**
AZOTE **8588**
AZOTITE **11725**
AZOTITE **9691**
AZOTITE **8586**
AZOTYLE **8591**
AZURITE **886**
BAC **13228**
BAC D'ALIMENTATION **5075**
BAC DE BATTERIE **1067**
BACHE **13228**
BAGASSE **924**
BAGUE **1773**
BAGUE **10600**
BAGUE **9379**
BAGUE AMOVIBLE **7743**
BAGUE COLLECTRICE **11618**
BAGUE CONIQUE **2945**
BAGUE D'ARRET **7743**
BAGUE D'ARRET **12904**
BAGUE D'ARRET EN DEUX PIECES **11939**
BAGUE D'ETANCHEITE **9007**
BAGUE D'EXCENTRIQUE **4435**
BAGUE DE BUTEE **12904**
BAGUE DE CALIBRE **2721**
BAGUE DE CENTRAGE **2154**
BAGUE DE CENTRAGE **10413**
BAGUE DE FOND **9000**
BAGUE DE FOND **1565**
BAGUE DE FREINAGE **1541**
BAGUE DE GARNITURE **9007**
BAGUE DE GRAISSAGE **8774**
BAGUE DE GRAISSAGE **8790**
BAGUE DE GUIDAGE **6163**
BAGUE DE PISTON A JOINT EN BISEAU **9381**

BAGUE DE PISTON AVEC FENTE D'EQUERRE **9380**
BAGUE DE PISTON AVEC JOINT (FENTE) A RECOUVREMENT **9382**
BAGUE DE PISTON FENDUE **11944**
BAGUE DE PRESSION **9830**
BAGUE DE RACCORD **2961**
BAGUE DE RETENUE **10515**
BAGUE DE ROULEMENT **964**
BAGUE DE SERRAGE **11391**
BAGUE EN CAOUTCHOUC **10823**
BAGUE EN PLOMB **7470**
BAGUE EN TERRE GLAISE **2468**
BAGUE EXTERIEURE D'UN ROULEMENT A BILLES **8898**
BAGUE INTERIEURE D'UN ROULEMENT A BILLES **6953**
BAGUE LARMIER **13879**
BAGUE PERIPHERIQUE **1232**
BAGUE SOUDEE **13771**
BAGUE TARAUDEE DU CALIBRE DE FILETAGE **11050**
BAGUETTE **10660**
BAGUETTE DE RETENUE **10516**
BAGUETTE EN VERRE **12260**
BAILLEMENT **5806**
BAIN **1062**
BAIN **3967**
BAIN BLEU **1369**
BAIN D'HUILE **8745**
BAIN D'HYDROXYDE DE SODIUM **2117**
BAIN DE BLANCHIMENT **8059**
BAIN DE BRILLANTAGE **1608**
BAIN DE BRILLANTAGE **1564**
BAIN DE BRILLANTAGE **1609**
BAIN DE CEMENTATION **1994**
BAIN DE CHLORURE DE MERCURE **1369**
BAIN DE DECALAMINAGE **3790**
BAIN DE DECAPAGE **9289**
BAIN DE DECAPAGE **111**
BAIN DE DECAPAGE **12382**
BAIN DE DECAPAGE A L'ACIDE **9290**
BAIN DE DECAPAGE MAT **8059**
BAIN DE DECAPAGE PRELIMINAIRE **2476**
BAIN DE DECOMPOSITION DE L ELECTROLYTE **7538**
BAIN DE DEROUILLAGE **10872**
BAIN DE FUSION **9961**
BAIN DE FUSION **8137**
BAIN DE MORDANCAGE **9289**
BAIN DE PASSIVATION **9103**
BAIN DE PLOMB **7454**
BAIN DE SABLE **10918**
BAIN DE SEL **10905**
BAIN DE TREMPE **10096**
BAIN DE TREMPE **6294**

BAIN DE VAPEUR **12157**
BAIN-MARIE **13641**
BAINITE **925**
BAIONNETTE **1074**
BAISSE DE TEMPERATURE **5022**
BALAI AU CHARBON **1946**
BALAI ELECTRIQUE **6534**
BALANCE **931**
BALANCE **3660**
BALANCE A FLEAU **1095**
BALANCE A RESSORT **11976**
BALANCE D'ANALYSE **2294**
BALANCE DE PRECISION **9764**
BALANCE DE TORSION **13052**
BALANCE DU COMMERCE EXTERIEUR **13099**
BALANCE HYDROSTATIQUE **6725**
BALANCIER A VIS **11040**
BALATA **943**
BALAYAGE **10992**
BALAYER **10990**
BALLANT D'UNE COURROIE **13825**
BALLAST **969**
BALLASTAGE **971**
BALLE (DE MARCHANDISE) **945**
BALLON (CHIM.) **5370**
BALLON A FOND PLAT **9142**
BALLON ORDINAIRE **10798**
BALOURD **13341**
BAMBOU **972**
BAN DE TRACTION **7846**
BANC **1167**
BANC A RECTIFIER **12320**
BANC A TREFILER LES GROS FILS **1727**
BANC D'ESSAI **12773**
BANC D'ETIRAGE **10023**
BANC D'ETIRAGE **4218**
BANC DE CONTROLE **12773**
BANC DE MENUISIER **7234**
BANC PHOTOMETRIQUE **9268**
BANC POUR VERIFIER LES ENGRENAGES **5902**
BANC REFROIDISSEUR **6617**
BANDAGE DE ROUE **13322**
BANDAGES DE ROUE **12220**
BANDE **11983**
BANDE **974**
BANDE A JUMEAUX **570**
BANDE A TUBES **11527**
BANDE ARRASIVE **18**
BANDE D'ABSORPTION **45**
BANDE D'ACIER **12197**
BANDE D'AMIANTE **754**

BANDE D'INTERFERENCE **7036**
BANDE DE CUIVRE **3131**
BANDE DE DEFORMATION **3706**
BANDE DE FERRITE LIBRE **5105**
BANDE DE FREIN **1544**
BANDE DE PLIAGE **7314**
BANDE DE RECHAPAGE **1892**
BANDE DE RECOUVREMENT COUVRE-JOINT **1783**
BANDE DE ROULEMENT **13172**
BANDE MAGNETIQUE **7919**
BANDE PERFOREE **10005**
BANDELETTE **5375**
BANDER UN RESSORT **10032**
BANDES **980**
BANDES DE GLISSEMENT **11613**
BANDES DE METAL DE RECOUVREMENT **1928**
BANDES DE NEUMANN **8540**
BANDES GALVANISEES **5788**
BANDES NOIRES **1289**
BANDES REVELEES PAR ATTAQUE CHIMIQUE **4840**
BARAQUE DE CHANTIER **2987**
BARAQUE DE CHANTIER **11519**
BARAQUE DE VESTIAIRE **2247**
BARBE **1764**
BARBOTEUR POUR LAVAGES **5845**
BARBOTIN **2217**
BARBURE **1764**
BAREME **8052**
BARGE **991**
BARGES DE MANUTENTION **3787**
BARGES DE POSE **7440**
BARITEL **1731**
BARN **1006**
BAROMETRE **1008**
BAROMETRE A CUVETTE **2440**
BAROMETRE A MERCURE **8150**
BAROMETRE A SIPHON **11515**
BAROMETRE ANEROIDE **504**
BAROMETRE ENREGISTREUR **1007**
BAROMETRE METALLIQUE **504**
BAROMETRIQUE **1009**
BAROSCOPE **1011**
BARRAGE **13753**
BARRAGE D'UNE VALLEE **3597**
BARRAGE DE SORTIE **8904**
BARRE **10660**
BARRE **10663**
BARRE **981**
BARRE **986**
BARRE A NOYAU FUSIBLE **3168**
BARRE A OEILLETS **5551**

BARRE A SECTION DECROISSANTE **12670**
BARRE COLLECTRICE **1770**
BARRE D'ACCOUPLEMENT **12231**
BARRE D'ACCOUPLEMENT **12915**
BARRE D'ACCOUPLEMENT **12916**
BARRE D'ALESAGE MICROMETRIQUE **8240**
BARRE D'ALESAGE MONOPIECE **11778**
BARRE D'ATTELAGE **4217**
BARRE D'EXCENTRIQUE **4431**
BARRE DE CONNEXION **2962**
BARRE DE CONTACT **2999**
BARRE DE CUIVRE **3113**
BARRE DE DEVERSOIR **4185**
BARRE DE DEVERSOIR **3593**
BARRE DE FER BRUT **8441**
BARRE DE GUIDAGE **6164**
BARRE DE METAL NOBLE **1730**
BARRE DE PIQUAGE **7333**
BARRE DE TORSION **13053**
BARRE EN ACIER CEMENTE **1328**
BARRE OMNIBUS **1770**
BARRE PLATE **5374**
BARRE SOUMISE AUX ESSAIS **12772**
BARRE STABILISATRICE **12539**
BARRE TRAVAILLANT A LA COMPRESSION **2855**
BARRE TRAVAILLANT A LA TENSION **12754**
BARREAU **10660**
BARREAU AIMANTE **985**
BARREAU DE GRILLE **5237**
BARREAU ENTAILLE **8675**
BARREAU-EPROUVETTE **12772**
BARREAUX VERGES DE FER **984**
BARRES DE FORAGE **1470**
BARRES ETIREES **4248**
BARRES HUIT-PANS EN ACIER ALLIE **375**
BARRES LAMINEES A FROID **2695**
BARRES POUR GRILLAGES **6069**
BARRETTE **5211**
BARRIERE DE COTTRELL **3224**
BARYTE **1020**
BARYTINE **1021**
BARYTITE **1021**
BARYUM **992**
BASALTE **1022**
BASALTIQUE **1023**
BASCULANT **6835**
BASCULE **931**
BASCULE **3660**
BASCULE AU DIXIEME (AU 10) **3660**
BASCULE BINAIRE **5429**
BASCULE POUR ALLIAGE **362**

BASCULE ROMAINE **12225**
BASCULE ROMAINE (A CURSEUR) **12225**
BASCULEUR DE WAGONS **1932**
BASE **1027**
BASE **1026**
BASE (CHIM.) **1024**
BASE (D'UN SOLIDE) **1025**
BASE D'UN FILET **1028**
BASE D'UN LOGARITHME **1033**
BASICITE **1055**
BASIQUE **1042**
BASSE PRESSION **7774**
BASSE TEMPERATURE **7775**
BASSE TENSION **7778**
BASSIN A VIDE **13451**
BASSIN DE FILTRATION **5178**
BASSIN REFROIDISSANT **3090**
BATI D'UNE MACHINE **5626**
BATIMENT ANNEXE **585**
BATIMENT D'USINE **5012**
BATIMENT DE GENERATEUR **1413**
BATIMENT DES CHAUDIERES **1413**
BATIMENT DES MACHINES **4753**
BATITURE **8279**
BATTAGE DE PIEUX **9306**
BATTEMENT **1119**
BATTEMENT D'OR **5994**
BATTEMENT DES METAUX **1120**
BATTERIE **4534**
BATTERIE D'ACCUMULATEURS **77**
BATTERIE D'ACCUMULATEURS **1066**
BATTERIE DE CHAUDIERES **1069**
BATTERIE DE PILES **9845**
BATTERIE-TAMPON **1687**
BATTEUSES **12884**
BATTITURES **6217**
BATTRE EN FORME **1482**
BAUME DE COPAHU **3101**
BAUXITE **1071**
BAVURE **1764**
BAVURE **5357**
BAVURE **8948**
BAVURE **10836**
BAVURE (D'UNE PIECE MOULEE) **5184**
BAVURE (DE SOUDURE) **8949**
BEANT D'UN JOINT **5806**
BEC (D'UN OUTIL) **8666**
BEC A GAZ **5811**
BEC BUNSEN **1741**
BEC DE COULEE **7632**
BEC DE COULEE DE LA POCHE **7361**

BEC DE LA POCHE DE COULEE 7363
BEC DE PASSAGE DU LAITIER 2399
BEC DE SOUFFLAGE 275
BEC DU CONVERTISSEUR 3072
BEC-D'ANE 3364
BECHER 1092
BEDANE 1919
BEDANE 3364
BEDANE DE MENUISIER 8398
BELIER HYDRAULIQUE 6694
BENEFICE DANS LE TRAVAIL 8094
BENJOIN 1203
BENZALDEHYDE 1198
BENZENE 1199
BENZIDINE 1201
BENZINE 1199
BENZOL 1204
BENZOLINE 7580
BENZOPHENONE 1205
BERCEAU 3288
BERCEAU 7816
BERCEAU 10877
BERCEAU A GALETS 10690
BERCEAU A GALETS 10694
BERCEAU A PATINS 11606
BERLINE 11128
BERLINE 10902
BERLINE 4 PORTES 5599
BERYLLIUM 1207
BETON 2894
BETON 12203
BETON COULE DANS DES TROUS 9714
BETON DAME 10193
BETON DE CIMENT 2135
BETON DE LAITIER 11565
BETON DE MACHEFER 11565
BETON MOULE 8426
BETONNAGE 2902
BETONNER 2895
BEURRE D'ANTIMOINE LIQUIDE 626
BEURRE DE COCO 2620
BIAIS 1223
BIAIS 1221
BIATOMIQUE 3881
BIBORATE DE SODIUM 1459
BICARBONATE D'AMMONIUM 459
BICARBONATE DE CALCIUM 1847
BICARBONATE DE MAGNESIUM 7883
BICARBONATE DE POTASSE 9676
BICARBONATE DE SOUDE 126
BICHLORURE D'ETAIN 12096

BICHLORURE DE MERCURE 8153
BICHROMATE DE POTASSIUM 9677
BIDON (POUR HUILEESSENCEECT.) 12961
BIEF D'AMONT 6319
BIEF D'AVAL 12596
BIELLE 2962
BIELLE D'ACCOUPLEMENT 3256
BIELLE DE COMMANDE 4317
BIER 2251
BIGORNE 1090
BIGORNE 1091
BILAN CALORIFIQUE 6337
BILAN THERMIQUE 6337
BILLAGE 1618
BILLE 1243
BILLE 950
BILLE POUR ROULEMENTS 1102
BILLES A POLIR 1759
BILLES POUR BRUNISSAGE 1759
BILLETTE 4371
BILLETTE 1243
BILLETTE DE LAITON 1563
BILLETTES EN ACIER ALLIE 370
BILLOT 633
BIOXYDE D'AZOTE 8581
BIOXYDE D'ETAIN 12098
BIOXYDE D'HYDROGENE 6716
BIOXYDE DE BARYUM 998
BIOXYDE DE MANGANESE 10047
BIOXYDE DE SODIUM 11727
BIPLAN 1261
BIREFRINGENT 1265
BISCUIT 1276
BISEAU 1221
BISEAUTAGE 1231
BISEAUTER 1222
BISECTRICE AIGUE 156
BISMUTH 1269
BISMUTHINE 1275
BISSECTER 1266
BISSECTION 1267
BISSECTRICE 1268
BISULFITE D'AMMONIAQUE 460
BISULFITE DE SODIUM 11718
BISULFURE D'ARSENIC 10262
BISULFURE D'ETAIN 12099
BIT 1249
BITARTRATE DE POTASSIUM 119
BITORD 12005
BITTERN 8405
BITUME SOLIDE 768

BITUMER **763**

BITUMINEUX **1281**

BLANC D'ARGENT **13834**

BLANC D'ESPAGNE **13841**

BLANC DE BALEINE **11882**

BLANC DE BARYTE **1003**

BLANC DE CERUSE **13834**

BLANC DE MEUDON **13841**

BLANC DE NEIGE **14043**

BLANC DE PLOMB **13834**

BLANC DE TROYES **13841**

BLANC DE ZINC **14043**

BLANC EBLOUISSANT **6823**

BLANC FIXE **1003**

BLANC INCANDESCENT **6823**

BLANC SOUDANT **13784**

BLENDE **1324**

BLEU **1375**

BLEU (CHAUDE) **1370**

BLEU DE BERLIN **1206**

BLEU DE CHYPRE **1376**

BLEU DE PRUSSE **1206**

BLEUIR L'ACIER **1368**

BLEUISSAGE **1377**

BLEUISSAGE **572**

BLOC (D'INFORMATIONS) **1332**

BLOC DE LA CROSSE **11587**

BLOC ERRATIQUE **4821**

BLOC OPTIQUE **11079**

BLOC-CYLINDRES **3564**

BLOCAGE **7704**

BLOCAGE **7711**

BLOCAGE AUTOMATIQUE (A) **11157**

BLOCAGE DES ECROUS **7707**

BLONDIN **1344**

BLOOM **1346**

BLOOMS EN ACIER AU CARBONE **1955**

BLOQUER UN ECROU **7700**

BLOQUER UNE CLAVETTE **7700**

BOBINAGE **2648**

BOBINAGE **13862**

BOBINE **2645**

BOBINE **4499**

BOBINE **1384**

BOBINE D'ALLUMAGE **6761**

BOBINE D'INDUCTION **6894**

BOBINE DE RUHMKORFF **10830**

BOCARD **12057**

BOIS **13913**

BOIS A BRULER **5258**

BOIS A COEUR POURRI **5610**

BOIS BLANC **11746**

BOIS BOUGE **1196**

BOIS CARIE **10768**

BOIS DE CHARPENTE **12937**

BOIS DE CHAUFFAGE **5258**

BOIS DE CONSTRUCTION **12937**

BOIS DE FER-BLANC **13833**

BOIS DE FIL **12939**

BOIS DE MAILLE **12938**

BOIS DE SABLE **13833**

BOIS DEBITE **10962**

BOIS DUR **6282**

BOIS EN GRUME **7723**

BOIS EN QUARTIERS **10076**

BOIS EN RONDINS **10803**

BOIS EQUARRI **12035**

BOIS FEUILLU **6301**

BOIS LEGER **11746**

BOIS LOURD **6282**

BOIS MI-PLAT **6210**

BOIS POUILLEUX **5610**

BOIS RESINEUX **10483**

BOIS SAIN **11820**

BOIS SANS NOEUDS **12940**

BOIS SECHE A L'AIR LIBRE **297**

BOIS TAILLE CONTRE LE FIL **4719**

BOIS TENDRE **11746**

BOIS VERMOULU **13977**

BOISSEAU D'UN ROBINET **11334**

BOISSEAU DE LA BOITE A ETOUPE **9008**

BOITE A BOURRAGE **9008**

BOITE A ETOUPE **12405**

BOITE A FEU **5238**

BOITE A FUSIBLES **5752**

BOITE A GANTS **5980**

BOITE A GARNITURES **9008**

BOITE A GRAISSE **6085**

BOITE A GRAISSE **877**

BOITE A HUILE **877**

BOITE A NOYAUX **3147**

BOITE A OUTILS **12997**

BOITE A RESSORT **11980**

BOITE A SOUPAPE **2229**

BOITE A VIDE **13435**

BOITE A VIDE **13438**

BOITE D'ESSIEU **877**

BOITE D'ESSIEU **876**

BOITE DE COMPAS **8051**

BOITE DE DISTRIBUTION **2229**

BOITE DE GRAISSAGE **877**

BOITE DE VITESSE **5890**

BOITE PROTECTRICE **9948**
BOITEMENT **7165**
BOITES DE CEMENTATION **1989**
BOITIER **6660**
BOITIER DE DIRECTION **12229**
BOITIER DE DISTRIBUTION **4086**
BOL **1430**
BOL DIARMENIE **8720**
BOLOMETRE **1431**
BOMBAGE **1889**
BOMBAGE **1723**
BOMBE **1720**
BOMBE **1722**
BOMBE CALORIMETRIQUE **1879**
BOMBEMENT **1720**
BOMBEMENT DE LA JANTE D'UNE POULIE **3398**
BOMBER **13628**
BONBONNE **1981**
BONDE DE FOND **1492**
BORATE DE SOUDE ANHYDRE **1459**
BORAX **1459**
BORD **4441**
BORD A SOUDER **6130**
BORD OU FLANC DE LA BANDE D'ABSORPTION **49**
BORD RABATTU **1083**
BORD TOMBE **1083**
BORDELAGE **1088**
BORE **1477**
BORNE **12763**
BORNE **12762**
BORNE D'ARRIVEE **6944**
BORNE DE RACCORDEMENT **12762**
BORNE DE RACCORDEMENT **1254**
BORT **1480**
BOSSAGE **1883**
BOSSAGE DE MISE A LA TERRE **4418**
BOSSE **1720**
BOSSE RENTRANTE **3760**
BOSSELAGE **4684**
BOTTE (DE TUBES OU PROFILES) **1736**
BOTTE DE FEUILLARD **1738**
BOTTELER **1737**
BOTTELEUSES MECANIQUES A TOUTE DENSITE **948**
BOUCHE D'ARROSAGE **6672**
BOUCHE D'EAU **6672**
BOUCHE D'INCENDIE **5244**
BOUCHER (SE) **2358**
BOUCHON **9529**
BOUCHON **3173**
BOUCHON **12285**
BOUCHON **1910**

BOUCHON **5758**
BOUCHON A L'EMERI **6146**
BOUCHON A VIS **11060**
BOUCHON DE CAOUTCHOUC **10824**
BOUCHON DE COULEE **12284**
BOUCHON DE JAUGE **5883**
BOUCHON DE LIEGE **3176**
BOUCHON DE RADIATEUR **10151**
BOUCHON DE REMPLISSAGE **5158**
BOUCHON DE VALVE **13464**
BOUCHON DE VAPEUR **13481**
BOUCHON DE VIDANGE **4208**
BOUCHON FEMELLE **11025**
BOUCHON FEMELLE **1910**
BOUCHON FILETE **11025**
BOUCHON FILETE **11060**
BOUCHON MALE **11060**
BOUCHON TARAUDE **11025**
BOUCHON VISSE **11060**
BOUCLE D'UNE CORDE **7742**
BOUCLE DE RETOUR **5080**
BOUCLIER **11342**
BOUDIN D'AMIANTE **749**
BOUDIN DE SOUDURE **1082**
BOUE **11609**
BOUE **8444**
BOUE **11655**
BOUE **11654**
BOUE D'ANODE **600**
BOUES **8444**
BOUEUX **8445**
BOUGIE **6419**
BOUGIE **1899**
BOUGIE D'ALLUMAGE **11841**
BOUGIE D'ALLUMAGE **11838**
BOUGIE DE PRECHAUFFAGE **6387**
BOUGIE DECIMALE **3663**
BOUILLIR **1401**
BOUILLONNEMENT **4461**
BOULE/SPHERE DU REGULATEUR **6015**
BOULON **1432**
BOULON A BASCULE **6509**
BOULON A CHARNIERE **6509**
BOULON A CROCHET **6569**
BOULON A ERGOT **5062**
BOULON A TETE A SIX PANS **6457**
BOULON A TETE CARREE **12033**
BOULON A TETE DE MARTEAU **12701**
BOULON A TIGE ENTIEREMENT FILETEE **12398**
BOULON A VIS **11021**
BOULON BRUT **1286**

BOULON D'ANCRAGE **501**
BOULON D'ASSEMBLAGE **771**
BOULON D'ASSEMBLAGE PROVISOIRE **12592**
BOULON D'ECLISSE **5269**
BOULON DE BLOCAGE **2281**
BOULON DE CARROSSERIE **2553**
BOULON DE CHARRONNAGE **2553**
BOULON DE FONDATION **5590**
BOULON DE LIAISON **2959**
BOULON DE SCELLEMENT **10172**
BOULON DE SCELLEMENT A BARBELURES **7209**
BOULON DECOLLETE **1607**
BOULON ETOQUIAU **2149**
BOULON FAIT A LA MACHINE **7840**
BOULON GRAISSEUR **7811**
BOULON LIBRE **1441**
BOULON MECANIQUE **7840**
BOULON NON FILETE **1302**
BOULON POUR COURROIES **1156**
BOULON PRISONNIER **12399**
BOULONNAGE **1443**
BOULONNER **1435**
BOULONNERIE **1443**
BOULONS A TETE PLATE **5386**
BOURNONITE **1502**
BOURRAGE **6445**
BOURRAGE **7247**
BOURRAGE **9004**
BOURRAGE **9009**
BOURRAGE DE TIGE DE PISTON **9386**
BOURRAGE PLASTIQUE ETANCHE **11078**
BOURRE DE COCO **2654**
BOURRELET D'ETANCHEITE **13854**
BOURRELET D'UNE ROUE **5331**
BOURRELET EN TERRE GLAISE **2468**
BOURRER **6153**
BOURSOUFLURES **1330**
BOUSSOLE **2799**
BOUT ANODIQUE **591**
BOUT D'ELECTRODE **4578**
BOUT DE TUBE FERME **1303**
BOUT FEMELLE D'UN TUYAU **11710**
BOUT MALE D'UN TUYAU **11901**
BOUTEROLLE A MAIN **10626**
BOUTEROLLE A MANCHE OU A OEIL **4979**
BOUTISSE **6324**
BOUTON D'INTERRUPTEUR **9797**
BOUTON DE MANIVELLE **3302**
BOUTON DE TENSION **12755**
BOUTON DE VILEBREQUIN **9319**
BOUTON ELECTRIQUE **9797**

BOUTON MARCHE/ARRET **12115**
BOUTON-POUSSOIR **9797**
BOUVET A FOURCHEMENT **8047**
BOUVET FEMELLE **6139**
BOUVET MALE **12993**
BOUVETAGE **12991**
BOUVETER **12988**
BOYAU **6614**
BRABANT (TRAINEES OU PORTEES) **9526**
BRACON (ELEMENT DE CHARPENTE METALLIQUE) **1516**
BRACONS DE CHARPENTE **1520**
BRAI GRAS NATUREL **1280**
BRAI LIQUIDE **2570**
BRAI SEC **2737**
BRAI SEC MINERAL **2569**
BRAISETTE **2659**
BRAME **11551**
BRAME (TRAIN A BRAMES) **11552**
BRAMES (CINTRAGE DES) **11553**
BRAN DE SCIE **10953**
BRANCARD **5625**
BRANCARD DE CREUSET **3403**
BRANCHE **11845**
BRANCHE D'UN FER CORNIERE **13728**
BRANCHE D'UNE COURBE **1555**
BRANCHE DE LA POINTE TRACANTE DU PLANIMETRE **10662**
BRANCHE PORTE-AIGUILLE D'UN COMPAS **8521**
BRANCHE PORTE-CRAYON **9165**
BRANCHE TIRE-LIGNE D'UN COMPAS **9160**
BRANCHEMENT **1554**
BRANCHEMENTS SOUTERRAINS **1744**
BRANCHER (SE) **1556**
BRAS D'ARTICULATION **6505**
BRAS D'UN VOLANT **705**
BRAS D'UNE POULIE **705**
BRAS DE L'ELECTRODE **6592**
BRAS DE LEVIER **7531**
BRAS DE LEVIER DE LA FORCE **706**
BRAS DE MANIVELLE **3301**
BRAS DE MANIVELLE **3296**
BRAS DE POTENCE **3619**
BRAS DE SUSPENSION **3050**
BRAS DROIT D'UNE POULIE **12302**
BRAS DYNAMOMETRIQUE **13049**
BRAS OSCILLANT **12549**
BRAS PARABOLIQUE D'UNE POULIE **3497**
BRAS RADIAL **10128**
BRAS/BRANCHE DE REGULATEUR **6014**
BRASAGE **11762**
BRASAGE **1576**

BRASAGE AU BAIN SALIN **10911**
BRASAGE AU CHALUMEAU **5810**
BRASAGE AU FOUR **5740**
BRASAGE AU TREMPE BRASAGE PAR IMMERSION **2304**
BRASAGE DUR ELECTRIQUE **4498**
BRASAGE FORT A L'ARC **666**
BRASAGE PAR INDUCTION **6893**
BRASAGE PAR INDUCTION PROGRESSIVE **9908**
BRASAGE PAR RESISTANCE **10487**
BRASEMENT **1576**
BRASER **1573**
BRASERO **3821**
BRASSURE (D'UNE ROUE) **13814**
BRASURE **6279**
BRASURE **1576**
BRASURE A FROID **2705**
BRASURE A L'ALLIAGE D'ARGENT **11460**
BREAK **12141**
BREVET D'INVENTION **9112**
BREVET ETRANGER **5540**
BREVETE **9123**
BREVETER **9113**
BREVETS D'INVENTION (BUREAU DES) **9120**
BREVETS D'INVENTION (LOI SUR LES) **9116**
BRIDE **5323**
BRIDE **13773**
BRIDE **7817**
BRIDE A COLLET RABATTU **9803**
BRIDE A FACE SURELEVEE **10183**
BRIDE A MANDRINER **10699**
BRIDE A MOYEU **6663**
BRIDE A REDUCTION **10350**
BRIDE A SOUDER (EN BOUT A L'INTERIEUR) **13794**
BRIDE A VISSER **11055**
BRIDE ANGULAIRE **512**
BRIDE BRASEE **11758**
BRIDE BRUTE DE FONTE **10776**
BRIDE D'ACCOUPLEMENT **3255**
BRIDE DE CHAPEAU **1452**
BRIDE DE FOULOIR **5955**
BRIDE DE PRESSE-ETOUPE **9002**
BRIDE DE RECOUVREMENT **1304**
BRIDE DE TUBE **9351**
BRIDE FIXE **5039**
BRIDE FOLLE **7746**
BRIDE MANDRINEE **10672**
BRIDE OBTURATRICE **1304**
BRIDE OVALE **8918**
BRIDE PLATE **5382**
BRIDE PLEINE **1326**
BRIDE PLEINE **1304**

BRIDE RAPPORTEE **1708**
BRIDE RIVETEE **10638**
BRIDE RONDE **10797**
BRIDE RONDELLE DE TUBE **9351**
BRIDE TARAUDEE **12866**
BRIDE TARAUDEE **11064**
BRIDE TOURNANTE **7398**
BRIDE TOURNEE **5003**
BRIDE VENUE DE FONTE **2063**
BRIDE VISSEE **11063**
BRIDE-DIAMETRE DE PERCAGE **1439**
BRIDER DES TUYAUX **7243**
BRIDES D'USINAGE **13013**
BRIDES DE MORTAISAGE **7289**
BRILLANTER **1756**
BRIN CONDUCTEUR **4316**
BRIN CONDUIT **5516**
BRIN COURANT **10851**
BRIN D'UNE CHAINE SANS FIN **11413**
BRIN D'UNE COURROIE **11413**
BRIN DE CABLE **10740**
BRIN DORMANT **12092**
BRIN GARANT DU CABLE **5023**
BRIN LACHE **5516**
BRIN MENANT **4316**
BRIN MENE **5516**
BRIN MOTEUR **4316**
BRIN MOU **5516**
BRIN SORTANT (D UNE COURROIE SANS FIN) **5516**
BRIN TENDU (D'UNE COURROIE SANS FIN) **4316**
BRIQUE **1596**
BRIQUE ABRASIVE **19**
BRIQUE CIRCULAIRE, ARRONDIE, CINTREE **13738**
BRIQUE CREUSE **245**
BRIQUE CRUE **7692**
BRIQUE CUITE **1760**
BRIQUE DE CHROMITE **2371**
BRIQUE DE PAREMENT **5004**
BRIQUE DE POCHE **7359**
BRIQUE DINAS **5803**
BRIQUE DURE **6273**
BRIQUE DURE ET TRES CUITE **2500**
BRIQUE EMAILLEE **5970**
BRIQUE EN LIEGE **3174**
BRIQUE EN TOURBE **9148**
BRIQUE PILEE **9724**
BRIQUE PLEINE **11780**
BRIQUE PROFILEE **8425**
BRIQUE REFRACTAIRE **5239**
BRIQUE REFRACTAIRE **5252**
BRIQUE REFRACTAIRE ALUMINEUSE **5254**

BRIQUE SECHEE A L'AIR **7692**

BRIQUE SILICEUSE **11435**

BRIQUE VERNISSEE **5970**

BRIQUES CREUSES **1697**

BRIQUETAGE **1602**

BRIQUETTE **1623**

BRIQUETTE DE FERRO ALLIAGE **1622**

BRIQUETTE DE MINERAI **8860**

BRIS **10861**

BRIS D'UN TUBE **9353**

BRISE **10397**

BROCHAGE **1637**

BROCHE **11904**

BROCHE **4260**

BROCHE **12985**

BROCHE (FRITTAGE) **3162**

BROCHE CONIQUE **2009**

BROCHE D'ASSEMBLAGE **4261**

BROCHE POUR ASSEMBLAGE PROVISOIRE DES TOLES **12592**

BROMARGURE **1645**

BROMARGYRITE **1645**

BROMATE **1646**

BROME **1649**

BROMITE **1645**

BROMURE **1648**

BROMURE DE POTASSIUM **9678**

BROMYRITE **1645**

BRONZAGE **1664**

BRONZE **1650**

BRONZE **1658**

BRONZE (DE) **216**

BRONZE A CANON **6178**

BRONZE ALPHA **1656**

BRONZE ARCHITECTURAL **692**

BRONZE AU BERYLLIUM **1208**

BRONZE AU MANGANESE **7967**

BRONZE AU NICKEL **8560**

BRONZE AU PLOMB **7456**

BRONZE AU TUNGSTENE **13263**

BRONZE AU ZING **10326**

BRONZE D'ALUMINIUM **420**

BRONZE D'ALUMINIUM **58**

BRONZE D'ALUMINIUM CUPRO-ALUMINIUM **434**

BRONZE D'ART **12147**

BRONZE D'OR **5995**

BRONZE DE CANONS **197**

BRONZE DE CLOCHE **1134**

BRONZE DE COUSSINET **1660**

BRONZE DE NICKEL ET ALUMINIUM **8572**

BRONZE DE QUALITE COMMERCIALE **1657**

BRONZE DE RESSORT **1661**

BRONZE DES MONNAIES **8183**

BRONZE DU COMMERCE **1657**

BRONZE EN POUDRE **1652**

BRONZE ORNEMENTAL **1663**

BRONZE PHOSPHOREUX **9246**

BRONZE PLASTIQUE **1660**

BRONZE POUR MIROIRS DE TELESCOPES **11867**

BRONZE POUR VISSERIE **1659**

BRONZE RESISTANT AUX ACIDES **1655**

BRONZE SILICEUX **11444**

BRONZE STATUAIRE **12147**

BRONZE STATUAIRE **1662**

BRONZER **1651**

BROSSE **9019**

BROSSE A LIMES **5150**

BROSSE CIRCULAIRE A POLIR **2425**

BROSSE EN FILS METALLIQUES **13884**

BROSSE METALLIQUE **13884**

BROSSE METALLIQUE **11014**

BROSSE RACLETTE POUR TUBES **13247**

BROUETTE **13812**

BROUILLARD METALLIQUE **8182**

BROYAGE **6117**

BROYAGE **8287**

BROYAGE FIN (FRITTAGE) **6106**

BROYER **6103**

BROYEUR **3419**

BROYEUR A BOULETS **959**

BROYEUR A CYLINDRES **10687**

BRUIT **8608**

BRULEMENT DU FER **1754**

BRULER **1745**

BRULER LE FER **1746**

BRULER SANS FLAMME **11688**

BRULEUR **1749**

BRULEUR A GAZ **5811**

BRULEUR BUNSEN **1741**

BRULURE **1750**

BRUNI **1755**

BRUNIR **1756**

BRUNISSAGE **1758**

BRUNISSOIR **1757**

BRUNISSURE **1755**

BRUT **3406**

BRUT DE COULEE **739**

BRUT DE FORGE **742**

BRUT DE LAMINAGE **10778**

BRUT DE LAMINAGE **744**

BRUT DE TREMPE **743**

BUCHE **1243**

BUEE **1404**

BUFFLE **1685**

BUIS **1507**

BULBE **1711**

BULLAGE **1672**

BULLAGE **5703**

BULLE D'AIR **1670**

BULLE DE GAZ **1671**

BUREAU D'ETUDE **3796**

BUREAU D'ETUDES **3800**

BUREAU DE DESSIN **4198**

BUREAU DE DESSIN **4238**

BURETTE **1743**

BURETTE A HUILE **8749**

BURETTE A PISTON **9994**

BURETTE DE GRAISSAGE **8749**

BURIN **5377**

BURIN **2349**

BURINAGE **2347**

BURINER **2344**

BUSE **12966**

BUSE **8685**

BUSE DE COUPE **3537**

BUTANE **1775**

BUTEE **57**

BUTEE **12902**

BUTEE **12278**

BUTEE **898**

BUTEE A BILLES **12901**

BUTEE A BILLES **966**

BUTEE A ROULEAUX (A GALETS) **10691**

BUTEE D'UN ARBRE **11384**

BUTEE D'UNE SOUPAPE **13461**

BUTEE DE DEBRAYAGE **2549**

BUTEE REGLABLE **187**

BUTOIR **1686**

BUTOIR **12278**

BUTOIR DE PARE-CHOCS **8956**

BUTTEUSES POUR POMMES DE TERRE **9702**

BUTYLENE **1816**

BUVARD **1350**

BY-PASS **1821**

BY-PASS **1818**

CABESTAN **1929**

CABESTAN A BRAS **6229**

CABINE AVANCEE **1824**

CABINE DE CAMION **13229**

CABLAGE **7389**

CABLAGE **1828**

CABLAGE **10739**

CABLAGE ALTERNATIF **10414**

CABLE **10730**

CABLE A CABLAGE ALBERT **309**

CABLE A CABLAGE LANG OU ALBERT **7390**

CABLE A GAINE METALLIQUE **8177**

CABLE A SECTION CARREE **12028**

CABLE A SECTION DECROISSANTE **12661**

CABLE A SECTION TRIANGULAIRE **13194**

CABLE A TORONS TRIANGULAIRES **13195**

CABLE AERIEN **8938**

CABLE ANTIGIRATOIRE **8646**

CABLE ARME **713**

CABLE CLOS **2513**

CABLE CONDUCTEUR TOUT-ALUMINIUM **343**

CABLE CONIQUE **12661**

CABLE D'ASCENSEUR **7546**

CABLE D'EXTRACTION **6520**

CABLE DE CHANVRE **6443**

CABLE DE COMMANDE **3051**

CABLE DE DISTRIBUTION **4079**

CABLE DE LEVAGE **6520**

CABLE DE MASSE **4411**

CABLE DE TRANSMISSION **4314**

CABLE DIMINUE **12661**

CABLE ELECTRIQUE **1825**

CABLE EN ACIER **12215**

CABLE EN ACIER **12201**

CABLE EN COTON **3221**

CABLE EN FILIN **12215**

CABLE EN FILS DE FER **7147**

CABLE EN GRELIN **12201**

CABLE ISOLE AU P.V.C. **9472**

CABLE METALLIQUE **13899**

CABLE METALLIQUE A FILS PARALLELES **7390**

CABLE METALLIQUE A TORONS MEPLATS **5407**

CABLE METALLIQUE EN FILS FINS **5207**

CABLE METALLIQUE EN GROS FILS **2579**

CABLE METALLIQUE TORDU ALTERNATIF **10415**

CABLE PLAT **5393**

CABLE PRINCIPAL **5081**

CABLE ROND **10801**

CABLE SANS FIN **4733**

CABLE TELEDYNAMIQUE **12705**

CABLE TORDU A DROITE **10587**

CABLE TORDU A GAUCHE **7511**

CABLE VOLANT **7264**

CABLER **13303**

CABLES **12739**

CABLES D'ALIMENTATION **12484**

CABLES INJECTES **1446**

CABROUET **947**

CACHE-ENTREE **4835**

CADMIAGE **1832**
CADMIUM **1831**
CADRAN DIVISE **3837**
CADRAN DU BOIS **12108**
CADRAN GRADUE **6027**
CADRANURE **12108**
CADRE **5624**
CADRE A CREMAILLERES **4150**
CADRE DENTE **4150**
CAESIUM **1833**
CAGE A BILLES **957**
CAGE A CYLINDRE VERTICAUX CYLINDRE DE REFOULEMENT **4456**
CAGE D'UN ROULEMENT A GALETS **12040**
CAGE DE CYLINDRES **10704**
CAGE DE FARADAY **5036**
CAGE DE LAMINOIR **12065**
CAGE DE ROULEMENT A BILLES **955**
CAGE DEGROSSISEURS **10787**
CAGE DEGROSSISSEUSE **9622**
CAGE FINISSEURS **5225**
CAGE FINISSEUSE **5219**
CAGE MEDIANE **7098**
CAGE PROTECTRICE **9948**
CAGE REFOULEUSE **4454**
CAHIER DES CHARGES **11855**
CAILLEBOTIS **6101**
CAILLOU **9152**
CAISSE DE CEMENTATION **3075**
CAISSE DE RECUIT **563**
CAISSE DE RECUIT **567**
CAISSE OU CORPS D'UN EVAPORATEUR **4870**
CAISSON **2798**
CAISSON (TOIT A...) **9621**
CAISSON DE SOUTIRAGE **4227**
CAISSONS ETANCHES **7658**
CALAGE **13740**
CALAGE D'UNE CLAVETTE **4310**
CALAGE DE LA DISTRIBUTION **12952**
CALAMINE **1836**
CALANDRE DE RADIATEUR **10152**
CALCAIRE **10969**
CALCAIRE **7586**
CALCAIRE CARBONIFERE **1969**
CALCAIRE COQUILLIER **8487**
CALCAIRE D'EAU DOUCE **5664**
CALCAIRE GROSSIER **5664**
CALCAIRE JURASSIQUE **7267**
CALCAIRE OOLITHIQUE **8801**
CALCAIRE PISOLITHIQUE **8801**
CALCAIRE SILICEUX **11451**

CALCINATION **1839**
CALCINATION DES MINERAIS **1840**
CALCINE **2272**
CALCINER LES MINERAIS **1841**
CALCITE **1845**
CALCITE **1844**
CALCIUM **1846**
CALCUL **1860**
CALCUL **3794**
CALCUL APPROXIMATIF **652**
CALCUL DIFFERENTIEL **3908**
CALCUL GRAPHIQUE **6060**
CALCUL INFINITESIMAL **6912**
CALCUL INTEGRAL **7011**
CALCUL RAPIDE APPROCHE **10772**
CALCULER **3795**
CALCULER **1857**
CALE **13731**
CALE **13739**
CALE D'EPAISSEUR **13731**
CALE D'EPAISSEUR **11352**
CALE DE RATTRAPAGE **185**
CALER **7277**
CALER SUR... **7279**
CALES D'USURE **13711**
CALES DE COMPARAISON **2796**
CALES MAGNETIQUES DE FIXATION **7897**
CALES-ETALONS **5873**
CALIBRAGE **2651**
CALIBRAGE **1867**
CALIBRAGE **11524**
CALIBRE **5871**
CALIBRE A COULISSE AVEC REGLE DE PROFONDEUR **1871**
CALIBRE A COULISSE DE PROFONDEUR **13538**
CALIBRE A COULISSE DE PROFONDEUR A FIXATION MICROMETRIQUE **13539**
CALIBRE A DEBIT **5455**
CALIBRE D'ALESAGE **10661**
CALIBRE D'ANGLES **513**
CALIBRE D'EPAISSEUR **5086**
CALIBRE D'UNE TOLE **9487**
CALIBRE DE CONICITE NORMAL **8393**
CALIBRE DE FILETAGE **11047**
CALIBRE DE HAUTEUR DE PRECISION **9768**
CALIBRE DE MATRICE **6532**
CALIBRE DE PERCAGE **4267**
CALIBRE DE PROFILAGE **11123**
CALIBRE DE PROFONDEUR **3781**
CALIBRE DE REGLAGE DE SOUPAPE **13473**
CALIBRE DE TOLERANCE **3905**
CALIBRE DE TRACAGE **8016**

CALIBRE EN FER A CHEVAL 6608
CALIBRE FEMELLE LISSE 9431
CALIBRE FIXE 5293
CALIBRE LIMITE (OU DE TOLERANCE) 4584
CALIBRE MALE CONIQUE 12656
CALIBRE MALE ET FEMELLE 3581
CALIBRE NORMAL 12076
CALIBRE PASSE-PASSE PAS 5991
CALIBRE POUR PROFILS 8422
CALIBRE POUR TUBES 2027
CALIBRER 1861
CALIBRES 5769
CALIBRES FEMELLES CONIQUES 12659
CALIBRES MALES CYLINDRIQUES 3587
CALIBRES MALES POUR FILETAGES 12862
CALIBRES MICROMETRIQUES 8251
CALIBRES MICROMETRIQUES D'INTERIEUR 6979
CALIBRES TELESCOPIQUES 12713
CALIBRES-MACHOIRES 11693
CALICHE 2330
CALOMEL 8159
CALORIE ANGLAISE 1625
CALORIE GRAMME-DEGRE 6045
CALORIE KILOGRAMME-DEGRE 7298
CALORIFUGE 6998
CALORIFUGE 8628
CALORIFUGE 6352
CALORIFUGEAGE 6639
CALORIFUGEAGE 7623
CALORIMETRE 1877
CALORIMETRE A CONDENSATION 12160
CALORIMETRE A EAU 13643
CALORIMETRE A FUSION DE LA GLACE 6749
CALORIMETRIE 1881
CALORIMETRIQUE 1878
CALORISATION 1882
CALOTTE (D'UNE SPHERE) 1911
CALOTTE SPHERIQUE 11887
CALOTTE SPHERIQUE 11137
CALQUAGE 13082
CALQUE 13082
CALQUER 13077
CALQUEUR 13078
CAMBOUIS 879
CAMBOUIS 3308
CAMBOUIS 6063
CAMBRAGE 1889
CAME 6329
CAME 4442
CAME 13878
CAME A DISQUE 9482

CAME A PLATEAU 9482
CAME A PLATEAU COURBE 11913
CAME A RAINURE 6134
CAME A RAINURE 3270
CAME EXCENTRIQUE BOSSE 1883
CAME OSCILLANTE 8878
CAMES-DISQUES 9497
CAMION 7752
CAMION A BENNE BASCULANTE 4375
CAMION A RIDELLES 7753
CAMION AUTOMOBILE 8414
CAMION BACHE 12686
CAMION POIDS LOURD 6405
CAMIONNETTE 7563
CAMIONS TRANSPALETTES 9025
CAMPHRE 1893
CANAL 2248
CANAL D'ARRIVEE 6319
CANAL DE COULEE 10844
CANAL DE COULEE 7436
CANAL DE DECHARGE 12596
CANAL DE FUITE 12596
CANAL DE TROP-PLEIN 6185
CANAL DECOUVERT 8807
CANALISATION 9339
CANALISATION 13354
CANALISATION 13666
CANALISATION D'AIR 281
CANALISATION DE SECOURS 4690
CANALISATION DE VAPEUR 12178
CANALISATION ELECTRIQUE 13353
CANALISATION ELECTRIQUE 4523
CANALISATION EN MER OUVERTE 11080
CANAUX DE VENT 247
CANIVEAU 2251
CANIVEAU (SOUDURE) 13351
CANNEL-COAL 1901
CANNELE 5484
CANNELURE 4372
CANNELURE 7282
CANNELURE 4455
CANNELURE EMBOITEE 2526
CANNELURE FERMEE 2526
CANNELURES FERMEES 2527
CANON A ELECTRONS 4628
CANON ANTI-AERIEN 610
CAOLIN 2340
CAOUTCHOUC 10825
CAOUTCHOUC 10816
CAOUTCHOUC AVEC TOILE INTERCALEE 6973
CAOUTCHOUC BRUT 3413

CAOUTCHOUC D'ETANCHEITE 13722
CAOUTCHOUC DE PARA 9037
CAOUTCHOUC DURCI 2363
CAOUTCHOUC VIERGE 3413
CAOUTCHOUC VULCANISE 13610
CAPACITE 1917
CAPACITE (ELECTR.) 1917
CAPACITE CALORIFIQUE 41
CAPACITE CALORIFIQUE 6339
CAPACITE CALORIFIQUE 6335
CAPACITE CALORIQUE 6342
CAPACITE D'AMORTISSEMENT 3605
CAPACITE DE CHARGE 7682
CAPACITE DE CHARGE 2011
CAPACITE DE DEFORMATION 3705
CAPACITE THERMIQUE 12803
CAPILLARITE 1920
CAPOT 1451
CAPOT DE VOITURE 6566
CAPOT EN TOLE 11310
CAPOT-MOTEUR 1451
CAPOTE PLIANTE 2719
CAPSULE 4864
CAPTAGE DES GAZ SANS COMBUSTION 5841
CAPTAGE DU GAZ ET DEPOUSSIERAGE 4392
CAPTEUR 13108
CAPTEUR DE POSITION 9649
CAPUCHON 1910
CAR A LINGOTS (TRANSPORT DE LINGOT) 6920
CARACTERE 2255
CARACTERISTIQUE 9075
CARACTERISTIQUE 2256
CARACTERISTIQUE D'UN LOGARITHME 2257
CARACTERISTIQUES CHIMIQUES 2295
CARACTERISTIQUES MECANIQUES 12746
CARACTERISTIQUES MECANIQUES 8109
CARBOLINEUM 1939
CARBONADO 1967
CARBONATE 1968
CARBONATE ACIDE DE SODIUM 126
CARBONATE D'AMMONIUM 461
CARBONATE DE BARYUM 996
CARBONATE DE CHAUX 1849
CARBONATE DE CUIVRE 3470
CARBONATE DE LITHIUM 7665
CARBONATE DE MAGNESIE 7884
CARBONATE DE SODIUM 11713
CARBONATE DE SOUDE ANHYDRE 551
CARBONATE DE ZINC 14037
CARBONATE NATUREL DE BARYTE 13911
CARBONATE NEUTRE DE POTASSE 9679

CARBONE 1966
CARBONE 1940
CARBONE A L'ETAT GRAPHITOIDE 6059
CARBONE COMBINE 5289
CARBONE COMBINE 1962
CARBONE COMBINE AU FER 2767
CARBONE DE REVENU 1965
CARBONE DISSOUS 1963
CARBONE NON COMBINE AU FER 13345
CARBONISATION 1975
CARBONISATION 1974
CARBONISATION 1970
CARBONISATION DE LA HOUILLE 7991
CARBONISATION INCOMPLETE 7776
CARBONISE 2272
CARBONISER 1971
CARBONITRURATION 1972
CARBONYLE 1939
CARBONYLE 1976
CARBONYLE DE FER 7140
CARBONYLE DE NICKEL 8561
CARBORUNDUN 1980
CARBURANT 9222
CARBURANT ANTI-DETONNANT 613
CARBURANTS 2024
CARBURANTS 1988
CARBURATEUR 1984
CARBURATEUR 1982
CARBURATEUR A DOUBLE CORPS 4360
CARBURATEUR INVERSE 4189
CARBURATION DU FER 1985
CARBURE 1933
CARBURE 1848
CARBURE D'HYDROGENE 6705
CARBURE DE BARYUM 995
CARBURE DE BORE 1479
CARBURE DE FER 7138
CARBURE DE SILICIUM 11445
CARBURE DE SILICUM 1980
CARBURER LE FER 1986
CARCASSE 5624
CARDAN 6573
CARDAN 5567
CARDAN 1997
CARDE 5150
CARDIOIDE 2001
CARIE 10770
CARNEAU 5467
CARNEAU 13418
CARNEAU A GAZ 5821
CARNEAU DE FUMEE 5743

CARNEAU DE RACCORDEMENT **5034**
CARNEAU INTERIEUR D'UNE CHAUDIERE **1412**
CAROTTE **3142**
CARRE **12025**
CARRE (DEUXIEME PUISSANCE) **12013**
CARRE (GEOM.) **12013**
CARRE (POUR CLEF) **9533**
CARRE D'ENTRAINEMENT **4288**
CARREAU **12019**
CARREAU (CONSTR.) **12933**
CARREE **12019**
CARRELET **12019**
CARRES EN ACIER ALLIE **379**
CARRIERE **10073**
CARROSSAGE **1889**
CARROSSERIE **1385**
CARROSSERIE PORTANTE **13381**
CARTE **1999**
CARTE PERFOREE **10002**
CARTER **2026**
CARTER D'EMBRAYAGE **2546**
CARTER D'EMBRAYAGE **1133**
CARTER D'ENGRENAGE **5890**
CARTER D'ENGRENAGES **13815**
CARTER D'HUILE (INFERIEUR) **8782**
CARTER D'HUILE (INFERIEUR) **8767**
CARTER DE CHAINES **2204**
CARTER DE DISTRIBUTION **12947**
CARTER DE POMPE D'INJECTION **5712**
CARTER DE REGULATEUR **6016**
CARTER-MOTEUR **3306**
CARTON **1381**
CARTON BITUME OU ASPHALTE **10712**
CARTON D'AMIANTE **747**
CARTON LEGER **1999**
CARTOUCHE **2015**
CARTOUCHE FILTRANTE **5175**
CASCADE **2018**
CASE **2798**
CASEINE **2025**
CASQUE **6432**
CASSANT **1627**
CASSANT A CHAUD **6658**
CASSANT A CHAUD **10337**
CASSE-COKE **2658**
CASSE-FER **631**
CASSETTE **2028**
CASSINOIDE **2029**
CASSITERITE **12959**
CASSURE **5620**
CASSURE **5617**

CASSURE CEROIDE **11936**
CASSURE CONCHOIDALE **2893**
CASSURE CROCHUE **6195**
CASSURE FIBREUSE **5138**
CASSURE GRENUE **6051**
CASSURE HACHEE **6195**
CASSURE INEGALE **13357**
CASSURE INTERGRANULAIRE **7042**
CASSURE LAMELLAIRE **7378**
CASSURE NETTE **11683**
CASSURE SOYEUSE **11454**
CASSURE TERREUSE **4421**
CASSURE TRANSCRISTALLINE **13117**
CASSURE UNIE **4871**
CATALYSE **2090**
CATALYSEUR **2091**
CATALYTIQUE **2092**
CATAPHORESE **4639**
CATARACTE **3611**
CATEGORIE DE FER **2458**
CATEGORIE SORTE D'ACIER **2459**
CATHETOMETRE **2096**
CATHODE **2097**
CATHODE **2106**
CATHODE DE REDRESSEUR **10316**
CATHOLYTE **2111**
CATION **2112**
CATION **9657**
CAUSE D'ERREUR **11825**
CAUSTIQUE **4847**
CAUSTIQUE (OPT.) **2116**
CAUSTIQUE PAR REFLEXION **2089**
CAUSTIQUE PAR REFRACTION **3829**
CAVALIER **1735**
CAVALIER **12106**
CAVALIER DE FOND **12104**
CAVALIERS **3946**
CAVET **2122**
CAVITATION **2123**
CAVITE **7073**
CAVITE **2125**
CAVITE DE CELLULE **2127**
CAVITE DU MOULE **8353**
CAVITE EN V **5271**
CEDER **5949**
CEINTURE DE SECURITE **11096**
CEINTURE DE SECURITE **10882**
CELLULE (OU CUVE) ELECTROLYTIQUE **4588**
CELLULE AU MERCURE **8165**
CELLULE PHOTO-CONDUCTRICE **9261**
CELLULE PHOTO-ELECTRIQUE **9260**

CELLULE PHOTO-EMISSIVE **9273**
CELLULE PHOTOVOLTAIQUE **9277**
CELLULOID **2130**
CELLULOSE **2131**
CELLULOSE NITREE **6176**
CEMENT **2022**
CEMENT **2139**
CEMENT DE CUIVRE **2136**
CEMENTATION **2021**
CEMENTATION **2141**
CEMENTATION (DES BARRES DE FER) **2142**
CEMENTATION A L'AZOTE **5818**
CEMENTATION A LA POUDRE **8992**
CEMENTATION AU PRUSSIATE DE POTASSE **9670**
CEMENTATION AU PRUSSIATE JAUNE **9670**
CEMENTATION BRILLANTE **9959**
CEMENTATION DES PIECES EN ACIER DOUX **2023**
CEMENTATION EN BAIN **7646**
CEMENTATION EN CAISSE **1996**
CEMENTATION HOMOGENE **1993**
CEMENTATION PAR GAZ **1992**
CEMENTATION PAR LE CARBONE **1987**
CEMENTATION PAR LE CHROME **2390**
CEMENTATION PARTIELLE **2023**
CEMENTATION SELECTIVE **1995**
CEMENTATION SELECTIVE **11143**
CEMENTER **2134**
CEMENTITE **2147**
CEMENTS **2024**
CENDRE **760**
CENDRE **2397**
CENDRES **760**
CENDRES VOLANTES **5493**
CENDRIER **759**
CENTIMETRE **2158**
CENTIMETRE CARRE **12016**
CENTIMETRE CUBE **3449**
CENTRAGE **2187**
CENTRALE ELECTRIQUE **4512**
CENTRALE HYDRAULIQUE **6692**
CENTRE **2148**
CENTRE **2162**
CENTRE A COMPAS **6594**
CENTRE CUBIQUE **1399**
CENTRE D'OSCILLATION **2170**
CENTRE DE COURBURE **2167**
CENTRE DE GRAVITE **2168**
CENTRE DE ROTATION **5717**
CENTRE DE ROTATION **2169**
CENTRE DE ROUE EN ACIER FONDU **2056**
CENTRE DE SURFACE **2191**

CENTRE DU RIVET **2171**
CENTRE INSTANTANE DE ROTATION **6989**
CENTRER **2163**
CERAMIQUE **2193**
CERCLE **2401**
CERCLE ANNUEL **586**
CERCLE CIRCONSCRIT **2435**
CERCLE DE BASE **1030**
CERCLE DE COURBURE **2402**
CERCLE DE CREUX **10723**
CERCLE DE FOND **10723**
CERCLE DE PIED **10717**
CERCLE DE PIED **10723**
CERCLE DE POINT NEUTRE (CALCUL DES CHARPENTES DE
   RESISTANCE SOUS PRESSION) **8604**
CERCLE DE TETE **3396**
CERCLE DE TETE **166**
CERCLE DE TETE **167**
CERCLE DECRIT PAR LA MANIVELLE **3298**
CERCLE DES TROUS DE BOULONS **1433**
CERCLE DIVISE **6026**
CERCLE ELASTIQUE **11990**
CERCLE EXINSCRIT **4834**
CERCLE EXTERIEUR **166**
CERCLE FIXE **5290**
CERCLE GRADUE **6026**
CERCLE INSCRIT **6963**
CERCLE MOBILE **5921**
CERCLE PRIMITIF **1030**
CERCLE PRIMITIF **9404**
CERCLE PRIMITIF DE DENTURE **9403**
CERCLE ROULANT **5921**
CERCLES CONCENTRIQUES **2890**
CERCLES EXCENTRIQUES **4426**
CERESINE **2194**
CERIUM **2195**
CERMET **2192**
CERTIFICAT D'EPREUVE **12775**
CERTIFICAT DE RECEPTION **70**
CERUSE **13834**
CERUSSITE **2196**
CESIUM **2197**
CESIUM **1833**
CESSION DE CHALEUR **3735**
CHABOTTE D'ENCLUME **633**
CHAINE **2201**
CHAINE A ETANCONS **12403**
CHAINE A MAILLES ETANCONNEES **12403**
CHAINE A MAILLES SOUDEES **13770**
CHAINE A MAILLONS COURTS **11373**
CHAINE A MAILLONS LONGS **7730**

CHAINE A MAILLONS SERRES 11373
CHAINE A ROULEAUX 10684
CHAINE CALIBREE 9412
CHAINE D'UN TISSU 13627
CHAINE DE BICYCLETTE 1239
CHAINE DE DISTRIBUTION 12950
CHAINE DE DISTRIBUTION 12948
CHAINE DE LEVAGE 6518
CHAINE DE TRANSMISSION 4306
CHAINE DE TRANSMISSION 4305
CHAINE DE VAUCANSON 6570
CHAINE DECOUPEE 12058
CHAINE GALLE 12002
CHAINE MOTRICE 4306
CHAINE ORDINAIRE 8855
CHAINE PLATE 1336
CHAINE SANS FIN 4732
CHAINE SILENCIEUSE 11432
CHAINE-CABLE 6518
CHAINE-CABLE A ETAIS 12403
CHAINES CALIBREES 1862
CHAINES DE MARINE 8004
CHAINES DE NEIGE 11696
CHAINETTE (GEOM.) 2095
CHAINON 7627
CHAISE D'APPLIQUE 13618
CHAISE DE PALIER 1103
CHAISE EN BOUT 4726
CHAISE EN U 9166
CHAISE PENDANTE A DEUX JAMBES 9166
CHAISE SUR LE SOL 12064
CHAISE-CONSOLE 13618
CHALCOPYRITE 3126
CHALCOSINE 3118
CHALEUR 6333
CHALEUR ATOMIQUE 802
CHALEUR BLANCHE 6823
CHALEUR D'ABSORPTION 6354
CHALEUR D'EVAPORATION 6359
CHALEUR D'OXYDATION 6361
CHALEUR DE CARBURATION 6355
CHALEUR DE COMBINAISON 9656
CHALEUR DE COMBUSTION 6356
CHALEUR DE DECOMPOSITION 6357
CHALEUR DE DECOMPOSITION 8528
CHALEUR DE DISSOCIATION 6358
CHALEUR DE FORGE 2744
CHALEUR DE FORMATION 9656
CHALEUR DE FUSION 6360
CHALEUR DE REDUCTION 6362
CHALEUR DE SUBLIMATION 6363

CHALEUR DE TRANSFORMATION 6364
CHALEUR DE VAPORISATION 6366
CHALEUR DEGAGEE 6370
CHALEUR DEGAGEE DANS UNE REACTION 6348
CHALEUR EFFECTIVE 4457
CHALEUR EMMAGASINEE 10411
CHALEUR INTERNE 7070
CHALEUR LATENTE 7415
CHALEUR LATENTE 6365
CHALEUR LATENTE DE VAPORISATION 7416
CHALEUR PERDUE 13635
CHALEUR PRODUITE PAR FROTTEMENT 6343
CHALEUR RADIANTE 10140
CHALEUR RAYONNANTE 10140
CHALEUR ROUGE 10329
CHALEUR SENSIBLE 11184
CHALEUR SOUDANTE 13784
CHALEUR SPECIFIQUE 11852
CHALEUR SPECIFIQUE MOYENNE 8076
CHALEUR UTILE 4457
CHALUMEAU 1355
CHALUMEAU 3539
CHALUMEAU 13042
CHALUMEAU 5307
CHALUMEAU A BOUCHE 8433
CHALUMEAU A DECOUPER 3216
CHALUMEAU A OXYGENE COMPRIME 2848
CHALUMEAU COUPEUR 5312
CHALUMEAU DE CHAUFFAGE 6399
CHALUMEAU DE PRECHAUFFAGE 6391
CHALUMEAU DEROUILLEUR 5311
CHALUMEAU ELECTRIQUE 4496
CHALUMEAU OXHYDRIQUE 8986
CHAMBRAGE 3236
CHAMBRAGE 3242
CHAMBRE A AIR 6955
CHAMBRE A AIR 13246
CHAMBRE A EAU D'UNE CHAUDIERE 13677
CHAMBRE A VIDE 13440
CHAMBRE A VIDE 13449
CHAMBRE D'EXPLOSIONS D'UN MOTEUR 2778
CHAMBRE D'IONISATION 7125
CHAMBRE DE COMBUSTION 2776
CHAMBRE DE COMBUSTION 2856
CHAMBRE DE COMBUSTION D'UN FOYER 2777
CHAMBRE DE DISTRIBUTION 2229
CHAMBRE DE TURBULENCE 12551
CHAMBRE DE VAPEUR D'UNE CHAUDIERE 12185
CHAMBRE DE VENT 241
CHAMBRE NOIRE 9264
CHAMOTTE 2235

CHAMP DE VISION **5143**
CHAMP ELECTRIQUE **4505**
CHAMP ELECTRIQUE **4504**
CHAMP MAGNETIQUE **7901**
CHAMP VISUEL **5143**
CHANCES D'INCENDIE **3608**
CHANDELLE **1899**
CHANDELLE (ETAI VERTICAL) **11365**
CHANFREIN **1902**
CHANFREIN **2231**
CHANFREIN (SOUDURE) EN K **1793**
CHANFREIN D'ENTREE **516**
CHANFREIN EN DEMI-V **1800**
CHANFREINAGE **3234**
CHANFREINAGE **2233**
CHANFREINAGE **1231**
CHANFREINER **2230**
CHANGEMENT D'ETAT **2238**
CHANGEMENT D'ETAT NON REVERSIBLE **7170**
CHANGEMENT D'ETAT PHYSIQUE **9280**
CHANGEMENT D'ETAT REVERSIBLE **10539**
CHANGEMENT DE DIRECTION **2236**
CHANGEMENT DE FORME **3707**
CHANGEMENT DE MULTIPLICATION **2245**
CHANGEMENT DE PRESSION **2237**
CHANGEMENT DE TEMPERATURE **2239**
CHANGEMENT DE VITESSE **2245**
CHANGEMENT DE VOIE **12556**
CHANGEMENT DE VOLUME **2240**
CHANGEMENT DU SENS D'UN COURANT **2246**
CHANGER LE SENS DU COURANT **2241**
CHANGEUR D'OUTILS **13000**
CHANTIER **12265**
CHANTIER DE COULEE **2071**
CHANTOURNAGE **10167**
CHANVRE **6439**
CHANVRE DE MANILLE **7980**
CHAPE **13172**
CHAPE **5567**
CHAPE DE LA TETE DE BIELLE **12345**
CHAPE DE POULIE **1333**
CHAPE DU BATTANT PORTE-OUTIL **2454**
CHAPE-FOURCHETTE **5568**
CHAPEAU **1910**
CHAPEAU **1454**
CHAPEAU **1451**
CHAPEAU **1912**
CHAPEAU A ARCADE **14015**
CHAPEAU DE LA BOITE A ETOUPE **5953**
CHAPEAU DE PALIER **1104**
CHAPELLE **2229**

CHAPITEAU D'UNE COLONNE **1926**
CHARBON **2558**
CHARBON ANIMAL **554**
CHARBON DE BOIS **13914**
CHARBON DE CORNUE **5812**
CHARBON DE FORGE **5543**
CHARBON DE TERRE **2558**
CHARBON FOSSILE **2558**
CHARBON GRAPHITIQUE **9851**
CHARBON GROS **7826**
CHARBON PULVERISE **9984**
CHARBONNAGE **2566**
CHARBONNER **1971**
CHARGE **2259**
CHARGE **12367**
CHARGE **12331**
CHARGE **4954**
CHARGE **7679**
CHARGE ADMISE **9215**
CHARGE ADMISSIBLE **13956**
CHARGE CENTRALE **2161**
CHARGE CHIMIQUE **5157**
CHARGE D'EAU **6323**
CHARGE D'ESSAI **12781**
CHARGE DE FORGEAGE **9823**
CHARGE DE LA VITESSE **13515**
CHARGE DE NEIGE **11695**
CHARGE DE RUPTURE **13331**
CHARGE DE RUPTURE **12749**
CHARGE DE RUPTURE **1585**
CHARGE DE RUPTURE AU FLAMBAGE **3338**
CHARGE DE SECURITE **10879**
CHARGE DUE AU VENT **13857**
CHARGE ELECTRIQUE **2259**
CHARGE EXCENTRIQUE **4428**
CHARGE HYDROSTATIQUE **7652**
CHARGE INCOMPLETE **7562**
CHARGE INFERIEURE A LA NORMALE **7562**
CHARGE INTERMITTENTE **7672**
CHARGE ISOLEE **7183**
CHARGE LIQUIDE **7647**
CHARGE MAXIMUM **8065**
CHARGE MOBILE **10706**
CHARGE MORTE **3631**
CHARGE NEUTRE **13842**
CHARGE NORMALE **8661**
CHARGE NORMALE **8655**
CHARGE PAR UNITE DE SURFACE **13371**
CHARGE PERIODIQUE **6812**
CHARGE PERMANENTE **3633**
CHARGE PRATIQUE **10879**

CHARGE RECTANGULAIRE **10303**
CHARGE REPARTIE **4078**
CHARGE ROULANTE **10706**
CHARGE SANS CHOCS **7685**
CHARGE STATIQUE **12138**
CHARGE TOTALE **13062**
CHARGE TRAPEZOIDALE **13150**
CHARGE TRIANGULAIRE **13190**
CHARGE UNIFORMEMENT REPARTIE **13366**
CHARGE UTILE **9139**
CHARGE UTILE **13425**
CHARGE VARIABLE **7672**
CHARGE VIVE **7671**
CHARGE-LIMITE D'ELASTICITE **1586**
CHARGEMENT **2263**
CHARGEMENT **5262**
CHARGEMENT **2269**
CHARGEMENT D'UN ACCUMULATEUR **2265**
CHARGEMENT D'UN FOUR METALLURGIQUE **2264**
CHARGEMENT PAR CONVOYEUR A BANDE **1141**
CHARGER **5156**
CHARGER (RESISTANCE DES MATERIAUX) **7680**
CHARGER UN ACCUMULATEUR **2261**
CHARGER UN FOUR METALLURGIQUE **2260**
CHARGER UN FOYER **12269**
CHARGEUR **8110**
CHARGEUR (OU CHARGEUSE) POUR FOURRAGE VERT **7687**
CHARGEURS DE FUMIER **7689**
CHARGEURS DE SACS ET DE BETTERAVES **7688**
CHARIOT **4116**
CHARIOT (POUR MANUTENTION D'ATELIER ET MAGASIN)
  13230
CHARIOT D'ECRASEMENT **8436**
CHARIOT D'UN PONT-ROULANT **13155**
CHARIOT DE SUSPENSION POUR PIECES A TREMPER **9505**
CHARIOT PORTE-PALAN **13155**
CHARIOT TRANSVERSAL **3379**
CHARIOTAGE **12668**
CHARIOTAGE D'UNE SURFACE CYLINDRIQUE **13164**
CHARIOTAGE LONGITUDINAL **13164**
CHARIOTAGE TRANSVERSAL **12522**
CHARIOTER UNE SURFACE CYLINDRIQUE **13162**
CHARIOTS A BRAS **13231**
CHARNIERE **6505**
CHARNIERE (MALE) **6508**
CHARNIERE TYPE PENTURE **12347**
CHARNIERE UNIVERSELLE **6573**
CHARPENTE **5624**
CHARPENTE **12942**
CHARPENTE **5630**
CHARPENTE A CROISILLONS **3360**

CHARPENTE A FERMES **13244**
CHARPENTE A POTEAUX **2759**
CHARPENTE A POTEAUX **2754**
CHARPENTE AUTOPORTANTE **11161**
CHARPENTE METALLIQUE **12208**
CHARPENTE METALLIQUE **7152**
CHARPENTERIE **2006**
CHARPENTIER **2005**
CHARRETTE **2013**
CHARRETTE A BRAS **6230**
CHARRUES **9526**
CHARRUES A DISQUES (TRAINEES OU PORTEES) **9523**
CHARRUES BILLONEUSES **9527**
CHARRUES PORTEES **9524**
CHARRUES RIGOLEURSE **8344**
CHARRUES TRAINEES **9525**
CHASSE **9516**
CHASSE **2066**
CHASSE (ANGLE DE) **2067**
CHASSE (OUTIL DE FORGERON) **11218**
CHASSE A BISEAU **2232**
CHASSE A PARER **5410**
CHASSE-CLAVETTE **7278**
CHASSE-CLOU **12127**
CHASSE-GOUPILLE **9326**
CHASSE-POINTE **7332**
CHASSE-RIVET **10626**
CHASSER LES RIVETS **3511**
CHASSER UNE CLAVETTE **4295**
CHASSIS **5624**
CHASSIS **2275**
CHASSIS DE FENETRE **13865**
CHASSIS DE MOULAGE **5369**
CHASSIS POUR BLEUS **9874**
CHASSIS-CAISSON **1511**
CHASSIS-TENDEUR **12756**
CHATEAU D'EAU **13680**
CHAUDE **10329**
CHAUDE **5981**
CHAUDE ROUGE-CERISE **2326**
CHAUDE SUANTE **13784**
CHAUDIERE **1406**
CHAUDIERE A BASSE PRESSION **7782**
CHAUDIERE A FAIBLE VOLUME **11663**
CHAUDIERE A FOYER INTERIEUR **5468**
CHAUDIERE A GRAND VOLUME **7412**
CHAUDIERE A HAUTE PRESSION **6487**
CHAUDIERE A MOYEN VOLUME **8126**
CHAUDIERE A MOYENNE PRESSION **8127**
CHAUDIERE A TUBE FOYER **5468**
CHAUDIERE A TUBES D'EAU **13682**

CHAUDIERE A TUBES DE FUMEE **5249**
CHAUDIERE A VAPEUR **12158**
CHAUDIERE A VAPORISATION RAPIDE **13682**
CHAUDIERE CYLINDRIQUE **3578**
CHAUDIERE ELECTRIQUE **4497**
CHAUDIERE HORIZONTALE **6579**
CHAUDIERE IGNITUBULAIRE **5249**
CHAUDIERE INDUSTRIELLE **12143**
CHAUDIERE INEXPLOSIBLE **13682**
CHAUDIERE MARINE **8003**
CHAUDIERE MULTITUBULAIRE **13682**
CHAUDIERE TUBULAIRE **5249**
CHAUDIERE TYPE LOCOMOTIVE **7714**
CHAUDIERE VERTICALE **13544**
CHAUDRONNERIE **3136**
CHAUDRONNERIE EN FER **1423**
CHAUDRONNIER EN CUIVRE **3135**
CHAUDRONNIER EN FER **1415**
CHAUDRONNIER EN TOLE **1415**
CHAUFFAGE **6390**
CHAUFFAGE **5259**
CHAUFFAGE A AIR CHAUD **6616**
CHAUFFAGE A AIR CHAUD **268**
CHAUFFAGE A HUILE LOURDE **1752**
CHAUFFAGE AU CHARBON **2564**
CHAUFFAGE AU CHARBON PULVERISE **1753**
CHAUFFAGE AU GAZ **1751**
CHAUFFAGE AU NAPHTE **1752**
CHAUFFAGE AU PETROLE **1752**
CHAUFFAGE AU-DESSUS DU POINT DE FUSION **3634**
CHAUFFAGE CENTRAL **2160**
CHAUFFAGE DE L'AIR **268**
CHAUFFAGE DES BATIMENTS **6395**
CHAUFFAGE ELECTRIQUE **4513**
CHAUFFAGE MECANIQUE **8111**
CHAUFFAGE PAR INDUCTION **6897**
CHAUFFAGE PAR L'EAU CHAUDE **6648**
CHAUFFAGE PAR L'EAU CHAUDE A BASSE PRESSION **7784**
CHAUFFAGE PAR L'EAU CHAUDE A HAUTE PRESSION **6489**
CHAUFFAGE PAR LA VAPEUR **12167**
CHAUFFAGE PAR LA VAPEUR A BASSE PRESSION **7787**
CHAUFFAGE PAR LA VAPEUR A HAUTE PRESSION **6493**
CHAUFFAGE PAR LA VAPEUR A UNE PRESSION INFERIEURE
  A LA PRESSION ATMOSPHERIQUE **8318**
CHAUFFAGE PARTIEL **7696**
CHAUFFAGE PREALABLE **9776**
CHAUFFAGE SELECTIF **3915**
CHAUFFAGE SUPPLEMENTAIRE **2904**
CHAUFFE **2327**
CHAUFFE **6390**
CHAUFFE A BLANC **6824**

CHAUFFE AU BLEU **1371**
CHAUFFE AU ROUGE **10330**
CHAUFFE AU ROUGE CLAIR **1614**
CHAUFFE AU ROUGE SOMBRE **4368**
CHAUFFE EAU A GAZ **5846**
CHAUFFER **5235**
CHAUFFER AU CONTACT DE L'AIR **6351**
CHAUFFER EN PRESENCE **6351**
CHAUFFER EN VASE CLOS **6367**
CHAUFFER PREALABLEMENT **9774**
CHAUFFERIE **1418**
CHAUFFERIE **12270**
CHAUFFEUR **12272**
CHAUFFEUR **12271**
CHAUX **7582**
CHAUX ANHYDRE **1854**
CHAUX ETEINTE **1851**
CHAUX FLUATEE **5479**
CHAUX GRASSE **5046**
CHAUX HYDRATEE **1851**
CHAUX HYDRAULIQUE **6689**
CHAUX MAIGRE **9623**
CHAUX VIVE **1854**
CHEF D'EQUIPE **7482**
CHEF DE CHANTIER **5145**
CHEF DE CHANTIER **11520**
CHEF DE FABRICATION, DE SERVICE, D'ATELIER **13964**
CHELATION **11195**
CHEMIN **4063**
CHEMIN DE FER A VOIE ETROITE **9639**
CHEMIN DE FER FUNICULAIRE **1826**
CHEMIN DE ROULEMENT **1111**
CHEMIN DE ROULEMENT **7606**
CHEMIN DE ROULEMENT (POUR ECHELLE ROULANTE DE
  TOIT) **10857**
CHEMIN DE ROULEMENT DES BILLES **965**
CHEMIN DE ROULEMENT DES GALETS **10689**
CHEMINEE **2339**
CHEMISE **3274**
CHEMISE **12168**
CHEMISE **11303**
CHEMISE **7623**
CHEMISE D'EAU **13655**
CHEMISE DE CYLINDRE **1465**
CHEMISE DU CYLINDRE **3571**
CHEMISE DU FOUR **5749**
CHEMISE EN BOIS **13927**
CHEMISE HUMIDE **13804**
CHEMISE SECHE **4348**
CHEVAL EFFECTIF **151**
CHEVAL NOMINAL **6873**

CHEVAL-HEURE **6603**
CHEVAL-VAPEUR **6602**
CHEVAUCHEMENT **8951**
CHEVAUCHER **8950**
CHEVILLE **9315**
CHEVILLE **4180**
CHEVILLE **7276**
CHEVILLE D'ASSEMBLAGE **3215**
CHEVILLE EN BOIS **13179**
CHEVRE A TROIS PIEDS **5945**
CHEVRON **4144**
CHICANE **921**
CHICANE **922**
CHIEN **2094**
CHIFFON A NETTOYER **11952**
CHIFFON D'ESSUYAGE **11952**
CHIFFRE **3940**
CHIFFRE DE DURETE BRINELL **1619**
CHIFFRES ET LETTRES A CHAUD/FROID **1560**
CHIGNOLE **9732**
CHIGNOLE **6258**
CHIMIE **2322**
CHIMIE ANALYTIQUE **495**
CHIMIE APPLIQUEE **648**
CHIMIE INDUSTRIELLE OU TECHNOLOGIQUE **9286**
CHIMIE MINERALE OU INORGANIQUE **6958**
CHIMIE ORGANIQUE **8865**
CHIMIE PHYSIQUE **9281**
CHIMIQUE **2290**
CHIMIQUEMENT COMBINE **2317**
CHIMIQUEMENT PUR **2319**
CHIMISTE **2321**
CHINA-GRASS **10192**
CHLORALUM **421**
CHLORATE **2351**
CHLORATE DE POTASSIUM **9680**
CHLORE **2354**
CHLORHYDRATE D'AMMONIAQUE **462**
CHLOROFORME **2356**
CHLORURATION **2353**
CHLORURE **2352**
CHLORURE CHROMEUX **2391**
CHLORURE CHROMIQUE **2381**
CHLORURE CUIVREUX **3478**
CHLORURE CUIVRIQUE **4853**
CHLORURE D'ACETYLE **92**
CHLORURE D'ALUMINIUM **421**
CHLORURE D'AMMONIUM **462**
CHLORURE D'ARGENT **11461**
CHLORURE D'ETHYLE **4849**
CHLORURE D'OR **823**

CHLORURE DE BARYUM **997**
CHLORURE DE CALCIUM **1850**
CHLORURE DE CHAUX **1317**
CHLORURE DE COBALT **2609**
CHLORURE DE LITHIUM **7666**
CHLORURE DE MAGNESIUM **7885**
CHLORURE DE METHYLE **8221**
CHLORURE DE PLOMB **7458**
CHLORURE DE POTASSE SOLUTION DE **9688**
CHLORURE DE POTASSIUM **9681**
CHLORURE DE SODIUM **2790**
CHLORURE DE SOUFRE **12451**
CHLORURE DE ZINC **14038**
CHLORURE DOUBLE D'ETAIN ET D'AMMONIUM **470**
CHLORURE FERREUX **5121**
CHLORURE FERRIQUE **5100**
CHLORURE MANGANEUX **7972**
CHLORURE MERCUREUX **8159**
CHLORURE MERCURIQUE **8153**
CHLORURE NICKELEUX **8562**
CHLORURE PLATINEUX **9509**
CHLORURE PLATINIQUE **9507**
CHLORURE STANNEUX **12100**
CHLORURE STANNIQUE **12096**
CHOC **11359**
CHOC **6783**
CHOC **4529**
CHOC DE LA SOUPAPE **6224**
CHOC EN ARRIERE **10286**
CHOC EN RETOUR **10286**
CHOC THERMIQUE **12812**
CHOMAGE **7092**
CHOMAGE MAT **4366**
CHROMAGE ELECTROLYTIQUE **2388**
CHROMATATION **2367**
CHROMATE **2366**
CHROMATE DE PLOMB **2375**
CHROMATE DE ZINC **14039**
CHROMATE NEUTRE DE POTASSIUM **9682**
CHROMATIQUE **2368**
CHROME **2385**
CHROME A FOUR **5741**
CHROMITE **2384**
CHROMITE **2373**
CHRONOMETRE A STOP **12283**
CHRYSOLITHE **2392**
CHRYSOTILE **11205**
CHUTE **3356**
CHUTE (DE TENSION) CATHODIQUE **2100**
CHUTE (RESTE) DE TOLE **9493**
CHUTE ANODIQUE **595**

CHUTE DE POTENTIEL **13594**
CHUTE DE PRESSION **9815**
CHUTE DE TEMPERATURE **12724**
CHUTE DE TENSION **13594**
CHUTE DE VITESSE **3677**
CHUTE LIBRE **5633**
CHUTE LIBRE (A) **5024**
CIBLE **12682**
CIMENT **2133**
CIMENT A PRISE LENTE **11648**
CIMENT A PRISE RAPIDE **10107**
CIMENT ARME **12203**
CIMENT DE LAITIER **11564**
CIMENT DE MACHEFER **11564**
CIMENT DE PORTLAND **9644**
CIMENT DURCISSANT A L'AIR **302**
CIMENT POUZZOLANIQUE **9749**
CIMENT REFRACTAIRE **5255**
CIMENT REFRACTAIRE AU SILICATE D'ALUMINIUM **443**
CIMENT ROMAIN **10707**
CIMENTATION (RECOUVREMENT D'UNE COUCHE DE CIMENT) **2146**
CIMENTER (COUVRIR D'UNE COUCHE DE CIMENT) **2134**
CINABRE **2400**
CINEMATIQUE **7305**
CINEMATIQUE **7306**
CINETIQUE **7307**
CINETIQUE **7310**
CINTRAGE **1182**
CINTRAGE **5579**
CINTRAGE A FROID **2668**
CINTRAGE D'UN TUBE **9341**
CINTRE **3494**
CINTRE DE DILATATION **4913**
CINTRER **1173**
CINTRER UN TUYAU **1175**
CIRCLIP **11990**
CIRCONFERENCE **2404**
CIRCONFERENCE CIRCONSCRITE **2435**
CIRCONFERENCE D'ECHANFREINEMENT **166**
CIRCONFERENCE D'EVIDEMENT **10723**
CIRCONFERENCE INSCRITE **6963**
CIRCONFERENCE PRIMITIVE **9404**
CIRCUIT DE FREINAGE **1545**
CIRCUIT DE REFROIDISSEMENT **3093**
CIRCUIT ELECTRIQUE **2406**
CIRCUIT IMPRIME (C.I.) **9871**
CIRCULATION **2433**
CIRCULATION AU CYANURE **3549**
CIRCULATION D'EAU **2434**
CIRCULER **2432**

CIRE **13702**
CIRE D'ABEILLES **1129**
CIRE DE CARNAUBA **2003**
CIRE DE LIGNITE **7579**
CIRE DE PARAFFINE **7579**
CIRE DE PARAFFINE **9045**
CIRE DU JAPON **7213**
CIRE FOSSILE **4413**
CIRE MINERALE **2194**
CISAILLAGE **11290**
CISAILLAGE **10507**
CISAILLE **11297**
CISAILLE **11279**
CISAILLE A DECOUPER LA CHUTE **3358**
CISAILLE A EBOUTER **12037**
CISAILLE A PORTE A FAUX **5495**
CISAILLE A TOLES **9494**
CISAILLE BASCULANTE **10658**
CISAILLE CIRCULAIRE **2403**
CISAILLE CIRCULAIRE **2429**
CISAILLE MECANIQUE **11292**
CISAILLE POUR FERS A BETON **2897**
CISAILLE VOLANTE **5495**
CISAILLEMENT **11290**
CISAILLEMENT **11288**
CISAILLEMENT **11279**
CISAILLER **1078**
CISAILLER (COUPER AVEC DES CISAILLES) **11280**
CISAILLES **11297**
CISAILLES A BRAMES **11554**
CISAILLES A GUILLOTINE **6170**
CISAILLES A MAIN **6251**
CISAILLES ARTICULEES **856**
CISAILLES D'ETABLI **12264**
CISAILLES POUR BILLETTES **1246**
CISAILLES POUR BILLETTES ET LARGETS **1244**
CISAILLES ROTATIVES **10759**
CISAILLEUSE **11292**
CISEAU **2349**
CISEAU A DOUILLE **11704**
CISEAU A FROID **2672**
CISEAU A FROID **2717**
CISEAU POUR MENUISIER **9080**
CISSOIDE **2438**
CITER **10113**
CITERNE **2439**
CITRATE DE SODIUM **11719**
CLAME D'ASSEMBLAGE **7280**
CLAMEAU **12105**
CLAPET **5353**
CLAPET **13453**

CLAPET **970**
CLAPET **2617**
CLAPET (DE ROBINET) **4016**
CLAPET ANTIGEL **8633**
CLAPET D'ASPIRATION **12436**
CLAPET D'UNE SOUPAPE **13459**
CLAPET DE DECHARGE **10441**
CLAPET DE FOND **5523**
CLAPET DE PIED **5523**
CLAPET DE REFOULEMENT **3739**
CLAPET DE RETENUE **2284**
CLAPET DE RETENUE **2283**
CLAPET DE RETENUE A BATTANT **12545**
CLAPET DE RETENUE A SOUPAPE **7543**
CLAPET DE SOUPAPE **9024**
CLAPET INVERSEUR **10553**
CLAPET-CREPINE **5523**
CLARIFICATION **2455**
CLARIFIER **2456**
CLASSEMENT **2460**
CLAVETAGE **13740**
CLAVETAGE **7288**
CLAVETER **7279**
CLAVETER **7277**
CLAVETTE **7276**
CLAVETTE **5211**
CLAVETTE **13739**
CLAVETTE A FRICTION **6542**
CLAVETTE A MENTONNET **5939**
CLAVETTE A MEPLAT **5388**
CLAVETTE A TALON **5939**
CLAVETTE A TETE **5939**
CLAVETTE CONIQUE **12665**
CLAVETTE CREUSE **6542**
CLAVETTE D'ARRET **3213**
CLAVETTE DE SERRAGE **185**
CLAVETTE DEMI-RONDE **13929**
CLAVETTE EN COIN **3215**
CLAVETTE ET CONTRE-CLAVETTE **5938**
CLAVETTE EVIDEE **6542**
CLAVETTE LONGITUDINALE **7276**
CLAVETTE NOYEE **12458**
CLAVETTE PLATE **5388**
CLAVETTE POSEE A PLAT **5388**
CLAVETTE TANGENTIELLE **12618**
CLAVETTE TRANSVERSALE **3215**
CLAVETTE WOODRUFF **13929**
CLAVETTE-CLEF **3212**
CLAVETTE-DISQUE **13929**
CLE **3212**
CLE A DOUILLE **11709**

CLE DYNAMOMETRIQUE **13049**
CLE EN MAIN (AFFAIRE) **13284**
CLEF **3212**
CLEF A CLIQUET **10224**
CLEF A CROCHET **1822**
CLEF A DOUILLE **1515**
CLEF A ECROUS **11833**
CLEF A FOURCHE **7284**
CLEF A FOURCHE DOUBLE **4167**
CLEF A FOURCHE SIMPLE **11499**
CLEF A GRIFFES EN BOUT **9322**
CLEF A MOLETTE **2551**
CLEF A TUBES **9366**
CLEF AJUSTABLE **11052**
CLEF ANGLAISE **11052**
CLEF COUDEE **1195**
CLEF CROCODILE **9366**
CLEF D'ESSIEU **883**
CLEF DE CALIBRE **7284**
CLEF DE CALIBRE DOUBLE **4167**
CLEF DE CALIBRE SIMPLE **11499**
CLEF DE CHAPEAU **883**
CLEF DE MANOEUVRE (POUR ROBINET A BOISSEAU) **2618**
CLEF DE ROBINET **9532**
CLEF EN BOUT **1515**
CLEF TUBULAIRE **1515**
CLEF TUBULAIRE COURBE **509**
CLICHE PHOTOGRAPHIQUE **8524**
CLICHE TYPOGRAPHIQUE **1340**
CLIGNOTANTS **5366**
CLIGNOTANTS **3989**
CLIMATISEUR **250**
CLINQUANT D'EPAISSEUR **7621**
CLIQUET **2496**
CLIQUET (MECANISME POUR TRANSFORMER UN MOUVEMENT
 ALTERNATIF EN MOUVEMENT CONTINU) **10218**
CLIQUET A RESSORT **11987**
CLIQUET D'ARRET **2094**
CLIQUET DE FROTTEMENT **5675**
CLIQUET POUR PERCER **10219**
CLIVAGE **2487**
CLOISON **4110**
CLOISON **1726**
CLOISON **3870**
CLOISON PARE-FEU **5251**
CLOQUAGE **1330**
CLOTURE **5094**
CLOU **8491**
CLOU A BATEAU **1382**
CLOU A CROCHET **2457**
CLOU DECOUPE **7841**

CLOU FONDU 2051
CLOU FORGE 13990
CLOU SEMENCE 12588
CLOUER 8492
CLOUERE 6325
CLOUIERE 6325
CLOUS DE CARROSSERIE 2554
CLOUS DE FER A CHEVAL 6601
CLOUTERE 6325
CLOUTIERE 6325
CLOUURE 10639
CLOUURE SIMPLE 11487
CLOUVIERE 6325
COAGULATION 2556
COAGULER (SE) 2555
COAGULUM 2557
COALESCENCE 2572
COAXIAL 2605
COBALT 2607
COBALT-CARBONYLE 2608
COBALTINE 2614
CODAGE 4716
CODE BINAIRE DECIMAL 1248
CODE DE VITESSE D'AVANCE 5085
CODE DECIMAL 3661
CODEUR NUMERIQUE 3943
COEFFICIENT 2621
COEFFICIENT BINOMIAL 1257
COEFFICIENT CALORIFIQUE (OU DE CONDUCTIVITE THERMIQUE) 7269
COEFFICIENT D'ABSORPTION APPARENT 46
COEFFICIENT D'ABSORPTION MASSIQUE 8030
COEFFICIENT D'ECOULEMENT 2626
COEFFICIENT D'EFFET THERMOELECTRIQUE COUPLE THERMOELECTRIQUE 2637
COEFFICIENT D'ELASTICITE DE CISAILLEMENT 5974
COEFFICIENT DE CONDUCTIBILITE CALORIFIQUE 6373
COEFFICIENT DE CONDUCTIBILITE CALORIFIQUE 2622
COEFFICIENT DE CONTRACTION 2623
COEFFICIENT DE CORROSION 2624
COEFFICIENT DE CORROSION ANODIQUE 594
COEFFICIENT DE DIFFUSION 3931
COEFFICIENT DE DILATATION 2628
COEFFICIENT DE DILATATION 2636
COEFFICIENT DE DILATATION CUBIQUE 2625
COEFFICIENT DE DILATATION LINEAIRE 2630
COEFFICIENT DE DILATATION SUPERFICIELLE 2635
COEFFICIENT DE DILATATION THERMIQUE 12809
COEFFICIENT DE DILATION 4914
COEFFICIENT DE FROTTEMENT 2629
COEFFICIENT DE FROTTEMENT 5670

COEFFICIENT DE GLISSEMENT 2634
COEFFICIENT DE PROPORTIONNALITE 2631
COEFFICIENT DE REGULARITE 3719
COEFFICIENT DE RESISTANCE 2632
COEFFICIENT DE RESISTANCE A L'ALLONGEMENT 8340
COEFFICIENT DE RESISTIVITE ELECTRIQUE 2627
COEFFICIENT DE RETRAIT 477
COEFFICIENT DE ROULEMENT 2633
COEFFICIENT DE SECURITE 10885
COEFFICIENT DE SECURITE 5009
COEFFICIENT DE SENSIBILITE 3723
COEFFICIENT DE TRAINEE 4200
COEFFICIENT DE VITESSE 2638
COEFFICIENT DIFFERENTIEL 3909
COERCITIF 2640
COFFRAGE GLISSANT (POUR BETON) 11616
COFFRE 1458
COFFRE 13239
COGNEE 11572
COGNEMENT 7330
COHERENCE 2643
COHESION 2644
COIN 185
COIN 13739
COIN 13731
COIN D'EPAISSEUR 185
COIN DU RABOT 13733
COIN PLAT 5400
COIN POUR LE RATTRAPAGE DU JEU 185
COINCEMENT 7212
COINCEMENT DE LA CORDE DANS LA GORGE 1278
COINCER (SE) 7210
COKE 2655
COKE D'USINE A GAZ 5849
COKE DE CUBILOT 5595
COKE DE FOUR 2662
COKE DE GAZ 5849
COKE DENSE 2792
COKE METALLURGIQUE 2662
COKEFACTION 2666
COKEFACTION 1975
COKEFICATION 2666
COKEFIER 2656
COL 8512
COL DE CYGNE 1194
COL DE CYGNE 6009
COL DE CYGNE 12535
COLCOTHAR 7148
COLLAGE 7351
COLLAGE 5987
COLLE 5984

COLLE A NOYAUX 3151
COLLE D'AMIDON 12112
COLLE DE POISSON 7174
COLLE DES PEAUX 11539
COLLE FORTE DES OS 1449
COLLE POUR COURROIES 2140
COLLECTEUR 11618
COLLECTEUR 10839
COLLECTEUR 7979
COLLECTEUR D'ADMISSION 7007
COLLECTEUR D'ADMISSION 6945
COLLECTEUR D'ECHAPPEMENT 4893
COLLECTEUR DE GAZ 5842
COLLER 5985
COLLERETTE 9803
COLLET 12904
COLLET D'UN ARBRE 11384
COLLET DE T.H. 7998
COLLET DE TROU D'HOMME 7978
COLLIER 2448
COLLIER 2720
COLLIER A VIS POUR TUYAUX FLEXIBLES 6611
COLLIER D'EXCENTRIQUE 4435
COLLIER DE FIXATION (POUR TUBE) 9344
COLLIER DE FREIN 1544
COLLIER DE SERRAGE (POUR TUYAU) 6610
COLLIER DE TUBE 13324
COLLIER POSE A CHAUD 11391
COLLIER POUR ARBRES VERTICAUX 13547
COLLIER POUR SCELLEMENT DANS LES MURS 9345
COLLIER POUR TUBES 9345
COLLIMATER 2730
COLLIMATEUR 2732
COLLIMATION 2731
COLLISION (DE NEUTRONS AVEC DES ATOMES) 7327
COLLODION 2733
COLLOIDAL 2735
COLLOIDE 2734
COLMATAGE 2510
COLONNE 2748
COLONNE BAROMETRIQUE 2751
COLONNE D'EAU 2752
COLONNE DE DIRECTION 12228
COLONNE DE MANOEUVRE 5448
COLONNE DE MERCURE 2751
COLONNE LIQUIDE 2750
COLONNE MONTANTE 2753
COLONNE MONTANTE 12093
COLONNE MONTANTE 10614
COLONNE PERDUE 7620
COLONNE PERDUE A FENTES 11639

COLONNE PERDUE A TROUS 9187
COLONNE PERDUE FORABLE 4272
COLONNE PERDUE NON-PERFOREE 1306
COLOPHANE 2737
COLORANT 4397
COLORANT 9303
COLORATION 2740
COLORATION THERMIQUE 6371
COMBINAISON (MATH.) 2761
COMBINAISON ALIPHATIQUE 326
COMBINAISON CHIMIQUE 2300
COMBURANT 277
COMBUSTIBILITE 2773
COMBUSTIBLE 2772
COMBUSTIBLE 5705
COMBUSTIBLE ARTIFICIEL 732
COMBUSTIBLE DE TRES BONNE QUALITE 6484
COMBUSTIBLE EN GROS MORCEAUX 7827
COMBUSTIBLE FOSSILE 5586
COMBUSTIBLE GAZEUX 5854
COMBUSTIBLE LIQUIDE 7650
COMBUSTIBLE NATUREL 8503
COMBUSTIBLE NECESSAIRE 3745
COMBUSTIBLE SOLIDE 11785
COMBUSTION 2774
COMBUSTION 11956
COMBUSTION COMPLETE 2779
COMBUSTION INCOMPLETE 6845
COMBUSTION LENTE 11644
COMBUSTION NUCLEAIRE 1747
COMBUSTION PARFAITE 9184
COMBUSTION VIVE 10205
COMMAMDE ELECTRIQUE 4503
COMMAMDE PAR VIS MOLETEE 7337
COMMANDE 4297
COMMANDE 3048
COMMANDE 4309
COMMANDE A DISTANCE 10448
COMMANDE A LA MAIN 6233
COMMANDE A LA MAIN (A) 6241
COMMANDE ADAPTIVE 161
COMMANDE CONTINUE 3032
COMMANDE DE CONTOURNAGE 3039
COMMANDE DE DEMARREUR 12117
COMMANDE DE POSITIONNEMENT 9650
COMMANDE DIRECTE PAR CALCULATEUR 3981
COMMANDE DIRECTE PAR CALCULATEUR 3985
COMMANDE DIRECTE PAR CALCULATEUR 2875
COMMANDE ELASTIQUE 8642
COMMANDE HYDRAULIQUE 6683
COMMANDE INDIVIDUELLE 6886

COMMANDE MECANIQUE **9734**
COMMANDE MECANIQUE **9654**
COMMANDE MECANIQUE **8119**
COMMANDE PAR ARBRE DE TRANSMISSION **7609**
COMMANDE PAR CABLE **10735**
COMMANDE PAR CHAINE **2206**
COMMANDE PAR CHAINE **2208**
COMMANDE PAR COURROIE **4319**
COMMANDE PAR ENGRENAGE **5892**
COMMANDE PAR ENGRENAGE CONIQUE **1230**
COMMANDE PAR ENGRENAGE DROIT **12010**
COMMANDE PAR ENGRENAGES **13025**
COMMANDE PAR GROUPES **6150**
COMMANDE PAR MOTEUR **8119**
COMMANDE PAR MOTEUR **9734**
COMMANDE PAR ROUE DENTEE **13025**
COMMANDE PARAXIALE **12314**
COMMANDE PNEUMATIQUE **2842**
COMMANDE POINT-A-POINT **9572**
COMMANDE POSITIVE **9654**
COMMANDER **3049**
COMMANDER **4289**
COMMERCIALE **12141**
COMMETTAGE **10739**
COMMETTRE LES TORONS **13303**
COMMOTION **4529**
COMMUTATEUR **2244**
COMMUTATEUR DES BOUGIES DE PRECHAUFFAGE **6389**
COMMUTATION **2246**
COMMUTATRICE **3071**
COMMUTER **2241**
COMPACT TERMINE **8551**
COMPARAISON DE DIMENSIONS **7445**
COMPARATEUR **2794**
COMPARATEUR A CADRAN POUR ALESAGES **3838**
COMPARATEUR A LAMES **10365**
COMPARATEUR ELECTRIQUE **4535**
COMPARATEUR ELECTRONIQUE **4632**
COMPARATEUR INTERFERENTIEL **7038**
COMPARATEUR MECANIQUE **8097**
COMPARATEUR MICROMETRIQUE POUR FILETAGES **11048**
COMPARATEUR OPTICO-MECANIQUE **8118**
COMPARATEUR OPTIQUE **8842**
COMPARATEURS **2795**
COMPARTIMENT **2798**
COMPARTIMENT MOTEUR **4757**
COMPAS **4100**
COMPAS **9022**
COMPAS A CHEVEU **6201**
COMPAS A COULISSE **13537**
COMPAS A CREMAILLERE **10122**

COMPAS A PIECES DE RECHANGE **2805**
COMPAS A POINTES SECHES **4101**
COMPAS A PORTE-CRAYON **2804**
COMPAS A QUART DE CERCLE **13876**
COMPAS A TIRE-LIGNE **2803**
COMPAS A TROIS BRANCHES **13187**
COMPAS A VERGE **13105**
COMPAS A VERGE **1096**
COMPAS D'EPAISSEUR **1873**
COMPAS D'EPAISSEUR **8912**
COMPAS DE CALIBRE **1874**
COMPAS DE DIAMETRE INTERIEUR **6975**
COMPAS DE PRECISION **6201**
COMPAS DE REDUCTION **9937**
COMPAS DROIT A POINTES **4101**
COMPAS-BALUSTRE **1503**
COMPATIBILITE **2806**
COMPENSATEUR **2808**
COMPENSATEUR DE DILATATION **4913**
COMPENSATION **2809**
COMPENSER **2807**
COMPENSER **930**
COMPENSER LE JEU D'UN COUSSINET **12603**
COMPLEMENT D'UN ANGLE **2810**
COMPLEMENT DIAMETRAL **3851**
COMPOSANT **2815**
COMPOSANT **2981**
COMPOSANT (CHIM.) **2983**
COMPOSANT D'UN ALLIAGE **2982**
COMPOSANTE **2817**
COMPOSANTE AXIALE **859**
COMPOSANTE RADIALE **10132**
COMPOSE INTERMETALLIQUE **7052**
COMPOSE ISOLANT **6997**
COMPOSES AROMATIQUES **716**
COMPOSITION **2823**
COMPOSITION CHIMIQUE **2299**
COMPOSITION CHIMIQUE **2827**
COMPOSITION DE DEGRAISSAGE **3713**
COMPOSITION DE POLISSAGE A LA MEULE **1692**
COMPOSITION DES FORCES **2825**
COMPOSITION GRANULOMETRIQUE **9090**
COMPOSITION POUR COUSSINET **1106**
COMPRESSEUR **2874**
COMPRESSEUR A GRAISSE **11982**
COMPRESSEUR A PISTON **9374**
COMPRESSEUR CENTRIFUGE **2178**
COMPRESSEUR COMPOUND **2836**
COMPRESSEUR D'AIR **249**
COMPRESSEUR DE RESERVE **12069**
COMPRESSEUR ETAGE **2836**

COMPRESSEUR HUMIDE **13801**
COMPRESSEUR HYDRAULIQUE **6680**
COMPRESSEUR SEC **4345**
COMPRESSIBILITE **2852**
COMPRESSIBILITE **2851**
COMPRESSIBLE **2853**
COMPRESSION **2854**
COMPRESSION **9806**
COMPRESSION **11355**
COMPRESSION **10028**
COMPRESSION DANS LA DIRECTION DES FIBRES **10021**
COMPRESSION DE L'AIR **2858**
COMPRESSION NORMALEMENT A LA DIRECTION DES FIBRES **10022**
COMPRIME **2791**
COMPRIMER **2837**
COMPRIMER UN RESSORT **2838**
COMPTE-GOUTTES **4321**
COMPTE-SECONDES **12946**
COMPTE-SECONDES **12283**
COMPTE-TOURS **12587**
COMPTE-TOURS **10556**
COMPTEUR **8214**
COMPTEUR **3229**
COMPTEUR A CHIFFRES **3559**
COMPTEUR A EAU **13661**
COMPTEUR A GAZ **5829**
COMPTEUR A POINTAGES **12946**
COMPTEUR D'EAU VENTURI **13532**
COMPTEUR D'ELECTRICITE **4553**
COMPTEUR D'ERREURS **4830**
COMPTEUR D'ERREURS **4823**
COMPTEUR DE TOURS **10556**
COMPTEUR DE VAPEUR **12174**
COMPTEUR DE VITESSE **11877**
COMPTEUR GEIGER **5911**
COMPTEUR VENTURI **13532**
COMPTEURS A FENETRES **3559**
CONCASSAGE **1588**
CONCASSAGE **2786**
CONCASSER **1580**
CONCASSEUR **4045**
CONCASSEUR **3419**
CONCASSEUR A CHARBON **2559**
CONCASSEUR A COKE **2658**
CONCASSEUR A MACHOIRES **12273**
CONCAVE **2877**
CONCAVITE **2880**
CONCENTRATION **2886**
CONCENTRATION D'IONS **7122**
CONCENTRATION D'UNE SOLUTION PAR EVAPORATION **2885**

CONCENTRATION DE LA CONTRAINTE **12360**
CONCENTRATION DES IONS HYDROGENE **6715**
CONCENTRE **2881**
CONCENTRER **2882**
CONCENTRER (SE) **1122**
CONCENTRICITE **2891**
CONCEPTION **3794**
CONCEPTION STANDARD **12073**
CONCESSIONNAIRE DE LICENCE **7540**
CONCHOIDE **2892**
CONCRETION CALCAIRE **1837**
CONDENSAT **2905**
CONDENSATEUR **1916**
CONDENSATEUR (ELECTR.) **2908**
CONDENSATEUR D'ARRET **1342**
CONDENSATEUR DE BLOCAGE **1342**
CONDENSATEUR ELECTROLYTIQUE **4590**
CONDENSATEUR-ETALON **1868**
CONDENSATION **2906**
CONDENSER (SE) **2907**
CONDENSEUR (CHIM.) **2908**
CONDENSEUR (DE VAPEUR) **1916**
CONDENSEUR (OPT.) **2908**
CONDENSEUR A EVAPORATION **4868**
CONDENSEUR BAROMETRIQUE **1010**
CONDENSEUR PAR EJECTION **4470**
CONDENSEUR PAR INJECTION **7220**
CONDENSEUR PAR MELANGE **3988**
CONDENSEUR PAR SURFACE **12499**
CONDENSEUR PAR SURFACE AU MOYEN DE L'AIR **296**
CONDENSEUR PAR SURFACE AU MOYEN DE L'EAU **13684**
CONDITION CRITIQUE DE DEFORMATION **5935**
CONDITIONNEMENT **8995**
CONDITIONS D'EMPLOI **11208**
CONDITIONS DE LIVRAISON **3737**
CONDITIONS DE SERVICE **8834**
CONDITIONS DE SERVICE **13949**
CONDUCTANCE MOL(ECUL)AIRE **8350**
CONDUCTEUR (DE LA CHALEUR DE L'ELECTRICITE) **2918**
CONDUCTEUR (ELECTR.) **2921**
CONDUCTEUR D'ELECRICITE **2922**
CONDUCTEUR DE LA CHALEUR **6340**
CONDUCTEUR DE MACHINE **4765**
CONDUCTEUR DE RESEAU **4079**
CONDUCTEUR ELECTRIQUE **4537**
CONDUCTEUR PARCOURU **7670**
CONDUCTEUR SANS COURANT **3627**
CONDUCTEUR SANS TENSION **3627**
CONDUCTIBILITE **2919**
CONDUCTIBILITE ELECTRIQUE **4536**
CONDUCTIBILITE EQUIVALENTE **4797**

CONDUCTIBILITE THERMIQUE CALORIFIQUE **2920**
CONDUCTION **2916**
CONDUCTION DE LA CHALEUR **2917**
CONDUCTIVITE **2919**
CONDUCTIVITE THERMIQUE **12804**
CONDUCTIVITE THERMO-ELECTRIQUE **10044**
CONDUIRE (L'ELECTRICITELA CHALEUR) **2913**
CONDUIT ANNULAIRE DE VENT CHAUD **1774**
CONDUIT DE PRESSION **9827**
CONDUITE **12178**
CONDUITE **12226**
CONDUITE **2845**
CONDUITE **9357**
CONDUITE **13666**
CONDUITE (D'UNE MACHINE) **814**
CONDUITE ASCENDANTE **756**
CONDUITE CIRCULAIRE **2418**
CONDUITE D'AIR **281**
CONDUITE D'ARRIVEE DU COMBUSTIBLE **12489**
CONDUITE D'EAU CHAUDE **6649**
CONDUITE D'EAU FORCEE **9840**
CONDUITE D'EAU SOUS PRESSION **9840**
CONDUITE DE DERIVATION **1818**
CONDUITE DE GAZ **5834**
CONDUITE DE REFOULEMENT **4022**
CONDUITE DE VENT **1314**
CONDUITE DE VENT **1367**
CONDUITE DE VENT CHAUD **6651**
CONDUITE DESCENDANTE **3791**
CONDUITE EN FONTE **2062**
CONDUITE EN MER **11080**
CONDUITE INTERIEURE **10902**
CONDUITE LIBRE **8807**
CONDUITE POUR HAUTE PRESSION **6490**
CONDUITE SOUTERRAINE **13354**
CONDUITE VERTICALE **2753**
CONE (GEOM.) **2924**
CONE A BASE CIRCULAIRE **2409**
CONE A ETAGES **11871**
CONE AUTO-DEMONTABLE **11159**
CONE CIRCULAIRE **2409**
CONE COMPLEMENTAIRE **5918**
CONE D'ATTAQUE **11871**
CONE DE DEFLECTION (DANS RESERVOIR) **11924**
CONE DE FRICTION **5671**
CONE DE FRICTION **1225**
CONE DE PIED **10718**
CONE DE REDUCTION **10358**
CONE DE TRANSMISSION **11871**
CONE DROIT **10578**
CONE ET CONTRE-CONE **404**

CONE GALOPIN **2941**
CONE GENERATEUR **5918**
CONE LISSE **12669**
CONE LUMINEUX **2932**
CONE OBLIQUE **8709**
CONE PRIMITIF **9406**
CONE PYROMETRIQUE **10050**
CONE PYROMETRIQUE (DE SEGER) **11134**
CONE TRONQUE **13235**
CONE VERTICAL **10578**
CONFECTION **8028**
CONFECTION **10462**
CONFOCAL **2933**
CONGE **5161**
CONGE **8512**
CONGE DE RACCORDEMENT **6537**
CONGELATION DU FOUR **5655**
CONGLOMERAT **2934**
CONGRUENCE **2937**
CONICITE **2948**
CONICITE **12646**
CONICITE D'UNE CLAVETTE **12654**
CONIQUE **2940**
CONJONCTEUR-DISJONCTEUR **7949**
CONJUGUE **2952**
CONSERVABILITE **4385**
CONSERVATION DE L'ENERGIE **2967**
CONSERVATION DU BOIS **9790**
CONSISTANCE **2968**
CONSISTANCE D'UNE HUILE **1395**
CONSOLE **13618**
CONSOLE **1522**
CONSOLIDATION DES BOULONS **7707**
CONSOMMATION **13710**
CONSOMMATION **2995**
CONSOMMATION D'ENERGIE **2876**
CONSOMMATION DE COMBUSTIBLE **2994**
CONSOMMATION DE COURANT **2993**
CONSTANTAN **2980**
CONSTANTE **2969**
CONSTANTE CALORIFIQUE **6341**
CONSTANTE CAPILLAIRE **1923**
CONSTANTE D'ABSORPTION **47**
CONSTANTE D'EQUILIBRE **4787**
CONSTANTE D'UN GAZ **5815**
CONSTANTE D'UNE CELLULE ELECTROLYTIQUE **2128**
CONSTANTE DE RADIATION **10144**
CONSTANTE DE RADIATION APPARENTE **643**
CONSTANTE DE TEMPS **12944**
CONSTANTE DIELECTRIQUE **3899**
CONSTANTES D'ELASTICITE **4475**

CONVERSION 3067
CONVERTISSEUR 3071
CONVERTISSEUR BESSEMER 1212
CONVEXE 3076
CONVEXION 3060
CONVEXITE 3396
CONVEXITE 1720
CONVOYEUR 1675
COORDONNEES 3100
COORDONNEES CARTESIENNES 2014
COORDONNEES COURANTES 3487
COORDONNEES CYLINDRIQUES 3580
COORDONNEES DANS L'ESPACE 13203
COORDONNEES OBLIQUES 8710
COORDONNEES ORTHOGONALES 10301
COORDONNEES PLANES 698
COORDONNEES POLAIRES 9578
COORDONNEES RECTANGULAIRES 10301
COPAHU 3101
COPAL 3102
COPALE 3102
COPEAU 2346
COPEAU 2343
COPEAU 12536
COPEAU ADHERENT 1707
COPEAU DE BURINAGE 2343
COPEAUX 3543
COPEAUX 11278
COPEAUX (DE BOIS) 13919
COPEAUX DE FIBRES DE BOIS 13923
COPEAUX DE FORAGE 1476
COPEAUX METALLIQUES 2348
COPIE (PHOT.) 9870
COPPERWELD 3137
COQUE 7313
COQUILLE 3883
COQUILLE 2331
COQUILLE 11303
COQUILLE 9207
COQUILLE DE COUSSINET 1113
COQUILLE DE SECHAGE 3148
COQUILLE INFERIEURE 1487
COQUILLE SUPERIEURE 13030
CORBEAU 3140
CORBEILLE DE TREMPAGE 3976
CORDAGE 10745
CORDE 10730
CORDE A PIANO 9287
CORDE D'UN CERCLE 2364
CORDE DE CHANVRE 6443
CORDE EN MANILLE 7981

CORDE GOUDRONNEE 12688
CORDE NOIRE 12688
CORDE NON GOUDRONNEE 13839
CORDE TENDUE 12923
CORDE TRESSEE 1527
CORDE-COURROIE 4314
CORDON D'AMIANTE 751
CORDON D'ECRASEMENT 3382
CORDON D'UN CABLE 12338
CORDON DE SOUDURE 13756
CORDON DE SOUDURE 1082
CORDON DE SOUDURE PLAT 5401
CORDON SOUDE RECTILIGNE 12376
CORDON SUPPORT (A L'ENVERS) 904
CORINDON 4699
CORINDON 3201
CORNE D'EMBOUTISSAGE 4408
CORNE D'ENCLUME 1090
CORNE RAPEE 6595
CORNIERE 506
CORNIERE 514
CORNIERE A AILES EGALES 4781
CORNIERE A AILES INEGALES 13356
CORNIERE DE PIED DE BAC 1490
CORNIERE DE TETE(D'UN RESERVOIR) 13028
CORNIERE SUPPORT 3058
CORNIERES A BOUDIN 1713
CORNIERES DE FER AU PAQUET 7137
CORNIERES EN ACIER ALLIE 369
CORNUE 10523
CORNUE TUBULEE 10525
CORPS 1385
CORPS AU REPOS 1386
CORPS CREUX 6538
CORPS CYLINDRIQUE D'UNE CHAUDIERE 1419
CORPS D'UNE CLAVETTE 11262
CORPS D'UNE SOUPAPE 13454
CORPS DE L'ESTAMPE 3885
CORPS DE LA BIELLE 11261
CORPS DE POMPE 1015
CORPS DE REVOLUTION 11790
CORPS DU MARTEAU 6317
CORPS DU PALIER 9313
CORPS DU PISTON 9373
CORPS DU RIVET 11263
CORPS ECLAIRE 6773
CORPS EN MOUVEMENT 1392
CORPS ETRANGERS 6813
CORPS FLOTTANT 5436
CORPS GAZEUX 5808
CORPS GRIS 6083

CORPS LIQUIDE **7641**
CORPS LUMINEUX **11147**
CORPS MOBILE **1392**
CORPS NOIR **1285**
CORPS PLASTIQUE **9467**
CORPS RIGIDE **10588**
CORPS SIMPLE **4654**
CORPS SOLIDE **11775**
CORRECTEUR D'AVANCE **11837**
CORRECTION **3181**
CORRECTION D'OUTIL **3521**
CORRECTION DE POSITION D'OUTIL **13007**
CORRECTION DE VITESSE D'AVANCE **5079**
CORRIGER L'ACIER PAR UN REVENU **6808**
CORRODANT **3197**
CORRODER **3182**
CORROSIF **3197**
CORROSIF **3196**
CORROSIF **4847**
CORROSION **3183**
CORROSION ATMOSPHERIQUE **795**
CORROSION AVEC DEGAGEMENT D'HYDROGENE **6718**
CORROSION CATHODIQUE **2107**
CORROSION CHIMIQUE **2302**
CORROSION DE LA PILE DE CONCENTRATION **2888**
CORROSION DE SOUDURE **13759**
CORROSION DES FACES EN CONTACT **5667**
CORROSION DUE A DES DEPOTS **3774**
CORROSION FISSURANTE **3185**
CORROSION FISSURANTE **7041**
CORROSION GALVANIQUE **5783**
CORROSION HUMIDE **13802**
CORROSION INTERCRISTALLINE **7030**
CORROSION INTERCRISTALLINE **7041**
CORROSION INTERCRISTALLINE DES LAITONS 70/30 **11093**
CORROSION INTERGRANULAIRE **7030**
CORROSION LOCALE **7695**
CORROSION LOCALISEE **9418**
CORROSION PAR CONTACT **3002**
CORROSION PAR COURANT VAGABOND **12350**
CORROSION PAR FROTTEMENT **5666**
CORROSION PAR INFILTRATION STRATIFIEE **7443**
CORROSION SOUS EROSION **6799**
CORROSION SOUS TENSION **12361**
CORROSION SOUS TENSION **3195**
CORROSION SOUS-JACENTE **13352**
COSECANTE **3202**
COSINUS **3203**
COSMOLINE **13508**
COSSE **12848**
COSSE **1827**

COSSE OVALE **6330**
COSSES **12765**
COTANGENTE **3211**
COTATION ABSOLUE **3097**
COTATION ABSOLUE **29**
COTATION RELATIVE **6854**
COTE **3956**
COTE CHAIR DU CUIR **5414**
COTE D INTRODUCTION **4774**
COTE D'ENTREE **4774**
COTE D'UN ANGLE **7512**
COTE D'UN POLYGONE **11414**
COTE DE BASE **1053**
COTE DE L'ANGLE DROIT **9218**
COTE DU VENT **13721**
COTE EXTERIEUR **8911**
COTE FLEUR DU CUIR **6038**
COTE INTERIEUR **6974**
COTE INTERIEUR D'UNE COURROIE **10853**
COTE POIL DU CUIR **6038**
COTE SOUS LE VENT **7505**
COTES DE RACCORDEMENT **2793**
COTON **3217**
COTON A NETTOYER **3223**
COTONPOUDRE **6176**
COUCHE **5166**
COUCHE **7442**
COUCHE **254**
COUCHE **10832**
COUCHE **9096**
COUCHE (DE PROTECTION) ANODIQUE **599**
COUCHE ANNUELLE **586**
COUCHE D'AIR **2124**
COUCHE D'OXYDE DE CUIVRE **5246**
COUCHE D'OXYDE FERREUX **12443**
COUCHE DE BASE **9864**
COUCHE DE BEILBY **1130**
COUCHE DE CHROMATE **2591**
COUCHE DE COULEUR A L'HUILE **2582**
COUCHE DE DEMI-ATTENUATION **6211**
COUCHE DE FOND **9857**
COUCHE DE LAQUE **2581**
COUCHE DE PHOSPHATE **9245**
COUCHE DES FIBRES INVARIABLES **8544**
COUCHE INTERMEDIAIRE DECARBUREE **1005**
COUCHE PRIMAIRE **9857**
COUCHE PROTECTRICE **9793**
COUCHE PROTECTRICE **9950**
COUCHE SUPERFICIELLE **11544**
COUCHE SUPERFICIELLE **2019**
COUCHE SUPERFICIELLE (DE PEINTURE) **5217**

COUCHES **2590**
COUCHETTE **1739**
COUDE **1194**
COUDE **1172**
COUDE **4913**
COUDE **4493**
COUDE A 180 **13323**
COUDE AU 1/2 **4467**
COUDE D'EQUERRE **7320**
COUDE D'UN ARBRE **6976**
COUDE EN U **13323**
COUDE ROND **1180**
COUDE ROND A ANGLE DROIT **10074**
COUDE ROND AU 1/4 **10074**
COUDER UN TUYAU **1175**
COULABILITE **2064**
COULAGE **2068**
COULAGE (D'UN LIQUIDE) **7492**
COULAGE DE CIMENT **6154**
COULEE **5592**
COULEE **8134**
COULEE **6333**
COULEE **12703**
COULEE **9715**
COULEE **10844**
COULEE **10845**
COULEE **2068**
COULEE (LOT DE...) **2076**
COULEE (OPERATION) **2070**
COULEE CENTRIFUGE **2176**
COULEE CONTINUE **3019**
COULEE CONTINUE **3029**
COULEE DIRECTE **13031**
COULEE EN COQUILLE **2334**
COULEE EN COQUILLE **9208**
COULEE EN FOSSE **9394**
COULEE EN SABLE **10922**
COULEE EN SABLE SEC **4353**
COULEE EN SOURCE **1488**
COULEE RATEE **8732**
COULEE SOUS PRESSION **9812**
COULEE SOUS PRESSION **3886**
COULEE SOUS PRESSION PAR GRAVITE **6079**
COULEE TRANQUILLE **4388**
COULEE TRANQUILLE (PROCEDE DURVILLE) **10109**
COULER **6153**
COULER **2030**
COULEUR **7362**
COULEUR **9303**
COULEUR **12732**
COULEUR **9015**

COULEUR (PHYS.) **2741**
COULEUR CERISE **2326**
COULEUR COMPLEMENTAIRE **2812**
COULEUR COMPOSEE **11103**
COULEUR D'INTERFERENCE **7037**
COULEUR OPAQUE **1389**
COULEUR SIMPLE **9847**
COULEURS SPECTRALES **11859**
COULIS **6152**
COULIS **10406**
COULIS REFRACTAIRE **6125**
COULIS REFRACTAIRE **6126**
COULISSE **11633**
COULISSE ET MANIVELLE **11631**
COULISSEAU **10188**
COULISSEAU **11585**
COULISSEAU D'OBTURATION **5160**
COULISSEAU MOBILE **7625**
COULISSER **11586**
COULOIR **2395**
COULOMB **3225**
COULOMBMETRE **452**
COULURE **7489**
COULURE POCHE **10892**
COULURES **10895**
COULURES **10856**
COUP **12375**
COUP DE BELIER **13653**
COUP DE PISTON **9389**
COUP DE SOUDURE POINT D'AMORCAGE DE L'ARC **677**
COUPAGE **3523**
COUPAGE A CHAUD **6622**
COUPAGE A L'ACETYLENE **94**
COUPAGE A L'ARC AU CARBONE **1942**
COUPAGE A L'ARC METALLIQUE **8175**
COUPAGE A L'AUTOGENE **5817**
COUPAGE A LA FLAMME **5318**
COUPAGE AU CHALUMEAU **5817**
COUPE **1309**
COUPE **9898**
COUPE **3503**
COUPE **11122**
COUPE **12511**
COUPE **7717**
COUPE **3523**
COUPE A LA MAIN **6232**
COUPE ANGULAIRE **539**
COUPE CHARDONS ET FOUGERES **13741**
COUPE D'UN DESSIN **11112**
COUPE INFERIEURE A LA COTE **13351**
COUPE LONGITUDINALE **7738**

COUPE MICROGRAPHIQUE **8261**
COUPE PAR ELECTRO-EROSION **4559**
COUPE TRANSVERSALE **3377**
COUPE TRANSVERSALE **11111**
COUPE-CIRCUIT DE SURETE **10886**
COUPE-FIL **13886**
COUPE-RACINES **10715**
COUPE-RAVES **7272**
COUPE-TUBE **13248**
COUPE-TUBES **9348**
COUPELLATION **3463**
COUPELLE DE RESSORT **11988**
COUPER (GEOM.) **3504**
COUPER (SE) (GEOM.) **7094**
COUPER AU TRANCHET **3515**
COUPER LE COURANT **7090**
COUPEROSE BLANCHE **14046**
COUPEROSE BLEUE **1376**
COUPEROSE VERTE **6093**
COUPLAGE **3254**
COUPLAGE EN QUANTITE **9051**
COUPLAGE EN SERIE **11199**
COUPLAGE EN TENSION **11199**
COUPLE **3252**
COUPLE **13048**
COUPLE DE DEMARRAGE **12129**
COUPLE DE ROUE ET VIS TANGENTE **13974**
COUPLE DE ROUES HELICOIDALES **11916**
COUPLE DE SERRAGE **12931**
COUPLE ELECTRO-CHIMIQUE **5782**
COUPLE METRE **13050**
COUPLE MOTEUR **13048**
COUPLE REACTION **8374**
COUPLE RESISTANT **8374**
COUPLE THERMO-ELECTRIQUE **12827**
COUPLE THERMOELECTRIQUE **12821**
COUPURE **3278**
COUPURE DE COURANT **8734**
COURANT **3486**
COURANT A BASSE TENSION **7777**
COURANT A FILETS PARALLELES **9055**
COURANT A HAUTE FREQUENCE **6468**
COURANT A HAUTE TENSION **6500**
COURANT ALTERNATIF **1**
COURANT ALTERNATIF **406**
COURANT ALTERNATIF HAUTE FREQUENCE **3**
COURANT CONTINU **3023**
COURANT D'AIR **253**
COURANT D'AIR FORCE **1315**
COURANT D'INDUCTION **6889**
COURANT D'UN PALAN **10851**

COURANT DE CHARGE **2266**
COURANT DE DISPERSION **7490**
COURANT DE FAIBLE INTENSITE **13705**
COURANT DE FOUCAULT **4438**
COURANT DE FUITE **7490**
COURANT DE GRANDE INTENSITE **6401**
COURANT DE LA CHAMBRE D'IONISATION **7126**
COURANT DIPHASE **13314**
COURANT DU BAIN **1064**
COURANT ELECTRIQUE **4500**
COURANT ELECTRIQUE ENERGIE **4525**
COURANT FORCE **5534**
COURANT FORT **6401**
COURANT GAZEUX **5816**
COURANT LIBRE **5634**
COURANT MONOPHASE **11501**
COURANT POLYPHASE **8462**
COURANT SECONDAIRE DE SOUDAGE **11110**
COURANT THERMOELECTRIQUE **12824**
COURANT TRIPHASE **12877**
COURANT TURBULENT **13274**
COURANT VAGABOND **12349**
COURANTS CROISES **3363**
COURANTS DE FOUCAULT **4440**
COURANTS DE MEME SENS **13360**
COURANTS DE SENS CONTRAIRE **3230**
COURBAGE **1182**
COURBE **3496**
COURBE A DOUBLE COURBURE **13307**
COURBE A FAIBLE RAYON **10102**
COURBE A GRAND RAYON **5378**
COURBE A PETIT RAYON **10102**
COURBE ADIABATIQUE **179**
COURBE APLATIE **5378**
COURBE CHARGE-ALLONGEMENT **12366**
COURBE D'UN DIAGRAMME **3495**
COURBE DANS L'ESPACE **13307**
COURBE DE CONTACT **675**
COURBE DE DETENTE **4916**
COURBE DE REFROIDISSEMENT **3087**
COURBE DE VITESSE **13513**
COURBE EFFORT-DEFORMATION **12365**
COURBE FERMEE **2520**
COURBE GAUCHE **13307**
COURBE ISOBARIQUE **7177**
COURBE LIMITE **1460**
COURBE PLANE **9435**
COURBE POLYTROPIQUE **9618**
COURBE RAIDE **10102**
COURBE ROULANTE **10697**
COURBE SINUEUSE **11476**

COURBE VIVE 10102
COURBER 1173
COURBER UN TUYAU 1175
COURBES DE TEMPERATURE 12723
COURBURE 3494
COURONNE CIRCULAIRE 589
COURONNE D'UNE ROUE 10596
COURONNE DE FIL 13898
COURONNE DENTEE 3399
COURONNE DENTEE 13023
COURONNE DIRECTRICE D'UNE TURBINE 6168
COURONNE FIXE 6168
COURONNE MOBILE 13823
COURONNE SUPPORT DE PLATEAU 12492
COURONNE-SUPPORT DE PLATEAUX 13170
COURONNEMENT 1604
COURROIE 1140
COURROIE (EPAISSEUR D'UNE) 1165
COURROIE A PLUSIEURS EPAISSEURS 8463
COURROIE A X PLIS 1166
COURROIE COLLEE 2145
COURROIE COUSUE 7350
COURROIE CROISEE 3387
COURROIE DE COMMANDE 1163
COURROIE DE SECTION TRIANGULAIRE 13511
COURROIE DE VENTILATEUR 5029
COURROIE DEMI-CROISEE 6207
COURROIE DOUBLE 4128
COURROIE DROITE 8804
COURROIE EN ACIER 12200
COURROIE EN BALATA 944
COURROIE EN BOYAUX 6183
COURROIE EN CAOUTCHOUC 10817
COURROIE EN CHANVRE 1907
COURROIE EN COTON 3218
COURROIE EN CRIN 6200
COURROIE EN CUIR 7495
COURROIE EN CUIR 7503
COURROIE EN CUIR A MAILLONS 7496
COURROIE EN FIBRES TEXTILES 4981
COURROIE EN LANIERES DE CUIR TRAVAILLANT SUR CHAMP 7376
COURROIE EN POILS 6200
COURROIE EN POILS DE CHAMEAU 1891
COURROIE EN TISSU CAOUTCHOUTE 10817
COURROIE EN TISSU TUBULAIRE 5397
COURROIE EN X EPAISSEURS 1166
COURROIE MOTRICE 1163
COURROIE MULTIPLE 8463
COURROIE OUVERTE 8804
COURROIE RENVERSEE 3387

COURROIE SIMPLE 11477
COURROIE TISSEE 13983
COURROIE TORDUE 6207
COURROIE TRAPEZOIDALE 13511
COURROIE TRAPEZOIDALE 2925
COURROIE TRAPEZOIDALE 13431
COURROIE TRIPLE 12878
COURSE 12387
COURSE A VIDE (D'UNE MACHINE-OUTIL) 10529
COURSE ALLER 5585
COURSE ASCENDANTE 13409
COURSE AVANT 5585
COURSE COMPLETE (ALLER ET RETOUR) DU PISTON 13406
COURSE D'ASPIRATION 12435
COURSE D'ECHAPPEMENT 4898
COURSE DE COMBUSTION 2775
COURSE DE COMPRESSION 2864
COURSE DE DETENTE 2775
COURSE DE L'EXCENTRIQUE 12894
COURSE DE LA CAME 12894
COURSE DE REFOULEMENT 9834
COURSE DE RETOUR 10530
COURSE DE TRAVAIL 3536
COURSE DESCENDANTE DU PISTON 4187
COURSE DIRECTE DU PISTON 5585
COURSE DU PISTON 7519
COURSE MONTANTE DU PISTON 13409
COURSE RETROGRADE DU PISTON 10530
COURSE TRANSVERSALE DE LA TABLE 12585
COURSE UTILE (D'UNE MACHINE-OUTIL) 3536
COURSE VERTICALE 13549
COURT-CIRCUIT 11372
COURT-CIRCUIT 11369
COUSSIN D'AIR 254
COUSSIN PNEUMATIQUE 256
COUSSINET 1100
COUSSINET A BILLES 954
COUSSINET A CANNELURES 12903
COUSSINET A ROTULE 12563
COUSSINET AUTO-GRAISSEUR 11158
COUSSINET DE PALIER 7941
COUSSINET DE PIED DE BIELLE 3390
COUSSINET DE TETE DE BIELLE 1242
COUSSINET EN COQUILLES 1572
COUSSINET EN DEUX PIECES 1572
COUSSINET EN UNE SEULE PIECE 1771
COUSSINET OVALISE PAR L'USURE 13980
COUSSINET USE 13980
COUSSINETS DE FILIERE 3896
COUTEAU 7325
COUTEAU 7323

COUTEAU CIRCULAIRE **2417**
COUVER **11688**
COUVERCLE **3269**
COUVERCLE **12639**
COUVERCLE A CHARNIERE **6510**
COUVERCLE A GLISSIERE **11604**
COUVERCLE D'UN ROBINET **3266**
COUVERCLE D'UNE SOUPAPE **3266**
COUVERCLE DU COFFRE **13240**
COUVERCLE DU CYLINDRE **3567**
COUVERCLE DU PALIER **1912**
COUVERCLE DU PISTON **7266**
COUVERCLE PROTECTEUR **9944**
COUVERTE **5968**
COUVERTE **615**
COUVERTURE DE SOL **5450**
COUVRE CULBUTEURS **3567**
COUVRE-OBJET **3265**
COUVRE-ROUES **13815**
COVARIANCE **3261**
CRABOTS ENCLENCHEURS/DECLENCHEURS **13214**
CRACHEMENT **11842**
CRACKING **3285**
CRAFE **2224**
CRAIE **13842**
CRAIE DE BRIANCON **11703**
CRAIE ROUGE **8720**
CRAMPON **12105**
CRAN **8672**
CRAPAUDINE **12241**
CRAPAUDINE **11905**
CRAPAUDINE A BILLES **11890**
CRAPAUDINE A PIVOT ANNULAIRE **2722**
CRAQUAGE OU CRACKING **3284**
CRAQUELURE SAISONNIERE **11092**
CRAQUELURE SUPERFICIELLE **3319**
CRASSE **4331**
CRATERE **3317**
CRAYON **7478**
CRAYON (D'UN APPAREIL ENREGISTREUR) **9161**
CRAYON DE COULEUR **2745**
CREMAILLERE **10119**
CREMAILLERE **13022**
CREMAILLERE ET PIGNON **10120**
CREME DE TARTRE **119**
CREOSOTAGE **3332**
CREOSOTE **3331**
CREPINE **12332**
CREPINE D'HUILE **8770**
CREPINE PREGRAVILLONNEE **9783**
CRETE D'UN DEVERSOIR **13754**

CRETE DE L'ONDE **3334**
CREUSAGE **11030**
CREUSEMENT D'UN FILET **11030**
CREUSER UN FILER **11046**
CREUSET **6331**
CREUSET **8142**
CREUSET (DE HAUT FOURNEAU) **3401**
CREUSET EN GRAPHITE **6064**
CREUSET EN PLATINE **9512**
CREUSET EN PORCELAINE **9626**
CREUX **2125**
CREUX DE LA DENT **6548**
CREUX DE LA SPHERE **13326**
CREUX ENTRE LES ONDES **6543**
CREUX SOUS LE PRIMITIF **3681**
CREVASSE **3278**
CRIBLAGE **11424**
CRIBLE **10572**
CRIBLE **11422**
CRIBLE A SECOUSSES **1733**
CRIBLE OSCILLANT **11252**
CRIBLE POUR LA TERRE **11628**
CRIBLE POUR LE SABLE **11628**
CRIBLER **11423**
CRIC **10121**
CRIC A LEVIER **7533**
CRIC A VIS **11035**
CRIC MECANIQUE **11035**
CRIN **12594**
CRIN DE CHEVAL **6604**
CRINS **7965**
CRIQUAGE **3284**
CRIQUE **3278**
CRIQUE **10097**
CRIQUE **2278**
CRIQUE A CHAUD **5240**
CRIQUE A CHAUD **6621**
CRIQUE DANS LE BOIS **11250**
CRIQUE DE CLIVAGE **11615**
CRIQUE DE RETRAIT **6645**
CRIQUE DE TENSION **5304**
CRIQUE DE VIEILLISSEMENT **11092**
CRIQUES D'ATTAQUE CHIMIQUE **4841**
CRISTAL **6030**
CRISTAL (MIN.) **3424**
CRISTAL (VERRE) **3428**
CRISTAL BASALTIQUE **2757**
CRISTAL CUBIQUE **3450**
CRISTAL DE ROCHE **10652**
CRISTAL HEMIEDRIQUE **6436**
CRISTAL HEMIMORPHE **6437**

CRISTAL HEXAGONAL **6459**
CRISTAL HOLOEDRE **6552**
CRISTAL IDIOMORPHE **6751**
CRISTAL INHOMOGENE **3169**
CRISTAL MONOCLINIQUE **8384**
CRISTAL OSCILLATEUR **3432**
CRISTAL RHOMBOEDRIQUE **10565**
CRISTALLIN **8688**
CRISTALLIN **3433**
CRISTALLISABILITE **3436**
CRISTALLISABLE **3437**
CRISTALLISATION **3438**
CRISTALLISATION PRIMAIRE **9849**
CRISTALLISER **3439**
CRISTALLITE **3440**
CRISTALLOGRAMME **3441**
CRISTALLOGRAPHIE **3442**
CRISTALLOIDE **3443**
CRISTAUX **119**
CRISTAUX DE SOUDE **11711**
CRISTAUX EQUIAXES **4784**
CRISTAUX MIXTES **8329**
CRISTAUX ORTHORHOMBIQUES **8874**
CROC **4974**
CROCHET **5772**
CROCHET **6568**
CROCHET A NOYAUX **3152**
CROCHET A OEIL **4974**
CROCHET A QUEUE DE COCHON **9301**
CROCHET A TETE DE BELIER **10191**
CROCHET A VIS **11057**
CROCHET D'ATTELAGE **4217**
CROCHET DE CHAINE **3822**
CROCHET DE LEVAGE **7552**
CROCHET DE SURETE **10888**
CROCHET DE SUSPENSION **12527**
CROCHET DOUBLE **10191**
CROCHET PIVOTANT **12564**
CROCHET POUR TUBES **9354**
CROCHETS **11902**
CROCHETS DE TRACTION **6307**
CROISILLONS DE CHARPENTE **1520**
CROISSANCE **6155**
CROISSANCE ANORMALE DES CRISTAUX **7**
CROISSANCE DES GRAINS **6035**
CROISSANT **7829**
CROIX **3359**
CROIX A QUATRE DIRECTIONS **3368**
CROQUER **11529**
CROQUIS **11532**
CROQUIS **11528**

CROQUIS DE PRINCIPE **3797**
CROQUIS SCHEMATIQUE **3836**
CROSSETTE **3388**
CROUTE **11624**
CROUTE **3421**
CROUTE **9155**
CROUTE DE LA FONTE **11538**
CROUTE DE LAMINAGE **11544**
CROWN-GLASS **3397**
CRUE **586**
CRYOHYDRATE **3422**
CRYOLITHE **3423**
CUBAGE **13598**
CUBAGE (EVALUATION EN UNITES CUBES) **3444**
CUBE (GEOM.) **3445**
CUBER **3446**
CUBILOT **3466**
CUBILOT **3464**
CUBIQUE A FACE CENTREE **4991**
CUCURBITE **1407**
CUIR **7494**
CUIR BRULE **2272**
CUIR BRUT **10237**
CUIR CHROME **2377**
CUIR DE BOEUF **3275**
CUIR DE VACHE **3275**
CUIR DU CROUPON **1780**
CUIR EMBOUTI **3458**
CUIR FORT **11770**
CUIR TANNE A L'ECORCE DE CHENE **8704**
CUIR VERNI **9117**
CUIR VERT **10237**
CUIVRAGE **3125**
CUIVRE (CUIVRE ROUGE) **3110**
CUIVRE A BALAI **1669**
CUIVRE A OXYDE CUIVREUX **13069**
CUIVRE A SOUFFLURES **1329**
CUIVRE AFFINE **2073**
CUIVRE AFFINE **13070**
CUIVRE AFFINE AU FEU **5250**
CUIVRE ANODIQUE **593**
CUIVRE ARGENTIFERE **11468**
CUIVRE ARSENICAL **725**
CUIVRE CEMENTATOIRE **2136**
CUIVRE DE BRASAGE **1575**
CUIVRE DE CEMENT **2136**
CUIVRE DESOXYDE **3764**
CUIVRE DU LAC SUPERIEUR **7370**
CUIVRE ELECTROLYTIQUE **2099**
CUIVRE ELECTROLYTIQUE **4591**
CUIVRE ELECTROLYTIQUE A OXYDE CUIVREUX **4610**

CUIVRE ELECTROLYTIQUE COALESCE **2571**
CUIVRE EN FEUILLES **11308**
CUIVRE EXEMPT D'OXYGENE A HAUTE CONDUCTIVITE **8983**
CUIVRE GRIS **12796**
CUIVRE JAUNE **1562**
CUIVRE NATIF **7370**
CUIVRE NOIR BRUT **1287**
CUIVRE OXYDULE **3476**
CUIVRE PANACHE **13501**
CUIVRE PHOSPHOREUX **9250**
CUIVRE PHOSPHOREUX **9253**
CUIVRE POUR GRAVURE **4767**
CUIVRE PYRITEUX **3126**
CUIVRE REGENERE **2136**
CUIVRE ROSETTE **10748**
CUIVRE SILICEUX **11446**
CUIVRE SUFURE GRIS **3118**
CUIVRE VITREUX **3118**
CUIVRER **3124**
CULASSE **3569**
CULASSE D'UN AIMANT **14017**
CULBUTEUR **10654**
CULBUTEUR DE LINGOTS **7982**
CULEE **57**
CULOT D'UNE CRAPAUDINE **12905**
CULOT DE BOUGIE **11839**
CULOTTE **1595**
CULTIVATEURS **3457**
CUNEIFORME **13737**
CUPRO-BERYLLIUM **1209**
CUPRO-NICKEL **3477**
CURAGE **11011**
CURCUMINE **3481**
CURER **11008**
CURETTE **11009**
CURSEUR **7625**
CURSEUR D'UNE BALANCE **13160**
CURVILIGNE **3499**
CURVIMETRE **3501**
CUVE **13228**
CUVE **12046**
CUVE A NIVEAU CONSTANT **5431**
CUVE DE DECANTATION **11229**
CUVE DE FOUR **9029**
CUVE DE TREMPAGE **3973**
CUVE DE TREMPE **10090**
CUVETTE **11073**
CUVETTE D'EGOUTTAGE **4283**
CUVETTE DE PLATEAUX **11074**
CUVETTE DE RETENTION D'UN RESERVOIR **12629**
CYANATE **3544**

CYANOFERRURE DE POTASSIUM **9685**
CYANOGENE **3551**
CYANOTYPE **5119**
CYANURE **3547**
CYANURE D'ARGENT **11462**
CYANURE D'OR **5997**
CYANURE DE CUIVRE **3115**
CYANURE DE POTASSIUM **9683**
CYANURE DE SODIUM **11720**
CYANURE DE ZINC **14041**
CYANURE LIBRE **5632**
CYCLE **3554**
CYCLE **3553**
CYCLE AUTOMATIQUE **839**
CYCLE DE CARNOT **2004**
CYCLE DE TRAVAIL **3553**
CYCLE FIXE **5292**
CYCLIQUE **3555**
CYCLOIDE **3557**
CYCLOIDE RACCOURCIE **3493**
CYCLOIDE RALLONGEE **9918**
CYCLONE PULVERISATEUR **3560**
CYLINDRAGE **13164**
CYLINDRE **1015**
CYLINDRE **10671**
CYLINDRE (GEOM.) **3561**
CYLINDRE A AIR **2841**
CYLINDRE A BASE CIRCULAIRE **2411**
CYLINDRE A BASE ELLIPTIQUE **4675**
CYLINDRE A BASSE PRESSION **7783**
CYLINDRE A CINTRER **1186**
CYLINDRE A FREIN **1532**
CYLINDRE A HAUTE PRESSION **6488**
CYLINDRE A RAINURE **3579**
CYLINDRE A VAPEUR **12162**
CYLINDRE A VIDE **4373**
CYLINDRE CANNELE **5486**
CYLINDRE CANNELE **6136**
CYLINDRE CANNELE **11206**
CYLINDRE CREUX **6540**
CYLINDRE D'ALIMENTATION **5074**
CYLINDRE D'EQUILIBRAGE **938**
CYLINDRE D'EQUILIBRAGE **939**
CYLINDRE DE CINTRAGE **1186**
CYLINDRE DE LEVAGE **7550**
CYLINDRE DE POMPE **9992**
CYLINDRE DE TRAVAIL **13954**
CYLINDRE DROIT **10579**
CYLINDRE EN PORTE-A-FAUX **8935**
CYLINDRE ENTRAINEUR **5074**
CYLINDRE GRADUE **3582**

CYLINDRE HYPERBOLIQUE **6736**
CYLINDRE LISSE **5391**
CYLINDRE MOTEUR **13954**
CYLINDRE OBLIQUE **8712**
CYLINDRE PARABOLIQUE **9039**
CYLINDRE RUGUEUX **10174**
CYLINDRE TOURNANT D'UN ENREGISTREUR **10290**
CYLINDREE **2262**
CYLINDREE **3448**
CYLINDREE **3566**
CYLINDRER **13162**
CYLINDRES DE SUPPORT **903**
CYLINDRES TRANSVERSAUX **1643**
CYLINDRIQUE **3577**
DALLAGE **9136**
DALLE **11551**
DALLE **5302**
DALLE DE BETON (SUPPORT D'UN RESERVOIR) **2900**
DALLE SUPPORT / SUR PIEUX **9305**
DAMAGE **10195**
DAME **12611**
DAME-JEANNE **1981**
DAMER **10189**
DAMMAR **3598**
DAMPER **3602**
DANGER D'INCENDIE **3608**
DANGEREUX **6503**
DANS LE SENS DES AIGUILLES D'UNE MONTRE **2506**
DARD **6951**
DARD DE CHALUMEAU **8975**
DARD DE LA FLAMME **2923**
DARD INTERIEUR **6952**
DASHPOT **3611**
DASHPOT A AIR **256**
DATE DE L'ACCORD DU BREVET **3617**
DATE DU DEPOT DE LA DEMANDE DE BREVET **3616**
DEBIT **10069**
DEBIT **9730**
DEBIT **8908**
DEBIT **5459**
DEBIT **8896**
DEBIT A LA SCIE DE LONG **1058**
DEBIT AU COIN DU BOIS **10075**
DEBIT D'UNE POMPE **10228**
DEBIT NORMAL **8658**
DEBIT TOTAL **13064**
DEBITER AUX DIMENSIONS VOULUES **3509**
DEBITMETRE **5458**
DEBITMETRE **13604**
DEBLOCAGE **11560**
DEBOUCHAGE **12675**

DEBOURBAGE **7537**
DEBRAYAGE **4028**
DEBRAYAGE **12898**
DEBRAYAGE **2548**
DEBRAYER **10034**
DECAGONE **3644**
DECAGRAMME **3645**
DECALAGE **541**
DECALAGE (ACTION D'OTER LES CALES) **4312**
DECALAGE A DROITE **10583**
DECALAGE A GAUCHE **7507**
DECALAGE DE PHASE **9237**
DECALAGE DES PHASES **9234**
DECALAGE DU POINT D'ORIGINE **14026**
DECALAGE DU POINT D'ORIGINE **14031**
DECALAGE EN ARRIERE **7368**
DECALAGE EN AVANCE **7483**
DECALAMINAGE **10979**
DECALAMINAGE A L'EAU **13647**
DECALAMINAGE A LA FLAMME **5313**
DECALAMINAGE AU CHALUMEAU **5310**
DECALCOMANIE **13109**
DECALE **8731**
DECALE **8730**
DECALER **4295**
DECALESCENCE **3646**
DECALITRE **3647**
DECALQUE **13082**
DECAMETRE **3648**
DECANTATION **3650**
DECANTATION **4681**
DECANTATION **11228**
DECANTER **3649**
DECAPAGE **11006**
DECAPAGE A FROID **2707**
DECAPAGE A L'ACIDE **9288**
DECAPAGE ANODIQUE **601**
DECAPAGE AU BAIN ACIDULE **112**
DECAPAGE AU GAZ **5832**
DECAPAGE CATHODIQUE **2103**
DECAPAGE CHIMIQUE **9288**
DECAPAGE ELECTROLYTIQUE **4603**
DECAPAGE MAT **3628**
DECAPAGE PAR GRENAILLAGE **11381**
DECAPER **11005**
DECARBONISATION **3651**
DECARBONISER **3652**
DECARBURATION **3655**
DECARBURATION DU FER **3653**
DECARBURER LE FER **3654**
DECAUVILLE **9639**

DECELERATION AUTOMATIQUE **841**
DECENTRE **8730**
DECHARGE **4024**
DECHARGE **4017**
DECHARGE **2395**
DECHARGE D'UN ACCUMULATEUR **4025**
DECHARGE ELECTRIQUE **4502**
DECHARGER **4018**
DECHARGER UN ACCUMULATEUR **4019**
DECHARGEURS A GRIFFES (MAT. ET GRUE) **6310**
DECHEANCE D'UN BREVET **7408**
DECHETS **13637**
DECHIRER (SE) **10859**
DECHIRURE **3278**
DECIGRAMME **3657**
DECILITRE **3658**
DECIMALE **3659**
DECIMETRE **3666**
DECIMETRE CARRE **12018**
DECIMETRE CUBE **3451**
DECLAVETAGE **7751**
DECLAVETER **7750**
DECLENCHEMENT **4031**
DECLENCHER **4030**
DECLINAISON MAGNETIQUE **3668**
DECLIQUETAGE **4031**
DECLIQUETER **4030**
DECLIVITE **3792**
DECOCHAGE **7328**
DECOCHAGE **9554**
DECOCTION **3669**
DECODEUR **3670**
DECOLLEMENT **1354**
DECOLLEMENT **9156**
DECOLLETAGE **3533**
DECOLLETER **3510**
DECOLORATION **5014**
DECOMPOSABLE **3671**
DECOMPOSER **3672**
DECOMPOSER (SE) **3673**
DECOMPOSITION **10769**
DECOMPOSITION (CHIM.) **3674**
DECOMPOSITION DE LA LUMIERE **3676**
DECOMPOSITION DES FORCES **10505**
DECOMPOSITION ELECTROLYTIQUE **4596**
DECOUPAGE **10012**
DECOUPAGE **3533**
DECOUPAGE A L'EMPORTE-PIECE **6514**
DECOUPAGE A LA MACHINE **7842**
DECOUPAGE A LA POUDRE **9719**
DECOUPAGE AUTOGENE **832**

DECOUPER **9998**
DECOUPER **3510**
DECOUPER A L'AUTOGENE **3505**
DECOUPER AU CHALUMEAU **3505**
DECOUPER DES TOLES **3513**
DECOUPOIR **6545**
DECRIQUER (AU CHALUMEAU) **10993**
DECROISSANCE **3656**
DEDOUBLEMENT **11950**
DEDOUBLER **11946**
DEFAILLANCE PRECOCE **6908**
DEFAUT **5411**
DEFAUT **3691**
DEFAUT D'ALIGNEMENT OU DENIVELLATION (D'UN JOINT SOUDE) **8737**
DEFAUT DE COULAGE **5056**
DEFAUT DE FABRICATION **3693**
DEFAUT DE MATIERE **3692**
DEFAUT DE SURFACE **12502**
DEFAUTS DE LA SOUDURE **13760**
DEFAUTS DE PEINTURE **9021**
DEFECTUEUX **5057**
DEFLECTEUR **921**
DEFLECTEUR **13531**
DEFLECTEUR **3702**
DEFLECTION **3699**
DEFONCEMENT **3760**
DEFORMABILITE **3705**
DEFORMATION **12322**
DEFORMATION **3707**
DEFORMATION **13629**
DEFORMATION (FLECHE) **3698**
DEFORMATION CRITIQUE **3350**
DEFORMATION ELASTIQUE **4485**
DEFORMATION ELASTIQUE **4478**
DEFORMATION EN GENOU **7315**
DEFORMATION PAR ENROULEMENT **13629**
DEFORMATION PAR FLEXION **5426**
DEFORMATION PAR TRACTION **12744**
DEFORMATION PERMANENTE **9202**
DEFORMATION PERMANENTE **9210**
DEFORMATION PERMANENTE **3337**
DEFORMATION PERMANENTE DE L'AME D'UNE POUTRE **13727**
DEFORMATION PLASTIQUE **9469**
DEFORMATION PLASTIQUE **9471**
DEFORMATION PRINCIPALE **2000**
DEFORMATION PROGRESSIVE **6853**
DEFORMATIONS EN DEMI-LUNE DU CORDON DE SOUDURE **5212**
DEFOURNEMENT PAR EXTRACTEUR A BRAS **4471**

DEGAGEMENT 910
DEGAGEMENT DE CHALEUR 4874
DEGAGEMENT DE CHALEUR 10274
DEGAGEMENT DE GAZ 3818
DEGAGER 893
DEGAUCHIR 12318
DEGAUCHISSAGE 12319
DEGAUCHISSAGE A FROID 2691
DEGAUCHISSEMENT 12319
DEGAUCHISSEUSE 12515
DEGAZAGE 3710
DEGAZAGE DE L'EAU 3623
DEGAZER 10455
DEGIVREUR 3708
DEGORGEMENT 1321
DEGORGEMENT 5728
DEGORGEOIR 5729
DEGOULINAGE 10892
DEGOURDISSAGE (PRECHAUFFAGE AVANT SOUDAGE) 6260
DEGRAISSAGE 2474
DEGRAISSAGE 3712
DEGRAISSAGE A LA VAPEUR 13486
DEGRAISSAGE D'UN METAL 11006
DEGRAISSAGE ELECTROLYTIQUE 4589
DEGRAISSER UN METAL 11005
DEGRAISSEUR AU SOLVANT 2473
DEGRAS 3711
DEGRE 12239
DEGRE 3714
DEGRE 6021
DEGRE BAUME 3715
DEGRE CENTIGRADE 3716
DEGRE D'ENCROUTEMENT 4683
DEGRE D'EXACTITUDE 3721
DEGRE D'USINAGE 7866
DEGRE DE DURETE 3720
DEGRE DE FIN D'UN ALLIAGE 5209
DEGRE DE PRECISION 3721
DEGRE DE SATURATION 3722
DEGRE DE TREMPE DE L'ACIER 6299
DEGRE ENGLER 3717
DEGRE FAHRENHEIT 3718
DEGRE HYDROTIMETRIQUE 6298
DEGRE HYGROMETRIQUE DE L'AIR 10433
DEGRE REAUMUR 3724
DEGRE TWADDELL 3725
DEGRES DE LIBERTE 3726
DEGROSSIR 10775
DEGROSSISSAGE 10786
DEGROSSISSAGE A LA MEULE 10777
DEJETER (SE) 13628

DELARDAGE 12646
DELARDAGE D'UNE TOLE 12671
DELAVAGE 13632
DELAYER 12718
DELIQUESCENCE 3733
DELIQUESCENT 3734
DELIVRANCE D'UN BREVET 6049
DEMAGNETISEURS 3744
DEMANDE 646
DEMANDER UN BREVET 651
DEMANDEUR D'UN BREVET 645
DEMARIEUSE POUR BETTERAVES A SUCRE 12438
DEMARRAGE 12118
DEMARRAGE D'UN MOTEUR 12122
DEMARRAGE DOUX 11685
DEMARRAGE SANS A-COUPS 11685
DEMARRER 12113
DEMARREUR 3310
DEMARREUR 12116
DEMARREUR 8418
DEMARREUR AUTOMATIQUE 828
DEMI-AXE 11167
DEMI-CELLULE 6204
DEMI-CERCLE 11179
DEMI-CERCLE 12261
DEMI-CIRCULAIRE 11169
DEMI-COUSSINET INFERIEUR 1487
DEMI-COUSSINET SUPERIEUR 13030
DEMI-COUSSINETS 11947
DEMI-MANCHON 6205
DEMI-RONDS TREFILES 4249
DEMI-SPHERE 6438
DEMODULATEUR 8338
DEMODULATION 10312
DEMOISELLE 12611
DEMONTABLE 4027
DEMONTAGE 12604
DEMONTAGE D'UNE CLAVETTE 7751
DEMONTE 7329
DEMONTER 12600
DEMONTER UN BOULON 13395
DEMOULAGE 9554
DEMOULAGE 11251
DEMOULER 4468
DEMOULEUSE 4237
DEMULTIPLICATION 5908
DEMULTIPLICATION 10357
DENATURANT 3747
DENDRITE 3748
DENDRITE 9331
DENICKELAGE 3751

DEROUILLEMENT 10826
DEROUILLER 10815
DEROULEMENT 13405
DEROULER 13404
DEROULEUSE DE FIL 13903
DERRICK 3788
DES DE FRAISAGE 8290
DESABLAGE 2478
DESABLER 2469
DESACTIVATION 3789
DESAGREGATION 4000
DESAGREGER 3999
DESAGREGER (SE) 4043
DESAIMANTATION 3742
DESAIMANTER 3743
DESAIMANTER (SE) 1123
DESARGENTAGE 3802
DESARGENTER 3801
DESASSEMBLAGE 12604
DESASSEMBLER 12600
DESAXAGE 4055
DESAXE 8893
DESAXE 8730
DESCENDRE LA COURROIE 12895
DESCENTE D'UN FARDEAU 7797
DESCENTE DE COULEE 4186
DESCENTE DE COULEE 12003
DESCENTE DU PISTON 4187
DESCRIPTION D'UN BREVET 11856
DESEMBRAYAGE 12898
DESEMBRAYER 10034
DESEMBUAGE 3704
DESENGRENAGE 12898
DESENGRENE 8892
DESENGRENER 10034
DESEQUILIBRE 13341
DESHUILEUR DE VAPEUR 8779
DESHYDRATATION 3728
DESHYDRATER 3727
DESHYDRATER (SE) 3727
DESINCRUSTANT CURATIF 4041
DESINCRUSTANT PREVENTIF 10978
DESINCRUSTATION 10987
DESINCRUSTER 10971
DESINFECTANT 4042
DESINTEGRATION 4044
DESINTEGRATION 3418
DESINTEGRATION 4000
DESINTEGRER 3999
DESIONISATION 3729
DESOXYDANT 3765

DESOXYDATION 3766
DESOXYDATION 3762
DESOXYDER 3763
DESOXYGENATION 3762
DESOXYGENER 3763
DESSECHAGE/DESSICCATION DES BOIS A L'AIR LIBRE
   11095
DESSECHER 4338
DESSERRAGE 11560
DESSERRER (SE) 11559
DESSERRER UN ECROU 11558
DESSERRER UN FREIN 10436
DESSERRER UNE VIS 11558
DESSICCATEUR 3804
DESSICCATION 4359
DESSIN 4229
DESSIN 9430
DESSIN (ART DU DESSINATEUR) 4246
DESSIN (REPRESENTATION GRAPHIQUE) 4245
DESSIN A L'ECHELLE 10983
DESSIN A MAIN LEVEE 5648
DESSIN AU CRAYON 7467
DESSIN AU TRAIT 8905
DESSIN COTE 3959
DESSIN D'ARCHITECTURE 693
DESSIN DE DETAIL 3809
DESSIN DE MACHINE 7843
DESSIN EN COUPE 11122
DESSIN EN GRANDEUR NATURELLE 5724
DESSIN GEOMETRIQUE 5928
DESSIN INDUSTRIEL 8099
DESSIN LAVE 12965
DESSIN LINEAIRE 8905
DESSIN NATURE 5724
DESSIN OMBRE 11235
DESSIN PASSE A L'ENCRE 6941
DESSINATEUR 4213
DESSINER 4195
DESSINER A GRANDS TRAITS 11529
DESSOUDAGE 13400
DESSOUDER 13399
DESSUS D'UN FILET 12455
DESTRUCTIBILITE 6
DETAIL DE CONSTRUCTION 3810
DETALONNAGE 1387
DETARTRAGE 10987
DETARTRANT 10978
DETARTRER 10971
DETECTEUR DE CRIQUES ET FELURES 3280
DETENDEUR 10420
DETENDEUR 10355

DETENDRE (SE) **4902**
DETENTE **10435**
DETENTE D UN GAZ **4921**
DETENTE DE LA VAPEUR **4921**
DETENTE DES GAZ **4920**
DETERGENT **3813**
DETERIORE **5057**
DETERMINANT **3814**
DETERMINANT MINEUR **8315**
DETERMINATION ANALYTIQUE **496**
DETERMINATION EXPERIMENTALE **4929**
DETONATION **7330**
DETREMPAGE **7549**
DETREMPE DE L'ACIER **12733**
DETREMPER L'ACIER **12719**
DEVELOPPABLE **3817**
DEVELOPPABLE **3816**
DEVELOPPANTE **7117**
DEVELOPPANTE DU CERCLE **7118**
DEVELOPPEE **4873**
DEVELOPPER (GEOM.) **3815**
DEVERSEMENT **4376**
DEVERSER (SE) **13628**
DEVERSOIR **13753**
DEVERSOIR **8931**
DEVERSOIR DES PLATEAUX **4188**
DEVIATION **3819**
DEVIATION **3699**
DEVIATION **3380**
DEVIATION LATERALE **7419**
DEVIATION NULLE **14024**
DEVIDOIR **13898**
DEVIS **4838**
DEVISSAGE **13396**
DEVISSER **13395**
DEVITRIFICATION **3823**
DEXTRINE **3825**
DEXTROGYRE **3827**
DEXTRORSUM **2508**
DIABASE **3828**
DIABLE **947**
DIACAUSTIQUE **3829**
DIAGMAGNETISME **3848**
DIAGONAL **3831**
DIAGONALE **3831**
DIAGRAMME **3834**
DIAGRAMME **7717**
DIAGRAMME D'EQUILIBRE **4788**
DIAGRAMME D'INDICATEUR **6881**
DIAGRAMME D'OUVERTURE ET DE FERMETURE D'UNE
    SOUPAPE **13457**

DIAGRAMME DE DIFFRACTION A RAYONS X **3441**
DIAGRAMME DE DIFFRACTION DES RAYONS X. **13997**
DIAGRAMME DE DIFFRACTION RX **3926**
DIAGRAMME DE DISTRIBUTION **11593**
DIAGRAMME DE LAUE **7435**
DIAGRAMME DE PHASES **2984**
DIAGRAMME DE RX **9128**
DIAGRAMME DE TRAVAIL **6881**
DIAGRAMME DES EFFORTS **5526**
DIAGRAMME DES FORCES **5526**
DIAGRAMME DES FORCES **3835**
DIAGRAMME DES RESISTANCES **10489**
DIAGRAMME DES VITESSES **13514**
DIAGRAMME ENTROPIQUE **6346**
DIAGRAMME FER-CARBURE **7139**
DIAGRAMME RANKINISE **10203**
DIAGRAMME TOTALISE **10203**
DIALYSE **3846**
DIALYSER **3844**
DIALYSEUR **3845**
DIAMAGNETIQUE **3847**
DIAMANT **3862**
DIAMETRE **3849**
DIAMETRE (DU CERCLE) DE PERCAGE **3853**
DIAMETRE A FLANC DE FILET **9407**
DIAMETRE A FOND DE FILET **3850**
DIAMETRE ADMIS AU-DESSUS DU BANC **12547**
DIAMETRE CONJUGUE **2954**
DIAMETRE D'ALESAGE **3855**
DIAMETRE D'UN CABLE **10733**
DIAMETRE D'UN FIL **3860**
DIAMETRE D'UN TUYAU **3858**
DIAMETRE DE LA BARRE DE FER D'UN MAILLON **3856**
DIAMETRE DE LA POINTE DU FORET **4269**
DIAMETRE DE PERCAGE **1438**
DIAMETRE DE PIED **10719**
DIAMETRE DE TROU DE BOULON **3854**
DIAMETRE DU BOULON **1434**
DIAMETRE DU NOYAU **3850**
DIAMETRE DU RIVET **3859**
DIAMETRE DU TROU **3857**
DIAMETRE DU TROU DE BOULON **1440**
DIAMETRE EXTERIEUR **3852**
DIAMETRE EXTERIEUR **8913**
DIAMETRE EXTERIEUR **4958**
DIAMETRE INTERIEUR **8314**
DIAMETRE INTERIEUR **6977**
DIAMETRE INTERIEUR **7071**
DIAMETRE INTERIEUR **3850**
DIAMETRE INTERIEUR D'UN TUYAU **1464**
DIAMETRE NOMINAL **8611**

DIAMETRE PRIMITIF **9407**
DIAMETRE PRIMITIF **3861**
DIAMETRE UTILE **13423**
DIAPHANE **3869**
DIAPHANEITE **3868**
DIAPHRAGME **4110**
DIAPHRAGME **3870**
DIAPHRAGME (OPT.) **3871**
DIAPHRAGME ELASTIQUE **4479**
DIAPHRAGME EN CAOUTCHOUC **10819**
DIAPHRAGME EN CUIR **7498**
DIAPHRAGME IRIS **7132**
DIAPHRAGMER **3877**
DIATHERMANE **3879**
DIATHERMANEITE **3878**
DIATHERMANSIE **3878**
DIATHERMIQUE **3879**
DIATOMIQUE **3881**
DIBASIQUE **3882**
DIEDRE **3944**
DIELECTRIQUE **3898**
DIESEL **3901**
DIFFERENCE **3902**
DIFFERENCE DE POTENTIEL **9709**
DIFFERENCE DE POTENTIEL AUX BORNES **12764**
DIFFERENCE DE PRESSION **9814**
DIFFERENCE DE TEMPERATURE **3904**
DIFFERENCIATION **3923**
DIFFERENCIER **3922**
DIFFERENTIEL **3913**
DIFFERENTIEL **3906**
DIFFRACTION **3924**
DIFFRACTION **10387**
DIFFUSEUR **3928**
DIFFUSEUR OU BUSE **2359**
DIFFUSIBILITE **3936**
DIFFUSION **3929**
DIFFUSION **10994**
DIFFUSION DE LA LUMIERE **3933**
DIFFUSION DES GAZ **3932**
DIGESTEUR **829**
DIGESTION (CHIM.) **3939**
DILATABILITE **3947**
DILATABLE **3948**
DILATATION **3949**
DILATATION CUBIQUE **13603**
DILATATION D'UN METAL **4919**
DILATATION LATERALE **7421**
DILATATION LINEAIRE **7612**
DILATATION SOUS L'INFLUENCE DE LA CHALEUR **4948**
DILATATION SUPERFICIELLE **12466**

DILATATION THERMIQUE **12808**
DILATOMETRE **3950**
DILUANT **3951**
DILUER **3952**
DILUTION D'UNE SOLUTION **3954**
DIMENSION (A DEUX) **8729**
DIMENSION (A UNE) **8727**
DIMENSION INFERIEURE AUX PRESCRIPTIONS **13349**
DIMENSION NOMINALE **8612**
DIMENSION NOMINALE **8615**
DIMENSION NORMALE **12071**
DIMENSION REELLE **152**
DIMENSIONS **3960**
DIMENSIONS (A TROIS) **8728**
DIMENSIONS D'ENCOMBREMENT **7481**
DIMENSIONS HORS-TOUT **8922**
DIMENSIONS NOMINALES **11857**
DIMETHYLCETONE **90**
DIMETHYLE **4848**
DIMINUTION **3677**
DIMINUTION DE LA SECTION TRANSVERSALE **10362**
DIODE **3964**
DIORITE **3966**
DIOXYDE DE PLOMB **7461**
DIPHENYLAMINE **3974**
DIRECTEUR **3995**
DIRECTEUR D'USINE **5915**
DIRECTION ASSISTEE **9743**
DIRECTION D'UNE FORCE **3990**
DIRECTION DE MOUVEMENT **3991**
DIRECTION DES FIBRES **3993**
DIRECTION DES FIBRES **5133**
DIRECTION DES USINES, DES ATELIERS, DU TRAVAIL **13963**
DIRECTION/ORGANISATION SCIENTIFIQUE DES ATELIERS **11001**
DIRECTRICE **3996**
DISCRIMINANT **4029**
DISJONCTEUR **2407**
DISJONCTEUR **3516**
DISLOCATION **4047**
DISLOCATION-COIN **4443**
DISLOCATION-VIS (EN HELICE) **11031**
DISPERSANT **4050**
DISPERSEUR DE LIQUIDES **809**
DISPERSION **4051**
DISPERSION ELECTROMAGNETIQUE **4616**
DISPOSITIF **3820**
DISPOSITIF A RETIRER **4239**
DISPOSITIF ACCESSOIRE **5766**
DISPOSITIF BIELLE ET MANIVELLE **1094**

DISPOSITIF D'ARRET **12278**

DISPOSITIF D'ECLAIRAGE **6774**

DISPOSITIF DE BLOCAGE **7705**

DISPOSITIF DE CHANGEMENT DE SENS DE MARCHE PAR COURROIE **1154**

DISPOSITIF DE CHANGEMENT DE SENS DE MARCHE PAR ENGRENAGE CONIQUE **10545**

DISPOSITIF DE CHANGEMENT DE SENS DE MARCHE PAR ENGRENAGE DROIT **10547**

DISPOSITIF DE CHANGEMENT DE SENS DE MARCHE PAR ROUE DE FRICTION **10546**

DISPOSITIF DE CHANGEMENT DE SENS DE MARCHE POUR TRANSMISSION **10542**

DISPOSITIF DE CONTROLE **3052**

DISPOSITIF DE CONTROLE DE PRESSION **9829**

DISPOSITIF DE COUPE **3525**

DISPOSITIF DE DEPART A FROID **12342**

DISPOSITIF DE FIXATION **7227**

DISPOSITIF DE PRECONTRAINTE **9781**

DISPOSITIF DE REGLAGE **10418**

DISPOSITIF DE REPERAGE DES RIVES **4444**

DISPOSITIF DE SECURITE **10881**

DISPOSITIF DE SERRAGE **7227**

DISPOSITIF DE SURETE **10881**

DISPOSITIF POULIE ET COURROIE **1144**

DISPOSITIF PROTECTEUR **10881**

DISPOSITION **4056**

DISPOSITION GENERALE **5914**

DISQUE **4015**

DISQUE **3837**

DISQUE (DE CLAPET) **4014**

DISQUE A CAMES **13221**

DISQUE A COUPER **3542**

DISQUE D' EXCENTRIQUE **4434**

DISQUE DE COUPER **3542**

DISQUE DE RUPTURE **10887**

DISQUE DE RUPTURE **10860**

DISQUE EN BUFFLE **1684**

DISQUE EN DRAP **10173**

DISQUE EN LISIERE DE TISSU **4116**

DISQUE POLISSEUR **1693**

DISQUE-SCIE **2427**

DISQUE-VOLANT **4004**

DISQUES ABRASIFS **26**

DISSOCIATION **4058**

DISSOCIATION ELECTROLYTIQUE **4595**

DISSOLUTION **4060**

DISSOLUTION (ACTION) (CHIM.) **4059**

DISSOLUTION ELECTROLYTIQUE **4581**

DISSOLVANT **11814**

DISSOUDRE **11813**

DISSOUDRE **4061**

DISSOUDRE (SE) **4061**

DISSYMETRIE **779**

DISSYMETRIQUE **778**

DISTANCE **4063**

DISTANCE D'AXE EN AXE **4065**

DISTANCE DE CENTRE A CENTRE DES RIVETS **9409**

DISTANCE DES RIVETS DE CENTRE A CENTRE EN LIGNE CONTINUE **7736**

DISTANCE DU CENTRE DU RIVET AU BORD DE LA TOLE **4066**

DISTANCE ENTRE LES BRAS **6596**

DISTANCE ENTRE POINTES **4065**

DISTANCE FOCALE **5500**

DISTANCE FOCALE **5503**

DISTANCE POLAIRE **9579**

DISTANCE RETICULAIRE **7087**

DISTILLATION **4071**

DISTILLATION (ACTION) **4072**

DISTILLATION A LA VAPEUR D'EAU **4073**

DISTILLATION AVEC CRACKING **3285**

DISTILLATION AVEC DECOMPOSITION **3805**

DISTILLATION CONTINUE **3024**

DISTILLATION DANS LE VIDE **4075**

DISTILLATION DE LA HOUILLE **4074**

DISTILLATION DES COMBUSTIBLES A BASSE TEMPERATURE **7776**

DISTILLATION FRACTIONNEE **5612**

DISTILLATION SECHE **4346**

DISTILLER **4070**

DISTORSION **4077**

DISTRIBUTEUR **4086**

DISTRIBUTEUR **13453**

DISTRIBUTEUR **4084**

DISTRIBUTEUR **5353**

DISTRIBUTEUR **11592**

DISTRIBUTEUR D'ALLUMAGE **6762**

DISTRIBUTEUR GLISSANT A ROBINET **10760**

DISTRIBUTEUR OSCILLANT **10760**

DISTRIBUTEURS D'ENGRAIS **4087**

DISTRIBUTEURS DE CHAUX **4089**

DISTRIBUTEURS VEHICULES POUR ATTACHEMENT **4088**

DISTRIBUTION **13471**

DISTRIBUTION A DISTANCE DE L'ENERGIE **13135**

DISTRIBUTION D'AIR **281**

DISTRIBUTION D'AIR COMPRIME **2845**

DISTRIBUTION D'EAU **13666**

DISTRIBUTION D'EAU **13678**

DISTRIBUTION DE FORCE MOTRICE **9744**

DISTRIBUTION DE GAZ **5834**

DISTRIBUTION DE LA TEMPERATURE **4083**

DISTRIBUTION DES RIVETS **717**

DISTRIBUTION GAUSSIENNE **5887**
DISTRIBUTION PAR COULISSE **7626**
DISTRIBUTION PAR SOUPAPE **13460**
DISTRIBUTION PAR TIROIR **11594**
DISTRIBUTION PUBLIQUE D'ENERGIE ELECTRIQUE **12487**
DIVALENT **3881**
DIVERGENCE **4092**
DIVERGENT **4093**
DIVERGER **4091**
DIVIDENDE **4099**
DIVISER (MATH.) **4097**
DIVISER EN DEUX **1266**
DIVISEUR **4111**
DIVISEUR **4100**
DIVISEUR DE PRECISION **9769**
DIVISEUR-AERATEUR **209**
DIVISION **10981**
DIVISION (MATH.) **4107**
DIVISION D'UNE LIGNE EN MOYENNE ET EXTREME RAISON **8123**
DIVISION DU CERCLE **4108**
DIVISION DU TRAVAIL **12410**
DODECAEDRE **4112**
DOIGT **5211**
DOIGT **4301**
DOIGT D'ALLUMEUR **4085**
DOIGT D'ENCLIQUETAGE **2493**
DOIGT D'ENCLIQUETAGE **2094**
DOIGT DE DECLENCHEMENT **10437**
DOIGT DE RETENUE **2094**
DOIGT INCLINE **1887**
DOLERITE **4114**
DOLOMIE **4117**
DOLOMITE **4117**
DOMAINE **11845**
DOMAINE DE TRANSFORMATION **13113**
DOME DE PRISE DE VAPEUR **12163**
DONNEE **5951**
DONNEES **3612**
DONNEES JUSTIFICATIVES **12494**
DONNEES NUMERIQUES **8696**
DONNEES PRINCIPALES **7481**
DONNER LA VOIE A UNE SCIE **11224**
DONNER LE FIL A UN INSTRUMENT TRANCHANT **6790**
DORAGE **6002**
DORER **5941**
DORMANT (DU CABLE D'UN PALAN) **12092**
DORURE **5942**
DORURE AU FEU **5243**
DOS **890**
DOS DE LA FACE DE BRIDE **5326**

DOSAGE **10064**
DOSAGE **1325**
DOSAGE **3011**
DOSAGE DU CARBONE **1947**
DOSAGE DU MELANGE **9934**
DOSER **1323**
DOSSIER **917**
DOSSIER **12012**
DOSSIER TECHNIQUE **3614**
DOUBLAGE **2447**
DOUBLAGE **7623**
DOUBLE **5999**
DOUBLE BRIDE **4139**
DOUBLE D'OR **10673**
DOUBLE DECIMETRE **4135**
DOUBLE DECOMPOSITION **4136**
DOUBLE EFFET (A) **4163**
DOUBLE FOND **5026**
DOUBLE HARNAIS **4142**
DOUBLE PAROI (A) **4170**
DOUBLE REFRACTION **4151**
DOUBLE TAILLE D'UNE LIME **4134**
DOUBLE TRAIN D'ENGRENAGES **4142**
DOUBLER **7598**
DOUBLET **4172**
DOUBLURE **7623**
DOUBLURE **11926**
DOUILLE **1773**
DOUILLE **11576**
DOUILLE A RESSORT **11979**
DOUILLE D'ARCADE **14016**
DOUILLE DE REGLAGE **194**
DOUILLE DE SERRAGE **2728**
DOUILLE POUR POULIE FOLLE **7748**
DOUILLE-FOULOIR (DE PRESSE-ETOUPE) **9001**
DOUVE **12149**
DRAINAGE **4210**
DRAINER **4204**
DRAINER UNE TERRE HUMIDE **4209**
DRAPERIES/COULURES EN FESTONS **3492**
DRESSAGE **12319**
DRESSAGE **12720**
DRESSAGE (ACTION DE RENDRE PLAN) **13234**
DRESSER (RENDRE PLAN) **13232**
DRESSEUSE A GALETS **10685**
DRILLE **689**
DROITE **12310**
DROITS DU BREVETE **9121**
DUCTILE **4361**
DUCTILITE **4363**
DUCTILITE LONGITUDINALE **7735**

DUDGEONNAGE **4908**
DURABILITE **4385**
DURABLE **4386**
DURALUMIN **4387**
DURCI (NON-) **13359**
DURCIR **6286**
DURCIR **6287**
DURCISSEMENT **6291**
DURCISSEMENT **6292**
DURCISSEMENT **3483**
DURCISSEMENT A COEUR **5721**
DURCISSEMENT A COEUR **12892**
DURCISSEMENT PAR PRECIPITATION **9762**
DURCISSEMENT PAR TREMPE **10089**
DURCISSEMENT PAR VIEILLISSEMENT **223**
DURCISSEMENT STRUCTURAL **224**
DURCISSEUR (ALLIAGE) **6290**
DUREE **7541**
DUREE D'UNE OSCILLATION **12945**
DUREE DU BREVET **12761**
DUREE DU CYCLE DE SOUDAGE **13788**
DURETE **6297**
DURETE A CHAUD **6629**
DURETE A CHAUD **10328**
DURETE APRES REVENU **10520**
DURETE BRINELL **1619**
DURETE DE L'EAU **6298**
DURETE DE MEULAGE **14**
DURETE PERMANENTE DE L'EAU **9203**
DURETE SHORE **11003**
DURETE TEMPORAIRE DE L'EAU **12735**
DURETE VICKERS **13566**
DURETE VICKERS **6860**
DURETE VICKERS A LA PYRAMIDE **10039**
DURITE **6609**
DYNAMIQUE **4399**
DYNAMIQUE **4400**
DYNAMITE **4401**
DYNAMO **4402**
DYNAMO **5924**
DYNAMO **9867**
DYNAMO **407**
DYNAMO **3021**
DYNAMO COMPOUND **3228**
DYNAMO EN DERIVATION **11404**
DYNAMO EN SERIE **11201**
DYNAMO-SHUNT **11404**
DYNAMOGRAPHE **10291**
DYNAMOMETRE **7846**
DYNAMOMETRE **4403**
DYNAMOMETRE D'ABSORPTION **55**

DYNAMOMETRE D'ABSORPTION **48**
DYNAMOMETRE DE TRANSMISSION **13127**
DYNAMOMETRE DIFFERENTIEL **3911**
DYNAMOMETRE-FREIN **55**
DYNAMOMETRIQUE **4404**
DYNE **4405**
DYNE-CENTIMETRE **4818**
DYNODE **4406**
DYSPROSIUM **4407**
EAU **13640**
EAU A SOUDER **7293**
EAU ACIDULEE **134**
EAU AMMONIACALE DU GAZ **456**
EAU BOUEUSE **8446**
EAU BOUILLANTE **1428**
EAU CALCAIRE **2227**
EAU CALCAIRE **6281**
EAU CHIMIQUE **2318**
EAU COURANTE **5466**
EAU D'ALIMENTATION **5077**
EAU D'AMONT **6322**
EAU D'AVAL **12597**
EAU D'HYDRATATION **6675**
EAU D'INFILTRATION **9181**
EAU DE BARYTE **1019**
EAU DE BOISSON **4281**
EAU DE CHAUX **7585**
EAU DE CONDENSATION **13662**
EAU DE CONSTITUTION **2318**
EAU DE CRISTALLISATION **13663**
EAU DE DISTRIBUTION **13075**
EAU DE LABARRAQUE (SOLUTION DE CHLORURE DE SOUDE) **4422**
EAU DE MER **11072**
EAU DE PUITS **13796**
EAU DE PUITS DE MINE **9397**
EAU DE REFRIGERATION **3095**
EAU DE REFROIDISSEMENT **3095**
EAU DE RIVIERE **10620**
EAU DE SAVON **11702**
EAU DE SOURCE **11996**
EAU DE SURFACE **12520**
EAU DISTILLEE **4076**
EAU DORMANTE **12050**
EAU DOUCE **5663**
EAU DU JOUR **12520**
EAU DURE **6281**
EAU EXEMPTE DE MICROBES **13650**
EAU FLUVIALE **10620**
EAU FORTE **8580**
EAU FORTE **657**

EAU HYGROMETRIQUE **6731**
EAU INDUSTRIELLE **13649**
EAU LOURDE **6414**
EAU MERE **8405**
EAU METEORIQUE **8213**
EAU ORDINAIRE **10240**
EAU OXYGENEE **6716**
EAU PEU CHARGEE **11745**
EAU PLUVIALE **10179**
EAU POTABLE **4281**
EAU PURE **11745**
EAU REGALE **658**
EAU SALEE **1617**
EAU SALINE **10910**
EAU SAUMATRE **1524**
EAU SAVONNEUSE **11702**
EAU SOUS PRESSION **9841**
EAU STAGNANTE **12050**
EAU STERILISEE **13650**
EAU SUPERFICIELLE **12520**
EAU SURCHAUFFEE **12470**
EAUX D'EGOUTS **11230**
EAUX RESIDUAIRES **6902**
EAUX SOUTERRAINES **6147**
EBARBAGE **5129**
EBARBAGE **11689**
EBARBAGE **2504**
EBARBAGE **2478**
EBARBAGE **13209**
EBARBAGE **13210**
EBARBAGE **3642**
EBARBAGE A CHAUD **6647**
EBARBAGE A FROID **2711**
EBARBAGE DE TOLE EN BISEAU **9496**
EBARBE **1764**
EBARBEMENT **13210**
EBARBER **13206**
EBARBER **5364**
EBARBER **2469**
EBARBER UNE TOLE **13205**
EBARBEUR **4254**
EBARBURE **1764**
EBAUCHAGE **1341**
EBAUCHAGE **10786**
EBAUCHAGE **1583**
EBAUCHAGE **10785**
EBAUCHE **1301**
EBAUCHE **11651**
EBAUCHE A PLUSIEURS CONSTITUANTS **2828**
EBAUCHE DE COMPACT **6090**
EBAUCHE DE FORGEAGE PREALABLE **1343**

EBAUCHE POUR PRESSE A FILER **4971**
EBAUCHER **10775**
EBAUCHES **1349**
EBAVURAGE **3642**
EBAVURAGE **1766**
EBAVURAGE **13210**
EBAVURER **5364**
EBENE **4423**
EBONITE **2363**
EBOUTAGE **3357**
EBRANLEMENTS **13559**
EBULLITION BAS (A POINT D') **1426**
EBULLITION ELEVE (A POINT D') **1425**
EBULLITION PREABLE **9776**
EBULLITION PREALABLE **1429**
ECAILLAGE **10985**
ECAILLAGE **11830**
ECAILLAGE **5306**
ECAILLAGE **9156**
ECAILLAGE **11339**
ECAILLE **10968**
ECAILLE **10964**
ECAILLE **288**
ECAILLEMENT **4889**
ECAILLER (S') **10975**
ECAILLEUSE **11936**
ECART **3819**
ECART MOYEN **8074**
ECART TYPE **12074**
ECARTEMENT DE LA VOIE (CH. DE FER) **5871**
ECARTEMENT DES ESSIEUX **13813**
ECARTEMENT DES MAILLES **11523**
ECARTEMENT DES PALIERS **4064**
ECARTEMENT DES PLATEAUX **13169**
ECARTEUR **11973**
ECHAFAUDAGE **10965**
ECHAFAUDAGE VOLANT **7260**
ECHANCRURE **8672**
ECHANGE DE CHALEUR **4884**
ECHANGEUR DE CHALEUR **6349**
ECHANGEUR THERMIQUE **6349**
ECHANTILLON **10915**
ECHANTILLONNAGE **10916**
ECHAPPEMENT (DES GAZ D'UN MOTEUR) **4890**
ECHAPPEMENT DE GAZ **4833**
ECHAPPEMENT DE VAPEUR **4833**
ECHAPPEMENT LIBRE **8627**
ECHAPPER (S') **4832**
ECHAUFFEMENT DES COUSSINETS **6400**
ECHAUFFEMENT INTERNE **6338**
ECHAUFFEMENT RAPIDE **5362**

ECHAUFFER **6334**
ECHELLE **10981**
ECHELLE **10974**
ECHELLE **10968**
ECHELLE (MOYEN DE COMPARAISON) **10982**
ECHELLE ANNULAIRE **3841**
ECHELLE CENTIGRADE **2157**
ECHELLE CIRCULAIRE MICROMETRIQUE **8253**
ECHELLE D'UN PLAN **10973**
ECHELLE DE CELSIUS **2157**
ECHELLE DE DURETE DE MOHS **8346**
ECHELLE DE PROPORTION **3833**
ECHELLE DE REAUMUR **10267**
ECHELLE FAHRENHEIT **5019**
ECHELLE MOBILE **11607**
ECHELLE REDUITE **10343**
ECHELLE THERMOMETRIQUE **12835**
ECHELLE VERNIER **13541**
ECHELONNE **12246**
ECLABOUSSURE **11842**
ECLABOUSSURES DE METAL **11929**
ECLAIRAGE **7574**
ECLAIRAGE ARTIFICIEL **733**
ECLAIRAGE AU GAZ **5828**
ECLAIRAGE DE SECOURS **4688**
ECLAIRAGE DE SECOURS **4689**
ECLAIRAGE ELECTRIQUE **4518**
ECLAIRAGE EXTERIEUR **8897**
ECLAIRAGE INTERIEUR **6887**
ECLAIRAGE NATUREL **8505**
ECLAIRAGE PROVISOIRE **4688**
ECLAT (MIN.) **7830**
ECLAT ADAMANTIN **158**
ECLAT LUMINEUX **1616**
ECLAT METALLIQUE **8195**
ECLAT NACRE **9146**
ECLAT SOYEUX **11455**
ECLAT VITREUX **13583**
ECLATEMENT **1767**
ECLATEMENT **1768**
ECLATEMENT D'UN VOLANT **1769**
ECLATEMENT D'UNE POULIE **1769**
ECLATEMENT DU BOIS **11948**
ECLISSE **1783**
ECLISSES **5270**
ECLUSE **7698**
ECOINE **11481**
ECONOMIE DANS LE TRAVAIL **10947**
ECONOMIE DE COMBUSTIBLE **5707**
ECONOMIE DE FORCE MOTRICE **10948**
ECONOMIE DE MAIN-D'OEUVRE **10947**

ECONOMIE DE TEMPS **10949**
ECONOMIE EN EQUILIBRE **932**
ECONOMIE SAINE **932**
ECONOMISEUR (DE CARBURANT) **4437**
ECOUANE **11481**
ECOUENNE **11481**
ECOULEMENT **4210**
ECOULEMENT **5454**
ECOULEMENT D'EAU **8899**
ECOULEMENT DU MATERIAU **5465**
ECOULER (S') **4832**
ECRAN **6252**
ECRAN **11018**
ECRAN **11346**
ECRAN (OPT.) **11019**
ECRAN DE PROJECTION **13568**
ECRAN DE PROTECTION **11342**
ECRAN DE SOUDAGE **5000**
ECRAN FLUORESCENT **5477**
ECRAN LUMINEUX **5477**
ECRAN RENFORCATEUR **7017**
ECRASEMENT **3420**
ECRASEMENT **7265**
ECRASEMENT DES RIVETS **2490**
ECRASER **9712**
ECRASER **7263**
ECRASER LES RIVETS **2489**
ECRITURE CHINOISE **2341**
ECRITURE DROITE **6243**
ECRITURE PERPENDICULAIRE **6243**
ECRITURE RONDE **10799**
ECROU **8698**
ECROU **11056**
ECROU A CHAPEAU **1512**
ECROU A CRENEAUX **2085**
ECROU A EMBASE **5342**
ECROU A ENCOCHES **6533**
ECROU A ENTAILLES **2085**
ECROU A FENETRES **2085**
ECROU A FENTE **11942**
ECROU A OREILLES **12911**
ECROU A RACCORD **11061**
ECROU A SIX PANS **6455**
ECROU A TROUS **6533**
ECROU BORGNE **4118**
ECROU BORGNE **1512**
ECROU BRUT DE FORGE **1290**
ECROU CARRE **12024**
ECROU CARRE A SOUDER **1307**
ECROU CRENELE **2085**
ECROU CREUX **1512**

ECROU D'ARRET 10514
ECROU D'OPERCULE 11907
ECROU DE BLOCAGE 7702
ECROU DE CLAPET 4009
ECROU DE DOUILLE D'ARCADE 14018
ECROU DE PRESSE-ETOUPE 9005
ECROU DE RAPPEL 13369
ECROU DE REGLAGE 191
ECROU DE SERRAGE 2453
ECROU DE SURETE 7702
ECROU DE VOLANT 6268
ECROU DECOLLETE 1612
ECROU EMBOUTI 8700
ECROU FENDU 11942
ECROU HEXAGONAL (6-PANS) A EMBASE 5340
ECROU MOBILE 13159
ECROU MOLETE 8280
ECROU NON TARAUDE 1308
ECROU PAPILLON 12911
ECROUIR 2684
ECROUIR 4219
ECROUISSAGE 6219
ECROUISSAGE 2685
ECROUISSAGE 13715
ECROUISSAGE 2713
ECROUISSAGE 4235
ECROUISSAGE (DEFORMATION) 12326
ECROUTAGE 10988
ECROUTAGE 9156
ECROUTAGE DES PIECES COULEES 2478
ECROUTER LES PIECES COULEES 2469
ECUME 4331
ECUME DE GRAPHITE 7317
ECUMOIRE 11536
EFFACER 4799
EFFERVESCENCE 4461
EFFET 9730
EFFET BAUSCHINGER 1070
EFFET D'ANODE 596
EFFET D'EGALISATION 7528
EFFET DE CLIQUET 10220
EFFET DE LIGNE DE TRANSMISSION 7729
EFFET DE MASSE 8027
EFFET DE NETTOYAGE 10017
EFFET DE PEAU D'ORANGE 8851
EFFET DE PINCEMENT 9329
EFFET DE SOUPAPE 13458
EFFET LEONARD 7523
EFFET SIMPLE (A) 11493
EFFET THERMOELECTRIQUE 12830
EFFET UTILE 4463

EFFICACITE ENERGETIQUE 4746
EFFLORESCENCE 4464
EFFORT 12331
EFFORT 5532
EFFORT ADMISSIBLE 10880
EFFORT D'INERTIE 6907
EFFORT DANS LA FIBRE 5134
EFFORT DE CISAILLEMENT 11295
EFFORT DE CISAILLEMENT PAR UNITE DE SECTION 11296
EFFORT DE COMPRESSION 2863
EFFORT DE COMPRESSION 2869
EFFORT DE COMPRESSION PAR UNITE DE SECTION 2871
EFFORT DE COMPRESSION SUR PIECES LONGUES 3339
EFFORT DE FLAMBAGE 3339
EFFORT DE FLAMBAGE PAR UNITE DE SECTION 3341
EFFORT DE FLEXION 1187
EFFORT DE FLEXION 1191
EFFORT DE FLEXION PAR UNITE DE SECTION 1193
EFFORT DE FLUAGE 5460
EFFORT DE RUPTURE 5618
EFFORT DE TENSION 12373
EFFORT DE TORSION 13056
EFFORT DE TORSION PAR UNITE DE SECTION 13058
EFFORT DE TRACTION 13089
EFFORT DE TRACTION 12747
EFFORT DE TRACTION PAR UNITE DE SECTION 12750
EFFORT EXTERIEUR 7679
EFFORT EXTERIEUR 4954
EFFORT INTERIEUR MOLECULAIRE 12367
EFFORT MACROSCOPIQUE 7874
EFFORT PAR UNITE DE SECTION 12367
EFFORT TANGENTIEL 12624
EFFORT TANGENTIEL DU FREIN 1551
EFFORT TRANCHANT 11295
EFFORT TRANSVERSAL 1191
EFFORTS DE REFROIDISSEMENT 3091
EFFORTS DUS AU FORGEAGE 5564
EFFRITEMENT 3418
EFFRITER (S') 4043
EFFUSION 4465
EFFUSION DES GAZ 4466
EGRISE 3863
EJECTER 4468
EJECTEUR 12172
EJECTEUR 4472
EJECTEUR 4469
EJECTEUR HYDRAULIQUE 13658
EJECTEUR PNEUMATIQUE 271
EJECTEUR PNEUMATIQUE 9554
EJECTO-CONDENSEUR 4470
ELABORATION DE L'ACIER 12214

ELANCEMENT 11580
ELANCEMENT D'UNE POUTRE 11581
ELARGIR 5356
ELARGIR AU MANDRIN 4262
ELARGIR UN TUBE AU MANDRIN 4903
ELARGISSEMENT 1723
ELASTICITE 4487
ELASTICITE DE CISAILLEMENT 13144
ELASTICITE DE COMPRESSION 4488
ELASTICITE DE FLEXION 4489
ELASTICITE DE SUITE 4474
ELASTICITE DE TORSION 4490
ELASTICITE DE TRACTION 4491
ELASTICITE RESIDUELLE 10472
ELASTICITE TRANSVERSALE 13144
ELASTIQUE 4473
ELECTRE 4653
ELECTRICIEN 4551
ELECTRICITE 4552
ELECTRICITE ATMOSPHERIQUE 796
ELECTRICITE INDUSTRIELLE 4541
ELECTRICITE NEGATIVE 8526
ELECTRICITE POSITIVE 9655
ELECTRIFICATION 4554
ELECTRIFIER 4555
ELECTRISER (S') 1124
ELECTRO-AIMANT 4567
ELECTRO-AIMANT 4611
ELECTRO-AIMANT PORTEUR 7554
ELECTRO-CHIMIE 4573
ELECTRO-CORINDON 5753
ELECTRO-EROSION 4565
ELECTRO-FORMAGE 4557
ELECTRO-METALLURGIE 4621
ELECTRO-OSMOSE 4638
ELECTROANALYSE 4657
ELECTROANALYSE 4570
ELECTROCHIMIE 4640
ELECTROCHIMIE 4561
ELECTROCHIMIQUE 4560
ELECTRODE 4574
ELECTRODE 13789
ELECTRODE A CONTACT 3001
ELECTRODE AU CALOMEL 1876
ELECTRODE BIPOLAIRE 1233
ELECTRODE BLINDEE 11299
ELECTRODE COMPOSITE 2819
ELECTRODE DE CHARBON 1950
ELECTRODE DE VERRE 5959
ELECTRODE ENROBE 5492
ELECTRODE ENROBEE 11344

ELECTRODE ENROBEE 3271
ELECTRODE ENROBEE 2584
ELECTRODE FUSIBLE 2990
ELECTRODE METALLIQUE 8180
ELECTRODE NEGATIVE 2106
ELECTRODE NON-CONSOMMABLE 8618
ELECTRODE NUE 989
ELECTRODE POSITIVE 602
ELECTRODYNAMIQUE 4563
ELECTRODYNAMOMETRE 4564
ELECTROEROSION 4531
ELECTROLYSE 4586
ELECTROLYSER 4585
ELECTROLYTE 4587
ELECTROLYTE IMPUR 5588
ELECTROMAGNETIQUE 4612
ELECTROMAGNETISME 4620
ELECTROMOTEUR 4623
ELECTRON 4624
ELECTRON POSITIF 9662
ELECTRONEGATIF 4631
ELECTRONIQUE 4637
ELECTROPHORESE 4639
ELECTROPOSITIF 4641
ELECTROSTATIQUE 4568
ELECTROTECHNIQUE 4648
ELECTROTECHNIQUE 4647
ELECTROTHERMIQUE 4650
ELECTROTYPIE 4569
ELECTRUM 4653
ELEMENT 8145
ELEMENT 4654
ELEMENT 7852
ELEMENT A DEUX LIQUIDES 2887
ELEMENT ACIDIFICATEUR 115
ELEMENT ASYMETRIQUE 777
ELEMENT BASIQUE 1038
ELEMENT D'ACCUMULATEUR 78
ELEMENT D'ADDITION 168
ELEMENT D'APPOINT 812
ELEMENT D'UN RECIPIENT A PRESSION 11117
ELEMENT DE CHARPENTE INCORPORE 1704
ELEMENT DE PILE THERMOELECTRIQUE 12821
ELEMENT GALVANIQUE 5782
ELEMENT OU PARTIE TRONCONIQUE 13121
ELEMENT PRO-EUTECTIQUE 9848
ELEMENT RAIDISSEUR 12256
ELEMENT REDRESSEUR 4549
ELEMENT STRUCTURAL 12391
ELEMENTS A COUVRE-JOINTS 11932
ELEMENTS D'ADDITION 395

ELEMENTS DE REDUCTION TRONCONIQUES **2950**
ELEMI **4656**
ELEVATEUR **4472**
ELEVATEUR (U.S. = ASCENSEUR) **4660**
ELEVATEUR A GODETS **1676**
ELEVATEURS DE FOIN ET DE PAILLE **4661**
ELEVATEURS DE GRAINS/AERO ENGRANGEURS **6034**
ELEVATION **4659**
ELEVATION CAPILLAIRE **1925**
ELEVER A UNE PUISSANCE **10180**
ELEVER UN FARDEAU **7542**
ELEVER UNE PERPENDICULAIRE **4806**
ELEVER UNE VERTICALE **4806**
ELIMINATEUR D'IMPURETES **10995**
ELIMINATION **4667**
ELIMINATION (MATH.) **4664**
ELIMINATION DES BUEES **13485**
ELIMINATION DES DEFAUTS DE SURFACE AU CHALUMEAU
  8976
ELIMINATION DES DEFAUTS SUPERFICIELS **3793**
ELIMINATION DU FER DE L'EAU **10454**
ELIMINER **4663**
ELIMINER (MATH.) **4662**
ELINGUE **11610**
ELINVAR **4668**
ELLIPSE **4669**
ELLIPSE DE CASSINI **2029**
ELLIPSOGRAPHE **4670**
ELLIPSOIDE **4671**
ELLIPSOIDE DE REVOLUTION **4672**
ELLIPSOIDE DE ROTATION **4672**
ELONGATION DE L'AIGUILLE **3699**
EMAIL **4709**
EMAILLAGE **4712**
EMAILLER **4710**
EMANATION **10159**
EMBALLAGE **8999**
EMBALLAGE MOULANT **11540**
EMBALLEMENT D'UN MOTEUR **10118**
EMBALLER **10117**
EMBALLER **8991**
EMBALLER (S') **10117**
EMBASE **1040**
EMBASE **2724**
EMBASE D'UN PIVOT CANNELE **12904**
EMBOITAGE **5281**
EMBOITEMENT **5281**
EMBOITEMENT DOUBLE (M ET F) **12990**
EMBOITEMENT SIMPLE (M ET F) **7957**
EMBOITER **5277**
EMBOUCHURE **12966**

EMBOUCHURE **196**
EMBOUCHURE CONVERGENTE **3066**
EMBOUCHURE DIVERGENTE **4096**
EMBOUTIR **5324**
EMBOUTISSAGE **5345**
EMBOUTISSAGE **8189**
EMBOUTISSAGE **4040**
EMBOUTISSAGE EN COUPE DE FILS **3469**
EMBOUTISSAGE PROFOND **3469**
EMBOUTISSAGE PROFOND **7158**
EMBOUTISSAGE PROFOND **3683**
EMBOUTISSEUSE **5346**
EMBRANCHEMENT INDUSTRIEL **13966**
EMBRAYAGE **12897**
EMBRAYAGE **2544**
EMBRAYAGE **3254**
EMBRAYAGE **2461**
EMBRAYAGE A CONE **2927**
EMBRAYAGE A CONES **2930**
EMBRAYAGE A DENTS DE SCIE **10960**
EMBRAYAGE A DENTURE CONIQUE **6839**
EMBRAYAGE A DENTURE DROITE **12313**
EMBRAYAGE A DISQUE **4002**
EMBRAYAGE A DISQUES **4046**
EMBRAYAGE A DISQUES **9486**
EMBRAYAGE A FRICTION **11614**
EMBRAYAGE A FRICTION **5672**
EMBRAYAGE A GRIFFES **4113**
EMBRAYAGE A LAMES **9486**
EMBRAYAGE A ROUE LIBRE **8957**
EMBRAYAGE CENTRIFUGE **2177**
EMBRAYAGE HYDRAULIQUE **6681**
EMBRAYAGE MAGNETIQUE **7934**
EMBRAYAGE PROGRESSIF A FRICTION **11619**
EMBRAYER **10033**
EMBRYON **4686**
EMBUS **11289**
EMERI **4692**
EMERI **4699**
EMERILLON **12561**
EMETTRE DES RAYONS **4703**
EMEULAGE **1691**
EMISSION **4700**
EMMAGANISER **12294**
EMMAGASINAGE **12288**
EMMAGASINAGE **12299**
EMMAGASINER DE L'ENERGIE **12295**
EMMAGASINER LA CHALEUR **12296**
EMMANCHEMENT **10847**
EMMANCHEMENT **11389**
EMMANCHEMENT **4307**

EMMANCHEMENT 5275
EMMANCHEMENT 1074
EMMANCHEMENT 3254
EMMANCHEMENT 11600
EMMANCHEMENT 10025
EMMANCHEMENT A CHAUD 11398
EMMANCHEMENT A CONE 12649
EMMANCHEMENT A LA PRESSE 5527
EMMANCHEMENT CONIQUE 12649
EMMANCHEMENT COULISSANT 11600
EMMANCHEMENT DE TUBES 5281
EMMANCHEMENT GRAS 11600
EMMANCHER 11390
EMOUDRE 11273
EMOULAGE 11275
EMOUSSAGE 5972
EMPAQUETAGE 1772
EMPECHER 6926
EMPILAGE DU BOIS 9309
EMPILER LE BOIS 9307
EMPLACEMENT A BATIR 1699
EMPOIS 12112
EMPORTE-PIECE 10000
EMPORTE-PIECE CYLINDRIQUE 6545
EMPORTE-PIECE POUR COURROIE 1153
EMPORTER SUR...(L') 8928
EMPREINTE 10465
EMPREINTE 6806
EMPREINTE DE DURETE 6859
EMULSEUR 273
EMULSION 4708
EMULSIONNER 4707
ENANTIOTROPIQUE 4713
ENCASTRE AUX DEUX EXTREMITES 5284
ENCASTRE PAR UNE EXTREMITE 5285
ENCASTREMENT 4310
ENCASTREMENT 5300
ENCASTRER DANS LE BETON 4682
ENCEINTE A VIDE 13440
ENCLANCHEMENT 4751
ENCLENCHEMENT 4751
ENCLENCHER 4749
ENCLIQUETAGE 7705
ENCLIQUETAGE A FROTTEMENT 5676
ENCLIQUETAGE A ROCHET 9138
ENCLIQUETAGE DOUBLE 2494
ENCLIQUETAGE SILENCIEUX 11433
ENCLIQUETER 4749
ENCLUME 630
ENCOCHAGE 8676
ENCOCHE 8672

ENCOMBREMENT 11829
ENCRASSEMENT 2510
ENCRE DE CHINE 6871
ENCRER 6940
ENDOMMAGE 5057
ENDOSMOSE 4736
ENDOTHERMIQUE 4737
ENDUCTION 12529
ENDUIRE 1909
ENDUIRE 9014
ENDUIRE (MACON.) 9464
ENDUIT 2583
ENDUIT (MACON.) 9463
ENDUIT DE FINITION 3264
ENDUIT POUR LINGOTIERES 8355
ENDUIT POUR NOYAUX 3171
ENDURANCE 4738
ENDURCISSEMENT AU CYANURE 3549
ENERGIE 3746
ENERGIE 4747
ENERGIE 4539
ENERGIE 4745
ENERGIE ACTUELLE 7308
ENERGIE CALORIFIQUE 6344
ENERGIE CHIMIQUE 2305
ENERGIE CINETIQUE 7308
ENERGIE D'ACTIVATION 147
ENERGIE D'ECOULEMENT 4748
ENERGIE DE RAYONNEMENT 10139
ENERGIE DU CHOC 12757
ENERGIE LATENTE 9710
ENERGIE MAGNETIQUE 7900
ENERGIE MECANIQUE 8101
ENERGIE POTENTIELLE 9710
ENERGIE THERMIQUE 6344
ENFLAMMER (S') 6757
ENFONCER UN CLOU 4292
ENFONCER UN COIN 4293
ENGAGE DANS UNE MACONNERIE 1600
ENGAGER 4750
ENGIN 6519
ENGORGEMENT 2362
ENGORGER (S') 2358
ENGRENAGE 5889
ENGRENAGE 13021
ENGRENAGE A COIN 13736
ENGRENAGE A DENTS RECROISEES 12247
ENGRENAGE A DENTURE CROISEE 12247
ENGRENAGE A DENTURE SPIRALE 11534
ENGRENAGE A DEVELOPPANTE DE CERCLE 7119
ENGRENAGE A DOUBLES CHEVRONS 13176

ENGRENAGE A FRICTION **5674**
ENGRENAGE A FUSEAUX **9325**
ENGRENAGE A LANTERNE **9325**
ENGRENAGE A POINT **9325**
ENGRENAGE A VIS SANS FIN **13974**
ENGRENAGE BALADEUR **11598**
ENGRENAGE BRUT DE FONTE **10773**
ENGRENAGE CONIQUE **1227**
ENGRENAGE CYCLOIDAL **3558**
ENGRENAGE CYLINDRIQUE **12008**
ENGRENAGE D'ANGLE **1227**
ENGRENAGE DE FORCE **6402**
ENGRENAGE DE GRANDE FATIGUE **6402**
ENGRENAGE DE TRANSMISSION **13128**
ENGRENAGE DEMULTIPLICATEUR **11873**
ENGRENAGE DIFFERENTIEL **3913**
ENGRENAGE DROIT **12008**
ENGRENAGE ELLIPTIQUE **4677**
ENGRENAGE EPICYCLIQUE **9446**
ENGRENAGE EPICYCLOIDAL **4780**
ENGRENAGE EXTERIEUR **4956**
ENGRENAGE HELICOIDAL **6425**
ENGRENAGE HYPERBOLOIDE **11534**
ENGRENAGE HYPOCYCLOIDAL **6744**
ENGRENAGE INTERIEUR **7058**
ENGRENAGE INTERMEDIAIRE **7050**
ENGRENAGE PAR ROUE DENTEE ET CREMAILLERE **10120**
ENGRENAGE TAILLE **3508**
ENGRENAGES **5907**
ENGRENAGES A CHEVRONS **6448**
ENGRENAGES CONCOURANTS **8323**
ENGRENAGES HYPOIDES **6745**
ENGRENAGES INTERIEURS **7067**
ENGRENEMENT **4752**
ENGRENEMENT **5910**
ENGRENER **4750**
ENGRENER ENSEMBLE **1077**
ENJOLIVEUR DE ROUE **6662**
ENJOLIVEURS **10531**
ENLEVEMENT **4667**
ENLEVEMENT DE COUCHES SUPERFICIELLES **3807**
ENLEVEMENT DE LA COUCHE SUPERFICIELLE **2912**
ENLEVEMENT DES BAVURES **13210**
ENLEVEMENT PAR EMOULAGE **13**
ENLEVER (OU USER) EN EMOULANT **10**
ENLEVER DE LA CHALEUR **4663**
ENLEVER LES BAVURES **13206**
ENREGISTRER **10288**
ENREGISTREUR **10289**
ENREGISTREUR A DEROULEMENT CONTINU **12380**
ENREGISTREUR A DIAGRAMME CIRCULAIRE **2408**

ENREGISTREUR A TAMBOUR **3573**
ENRICHISSEMENT DU MINERAI **8861**
ENROBAGE **8999**
ENROBAGE **2587**
ENROBAGE BASIQUE **7584**
ENROULE A DROITE **2508**
ENROULE A GAUCHE **3239**
ENROULEMENT **2650**
ENROULEMENT **13862**
ENROULER **13853**
ENROULEUR DE COURROIE LENIX LENEVEU **13750**
ENROULEUSE **13903**
ENSEMBLE DE COMMANDE A VITESSE VARIABLE **13492**
ENSILEUSES **11430**
ENSILEUSES **3522**
ENTAILLE **8668**
ENTAILLE **8672**
ENTAILLER **8554**
ENTARTRAGE **6856**
ENTARTRE **6857**
ENTONNOIR **5738**
ENTONNOIR **12003**
ENTONNOIR DE COULEE **9716**
ENTOURER UN TUYAU **3263**
ENTOURER UNE CHAUDIERE DE MACONNERIE **11215**
ENTRAINEMENT **4303**
ENTRAINEMENT (PAR FROTTEMENT) **4304**
ENTRAINER PAR FROTTEMENT **4291**
ENTRAINEUR **4301**
ENTRAXE **4065**
ENTRAXE **2165**
ENTREE **6943**
ENTREE D'EAU **6915**
ENTREE DES DONNEES PAR BANDE **12643**
ENTREE EN PARALLELE **9054**
ENTREFER **262**
ENTREFER D'UN AIMANT **263**
ENTREPOSER **12294**
ENTREPRENEUR DE BATIMENT **1698**
ENTRETIEN **7946**
ENTRETOISAGE EN TUBES **13258**
ENTRETOISE **12151**
ENTRETOISE **4068**
ENTRETOISE **4067**
ENTRETOISE **3388**
ENTRETOISE **9489**
ENTRETOISE (CHARPENTE) **1516**
ENTRETOISE D'UN CHAINON **12402**
ENTRETOISEMENT **12153**
ENTRETOISER **12150**
ENTROPIE **4773**

ENVASEMENT D'UNE CONDUITE **11457**
ENVELOPPE **11303**
ENVELOPPE **7986**
ENVELOPPE **3274**
ENVELOPPE (GEOM.) **4776**
ENVELOPPE DE TUYAUX **7369**
ENVELOPPE DE VAPEUR **12168**
ENVELOPPE PROTECTRICE **9948**
ENVELOPPER UN TUYAU **3263**
ENVOYER DE LA CHALEUR **12485**
EPAISSEUR **13731**
EPAISSEUR **12843**
EPAISSEUR A L'ARC **678**
EPAISSEUR CIRCULAIRE DE LA DENT **2431**
EPAISSEUR D'UN FIL **3860**
EPAISSEUR D'UNE SOUDURE BOUT A BOUT **1786**
EPAISSEUR D'UNE SOUDURE EN ANGLE **12886**
EPAISSEUR D'UNE TOLE **12845**
EPAISSEUR DE DENT A LA CORDE **2365**
EPAISSEUR DE FILM **5170**
EPAISSEUR DE LA DENT **12846**
EPAISSEUR DE LA PAROI (DU METAL) D'UN TUBE **12844**
EPAISSEUR DE LA SOUDURE **12885**
EPAISSISSEMENT **1701**
EPAISSISSEMENT (PEINTURE) **7675**
EPANDEURS DE FUMIER **7995**
EPANOUISSEMENT D'UN TUBE **4768**
EPAULEMENT **7321**
EPAULEMENT **11383**
EPAULEMENT D'UN ARBRE **11384**
EPAULEMENT D'UNE ROUE D'ENGRENAGE **11400**
EPICYCLOIDE **4779**
EPISSURE **11930**
EPISSURE (ACTION) **11934**
EPISSURE (RESULTAT) **11933**
EPISSURER **11931**
EPONGE DE FER **7151**
EPONGE DE PLATINE **11955**
EPREUVE **12792**
EPREUVE **9870**
EPREUVE **9921**
EPREUVE **12770**
EPREUVE A CHAUD DE LA CHAUDIERE **12182**
EPREUVE A L'AIR COMPRIME **2846**
EPREUVE DE PRESSION **9836**
EPREUVE DE RESISTANCE **12356**
EPREUVE HYDRAULIQUE **6695**
EPREUVE NEGATIVE **8524**
EPREUVE OFFICIELLE **8740**
EPROUVETTE **12784**
EPROUVETTE **12780**

EPROUVETTE **12772**
EPROUVETTE **12788**
EPROUVETTE (CHIM) **12791**
EPROUVETTE (MORCEAU D'ESSAI) **12789**
EPROUVETTE AVEC ENTAILLE **8675**
EPROUVETTE DE METAL **11858**
EPROUVETTE DE METAL DE BASE **1039**
EPROUVETTE DU METAL DEPOSE PUR **347**
EPURATEUR **10020**
EPURATION **13638**
EPURATION CHIMIQUE DES EAUX D'ALIMENTATION **13676**
EPURATION DES GAZ **5839**
EPURATION DES GAZ **10018**
EPURE D'UNE PIECE MECANIQUE **11363**
EPURE DE DISTRIBUTION **13456**
EPURER LES EAUX D'ALIMEMTATION PAR VOIE CHIMIQUE **11748**
EQUARRISSOIR **1636**
EQUATION **4782**
EQUATION ALGEBRIQUE **322**
EQUATION BICARREE **1262**
EQUATION BINOME **1259**
EQUATION BIQUADRATIQUE **1262**
EQUATION CUBIQUE **3452**
EQUATION D'ETAT **4783**
EQUATION DIFFERENTIELLE **3912**
EQUATION DIFFERENTIELLE PARTIELLE **9086**
EQUATION DU CINQUIEME DEGRE **10112**
EQUATION DU PREMIER DEGRE **7616**
EQUATION DU QUATRIEME DEGRE **1262**
EQUATION DU SECOND DEGRE **10057**
EQUATION DU TROISIEME DEGRE **3452**
EQUATION HOMOGENE **6557**
EQUATION LINEAIRE **7616**
EQUATION QUADRATIQUE **10057**
EQUATION QUINTIQUE **10112**
EQUATION TRANSCENDANTE **13107**
EQUATION TRINOME **13213**
EQUERRE **514**
EQUERRE **1131**
EQUERRE **12014**
EQUERRE (D') **10577**
EQUERRE A DESSIN **11223**
EQUERRE A EPAULEMENT A CHAPEAU **13245**
EQUERRE A ONGLET **8325**
EQUERRE A SIX PANS **6456**
EQUERRE A TRACER LES CENTRES **2173**
EQUERRE D'ABLOCAGE **533**
EQUERRE D'ASSEMBLAGE **510**
EQUERRE DOUBLE **12581**
EQUILIBRAGE **942**

EQUILIBRAGE **935**
EQUILIBRAGE D'UNE SOUPAPE **936**
EQUILIBRAGE DES MASSES **941**
EQUILIBRATION **942**
EQUILIBRE **929**
EQUILIBRE **4793**
EQUILIBRE **4786**
EQUILIBRE CHIMIQUE **2307**
EQUILIBRE INDIFFERENT **6882**
EQUILIBRE INSTABLE **13401**
EQUILIBRE STABLE **12045**
EQUILIBRER **930**
EQUILIBRER (S') **1076**
EQUIPE D'OUVRIERS **11348**
EQUIPEMENT ELECTRIQUE **4542**
EQUIPEMENT POUR NIVELER LE SOL **4414**
EQUIPEMENT SANITAIRE **10931**
EQUIVALENCE **4795**
EQUIVALENT **4796**
EQUIVALENT CALORIFIQUE DU TRAVAIL **6347**
EQUIVALENT CHIMIQUE **2308**
EQUIVALENT DU CONE PYROMETRIQUE **10051**
EQUIVALENT ELECTROCHIMIQUE **4571**
EQUIVALENT MECANIQUE DE LA CHALEUR **8104**
ERBIUM **4804**
ERG **4818**
EROSION **4820**
ERREUR **4822**
ERREUR D'APPRECIATION **4827**
ERREUR D'ETALONNAGE **1869**
ERREUR D'OBSERVATION **4829**
ERREUR DE CALCUL **4824**
ERREUR DE COLLIMATION **4828**
ERREUR DE CONCEPTION DU CONSTRUCTEUR **4825**
ERREUR DE LECTURE **4826**
ERREUR DE MESURE **8089**
ERREUR PARALLACTIQUE **9048**
ERUGINEUX **219**
ERYTHROSINE **4831**
ESCALIER **12053**
ESCALIER DROIT **5641**
ESCALIER DROIT **5646**
ESCALIER HELICOIDAL **11919**
ESCARBILLE **2659**
ESCARBILLES **5469**
ESPACE CREUX **2125**
ESPACE MORT **2482**
ESPACE NEUTRE **2482**
ESPACE NUISIBLE **2482**
ESPACEMENT **4063**
ESPECE DE BOIS **13503**

ESPECE DE FER **2458**
ESPERANCE MATHEMATIQUE **4927**
ESPRIT DE BOIS **13920**
ESPRIT DE SEL **6706**
ESPRIT DE SEL DENATURE **7293**
ESQUILLEUSE **11936**
ESSAI **9921**
ESSAI **12792**
ESSAI **12770**
ESSAI (OU CONTROLE) AU BALAI ELECTRIQUE **6535**
ESSAI A FROID **2709**
ESSAI A HAUTE TEMPERATURE **6499**
ESSAI A L'ETINCELLE **11840**
ESSAI A LA COMPRESSION **2872**
ESSAI A LA FLEXION **1189**
ESSAI A LA GOUTTE **11958**
ESSAI A LA PERFORATION **10010**
ESSAI A LA TORSION **13059**
ESSAI A LA TRACTION **12751**
ESSAI ACOUSTIQUE **11815**
ESSAI AU BANC **1170**
ESSAI AU BROUILLARD SALIN **10909**
ESSAI AU FLAMBAGE **3342**
ESSAI AU FREIN **1546**
ESSAI AUX CHOCS REPETES **5049**
ESSAI AUX ULTRA-SONS **12480**
ESSAI BRINELL **1620**
ESSAI D'APLATISSEMENT **5409**
ESSAI D'ECRASEMENT **4374**
ESSAI D'ECRASEMENT **7262**
ESSAI D'ECRASEMENT **11949**
ESSAI D'ECRASEMENT **13416**
ESSAI D'ECRASEMENT SUR CYLINDRE DE BETON MASSIF **1561**
ESSAI D'EMBOUTISSAGE **10184**
ESSAI D'EMBOUTISSAGE ERICKSEN **8797**
ESSAI D'ENDURANCE **5049**
ESSAI D'ESTAMPAGE **12061**
ESSAI D'USURE **13714**
ESSAI DE CHARGE CONSTANTE **2974**
ESSAI DE CHOC (CHUTE) **4328**
ESSAI DE CHOC DE CHARPY (SUR EPROUVETTE ENTAILLEE) **2270**
ESSAI DE CHOC ET BALLOTTEMENT **7249**
ESSAI DE CHOC SUR BARREAU ENTAILLE **8673**
ESSAI DE CHOC SUR BARREAUX NON ENTAILLES **4328**
ESSAI DE CHOC SUR ENTAILLE **1190**
ESSAI DE CHUTE DE POIDS **4330**
ESSAI DE CINTRAGE **1189**
ESSAI DE CINTRAGE A FROID **2667**
ESSAI DE CISAILLEMENT **11949**

ESSAI DE CISAILLEMENT **11287**

ESSAI DE COMPRESSION **2865**

ESSAI DE COMPRESSION **13416**

ESSAI DE COULABILITE **2065**

ESSAI DE DEPERDITION THERMIQUE **1403**

ESSAI DE DURETE **3865**

ESSAI DE DURETE **6300**

ESSAI DE DURETE A LA LIME **5152**

ESSAI DE DURETE PAR EMPREINTE DE BILLE **6861**

ESSAI DE DURETE ROCKWELL **10659**

ESSAI DE FATIGUE / D'ENDURANCE **4743**

ESSAI DE FLEXION ALTERNE **12774**

ESSAI DE FLEXION ALTERNEE **10535**

ESSAI DE FLEXION CONSTANTE **2972**

ESSAI DE FLEXION DE LA FACE DE LA SOUDURE **1178**

ESSAI DE FLEXION ESSAI DE PLIAGE **1177**

ESSAI DE FLEXION PAR CHOC **4328**

ESSAI DE FLEXION PAR CHOC SUR BARREAUX ENTAILLES **1190**

ESSAI DE FLEXION SUR EPROUVETTE ENTAILLEE **8555**

ESSAI DE FLEXION TOURNANTE **13310**

ESSAI DE FORAGE **1475**

ESSAI DE FORGEAGE **5545**

ESSAI DE LA CHAUDIERE **12182**

ESSAI DE LA CHAUDIERE SOUS PRESSION **1421**

ESSAI DE MACRO-ATTAQUE **3684**

ESSAI DE MANDRINAGE **4911**

ESSAI DE MICRODURETE **8246**

ESSAI DE PERCAGE **1475**

ESSAI DE PLIAGE **1189**

ESSAI DE PLIAGE **4173**

ESSAI DE PLIAGE A CHAUD **6618**

ESSAI DE PLIAGE A FROID **2667**

ESSAI DE PLIAGE A L'ENVERS **1179**

ESSAI DE POINCONNAGE **10010**

ESSAI DE RABATTEMENT **5347**

ESSAI DE REBONDISSEMENT BRUSQUE **7254**

ESSAI DE RECEPTION **71**

ESSAI DE RESILIENCE **8673**

ESSAI DE RESISTANCE A CHAUD **6646**

ESSAI DE RESISTANCE A LA FATIGUE **5054**

ESSAI DE RESISTANCE AU CHAUD BLEU **1372**

ESSAI DE RESISTANCE AU CHAUD ROUGE **10331**

ESSAI DE RESISTANCE AU CHOC **6786**

ESSAI DE RUPTURE **5619**

ESSAI DE RUPTURE AU CHOC **6786**

ESSAI DE SOUDABILITE **13792**

ESSAI DE TORSION **13306**

ESSAI DE TRACTION **12748**

ESSAI DES FILS AU PLIAGE **12774**

ESSAI DES MATERIAUX **12782**

ESSAI DESTRUCTIF **3806**

ESSAI HYDROSTATIQUE **6727**

ESSAI IZOD **7200**

ESSAI JOMINY **7254**

ESSAI NON DESTRUCTIF **9924**

ESSAI NON DESTRUCTIF **8630**

ESSAI PAR CHOC **6786**

ESSAI PAR EMPREINTE DE BILLE **1618**

ESSAI PAR IMMERSIONS ET EMERSIONS ALTERNEES **405**

ESSAI PAR LA BILLE DE BRINELL **1618**

ESSAI PAR REBONDISSEMENT DES BILLES (APPAREIL SHORE) **11367**

ESSAI PHYSIQUE **9284**

ESSAI PRELIMINAIRE **9779**

ESSAI SCLEROMETRIQUE **11017**

ESSAIS D'USURE **16**

ESSAIS MECANIQUES **8112**

ESSENCE **9222**

ESSENCE **5864**

ESSENCE CRAQUEE **10647**

ESSENCE D'AMANDES AMERES **1198**

ESSENCE DE BOIS **13503**

ESSENCE DE MIRBANE **8587**

ESSENCE DE POIRE **487**

ESSENCE DE TEREBENTHINE **8765**

ESSENCE DE TEREBENTHINE RUSSE **10862**

ESSENCE MINERALE **9222**

ESSENCE MINERALE BRUTE **3408**

ESSIEU **884**

ESSIEU ARRIERE (NON-MOTEUR) **10265**

ESSIEU AVANT **5691**

ESSIEU FIXE **3624**

ESSIEU MOTEUR **4290**

ESSIEU OU ARBRE **875**

ESSIEU PORTEUR **2010**

ESSIEU-FLOTTANT **5435**

ESSORER PAR FORCE CENTRIFUGE **6702**

ESSOREUSE CENTRIFUGE **6703**

ESSUIE-GLACE **13871**

ESTAMPAGE **12532**

ESTAMPAGE **12534**

ESTAMPAGE **2651**

ESTAMPAGE **12059**

ESTAMPAGE A CHAUD **6632**

ESTAMPAGE A FROID **2673**

ESTAMPE **3883**

ESTAMPE A DEGROSSIR **5726**

ESTAMPE A ETIRER **5726**

ESTAMPE D'EMBOUTISSAGE **5580**

ESTAMPER **12056**

ESTAMPER **12530**

ESTAMPES FERMEES **2521**
ESTAMPEUSE **12060**
ESTIMATION **4838**
ETABLI **7234**
ETABLI **1167**
ETABLIR UN CONTACT **7950**
ETABLISSEMENT DE CONSTRUCTON DE MACHINES **4764**
ETABLISSEMENT DE FABRICATION **5013**
ETAGE DE PRESSION **9833**
ETAGE DE VITESSE **13523**
ETAI **12402**
ETAI **9925**
ETAIEMENT **12153**
ETAIN **12953**
ETAIN A SOUDER A FONDANT ACIDE **110**
ETAIN CHIMIQUEMENT PUR **2324**
ETAIN DE BANCA **973**
ETAIN DE MALACCA **12335**
ETAIN DUR **9230**
ETAIN EN FEUILLES **11326**
ETAIN EN GRENAILLES/EN GRAINS **6042**
ETAIN EN SAUMONS **1339**
ETALON **5768**
ETALON **12077**
ETALON **12076**
ETALON HEFNER **6419**
ETALON NORMALISE **12082**
ETALON PHOTOMETRIQUE **9270**
ETALON VIOLLE **13569**
ETALONNAGE **5886**
ETALONNAGE **1867**
ETALONNAGE (D'UN RESERVOIR) **1863**
ETALONNAGE D'UNE JAUGE **5770**
ETALONNAGE D'UNE JAUGE **5874**
ETALONNEMENT **5886**
ETALONNER **5872**
ETALONS **5769**
ETAMAGE **12963**
ETAMER **12954**
ETAMPAGE **12059**
ETAMPAGE A CHAUD **12534**
ETAMPAGE A CHAUD **6643**
ETAMPE **12531**
ETAMPE **1497**
ETAMPE **13036**
ETAMPER **12056**
ETAMPER A CHAUD **12530**
ETAMPEUSE **12060**
ETANCHE **12920**
ETANCHE (A LA VAPEUR) **12193**
ETANCHE A L'AIR **303**

ETANCHE A L'EAU **13688**
ETANCHE A L'EAU **13686**
ETANCHE AUX GAZ **5853**
ETANCHEITE **12932**
ETANCHEITE **11081**
ETANCHEITE (MANQUE D') **7491**
ETANCHEMENT **7954**
ETANCHER **7953**
ETANCON **12402**
ETANCONNER **12150**
ETAT BRUT **740**
ETAT CRITIQUE **3349**
ETAT D'EQUILIBRE **12132**
ETAT DE COLISAGE **9003**
ETAT DE LIVRAISON (A L') **741**
ETAT DE SERVICE (EN) **6819**
ETAT DE SOUDAGE (A L') **745**
ETAT DE SURFACE **12500**
ETAT DE SURFACE **12504**
ETAT DE SURFACE **5213**
ETAT DESCRIPTIF DU CONSTRUCTEUR **7993**
ETAT FINAL **5190**
ETAT GAZEUX **5857**
ETAT HYGROMETRIQUE DE L'AIR **6666**
ETAT INITIAL **6930**
ETAT LIBRE (A L') **6818**
ETAT LIQUIDE **7657**
ETAT NATIF (A L') **5647**
ETAT PATEUX **8488**
ETAT SOLIDE **11795**
ETAU **13579**
ETAU **13561**
ETAU A CHANFREIN **13565**
ETAU A MAIN **6254**
ETAU A PIED **6511**
ETAU A SERRAGE PARALLELE **9066**
ETAU A TUBES **13256**
ETAU A TUBES **9363**
ETAU D'AFFUTAGE POUR SCIES **10955**
ETAU D'ETABLI **1171**
ETAU DE TUYAUTEUR **13256**
ETAU DU NORD **6511**
ETAU INCLINABLE **535**
ETAU ORDINAIRE **6511**
ETAU ORIENTABLE **12562**
ETAU PIVOTANT **12562**
ETAU SIMPLE **9429**
ETAU UNIVERSEL **13389**
ETAU-LIMEUR **11269**
ETAUX-LIMEUR **11267**
ETAYAGE **12153**

ETAYEMENT 12153
ETAYER 12150
ETEINDRE LA CHAUX 11568
ETENDRE UNE SOLUTION 3952
ETENDUE DE L'ECHELLE D'UN INSTRUMENT 10201
ETHANE 4848
ETHENE 4851
ETHER ACETIQUE 88
ETHER AZOTIQUE 4850
ETHER CHLORHYDRIQUE 4849
ETHER COMPOSE 4837
ETHER DE PETROLE 9228
ETHER NITRIQUE 4850
ETHER ORDINAIRE 8853
ETHER OXYDE METHYLIQUE 8222
ETHER SULFURIQUE 8853
ETHYLAL 84
ETHYLENE 4851
ETINCELAGE ELECTRIQUE 4531
ETINCELLE 11836
ETIRABILITE 4228
ETIRAGE 4230
ETIRAGE A CHAUD 6624
ETIRAGE A FROID 2676
ETIRAGE A FROID 4230
ETIRAGE A FROID DE TUBES COULES SANS SOUDURE 2053
ETIRAGE BRILLANT 4347
ETIRAGE DES TUBES 13249
ETIRAGE EN FIL 13888
ETIRE 4247
ETIRE A FROID 2677
ETIRER A CHAUD 6623
ETIRER A FROID 2675
ETIRER EN FIL 4220
ETIRER LES TUBES 4224
ETOFFE CAOUTCHOUTEE 10818
ETOFFE FILTRANTE 5176
ETOUPE 13073
ETOUPE DE CHANVRE 6444
ETOUPE DE LIN 5413
ETRANGLEMENT 12891
ETRANGLEMENT DE LA VEINE LIQUIDE 3044
ETRANGLER 12889
ETRE EN EBULLITION 1401
ETRE SOUMIS A UN EFFORT DE CISAILLEMENT 1078
ETRIER 2720
ETRIER 12261
ETRIER 13324
ETRIER 2450
ETUDE 3794
ETUDE 3799

ETUDE DE SOL 11752
ETUDE DE TERRAIN 11752
ETUDIER 3795
ETUVE 3158
ETUVE 4353
ETUVE 4256
ETUVE 4358
ETUVE A SECHAGE RAPIDE 5360
EUDIOMETRE 4852
EUROPIUM 4854
EUTECTIQUE 4855
EUTECTOIDE 4859
EVACUATION DE L'AIR D'UNE CONDUITE 4665
EVACUATION DE LA VAPEUR D'EBULLITION 1405
EVACUATION DES EAUX RESIDUAIRES 10453
EVACUATION DES POUSSIERES 10452
EVACUATION DU LAITIER 3803
EVACUER LES GAZ 10455
EVANT 10614
EVAPORABLE 4861
EVAPORATION 4867
EVAPORATION A L'AIR LIBRE 8502
EVAPORER 4862
EVAPORER (S') 4862
EVAPORER (S') A L'AIR LIBRE 4863
EVENT 269
EVENT 13528
EVENT AUTOMATIQUE 1320
EVENT D'ETANCHEITE 10597
EVENT LIBRE 5642
EVIDEMENT 8831
EVIDEMENT 10279
EVOLUTE 4873
EXAMEN 4876
EXAMEN 12792
EXAMEN A FROID 2687
EXAMEN A L'ULTRAMICROSCOPE 13336
EXAMEN MAGNETOGRAPHIQUE 7931
EXAMEN MICROSCOPIQUE 7114
EXAMEN MICROSCOPIQUE 8260
EXAMEN PAR RESSUAGE 7656
EXAMEN RADIOGRAPHIQUE 10162
EXAMEN RADIOGRAPHIQUE 13993
EXCENTRATION 10892
EXCENTRE 4424
EXCENTRICITE 4436
EXCENTRIQUE 4433
EXCENTRIQUE 4425
EXCENTRIQUE 4424
EXCENTRIQUE A ONDES 4442
EXCENTRIQUE CIRCULAIRE A COLLIER 4433

EXCENTRIQUE EN COEUR **6329**
EXCENTRIQUE PLAN **8805**
EXCES **4878**
EXCES DE TRAVAIL **4880**
EXCITATION **4885**
EXCITATION SEPAREE **11190**
EXCITATRICE **4888**
EXCITER **4887**
EXECUTION D'UNE GORGE **8515**
EXEMPT D'ACIDE **5635**
EXEMPT DE CARBONE **1961**
EXEMPT DE CENDRES **5636**
EXERCER UN EFFORT SUR UN CORPS **12323**
EXFOLIATION **11328**
EXHAUSTEUR **4900**
EXIGENCES D'USINAGE **4987**
EXISTENCE **12263**
EXOTHERMIQUE **4901**
EXPERIENCE **12792**
EXPERT **4930**
EXPERTISE **4931**
EXPIRATION D'UN BREVET **4932**
EXPLOITATION **8837**
EXPLOITATION D'UNE INVENTION **13948**
EXPLOITATION MINIERE **8311**
EXPLOITER UNE INVENTION **13938**
EXPLOSER **4933**
EXPLOSIF **4937**
EXPLOSION **4935**
EXPLOSION D'UNE CHAUDIERE **1410**
EXPOSANT **4939**
EXPOSANT D'ECOULEMENT **4940**
EXPOSITION **4944**
EXPRESSION FRACTIONNAIRE **6807**
EXPRIMER **12039**
EXSUDATION **12540**
EXTENSION **4946**
EXTENSOMETRE **4949**
EXTENSOMETRE **12325**
EXTINCTION DE LA CHAUX **11569**
EXTRACTEUR **4900**
EXTRACTEUR (CHIM.) **4967**
EXTRACTEURS DE CONES **2929**
EXTRACTION DE L'HUILE **8777**
EXTRACTION DE L'OR PAR CYANURATION **3546**
EXTRACTION DE LA HOUILLE **13877**
EXTRACTION DES METAUX **4966**
EXTRACTION ELECTROLYTIQUE **4582**
EXTRACTION ELECTROLYTIQUE **4652**
EXTRAIRE **12039**
EXTRAIT **4964**

EXTRAIT MUCILAGINEUX **8440**
EXTREMITE **9384**
EXTREMITE **6314**
EXTREMITE DE LA BIELLE ARTICULEE AU PISTON **3389**
EXTREMITE DU BERCEAU **6597**
EXTREMITE LIBRE **5023**
EXTRUSION **4970**
EXTRUSION PAR CHOC **6784**
F E M **4622**
FABRICANT **7992**
FABRICATION **7989**
FABRICATION **8028**
FABRICATION **10462**
FABRICATION D'OUTILS A TAILLANTS **13006**
FABRICATION DES AGGLOMERES **7990**
FABRICATION DES MACHINES **8103**
FABRICATION DU COKE **7991**
FABRICATION EN GRANDES SERIES **7727**
FABRIQUE **5013**
FABRIQUE **5730**
FABRIQUER **7948**
FACE CRISTALLINE **3427**
FACE D'APPUI **1026**
FACE D'ATTAQUE **4989**
FACE D'UN TROU **13624**
FACE DE BRIDE **5327**
FACE DE LA DENT **4997**
FACE DE REFERENCE **8043**
FACES CENTREES (A) **5002**
FACES CENTREES (A) **4729**
FACILE A TRAVAILLER **13947**
FACILITE DU TRAVAIL **13946**
FACONNAGE **5581**
FACONNAGE **4983**
FACONNAGE **11268**
FACONNAGE A CHAUD **6627**
FACONNEMENT DES TETES A FROID **2686**
FACONNER **5571**
FACONNER UNE TETE DE RIVET **2516**
FACTEUR **5008**
FACTEUR DE FORME **5572**
FACTEUR DE PUISSANCE **9735**
FACTEUR DE SECURITE **5009**
FACTEUR DE VIBRATION **11981**
FACULTE DE PRENDRE BIEN LA TREMPE **6296**
FAIBLE FREQUENCE **7771**
FAIENCAGE **2286**
FAIRE CORRESPONDRE LES TROUS DE RIVETS **9969**
FAIRE DES ESSAIS **4928**
FAIRE DIGERER **3937**
FAIRE PRISE **6286**

FAIRE PRISE **11214**

FAIRE PRISE **6287**

FAIRE REVENIR **12717**

FAIRE REVENIR **12719**

FAISCEAU **1093**

FAISCEAU CONCENTRE **5504**

FAISCEAU D'ELECTRONS **4625**

FAISCEAU DE COURBES **5028**

FAISCEAU DE FILS **7452**

FAISCEAU DE RAYONS **9164**

FAISCEAU DE RAYONS X **13995**

FAISCEAU ELECTRONIQUE **1093**

FAISCEAU FOCALISE **5504**

FAISCEAU LUMINEUX **9162**

FAISCEAU TUBULAIRE **8538**

FALSIFICATION **205**

FALSIFIER **204**

FANEUSES **12696**

FARAD **5035**

FARINAGE **2226**

FARINE FOSSILE **7292**

FATIGUE **5047**

FATIGUE D'UN MATERIEL **5051**

FATIGUE PAR CONTACTS DE FROTTEMENT **2200**

FATIGUE PAR CORROSION **3187**

FATIGUE PAR L'HYDROGENE SULFURE **12442**

FATIGUER UN MATERIEL **8962**

FAUCHEUSES A FOURRAGES **5525**

FAUCHEUSES A TRACTEUR-TRAINEES/MONTEES **8439**

FAUSSE **1197**

FAUSSE BRIDE **1304**

FAUSSE EQUERRE **1228**

FAUSSE GALENE **1324**

FAUTE D'EXECUTION **3693**

FAUX FOND **5026**

FAUX NOYAU **5027**

FAUX-ROND **8894**

FECULE **12111**

FEEDER **5081**

FELDSPATH **5087**

FELDSPATH **5089**

FELURE **2278**

FELURE A FROID **2674**

FENDAGE **11622**

FENDILLEMENT **5274**

FENDILLEMENT DES TROUS DE RIVET **3286**

FENDILLEMENT SUR LES BORDS **3281**

FENDILLER (SE) **3279**

FENETRE **13864**

FENTE **11937**

FENTE **11629**

FENTE **3278**

FENTE SUPERFICIELLE **6834**

FER **7133**

FER (MOULURE) A VITRAGES **10937**

FER A BARREAUX DE GRILLE **6098**

FER A BOUDIN **1716**

FER A BOUDIN A PATIN **1719**

FER A BRANCARDS **2250**

FER A CHEVAL (EN) **6606**

FER A GORGE **2248**

FER A GRAIN FIN **2511**

FER A GRAIN SERRE **2511**

FER A GROS GRAIN **8819**

FER A MAIN COURANTE **6265**

FER A NERF **5139**

FER A SOUDER **11765**

FER A SOUDER **11764**

FER A SOUDER A CHAUFFAGE AUTOMATIQUE **11155**

FER A SOUDER A TETE CARREE **2350**

FER A SOUDER AU GAZ **5824**

FER A SOUDER DROIT **9573**

FER A SOUDER ELECTRIQUE **4530**

FER ACIEREUX **12224**

FER AIGRE **2702**

FER ARMCO **711**

FER BRULE **1762**

FER BRUT COULE EN COQUILLE **2333**

FER CARRE **12015**

FER CASSANT **11368**

FER CASSANT A CHAUD **10338**

FER CASSANT A FROID **2702**

FER CAVALIER **6605**

FER CORNIERE **514**

FER CORNIERE A ANGLE ARRONDI **10794**

FER CORNIERE A ANGLE VIF **11271**

FER CORNIERE A BOUDIN **1712**

FER CORROYE **11490**

FER DANS LE BROYEUR **13106**

FER DE CARBONYLE **1977**

FER DE COULEUR **10338**

FER DE MASSE **5553**

FER DE RIBLONS **5553**

FER DELTA **3740**

FER DEMI-ROND CREUX **6541**

FER DEMI-ROND PLEIN **11786**

FER DOUBLE CORROYE **4156**

FER DOUX **13988**

FER DU RABOT **9436**

FER EBAUCHE **8441**

FER EN BANDES **5375**

FER EN BARREAUX EN VERGES **984**

FER EN CROIX 3372
FER EN CROIX 3381
FER EN DOUBLE T 6194
FER EN E 2250
FER EN I 6194
FER EN L 514
FER EN QUART DE ROND 9241
FER EN T 12698
FER EN T A BOUDIN 1719
FER EN T A LARGES AILES 1640
FER EN U 2250
FER EN U 2248
FER EN Z 14023
FER FEUILLARD 6575
FER FONDU 6921
FER FORGE 13988
FER FORT 11490
FER FORT SUPERIEUR 4156
FER GAMMA 5795
FER HEXAGONAL 6454
FER KAHLBAUM 7270
FER LAMINE 10676
FER LAMINE PROFILE 10675
FER MAIN-COURANTE 6245
FER MALLEABLE ANORMAL 9
FER MARCHAND 8148
FER METIS 10338
FER NERVEUX 5139
FER NOIR 1296
FER OLIVE 8916
FER OXYDULE 7927
FER PLAT 5375
FER PLAT A BOUDIN 1714
FER PLAT A BOUDIN 5376
FER POUR CLOTURE 1086
FER POUR TABLIERS DE PONT 7112
FER PROFILE FACONNE 11119
FER PUDDLE 9962
FER ROND 10795
FER SOUDE 13766
FER SPATHIQUE 11421
FER SPECULAIRE 11865
FER TENACE 13068
FER VIRGINAL 346
FER ZORES 7112
FER-BLANC 12957
FERAILLES DE PRODUCTION PROPRE 6554
FERBLANTIER 12964
FERME 6447
FERMENT 5097
FERMENTATION 5098

FERMENTER 10036
FERMER L'INTRODUCTION 11407
FERMER LE CIRCUIT ELECTRIQUE 2515
FERMES 13243
FERMETURE A BAIONNETTE 1074
FERMETURE A GRENOUILLERE 7336
FERMETURE DE LA SOUPAPE 2537
FERMOIR 5264
FERRAILLE 5126
FERRAILLE 11007
FERRAILLES DE PROTECTION PROPRE 4124
FERRICYANURE DE POTASSIUM 9684
FERRITE 5104
FERRO-ALUMINIUM 5106
FERRO-BORE 5107
FERRO-CHROME 5108
FERRO-MANGANESE 5110
FERRO-MOLYBDENE 5111
FERRO-NICKEL 5112
FERRO-SILICIUM 5113
FERRO-TITANE 5114
FERRO-TUNGSTENE 5115
FERRO-VANADIUM 5116
FERROALLIAGES 5117
FERROCYANURE DE POTASSIUM 9685
FERROMAGNETIQUE 5118
FERRUGINEUX 5127
FERRURE 7159
FERS A BETON 12198
FERS D'HUISSERIE 4122
FERS FORGES 5566
FEU D'OXYDATION 8636
FEU DE DEPASSEMENT 9102
FEU DE FORGE 5548
FEU DE REDUCTION 7820
FEU NU 8813
FEUILLARD 12378
FEUILLARD 12343
FEUILLARD (TOLE MINCE) 12216
FEUILLARD DE FER 6575
FEUILLARD INOX 10875
FEUILLARD-SUPPORT 902
FEUILLARDS 980
FEUILLARDS D'ACIER POUR EMBOUTISSAGE PROFOND 3688
FEUILLARDS EN ACIER ALLIE 371
FEUILLE 5508
FEUILLE D'ALUMINIUM 436
FEUILLE D'ETAIN 12955
FEUILLE D'OR 6000
FEUILLE D'UN RESSORT 9492
FEUILLE DE CAOUTCHOUC (NEOPRENE) 11329

FEUILLE DE CLINQUANT 7621
FEUILLE DE METAL 5508
FEUILLE DE PLACAGE 13524
FEUILLE DE PLOMB 7462
FEUILLE DE TOLE 11313
FEUILLE DE ZINC 11327
FEUILLE METALLIQUE 8184
FEUILLE METALLIQUE 11315
FEUILLE MINCE DE CUIVRE 3117
FEUILLE MINCE DE METAL 5509
FEUILLERET 10115
FEUILLET 13524
FEUILLETAGE (DEDOUBLEMENT DES BORDS) DE TOLE 7382
FEUILLURE 10269
FEUTRE 5090
FEUTRE ASPHALTE 765
FEUTRE D'AMIANTE 750
FEUX ARRIERE 12595
FEUX DE DIRECTION 3989
FEUX DE RECUL 908
FIABILITE 10439
FIBRE 13609
FIBRE 5131
FIBRE 5136
FIBRE DE COCO 2654
FIBRE DE PAILLE 12348
FIBRE DU CANADA 1898
FIBRE NEUTRE 8542
FIBRE NEUTRE 2151
FIBRE TEXTILE 12799
FIBRE VULCANISEE 13609
FIBRES (DANS LE SENS) 6031
FIBREUX 5137
FICHE D'ESSAI 12785
FICHE DE PRISE DE COURANT 9529
FICHES TECHNIQUES 3615
FIDELITE DE SERVICE 10439
FIGURE 6778
FIGURE D'INTERFERENCE 7039
FIGURE DE CORROSION 4845
FIGURES D'ATTAQUE A L'ACIDE 4842
FIL (BAGUETTE) A SOUDER 13793
FIL (CROISSANCE EN FILAMENT) 13828
FIL A PLOMB 9539
FIL A SOUDER 13789
FIL BARBELE 987
FIL CLAIR 13340
FIL CONDUCTEUR 2922
FIL D'ACIER 12221
FIL D'ACIER POUR CORDES A PIANO 9287
FIL D'ALUMINIUM 428

FIL D'AMIANTE 749
FIL D'ARCHAL 1571
FIL DE BRONZE 1654
FIL DE BRONZE PHOSPHOREUX 5607
FIL DE CARET 10744
FIL DE CUIVRE 3133
FIL DE CUIVRE 3106
FIL DE CUIVRE PLAT 3116
FIL DE FER 7155
FIL DE FER CLAIR CRU 1610
FIL DE FER RECUIT NOIR 1284
FIL DE LAITON 1571
FIL DE LIGATURE 1255
FIL DE PLATINE 9513
FIL DE PLOMB 7477
FIL DE ZINC 14048
FIL ELECTRIQUE 2922
FIL ELLIPTIQUE 4678
FIL EN VERGE POUR TREFILERIE 13882
FIL ETAME 12962
FIL ETIRE 4252
FIL ETIRE A FROID 2715
FIL FIN 5205
FIL GAINE 11300
FIL GAINE DE PLASTIQUE 9468
FIL GALVANISE 5786
FIL GALVANISE 12962
FIL GUIPE 3219
FIL HEXAGONAL 6462
FIL ISOLE 6996
FIL LAMINE 10678
FIL LAMINE DE LAITON 1568
FIL MACHINE DE LAITON 1568
FIL MACHINE EN ACIER ALLIE 376
FIL MEPLAT 5402
FIL METALLIQUE 13880
FIL METALLIQUE FIN 12849
FIL MINCE 5205
FIL MONOCRISTALLIN 13828
FIL NU 990
FIL OVALE 4678
FIL PLAT 5402
FIL PLAT EN CUIVRE 3116
FIL PLOMBE 7459
FIL POISSE 13704
FIL POUR CABLE 10743
FIL POUR RESISTANCES ELECTRIQUES 10501
FIL POUR RESSORT 11997
FIL PROFILE 11847
FIL RECOUVERT 3272
FIL RECTANGULAIRE 5402

FIL RECUIT **561**
FIL RECUIT BLANC **1606**
FIL ROND **10804**
FIL TELEGRAPHIQUE **12708**
FIL TELEPHONIQUE **12711**
FIL TREFILE **4252**
FIL TRESSE **1528**
FIL TRIANGULAIRE **13196**
FIL ZINGUE **5786**
FILAGE **4970**
FILAGE **4230**
FILAGE A FROID **4970**
FILAGE AVEC REMONTEE DE MATIERES **6884**
FILAGE DIRECT **3984**
FILAMENT **6386**
FILAMENT DE QUARTZ **10078**
FILASSE DE CHANVRE **6439**
FILET **5092**
FILET **11045**
FILET (AGE) INTERIEUR **7078**
FILET A DROITE **10585**
FILET A GRAND DIAMETRE **2575**
FILET A PAS ALLONGE **10105**
FILET A PAS FIN **5201**
FILET A PAS RAPIDE **10105**
FILET A PETIT PAS **11647**
FILET CARRE **12030**
FILET EXTERIEUR **4960**
FILET FIN **5204**
FILET FLUIDE **12351**
FILET FOIRE **13981**
FILET JARETE **13981**
FILET METRIQUE INTERNATIONAL **7083**
FILET PROTECTEUR **6157**
FILET ROND **10802**
FILET TRAPEZOIDAL **1814**
FILET TRIANGULAIRE **545**
FILETAGE **12857**
FILETAGE **12868**
FILETAGE **11045**
FILETAGE **3535**
FILETAGE (OPERATION) **11028**
FILETAGE CONIQUE **12666**
FILETAGE CYLINDRIQUE **9065**
FILETAGE EXTERIEUR **7955**
FILETAGE EXTERIEUR **4961**
FILETAGE FEMELLE **5092**
FILETAGE INTERIEUR **5092**
FILETAGE MALE **7955**
FILETER **3512**
FILETEUSE **11066**

FILIERE **4223**
FILIERE **3883**
FILIERE A COUSSINETS **12266**
FILIERE A TREFILER **13887**
FILIERE A TRUELLE **11039**
FILIERE A TUBES **1121**
FILIERE DE TREFILAGE **4223**
FILIERE RONDE MONOBLOC **11781**
FILIERE SIMPLE **11039**
FILIERES **189**
FILM **5166**
FILM ANTI-MOUILLANT **13648**
FILM RIGIDE **8993**
FILTRAGE **5182**
FILTRAT **5181**
FILTRATION **5182**
FILTRE **5172**
FILTRE **12333**
FILTRE A AIR **260**
FILTRE A AIR A BAIN D'HUILE **8746**
FILTRE A CARBURANT **5708**
FILTRE A COKE **2660**
FILTRE A COMBUSTIBLE **5708**
FILTRE A GRAVIER **6074**
FILTRE A HUILE **8757**
FILTRE A SABLE **10923**
FILTRE PREPARATOIRE **5265**
FILTRE-PRESSE **5180**
FILTRER **5173**
FILTRES SYMETRIQUES **933**
FIN DE BANDE **4725**
FIN DE BLOC **4730**
FIN DE LIGNE **4731**
FIN DE PROGRAMME **4724**
FINESSE **5208**
FINI SATINE **10938**
FINIR **5214**
FINISSAGE **5223**
FINISSAGE **5213**
FINISSAGE **11682**
FINISSAGE A CHAUD **6655**
FINISSAGE A FROID **2679**
FINISSAGE A LA BROSSE METALLIQUE **11016**
FINISSAGE A LA MACHINE **7844**
FINISSAGE A LA MEULE **5224**
FINISSAGE AU LAMINOIR **8278**
FINISSAGE AU TONNEAU **1014**
FINISSAGE DIMENSIONNEL **11524**
FINITION **11682**
FINITION A LA FRAISE-MERE **6513**
FINITION PARTIELLE **11960**

FISSURAGE 3284
FISSURATION 3284
FISSURATION PAR CORROSION SOUS TENSION 12364
FISSURE 3278
FISSURE 2278
FISSURE 4377
FISSURE (MICROFISSURE) 3283
FISSURE CAPILLAIRE 3318
FISSURE CAPILLAIRE/MICROCRIQUE 6202
FISSURE CRIQUE DE LAMINAGE 7383
FISSURE DE FATIGUE / D'ENDURANCE 4739
FISSURE DE TREMPE 10097
FISSURE PAR FATIGUE 5048
FISSURE SOUS CORDON 13350
FISSURE SUPERFICIELLE 12501
FIXAGE 5044
FIXATION 5044
FIXATION 5416
FIXATION D'UN COIN DE SERRAGE 7704
FIXATION D'UNE EPROUVETTE 5300
FIXATION RIGIDE 10589
FIXER 5041
FLACON DESSECHANT 4308
FLACON LAVEUR 5845
FLACON SECHEUR 4308
FLAMBAGE 3343
FLAMBEMENT 3343
FLAMME CARBURANTE 1990
FLAMME D'HYDROGENE 6714
FLAMME ECLAIRANTE 7820
FLAMME FULIGINEUSE 11678
FLAMME INCOLORE 8636
FLAMME NEUTRE 8543
FLAMME NORMALE 8543
FLAMME OXYDANTE 8636
FLAMME OXYDANTE 8975
FLAMME OXYDANTE 4883
FLAMME REDUCTRICE 7820
FLAMME REDUCTRICE 4882
FLAMME REDUCTRICE 10349
FLANC 5348
FLANC DE BASE 10720
FLANC DE DENT 13018
FLANC DE LA DENT 5351
FLANC DE RACINE 10721
FLANC DE SAILLIE 167
FLANC DU FILET 5350
FLANELLE 5352
FLASQUE 5323
FLASQUE D'UN ARBRE COUDE 3304
FLASQUE D'UNE CHAINE A ROULEAUX 11416

FLEAU DE BALANCE 10972
FLECHE 5426
FLECHE 3698
FLECHE 10613
FLECHE 719
FLECHE (MEC.) 3701
FLECHE DE COTE 3395
FLECHE DE GRUE 3291
FLECHE MAXIMUM 8064
FLECHE PRINCIPALE D'UNE GRUE 7226
FLECHIR 1174
FLECHIR 10893
FLECHISSEMENT 10892
FLECHISSEMENT 10894
FLECHISSEMENT 3700
FLEUR DE SOUFRE 5464
FLEUR DU CUIR 6038
FLEURET ELECTRIQUE 4556
FLEXIBILITE 5415
FLEXIBLE 9518
FLEXION 5422
FLEXION 3698
FLEXION 1182
FLEXION A CHAUD 6619
FLEXION EFFECTUE SUR DES ACIERS TRAITES AVEC DES TREMPES (ESSAI DE...) 10088
FLEXION ELASTIQUE 4477
FLEXION TRANSVERSALE 3700
FLINT 5427
FLINT-GLASS 5428
FLOCON 5303
FLOCULATION 5441
FLOTTABILITE 1742
FLOTTATION 344
FLOTTATION 5452
FLOTTATION ANIONIQUE 557
FLOTTATION SELECTIVE 11144
FLOTTEMENT D'UNE COURROIE 13825
FLOTTEUR 5430
FLOTTEUR (DE JAUGE) 5876
FLOTTEUR DE STOCKAGE ET D'ACCOSTAGE 10217
FLUAGE (METAUX) 3324
FLUAGE A FROID 2680
FLUAGE PLASTIQUE 9471
FLUAGE SECONDAIRE 11104
FLUAGE TRANSITOIRE 13118
FLUAGE VISQUEUX 13578
FLUCTUATIONS DE VITESSE 13499
FLUIDE 7641
FLUIDE DE COUPE 3530
FLUIDE DE REFROIDISSEMENT 3083

FLUIDE GAZEUX **5808**
FLUIDE LIQUIDE **7641**
FLUIDES DE COUPE CHIMIQUES **2303**
FLUIDES DE COUPE SYNTHETIQUES **2303**
FLUIDITE **5474**
FLUIDITE **5463**
FLUOR **5478**
FLUORESCENCE **5476**
FLUORINE **5479**
FLUORURE D'AMMONIUM **463**
FLUORURE DE BARYUM **999**
FLUORURE DE CALCIUM **5479**
FLUORURE DE LITHIUM **7667**
FLUORURE DE MAGNESIUM **7886**
FLUORURE DE POTASSIUM **9686**
FLUORURE DE SODIUM **11721**
FLUORURE DE STRONTIUM **12390**
FLUSSPATH **5479**
FLUX BASIQUE **1046**
FLUX LUMINEUX **7821**
FLUX MAGNETIQUE **7903**
FOIE DE SOUFRE **7674**
FOISONNER **12542**
FOLIUM DE DESCARTES **5513**
FONCTION (MATH.) **5734**
FONCTION AUXILIAIRE **8319**
FONCTION AUXILIAIRE **854**
FONCTION CIRCULAIRE **2413**
FONCTION CONTINUE **3025**
FONCTION CYCLIQUE **2413**
FONCTION DE (EN) **13542**
FONCTION DE DISTRIBUTION **9881**
FONCTION DE REPARTITION COMPOSEE **7239**
FONCTION EXPONENTIELLE **4942**
FONCTION GAMMA **5794**
FONCTION GENERATRICE (DES MOMENTS) **5919**
FONCTION HYPERBOLIQUE **6737**
FONCTION OUTIL **13001**
FONCTION PREPARATOIRE **9786**
FONCTIONNEMENT **10850**
FONCTIONNEMENT **8837**
FONCTIONNEMENT CONTINU **3038**
FONCTIONNEMENT INTERMITTENT **7056**
FONCTIONNEMENT SUR (DE) **10440**
FOND **6314**
FOND **1028**
FOND BOMBE **4036**
FOND D'UN BALLON **6316**
FOND DE FILET **12864**
FOND DE POCHE **1812**
FOND DU CYLINDRE **1494**

FOND EMBOUTI **4038**
FOND EN ANSE DE PANIER **13045**
FOND TORICONIQUE **13044**
FONDANT **5475**
FONDANT **5489**
FONDANT ACIDE **114**
FONDANT POUR SOUDER **5490**
FONDATION **5589**
FONDATION **2896**
FONDATION D'UNE MACHINE **7848**
FONDATION EN BRIQUES **1599**
FONDATION EN MACONNERIE **1599**
FONDERIE **5593**
FONDERIE **5592**
FONDERIE (ART) **5592**
FONDERIE D'ACIER **12207**
FONDERIE DE CUIVRE **1566**
FONDERIE DE FER/DE FONTE **7142**
FONDERIE SUR MODELES **7230**
FONDERIES (ETABLISSEMENT) **5597**
FONDEUR **5591**
FONDRE **8135**
FONDRE **2030**
FONDRE (CHIM.) **4061**
FONDRE (LE COUPE-CIRCUIT FOND) **1352**
FONDRE (LE COUSSINET FOND) **10837**
FONTE **2068**
FONTE **2034**
FONTE **7157**
FONTE (DURCIE) **2336**
FONTE A GRANDE RESISTANCE **2042**
FONTE A GRAPHITE SPEROIDAL **4362**
FONTE A GRAPHITE SPHEROIDAL **2045**
FONTE A GRAPHITE SPHEROIDAL **11895**
FONTE A GRAPHITE SPHEROIDAL **11894**
FONTE A L'AIR CHAUD **6620**
FONTE A L'AIR FROID **2669**
FONTE A SOUFFLURES **6563**
FONTE ACIEREE **11177**
FONTE ALLIEE **363**
FONTE ALLIEE **2037**
FONTE AU CHARBON DE BOIS **2258**
FONTE AU COKE (DE HAUT-FOURNEAU) **2657**
FONTE AU COKE BOIS **2664**
FONTE AU FOUR ELECTRIQUE **4510**
FONTE BESSEMER **1213**
FONTE BESSEMER **117**
FONTE BLANCHE **13830**
FONTE BLANCHE **2049**
FONTE BRUTE **9299**
FONTE BRUTE BESSEMER **105**

FONTE CRUE **9299**
FONTE D'ACIER **2055**
FONTE D'AFFINAGE **5544**
FONTE DE COQUILLE **2038**
FONTE DE FER **2034**
FONTE DE MOULAGE **5596**
FONTE DE TRANSITION **8733**
FONTE DUCTILE **11894**
FONTE DUCTILE **2045**
FONTE DUCTILE **2040**
FONTE ELECTRIQUE **4509**
FONTE ELECTRIQUE **4550**
FONTE EN CHAMBRE FROIDE **2671**
FONTE EN COQUILLE **2334**
FONTE EN COQUILLE **2335**
FONTE GRAPHITIQUE **1292**
FONTE GRISE **2041**
FONTE GRISE **5596**
FONTE GRISE **6094**
FONTE HEMATITE **6435**
FONTE MALLEABLE **7959**
FONTE MALLEABLE **2043**
FONTE MALLEABLE DE CUBILOT **3467**
FONTE MALLEABLE PERLITIQUE **2047**
FONTE MECANIQUE **7863**
FONTE MECANIQUE **4763**
FONTE NODULAIRE **8607**
FONTE NODULAIRE **11894**
FONTE PERLITIQUE **9200**
FONTE PERLITIQUE **2046**
FONTE RESISTANTE A L'USURE **2048**
FONTE RESISTANTE A LA CORROSION **2039**
FONTE SANS RETASSURES **2075**
FONTE SANS TENSIONS INTERNES **2074**
FONTE SPECULAIRE **11899**
FONTE THOMAS **1049**
FONTE TREMPEE **2038**
FONTE TRUITEE **2044**
FONTE TRUITEE **8420**
FORAGE **4277**
FORAGE **1468**
FORAGE D'ESSAI **12778**
FORAGE EN PROFONDEUR **3780**
FORCE **12352**
FORCE **6907**
FORCE **5532**
FORCE ASCENSIONNELLE **1742**
FORCE ATTRACTIVE **5529**
FORCE CENTRIFUGE **2180**
FORCE CENTRIPETE **2190**
FORCE COERCITIVE **2641**

FORCE COERCITIVE **2639**
FORCE COMPOSANTE **2816**
FORCE CONSTANTE **2973**
FORCE CONTRE-ELECTROMOTRICE **3231**
FORCE D'UN RESSORT **5528**
FORCE DE CISAILLEMENT **11291**
FORCE DE COHESION **2644**
FORCE DE COMPRESSION **2868**
FORCE DE L'HOMME **9738**
FORCE DE TORSION **13055**
FORCE DE TRACTION **12745**
FORCE DU FROTTEMENT **5683**
FORCE ELASTIQUE D'UN GAZ **4480**
FORCE ELECTROMOTRICE **4622**
FORCE ELECTROMOTRICE THERMIQUE **12807**
FORCE ELECTROSTATIQUE **4642**
FORCE EXTERIEUR **4954**
FORCE HYDRAULIQUE **13668**
FORCE IMPULSIVE **6811**
FORCE INTERIEURE **7064**
FORCE MAGNETIQUE **7904**
FORCE MAGNETISANTE **7925**
FORCE MAGNETOMOTRICE **7932**
FORCE MAGNETOMOTRICE SPECIFIQUE **7929**
FORCE MOLECULAIRE **8360**
FORCE MOTRICE **8409**
FORCE MOTRICE **9733**
FORCE NECESSAIRE **3746**
FORCE POLAIRE **9596**
FORCE REACTIVE **10248**
FORCE REPULSIVE **5530**
FORCE RESISTANCE **10484**
FORCE TANGENTIELLE **12622**
FORCE THERMOELECTRIQUE **12831**
FORCE UNIQUE **11483**
FORCE VIVE **7308**
FORCER UNE CLAVETTE DANS SA RAINURE **4293**
FORCES CONCOURANTES **2903**
FORCES DE MEME SENS **5538**
FORCES OPPOSEES EN DIRECTION **5537**
FORCES PARALLELES **9056**
FORE **4275**
FORER **4264**
FORER PREALABLEMENT **10771**
FORET **4271**
FORET **4263**
FORET A CANALISATION D'HUILE **8762**
FORET A CANON **6177**
FORET A CENTRE **2164**
FORET A CENTRER **2763**
FORET A COUTURE DROITE **12304**

FORET A CUILLER **11957**
FORET A FRAISER **3245**
FORET A HELICE **13304**
FORET A HELICE SERREE **6470**
FORET A LANGUE D'ASPIC **5380**
FORET A MISE EN CARBURE **1936**
FORET A TETON **2164**
FORET A TETON CYLINDRIQUE **9316**
FORET ALESEUR **3241**
FORET ALESEUR **3149**
FORET AMERICAIN **13304**
FORET CHAMPIGNON **3245**
FORET ELECTRIQUE **4556**
FORET HELICOIDAL **13304**
FORET HELICOIDE **13304**
FORET OU MECHE A CANON **3563**
FORET POUR PERCAGE PROFOND **3686**
FOREUR **4276**
FOREUSE **4279**
FOREUSE A COLONNE **9312**
FORGE **5548**
FORGE **5547**
FORGE (USINE) **7160**
FORGE A LA MAIN **6261**
FORGE D'UNE PIECE DANS LA MASSE **5552**
FORGE DE MARECHALE **5548**
FORGE PORTATIVE **9637**
FORGE VOLANTE **9637**
FORGEABILITE **7958**
FORGEABLE **7960**
FORGEAGE **5555**
FORGEAGE **5557**
FORGEAGE A CHAUD **6626**
FORGEAGE A FROID **2681**
FORGEAGE A LA MACHINE **7847**
FORGEAGE A LA MAIN **11669**
FORGEAGE A LA PRESSE **9807**
FORGEAGE A LA PRESSE **9798**
FORGEAGE PAR LAMINAGE **10668**
FORGER **5542**
FORGER A LA PRESSE **9800**
FORGERON **11671**
FORGERON A MAIN **6225**
FORMAGE **5579**
FORMAGE A CHAUD **6627**
FORMALDEHYDE **5573**
FORMAT **5574**
FORMAT A ADRESSES **13935**
FORMAT A ADRESSES DE BLOCS **1334**
FORMAT A BLOC FIXE **5288**
FORMAT A BLOC VARIABLE **13489**

FORMAT A SEQUENCE FIXE **5297**
FORMAT A TABULATION **12583**
FORMATION **5579**
FORMATION D'INCRUSTATIONS **6856**
FORMATION D'UNE CROUTE **1927**
FORMATION DE CIRQUES **2286**
FORMATION DE FISSURES **644**
FORMATION DE GERMES **5934**
FORMATION DE HALO **6203**
FORMATION DE LA FUMEE **11675**
FORMATION DE LA MOISISSURE **5575**
FORMATION DE PEAUX **11545**
FORMATION DE PONT **1604**
FORMATION DE ROUILLE **10865**
FORMATION DES TACHES **11961**
FORMATION DU MODELE **3041**
FORME **11264**
FORMENE **8215**
FORMER LA TETE DU RIVET **2516**
FORMOL **5573**
FORMULE **5582**
FORMULE APPROXIMATIVE **654**
FORMULE BRUTE **4704**
FORMULE CHIMIQUE **2310**
FORMULE D'APPROXIMATION **654**
FORMULE DE CONSTITUTION **5583**
FORTE PRESSION **6473**
FOSSE A BATTITURES **10977**
FOSSE DU VOLANT **5498**
FOUETTEMENT D'UNE COURROIE **13825**
FOULOIR A MAIN **6246**
FOULOIR PNEUMATIQUE **286**
FOUR **5739**
FOUR **8920**
FOUR **12300**
FOUR **11667**
FOUR A ARC **4495**
FOUR A ARC A RESISTANCE **10485**
FOUR A ARC ELECTRIQUE **668**
FOUR A ARC ELECTRIQUE A ELECTRODES DE CHARBON **670**
FOUR A ARC INDIRECT **669**
FOUR A ATMOSPHERE CONTROLEE **3055**
FOUR A BAIN DE SEL **10912**
FOUR A BANDE **1148**
FOUR A CHARGER **1060**
FOUR A CHARIOTER **11603**
FOUR A CHAUFFER LES RIVETS **10623**
FOUR A CLOCHE **1132**
FOUR A CLOCHE **1135**
FOUR A COKE **2661**

FOUR A CORNUE **10524**
FOUR A CREUSET **9669**
FOUR A CREUSET **4597**
FOUR A CUVE **11241**
FOUR A CYANURATION **3550**
FOUR A FONDRE **11665**
FOUR A FRITTER **11512**
FOUR A HAUTE FREQUENCE **6469**
FOUR A INDUCTION **6895**
FOUR A INDUCTION A BASSE FREQUENCE **7772**
FOUR A MANCHE **11241**
FOUR A MOUFLE **8447**
FOUR A NITRURATION **8584**
FOUR A POULIES **9975**
FOUR A RECHAUFFER EN PAQUETS **9308**
FOUR A RECUIRE **564**
FOUR A RECUIRE **10426**
FOUR A REGENERATION **10321**
FOUR A REGENERATION **10410**
FOUR A REVERBERE **10532**
FOUR A SECOUSSES **10024**
FOUR A SOIE **6332**
FOUR A SOUDER **11763**
FOUR CHAUFFE AU GAZ **5851**
FOUR CONTINU **3026**
FOUR D'AFFINAGE **10379**
FOUR D'EGALISATION **11698**
FOUR D'HORLOGER **13639**
FOUR DE CALCINATION **1843**
FOUR DE CALCINATION **7295**
FOUR DE CARBURISATION **1991**
FOUR DE CEMENTATION **3074**
FOUR DE CHAUFFERIE **5747**
FOUR DE FUSION **8138**
FOUR DE GRILLAGE **10649**
FOUR DE MAINTIEN **6527**
FOUR DE PREMIER ALLIAGE **3607**
FOUR DE PUDDLAGE **9965**
FOUR DE RECUIT **564**
FOUR DE REVENU **564**
FOUR ELECTRIQUE **4507**
FOUR ELECTRIQUE A ARC DIRECT **671**
FOUR ELECTRIQUE A ARC DIRECT **667**
FOUR ELECTRIQUE A INDUCTION **4515**
FOUR ELECTRIQUE A RESISTANCE **10490**
FOUR INDUSTRIEL **6900**
FOUR MARTIN **8815**
FOUR MARTIN **8829**
FOUR METALLURGIQUE **8207**
FOUR NON CONTINU **1060**
FOUR POUSSANT **10030**

FOUR PRECHAUFFEUR **9777**
FOUR REDUCTEUR **10359**
FOUR REVERBERE **261**
FOUR STATIONNAIRE FIXE **12142**
FOURCHE DE DEBRAYAGE **12346**
FOURCHE GUIDE-COURROIE **12346**
FOURCHETTE **12346**
FOURGON **13474**
FOURNEAU **12300**
FOURNEAU **5739**
FOURNEAU A BRASER **1577**
FOURNEAU A CALEBASSE **3402**
FOURNEAU A CREUSET **3402**
FOURNEAU A LA WILKINSON **3466**
FOURNEAU D'ACHESON **97**
FOURNEAU DE FUSION **11667**
FOURNIR DE NOUVELLES CALORIES **12485**
FOURNITURE DE CHALEUR **12488**
FOURREAU **11576**
FOURRURE **7623**
FOURRURE DUN PALIER **7624**
FOURRURE EN BOIS **13927**
FOYER (OPT. ET GEOM.) **5502**
FOYER A CHARBON **2565**
FOYER A CHARBON PULVERISE **5746**
FOYER A GAZ **5744**
FOYER A HOUILLE **2565**
FOYER A HUILE LOURDE **5745**
FOYER A PETROLE **5745**
FOYER A RECUPERATION INTERMITTENTE **10410**
FOYER AUTOMATIQUE **8110**
FOYER DE CHAUDIERE **5747**
FOYER FIN (TUBE DE RAYONS X) **8239**
FOYER FUMIVORE **11674**
FOYER REEL **10258**
FOYER VIRTUEL IMAGINAIRE **13571**
FRACTION **5613**
FRACTION DE DISTILLATION **5611**
FRACTION DECIMALE **3662**
FRACTION DECIMALE PERIODIQUE **9192**
FRACTION DECIMALE TERMINEE **5231**
FRACTION ORDINAIRE **13611**
FRACTION PROPRE **9932**
FRACTIONNER **5615**
FRACTURE **10858**
FRAGILE **5621**
FRAGILISATION A CHAUD **6625**
FRAGILITE **4685**
FRAGILITE **5622**
FRAGILITE **11378**
FRAGILITE **1628**

FRAGILITE (NON-) **13342**
FRAGILITE A CHAUD **10339**
FRAGILITE A CHAUD **6641**
FRAGILITE A FROID **2703**
FRAGILITE AU BLEU **1631**
FRAGILITE AU ZINGAGE **5790**
FRAGILITE CAUSTIQUE **2118**
FRAGILITE D'ENTAILLE **1632**
FRAGILITE D'ENTAILLE **8669**
FRAGILITE DE DECAPAGE **109**
FRAGILITE DE DECAPAGE **6713**
FRAGILITE DE REVENU **1634**
FRAGILITE DU FER **1629**
FRAGILITE PAR CORROSION **3186**
FRAGILITE PAR DECAPAGE **1630**
FRAGILITE PAR PENETRATION DE SOUDURE **11757**
FRAGILITE RHEOTROPIQUE **1633**
FRAGMENTATION DE GRAINS **5623**
FRAGMENTS MENUS **8267**
FRAIS **7346**
FRAIS D'EMBALLAGE **3205**
FRAIS D'ENTRETIEN **7945**
FRAIS D'EXPLOITATON **13950**
FRAIS DE CONSOMMATION **8049**
FRAIS DE FABRICATION **3207**
FRAIS DE FORCE MOTRICE **3206**
FRAIS DE MANUTENTION **3204**
FRAIS DE MATIERE **8049**
FRAIS DE PREMIER ETABLISSEMENT **9858**
FRAIS DE PRODUCTION **3207**
FRAIS DE REPARATION **3208**
FRAIS DE TRANSPORT **3209**
FRAIS GENERAUX **8798**
FRAISAGE **3234**
FRAISAGE **8287**
FRAISAGE CLASSIQUE **3061**
FRAISAGE D'UN TROU DE RIVET **3247**
FRAISAGE DES CONTOURS **3040**
FRAISAGE DES MATRICES **3892**
FRAISAGE EN AVALANT **2497**
FRAISAGE EN DESCENDANT **4182**
FRAISAGE EN DUPLEX **12301**
FRAISAGE EN REMONTANT **13408**
FRAISAGE EN TRAIN **5800**
FRAISAGE RECIPROQUE **10282**
FRAISE **8289**
FRAISE **3519**
FRAISE A DENTS DEGAGEES **905**
FRAISE A DENTS DEPOUILLEES **905**
FRAISE A DENTS RAPPORTEES **6971**
FRAISE A DENTURE DROITE **12307**

FRAISE A DENTURE HELICOIDALE **11917**
FRAISE A DISQUE **8496**
FRAISE A LAMES RAPPORTEES **6971**
FRAISE A PROFIL CONSTANT **905**
FRAISE A QUEUE **11259**
FRAISE A SURFACER **4722**
FRAISE A TAILLER LES ENGRENAGES **5898**
FRAISE ANGULAIRE **543**
FRAISE AXIALE **4446**
FRAISE CONIQUE **543**
FRAISE CYLINDRIQUE **13848**
FRAISE D'EXTERIEUR POUR TUBES **4951**
FRAISE D'INTERIEUR POUR TUBES **7059**
FRAISE DE FACE **4722**
FRAISE DE FORME **9899**
FRAISE DISQUE **11412**
FRAISE EBARBEUSE **1765**
FRAISE EBARBEUSE FEMELLE **4951**
FRAISE EBARBEUSE MALE **7059**
FRAISE EN BOUT **4722**
FRAISE FRONTALE **4722**
FRAISE MERE **13971**
FRAISE POUR LOGEMENT DE TETES DE VIS **3245**
FRAISE POUR RAINURES **11634**
FRAISE POUR TAILLER DES ENGRENAGES HELICOIDAUX
  13972
FRAISE PROFILEE **9899**
FRAISE RADIALE **4722**
FRAISE RECTILIGNE **13848**
FRAISE-MERE **6512**
FRAISE-SCIE **8193**
FRAISER **8276**
FRAISER UN TROU DE RIVET **3246**
FRAISER UN TROU DE VIS **3246**
FRAISES A COUTEAUX **11632**
FRAISES A DEUX TAILLES EN BOUT **4992**
FRAISES A RAINURE WOODRUFF **13930**
FRAISES A RAINURES **11632**
FRAISES A UNE TAILLE LATERALE **6206**
FRAISES ARBREES **663**
FRAISES CONIQUES **540**
FRAISES EN BOUT **4723**
FRAISES EN BOUT ALESEES **11337**
FRAISES EN BOUT PLEINES **11783**
FRAISES POUR QUEUE D'ARONDE **4176**
FRAISES POUR RAINURES EN T **12582**
FRAISES SIMPLES POUR TRAVAUX LOURDS **6404**
FRAISES SIMPLES A HELICE RAPIDE **6471**
FRAISES-MERES **5577**
FRAISEUR **8292**
FRAISEUSE **8291**

FRAISEUSE A BANC FIXE **5298**
FRAISEUSE A COMMANDE PAR DISPOSITIF DE COPIAGE **13080**
FRAISEUSE A CONSOLE ET COLONNE **7319**
FRAISEUSE A COPIER **9900**
FRAISEUSE A PORTIQUE **9642**
FRAISEUSE A TAILLER LES ENGRENAGES **5894**
FRAISEUSE CIRCULAIRE **2420**
FRAISEUSE DUPLEX A GRAND RENDEMENT **4379**
FRAISEUSE HORIZONTALE **6589**
FRAISEUSE MULTIPLE **8471**
FRAISEUSE RADIALE A COULISSE **10190**
FRAISEUSE UNIVERSELLE **13387**
FRAISEUSE VERTICALE **13548**
FRAISEUSE-FACONNEUSE **9900**
FRAISEUSE-RABOTEUSE **9460**
FRAISSAGE D'UN TROU A VIS **3247**
FRANCHISE (DELAI AVANT) **6020**
FRANGE D'INTERFERENCE **7040**
FRAPPE-DEVANT **11573**
FRAPPEUR DE POINTES **13891**
FRAUDE **205**
FREIN **1529**
FREIN **5031**
FREIN A AIR COMPRIME **2840**
FREIN A BANDE **5681**
FREIN A CHAINE **2203**
FREIN A COLLIER **5681**
FREIN A CONE **2926**
FREIN A CORDE **10732**
FREIN A COURANTS DE FOUCAULT **4439**
FREIN A DEPRESSION **13436**
FREIN A DEUX MACHOIRES **4129**
FREIN A DEUX SABOTS **4129**
FREIN A DISQUE **8454**
FREIN A DISQUE **4001**
FREIN A ENROULEMENT **5681**
FREIN A FRICTION **5669**
FREIN A GENOUILLERE **12978**
FREIN A HUILE **8755**
FREIN A LAME **8454**
FREIN A LEVIER **7532**
FREIN A LEVIER ET CONTREPOIDS **3638**
FREIN A MAIN **6227**
FREIN A MAIN **9082**
FREIN A PEDALE **5519**
FREIN A PIED **5519**
FREIN A RUBAN **5681**
FREIN A SABOT **1335**
FREIN A TAMBOUR **4334**
FREIN A VAPEUR **12159**

FREIN A VIDE **13436**
FREIN A VIS **12344**
FREIN CENTRIFUGE **2175**
FREIN D'ECROU **8701**
FREIN DE RETENUE **7681**
FREIN DE SECOURS **4687**
FREIN DE SECURITE **10883**
FREIN DE STATIONNEMENT **9082**
FREIN DIFFERENTIEL **3907**
FREIN DYMAMOMETRIQUE DE PRONY **9920**
FREIN ELECTROMAGNETIQUE **4613**
FREIN HYDRAULIQUE **8755**
FREIN HYDRAULIQUE **6678**
FREIN HYDRAULIQUE **6709**
FREIN HYDROPNEUMATIQUE **6678**
FREIN MAGNETIQUE **7896**
FREIN-MOTEUR **4760**
FREINAGE **1549**
FREINAGE INITIAL **6928**
FREINER **1530**
FREINS ASSISTES **9731**
FREQUENCE **5661**
FREQUENCE EMPIRIQUE **5662**
FRETTAGE **11398**
FRETTE **7203**
FRETTE **2931**
FRETTE **1252**
FRETTE POSEE A CHAUD **11391**
FRETTER A CHAUD **11390**
FRICTION **5668**
FRICTION INTERNE **7065**
FRISAGE **13985**
FRITTAGE **5690**
FRITTAGE **11511**
FRITTAGE CONTINU **3014**
FRITTAGE SOUS PRESSION **9799**
FRITTAGE SOUS VIDE **13448**
FRITTER **11509**
FRONT DE CRISTALLISATION **11798**
FRONT DE SOLIDIFICATION **11798**
FROTTAGE **8450**
FROTTEMENT **5668**
FROTTEMENT AU DEMARRAGE **5678**
FROTTEMENT AU DEPART **5678**
FROTTEMENT CINETIQUE **7309**
FROTTEMENT DANS LES PALIERS **1105**
FROTTEMENT DE GLISSEMENT **11601**
FROTTEMENT DE ROULEMENT **10700**
FROTTEMENT DES TOURILLONS **7258**
FROTTEMENT EN MARCHE **5677**
FROTTEMENT INTERIEUR DES LIQUIDES **5473**

FROTTEMENT INTERNE **7065**
FROTTEMENT PENDANT LE MOUVEMENT **5677**
FROTTEMENT STATIQUE **12134**
FROTTEUR **8451**
FUEL **5714**
FUIR **7487**
FUIT (QUI) **7493**
FUITE **7488**
FUITE **7492**
FUITE AUX JOINTS **7486**
FULMICOTON **6176**
FULMINATE D'ARGENT **11463**
FULMINATE DE MERCURE **8154**
FUMEE **11672**
FUMEES **5732**
FUMEES INDUSTRIELLES **6899**
FUMEES ROUSSES **10327**
FUMIVORE **11673**
FUNICULAIRE **1826**
FUSEAU D'UN ENGRENAGE A LANTERNE **9323**
FUSEAU SPHERIQUE **11886**
FUSEE D'ESSIEU **882**
FUSEE ECLAIRANTE **7573**
FUSEES **5443**
FUSEES **5434**
FUSIBILITE **5755**
FUSIBLE **5751**
FUSIBLE **5758**
FUSIBLE **5756**
FUSIBLE DES COUPE-CIRCUIT **10886**
FUSION **11666**
FUSION **5760**
FUSION **8134**
FUSION **8136**
FUSION CONGRUENTE **2938**
FUSION EUTECTIQUE **4856**
FUSION IGNEE **5764**
FUSION PAR ZONE **14050**
FUSION SOUS VIDE **13444**
FUT D'UNE COLONNE **11243**
FUT DE CYLINDRE **3562**
FUT DE RABOT **12262**
FUT DU RIVET **11263**
GABARIT **9128**
GABARIT **12734**
GABARIT **5768**
GABARIT DE PERCAGE **4267**
GABARIT DE SOUDAGE **13785**
GABARIT POUR OUTILS A FILETER **12865**
GABARITS **5769**
GABARITS DE PERCAGE **4268**

GABBRO **5765**
GACHER **12718**
GADOLINIUM **5767**
GAILLETINS **2328**
GAILLETTE **2606**
GAILLETTERIE **2606**
GAIN **5773**
GAIN DE TEMPS **10949**
GAIN DE TRAVAIL **8094**
GAINE **7986**
GALBE **3396**
GALENE **5775**
GALET **1895**
GALET **10671**
GALET **12334**
GALET **10679**
GALET CONIQUE **12660**
GALET CYLINDRIQUE **3589**
GALET DE CONTACT **3008**
GALET DE FRICTION **5679**
GALET DE GUIDAGE **6161**
GALET DE POUSSOIR **1885**
GALET DE POUSSOIR **12674**
GALET DE ROULEMENT **10844**
GALET DE ROULEMENT **10854**
GALET POUR ROULEMENTS **1112**
GALET TENDEUR **4910**
GALET-GUIDE **6161**
GALETAGE **1758**
GALETS (GEOL.) **11354**
GALETS (POUR MISE SUR CHAMP) **10695**
GALETTE **9011**
GALLIUM **5779**
GALLON (MESURE ANGLAISE) **5781**
GALVANISATION **5787**
GALVANISATION **5789**
GALVANISATION **14044**
GALVANISATION (OU ZINGAGE) ELECTROLYTIQUE **4583**
GALVANISATION A CHAUD **6653**
GALVANISATION A FROID **2683**
GALVANISATION EN CONTINU **3027**
GALVANISATION PAR TREMPE **6628**
GALVANISATION PARTIELLE **9076**
GALVANISER **5784**
GALVANISER **366**
GALVANOMETRE **5791**
GALVANOPLASTIE **4640**
GALVANOPLASTIE **4651**
GALVANOPLASTIE AU TONNEAU **1016**
GALVANOPLASTIQUE **5792**
GALVANOSTEGIE **4640**

GAMMA-GRAPHIE **5798**
GAMME **10200**
GAMME DE DURETE **10974**
GAMME DE FABRICATION **7994**
GAMME DE TEMPERATURE **12730**
GAMME DES OPERATIONS **2222**
GANGUE **5802**
GANNISTER **5803**
GARANT (D'UN PALAN) **5023**
GARDE AU SOL **6143**
GARDE-BOUE **11927**
GARDE-CORPS **6259**
GARNIR DE MATIERE DE JOINT **8990**
GARNISSAGE **7623**
GARNISSAGE DU FOUR **5748**
GARNISSAGE REFRACTAIRE ACIDE **123**
GARNITURE **9386**
GARNITURE **7347**
GARNITURE **9004**
GARNITURE **4482**
GARNITURE **7620**
GARNITURE **9009**
GARNITURE ANTI-FRICTION **13837**
GARNITURE D'EMBRAYAGE **2545**
GARNITURE DE BOITE A BOURRAGE **12406**
GARNITURE DE FREIN **1538**
GARNITURE DE JOINT **8998**
GARNITURE DE METAL ANTIFRICTION **7624**
GARNITURE DE PISTON **9375**
GARNITURE DE PRESSE-ETOUPE **12406**
GARNITURE DE SOUPAGE **8998**
GARNITURE EN CHANVRE **6445**
GARNITURE EN CUIR **7502**
GARNITURE EN CUIR EMBOUTI **3459**
GARNITURE EN TRESSE **1526**
GARNITURE INTERIEURE DU CYLINDRE **3571**
GARNITURE METALLIQUE **8196**
GARNITURE MONOLITHIQUE **8385**
GARNITURE SOUPLE **11739**
GARNITURES **5283**
GARNITURES DE CHAUDIERES **1411**
GAUCHIR **13628**
GAUCHISSEMENT **13629**
GAUCHISSEMENT LATERAL **7418**
GAZ **5808**
GAZ A L'AIR **264**
GAZ A L'EAU **13651**
GAZ AMMONIAC **453**
GAZ CARBONIQUE **1948**
GAZ CARBONIQUE LIQUEFIE **7645**
GAZ COMBUSTIBLE **5709**

GAZ COMPRIME **2847**
GAZ D'ECHAPPEMENT **4891**
GAZ D'ECLAIRAGE **6775**
GAZ D'HUILE **8758**
GAZ DE CHAUFFAGE **5709**
GAZ DE FOUR A COKE **2663**
GAZ DE GAZOGENE **9890**
GAZ DE HAUT FOURNEAU **1313**
GAZ DE LA COMBUSTION **5470**
GAZ DE PETROLE LIQUEFIE **7638**
GAZ DE PROTECTION **6905**
GAZ DES MARAIS **8215**
GAZ DU FOYER **5470**
GAZ INCANDESCENT **6827**
GAZ INERTE **6905**
GAZ LIQUEFIE **7342**
GAZ LIQUEFIE **7637**
GAZ LUMIERE **6775**
GAZ MOND **8378**
GAZ NATUREL **8504**
GAZ OLEFIANT **4851**
GAZ OXHYDRIQUE **8985**
GAZ PARFAIT **9185**
GAZ PAUVRE **4194**
GAZ PERDU **13634**
GAZ POUR ACTIONNER LES MOTEURS **9736**
GAZ PUANT **6717**
GAZ RICHE **8758**
GAZE **5888**
GAZE METALLIQUE **13893**
GAZEIFIABLE **5860**
GAZEIFICATION **5865**
GAZEIFICATION **5861**
GAZEIFIER **5862**
GAZOGENE **5838**
GAZOGENE A ASPIRATION **12430**
GAZOGENE A VENT SOUFFLE **9819**
GAZOLE **5831**
GAZOLINE **9228**
GAZOLINE **9222**
GAZOMETRE **5859**
GEL **5912**
GELATINE PURE **5913**
GELIVURE DU BOIS **5700**
GENER MUTUELLEMENT (SE) **5587**
GENERATEUR A GAZ **5838**
GENERATEUR A VAPEUR **12169**
GENERATEUR DE VAPEUR **12158**
GENERATEUR ELECTROSTATIQUE **4643**
GENERATION DE VAPEUR **5923**
GENERATRICE **9867**

GENERATRICE 407
GENERATRICE 3021
GENERATRICE (GEOM.) 5925
GENERATRICE A COURANT TRIPHASE 12875
GENERATRICE A GAZ 5850
GENERATRICE A VAPEUR 12191
GENERATRICE DE COURANT 5924
GENERATRICE DE SOUDAGE POUR COURANT CONSTANT 2970
GENERATRICE HYDROELECTRIQUE 6700
GENIE CIVIL 2444
GENOU 953
GENOUILLERE 12546
GENOUILLERE 12979
GEOMETRIE 5931
GEOMETRIE DU JOINT 7240
GEOMETRIE DU TRAIN AVANT 5696
GEOMETRIQUE 5927
GERCE 3282
GERCE 11250
GERCE 3278
GERCURE 3282
GERCURE 5304
GERCURE 5273
GERCURE DU BOIS 11250
GERCURE SUPERFICIELLE 12501
GERMANIUM 5933
GERME 4686
GERME (DE CRISTAL) 8688
GERMINATION 8687
GERMINATION 5934
GEYSER (NATURE) 5936
GIBET 5945
GICLEUR 7218
GICLEUR 196
GICLEUR DE RALENTI 11646
GICLEUR PRINCIPAL 7942
GIFFARD 6938
GIROUETTE 13860
GISEMENT MINERAL 8303
GITE MINIER 8303
GIVRAGE 5702
GLACAGE 5972
GLACE A REFRACTION 10396
GLACE DEPOLIE (PHOT.) 6145
GLACE DU TIROIR 4996
GLACE FONDANTE 8139
GLACE RONDE 2414
GLACER 5969
GLACURE 5968
GLACURE 3264

GLACURE 5972
GLAISE 7691
GLISSEMENT 5973
GLISSEMENT 11611
GLISSEMENT 11595
GLISSEMENT DE LA COURROIE 11620
GLISSEMENT TRANSVERSAL 11288
GLISSER 11586
GLISSER 11612
GLISSETTE 5976
GLISSIERE 6160
GLISSIERE 11588
GLISSIERE 2395
GLISSIERE (DE FENETRE) 13868
GLISSIERE CYLINDRIQUE 3584
GLISSIERE DE CROSSE 6169
GLISSIERE PLATE 5384
GLISSOIR 11587
GLISSOIRE 11585
GLOBE 11883
GLOBERTITE 7880
GLOBULISATION 11896
GLU MARINE 8006
GLUCINIUM 1210
GLUCOSE 6058
GLYCERINE 5988
GNEISS 5989
GODET 1674
GODET (DU DESSINATEUR) 10946
GODET DE GRAISSAGE 8753
GODET EN VERRE A TIGE 8520
GODET GRAISSEUR 8753
GODET HUILEUR 8753
GOMME A CRAYON 9163
GOMME A ENCRE 6939
GOMME ADRAGANTE 6174
GOMME ARABIQUE 6172
GOMME DAMMAR 3598
GOMME DE SUMATRA 6184
GOMME ELASTIQUE 10825
GOMME GETTANIA 6184
GOMME PLASTIQUE 6184
GOMME-GRATTOIR 4801
GOMME-GUTTE 5793
GOMME-LAQUE 11338
GOMME-LAQUE EN BRANCHES 12254
GOMME-LAQUE EN GRAINS 11131
GOMME-RESINE 6173
GONFLEMENT 1720
GONFLEMENT 6155
GONFLEMENT 1331

GONFLEMENT (PAR LES GAZ DE FISSION) **12543**
GONFLEMENT DE LA FONTE **2061**
GONFLEMENT DU BOIS **12544**
GONFLER **12542**
GONFLER (SE) **1721**
GONIOMETRE **6004**
GORGE **6127**
GORGE **5483**
GORGE **2122**
GORGE D'UNE POULIE **6131**
GORGE DE ROULEMENT **13085**
GOUDRON **12679**
GOUDRON DE BOIS **13921**
GOUDRON DE GAZ **2570**
GOUDRON DE HOUILLE **2570**
GOUDRON DE LIGNITE **1667**
GOUDRON DE SCHISTE **11255**
GOUDRON VEGETAL **13921**
GOUDRONNAGE **12689**
GOUDRONNER **12680**
GOUGE DE FORGERON **6546**
GOUGE DE MENUISIER **6010**
GOUGE TRIANGULAIRE **13432**
GOUGEAGE **6011**
GOUJON **12399**
GOUJON **4181**
GOUJON **12397**
GOUJON AIGUILLE **9315**
GOUJON D'ASSEMBLAGE D'UNE CHAINE A ROULEAUX **9320**
GOUJON DE CENTRAGE **1513**
GOUJON EN BOIS **4180**
GOUJURE **5483**
GOULOTTE **2395**
GOUPILLE **9159**
GOUPILLE **9315**
GOUPILLE **13986**
GOUPILLE A TETE **2492**
GOUPILLE CONIQUE **12655**
GOUPILLE DE BLOCAGE **11220**
GOUPILLE DE CHASSIS **5372**
GOUPILLE DE CISAILLEMENT **11294**
GOUPILLE DE CISAILLEMENT **11283**
GOUPILLE DE CLAVETAGE **11220**
GOUPILLE DE RETENUE **11220**
GOUPILLE DE RETENUE **6522**
GOUPILLE FENDUE **11943**
GOUPILLE FENDUE **11940**
GOUSSET **6182**
GOUSSET **3058**
GOUTIERE A SECOUSSES **11253**
GOUTTE **4282**

GOUTTE FROIDE **2706**
GOUTTIERE **2395**
GOUTTIERE **6185**
GOUTTIERE **4285**
GRADE **6021**
GRADIN **12239**
GRADINS (EN) **12246**
GRADINS (A) **12246**
GRADUATION **10981**
GRADUER **6025**
GRAIN **12395**
GRAIN **12905**
GRAIN **6030**
GRAIN (MESURE ANGLAISE) **6029**
GRAIN ABRASIF **21**
GRAIN D'ORGE **9571**
GRAINS CONTENUS DANS LA FONTE **3998**
GRAINS DE CONTACT (RUPTEUR) **3000**
GRAINS POUR GAZOGENES **9140**
GRAISSAGE **7814**
GRAISSAGE **7803**
GRAISSAGE **7808**
GRAISSAGE **7812**
GRAISSAGE (ENDUCTION A LA GRAISSE) **6087**
GRAISSAGE A BAGUE **10601**
GRAISSAGE A CHAINETTE **2212**
GRAISSAGE A L'HUILE **8791**
GRAISSAGE A LA GRAISSE CONSISTANCE **6087**
GRAISSAGE A POMPE ET CIRCULATION D'HUILE **5442**
GRAISSAGE AU SUIF **6087**
GRAISSAGE AUTOMATIQUE **848**
GRAISSAGE COMPTE-GOUTTES **4287**
GRAISSAGE CONTINU **3012**
GRAISSAGE INTERMITTENT **9195**
GRAISSAGE MECANIQUE **848**
GRAISSAGE PAR BARBOTAGE **11925**
GRAISSAGE PAR COMPTE-GOUTTES **4284**
GRAISSAGE PAR MECHE **13847**
GRAISSAGE PAR OLEO-COMPRESSEUR **7809**
GRAISSAGE SOUS PRESSION **9818**
GRAISSAGE SOUS PRESSION **9824**
GRAISSAGE SOUS PRESSION A CIRCULATION CONTINUE **5442**
GRAISSE **5045**
GRAISSE AMELIOREE **6063**
GRAISSE CONSISTANTE **6275**
GRAISSE D'ETIRAGE **4233**
GRAISSE DE LUBRIFICATION **7804**
GRAISSE DE VOITURE **879**
GRAISSE DES TANNEURS **3711**
GRAISSE GRAPHITEE **6063**

GRAISSE MINERALE **13508**
GRAISSER **7801**
GRAISSER (ENDUIRE DE GRAISSE) **6084**
GRAISSER A L'HUILE **8744**
GRAISSEUR **6086**
GRAISSEUR **7812**
GRAISSEUR A DEBIT VISIBLE **11425**
GRAISSEUR A EPINGLETTE **8520**
GRAISSEUR A MECHE **11516**
GRAISSEUR A SOUPAPE **13467**
GRAISSEUR A TIGE **8520**
GRAISSEUR AUTOMATIQUE **851**
GRAISSEUR CENTRIFUGE **2182**
GRAISSEUR COMPTE-GOUTTES **4323**
GRAISSEUR MOLLERUP **8105**
GRAISSEUR STAUFFER **12148**
GRAMME **6046**
GRAND AXE D'UNE ELLIPSE **7947**
GRAND DEBIT DU BOIS **1058**
GRAND DIAMETRE DU FILET **3852**
GRAND LEVIER **3394**
GRANDE CALORIE **7298**
GRANDE COURONNE DE DIFFERENTIEL **3910**
GRANDE PENTE DE COUPE EFFECTIVE **7411**
GRANDES MAILLES (A) **2578**
GRANDEUR **7939**
GRANDEUR **10065**
GRANDEUR CHERCHEE **10469**
GRANDEUR D'EXECUTION (EN) **5723**
GRANDEUR DE COMPARAISON **12079**
GRANDEUR NATURELLE (EN) **5723**
GRANIT **6047**
GRANULAIRE **6050**
GRANULATION **1279**
GRANULATION **6057**
GRANULATION **6043**
GRANULATION SUPERFICIELLE **9314**
GRANULOMETRIE **6040**
GRANULOMETRIE **6032**
GRAPHIQUE **7717**
GRAPHIQUE **3834**
GRAPHIQUE DE LA RESISTANCE MECANIQUE **12366**
GRAPHITE **6062**
GRAPHITE **9541**
GRAPHITE DEFLOCULE **3703**
GRAPHITE EN PAILLETTES **10989**
GRAPHITE LAMELLAIRE **5305**
GRAPHITISANT **6067**
GRAPHITISATION **6066**
GRAPHOSTATIQUE **6061**
GRATTAGE **4802**

GRATTAGE **11011**
GRATTAGE (RESULTAT) **4803**
GRATTER **4799**
GRATTER **11008**
GRATTOIR **11009**
GRATTOIR DE BUREAU **4800**
GRATTOIR DE DESSINATEUR **4800**
GRAUWACKE **6095**
GRAVIER **6073**
GRAVITE **6078**
GRAVURE CATHODIQUE **2110**
GRELIN **1829**
GRELIN COMPOSE DE QUATRE AUSSIERES **11401**
GRENAILLAGE **11380**
GRENAILLAGE (FIN) **6124**
GRENAILLAGE (GROS) **11382**
GRENAILLE **25**
GRENAILLE **11379**
GRENAILLE DE CUIVRE **3128**
GRENAILLE DE NICKEL **8568**
GRENAILLE DE PLOMB **7472**
GRENAILLE DE ZINC **8404**
GRENAILLE FINE **6123**
GRENAT **5807**
GRES (GEOL.) **10928**
GRES BIGARRE **13502**
GRES CERAME **12276**
GRES ROUGE **10336**
GRESILLON DE COKE **11657**
GRIFFE **11013**
GRIFFE **2464**
GRIFFE DE FORGERON **6571**
GRIGNOTAGE **8552**
GRIGNOTEUSE **8553**
GRILLAGE **6071**
GRILLAGE **6096**
GRILLAGE **1839**
GRILLAGE DES MINERAIS **10650**
GRILLAGE DU MINERAI **8862**
GRILLAGE ONDULE **3200**
GRILLAGE TOTAL **3636**
GRILLE **6070**
GRILLE **6096**
GRILLE A BARREAUX MOBILES **10655**
GRILLE A CHAINE **2209**
GRILLE A ETAGES **12248**
GRILLE A GRADINS **12248**
GRILLE A GRAVIER **6075**
GRILLE A SECOUSSES **10655**
GRILLE DE DECOCHAGE **7334**
GRILLE HORIZONTALE **6584**

GRILLE INCLINEE **6837**
GRILLE OSCILLANTE **10655**
GRILLE ROTATIVE **10559**
GRILLE TOURNANTE **10559**
GRILLER LES MINERAIS **10648**
GRIMPEMENT D'UNE SOLUTION **3330**
GRIPPAGE **11141**
GRIPPAGE **5778**
GRIPPEMENT **11141**
GRIPPER **11140**
GROS FILET **2576**
GROS GRAIN **2573**
GROSSE CHAUDRONNERIE **1423**
GROSSE POUDRE **2580**
GROSSE TETE DE BIELLE **3299**
GROSSEUR DE GRAIN MINIMALE **8317**
GROSSEUR DE LA PARTICULE **9089**
GROSSEUR DES GRAINS D'ABRASIF **23**
GROSSEUR DU GRAIN **11522**
GROSSEUR DU GRAIN **6039**
GROSSISSEMENT **7937**
GROSSISSEMENT **7935**
GROSSISSEMENT D'UNE LENTILLE **7938**
GROSSISSEMENT DU GRAIN **6035**
GROUPE (DE MACHINES ELECTRIQUES) **11213**
GROUPE DE MACHINES **6151**
GROUPE MOTEUR-GENERATEUR **8413**
GROUPEMENT **9051**
GROUPEMENT **11199**
GROUPER EN QUANTITE **2956**
GROUPER EN SERTIE **2957**
GROUPES ELECTROGENES **4524**
GRUE **3290**
GRUE A FLECHE **7224**
GRUE A PORTIQUE **9641**
GRUE A POTENCE **13623**
GRUE DERRICK **3788**
GRUE FIXE **5291**
GRUE MOBILE **13161**
GRUE PIVOTANTE **11583**
GRUE ROULANTE **13161**
GRUE STATIONNAIRE **5291**
GRUES AGRICOLES **3294**
GRUGEAGE **3107**
GRUGEOIR **8677**
GRUMELAGE **1279**
GUEUSE **951**
GUEUSE **9302**
GUEUSE **9298**
GUIDAGE **6160**
GUIDE **6160**

GUIDE A REFENDRE **10610**
GUIDE DE CENTRAGE **13087**
GUIDE DE LA CROSSE **6169**
GUIDE DE LA TIGE DU PISTON **9385**
GUIDE DE SOUPAPE **13462**
GUIDE DU MOUVEMENT RECTILIGNE **12315**
GUIDE-COURROIE **1150**
GUIDE-RUBAN INFERIEUR **7796**
GUILLAUME **9522**
GUILLAUME DE COTE **11417**
GYPSE **6190**
GYROSCOPE **6191**
HACHE **857**
HACHE PAILLES **2199**
HACHEREAU **6305**
HACHERON **6305**
HACHETTE **6305**
HACHURE **6306**
HACHURER **6304**
HAFNIUM **6199**
HALLE **1072**
HALLE DE COULEE **2071**
HALLE DE COULEE **2077**
HALLE DE MONTAGE **770**
HALLE DE MONTAGE **773**
HALO **6213**
HALOGENE **6214**
HANGAR **11302**
HAPPE **7548**
HARNAIS D'ENGRENAGES **7050**
HARNAIS D'ENGRENAGES **5907**
HAUBAN **6187**
HAUBANS DE SPHERE **12919**
HAUSSE **10611**
HAUT FOURNEAU **1311**
HAUT VOLTAGE **6479**
HAUTE FREQUENCE **6467**
HAUTE PERSSION **6473**
HAUTE TEMPERATURE **6476**
HAUTE TENEUR (A) **6483**
HAUTE TENSION **6478**
HAUTE TENSION **6479**
HAUTEUR **6420**
HAUTEUR D'ACTION **13951**
HAUTEUR D'ASCENSION **10611**
HAUTEUR D'ASPIRATION D'UNE POMPE **12432**
HAUTEUR D'ELEVATION D'UNE POMPE **13061**
HAUTEUR DE BUSE **8680**
HAUTEUR DE CHUTE **6421**
HAUTEUR DE CONSTRUCTION **8924**
HAUTEUR DE DENT **13845**

HAUTEUR DE DENT **13017**
HAUTEUR DE FACE **164**
HAUTEUR DE LA TETE D'UNE DENT **164**
HAUTEUR DE LEVAGE D'UNE GRUE **7551**
HAUTEUR DE REFOULEMENT D'UNE POMPE **4021**
HAUTEUR DU FLANC **3681**
HAUTEUR DU PIED DE DENT **3681**
HAUTEUR SOUS TRAVERSE **6423**
HAUTEUR TOTALE **8924**
HAUTEUR TOTALE DE LA DENT **3785**
HAYON **891**
HECTOLITRE **6415**
HECTOWATT **6416**
HECTOWATT-HEURE **6417**
HELIANTHINE **8223**
HELICE **6430**
HELICE PROPULSIVE **11041**
HELICO-MELANGEUR **9931**
HELICOIDE **6428**
HELIUM **6429**
HEMATITE **6434**
HEMATITE BRUNE **7596**
HEMATITE ROUGE **11865**
HEMICELLULE **6204**
HEMISPHERE **6438**
HERMETIQUE **303**
HERMETIQUEMENT CLOS **6447**
HERMINETTE **207**
HERSES (A CHAINEHERSES A ZIG-ZAGHERSES FLEXIBLE) **6303**
HETEROGENE **6451**
HETEROGENEITE **3170**
HEURES SUPPLEMENTAIRES **8964**
HEURT **11359**
HEXAEDRE **6463**
HEXAGONE **6453**
HEXAGONES EN ACIER ALLIE **374**
HOLMIUM **6551**
HOMOGENE **6555**
HOMOGENEISATION **6561**
HOMOGENEITE D'UN MATERIEL **13364**
HOMOGENEITE D'UNE SOUDURE **11821**
HONAGE AU JET DE VAPEUR **7653**
HORIZONTAL **6578**
HORLOGE POINCONNEUSE **12946**
HORNBLENDE **6598**
HORS D'AXE **8731**
HORS D'EQUERRE **8895**
HOTTE **1455**
HOUES (A TRACTEUR TRAINEE OU MONTEE) **6515**
HOUILLE **2558**

HOUILLE A COURTE FLAMME **11371**
HOUILLE A GAZ **5813**
HOUILLE A LONGUE FLAMME **7724**
HOUILLE ANTHRACITEUSE **11165**
HOUILLE COLLANTE **1835**
HOUILLE ECLATANTE **609**
HOUILLE GRASSE **1282**
HOUILLE MAIGRE **11168**
HOUILLE MAIGRE A COURTE FLAMME **11165**
HOUILLE MARECHALE **5543**
HOUILLE MENUE **11557**
HOUILLE NON COLLANTE **8626**
HOUILLE SECHE A LONGUE FLAMME **4344**
HOUILLE TOUT-VENANT **10841**
HOUILLERE **2566**
HUILAGE **8791**
HUILE **8743**
HUILE A ANTHRACENE **608**
HUILE A NOYAUX **3157**
HUILE ANIMALE **555**
HUILE BRUTE DE PETROLE **3412**
HUILE COMPOUND **2832**
HUILE D'ECLAIRAGE **7275**
HUILE D'EGOUTTAGE **13636**
HUILE D'OEILLETTE **9624**
HUILE DE BALEINE **13103**
HUILE DE BLANC DE BALEINE **11881**
HUILE DE CAMPHRE **1894**
HUILE DE CHENEVIS **6446**
HUILE DE COCO **2620**
HUILE DE COLZA **10204**
HUILE DE COTON **3222**
HUILE DE COUPE **3534**
HUILE DE FLUXAGE **5491**
HUILE DE GOUDRON **12681**
HUILE DE GOUDRON DE HOUILLE **2568**
HUILE DE GOUDRON DE LIGNITE **1668**
HUILE DE GRAISSAGE **7805**
HUILE DE GRAISSE **7409**
HUILE DE LIN **7631**
HUILE DE LIN A VERNIS **8786**
HUILE DE LIN BOUILLIE **8786**
HUILE DE NAVETTE **10204**
HUILE DE NETTOYAGE **2477**
HUILE DE PALME **9027**
HUILE DE PIED DE BOEUF **1450**
HUILE DE PIED DE MOUTON **1450**
HUILE DE PIERRE **9229**
HUILE DE PIN **10749**
HUILE DE POISSON **13103**
HUILE DE POMMES DE TERRE **5754**

HUILE DE RESINE 10749
HUILE DE RICIN 2086
HUILE DE SAINDOUX 7409
HUILE DE SCHISTE 11256
HUILE DE SESAME 11212
HUILE DE TREMPE 10099
HUILE DE TREMPE 6295
HUILE DE VITRIOL 12453
HUILE EMULSIONNEE 4706
HUILE EN EXCES 13636
HUILE ENTHRACENIQUE 608
HUILE EPUREE 10373
HUILE ESSENTIELLE 4836
HUILE FIXE 5294
HUILE GRAPHITEE 6063
HUILE GRASSE 5294
HUILE H.D. 6403
HUILE LAMPANTE DE PETROLE 7275
HUILE LEGERE 7570
HUILE LOURDE 6408
HUILE LOURDE (POUR FORCE MOTRICE) 5714
HUILE LOURDE DE PARAFFINE 9044
HUILE LUBRIFIANTE 7805
HUILE MINERALE 8304
HUILE MINERALE BLONDE 2480
HUILE MINERALE BRUNE 1291
HUILE MINERALE POUR BROCHES 11908
HUILE MIXTE 2832
HUILE MOYENNE 8125
HUILE PARAFFINE 9044
HUILE POUR CYLINDRES 3572
HUILE POUR FORER 4280
HUILE POUR ISOLEMENT 6999
HUILE POUR MACHINES 4755
HUILE POUR MOTEUR 8415
HUILE PROJETEE 11928
HUILE RANCE 10197
HUILE SICCATIVE 4357
HUILE SOLAIRE 11753
HUILE SOLUBLE 11810
HUILE SPECIALE 6403
HUILE VEGETALE 13512
HUILE VERSEE 11928
HUILE VOLATILE 4836
HUILER 8744
HUILES BLANCHES 13838
HUILES CONDITIONNEES 8996
HUMECTATION 13807
HUMECTER 8347
HUMIDIFIANT 6664
HUMIDIFICATION 8348

HUMIDITE ABSOLUE DE L'AIR 30
HUMIDITE CRITIQUE 3346
HUMIDITE DE L'AIR 6666
HUMIDITE DE LA VAPEUR 13806
HUMIDITE RELATIVE 10432
HUMIDITE RELATIVE DE L'AIR 10433
HUMUS 6667
HYDRATATION DE LA CHAUX 11569
HYDRATE 6674
HYDRATE 6673
HYDRATE D'ETHYLE 313
HYDRATE DE BARYTE 1000
HYDRATE DE BARYUM 1000
HYDRATE DE CARBONE 1937
HYDRATE DE CHAUX 1851
HYDRATE DE MAGNESIE 7887
HYDRATE DE METHYLE 13920
HYDRATE DE POTASSIUM 9687
HYDRATE DE SODIUM 11722
HYDRATE DE VINYLE 84
HYDRATE STANNEUX 12101
HYDRAULIQUE 6676
HYDRAULIQUE 6698
HYDROCARBURE 6705
HYDROCARBURES DE LA SERIE AROMATIQUE 2518
HYDROCARBURES DE LA SERIE GRASSE 8806
HYDROCARBURES NONSATURES (HYDROCARBURES
  ETHYLENIQUESHYDROCARBURES ACETYLENIQUES) 13393
HYDROCARBURES PARAFFINIQUES 10940
HYDROCARBURES SATURES 10940
HYDROCHLORATE 462
HYDRODYNAMIQUE 6710
HYDRODYNAMIQUE 6708
HYDROFUGE 8349
HYDROFUGE 13685
HYDROGENE 6712
HYDROGENE ARSENIE 730
HYDROGENE ATOMIQUE 150
HYDROGENE BICARBONE 4851
HYDROGENE SULFURE 6717
HYDROLYSE 6719
HYDROMETALLURGIE 6721
HYDROPHILE 6724
HYDROSTATIQUE 6728
HYDROXYDE 6673
HYDRURE D'ACETYLE 84
HYDRURE D'ETHYLE 4848
HYDRURE DE BENZYLE 12983
HYDRURE DE BUTYLE 1775
HYDRURE DE CRESYLE 12983
HYDRURE DE PROPYLE 9927

HYDRURE METHYLIQUE **8215**
HYGROMETRE **6729**
HYGROMETRICITE **6732**
HYGROMETRIQUE **6730**
HYPERBOLE **6734**
HYPERBOLIQUE **6735**
HYPERBOLOIDE A DEUX NAPPES **6740**
HYPERBOLOIDE A UNE NAPPE **6739**
HYPEREUTECTOIDE **6733**
HYPOAZOTIDE **8591**
HYPOAZOTIQUE **8591**
HYPOCYCLOIDE **6743**
HYPOEUTECTOIDE **6741**
HYPOSULFITE **12852**
HYPOSULFITE DE PLOMB **7474**
HYPOSULFITE DE SODIUM **11736**
HYPOTENUSE **6746**
HYPOTHESES DE BASE **12125**
HYSTERESIS **6747**
ICHTYOCOLLE **7174**
IGNIFUGE **5257**
IGNITION **6760**
ILLINIUM **6771**
ILLUMINATION **7574**
ILLUSTRATION **6778**
IMAGE **6779**
IMAGE NETTE **2479**
IMAGE REELLE **10259**
IMAGE VIRTUELLE **13572**
IMBIBER **6802**
IMBIBITION **6805**
IMBRULE **13344**
IMBRULES **6765**
IMCOMBUSTIBLE **6843**
IMMERGER **3968**
IMMERSION **3967**
IMMERSION **3975**
IMMERSION **3978**
IMMOBILISATION DES ECROUS ET BOULONS **7707**
IMPEDANCE **6794**
IMPERFECTIONS DE SURFACE **6464**
IMPERMEABILISATION **13691**
IMPERMEABILISER **13689**
IMPERMEABILITE **6797**
IMPERMEABILITE A L'EAU **6796**
IMPERMEABLE **13671**
IMPERMEABLE **6798**
IMPERMEABLE A L'HUMIDITE **8349**
IMPERMEABLE A LA LUMIERE **7572**
IMPONDERABLE **6800**
IMPREGNATION **6805**

IMPREGNATION DU BOIS AU SUBLIME CORROSIF PAR LE PROCEDE KYAN **7340**
IMPREGNER **6802**
IMPRIMANTE **9872**
IMPULSION **9979**
IMPULSION **6809**
IMPURETE **6814**
IMPURETES **6813**
INACCESSIBILITE **6820**
INACCESSIBLE **6821**
INALTERABLE **13723**
INATTAQUABILITE AUX ACIDES **10496**
INATTAQUABLE A L'EAU **13671**
INATTAQUABLE A L'HUILE **8792**
INATTAQUABLE AUX ACIDES **128**
INATTAQUABLE AUX INTEMPERIES **13723**
INCANDESCENCE **13832**
INCANDESCENCE **6822**
INCANDESCENT **5983**
INCASSABLE **13343**
INCIDENCE **6831**
INCLINABLE **6835**
INCLINAISON **6836**
INCLINAISON **11627**
INCLINAISON **1065**
INCLUSION **6840**
INCLUSION ALLONGEE **6841**
INCLUSION DE GAZ **5836**
INCLUSION DE SCORIE **4772**
INCLUSION GAZEUSE **5855**
INCLUSION SECONDAIRE **11106**
INCLUSIONS NON-METALLIQUES **8620**
INCOMBUSTIBILITE **6842**
INCOMPRESSIBILITE **6846**
INCOMPRESSIBLE **6847**
INCONGELABILITE **8634**
INCONGELABLE **8635**
INCONNUE **13391**
INCRUSTATION DE CALAMINE **5367**
INCRUSTATIONS DES CHAUDIERES **10969**
INCRUSTE **6857**
INDEPENDANTE **11190**
INDICATEUR **6879**
INDICATEUR **5768**
INDICATEUR **3837**
INDICATEUR A CADRAN **3840**
INDICATEUR A REFRACTION **10395**
INDICATEUR D'OUVERTURE **13153**
INDICATEUR DE CHARGE **2267**
INDICATEUR DE NIVEAU **13652**
INDICATEUR DE NIVEAU **7526**

INSTALLER LA COURROIE **11357**
INSTITUT BRITANNIQUE DE NORMALISATION **887**
INSTRUCTION **6992**
INSTRUCTION **6991**
INSTRUCTION DE SAUT **11546**
INSTRUCTIONS **8835**
INSTRUCTIONS RELATIVES AU TRAVAIL **11209**
INSTRUMENT A CENTRER **2166**
INSTRUMENT APERIODIQUE **635**
INSTRUMENT D'OPTIQUE **8844**
INSTRUMENT DE MESURE **8090**
INSTRUMENT DE VERIFICATION **5871**
INSTRUMENT ENREGISTREUR **10289**
INSTRUMENT POUR LA DETERMINATION DE LA VITESSE D'UN COURANT **3490**
INSTRUMENTS DE MATHEMATIQUES **8051**
INTEGRALE **7009**
INTEGRALE DEFINIE **3697**
INTEGRALE DOUBLE **4145**
INTEGRALE ELLIPTIQUE **4676**
INTEGRALE INDEFINIE **6858**
INTEGRALE LINEAIRE **7600**
INTEGRALE MULTIPLE **8465**
INTEGRALE SINGULIERE **11471**
INTEGRALE TRIPLE **13215**
INTEGRAPHE **7013**
INTEGRATEUR **7016**
INTEGRATION **7015**
INTEGRER **7014**
INTENSITE **7018**
INTENSITE D' ECLAIREMENT **6777**
INTENSITE D'UN COURANT **7021**
INTENSITE D'UNE FORCE **7939**
INTENSITE DE CHAMP **4506**
INTENSITE DE LA PESANTEUR **2975**
INTENSITE DE POLE **9596**
INTENSITE DE RADIATION **7019**
INTENSITE DU CHAMP **5144**
INTENSITE DU COURANT **478**
INTENSITE DU COURANT DE SOUDAGE **13780**
INTENSITE EN BOUGIES **1900**
INTENSITE LUMINEUSE **7822**
INTENSITE LUMINEUSE **7019**
INTENSITE LUMINEUSE PAR UNITE DE SURFACE **7104**
INTERCALATION **7023**
INTERCALER **7022**
INTERCEPTER DES RAYONS **7024**
INTERCHANGEABILITE **7025**
INTERCHANGEABILITE DE FABRICATION **7028**
INTERCHANGEABLE **7026**
INTERFACE **7032**

INTERFERENCE **7035**
INTERPOLATEUR **3995**
INTERPOLATION **7089**
INTERPOLATION CIRCULAIRE **2416**
INTERPOLATION LINEAIRE **7613**
INTERPOLATION PARABOLIQUE **9040**
INTERPOLER **7088**
INTERROMPRE LE CIRCUIT ELECTRIQUE **7090**
INTERRUPTEUR **12553**
INTERRUPTEUR A ACTION INSTANTANEE **11690**
INTERRUPTEUR-CONJONCTEUR **12553**
INTERRUPTEUR-DISJONCTEUR **12553**
INTERRUPTION **11546**
INTERRUPTION DE PASSAGE DU COURANT ELECTRIQUE **7093**
INTERRUPTION DE SERVICE **7092**
INTERSTICE **262**
INTERVALLE **7101**
INTERVALLE DE FUSION **8143**
INTERVALLE DE SOLIDIFICATION **11801**
INTERVALLE DE SOLIDIFICATION **5658**
INTERVALLE LIBRE **278**
INTRODUCTION D'AIR **12486**
INTRODUCTION MANUELLE DES DONNEES **7987**
INVAR **7105**
INVARIANT **7106**
INVENTEUR **7108**
INVENTION **7107**
INVENTION BREVETABLE **9122**
INVERSEUR DE MARCHE **10548**
INVERSEUR-PHARE-CODE **3963**
INVERSION (MATH.) **7110**
INVERSION DE MOUVEMENT **10552**
INVERTIR **2241**
IODE **7120**
IODURE D'ARGENT **11464**
IODURE DE MERCURE **8155**
IODURE DE POTASSIUM **9689**
ION **7121**
ION HYDROGENE **2112**
IONISATION **7124**
IONISATION **4596**
IONOGENE **7129**
IRIDIUM **7131**
IRRADIATION **7161**
IRRADIATION **10241**
IRREGULARITE DE SURFACE **12503**
IRREGULARITE DU MOUVEMENT **7167**
IRREGULIER **7163**
IRREVERSIBILITE **7168**
IRREVERSIBLE **7169**

ISOBARE **7176**
ISOCHRONE **7179**
ISOCHRONISME **7178**
ISOLANT **7006**
ISOLANT **3898**
ISOLANT **6998**
ISOLATEUR **7006**
ISOLATION **7003**
ISOLATION **7184**
ISOLATION **12287**
ISOLATION A FROID **2700**
ISOLATION CONTRE LE BRUIT **7004**
ISOLATION CONTRE LE SON **7004**
ISOLATION ELECTRIQUE **4545**
ISOLATION THERMIQUE **7003**
ISOLATION THERMIQUE **12810**
ISOLEMENT **7003**
ISOLER **6994**
ISOMERE **7185**
ISOMERIE **357**
ISOMERIE **7186**
ISOMORPHE **7190**
ISOMORPHIE **7189**
ISOMORPHISME **7191**
ISOPLERE **7187**
ISOTOPES **7197**
ISOTROPE **7198**
ITINERAIRE MARITIME **10810**
IVOIRE **7199**
JAMBE DE FORCE **12396**
JANTE **10592**
JANTE (COURONNE) D'UNE ROUE D'ENGRENAGE **10594**
JANTE A BASE CREUSE **4322**
JANTE D'UNE POULIE **10595**
JANTE D'UNE ROUE **5088**
JANTE DE ROUE **13818**
JANTE DE VOLANT **10593**
JAQUETTE **8354**
JAUGE **5768**
JAUGE **5871**
JAUGE (NON- VARIABLE) **8625**
JAUGE ANGLAISE **12089**
JAUGE AVEC RENVOI A BAIN D'HUILE **8747**
JAUGE D'AJUSTAGE **11225**
JAUGE D'EXTERIEUR **5091**
JAUGE D'EXTERIEUR **4955**
JAUGE D'HUILE **8764**
JAUGE D'HUILE **3979**
JAUGE D'INTERIEUR **7066**
JAUGE DE BIRMINGHAM POUR FILS METALLIQUES **12089**
JAUGE DE CONTRAINTE (A FIL RESISTANT) **12329**

JAUGE DE CONTRAINTE A RESISTANCE **10494**
JAUGE DE CONTROLE **3053**
JAUGE DE CONTROLE **12779**
JAUGE DE FABRICATION **13967**
JAUGE DE FABRICATION **13955**
JAUGE DE FILETAGE **2760**
JAUGE DE FILETAGE **11038**
JAUGE DE PRECISION **9766**
JAUGE DE RECEPTION **10015**
JAUGE DE REFERENCE **10366**
JAUGE DE REFERENCE **8035**
JAUGE DE REVISION **5011**
JAUGE DE TREFILERIE **13892**
JAUGE DEMONTABLE **4048**
JAUGE LIMITE DE TOLERANCE **3903**
JAUGE MACHOIRE NORMALE **12088**
JAUGE MAGNETIQUE **7905**
JAUGE MICROMETRIQUE **8256**
JAUGE NORMALE - TAMPON FILETE **12861**
JAUGE NORMALE BAGUE FILETEE **12863**
JAUGE PNEUMATIQUE **283**
JAUGE POUR FILS METALLIQUES **13892**
JAUGE POUR RAINURE DE CLAVETTE **7283**
JAUGE POUR TOLES **11311**
JAUGE TAMPON NORMALE **12087**
JAUGE-FOURCHE A TOLERANCES **4148**
JAUGE-MACHOIRE A UNE BRANCHE POUR COTES 'MINI' ET 'MAXI' **6607**
JAUGE-MACHOIRE POUR FILETS **10693**
JAUGE-MACHOIRE REGLABLE **186**
JAUGE-TAMPON AU CONE METRIQUE **8230**
JAUGE-TAMPON AU CONE MORSE **8392**
JAUGE-TAMPON CONIQUE **7956**
JAUGE-TAMPON CONIQUE **7076**
JAUGE-TAMPON CONIQUE **12656**
JAUGE-TAMPON CONIQUE AVEC TETON **12657**
JAUGE-TAMPON DOUBLE A LIMITES **4137**
JAUGEAGE **1867**
JAUGEAGE A DISTANCE **10449**
JAUGER **1861**
JAUGES **5769**
JAUGES POUR FILETAGES **12860**
JAUNE DE CHROME **2375**
JAUNE DE ZINC **14039**
JET COMPACT **11787**
JET D'AIR **242**
JET D'EAU **13656**
JET DE COULEE **12004**
JET DE VAPEUR **12170**
JET LIQUIDE **7221**
JETER DE LA FLAMME **5359**

JEU **11213**
JEU **13873**
JEU **358**
JEU **2483**
JEU (MOUVEMENT) **9516**
JEU A FOND DE FILET **12858**
JEU A FOND DES DENTS **1489**
JEU A LA CRETE **3335**
JEU AXIAL **4727**
JEU D'ENGRENAGES **5901**
JEU D'UNE SOUPAPE **13154**
JEU DE COIFFAGE **358**
JEU DE COIFFAGE **2483**
JEU DE LABORATOIRE **7345**
JEU DE MONTAGE **9516**
JEU DE POIDS **11219**
JEU DES SOUPAPES **12673**
JEU DU FOND DE LA DENT **1489**
JEU DU PISTON **7519**
JEU ENTRE DENTS **914**
JEU ENTRE LES POINTS DE CONTACT DES DENTS **5349**
JEU ETALON **8036**
JEU INUTILE **916**
JEU LATERAL DES ARBRES **4735**
JEU NUISIBLE **916**
JEU PERNICIEUX **916**
JEU TOTAL PERDU **13063**
JEU UTILE **9516**
JOINT **1500**
JOINT **5863**
JOINT **6127**
JOINT **7247**
JOINT (DE RETENUE) D'HUILE **8776**
JOINT (POINT DE JONCTION) **7238**
JOINT A BAGUE DE CAOUTCHOUC **10820**
JOINT A BAGUE DE PLOMB MATEE POUR TUYAUTERIES **7464**
JOINT A CARDAN **13385**
JOINT A CHICANE **7347**
JOINT A DOUBLE TOURNEVIS **8794**
JOINT A ECLISSES **1782**
JOINT A EMBOITEMENT **11900**
JOINT A INSERTION **3159**
JOINT A LABYRINTHE **7347**
JOINT A RECOUVREMENT **7397**
JOINT A RECOUVREMENT SOUDE D'UN SEUL COTE **11504**
JOINT A RECOUVREMENT SOUDE DES DEUX COTES **4171**
JOINT A ROTULE **953**
JOINT ABOUTE SOUDE D'UN SEUL COTE **1779**
JOINT ABOUTE SOUDE DES DEUX COTES **1777**
JOINT ANNULAIRE **10604**

JOINT AU MASTIC DE FONTE **10867**
JOINT AUTOCLAVE **7963**
JOINT BOUT A BOUT **1776**
JOINT COLLE **5986**
JOINT COULISSANT **11617**
JOINT D'ETANCHEITE **5863**
JOINT D'OLDHAM **8794**
JOINT DE CARDAN **6573**
JOINT DE CULASSE **3570**
JOINT DE DILATATION **4918**
JOINT DE SEPARATION D'UN PALIER **7241**
JOINT DE SOUDURE EN J FERMEE **1790**
JOINT DE SOUDURE PAR BOSSAGES **9915**
JOINT DE TUBES **9347**
JOINT EN ANGLE EXTERIEUR **3179**
JOINT ETANCHE **12921**
JOINT FUSIBLE **2991**
JOINT HYDRAULIQUE **13675**
JOINT METALLOPLASTIQUE **3134**
JOINT PLASTIQUE **4482**
JOINT PLASTIQUE **11078**
JOINT POUR TUYAUTERIE **9355**
JOINT RIGIDE **10589**
JOINT SECONDAIRE D'ETANCHEITE **13720**
JOINT SOUDE **13767**
JOINT SOUDE **13774**
JOINT SOUDE **13776**
JOINT SOUDE **7237**
JOINT SOUDE A DOUBLE CLIN **4141**
JOINT SOUDE A SIMPLE CLIN **5720**
JOINT SOUPLE **5416**
JOINT SPHERIQUE **953**
JOINT SUR TRANCHES **4445**
JOINT TORIQUE **8703**
JOINT UNIVERSEL **6573**
JONC D'ARRET **11692**
JONC DES INDES **10236**
JONCTION **3254**
JONCTION BOUT A BOUT **1778**
JONCTION DES COURROIES **1151**
JONCTION DES COURROIES PAR COLLAGE **2143**
JONCTION DES COURROIES PAR LANIERES EN CUIR **7348**
JONCTION DES COURROIES PAR RIVET **10555**
JONCTIONNEMENT DES COURROIES **1142**
JOUE D'UN RABOT **5093**
JOUE D'UNE CHAINE A ROULEAUX **11416**
JOUE D'UNE POULIE **5332**
JOUE D'UNE ROUE D'ENGRENAGE **11400**
JOUE DE CINTRAGE **1183**
JOUE DE CINTRAGE **1529**
JOUE DE CONTACT **3003**

JOUE DU COUSSINET D'UNE BUTEE **2723**
JOULE **7255**
JOUR **3616**
JOUR DE LA SIGNATURE DU BREVET **3617**
JOURNEE DE TRAVAIL **11347**
JUMELLE **2491**
JUMELLE DE RESSORT **11991**
JUMELLES **11233**
JUPE **11547**
JUPE (D'UN RESERVOIR) **11549**
JUTE **7268**
KAOLIN **2340**
KAOLIN (POUR LES PNEUS) **13842**
KEROSENE **7275**
KIESEGUHR **7292**
KILODYNE **7296**
KILOGRAMME **7297**
KILOGRAMMETRE **7299**
KILOMETRE **7300**
KILOVOLT **7301**
KILOVOLT-AMPERE **7302**
KILOWATT **7303**
KILOWATT-HEURE **7304**
KIP = 1000 LIVRES = 453,59 KG **7316**
KLAXON **6592**
KLINGERITE **7318**
KRYPTON **7339**
LABILE **7343**
LABORATOIRE **7344**
LACET **7741**
LACUNE **6530**
LACUNE **13585**
LACUNE **13433**
LACUNES (AFNOR) **6536**
LAINE (DE MOUTON) **13932**
LAINE DE BOIS **13923**
LAINE DE VERRE **5966**
LAINE MINERALE **11566**
LAIT DE CHAUX **8274**
LAIT DE CIMENT **6152**
LAITIER **11561**
LAITIER **1312**
LAITIER **2397**
LAITIER GRANULE **6055**
LAITON **1562**
LAITON **14008**
LAITON A FORGER **5558**
LAITON A QUALITE D'EMBOUTISSAGE **3690**
LAITON ALPHA **399**
LAITON ALPHA-BETA **403**
LAITON ALPHA-BETA **8511**

LAITON AU MANGANESE **7967**
LAITON AU NICKEL **8559**
LAITON BETA **1216**
LAITON COULE **2031**
LAITON D'ALUMINIUM **419**
LAITON D'ETIRAGE **4232**
LAITON DE BASE **8056**
LAITON DE DECOLLETAGE **5645**
LAITON DE FAIBLE ALLIAGE **7768**
LAITON DE L'AMIRAUTE **8511**
LAITON DE MARINE **198**
LAITON DE QUALITE **6480**
LAITON DE RESSORT **11978**
LAITON EN FEUILLES **11307**
LAITON POUR CARTOUCHES **2016**
LAITON POUR HORLOGES **2505**
LAITON ROUGE **2824**
LAITON ROUGE **10325**
LAITON SILICEUX **11443**
LAMAGE **11960**
LAMBAGE (DE TOLES) **1681**
LAME **1299**
LAME D'ACCUMULATEUR **81**
LAME D'UN OUTIL TRANCHANT **1300**
LAME D'UN RESSORT **9492**
LAME D'UNE CISAILLE **11282**
LAME DE CUIVRE **3131**
LAME DE RESSORT **11977**
LAME DE SCIE **10952**
LAME DE VERRE **11589**
LAME POUR SCIE A METAUX **6197**
LAME TRANCHANTE **1300**
LAMELLAIRE **7372**
LAMELLE **7371**
LAMELLE **5303**
LAMINAGE **10696**
LAMINAGE A CHAUD **6656**
LAMINAGE A CHAUD **6638**
LAMINAGE A FAIBLE PRESSION **9330**
LAMINAGE A FROID **2698**
LAMINAGE CONTINU **3035**
LAMINAGE DE LINGOTS **2642**
LAMINAGE EN PAQUET **8994**
LAMINAGE EN SANDWICH **10929**
LAMINAGE FINISSEUR **5215**
LAMINAGE STRATIFIE **10929**
LAMINAGE TRANSVERSAL **3384**
LAMINE **7375**
LAMINE A FROID **2716**
LAMINER **10665**
LAMINER A CHAUD **6636**

LAMINER A FROID **2694**
LAMINOIR **10701**
LAMINOIR A BILLETTES **1245**
LAMINOIR A BLOOMS **1348**
LAMINOIR A BRAMES **11555**
LAMINOIR A FORGER **5563**
LAMINOIR A POLIR **9456**
LAMINOIR A SIX CYLINDRES **2543**
LAMINOIR A TUBES **9358**
LAMINOIR FINISSEUR **5222**
LAMINOIR POUR TOLES FORTES **6410**
LAMINOIR REDUCTEUR **11508**
LAMINOIR SENDZIMIR **11181**
LAMINOIR TRANSVERSAL **1642**
LAMINOIR UNIVERSEL **13386**
LAMPE **13453**
LAMPE **7384**
LAMPE A ACETYLENE **95**
LAMPE A ALCOOL **11923**
LAMPE A ARC **672**
LAMPE A ESSENCE MINERALE **9224**
LAMPE A FILAMENT DE CARBONE **1951**
LAMPE A FILAMENT DE TUNGSTENE **13264**
LAMPE A FILAMENT METALLIQUE **8181**
LAMPE A GAZ **6828**
LAMPE A GAZ D'ECLAIRAGE **5826**
LAMPE A INCANDESCENCE **6828**
LAMPE A INCANDESCENCE **6825**
LAMPE A PETROLE **9227**
LAMPE A SOUDER **11766**
LAMPE A VAPEUR DE MERCURE **8167**
LAMPE A VAPEUR DE MERCURE EN QUARTZ **10080**
LAMPE AU NEON **8536**
LAMPE DIODE **3965**
LAMPE ELECTRIQUE **4516**
LANCE **8679**
LANCE A OXYGENE **8981**
LANCEMENT D'UN MOTEUR **12130**
LANCER UN MOTEUR **12114**
LANCEUR **9336**
LANGAGE DE PROGRAMMATION ADAPT **159**
LANGAGE DE PROGRAMMATION APT **656**
LANGUE D'ASPIC **5380**
LANGUETTE **5059**
LANGUETTE ET RAINURE **12989**
LANIERE POUR COURROIES **7500**
LANOLINE **7391**
LANTERNE **7393**
LANTERNE D'UNE VANNE **7392**
LANTERNE DE PRESSE-ETOUPE **7392**
LANTHANE **7394**

LAQUE **7354**
LAQUE DE BRONZE **6001**
LAQUE EN ECAILLES **11338**
LAQUE EN FEUILLES **11338**
LAQUE EN PLAQUES **11338**
LAQUE PLATE **11338**
LAQUER **7353**
LARDONS **5940**
LARGET **11306**
LARGETS EN ACIER ALLIE **378**
LARGETS EN ACIER AU CARBONE **1956**
LARGEUR **13850**
LARGEUR D'UNE SOUDURE **12887**
LARGEUR D'USINAGE **7867**
LARGEUR DE CONSTRUCTION **8926**
LARGEUR DE LA SOUDURE DE BASE **10727**
LARGEUR INTERIEURE D'UN MAILLON **6982**
LARGEUR TOTALE **8926**
LARGEUR UTILE DE LAMINAGE **13427**
LATEX **7425**
LATTE **7426**
LAVAGE (DU SABLE) **4681**
LAVE **7437**
LAVE-GLACE **13870**
LAVER (LES MINERAIS) **2550**
LAVER A GRANDE EAU **10607**
LAVER UN DESSIN **2742**
LAVEUSES ET TRIEURS DE RACINES ET DE FRUITS **5704**
LAVIS **12965**
LECTEURS OPTIQUES **8847**
LECTURE **10254**
LEDEBURITE **7504**
LEGENDE **7513**
LEIN (GEOL.): CIMENT **1253**
LEMNISCATE **2029**
LEMNISCATE **7514**
LENTILLE **7522**
LENTILLE A BORD EPAIS **4095**
LENTILLE A BORD MINCE **2910**
LENTILLE BICONCAVE **1237**
LENTILLE BICONVEXE **1238**
LENTILLE CONVERGENTE **2909**
LENTILLE CONVERGENTE **2910**
LENTILLE DE CONCENTRATION DES ELECTRONS **4626**
LENTILLE DE SOUDURE **13764**
LENTILLE DIVERGENTE **4095**
LENTILLE DIVERGENTE **9914**
LENTILLE PLAN-CONCAVE **9458**
LENTILLE PLAN-CONVEXE **9459**
LESSIVAGE **7677**
LESSIVAGE **7446**

LESSIVE **7835**
LESSIVE **2120**
LESSIVE DE POTASEE **9671**
LESSIVE DE SOUDE **11712**
LESSIVER **7676**
LESSIVEUR **3938**
LEVAGE **6553**
LEVE-GLACE **13867**
LEVE-ROUE **10121**
LEVEE **10611**
LEVEE DE CAME **1884**
LEVEE DE LA SOUPAPE **7544**
LEVEE DE TERRE **3945**
LEVER **6517**
LEVIER **7530**
LEVIER A CLIQUET **10223**
LEVIER A CONTREPOIDS **13751**
LEVIER A COULISSE **11638**
LEVIER A DEUX BRAS **4147**
LEVIER A FOURCHE **5569**
LEVIER A GENOUILLERE **12979**
LEVIER A PEDALE **5521**
LEVIER A POIGNEE **6239**
LEVIER A SONNETTE **1131**
LEVIER A TROIS BRAS **4131**
LEVIER ARTICULE **12979**
LEVIER COUDE **1131**
LEVIER D'ATTAQUE DE DIRECTION **12227**
LEVIER D'AVANCE AUTOMATIQUE **842**
LEVIER D'INVERSION DE L'AVANCE **5073**
LEVIER DE CHANGEMENT DE MARCHE **10549**
LEVIER DE COMMANDE **9414**
LEVIER DE DEBRAYAGE **4035**
LEVIER DE DECLENCHEMENT **10438**
LEVIER DE DEMARRAGE **12120**
LEVIER DE DESEMBRAYAGE **4035**
LEVIER DE FREIN **1537**
LEVIER DE MANOEUVRE **8833**
LEVIER DE PURGE LIBRE **7553**
LEVIER DE SOUPAPE **13465**
LEVIER DE VIS **12985**
LEVIER DOUBLE **4147**
LEVIER DROIT **12306**
LEVIER OSCILLANT **10656**
LEVIER ROULANT **10686**
LEVIER SELECTEUR DES AVANCES **5065**
LEVIGATION **4681**
LEVIGATIONS **7537**
LEVOGYRE **7365**
LEVRES DE COUPES **7633**
LIAISON **5416**

LIAISON RIGIDE **10589**
LIAISON SOUDEE **13772**
LIAISONS DE COURROIE **1142**
LIANT **1253**
LIANT **1445**
LIANT **1250**
LIANT DE NOYAUTAGE **3145**
LIANT DURCISSANT A L'AIR **301**
LIANT SEC **4342**
LIBER **1056**
LIBERANT (SE) (CHIM.) **8497**
LICENCE D'EXPLOITATION **7539**
LIEGE **3173**
LIEU D'ETABLISSEMENT **9423**
LIEU D'INSTALLATION **9423**
LIEU DE MONTAGE **9423**
LIEU DE TRAVAIL **13943**
LIEU GEOMETRIQUE **7715**
LIGNE **7597**
LIGNE **4523**
LIGNE **10812**
LIGNE **179**
LIGNE A HAUTE TENSION **6501**
LIGNE AERIENNE **212**
LIGNE AERODYNAMIQUE **12351**
LIGNE ATMOSPHERIQUE **798**
LIGNE BRISEE **7166**
LIGNE CONDUCTRICE **4523**
LIGNE COURBE DE VOLUME CONSTANT **7187**
LIGNE COURBE ISOTHERMIQUE **7193**
LIGNE D'ADMISSION DE LA VAPEUR **199**
LIGNE D'ARBRES **7607**
LIGNE D'ECHAPPEMENT DE LA VAPEUR **4892**
LIGNE D'ECLAIRAGE **4519**
LIGNE D'ECOULEMENT **5457**
LIGNE D'ENGRENEMENT **7602**
LIGNE D'INSPECTION **6986**
LIGNE D'INTERSECTION **7605**
LIGNE DE COLLIMATION **7601**
LIGNE DE COMPRESSION **2857**
LIGNE DE COTE **3957**
LIGNE DE FERMETURE **2536**
LIGNE DE FORCE **7603**
LIGNE DE FUITE **324**
LIGNE DE JOINT **9093**
LIGNE DE JONCTION DES TOLES **11084**
LIGNE DE NULLE PRESSION (DU DIAGRAMME D'INDICATEUR) **14029**
LIGNE DE RIVETS **10813**
LIGNE DE ROULEAUX D'AMENEE **4775**
LIGNE DE SECOURS **4690**

LIGNE DE TRANSMISSION **7607**
LIGNE DE TRANSPORT A GRANDE DISTANCE **4521**
LIGNE DE TRANSPORT DE FORCE **4526**
LIGNE DE TRUSQUINAGE **7604**
LIGNE DROITE **12310**
LIGNE DU LIQUIDUS **7661**
LIGNE HORIZONTALE **6585**
LIGNE ISENTROPIQUE **7173**
LIGNE ISOCLINE **7181**
LIGNE ISODYNAME **7182**
LIGNE ISODYNAMIQUE **7182**
LIGNE NEUTRE **8545**
LIGNE PLEINE **5725**
LIGNE POINTILLEE **4125**
LIGNE POLYGONALE **9611**
LIGNE PONCTUEE **4125**
LIGNE SOUTERRAINE **13353**
LIGNE UN TRAIT **2205**
LIGNES D'HARTMANN **7815**
LIGNES DE LUDERS **7815**
LIGNES DE LUDERS **7622**
LIGNES DE LUDERS **12371**
LIGNES LIMITES **5689**
LIGNEUX **7578**
LIGNINE **7577**
LIGNITE **7578**
LIGNITE NOIR **9405**
LIGNITE PARFAIT **1665**
LIGNONE **7577**
LIGNOSE **7577**
LIGROINE **7580**
LIMAGE **5153**
LIMAILLE **5155**
LIMBE **6026**
LIME **5148**
LIME A ALUMINIUM **422**
LIME A BARRETTES **1903**
LIME A BISEAU **1903**
LIME A COTES LISSES **10878**
LIME A DOUBLE TAILLE **4133**
LIME A GROSSE TAILLE **10774**
LIME A LOSANGE **11623**
LIME A MAIN **6234**
LIME A SCIES **10954**
LIME A TAILLE BATARDE **1057**
LIME A TAILLE CROISEE **4133**
LIME A TAILLE DEMI-DOUCE **11100**
LIME A TAILLE DOUCE **11680**
LIME A TAILLE MOYENNE **8266**
LIME A TAILLE RUDE **10774**
LIME A TAILLE TRES DOUCE **12467**

LIME AIGUILLE **8519**
LIME CARREE **12019**
LIME D'ENTREE **13625**
LIME DEMI-RONDE **6208**
LIME FEUILLE DE SAUGE **3392**
LIME FINE **11680**
LIME FORTE **10774**
LIME LARGE **9067**
LIME OLIVE **13041**
LIME PLATE **5381**
LIME PLATE MAIN **9067**
LIME PLATE POINTUE **5390**
LIME POINTUE **12648**
LIME QUATRE-QUARTS **12019**
LIME QUEUE DE RAT **10216**
LIME RONDE **10796**
LIME ROTATIVE **10755**
LIME TIERS-POINT **13189**
LIME TIERS-POINT **12870**
LIME TROIS-QUARTS **13189**
LIME 4/4 **12019**
LIME-COUTEAU **7324**
LIME-FRAISE **1390**
LIMER **5149**
LIMITE **7587**
LIMITE (CONVENTIONNELLE) DE FLUAGE **3325**
LIMITE (D'ELASTICITE) PROPORTIONNELLE **9936**
LIMITE D'ABSORPTION **50**
LIMITE D'ACCEPTATION **69**
LIMITE D'ALLONGEMENT **9923**
LIMITE D'ECOULEMENT **5456**
LIMITE D'ECOULEMENT **14011**
LIMITE D'ECOULEMENT **14012**
LIMITE D'ECRASEMENT **2873**
LIMITE D'ELASTICITE **4481**
LIMITE D'ELASTICITE PROPORTIONNELLE **7590**
LIMITE D'ENDURANCE **4741**
LIMITE D'ENDURANCE OU DE FATIGUE **5050**
LIMITE D'ERREURS **7589**
LIMITE DE FATIGUE **5053**
LIMITE DE FATIGUE PAR CORROSION **3188**
LIMITE DE REACTION **10249**
LIMITE DE SATURATION **7591**
LIMITE DE TEMPERATURE **7592**
LIMITE ELASTIQUE **14011**
LIMITE ELASTIQUE **14012**
LIMITE ELASTIQUE **14013**
LIMITE ELASTIQUE **4481**
LIMITE ELASTIQUE **5456**
LIMITE ELASTIQUE **7588**
LIMITE ELASTIQUE CONVENTIONNELLE **9923**

LIMITE ELASTIQUE CONVENTIONNELLE OU APPARENTE 9922
LIMITE ELASTIQUE MINIMALE GARANTIE **8308**
LIMITE TARDIVE **14011**
LIMITE/JOINT DE GRAIN **6033**
LIMITER **6926**
LIMITES DE TOLERANCE **7595**
LIMITEUR OU REGULATEUR DE DEBIT **4879**
LIMON **12377**
LIMON **12052**
LIMONITE **7596**
LIMONITE ARGILEUSE **2467**
LIMOUSINE **11521**
LIN **5412**
LIN DE LA NOUVELLE-ZELANDE **8550**
LINGOT **6919**
LINGOT **2069**
LINGOT BRULE **1748**
LINGOT D'ACIER **12210**
LINGOT DE CUIVRE **3120**
LINGOT DE DEPART **1834**
LINGOT ENTAILLE **8674**
LINGOT EPROUVETTE **12777**
LINGOT PLAT DE CUIVRE **3129**
LINGOT-EPROUVETTE (EN FORME DE QUILLE DE NAVIRE) 7273
LINGOTIERE **6923**
LINGOTIERE **8422**
LINOLEUM **7630**
LIQUATION **7635**
LIQUEFACTION **5764**
LIQUEFACTION D'UN GAZ **7636**
LIQUEFIABLE **7639**
LIQUEFIABLE **5756**
LIQUEFIER **8135**
LIQUEFIER UN GAZ **7640**
LIQUETURE **8421**
LIQUEUR NORMALE **12084**
LIQUEUR TITREE **12084**
LIQUIDE **7642**
LIQUIDE **7641**
LIQUIDE CAUSTIQUE **2119**
LIQUIDE D'EXTRACTION **4965**
LIQUIDE DE DECAPAGE **7293**
LIQUIDE FILTRE **5181**
LIQUIDE OBTURATEUR **11082**
LIQUIDE PENETRANT **9169**
LIQUIDE SURFONDU **12464**
LIQUIDES **7660**
LIQUIDUS **7661**
LIRE **12692**

LIRE **10252**
LISSAGE **11686**
LISSAGE **5408**
LISSE **13035**
LISSER **11679**
LIT-PLACARD **1739**
LITHARGE **7663**
LITHIUM **7664**
LITRE **7669**
LIVRE-PIED **5522**
LIXIVIATION **7677**
LIXIVIER **7676**
LOCOMOBILE **9640**
LOCOMOTION **7712**
LOCOMOTION A BENZINE **9225**
LOCOMOTIVE **7713**
LOCOMOTIVE A AIR COMPRIME **2844**
LOCOMOTIVE A ESSENCE **9225**
LOCOMOTIVE A VAPEUR **12173**
LOCOMOTIVE ELECTRIQUE **4520**
LOCOMOTIVE ELECTRIQUE A ACCUMULATEURS **79**
LOCOMOTIVE SANS FOYER **5256**
LOESS **7716**
LOGARITHME **7718**
LOGARITHME DECIMAL **2788**
LOGARITHME HYPERBOLIQUE **8509**
LOGARITHME NATUREL **8509**
LOGARITHME VULGAIRE **2788**
LOGEMENT **6660**
LOGEMENT D'UNE CLAVETTE **7282**
LOI BINOMIALE **1258**
LOI DE BRAGG **1525**
LOI DE HOOKE **6574**
LOI DES VALEURS EXTREMES **4968**
LOI EXPONENTIELLE **4941**
LONGERON **5627**
LONGRINE **7734**
LONGUE-VIS **7726**
LONGUEUR **7516**
LONGUEUR D'ARC **7517**
LONGUEUR D'ONDE **13700**
LONGUEUR D'ONDE MINIMALE **8309**
LONGUEUR D'ONDES CRITIQUE **10072**
LONGUEUR DE CONSTRUCTION **8925**
LONGUEUR DE LA FLECHE **7225**
LONGUEUR DE REFERENCE **5878**
LONGUEUR DE RUPTURE **1584**
LONGUEUR DU BRAS DE LEVIER **7536**
LONGUEUR DU CORDON **7520**
LONGUEUR FOCALE **5500**
LONGUEUR LIBRE **7518**

LONGUEUR TOTALE **8925**

LONGUEUR UTILE **13424**

LONGUEUR/LARGEUR DE LA DENT **1579**

LOPIN **1301**

LOQUET **7414**

LOSANGE **10568**

LOT **7766**

LOT DE COULEE **1059**

LOUCHE **1378**

LOUCHE **1347**

LOUCHISSEMENT **2541**

LOUP **10896**

LOUP **11550**

LOUPE **951**

LOUPE **7936**

LOUPE (MET.) **1345**

LUBRIFIANT **1250**

LUBRIFIANT **7800**

LUBRIFIANT DE MOULE **9092**

LUBRIFIANT DE RODAGE **6112**

LUBRIFICATION **7803**

LUBRIFICATION **7808**

LUBRIFIER **7801**

LUMEN **7818**

LUMIERE **7560**

LUMIERE (D'UN CYLINDRE A VAPEUR) **12176**

LUMIERE A INCANDESCENCE **6826**

LUMIERE ANODIQUE **9660**

LUMIERE ARTIFICIELLE **734**

LUMIERE D'ADMISSION **200**

LUMIERE D'ECHAPPEMENT **4895**

LUMIERE D'UN RABOT **8434**

LUMIERE DE DRUMMOND **8987**

LUMIERE DE L'ARC VOLTAIQUE **673**

LUMIERE DIFFUSE **3955**

LUMIERE DIFFUSE **3927**

LUMIERE DU GAZ **5827**

LUMIERE DU GAZ A INCANDESCENCE **6829**

LUMIERE DU JOUR **3622**

LUMIERE ELECTRIQUE **4517**

LUMIERE OXHYDRIQUE **8987**

LUMIERE POLARISEE **9585**

LUMIERE REFLECHIE **10383**

LUMINESCENCE **7819**

LUNETTE A SUIVRE **5514**

LUNETTE ARRIERE **899**

LUNETTE D'APPROCHE **10398**

LUNETTE DE CALIBRE **2721**

LUNETTE FIXE **12154**

LUNETTE MOBILE **5514**

LUNETTE VERIFICATRICE **2721**

LUNETTES D'ALESAGE **1474**

LUNETTES DE PROTECTION **5992**

LUT **8038**

LUTECIUM **7833**

LUTER **2134**

LUTTER **7831**

LUX **7834**

LYRE (D'UN TUYAU) **7832**

LYRE DE COMPENSATION **4913**

LYRE DE DILATATION **4913**

MACERATION **7837**

MACERER **7836**

MACHEFER (FOND) **2500**

MACHINE **6345**

MACHINE A ACCOUPLEMENT DIRECT **3982**

MACHINE A AFFUTER LES OUTILS **13002**

MACHINE A ALESER **1471**

MACHINE A ALESER EN CREUX **1013**

MACHINE A BOBINER LES INDUITS **709**

MACHINE A BOIS **13925**

MACHINE A BORDER **3323**

MACHINE A BOUVETER **12992**

MACHINE A BROYER **3419**

MACHINE A CALCULER **1859**

MACHINE A CALIBRER **1865**

MACHINE A CANNELER **6138**

MACHINE A CENTRER ET A DRESSER **2188**

MACHINE A CHAMBRE FROIDE **2670**

MACHINE A CHANFREINER LES TOLES **9484**

MACHINE A CHANTOURNER **5665**

MACHINE A CHARGER LES FOURS **2268**

MACHINE A CINTRER **1184**

MACHINE A CINTRER LES TOLES **9480**

MACHINE A CISAILLER **11292**

MACHINE A CISAILLER LES FERS ET PROFILES **7136**

MACHINE A CISAILLER LES METAUX EN FEUILLES OU EN BANDES **11323**

MACHINE A COLONNE D'EAU **13670**

MACHINE A COMMANDE INDEPENDANT **6863**

MACHINE A COMMANDE PAR CABLE **10736**

MACHINE A COMMANDE PAR COURROIE **1145**

MACHINE A COMMANDE PAR ENGRENAGE **5893**

MACHINE A CONCASSER **3419**

MACHINE A COUPER LES TUBES **9349**

MACHINE A COURANT ALTERNATIF **407**

MACHINE A COURANT CONTINU **3021**

MACHINE A COURBER **1391**

MACHINE A CYLINDRE OSCILLANT **8880**

MACHINE A DEBOURRER **3153**

MACHINE A DECOUPER **6012**

MACHINE A DECOUPER AUTOGENE **833**

MACHINE A DEGAUCHIR **12515**
MACHINE A DENOYAUTER **3153**
MACHINE A DETENTE **4924**
MACHINE A DEUX CYLINDRES **13313**
MACHINE A DIVISER **4102**
MACHINE A DIVISER LES CERCLES **4103**
MACHINE A DIVISER LES LIGNES DROITES **4104**
MACHINE A DOUBLE EFFET **4165**
MACHINE A DOUBLE EXPANSION **4168**
MACHINE A DRESSER **6113**
MACHINE A DRESSER **12372**
MACHINE A DRESSER ET A SURFACER **9442**
MACHINE A DRESSER LES ARBRES **11245**
MACHINE A ECRASER **3419**
MACHINE A ELARGIR LES TUBES **9352**
MACHINE A EMBOUTIR **5346**
MACHINE A ENROULER LES RESSORTS A BOUDIN **13901**
MACHINE A ENTAILLER LES TOLES **11319**
MACHINE A EQUICOURANT **12407**
MACHINE A EQUILIBRER **940**
MACHINE A ESSAYER **12793**
MACHINE A ESSAYER LES HUILES **8785**
MACHINE A ESSAYER LES RESSORTS ET LES FILS **11975**
MACHINE A ETIRER LES BARRES ET LES TUBES **982**
MACHINE A EXPANSION **4924**
MACHINE A FAIBLE COURSE **11374**
MACHINE A FAIRE LES BOURRELETS **1089**
MACHINE A FAIRE LES CHAINES EN FIL DE FER **13885**
MACHINE A FAIRE LES FEUILLES DE PLACAGE **13525**
MACHINE A FAIRE LES GRILLAGES **13897**
MACHINE A FAIRE LES JOINTS **7246**
MACHINE A FAIRE LES RAINURES ET LANGUETTES **12992**
MACHINE A FAIRE LES TENONS **12742**
MACHINE A FAIRE LES TENONS EN QUEUE D'ARONDE **4179**
MACHINE A FIL BARBELE **988**
MACHINE A FILETER **12869**
MACHINE A FILETER ET TARAUDER **11066**
MACHINE A FORER **13182**
MACHINE A FORER **4279**
MACHINE A FORER ET ALESER LES TROUS PROFONDS **3687**
MACHINE A FORGER **5561**
MACHINE A FRAISER **8291**
MACHINE A FRAISER A CONSOLE **7322**
MACHINE A FRAISER LES RAINURES **7281**
MACHINE A FRAISER LES VIS **11036**
MACHINE A FROID **10407**
MACHINE A GLACE **6750**
MACHINE A GRAND DEBIT **6466**
MACHINE A GRAND RENDEMENT **6466**

MACHINE A GRANDE COURSE **7731**
MACHINE A GRANDE VITESSE **6496**
MACHINE A GRATTER **11012**
MACHINE A HAUTE PRESSION **6492**
MACHINE A INJECTION PNEUMATIQUE **299**
MACHINE A LIMER **5154**
MACHINE A MANCHONNAGE DIRECT **3982**
MACHINE A MANDRINER LES BRIDES **5333**
MACHINE A MARCHE LENTE **7789**
MACHINE A MARCHE RAPIDE **6496**
MACHINE A MESURER WHITWORTH **13843**
MACHINE A METAUX **8178**
MACHINE A MEULER **6113**
MACHINE A MEULER A AXE HORIZONTAL **6588**
MACHINE A MEULER LES NOYAUX **3150**
MACHINE A MEULER SANS POINTES **2174**
MACHINE A MOLETER LES CYLINDRES D'IMPRIMERIE **9873**
MACHINE A MORTAISER **11643**
MACHINE A MOULER **8358**
MACHINE A MOULER **8429**
MACHINE A MOULER A SECOUSSES **7250**
MACHINE A MOULER A SECOUSSES AVEC PLAQUE REVERSIBLE **7251**
MACHINE A MOULER A SECOUSSES ET A SERRAGE COMBINES AVEC DEMOULAGE **7253**
MACHINE A MOULER LES NOYAUX **3154**
MACHINE A MOULER RAPIDE A SECOUSSES **7215**
MACHINE A MOULURER **8429**
MACHINE A MULTIPLE EXPANSION **8477**
MACHINE A PERCER **9636**
MACHINE A PERCER **4279**
MACHINE A PERCER A BROCHES MULTIPLES **8470**
MACHINE A PERCER LES TROUS A ANGLE **3178**
MACHINE A PERCER LES TUYERES **8681**
MACHINE A PERCER RADIALE **10129**
MACHINE A PERCER RAILS ET ECLISSES **10176**
MACHINE A PERCER SENSITIVE **11186**
MACHINE A PERCER VERTICALE **13552**
MACHINE A PETITE VITESSE **7789**
MACHINE A PISTON **10285**
MACHINE A PLANER ET DRESSER LES TOLES **11316**
MACHINE A PLANER LES TOLES **9485**
MACHINE A PLEINE PRESSION **12165**
MACHINE A PLIER LES FILS **13883**
MACHINE A POINCONNER **10009**
MACHINE A POINCONNER ET A CISAILLER **10007**
MACHINE A POINTER **7228**
MACHINE A POINTER PAR ETINCELAGE EN COORDONNEES RECTANGULAIRES **10300**
MACHINE A POLIR **9603**
MACHINE A PONCER AU PAPIER DE VERRE **10926**

MACHINE A PRESSION ORDINAIRE **7786**

MACHINE A PROFILER A GALETS MULTIPLES **8457**

MACHINE A PROFILER LES JANTES DE ROUES **13822**

MACHINE A QUADRUPLE EXPANSION **10059**

MACHINE A RABOTER **9451**

MACHINE A RABOTER **9452**

MACHINE A RABOTER A FOSSE **9395**

MACHINE A RABOTER A TABLE MOBILE **2007**

MACHINE A RABOTER A UN SEUL MONTANT **8821**

MACHINE A RABOTER LATERALE **11415**

MACHINE A RABOTER OUVERTE SUR LE COTE **8821**

MACHINE A RABOTER TIRANT LES BOIS D'EPAISSEUR **12847**

MACHINE A RAINER **6138**

MACHINE A RECTIFIER A LA MEULE **6113**

MACHINE A RECTIFIER L'INTERIEUR DES CYLINDRES **3568**

MACHINE A RECTIFIER LES PIECES CYLINDRIQUES **3583**

MACHINE A RECTIFIER LES SURFACES INTERIEURES **7068**

MACHINE A RECTIFIER LES SURFACES PLANES **12505**

MACHINE A REDRESSER **12320**

MACHINE A REFOULER **13415**

MACHINE A REFOULER **5561**

MACHINE A REPRODUIRE **3139**

MACHINE A RETREINDRE **12533**

MACHINE A RETROUSSER **5346**

MACHINE A RIVER **10645**

MACHINE A RIVETER **10642**

MACHINE A RODER **6565**

MACHINE A ROULER **1186**

MACHINE A SCIER **9747**

MACHINE A SCIER ALTERNATIVE **6198**

MACHINE A SERRAGE PAR SECOUSSE ET PRESSION SANS DEMOULAGE **7252**

MACHINE A SIMPLE EFFET **11495**

MACHINE A SIMPLE EXPANSION **11500**

MACHINE A SOUDER **13786**

MACHINE A SOUFFLER LES NOYAUX **3146**

MACHINE A TAILLER LES ENGRENAGES **5891**

MACHINE A TARAUDER ET FILETER LES TUBES **9361**

MACHINE A TRAVAILLER LES FILS METALLIQUES **13904**

MACHINE A TREFILER **13889**

MACHINE A TRIPLE EXPANSION **13217**

MACHINE A TROIS CYLINDRES **12873**

MACHINE A TRONCONNER **3541**

MACHINE A VAPEUR **12164**

MACHINE A VAPEUR DEMI-FIXE **11175**

MACHINE A VAPEUR FIXE **12144**

MACHINE A VAPEUR MURALE **13620**

MACHINE A VAPEUR REVERSIBLE **10543**

MACHINE A VAPEUR SURCHAUFFEE **4759**

MACHINE A VISSER **11066**

MACHINE ALTERNATIVE A SCIER **5629**

MACHINE ARITHMETIQUE **1859**

MACHINE AU JET DE SABLE **10920**

MACHINE AUXILIAIRE **853**

MACHINE AVEC CONDENSATION **2911**

MACHINE COMPOUND **2829**

MACHINE D'ESSAI D'EMBOUTISSAGE D'ERICHSEN **4819**

MACHINE D'ESSAI DE TRACTION **12752**

MACHINE D'ESSAIS UNIVERSELLE **13388**

MACHINE DE MANIPULATION **7982**

MACHINE DE PRECISION **9770**

MACHINE DE RESERVE **12070**

MACHINE DE SOUDAGE A L'ARC **680**

MACHINE DE SOUDAGE DOUBLE POINT **4382**

MACHINE DE TRAVAIL **7839**

MACHINE ELECTRIQUE **4522**

MACHINE FRIGORIFIQUE **10407**

MACHINE HORIZONTALE **6582**

MACHINE HYDRAULIQUE **6691**

MACHINE ISOLEE **11151**

MACHINE MAGNETO-ELECTRIQUE **7930**

MACHINE MARINE **8005**

MACHINE MONOCYLINDRIQUE **11498**

MACHINE MOTRICE **9859**

MACHINE MULTIPLE **8456**

MACHINE POUR ESSAIS DE FLEXION **7845**

MACHINE POUR TRAVAUX SPECIAUX **11486**

MACHINE RECEPTRICE **7839**

MACHINE ROTATIVE **10761**

MACHINE SANS CONDENSATION **8627**

MACHINE SOUFFLANTE **1363**

MACHINE STUMPF **12407**

MACHINE TANDEM **12613**

MACHINE UNIVERSELLE A RECTIFIER LES SURFACES DE REVOLUTION **13383**

MACHINE VERTICALE **13546**

MACHINE-OUTIL **7859**

MACHINE-OUTIL A TRAVAILLER LE BOIS **13925**

MACHINE-OUTIL A TRAVAILLER LES METAUX **8178**

MACHINERIE **7862**

MACHINES A BOIS **13931**

MACHINES A COUDER **7232**

MACHINES A ETETER LES BETTERAVES **12441**

MACHINES A PERFORER LES TOLES **11317**

MACHINES A PLANER LES TOLES **11318**

MACHINES A TAILLER LES ENGRENAGES DROITS ET HELICOIDAUX **12006**

MACHINES A TRAIRE **8275**

MACHINES A TRAVAILLER LA TOLE **11320**

MACHINES DE LAITERIE **3595**

MACHINES POUR COUPER LES HAIES **6418**

MACHINISTE **4765**
MACHOIRE **13564**
MACHOIRE (D'ETAU) **6122**
MACHOIRE DE BLOCAGE **13945**
MACHOIRE DE CLEF **7216**
MACHOIRE DE CROCODILE **9366**
MACHOIRE DE FREIN **1536**
MACHOIRE FIXE D'UN ETAU **5299**
MACHOIRE MOBILE D'UN ETAU **8437**
MACLAGE **13302**
MACLE **13300**
MACLE DE DEFORMATION **8113**
MACONNER **1597**
MACONNERIE **1601**
MACONNERIE **8025**
MACONNERIE EN BRIQUES **1602**
MACONNERIE EN MOELLONS **10828**
MACONNERIE EN PIERRES DE TAILLE **762**
MACROGRAPHIE **7871**
MACROPHOTOGRAPHIE **9266**
MACROSCOPIQUE **7872**
MACROSEGREGATION **7876**
MACROSTRUCTURE **7869**
MADRIER **9457**
MAGASIN **12297**
MAGASINAGE **12288**
MAGASINIER **12298**
MAGNALIUM **7878**
MAGNESIE **7879**
MAGNESIE CALCINEE **1842**
MAGNESIE EN POUDRE **9725**
MAGNESITE **7880**
MAGNESITE MORTE **3639**
MAGNESIUM **7882**
MAGNETIQUE **7893**
MAGNETISABLE **7922**
MAGNETISATION **7928**
MAGNETISATION **7906**
MAGNETISME **7926**
MAGNETISME LATENT **7417**
MAGNETISME LIBRE **5640**
MAGNETISME PERMANENT **9206**
MAGNETISME REMANENT **10445**
MAGNETISME RESIDUEL **10445**
MAGNETITE **7927**
MAGNETO **7930**
MAGNETOPHONE **12644**
MAGNETOSCOPIE **7877**
MAGNETOSCOPIE **7910**
MAGNETOSTRICTION **7933**
MAILLE **9032**

MAILLE (D'UN CRIBLE) **8169**
MAILLE D'UN TAMIS **8170**
MAILLE D'UNE CHAINE **7627**
MAILLE D'UNE TOILE METALLIQUE **8170**
MAILLE RECTANGULAIRE **10304**
MAILLECHORT **8569**
MAILLECHORT **5932**
MAILLES SERREES (A) **5206**
MAILLET **13926**
MAILLET EN BOIS **7962**
MAILLON **7627**
MAILLON DE CHAINE **2211**
MAIN-D'OEUVRE **13961**
MAINTIEN **6526**
MAINTIEN EN TEMPERATURE **6528**
MAITRE A DANSER **6975**
MAITRE A DANSER **1870**
MAITRE-CYLINDRE **8037**
MAL VENUE **8321**
MALACHITE **6089**
MALAXER (DU MORTIERDU CIMENT) **12718**
MALAXEUR **8334**
MALAXEUR **9966**
MALAXEUR **8449**
MALLE **13239**
MALLE **1458**
MALLEABILISATION **7961**
MALLEABILISATION DE LA FONTE **13175**
MALLEABILISER LA FONTE **7952**
MALLEABILITE **7958**
MALLEABLE **7960**
MAMELON **8578**
MAMELON A DEUX FILETS **4149**
MANCHE **6262**
MANCHE D'UN MARTEAU **11244**
MANCHE D'UN OUTIL **6263**
MANCHE D'UNE HACHE **6433**
MANCHETTE **11576**
MANCHETTE D'UN TAMPON **11548**
MANCHON **3258**
MANCHON **1773**
MANCHON A BANDE DE COURROIE **7349**
MANCHON A BOULONS NOYES **2449**
MANCHON A CLIQUETS **10221**
MANCHON A COQUILLES **2449**
MANCHON A FRETTES **11577**
MANCHON A LANIERES DE CUIR **7501**
MANCHON A VIS **11026**
MANCHON ARTICULE UNIVERSEL **6572**
MANCHON D'ACCOUPLEMENT **3257**
MANCHON D'ACCOUPLEMENT CYLINDRIQUE **2449**

MANCHON D'ACCOUPLEMENT SELLERS **11164**

MANCHON D'EMBRAYAGE **11599**

MANCHON DE CENTRAGE **2087**

MANCHON DE DILATATION **4915**

MANCHON DE MANIVELLE A BRAS **3300**

MANCHON DE RACCORD **11705**

MANCHON DE REDUCTION D'ALESAGE **3962**

MANCHON DE SURETE **10884**

MANCHON DE T.H. COLLET DE T.H. **7998**

MANCHON ELASTIQUE **4476**

MANCHON FEMELLE **11061**

MANCHON FLEXIBLE **5417**

MANCHON MOBILE **11599**

MANCHON PORTE-MECHE **4266**

MANCHON POUR RACCORDER DEUX TUYAUX COUPES **4158**

MANCHON PROTECTEUR **9942**

MANCHON REDUCTEUR **10351**

MANCHON TARAUDE **11061**

MANCHON/DOUILLE DU REGULATEUR **6017**

MANCHONNAGE **3254**

MANDRIN **2393**

MANDRIN **2727**

MANDRIN **4260**

MANDRIN **11904**

MANDRIN A ELARGIR LES TUBES **13250**

MANDRIN A QUATRE MORS INDEPENDANTS **5604**

MANDRIN A QUEUE CONIQUE **12667**

MANDRIN CONIQUE **12651**

MANDRIN DE MONTAGE **7964**

MANDRIN DE MONTAGE EXPANSIBLE **4907**

MANDRIN DE MONTAGE LISSE **9425**

MANDRIN DE REPRISE **7964**

MANDRIN DE SERRAGE **6121**

MANDRIN DE TOUR **2393**

MANDRIN ET PINCES PORTE-FRAISE **2729**

MANDRIN EXPANSIBLE JACOBS **7208**

MANDRIN FILETE **12867**

MANDRIN FLOTTANT **5438**

MANDRIN LISSE **11789**

MANDRIN PNEUMATIQUE **279**

MANDRIN UNIVERSEL **13382**

MANDRIN UNIVERSEL A TROIS MORS **12874**

MANDRINER UN TROU DE RIVET **4262**

MANDRINER UN TUBE **4903**

MANDRINS AUTO-ENTRAINEURS **11152**

MANDRINS EXPANSIBLES **4906**

MANDRINS HYDRAULIQUES **6679**

MANDRINS MAGNETIQUES **7898**

MANDRINS MECANIQUES **8095**

MANDRINS PNEUMATIQUES **9550**

MANEGE **1731**

MANETON **3302**

MANETON **3300**

MANETON **3311**

MANETON **9319**

MANETON DE VILEBREQUIN **3311**

MANETTE DE COMMANDE **7706**

MANGANESE **7966**

MANGANESE ELECTROLYTIQUE **4599**

MANGANESIFERE **7971**

MANIABLE **6269**

MANIABLE (PEU) **13402**

MANILLE **7980**

MANILLE D'ASSEMBLAGE **11232**

MANIVELLE **3295**

MANIVELLE A COULISSE **11631**

MANIVELLE A FAIBLE COURSE **11375**

MANIVELLE A GRANDE COURSE **7732**

MANIVELLE A MAIN **6231**

MANIVELLE A MANCHON **3305**

MANIVELLE AMOVIBLE **10450**

MANIVELLE COUDEE A LA FORGE **5550**

MANIVELLE D'INDEXAGE **6865**

MANIVELLE DU VILEBREQUIN **6976**

MANIVELLE EN BOUT **8947**

MANIVELLE EN PLUSIEURS PIECES **1706**

MANIVELLE EN PORTE-A-FAUX **8947**

MANIVELLE FORGEE **5550**

MANIVELLE FRONTALE **8947**

MANIVELLE VENUE DE FORGE **5550**

MANO-VACUOMETRE **2830**

MANOEUVRE **12226**

MANOEUVRE DE SERVICE **13398**

MANOEUVRE DU FREIN **1549**

MANOEUVRER **3049**

MANOMETRE **9820**

MANOMETRE **9821**

MANOMETRE A AIR COMPRIME **2534**

MANOMETRE A AIR LIBRE **8827**

MANOMETRE A MERCURE **8151**

MANOMETRE A PLAQUE **3873**

MANOMETRE A RESSORT **11989**

MANOMETRE A TUBE **1501**

MANOMETRE BOURDON **1501**

MANOMETRE DIFFERENTIEL **3921**

MANOMETRE ENREGISTREUR **10292**

MANOMETRE ETALON **12080**

MANOMETRE METALLIQUE **11989**

MANOMETRE POUR DETERMINATIONS ANEMOMETRIQUES
4211

MANOSTAT **9835**

MANQUE **11537**

MANQUE DE FUSION **6782**
MANQUE DE MATIERE **13348**
MANQUE DE PLANEITE **1506**
MANQUE DE SOUPLESSE **1628**
MANQUES (AFNOR) **6536**
MANTISSE **7985**
MANUEL DE FABRICATION **9885**
MANUFACTURABLE **12802**
MANUFACTURE **5013**
MANUFACTURER **7948**
MANUSCRIT **7996**
MANUTENTION MECANIQUE **8106**
MANUTENTIONNAIRE **12298**
MAQUETTE **4370**
MARBRE **12321**
MARBRE **6048**
MARBRE **6114**
MARBRE **8001**
MARBRE A DRESSER **12516**
MARCHE **10850**
MARCHE **12239**
MARCHE **13172**
MARCHE A VIDE **6752**
MARCHE A VIDE **8598**
MARCHE ARRIERE **10846**
MARCHE AVANT **10848**
MARCHE BRUYANTE **8610**
MARCHE DOUCE **7567**
MARCHE EN ARRIERE DU PISTON **10530**
MARCHE EN AVANT DU PISTON **5585**
MARCHE EN CONTRE-PRESSION **894**
MARCHE IRREGULIERE **7165**
MARCHE LOURDE **6412**
MARCHE REGULIERE **11681**
MARCHE SILENCIEUSE **8609**
MARCHER A CHARGE INCOMPLETE **1080**
MARCHER A GRANDE VITESSE **10833**
MARCHER A PETITE VITESSE **10834**
MARCHER A PLEINE CHARGE **1079**
MARCHER A VIDE **1081**
MARCHES EN CAILLEBOTIS **6102**
MARMITE AUTOCLAVE **829**
MARMITE DE PAPIN **829**
MARNE **8019**
MARNE ARGILEUSE **699**
MARNE CALCAIRE **2225**
MARNE SILICEUSE **11439**
MARQUAGE **8015**
MARQUAGE DES TOLES AU POINCON **3894**
MARQUE DE COMMERCE **13100**
MARQUE DE FABRIQUE **13100**

MARQUE DE L'ESTAMPE **3888**
MARQUE REPERE DE NIVEAU D'EAU **13660**
MARQUER **8008**
MARQUER **8010**
MARQUER UN REPERE AU POINTEAU **8011**
MARQUES DE MEULAGE **15**
MARQUES DE VIBRATION **2276**
MARQUES SUPERFICIELLES EN FORME DE V **2282**
MARTEAU **6215**
MARTEAU A AIR COMPRIME **9553**
MARTEAU A DENT **2462**
MARTEAU A DEUX TETES **4169**
MARTEAU A DEVANT **12308**
MARTEAU A EBARBER LA FONTE **10986**
MARTEAU A FRAPPER DEVANT **11573**
MARTEAU A MAIN **6235**
MARTEAU A MATER **1872**
MARTEAU A PANNE BOMBEE **960**
MARTEAU A PANNE FENDUE **2462**
MARTEAU A PANNE RONDE **962**
MARTEAU A PIQUER LES CHAUDIERES **10986**
MARTEAU A PLANER **9455**
MARTEAU A RIVER **10644**
MARTEAU A RIVER PNEUMATIQUE **9555**
MARTEAU A SUAGE **3321**
MARTEAU A TETE PLASTIQUE **11747**
MARTEAU A VAPEUR **12166**
MARTEAU ARRONDI **960**
MARTEAU D'AJUSTEUR **5279**
MARTEAU D'OUTILLEUR **13014**
MARTEAU DE FORGE **5559**
MARTEAU DE FORGE DOUBLE CORPS **4140**
MARTEAU DE TAPISSIER AIMANTE **7921**
MARTEAU MECANIQUE **5559**
MARTEAU MECANIQUE A AIR COMPRIME **9555**
MARTEAU PNEUMATIQUE **9551**
MARTEAU PNEUMATIQUE **9553**
MARTEAU-PELIN A PLANCHE **1380**
MARTEAU-PILON **4326**
MARTEAU-PILON **4325**
MARTEAU-PILON **4327**
MARTEAU-PILON **6080**
MARTEAU-PILON **9552**
MARTEAU-PILON **12166**
MARTEAU-PIQUEUR **258**
MARTEAU-RIVOIR **10644**
MARTELAGE **6223**
MARTELAGE **9157**
MARTELAGE A FROID **2685**
MARTELE **6220**
MARTELER **6216**

MARTELER A FROID **2684**
MARTELURES **6217**
MARTENSITE **8022**
MARTENSITE ACICULAIRE **103**
MASQUE **6252**
MASQUE DE SOUDAGE **6432**
MASSE **3119**
MASSE **11572**
MASSE **12611**
MASSE (MEC.) **8026**
MASSE (TERRE) **4409**
MASSE D'ACIER **951**
MASSE EN ROTATION **10561**
MASSE PRINCIPALE **8055**
MASSE TOURNANTE **10561**
MASSELOTTE **11506**
MASSELOTTE **3629**
MASSELOTTE A NOYAU ATMOSPHERIQUE **799**
MASSES POLAIRES **9593**
MASSETTE EN CUIVRE **3119**
MASSETTE EN PLOMB **7463**
MASSICOT **8031**
MASSIF DE FONDATION **5589**
MASSIF EN BETON **2896**
MASTIC **8038**
MASTIC **10037**
MASTIC A L'HUILE **8771**
MASTIC A LA RESINE **10482**
MASTIC A PIERRE **12274**
MASTIC AU MINIUM **10332**
MASTIC D'ASPHALTE **764**
MASTIC DE FER **7141**
MASTIC DES FONTAINIERS **13797**
MASTIC DES VITRIERS **5971**
MASTIC EN GRAINS **8039**
MASTIC EN LARMES **8039**
MASTIC POUR FONTE **7141**
MASTIC RESINEUX **10482**
MASTIC ROUGE **10332**
MASTICAGE **2146**
MASTIQUER **2134**
MASURIUM **8040**
MATAGE **2115**
MATELAS **254**
MATELAS D'HUILE **8754**
MATER **2113**
MATERIAUX REFRACTAIRES **10466**
MATERIEL D'IRRIGATION **7172**
MATERIEL DE CONSTRUCTION **8050**
MATERIEL DE MONTAGE **4810**
MATERIEL DE MONTAGE **4809**

MATHEMATIQUE **8053**
MATIERE **12419**
MATIERE A AIGUISER **6107**
MATIERE A POLIR **9601**
MATIERE A POLIR **11**
MATIERE A POLIR **22**
MATIERE ABRASIVE **9601**
MATIERE AGGLUTINANTE **1253**
MATIERE CALORIFUGE POUR CHAUDIERES **1408**
MATIERE CHARGEANTE **7690**
MATIERE COLORANTE **9303**
MATIERE COLORANTE A BASE DE CHROME **2387**
MATIERE COLORANTE A BASE DE PLOMB **7466**
MATIERE COLORANTE DERIVEE DU GOUDRON DE HOUILLE **2567**
MATIERE COLORANTE MINERALE **4412**
MATIERE DE MELANGE **228**
MATIERE DE REMPLISSAGE **5164**
MATIERE EN FUSION **8365**
MATIERE ETOUFFANT LE BRUIT **3641**
MATIERE FIBREUSE **5140**
MATIERE FILTRANTE **5179**
MATIERE IMPERMEABLE AU SON **3641**
MATIERE INORGANIQUE **6957**
MATIERE ISOLANTE **6998**
MATIERE ORGANIQUE **8866**
MATIERE POUR JOINT **7247**
MATIERE PREMIERE **10239**
MATIERE PROTECTRICE **9952**
MATIERES EN DISSOLUTION **8061**
MATIERES EN SUSPENSION **8062**
MATIERES PLASTIQUES **9476**
MATOIR **2114**
MATRAS **10277**
MATRICAGE **12059**
MATRICAGE **2651**
MATRICAGE **12532**
MATRICAGE (DEFAUT DE SURFACE DE TOLE) **3894**
MATRICE **3883**
MATRICE **8055**
MATRICE A DESCENTE COMMANDEE **4193**
MATRICE A DISQUE **1310**
MATRICE A ETIRER **4234**
MATRICE AMOVIBLE **6966**
MATRICE COMPOSITE **2818**
MATRICE D'ESTAMPAGE **2652**
MATRICE DE DESSOUS **1497**
MATRICE DE DESSUS **13036**
MATRICE ETAMPE ESTAMPE DE FORGERON **3897**
MATRICE FINISSEUSE **5219**
MATRICE FLOTTANTE **5437**

MATRICE NEGATIVE **8529**
MATRICE OUVERTE **8810**
MATRICE RONDE ET DIVISEE CAGE REFOULEUSE **4451**
MATRICER **12056**
MATTE **11878**
MATTE **8058**
MATTE DE CUIVRE **3121**
MATTE DE NICKEL **8563**
MAUVAIS COMBUSTIBLE **7780**
MAZOUT **6408**
MAZOUT **8041**
MAZOUT **5714**
MECANICIEN **4765**
MECANICIEN-CONSTRUCTEUR **4214**
MECANIQUE **8121**
MECANIQUE **8093**
MECANIQUE (ESSAI) **12743**
MECANIQUE DES CORPS LIQUIDES **6720**
MECANISME **8122**
MECANISME **3913**
MECANISME A CAME **1886**
MECANISME A MANIVELLE **1094**
MECANISME A MANIVELLE EXCENTRIQUE **4427**
MECANISME A VIS SANS FIN **13974**
MECANISME D'ARRET **7709**
MECANISME D'ARRET DES ECROUS **8701**
MECANISME D'EMBRAYAGE **4034**
MECANISME D'INDEXAGE **6869**
MECANISME D'INDEXAGE POUR CREMAILLERE **10123**
MECANISME DE CHANGEMENT DE MARCHE **10544**
MECANISME DE CHANGEMENT DE VITESSE **2242**
MECANISME DE COMMANDE **4311**
MECANISME DE COMMANDE **4309**
MECANISME DE DEBRAYAGE **4034**
MECANISME DE DEBRAYAGE DE LA COURROIE **1157**
MECANISME DE REPRISE DES JEUX **915**
MECANISME DE RETENUE EN POSITION D'OUVERTURE **6525**
MECANISME DE ROUE LIBRE **5652**
MECANISME DES AVANCEMENTS **5067**
MECANISME POUR RETOUR RAPIDE **10106**
MECHE **11957**
MECHE **4271**
MECHE **4263**
MECHE (D'UNE LAMPED'UN GRAISSEUR) **13846**
MECHE A CENTRE **2164**
MECHE A COUPE RAPIDE **10206**
MECHE A COUTEAUX RENVERSES **4160**
MECHE A TROIS POINTES **2164**
MECHE DE GRAISSAGE **7807**
MECHE EN LISIERE DE DRAP **2538**
MECHE FACON SUISSE **11023**

MECHE PLATE **5380**
MECHE STYRIENNE **11023**
MECHE TORSE **4160**
MEDIANE **8124**
MEGADYNE **8129**
MEGAVOLT **8130**
MEGERG **8131**
MEGOHM **8132**
MELANGE **8327**
MELANGE **1325**
MELANGE (ACTION) **8336**
MELANGE (CHIMIQUE) **8336**
MELANGE CARBURE **8337**
MELANGE DE CARBURES FRITTES **2192**
MELANGE DE DIVERSES FRACTIONS DE POUDRES D'UNE
  MEME SUBSTANCE **1325**
MELANGE DE GRAIN SANS GRAINS INTERMEDIAIRES **5805**
MELANGE DETONANT **4938**
MELANGE EUTECTIQUE **4857**
MELANGE EXPLOSIF **4938**
MELANGE GAZEUX **5856**
MELANGE INTIME **7102**
MELANGE REFRIGERANT **5654**
MELANGE SULFONITRIQUE **8595**
MELANGE TONNANT **4938**
MELANGER **8328**
MELANGER **1323**
MELANGEUR **8334**
MELANGEUR **8331**
MELANGEUR A SOLE PLATE **5387**
MELANGEURS **8332**
MELAPHYRE **8133**
MELINOSE **13991**
MEMBRANE **3870**
MEMBRANE **4479**
MEMBRANE D'ECLATEMENT **10860**
MEMBRURE **8145**
MEMOIRE **8146**
MEMOIRE DESCRIPTIF DU BREVET **11856**
MEMOIRE INTERMEDIAIRE **1689**
MEMOIRE TEMPORAIRE **12736**
MENEUR **4316**
MENISQUE CONVERGENT **8147**
MENTONNET **12672**
MENTONNET D'UNE CLAVETTE **6318**
MENUISERIE **7236**
MENUISERIE (ATELIER) **7235**
MENUISERIE DE MODELES (ART) **9131**
MENUISERIE DE MODELES (ATELIER) **9132**
MENUISIER **7233**
MENUS DE HOUILLLE **11557**

MER (EN) **8738**
MERCURE **8168**
MESURAGE **8088**
MESURAGE EXACT **4875**
MESURAGE PRECIS **4875**
MESURE **8077**
MESURE COMPARATIVE **2797**
MESURE DE CAPACITE **8079**
MESURE DE CONTROLE **12077**
MESURE DE L'ARC INTERCEPTE **2419**
MESURE DE L'ELASTICITE **8340**
MESURE DE LA TEMPERATURE **8086**
MESURE DE SURFACE **8081**
MESURE DE VOLUME **8082**
MESURE ETALON **12077**
MESURE LINEAIRE **8080**
MESURE METRIQUE **8227**
MESURER **8078**
MESURER LA PUISSANCE DES FREINS **8083**
MESURES DE SECURITE **10889**
METACENDRE **8172**
METAL **8173**
METAL **13836**
METAL A GRENAILLES **6054**
METAL A HAUTE DENSITE **6482**
METAL ALLIAGE DELTA **3741**
METAL ANGLAIS **9230**
METAL ANTI-FRICTION **13836**
METAL ANTIFRICTION **619**
METAL ANTIFRICTION **888**
METAL AUTRE QUE LE FER **8632**
METAL BLANC **888**
METAL BLANC DEFLOCULE **13835**
METAL BLANC DUR **6276**
METAL BRITANNIQUE **1624**
METAL BRUT **3410**
METAL CONSTITUANT **2982**
METAL COULE **2050**
METAL D'ACCUMULATEUR **80**
METAL D'APPORT **3775**
METAL D'APPORT **5159**
METAL DE BASE **8057**
METAL DE BASE **1032**
METAL DE BASE **12422**
METAL DE BASE **9079**
METAL DE CUBILOT **3468**
METAL DE L'ELECTRODE **4576**
METAL DE RECHARGE **3775**
METAL DE RECUPERATION **11107**
METAL DE SOUDURE **13763**
METAL DEPLOYE **4904**

METAL DEPOSE **13761**
METAL DUR **6277**
METAL EN ETAT DE FUSION **6631**
METAL EN FUSION **8365**
METAL EN FUSION **6631**
METAL FONDU APPLIQUE **3776**
METAL FONDU EN COQUILLE SANS DEFORMATION **2337**
METAL INVAR **7105**
METAL LEGER **7779**
METAL LEGER **7564**
METAL LOURD **6407**
METAL MARTELLE **6221**
METAL MONEL **8379**
METAL MOU **11738**
METAL MUNTZ **8486**
METAL NOBLE EN BARRES **1729**
METAL NOBLE OU PRECIEUX **8602**
METAL NON PRECIEUX **1031**
METAL POLYCRISTALLIN **9608**
METAL POUR COUSSINET **1106**
METAL POUR MATRICES **3889**
METAL PREAFFINE **1366**
METAL PRECIEUX **9756**
METAL PRECIEUX **8603**
METAL PROPRE A LA FABRICATION DES CARACTERES
   D'IMPRIMERIE **13319**
METAL PULVERISE **8185**
METAL STRATIFIE **7379**
METAL SUPPORT **909**
METAL TENDRE **11738**
METAL VIERGE **13570**
METAL VIERGE OU NAISSANT **9853**
METALLIFERE **8197**
METALLISATION **8201**
METALLISATION **8190**
METALLISATION AU PISTOLET **5361**
METALLISATION AU ZINC **14045**
METALLISATION SOUS VIDE **13445**
METALLISATION SOUS VIDE **13487**
METALLOGRAPHIE **8203**
METALLOGRAPHIE EN COULEURS **2738**
METALLOGRAPHIQUE **8202**
METALLOIDE **8640**
METALLOIDE **8205**
METALLURGIE **8208**
METALLURGIE D'EXTRACTION **9886**
METALLURGIE DES POUDRES **9720**
METALLURGIE DU FER **8209**
METALLURGIE DU FER **5123**
METALLURGIE MECANIQUE **8107**
METALLURGIE PHYSIQUE **9282**

METALLURGIE SOUS VIDE 13446

METALLURGIQUE 8206

METASTABLE 8212

METAUX ALCALIN-TERREUX 335

METAUX ALCALINS 329

METAUX DE CONSTRUCTION (ORDINAIRES) 2988

METAUX UTILISES COMME CONTACT ELECTRIQUE 4538

METHANAL 5573

METHANE 8215

METHODE 8216

METHODE 8220

METHODE AU TREMPE 3977

METHODE D'IONISATION 7127

METHODE DE BRAGG 1525

METHODE DE CRISTAL OSCILLANT 8879

METHODE DE DEBYE-SCHERRER 3643

METHODE DE LAVE 7438

METHODE DE MESURE 8219

METHODE DE TRAVAIL 8220

METHODE DES POUDRES 9721

METHODE DITE A TROIS FILS 12883

METHODE STROBOSCOPIQUE 12386

METHYLBENZENE 12983

METHYLENE 13920

METHYLORANGE 8223

METRE 8226

METRE (UNITE DE MESURE) 8225

METRE A RUBAN 8092

METRE CARRE 12022

METRE COURANT 10849

METRE CUBE 3455

METRE DROIT 8226

METRE ETALON 12081

METRE PLIANT 5512

METRE RIGIDE 8226

METTRE A DIGERER 3937

METTRE A LA TERRE 4410

METTRE DES CONTREFICHES 1517

METTRE EN CIRCUIT 12555

METTRE EN EXPLOITATION UNE INVENTION 13938

METTRE EN MARCHE 12114

METTRE EN MOUVEMENT 4289

METTRE EN PLACE LA COURROIE 11357

METTRE EN ROUTE UNE MACHINE 12114

METTRE EN TRAIN 12114

METTRE HORS CIRCUIT 12554

METTRE UNE MACHINE HORS SERVICE 11406

MEUBLE A POLIR 9604

MEULAGE 11275

MEULAGE 11689

MEULAGE 6106

MEULAGE A LA MACHINE 7849

MEULE 6115

MEULE 9604

MEULE D'AFFUTAGE 6118

MEULE D'EMERI 4698

MEULE D'EMERI 20

MEULE DE MOULIN 8298

MEULE DIAMANTEE 3867

MEULE EN GRES 6118

MEULE FLEXIBLE 4116

MEULE VERTICALE 4450

MEULE-DIAMANT 3864

MEULER 11273

MEULES 26

MEULEUSE 6113

MEULIERE 8299

MI-POSITION 8265

MICA 8231

MICANITE 8233

MICASCHISTE 8232

MICRO POUCE 8248

MICRO-PALPEUR DE CONICITE 12652

MICROAMPERE 8238

MICROAMPEREMETRE 8237

MICROBE PATHOGENE 9126

MICROCRIQUE 3318

MICRODURETE 8245

MICROFARAD 8242

MICROFISSURE 1263

MICROFISSURES 8234

MICROGRAPHIE 8244

MICROGRAPHIE 8258

MICROGRAPHIQUE 8243

MICROHM 8247

MICROMETRE 8249

MICROMETRE 1875

MICROMETRE A CADRAN 6877

MICROMETRE A VERNIER 13540

MICROMETRE DE CONICITE 12652

MICROMETRE DE PROFONDEUR 8252

MICROMETRE METRIQUE 8228

MICRON 8257

MICROPHOTOGRAMME 9272

MICROSCOPE 8259

MICROSCOPE BINOCULAIRE 1256

MICROSCOPE ELECTRONIQUE 4629

MICROSCOPE POLARISANT 9583

MICROSEGREGATION 3172

MICROSTRUCTURE 8262

MICROTOME 8263

MICROVOLT 8264

MIGRATION **8268**
MIGRATION **1321**
MIGRATION D'IONS **7123**
MILIEU CONDUCTEUR **2914**
MILIEU DE TREMPE **10098**
MILIEU REFRACTIF **10401**
MILIEU TRANSPARENT **13139**
MILLE (ANGLAIS) **8272**
MILLIAMPERE **8284**
MILLIAMPERE/METRE **8283**
MILLIEME DE MILLIMETRE **8257**
MILLIGRAMME **8285**
MILLIMETRE **8286**
MILLIMETRE CARRE **12023**
MILLIMETRE CUBE **3456**
MILLIVOLT **8294**
MINCE COUCHE D'HUILE DE GRAISSAGE **5167**
MINE **8300**
MINE DE HOUILLE **2566**
MINE DE PLOMB **9542**
MINE METALLIQUE **8198**
MINERAI **8859**
MINERAI BRIQUETTE **8860**
MINERAI D'ARGENT **11625**
MINERAI D'ETAIN **12956**
MINERAI DE BASSE TENEUR **7781**
MINERAI DE CUIVRE **3122**
MINERAI DE FER **7143**
MINERAI DE FER MAGNETIQUE **7927**
MINERAI DE HAUTE TENEUR **6485**
MINERAI DE NICKEL **8564**
MINERAI DE PLOMB **7465**
MINERAI DE ZINC **14042**
MINERAI PAUVRE **7781**
MINERAI RICHE **6485**
MINERAIS DE MANGANESE FERREUX **5122**
MINERAIS ET FONDANTS **8863**
MINERAL **8301**
MINETTE **8305**
MINIUM **10333**
MINUENDE **8316**
MINUTERIE **3843**
MIROIR CONCAVE **2879**
MIROIR CONVEXE **3078**
MIROIR PARABOLIQUE **9041**
MIROIR PLAN **9437**
MIROIR SPHERIQUE **11888**
MISCHMETAL **8320**
MISE A LA TERRE **6149**
MISE A LA TERRE **4417**
MISE AU POINT **8419**

MISE AU POINT D UN INSTRUMENT D OPTIQUE **5505**
MISE AU POINT D'UN OUTIL **1621**
MISE DE CONTREFICHES **1521**
MISE EN BOTTES **1772**
MISE EN CIRCUIT **12560**
MISE EN MARCHE **12130**
MISE EN MARCHE **12118**
MISE EN PLACE D'UNE COURROIE **1160**
MISE EN ROUTE D'UNE MACHINE **12130**
MISE EN SERVICE **12118**
MISE EN TRAIN **12130**
MISE EN VOIE DES DENTS D'UNE SCIE **10958**
MISE HORS CIRCUIT **12559**
MISE HORS SERVICE D'UNE MACHINE **11410**
MISES BRASEES **1574**
MITRAILLE **11379**
MITRAILLE **11007**
MITRAILLE **5553**
MITRAILLES PAQUETEES **5016**
MOBILE **9635**
MODE D'ATTAQUE **8217**
MODE D'EMPLOI **6992**
MODE OPERATOIRE **8220**
MODELAGE **9135**
MODELE **9133**
MODELE **9128**
MODELE A CIRE PERDUE **7116**
MODELE A ENTONNOIRS **5868**
MODELE DEMONTABLE **7744**
MODELE EN CIRE **13703**
MODELE EN DEUX **3105**
MODELE EN PLATRE **6189**
MODELE ETALON **8032**
MODELE POUR FONTE **9129**
MODELE SQUELETTE **11526**
MODELE STANDARD **12073**
MODEM **8338**
MODERATEUR **10421**
MODULATEUR **8338**
MODULE **8339**
MODULE D'ELASTICITE **8341**
MODULE D'ELASTICITE **4483**
MODULE D'ELASTICITE TRANVERSALE **5974**
MODULE D'INERTIE **11115**
MODULE DE CISAILLEMENT **8342**
MODULE DE COULOMB **3226**
MODULE DE DENTURE **3861**
MODULE DE GLISSEMENT **8342**
MODULE DE RUPTURE **8343**
MOELLE DU BOIS **9413**
MOELLON **10827**

MOELLON EN LAITIER **11563**
MOISISSURE **8271**
MOISSONNEUSES ET M. LIEUSES **1251**
MOISSONNEUSES-BATTEUSES **2766**
MOLECULE **8364**
MOLECULE-GRAMME **6044**
MOLETAGE **7338**
MOLETAGE **8287**
MOLETAGE D'UNE TETE DE VIS **8288**
MOLETTE (OU GALET) DE SOUDAGE **2412**
MOLETTE DE LA CAME **1895**
MOLETTER UNE TETE DE VIS **8277**
MOLYBDENE **8367**
MOLYBDENITE **8366**
MOMENT BASCULANT **12936**
MOMENT CENTRIFUGE **2183**
MOMENT D'INERTIE **8372**
MOMENT D'UNE FORCE **8370**
MOMENT DE FLEXION **1185**
MOMENT DE LA QUANTITE DE MOUVEMENT **8373**
MOMENT DE RENVERSEMENT **8966**
MOMENT DE STABILITE **8375**
MOMENT DU COUPLE **13308**
MOMENT DU FROTTEMENT **8371**
MOMENT FLECHISSANT **1185**
MOMENT RESISTANT **8374**
MOMENT STATIQUE **12139**
MOMENT TORDANT **13308**
MONEL **8379**
MONNAIES DE CUIVRE **3114**
MONNAYAGE **2651**
MONOATOMIQUE **8377**
MONOBLOC **11079**
MONOCRISTAL **11480**
MONOSULFURE DE CALCIUM **1852**
MONOTRON **8387**
MONOTROPE **8388**
MONTAGE **9051**
MONTAGE **11199**
MONTAGE **2986**
MONTAGE **5275**
MONTAGE **12493**
MONTAGE **5301**
MONTAGE **4815**
MONTAGE **4813**
MONTAGE **4808**
MONTAGE A BLANC **13184**
MONTAGE A LA PRESSE **5527**
MONTAGE A LA PRESSE A CHAUD **11389**
MONTAGE A VIS **11067**
MONTAGE BLOQUE **4307**

MONTAGE D'USINAGE BATI MANNEQUIN **4985**
MONTAGE DE CALES-ETALONS **5884**
MONTAGE DE LA COURROIE **1160**
MONTAGE EN PARALLELE **9051**
MONTAGE EN SURFACE **9051**
MONTAGE GLISSANT **11600**
MONTAGE LACHE **10025**
MONTAGE SUR CHANTIER **8800**
MONTAGE TOURNANT **10847**
MONTAGE ULTERIEUR **10527**
MONTANT **10615**
MONTANT **12063**
MONTANT **12062**
MONTANT D'UNE MACHINE **12078**
MONTANT DE GARDE-CORPS **6266**
MONTANT DE PORTE **6507**
MONTANT DOUBLE **4159**
MONTANT JUMELE **4159**
MONTANT LATERAL D'ENTREE **7211**
MONTANT PORTE-OUTIL **13010**
MONTE SIMULTANEMENT OU EN CONTINUITE **4807**
MONTE-CHARGE **6007**
MONTE-COURROIE **1159**
MONTE-JUS **8389**
MONTEE DE LA COURROIE **2499**
MONTEE DU PISTON **13409**
MONTEE EN TEMPERATURE **10231**
MONTER **4805**
MONTER **12491**
MONTER **6517**
MONTER A VIS **11037**
MONTER EN PARALLELE **2956**
MONTER EN TENSION **2957**
MONTER LA COURROIE **11357**
MONTER UN CALQUE **8431**
MONTEUR **4816**
MONTEUR-ELECTRICIEN **4543**
MONTRE DE SEGER **11134**
MONTRE FUSIBLE **11134**
MONTURE D'ESTAMPE A GUIDAGE A COLONNES **3890**
MONTURE D'UNE FILIERE A COUSSINETS **3895**
MORD D'UN ETAU **13564**
MORDACHE A CHARNIERE **13562**
MORDACHE D'UN ETAU **13563**
MORDACHES **6120**
MORDANT **8391**
MORDANT **3196**
MORFIL **13890**
MORS **13564**
MORS DOUX POUR MANDRINS **2394**
MORTAISAGE **11641**

MORTAISE **8401**
MORTAISE A CLAVETTE **7282**
MORTAISE A EXTREMITES ARRONDIES **8714**
MORTAISER **11630**
MORTAISEUSE **11643**
MORTAISEUSE A BOIS **8402**
MORTAISEUSES **11640**
MORTIER (A CONSTRUIRE) **8394**
MORTIER (VASE EN BRONZE) **8396**
MORTIER (VASE EN PORCELAINEEN AGATE) **8395**
MORTIER AERIEN **8854**
MORTIER DE CIMENT **2137**
MORTIER HYDRAULIQUE **6690**
MORTIER REFRACTAIRE **10406**
MOT **13934**
MOTEUR **9859**
MOTEUR A AIR CHAUD **6615**
MOTEUR A AIR COMPRIME **2843**
MOTEUR A ALCOOL **312**
MOTEUR A BASSE TENSION **7791**
MOTEUR A CARBURANT **8756**
MOTEUR A CARTER **4715**
MOTEUR A CHAMP TOURNANT **12876**
MOTEUR A COMBUSTION INTERNE **7061**
MOTEUR A COMBUSTION INTERNE **7062**
MOTEUR A COURANT ALTERNATIF **408**
MOTEUR A COURANT CONTINU **3022**
MOTEUR A CYLINDRES JUMELES **13301**
MOTEUR A DEUX TEMPS **13312**
MOTEUR A ESSENCE **5819**
MOTEUR A ESSENCE DE GRANDE PUISSANCE **7410**
MOTEUR A ESSENCE DE PETROLE **9223**
MOTEUR A EXPLOSIONS **7061**
MOTEUR A GAZ DE FAIBLE PUISSANCE **11658**
MOTEUR A HAUTE TENSION **6502**
MOTEUR A PETROLE **9046**
MOTEUR A PISTONS ROTATIFS **10754**
MOTEUR A QUATRE TEMPS **5603**
MOTEUR A QUATRE TEMPS **5600**
MOTEUR A VAPEUR **12164**
MOTEUR A VENT **13855**
MOTEUR A 2 CYLINDRES OPPOSES HORIZONTAUX **5398**
MOTEUR A 4 CYLINDRES OPPOSES HORIZONTAUX **5383**
MOTEUR ASYNCHRONE **783**
MOTEUR BLINDE OU CUIRASSE **4715**
MOTEUR BLINDE VENTILE **13529**
MOTEUR COMPOUND **2835**
MOTEUR DE DEMARRAGE **12121**
MOTEUR DE FAIBLE PUISSANCE **11661**
MOTEUR DE RESERVE **12070**
MOTEUR DIESEL **2866**

MOTEUR DIESEL **3901**
MOTEUR ELECTRIQUE **4623**
MOTEUR EN DERIVATION **11405**
MOTEUR EN ETOILE **10135**
MOTEUR EOLIEN PNEUMATIQUE **13855**
MOTEUR HYDRAULIQUE **6691**
MOTEUR MARIN **8005**
MOTEUR NON PROTEGE **8818**
MOTEUR OUVERT **8818**
MOTEUR POLYCYLINDRIQUE **8453**
MOTEUR PROTEGE **9941**
MOTEUR REFROIDI PAR AIR **251**
MOTEUR REVERSIBLE **10540**
MOTEUR SEMI-DIESEL **11171**
MOTEUR SOUS TENSION **7686**
MOTEUR SYNCHRONE **12571**
MOTEUR THERMIQUE **6345**
MOTEUR TRIPHASE **12876**
MOTEUR-GENERATEUR **8413**
MOTEUR-SERIE **11202**
MOTEUR-SHUNT **11405**
MOTEURS ELECTRIQUES **4524**
MOTEURS FIXES ET MOBILES **4766**
MOTOCULTEURS A FRAISES ROTATIVES **10753**
MOTTE DE TOURBE **9151**
MOUCHETTE **6544**
MOUFLE **9972**
MOUFLE A CORDE **10731**
MOUFLES **9973**
MOUILLAGE **8390**
MOUILLAGE **8348**
MOUILLER **8347**
MOULAGE **2068**
MOULAGE (REPRODUCTION A L'AIDE D'UN MOULE) **8428**
MOULAGE A CIRE PERDUE **7115**
MOULAGE A DECOUVERT **8820**
MOULAGE A LA TABLE **1169**
MOULAGE DE PRECISION **9765**
MOULAGE EN CHASSIS **1508**
MOULAGE EN COQUILLE **11333**
MOULAGE EN FOSSE **5445**
MOULAGE EN SABLE ETUVE **4354**
MOULAGE EN SABLE SEC **4354**
MOULAGE ETUVE **4354**
MOULAGE MECANIQUE **7851**
MOULAGE SORTIE EN LINGOTIERE **12703**
MOULAGE SOUS PRESSION **9812**
MOULAGE SOUS VIDE **13439**
MOULAGE SUR LE SOL **5445**
MOULAGES D'ACIER ALLIE **372**
MOULE **3883**

MOULE **5570**
MOULE **8352**
MOULE (FOND.) **8423**
MOULE A EMPREINTES MULTIPLES **2762**
MOULE DE DESSOUS **8678**
MOULE EN PLATRE **9465**
MOULE EN PLATRE **6188**
MOULE EN SABLE **10924**
MOULE EN TERRE GLAISE **7693**
MOULE METALLIQUE **3883**
MOULE MULTIPLE **8466**
MOULE NON-METALLIQUE **8650**
MOULER **8424**
MOULER **2030**
MOULEUR **8427**
MOULIN A BOCARDS **12057**
MOULIN A MEULES VERTICALES **4450**
MOULIN A VENT **13855**
MOULINET A PALETTES **5031**
MOULINET DYNAMOMETRIQUE **5031**
MOULINET HYDRAULIQUE (POUR LA MESURE DE LA VITESSE D'ECOULEMENT D'UN COURANT D'EAU) **11027**
MOULINS/APLATISSEURS **8296**
MOULINS/BROYEUR **8297**
MOULINS/COMBINE **8295**
MOULURES FILEES **4969**
MOUSQUETON **11691**
MOUSSAGE **4461**
MOUSSAGE **1331**
MOUSSE **5499**
MOUSSE A CELLULES FERMEES **2517**
MOUSSE DE PLATINE **9514**
MOUSSE DE PLATINE **11955**
MOUSSELINE **2288**
MOUTON **4325**
MOUTON **4326**
MOUTON A CHUTE LIBRE **4327**
MOUTON A PLANCHE **1380**
MOUTON-PENDULE CHARPY **2271**
MOUTON-PENDULE DE CHARPY **6787**
MOUVEMENT **8122**
MOUVEMENT **3913**
MOUVEMENT **12456**
MOUVEMENT **8438**
MOUVEMENT **8406**
MOUVEMENT A COMMANDE ELASTIQUE **8629**
MOUVEMENT A COMMANDE MECANIQUE **2985**
MOUVEMENT A COMMANDE POSITIVE **2985**
MOUVEMENT ABSOLU **32**
MOUVEMENT ACCELERE **60**
MOUVEMENT ALTERNATIF **10283**

MOUVEMENT ALTERNATIF **13725**
MOUVEMENT ANGULAIRE **544**
MOUVEMENT APERIODIQUE **636**
MOUVEMENT CIRCULAIRE **2421**
MOUVEMENT CONTINU **3031**
MOUVEMENT CURVILIGNE **3500**
MOUVEMENT D'AVANCE **5584**
MOUVEMENT D'HORLOGERIE **2509**
MOUVEMENT DANS LE SENS RETROGRADE **919**
MOUVEMENT DE GLISSEMENT **11605**
MOUVEMENT DE RECUL **919**
MOUVEMENT DE ROTATION **10766**
MOUVEMENT DE ROULEMENT **10702**
MOUVEMENT DE TRANSLATION **8407**
MOUVEMENT DE VA-ET-VIENT **10283**
MOUVEMENT DISCONTINU **7053**
MOUVEMENT EN SENS CONTRAIRE **3243**
MOUVEMENT GYROSCOPIQUE **6193**
MOUVEMENT HELICOIDAL **6426**
MOUVEMENT INTERMITTENT **7053**
MOUVEMENT MOTEUR **6809**
MOUVEMENT NON UNIFORME **8647**
MOUVEMENT NON UNIFORMEMENT ACCELERE **8648**
MOUVEMENT NON UNIFORMEMENT RETARDE **8649**
MOUVEMENT OSCILLATOIRE **8881**
MOUVEMENT PENDULAIRE **6302**
MOUVEMENT PENDULAIRE **10283**
MOUVEMENT PERIODIQUE **9194**
MOUVEMENT PLANETAIRE **9447**
MOUVEMENT RECTILIGNE **10320**
MOUVEMENT RELATIF **10434**
MOUVEMENT RETARDE **10518**
MOUVEMENT ROTATOIRE **10766**
MOUVEMENT SACCADE **7053**
MOUVEMENT SPHERIQUE **11889**
MOUVEMENT UNIFORME **13362**
MOUVEMENT UNIFORMEMENT ACCELERE **13365**
MOUVEMENT UNIFORMEMENT RETARDE **13367**
MOUVEMENT VARIE **13490**
MOYEN **8072**
MOYEN D'HOMOGENEISATION **3754**
MOYENNE ARITHMETIQUE **703**
MOYENNE GEOMETRIQUE **5929**
MOYEU **6661**
MOYEU **1481**
MOYEU D'UNE BRIDE **5329**
MOYEU D'UNE SEULE PIECE **11779**
MOYEU DE LA MANIVELLE **3297**
MOYEU EN DEUX PIECES **11938**
MUCILAGE **8440**
MULTIPLICANDE **8479**

MULTIPLICATEUR **8482**
MULTIPLICATEUR DE VITESSE **8921**
MULTIPLICATION **8480**
MULTIPLICATION **5909**
MULTIPLIER **8483**
MUNITION **474**
MURIATE D'AMMONIAQUE **462**
MUSCHELKALK **8487**
NACELLE **1383**
NACELLE DE PEINTRE **9017**
NAISSANT **8497**
NAPHTALENE **8494**
NAPHTALINE **8494**
NAPHTE **8493**
NAPHTOL **8495**
NAPHTOSCHISTE **8780**
NAPPE D'EAU SOUTERRAINE **6147**
NATIF **5647**
NECROSE **10770**
NEGATIVATION **4607**
NEODYME **8533**
NEODYMIUM **8534**
NEON **8535**
NERVURE **10569**
NERVURE D'UN TUBE **10570**
NETTETE D'UNE IMAGE **11276**
NETTOYAGE ALCALIN **334**
NETTOYAGE AU TAMBOUR **4335**
NETTOYAGE CATHODIQUE **2098**
NETTOYAGE DU METAL **2474**
NETTOYAGE MECANIQUE **8096**
NETTOYAGE PAR IMMERSION **11697**
NETTOYAGE PAR JETS D'AIR **244**
NETTOYANT ALCALIN **2472**
NETTOYER AU CHALUMEAU **5309**
NEUTRALISATION **8548**
NEUTRALISER **8549**
NEZ **8667**
NEZ D'UNE CLAVETTE **6318**
NICHE **10279**
NICHE **13617**
NICHE D'AGITATEUR **237**
NICKEL **8556**
NICKEL AU CARBONYLE **1978**
NICKEL ELECTROLYTIQUE **4600**
NICKELAGE **8567**
NICKELER **8566**
NIOBIUM **8576**
NIOBIUM **2747**
NITRATE **8579**
NITRATE **10903**

NITRATE D'AMMONIUM **465**
NITRATE D'ARGENT **7828**
NITRATE D'ETHYLE **4850**
NITRATE DE BISMUTH **1271**
NITRATE DE CUIVRE **3472**
NITRATE DE MERCURE **8156**
NITRATE DU CHILI **2330**
NITRATE MERCUREUX **8160**
NITRE **10903**
NITRITE **8586**
NITRITE D'AMMONIUM **466**
NITRITE DE CUIVRE **3473**
NITRITE DE POTASSE **9691**
NITRITE DE SOUDE **11725**
NITROBENZENE **8587**
NITROBENZINE **8587**
NITROCELLULOSE **6176**
NITROGENE **8588**
NITROGLYCERINE **8593**
NITROMETRE **8594**
NITRUM FLAMMANS **465**
NITRURATION **8589**
NITRURATION GAZEUSE **5830**
NIVEAU **6021**
NIVEAU **7525**
NIVEAU A BULLE D'AIR **13259**
NIVEAU A FIL **9540**
NIVEAU BAROMETRIQUE **6422**
NIVEAU D'EAU **13652**
NIVEAU D'EAU (HAUTEUR D'UN LIQUIDE) **13659**
NIVEAU DE L'EAU SOUTERRAINE **6148**
NIVEAU DE LA MER **11070**
NIVEAU SPHERIQUE **2430**
NODULAIRE **8606**
NOEUD **7313**
NOEUD COULANT **8651**
NOEUD D'OSCILLATION **8605**
NOEUD D'UNE COURBE **7741**
NOEUD DE VIBRATION **8605**
NOIR A NOIRCIR **7386**
NOIR ANIMAL **1447**
NOIR D'IVOIRE **1447**
NOIR DE FONDERIE **1297**
NOIR DE FUMEE **7386**
NOIR DE PLATINE **9511**
NOISETTES **2328**
NOIX D'ENTRAINEMENT POUR CHAINES-CABLES **11662**
NOIX DE ROBINET **9532**
NOIX POUR CHAINES **11662**
NOM COMMERCIAL **13101**
NOMBRE A SOUSTRAIRE **12426**

NOMBRE ATOMIQUE 803
NOMBRE COMMENSURABLE 10235
NOMBRE COMPLEXE 2813
NOMBRE COMPOSE 2820
NOMBRE D'AMPERES 478
NOMBRE D'UN LOGARITHME 8689
NOMBRE DE CALORIES DEGAGEES DANS UNE REACTION 6348
NOMBRE DE DENTS 8691
NOMBRE DE FILETS 8692
NOMBRE DE TOURS 8690
NOMBRE DE VOLTS 13593
NOMBRE DECIMAL 3664
NOMBRE ENTIER 7012
NOMBRE FRACTIONNAIRE 5613
NOMBRE IMAGINAIRE 6781
NOMBRE IMPAIR 8725
NOMBRE IRRATIONNEL/INCOMMENSURABLE 7162
NOMBRE PAIR 4872
NOMBRE PREMIER 9860
NOMBRE RATIONNEL 10235
NOMBRE REEL 10260
NOMENCLATURE 9095
NOMENCLATURE DES PIECES 7662
NON COMBINE 6818
NON CONDUCTEUR 3898
NON-FERREUX 8631
NON-METALLIQUE 8639
NON-REVERSIBILITE 7168
NON-REVERSIBLE 7169
NORIA 8960
NORIA 1676
NORMALE (GEOM.) 8652
NORMALE D'UNE COURBE 8662
NORMALISATION 12091
NORMALISER 12090
NORME 12071
NOYAU 3142
NOYAU D'UNE GARNITURE 3155
NOYAU D'UNE VIS 1394
NOYAU DE LA SECTION 3156
NOYAU ETUVE 927
NOYAU EXTERIEUR 5027
NOYAU FILTRE 3164
NOYAU MAGNETIQUE 3142
NOYER UNE TETE DE RIVET 3246
NOYER UNE TETE DE VIS 3246
NUANCAGE 5434
NUANCAGE 5443
NUANCAGE 11453
NUANCE/QUALITE DE L'ACIER 6022

NUISANCE PAR LA FUMEE 11676
NUMERATEUR 8693
NUMERIQUE 3941
NUMERO D'OPERATION EN COURS 8838
NUMERO D'ORDRE DE LA FABRICATION 11197
NUMERO DE CONSTRUCTION 4986
NUMERO DE JAUGE D'UN FIL 5881
NUMERO DE JAUGE D'UNE TOLE 5880
NUMERO DE SEQUENCE 11193
NUTATION 8702
OBELISQUE 8705
OBJECTIF 8708
OBJECTIF 12682
OBJET A EXAMINER AU MICROSCOPE SOUS FORME DE COUPE MINCE 11113
OBLIQUITE 1065
OBLONG 8715
OBSERVATION MACROSCOPIQUE A L'OEIL NU 7873
OBSIDIENNE 8716
OBSTRUCTION 2362
OBSTRUER (S') 2358
OBTURATEUR 10760
OBTURATEUR 9529
OBTURATEUR 13453
OBTURATEUR A LEVEE ANGULAIRE 5353
OBTURATEUR GLISSANT 11592
OBTURER L'ADMISSION DANS UNE TUYAUTERIE 11407
OCCLUSION (CHIM.) 8719
OCRE 8720
OCRE JAUNE 2467
OCTAEDRE 8722
OCTANGLE 8721
OCTANT 8724
OCTOGONE 8721
OCULAIRE 4977
OCULAIRE-MICROMETRE 8254
ODONTOGRAPHE 8726
OEIL 4980
OEIL D'UN MARTEAU 4975
OEIL D'UNE HACHE 11231
OEIL DE CHAT 2088
OEIL DE LA BIELLE 4976
OEILLET 4980
OHM 8742
OLIGISTE 11865
OLIVES 8916
OLIVINE 9243
OLIVINE 8796
OMBRE 11236
OMBRE PORTEE 13339
OMBRER 11234

ONCTUOSITE DU LUBRIFIANT 7813

ONDE 13699

ONDE CALORIFIQUE 6379

ONDE LUMINEUSE 7571

ONDE SONORE 11819

ONDE STATIONNAIRE 12145

ONDULATION 13701

ONDULATIONS 11419

ONDULATIONS 2619

ONDULATIONS DUES AU GAZ 5823

ONDULER 3336

ONGLET 8326

ONGLET CYLINDRIQUE 13358

ONGLET SPHERIQUE 11892

OPACITE 8802

OPAQUE 8803

OPERATION 9096

OPERATIONS A LA CHAINE 3034

OPERATIONS PRELIMINAIRES 9787

OPERATIONS PREPARATOIRES 9787

OPERCULE DOUBLE 4162

OPERCULE MONOBLOC (COIN) 11796

OPERCULE PARALLELE A SERRAGE MECANIQUE 9053

OPERCULE PARALLELE A SERRAGE PAR RESSORT 9052

OPERCULE SIMPLE 11492

OPHITE 11204

OPTIQUE 8849

OPTIQUE 8840

OR 5993

OR BATTU 6000

OR BLANC 13831

OR DE MONNAIE 5996

OR ELECTROLYTIQUE 4598

OR FIN 5198

OR MUSIF 12099

OR VERT 6092

ORDONNEE 8858

ORDRE 2780

ORDRE D'ALLUMAGE 5260

ORDRE D'INJECTION 6936

ORDRE DE FONCTIONNEMENT 13626

ORDRE DE GRANDEUR 8852

ORDURES 13074

OREILLE 7817

ORFEVRERIE 6003

ORGANE 7852

ORGANE 7852

ORGANE DE COMMANDE 3057

ORGANE DE COMMANDE 4309

ORGANE DE FIXATION 5042

ORGANE DE LUBRIFICATION 7812

ORGANE DE MACHINE 4756

ORGANE TRAVAILLANT 13953

ORGANES DE TRANSMISSION 13126

ORGANISATION DES USINES 6901

ORIENTATION 8867

ORIENTATION DES CRISTAUX 3994

ORIENTATION PREFERENTIELLE 9771

ORIENTE AU HASARD 10199

ORIFICE 9633

ORIFICE 6530

ORIFICE D'ADDUCTION 6947

ORIFICE D'ADMISSION 6947

ORIFICE D'ARRIVEE 6947

ORIFICE D'ECHAPPEMENT 8902

ORIFICE D'ECOULEMENT 8902

ORIFICE D'ENTREE 6947

ORIFICE D'INTRODUCTION 6947

ORIFICE DE DECHARGE 8902

ORIFICE DE GRAISSAGE 8761

ORIFICE DE PASSAGE 9634

ORIFICE DE PASSAGE 9100

ORIFICE DE SORTIE 8902

ORIFICE NOMINAL 8615

ORIFICES A BRIDES 5339

ORIFICES A SOUDER (EN BOUTA L'INTERIEUR) 1810

ORIFICES FILETES (MALES) 11054

ORIFICES TARAUDES (FEMELLES) 11053

ORIGINE DES COORDONNEES 8868

ORPIMENT 728

OS PULVERISES 1448

OSCILLATION 8884

OSCILLATION D'UN PENDULE 12548

OSCILLATION D'UN RESSORT 13558

OSCILLATION ELECTRIQUE 4547

OSCILLATION LIBRE 5643

OSCILLATION LONGITUDINALE 7739

OSCILLATION TRANSVERSALE 13146

OSCILLATIONS DU REGULATEUR 6671

OSCILLATIONS FORCEES 5535

OSCILLER 8877

OSCILLOGRAPHE 8885

OSCILLOSCOPE 8886

OSMIRIDIUM 8888

OSMIUM 8889

OSMOSE 8890

OSSATURE 5630

OSSEINE 1449

OUATE 13612

OUTIL 12994

OUTIL A DEGROSSIR 10791

OUTIL A ETIRER 4244

OUTIL A FINIR **5228**
OUTIL A MAIN **6253**
OUTIL A MAIN DE MARTELAGE **1734**
OUTIL A SIMPLE EFFET **11496**
OUTIL A TAILLANT **3538**
OUTIL A TREFILER **4244**
OUTIL A TREPANER **5494**
OUTIL COUPANT **3538**
OUTIL COUPANT (D'UNE MACHINE OUTIL) **13011**
OUTIL DE ROULAGE **3485**
OUTIL DE TOUR **13292**
OUTIL DIAMANTE **3866**
OUTIL PNEUMATIQUE **9556**
OUTIL TRANCHANT **3538**
OUTILLAGE **13015**
OUTILLAGE **4794**
OUTILLAGE DE MONTAGE **4812**
OUTILLAGE DU DESSINATEUR **4236**
OUTILLAGE DU MONTEUR **4817**
OUTILLAGE EN ECHINE **1143**
OUTILS A MISES EN CARBURE **2144**
OUTILS EN ACIER RAPIDE **6475**
OUTILS ET INSTRUMENTS DE DESSIN **4236**
OUTREMER **13334**
OUVERTURE **7963**
OUVERTURE **12976**
OUVERTURE DE CLEF **11832**
OUVERTURE DE LA SOUPAPE **8832**
OUVRAGE DE MACONNERIE **8025**
OUVRAGER **13937**
OUVRAGES EN BETON **2901**
OUVRIER **13960**
OUVRIER DE MACHINE -OUTIL **7868**
OUVRIER EXPERIMENTE **3289**
OUVRIER FONDEUR **5591**
OUVRIER MECANICIEN **7868**
OUVRIER MODELEUR **9130**
OUVRIER MOULEUR **8427**
OUVRIER NON SPECIALISE **13398**
OUVRIER PROFESSIONNEL **3289**
OUVRIER RABOTEUR **9443**
OUVRIER RIVEUR **10641**
OUVRIER TREMPEUR **13008**
OUVRIR LE CIRCUIT ELECTRIQUE **8826**
OVALE **4669**
OVALISATION D'UN COUSSINET **13719**
OXALATE **8968**
OXALATE ACIDE DE POTASSIUM **118**
OXALATE D'AMMONIAQUE **467**
OXALATE DE CALCIUM **1853**
OXALATE DE SODIUM **11726**

OXALATE FERREUX **5124**
OXALATE NEUTRE DE POTASSIUM **9692**
OXIDATION ANODIQUE A L'ACIDE CHROMIQUE **2380**
OXYCOUPAGE **5817**
OXYCOUPAGE **8977**
OXYCOUPAGE **7387**
OXYDABLE **8972**
OXYDANT **8974**
OXYDATION **8970**
OXYDATION **8984**
OXYDATION (OU CORROSION) SECHE **4350**
OXYDATION A TEMPERATURE ELEVEE **6498**
OXYDATION ANODIQUE TRAITEMENT ANODIQUE **605**
OXYDATION ELECTROLYTIQUE **4601**
OXYDATION PAR FROTTEMENT **5667**
OXYDE **8971**
OXYDE AZOTEUX **8597**
OXYDE AZOTIQUE **8581**
OXYDE CUIVREUX **3479**
OXYDE CUIVRIQUE **3474**
OXYDE CUIVRIQUE HYDRATE **3471**
OXYDE D'ALUMINIUM **413**
OXYDE D'ETHYLE **8853**
OXYDE D'ETHYLIDENE **84**
OXYDE DE BISMUTH **1272**
OXYDE DE CALCIUM **1854**
OXYDE DE CARBONE **1952**
OXYDE DE FER **5102**
OXYDE DE FER **7144**
OXYDE DE MAGNESIUM **7879**
OXYDE DE MERCURE **8157**
OXYDE DE METHYLE **8222**
OXYDE DE METHYLENE **5573**
OXYDE DE NICKEL **8574**
OXYDE DE SODIUM **11724**
OXYDE DE ZINC **14043**
OXYDE FERREUX **5125**
OXYDE FERRIQUE **7148**
OXYDE MERCUREUX **8161**
OXYDE MERCURIQUE **8157**
OXYDE PUCE **7461**
OXYDE SALIN DE MANGANESE **13207**
OXYDE SALIN DE PLOMB **10333**
OXYDE STANNEUX **12102**
OXYDE STANNIQUE **12098**
OXYDE SULFUREUX **12450**
OXYDER **8973**
OXYDER (S') **12685**
OXYDER (S') **8973**
OXYGENATION **8984**
OXYGENE **8980**

OZOKERITE **4413**
OZONE **8989**
PAIE **13613**
PAIEMENT **13613**
PAILLE **5056**
PAILLE **8279**
PAILLE **3283**
PAILLE DE FER **6217**
PAILLE DE FER **11953**
PAILLON D'ARGENT **11466**
PAILLON D'ETAIN **12955**
PAIRE **3252**
PAIRE DE CONES LISSES **404**
PALAN **6516**
PALAN **9973**
PALAN A CHAINE **2213**
PALAN A CORDE **10741**
PALAN A ENGRENAGE DROIT **12011**
PALAN A VIS SANS FIN **13973**
PALAN A VIS TANGENTE **13973**
PALAN DIFFERENTIEL **13798**
PALAN WESTON **13798**
PALES DE VENTILATEUR **5030**
PALETTE D'UNE ROUE HYDRAULIQUE **1677**
PALETTISATION **9026**
PALIER **9502**
PALIER **1117**
PALIER **1100**
PALIER **7388**
PALIER **12563**
PALIER A CHAISE SUR LE SOL **9154**
PALIER A CHARGE AXIALE **12902**
PALIER A CHARGE RADIALE/TRANSVERSALE **7257**
PALIER A COUSSINETS LISSES **9424**
PALIER A GALETS (A ROULEAUX) **10680**
PALIER A PATIN **9543**
PALIER A POTENCE **6321**
PALIER A ROULEMENT A BILLES **954**
PALIER A TOURILLON SPHERIQUE **952**
PALIER AUTO-REGULATEUR **11150**
PALIER AUXILIAIRE **7044**
PALIER AVEC PLAN DE SEPARATION INCLINE **508**
PALIER CONSOLE **1523**
PALIER COULISSANT **11596**
PALIER D'EXTREMITE **4720**
PALIER D'UN TOURILLON **11905**
PALIER DE BUTEE **12902**
PALIER DE BUTEE A BILLES **12900**
PALIER DE BUTEE A CANNELURES **12903**
PALIER DE BUTEE A UNE SEULE EMBASE **11479**
PALIER DE L'ARBRE DE COUCHE **3303**

PALIER DE L'ARBRE MANIVELLE **3303**
PALIER DE PIED; CRAPAUDINE **5524**
PALIER DE VILEBREQUIN **3314**
PALIER DE VILEBREQUIN **3303**
PALIER EN COQUILLES **4098**
PALIER EN DEUX PARTIES **4098**
PALIER EN UNE SEULE PIECE **11788**
PALIER FERME **11777**
PALIER GRAISSEUR A BAGUE **10602**
PALIER INTERMEDIAIRE **13547**
PALIER INTERMEDIAIRE **7048**
PALIER MURAL **1523**
PALIER PENDANT **6321**
PALIER PENDANT FERME **6270**
PALIER PENDANT OUVERT **6271**
PALIER SECONDAIRE **7044**
PALIER VERITICAL **5524**
PALIER-CONSOLE SUR COLONNE **9663**
PALIERS DE VILEBREQUIN **7941**
PALLADIUM **9023**
PALMER **11034**
PALMER **8249**
PALMER **1875**
PALPEUR **9882**
PALPEUR **11090**
PALPEUR A CADRAN **3842**
PAN D'UN ECROU **4994**
PANABASE **12796**
PANACHE **4777**
PAND D'UNE CLEF **7216**
PANNAGE **9157**
PANNE **1581**
PANNE A PIED DE BICHE **2463**
PANNE BOMBEE **963**
PANNE DU MARTEAU **9031**
PANNE EN LONG **12308**
PANNE EN TRAVERS **3369**
PANNE FENDUE **2463**
PANNE RONDE **961**
PANNE SPHERIQUE **963**
PANNEAU **9032**
PANNERESSE **12369**
PANTOGRAPHE **9033**
PAPIER A DESSIN **4240**
PAPIER A FILTRER **5177**
PAPIER A L'ACETATE DE PLOMB **7449**
PAPIER A LA PHENOIPHTALEINE **9239**
PAPIER ALBUMINE **311**
PAPIER AU FERRO-PRUSSIATE **5119**
PAPIER AU ROUGE CONGO **2936**
PAPIER BLEU POUR PHOTOCALQUE **5119**

PAPIER BUVARD **1350**
PAPIER CALQUE **13084**
PAPIER CARMIN **2002**
PAPIER CHERCHEUR DE POLES **9594**
PAPIER CONTINU **3013**
PAPIER CYANOFER **5119**
PAPIER D' EMERI **4696**
PAPIER D'AMIANTE **752**
PAPIER D'ANCHUSINE **342**
PAPIER D'EMBALLAGE **13984**
PAPIER D'INDICATEUR **6880**
PAPIER D'ORCANETINE **342**
PAPIER DE CURCUMA **13275**
PAPIER DE SOIE **12969**
PAPIER DE TOURNESOL **7668**
PAPIER DE VERRE **5961**
PAPIER GOMME/LAQUE **13506**
PAPIER HUILE **8788**
PAPIER IODOAMIDONNE **9690**
PAPIER JOSEPH **12969**
PAPIER MACHE **9036**
PAPIER METALLISE **8199**
PAPIER PELURE **12969**
PAPIER PHOTOCALQUE **9257**
PAPIER POLE **9594**
PAPIER POURRI **9036**
PAPIER PRUSSIATE **5119**
PAPIER QUADRILLE **11530**
PAPIER REACTIF **12783**
PAPIER SABLE **10925**
PAPIER TRANSPARENT **13084**
PAPIER-PARCHEMIN **9078**
PAPILLON **12888**
PAPILLON DE REGLAGE **12890**
PAPYRINE **9078**
PAQUET **5017**
PAQUET DE FER A SOUDER **5015**
PAQUETAGE **5018**
PARABOLE **9038**
PARABOLOIDE DE REVOLUTION **9042**
PARACHEVEMENT **11682**
PARACHEVEMENT **5220**
PARACHEVER **5214**
PARAFFINE **9043**
PARAFOUDRE **7575**
PARALLAXE **9047**
PARALLELE **9050**
PARALLELE **9049**
PARALLELEPIPEDE **9068**
PARALLELEPIPEDE DROIT **10581**
PARALLELEPIPEDE OBLIQUE **8713**

PARALLELEPIPEDE RECTANGLE **10305**
PARALLELISME **9069**
PARALLELOGRAMME **9070**
PARALLELOGRAMME ARTICULE **9057**
PARALLELOGRAMME DES FORCES **9071**
PARALLELOGRAMME DES VITESSES **9072**
PARAMAGNETIQUE **9073**
PARAMAGNETISME **9074**
PARAMETRE **9075**
PARAMETRES DE SOUDAGE **13787**
PARATONNERRE **7576**
PARC COUVERT **11302**
PARC DE STOCKAGE D'HYDROCARBURE **12627**
PARC NON COUVERT **12265**
PARCHEMIN **9077**
PARCHEMIN VEGETAL **9078**
PARCOURS **10810**
PARCOURS **4063**
PARE-BOUE **8443**
PARE-BRISE **13872**
PARE-CHOC **1732**
PARE-ETINCELLES **3109**
PARE-PIERRE **12275**
PARER **13204**
PAROI DE LA BOITE DE REFROIDISSEMENT **8381**
PAROI EPAISSE (A) **12841**
PAROI FRONTALE **4728**
PAROI MINCE (A) **12851**
PAROI PROTECTRICE **9953**
PARPAINGS **1697**
PART CENTRALE **2287**
PARTAGER **1266**
PARTICULE **9088**
PARTICULE ACICULAIRE **8517**
PARTICULE ALPHA **400**
PARTICULE BETA **1217**
PARTICULES COLLOIDALES **2736**
PARTICULES DE METAUX NOBLES **3548**
PARTICULES ENROBEES **2585**
PARTIE CONSTITUANTE VOLATILE **13586**
PARTIE DE DESSOUS (DE MOULE) **4199**
PARTIE DE DESSUS **3104**
PARTIE DEMONTABLE **7745**
PARTIE FIXE **3885**
PAS **12239**
PAS A DROITE **10585**
PAS A GAUCHE **7508**
PAS ANGLAIS **13844**
PAS CIRCONFERENTIEL **2423**
PAS D'UNE CHAINE **6978**
PAS DE FILETAGE **9399**

PAS DE L'ENGRENAGE 9408
PAS DE L'HELICE 9410
PAS DE VIS 9410
PAS DE VIS 11045
PAS DE VIS 12857
PAS DIAMETRAL 3861
PAS LINEAIRE 7615
PAS POUR TUBES 9360
PAS POUR TUBES A GAZ 5843
PAS SYSTEME BRIGGS POUR TUBES 1605
PAS SYSTEME INTERNATIONAL A BASE METRIQUE (S.I.) 7084
PAS SYSTEME LOWENHERZ 7792
PAS SYSTEME SELLERS 451
PAS SYSTEME THURY 12552
PAS WHITWORTH 13844
PASSAGE A L'ENCRE 6942
PASSAGE DE BROCHE 11906
PASSAGE DE LA CHALEUR 9098
PASSAGE DE LA COURROIE D'UNE POULIE SUR L'AUTRE 1158
PASSAGE DE LA LUMIERE 9099
PASSE 9096
PASSE 10832
PASSE A VIDE 4372
PASSE DE DECALAMINAGE (LAMINAGE) 11541
PASSE DE DEGAUCHISSAGE 11543
PASSE DE DRESSAGE 11543
PASSE DE FINISSAGE 11541
PASSE DE FINISSAGE 11542
PASSE HUMIDE 13803
PASSE LARGE 13724
PASSE REFOULEUSE 4455
PASSE-COURROIE 1159
PASSE-PARTOUT 3365
PASSER A L'ENCRE 6940
PASSER A L'ETAT LIQUIDE 8135
PASSERELLE 5776
PASSERELLE 13616
PASSERELLE DE LIAISON 2964
PASSIVANT 9106
PASSIVATION 9105
PASSIVITE 9107
PASTEL 2745
PASTILLEUSE 9310
PATE 12112
PATE A PAPIER 9035
PATE A POLIR 2827
PATE A SOUDER 9108
PATE A SURFACER ET A RODER 6108
PATE CHIMIQUE DE BOIS 2316

PATE MECANIQUE DE BOIS 8116
PATENTAGE 9124
PATENTAGE A L'AIR 280
PATIN 11587
PATIN D'UN PALIER 1034
PATINAGE DE LA COURROIE 11620
PATINAGE DES ROUES 11621
PATINE 9127
PATINER 11612
PATRON 12734
PATTE 7817
PATTE D'ARAIGNEE 7810
PATTE D'ARAIGNEE 8759
PATTES DE LEVAGE PAR VERINS 7206
PAVE 9136
PAVE (PIERRE) 9137
PEAU D'ORANGE 6008
PEAU D'ORANGE (AFNOR) 9134
PEAU DE BUFFLE 1685
PEAU DE CHAMOIS 2234
PEAU DE CHIEN 11270
PEAU DE CROCODILE 348
PEAU DE NETTOYAGE 2234
PEAU DE PORC 9300
PECHBLENDE 9411
PECHURANE 9411
PEDALE 13174
PEDALE D'EMBRAYAGE 2547
PEIGNE 12384
PEIGNE A FILETER 4952
PEIGNE A TARAUDER 7060
PEIGNE POUR L'EXTERIEUR 4952
PEIGNE POUR L'INTERIEUR 7060
PEIGNE POUR LES PAS DE VIS 2274
PEINTURAGE 9018
PEINTURE 9018
PEINTURE 9013
PEINTURE 2583
PEINTURE A L'EAU 13664
PEINTURE A L'HUILE 8766
PEINTURE AU PISTOLET 9020
PEINTURE AU PISTOLET 6179
PEINTURE PAR PULVERISATION AU PISTOLET 11965
PEINTURE SOUS-JACENTE 9864
PEINTURE VERNISSANTE 4711
PEINTURER 9014
PEINTURES 2599
PELLE 11385
PELLETER 11386
PELLICULE 5171
PELLICULE 11544

PELLICULE **5166**
PELLICULE D'EAU **5169**
PELLICULE D'OXYDE **5168**
PELLICULE EN BOBINE **10667**
PELLICULE GRASSE **5167**
PELLICULE PASSIVANTE **9104**
PELLICULE RIGIDE **8993**
PENDANT **6321**
PENDULE **9167**
PENDULE CIRCULAIRE **2422**
PENDULE COMPENSATEUR **2808**
PENDULE COMPENSE **2808**
PENETRATION **9170**
PENETRATION **11507**
PENETRATION A FROID **2689**
PENETRATION DE LA CHALEUR **6368**
PENETROMETRE **10061**
PENETROMETRE **9171**
PENICHE **991**
PENOMBRE **9177**
PENTACHLORURE D'ANTIMOINE **624**
PENTACHLORURE DE PHOSPHORE **9254**
PENTAEDRE **9175**
PENTAGONE **9172**
PENTANE NORMAL **9176**
PENTASULFURE D'ANTIMOINE **625**
PENTE **3792**
PENTE **6836**
PENTE **11627**
PENTE D'UNE CLAVETTE **12654**
PERCAGE **1468**
PERCAGE **4277**
PERCAGE **12675**
PERCAGE **10006**
PERCAGE PAR JET DE FLAMME **7222**
PERCARBONATE DE POTASSE **9693**
PERCEE **1589**
PERCER **4264**
PERCER **10771**
PERCER DES TROUS **9186**
PERCEUR -MECANICIEN **4276**
PERCEUSE **9296**
PERCEUSE **4279**
PERCEUSE **1472**
PERCEUSE A BRAS ARTICULE **12550**
PERCEUSE A COLONNE **2755**
PERCEUSE A COLONNE **9312**
PERCEUSE A GRANDE VITESSE **6495**
PERCEUSE A TETES MULTIPLES **5799**
PERCEUSE D'ETABLI **1168**
PERCEUSE DE CHAUDIERES **1409**

PERCEUSE ELECTRIQUE **9732**
PERCEUSE HORIZONTALE **6581**
PERCEUSE MULTIBROCHE EN LIGNE **8460**
PERCEUSE PORTATIVE **9636**
PERCEUSE POUR BOITE D'ESSIEU **878**
PERCEUSE RADIALE **10134**
PERCEUSE SUR BATI OU MONTANT **2755**
PERCEUSE SUR COLONNE STANDARD **13412**
PERCEUSE-ALESEUSE A COORDONNEES **3098**
PERCEUSES RIGIDES **10590**
PERCHLORATE DE POTASSE **9694**
PERCOIR POUR TOLE **9997**
PERCUSSION **11359**
PERD (QUI) **7493**
PERFORATION PAR DECHARGE DISRUPTIVE **10013**
PERFORER **9186**
PERFOREUSE **10001**
PERIDOT **9243**
PERIDOT **8796**
PERIMETRE **9190**
PERIODE DE DEMARRAGE OU DE MISE EN MARCHE D'UNE USINE **12123**
PERIODICITE **9196**
PERIODIQUE **9191**
PERIPHERIE DU CERCLE **2404**
PERLASSE **9143**
PERLE **9146**
PERLE DE SOUDURE **11842**
PERLITE **9144**
PERMALLOY **9201**
PERMANGANATE DE POTASSE **9695**
PERMEABILITE **9211**
PERMEABILITE MAGNETIQUE **7911**
PERMEABLE **9213**
PERMEABLE A L'AIR **9212**
PERMEAMETRE **9214**
PERMIS DE CONDUIRE **4302**
PERMUTATION **9216**
PERMUTATION CYCLIQUE **3556**
PEROXYDE **7148**
PEROXYDE D'AZOTE **8591**
PEROXYDE DE MANGANESE **10047**
PEROXYDE HYDRATE DE FER **5101**
PERPENDICULAIRE **9217**
PERPENDICULARITE **9219**
PERSONNEL OUVRIER **13962**
PERSPECTIF **9220**
PERSPECTIVE **9220**
PERSPECTIVE **1264**
PERSPECTIVE AXONOMETRIQUE **885**
PERSPECTIVE LINEAIRE **7614**

PERSULFATE D'AMMONIAQUE **468**

PERTE **13594**

PERTE **7754**

PERTE A VIDE **8599**

PERTE ADDITIONNELLE **170**

PERTE AU FEU **7761**

PERTE AU FEU **7341**

PERTE AU ROUGE **7341**

PERTE D'ENERGIE **7758**

PERTE DE CHALEUR **6353**

PERTE DE CHALEUR **7760**

PERTE DE CHALEUR PAR RAYONNEMENT **10145**

PERTE DE CHARGE **6315**

PERTE DE CHARGE **7759**

PERTE DE FROTTEMENT **7755**

PERTE DE POIDS **7764**

PERTE DE PRESSION **7763**

PERTE DE PUISSANCE **7762**

PERTE DE TRAVAIL **7765**

PERTE DUE AU FROTTEMENT **5685**

PERTE PAR CHOCS **7757**

PERTE PAR EVAPORATION **4866**

PERTE PAR FUITES **7756**

PERTE PAR REFLEXION **10389**

PERTE PAR REMPLISSAGE **5163**

PERTE PAR RESPIRATION **1594**

PERTE THERMIQUE **6353**

PERTES AU FEU **6763**

PERTES DE FUSION **8140**

PERTES DE MAGNETISATION **7907**

PERTES PAR CRACHEMENT **11843**

PERTES THERMOGENES **6381**

PERTURBATION **1581**

PESAGE **13746**

PESANTEUR **6078**

PESE-ACIDES **89**

PESE-ALCOOLS **316**

PESE-LIQUEURS **6722**

PESE-SELS **10901**

PESEE **13746**

PESER **13744**

PESON A RESSORT **11976**

PETIT AXE D'UNE ELLIPSE **8313**

PETIT DIAMETRE DU FILET **3850**

PETIT METAL **8632**

PETITE CALORIE **6045**

PETREOLINE **13508**

PETROLE **9226**

PETROLE **9222**

PETROLE **9229**

PETROLE BRUT **3411**

PETROLE BRUT **3412**

PETROLE BRUT SULFUREUX **11822**

PETROLE BRUT TOPPE **13040**

PETROLE LAMPANT **7275**

PETROLEINE **13508**

PH. **9231**

PHARE **6326**

PHARE **6327**

PHARE DE BROUILLARD **5506**

PHARE ENCASTRE **5480**

PHASE **9232**

PHASE CONTINUE **3033**

PHASE DISPERSEE **4049**

PHASE DISPERSIVE **3033**

PHENOL **1938**

PHENOLPHTALEINE **9238**

PHENOMENE **9240**

PHENOMENE LUMINEUX **7566**

PHENYLAMINE **552**

PHONAUTOGRAPHE **9242**

PHOSPHATATION **2600**

PHOSPHATE **9244**

PHOSPHATE D'AMMONIUM **469**

PHOSPHATE DE CALCIUM **1855**

PHOSPHATE DE MAGNESIE **8656**

PHOSPHATE DE POTASSIUM **9696**

PHOSPHATE DE SODIUM **11728**

PHOSPHATE DE SODIUM ET D'AMMONIUM **8241**

PHOSPHATE TRICALCIQUE **1855**

PHOSPHORE **9252**

PHOSPHORE BLANC **14009**

PHOSPHORE ROUGE **10334**

PHOSPHORESCENCE **9247**

PHOSPHORESCENT **9248**

PHOTO-DIODE **9258**

PHOTOCALQUE **9276**

PHOTOCOPIE **9276**

PHOTOGRAPHIE **9263**

PHOTOGRAPHIE **9265**

PHOTOGRAPHIER **9262**

PHOTOMETRE **9267**

PHOTOMETRIE **9271**

PHOTOMETRIQUE **9269**

PHOTOMULTIPLICATEUR **9274**

PHOTON **9275**

PHTALEINE DU PHENOL **9238**

PHYLLITE **9278**

PHYSIQUE **9285**

PHYSIQUE **9279**

PIECE **9083**

PIECE A TRAVAILLER **9293**

PIECE BRUTE **1301**
PIECE D'AJUSTAGE **9006**
PIECE D'APPUI **10876**
PIECE D'ECARTEMENT **4068**
PIECE D'USINAGE **9293**
PIECE DE COMPENSATION **4922**
PIECE DE FONDERIE **2079**
PIECE DE FONTE MOULEE **2079**
PIECE DE FORGE **5556**
PIECE DE GROSSE FORGE **6406**
PIECE DE MACHINE **7852**
PIECE DE MOULAGE CREUSE **6539**
PIECE DE RECHANGE **11834**
PIECE DE REFERENCE **8034**
PIECE DETACHEE **3811**
PIECE DETACHEE **11484**
PIECE FACONNEE **11265**
PIECE FINIE **5730**
PIECE FORGEE **5556**
PIECE FORGEE A EBAVURAGE EXCESSIF **8322**
PIECE FORGEE A LA MAIN **11669**
PIECE FORGEE AU MARTEAU **11670**
PIECE INTERCHANGEABLE **11834**
PIECE MATRICEE **4324**
PIECE MECANIQUE **7852**
PIECE MOULEE **2069**
PIECE RAPPORTEE **6965**
PIECE RAPPORTEE **6970**
PIECE TRAVAILLEE A LA FORGE **5556**
PIECES COULEES (OU MOULEES) **2080**
PIECES DEFECTUEUSES **3696**
PIECES DETACHEES (EN) **7329**
PIECES MOULEES EN FONTE GRISE **2083**
PIECES MOULEES RESISTANTES AUX TEMPERATURES
   ELEVEES **2084**
PIECES MOULEES(OU COULEES) RESISTANT A LA CORROSION
   2082
PIED **1041**
PIED **13537**
PIED A COULISSE **11597**
PIED A COULISSE A CADRAN **3986**
PIED A COULISSE A VERNIER POUR DENTS DE PIGNON
   5900
PIED A DENTURE **5899**
PIED A PROFONDEUR **3781**
PIED ANGLAIS **5518**
PIED CARRE ANGLAIS **12020**
PIED CUBE ANGLAIS **3453**
PIED CUBE NORMALISE **12072**
PIED DE BIELLE **3389**
PIED DE DENT **10724**

PIED DE LA BIELLE A FOURCHETTE **5568**
PIED DE LA DENT **10725**
PIEGE A CAILLOUX **9153**
PIERRAILLE **1644**
PIERRE A AFFILER **13824**
PIERRE A AFFUTER **13824**
PIERRE A AIGUISER **13824**
PIERRE A BATIR **1700**
PIERRE A BRIQUET **5427**
PIERRE A FUSIL **5427**
PIERRE A HUILE **8781**
PIERRE A PLATRE **6190**
PIERRE ARTIFICIELLE **737**
PIERRE D'AIMANT **8506**
PIERRE DE PAREMENT **5004**
PIERRE DE SAVON **10935**
PIERRE DE TAILLE **761**
PIERRE DE VERRE **5957**
PIERRE DE VIN **119**
PIERRE INFERNALE **7828**
PIERRE MEULIERE **8299**
PIERRE NATURELLE **8507**
PIERRE PONCE **9988**
PIERRE PONCE **9986**
PIERRES CONCASSEES **1644**
PIETEMENT **8679**
PIEZO-ELECTRICITE **9297**
PIEZOMETRE **12067**
PIGNON **5889**
PIGNON **9444**
PIGNON **9335**
PIGNON A FUSEAUX **9324**
PIGNON CONIQUE **1226**
PIGNON DE CHAINE **9402**
PIGNON DE DISTRIBUTION **12949**
PIGNON DE VILLEBREQUIN **3315**
PIGNON EN CUIR VERT **10238**
PILE **9304**
PILE DE VOLTA **13596**
PILE ELECTRIQUE **2126**
PILE GALVANIQUE **5782**
PILE HYDRO-ELECTRIQUE **5782**
PILE LOCALE **7694**
PILE PHOTO-ELECTRIQUE **9259**
PILE PRIMAIRE **9846**
PILE SECHE **4343**
PILE SECONDAIRE **4534**
PILE THERMIQUE **12820**
PILE THERMO-ELECTRIQUE **12836**
PILE THERMOELECTRIQUE **12820**
PILE VOLTAIQUE **13596**

PILER 9712
PILETTE MECANIQUE 1732
PILIER 9311
PILON DE BOCARD 12055
PILON DE MORTIER 9221
PIMELEINE 13508
PIN NOIR D'AUTRICHE 1293
PINACOIDE 9327
PINCE 3394
PINCE (D'UN APPAREIL DE LEVAGE) 7557
PINCE A CREUSET 3405
PINCE A SOUDER 11767
PINCE COUPANTE 3532
PINCE COUPANTE DIAGONALE 3832
PINCE DE FORGERON 1298
PINCE DE JET D'EAU 4286
PINCE DE MOHR 8345
PINCE EMPORTE-PIECE 10000
PINCE MOTORISTE 2764
PINCE PLATE 5403
PINCE RONDE 10805
PINCEAU 9019
PINCEAU 2743
PINCEAU LUMINEUX 9162
PINCEMENT 7212
PINCEMENT 12975
PINCES (S) 12987
PINCES A GAZ 5835
PINCES A TUBES 9362
PINCETTE 5536
PINT 9337
PION DE CENTRAGE 2153
PIPETTE 9368
PIPETTE A CYLINDRE 1718
PIPETTE DROITE GRADUEE 6028
PIQUAGE 9629
PIQUAGE AU MARTEAU 2347
PIQUEE 6333
PIQURE 9333
PIQURE 9393
PIQURE 10171
PIQURE DE CORROSION 3190
PIQURE DE LAMINAGE 9417
PIQURES 9416
PIQURES 9334
PIQURES DE CORROSION 4843
PISSETTE 13633
PISTE 13085
PISTE D'ESSAI 12794
PISTOLET 5660
PISTOLET A PROJETER 11962

PISTOLET DE PULVERISATION 11962
PISTOLET DE SOUDAGE PAR POINTS 6180
PISTOLETS A GOUJONS 12400
PISTON 9370
PISTON A AIR 282
PISTON A CLAPET 1673
PISTON A CONTRE-TIGE 9391
PISTON A DOUBLE EFFET 4164
PISTON A DOUBLE TIGE 9391
PISTON A EAU 13667
PISTON A SIMPLE EFFET 11494
PISTON A TIGE UNIQUE 9392
PISTON A VAPEUR 12179
PISTON CHAUFFE 6384
PISTON CONIQUE 2944
PISTON DE DEUX PIECES 1709
PISTON DE FERMETURE DU MOULE 2535
PISTON DE LEVAGE 7555
PISTON DE MOTEUR HYDRAULIQUE 13667
PISTON DE POMPE 9995
PISTON DE VERIN 10188
PISTON DISTRIBUTEUR 9390
PISTON ELEVATOIRE 1673
PISTON EVIDE 1514
PISTON OSCILLANT 8882
PISTON PLEIN 11791
PISTON PLONGEUR 9545
PISTON RAMSBOTTOM 10196
PISTON REFROIDI 3084
PISTON ROTATIF 10756
PISTON SANS TIGE 13241
PISTON SUEDOIS 10196
PIT A TREMPER 11699
PITON 4973
PITON-VIS 11057
PIVOT 9315
PIVOT 9419
PIVOT 9421
PIVOT 9420
PIVOT (D'UN ARBRE VERTICAL) 9421
PIVOT ANNULAIRE 10606
PIVOT D'UNE CHAINE A ROULEAU 9320
PIVOT SPHERIQUE 11893
PIVOTANT 9422
PLACAGE 9503
PLACAGE 13524
PLACAGE 2447
PLACAGE A LA FLAMME 5317
PLACAGE AU LAITON 1567
PLACAGE MECANIQUE 8108
PLACE FOU (SUR L'ARBRE) 7749

PLAFONNIER 10709
PLAFONNIER 4119
PLAGE D'OXYDE 5367
PLAN 4229
PLAN 3794
PLAN 9433
PLAN 11124
PLAN 9426
PLAN 9430
PLAN AUTOMOTEUR 6081
PLAN D'ACCOLEMENT (MACLE) 2826
PLAN D'ENSEMBLE 8905
PLAN D'EXECUTION 4984
PLAN D'EXECUTION (ATELIER) 11361
PLAN D'INTERSECTION 7095
PLAN DE CHARGE 7683
PLAN DE CISAILLEMENT 11284
PLAN DE CLIVAGE 2488
PLAN DE CRISTAL 3427
PLAN DE DISPOSITION 5914
PLAN DE GLISSEMENT 5975
PLAN DE JOINT 9093
PLAN DE LA TOLERIE 9504
PLAN DE MASSE 8905
PLAN DE MONTAGE 4814
PLAN DE MONTAGE 772
PLAN DE PROJECTION 9438
PLAN DE REFRACTION 9439
PLAN DE RUPTURE 3373
PLAN DE SYMETRIE 9440
PLAN DEFINITIF 5185
PLAN DETAILLE 3812
PLAN FOCAL 5501
PLAN INCLINE 6081
PLAN INCLINE 6838
PLAN OPTIQUE 8843
PLAN OSCULATEUR 8887
PLAN PRISMATIQUE 9877
PLAN RETICULAIRE 804
PLAN RETICULAIRE 7432
PLANAGE 9118
PLANAGE 9454
PLANCHE 9032
PLANCHE A DESSIN 4231
PLANCHE DU MARTEAU 4993
PLANCHE FORTE 9457
PLANCHE MINCE 1379
PLANCHER COUVERTURE DE PLANCHER 5450
PLANCHER DE VOITURE 5446
PLANE 4221
PLANER 5405

PLANETAIRE 9445
PLANEUSE 9451
PLANEUSE 12372
PLANIMETRAGE 8087
PLANIMETRE 9448
PLANIMETRER 8084
PLANIMETRIE 9449
PLANTEUSES DE POMMES DE TERRE 9705
PLAQUAGE AU NICKEL 8567
PLAQUE 9477
PLAQUE 9032
PLAQUE CIRCULAIRE 4999
PLAQUE D'ACCUMULATEUR 81
PLAQUE D'ANCRAGE 13621
PLAQUE D'APPUI 1127
PLAQUE D'ASSISE 1037
PLAQUE DE BASE 1037
PLAQUE DE BLINDAGE 712
PLAQUE DE CONSCIENCE (DE VILLEBREQUIN) 1590
PLAQUE DE CUIVRE 3123
PLAQUE DE DAME 3596
PLAQUE DE FOND 1037
PLAQUE DE FOND 1486
PLAQUE DE FONDATION 1037
PLAQUE DE GARDE 6593
PLAQUE DE GARDE 6160
PLAQUE DE LIEGE 3175
PLAQUE DE LIEGE AGGLOMERE 9802
PLAQUE DE RECOUVREMENT 3267
PLAQUE DE REDRESSEUR 10315
PLAQUE DE RENFORT 12257
PLAQUE DE TOURBE 9150
PLAQUE DE VERRE 5963
PLAQUE EN FEUTRE 11322
PLAQUE FRONTALE 4999
PLAQUE METALLIQUE 11315
PLAQUE MODELE 1710
PLAQUE MODELE DOUBLE FACE 8044
PLAQUE NEGATIVE 8530
PLAQUE PHOTOGRAPHIQUE 11187
PLAQUE POSITIVE 9658
PLAQUE SENSIBLE 11187
PLAQUE TOURNANTE 13295
PLAQUE TUBULAIRE 13254
PLAQUE-MODELE 2052
PLAQUER 9479
PLAQUES COLLECTIVES 2726
PLAQUETTE 9498
PLAQUETTE 11075
PLAQUETTES UNISERVICE 12896
PLASTICITE 3705

PLASTICITE **9474**
PLASTIFIANT **9475**
PLASTRON **1590**
PLAT DE MAINTIEN **13029**
PLAT-SUPPORT **902**
PLATE **8993**
PLATE-FORME A DIVISER **4109**
PLATE-FORME EN PORTE-A-FAUX **13875**
PLATEAU **13167**
PLATEAU A AIMANTS PERMANENTS **9205**
PLATEAU A DIVISIONS **4109**
PLATEAU D'ACCOUPLEMENT **3255**
PLATEAU D'UNE BALANCE **10976**
PLATEAU DE DEMOTTAGE **2012**
PLATEAU DE FOUR **6512**
PLATEAU DE FRICTION **5673**
PLATEAU DIVISE **4109**
PLATEAU DIVISEUR **4109**
PLATEAU DIVISEUR **4106**
PLATEAU TOURNANT **10762**
PLATEAU-MANIVELLE **4003**
PLATEAUX CIRCULAIRES **2424**
PLATEAUX DIVISEURS **6870**
PLATEAUX MAGNETIQUES **7912**
PLATEFORME **5776**
PLATEFORME DE JAUGEAGE **5885**
PLATELAGE **6101**
PLATELAGE **5449**
PLATELAGE **5451**
PLATINE **11306**
PLATINE (METAL) **9510**
PLATINE (PETITE PLAQUE METALLIQUE) **4174**
PLATINE DE FIXATION **9499**
PLATINE IRIDIE **9515**
PLATINITE **9508**
PLATINOCYANURE DE BARYUM **1002**
PLATRE **9466**
PLATS **5375**
PLATS EN ACIER ALLIE **373**
PLEINE CHARGE **5722**
PLI **4408**
PLI **5511**
PLI **3320**
PLI CHARNIERE **6506**
PLIAGE **1182**
PLIAGE A CHAUD **6619**
PLIAGE A FROID **2668**
PLIER **1173**
PLIEUSE **1181**
PLIEUSE **11087**
PLIEUSES **4453**

PLIS **1680**
PLISSER **3336**
PLOMB **7447**
PLOMB **9539**
PLOMB (DU FIL A PLOMB) **9538**
PLOMB (SCEAU EN PLOMB) **7480**
PLOMB AIGRE **621**
PLOMB ANTIMONIAL **621**
PLOMB ANTIMONIEUX **6276**
PLOMB D'OEUVRE **3409**
PLOMB DE SURETE **10886**
PLOMB DOUX **8014**
PLOMB DURCI **6276**
PLOMB EN FEUILLES **7462**
PLOMB EN FEUILLES **11314**
PLOMB FUSIBLE **10886**
PLOMB FUSIBLE **5758**
PLOMB IMPUR **1029**
PLOMB MARCHAND **8014**
PLOMB NON RAFFINE **1029**
PLOMB PUR **2311**
PLOMB SPONGIEUX **11954**
PLOMB SULFURE **5775**
PLOMBAGINE **9542**
PLOMBAGINE **9541**
PLOMBE **7479**
PLOMBER (MARQUER D'UN SCEAU EN PLOMB) **11077**
PLOMBERIE **7484**
PLONGER **3968**
PLONGEUR **9545**
PLONGEUR (DE THERMOMETRE) **1711**
PLOT **3009**
PLUME **9161**
PLUME A DESSIN **7999**
PLUME DE RONDE **10806**
PNEU **13322**
PNEU A CARCASSE RADIALE **10126**
PNEU A CLOUS **12404**
PNEU A CRAMPONS **12404**
PNEUMATIQUE **9549**
PNEUMATIQUE **9558**
PNEUS A FLANCS BLANCS **13840**
POCHE A MAIN A MANCHE **6250**
POCHE A QUENOUILLE **1496**
POCHE BRIQUETEE **7618**
POCHE DE COULEE **7356**
POCHE DE COULEE A ANSE **11258**
POCHE DE COULEE A MAIN **6238**
POCHE DE COULEE A RENVERSEMENT **12935**
POCHE DE GRUE DE COULEE **1728**
POCHE DE GRUE DE COULEE **3292**

POINT MORT HAUT (P.M.H.) 13411
POINT MORT HAUT (P.M.H) 13032
POINT TRIPLE 13216
POINT ZERO 14027
POINTE A TRACER 11068
POINTE DE L'ELECTRODE 4579
POINTE DE PARIS 13895
POINTE DE RECHANGE 3006
POINTE TRACANTE DU PLANIMETRE 13079
POINTEAU 8522
POINTEAU DE CENTRAGE 2172
POINTES ABRASIVES 24
POINTES POUR FONDERIE 5598
POINTS BRILLANTS 6472
POIRE DE VITESSE 11871
POISSAGE 12591
POIX 9400
POIX SECHE 2737
POLAIRE 9577
POLARISATION 9582
POLARISATION 9589
POLARISATION (D'UN ELECTRODE) PAR CHUTE DE
  CONCENTRATION 2889
POLARISATION ANODIQUE 604
POLARISATION CATHODIQUE 2108
POLARISATION ELECTROLYTIQUE 4604
POLARISCOPE 9584
POLARISEUR 9586
POLARITE 9587
POLAROGRAPHIE 9590
POLE 9592
POLE (ELECTR.) 9591
POLE D'UN AIMANT 9595
POLE MAGNETIQUE 7913
POLE NEGATIF 8531
POLE NORD D'UN AIMANT 8665
POLE POSITIF 9659
POLE SUD D'UN AIMANT 11827
POLES DE NOM CONTRAIRE 13392
POLES DISSEMBLABLES 13392
POLES SEMBLABLES 7581
POLI (QUI PREND BIEN LE) 9599
POLI (SUSCEPTIBLE D'UN BEAU) 9599
POLI (SUSCEPTIBLE DE PRENDRE UN) 9599
POLI PARFAIT 1755
POLIR 9598
POLIR A L'EMERI 4694
POLIR BRILLANT 1756
POLISSAGE 9600
POLISSAGE 9605
POLISSAGE 5408

POLISSAGE A BILLES 956
POLISSAGE A L'EMERI 4695
POLISSAGE A LA MAIN 6228
POLISSAGE AU TAMBOUR 1012
POLISSAGE DE LA SURFACE 3527
POLISSAGE ELECTROLYTIQUE 4605
POLISSOIR 1686
POLONIUM 9606
POLYCHROME 9607
POLYEDRE 9612
POLYGONE 9609
POLYGONE ARTICULE 5737
POLYGONE CIRCONSCRIT 2436
POLYGONE DES FORCES 9610
POLYGONE ETOILE 12107
POLYGONE FUNICULAIRE 5737
POLYGONE INSCRIT 6964
POLYGONE IRREGULIER 7164
POLYGONE REGULIER 10416
POLYMERE 9613
POLYMERIE 9616
POLYMERISATION 9614
POLYMERISER (SE) 9615
POLYMORPHISME 9617
POLYSULFURE DE POTASSIUM 9697
POMME D'ARROSOIR 10746
POMPAGE 6670
POMPE 9991
POMPE A ACIDE 121
POMPE A AIR 284
POMPE A AIR HUMIDE 13799
POMPE A AIR SEC 4340
POMPE A BRAS 6244
POMPE A DIAPHRAGME 3874
POMPE A ENGRENAGE 5895
POMPE A HUILE 8769
POMPE A MAIN 6244
POMPE A MENBRANE 3874
POMPE A MERCURE 8163
POMPE A PALETTES 13478
POMPE A PALETTES ROTATIVES 10758
POMPE A PISTON 9378
POMPE A PISTON A MOUVEMENT RECTILIGNE ET
  ALTERNATIF 10284
POMPE A PISTON OSCILLANT 11176
POMPE A PISTON PLONGEUR 9546
POMPE A ROTOR 10758
POMPE A VAPEUR 12183
POMPE A VIDE 284
POMPE ASPIRANTE 12434
POMPE CENTRIFUGE 13269

POMPE CENTRIFUGE 2184
POMPE D'AIR 249
POMPE D'ALIMENTATION 5715
POMPE D'ALIMENTATION 5071
POMPE D'ALIMENTATION 5713
POMPE D'EPREUVE 12787
POMPE D'INJECTION 5711
POMPE DE GRAISSAGE 8769
POMPE DE REPRISE 61
POMPE DUPLEX 4381
POMPE ELEVATOIRE 7545
POMPE FOULANTE 5531
POMPE HELICOIDALE 861
POMPE HELICOIDALE 11042
POMPE HYDRAULIQUE 8769
POMPE PNEUMATIQUE 284
POMPE PORTATIVE 9638
POMPE POUR PUITS PROFONDS 3689
POMPE ROTATIVE 10758
POMPE VOLUMETRIQUE 9653
POMPES 9996
PONCAGE 9990
PONCE 9988
PONCER 9987
PONCEUSE 4010
PONCEUSE 4006
PONDERABLE 9619
PONT 3667
PONT A BASCULE 13745
PONT ARRIERE (MOTEUR) 10265
PONT DE CHAUFFE 5308
PONT DE WHEATSTONE 13809
PONT ROULANT 13156
PONT ROULANT DEMOULEUR 12383
PONT SUSPENDU 12528
PONT TRANSBORDEUR 13141
PONT-ROULANT D'USINE 8942
PONT-ROULANT DE CHANTIER/DE GARE 5804
POQUETTE 1327
PORCELAINE 9625
PORE 9627
POREUX 9630
POROGENE 9628
POROSITE 9629
POROSITE A COMMUNICATION INTERNE 7029
POROSITES SUPERFICIELLES 9178
PORPHYRE 9632
PORPHYRE QUARTZIFERE 10081
PORT DE DECHARGEMENT (DE NAVIRES MARCHANDS) 8736
PORTANCE DU SOL 11751
PORTE 891

PORTE A INCANDESCENCE 6824
PORTE AU ROUGE CERISE 2327
PORTE AU ROUGE NAISSANT 1288
PORTE DU FOYER 5242
PORTE OUTILS 13012
PORTE-A-FAUX 8934
PORTE-A-FAUX 8936
PORTE-ALESOIRS 10264
PORTE-CLAPET 4008
PORTE-COURROIE 1159
PORTE-CRAYON D'UN COMPAS 9165
PORTE-ELECTRODE 4575
PORTE-ESTAMPE 3884
PORTE-ESTAMPES 8033
PORTE-GICLEUR 7219
PORTE-INJECTEUR 8682
PORTE-MICROMETRE 8250
PORTE-MOUSQUETON 11691
PORTE-OBJET 12047
PORTE-OUTIL 13003
PORTE-OUTIL A PLUSIEURS OUTILS 5801
PORTE-OUTILS 13005
PORTE-POINCONS 8033
PORTEE 7256
PORTEE (DISTANCE ENTRE DEUX POINTS D'APPUI) 11831
PORTEE D'ASSEMBLAGE DRESSEE (D'UNE PIECE DE FONDERIE) 5007
PORTEE D'UN ARBRE 1107
PORTEE D'UN MANETON 1116
PORTEE DE MODELE 3161
PORTEE DU SON 10202
PORTER AU ROUGE 5982
PORTES-BILLONNEUSES 10573
PORTIERE ARRIERE 891
PORTIQUE 9641
PORTIQUE DE ROULEMENT 13224
POSE D'UN FIL CONDUCTEUR 13909
POSE DE PLAQUETTES DE COUPE METAL DUR 6278
POSE DE TUBES 9356
POSE DES RIVETS 6972
POSER 11390
POSER DES TUYAUX 7441
POSER LES RIVETS 6967
POSER UN FIL CONDUCTEUR 13881
POSITION 3940
POSITION AU POINT MORT 9646
POSITION BINAIRE 1249
POSITION D'EQUILIBRE 9647
POSITION DE DEPART 12126
POSITION DE LA MANIVELLE 9645
POSITION DE MILIEU 8265

POSITION DE REPOS 9648
POSITION DE ZERO 14028
POSITION DU PISTOLET DE SOUDAGE 13782
POSITION FINALE 5187
POSITION INTERMEDIAIRE 8265
POST-PROCESSEUR 9665
POSTCHAUFFAGE 9668
POSTE 8839
POSTE D'ESSAI 12773
POSTE DE SOUDAGE A COURANT ALTERNATIF 2
POSTE DE SOUDAGE A COURANTS ALTERNATIF ET CONTINU
  59
POSTE RECEPTEUR 10278
POT A FEU 3821
POT A RECUIRE 566
POT D'ECHAPPEMENT 11431
POT D'ECHAPPEMENT 8448
POTASSE 9679
POTASSE AZOTEUSE 9691
POTASSE CAUSTIQUE 9687
POTASSE NITREUSE 9691
POTASSIUM 9672
POTEAU 12062
POTEAU 2748
POTEAU AVEC LES DEUX EXTREMITES GUIDEES 2756
POTEAU DE LIGNE ELECTRIQUE 7608
POTEE D'EMERI 5453
POTEE D'ETAIN 12098
POTENCE 3618
POTENTIEL 9708
POTENTIEL D'ELECTRODE 4577
POTENTIEL DE CONTACT 3007
POTENTIOMETRE 9711
POUCE ANGLAIS 6830
POUCE CARRE ANGLAIS 12021
POUCE CUBE ANGLAIS 3454
POUDINGUE 8490
POUDRAGE AU SULFURE 4394
POUDRE ABRASIVE 6110
POUDRE ALLIEE 367
POUDRE CARBURANTE 2139
POUDRE D'EMERI 4697
POUDRE D'OS 1448
POUDRE DE BOIS 13916
POUDRE DE BRONZE 1652
POUDRE DE CARBONYLE 1979
POUDRE DE CHARBON 9984
POUDRE DE DIAMANT 3863
POUDRE DE FER 7145
POUDRE DE LIEGE 6144
POUDRE DE PONCE 9727

POUDRE DE ZINC OXYDE 1374
POUDRE DENTRIDIQUE 3749
POUDRE GLOBULAIRE 5979
POUDRE GROSSIEREMENT BROYEE 2580
POUDRE IMPALPABLE 6789
POUDRE NOIRE 6181
POUDREUSES PORTATIVES 11969
POUDREUSES PULVERISATEURS 11971
POUDREUSES SOUFREUSES 11970
POULIE 11301
POULIE (APPAREIL DE LEVAGE) 9971
POULIE A BRAS COURBES 9976
POULIE A BRAS DROITS 9977
POULIE A BRAS PARABOLIQUES 9976
POULIE A BRASSURE DOUBLE 4127
POULIE A CABLE 10742
POULIE A CHAINE 2217
POULIE A CORDE 975
POULIE A COURROIE 1152
POULIE A EMPREINTES 2217
POULIE A EXPANSION 4909
POULIE A GORGE 6140
POULIE A GORGE 6135
POULIE A GRIFFES 2503
POULIE A JANTE PERFOREE 9188
POULIE AVEC JOUES 5336
POULIE AVEC REBORDS 5336
POULIE BOMBEE 3400
POULIE COMMANDEE 4298
POULIE CONDUCTRICE 4313
POULIE CONDUITE 4298
POULIE CYLINDRIQUE 5404
POULIE D' EXCENTRIQUE 4434
POULIE D'ATTAQUE 4313
POULIE DE COMMANDE 4313
POULIE DE FREIN 1548
POULIE DE RENVOI 6161
POULIE DE RENVOI 10551
POULIE DE TENSION 12334
POULIE DE VENTILATEUR 5032
POULIE DROITE 5404
POULIE EN CARTON 9034
POULIE EN DEUX PIECES 11945
POULIE EN TOLE EMBOUTIE 9805
POULIE EN UNE PIECE 11792
POULIE ENTRAINEE 4298
POULIE ETAGEE 11871
POULIE EXTENSIBLE 4909
POULIE FENDUE 11945
POULIE FIXE 5040
POULIE FIXE (D'UN PALAN) 5287

POULIE FOLLE **6754**
POULIE FOLLE **7747**
POULIE MAL TOURNEE **9974**
POULIE MENANTE **4313**
POULIE MENEE **4298**
POULIE MOBILE (D'UN PALAN) **8435**
POULIE MOTRICE **4313**
POULIE POUR CABLE DE TRANSMISSION **10742**
POULIE RECEPTRICE **4298**
POULIE-CONE **12669**
POULIE-GUIDE **6161**
POULIE-TAMBOUR **4333**
POULIE-VOLANT **1147**
POULIES MOUFLEES **9973**
POUPEE (DE TOUR) **7964**
POUPEE (ELECTR.) **12762**
POUPEE DE PERCAGE **4278**
POUPEE DE TOUR **6328**
POUPEE FIXE **6328**
POUPEE MOBILE **12599**
POURCENTAGE DE CENDRES **758**
POURCENTAGE DE CHARBON **1965**
POURRIR **10036**
POURRITURE (ETAT) **10035**
POURRITURE DU BOIS **10770**
POUSEE DE BAS EN HAUT **1742**
POUSSEE AU SOMMET **12908**
POUSSEE AXIALE **4734**
POUSSEE DES TERRES **12907**
POUSSEE DU VENT **13858**
POUSSEE HORIZONTALE **6590**
POUSSEE LONGITUDINALE **4734**
POUSSIER DE COKE **2659**
POUSSIER DE HOUILLE **2563**
POUSSIER DE TOURBE **9726**
POUSSIERE **4389**
POUSSIERE D'OR **5998**
POUSSIERE DE GUEULARD **5469**
POUSSIERES DE GAZ DE HAUT FOURNEAU **5469**
POUSSOIR **10029**
POUSSOIR **9797**
POUSSOIR **12672**
POUSSOIR A GALET **10692**
POUSSOIR DE SOUPAPE **13466**
POUTRE **1099**
POUTRE **1093**
POUTRE **5946**
POUTRE A LARGES AILES **1639**
POUTRE AU VENT **13856**
POUTRE COMPOSEE **9488**
POUTRE COMPOSEE **9490**

POUTRE CONTINUE **3018**
POUTRE DE PONT ROULANT **13157**
POUTRE EN BOIS **949**
POUTRE EN FER **12199**
POUTRE EN PORTE-A-FAUX **1904**
POUTRE EN PORTE-A-FAUX **1905**
POUTRE EN TREILLIS **7431**
POUTRE ENCASTREE **6968**
POUTRE ENCASTREE **5286**
POUTRE ENCASTREE **1702**
POUTRE ENCASTREE A SES DEUX EXTREMITES **1703**
POUTRE ENCASTREE A UNE EXTREMITE **1904**
POUTRE ENCASTREE A UNE EXTREMITE ET REPOSANT A
L'AUTRE SUR UN APPUI **1097**
POUTRE ENROBEE **4714**
POUTRE INCORPOREE **7010**
POUTRE MIXTE OU COMPOSEE **2831**
POUTRE NON ENTRETOISEE **1905**
POUTRE RAIDISSEUSE **13856**
POUTRE RAYONNANTE **10131**
POUTRE REPOSANT SOUTENUE LIBREMENT SUR DEUX APPUIS
1098
POUTRELLE **1093**
POUTRELLE **986**
POUTRELLE A LARGES AILES **13849**
POUTRELLE EN I **6748**
POUTRELLE EN U **2249**
POUTRELLES A LARGES AILES **1641**
POUVOIR ABSORBANT **42**
POUVOIR ADHESIF **176**
POUVOIR CALORIFIQUE **6378**
POUVOIR CALORIFIQUE **6392**
POUVOIR COUVRANT **3273**
POUVOIR DISSOLVANT **4062**
POUVOIR ECLAIRANT **6776**
POUVOIR EMISSIF **4701**
POUVOIR EMISSIF MONOCHROMATIQUE **8382**
POUVOIR GROSSISSANT **7937**
POUVOIR ISOLANT **7000**
POUVOIR LUBRIFIANT **7806**
POUVOIR OPACIFIANT INSUFFISANT **6119**
POUVOIR RAYONNANT **4702**
POUVOIR REFLECHISSANT **10393**
POUVOIR REFLECHISSANT **10392**
POUVOIR REFRINGENT **10403**
POUZZOLANE **9748**
POZZOLANE **9748**
PRASEODYME **9750**
PRATIQUER UNE SAIGNEE DANS UNE PIECE **8554**
PRE-ALLIAGE **5594**
PRE-MODELE **4370**

PRECAUTIONS **9753**
PRECESSION **9754**
PRECHARGE **9782**
PRECHAUFFAGE **9776**
PRECIPITABLE **9757**
PRECIPITANT **9758**
PRECIPITATION **9761**
PRECIPITATION **9091**
PRECIPITATION DE CARBURE **1934**
PRECIPITE **9759**
PRECIPITER **9760**
PRECIPITER (SE) **9760**
PRECISION **9763**
PRECISION **82**
PRECISION D'USINAGE **83**
PRECONTRAINTE **9782**
PRECONTRAINTE DU BETON **2899**
PREFILTRE **9751**
PREFORMAGE **9773**
PREFORMAGE **1341**
PREFORME **9772**
PREFRITTAGE **9794**
PREMIER FILTRE **5265**
PREMIERE COUCHE DE PEINTURE **9864**
PREMIERE TETE DE RIVET **5266**
PREPARATION **9784**
PREPARATION D'UNE SOLUTION **9785**
PREPARATION DES BORDS **4449**
PREPARATION DES MINERAIS **8861**
PREPARER UNE SOLUTION **9788**
PRESELECTEUR **9789**
PRESERVANT DE LA ROUILLE **10868**
PRESERVATIF CONTRE LA ROUILLE **10869**
PRESSAGE **9806**
PRESSAGE A FROID **2690**
PRESSE **9795**
PRESSE A CALIBRER **11525**
PRESSE A COIN **13734**
PRESSE A COLLER **2451**
PRESSE A COMPRIMER LES POUDRES **9718**
PRESSE A EBARBER **13211**
PRESSE A EXCENTRIQUE **4430**
PRESSE A FILER **4972**
PRESSE A FORGER **5562**
PRESSE A FORGER HYDRAULIQUE **6687**
PRESSE A GENOUILLERES **12980**
PRESSE A LEVIER **7534**
PRESSE A PASTILLES **9310**
PRESSE A VIS **11040**
PRESSE A VIS **2452**
PRESSE BRINELL **1553**

PRESSE DE COULEE **3370**
PRESSE DE MENUISIER **2451**
PRESSE HYDRAULIQUE **6693**
PRESSE HYDRAULIQUE A COL DE CYGNE **8828**
PRESSE HYDRAULIQUE A EMBOUTIR **6682**
PRESSE HYDRAULIQUE A MARTEAU-PILON **6687**
PRESSE MECANIQUE A ARCADE **4157**
PRESSE MECANIQUE A BATI COL DE CYGNE **8814**
PRESSE MECANIQUE A CALIBRER ET A MATRICER **1864**
PRESSE MECANIQUE A FRAPPER **9182**
PRESSE MECANIQUE A MONTANT **12316**
PRESSE MECANIQUE A PLATEAU REVOLVER **3839**
PRESSE MONETAIRE **2653**
PRESSE PLIEUSE **9481**
PRESSE POUR TOLES FORTES **6411**
PRESSE ROTATIVE **10757**
PRESSE-ETOUPE **5953**
PRESSE-ETOUPE **12405**
PRESSES TRANSFERT A POINCONS MULTIPLES **8461**
PRESSION **13434**
PRESSION **9808**
PRESSION (CHAUD.) **6314**
PRESSION ABSOLUE **33**
PRESSION ABSOLUE **4459**
PRESSION ATMOSPHERIQUE **794**
PRESSION BAROMETRIQUE **794**
PRESSION CONSTANTE **2976**
PRESSION CRITIQUE **3348**
PRESSION D'EAU **13669**
PRESSION D'EPREUVE **12786**
PRESSION D'ESSAI **12786**
PRESSION DANS UNE CHAUDIERE **1417**
PRESSION DE COMPRESSION **2859**
PRESSION DE LA VAPEUR **12181**
PRESSION DE REGIME **13957**
PRESSION DE REGLAGE **9832**
PRESSION DE SATURATION **10944**
PRESSION DE SERVICE **13957**
PRESSION DE X CM DE MERCURE **794**
PRESSION DU COURANT **9825**
PRESSION DU GAZ **5837**
PRESSION DU VENT **13858**
PRESSION DUE AU FROTTEMENT **9816**
PRESSION EN KILOGRAMMES PAR UNITE DE SURFACE **11853**
PRESSION EXPLOSIVE **4936**
PRESSION FINALE **5188**
PRESSION HYDRAULIQUE **13669**
PRESSION HYDROSTATIQUE **6726**
PRESSION INITIALE **6929**
PRESSION INTERIEURE DANS UNE CONDUITE **9822**

PRESSION LATERALE **7422**
PRESSION LIMITE **7593**
PRESSION MANOMETRIQUE **9809**
PRESSION MANOMETRIQUE **5771**
PRESSION MOYENNE **8075**
PRESSION NOMINALE **8613**
PRESSION NORMALE **8657**
PRESSION OSMOTIQUE **8891**
PRESSION PARTIELLE **9087**
PRESSION REELLE **4459**
PRESSION RELATIVE **5882**
PRESSION STATIQUE **12136**
PRESSION STATIQUE **12137**
PRESSION SUPERFICIELLE **12517**
PRESSION SUPERIEURE A LA PRESSION AUTORISEE **8955**
PRESSION SUR LE PISTON **9377**
PRESSION SUR LES ARETES **9826**
PRESSION SUR LES SURFACES D'APPUI **1110**
PRESSION TOTALE **13065**
PRESSION VARIABLE **13491**
PRESSION/REACTION SUR LES TOURILLONS **7259**
PRESSSPAHN **9801**
PRIMAGE (ENTRAINEMENT DE L'EAU PAR LA VAPEUR) **9863**
PRIMEROSE SOLUBLE **4831**
PRISE **3483**
PRISE (EN) **8171**
PRISE (DU CIMENT) **3484**
PRISE D'AIR **289**
PRISE D'AIR **270**
PRISE D'UN ECHANTILLON **10916**
PRISE DE FORCE **9745**
PRISE DE POUSSIERES **3997**
PRISE DIRECTE **3983**
PRISE DU CIMENT **11226**
PRISMATIQUE **9879**
PRISMATOIDE **9880**
PRISME **9875**
PRISME A SECTION OBLIQUE **9876**
PRISME D'EGALE RESISTANCE **1396**
PRISME HEXAGONAL **6460**
PRISME PENTAGONAL **9173**
PRISME POLARISATEUR **9586**
PRISME QUADRANGULAIRE **10055**
PRISME TRIANGULAIRE **13192**
PRISONNIER **12399**
PRISONNIER **11220**
PRIX A LA JOURNEE **3594**
PRIX DE MAIN-D'OEUVRE **7346**
PRIX DE REVIENT **3210**
PROCEDE **9883**

PROCEDE 'UNIONMELT' **684**
PROCEDE A L'OXYGENE **8982**
PROCEDE A PARTIR DES PRODUITS RESULTANT DE LA DECOMPOSITION DE L'HYDRURE. **6699**
PROCEDE ACIDE **120**
PROCEDE AU CREUSET **3404**
PROCEDE BASIQUE **1050**
PROCEDE BASIQUE **1043**
PROCEDE BAYER **1073**
PROCEDE BESSEMER **1214**
PROCEDE BESSEMER AU CONVERTISSEUR A GARNISSAGE ACIDE **106**
PROCEDE BETTS **1220**
PROCEDE CHAPMAN **2254**
PROCEDE CHIMIQUE **2291**
PROCEDE D'AFFINAGE **10380**
PROCEDE D'AMALGAMATION **448**
PROCEDE D'USINAGE **9887**
PROCEDE DE FABRICATION **8218**
PROCEDE DE FINITION CHIMIQUE DE SURFACE **2309**
PROCEDE DE MOULAGE AU MOYEN DU SAC EN CAOUTCHOUC **923**
PROCEDE DE MOULAGE INVERSE **11656**
PROCEDE DE REVETEMENT EN BAIN CHAUD **6659**
PROCEDE DUPLEX **4380**
PROCEDE FOUR KALDO **7271**
PROCEDE HANSGIRG **6272**
PROCEDE OLD **8793**
PROCEDE PAR ASPERSION **11972**
PROCEDE SIEMENS-MARTIN **8830**
PROCEDE THOMAS **1043**
PROCEDES CHIMIQUES DE TRAITEMENT DE SURFACE **2314**
PROCESSEUR **9889**
PROCESSUS **9883**
PROCESSUS DE RECHARGEMENT **1696**
PRODUCTION **10462**
PRODUCTION **8028**
PRODUCTION D'ENERGIE **5922**
PRODUCTION D'ENERGIE **9737**
PRODUCTION DE CHALEUR **9896**
PRODUCTION DE COURANT **9894**
PRODUCTION DU FROID **9895**
PRODUCTION EN SERIE **11198**
PRODUCTIVITE **9897**
PRODUIRE (DE LA VAPEURDES GAZ) **5916**
PRODUIRE UN COURANT ELECTRIQUE **5917**
PRODUIT **9892**
PRODUIT **9784**
PRODUIT (MATH.) **9891**
PRODUIT ANTI-ROUILLE **10870**
PRODUIT ANTI-ROUILLE **3189**

PRODUIT ANTI-ROUILLE **10866**
PRODUIT DE COMBUSTION **9893**
PRODUIT DE DISTILLATION **4071**
PRODUIT DE NETTOYAGE **2471**
PRODUIT DE RAFFINAGE **10374**
PRODUIT DE REVETEMENT **2589**
PRODUIT FINAL **5189**
PRODUIT FINI **5730**
PRODUIT FINI / PLAT / LONG **5218**
PRODUIT FORMANT DES POROSITES **9628**
PRODUIT HYDROFUGE **13674**
PRODUIT INTERMEDIAIRE **7043**
PRODUIT MANUFACTURE **7989**
PRODUIT PRIMAIRE **9852**
PRODUIT SCALAIRE **10966**
PRODUIT SECONDAIRE **1820**
PRODUIT SEMI-OUVRE **11174**
PRODUIT SPECIAL **11849**
PRODUIT VECTORIEL **13509**
PRODUITS CHIMIQUES **2320**
PRODUITS DE NETTOYAGE ET DE POLISSAGE **2475**
PRODUITS DE PREMIERE QUALITE **9862**
PRODUITS EN ACIER A RAIL **10177**
PROFIL **9898**
PROFIL **8906**
PROFIL **12734**
PROFIL **7717**
PROFIL (A DROITEA GAUCHE) **11420**
PROFIL D'UNE DENT **13019**
PROFIL DE DENTURE **13020**
PROFIL EN LONG **7738**
PROFIL NORMAL **12083**
PROFIL TRANSVERSAL **3377**
PROFILAGE **10668**
PROFILAGE A FROID **2682**
PROFILE **11119**
PROFILE EN T **12700**
PROFILE MI-ROND **6209**
PROFILE(S) **11114**
PROFILER **5571**
PROFILES LOURDS **6413**
PROFILES MARCHANDS **8149**
PROFONDEUR **3779**
PROFONDEUR D'ENGRENEMENT **13952**
PROFONDEUR DE COUPE **3782**
PROFONDEUR DE COUPE **3526**
PROFONDEUR DE LA COUCHE CEMENTEE **2020**
PROFONDEUR DU COL DE CYGNE **3783**
PROFONDEUR DU FILET **3784**
PROGRAMMATION **9904**
PROGRAMMATION AUTOMATIQUE **850**

PROGRAMMATION DE PIECE **9084**
PROGRAMMATION MANUELLE DE PIECE **6242**
PROGRAMME **9902**
PROGRAMME D'ORDINATEUR **10811**
PROGRAMME DE FABRICATION **7994**
PROGRAMME DIAGNOSTIQUE **3830**
PROGRESSION **11203**
PROGRESSION ARITHMETIQUE **704**
PROGRESSION GEOMETRIQUE **5930**
PROJECTEUR **6326**
PROJECTEUR D'OMBRE **11238**
PROJECTEUR DE LUMIERE **11091**
PROJECTION **11968**
PROJECTION **9912**
PROJECTION CENTRALE **10137**
PROJECTION DIMETRIQUE **3961**
PROJECTION GNOMOMIQUE **5990**
PROJECTION HORIZONTALE **11124**
PROJECTION ISOMETRIQUE **7188**
PROJECTION OBLIQUE **2501**
PROJECTION ORTHOGONALE **8870**
PROJECTION PARALLELE **9058**
PROJECTION TRIMETRIQUE **13208**
PROJECTION VERTICALE **4659**
PROJET **3794**
PROJETER **9909**
PROJETER DANS... **6934**
PROLONGATION DE DUREE D'UN BREVET **9919**
PROMETHEUM **6771**
PROPAGATION D'ONDES **9926**
PROPANE **9927**
PROPORTION **3011**
PROPORTION **9934**
PROPORTION (MATH.) **9933**
PROPORTION D'EAU **13645**
PROPORTION DE CARBONE **1947**
PROPORTIONNALITE **9938**
PROPORTIONNEL **9935**
PROPORTIONS DEFINIES **2771**
PROPRIETE COUVRANTE **1393**
PROPRIETE DE COUVRIR D'UNE COULEUR **1393**
PROPRIETE REFRACTAIRE **10405**
PROPRIETES ANTI-FRICTION **617**
PROPRIETES CHIMIQUES **2295**
PROPRIETES PHYSIQUES **9283**
PROPULSEUR **11041**
PROPULSION A 4 ROUES MOTRICES **5601**
PROPYLENE **9939**
PROSPERITE ECONOMIQUE D'UNE ENTREPRISE **9462**
PROTACTINIUM **9940**
PROTECTION **9943**

PROTECTION 11346
PROTECTION CATHODIQUE 2109
PROTECTION CONTRE LA CORROSION 3191
PROTECTION CONTRE LA ROUILLE 9945
PROTECTION CONTRE LA ROUILLE 10871
PROTECTION DE LA PROPRIETE INDUSTRIELLE 9946
PROTOCARBURE D'HYDROGENE 8215
PROTOCHLORURE D'ETAIN 12100
PROTOCHLORURE DE MERCURE 8159
PROTON 9954
PROTOSULFURE 627
PROTOTYPE 9955
PROTOXYDE D'AZOTE 8597
PROTOXYDE DE BARYUM 1020
PROTOXYDE DE FER 5125
PROTOXYDE DE MANGANESE 7973
PROTOXYDE DE MERCURE 8161
PROTOXYDE DE NICKEL 8565
PROTOXYDE DE PLOMB 7663
PRUSSIATE JAUNE 9685
PRUSSIATE ROUGE 9684
PSYCHROMETRE 9960
PUDDLER 9964
PUISSANCE 9730
PUISSANCE 7308
PUISSANCE 1917
PUISSANCE (MATH.) 9729
PUISSANCE ABSORBANTE 42
PUISSANCE ABSORBEE 4747
PUISSANCE AU FREIN 4458
PUISSANCE AU FREIN 1540
PUISSANCE AU FREIN 1535
PUISSANCE AU FREIN EN CHEVAUX 151
PUISSANCE CALORIFIQUE 6392
PUISSANCE DE FREINAGE 1551
PUISSANCE DE LEVAGE D'UN AIMANT 13090
PUISSANCE EFFECTIVE 1540
PUISSANCE EFFECTIVE 4458
PUISSANCE EFFECTIVE EN CHEVAUX 151
PUISSANCE FISCALE 10233
PUISSANCE INDIQUEE 6875
PUISSANCE INDIQUEE (OU FISCALE) 6874
PUISSANCE MAXIMUM 8066
PUISSANCE MOTRICE 8409
PUISSANCE NOMINALE 10233
PUISSANCE RECUEILLIE 4747
PUISSANCE REQUISE 3746
PUISSANCE/ALIMENTATION 4546
PUITS (D'UNE MINE) 9398
PUITS A GRANULATION DU LAITIER 6056
PUITS D'AERATION 291

PULSOMETRE 9980
PULVERISABLE 9981
PULVERISATEUR 11967
PULVERISATEUR 809
PULVERISATION 6117
PULVERISATION 811
PULVERISATION 2786
PULVERISATION 2226
PULVERISATION 11968
PULVERISATION 9985
PULVERISATION D'UN LIQUIDE 810
PULVERISE 9983
PULVERISER 9982
PULVERISER 6103
PULVERISER (UN LIQUIDE) 808
PULVERISEURS A DISQUES 4007
PULVERULENT 9728
PUNAISE 4242
PURE 9932
PURGE 4206
PURGER 4204
PURGEUR 10020
PURGEUR 1318
PURGEUR D'EAU CONDENSEE 12188
PURGEUR D'EAU CONDENSEE A DILATATION 12839
PURGEUR D'EAU CONDENSEE A FLOTTEUR 5432
PURIFICATION ANODIQUE 592
PURIFICATION DE L'EAU 10019
PURIFICATION DES EAUX RESIDUAIRES 13638
PURIFIE 10372
PURIFIER LES EAUX D'ALIMENTATION PAR VOIE CHIMIQUE 11748
PUSTULE 1327
PUSTULES 1330
PUTREFACTION (ACTION) 10035
PUTREFIER (SE) 10036
PYCNOMETRE 11851
PYRAMIDE 10038
PYRAMIDE PENTAGONALE 9174
PYRAMIDE QUADRANGULAIRE 12026
PYRAMIDE TRIANGULAIRE 13193
PYRIDINE 10041
PYRITE CUIVREUSE 3126
PYRITE JAUNE 7146
PYRITE MAGNETIQUE 7914
PYRITE MARTIALE 7146
PYROELECTRICITE 10043
PYROELECTRIQUE 10042
PYROGALLOL 10045
PYROMETALLURGIE 10048
PYROMETRE 10049

PYROMETRE A DISTANCE **12714**
PYROMETRE A GRAPHITE **6065**
PYROMETRE A RESISTANCE **10495**
PYROMETRE ELECTRIQUE **10495**
PYROMETRE OPTIQUE **8845**
PYROMETRE THERMOELECTRIQUE **12825**
PYROMETRIE **10052**
PYROMETRIE OPTIQUE **8846**
PYROPHOSPHATE DE SOUDE **11730**
PYROXYLE **6176**
PYROXYLINE **6176**
PYRRHOTINE **7914**
QUADRANGLE **10058**
QUADRANT **10056**
QUADRATIQUE **12795**
QUADRATURE **12036**
QUADRILATERE **10058**
QUADRILATERE **10055**
QUADRILATERE **12026**
QUADRILATERE INSCRIPTIBLE **6961**
QUADRILLAGE **6099**
QUADRIVALENT **12798**
QUALITE LUBRIFIANTE **7806**
QUANTITE **10065**
QUANTITE CONNUE **7335**
QUANTITE D'ELECTRICITE **10067**
QUANTITE DE CHALEUR **10066**
QUANTITE DE MOUVEMENT **10068**
QUANTITE DE MOUVEMENT **8376**
QUANTITE INCONNUE **13391**
QUANTITE MESUREE **10070**
QUANTITE RECIPROQUE **10280**
QUARTIER **10076**
QUARTZ **10077**
QUARTZITE **10083**
QUATRE-FEUILLES **10086**
QUATRIEME PUISSANCE **5608**
QUESTIONNAIRE **10101**
QUEUE **11257**
QUEUE CONIQUE **12664**
QUEUE D'ARONDE **4175**
QUEUE D'ARONDE (A) **4177**
QUEUE D'HIRONDE **4175**
QUEUE D'UN OUTIL **12614**
QUEUE DE DISTILLATION **7413**
QUEUE DE LA TIGE DU PISTON **9384**
QUEUE DE ROBINET **12236**
QUI PEUT SE COULER **1914**
QUINCAILLERIE **11660**
QUINCONE (EN) **12048**
QUINHYDRONE (HEMI-CELLULE A LA) **10110**

QUINTAL (50 KGS ENV.) **6668**
QUINTAL METRIQUE **10111**
QUOTIENT **10114**
QUOTIENT **10234**
QUOTIENT DIFFERENTIEL **3916**
RABATTEMENT **1088**
RABATTEMENT DE LA COLLERETTE DES TUBES **1088**
RABATTRE LA COLLERETTE DES TUBES **1084**
RABATTRE LES BORDS **1084**
RABBATRE L'EXCES DE LA TIGE DU RIVET **2489**
RABOT **9432**
RABOT A CONTRE-FER **4146**
RABOT A REPASSER **11687**
RABOT CINTRE **2801**
RABOT DENTE A DENTS **13027**
RABOT ROND **10800**
RABOTAGE **9450**
RABOTEMENT **9450**
RABOTER **9434**
RABOTEUR **9443**
RABOTEUSE **9452**
RABOTEUSE **4871**
RABOTEUSE **9443**
RABOTEUSE-FRAISEUSE **9453**
RACCAGNAC **10219**
RACCORD **5280**
RACCORD **13368**
RACCORD **3254**
RACCORD **2965**
RACCORD (ELECTR.) **2966**
RACCORD A ECROUS **1442**
RACCORD A EMBOITEMENT POUR SOUDURE **11708**
RACCORD A SOUDER **13781**
RACCORD ACIER FORGE **5549**
RACCORD CONIQUE **12647**
RACCORD CONIQUE **10352**
RACCORD FILETE **11061**
RACCORD MALE **8578**
RACCORD POUR TUBES **9350**
RACCORD POUR TUYAUX FLEXIBLES **6612**
RACCORD T **1559**
RACCORD TROIS PIECES **13369**
RACCORD VISSE **11058**
RACCORD-UNION **13369**
RACCORDEMENT (ELECTR.) **2958**
RACCORDEMENT DE TUYAUX **7248**
RACCORDER **2955**
RACCORDER DES TUYAUX **7244**
RACCOURCISSEMENT **11376**
RACCOURCISSEMENT D'UN CABLE **11377**
RACCOURCISSEMENT D'UNE COURROIE **11377**

RACHETER L'USURE 12603
RACINE 1028
RACINE 10720
RACINE (MATH.) 10714
RACINE BIQUADRATIQUE 5609
RACINE CARREE 12027
RACINE CINQUIEME 5147
RACINE CUBIQUE 3447
RACINE DE LA DENT 1036
RACINE DE LA SOUDURE 10726
RACLAGE 11011
RACLER 11008
RACLETTE 11009
RACLOIR 11009
RADIAL 10125
RADIAN 10138
RADIATEUR 10148
RADIATEUR A AILETTES 5233
RADIATEUR ALPHA 401
RADIATEUR DE CHAUFFAGE 10150
RADIATEUR DE REFROIDISSEMENT 10149
RADIATEUR NID D'ABEILLES 2129
RADIATION 10143
RADICAL 10154
RADIO DES NOEUDS DE SOUDURE 7097
RADIO-ACTIF 10157
RADIO-ACTIVITE 10160
RADIO-CRISTALLOGRAPHIE 13996
RADIOGRAPHIE 10664
RADIOGRAPHIE 10161
RADIOGRAPHIE 10163
RADIOLOGIE 10164
RADIOMETALLOGRAPHIE 10155
RADIOTELEGRAPHIE 13907
RADIOTELEPHONIE 13908
RADIUM 10165
RADON 10159
RAFFINAGE 10377
RAFFINAGE D'UN METAL 10378
RAFFINAGE SOUS VIDE 13447
RAFFINE 10372
RAFFINERIE 10375
RAIDEUR 10591
RAIDEUR D'UNE CORDE 12258
RAIDEUR D'UNE COURROIE 12258
RAIDISSEUR DE POUTRES EN TOLE 9489
RAIDISSEUR INTERMEDIAIRE 7047
RAIE 7597
RAIE DU SODIUM 11723
RAIES DE FRAUNHOFER 5631
RAIL 10175

RAIL DE CONTACT 1770
RAIL DE PONT ROULANT 3293
RAIL-GUIDE 6162
RAIL-TENDEUR 12756
RAILS A BOUDIN 1715
RAINAGE 6137
RAINER 6128
RAINER ET LANGUETTER 12988
RAINEUSE 6138
RAINURAGE 6137
RAINURAGE A LA FLAMME 5822
RAINURE 6127
RAINURE 7282
RAINURE ANNULAIRE 2415
RAINURE CIRCULAIRE 2415
RAINURE DE CLAVETAGE 7282
RAINURE DE CLAVETTE 7291
RAINURER 6128
RAISON DIRECTE (EN) 6815
RAISON INVERSE (EN) 6816
RALENTI 6755
RALENTI ACCELERE 5038
RALENTIR LA MARCHE D'UNE MACHINE 11645
RALENTISSEMENT DE LA MARCHE 11649
RALLONGE 4947
RALLONGE D'UN COMPAS 7521
RALLONGE DE MANCHE 4945
RAMASSE FOIN 12541
RAMASSE-BALLES 946
RAMASSEUSES DE FOURRAGES 6091
RAMIE 10192
RAMPAGE 2437
RAMPE 757
RANG DE RIVETS 10813
RANKINISATION DES DIAGRAMMES 2770
RANKINISER 2765
RAPE 10215
RAPIECAGE 9109
RAPIECAGE 9111
RAPIECEMENT (ACTION) 9111
RAPIECEMENT (RESULTAT) 9109
RAPIECER 9110
RAPPORT 10234
RAPPORT : LIMITE DE FATIGUE 4742
RAPPORT D'ABSORPTION 53
RAPPORT D'ELANCEMENT 11582
RAPPORT D'ENGRENAGE 5897
RAPPORT DE COMPRESSION 2860
RAPPORT DE CONVEXITE 3079
RAPPORT DE DENSITE 3759
RAPPORT DE DISTRIBUTION DU METAL 3777

RAPPORT DE DISTRIBUTION DU METAL **8179**
RAPPORT DE REDUCTION **10360**
RAPPORT DE REFLEXION **10391**
RAPPORT DES BRAS DE LEVIER **7536**
RAPPORT DES VITESSES **13522**
RAPPORT VOLUMETRIQUE **2860**
RAPPORTE **1705**
RAPPORTEUR **9956**
RAPPORTEUR A FORME D'EQUERRE **10308**
RAPPORTEUR CERCLE ENTIER **2426**
RAPPORTEUR DEMI-CERCLE **11180**
RAPPORTEUR FIXE **10366**
RAPPORTS D'ENGRENAGES **5896**
RAREFACTION DE L'AIR **10212**
RAREFACTION D'UN GAZ **10211**
RAREFIER UN GAZ **10214**
RATE D'UN MOTEUR **5021**
RATEAUX **6312**
RATEAUX A DEVERSEMENT LATERAL **11411**
RATEAUX-AMMEULONNEURS **1683**
RATELEUSES RAMASSEUSES **6311**
RATTRAPAGE DE JEU **13403**
RATTRAPAGE DU JEU **12606**
RATTRAPER LE JEU D'EMBRAYAGE **12603**
RAYON **10166**
RAYON CALORIFIQUE **6369**
RAYON D'ECARTEMENT ENTRE LES BORDS **6132**
RAYON D'ECARTEMENT ENTRE LES BORDS **10728**
RAYON DE COURBURE **10168**
RAYON DE COURBURE **1176**
RAYON DE GIRATION **10169**
RAYON DE ROTATION **10169**
RAYON EMERGENT **4691**
RAYON INCIDENT **6832**
RAYON LUMINEUX **7824**
RAYON MEDULLAIRE **8128**
RAYON POLAIRE **9580**
RAYON REFLECHI **10384**
RAYON REFRACTE **10397**
RAYON VECTEUR **10170**
RAYONNEMENT **10143**
RAYONNEMENT DE LA CHALEUR **12811**
RAYONNEMENT DE LA CHALEUR **10146**
RAYONNEMENT HETEROGENE **6450**
RAYONNEMENT HOMOGENE **6559**
RAYONNEMENT IRRADIATION DE LA LUMIERE **10147**
RAYONNEMENT MONOCHROMATIQUE **8383**
RAYONNER **10141**
RAYONS **10242**
RAYONS ACTINIQUES **140**
RAYONS ALPHA **402**

RAYONS BETA **1218**
RAYONS CATHODIQUES **2104**
RAYONS D'UNE ROUE **13811**
RAYONS DE LENARD **7515**
RAYONS GAMMA **5796**
RAYONS INFRA-ROUGES **7103**
RAYONS ULTRA-ROUGES **13332**
RAYONS ULTRA-VIOLETS **13333**
RAYONS X **14003**
REA D'UN MOUFLE **11301**
REACTANCE **10247**
REACTEUR **10251**
REACTEUR DE RADIOCHIMIE **2323**
REACTIF **10257**
REACTIF **4844**
REACTIF D'ATTAQUE A L'ACIDE **4847**
REACTIF INDICATEUR **6879**
REACTION **10248**
REACTION ACIDE **122**
REACTION ALCALINE **337**
REACTION CHIMIQUE **2312**
REACTION DES APPUIS **1110**
REACTION ELECTROLYTIQUE IRREVERSIBLE **7171**
REACTION EUTECTOIDE **4860**
REACTION MONOTECTOIDE **8386**
REACTION PERITECTIQUE **9199**
REALESAGE **10271**
REALESER **10270**
REALGAR **10262**
REALISER **7950**
REBONDIMETRE **11366**
REBONDIR **10273**
REBONDISSEMENT **10272**
REBORD **4441**
REBORD **5332**
REBUT **8321**
REBUT **10431**
REBUTS DE FABRICATION **3696**
RECALAGE (D'UN COUSSINET) **12606**
RECALESCENCE **10274**
RECARBURANT **10276**
RECARBURATION **1953**
RECARBURATION DU FER **10275**
RECEPTEUR HYDRAULIQUE **13692**
RECEPTION **67**
RECEPTION (DE MACHINES) **68**
RECEVOIR UNE FORCE **12601**
RECHANGE D'UN ORGANE DE MACHINE **10464**
RECHANGER **10463**
RECHARGEMENT D'UNE SOUDURE **13762**
RECHARGEMENT DUR **6283**

RECHAUFFAGE 10425
RECHAUFFEMENT 10425
RECHAUFFER 10424
RECHAUFFER 6334
RECHAUFFEUR 6385
RECHAUFFEUR 9775
RECHAUFFEUR D'AIR 287
RECHAUFFEUR D'EAU D'ALIMENTATION 5078
RECHERCHE APPLIQUEE 649
RECHERCHE FONDAMENTALE 5735
RECHERCHE PURE 1052
RECHERCHE PURE 10016
RECIPIENT 3010
RECIPIENT (D'UNE CORNUE) 10277
RECIPIENT PLAT 9029
RECIPIENT SOUS PRESSION 9839
RECOLTEUSES DE BETTERAVES 12439
RECOLTEUSES DE POIS 9141
RECOLTEUSES DE POMMES DE TERRE 9704
RECOUVREMENT 8951
RECOUVREMENT 8948
RECOUVREMENT 2590
RECOUVREMENT D'UN METAL PAR VOIE ELECTRO-CHIMIQUE 4562
RECOUVREMENT DE TOLES 9501
RECOUVREMENT DE TUYAUX 7369
RECOUVREMENT DU TIROIR 7399
RECOUVREMENT PAR DECOMPOSITION D'UN GAZ 13482
RECRISTALLISATION 10297
RECTANGLE 10299
RECTANGLE 10310
RECTIFICATION 11275
RECTIFICATION 10511
RECTIFICATION 6106
RECTIFICATION (CHIM.) 10311
RECTIFICATION A L'ALESOIR 10271
RECTIFICATION DE SURFACE 12506
RECTIFICATION DEMI-ONDE 6212
RECTIFICATION EXTERIEURE 4957
RECTIFICATION INTERIEURE 7069
RECTIFICATION INTERIEURE 6564
RECTIFICATION SANS POINTE(S) 2156
RECTIFIER 11273
RECTIFIER (CHIM.) 10318
RECTIFIER AVEC L'ALESOIR 10270
RECTIFIEUR 6105
RECTIFIEUSE POUR SURFACE PLANES 12507
RECTIFIEUSE POUR SURFACES DE REVOLUTION EXTERIEURES 4953
RECTIFIEUSE POUR SURFACES DE REVOLUTION SANS CENTRE 2155

RECTILIGNE 10319
RECUEIL DE DONNEES 3613
RECUIRE L'ACIER 560
RECUIT 562
RECUIT A GROSGRAIN 5719
RECUIT A LA FLAMME 5319
RECUIT AU CHALUMEAU 575
RECUIT AU FOUR ELECTRIQUE 4508
RECUIT BLANC 573
RECUIT COMPLET 576
RECUIT CONTINU 574
RECUIT DE DETENTE 584
RECUIT DE DETENTE 9667
RECUIT DE DETENTE 12044
RECUIT DE DETENTE 12362
RECUIT DE L'ACIER 581
RECUIT DE MISE EN SOLUTION 11812
RECUIT DE NORMALISATION 8664
RECUIT DE RELAXATION 583
RECUIT DE RELAXATION 12363
RECUIT ECHELONNE 12240
RECUIT EN CAISSE 5371
RECUIT EN NOIR 571
RECUIT INTERMEDIAIRE 577
RECUIT INTERMEDIAIRE D'USINAGE 9884
RECUIT INTERMEDIAIRE ENTRE DEUX ETIRAGES 7085
RECUIT INVERSE 578
RECUIT ISOTHERMIQUE 579
RECUIT PERIODIQUE 582
RECUIT SELECTIF 580
RECUIT SELECTIF 11142
RECUL 919
RECUPERATEUR 10412
RECUPERATEUR 10322
RECUPERATION 10296
RECUPERER 10295
REDONDANCE 10364
REDRESSAGE 12319
REDRESSEMENT 10312
REDRESSER 12318
REDRESSEUR (DE TUBES) ABRAMSEN 12
REDRESSEUR (OU DETECTEUR) ELECTROLYTIQUE 4606
REDRESSEUR A VAPEUR DE MERCURE 8164
REDRESSEUR DE COURANT 10314
REDRESSEUR ELECTROCHIMIQUE 4572
REDRESSEUR ELECTRONIQUE 4634
REDUCTEUR 10345
REDUCTEUR 10346
REDUCTEUR DE PRESSION 10355
REDUCTEUR DE VITESSE 11872
REDUCTEUR DE VITESSE 11873

REDUCTION **13121**
REDUCTION (CHIM.) **10356**
REDUCTION CONICITE **4196**
REDUCTION DE LA SECTION TRANSVERSALE **10362**
REDUCTION DE TREMPE **4**
REDUCTION DE VITESSE **10363**
REDUCTION DES EMPLACEMENTS **5774**
REDUCTION ELECTROLYTIQUE **4607**
REDUCTION FEMELLE-FEMELLE **10348**
REDUCTION MALE-FEMELLE **1773**
REDUCTRICE **10345**
REDUIRE (CHIM.) **10342**
REDUIRE EN POUDRE **6103**
REDUIRE LA VITESSE **3678**
REDUIRE PAR LAMINAGE **10666**
REEMPLOI **10246**
REFLECHIR **10382**
REFLECTEUR **10394**
REFLETER **10382**
REFLEXION **10387**
REFLEXION DE LA LUMIERE **10390**
REFLEXION EN RETOUR **896**
REFLEXION TOTALE **13060**
REFONDRE **10446**
REFONTE **10447**
REFOULAGE **7265**
REFOULEMENT **7265**
REFOULEMENT **4452**
REFOULEMENT **13414**
REFOULEMENT A FROID **2686**
REFOULEMENT AVEC CHAUFFAGE PAR EFFET JOULE **13417**
REFOULER **7263**
REFRACTAIRE **5257**
REFRACTAIRE PLASTIQUE **9473**
REFRACTAIRES BASIQUES **1051**
REFRACTION **10399**
REFRACTION DE LA LUMIERE **10400**
REFRACTOMETRE **10404**
REFRIGERANT **10409**
REFRIGERANT **3083**
REFRIGERANT DE CENDRES **2398**
REFRIGERANT DE MEULAGE **6109**
REFRIGERANT POUR LES EAUX DE CONDENSATION **3089**
REFRIGERATEUR **10409**
REFRIGERATION **3085**
REFRIGERER **3081**
REFRINGENT **10402**
REFRINGENT **10401**
REFROIDI PAR L'EAU **13683**
REFROIDIR **2332**
REFROIDIR (SE) **3081**

REFROIDIR AU DESSOUS DU POINT DE CONGELATION **12463**
REFROIDIR AU-DESSOUS DE LA TEMPERATURE DE CONDENSATION **12463**
REFROIDIR BRUSQUEMENT L'ACIER **10091**
REFROIDISSEMENT **3093**
REFROIDISSEMENT **3085**
REFROIDISSEMENT (SOUS-) **13347**
REFROIDISSEMENT A L'AIR **252**
REFROIDISSEMENT A L'EAU **13646**
REFROIDISSEMENT A L'HUILE **8752**
REFROIDISSEMENT BRUSQUE DE L'ACIER **10100**
REFROIDISSEMENT COMMANDE **3056**
REFROIDISSEMENT D'EAU (A) **13683**
REFROIDISSEMENT DU FOUR **5742**
REFROIDISSEMENT LENT **4230**
REFROIDISSEMENT PAR CHEMISE D'EAU **7204**
REFROIDISSEUR CIRCULAIRE **10752**
REFROIDISSEUR D'HUILE **8751**
REFROIDISSEUR DU TROU DE LAITIER **8380**
REFROIDISSEUR INTERNE **7046**
REFUS DE CRIBLAGE **9548**
REFUS DE RECEPTION **10431**
REGARD **11427**
REGARD **9158**
REGENERATEUR **10412**
REGENERATION DES PIECES DECARBURISEES **1953**
REGIME **10558**
REGIME **8857**
REGIME D'UTILISATION **4395**
REGIME DE ROTATION **10558**
REGIME DE VITESSE **8663**
REGION HOUILLERE **2562**
REGISTRE DE FUMEE **3603**
REGISTRE DE TIRAGE **3603**
REGLABLE **183**
REGLAGE **195**
REGLAGE CONSECUTIF **10256**
REGLAGE DE L'ALLUMAGE **6768**
REGLAGE DE LA PRESSION **10419**
REGLAGE PAR VIS **190**
REGLE A CALCUL **11591**
REGLE A RETRAIT **3047**
REGLE A TRACER DES PARALLELES **9059**
REGLE COURBE **5660**
REGLE DE DESSINATEUR **10831**
REGLE DE JAUGE **12303**
REGLE DES PHASES **9236**
REGLE DIVISEE DE PRECISION **9767**
REGLE GRADUEE **8085**
REGLEMENTS D'USINE **11209**

REGLER **182**
REGLES DE MORTAISAGE **7290**
REGLES DE SERVICE **8835**
REGULARITE CYCLIQUE **3719**
REGULARITE DU MOUVEMENT **10417**
REGULATEUR **6013**
REGULATEUR **12890**
REGULATEUR A BOULES **2185**
REGULATEUR A BRAS CROISES **3386**
REGULATEUR A DEPRESSION **12431**
REGULATEUR A DOUILLE **11578**
REGULATEUR A FORCE CENTRIFUGE **2185**
REGULATEUR A MANCHON **11578**
REGULATEUR A MASSE CENTRALE **13752**
REGULATEUR A RESSORT **11984**
REGULATEUR ASTATIQUE **775**
REGULATEUR AUTOMATIQUE **843**
REGULATEUR CENTRIFUGE **2181**
REGULATEUR CONIQUE **2943**
REGULATEUR D'INTENSITE **3491**
REGULATEUR D'INTENSITE **7020**
REGULATEUR D'INTENSITE **13595**
REGULATEUR DE COURANT **13595**
REGULATEUR DE COURANT **3491**
REGULATEUR DE DEBIT **13600**
REGULATEUR DE DEBIT **8909**
REGULATEUR DE DEBIT DE VENT **3465**
REGULATEUR DE NIVEAU **7654**
REGULATEUR DE PUISSANCE **8909**
REGULATEUR DE TEMPERATURE **12840**
REGULATEUR DE TENSION **13595**
REGULATEUR DE VITESSE **10421**
REGULATEUR DE VITESSE **11874**
REGULATEUR DE WATT **2185**
REGULATEUR ELECTRIQUE **4527**
REGULATEUR ELECTRONIQUE **4635**
REGULATEUR FARCOT **3386**
REGULATEUR ISOCHRONE **7180**
REGULATEUR PENDULE **9168**
REGULATEUR PLAN **11242**
REGULATEUR PORTER **13752**
REGULATEUR PSEUDO-ASTATIQUE **9957**
REGULATEUR STATIQUE **12135**
REGULATEUR SUR L'ARBRE **11242**
REGULE **10422**
REGULE **889**
REGULE D'ANTIMOINE **10423**
REHAUSSE ....DE TOIT FLOTTANT **11331**
RELAMINAGE **10470**
RELAXATION **10435**
RELEVE **7717**

RELEVER UN DIAGRAMME **9520**
RELEVES DE TEMPERATURE **12725**
RELIER (ELECTR.) **2955**
RELIER A LA TERRE (ELECTR.) **4410**
RELUCTANCE **10444**
REMANENCE **10445**
REMANENCE **10473**
REMBLAI **900**
REMBLAI HYDRAULIQUE **6686**
REMBLAYAGE **900**
REMETTRE EN MARCHE **10508**
REMISE A ZERO **14032**
REMISE A ZERO **14030**
REMISE EN ETAT **10287**
REMISE EN MARCHE **10509**
REMONTAGE **10245**
REMONTER **10244**
REMORQUES **13102**
REMOULEUR **6105**
REMOUS **13606**
REMPLACER **10463**
REMPLIR **5156**
REMPLISSAGE **2269**
REMUER A LA PELLE **13283**
RENDEMENT **14010**
RENDEMENT **4463**
RENDEMENT ANODIQUE **597**
RENDEMENT CATHODIQUE **2101**
RENDEMENT COMMERCIAL **2781**
RENDEMENT ECONOMIQUE **2781**
RENDEMENT ELECTROTHERMIQUE **4649**
RENDEMENT EN COURANT **3489**
RENDEMENT FINAL **5194**
RENDEMENT HYDRAULIQUE **6684**
RENDEMENT INDIQUE **6872**
RENDEMENT INDUSTRIEL **2781**
RENDEMENT MANOMETRIQUE **7984**
RENDEMENT MECANIQUE **8100**
RENDEMENT ORGANIQUE **8100**
RENDEMENT QUANTIQUE **10071**
RENDEMENT THERMIQUE **12806**
RENDEMENT TOTAL **8927**
RENDEMENT VOLUMETRIQUE **13602**
RENDRE ALCALIN **330**
RENDRE ETANCHE **7953**
RENDRE RUGUEUX **10783**
RENFERMANT DE LA RESINE **10479**
RENFORCEMENT **12359**
RENFORCEMENT DE LA SOUDURE **10427**
RENFORCEMENT EN TUBES **13258**
RENFORCEMENT PAR DES NERVURES **12152**

RENFORCER **12358**
RENFORT **10569**
RENFORT **10429**
RENFORT **1516**
RENFORT **12359**
RENIFLARD **1591**
RENIFLARD **3307**
RENIFLARD **11694**
RENONCIATION **13615**
RENOUVELER UN ORGANE DE MACHINE **10457**
RENOUVELLEMENT D'UN ORGANE DE MACHINE **10458**
RENTABILITE **9462**
RENTREE D'AIR **4771**
RENTREE DE FLAMME **5365**
RENVERSEMENT (CALCUL SOUS SEISMES) **8965**
RENVERSEMENT DE LA MARCHE D'UNE MACHINE **10541**
RENVERSER LE SENS DE LA MARCHE **10534**
RENVOI A COURROIE **7045**
RENVOI A ENGRENAGES **7050**
RENVOI A LA CHASSE **7231**
RENVOI D'ANGLE **1149**
RENVOI DE MOUVEMENT **3244**
RENVOI DE PLAFOND **8940**
RENVOI FIXE AU SOL **5444**
RENVOI FIXE CONTRE UN MUR **13619**
RENVOI MURAL **13619**
REPARATION **10459**
REPARER **10460**
REPARTITION D'UNE CHARGE **4081**
REPARTITION DES CHARGES **7684**
REPARTITION DES MASSES **4082**
REPARTITION DU TRAVAIL **13939**
REPASSER **11273**
REPERAGE **8015**
REPERAGE PAR COORDONNEES **3099**
REPERE **8012**
REPERE **5879**
REPERER **8008**
REPERER A L'AIDE D'UN POINTEAU **8011**
REPERES DE CALAGE **12951**
REPIQUEUSES **11133**
REPLI **7395**
REPLI (TRAVAIL DES TOLES) **5511**
REPLIURE DE LAMINAGE **7395**
REPLIURE DE LAMINAGE **5510**
REPOS **7100**
REPOSANT **8432**
REPOSANT LIBREMENT **5650**
REPOSANT LIBREMENT SUR DEUX APPUIS **11474**
REPOUSSAGE **8189**
REPOUSSE **8189**

REPRESENTATION CONFORME **8872**
REPRESENTATION GEOMETRIQUE **5926**
REPRISES VISIBLES **7407**
REPRODUCTION **8481**
REPRODUCTION (EN GRAND) **10970**
REPRODUCTION (RESULTAT) **10467**
REPRODUCTION AU FERRO-PRUSSIATE EN BLANC SUR FOND
BLEU **1375**
REPRODUCTION AU FERROPRUSSIATE EN BLEU SUR FOND
BLANC **1373**
REPRODUCTION EN PLOMB FONDU **7457**
REPRODUCTION PHOTOGRAPHIQUE **9263**
REPRODUIRE **8484**
REPULSION **10468**
REPULSION MAGNETIQUE **7915**
REQUETE DE BREVET **646**
RESEAU **7940**
RESEAU A GRANDE DISTANCE **7728**
RESEAU CRISTALLIN **7429**
RESEAU D'ELECTRIFICATION **6096**
RESEAU DE CONDUCTEURS ELECTRIQUES **4080**
RESEAU DE DIFFRACTION **3925**
RESEAU DE DISTRIBUTION ELECTRIQUE **4080**
RESEAU DE REFLEXION **10388**
RESEAU DE TRANSLATION **13120**
RESEAU DE TUYAUX **12579**
RESEAU DE WULFF (STEREOGRAPHIQUE) **13992**
RESEAU RECIPROQUE **10281**
RESEAU SPATIAL **11828**
RESERVE (DE) **12068**
RESERVOIR **10471**
RESERVOIR **12625**
RESERVOIR **1740**
RESERVOIR **5716**
RESERVOIR A AIR **294**
RESERVOIR A DOUBLE PAROI **4161**
RESERVOIR A EAU **13679**
RESERVOIR A FOND CONCAVE **12631**
RESERVOIR A FOND CONVEXE **12632**
RESERVOIR A PETROLE BRUT **3407**
RESERVOIR A TOIT BOMBE **4121**
RESERVOIR A TOIT CONIQUE AUTOPORTANT **11160**
RESERVOIR A TOIT FLOTTANT **5439**
RESERVOIR A TOIT SURBAISSE **13047**
RESERVOIR A TOIT SUSPENDU **12525**
RESERVOIR BALLON **6580**
RESERVOIR CALORIFUGE **6995**
RESERVOIR D'ESSENCE **5716**
RESERVOIR D'HUILE **8773**
RESERVOIR DE CARBURANT **5716**
RESERVOIR DE DECANTATION **11229**

RESERVOIR DE GNL **7678**
RESERVOIR DE JAUGEAGE **12790**
RESERVOIR DE LIQUIDE DE FREIN **1534**
RESERVOIR DE RECETTE **10843**
RESERVOIR SURELEVE **4492**
RESERVOIR TAMPON **1592**
RESERVOIRS DE STOCKAGE **12292**
RESIDU **10475**
RESIDU DE CALCINATION **6765**
RESIDU DE LA COMBUSTION **10476**
RESIDUS **12598**
RESILENCE **10477**
RESILIENCE **8671**
RESILIENCE **13071**
RESILIENCE **6785**
RESILIENCE **6788**
RESILIENCE **1918**
RESINE **10478**
RESINE DE VERNIS **10927**
RESINE ELEMI **4656**
RESINIFICATION **10480**
RESINIFIER (SE) **10481**
RESISTANCE **12352**
RESISTANCE **13327**
RESISTANCE **10484**
RESISTANCE **10502**
RESISTANCE A L'ABRASION **11015**
RESISTANCE A L'ALLONGEMENT **13331**
RESISTANCE A L'ECRASEMENT **13328**
RESISTANCE A L'ETAT DE RECUIT **569**
RESISTANCE A L'EXTENSION **13331**
RESISTANCE A L'USURE **13717**
RESISTANCE A L'USURE **10499**
RESISTANCE A LA COMPRESSION **2870**
RESISTANCE A LA COMPRESSION **13328**
RESISTANCE A LA CORROSION **3194**
RESISTANCE A LA DEFORMATION **10521**
RESISTANCE A LA FATIGUE **5052**
RESISTANCE A LA FLEXION **5424**
RESISTANCE A LA FLEXION TRANSVERSALE **1192**
RESISTANCE A LA RUPTURE **13327**
RESISTANCE A LA RUPTURE TRANSVERSALE **13330**
RESISTANCE A LA TORSION **13057**
RESISTANCE A LA TRACTION **12749**
RESISTANCE A LA TRACTION **13331**
RESISTANCE APPARENTE **6794**
RESISTANCE AU CHOC **6785**
RESISTANCE AU CHOC **1918**
RESISTANCE AU CINTRAGE **1188**
RESISTANCE AU CISAILLEMENT **13330**
RESISTANCE AU CISAILLEMENT **11285**

RESISTANCE AU DEMARRAGE **12128**
RESISTANCE AU FLAMBAGE **3340**
RESISTANCE AU FLAMBAGE **1682**
RESISTANCE AU FLUAGE **3328**
RESISTANCE AU FLUAGE POUR UNE DUREE FINIE **3327**
RESISTANCE AU FREINAGE **1552**
RESISTANCE AU GLISSEMENT **10498**
RESISTANCE AU GLISSEMENT TRANSVERSAL **13330**
RESISTANCE AU MOUVEMENT **10497**
RESISTANCE AU PLIAGE **1188**
RESISTANCE AU RACCOURCISSEMENT **13328**
RESISTANCE AU ROULEMENT **10703**
RESISTANCE AUX CHOCS REPETES **12353**
RESISTANCE AUX VARIATIONS DE TEMPERATURE OU AU CHOC THERMIQUE **12813**
RESISTANCE COMPLEXE **2822**
RESISTANCE COMPOSEE **2822**
RESISTANCE D'ISOLEMENT **7005**
RESISTANCE D'UN CONDUCTEUR **10491**
RESISTANCE DANS LA MARCHE A VIDE **8600**
RESISTANCE DE BOUGIE DE PRECHAUFFAGE **6388**
RESISTANCE DE L'AIR **10492**
RESISTANCE DE RUPTURE PAR FRACTION **4742**
RESISTANCE DES MATERIAUX **12355**
RESISTANCE DIELECTRIQUE OU DISRUPTIVE **3900**
RESISTANCE DISRUPTIVE **4057**
RESISTANCE DU FROTTEMENT **5686**
RESISTANCE EFFECTIVE **10261**
RESISTANCE ELECTRIQUE **4528**
RESISTANCE INTERNE **7072**
RESISTANCE MAGNETIQUE **10444**
RESISTANCE MOL(ECUL)AIRE **8351**
RESISTANCE PASSIVE **10497**
RESISTANCE PRIMITIVE **9854**
RESISTANCE SPECIFIQUE **4548**
RESISTANCE SPECIFIQUE ELECTRIQUE **10503**
RESISTANCE SUIVANT LA TANGENTE **9197**
RESISTANCE UTILE **13426**
RESISTANCE VIVE **10477**
RESISTANCE-LIMITE D'ENDURANCE **5053**
RESISTIVITE **10503**
RESISTIVITE **4548**
RESISTIVITE EQUIVALENTE **4798**
RESOLUTION **10504**
RESONANCE **10506**
RESOUDRE **11813**
RESPIRATION RENIFLARD **1593**
RESSAUT **9917**
RESSORT **11974**
RESSORT A BOUDIN **6427**
RESSORT A BOUDIN CYLINDRIQUE **3585**

RESSORT A LAME 9495

RESSORT A LAME RECTANGULAIRE A PROFIL PARABOLOIDE 10307

RESSORT A LAMES 7485

RESSORT A LAMES ETAGEES 2834

RESSORT A PLUSIEURS LAMES 2834

RESSORT AMORTISSEUR DES CHOCS 1688

RESSORT ANTAGONISTE 11974

RESSORT BANDE 12641

RESSORT CHARGE 11983

RESSORT CONIQUE 2942

RESSORT CONIQUE A LAME PLATE 13605

RESSORT CONIQUE A SECTION RECTANGULAIRE 13605

RESSORT D'INJECTEUR 8686

RESSORT DE COMPRESSION 2862

RESSORT DE FLEXION 11993

RESSORT DE FLEXION A ENROULEMENT 2649

RESSORT DE FLEXION DROIT 12312

RESSORT DE PISTON AVEC CUVETTE 9547

RESSORT DE RAPPEL 10526

RESSORT DE REGULATEUR 6019

RESSORT DE SOUPAPE 13470

RESSORT DE SUSPENSION 1114

RESSORT DE TORSION 11994

RESSORT DE TRACTION 12759

RESSORT DE VOITURE 2008

RESSORT DIAPHRAGME 3875

RESSORT EN C 1823

RESSORT EN HELICE 6427

RESSORT HELICOIDAL 2647

RESSORT RECTANGULAIRE A LAME PLATE 10306

RESSORT RECTANGULAIRE A LAMES SUPERPOSEES 7380

RESSORT SPIRALE 5396

RESSORT TARE 1866

RESSORT TENDU 11983

RESSORT TRAPEZOIDAL 13151

RESSORT TRIANGULAIRE 13191

RESSORT TRIANGULAIRE A LAMES SUPERPOSEES 7381

RESSORT TRONCONIQUE 13236

RESTAURATION 10296

RESULTANTE 10513

RESULTAT D'UNE EPREUVE 10512

RETAILLAGE DES LIMES 10324

RETAILLER LES LIMES 10323

RETARD 3730

RETARD A L'EBULLITION 10517

RETARD A LA FERMETURE DE LA SOUPAPE 10519

RETARD DE LIVRAISON 3731

RETARD DE SOUDAGE 12038

RETARDER 7366

RETARDER 6926

RETASSEMENT 11394

RETASSURE 11393

RETASSURE 11505

RETASSURE 2124

RETASSURE 9340

RETASSURE 7073

RETENDRE UNE COURROIE 12602

RETENUE D'EAU 10182

RETICULE 3383

RETIRURE 4230

RETOMBER (CLIQUET) 4320

RETOUCHE 11682

RETOUCHE DE REGLAGE 10256

RETOUCHER 5214

RETOUCHER LE REGLAGE 10255

RETOUR 10530

RETOUR A LA TEMPERATURE AMBIANTE 3096

RETOUR DE FLAMME 906

RETOUR DE LA FLAMME 892

RETOUR RAPIDE 10106

RETRACTION 11397

RETRACTION 2437

RETRAIT 477

RETRAIT 2437

RETRAIT 3043

RETRAIT 11392

RETRAIT ANGULAIRE 542

RETRAIT AU MOULE 1322

RETRAIT CONTRARIE 6504

RETRAIT DU BOIS 11396

RETRAIT THERMIQUE 12805

RETRANCHER 12424

RETRECI 3042

RETRECIR (SE) 11388

RETRECISSEMENT 3045

RETRECISSEMENT 11397

RETROUSSEMENT DES BORDS DES TOLES 1088

RETROUSSER LES BORDS DES TOLES 1084

RETROVISEUR 10266

REUNION 3254

REUNION DE COURROIES 1151

REUNION DES COURROIE ATTACHE DES COURROIES 1142

REUTILISATION 10246

REVENU 12733

REVENU 12731

REVENU 4230

REVENU-FLUAGE POSTERIEUR 221

REVERSIBILITE 10537

REVERSIBLE 10538

REVERSION 10554

REVETEMENT 2590

REVETEMENT 2587
REVETEMENT 7623
REVETEMENT (ENDUIT) ANODIQUE 599
REVETEMENT ANODIQUE 2588
REVETEMENT ANTI-ROUILLE 2601
REVETEMENT AVEC VERNIS FIN 7214
REVETEMENT CALORIFUGE 7367
REVETEMENT CATHODIQUE 2102
REVETEMENT CHIMIQUE 2297
REVETEMENT D'ALLIAGE 364
REVETEMENT D'ANTIFRICTION 7624
REVETEMENT D'ETAIN 2602
REVETEMENT D'OXYDE 2598
REVETEMENT D'UN CABLE 715
REVETEMENT EN BOIS 13915
REVETEMENT EXTERIEUR 3274
REVETEMENT FONDU 2032
REVETEMENT GALVANIQUE AU TAMBOUR 1016
REVETEMENT INTERIEUR 7623
REVETEMENT METALLIQUE 8194
REVETEMENT METALLIQUE 2596
REVETEMENT METALLIQUE PAR GALVANOPLASTIE 9503
REVETEMENT METALLISE 8200
REVETEMENT NON-METALLIQUE 2597
REVETEMENT NON-METALLIQUE 8641
REVETEMENT PAR COMPOSITION KB TI 7583
REVETEMENT PAR PEINTURE 2595
REVETEMENT PROTECTEUR 9949
REVETEMENT VITREUX 2603
REVETEMENTS DE DIFFUSION 3930
REVETEMENTS DE PROTECTION 9792
REVETEMENTS ELECTROLYTIQUES OU GALVANOPLASTIQUES 2593
REVETEMENTS EMAILLES 2594
REVETEMENTS PAR TREMPE 2592
REVETIR 7598
REVISER 8937
REVOIR 8937
REVOLUTION 10557
REVOLUTION DU FILET 13279
REVOLUTIONS PAR MINUTE 10814
RHEOLOGIE 10562
RHEOSTAT 10563
RHODIUM 10564
RHOMBE 10568
RHOMBOEDRE 10566
RHOMBOIDE 10567
RIBLONS DE FER 11007
RIDAGE 11399
RIFLARD 7201
RIFLOIR 10575

RIGIDITE 10591
RIGIDITE A LA FLEXION 5423
RIGOLE 6185
RIGOLE 2251
RIGOLE DE GRAISSAGE 7810
RINCAGE 10608
RINCER 10607
RINGARD 10116
RINGARDER 9575
RISQUE D'INCENDIE 3608
RIVE 4441
RIVE CISAILLEE 3506
RIVE DE TOLE 9483
RIVER 2516
RIVER A CHAUD 6634
RIVER A FROID 2692
RIVET 10621
RIVET A TETE FRAISEE 10635
RIVET A TETE AFFLEUREE 10635
RIVET A TETE CHANFREINEE 2946
RIVET A TETE CONIQUE 2946
RIVET A TETE CYLINDRIQUE 10632
RIVET A TETE FRAISEE AVEC BOMBE 3249
RIVET A TETE GOUTTE DE SUIF 10633
RIVET A TETE HEMISPHERIQUE 10637
RIVET A TETE NOYEE 10635
RIVET A TETE PERDUE 10635
RIVET A TETE PLATE 10634
RIVET A TETE RONDE 10636
RIVET A TETE TRONCONIQUE 9030
RIVET AU BORD DE LA TOLE 10622
RIVET D'ETANCHEITE 12922
RIVET DE CHAINE 2214
RIVET POUR COURROIES 1155
RIVETAGE 10643
RIVETAGE A LA MAIN 6248
RIVETAGE A RIVET SANS TETE 9321
RIVETAGE A TIGE 9321
RIVETAGE AU MARTEAU 6248
RIVETAGE MECANIQUE 7854
RIVETAGE POUR CONSTRUCTIONS METALLIQUES 9740
RIVETE EN SPIRALE 11918
RIVETER 10629
RIVETS 10646
RIVETS 1241
RIVETS (DISTANCE DES) DE CENTRE A CENTRE D'UNE LIGNE A L'AUTRE 13145
RIVEUR 10641
RIVEUSE 10642
RIVOIR 10644
RIVURE 10628

RIVURE **10639**
RIVURE A BANDE DE RECOUVREMENT **1781**
RIVURE A CHAUD **6635**
RIVURE A COUVRE-JOINT SIMPLE **11478**
RIVURE A DEUX COUPES **4155**
RIVURE A DEUX COUVRE-JOINTS **4130**
RIVURE A DEUX RANGS **4152**
RIVURE A DOUBLE COUVRE-JOINT **4130**
RIVURE A FROID **2693**
RIVURE A PLAT-JOINT **11478**
RIVURE A RECOUVREMENT **7400**
RIVURE A SIMPLE CLOUURE **11487**
RIVURE A TROIS COUPES **13178**
RIVURE A TROIS RANGS **13177**
RIVURE A UN RANG **11487**
RIVURE A UNE COUPE **11489**
RIVURE A UNE SEULE BANDE DE RECOUVREMENT **11478**
RIVURE D'ASSEMBLAGE DE FORCE **9740**
RIVURE DE FORCE **9740**
RIVURE DOUBLE **4152**
RIVURE EN CHAINE **2215**
RIVURE EN LOSANGE **7799**
RIVURE EN QUINCONCE **14034**
RIVURE SIMPLE **11487**
RIVURE TRIPLE A TRIPLE CLOUURE **13177**
ROBE **11330**
ROBE DU RESERVOIR **12630**
ROBINET **12639**
ROBINET **11592**
ROBINET **2617**
ROBINET A BEC COURBE **1234**
ROBINET A BOISSEAU **8997**
ROBINET A BOISSEAU **9535**
ROBINET A BOISSEAU LUBRIFIE **7802**
ROBINET A BOISSEAU RENVERSE **7111**
ROBINET A BOISSEAU SPHERIQUE (A BOULE) **968**
ROBINET A BOURRAGE **5954**
ROBINET A CLEF **9530**
ROBINET A DEUX VOIES **13317**
ROBINET A EAU **13644**
ROBINET A FLOTTEUR **5433**
ROBINET A GAZ **5814**
ROBINET A MANOEUVRE RAPIDE **10104**
ROBINET A MEMBRANE **3876**
ROBINET A MOTEUR **8417**
ROBINET A POINTEAU (A AIGUILLE) **8522**
ROBINET A QUATRE VOIES **5606**
ROBINET A QUATRE VOIES **5605**
ROBINET A SOUFFLET **1138**
ROBINET A SOUPAPE **5977**
ROBINET A SOUPAPE DOUBLE **4154**

ROBINET A SOUPAPE SIMPLE **11488**
ROBINET A TROIS VOIES **12881**
ROBINET A TROIS VOIES **12882**
ROBINET A VIS DE PRESSION **11033**
ROBINET D'AIR **293**
ROBINET D'ALIMENTATION **5066**
ROBINET D'ANGLE **511**
ROBINET D'ARRET **12281**
ROBINET D'ARRET **11408**
ROBINET D'EQUERRE **534**
ROBINET D'EXTRACTION **1359**
ROBINET D'ISOLEMENT **845**
ROBINET DE BAIGNOIRE **1063**
ROBINET DE BATTERIE **1068**
ROBINET DE BOUTEILLE (DE GAZ COMPRIME) **3575**
ROBINET DE DISTRIBUTEUR **8467**
ROBINET DE DOUCHE **11387**
ROBINET DE FOND DE CUVE **1493**
ROBINET DE JAUGE **12776**
ROBINET DE JAUGE **5875**
ROBINET DE LAVABO **13630**
ROBINET DE PRISE DE VAPEUR **12187**
ROBINET DE PUISAGE **1235**
ROBINET DE RADIATEUR **10153**
ROBINET DE VIDANGE **1357**
ROBINET DE VIDANGE **4205**
ROBINET DISTRIBUTEUR TOURNANT **10760**
ROBINET DROIT **5977**
ROBINET ELECTROMAGNETIQUE **11774**
ROBINET MELANGEUR **8335**
ROBINET PURGEUR **4205**
ROBINET REGULATEUR A MEMBRANE **3872**
ROBINET SANS PRESSE-ETOUPE **9010**
ROBINET-VALVE A SOUPAPE **13455**
ROBINET-VANNE **11653**
ROBINETTERIE **1566**
ROCHE **10651**
ROCHE CLIVABLE **2486**
ROCHE-MERE **5802**
ROCHET **10225**
ROCHET **1061**
RODAGE **6106**
RODAGE **10855**
RODAGE **7407**
RODAGE (D'UNE VOITURE) **1587**
RODAGE A L'EMERI **6111**
RODAGE AU LIQUIDE **6564**
RODER A L'EMERI **6104**
RODER UN MOTEUR **10835**
ROGNAGE **13209**
ROGNER **13204**

ROGNOIR **11009**
ROGNURES DE CUIR **7497**
ROGNURES DE FER **11007**
ROGNURES DE LIEGE **6144**
ROGNURES DE ZINC **14040**
RONCE ARTIFICIELLE **987**
RONDELLE **13631**
RONDELLE AUTOTARAUDEUSE **11163**
RONDELLE BELLEVILLE **3461**
RONDELLE BRASEE **11758**
RONDELLE D'AILE **5096**
RONDELLE D'EPAISSEUR **11351**
RONDELLE D'UN ARBRE **2724**
RONDELLE DE BLOCAGE **7703**
RONDELLE DE SERRAGE **7703**
RONDELLE DECORATIVE **5230**
RONDELLE ELASTIQUE **4486**
RONDELLE EN CAOUTCHOUC **10823**
RONDELLE FREIN **7703**
RONDELLE GOUPILLEE **7743**
RONDELLE GROWER **11995**
RONDELLE PLATE **5399**
RONDELLE PROTECTRICE **4835**
RONDELLE SOUDEE **13773**
RONDIN **10803**
RONDS EN ACIER ALLIE **377**
ROSE DES VENTS **2800**
ROTATION **10765**
ROTATION A DROITE **2507**
ROTATION A GAUCHE **3238**
ROTIN **10236**
ROTOR **10767**
ROTOR **6795**
ROTULE **7336**
ROUAGE **13821**
ROUE **13810**
ROUE **13692**
ROUE **9402**
ROUE A AUBES **13479**
ROUE A AUGETS **8960**
ROUE A CENTRE PLEIN **4012**
ROUE A CHEVRONS **4143**
ROUE A DENTS EN BOIS **8400**
ROUE A DENTS TAILLEES **3508**
ROUE A DENTURE BRUTE **10773**
ROUE A DENTURE INTERIEURE **7079**
ROUE A DISQUE **4012**
ROUE A EMPREINTES **2217**
ROUE A MEULER **6115**
ROUE A PALETTES, A AILETTES **13479**
ROUE A RAYONS **11951**

ROUE A ROCHET **10225**
ROUE A ROCHET **2495**
ROUE A VIS SANS FIN **13975**
ROUE A VOILE **4012**
ROUE A VOILE BOMBE **4037**
ROUE A VOILE DROIT **5379**
ROUE ATMOSPHERIQUE **13861**
ROUE CENTRALE D'UN TRAIN PLANETAIRE **12457**
ROUE CONDUCTRICE **4318**
ROUE CONDUITE **4300**
ROUE CONIQUE **1229**
ROUE CYLINDRIQUE **3591**
ROUE D'ANGLE **1229**
ROUE D'ENGRENAGE **5903**
ROUE D'ENGRENAGE AVEC JOUES **11402**
ROUE D'ENGRENAGE BRUTE DE FONDERIE **10773**
ROUE D'ENGRENAGE CYLINDRIQUE **12009**
ROUE D'ENGRENAGE DROITE **12009**
ROUE D'ENGRENAGE EPAULEE **11402**
ROUE D'UNE TURBINE **13823**
ROUE DE CHAMP **3399**
ROUE DE COMMANDE **4318**
ROUE DE COTE (HYDRAULIQUE) **7769**
ROUE DE DIFFERENTIEL **3917**
ROUE DE FRICTION **5682**
ROUE DE FRICTION (A JANTE EN FORME DE COIN) **13732**
ROUE DE FRICTION CONIQUE **1225**
ROUE DE FRICTION CYLINDRIQUE **12007**
ROUE DE POITRINE **6465**
ROUE DE RECHANGE **2243**
ROUE DE SECOURS **2243**
ROUE DE SECOURS **11835**
ROUE DE TRANSPORT **6754**
ROUE DENTEE **5903**
ROUE DENTEE A CLIQUET **10225**
ROUE DENTEE INTERROMPUE **7975**
ROUE DROITE **3591**
ROUE ELLIPTIQUE **4677**
ROUE ETOILEE **12109**
ROUE HELICOIDALE **6425**
ROUE INTERMEDIAIRE **6754**
ROUE LIBRE **5644**
ROUE LIBRE **5651**
ROUE MENANTE **4318**
ROUE MENEE **4300**
ROUE MOTRICE **4318**
ROUE PAR-DESSOUS **13355**
ROUE PAR-DESSUS **8960**
ROUE PLANETAIRE **9444**
ROUE POLISSEUSE **9604**
ROUE PORTEUSE **10854**

ROUE RECEPTRICE 4300
ROUE SATELLITE 9444
ROUES D'ASSORTIMENT 7027
ROUES DE SERIE 7027
ROUGE 10329
ROUGE A POLIR 7148
ROUGE ANGLAIS 7148
ROUGE AU FEU 10330
ROUGE CERISE 2326
ROUGE CLAIR 1613
ROUGE CONGO 2935
ROUGE D'ANGLETERRE 3355
ROUGE DE CHROME 2374
ROUGE DE PRUSSE 7148
ROUGE NAISSANT 1295
ROUGE SOMBRE 4367
ROUGE VIF 1613
ROUILLE 10873
ROUILLE 10335
ROUILLE 10863
ROUILLER (SE) 10864
ROULANT 9635
ROULEAU 12334
ROULEAU 12334
ROULEAU 10671
ROULEAU 10679
ROULEAU 2645
ROULEAU 1112
ROULEAU CONIQUE 12660
ROULEAU D'UNE CHAINE 10688
ROULEAU RENFLE 1017
ROULEAUX 10695
ROULEAUX DE POLISSAGE DRAPES 2539
ROULEMENT 954
ROULEMENT 10696
ROULEMENT 1117
ROULEMENT A AIGUILLES 8518
ROULEMENT A BILLES 954
ROULEMENT A BILLES 618
ROULEMENT A BILLES A CHARGE AXIALE 12901
ROULEMENT A BILLES A CHARGE RADIALE 10130
ROULEMENT A BILLES A CHARGE RADIALE ET AXIALE
   COMBINEE 538
ROULEMENT A BILLES A POUSSEE AXIALE ET LATERALE
   COMBINEE 538
ROULEMENT A GALETS 618
ROULEMENT A GALETS (A ROULEAUX) 10680
ROULEMENT A GALETS CONIQUES 10683
ROULEMENT A GALETS CYLINDRIQUES 10682
ROULEMENT A POUSSEE LATERALE 10130
ROULEMENT A ROTULE SUR ROULEAUX 10681

ROULEMENT LISSE 9424
ROULER 10665
ROULETTE 10697
ROULETTE D'ARPENTAGE 8092
ROULETTE DU PLANIMETRE 13817
ROULIS 12538
ROULURE DU BOIS 3460
RUBAN 12641
RUBAN 974
RUBAN 8092
RUBAN CAOUTCHOUTE 10822
RUBAN D'EMERI 18
RUBAN DE FREIN 1544
RUBAN ISOLANT 7001
RUBAN METALLIQUE DE PROTECTION 3268
RUBANS 980
RUBIDIUM 10829
RUBRIQUE 10340
RUGOSIMETRE 12509
RUGOSIMETRE ENREGISTREUR 12498
RUGOSITE 10792
RUGOSITE DE SURFACE 12518
RUPTURE 5617
RUPTURE 3278
RUPTURE 10858
RUPTURE 10861
RUPTURE 9353
RUPTURE A CHAUD 6621
RUPTURE BRUSQUE DU CORPS D'UNE VIS 13309
RUPTURE D'UN TUYAU 1768
RUPTURE DE FATIGUE / D'ENDURANCE 4740
RUPTURE DE LA COUCHE DE SCORIES 5354
RUPTURE DES TETES DE RIVET 11999
RUPTURE DU RIVET PAR CISAILLEMENT 11293
RUPTURE PAR FLEXION 5488
RUTHENIUM 10874
S'AGGLOMERER 11509
S'ELARGIR 5356
S'ENGRENER 1077
S'ETENDRE 5356
SABLAGE 1316
SABLAGE 10921
SABLE 10917
SABLE A GRAIN FIN 5199
SABLE A GROS GRAIN 2574
SABLE A NOYAUX 3163
SABLE DE CARRIERE 9396
SABLE DE CONTACT 5006
SABLE DE COUVERTURE 912
SABLE DE FONDERIE 8430
SABLE DE MER 11071

SABLE DE MOULAGE **8430**
SABLE DE RIVIERE **10619**
SABLE ISOLANT **9094**
SABLE QUARTZEUX **10082**
SABLE SEC **4255**
SABLER **10919**
SABLEUSE **10920**
SABOT **11360**
SABOT DE FREIN **1543**
SABOT DE FREIN **1531**
SABOT MAGNETIQUE **8188**
SABOT POUR DISPOSITIF DE LEVAGE **7207**
SABOTS D'ANIMAUX RAPES **6567**
SAC A OUTILS **12996**
SAGETTE **10613**
SAIGNEE **7274**
SAIGNEE **1321**
SAIGNER **8554**
SAILLANT **9910**
SAILLANT **4301**
SAILLIE **9913**
SAILLIE **9917**
SAILLIE **2093**
SAILLIE SUR LE PRIMITIF **164**
SALAIRE **10897**
SALAIRE **3594**
SALAIRE **13613**
SALAIRE A LA TACHE **9295**
SALAIRE AUX PIECES **9295**
SALLE DE CHAUFFERIE **1418**
SALLE DES CHAUDIERES **1418**
SALLE DES MACHINES **4757**
SALLE DES MACHINES **7855**
SALPETRE **2330**
SALPETRE **10903**
SAMARIUM **10914**
SANDARAQUE **10927**
SANG-DRAGON **4203**
SANGUINE **8720**
SANS FROTTEMENT **5688**
SANS SCORIES **5638**
SAPONIFIABLE **10932**
SAPONIFICATION **10933**
SAPONIFIER **10934**
SAPONITE **10935**
SARCLEUSES **13742**
SAS **11422**
SAS A AIR **274**
SASSER **11423**
SATELLITE **9444**
SATELLITE **9447**

SATELLITE **6168**
SATELLITE DE DIFFERENTIEL **3918**
SATURATION **10943**
SATURATION MAGNETIQUE **7916**
SATURER **10939**
SAUMURE **1617**
SAUT DE BLOC **1337**
SAUTERELLE **1228**
SAUTEUSE **5665**
SAVON **11701**
SAVON BLANC ORDINAIRE **3482**
SAVON DE MARSEILLE **8020**
SAVON DE POTASSE **11740**
SAVON MOU **11740**
SAVON NOIR **11740**
SAVON VERT **11740**
SCALAIRE **10967**
SCANDIUM **10991**
SCELLER DE CIMENT **6153**
SCHEELIN CALCAIRE **10997**
SCHEELITE **10997**
SCHEMA **3836**
SCHISTE **10999**
SCHISTE BITUMINEUX **8780**
SCIAGE **10961**
SCIAGE A CHAUD **6657**
SCIAGE A FROID **2699**
SCIAGE CONTRE FIL **4718**
SCIAGE EN LONG **1058**
SCIAGE EN TRAVERS **4718**
SCIAGE HOLLANDAIS **10075**
SCIAGE PAR FRICTION **5680**
SCIAGE PAR RAYONNEMENT **10075**
SCIAGE PARALLELE **1058**
SCIAGE VERTICAL **4718**
SCIE **10950**
SCIE A ARASER **12741**
SCIE A BOIS **13917**
SCIE A CADRE **5629**
SCIE A CHAINETTE **2216**
SCIE A CHANTOURNER **1505**
SCIE A CHASSIS **5629**
SCIE A CHAUD **6630**
SCIE A CHEVILLES **12741**
SCIE A DECOUPER **5665**
SCIE A DOS **897**
SCIE A FROID **2688**
SCIE A GRUGER **3108**
SCIE A GUICHET **2802**
SCIE A LAME SANS FIN **978**
SCIE A MAIN **6249**

SCIE A METAUX **6196**
SCIE A METAUX **8186**
SCIE A METAUX A POIGNEE PISTOLET **9369**
SCIE A MONTURE METALLIQUE **8186**
SCIE A MORTAISER **8193**
SCIE A PLACAGE **13526**
SCIE A REFENDRE **5628**
SCIE A RUBAN **977**
SCIE A RUBAN **978**
SCIE A TENONS **12741**
SCIE A TRONCONNER **3365**
SCIE CIRCULAIRE **2427**
SCIE CIRCULAIRE **2428**
SCIE DE LONG **10609**
SCIE EGOINE **11659**
SCIE MECANIQUE **9747**
SCIE OSCILLANTE **4337**
SCIE PASSE-PARTOUT **3365**
SCIE POUR METAUX AU ROUGE **6630**
SCIER **10951**
SCIERIE (USINE) **10956**
SCIERIE A MOUVEMENT ALTERNATIF **5629**
SCIERIE MECANIQUE **9747**
SCIES A BUCHES **10963**
SCINDER **11946**
SCIURE DE BOIS **10953**
SCLEROMETRE **11002**
SCLEROSCOPE **11366**
SCORIE **11561**
SCORIE **2397**
SCORIE DES HAUTS-FOURNEAUX **1312**
SCORIE EMPRISONNEE **11567**
SCORIES BASIQUES **1054**
SCORIES DE FORGE **6217**
SCORIFICATION **11004**
SEAU **1674**
SECANTE **11099**
SECHAGE **928**
SECHAGE **4359**
SECHER **4338**
SECHER LES BOIS A L'AIR LIBRE **11094**
SECHEUR **4356**
SECHOIR **4355**
SECHOIR **4256**
SECHOIRS A FOURRAGES **4259**
SECHOIRS A GRAINS **4258**
SECOND FILTRE **11101**
SECOUSSE **11359**
SECOUSSE ELECTRIQUE **4529**
SECTEUR **2782**
SECTEUR CIRCULAIRE **11126**

SECTEUR D'UN CERCLE **11126**
SECTEUR DENTE **13024**
SECTEUR GRADUE **6867**
SECTEUR SPHERIQUE **11127**
SECTION **11111**
SECTION **11122**
SECTION (DROITE OU OBLIQUE) D'UNE POUTRE **11116**
SECTION A NERVURES **6550**
SECTION ANNULAIRE **587**
SECTION CARREE **12017**
SECTION CHARGEE **4460**
SECTION CIRCULAIRE **2410**
SECTION CONIQUE **2940**
SECTION CONIQUE **13735**
SECTION DANGEREUSE **3609**
SECTION DE L'ORIFICE D'ECOULEMENT **4020**
SECTION DE PASSAGE **3374**
SECTION DE PASSAGE D'UNE SOUPAPE **11121**
SECTION DE RUPTURE **3373**
SECTION DE RUPTURE **5620**
SECTION DROITE **8654**
SECTION EFFICACE **3371**
SECTION ELLIPTIQUE **4674**
SECTION ETOILEE **12110**
SECTION LIBRE **3374**
SECTION MACROSCOPIQUE **7875**
SECTION MI-RONDE **11170**
SECTION NORMALE **8654**
SECTION OBLIQUE **8711**
SECTION OVALE **8917**
SECTION PRINCIPALE **9868**
SECTION RECTANGULAIRE **10302**
SECTION RECTANGULAIRE CREUSE **6550**
SECTION SEMI-CIRCULAIRE **11170**
SECTION SOUMISE A UN EFFORT DE COMPRESSION **3375**
SECTION SOUMISE A UN EFFORT DE TRACTION **3376**
SECTION TRANSVERSALE **3378**
SECTION TRAPEZOIDALE **13183**
SECTION TRIANGULAIRE **13188**
SECTION UTILE **13422**
SECURITE DE BON FONCTIONNEMENT **10439**
SEDIMENT **3773**
SEDIMENTATION **11228**
SEGMENT **11135**
SEGMENT D'ANNEAU CIRCULAIRE **10603**
SEGMENT D'ETANCHEITE **2867**
SEGMENT D'UN CERCLE **11136**
SEGMENT D'UNE DROITE **11118**
SEGMENT DE COMPRESSION **2861**
SEGMENT DE FEU **13034**
SEGMENT DE FREIN **1542**

SEGMENT DE PISTON **9379**
SEGMENT DE PISTON EXCENTRE **4429**
SEGMENT DE TOLE (D'INDUIT) **3160**
SEGMENT RACLEUR **8750**
SEGMENT RACLEUR **8775**
SEGMENT SPHERIQUE A DEUX BASES **14051**
SEGMENT SPHERIQUE A UNE BASE **11137**
SEGREGATION **11139**
SEGREGATION **3172**
SEGREGATION **7635**
SEGREGATION MAJEURE **6925**
SEGREGATION MAJEURE **8660**
SEGREGATION MINEURE **8235**
SEL (CHIM.) **10904**
SEL ALCALIN **338**
SEL AMMONIAC **462**
SEL ANGLAIS **7889**
SEL COMMUN **2790**
SEL D'EPSOM **7889**
SEL D'ETAIN **12100**
SEL D'OSEILLE **10907**
SEL DE GLAUBER **5967**
SEL DE PHOSPHORE **8241**
SEL DE SEDLITZ **7889**
SEL DE SEIGNETTE **11729**
SEL DE SOUDE **551**
SEL DOUBLE **4153** ⁻
SEL GEMME **10653**
SEL MARIN **2790**
SEL MICROCOSMIQUE **8241**
SEL NEUTRE **8546**
SEL SOLVAY **455**
SEL VOLATIL D'ANGLETERRE **461**
SELENIUM **11146**
SELF-INDUCTION **11156**
SELS AUGMENTANT LA CONDUCTIBILITE D'UNE SOLUTION **2915**
SELS DE POLISSAGE **1690**
SELS POUR BAINS **9506**
SEMELLE **1127**
SEMELLE D'UN RABOT **11771**
SEMELLE D'UNE POUTRE **5947**
SEMELLE DE PALIER **11772**
SEMENCE **12588**
SEMI-DIESEL **11171**
SEMI-POINTEUSE **11172**
SEMI-REMORQUE **11178**
SEMOIRS A LA VOLEE **11129**
SEMOIRS POUR PORTES OUTILS **11132**
SENESTRORSUM **3239**
SENS D'UNE FORCE **11182**

SENS DE LA MARCHE **11183**
SENS DE LAMINAGE **10698**
SENS DE ROTATION **3992**
SENS DU MOUVEMENT **11183**
SENS INVERSE DES AIGUILLES D'UNE MONTRE (EN) **3237**
SENSIBILITE A L'EFFET D'ENTAILLE **8670**
SENSIBILITE A L'EPAISSEUR **8027**
SENSIBILITE D'UN INSTRUMENT **11188**
SEPARATEUR D'EAU **12188**
SEPARATEUR D'HUILE **8778**
SEPARATEUR DE VAPEUR **12184**
SEPARATION (CHIM.) **11191**
SEPARATION D'UN MELANGE **11192**
SEPARATION ELECTROLYTIQUE **4602**
SEPARATION ELECTROMAGNETIQUE **4618**
SEPARATION ELECTROSTATIQUE **4646**
SEPARATION MAGNETIQUE **7917**
SEPARATION PNEUMATIQUE **290**
SEPARER (CHIM.) **11189**
SEPARER (SE) **9760**
SEQUENCE DES CALIBRES **9097**
SEQUENTIEL **11194**
SERGENT **2452**
SERIE **11203**
SERIE **11213**
SERIE **11219**
SERIE (EN) **11196**
SERIE BINOMIALE **1260**
SERIE CONVERGENTE **3065**
SERIE DIVERGENTE **4094**
SERIE EXPONENTIELLE **4943**
SERIE FINIE **5232**
SERIE HOMOLOGUE **6562**
SERIE INFINIE **6909**
SERIE LOGARITHMIQUE **7719**
SERINGUE A HUILE **8784**
SERPENTIN **2645**
SERPENTIN **9346**
SERPENTIN CHAUFFE PAR LA VAPEUR **12161**
SERPENTIN DE REFROIDISSEMENT **13978**
SERPENTINE **11204**
SERRAGE **11355**
SERRAGE **10194**
SERRAGE D'UN COIN **12928**
SERRAGE DE CLAVETTE **12653**
SERRAGE DU FREIN **1549**
SERRAGE PAR COIN **12929**
SERRAGE PAR ECROUS **12930**
SERRE **10194**
SERRE-BARRES **12981**
SERRE-FLANS **3887**

SERRE-JOINTS **2451**

SERRE-JOINTS **2452**

SERRER A BLOC UN ECROU **12927**

SERRER A FOND UN ECROU **12927**

SERRER A VIS **11022**

SERRER LA BOITE A ETOUPE **12926**

SERRER LE FREIN **650**

SERRER UN COIN **12924**

SERRER UN ECROU **12925**

SERRER UNE POULIE SUR L'ARBRE **4222**

SERRER UNE ROUE **4222**

SERRER UNE VIS **12925**

SERRURE DE PORTIERE **4123**

SERTISSAGE DES TUYAUX AU DUDGEON **4908**

SERTISSEUSE **5346**

SERTISSEUSE **11088**

SERVICE **8837**

SERVICE CONTINU **3038**

SERVICE CONTINU **9209**

SERVICE INTERMITTENT **7056**

SERVO-DIRECTION **9743**

SERVO-FREIN **1457**

SERVO-FREINS **9731**

SERVO-MECANISME **11210**

SERVO-MOTEUR **11211**

SERVOMOTEUR DE DEMARRAGE **1018**

SERVOMOTEUR DE LANCEMENT **1018**

SESQUICHLORURE DE CHROME **2381**

SESQUIOXYDE DE CHROME **2382**

SESQUIOXYDE DE COBALT **2611**

SESQUIOXYDE DE FER **7148**

SESQUIOXYDE DE MANGANESE **7970**

SHERARDISATION **11341**

SHERARDISER **11340**

SHIMMY **11353**

SHUNT **11403**

SICCATIF **4257**

SICCATIF **4256**

SIDEROSE **11421**

SIDEROXYLE **13833**

SIDERURGIE **8209**

SIEGE **11098**

SIEGE A REGLAGE AUTOMATIQUE **9741**

SIEGE BAQUET **1678**

SIEGE DE CORPS DE VANNE **1397**

SIEGE DE SOUPAPE **11097**

SIEGE DE SOUPAPE **13468**

SIEGE SUSPENDU **1383**

SIEGES OBLIQUES **12663**

SIEGES PARALLELES **9063**

SIFFLET **13829**

SIFFLET A VAPEUR **12190**

SIFFLET AVERTISSEUR **307**

SIFFLET D'ALARME **307**

SIGNAL ACOUSTIQUE **816**

SIGNAL D'ALARME **306**

SIGNAL OPTIQUE **13580**

SIGNE **11429**

SIGNE NEGATIF **8532**

SIGNE POSITIF **9661**

SILENCIEUX **11431**

SILENCIEUX **8448**

SILEX PYROMAQUE **5427**

SILICATE **11437**

SILICATE D'ALUMINE **425**

SILICATE DE FER **7149**

SILICATE DE MAGNESIUM **7888**

SILICATE DE POTASSE **9698**

SILICATE DE SOUDE **11731**

SILICE **11436**

SILICE **11434**

SILICE FARINEUSE **7292**

SILICEUX **10082**

SILICIUM **11442**

SILICIURE DE CARBONE **1980**

SILICO-SPIEGEL **11440**

SILICONE **11456**

SIMILITUDE **11470**

SIMITHSONITE **1836**

SINISTRORSUM **3239**

SINUOIDE **11513**

SINUS **11475**

SIPHON **11517**

SIPHON (DE MANOMETRE) **11514**

SIRENE D'ALARME **11518**

SOCLE **1026**

SOCLE EN FONTE **1037**

SODIUM **11714**

SOIE **11452**

SOIE **12614**

SOIE ARTIFICIELLE **736**

SOL AQUIFERE **13642**

SOLE **6331**

SOLE ACIDE **108**

SOLE DE HAUT-FOURNEAU **9011**

SOLE ET GARNISSAGE BASIQUES **1045**

SOLENOIDE **11774**

SOLENOIDE **11773**

SOLIDE **11775**

SOLIDE ELASTIQUE **4484**

SOLIDIFICATION **6292**

SOLIDIFICATION **5653**

SOLIDIFICATION 11797
SOLIDIFICATION D'UN LIQUIDE 11799
SOLIDIFICATION DE LA MASSE FONDUE DANS LA POCHE
 7360
SOLIDIFIER (SE) A L'AIR (CIMENT) 6286
SOLIDIFIER (SE) DANS L'EAU (CIMENT) 6287
SOLIDIFIER (SE) PAR REFROIDISSEMENT 11803
SOLIDUS 11805
SOLLICITATION 12331
SOLLICITATIONS 7679
SOLUBILISER 10456
SOLUBILITE 11806
SOLUBLE 11807
SOLUBLE DANS L'EAU 11809
SOLUTION 11811
SOLUTION ACIDE 9290
SOLUTION ADHERENTE 4201
SOLUTION ALCALINE 339
SOLUTION ALCOOLIQUE 315
SOLUTION AMMONIACALE 457
SOLUTION AMMONIACALE 660
SOLUTION ANODIQUE 606
SOLUTION AQUEUSE 661
SOLUTION CONCENTREE 2883
SOLUTION DE POTASSE CAUSTIQUE 9671
SOLUTION DE SOUDE CAUSTIQUE 2120
SOLUTION DILUEE 3953
SOLUTION ENTRAINEE 4202
SOLUTION ETENDUE 3953
SOLUTION FAIBLE 13706
SOLUTION FORTE 2883
SOLUTION GRAPHIQUE 6060
SOLUTION NON SATUREE 13394
SOLUTION PAR LE CALCUL 8694
SOLUTION SALINE 10908
SOLUTION SATUREE 10941
SOLUTION SOLIDE 11794
SOLUTION SURSATUREE 12478
SOLVANT 11814
SOLVANT IONOGENE 7130
SOLVANTS DE DECAPAGE 12381
SOLVANTS POUR CORROSIFS 3192
SOMME 12454
SOMMET 12455
SOMMET 638
SOMMET D'UN POLYGONE 13543
SOMMET DE LA DENT 12514
SOMMET DU FILET 3333
SONDAGE 12778
SONDE DE MOULEUR 8357
SONDEUSE SUR CAMION 13614

SONNERIE 1137
SOPHISTICATION 205
SORBITE 11817
SORTIE 8900
SORTIE DE COULEE (TROU) 12645
SORTIE EN LINGOTIERE 9717
SOUBASSEMENT D'UNE COLONNE 1035
SOUDABILITE 13768
SOUDABLE 13769
SOUDAGE 11762
SOUDAGE 13767
SOUDAGE 13779
SOUDAGE (A L'ARC) ELECTRIQUE 4533
SOUDAGE A DROITE 907
SOUDAGE A ENTAILLE 11636
SOUDAGE A FROID 2712
SOUDAGE A L'ACETYLENE 96
SOUDAGE A L'ARC A COURANT ALTERNATIF 685
SOUDAGE A L'ARC A COURANT CONTINU 686
SOUDAGE A L'ARC AVEC ELECTRODE AU CARBONE 1943
SOUDAGE A L'ARC ELECTRIQUE 8176
SOUDAGE A L'ARC EN ATMOSPHERE INERTE 683
SOUDAGE A L'ARC METALLIQUE SANS GAZ PROTECTEUR
 13397
SOUDAGE A L'ARC METALLIQUE SOUS GAZ PROTECTEUR
 11345
SOUDAGE A L'ARC METALLIQUE SOUS PROTECTION DE GAZ
 INERTE 11428
SOUDAGE A L'ARC SOUS PROTECTION GAZEUSE 5852
SOUDAGE A L'ARC SUBMERGE 12414
SOUDAGE A L'ARC SUBMERGE 684
SOUDAGE A LA CAROTTE 10027
SOUDAGE A LA CAROTTE 9576
SOUDAGE A LA FORGE 5546
SOUDAGE A LA FORGE 6218
SOUDAGE A PASSES MULTIPLES 8455
SOUDAGE AERO-ACETYLENIQUE 295
SOUDAGE ALUMINOTHERMIQUE PAR PRESSION 9837
SOUDAGE ARCATOM (A L'ARC PROTEGEA L'HYDROGENE
 ATOMIQUE) 801
SOUDAGE AU CONTACT 682
SOUDAGE AU PLAFOND 7113
SOUDAGE AUTOGENE 835
SOUDAGE AUTOGENE 8979
SOUDAGE AUTOGENE PAR PRESSION 8978
SOUDAGE AUTOMATIQUE 849
SOUDAGE BOUT A BOUT (PAR RESISTANCE) 1805
SOUDAGE CONTINU 3016
SOUDAGE CONTINU PAR RECOUVREMENT 7401
SOUDAGE DE FILS EN CROIX 3385
SOUDAGE DE PLOMB 7476
SOUDAGE DE RENFORCEMENT 10428

SOUDAGE DE TROUS DE RIVETS **9537**

SOUDAGE DES RIVETS **10631**

SOUDAGE EN ANGLE **5162**

SOUDAGE EN ATELIER **11362**

SOUDAGE EN BOUT PAR ETINCELAGE **1806**

SOUDAGE EN BOUT PAR RESISTANCE **1807**

SOUDAGE EN BOUT PAR RESISTANCE **10488**

SOUDAGE EN CONGE **5162**

SOUDAGE EN LIGNE CONTINUE A LA MOLETTE **11086**

SOUDAGE EN PAS DE PELERIN **918**

SOUDAGE EN SPIRALE **11921**

SOUDAGE ETANCHE **11076**

SOUDAGE INTERIEUR **7080**

SOUDAGE LONGITUDINAL **7737**

SOUDAGE MAL FAIT **920**

SOUDAGE MECANIQUE (AUTOMATIQUE) **8115**

SOUDAGE PAR ALUMINOTHERMIE **12819**

SOUDAGE PAR BOMBARDEMENT D'ELECTRONS **4630**

SOUDAGE PAR BOMBARDEMENT ELECTRONIQUE **4627**

SOUDAGE PAR BOSSAGES **9916**

SOUDAGE PAR BOSSAGES MULTIPLES **8468**

SOUDAGE PAR ETINCELAGE **5358**

SOUDAGE PAR FUSION **5762**

SOUDAGE PAR FUSION **8643**

SOUDAGE PAR INDUCTION **4619**

SOUDAGE PAR PERCUSSION A CONDENSATEUR **4645**

SOUDAGE PAR PERCUSSION ELECTROMAGNETIQUE **4617**

SOUDAGE PAR POINTS A RECOUVREMENT **7402**

SOUDAGE PAR POINTS AU PISTOLET **9576**

SOUDAGE PAR POINTS EN SERIE PAR RESISTANCE **11200**

SOUDAGE PAR POINTS MULTIPLES **8464**

SOUDAGE PAR POINTS PAR RESISTANCE **10493**

SOUDAGE PAR PRESSION **9842**

SOUDAGE PAR PULSATION **9978**

SOUDAGE PAR RAPPROCHEMENT **1805**

SOUDAGE PAR RECOUVREMENT **7403**

SOUDAGE PAR RESISTANCE **10500**

SOUDAGE PAR ULTRASONS **13338**

SOUDAGE PROVISOIRE PAR POINTS DE POINTAGE **12589**

SOUDAGE SEMI-AUTOMATIQUE A L'ARC **11166**

SOUDAGE SOUS POUDRE **9722**

SOUDAGE T.I.G (A L'ARC DE TUNGSTENE SOUS GAZ INERTE) **6906**

SOUDAGE TENDRE **12958**

SOUDAGE-FINITION **5216**

SOUDE **11713**

SOUDE A L'AMMONIAQUE **455**

SOUDE AZOTEUSE **11725**

SOUDE CAUSTIQUE **11722**

SOUDE LEBLANC **12445**

SOUDE NITREUSE **11725**

SOUDE SOLVAY **455**

SOUDER **11756**

SOUDER **13755**

SOUDER A L'AUTOGENE **13758**

SOUDER A L'ETAIN **11741**

SOUDER EN BISEAU **7404**

SOUDER FORT **1573**

SOUDER PAR CONTACT **1785**

SOUDER PAR ENCOLLAGE **1785**

SOUDEUR **13778**

SOUDURE **13776**

SOUDURE **13767**

SOUDURE **13779**

SOUDURE **11083**

SOUDURE **11762**

SOUDURE (ASSEMBLAGE) EN T **1801**

SOUDURE (ASSEMBLAGE) EN T **1795**

SOUDURE (COMPOSITION FUSIBLE) **11755**

SOUDURE (TRAVAIL FAIT EN SOUDANT AVEC INTERPOSITION D'UN ALLIAGE) **11761**

SOUDURE A ENTAILLE **11635**

SOUDURE A FROID **2712**

SOUDURE A L'ARC ELECTRIQUE **681**

SOUDURE A L'ARGENT **11466**

SOUDURE A L'ECRASEMENT **8024**

SOUDURE A L'ETAIN **11742**

SOUDURE A L'ETAIN **11743**

SOUDURE A PAS DE PELERIN **901**

SOUDURE ALUMINOTHERMIQUE **12818**

SOUDURE AU BISMUTH **1274**

SOUDURE AU CHALUMEAU **11768**

SOUDURE AU FER A SOUDER **11769**

SOUDURE AU PLAFOND **8944**

SOUDURE AUTOGENE **834**

SOUDURE AUTOGENE **831**

SOUDURE AUTOGENE **5847**

SOUDURE AVEC APPORT DE FER-THERMIT **12818**

SOUDURE BIEN FAITE **6005**

SOUDURE BOUT A BOUT **1784**

SOUDURE BOUT A BOUT **1809**

SOUDURE BOUT A BOUT PAR RESISTANCE **1799**

SOUDURE BOUT A BOUT PAR RESISTANCE **13413**

SOUDURE BOUT A BOUT SANS SUREPAISSEUR **5481**

SOUDURE CLAIRE **7475**

SOUDURE CONCAVE **1792**

SOUDURE CONTINE **11086**

SOUDURE CONTINUE **3037**

SOUDURE D'ANGLE A CORDON PLAT **12075**

SOUDURE D'ANGLE A RANGEES ALTERNEES SYMETRIQUES **2210**

SOUDURE D'ANGLE CONCAVE **7561**

SOUDURE D'ANGLE CONCAVE **2878**
SOUDURE D'ANGLE CONVEXE **3077**
SOUDURE D'ANGLE DISCONTINUE A RANGEES ALTERNEES **12049**
SOUDURE D'ANGLE OUVERTE **8809**
SOUDURE D'UN TUBE **11085**
SOUDURE DE BORD **6133**
SOUDURE DE BORD EN DOUBLE U FERMEE **2524**
SOUDURE DE BORD EN DOUBLE-V SANS ECARTEMENT **2525**
SOUDURE DE BORD EN U FERMEE **2530**
SOUDURE DE BORD EN V SANS ECARTEMENT **2531**
SOUDURE DEPOSEE A LA MAIN **7988**
SOUDURE DISCONTINUE **7055**
SOUDURE DOUBLE FERMEE **1788**
SOUDURE DOUBLE J FERMEE **2523**
SOUDURE DOUBLE OUVERTE **8812**
SOUDURE EN ANGLE A PLAT **6583**
SOUDURE EN BISEAU **7405**
SOUDURE EN BOUCHON **9536**
SOUDURE EN CONGE **7561**
SOUDURE EN CONGE **2878**
SOUDURE EN CORNICHE **6591**
SOUDURE EN DEMI-V AVEC ECARTEMENT **1797**
SOUDURE EN DEMI-V AVEC ECARTEMENT **8822**
SOUDURE EN DEMI-V SANS ECARTEMENT **1789**
SOUDURE EN DEMI-V SANS ECARTEMENT **2528**
SOUDURE EN I AVEC ECARTEMENT **8824**
SOUDURE EN I SANS ECARTEMENT **2532**
SOUDURE EN I SANS ECARTEMENT DES BORDS **1791**
SOUDURE EN J AVEC ECARTEMENT **1798**
SOUDURE EN J AVEC ECARTEMENT **8823**
SOUDURE EN J SANS ECARTEMENT **2529**
SOUDURE EN K AVEC ECARTEMENT **1796**
SOUDURE EN K AVEC ECARTEMENT **8811**
SOUDURE EN K SANS ECARTEMENT **1787**
SOUDURE EN K SANS ECARTEMENT **2522**
SOUDURE EN PASSES SUPERPOSEES **13757**
SOUDURE EN POSITION A PLAT **4190**
SOUDURE EN POSITION PLAFOND **8941**
SOUDURE EN POSITION VERTICALE DESCENDANTE **4192**
SOUDURE EN POSITON VERTICALE MONTANTE **13419**
SOUDURE EXECUTEE SUR CHANTIER **5146**
SOUDURE FORTE **1576**
SOUDURE FORTE (COMPOSITION FUSIBLE) **6279**
SOUDURE FORTE PAR IMMERSION **3969**
SOUDURE HORIZONTALE **5948**
SOUDURE INTERMITTENTE **7054**
SOUDURE JAUNE **11880**
SOUDURE LINEAIRE DE FRACTION **1085**
SOUDURE MAL FAITE **920**
SOUDURE MONOPASSE **11485**

SOUDURE OXHYDRIQUE **8988**
SOUDURE OXY-ÁCETYLENIQUE **8979**
SOUDURE PAR ENCOLLAGE **1802**
SOUDURE PAR ETINCELAGE **1794**
SOUDURE PAR PERCUSSION **9183**
SOUDURE PAR POINTS **11959**
SOUDURE PAR RAPPROCHEMENT **1784**
SOUDURE PAR RAPPROCHEMENT **1802**
SOUDURE PAR RECOUVREMENT **7405**
SOUDURE PORTANTE **12357**
SOUDURE POUSSEE VERS LA GAUCHE **5539**
SOUDURE PROVISOIRE PAR POINTS **12590**
SOUDURE RESISTANTE A LA PRESSION **9838**
SOUDURE SUR BORDS DROITS **12032**
SOUDURE SUR BORDS RELEVES **5335**
SOUFFLAGE MAGNETIQUE DE L'ARC **665**
SOUFFLERIE **1361**
SOUFFLERIE **242**
SOUFFLERIE DE VAPEUR **12172**
SOUFFLET **1139**
SOUFFLET DE FORGE **1139**
SOUFFLETTE **272**
SOUFFLETTE A AIR **276**
SOUFFLETTES **9796**
SOUFFLURE **7523**
SOUFFLURE **1353**
SOUFFLURE **1351**
SOUFFLURE **5825**
SOUFFLURE **5836**
SOUFFLURES **1362**
SOUFFLURES **1330**
SOUFFLURES **11903**
SOUFFRE DORE D'ANTIMOINE **625**
SOUFRE **12449**
SOUFRE AMORPHE **476**
SOUFRE CRISTALLISE **3435**
SOUFRE EN CANON **12255**
SOUFRE SUBLIME **5464**
SOULEVEMENT D'UN FARDEAU **7559**
SOULEVEMENT DE LA SOUPAPE **7544**
SOULEVER **7542**
SOULEVEUSES DE BETTERAVES **12440**
SOULEVEUSES DE RACINES **10722**
SOUMETTRE UN CORPS A UNE CHARGE **12323**
SOUMISSION A UNE PRE-TENSION **830**
SOUPAPE **13453**
SOUPAPE **13472**
SOUPAPE **2617**
SOUPAPE **2284**
SOUPAPE **12436**
SOUPAPE A AIR **259**

SOUPAPE A BOULET **968**

SOUPAPE A CHARNIERE **5353**

SOUPAPE A CLAPET **5353**

SOUPAPE A CLAPET **7547**

SOUPAPE A CLOCHE **1136**

SOUPAPE A CLOCHE INTERIEURE **11579**

SOUPAPE A COMMANDE HYDRAULIQUE **6697**

SOUPAPE A COMMANDE PNEUMATIQUE **9557**

SOUPAPE A DEUX VOIES **13318**

SOUPAPE A DISQUE **4011**

SOUPAPE A DOUBLE SIEGE **4166**

SOUPAPE A FLOTTEUR **5433**

SOUPAPE A MANCHON **11579**

SOUPAPE A QUADRUPLE SIEGE **10060**

SOUPAPE A SIEGE CONIQUE **2951**

SOUPAPE A SIEGE PLAN **4011**

SOUPAPE A SIMPLE SIEGE **11497**

SOUPAPE A TROIS VOIES **12882**

SOUPAPE ANNULAIRE **10605**

SOUPAPE AUTOMATIQUE **852**

SOUPAPE CASSE-VIDE **13437**

SOUPAPE CHAMPIGNON **7547**

SOUPAPE COMMANDEE **8120**

SOUPAPE D EMISSION **4899**

SOUPAPE D'ADMISSION **6949**

SOUPAPE D'ADMISSION **201**

SOUPAPE D'ALIMENTATION **5076**

SOUPAPE D'ARRET **12282**

SOUPAPE D'ARRET DE VAPEUR **12186**

SOUPAPE D'ECHAPPEMENT **4899**

SOUPAPE D'ECHAPPEMENT **797**

SOUPAPE D'EQUERRE **534**

SOUPAPE D'EVACUATION **1358**

SOUPAPE D'INJECTION **6937**

SOUPAPE DE CORNOUAILLES **1136**

SOUPAPE DE DECHARGE **1358**

SOUPAPE DE DECHARGE **10441**

SOUPAPE DE DERIVATION **1819**

SOUPAPE DE DETENTE **4926**

SOUPAPE DE PRISE DE VAPEUR **12187**

SOUPAPE DE REDUCTION **10355**

SOUPAPE DE REFOULEMENT **3739**

SOUPAPE DE SECURITE **1593**

SOUPAPE DE SURETE **10891**

SOUPAPE DE SURETE A CHARGE DIRECTE **3632**

SOUPAPE DE SURETE A CHARGE DIRECTE A RESSORT **3987**

SOUPAPE DE SURETE A CONTRE-POIDS **13749**

SOUPAPE DE SURETE A GRANDE LEVEE **10890**

SOUPAPE DE SURETE A LEVIER ET A RESSORT **11985**

SOUPAPE DE SURETE A LEVIER ET CONTREPOIDS **7535**

SOUPAPE DE SURETE A RESSORT **11998**

SOUPAPE DE SURETE A RESSORT **11986**

SOUPAPE DE TROP PLEIN **8933**

SOUPAPE DE VIDANGE **1358**

SOUPAPE DESMODROMIQUE **852**

SOUPAPE DROITE **12317**

SOUPAPE ELECTROMAGNETIQUE **11774**

SOUPAPE EQUILIBREE **934**

SOUPAPE ETAGEE **8475**

SOUPAPE MULTIPLE **8476**

SOUPAPE PRESSION-DEPRESSION **9843**

SOUPAPE UNIQUE **11491**

SOUPAPE-BILLE **968**

SOUPAPES EN TETE **8943**

SOUPLESSE **5415**

SOURCE D'ELECTRICITE **11823**

SOURCE D'ENERGIE **11824**

SOURCE D'ERREURS **11825**

SOURCE DE COURANT A TENSION CONSTANTE (POUR SOUDAGE) **2979**

SOURCE LINEAIRE **7599**

SOURCE LUMINEUSE **7568**

SOURCE LUMINEUSE **11826**

SOUS FORME DE VAPEUR **6817**

SOUS-COUCHE (D'OXYDE) **12416**

SOUS-JOINT **12408**

SOUS-JOINT A FAIBLE DESORIENTATION **7767**

SOUS-LISSE **7795**

SOUS-PRODUIT **1820**

SOUS-PROGRAMME **12415**

SOUS-SOLEUSES **8344**

SOUS-SOLEUSES ET CHARRUES RIGOLEUSES **4090**

SOUS-STRUCTURE **12423**

SOUSTRACTION **12425**

SOUSTRACTION DE LA CHALEUR **4666**

SOUSTRAIRE **12424**

SOUSTRAIRE DE LA CHALEUR **13910**

SOUTE A CHARBON **2561**

SOUTE A CHARBON **1740**

SOUTE A MAZOUT **5706**

SOUTENEMENT **12153**

SOUTENENENT **12493**

SOUTENIR **12150**

SOUTENIR LE TAS **6524**

SOUTENU LIBREMENT SUR DEUX APPUIS **11474**

SOUTES A COMBUSTIBLE **8748**

SOUTIRAGE-VIDANGE **4225**

SPATH CALCAIRE **1845**

SPATH FLUOR **5479**

SPATH PESANT **1021**

SPATULE **11844**

SPECIALITE **11845**
SPECIALITE PHARMACEUTIQUE **9119**
SPECIMEN **10915**
SPECTRE **11863**
SPECTRE CANNELE **5487**
SPECTRE CONTINU **3036**
SPECTRE D'ABSORPTION **54**
SPECTRE DES RAYONS X **14001**
SPECTRE DISCONTINU **7610**
SPECTRE MAGNETIQUE **7902**
SPECTRE PRODUIT PAR PRISME **9878**
SPECTRE SECONDAIRE **11108**
SPECTRE SOLAIRE **11754**
SPECTROGRAPHE **11860**
SPECTROMETRE **11862**
SPECTROMETRE A RAYONS X **14000**
SPECTROMETRE DE MASSE **8029**
SPECTROSCOPE **11862**
SPECTROSCOPIE COLORIMETRIQUE **2739**
SPECTROSCOPIE DE LA FLAMME **5320**
SPERMACETI **11882**
SPHALERITE **1324**
SPHERE **11883**
SPHERE CREUSE **6549**
SPHERIQUE **11884**
SPHEROIDAL **8606**
SPHEROIDISATION **11896**
SPHEROMETRE **11897**
SPIEGEL **11898**
SPIEGEL **11899**
SPIEGELEISEN **11899**
SPIRALE **11912**
SPIRALE ARCHIMEDIENNE **690**
SPIRALE D'ARCHIMEDE **690**
SPIRALE HYPERBOLIQUE **6738**
SPIRALE LOGARITHMIQUE **7720**
SPIRE **2646**
STABILISATEUR **12042**
STABILISATEUR AUTOMATIQUE **847**
STABILISATION **12043**
STABILITE **12041**
STABILITE **4385**
STABILITE (CHIM.) **2313**
STABILITE A SEC **4352**
STABILITE DIMENSIONNELLE **3958**
STADE **12239**
STALAGMOMETRIE **12054**
STAND DE L'OUVRIER **8839**
STANDARD **12071**
STANDARD TELEPHONIQUE **12557**
STANDARDISATION **12091**

STANDARDISER **12090**
STANDARDS POUR TUYAUTERIES **9359**
STANNATE **12095**
STANNATE DE SODIUM **11732**
STARTER **2357**
STARTER AUTOMATIQUE **838**
STATION CENTRALE **9742**
STATION GENERATRICE **9742**
STATIONNER **12294**
STATIQUE **12133**
STATIQUE **12140**
STATIQUE GRAPHIQUE **6061**
STATOR **12146**
STAUFFER **12148**
STEARINE **12195**
STEATITE **11703**
STEATITE **5659**
STELLITE **12234**
STEREOMETRIE **12251**
STEREOMETRIQUE **12250**
STERILISATION DE L'EAU **12252**
STIBINE **622**
STOCK **12263**
STOCKER **12294**
STOECHIOMETRIE **12268**
STOECHIOMETRIQUE **12267**
STRAPONTIN **7261**
STRATE **7444**
STRATIFICATION **7382**
STRICTION **7424**
STRICTION **8515**
STRICTION **3043**
STRICTION **8516**
STRICTION **3045**
STRICTION **10361**
STRIE **5484**
STRIE **5486**
STRIE **11206**
STRIE **5061**
STRIES **10576**
STROBOSCOPE **12385**
STRONTIUM **12389**
STROPHOIDE **7722**
STRUCTURE **12394**
STRUCTURE **12395**
STRUCTURE **12800**
STRUCTURE A GRAINS FINS **5200**
STRUCTURE A GROS GRAINS **7869**
STRUCTURE BASALTIQUE **2758**
STRUCTURE BETA **1219**
STRUCTURE COMPACTE (A) **2514**

STRUCTURE CRISTALLINE **3430**
STRUCTURE DENDRITIQUE **3750**
STRUCTURE DES ALLIAGES **2058**
STRUCTURE DU COEUR **3165**
STRUCTURE DU GRAIN **6041**
STRUCTURE DUE A LA DEFORMATION PLASTIQUE **5461**
STRUCTURE FIBREUSE **5135**
STRUCTURE FINE **5210**
STRUCTURE FINE (R.X.) **5203**
STRUCTURE GAMMA **5797**
STRUCTURE GRANULAIRE **6052**
STRUCTURE HEXAGONALE COMPACTE **6458**
STRUCTURE LAMELLAIRE **7373**
STRUCTURE MOLECULAIRE **8361**
STRUCTURE MOSAIQUE **8403**
STRUCTURE RETICULAIRE **8539**
STRUCTURE ZONALE **979**
STYLET **9161**
STYLET TRACEUR **9161**
SUAGE **3322**
SUBDIVISION **12409**
SUBLIMATION **12412**
SUBLIME CORROSIF **8153**
SUBLIMER **12411**
SUBMERGER **12413**
SUBSTANCE **12419**
SUBSTANCE ACTIVATRICE **4744**
SUBSTANCE ACTIVATRICE **148**
SUBSTANCE ADHESIVE **177**
SUBSTANCE AGGLUTINANTE **1253**
SUBSTANCE ANTI-INCRUSTANTE **10978**
SUBSTANCE CALORIFUGE **6352**
SUBSTANCE CORROSIVE **3197**
SUBSTANCE EN POUDRE FINE **6789**
SUBSTANCE EPAISSISSANTE **12842**
SUBSTANCE FINEMENT BROYEE **6789**
SUBSTANCE HOMOGENE **6558**
SUBSTANCE USANTE **6107**
SUBSTANCES ETRANGERES **6813**
SUBSTANCES PHOSPHORESCENTES **9251**
SUBSTANCES RADIOACTIVES **10158**
SUBSTITUTION (MATH.) **12421**
SUCCEDANE **12420**
SUCESSION DES OPERATIONS DE SOUDAGE **13790**
SUCETTE **12332**
SUCRE DE RAISIN **6058**
SUIE **11816**
SUIF **12609**
SUIF DES BOEUFS **1128**
SUIF VEGETAL DE CHINE **2342**
SUINT **13933**

SUINTEMENT **12540**
SUINTEMENT **7488**
SUITE **2222**
SUITE **11203**
SULFATE **12444**
SULFATE ACIDE DE SOUDE **11717**
SULFATE ANHYDRE DE CHAUX **549**
SULFATE D'ALUMINIUM **426**
SULFATE D'AMMONIUM **471**
SULFATE D'ARGENT **11467**
SULFATE DE BARYUM **1003**
SULFATE DE CALCIUM **1856**
SULFATE DE CALCIUM **6190**
SULFATE DE COBALT **2615**
SULFATE DE CUIVRE **3475**
SULFATE DE FER **6093**
SULFATE DE MAGNESIUM **7889**
SULFATE DE NICKEL **8571**
SULFATE DE PLOMB **7473**
SULFATE DE POTASSE **9699**
SULFATE DE ZINC **14046**
SULFATE DOUBLE DE NICKEL ET D'AMMONIAQUE **8558**
SULFATE FERRIQUE **5103**
SULFATE MANGANEUX **7974**
SULFATE MERCUREUX **8162**
SULFATE MERCURIQUE **8158**
SULFATE NEUTRE DE SODIUM **11733**
SULFHYDRATE D'AMMONIUM **464**
SULFITE **12448**
SULFITE ACIDE **11718**
SULFITE DE SOUDE **11734**
SULFOCYANURE DE POTASSIUM **9701**
SULFURE **12446**
SULFURE D'AMMONIUM **472**
SULFURE DE BARYUM **1004**
SULFURE DE CARBONE **1964**
SULFURE DE FER **12447**
SULFURE DE POTASSE **9700**
SULFURE DOREE **625**
SULFURE EN CHAINE **2221**
SULFURE FERREUX/FERRIQUE **7153**
SULFURE ROUGE DE MERCURE **13535**
SULFURE STANNEUX **12103**
SULFURE STANNIQUE **12099**
SUPER-RESEAU **12475**
SUPERFICIE DE LA SOUDURE **4998**
SUPERPOSITION DE VIBRATIONS **12476**
SUPPLEMENT D'UN ANGLE **12482**
SUPPORT **12490**
SUPPORT **1041**
SUPPORT **6586**

SUPPORT 57
SUPPORT 3288
SUPPORT 1522
SUPPORT (CONSTR.) 1118
SUPPORT A COUTEAU 7325
SUPPORT D'AME SUPPORT DE NOYAU 2253
SUPPORT D'INFORMATION 12289
SUPPORT D'INFORMATION D'ENTREE 6960
SUPPORT DE FUSEE 11910
SUPPORT DE FUSEE 12230
SUPPORT DE PIECE A ESTAMPER 1305
SUPPORT METALLIQUE DE NOYAU 662
SUPPORT MOTEUR 4754
SUPPORTER 12491
SUPPRESSION 1337
SUPPRESSION DE LA FUMEE 11677
SUPPRESSION DES ZEROS 14033
SURCHARGE 8963
SURCHARGE 8953
SURCHARGER 8954
SURCHARGER UN MATERIEL 8962
SURCHAUFFAGE 12473
SURCHAUFFAGE 12460
SURCHAUFFE 8946
SURCHAUFFE 12473
SURCHAUFFER 12468
SURCHAUFFEUR 12472
SURCOMPRESSEUR 1456
SURCOUCHE 8952
SURCROIT DE DIMENSION 8961
SUREPAISSEUR 8948
SUREPAISSEUR 1695
SUREPAISSEUR D'USINAGE 358
SUREPAISSEUR D'USINAGE 7865
SURFACAGE 12522
SURFACE 12521
SURFACE 12496
SURFACE 2019
SURFACE 694
SURFACE 11120
SURFACE CAUSTIQUE 2121
SURFACE CHARGEE 12368
SURFACE COURBE 3498
SURFACE D'AFFLEUREMENT 5058
SURFACE D'APPUI 1115
SURFACE D'ENGRENEMENT 696
SURFACE DE CHAUFFE 6398
SURFACE DE CONTACT 2998
SURFACE DE CONTACT 12510
SURFACE DE GLISSEMENT 11608
SURFACE DE GLISSEMENT 11585

SURFACE DE GRILLE 6068
SURFACE DE GUIDAGE 6166
SURFACE DE L'EMPREINTE 12512
SURFACE DE LA CIBLE (OFFERTE AUX RAYONNEMENTS) 12683
SURFACE DE LA COUPE 12511
SURFACE DE LIQUATION 7634
SURFACE DE NIVEAU 7527
SURFACE DE PORTEE D'UN TOURILLON 1116
SURFACE DE REFRIGERATION 3092
SURFACE DE REFROIDISSEMENT 3092
SURFACE DE REVOLUTION 12513
SURFACE DE ROULEMENT 10705
SURFACE DE ROULEMENT D'UNE ROUE 13173
SURFACE DEVALOPPABLE 3817
SURFACE DIFFUSIVE 3935
SURFACE EQUIPOTENTIELLE 7527
SURFACE FILTRANTE SATUREE 2360
SURFACE FROTTANTE 5687
SURFACE GAUCHE 11535
SURFACE HELICOIDALE 6428
SURFACE INTERIEURE D'UN TUYAU 6954
SURFACE LATERALE D'UN CONE 2949
SURFACE LATERALE D'UN CYLINDRE 3590
SURFACE MATE 8060
SURFACE METALLIQUE DECAPEE 2470
SURFACE METALLIQUE POLIE 1611
SURFACE MIROITANTE 11866
SURFACE PLANE 9441
SURFACE POLIE 6142
SURFACE PORTANTE 1101
SURFACE PORTANTE 1115
SURFACE RAYONNANTE 10142
SURFACE REFLECHISSANTE 10385
SURFACE REFROIDISSANTE 3092
SURFACE RUGUEUSE 10784
SURFACE RUGUEUSE 10779
SURFACE TRES POREUSE 6644
SURFACER 12497
SURFUSION 12465
SURMOULAGE 4383
SURPASSER 8928
SURPRESSEUR 1456
SURPRESSION 8955
SURREFROIDISSEMENT 12465
SURSATURATION 12461
SURSATURATION 12479
SURSATURER 12477
SURSOUFFLAGE BESSEMER 1211
SURSTRUCTURE 12475
SURVEILLANCE 12481

SURVEILLANCE DES CHAUDIERES **9193**
SURVEILLANT **8958**
SURVEILLER **12474**
SURVIEILLISSEMENT **8923**
SURVOLTAGE **4881**
SURVOLTAGE **8967**
SURVOLTEUR **1456**
SUSCEPTIBILITE MAGNETIQUE **7918**
SUSCEPTIBLE D'ETRE BREVETEE **9122**
SUSCEPTIBLE D'ETRE COULE **1914**
SUSCEPTIBLE D'ETRE LAMINE **1915**
SUSPENDU LIBREMENT **5649**
SUSPENDU SUR UN COUTEAU **6669**
SUSPENDU SUR UN PIVOT **8432**
SUSPENSION **12526**
SUSPENSION A COUTEAU **7326**
SUSPENSION A LA CARDAN **1998**
SUSPENSION A PIVOT **9570**
SUSPENSION A ROTULE **958**
SUSPENSION BIFILAIRE **1240**
SYENITE **12565**
SYMBOLE (CHIM.) **2315**
SYMBOLES DE SOUDAGE **13791**
SYMETRIE **12567**
SYMETRIQUE **12566**
SYNCHRO **12568**
SYNCHRONE **12570**
SYNCHRONISME **12569**
SYNTHESE **12572**
SYNTHETIQUE **12573**
SYNTONIE **12574**
SYPHON **11517**
SYSTEME A BOUCLE DE RETOUR **2533**
SYSTEME ANTI-DERAPANT **8645**
SYSTEME ARTICULE **7628**
SYSTEME C.G.S. **2159**
SYSTEME CENTIMETRE-GRAMME-SECONDE **2159**
SYSTEME DE COMMANDE EN BOUCLE OUVERTE **8817**
SYSTEME DE COMMANDE NUMERIQUE **8695**
SYSTEME DE COORDONNEES **12575**
SYSTEME DE COULEE ET D'ALIMENTATION **5870**
SYSTEME DE LENTILLES **12576**
SYSTEME DE MESURE **12578**
SYSTEME DE MESURE ABSOLUE **31**
SYSTEME DE PEINTURE A DEUX COMPOSANTS **13311**
SYSTEME DE POIDS **12580**
SYSTEME DECIMAL **3665**
SYSTEME DES ALLIAGES **381**
SYSTEME DES CRISTAUX **3431**
SYSTEME ENSEMBLE DE LEVIERS **12577**
SYSTEME METRIQUE **8229**

SYSTEME MULTIPLE **8472**
SYSTEME OBJECTIF **8708**
SYSTEME OCULAIRE **4977**
SYSTEME OPTIQUE **8848**
SYSTEME PYRAMIDAL **10040**
SYSTEME QUASI-BINAIRE **9958**
SYSTEME SIEMENS **10412**
SYSTEME TERNAIRE **12766**
SYSTEME TRICLINIQUE **13199**
T.H. **7997**
TABLE **12584**
TABLE A DESSIN **4243**
TABLE D'ENCLUME **4995**
TABLE D'EXAMEN **12773**
TABLE DE CONVERSION **3069**
TABLE DE DESSINATEUR **4243**
TABLE DE TRANSFORMATION **3069**
TABLE DES LOGARITHMES **7721**
TABLE DRESSEE **12516**
TABLEAU **2273**
TABLEAU (SERIE DE NOMBRES) **12584**
TABLEAU DE BORD **6993**
TABLEAU DE BORD **3610**
TABLEAU DE COMMANDE **3054**
TABLEAU DISTRIBUTEUR **12557**
TABLEAU GENERAL **12557**
TABLES DE FIXATION **8290**
TABLIER **5692**
TABLIER **655**
TABLIER DU MOTEUR **5251**
TACHE DE NUIT **8575**
TACHE PAR CRISTAUX DE SULFURE DE CUIVRE **3429**
TACHES D'EAU (DE PLUIE) **10178**
TACHYMETRE **10556**
TACHYMETRE **11869**
TACHYMETRE **12587**
TACHYMETRE ENREGISTREUR **12586**
TAILLAGE **3523**
TAILLAGE DES LIMES **3529**
TAILLANDERIE **13006**
TAILLE **3523**
TAILLE D'UNE LIME **3518**
TAILLE CROISEE **4134**
TAILLE CROISEE **4132**
TAILLE SIMPLE D'UNE LIME **11482**
TAILLER **3504**
TAILLER EN BIAIS **1222**
TAILLER EN SIFFLET **1222**
TAILLER LES LIMES **3507**
TAILLEUR DE LIMES **5151**
TAILLEUSE D'ENGRENAGES **5891**

TALC **5659**
TALC **12608**
TALC EN POUDRE **12607**
TALON **8667**
TALON D'UNE CLAVETTE **6318**
TALON DE CLAVETTE **7286**
TALUS **11627**
TAMBOUR **10290**
TAMBOUR **10671**
TAMBOUR **4333**
TAMBOUR **13817**
TAMBOUR **1532**
TAMBOUR **3579**
TAMBOUR A CHAINE **2202**
TAMBOUR D'ENROULEMENT **10737**
TAMBOUR DE FREIN **1533**
TAMBOUR DE NETTOTAGE **13261**
TAMBOUR POUR CHAINE-CABLE **2202**
TAMIS **11018**
TAMIS **10572**
TAMIS **11422**
TAMISAGE **11424**
TAMISER **11423**
TAMPON **9529**
TAMPON **1686**
TAMPON **12785**
TAMPON **9011**
TAMPON **9531**
TAMPON **3267**
TAMPON (COUVERCLE D'UN TAMPON) **3262**
TAMPON A ECHELONS DE TOLERANCE **9907**
TAMPON BOUCHON **1483**
TAMPON D'ALESAGE **11049**
TAMPON DE CONTROLE D'ALESAGE **6531**
TAMPON DE TROU D'HOMME **7977**
TAMPON ET BAGUE **3581**
TAMPON ET LUNETTE DE CALIBRE **3581**
TAMPON FILETE **12367**
TAMPON FILETE FEMELLE **2721**
TAMPON FILETE MALE **11049**
TAMPON OBTURATEUR **1326**
TAMPONS DE CONTROLE **5769**
TAN COMPRIME **12612**
TAN EN MOTTES **12612**
TAN EPUISE **12635**
TANGENTE **12615**
TANGENTE (GEOM.) **12616**
TANGENTE AU SOMMET **12617**
TANNATE **12633**
TANNATE DE SODIUM **11735**
TANNE AU CHROME **2377**

TANNIN **5780**
TANTALE **12634**
TAPAGE D'UNE POMPE **7331**
TAPURE **5304**
TAPURE **3278**
TAPURE DE L'ACIER TREMPE **3282**
TAPURE DE RETRAIT **11395**
TAQUET **12278**
TAQUET **2485**
TAQUET **4301**
TAQUET DE MISE A LA TERRE **4419**
TARAGE **5886**
TARAGE DE SOUPAPE **13469**
TARARES CRIBLEURS T. TRIEURS **3177**
TARAUD **12640**
TARAUD **11044**
TARAUD A LA MACHINE **7858**
TARAUD A MAIN **6257**
TARAUD CONIQUE **5268**
TARAUD CYLINDRIQUE **9534**
TARAUD DEMI-CONIQUE **11102**
TARAUD EBAUCHEUR **5268**
TARAUD FINISSEUR **9534**
TARAUD FINISSEUR **1499**
TARAUD INTERMEDIAIRE **11102**
TARAUDAGE **4277**
TARAUDAGE **12868**
TARAUDAGE **12678**
TARAUDAGE (INTERIEUR) **7078**
TARAUDER **12636**
TARAUDEUSE **11066**
TARAUDEUSE **12677**
TARAUDEUSE **7858**
TARER **5872**
TARER UNE SOUPAPE **11216**
TARIERE **818**
TARTRATE **12691**
TARTRATE EMETIQUE **9675**
TARTRATE D'AMMONIAQUE **473**
TARTRATE DE POTASSE ET D'ANTIMOINE **9675**
TARTRATE DE POTASSE ET DE SOUDE **11729**
TARTRE **119**
TARTRE DES CHAUDIERES **10969**
TARTRE STIBIE **9675**
TARTRIFUGE **4041**
TAS **632**
TAS **4116**
TAS AVEC CONTRE-BOUTEROLLE **11032**
TASSEAU **632**
TASSEMENT **11355**
TAUX **10226**

TAUX D'EVAPORATION **1402**
TAUX DE COMPRESSION **2860**
TAUX DE COURANT PRIMAIRE **9850**
TAUX DE TRAVAIL **13958**
TAXES DE BREVET **9115**
TE **12697**
TE A TROIS DIRECTIONS **12699**
TE DE DESSIN **12581**
TECHNETIUM **8040**
TECHNIQUE **12695**
TECHNIQUE **12694**
TEINTE DE RECUIT **12732**
TEINTE DE REVENU **12732**
TELE-THERMOMETRE **4069**
TELECOMMANDE **10448**
TELEGRAMME **1830**
TELEGRAPHE **12707**
TELEGRAPHIE **12709**
TELEGRAPHIE SANS FIL **13907**
TELEJAUGEAGE **10449**
TELEPHONE **12710**
TELEPHONIE **12712**
TELEPHONIE SANS FIL **13908**
TELESCOPE **10386**
TELLURE **12716**
TEMOIN D'ALLUMAGE **6769**
TEMPERATURE **12721**
TEMPERATURE ABAISSANTE **5025**
TEMPERATURE ABSOLUE **35**
TEMPERATURE AMBIANTE **12729**
TEMPERATURE AMBIANTE **10713**
TEMPERATURE CONSTANTE **2977**
TEMPERATURE CRITIQUE **3352**
TEMPERATURE CROISSANTE **10618**
TEMPERATURE D'AFFINAGE **10381**
TEMPERATURE D'ALLUMAGE **6767**
TEMPERATURE D'EBULLITION **1427**
TEMPERATURE D'EQUILIBRE **4791**
TEMPERATURE D'ETUDE **3798**
TEMPERATURE D'INFLAMMATION **12727**
TEMPERATURE D'INFLAMMATION **6767**
TEMPERATURE DE CALCUL **3798**
TEMPERATURE DE FINISSAGE **5227**
TEMPERATURE DE FORGEAGE **5565**
TEMPERATURE DE FORGEAGE **2744**
TEMPERATURE DE FORGEAGE **5560**
TEMPERATURE DE FUSION **6360**
TEMPERATURE DE L'AIR **12728**
TEMPERATURE DE LA PASSE INTERMEDIAIRE **7086**
TEMPERATURE DE RECRISTALLISATION **10298**
TEMPERATURE DE RECUIT **11700**

TEMPERATURE DE REDUCTION **10354**
TEMPERATURE DE REPERE **10369**
TEMPERATURE DE SATURATION **10945**
TEMPERATURE DE SERVICE **8836**
TEMPERATURE DE TRANSFORMATION **13114**
TEMPERATURE DE TRAVAIL **8836**
TEMPERATURE DE VAPORISATION **4865**
TEMPERATURE ELEVEE **4658**
TEMPERATURE EUTECTIQUE **4858**
TEMPERATURE FINALE **5191**
TEMPERATURE INITIALE **6931**
TEMPERATURE MAXIMUM **8068**
TEMPERATURE MINIMUM **8306**
TEMPERATURE MOYENNE **855**
TEMPERATURE NORMALE **10713**
TEMPERATURE ORDINAIRE DE LA SALLE **10713**
TEMPERATURE REDUITE **10344**
TEMPERATURE SUPERFICIELLE MAXIMALE **6654**
TEMPERATURES DE VIEILLISSEMENT **230**
TEMPS **12387**
TEMPS **12943**
TEMPS ACTIF **13410**
TEMPS D'ADMISSION **7008**
TEMPS D'ECOULEMENT **5459**
TEMPS D'ECOULEMENT DE COURANT **6382**
TEMPS D'INVERSION **10533**
TEMPS DE CHANGEMENT D'OUTIL **12999**
TEMPS DE COMPRESSION **2864**
TEMPS DE MAINTIEN DE L'EFFORT **6523**
TEMPS DE POSE **4944**
TEMPS DE POSITIONNEMENT **9651**
TEMPS DE REFROIDISSEMENT **3082**
TEMPS DE REPOS **8734**
TEMPS DE SOUDAGE EFFECTIF **6382**
TEMPS MORT **4191**
TEMPS MORT **6753**
TENACE **13067**
TENACITE **12737**
TENACITE **13071**
TENACITE A L'ENTAILLE **8671**
TENACITE DES BARREAUX ENTAILLES **12738**
TENACITE EXTREME **13327**
TENAILLE **9328**
TENAILLE **1298**
TENAILLE (S) **12987**
TENAILLE A CHANFREIN **13565**
TENAILLE A MORS COUPANTS **3532**
TENAILLE A RIVETS **10630**
TENAILLE A SOUDER **11767**
TENDEUR **12334**
TENDEUR A GALET **12334**

TENDEUR A VIS **13285**
TENDEUR DE CHAINE **2219**
TENDEUR DE CHAINE **2218**
TENDEUR DE COURROIES **1161**
TENDRE UN RESSORT **10032**
TENEUR **3011**
TENEUR EN CARBONE **1947**
TENEUR EN CENDRES **758**
TENEUR EN EAU **13645**
TENEUR EN SEL **10906**
TENIR (RIVETAGE) **6524**
TENIR LE COUP **6524**
TENON **12740**
TENON D'AGRAFAGE **4175**
TENSION **9808**
TENSION **12764**
TENSION **12753**
TENSION (VOLTAGE) DE L'ARC **676**
TENSION A VIDE **8808**
TENSION AUX ELECTRODES **9709**
TENSION CONSTANTE **2978**
TENSION CRITIQUE **3351**
TENSION D'EQUILIBRE D'UNE ELECTRODE **4789**
TENSION D'EQUILIBRE D'UNE REACTION **4790**
TENSION D'ETIRAGE **12370**
TENSION DE CISAILLEMENT **11286**
TENSION DE CLAQUAGE **1582**
TENSION DE COULAGE **2078**
TENSION DE COULEE **2081**
TENSION DE DECOMPOSITION **3675**
TENSION DE FLEXION **5425**
TENSION DE FORMATION **5576**
TENSION DE LA COURROIE **1162**
TENSION DE SERVICE **13957**
TENSION DE SERVICE DE L'ARC **679**
TENSION DE SOLUTION ELECTROLYTIQUE **4609**
TENSION DU COURANT **13593**
TENSION INTERNE **7074**
TENSION PRINCIPALE **9869**
TENSION SUPERFICIELLE **7033**
TENSION SUPERFICIELLE **12519**
TENSION THERMIQUE **12814**
TENSIONS INTERNES DANS LA FONTE **7075**
TENUE DE ROUTE EN COTE **2498**
TENUE DE ROUTE EN VIRAGE **3180**
TERBIUM **12760**
TEREBENTHINE **3415**
TEREBENTHINE DE VENISE **13527**
TERNIR (SE) **12685**
TERNISSEMENT **4369**
TERRAIN A BATIR **1699**

TERRAIN ARGILEUX **2466**
TERRAIN GRAVELEUX **6076**
TERRAIN SABLEUX **10930**
TERRASSEMENTS **4420**
TERRAZZO **12768**
TERRE **4409**
TERRE (ELECTR.) **4415**
TERRE A BRIQUES **1598**
TERRE A FOULON **5727**
TERRE CUITE **12767**
TERRE D'INFUSOIRES **7292**
TERRE DE DIATOMEES **3880**
TERRE DE PIPE **9367**
TERRE GLAISE **7691**
TERRE POURRIE **7292**
TERRE REFRACTAIRE **5241**
TERRE RUBRIQUE **10340**
TERRE VEGETALE **8423**
TERRES ALCALINES **336**
TERRES RARES **10210**
TEST D'AJUSTEMENT **6006**
TEST DE MCQUAID-EHN **8071**
TETE A SIX PANS D'UNE VIS **6461**
TETE CARREE D'UNE VIS **10309**
TETE CYLINDRIQUE D'UNE VIS **3574**
TETE D'EPINGLE **9317**
TETE D'UN BOULON **1436**
TETE D'UN MARTEAU **4993**
TETE D'UNE CLAVETTE **6318**
TETE D'UNE SOUPAPE **13463**
TETE D'UNE VIS **1436**
TETE DE BIELLE **2963**
TETE DE BIELLE A CAGE FERMEE **1509**
TETE DE BIELLE A CAGE OUVERTE **8007**
TETE DE DISTILLATION **5267**
TETE DE FRAISAGE (OU DE COUPE) **3531**
TETE DE JAUGE **5877**
TETE DE LA DENT **9569**
TETE DE LA TIGE D'EXCENTRIQUE **4432**
TETE DE PERCAGE MULTIBROCHES **8458**
TETE DE POTEAU **2749**
TETE DE RIVET **10624**
TETE DE SOUDAGE **13783**
TETE DE SOUDAGE UNIVERSELLE **13390**
TETE DU CYLINDRE **3569**
TETE FRAISEE **3250**
TETE GOUTTE DE SUIF D'UNE VIS **4673**
TETE MOLETEE D'UNE VIS **8281**
TETE MULTI-BROCHES **8459**
TETE NOYEE D'UNE VIS **3250**
TETE PLATE D'UNE VIS **5394**

TETE RONDE D'UNE VIS **1813**	THYRISTOR **12913**
TETES DE MOINEAU **2328**	TIGE **11257**
TETRACHLORURE **12096**	TIGE **10663**
TETRACHLORURE DE CARBONE **1960**	TIGE **10660**
TETRAEDRE **12797**	TIGE **12237**
TETRAIOD-FLUORESCEINE **4831**	TIGE A VIS EXTERIEURE **8915**
TETRATOMIQUE **12798**	TIGE A VIS INTERIEURE **6981**
TEXTURE **12800**	TIGE BILATERALE DE PISTON **9387**
TEXTURE **12395**	TIGE D'EXCENTRIQUE **4431**
TEXTURE CRISTALLINE **3434**	TIGE D'UN BOULON **11260**
TEXTURE FIBREUSE **5141**	TIGE D'UNE CLAVETTE **11262**
TEXTURE GRENUE **6053**	TIGE D'UNE VIS **11260**
TEXTURE HETEROGENE **6452**	TIGE DE CREMAILLERE **10124**
TEXTURE HOMOGENE **6560**	TIGE DE CULBUTEURS **10031**
TEXTURE LAMELLEUSE **7374**	TIGE DE GUIDAGE **6164**
THALLIUM **12801**	TIGE DE PISTON TRAVERSANTE **9387**
THEORIE ATOMIQUE **805**	TIGE DE PISTON UNILATERALE **11472**
THEORIE DE LA DISTORSION DU RESEAU **7434**	TIGE DE RACCORDEMENT **2962**
THEORIE DES CASSURES **5616**	TIGE DE SOUPAPE **12236**
THERMIT **12817**	TIGE DU CULBUTEUR **10026**
THERMITE DE FONTE **2036**	TIGE DU PISTON **9383**
THERMO-ELECTRICITE **12826**	TIGE DU RIVET **11263**
THERMO-ELECTRIQUE **12823**	TIGE DU THERMOMETRE **13255**
THERMOCHIMIE **12822**	TIGE FILETEE **11062**
THERMOCOUPLE **12827**	TIGE FILETEE DE CHAPEAU **1453**
THERMODYNAMIQUE **12828**	TIGE FIXE **8623**
THERMODYNAMIQUE **12829**	TIGE MONTANTE **10616**
THERMOGRAPHE **12832**	TIGE VERTICALE **6018**
THERMOGRAPHE **10293**	TIMBRAGE **9828**
THERMOMETALLURGIE **4349**	TIMBRE **5886**
THERMOMETRE **12726**	TIMBRE D'UN RESERVOIR (A. PRESSION) **8614**
THERMOMETRE **12833**	TIMBRE D'UNE SOUPAPE **9828**
THERMOMETRE A ALCOOL **11922**	TIMBRE-AVERTISSEUR **1137**
THERMOMETRE A GAZ **265**	TIMBRER **5872**
THERMOMETRE A MAXIMA ET MINIMA **8063**	TIMONERIE **12232**
THERMOMETRE A MERCURE **8152**	TIRAGE **4196**
THERMOMETRE CENTIGRADE **2186**	TIRAGE A SOUFFLERIE **8098**
THERMOMETRE DIFFERENTIEL **3920**	TIRAGE ARTIFICIEL **8098**
THERMOMETRE ENREGISTREUR **10293**	TIRAGE D'AIR **257**
THERMOMETRE ETALON **12085**	TIRAGE D'AIR DES NOYAUX **3166**
THERMOMETRE FAHRENHEIT **5020**	TIRAGE D'UNE CHEMINEE **4212**
THERMOMETRE GRADUE SUR TIGE **12235**	TIRAGE FORCE **5533**
THERMOMETRE REAUMUR **10268**	TIRAGE INDUIT **6890**
THERMOMETRE-FRONDE **13826**	TIRAGE NATUREL **8501**
THERMOSCOPE **12837**	TIRANT **12917**
THERMOSTAT **12838**	TIRANT D'ENLEVEMENT **5772**
THERMOSTAT **12722**	TIRANTS (DE SPHERE) **1520**
THIOSULFATE **12852**	TIRE-FOND **11043**
THIXOTROPIE **12854**	TIRE-LIGNE **4241**
THORIUM **12856**	TIRE-LIGNE A POINTILLER **4126**
THULIUM **12910**	TIREFONDS **11575**

TIRER **4215**
TIRETTE **5772**
TIROIR **11592**
TIROIR A COQUILLE **3592**
TIROIR A GRILLE **6100**
TIROIR A PISTON **9390**
TIROIR CYLINDRIQUE **9390**
TIROIR EQUILIBRE **9390**
TIROIR PLAT **9426**
TISSU CAOUTCHOUTE **10818**
TISSU D'AMIANTE **748**
TISSU IMPERMEABLE **13690**
TITANE **12970**
TITRAGE **12974**
TITRAGE **12968**
TITRE D'UNE SOLUTION **12354**
TITRER **12973**
TOBOGGAN **11914**
TOC D'ENTRAINEMENT **4301**
TOC LIMITANT LA COURSE **12278**
TOILE **7619**
TOILE A CALQUER **13083**
TOILE A VOILE **1906**
TOILE CAOUTCHOUTEE **10818**
TOILE D'AMIANTE **748**
TOILE D'ARAIGNEE **2616**
TOILE D'EMBALLAGE **6449**
TOILE D'UN VOLANT **4013**
TOILE D'UNE ROUE **4013**
TOILE DE JUTE **6449**
TOILE EMERI **4693**
TOILE METALLIQUE **5888**
TOILE METALLIQUE **13893**
TOILE TRANSPORTEUSE **1164**
TOILE VERREE **5958**
TOIT **3667**
TOIT **10708**
TOIT BOMBE **4120**
TOIT BOMBE **4039**
TOIT CONIQUE **2947**
TOIT FIXE **5296**
TOIT SURBAISSE **13046**
TOIT SUSPENDU **12524**
TOLE **11315**
TOLE **11298**
TOLE **11305**
TOLE (D'ACIER) ALUMINEE **441**
TOLE (POUR CONSTRUCTION) NAVALE **11356**
TOLE A GRAIN ORIENTE **6036**
TOLE A TROUS FORES **4274**
TOLE A TROUS POINCONNES **10003**

TOLE AJOUREE **9189**
TOLE BOMBEE **1679**
TOLE CANNELEE **2252**
TOLE D'ACIER **12213**
TOLE D'ACIER **11325**
TOLE D'USURE **13716**
TOLE DE BLINDAGE **11343**
TOLE DE CHAUDIERE **1416**
TOLE DE CONSTRUCTION **1109**
TOLE DE CUIVRE **3127**
TOLE DE FER **11309**
TOLE DE LAITON **1569**
TOLE DE RENFORT **10430**
TOLE DE RENFORT **12257**
TOLE DE RENFORT OU DOUBLANTE **11343**
TOLE DE ROBE **11335**
TOLE DE TRANSFORMATEUR **4544**
TOLE DECOUPEE A JOUR **9189**
TOLE DEFLECTRICE **921**
TOLE DOUBLANTE **10430**
TOLE DRESSEE **5395**
TOLE ELECTRIQUE **4544**
TOLE ELECTRIQUE **11448**
TOLE EMBOUTIE **1679**
TOLE EN SAILLIE **9911**
TOLE EPAISSE **9477**
TOLE ETAMEE **12957**
TOLE FACONNEE **11846**
TOLE FINE **11304**
TOLE FINE **12850**
TOLE FORTE **9477**
TOLE FORTE **6409**
TOLE GALVANISEE **5785**
TOLE GAUFREE **12769**
TOLE GAUFREE **2285**
TOLE LAMINEE **10674**
TOLE LAMINEE A CHAUD **6637**
TOLE LAMINEE A FROID **2696**
TOLE LARMEE **10181**
TOLE MARGINALE **11531**
TOLE MARGINALE **8002**
TOLE NOIRE **1294**
TOLE NOIRE **1296**
TOLE ONDULEE **3198**
TOLE PERFOREE **9189**
TOLE PLAQUEE **2445**
TOLE PLOMBEE **7460**
TOLE POUR EMBOUTISSAGE **9478**
TOLE POUR LES INDUITS DES DYNAMOS **708**
TOLE POUR REVETEMENT DE SOL **5447**
TOLE POUR TRANSFORMATEURS **13116**

TOLE PROFILEE **11846**	TORSION **13054**
TOLE STRIEE **10574**	TORSION **13051**
TOLE STRIEE **2325**	TORSION ALTERNATIVE **10414**
TOLE STRIEE **2285**	TORSION D'UNE CORDE **13305**
TOLE STRIEE **2252**	TORSION LANG **7389**
TOLE TECHNOLOGIQUE **10842**	TOTALISER LES DIAGRAMMES **2765**
TOLE TERNE **1296**	TOUCHER (GEOM.) **13066**
TOLE TRES MINCE **12593**	TOUPIE **9901**
TOLE ZINGUEE **5785**	TOUR **10557**
TOLE-SUPPORT **12495**	TOUR **7427**
TOLERANCE **12982**	TOUR (MACHINE-OUTIL) **13289**
TOLERANCE **358**	TOUR A ARBRE A CAMES **1897**
TOLERANCE **2483**	TOUR A ARBRE COUDE **3316**
TOLERANCE ADMISE **359**	TOUR A ARBRES **11246**
TOLERANCE D'USINAGE **359**	TOUR A BANDAGES **13321**
TOLERANCE LIMITE **7587**	TOUR A BOIS **13922**
TOLERANCES D'USINAGE **4988**	TOUR A BOULONS **12401**
TOLES (ORDRE DE MISE EN PLACE DES) **9500**	TOUR A BRIDES **5334**
TOLES DECOUPEES **11266**	TOUR A CHARIOT **11590**
TOLES ENROBEES **2586**	TOUR A CHARIOTER **11590**
TOLES ETIREES **4250**	TOUR A COMMANDES PAR VIS-MERE **7428**
TOLES MINCES EN ACIER ALLIE **380**	TOUR A COPIER **3138**
TOLUENE **12983**	TOUR A CYCLES AUTOMATIQUES **840**
TOLUOL **12983**	TOUR A CYLINDRER **9428**
TOMBAC **12984**	TOUR A DECOLLETER **11584**
TOMBAGE **1088**	TOUR A DECOLLETER **3540**
TOMBER EN DELIQUESCENCE **3732**	TOUR A DECOLLETER LES ECROUS **8699**
TOMBER LES BORDS **1084**	TOUR A DEGAGER **913**
TOMBEREAUX **2017**	TOUR A DEPOUILLER **913**
TONDEUSES A GAZON **7439**	TOUR A DETALONNER **10443**
TONNE (1000 KG) **12986**	TOUR A EBAUCHER **10788**
TONNEAU A POLIR **9602**	TOUR A EPROUVETTES **12771**
TONNEAU DE FINISSAGE **13261**	TOUR A ESSIEUX **881**
TONNEAU DE NETTOYAGE **13261**	TOUR A FILETER **12859**
TOPAZE **13038**	TOUR A FILETER **11029**
TOPOCHIMIE **13039**	TOUR A LINGOT **6922**
TORCHE **13042**	TOUR A METAUX **8191**
TORCHE A HELIUM **6424**	TOUR A OVALES **8919**
TORCHERE **5820**	TOUR A PEDALE **5520**
TORCHERE **5355**	TOUR A PLATEAUX **13171**
TORCHERE DE RAFFINERIE **10376**	TOUR A POINTE **2150**
TORE **13043**	TOUR A REPOUSSER ET A LISSER **11911**
TORE CIRCULAIRE **3588**	TOUR A REPRODUIRE **3138**
TORON **12336**	TOUR A REVOLVER **1931**
TORON D'UNE CORDE **12339**	TOUR AUTOMATIQUE **846**
TORON METALLIQUE **13902**	TOUR AUTOMATIQUE A BROCHES MULTIPLES **8469**
TORONNAGE **12340**	TOUR AUTOMATIQUE TRAVAILLANT EN BARRE **983**
TORONNER **12337**	TOUR D'UNE HELICE **13278**
TORONNEUSE **12341**	TOUR DE CRAQUAGE **3287**
TORONS EN ACIER POUR CABLES **12217**	TOUR DE FRACTIONNEMENT **5614**
TORS **13305**	TOUR DE REFROIDISSEMENT **3094**

TOUR DE REPRISE MULTIBROCHES **8485**
TOUR DE SPIRE **2646**
TOUR EN L'AIR **5005**
TOUR EN L'AIR **5001**
TOUR EN L'AIR A PLATEAU HORIZONTAL **1469**
TOUR FRONTAL (TOUR EN L'AIR) **5697**
TOUR MINUTE **10814**
TOUR MONOBROCHE AUTOMATIQUE A TOURELLE **13298**
TOUR PARALLELE **9428**
TOUR POUR FUSEE D'ESSIEU **880**
TOUR POUR METAUX LEGERS **7565**
TOUR POUR ROUES **13816**
TOUR REVOLVER **13297**
TOUR VERTICAL **13545**
TOURBE **9147**
TOURBE COMPRIMEE **9804**
TOURBE MENUE **9726**
TOURBE MOTTIERE **4364**
TOURBILLON **13606**
TOURBILLONNEMENT **13827**
TOURELLES PORTE-OUTILS **13004**
TOURIE **1981**
TOURILLON **7256**
TOURILLON **11893**
TOURILLON **9420**
TOURILLON A CANNELURES **12906**
TOURILLON A FOURCHETTE **3391**
TOURILLON CROCHET DE LEVAGE **7558**
TOURILLON D EXTREMITE **4721**
TOURILLON D'APPUI **9421**
TOURILLON D'ARTICULATION **7242**
TOURILLON DE BUTEE CANNELE **12906**
TOURILLON DE BUTEE POUR CRAPAUDINE **9421**
TOURILLON DE CROSSE **3391**
TOURILLON FRONTAL **4721**
TOURILLON INTERMEDIAIRE **8513**
TOURILLON SERVANT D'AXE DE ROTATION **13242**
TOURILLON-AXE POUR CHAPE-FOURCHETTE **3391**
TOURMALINE **13072**
TOURNAGE **13287**
TOURNAGE CONIQUE **12668**
TOURNAGE DES SURFACES BOMBEES **13288**
TOURNAGE DES SURFACES CONIQUES **12668**
TOURNAGE DES SURFACES FIGUREES **8869**
TOURNAGE DES SURFACES SPHERIQUES DES CORPS RONDS **13291**
TOURNE-A-GAUCHE **12638**
TOURNE-A-GAUCHE POUR AVOYER **10957**
TOURNE-A-GAUCHE POUR DONNER DU PAS/DE VOIE A UNE SCIE **10957**
TOURNER **10763**

TOURNER **13276**
TOURNER DES CORPS RONDS **13281**
TOURNER DES SURFACES BOMBEES **13277**
TOURNER DES SURFACES FIGUREES **13280**
TOURNER DES SURFACES OBLIQUES/RAMPANTES **13282**
TOURNER DES SURFACES SPHERIQUES **13281**
TOURNER FAUX-ROND **10840**
TOURNEUR **13286**
TOURNEVIS **13294**
TOURNURES **13293**
TOURS A LA MINUTE **8690**
TOUT EN METAUX FERREUX **345**
TOUT-VENANT **10841**
TRACAGE **8017**
TRACAGE DES TOLES **8018**
TRACE **8017**
TRACE **13076**
TRACE **3794**
TRACE D'UN DIAGRAMME **9521**
TRACER **8010**
TRACER **3795**
TRACER **4195**
TRACER GRAPHIQUEMENT **9519**
TRACER UN DIAGRAMME **9520**
TRACERET **11068**
TRACEUR **9161**
TRACEUR-MECANICIEN **8013**
TRACHYTE **13081**
TRACOIR **11068**
TRACTEUR **9859**
TRACTEUR **13097**
TRACTEUR **13091**
TRACTEUR AGRICOLE A CHENILLES **13094**
TRACTEUR AGRICOLE A ROUES **13095**
TRACTEUR AGRICOLE DEMI TRAC **13093**
TRACTEUR MARAICHERS **13096**
TRACTION **9970**
TRACTION AVANT **5699**
TRACTION DANS LA DIRECTION DES FIBRES **9968**
TRACTION DES VEHICULES **13088**
TRACTION NORMALEMENT A LA DIRECTION DES FIBRES **9967**
TRACTOIRE **13098**
TRACTRICE **13098**
TRADUCTION **13123**
TRAIN (DE LAMINOIRS) A TOLES FINES **11324**
TRAIN A BANDES **11321**
TRAIN BALADEUR **11602**
TRAIN CONTINU (DE LAMINOIR) **3030**
TRAIN D'ENGRENAGES **5901**
TRAIN DE LAMINAGE **10670**

TRAIN DE LAMINAGE A FROID **2708**
TRAIN DE LAMINOIR **10701**
TRAIN DE LAMINOIR **10669**
TRAIN DE ROULEAUX DE SORTIE **10838**
TRAIN DIFFERENTIEL **3913**
TRAIN EBAUCHEUR **1348**
TRAIN EBAUCHEUR **10789**
TRAIN EPICYCLOIDAL **4778**
TRAIN FINISSEUR **5229**
TRAIN PLANETAIRE **12456**
TRAIN PLANETAIRE **9446**
TRAIN REDUCTEUR **10353**
TRAIN(S) A BRAMES **11556**
TRAINEE D'AIR **266**
TRAIT D'UNE ECHELLE **8009**
TRAIT DE SCIE **3517**
TRAIT PONCTUE **4125**
TRAIT ZERO **14025**
TRAITEMENT **13959**
TRAITEMENT AU BICHROMATE **1236**
TRAITEMENT AU JET DE SABLE **2540**
TRAITEMENT DE DETENTE APRES SOUDAGE **9666**
TRAITEMENT DE RECUIT **568**
TRAITEMENT DES METAUX A FROID **2710**
TRAITEMENT DES MINERAIS **11666**
TRAITEMENT DIRECT **8799**
TRAITEMENT INDIRECT **8735**
TRAITEMENT ISOTHERME **7194**
TRAITEMENT MECANIQUE **8861**
TRAITEMENT PRELIMINAIRE **9780**
TRAITEMENT PREPARATOIRE **9780**
TRAITEMENT THERMIQUE **12815**
TRAITEMENT THERMIQUE **6375**
TRAITEMENT THERMIQUE **6377**
TRAITEMENT THERMIQUE SOUS VIDE **13443**
TRAITEMENT ULTERIEUR **12417**
TRAITER **13937**
TRAITER LES MINERAIS **11664**
TRAITER MECANIQUEMENT LES MINERAIS **4253**
TRAJECTOIRE **13104**
TRAJET SUIVI PAR UN FLUIDE **9125**
TRAJET, ECARTEMENT **4063**
TRAME **13743**
TRANCHANT **12995**
TRANCHANT D'UN OUTIL **3528**
TRANCHE **11217**
TRANCHE A CHAUD **6640**
TRANCHE A FROID **2701**
TRANCHER **3515**
TRANCHET D'ENCLUME **631**
TRANSBORDEUR **13141**

TRANSDUCTEUR **13108**
TRANSFERT DE CHALEUR **6372**
TRANSFERT PAR COURT-CIRCUIT **11370**
TRANSFERT PAR VEINE LIQUIDE **7659**
TRANSFORMATEUR **13115**
TRANSFORMATEUR AMPLIFICATEUR DE POTENTIEL **12245**
TRANSFORMATEUR DE POTENTIEL **13115**
TRANSFORMATEUR DEVOLTEUR **12244**
TRANSFORMATEUR REDUCTEUR DE POTENTIEL **12244**
TRANSFORMATEUR SURVOLTEUR **12245**
TRANSFORMATION **3067**
TRANSFORMATION ALLOTROPIQUE **355**
TRANSFORMATION CHIMIQUE **2296**
TRANSFORMATION CONGRUENTE **2939**
TRANSFORMATION D'ENERGIE **3070**
TRANSFORMATION DU FER EN ACIER **3068**
TRANSFORMATION EN COKE **2666**
TRANSFORMATION ISOTHERME **7196**
TRANSFORMER EN COKE **2656**
TRANSFORMER L'ENERGIE **13112**
TRANSISTOR **13119**
TRANSLATION **13158**
TRANSLATION **13123**
TRANSLATION PRIMITIVE **9865**
TRANSLUCIDE **13125**
TRANSLUCIDITE **13124**
TRANSMISSION **11248**
TRANSMISSION **6372**
TRANSMISSION **4294**
TRANSMISSION AUX ROUES **5186**
TRANSMISSION COURROIES DE CHASSE **3141**
TRANSMISSION D'ELECTRICITE **13133**
TRANSMISSION D'ENERGIE ELECTRIQUE **13129**
TRANSMISSION D'UN MOUVEMENT **13130**
TRANSMISSION DE CHALEUR **6374**
TRANSMISSION DE LA CHALEUR **13111**
TRANSMISSION DE PUISSANCE **13134**
TRANSMISSION DE PUISSANCE **9746**
TRANSMISSION FLEXIBLE **5420**
TRANSMISSION FUNICULAIRE **10734**
TRANSMISSION HYDRAULIQUE **5472**
TRANSMISSION INTERMEDIAIRE **3244**
TRANSMISSION PAR CABLE TELEDYNAMIQUE **12706**
TRANSMISSION PAR CABLES **10734**
TRANSMISSION PAR CHAINES **2207**
TRANSMISSION PAR CONES DE FRICTION **1224**
TRANSMISSION PAR CORDES COURROIES **10734**
TRANSMISSION PAR COURROIES **1144**
TRANSMISSION PAR ENGRENAGE **13131**
TRANSMISSION PAR FRICTION **5684**
TRANSMISSION PAR LEVIER **13132**

TREMIE **6577**
TREMPABILITE **6288**
TREMPE **6291**
TREMPE **2338**
TREMPE **10094**
TREMPE **3975**
TREMPE **3978**
TREMPE (NON-) **13359**
TREMPE A L'AIR **285**
TREMPE A L'AIR **267**
TREMPE A L'EAU **13672**
TREMPE A L'EAU **13654**
TREMPE A L'EAU SUIVIE DE REVENU **13673**
TREMPE A L'HUILE **8772**
TREMPE A L'HUILE **8760**
TREMPE AU BAIN DE SEL **10913**
TREMPE AU CHALUMEAU **5315**
TREMPE AU GAZ **5840**
TREMPE CONTINUE A CHAUD **3028**
TREMPE DE SURFACE **2021**
TREMPE DES METAUX **6293**
TREMPE DIFFERENTIELLE **3914**
TREMPE DOUCE **8760**
TREMPE ECHELONNEE **6633**
TREMPE EN BAIN CHAUD **8021**
TREMPE EN BAIN DE SEL **7651**
TREMPE EN ETAPES **824**
TREMPE EN SURFACE **2023**
TREMPE ET REVENU **10095**
TREMPE ET REVENU **10092**
TREMPE ETAGEE **12242**
TREMPE ETAGEE BAINITIQUE **824**
TREMPE ETAGEE MARTENSITIQUE **12243**
TREMPE INTERROMPUE **12249**
TREMPE INVERSE **7109**
TREMPE ISOTHERME **7195**
TREMPE LOCALISEE **3914**
TREMPE MARTENSITIQUE **8021**
TREMPE NEGATIVE **8527**
TREMPE PAR INDUCTION **6896**
TREMPE PAR NITRURATION **8582**
TREMPE PAR PULVERISATION **11966**
TREMPE PARTIELLE **11145**
TREMPE PARTIELLE **7697**
TREMPE PRODUITE PAR ETIRAGE A FROID **6274**
TREMPE SECONDAIRE **11105**
TREMPE SUPERFICIELLE **12508**
TREMPER **2332**
TREMPER (FONTE) **2332**
TREMPER UN METAL **6285**
TREMPEUR D'OUTILLAGE **13008**

TREPAN **4265**
TREPIDATIONS **13559**
TREPIED **13219**
TRES FLUIDE **5471**
TRES MOBILE **5471**
TRESSE **5863**
TRESSE **751**
TRESSE DE CHANVRE SUIFFEE **6442**
TRESSE EN CHANVRE **6441**
TRESSE EN COTON **3220**
TREUIL **13851**
TREUIL A BRAS **6256**
TREUIL A ENGRENAGE **10014**
TREUIL A MANIVELLE **6256**
TREUIL SIMPLE **13863**
TREUILS **13852**
TRI **11139**
TRIAGE **11524**
TRIAGE PAR COURANT GAZEUX **248**
TRIAGE PAR COURANT GAZEUX **4681**
TRIANGLE **13185**
TRIANGLE ACUTANGLE **157**
TRIANGLE DE FORCES **13186**
TRIANGLE DE LIAISON **2962**
TRIANGLE EQUILATERAL **4785**
TRIANGLE ISOCELE **7192**
TRIANGLE OBTUSANGLE **8718**
TRIANGLE POLAIRE **9581**
TRIANGLE RECTANGLE **10582**
TRIANGLE SCALENE **10984**
TRIANGLE SPHERIQUE **11891**
TRIATOMIQUE **13197**
TRIBASIQUE **13198**
TRIBOMURE DE PHOSPHORE **9255**
TRICHLORURE D'ANTIMOINE **626**
TRICHLORURE D'ARSENIQUE **724**
TRICHLORURE D'OR **821**
TRICHLORURE DE BISMUTH **1270**
TRICHLORURE DE PHOSPHORE **9256**
TRIEDRE **13202**
TRIEURS DE POMMES DE TERRE **9706**
TRIEUSES ELECTRONIQUES **11185**
TRIGONOMETRIE **13201**
TRIGONOMETRIQUE **13200**
TRINGLE **10663**
TRINGLE DE GUIDAGE **6164**
TRINGLERIE DE DIRECTION **12232**
TRINGLERIE DE FREIN **1539**
TRINITRINE **8593**
TRIPOLI **13220**
TRIPOLI SILICEUX **7292**

TRIRANTS 6186
TRISULFURE D'ANTIMOINE 627
TRISULFURE D'ARSENIC 728
TRITURATION 6117
TRITURER 13222
TRIVALENT 13197
TROCHOIDE 13223
TROLLEYBUS 13225
TROMMEL 13226
TROMPE 11518
TRONC DE CONE 13235
TRONC DE PARALLELEPIPEDE 13237
TRONC DE PYRAMIDE 13238
TRONCATURE D'UN FILET 5389
TRONCON D'ARBRE 9643
TRONCON DE TUYAU 9292
TRONCONNAGE 3533
TRONCONNAGE DU BOIS 4718
TRONCONNEMENT 3533
TRONCONNER 3510
TRONCONNEUSE 3541
TROOSTITE 13227
TROP PLEIN 8931
TROP-PLEIN 8932
TROU 6530
TROU A CRASSE 4332
TROU A MAIN 6236
TROU ALESE 4273
TROU BORGNE 3630
TROU D'AIR 13528
TROU D'AXE OU DE GOUPILLE 9332
TROU D'HOMME 7976
TROU D'HOMME 9620
TROU D'HOMME 7963
TROU D'HOMME 7997
TROU D'HOMME DE ROBE 11332
TROU D'HOMME DE TOIT 10710
TROU DE BOULON 1437
TROU DE BRAS 6236
TROU DE COULEE 12637
TROU DE COULEE 7287
TROU DE GOUPILLE 3214
TROU DE JAUGE 3971
TROU DE JAUGE COMBINE AVEC EVENT 3972
TROU DE RIVET 10625
TROU DE RIVET PERCE 4275
TROU DE RIVET POINCONNE 10004
TROU DE SONDAGE 1463
TROU DE SOUFFLAGE 6393
TROU DE VIDANGE 8442
TROU DE VISITE D'UN CAISSON 9620

TROU DES TOLES A ASSEMBLER 10625
TROU EN CUL DE SAC 3630
TROU FILETE 11056
TROU FORE 4273
TROU GRAISSEUR 8761
TROU OBLONG 8714
TROU TEMOIN 12715
TROU TRAVERSANT LA PIECE 12893
TROU VENU DE FONTE 2033
TROUBLE 2541
TROUEE 11831
TROUS DE BRIDE 5328
TROUS DE MANIPULATION 6264
TROUSSE (METALLURGIE) 5017
TROUSSEAU 12734
TRUSQUIN A MARBRE 11069
TRUSQUIN A POINTE 8016
TRUSQUIN D'ASSEMBLAGE 8399
TUBE 9339
TUBE 13257
TUBE A AILERONS 10571
TUBE A AILETTES 10571
TUBE A AILETTES EXTERIEURES 4963
TUBE A AILETTES INTERIEURES 7082
TUBE A BORD RABATTU 1087
TUBE A CATHODE CHAUDE 6652
TUBE A COLLET RABATTU 1087
TUBE A NERVURES LONGITUDINALES 7740
TUBE A NERVURES TRANSVERSALES 13147
TUBE A RACCORDS 9365
TUBE A RAPPROCHEMENT 1804
TUBE A RAYONS CATHODIQUES 2105
TUBE A RAYONS X 14002
TUBE A RAYONS-X 12684
TUBE A RECOUVREMENT 7406
TUBE BRASE 11760
TUBE BROYEUR 13252
TUBE CAPILLAIRE 13255
TUBE CAPILLAIRE 1921
TUBE CATHODIQUE 2105
TUBE COMPENSATEUR 4913
TUBE COUDE A DEUX BRANCHES COMMUNIQUANT ENTRE ELLES 13325
TUBE COULE DEBOUT 9343
TUBE COULE HORIZONTAL 9342
TUBE CYLINDRIQUE GRADUE 5960
TUBE D'ACIER 12212
TUBE D'ACIER (POUR GAZ LIQUEFIES) 12204
TUBE D'EAU 13681
TUBE D'ECHANGEUR THERMIQUE 6350
TUBE DE CHAUDIERE 1422

TUBE DE DERIVATION **1558**
TUBE DE DESCENTE **4183**
TUBE DE FORCE **13253**
TUBE DE FUMEE **5248**
TUBE DE PITOT **9415**
TUBE DE REDUCTION **10352**
TUBE DE VAPEUR **12177**
TUBE DE VENTURI **13532**
TUBE DROIT **12309**
TUBE DUDGEONNE **4905**
TUBE EN ACIER **12219**
TUBE EN ALUMINIUM **427**
TUBE EN BRONZE **1653**
TUBE EN GRES **12277**
TUBE EN LAITON **1570**
TUBE EN VERRE **5965**
TUBE ETIRE **11782**
TUBE ETIRE A FROID **2714**
TUBE FILETE **11059**
TUBE FLEXIBLE **6614**
TUBE FLEXIBLE AVEC ARMATURE EN FIL DE FER **714**
TUBE FLEXIBLE CERCLE EN ACIER **714**
TUBE FOYER **1412**
TUBE ISOLANT **7002**
TUBE LAMINE **10677**
TUBE LISSE **11684**
TUBE MANNESMANN **7983**
TUBE PLISSE **3199**
TUBE PROTEGE PAR GOUDRONNAGE **12687**
TUBE PROTEGE PAR UN RECOUVREMENT DE JUTE ASPHALTE
  766
TUBE PROTEGE PAR UNE PEINTURE **9016**
TUBE REDRESSEUR **10317**
TUBE RUGUEUX **10780**
TUBE SANS SOUDURE **11089**
TUBE SERVE **7740**
TUBE SOUDE **13775**
TUBE SOUDE **13777**
TUBE SOUDE A L'AUTOGENE **9364**
TUBE SOUDE BOUT A BOUT **1803**
TUBE SOUDE EN HELICE **11920**
TUBE SOUDE EN SPIRALE **11920**
TUBE SOUDE PAR RAPPROCHEMENT **1804**
TUBE SOUDE PAR RECOUVREMENT **7406**
TUBE TARAUDE **11059**
TUBES A NERVURES **10571**
TUBES ELECTRONIQUES **4636**
TUBES ETIRES A FROID **2678**
TUBES RAPPROCHES **2512**
TUBULURE **8679**
TUBULURE A BRIDE **5341**

TUBULURE A BRIDE **5337**
TUBULURE A COLLET **8683**
TUBULURE A COLLET **8514**
TUBULURE A COLLET DE RACCORDEMENT **5337**
TUBULURE AUTORENFORCEE **9012**
TUBULURE D ECHAPPEMENT **4894**
TUBULURE D'ADMISSION **6948**
TUBULURE D'ALIMENTATION **5084**
TUBULURE D'ARRIVEE **6946**
TUBULURE D'ENTREE **6946**
TUBULURE DE BRANCHEMENT **1557**
TUBULURE DE DEVERSEMENT **3736**
TUBULURE DE FOND **1495**
TUBULURE DE PURGE AVEC CUVETTE PERIPHERIQUE **4226**
TUBULURE DE SORTIE **8901**
TUBULURE DE SORTIE/D'ASPIRATION **4705**
TUBULURE DE TOIT **10711**
TUBULURE ENCASTREE **6969**
TUBULURE POUR PURGE SOUPLE **6613**
TUF **13260**
TUF CALCAIRE **1838**
TUF PONCEUX **9989**
TUF PORPHYRITIQUE **9631**
TUF VOLCANIQUE **13590**
TUILE **12934**
TUILE EN VERRE **5964**
TUNGSTATE DE SOUDE **11737**
TUNGSTENE **13262**
TURBINE **13267**
TURBINE **6795**
TURBINE A ACTION **6810**
TURBINE A ADMISSION PARTIELLE **9085**
TURBINE A ADMISSION TOTALE **5718**
TURBINE A AIR **13859**
TURBINE A BASSE PRESSION **7788**
TURBINE A DETENTE **4925**
TURBINE A GAZ **5844**
TURBINE A HAUTE PRESSION **6494**
TURBINE A IMPULSION **6810**
TURBINE A LIBRE DEVIATION **6810**
TURBINE A PRESSION INTERIEURE **10250**
TURBINE A REACTION **10250**
TURBINE A VAPEUR **12189**
TURBINE AMERICAINE **8330**
TURBINE AXIALE **862**
TURBINE DE POMPE CENTRIFUGE **6795**
TURBINE DE RETOUR D'HUILE **8763**
TURBINE HYDRAULIQUE **6696**
TURBINE MIXTE **8330**
TURBINE PARALLELE **862**
TURBINE RADIALE **10136**

TURBINE TANGENTIELLE **12621**
TURBO-COMPRESSEUR **13272**
TURBO-COMPRESSEUR **2178**
TURBO-GENERATEUR **13271**
TURBO-VENTILATEUR **13270**
TURBULENCE **13273**
TURC **11032**
TUYAU **9339**
TUYAU **4183**
TUYAU **13257**
TUYAU (FLEXIBLE) DE CAOUTCHOUC **10821**
TUYAU A AILETTES **5234**
TUYAU A BRIDES **5343**
TUYAU A EMBOITEMENT **11707**
TUYAU A GAZ **5833**
TUYAU A GAZ EN PAPIER IMPREGNE DE BITUME **6803**
TUYAU A JOINT SPHERIQUE **5418**
TUYAU A PRESSION **3736**
TUYAU A VAPEUR **12177**
TUYAU ADDUCTEUR **12489**
TUYAU BOUCHE **2361**
TUYAU CINTRE **1194**
TUYAU COLLECTEUR **2725**
TUYAU CYLINDRIQUE **3586**
TUYAU D'ALIMENTATION **5070**
TUYAU D'AMENEE D'ARRIVEE **12489**
TUYAU D'ASCENSION **10615**
TUYAU D'ASPIRATION **12433**
TUYAU D'ECHAPPEMENT **4894**
TUYAU D'ECHAPPEMENT DE VAPEUR **4897**
TUYAU D'ECOULEMENT **4207**
TUYAU D'EVACUATION **4207**
TUYAU DE CHAUFFAGE CENTRAL **6396**
TUYAU DE COMMUNICATION **2960**
TUYAU DE CONDUITE **13665**
TUYAU DE CONDUITE **9357**
TUYAU DE DEBIT **4207**
TUYAU DE DECHARGE **4207**
TUYAU DE DISTRIBUTION D'EAU **13665**
TUYAU DE PRISE D'EAU **12066**
TUYAU DE RACCORDEMENT **2960**
TUYAU DE REFOULEMENT **3736**
TUYAU DE TROP-PLEIN **8932**
TUYAU ELEVATOIRE **10615**
TUYAU EM PLOMB ETAME A L'INTERIEUR **7469**
TUYAU EN BOIS **13928**
TUYAU EN CIMENT **2898**
TUYAU EN CIMENT ARME **5109**
TUYAU EN CUIR **7499**
TUYAU EN CUIVRE **3132**
TUYAU EN ETAIN **12960**

TUYAU EN FER **13989**
TUYAU EN FONTE **2035**
TUYAU EN PLOMB **7468**
TUYAU EN POTERIE **4416**
TUYAU EN TERRE CUITE **4416**
TUYAU EN TOILE DE CHANVRE **1908**
TUYAU EN TOLE **11312**
TUYAU EN ZINC **14047**
TUYAU ENGORGE **2361**
TUYAU METALLIQUE FLEXIBLE **5419**
TUYAU METALLIQUE FLEXIBLE **5421**
TUYAU RIVE **10640**
TUYAU VERTICAL **10615**
TUYAUTAGE **9357**
TUYAUTERIE **9357**
TUYAUTERIE **3736**
TUYAUTERIE **12178**
TUYAUTERIE **2845**
TUYERE **8685**
TUYERE A VAPEUR **12175**
TUYERE CONVERGENTE **2769**
TUYERE D'AMENEE DE L'AIR SOUS PRESSION **13299**
TUYERE DE PULVERISATION **11963**
TUYERE DE PULVERISATION **11964**
TUYERE DE REFOULEMENT **3738**
TUYERE DIVERGENTE **3738**
TYPE **13320**
TYPE STANDARD **12086**
U DE MONTAGE **12388**
ULTRAMICROSCOPE **13335**
UNIFICATION **12091**
UNIFIER **12090**
UNION **13368**
UNION HOMOGENE **6556**
UNITE DE CHALEUR **12816**
UNITE DE LONGUEUR **13373**
UNITE DE LUMIERE **9270**
UNITE DE MASSE **13374**
UNITE DE MESURE **13375**
UNITE DE POIDS **13378**
UNITE DE POINCONNAGE **10011**
UNITE DE SURFACE **13372**
UNITE DE SURFACE NUCLEAIRE **1006**
UNITE DE TEMPS **13376**
UNITE DE TRAVAIL D'ENERGIE **13379**
UNITE DE VOLUME **13377**
UNITE THERMIQUE **12816**
UNITE THERMIQUE ANGLAISE **1626**
UNIVALENT **8377**
URANIUM **13420**
USAGE **13982**

VARIANTE 410
VARIATEUR DE VITESSE 13493
VARIATION 13495
VARIATION (MATH.) 13496
VARIATION BRUSQUE DE TEMPERATURE 12437
VARIATION DE TEMPERATURE 13498
VARIATION PROGRESSIVE DE TEMPERATURE/DE VITESSE 6024
VARIATIONS DE NIVEAU D'EAU 13500
VARIATIONS DE PRESSION 13497
VARIATIONS DE VITESSE 13499
VARLOPE 7245
VASE CLOS 4075
VASELINE 13508
VASES COMMUNICANTS 13553
VECTEUR 13510
VECTEUR UNITAIRE 13380
VEHICULE A BENNE BASCULANTE 12967
VEHICULE A TROIS ESSIEUX 12871
VEHICULE ARTICULE 13092
VEHICULE TOUS TERRAINS 3362
VEHICULE UTILITAIRE 2784
VENT 1315
VENT ENRICHI D'OXYGENE 4770
VENTILATEUR 5033
VENTILATEUR A FORCE CENTRIFUGE 2179
VENTILATEUR A HELICE 9929
VENTILATEUR A PISTON 9372
VENTILATEUR A PISTON ROTATIF 10751
VENTILATEUR ASPIRANT 12429
VENTILATEUR ASPIRANT ET SOUFFLANT 2768
VENTILATEUR DE REFROIDISSEMENT 3088
VENTILATEUR REFOULANT 9817
VENTILATEUR SOUFFLANT 9817
VENTILATEUR SYSTEME ROOT 10729
VENTILATION 13530
VENTILATION ARTIFICIELLE 738
VENTILATION NATURELLE 8508
VENTILATION PAR APPEL 13450
VENTILATION PAR PULSION 9517
VENTRE DE POISSON (A) 5272
VENU DE FONDERIE 2054
VENU DE FONTE 2054
VENUE A LA COULEE 2054
VERDET 13533
VERIFICATEUR (METR) 5871
VERIFICATION 13534
VERIFICATION DE LA COMPOSITION CHIMIQUE 2280
VERIFICATION DU PARALLELISME 13086
VERIFIER LES MESURES 2279
VERIN 11035

VERIN A BOUTEILLE 1485
VERIN A CHARIOT 13165
VERIN A CLIQUET 10222
VERIN A TREPIED 13218
VERIN HYDRAULIQUE 6688
VERMILLON 13535
VERMOULU 13976
VERNIER 13536
VERNIR 7353
VERNIS 13504
VERNIS A L'ALCOOL 7354
VERNIS A L'ESSENCE 13296
VERNIS A L'HUILE 8786
VERNIS A LA GOMME-LAQUE 10998
VERNIS AU BITUME 1283
VERNIS AU COPAL 3103
VERNIS DU JAPON 239
VERNIS JAPON 1283
VERNIS ZAPON 14022
VERNISSAGE 13505
VERRE 8708
VERRE 5956
VERRE A BOUTEILLES 1484
VERRE A FIL DE FER NOYE 13894
VERRE A GLACES 9491
VERRE A VITRES 13866
VERRE ARME 13894
VERRE BLANC DE BOHEME 1400
VERRE D'IENA 7217
VERRE DE QUARTZ 10079
VERRE DEPOLI 6145
VERRE DEPOLI 5701
VERRE FILE 5966
VERRE GROSSISSANT 7936
VERRE INCOLORE 2746
VERRE LIQUIDE 11808
VERRE OPAQUE 8273
VERRE PILE 5962
VERRE SOLUBLE 11808
VERRE SOUFFLE 1365
VERROU 1432
VERROUILLAGE 7710
VERROUILLER 7699
VERT DE SCHEELE 10996
VERT DE SCHWEINFURT 11000
VERT EMERAUDE 2372
VERT-DE-GRIS 220
VERT-DE-GRIS 13533
VERTICAL 13551
VERTICALE 9217
VIBRATION 8884

VIBRATION **13554**
VIBRATION **13146**
VIBRATION AMORTIE **3600**
VIBRATION DUE A DES EFFORTS DE FLEXION **13556**
VIBRATION DUE A DES EFFORTS DE TORSION **13557**
VIBRATION FONDAMENTALE **5736**
VIBRATION LUMINEUSE **7825**
VIBRATION MECANIQUE **8114**
VIBRATION NON AMORTIE **13346**
VIBRATION PROPRE **5736**
VIBRATION SONORE **11818**
VIBRATIONS **2277**
VIBRATIONS **13559**
VIBRATIONS DE L'ETHER **13560**
VIBRATIONS DUES A LA RESONANCE **8883**
VIBRER **8877**
VIBROCULTEURS **3457**
VICE **3692**
VICE DE CONCEPTION **3694**
VICE DE CONSTRUCTION **3693**
VICE DE FONCTIONNEMENT **3695**
VICE DE MATIERE **3680**
VIDANGE **4024**
VIDANGE D'UNE CHAUDIERE **1364**
VIDANGER **4204**
VIDE **13434**
VIDE ENTRE DEUX DENTS **6548**
VIDE ENTRE LES BARREAUX DE LA GRILLE **278**
VIDE ENTRE LES ONDES **6543**
VIDE LIBRE **2481**
VIDE POUSSE **6477**
VIDER **4018**
VIDER **4204**
VIEILLISSEMENT **225**
VIEILLISSEMENT ARTIFICIEL **226**
VIEILLISSEMENT ARTIFICIEL **731**
VIEILLISSEMENT COMPLET **231**
VIEILLISSEMENT CRITIQUE **3344**
VIEILLISSEMENT ECHELONNE **232**
VIEILLISSEMENT INTERROMPU **7091**
VIEILLISSEMENT NATUREL **233**
VIEILLISSEMENT NATUREL **8500**
VIEILLISSEMENT PAR DEFORMATION **12328**
VIEILLISSEMENT PAR DEFORMATION **12327**
VIEILLISSEMENT PAR LES EFFORTS **12324**
VIEILLISSEMENT PAR REFROIDISSEMENT RAPIDE **10087**
VIEILLISSEMENT PAR TRAVAIL A FROID **235**
VIEILLISSEMENT PROGRESSIF **9906**
VIEILLISSEMENT PROGRESSIF **234**
VIERGE (MIN.) **5647**
VILEBREQUIN **3313**

VILEBREQUIN (OUTIL A PERCER) **1519**
VILEBREQUIN A CONTREPOIDS **3235**
VILEBREQUIN A ENGRENAGE **5905**
VILEBREQUIN A PLUSIEURS COUDES **8478**
VILEBREQUIN A TROIS COUDES **12880**
VILEBREQUIN A VIS DE PRESSION **1518**
VINAIGRE RADICAL **86**
VIOLLE **13569**
VIRE-ANDAINS **12537**
VIROLE **5128**
VIROLE D'UNE CHAUDIERE **3260**
VIS **2453**
VIS **6457**
VIS **11021**
VIS A BOIS A TETE CARREE **2552**
VIS A BROCHE **1930**
VIS A DEUX FILETS **13315**
VIS A DROITE **10584**
VIS A FILET CARRE **12034**
VIS A FILET ROND **10807**
VIS A FILET TRAPEZOIDAL **1815**
VIS A FILET TRIANGULAIRE **547**
VIS A GAUCHE **7510**
VIS A LEVIER **1930**
VIS A METAUX **8187**
VIS A OREILLES **12912**
VIS A PLUSIEURS FILETS **8474**
VIS A TETE **1913**
VIS A TETE CARREE **12033**
VIS A TETE CYLINDRIQUE **5165**
VIS A TETE FENDUE **11637**
VIS A TETE NOYEE **3248**
VIS A TETE PLATE **5385**
VIS A TETE RONDE **11051**
VIS A TROIS FILETS **12879**
VIS A UN FILET **11502**
VIS AILEE **12912**
VIS AUTOTARAUDEUSE **11162**
VIS D'ARCHIMEDE **13970**
VIS D'ARRET **2453**
VIS DE BLOCAGE **7708**
VIS DE BUTEE **12909**
VIS DE PRESSION **193**
VIS DE PRESSION **9831**
VIS DE PURGE **1319**
VIS DE REGLAGE **192**
VIS DE RELEVAGE **7556**
VIS DE TENSION **12758**
VIS DE TRANSLATION **10852**
VIS DECOLLETEE **1607**
VIS DIFFERENTIELLE **2833**

VIS EGALISATRICE **7529**
VIS FILETEE A DROITE **10580**
VIS FILETEE A GAUCHE **7506**
VIS GLOBIQUE **5978**
VIS MERE **7471**
VIS METALLIQUE A BOIS **13918**
VIS MICROMETRIQUE **8255**
VIS OU BOULON DE FIXATION **5043**
VIS OU BOULON DE MONTAGE **5043**
VIS POINTEAU POUR ARRET DE BAGUES **11222**
VIS SANS FIN **13979**
VIS SANS FIN **13969**
VIS SANS TETE **6156**
VIS TANGENTE **12619**
VIS TRANSPORTEUSE **13970**
VIS VIOLON **12911**
VIS-MERE **6165**
VISCOSE **13575**
VISCOSIMETRE **13574**
VISCOSITE **13576**
VISITE DE LA CHAUDIERE **1414**
VISQUEUX **13577**
VISSAGE **11065**
VISSER **11022**
VISSER SUR **11037**
VISUALISATION **10253**
VITESSE **11868**
VITESSE **10226**
VITESSE **11875**
VITESSE A L'ARRIVEE **6950**
VITESSE A LA SORTIE **8903**
VITESSE ANGULAIRE **546**
VITESSE CIRCONFERENTIELLE **9198**
VITESSE CRITIQUE **3353**
VITESSE CROISSANTE **6852**
VITESSE D'ARRIVEE **6950**
VITESSE D'ECOULEMENT **5462**
VITESSE D'ECOULEMENT **8903**
VITESSE D'ECOULEMENT D'HUILE **10232**
VITESSE D'ENTREE **6950**
VITESSE D'INFLAMMATION **11870**
VITESSE DE BROCHE **11909**
VITESSE DE COMBUSTION **10227**
VITESSE DE CORROSION **3193**
VITESSE DE COUPE CONSTANTE **2971**
VITESSE DE CROISIERE **3417**
VITESSE DE DEFORMATION **12330**
VITESSE DE DEPOT **10229**
VITESSE DE FLUAGE **3326**
VITESSE DE FLUAGE **5459**
VITESSE DE FUSION **8144**

VITESSE DE LA LUMIERE **7569**
VITESSE DE LA LUMIERE **13517**
VITESSE DE PASSAGE **13518**
VITESSE DE PROPAGATION **13519**
VITESSE DE PROPAGATION DE LA FLAMME **10230**
VITESSE DE REGIME **8663**
VITESSE DE ROTATION **11876**
VITESSE DE SORTIE **8903**
VITESSE DE TRANSFORMATION **13521**
VITESSE DECROISSANTE **3679**
VITESSE DU COURANT **13516**
VITESSE DU FIL **13900**
VITESSE DU PISTON **9388**
VITESSE DU SON **13520**
VITESSE FINALE **5193**
VITESSE INITIALE **6933**
VITESSE LINEAIRE **7617**
VITESSE MAXIMUM **8067**
VITESSE NORMALE DE FONCTIONNEMENT **8663**
VITESSE PERIPHERIQUE **9198**
VITESSE SURMULTIPLIEE **8930**
VITESSE TANGENTIELLE **12623**
VITESSE UNIFORME CONSTANTE **13363**
VITESSE VARIEE **13494**
VITESSE VIRTUELLE **13573**
VITRAGE **5972**
VITRER **5969**
VITREUX **13582**
VITRIFICATION **13584**
VITRIOL **12453**
VITRIOL BLANC **14046**
VITRIOL BLEU **1376**
VITRIOL MARTIAL **6093**
VITRIOL VERT **6093**
VOIE **13085**
VOIE DECAUVILLE **9639**
VOIE FERREE **7606**
VOIE FERREE D'ATELIER **13965**
VOIE PORTATIVE **9639**
VOILE **3667**
VOILE **1378**
VOILE **6313**
VOILE **1347**
VOILER (SE) **13628**
VOITURE AUTOMOBILE **8411**
VOITURE DECAPOTABLE **3073**
VOLANT **13810**
VOLANT A CHAINE **2223**
VOLANT A CHAINE ADAPTABLE **2220**
VOLANT A GORGES **10738**
VOLANT A MAIN **6267**

VOLANT D'UNE SEULE PIECE 11784
VOLANT DE CHARIOTAGE 13163
VOLANT DE DIRECTION 12233
VOLANT DE MANOEUVRE 6255
VOLANT DENTE 5906
VOLANT EN DEUX SEGMENTS 5497
VOLANT EN DISQUE 4004
VOLANT EN PLUSIEURS SEGMENTS 11138
VOLANT ENGRENAGE 5906
VOLANT FENDU 11941
VOLANT-MOTEUR 5496
VOLANT-POULIE 1147
VOLATILISATION 13587
VOLATILISER 13588
VOLATILITE 13589
VOLET D'AIR 292
VOLET DE GAZ 12888
VOLT 13591
VOLT-AMPERE 13592
VOLT-COULOMB 7255
VOLTAGE 13593
VOLTAGE AUX BORNES 12764
VOLTAGE NORMAL 8659
VOLTAMETRE 3227
VOLTMETRE 13597
VOLUME 13598
VOLUME (POUR CENT DU) 9179
VOLUME ATOMIQUE 806
VOLUME CRITIQUE 3354
VOLUME DEPLACE (PHYS.) 4053
VOLUME ENGENDRE PAR LE DEPLACEMENT DU PISTON
    13599
VOLUME MOLECULAIRE 8362
VOLUME QUI PASSE AU TRAVERS D'UN ORIFICE PAR
    SECONDE 4023
VOLUME SPECIFIQUE 11854
VOUSSEAU 687
VOUSSOIR 687
VOUSSURE 3396
VOYAGEUR 7625
VOYANT ROND 2414
VRILLE 5944
VRILLES 9664
VUE 13567
VUE A VOL D'OISEAU 1264
VUE D'ENSEMBLE 9430
VUE DE BAS EN HAUT 1498
VUE DE COTE 11420
VUE DE FACE 5694
VUE DE FACE POSTERIEURE 4717
VUE DE HAUT EN BAS 13037

VUE ECLATEE 4934
VUE PAR-DESSOUS 1498
VULCANISATION 3484
VULCANISATION 13607
VULCANISER 13608
WAGON-CITERNE 12626
WAGON-RESERVOIR 12626
WAGONS CIGARES/POCHES 2072
WATT 13694
WATT-HEURE 13695
WATT-MINUTE 13696
WATT-SECONDE 13697
WATTMETRE 13698
WITHERITE 13911
WOLFRAM 13912
WULFENITE 13991
XENON 14004
XEROGRAPHIE 14005
XYLENE 14006
XYLOIDINE 6176
XYLOL 14006
YARD 14007
YARD CARRE 12031
YTTERBIUM 14019
YTTRIUM 14020
ZAMAK 14021
ZERO (D'UNE ECHELLE) 14027
ZERO ABSOLU 36
ZERO FLOTTANT 5440
ZINC 14035
ZINC (COMMERCIAL) 11879
ZINC BRUT 2785
ZINC DU COMMERCE 2785
ZINC EN FEUILLES 11327
ZINC REFONDU 10341
ZINC SILICATE 11438
ZINGAGE 2604
ZINGAGE 14044
ZINGAGE 5787
ZINGAGE 5789
ZINGAGE ELECTROCHIMIQUE ELECTROLYTIQUE 4566
ZINGUER 5784
ZIRCONIUM 14049
ZONE DE DIFFUSION 3934
ZONE DE FUSION 5761
ZONE DE FUSION 5763
ZONE DE GUINIER-PRESTON 6171
ZONE DE SOLIDIFICATION 5658
ZONE DE SURFACE DE SEPARATION 7034
ZONE INFLUENCEE PAR LA CHALEUR 6380
ZONE IONISEE 7128

ZONE NEUTRE **8547**   ZONE SPHERIQUE **14051**

DEUTSCHER INDEX

ABAKUS (WULFFSCHER) **13992**
ABBAUEN UND ERSETZEN **10451**
ABBEIZEN **11006**
ABBEIZEN (EIN METALL) **11005**
ABBILDUNG **6778**
ABBIMSEN **9990**
ABBIMSEN **9987**
ABBINDEN **11214**
ABBINDEN DES ZEMENTS **11226**
ABBLASEHAHN **1357**
ABBLASEN EINES KESSELS **1364**
ABBLASEN MIT PRESSLUFT **244**
ABBLASEVENTIL **1358**
ABBLASHAHN **272**
ABBLÄTTERN **10975**
ABBLÄTTERN **9156**
ABBLÄTTERUNG **11830**
ABBLÄTTERUNG **5306**
ABBLÄTTERUNG **11328**
ABBLÄTTERUNG **11339**
ABBLÄTTERUNG **10985**
ABBLÄTTERUNG **4889**
ABBLENDEN **3877**
ABBRAND **1747**
ABBRANDVERLUST **7761**
ABBRENNSCHWEISSEN **5358**
ABBRENNSTUMPFSCHWEISSEN **1806**
ABBRENNSTUMPFSCHWEISSUNG **1794**
ABDAMPF **4896**
ABDAMPF-ENTÖLER **8779**
ABDAMPFROHR **4897**
ABDAMPFSCHALE **4864**
ABDAMPFVORRICHTUNG **4869**
ABDECKLEISTEN **1928**
ABDECKMITTEL **615**
ABDECKPLATTE **3267**
ABDECKUNG **12287**
ABDICHTEN **7954**
ABDICHTEN **7953**
ABDICHTEN **7831**
ABDICHTUNG **11081**
ABDREHEN **13164**
ABDREHEN **13162**
ABDRUCK **10465**
ABDRÜCKSCHRAUBE **7556**
ABERRATION (CHROMATISCHE) **2369**
ABERRATION (SPHÄRISCHE) **11885**
ABERRATION DES LICHTES **5**
ABFALL **12536**
ABFÄLLE **13637**
ABFALLEISEN **5553**

ABFALLENDE **3356**
ABFALLSTOFFE **13637**
ABFASEN **2230**
ABFASEN **2233**
ABFASUNG **2231**
ABFLACHEN **5408**
ABFLACHUNG DES GEWINDES **5389**
ABFLUCHTUNG **325**
ABFLUSS **8900**
ABFLUSS **8899**
ABFLUSSGESCHWINDIGKEIT **8903**
ABFLUSSMENGE **4023**
ABFLUSSROHR **4207**
ABFORMEN **8428**
ABFORMEN **8424**
ABGAS **13634**
ABGEFLACHT **5406**
ABGENUTZT **13982**
ABGEPLATTET **5406**
ABGERUNDET **10794**
ABGESCHRECKTEN ZUSTAND (IM) **743**
ABGESTUFT **12246**
ABGIESSEN **3649**
ABGIESSEN **3650**
ABGRATEN **1766**
ABGRATEN **3642**
ABGRATEN **13209**
ABGRATEN **13206**
ABGRATEN **13210**
ABGRATEN (EIN BLECH) **13205**
ABGRATPRESSE **13211**
ABGUSS **9717**
ABGUSS **12703**
ABHANG **11627**
ABHÄNGIGKEIT VON (IN) **13542**
ABHEBEFORMMASCHINE **4237**
ABKANTEN **2233**
ABKANTEN **12671**
ABKANTEN **12646**
ABKANTEN **2230**
ABKANTMASCHINEN **4453**
ABKLÄREN **3650**
ABKLÄREN **3649**
ABKLOPPEN **2347**
ABKOCHUNG **3669**
ABKREIDEN **2226**
ABKÜHLEN **3085**
ABKÜHLEN **2332**
ABKÜHLEN (SICH) **3081**
ABKÜHLUNG MIT WASSERMANTEL **7204**
ABKÜHLUNGSKURVE **3087**

AMPERESEKUNDE **483**
AMPERESTUNDE **481**
AMPEREWINDUNG **480**
AMPEREZAHL **478**
AMPHOTER **484**
AMPLITUDE **486**
AMYLALKOHOL **488**
AMYLAZETAT **487**
AMYLOXYDHYDRAT **488**
ANAEROB **489**
ANALOG **490**
ANALYSATOR **492**
ANALYSE **493**
ANALYSE (CHEMISCHE) **2293**
ANALYSE (ELEKTROLYTISCHE) **4570**
ANALYSE (KOLORIMETRISCHE) **2739**
ANALYSE (MAGNETISCHE) **7894**
ANALYSE (QUALITATIVE) **10062**
ANALYSE (QUANTITATIVE) **10064**
ANALYSE (SELEKTIVE THERMISCHE) **3919**
ANALYSE AUF NASSEM WEGE **13800**
ANALYSE AUF TROCKENEM WEGE **4341**
ANALYSENWAAGE **2294**
ANALYSIEREN **491**
ANALYTISCH **494**
ANASTIGMAT **497**
ANBAU **585**
ANBRUCH **6834**
ÄNDERUNG (ALLMÄHLICHE DER TEMPERATUR) **6024**
ÄNDERUNG (PLÖTZLICHE) DER TEMPERATUR **12437**
ÄNDERUNG DER GESCHWIDIGKEIT **6024**
ANDREISSSCHABLONE **8016**
ANEMOMETER **503**
ANEROID BAROMETER **504**
ANFANGSDRUCK **6929**
ANFANGSGESCHWINDIGKEIT **6933**
ANFANGSKRIECHEN **6928**
ANFANGSLAGE **12126**
ANFANGSPUNKT DER KOORDINATEN **8868**
ANFANGSSTELLUNG **12126**
ANFANGSTEMPERATUR **6931**
ANFANGSWERT **6932**
ANFANGSZUSTAND **6930**
ANFEUCHTEN **8348**
ANFEUCHTEN **8347**
ANFEUERN (EINEN KESSEL) **5236**
ANFEUERN EINES KESSELS **5263**
ANFRESSEN **3182**
ANFRESSUNG **5778**
ANFRESSUNG **3183**
ANFÜHREN **10113**

ANGEL **9419**
ANGEL EINES WERKZEUGS **12614**
ANGENÄHERTER WERT **653**
ANGESCHRIEBENER KREIS **4834**
ANGETRIEBENE SCHEIBE **4298**
ANGREIFEN **647**
ANGREIFEN **3182**
ANGREIFLÖCHER **6264**
ANGRENZEND **181**
ANGRIFF **813**
ANGRIFF (INTERDENDRITISCHER) **7031**
ANGRIFF EINER KRAFT **647**
ANGRIFFSPUNKT **9561**
ANGSTRÖM **536**
ANGUSS **10845**
ANGUSS **12004**
ANHAFTEN **8042**
ANHALTEN (EINE MASCHINE) **12280**
ANHALTEN EINER MASCHINE **12286**
ANHALTSPUNKTE **12125**
ANHÄNGER **13102**
ANHÄNGEVORRICHTUNGEN FÜR WAGEN **4088**
ANHEIZEN **5236**
ANHEIZEN EINES KESSELS **5263**
ANHYDRID **548**
ANHYDRIT **549**
ANILIN **552**
ANILINÖL **553**
ANION **556**
ANISOTROP **558**
ANISOTROPIE **559**
ANKER **5590**
ANKER **500**
ANKER EINER GLEICHSTROMMASCHINE **707**
ANKER EINES MAGNETEN **710**
ANKERBLECH **708**
ANKERBOLZEN **5590**
ANKERDRAHT **6187**
ANKERPLATTE **13621**
ANKERPLATTE (IM FUNDAMENT) **502**
ANKERPLATZ **8390**
ANKERROSETTE **13621**
ANKERSCHNITT **3160**
ANKERSCHRAUBE **5590**
ANKERSCHRAUBE **501**
ANKERSEILE **6186**
ANKERUNG **499**
ANKERWICKELMASCHINE **709**
ANKERZUGSEILE **12919**
ANKERZUGSTANGE **12919**
ANKOMMENDER STROM **6844**

ANSCHLUSSROHR 2960
ANSCHLUSSSTUTZEN 1557
ANSCHMIEGUNGSWINKEL 517
ANSCHNITTSYSTEM 5870
ANSCHRAUBEN 11065
ANSCHRAUBEN 11022
ANSCHÜTTEN 971
ANSETZEN 1621
ANSETZEN (EINE LÖSUNG) 9788
ANSETZEN EINER LÖSUNG 9785
ANSETZEN VON SCHMUTZ 2510
ANSICHT 13567
ANSICHT (EXPLODIERTE) 4934
ANSTÄHLEN 12205
ANSTÄHLEN 12206
ANSTAUUNG 10182
ANSTELLEN EINES WERKZEUGES 1621
ANSTELLUNGSWINKEL 2484
ANSTIEG 757
ANSTIEG (MECH.) 10611
ANSTOSS 7327
ANSTOSSEND*181
ANSTREICHEN 12529
ANSTREICHEN 9018
ANSTREICHEN 9014
ANSTRICH 9013
ANSTRICH 9018
ANSTRICHFARBE 9015
ANSTRICHGONDEL 9017
ANSTRICHMÄNGEL 9021
ANSTRICHMITTEL 2589
ANTHRAZEN 607
ANTHRAZENÖL 608
ANTHRAZIT 609
ANTIFRIKTIONSMETALL 619
ANTIFRIKTIONSWIRKUNG 617
ANTIKATHODE 12682
ANTIKATHODE 611
ANTILOGARITHMUS 620
ANTIMAGNETISCH 614
ANTIMON 623
ANTIMONBLEI 6276
ANTIMONBLEI 621
ANTIMONBUTTER 626
ANTIMONCHLORÜR 626
ANTIMONGLANZ 622
ANTIMONIT 622
ANTIMONPENTACHLORID 624
ANTIMONPENTASULFID 625
ANTIMONREGULUS 10423
ANTIMONSULFÜR 627

ANTIMONSUPERCHLORID 624
ANTIMONTRICHLORID 626
ANTIMONTRISULFID 627
ANTISEPTIKUM 629
ANTISEPTISCH 628
ANTISEPTISCHES MITTEL 629
ANTREIBEN 4289
ANTREIBEN 4297
ANTRIEB 4303
ANTRIEB 4297
ANTRIEB (ELEKTRISCHER) 4503
ANTRIEB (HYDRAULISCHER) 6683
ANTRIEB (KRAFTSCHLÜSSIGER) 8642
ANTRIEB (MASCHINELLER) 9734
ANTRIEB (MECHANISCHER) 9734
ANTRIEB (ZWANGSLAÜFIGER) 9654
ANTRIEB EINER KRAFT 6809
ANTRIEB MIT DRUCKLUFT 2842
ANTRIEB VON HAND 6233
ANTRIEBART 8217
ANTRIEBMACHINE 9859
ANTRIEBMECHANISMUS 4309
ANTRIEBSCHEIBE 4313
ANTRIEBSCHRAUBE 11041
ANTRIEBSEIL 4314
ANTRIEBSELEMENTE 13126
ANTRIEBSPINDEL 4317
ANTRIEBSWELLE 4296
ANTRIEBVORRICHTUNG 4309
ANTRIEBWELLE 4315
ANWÄRMBRENNER 6399
ANWÄRMBRENNER 6391
ANWÄRMEN 6260
ANWÄRMER 6385
ANWÄRMHITZE (KURZE) 5362
ANWEISUNG 6991
ANZAPFUNGSTANK 4227
ANZEICHNEN 8015
ANZEICHNEN 8008
ANZEIGE 10253
ANZEIGE (DIGITALE) 8696
ANZEIGEBEREICH EINES INSTRUMENTS 10201
ANZEIGEEINRICHTUNG 12715
ANZEIGEVORRICHTUNG 13153
ANZIEHEN (DIE STOPFBÜCHSE) 12926
ANZIEHEN (EINE SCHRAUBE) 12925
ANZIEHEN (EINEN KEIL) 12924
ANZIEHEN (ZEMENT) 11214
ANZIEHEN EINES KEILES 12928
ANZIEHUNG 815
ANZIEHUNG (MAGNETISCHE) 7895

ARSENDISULFID 10262
ARSENIAT 720
ARSENID 726
ARSENIGS ÄUREANHYDRID 727
ARSENIGSÄURESALZ 729
ARSENIK (GELBES) 728
ARSENIK (WEISSER) 727
ARSENIT 729
ARSENKUPFER 725
ARSENMETALL 726
ARSENPENTOXYD 723
ARSENSÄURE 722
ARSENSÄUREANHYDRID 723
ARSENSÄURESALZ 720
ARSENSULFÜR 10262
ARSENSUPERSULFÜR 728
ARSENTRICHLORID 724
ARSENTRIOXYD 727
ARSENTRISULFID 728
ARSENWASSERSTOFF 730
ASBEST 746
ASBEST 755
ASBESTFILZ 750
ASBESTGEWEBE 748
ASBESTPAPIER 752
ASBESTPAPPE 747
ASBESTRING 753
ASBESTSCHNUR 749
ASBESTSTREIFEN 754
ASBESTZOPF 751
ASCHE 2397
ASCHE 760
ASCHE (KLEINE) 11904
ASCHEFREI 5636
ASCHENFALL 759
ASCHENGEHALT 758
ASCHENKÜHLER 2398
ASCHENRAUM 759
ASPHALT 768
ASPHALT-GOUDRON 1280
ASPHALT-TEER 1280
ASPHALTFILZ 765
ASPHALTIEREN 763
ASPHALTIEREN 767
ASPHALTIERTES ROHR 766
ASPHALTLACK 1283
ASPHALTMASTIX 764
ASPHALTROHR 6803
ASPHALTSTEINPULVER 9723
ASPIRATIONSVENTILATION 13450
ASPIRATOR 769

AST 1555
ASTATISCH 774
ASTERISMUS 776
ASTROIDE 5602
ASYMMETRIE 779
ASYMMETRISCH 778
ASYMPTOTE 780
ASYMPTOTISCH 781
ASYNCHRON 784
ASYNCHRONISMUS 782
ASYNCHRONMOTOR 783
ÄTHAN 4848
ÄTHANDISÄURE 8969
ÄTHER 8853
ATHERMAN 786
ÄTHERSCHWINGUNGEN 13560
ÄTHYLALDEHYD 84
ÄTHYLALKOHOL 313
ÄTHYLÄTHER 8853
ÄTHYLAZETAT 88
ÄTHYLCHLORID 4849
ÄTHYLEN 4851
ÄTHYLNITRAT 4850
ÄTHYLWASSERSTOFF 4848
ATLASGLANZ 11455
ATMOSPHÄRE 787
ATMOSPHÄRE (MASSEINHEIT) 788
ATMOSPHÄRE (PRÄPARIERTE) 789
ATMOSPHÄRE (VERDÜNNTE) 13441
ATMOSPHÄRILIEN 793
ATMUNGSVERLUST 1594
ATOM 800
ATOMGEWICHT 807
ATOMTHEORIE 805
ATOMVOLUMEN 806
ATOMWÄRME 802
ATOMZAHL 803
ÄTZALKALISCHE LÖSUNG 9671
ÄTZAMMONIAK 660
ÄTZBARYT 1000
ÄTZE 657
ÄTZEN 4839
ÄTZEN 4846
ÄTZEN (KATHODISCHES) 2110
ÄTZEND 3196
ÄTZFIGUR 4845
ÄTZFIGUREN 4842
ÄTZFLÜSSIGKEIT 4847
ÄTZGRÜBCHEN 9416
ÄTZKALI 9687
ÄTZKALILAUGE 9671

AUFSCHRAUBEN 13395
AUFSCHRUMPFEN 11390
AUFSCHRUMPFEN 11398
AUFSEHER 8958
AUFSICHT 12481
AUFSICHTSBILD 13037
AUFSPANNPLATTE 9499
AUFSPANNVORRICHTUNG 5301
AUFSPEICHERN 12294
AUFSPEICHERN 12299
AUFSPEICHERUNG VON ENERGIE 12290
AUFSPEICHREUNG VON VÄRME 12291
AUFSTAPELN (HOLZ) 9307
AUFSTAPELN DES HOLZES 9309
AUFSTECKFRÄSER 663
AUFSTECKREIBAHLE 11336
AUFSTECKSCHLÜSSEL 509
AUFSTELLEN 4805
AUFSTELLER 4816
AUFSTELLUNG 4808
AUFSTELLUNGSORT 9423
AUFTRAG 2587
AUFTRAGEN EINER SCHUTZSCHICHT 6792
AUFTRAGSCHWEISSUNG 13762
AUFTRAGSMETALL 3775
AUFTRAGSPROZESS 1696
AUFTRAGSSCHWEISSZUSATZ 8952
AUFTREIBEN (EIN ROHR) 4903
AUFTREIBEPROBE 4911
AUFTRIEB 1742
AUFWACHSEN 12543
AUFWALLEN 4461
AUFWÄRTSSCHWEISSUNG 13419
AUFWÄRTSTRANSFORMATOR 12245
AUFWEITEN 5356
AUFWEITEN (EIN ROHR) 4903
AUFWEITEPROBE 4911
AUFWEITUNG 1723
AUFWICKELN 13862
AUFWICKELN 13853
AUFWINDEN 6517
AUFWINDEN 13853
AUFWINDEN 6553
AUFWINDEN 13862
AUFWÖLBUNG 1723
AUFWULSTUNG 1723
AUFZEICHNEN 9519
AUFZEICHNEN 10288
AUFZEICHNEN EINES DIAGRAMMES 9521
AUFZIEHEN 7549
AUFZIEHEN (EIN RAD) 4222

AUFZIEHEN (EINE RIEMENSCHEIBE) 4222
AUFZUG 9101
AUFZUG 6007
AUFZUGSEIL 7546
AUGE 4980
AUGE (ANGEGOSSENES) 7817
AUGE EINES HAMMERS 4975
AUGENLAGER 11788
AUGENSTAB (GESCHMIEDETER) 5551
AUGIT 819
AUGITSYENIT 820
AURICHLORID 821
AURIPIGMENT 728
AUROCHLORID 823
AUS DEM GROBEN 10786
AUSBALANCIEREN 930
AUSBALANCIERUNG 942
AUSBAUCHEN (SICH) 1721
AUSBAUCHUNG 1723
AUSBAUCHUNG 1720
AUSBESSERN 10460
AUSBESSERUNG 10459
AUSBESSERUNGSKOSTEN 3208
AUSBEULEN 5405
AUSBEULUNG 1720
AUSBLASEHAHN 1357
AUSBLASEVENTIL 1358
AUSBLÜHUNG 4464
AUSBLUTEN 1321
AUSBOHREN 1468
AUSBOHREN 1462
AUSBOHRMASCHINE 1471
AUSBREITEPROBE 5409
AUSDEHNEN (SICH) 4902
AUSDEHNUNG 4679
AUSDEHNUNG (LINEAIRE) 7612
AUSDEHNUNG (RÄUMLICHE) 13603
AUSDEHNUNG DURCH WÄRME 4948
AUSDEHNUNG VON DAMPF ODER GAS 4921
AUSDEHNUNGSFÄHIG 3948
AUSDEHNUNGSFÄHIGKEIT 3947
AUSDEHNUNGSFUGE 4918
AUSDEHNUNGSHUB 2775
AUSDEHNUNGSKOEFFIZIENT 2636
AUSDEHNUNGSKOEFFIZIENT 4914
AUSDEHNUNGSKUPPLUNG 4915
AUSDEHNUNGSKUPPLUNG 11617
AUSDEHNUNGSSTÜCK 4922
AUSDEHNUNGSZAHL 2628
AUSDREHEN 1473
AUSDREHEN 1466

AUSLEGERARM 10128
AUSLEGERKRAN 7224
AUSLERGERLÄNGE 7225
AUSLÖSCHEBEL 10438
AUSLÖSEDAUMEN 10437
AUSLÖSEFINGER 10437
AUSLÖSER 2407
AUSLÖSUNGSKUPPLUNG 4032
AUSMASSE 3960
AUSMESSEN 8078
AUSNÜTZEN 13430
AUSNUTZEN (EINE ERFINDUNG) 13938
AUSNUTZUNG 13429
AUSNUTZUNG EINER ERFINDUNG 13948
AUSPRESSEN 12039
AUSPUFF 4890
AUSPUFFDAMPFMASCHINE 8627
AUSPUFFGAS 4891
AUSPUFFGERÄUSCHDÄMPFER 8448
AUSPUFFHUB 4898
AUSPUFFROHR 4894
AUSPUFFSAMMELROHR 4893
AUSPUFFTOPF 11431
AUSPUFFTOPF 8448
AUSPUFFVENTIL 797
AUSRADIEREN 4799
AUSRECHNEN 3795
AUSREIBER 1636
AUSRICHTEN 324
AUSRICHTEN 12319
AUSRICHTEN 323
AUSRICHTEN 12318
AUSRÜCKBARE KUPPLUNG 4033
AUSRÜCKEN 2548
AUSRÜCKEN 10034
AUSRÜCKEN 12898
AUSRÜCKHEBEL 4035
AUSRÜCKMUFFE 11599
AUSRÜCKUNG 4028
AUSRÜCKUNG 12898
AUSRÜCKVORRICHTUNG 4034
AUSRUNDUNG 6537
AUSRÜSTUNG 5283
AUSSCHALTEN 12559
AUSSCHALTEN 12898
AUSSCHALTEN 10034
AUSSCHALTEN (EINE ROHRLEITUNG) 11407
AUSSCHALTEN (ELEKTR.) 12554
AUSSCHALTER 2407
AUSSCHALTER (ELEKT.) 12553
AUSSCHALTUNG 12898

AUSSCHEIDEN (SICH) 9760
AUSSCHEIDUNG 9761
AUSSCHEIDUNGSHÄRTUNG 9762
AUSSCHLAG 486
AUSSCHLAGBILDUNG 11961
AUSSCHLAGEISEN 6545
AUSSCHLAGEN 9554
AUSSCHLAGEN 11251
AUSSCHLAGEN 7328
AUSSCHLAGPUNZE 6545
AUSSCHLAGROST 7334
AUSSCHLAGRÜTTLER 3153
AUSSCHLAGSTAHL 6545
AUSSCHLAGWINKEL 518
AUSSCHLEUDERN 6702
AUSSCHMELZMODELL 7116
AUSSCHNEIDEN 10012
AUSSCHNEIDEN 9998
AUSSCHNITT 11135
AUSSCHNITT 5866
AUSSCHUSS 3696
AUSSCHUSSWARE 3696
AUSSCHWIMMEN 5443
AUSSCHWIMMEN 5434
AUSSCHWITZUNG 12540
AUSSEN-RUNDSCHLEIF-MASCHINE 4953
AUSSEN-RUNDSCHLEIF-MASCHINE (SPITZENLOSE) 2155
AUSSENBELEUCHTUNG 8897
AUSSENDURCHMESSER 8913
AUSSENGETRIEBE 4956
AUSSENGEWINDE 4960
AUSSENGEWINDE 4961
AUSSENGEWINDE 7955
AUSSENGEWINDESCHNEIDEN 3512
AUSSENGEWINDESCHNEIDMASCHINE 12869
AUSSENHANDELSBILANZ 13099
AUSSENLEHRE 5091
AUSSENLEHRE 4955
AUSSENLUFT 450
AUSSENMASSE 8922
AUSSENPUTZ 3264
AUSSENRING EINES KUGELLAGERS 8898
AUSSENSCHLIFF 4957
AUSSENSEITE 8911
AUSSENSTRÄHLER 4952
AUSSENTASTER 8912
AUSSENVERZAHNUNG 4959
AUSSENWINKEL 4950
AUSSER BETRIEB SETZEN (EINE MASCHINE) 11406
AUSSER EINGRIFF 8892
AUSSERACHSIG 8893

AZETYLZELLULOSE 91
AZETYLZELLULOSE 2132
AZIDIMETRIE 131
AZURBLAU 13334
BACKEN EINES SCHRAUBENSCHLÜSSELS 7216
BACKEN EINES SCHRAUBSTOCKS 13564
BACKENFUTTER EINES SCHRAUBSTOCKS 13563
BACKSTEIN 1596
BAD 1062
BAD 3967
BAD (ELEKTROLYTISCHES) 4588
BADSTROM 1064
BADZEMENTIEREN 7646
BADZUSATZ 168
BADZUSÄTZE 9506
BAGASSE 924
BAGGER 4414
BAHN 13104
BAHN (ABSCHÜSSIGE) 2395
BAHN EINES HAMMERS 4993
BAHNSTEUERUNG 3039
BAINITE 925
BAJONETTVERSCHLUSS 1074
BALANCIEREN 935
BALATA 943
BALATAHARZ 943
BALATARIEMEN 944
BALKEN (EINGEBAUTER) 7010
BALKEN (HÖLZERNER) 949
BALKENWAAGE 1095
BALKENWERK 5630
BALLEN 945
BALLENKARREN 947
BALLHAMMER 2232
BALLIGDREHEN 13288
BALLIGDREHEN 13277
BALLIGKEIT 1506
BALLIGKEIT 3396
BALLON 1981
BAMBUSROHR 972
BAND 12641
BAND 974
BANDBREMSE 5681
BANDEISEN 6575
BÄNDER 980
BANDKETTE 12058
BANDKUPPKLUNG 7349
BANDMASS 8092
BANDNIETUNG 1781
BANDSÄGE 978
BANDSÄGEMASCHINE 977

BANDSCHARNIER 12347
BANDSCHREIBER 12380
BANDSEIL 5393
BANDSPANNUNGSANZEIGER 976
BANDSPEKTRUM 5487
BANDSTAHL 12216
BANDSTAHL 902
BANDSTAHL (WARMGEWALZTER) 371
BANDTRANSPORTEUR 1164
BANK 1167
BANK (OPTISCHE) 9268
BANKAZINN 973
BANKBOHRMASCHINE 1168
BANKFORMUNG 1169
BANKHAMMER 6235
BANKMEISSEL 2717
BANKSCHRAUBSTOCK 1171
BÄR 11550
BARETTFEILE 1903
BARIUM 992
BARIUM (KOHLENSAURES) 996
BARIUM (SCHWEFELSAURES) 1003
BARIUMALUMINAT 994
BARIUMAZETAT (ESSIGSAURES) 993
BARIUMCHLORID 997
BARIUMDIOXYD 998
BARIUMHYDROXYD 1000
BARIUMHYPEROXYD 998
BARIUMKARBID 995
BARIUMKARBONAT 996
BARIUMMONOXYD 1020
BARIUMNITRAT 1001
BARIUMOXYDHYDRAT 1000
BARIUMPLATINZYANÜR 1002
BARIUMSULFAT 1003
BARIUMSULFID 1004
BARIUMSUPEROXYD 998
BARN 1006
BAROGRAPH 1007
BAROMETER 1008
BAROMETERSTAND 6422
BAROMETRISCH 1009
BAROSKOP 1011
BART 1764
BARUIMFLUORID 999
BARYT 993
BARYT 1020
BARYT (KAUSTISCHER) 1000
BARYT (KOHLENSAURER) 13911
BARYT (SALPETERSAURER) 1001
BARYT (SCHWEFELSAURER) 1021

BARYTHYDRAT 1000
BARYTWASSER 1019
BARYTWEISS 1003
BASALT 1022
BASALTTUFF 1023
BASE 1026
BASE (CHEM.) 1024
BASIS (GEOM.) 1027
BASIS EINES GEWINDES 1028
BASIS EINES LOGARITHMUS 1033
BASISCH 1042
BASIZITÄT 1055
BASIZITÄT 340
BAST 1056
BASTARDFEILE 1057
BATTERIE 1066
BATTERIE 4534
BATTERIE (GALVANISCHE) 9845
BATTERIEKASTEN 1067
BATTERIESCHALTER 1068
BAU 12395
BAUART 13320
BAUART 3794
BAUBREITE 8926
BAUBRONZE 692
BAUCHSÄGE 3365
BAUEN 1694
BAUGERÜST 10965
BAUGLIED 8145
BAUGRUND 1699
BAUHÖHE 8924
BAUHOLZ 12937
BAUINGENIEUR 2443
BAUINGENIEURWESEN 2444
BAULÄNGE 8925
BAULÄNGE EINES KETTENGLIEDES 6978
BAULEITER 5145
BAUMATERIAL 8050
BAUMETALLE (GEWÖHNLICHE) 2988
BAUMWOLLE 3217
BAUMWOLLRIEMEN 3218
BAUMWOLLSAMENÖL 3222
BAUMWOLLSEIL 3221
BAUMWOLLZOPF 3220
BAUSCHINGER-EFFEKT 1070
BAUSTAHL 12392
BAUSTAHL 12393
BAUSTAHLBLECH 1109
BAUSTEIN 1700
BAUSTEINE 1697
BAUSTELLENBARACKE 2987

BAUSTELLENBUDE 11519
BAUSTELLENLEITER 11520
BAUSTELLENMONTAGE 8800
BAUSTELLENSCHWEISSNAHT 5146
BAUSTOFF 8050
BAUTEIL (EINGEBAUTES) 1704
BAUUNTERNEHMER 1698
BAUWERK 12395
BAUXIT 1071
BAUZEICHNUNG 693
BAYER-VERFAHREN 1073
BEANSPRUCHEN (EINEN KÖRPER) 12323
BEANSPRUCHT WERDEN (AUF DRUCK) 1075
BEANSPRUCHT WERDEN (AUF ZUG) 1075
BEANSPRUCHUNG 12331
BEANSPRUCHUNG 12367
BEANSPRUCHUNG 7679
BEANSPRUCHUNG (DYNAMISCHE) 4398
BEANSPRUCHUNG (KRITISCHE) 3351
BEANSPRUCHUNG (MAKROSKOPISCHE) 7874
BEANSPRUCHUNG (ZULÄSSIGE) 9215
BEANSPRUCHUNG (ZULÄSSIGE) 10880
BEANSPRUCHUNG (ZULÄSSIGE) 353
BEANSPRUCHUNG AUF BIEGUNG 1191
BEANSPRUCHUNG AUF DRUCK 2869
BEANSPRUCHUNG AUF KNICKUNG 3339
BEANSPRUCHUNG AUF SCHUB 11295
BEANSPRUCHUNG AUF ZUG 12747
BEARBEITBAR 13947
BEARBEITBARKEIT 7838
BEARBEITBARKEIT 13946
BEARBEITEN 13937
BEARBEITEN 13959
BEARBEITEN MIT SANDSTRAHL 10921
BEARBEITET 7861
BEARBEITUNG 7864
BEARBEITUNG 10786
BEARBEITUNG 13959
BEARBEITUNG (VOREINGESTELLTE) 9752
BEARBEITUNG AUS DEM ROHEN 10786
BEARBEITUNGSFÄHIG 13947
BEARBEITUNGSFLANSCHE 13013
BEARBEITUNGSSTUFE 7866
BEARBEITUNGSZUGABE 7865
BEARBEITUNGSZUGABE 358
BEAUFSICHTIGEN 12474
BEAUFSICHTIGUNG 12481
BECHERGLAS 1092
BECHERKABEL 1675
BECHERKETTE 1675
BECHERWERK 1676

BESTANDTEIL (ARBEITENDER) 13953
BESTANDTEIL (FLÜCHTIGER) 13586
BESTANDTEIL EINER LEGIERUNG 2982
BESTIMMUNG (ANALYTISCHE) 496
BESTIMMUNG (ELEKTROLYTISCHE) 4594
BESTIMMUNG (EXPERIMENTELLE) 4929
BESTIMMUNG (GRAVIMETRISCHE) 6077
BESTIMMUNG (KALORIMETRISCHE) 1880
BESTRAHLUNG 7161
BESTRAHLUNG 10241
BETAMESSING 1216
BETASTRAHLEN 1218
BETASTRUKTUR 1219
BETATEILCHEN 1217
BETON 2894
BETON (ARMIERTER) 12203
BETON (BEWEHRTER) 12203
BETON-(DECKEN-)PLATTE 2900
BETONARBEITEN 2901
BETONEISEN 12198
BETONEISENSCHERE 2897
BETONFUNDAMENT 2896
BETONIEREN 2895
BETONIEREN 2902
BETONROHR 2898
BETONVORSPANNUNG 2899
BETRIEB 8837
BETRIEB (AUSSETZENDER) 7056
BETRIEB (INTERMITTIERENDER) 7056
BETRIEB (UNUNTERBROCHENER) 3038
BETRIEBFOLGE 13626
BETRIEBSBEDINGUNGEN 11208
BETRIEBSBELASTUNG 10879
BETRIEBSDRUCK 13957
BETRIEBSFÄHIG 6819
BETRIEBSFÄHIGEM ZUSTAND (IN) 6819
BETRIEBSFÜHRER 13964
BETRIEBSFÜHRUNG 13963
BETRIEBSFÜHRUNG (WISSENCHAFTLICHE) 11001
BETRIEBSGESCHWINDIGKEIT 8663
BETRIEBSKOSTEN 13950
BETRIEBSKRAFT 8409
BETRIEBSLEITER 13964
BETRIEBSPAUSE 7100
BETRIEBSSICHER 10440
BETRIEBSSICHERHEIT 10439
BETRIEBSSPANNUNG (ELEKTR.) 8659
BETRIEBSSTOFF 8408
BETRIEBSSTÖRUNG 1581
BETRIEBSTEMPERATUR 8836
BETRIEBSUNFALL 75

BETRIEBSUNTERBRECHUNG 7092
BETRIEBSVERHÄLTNISSE 13949
BETRIEBSVERHÄLTNISSE 8834
BETRIEBSVORSCHRIFTEN 11209
BETRIEBSWASSER 13649
BETRIEBSZUSTAND (NORMALER) 8857
BETTS-VERFAHREN 1220
BEUGUNG 3924
BEUGUNGSGITTER 3925
BEULE 1720
BEWÄSSERUNGSANLAGEN 7172
BEWEGUNG 8438
BEWEGUNG 8406
BEWEGUNG (ABSOLUTE) 32
BEWEGUNG (APERIODISCHE) 636
BEWEGUNG (AUSSETZENDE) 7053
BEWEGUNG (BESCHLEUNIGTE) 60
BEWEGUNG (DREHENDE) 10766
BEWEGUNG (FORTSCHREITENDE) 8407
BEWEGUNG (GEGENLÄUFIGE) 3243
BEWEGUNG (GEGENSEITIGE) 10434
BEWEGUNG (GERADLINIGE) 10320
BEWEGUNG (GLEICHFÖRMIGE) 13362
BEWEGUNG (GLEICHMÄSSIG BESCHLEUNIGTE) 13365
BEWEGUNG (GLEICHMÄSSIG VERZÖGERTE) 13367
BEWEGUNG (GLEITENDE) 11605
BEWEGUNG (HIN-UND HERGEHENDE) 13725
BEWEGUNG (HIN-UND HERGEHENDE) 10283
BEWEGUNG (INTERMITTIERENDE) 7053
BEWEGUNG (KRAFTSCHLÜSSIGE) 8629
BEWEGUNG (KREISFÖRMIGE) 2421
BEWEGUNG (KRUMMLINIGE) 3500
BEWEGUNG (PERIODISCHE) 9194
BEWEGUNG (RELATIVE) 10434
BEWEGUNG (ROLLENDE) 10702
BEWEGUNG (RÜCKLÄUFIGE) 919
BEWEGUNG (RUCKWEISE) 7053
BEWEGUNG (SCHWINGENDE) 8881
BEWEGUNG (SENKRECHTE) 13549
BEWEGUNG (SPHÄRISCHE) 11889
BEWEGUNG (STETIGE) 3031
BEWEGUNG (STOSSWEISE) 7053
BEWEGUNG (UNGLEICHFÖRMIG BESCHLEUNIGTE) 8648
BEWEGUNG (UNGLEICHFÖRMIG VERZÖGERTE) 8649
BEWEGUNG (UNGLEICHFÖRMIGE) 8647
BEWEGUNG (UNTERBROCHENE) 7053
BEWEGUNG (VARIABLE) 13490
BEWEGUNG (VERÄNDERLICHE) 13490
BEWEGUNG (VERZÖGERTE) 10518
BEWEGUNG (ZWANGSLÄUFIGE) 2985
BEWEGUNG AUF EINER SCHRAUBENLINIE 6426

BINDERMASCHINEN ALLER ARTEN 948
BINOMIALKOEFFIZIENT 1257
BINOMIALREIHE 1260
BINOMIALVERTEILUNG 1258
BIRNE 7385
BISEKTRIX (SPITZE) 156
BISKUIT 1276
BISKUITGUT 1276
BISMUTIN 1275
BIT 1249
BITTERMANDELÖL 1198
BITTERSALZ 7889
BITTERSPAT 7880
BITUMEN 1280
BITUMENHALTIG 1281
BITUMINÖS 1281
BLANK SCHLEIFEN 9598
BLANKBEIZEN 11005
BLANKBEIZEN 11006
BLANKE STELLE 6536
BLANKETT 1301
BLANKGLÜHDRAHT 1606
BLANKGLÜHEN 573
BLANKSCHLEIFEN 9605
BLASE 1353
BLASE 1351
BLASEBALG 1139
BLASEBALGHAHN 1138
BLASEN 1330
BLASEN 11903
BLASEN 1362
BLASENBILDUNG 1330
BLASENBILDUNG 2437
BLASENBILDUNG 1672
BLASENDER LÜFTER 9817
BLASENSTAHL 2138
BLASFORM 13299
BLASIGE STELLE IM GUSS 1353
BLASLOCH 6393
BLASLÖTROHR 8433
BLASWIRKUNG (MAGNETISCHE) 665
BLATT 1299
BLATT (DESCARTESSCHES) 5513
BLÄTTERMAGNET 7377
BLATTFEDER 7485
BLATTFEDER 9495
BLATTFEDERWERK 2834
BLATTGOLD 6000
BLATTGOLDSCHLÄGEREI 5994
BLATTKOHLSCHNEIDER 7272
BLATTMETALL 5509

BLÄTTRIGE STRUKTUR 7373
BLATTZINN 12955
BLAU (BERLINER) 1206
BLAUBRENNE 1369
BLAUBRÜCHIGKEIT 1631
BLAUBRUCHPROBE 1372
BLAUDRUCKPAPIER 5119
BLAUGLÜHEN 572
BLAUGLUT 1370
BLAUPAUSE 1375
BLAUSÄURE 6707
BLAUSAURES KALI 9683
BLAUSTICHIGES EOSIN 4831
BLAUUNG 1377
BLAUWARM 1371
BLECH 11315
BLECH 11298
BLECH 11305
BLECH (EIN) ENTGRATEN 13205
BLECH (GALVANISIERTES) 5785
BLECH (GELOCHTES) 9189
BLECH (GERICHTETES) 5395
BLECH (GERIFFELTES) 2325
BLECH (GERIPPTES) 2325
BLECH (GEWALZTES) 10674
BLECH (GLATTES) 1296
BLECH (KALTGEWALZTES) 2696
BLECH (KORNORIENTIERTES) 6036
BLECH (PERFORIERTES) 9189
BLECH (PLATTIERTES) 2445
BLECH (SCHWARZES) 1294
BLECH (VERBLEITES) 7460
BLECH (VERZINKTES) 5785
BLECH (WARMGEWALZTES) 6637
BLECH (ZUGESCHNITTENES) 11266
BLECH MIT GEBOHRTEN LÖCHERN 4274
BLECH MIT GESTANZTEN LÖCHERN 10003
BLECH- ODER BANDMETALLSCHERMASCHINE 11323
BLECHABFALL 9493
BLECHABKANTPRESSE 9481
BLECHANHAFTUNG 8046
BLECHANREISSEN 8018
BLECHBEARBEITUNGSMASCHINEN 11320
BLECHBIEGEMASCHINE 9480
BLECHDICKE 9487
BLECHDICKE 12845
BLECHE (GESTRECKTE) 4250
BLECHE (UMHÜLLTE) 2586
BLECHE ANEINANDERLEGEN 8045
BLECHE AUS LEGIERTEN STÄHLEN 380
BLECHEINLEGEFOLGE 9500

BLECHKANNE 12961
BLECHKANTE 9483
BLECHKANTENABSCHRÄGUNG 9496
BLECHKANTENHOBELMASCHINE 9484
BLECHKONSTRUKTIONSZEICHNUNG 9504
BLECHLEHRE 11311
BLECHMANTEL 11310
BLECHNUMMER 5880
BLECHPERFORIERMASCHINEN 11317
BLECHPLANIERMASCHINEN 11318
BLECHRAND 9483
BLECHRAND 4441
BLECHRICHTMASCHINE 9485
BLECHRICHTMASCHINE 11316
BLECHRITZMASCHINE 11319
BLECHROHR 11312
BLECHSCHERE 9494
BLECHSCHERMASCHINE 9494
BLECHSCHUTZ 655
BLECHSTÄRKE 12845
BLECHSTREIFENWALZWERK 11321
BLECHTAFEL 11313
BLECHTRÄGER 9490
BLECHWALZSTRASSE 11324
BLEI 7447
BLEI (BASISCH) 13834
BLEI (CHROMSAURES) 2375
BLEI (KOHLENSAURES) 13834
BLEI (REINES) 2311
BLEI (SCHWEFESAURES) 7473
BLEI (UNREINLICHES) 1029
BLEI (UNTERSCHWEFLIGSAURES) 7474
BLEI UNRAFFINIERTES) 1029
BLEI-ZINNLOT 7475
BLEIABGUSS 7457
BLEIAKKUMULATOR 7455
BLEIAZETAT 7450
BLEIBAD 7454
BLEIBENDE HÄRTE DES WASSERS 9203
BLEIBLECH 11314
BLEIBRONZE 7456
BLEICHERDE 5727
BLEICHKALK 1317
BLEICHLORID 7458
BLEICHPULVER 1317
BLEICHROMAT 2375
BLEIDICHTUNG FÜR ROHRE 7464
BLEIDIOXYD 7461
BLEIDRAHT 7477
BLEIERZ 7465
BLEIESSIGSÄURE 7450

BLEIFARBE 7466
BLEIFOLIE 7462
BLEIGELB 8031
BLEIGLANZ 5775
BLEIGLÄTTE 7663
BLEIHAMMER 7463
BLEIKARBONAT 2196
BLEILEGIERUNG 7451
BLEILEGIERUNG 7453
BLEILOT 9540
BLEIMENNIGE 10333
BLEIOXYD 7663
BLEIOXYD (GELBES) 8031
BLEIPEROXYD 7461
BLEIRING 7470
BLEIROHR 7468
BLEISCHROT 7472
BLEISCHWAMM 11954
BLEISCHWEISSEN 7476
BLEISPAT 2196
BLEISTIFT 7478
BLEISTIFTEINSATZ FÜR ZIRKEL 9165
BLEISTIFTGUMMI 9163
BLEISTIFTZEICHNUNG 7467
BLEISTIFTZIRKEL 2804
BLEISULFAT 7473
BLEISUPEROXYD 7461
BLEITHIOSULFAT 7474
BLEIVERARBEITUNG 7484
BLEIWAAGE 9540
BLEIWEISS 13834
BLEIZUCKER 7450
BLEIZUCKERPAPIER 7449
BLENDE 3871
BLENDE 1324
BLENDUNG 1325
BLINDAUFKOHLEN 9959
BLINDBODEN 5026
BLINDFLANSCH 1304
BLINDFLANSCH 1326
BLINDKALIBER 4372
BLINDLOCH 3630
BLINDSTICH 4372
BLINDWALZE 4373
BLINDWERDEN 5507
BLINDWIDERSTAND 10247
BLINKLEUCHTE 5366
BLINKLICHTER 3989
BLITZABLEITER 7576
BLITZSCHUTZVORRICHTUNG 7575
BLITZTROCKNER 5360

BLOCK **6919**
BLOCK **1332**
BLOCK ERRATISCHER **4821**
BLOCKDREHBANK **6922**
BLÖCKE AUS KOHLENSTOFFSTAHL **1955**
BLOCKENDE **4730**
BLOCKFORM **6923**
BLOCKKETTE **1336**
BLOCKSEIGERUNG **8660**
BLOCKSEIGERUNG **6925**
BLOCKWAGEN **6920**
BLOCKWALZEN **2642**
BLOCKZINN **1339**
BLUTALBUMIN **11207**
BOCK **12490**
BOCKKRAN **5804**
BOCKLAGER **9154**
BOCKWINDE **13851**
BODEN (FALSCHER) **5026**
BODEN (FLACHGEWÖLBTER) **13045**
BODEN (GEWÖLBTER) **4036**
BODEN (WASSERFÜHRENDER) **13642**
BODEN ENTWÄSSERN **4209**
BODENABLAUF **1492**
BODENBEARBEITUNGSGERÄTE **3457**
BODENBELAG **5451**
BODENBLECH **5446**
BODENBRETT **5446**
BODENDÜSE **1495**
BODENFREIHEIT **6143**
BODENLÄNGSTRÄGER **5625**
BODENPLATTE **5449**
BODENSATZ **9759**
BODENSTEIN (BASISCHER) UND FUTTER (BASISCHES) **1045**
BODENUNTERSUCHUNG **11752**
BODENVENTIL **1493**
BODENVENTIL **5523**
BOGEN **664**
BOGENDACH **4039**
BOGENHÖHE **10613**
BOGENLAMPE **672**
BOGENLÄNGE **7517**
BOGENLICHT **673**
BOGENMASS **2419**
BOGENROHR **1180**
BOGENSÄGE **8186**
BOGENSÄGE **3108**
BOGENSTÜCK **1180**
BOGENZIRKEL **13876**
BOGENZIRKEL MIT GEZAHNTEM BOGEN **10122**
BOHLE **9457**

BOHR-LÜNETTEN **1474**
BOHRBRETT **1590**
BOHRDURCHMESSER **1439**
BOHREN **4277**
BOHREN **4264**
BOHREN **1468**
BOHREN (EINES GEWINDES) **12678**
BOHREN (GEWINDE) **12636**
BOHRER **4271**
BOHRER **4263**
BOHRER (ARBEITER) **4276**
BOHRER (GERADEGENUTETER) **12304**
BOHRER (HARTMETALLBESTÜCKTER) **1936**
BOHRERLEHRE **4267**
BOHRERSPITZENDURCHMESSER **4269**
BOHRFUTTER **4266**
BOHRHAMMER **258**
BOHRKERN **3142**
BOHRKNARRE **10219**
BOHRKÖPFE (MEHRSPINDLIGE) **8458**
BOHRKRONE **4265**
BOHRKURBEL **1518**
BOHRLEHRE **4267**
BOHRLOCH **4273**
BOHRLOCH (IM GESTEIN) **1463**
BOHRMASCHINE **1471**
BOHRMASCHINE **4279**
BOHRMASCHINE (FAHRBARE) (TRAGBARE) **9636**
BOHRMASCHINE (LIEGENDE) **6581**
BOHRMASCHINE (MEHRSPINDLIGE) **8470**
BOHRMASCHINE (STEHENDE) **13552**
BOHRMASCHINEN FÜR MASTEN **9664**
BOHRÖL **4280**
BOHRRATSCHE **10219**
BOHRSPÄNE **1476**
BOHRSPÄNE **11277**
BOHRSPINDELSTOCK **4278**
BOHRSTÄHLE FÜR GESTEINS-BOHRUNGEN **4270**
BOHRSTANGE **4265**
BOHRSTANGE (EINTEILIGE) **11778**
BOHRSTANGEN **1470**
BOHRTURM PONTON **3787**
BOHRUNG **3855**
BOHRUNG **4277**
BOHRUNG **6530**
BOHRUNG **1461**
BOHRUNG EINER RIEMENSCHEIBE **4978**
BOHRUNG EINES RADES **4978**
BOHRUNGSDURCHMESSER **3855**
BOHRUNGSLEHRE **6531**
BOHRUNGSMESSGERÄT MIT SKALA **3838**

BREMSGEWICHT 1547
BREMSHEBEL 1537
BREMSKLOTZ 1543
BREMSKLOTZ 1531
BREMSKRAFT 1551
BREMSKREIS 1545
BREMSLEISTUNG 1535
BREMSLEISTUNG 1540
BREMSLEISTUNG 4458
BREMSPFERDESTÄRKE (EFFEKTIVE) 151
BREMSRING 1541
BREMSSCHEIBE 1548
BREMSTROMMEL 1532
BREMSTROMMEL 1533
BREMSVERSUCH 1546
BREMSWIDERSTAND 1552
BREMSWIRKUNG 1550
BREMSZYLINDER 1532
BRENNBAR 2772
BRENNBARKEIT 2773
BRENNEBENE 5501
BRENNEN 1745
BRENNER 1749
BRENNER 13042
BRENNER 2848
BRENNFLÄCHE 2121
BRENNGAS 5709
BRENNGESCHWINDIGKEIT 10227
BRENNHÄRTEN 5315
BRENNHOLZ 5258
BRENNKEGEL 11134
BRENNLINIE 2116
BRENNLINIE DURCH REFLEXION 2089
BRENNLINIE DURCH REFRAKTION 3829
BRENNMITTEL 5705
BRENNOFEN 7295
BRENNPUNKT 5502
BRENNPUNKT (EINES ÖLES) 5247
BRENNPUNKT (REELLER) 10258
BRENNPUNKT (SCHEINBARER) 13571
BRENNPUNKT (VIRTUELLER) 13571
BRENNPUNKT (WIRKLICHER) 10258
BRENNPUNKTABSTAND 5503
BRENNPUTZEN 5309
BRENNPUTZEN (FLÄMMEN) 8976
BRENNSCHNEIDEN 7387
BRENNSCHNEIDEN 5817
BRENNSTOFF 5705
BRENNSTOFF (FESTER) 11785
BRENNSTOFF (FLÜSSIGER) 7650
BRENNSTOFF (FOSSILER) 5586

BRENNSTOFF (GASFÖRMIGER) 5854
BRENNSTOFF (HOCHWERTIGER) 6484
BRENNSTOFF (KÜNSTLICHER) 732
BRENNSTOFF (NATÜRIICHER) 8503
BRENNSTOFF (STÜCKIGER) 7827
BRENNSTOFF MINDERWERTIGER 7780
BRENNSTOFFBEDARF 3745
BRENNSTOFFERSPARNIS 5707
BRENNSTOFFVERBRAUCH 2994
BRENNSTOFFZUFÜHRUNG 5082
BRENNWEITE 5503
BRENNWEITE 5500
BRENZESSIGGEIST 90
BRETT 1379
BRETTFALLHAMMER 1380
BRETTSÄGE 10609
BRIGGS'SCHES GEWINDE 1605
BRIKETT 1623
BRIKETTIERTES ERZ 8860
BRIKETTIERUNG 7990
BRILLE (FESTSTEHENDE) 12154
BRILLE (LAUFENDE) 5514
BRINELL-HÄRTEPRÜFER 1553
BRINELL-PRESSE 1553
BRINELL'SCHER 1620
BRINELLHÄRTE 1619
BRINELLPROBE 1618
BRITANNIAMETALL 1624
BROM 1649
BROMARGYRIT 1645
BROMAT 1646
BROMID 1648
BROMIT 1645
BROMKALIUM 9678
BROMMETALL 1648
BROMSÄURE 1647
BROMSILBER 1645
BROMWASSERSTOFFSÄURE 6704
BRONZE 1650
BRONZE (SÄUREBESTÄNDIGE) 1655
BRONZEDRAHT 1654
BRONZELACK 6001
BRONZEN 216
BRONZEPULVER 1652
BRONZEROHR 1653
BRONZIEREN 1651
BRONZIEREN 1664
BRUCH 5620
BRUCH 10861
BRUCH 10858
BRUCH 3278

DÄMPFUNG 3606
DÄMPFUNG 3604
DÄMPFUNG (INNERE) 7065
DÄMPFUNGSFÄHIGKEIT 3605
DÄMPFUNGSVERMÖGEN 3605
DÄMPFUNGVORRICHTUNG (HYDRODYNAMISCHE) 6709
DAMPFVERBRAUCH 2995
DAMPFVERBRAUCHMESSER 12174
DAMPFWEG 12176
DARMSAITENRIEMEN 6183
DARSTELLUNG (GEOMETRISCHE) 5926
DATEN 3612
DATEN 8696
DATEN (NUMERISCH EINGEGEBENE) 3942
DATENEINGABE (MANUELLE) 7987
DATENREGISTER 3613
DATENTRÄGER 12289
DATENWORT 13934
DAUBE 12149
DAUER EINES PATENTRECHTES 12761
DAUERBELASTUNG 3633
DAUERBETRIEB 3038
DAUERBETRIEB 9209
DAUERBRUCH 4740
DAUERFESTIGKEIT 5052
DAUERFESTIGKEIT 4741
DAUERFESTIGKEITSVERHÄLTNIS 4742
DAUERFORM 9207
DAUERGRENZE 5050
DAUERHAFT 4386
DAUERHAFTIGKEIT 4385
DAUERHAFTIGKEIT 4738
DAUERLASTPRÜFUNG 2974
DAUERMAGNET 9204
DAUERMAGNET-SPANNPLATTE 9205
DAUERRISS 4739
DAUERSCHWINGFESTIGKEIT 5053
DAUERSPANNUNG 2978
DAUERSTANDGRENZE 3325
DAUERVERSUCH 5049
DAUERVERSUCH 4743
DAUMEN 1883
DAUMENKRAFT 6688
DAUMENSCHEIBE 13221
DAUMENWELLE 1888
DÄUMLING 13878
DEBYE-SCHERRER METHODE 3643
DECHSEL 207
DECKANSTRICH 5217
DECKE 3667
DECKEL 13033

DECKEL 3269
DECKEL 3266
DECKEL 1910
DECKELBILDUNG 1927
DECKELFLANSCH 1304
DECKELMANSCHETTE 11548
DECKENLAGER 6321
DECKENLEUCHTE 4119
DECKENLEUCHTE 10709
DECKENVORGELEGE 8940
DECKFÄHIGKEIT 3273
DECKFARBE 1389
DECKGLAS 3265
DECKKRAFT EINER FARBE 1393
DECKLACK 13505
DECKPLATTE 13033
DECKPLATTE 3267
DECKRING 1232
DECKSCHICHT 9950
DEFEKT 5411
DEFORMATIONSBAND 3706
DEGRAS 3711
DEHNBAR 4361
DEHNBARKEIT 4363
DEHNGRENZE 9923
DEHNUNG 3949
DEHNUNG 4946
DEHNUNG 4679
DEHNUNG (RELATIVE) 13370
DEHNUNGSDICHTUNG 4918
DEHNUNGSFUGE 4918
DEHNUNGSKURVE 12365
DEHNUNGSMESSER 4949
DEHNUNGSMESSER 12325
DEHNUNGSMESSSTREIFE 12329
DEHNUNGSROHR 4913
DEICH 3945
DEKAGRAMM 3645
DEKALESZENZ 3646
DEKALITER 3647
DEKAMETER 3648
DEKANTIEREN 3649
DEKANTIEREN 3650
DEKAPIERBAD 111
DEKAPIEREN 11005
DEKAPIEREN 9288
DEKAPIEREN EINES METALLES 11006
DEKLINATION 3668
DEKODIERER 3670
DEKUPIEREN 8552
DEKUPIERMASCHINE 6012

DELIQUESZENZ 3733
DELIQUESZIEREN 3732
DELTA-EISEN 3740
DELTAMETALL 3741
DEMIJOHN 1981
DEMODULATION 10312
DEMODULATOR 8338
DEMONTAGE 12604
DEMONTIEREN 12600
DENATURIERTER SPIRITUS 8224
DENATURIERUNGSMITTEL 3747
DENDRIT 3748
DENDRIT 9331
DENSIMETER 3755
DEPOLARISATION 3771
DEPOLARISATION 3768
DEPOLARISATOR 3772
DEPOLARISATOR 3770
DEPOLARISIEREN 3769
DERIVAT 3786
DERRICKKRAN 3788
DESINFEKTIONSMITTEL 4042
DESINTEGRATOR 4045
DESOXYDATION 3762
DESOXYDATION 3766
DESOXYDATIONSMITTEL 3765
DESOXYDIEREN 3763
DESTILLAT 4071
DESTILLATION 4072
DESTILLATION (FRAKTIONIERTE) 5612
DESTILLATION (STETIGE) 3024
DESTILLATION (STUFENWEISE) 5612
DESTILLATION (TROCKENE) 4346
DESTILLATION (UNTERBROCHENE) 5612
DESTILLATION IM VAKUUM 4075
DESTILLATION MIT WASSERDAMPF 4073
DESTILLATIONSERZEUGNIS 4071
DESTILLIERBLASE 1407
DESTILLIEREN 4070
DESTILLIEREN (STUFENWEISE) 5615
DESTILLIERKOLBEN 10523
DESTILLIERVORRICHTUNG 12259
DESTRUKTIVE DESTILLATION 3805
DETAILZEICHNUNG 3809
DETERMINANTE 3814
DEUTSCHES GASÖL 1668
DEXTRIN 3825
DEXTROSE 6058
DEZENTRIERT 8893
DEZIGRAMM 3657
DEZILITER 3658

DEZIMALBRUCH 3662
DEZIMALBRUCH (ENDLICHER) 5231
DEZIMALBRUCH (PERIODISCHER) 9192
DEZIMALE 3659
DEZIMALKERZE 3663
DEZIMALKODE 3661
DEZIMALSTELLE 3659
DEZIMALSYSTEM 3665
DEZIMALWAAGE 3660
DEZIMALZAHL 3664
DEZIMETER 3666
DIABAS 3828
DIAGNOSEPROGRAMM 3830
DIAGONAL 3831
DIAGONALE 3831
DIAGRAMM 7717
DIAGRAMM 3834
DIAGRAMM (RANKINISIERTES) 10203
DIAGRAMM AUFZEICHNEN (EIN) 9520
DIAGRAMME RANKINISIEREN 2765
DIAGRAMMKURVE 3495
DIAGRAMMLINIE 3495
DIAKAUSTISCHE LINIE 3829
DIALYSATOR 3845
DIALYSE 3846
DIALYSIEREN 3844
DIAMAGNETISCH 3847
DIAMAGNETISMUS 3848
DIAMANT 3862
DIAMANT (SCHWARZER) 1967
DIAMANTGLANZ 158
DIAMANTHÄRTEPROBE 3865
DIAMANTSCHLEIFSCHEIBE 3867
DIAMANTSCHLEIFSCHEIBE 3864
DIAMANTSTAUB 3863
DIAMANTWERKZEUG 3866
DIANTHIN 4831
DIAPHAN 3869
DIAPHANITÄT 3868
DIAPHRAGMA 3870
DIAPHRAGMA 3871
DIATHERMAN 3879
DIATHERMANITÄT 3878
DIATHERMANSIE 3878
DIATOMEENERDE 7292
DIATOMEENERDE 3880
DICHT 12920
DICHTE 3756
DICHTE (ELEKTRISCHE) 4501
DICHTE (MAGNETISCHE) 7899
DICHTEN 7954

DIVISOR 4111
DOCHT 13846
DOCHTÖLER 11516
DOCHTSCHMIERUNG 13847
DODBKAEDER 4112
DOLERIT 4114
DOLOMIT 4117
DOPPEL-J-NAHT MIT LUFTSPALT 8812
DOPPEL-J-NAHT OHNE LUFTSPALT 2523
DOPPEL-T-EISEN 6194
DOPPEL-U-FUGENNAHT OHNE LUFTSPALT 2524
DOPPELBACKENBREMSE 4129
DOPPELBODEN 5026
DOPPELBRECHEND 1265
DOPPELBRECHUNG 4151
DOPPELDECKER 1261
DOPPELFLANSCH 4139
DOPPELGÄRBSTAHL 4156
DOPPELGETRIEBE 4142
DOPPELHAKEN 10191
DOPPELHE 4147
DOPPELHOBEL 4146
DOPPELINTEGRAL 4145
DOPPELKALIBER 3905
DOPPELKEHLNAHT 4141
DOPPELKEIL 5938
DOPPELKLINKENGETRIEBE 2494
DOPPELKLOTZBREMSE 4129
DOPPELKOPFSCHIENEN 1715
DOPPELKRÜMMER 13323
DOPPELLASCHENNIETUNG 4130
DOPPELMUFFE 4158
DOPPELNAHT OHNE LUFTSPALT 1788
DOPPELNIPPEL 4149
DOPPELPFEILZAHNRAD 13176
DOPPELPUNKTSCHWEISSMASCHINE 4382
DOPPELRIEMEN 4128
DOPPELSALZ 4153
DOPPELSAÜLESCHMIEDEHAMMER 4140
DOPPELSCHLUSSDYNAMO 3228
DOPPELSCHLÜSSEL 4167
DOPPELSCHLUSSMOTOR 2835
DOPPELSCHNITTFEILE 4133
DOPPELSITZVENTIL 4154
DOPPELSITZVENTIL 4166
DOPPELSTÄNDER 4159
DOPPELSTÄNDER PRESSE (MECHANISCHE) 4157
DOPPELT GEKRÖPFTE KURBELWELLE 13316
DOPPELT KOHLENSAURES KALI 9676
DOPPELTCHROMSAURES KALI 9677
DOPPELTER ANSCHLAGWINKEL 12581

DOPPELTKOHLENSÄURE MAGNESIA 7883
DOPPELTWIRKEND 4163
DOPPELVERGASER 4360
DOPPELWANDIG 4170
DOPPELWANDIGER 4161
DOPPELWINKELZAHNRAD 13176
DOPPELZENTNER 10111
DÖPPER 10626
DOPPLUNG 7382
DORN 7964
DORN 11904
DORN 3162
DORN 4260
DORN (FESTER) 9425
DORN (FESTER) 11789
DOSENBAROMETER 504
DOSENLIBELLE 2430
DOUBLE 5999
DOUBLET 4172
DOWSONGAS 4194
DRACHENBLUT 4203
DRAHT 13880
DRAHT (BLANKER) 13340
DRAHT (BLANKER) 990
DRAHT (DREIKANTIGER) 13196
DRAHT (EIRUNDER) 4678
DRAHT (ELLIPTISCHER) 4678
DRAHT (FEINER) 12849
DRAHT (FEINGEZOGNER) 12849
DRAHT (GALVANISIERTER) 5786
DRAHT (GEFLOCHTENER) 1528
DRAHT (GEGLÜHTER) 561
DRAHT (GEWALZTER) 10678
DRAHT (GEZOGENER) 4252
DRAHT (HARTGEZOGENER) 13340
DRAHT (ISOLIERTER) 6996
DRAHT (KALTGEZOGENER) 2715
DRAHT (KUNSTSTOFFUMKLEIDETER) 9468
DRAHT (NACKTER) 990
DRAHT (NICHT ISOLIERTER ELEKTR.) 990
DRAHT (OVALER) 4678
DRAHT (RECHTECKIGER) 5402
DRAHT (RUNDER) 10804
DRAHT (UMHÜLLTER) 11300
DRAHT (UMKLÖPPELTER) 1528
DRAHT (UMMANTELTER) 3272
DRAHT (UNGEGLÜHTER) 13340
DRAHT (VERBLEITER) 7459
DRAHT (VERZINKTER) 5786
DRAHT (VERZINNTER) 12962
DRAHT UMFLOCHTENER 1528

DRAHTABSPULER 13898
DRAHTABSPULGESCHWINDIGKEIT 13900
DRAHTAUF-U.ABWICKELMASCHINE 13903
DRAHTBARREN 13882
DRAHTBIEGEMASCHINE 13883
DRAHTBÜRSTE 13884
DRAHTFEDERNWINDEMASCHINE 13901
DRAHTFLECHTMASCHINEN 13897
DRAHTGAZE 13893
DRAHTGEFLECHT 13896
DRAHTGEWEBE 13893
DRAHTGEWEBE 5888
DRAHTGLAS 13894
DRAHTHASPEL 13898
DRAHTKETTENMASCHINE 13885
DRAHTKLAMMER 12104
DRAHTLEHRE 13892
DRAHTLEHRE (ENGLISCHE) (B. W. G.) 12089
DRAHTLITZE 13902
DRAHTNAGEL 13895
DRAHTNETZ 13896
DRAHTNUMMER 5881
DRAHTÖSE 12106
DRAHTSCHNEIDER 13886
DRAHTSEIL 13899
DRAHTSEIL (FLACHLITZIGES) 5407
DRAHTSEIL IM GLEICHSCHLAG 309
DRAHTSEIL NACH ALBERTSCHLAG 7390
DRAHTSEILBAHN 213
DRAHTSEILHERSTELLUNGSMASCHINE 12341
DRAHTSTÄRKE 3860
DRAHTSTIFT 13895
DRAHTSTIFTSCHLAGMASCHINE 13891
DRAHTVERARBEITUNGSMASCHINE 13904
DRAHTZANGE (FLACHE) 5403
DRAHTZANGE (RUNDE) 10805
DRAHTZIEHBANK 4218
DRAHTZIEHEISEN 4223
DRAHTZIEHEN 4230
DRAHTZIEHEN 13888
DRAHTZIEHEN 13906
DRAHTZIEHEN 4220
DRAHTZIEHEREI 13905
DRAHTZIEHMASCHINE 13889
DRAHTZIEHSTEIN 13887
DRAHTZIEHSTEIN 189
DRAHTZWICKZANGE 3532
DRALL EINES SEILES 13305
DRAUFSICHTSBILD 13037
DREHACHSE 868
DREHAUTOMAT 846

DREHBANK 13289
DREHBANK 7427
DREHBANK (SELBSTTÄTIGE) 846
DREHBANKFUTTER 2393
DREHBAR 9422
DREHBEANSPRUCHUNG 13056
DREHBESCHLEUNIGUNG 537
DREHBEWEGUNG 10766
DREHEN (AUF DER DREHBANK) 13276
DREHEN (AUF DER DREHBANK) 13287
DREHEN (KEGELIG) 13282
DREHEN (KONISCH) 13282
DREHEN (SICH) 10763
DREHENDE WELLE (SICH) 10764
DREHER 13286
DREHEREI 13290
DREHFEILE 10755
DREHFUTTER-BACKEN 2394
DREHGESCHWINDIGKEIT 11876
DREHKOLBEN 10756
DREHKOLBENPUMPE 10758
DREHKRAN 11583
DREHLING 7393
DREHMASCHINE 1018
DREHMASCHINE MIT AUTOMATISCHEN ARBEITSABLÄUFEN
  840
DREHMASCHINE MIT FRONTBEDIENUNG 5697
DREHMELDER 12568
DREHMOMENT 13048
DREHMOMENT 13308
DREHMOMENTMESSER 13050
DREHMOMENTSCHRAUBENSCHLÜSSEL 13049
DREHPUNKT 2169
DREHPUNKT 5717
DREHRICHTUNG 3992
DREHROST 10559
DREHSCHEIBE 13295
DREHSCHIEBER 10760
DREHSCHWINGUNG 13557
DREHSINN 3992
DREHSPÄNE 13293
DREHSPANNUNG 13058
DREHSTAB 13053
DREHSTABSTABILISATOR 12539
DREHSTAHL 13292
DREHSTICHEL 13292
DREHSTROM 12877
DREHSTROMDYNAMO 12875
DREHSTROMMASCHINE 12875
DREHSTROMMOTOR 12876
DREHSTUHL 13639

DREHTISCH 10762
DREHUNG 10765
DREHUNG 13054
DREHUNGSELLIPSOID 4672
DREHUNGSFEDER 11994
DREHUNGSFESTIGKEIT 13057
DREHUNGSFLÄCHE 12513
DREHUNGSHALBMESSER 10169
DREHUNGSKÖRPER 11790
DREHUNGSPARABOLOID 9042
DREHUNGSWINKEL 528
DREHZAHL 8690
DREHZAHL 10814
DREHZAHL 10558
DREHZAHLMESSER 10556
DREHZAHLREGLER 11874
DREHZAHLREGLER 13493
DREHZAHLREGLER 6013
DREHZAPFEN 13242
DREHZAPFEN 7256
DREIATOMIG 13197
DREIBACKENFUTTER 12874
DREIBASISCH 13198
DREIBEIN 5945
DREIDIMENSIONAL 8728
DREIECK 11223
DREIECK (GEOM.) 13185
DREIECK (GLEICHSCHENKLIGES) 7192
DREIECK (GLEICHSEITIGES) 4785
DREIECK (RECHTWINKLIGES) 10582
DREIECK (SPHÄRISCHES) 11891
DREIECK (SPITZWINKLIGES) 157
DREIECK (STUMPFWINKLIGES) 8718
DREIECK (UNGLEICHSEITIGES) 10984
DREIECKFEDER 13191
DREIECKFEDER (GESCHICHTETE) 7381
DREIECKGEWINDE 545
DREIECKLAST 13190
DREIFACH-SCHWEFELANTIMON 627
DREIFACHEXPANSIONSMASCHINE 13217
DREIFADENMETHODE 12883
DREIFLACH 13202
DREIFUSSWINDE 13218
DREIKANT 13202
DREIKANTFEILE 13189
DREIKANTLITZENSEIL 13195
DREIKANTSEIL 13194
DREIPHASENDYNAMO 12875
DREIPHASENMOTOR 12876
DREIPHASENSTROM 12877
DREISCHENKELZIRKEL 13187

DREISTOFFSYSTEM 12766
DREIWEGESTÜCK 12699
DREIWEGEVENTIL 12882
DREIWEGHAHN 12881
DREIWEGSCHIEBER 3592
DREIWERTIG 13197
DREIZYLINDERMASCHINE 12873
DRESCHMASCHINEN 12884
DRESSIEREN 12720
DRESSIEREN 12319
DRESSIEREN 9157
DRILLBOHRER 689
DRILLBRETT 1590
DRILLUNG 532
DRITTE POTENZ 3445
DRITTE WURZEL 3447
DROSSELKLAPPE 2357
DROSSELKLAPPE 1811
DROSSELKLAPPE 12890
DROSSELKLAPPE 12888
DROSSELN 12889
DROSSELN 12891
DROSSELUNG 12891
DRUCK 10028
DRUCK 6314
DRUCK 9808
DRUCK (ABSOLUTER) 4459
DRUCK (ABSOLUTER) 33
DRUCK (ATMOSPHÄRISCHER) 794
DRUCK (GLEICHBLEIBENDER) 2976
DRUCK (HOHER) 6473
DRUCK (HYDROSTATISCHER) 6726
DRUCK (HYDROSTATISCHER) 7652
DRUCK (INNERER) 9822
DRUCK (KONSTANTER) 2976
DRUCK (KRITISCHER) 3348
DRUCK (MANOMETRISCHER) 5771
DRUCK (MITTLERER) 8075
DRUCK (NIEDERER) 7774
DRUCK (OSMOTISCHER) 8891
DRUCK (RUHENDER) 12137
DRUCK (SEITLICHER) 7422
DRUCK (SPEZIFISCHER) 11853
DRUCK (STATISCHER) 12137
DRUCK (STATISCHER) 12136
DRUCK (VERÄNDERLICHER) 13491
DRUCK (ZULÄSSIGER) 9828
DRUCK PARALLEL ZUR FASER 10021
DRUCK SENKRECHT ZUR FASER 10022
DRUCK-UND PLANIERMASCHINE 11911
DRUCKABFALL 9815

DRUCKÄNDERUNG 2237
DRUCKANSTIEG 6848
DRUCKBACKEN EINER SCHERE 11282
DRUCKBEANSPRUCHUNG 2863
DRUCKBEANSPRUCHUNG 2869
DRUCKBEGRENZUNGSVENTIL 10441
DRUCKBELASTUNG 9823
DRUCKDÜSE 3738
DRUCKELASTIZITÄT 4488
DRÜCKEN 8189
DRUCKER 9872
DRUCKFEDER 2862
DRUCKFESTIGKEIT 2870
DRUCKFESTIGKEIT 13328
DRUCKGASGENERATOR 9819
DRUCKGEFÄSS 9839
DRUCKGIESSVERFAHREN 3886
DRUCKGUSS 3886
DRUCKGUSS 9812
DRUCKGUSSLEGIERUNG 388
DRUCKHÖHE (HÖHE DER FLÜSSIGKEITSSÄULE) 6323
DRUCKHÖHE EINER PUMPE 4021
DRUCKHUB 9834
DRUCKKESSEL 829
DRUCKKLAPPE 3739
DRUCKKNOPF 9797
DRUCKKRAFT 2868
DRUCKLAGER 12902
DRUCKLEITUNG 3736
DRUCKLUFT 2839
DRUCKLUFTANTRIEB 2842
DRUCKLUFTBEHÄLTERGLIED 11117
DRUCKLUFTBREMSE 2840
DRUCKLÜFTER 9817
DRUCKLUFTFUTTER 279
DRUCKLUFTHAMMER 9551
DRUCKLUFTHAMMER 9553
DRUCKLUFTLEITUNG 2845
DRUCKLUFTLOKOMOTIVE 2844
DRUCKLUFTMOTOR 2843
DRUCKLUFTNIETHAMMER 9555
DRUCKLUFTNIETHAMMER 12400
DRUCKLUFTPROBE 2846
DRUCKLUFTPUMPE 273
DRUCKLÜFTUNG 9517
DRUCKLUFTZYLINDER 2841
DRUCKMESSER 9820
DRUCKMESSER 9821
DRUCKMINDERER 10355
DRUCKMINDERER 10420
DRUCKMINDERUNGSVENTIL 10355

DRUCKMINDERVENTIL 10355
DRUCKPROBE 9836
DRUCKPROBE 2865
DRUCKPUMPE 5531
DRUCKQUERSCHNITT 3375
DRUCKREGLER 9813
DRUCKREGLER 9829
DRUCKREGLER 10355
DRUCKREGLUNG 10419
DRUCKRING 9830
DRUCKRING 12904
DRUCKROHRLEITUNG 3736
DRUCKROHRLEITUNG 9827
DRUCKSCHALTER 9835
DRUCKSCHMIERUNG 9824
DRUCKSCHMIERUNG 9818
DRUCKSCHRAUBE 9831
DRUCKSCHRAUBE 193
DRUCKSCHWANKUNG 13497
DRUCKSINTERN 9799
DRUCKSPANNUNG 2871
DRUCKSTAB 2855
DRUCKSTEIGERUNG 6848
DRUCKSTOCK 1340
DRUCKSTUFE 9833
DRUCKTURBINE 6810
DRUCKUNTERSCHIED 9814
DRUCKVENTIL 3739
DRUCKVERLUST 6315
DRUCKVERLUST 7759
DRUCKVERLUST 7763
DRUCKVERSUCH 2872
DRUCKVOLUMENDIAGRAMM 6881
DRUCKWALZENRÄNDELMASCHINE 9873
DRUCKWASSER 9841
DRUCKWASSERANTRIEB 6683
DRUCKWASSERLEITUNG 9840
DRUCKWASSERPRESSE 6693
DRUCKWASSERPROBE 6695
DRUCKWASSERSCHMIEDEPRESSE 6687
DRUCKWASSERSPEICHER 6677
DRUCKZUG 5533
DRUCKZUNAHME 6848
DRUMMONDSCHES LICHT 8987
DÜBEL 7276
DÜBEL 12397
DÜBEL 4180
DUBLEE 5999
DUBLETT 4172
DUKTILITÄT 4363
DUKTILITÄTSPRÜFMASCHINE 3462

DÜSE EINES GEBLÄSES 8685
DÜSENBOHRMASCHINE 8681
DÜSENFEDER 8686
DÜSENHALTER 8682
DÜSENHÖHE 8680
DÜSENNADEL 8684
DÜSENTRÄGER 7219
DYN 4405
DYNAMIK 4400
DYNAMISCH 4399
DYNAMIT 4401
DYNAMO (IN SERIE GESCHALTETER) 11201
DYNAMOBLECH 708
DYNAMOELEKTRISCHE MASCHINE 5924
DYNAMOGRAPH 10291
DYNAMOMASCHINE 5924
DYNAMOMETER 4403
DYNAMOMETER MIT SCHREIBVORRICHTUNG 10291
DYNAMOMETRISCH 4404
DYNE 4405
DYNODE 4406
DYSPROSIUM 4407
E-EISEN 2250
EAU DE JAVEL 9688
EAU DE LABARRAQUE
   (NATRIUMHYPOCHLORITLÖSUNGCHLORSODALÖSUNG) 4422
EBENE (GEOM.) 9433
EBENE (PRISMATISCHE) 9877
EBENE (SCHIEFE) 6838
EBENENWINKEL 3944
EBENES VENTIL 4011
EBENHOLZ 4423
EBONIT 2363
ECKBLECH 6182
ECKBOHRMASCHINE 3178
ECKBOHRWINDE 5905
ECKE 4408
ECKE (EINSPRINGENDE) 10243
ECKE EINES POLYGONS 13543
ECKEN 7212
ECKEN 7210
ECKENBOHRER 5905
ECKENLINIE 3831
ECKNAHTVERBINDUNG MIT LUFTSPALT 8809
ECKNAHTVERBINDUNG OHNE LUFTSPALT 2522
ECKVENTIL 534
ECKVERBAND 2519
ECKVERBAND 4445
ECKWINKEL 510
EDELGAS 6905
EDELMETALL 9756

EDELMETALL 8603
EDELMETALL 8602
EDELMETALLBAR 1730
EDELMETALLBARREN 1729
EDELMETALLTEILCHEN 3548
EDELROST 220
EDELROST 9127
EDELSTAHL 12462
EFEUBLATTKURVE 2438
EFFEKT 9730
EFFEKT-KOEFFIZIENT (THERMOELEKTRISCHER) 2637
EFFLORESZENZ 4464
EFFUSION 4465
EFFUSION VON GASEN 4466
EGALISIERBANK 9428
EGALISIEREN 13164
EGALISIEREN 13162
EGGEN (GLEIDEREGGENZIGZAGEGGEN) 6303
EICHEN 5872
EICHEN 5886
EICHFEHLER 1869
EICHKONDENSATOR 1868
EICHMASS 5768
EICHMASS 12076
EICHUNG 5886
EICHUNG 1867
EICHUNG (EINES BEHÄLTERS) 1863
EIGENERREGUNG 11153
EIGENFEDERUNG 4487
EIGENGEWICHT 3637
EIGENLAST 3637
EIGENSCHAFTEN (MECHANISCHE) 8109
EIGENSCHAFTEN (PHYSIKALISCHE) 9283
EIGENSCHWINGUNG 5736
EIGENSPANNUNG 10474
EIMER 1674
EIN-UND AUSRÜCKKUPPLUNG 4033
EINATOMIG 8377
EINBAUTRÄGER 1702
EINBAUTRÄGER 6968
EINBETONIEREN 4682
EINBETONIERTER TRÄGER 4714
EINBEULUNG 3760
EINBEULVERSUCH 8797
EINBRANDKERBEN 13351
EINBRENNHÄRTUNG 9670
EINDAMPFEN EINER LÖSUNG 2885
EINDIMENSIONAL 8727
EINDRINGUNGSGRAD 4683
EINDRUCK 6806
EINDRUCKFLÄCHE 12512

EISENSPALT 262
EISENSPAT 11421
EISENSULFID 12447
EISENSULFID 7153
EISENTRÄGER 12199
EISENVITRIOL 6093
EISENWERK 7160
EISENZAHN 7154
EISENZYANPAPIER 5119
EISESSIG 5952
EISKALORIMETER 6749
EISKLUFT 5700
EISMASCHINE 6750
EISTEIN 3423
EIWEISS 310
EIWEISSPAPIER 311
EJEKTOR 4469
EJEKTOR 4472
EKONOMISER 4437
ELASTISCH 4473
ELASTIZITÄT 4487
ELASTIZITÄTSGRENZE 4481
ELASTIZITÄTSGRENZE 14013
ELASTIZITÄTSGRENZE 7588
ELASTIZITÄTSMASS 8340
ELASTIZITÄTSMODUL 8340
ELASTIZITÄTSMODUL 4483
ELASTIZITÄTSMODUL 8341
ELECTROLYTREINIGUNGSBAD 7538
ELEKTRIFIZIEREN 4555
ELEKTRIFIZIERUNG 4554
ELEKTRISCH WERDEN 1124
ELEKTRISCHE ZENTRALE 4512
ELEKTRISCHEN STROM ERZEUGEN 5917
ELEKTRISCHEN STROM UNTERBRECHEN (DEN) 7090
ELEKTRISCHER SAMMLER 4534
ELEKTRIZITÄT 4552
ELEKTRIZITÄT (ATMOSPHÄRISCHE) 796
ELEKTRIZITÄT (NEGATIVE) 8526
ELEKTRIZITÄT (POSITIVE) 9655
ELEKTRIZITÄTSMENGE 10067
ELEKTRIZITÄTSVERSORGUNG 12487
ELEKTRIZITÄTSWERK 4512
ELEKTRIZITÄTSZÄHLER 4553
ELEKTRO-OSMOSIS 4638
ELEKTROANALYSE 4657
ELEKTROANALYSE 4570
ELEKTROBLECH 4544
ELEKTROBLECH 11448
ELEKTROBOHRMASCHINE 4556
ELEKTROCHEMIE 4573

ELEKTROCHEMIE 4561
ELEKTROCHEMISCH 4560
ELEKTRODAMPFKESSEL 4497
ELEKTRODE 4574
ELEKTRODE (BLECHUMHÜLLTE) 11299
ELEKTRODE (DICKUMHÜLLTE) 11344
ELEKTRODE (NACKTE) 989
ELEKTRODE (NEGATIVE) 2106
ELEKTRODE (NICHTSCHMELZENDE) 8618
ELEKTRODE (POSITIVE) 602
ELEKTRODE (SCHMELZBARE) 2990
ELEKTRODE (UMHÜLLTE) 2584
ELEKTRODE (UMMANTELTE) 5492
ELEKTRODE (ZWEIPOLIGE) 1233
ELEKTRODENARBEITSFLÄCHE 4579
ELEKTRODENARM 6592
ELEKTRODENARMAUSLADUNG 12885
ELEKTRODENHALTER 4575
ELEKTRODENMETALL 4576
ELEKTRODENPOTENTIAL 4577
ELEKTRODENREST 4578
ELEKTRODENROLLE 2412
ELEKTRODENSPANNUNG 9709
ELEKTRODENSPITZE 4579
ELEKTRODENZANGE 4575
ELEKTRODYNAMISCH 4563
ELEKTRODYNAMOMETER 4564
ELEKTROEROSION 4565
ELEKTROGRENZLEHRE 4584
ELEKTROGUSSEISEN 4510
ELEKTROGUSSEISEN 4509
ELEKTROHARTLÖTEN 4498
ELEKTROINGENIEUR 4540
ELEKTROINGENIEURWESEN 4541
ELEKTROKUNSTKORUND 5753
ELEKTROLICHTBOGENOFEN 4495
ELEKTROLYSE 4586
ELEKTROLYSEUR 4588
ELEKTROLYSIEREN 4585
ELEKTROLYT 4587
ELEKTROLYT (VERBRAUCHTER) 5588
ELEKTROLYTGLEICHRICHTER 4606
ELEKTROLYTGOLD 4598
ELEKTROLYTKONDENSATOR 4590
ELEKTROLYTKUPFER 4591
ELEKTROLYTKUPFER 2099
ELEKTROLYTNICKEL 4600
ELEKTROLYTSILBER 4608
ELEKTROLYTZELLENKONSTANTE 2128
ELEKTROMAGNET 4567
ELEKTROMAGNET 4611

ERNEUERUNG EINES MASCHINENTEILS **10458**
ERNTEMASCHINEN FÜR HACKFRÜCHTE **10722**
EROSION **4820**
ERREGEN **4887**
ERREGERANODE **4886**
ERREGERMASCHINE **4888**
ERREGERWICKLUNG **6898**
ERREGUNG (PHYS.) **4885**
ERSATZ **12420**
ERSATZRAD **11835**
ERSATZRAD **2243**
ERSATZSTOFF **12420**
ERSATZTEIL **11834**
ERSCHÜTTERUNGEN **13559**
ERSTARREN **11797**
ERSTARREN **11803**
ERSTARRUNG **5653**
ERSTARRUNG **11797**
ERSTARRUNG DER SCHMELZE IN DER PFANNE **7360**
ERSTARRUNG EINER FLÜSSIGKEIT **11799**
ERSTARRUNGS-BEREICH **5658**
ERSTARRUNGS-INTERVALL **5658**
ERSTARRUNGSBEREICH **11801**
ERSTARRUNGSFRONT **11798**
ERSTARRUNGSPUNKT **5656**
ERSTARRUNGSPUNKT **11800**
ERSTARRUNGSPUNKT **11804**
ERSTARRUNGSSCHWINDUNG **11802**
ERSTFARBE **9847**
ERTEILUNG EINES PATENTS **6049**
ERTRAGSFÄHIGKEIT **9897**
ERUPTIONSABSPERRVORRICHTUNG **1360**
ERWÄRMEN **6334**
ERWÄRMEN **6390**
ERWÄRMUNG **6390**
ERWARTUNGSWERT **4927**
ERWEICHUNGSPUNKT **11750**
ERWEITERN (SICH) **5356**
ERWEITERTER TEIL (EINES ROHRES) **4768**
ERWEITERUNG **4768**
ERYTHROSIN **4831**
ERZ **8859**
ERZ (ARMES) **7781**
ERZ (REICHES) **6485**
ERZ UND SCHMELZMITTEL **8863**
ERZANREICHERUNG **8861**
ERZBERGWERK **8198**
ERZBRIKETT **8860**
ERZE AUFBEREITEN **4253**
ERZE GLÜHEN **1841**
ERZE RÖSTEN **10648**

ERZEUGEN **7948**
ERZEUGEN (DAMPF, GAS) **5916**
ERZEUGENDE **5925**
ERZEUGENDE FUNKTION **5919**
ERZEUGER **7992**
ERZEUGNIS **7989**
ERZEUGNIS **9892**
ERZEUGUNG **4983**
ERZGRUBE **8198**
ERZPRESSSTEIN **8860**
ERZRÖSTEN **8862**
ERZZIEGEL **8860**
ESELRÜCKENBOGEN **8741**
ESSE **2339**
ESSENKLAPPE **3603**
ESSIGGEIST **90**
ESSIGSÄURE **86**
ESSIGSÄUREANHYDRID **87**
ESSIGSÄUREÄTHER **88**
ESSIGSÄUREÄTHYLESTER **88**
ESSIGSAURES EISEN **7134**
ESSIGSAURES EISENOXYDUL **5120**
ESSIGSAURES KALI **9673**
ESSIGSAURES SALZ **85**
ESTER **4837**
ETAGENVENTIL **8475**
EUDIOMETER **4852**
EUROPIUM **4854**
EUTEKTIKUM **4855**
EUTEKTISCH **4855**
EUTEKTOID **4859**
EUTEKTOIDE REAKTION **4860**
EVOLUTE **4873**
EVOLVENTE **7117**
EVOLVENTENVERZAHNUNG **7119**
EXHAUSTOR **4900**
EXOTHERM **4901**
EXOTHERMISCH **4901**
EXPANDIEREN **4902**
EXPANSION **4921**
EXPANSIONSDAMPFMASCHINE **4924**
EXPANSIONSHUB **2775**
EXPANSIONSKURVE **4916**
EXPANSIONSREIBAHLE **188**
EXPANSIONSRIEMENSCHEIBE **4909**
EXPANSIONSTURBINE **4925**
EXPANSIONSVENTIL **4926**
EXPERIMENT **12792**
EXPERIMENTIEREN **4928**
EXPLODIEREN **4933**
EXPLOSION **4935**

EXPLOSION EINER SCHEIBE **1769**
EXPLOSION EINES KESSELS **1410**
EXPLOSION EINES SCHWUNGRADES **1769**
EXPLOSIONSDRUCK **4936**
EXPLOSIV **4937**
EXPLOSIVSTOFF **4937**
EXPONENT **4939**
EXPONENTIALFUNKTION **4942**
EXPONENTIALREIHE **4943**
EXPONENTIALVERTEILUNG **4941**
EXSIKKATOR **3804**
EXTRAKT **4964**
EXTRAKTION (ELEKTROLYTISCHE) **4582**
EXTRAKTION (ELEKTROLYTISCHE) **4652**
EXTRAKTIONSAPPARAT **4967**
EXTRAKTIONSMETALLURGIE **9886**
EXZENTER **4433**
EXZENTER **4425**
EXZENTER (GESCHLOSSENES) **3270**
EXZENTER (OFFENES) **8805**
EXZENTER (UNRUNDE SCHEIBEHUBSCHEIBE) **1883**
EXZENTERBÜGEL **4435**
EXZENTERHUB **12894**
EXZENTERPRESSE **4430**
EXZENTERRING **4435**
EXZENTERROLLE **1895**
EXZENTERSCHEIBE **4434**
EXZENTERSTANGE **4431**
EXZENTERSTANGENKOPF **4432**
EXZENTERTRIEBWERK **1886**
EXZENTRISCH **4424**
EXZENTRISCH **8730**
EXZENTRIZITÄT **4436**
FABRIK **5013**
FABRIKABWÄSSER **6902**
FABRIKANSCHLUSSGLEIS **13966**
FABRIKANT **7992**
FABRIKAT **7989**
FABRIKATION **7989**
FABRIKATIONSFEHLER **3693**
FABRIKATIONSGANG **2222**
FABRIKATIONSLÄNGE **8925**
FABRIKATIONSVERFAHREN **8218**
FABRIKAUFSEHER **8958**
FABRIKBAHN **13965**
FABRIKDIREKTOR **5915**
FABRIKGEBÄUDE **5012**
FABRIKMARKE **13100**
FABRIKNUMMER **11197**
FABRIKNUMMER **4986**
FABRIKORGANISATION **6901**

FABRIZIEREN **7948**
FACH **2798**
FACHGEBIET **11845**
FACHWERK **7433**
FACHWERKTRÄGER **7431**
FACKEL **13042**
FACKEL **5355**
FADENKREUZ **3383**
FADENKRISTALL **13828**
FADENZIEHEN **2616**
FAHLERZ **12796**
FAHRARM DES PLANIMETERS **10662**
FAHRBAR **9635**
FAHRBARER KRAN **13161**
FAHRBEREICH **3416**
FAHRBEWEGUNG **13158**
FAHRENHEITGRAD **3718**
FAHRENHEITSKALA **5019**
FAHRENHEITTHERMOMETER **5020**
FAHRFLÄCHE-LAUFBAHN **10857**
FAHRGESTELL **5624**
FAHRGESTELL **2275**
FAHRRADKETTE **1239**
FAHRSCHIENE **10175**
FAHRSTIFT DES PLANIMETERS **13079**
FAHRSTRAHL **10170**
FAHRT **10810**
FAHRTRICHTUNGSANZEIGER **3989**
FAKTOR (MATH.) **5008**
FALLBESCHLEUNIGUNG **63**
FALLE **7414**
FÄLLEN **9761**
FÄLLEN (EIN LOT) **4216**
FALLENDGIESSEN **13031**
FALLGEWICHT **4329**
FALLHAMMER **4326**
FALLHAMMER **4327**
FALLHAMMER **4325**
FALLHÖHE **6421**
FALLKLINKE **7414**
FÄLLMITTEL **9758**
FÄLLMITTEL **11130**
FALLROHR **4185**
FALLROHR **4183**
FALLROHR **3593**
FALLROHR UNTER SCHEIBEN **4188**
FALLSTROMVERGASER **4189**
FÄLLUNG (ELEKTROLYTISCHE) **4593**
FALLVERSUCH **4328**
FALLWERK **4326**
FALSCHLUFT **4771**

FÄLSCHUNG 205
FALTE 3320
FALTE 4408
FALTEN 1680
FALTEN VERFORMEN 3336
FALTENBALG 1139
FALTENBILDUNG 13985
FALTMASSSTAB 5512
FALTVERSUCH 4173
FALZ (BAUWESEN) 10269
FALZ (BLECHBEARBEITUNG) 5511
FALZBIEGEMASCHINE 11087
FALZHOBEL 10115
FALZMASCHINE 11088
FANGELEKTRODE 12682
FARAD 5035
FARADAYSCHER KÄFIG 5036
FARBANSTRICH 2583
FARBAUFTRAG 2599
FARBE 9303
FARBE (DECKENDE) 1389
FARBE (PHYS.) 2741
FARBMETALLOGRAPHIE 2738
FARBSPRITZVERFAHREN 6179
FARBSTIFT 2745
FARBSTOFF 4397
FARBSTOFF 9303
FARBÜBERZUG 2583
FÄRBUNG 2740
FASER 5136
FASER 5131
FASER (NEUTRALE) 8542
FASER (PARALLEL ZUR) 6031
FASERACHSE 5132
FASERIG 5137
FASERN (IN DER RICHTUNG) 6031
FASERRICHTUNG 5133
FASERRICHTUNG 3993
FASERSCHICHT (NEUTRALE) 8544
FASERSPANNUNG 5134
FASERSTOFF 5140
FASERSTOFFRIEMEN 4981
FASERSTRUKTUR 5135
FASSFLÄCHENKORROSION 5667
FASSONDRAHT 11847
FASSONDREHBANK 3138
FASSONEISEN 11119
FASSONFRÄSER 9899
FASSONROHR 9350
FASSPOLIEREN 1012
FASSUNGSVERMÖGEN 1917

FAULEN 10036
FÄULNIS 10035
FÄULNIS 10769
FÄULNIS DES HOLZES 10770
FÄULNIS ÜBERGEHEN (IN) 10036
FÄULNISVERHINDERNDES MITTEL 629
FÄULNISWIDRIG 628
FAUSTHAMMER 6235
FAUSTLEIER 1519
FAUSTSCHERE 856
FEDER 5059
FEDER 11974
FEDER (EINE) ZUSAMMENDRÜCKEN 2838
FEDER (GESCHICHTETE) 2834
FEDER (GESPANNTE) 11983
FEDER (ZUSAMMENGESETZTE) 2834
FEDER SPANNEN (EINE) 10032
FEDER UND NUT 12990
FEDER UND NUT 12989
FEDER-UND DRAHT-PRÜFMASCHINE 11975
FEDERBAROMETER 504
FEDERBLATT 11977
FEDERBLATT 9492
FEDERBOLZEN 2149
FEDERBRONZE 1661
FEDERDRAHT 11997
FEDERGEHÄNGE 11233
FEDERHARZ 10825
FEDERHOBEL 12993
FEDERKEIL 5059
FEDERKLINKE 11987
FEDERKRAFT 4487
FEDERKRAFTREGLER 11984
FEDERLASCHE 11991
FEDERMANOMETER 11989
FEDERMEMBRAN 4479
FEDERMESSING 11978
FEDERND 4473
FEDERPLATTE 11988
FEDERPUFFER 11980
FEDERRING 11990
FEDERRING 11995
FEDERRING 4486
FEDERROHR 4913
FEDERSCHMIERBÜCHSE 11982
FEDERSICHERHEITSVENTIL 11998
FEDERSTAHL 11992
FEDERTELLER 11988
FEDERWAAGE 11976
FEDERWAAGE 4403
FEDERWIRKUNG 4487

FERNSTEUERUNG 10448
FERNTHERMOMETER 12714
FERNTHERMOMETER 4069
FERNTRIEB 12706
FERNTRIEBSEIL 12705
FERNÜBERTRAGUNG VON ELEKTRISCHER ENERGIE 13133
FERRIAZETAT 5099
FERRICHLORID 5100
FERRIFERROZYANID 1206
FERRIHYDRAT 5101
FERRIHYDROXYD 5101
FERRISULFAT 5103
FERRIT 5104
FERRITBAND (FREIES) 5105
FERRIZYANKALIUM 9684
FERROALUMINIUM 5106
FERROAZETAT 5120
FERROBOR 5107
FERROCHLORID 5121
FERROCHROM 5108
FERROLEGIERUNGSBRIKETT 1622
FERROMAGNETISCH 5118
FERROMANGAN 5110
FERROMOLYBDÄN 5111
FERRONICKEL 5112
FERROOXALAT 5124
FERROOXYD 5125
FERROSILIZIUM 5113
FERROSULFAT 6093
FERROTITAN 5114
FERROVANADIUM 5116
FERROWOLFRAM 5115
FERROZYANKALIUM 9685
FERTIGBEARBEITUNG 11682
FERTIGBEARBEITUNG 5220
FERTIGBEARBEITUNG (MASCHINELLE) 7844
FERTIGER PRESSLING 8551
FERTIGERZEUGNIS 5218
FERTIGERZEUGNIS 5730
FERTIGGERÜST 5219
FERTIGGESENK 5219
FERTIGSCHLEIFEN 5224
FERTIGSCHNEIDER 1499
FERTIGSCHWEISSEN 5216
FERTIGSTRASSE 5229
FERTIGUNGSARBEIT 5223
FERTIGUNGSBEREICH 7994
FERTIGUNGSPROGRAMM 7994
FERTIGUNGSTEMPERATUR 5227
FERTIGUNGSUMFANG 7994
FERTIGWALZE 5225

FERTIGWALZEN 5215
FERTIGWALZWERK 5222
FEST ANZIEHEN (EINE SCHRAUBE) 12927
FESTBETTFRÄSMASCHINE 5298
FESTDACH 5296
FESTER AGGREGAT-ZUSTAND 11795
FESTER STEMPELTEIL 3885
FESTER ZYKLUS 5292
FESTFRESSEN 11141
FESTIGKEIT 12352
FESTIGKEIT 10484
FESTIGKEIT (ZUSAMMENGESETZTE) 2822
FESTIGKEIT DES MATERIALS 12355
FESTIGKEITS-PRÜFMASCHINE 12793
FESTIGKEITSEIGENSCHAFTEN 12746
FESTIGKEITSGUSS 2042
FESTIGKEITSPRÜFUNG 12356
FESTKÖRPER 11775
FESTPUNKT 5295
FESTSCHEIBE 5040
FESTSCHRAUBEN 11065
FESTSCHRAUBEN 11022
FESTSPANNEINRICHTUNG 13945
FESTSTAMPFEN 10189
FESTSTAMPFEN 10195
FESTSTELLBREMSE 9082
FESTSTELLBREMSE 9082
FESTSTELLHEBEL 7706
FESTSTELLSCHRAUBE 11222
FESTSTELLVORRICHTUNG 7705
FESTSTOFF (ELASTISCHER) 4484
FESTWERT 2969
FETT (CHEM.) 5045
FETT (KONSISTENTES) 6275
FETTGAS 8758
FETTKALK 5046
FETTKOHLE 1282
FETTLÖSUNGSMITTEL 2473
FETTSÄURE 5055
FETTSCHMIERUNG 6087
FEUCHTE LUFT 3599
FEUCHTEN 13807
FEUCHTIGKEIT (ABSOLUTE) DER LUFT 30
FEUCHTIGKEIT (HYGROSKOPISCHE) 6731
FEUCHTIGKEIT (KRITISCHE) 3346
FEUCHTIGKEIT (RELATIVE) 10432
FEUCHTIGKEITSGEHALT 30
FEUCHTIGKEITSGRAD 10433
FEUCHTIGKEITSKORROSION 13802
FEUCHTIGKEITSMESSER 6729
FEUERBESTÄNDIG 5257

FLACHFEILE 5390
FLACHFEILE 5381
FLACHFÜHRUNG 5384
FLACHGÄNGIGE SCHRAUBE 12034
FLACHGEWINDE 12030
FLACHHAMMER 5410
FLACHHERDMISCHER 5387
FLACHKABBER (GESCHLOSSENES) 2526
FLACHKEHLNAHT 12075
FLACHKEIL 5388
FLACHKEIL 5400
FLACHMEISSEL 2672
FLACHMEISSEL 5377
FLACHNAHT 5481
FLACHREGLER 11242
FLACHS 5412
FLACHSCHEIBE 5399
FLACHSCHIEBER 9426
FLACHSEIL 5393
FLACHSTAB 5374
FLACHSTAHL 5392
FLACHSWERG 5413
FLACHWALZE 5391
FLACHWULSTEISEN 5376
FLACHWULSTEISEN 1714
FLACHZANGE 5403
FLADERSCHNITT 1058
FLAK-GESCHÜTZ 610
FLAMME (AUFKOHLENDE) 1990
FLAMME (KARBURIERENDE) 1990
FLAMME (LEUCHTENDE) 7820
FLAMME (NICHTLEUCHTENDE) 8636
FLAMME (OXIDIERENDE) 8975
FLAMME (REDUZIERENDE) 10349
FLAMME (RUSSENDE) 11678
FLÄMMEN 10993
FLAMMEN-LÖSCHER 5321
FLAMMENDER SALPETER 465
FLAMMENENTZUNDERUNG 5310
FLAMMENKEGEL 2923
FLAMMENKEGEL (INNERER) 6952
FLAMMENLÖTUNG 11768
FLAMMENRÜCKSCHLAG 906
FLAMMENRÜCKSCHLAG 5365
FLAMMENSICHER 5322
FLAMMENSPEKTROSKOPIE 5320
FLAMMENWÄCHTER 5314
FLAMMENZÜNDUNG 5316
FLÄMMHOBELN 6011
FLAMMLÖTEN 5810
FLAMMOFEN 261

FLAMMOFEN 10532
FLAMMPLATTIERUNG 5317
FLAMMPUNKT 6764
FLAMMPUNKT 5368
FLAMMPUNKT 5363
FLAMMPUTZEN 5309
FLAMMROHR 1412
FLAMMROHRKESSEL 5468
FLANELL 5352
FLANGEABDECKUNG 3267
FLANKE 5348
FLANKE EINES GEWINDES 5350
FLANKENDURCHMESSER 9407
FLANKENKEHLNAHT 12702
FLANKENSPIEL (ZAHNRÄDERN) 5349
FLANKENSPIELRAUM 914
FLANKENWINKEL DES GEWINDES 530
FLANSCH 5323
FLANSCH (ANGEGOSSENER) 2063
FLANSCH (AUFGELÖTETER) 11758
FLANSCH (AUFGESCHRAUBTER) 11063
FLANSCH (AUFGESCHWEISSTER) 13773
FLANSCH (AUFGEWALZTER) 10672
FLANSCH (BEARBEITETER) 5003
FLANSCH (FESTER) 5039
FLANSCH (GLATTER) 5382
FLANSCH (LOSER) 7746
FLANSCH (OVALER) 8918
FLANSCH (ROHER) 10776
FLANSCH (RUNDER) 10797
FLANSCH (ÜBERHÖHTER) 10183
FLANSCH (UNBEARBEITETER) 10776
FLANSCH EINES EISENS 5330
FLANSCH EINES ROHRES 9351
FLANSCHANPASSSTÜCK 5325
FLANSCHBOHRUNGEN 5328
FLANSCHENANSCHLUSS 5339
FLANSCHENAUFWALZMASCHINE 5333
FLANSCHENDECKEL 1304
FLANSCHENDREHBANK 5334
FLANSCHENKUPPLUNG 5338
FLANSCHENROHR 5343
FLANSCHENVERBINDUNG 5344
FLANSCHENVERSCHRAUBUNG 5344
FLANSCHFLÄCHE 5327
FLANSCHNABE 5329
FLANSCHRING 1708
FLANSCHROHRSTUTZEN 8683
FLANSCHRÜCKFLÄCHE 5326
FLANSCHSTUTZEN 5337
FLANSCHWULSTEISEN 1719

FLASCHE 1333
FLASCHE (FESTE) 5287
FLASCHE (LOSE) 8435
FLASCHENGLAS 1484
FLASCHENSCHRAUBSTOCK 6511
FLASCHENWINDE 1485
FLASCHENZUG 9973
FLASCHENZUG 9972
FLATTERN 11353
FLATTERN 2277
FLATTERN DES RIEMENS 13825
FLECHTWERK 13180
FLEISCHSEITE DES LEDERS 5414
FLICKEN 9109
FLICKEN 9110
FLICKEN 9111
FLICKMASSE 5803
FLIEGEND ANGEORDNETE KURBEL 8947
FLIEGEND GELAGERT 8936
FLIEHKRAFT 2180
FLIEHKRAFTBESCHLEUNIGUNG 10127
FLIEHKRAFTKUPPLUNG 2177
FLIEHKRAFTREGLER 2185
FLIEHKRAFTREGLER 2181
FLIESE 12933
FLIESSARBEIT 3034
FLIESSEN 5454
FLIESSEN 13578
FLIESSEN (PLASTISCHES) 9471
FLIESSEN DES MATERIALS 5465
FLIESSFÄHIGKEIT 5463
FLIESSFIGUREN 12371
FLIESSGESCHWINDIGKEIT 5459
FLIESSGRENZE 5456
FLIESSGRENZE 14012
FLIESSGRENZE 14011
FLIESSKUNDE 10562
FLIESSLINIE 5457
FLIESSMARK 5457
FLIESSPAPIER 1350
FLIESSPRESSE 4972
FLIESSPRESSEN 4970
FLIESSSCHEIDE 8547
FLIESSTEXTUR 5461
FLIESSVERMÖGEN 2064
FLIESSZEIT 5459
FLINT 5427
FLINTGLAS 5428
FLOCKE 5303
FLOCKEN VON WEISSMETALL 13835
FLOCKENGRAPHIT 5305

FLOCKENGRAPHIT 3703
FLOTATION 5452
FLOTATION (ANIONISCHE) 557
FLÜCHTIGKEIT 13589
FLUCHTLINIE 324
FLUCHTLINIENTAFEL 8617
FLUCHTSCHLEUSE 8904
FLUGASCHE 5493
FLÜGEL EINES VENTILS 5060
FLÜGELBREMSE 5031
FLÜGELGEBLÄSE 10751
FLÜGELMUTTER 12911
FLÜGELPUMPE 13478
FLÜGELPUMPE 11176
FLÜGELRAD 13479
FLÜGELSCHRAUBE 12912
FLUOBARIUM 999
FLUOR 5478
FLUORAMMONIUM 463
FLUORESZENZ 5476
FLUORESZENZSCHIRM 5477
FLUORKALIUM 9686
FLUORKALZIUM 5479
FLUORMAGNESIUM 7886
FLUORNATRIUM 11721
FLUORWASSERSTOFFSÄURE 6711
FLURSÄULE 5448
FLUSS (BASISCHER) 1046
FLUSSEISEN 6921
FLUSSEISEN 11744
FLÜSSIG 7642
FLÜSSIGGAS 7342
FLÜSSIGGAS 7638
FLÜSSIGKEIT 7641
FLÜSSIGKEITEN 7660
FLÜSSIGKEITSBREMSE 6678
FLÜSSIGKEITSDÄMPFUNG 7649
FLÜSSIGKEITSDICHTE 3758
FLÜSSIGKEITSDRUCK 6726
FLÜSSIGKEITSGRAD 5474
FLÜSSIGKEITSREIBUNG 5473
FLÜSSIGKEITSSÄULE 2750
FLÜSSIGKEITSSTRAHL 7221
FLUSSMITTEL 5475
FLUSSMITTEL 5489
FLUSSMITTEL (SAURES) 114
FLUSSSAND 10619
FLUSSSÄURE 6711
FLUSSSPAT 5479
FLUSSSPATSÄURE 6711
FLUSSSTAHL 6924

FLUSSSTAHL 11744
FLUSSTAHL 1958
FLUSSWASSER 10620
FOKUS 5502
FOKUSSIEREN 5505
FOLIE 5508
FOLIE 5509
FOLIUM (KARTESISCHES) 5513
FÖRDERBAND 1164
FÖRDERBANDBELADUNG 1141
FÖRDERBANDOFEN 1148
FÖRDEREINRICHTUNG 3080
FÖRDERGURT 1164
FÖRDERHÖHE EINER PUMPE 13061
FÖRDERHÖHE EINES KRANS 7551
FÖRDERKARREN 13230
FÖRDERKETTE 1675
FÖRDERKOHLE 10841
FÖRDERKOSTEN 3204
FÖRDERMENGE 10228
FÖRDERPUMPE 5713
FÖRDERRINNE 2395
FÖRDERSCHNECKE 13970
FÖRDERSEIL 6520
FÖRDERUNG 8106
FÖRDERUNGSANLAGE (HEU UND STROH) 4661
FORM 3883
FORM 5570
FORM 8352
FORM 11264
FORM (NICHT-METALLISCHE) 8650
FORM GEBEN 5571
FORMALDEHYD 5573
FORMALIN 5573
FORMÄNDERUNG 3707
FORMÄNDERUNG (BLEIBENDE) 3337
FORMÄNDERUNG (BLEIBENDE) 9210
FORMÄNDERUNG (ELASTISCHE) 10477
FORMÄNDERUNG (ELASTISCHE) 4485
FORMÄNDERUNG (ELASTISCHE) 4478
FORMÄNDERUNGSVERMÖGEN 3705
FORMAT 5574
FORMBESTÄNDIGKEIT 3958
FORMBESTÄNDIGKEIT 10521
FORMBLECH 11846
FORMDRAHT 11847
FORMDREHBANK 3138
FORMEISEN 11119
FORMEL 5582
FORMEL (CHEMISCHE) 2310
FORMEN 8428

FORMEN 5571
FORMEN 5581
FORMEN NACH DEM GUSSSTÜCK 4383
FORMENHOHLRAUM 8353
FORMENTON 9470
FORMER 8427
FORMERSTIFTE 5598
FORMFAKTOR 5572
FORMFRÄSER 9899
FORMFRÄSER 5577
FORMFRÄSMASCHINE 9900
FORMGEBUNG 11268
FORMGEBUNG 5581
FORMGEBUNG 5579
FORMIERUNG 5579
FORMIERUNGSSPANNUNG 5576
FORMKALIBER 8422
FORMKASTEN 5369
FORMKASTENSTIFT 5372
FORMMASCHINE 8358
FORMMASCHINE 8429
FORMOBERTEIL 3104
FORMOL 5573
FORMSAND 8430
FORMSCHLICHTE 8355
FORMSCHWÄRZE 1297
FORMSTAHL 11114
FORMSTEIN 8425
FORMSTÜCK 11265
FORMSTÜCK FÜR ROHRLEITUNGEN 9350
FORMUNG AUF DEM BODEN 5445
FORMUNTERTEIL 4199
FORMVERSATZ 1322
FORMYLSÄURE 5578
FORSCHUNG (ANGEWANDTE) 649
FORTBEWEGUNG 7712
FORTFÜHRUNG DER WARME 3060
FORTPFLANZUNG VON WELLEN 9926
FORTPFLANZUNGSGESCHWINDIGKEIT 13519
FORTRÜCKUNG (ALLG.) 4053
FOTODIODE 9258
FOUCAULTSCHE STRÖME 4440
FOURDRINIERDRAHT 5607
FRACHT 3209
FRACHTKOSTEN 3209
FRAGEBOGEN 10101
FRAKTION (EINER DESTILLATION) 5611
FRAKTIONIEREN 5615
FRAKTIONIERTURM 5614
FRAKTIONIERTURM 13171
FRAKTOGRAPHIE 5616

FRÄSE 8289
FRÄSEN 8276
FRÄSEN 8287
FRÄSEN (GEGENLÄUFIGES) 3061
FRÄSEN (GLEICHLÄUFIGES) 2497
FRÄSER 3519
FRÄSER 8289
FRÄSER (ARBEITER) 8292
FRÄSER (EINFACHER) FÜR SCHWERE SCHNITTE 6404
FRÄSER FÜR WOODRUFFKEILE 13930
FRÄSER MIT EINGESETZTEN ZÄHNEN 6971
FRÄSER MIT GERADEN SCHNEIDEN 12307
FRÄSER MIT HINTERDREHTEN ZÄHNEN 905
FRÄSER MIT SCHRAUBENFÖRMIGEN SCHNEIDEN 11917
FRÄSEREI 8293
FRÄSERSPINDEL 3520
FRÄSKOPF 3531
FRÄSMASCHINE 8291
FRÄSMASCHINE (FÜHLERGESTEUERTE) 13080
FRÄSMASCHINE (HORIZONTALE) 6589
FRÄSMASCHINE (STEHENDE) 13548
FRÄSMASCHINE (WAAGERECHTE) 6589
FRÄSMASCHINE MIT MEHREREN SPINDELN 8471
FRÄSMASCHINE VERTIKALE 13548
FRÄSSTIFTE 8290
FRÄSVORRICHTUNG 1390
FRAUNHOFERSCHE LINIEN 5631
FREI AUFGEHÄNGT 5649
FREI AUFLIEGEND 11474
FREI AUFLIEGEND 5650
FREI GELAGERT 5650
FREIES FEUER 8813
FREIFALL 5633
FREIFALLEND 5024
FREIFALLHAMMER 6080
FREIFORMSCHMIEDESTÜCK 11670
FREIHANDZEICHNUNG 5648
FREIHÄNGEND 8936
FREIHEITSGRADE 3726
FREILÄNGE 7518
FREILAUF 5651
FREILAUFKUPPLUNG 10221
FREILAUFKUPPLUNG 8957
FREILAUFMECHANISMUS 5652
FREILAUFRAD 5644
FREILEITUNG 212
FREISCHWEBEND 8936
FREISCHWEBENDER KOLBEN 9391
FREITRÄGER 1905
FREITRÄGER 1904
FREIWERDEND (CHEM.) 8497

FREMDERREGUNG 11190
FREMDSTOFFE 6813
FREQUENZ 5661
FRESSEN 5778
FRESSEN 11141
FRESSEN 11140
FRIKTIONSSCHEIBE 5673
FRISCHDAMPF 7673
FRISCHEISEN 346
FRISCHEREIÖFEN 3607
FRISCHOFEN 10379
FRISCHUNGSTEMPERATUR 10381
FRISCHVERFAHREN 10380
FRONT FÜHRERHAUS 1824
FROSCHLAGER 11788
FROSCHPERSPEKTIVE 1498
FROSTBESTÄNDIG 8635
FROSTBESTÄNDIGKEIT 8634
FROSTRISS DES HOLZES 5700
FROSTSCHUTZMITTEL 612
FROSTSCHUTZMITTEL 5657
FROSTSCHUTZVENTIL 8633
FRÜHEINSTELLUNG 206
FRÜHZÜNDUNG 9778
FUCHS (EINER FEUERUNGSANLAGE) 13418
FUCHSSCHWANZ 11659
FUGE 11083
FUGE 7238
FUGE 6127
FUGE (DICHTE) 12921
FUGEINLAGE 7621
FÜGEMASCHINE 7246
FUGENFLANKE 6130
FUGENNAHT 6133
FUGENÖFFNUNGSWINKEL 6129
FUGENRADIUS 6132
FUGENRADIUS 10728
FÜHLER 11090
FÜHLER 1711
FÜHRERHAUS 13229
FÜHRERSCHEIN 4302
FÜHRUNG 6160
FÜHRUNGSARM 6505
FÜHRUNGSBAHN 6166
FÜHRUNGSLEISTE 5060
FÜHRUNGSRING 6163
FÜHRUNGSRIPPE 5060
FÜHRUNGSROLLE 6161
FÜHRUNGSSCHIENE 6162
FÜHRUNGSSCHUH 11587
FÜHRUNGSSTANGE 6164

FÜHRUNGSSTIFT 1513
FÜLLDICHTE 1724
FÜLLEN 5156
FÜLLEN 2269
FULLERERDE 5727
FÜLLHAHN 5066
FÜLLMETALL 5159
FÜLLSAND 912
FÜLLSTOFF 5157
FÜLLSTOFF 5164
FÜLLSTOFF (BAUW.) 228
FÜLLTRICHTER 6577
FÜLLUNG 2269
FÜLLVERLUST 5163
FUNDAMENT 5589
FUNDAMENT EINER MASCHINE 7848
FUNDAMENTALSCHWINGUNG 5736
FUNDAMENTANKER 501
FUNDAMENTBOLZEN 5590
FUNDAMENTPLATTE 2900
FUNDAMENTSCHRAUBE 5590
FUNDIERUNG 5589
FÜNFECK 9172
FÜNFFACH-SCHWEFELANTIMON 625
FÜNFFLACH 9175
FÜNFFLÄCHNER 9175
FUNKE 11836
FUNKENEROSION 4531
FUNKENEROSIONS-LEHRENBOHR-MASCHINE (IN RECHTECKIGEN
    KOORDINATEN ARBEITEND) 10300
FUNKENEROSIONSCHNITT 4559
FUNKENFÄNGER 3109
FUNKENPROBE 11840
FUNKTION (MATH.) 5734
FUNKTION (STETIGE) 3025
FURCHENZIEHER (2-3-UND 4-REIHIG) 10573
FURNIERHOBELMASCHINE 13525
FURNIERHOLZ 13524
FURNIERSÄGE 13526
FUSELÖL 5754
FUSS (ENGLISCHER) 5518
FUSS EINER SÄULE 1035
FUSSBODEN (BELAG) 5450
FUSSBODENBELAG 5450
FUSSBODENBELAGBLECH 5447
FUSSBODENPLATTE 5302
FUSSBODENVORGELEGE 5444
FUSSBREMSE 5519
FUSSGESTELL 1041
FUSSHEBEL 5521
FUSSKREIS 10717

FUSSKREIS 10723
FUSSLAGER 5524
FUSSLEISTE 12977
FUSSPFUND 5522
FUSSSCHRAUBE 7529
FUSSTRITTDREHBANK 5520
FUSSVENTIL 5523
FUSSVENTIL 5523
FUTTER 7623
FUTTER 10614
FUTTER (HYDRAULISCHE) 6679
FUTTER (MECHANISCHES) 8095
FUTTER (PNEUMATISCHE) 9550
FUTTER (SELBSTTÄTIGE) 11152
FÜTTERN 7623
G.C.S.-SYSTEM 2159
GABBRO 5765
GABELBOLZEN 2492
GABELGELENK 5567
GABELGIESSPFANNE 11258
GABELHEUWENDER 12696
GABELROHR 1595
GABELSCHLÜSSEL 9322
GABELZAPFEN 6159
GADOLINIUM 5767
GALENIT 5775
GALLESCHE KETTE 12002
GALLIUM 5779
GALLONE 5781
GALLUSGERBSÄURE 5780
GALLUSSÄURE 5777
GALMEI (EDLER) 1836
GALVANISCHER UEBERZUG 9503
GALVANISIEREN 366
GALVANISIEREN 5789
GALVANISIERUNG 4562
GALVANOMETER 5791
GALVANOPLASTIK 4557
GALVANOPLASTISCH 5792
GALVANOSTEGIE 4640
GAMMA-EISEN 5795
GAMMA-FUNKTION 5794
GAMMA-STRAHLEN 5796
GAMMA-STRAHLENBILD 5798
GAMMA-STRUKTUR 5797
GANG (DIREKTER) 3983
GANG (GERÄUSCHLOSER) 8609
GANG (GERÄUSCHVOLLER) 8610
GANG (HARTER) 6412
GANG (LEICHTER) 7567
GANG (RUHIGER) 11681

GANG (SCHWERER) **6412**
GANG (STOSSENDER) **6412**
GANG (STOSSFREIER) **7567**
GANG (TOTER) **13873**
GANG (TOTER) **916**
GANG (UNRUHIGER) **7165**
GANG (WEICHER) **7567**
GANG EINER MASCHINE **10850**
GANG EINES GEWINDES **13279**
GANG SETZEN (WIEDER IN) **10508**
GANGART **5802**
GANGGESTEIN **5802**
GANGHÖHE EINER SCHRAUBE **9410**
GANGMINERAL **5802**
GANGSPILL **6229**
GANGTIEFE EINER SCHRAUBE **3784**
GANGZAHL EINES GEWINDES **8692**
GANISTER **5803**
GANZ-ALUMINIUMLEITER **343**
GANZEISEN **345**
GANZFABRIKAT **5730**
GANZHOLZ **7723**
GÄRBSTAHL **11490**
GÄRFUTTERZUBEREITUNGSMASCHINEN **11430**
GARNITUR **11213**
GARSCHAUMGRAPHIT **7317**
GÄRSTOFF **5097**
GÄRTNEREI- **13096**
GÄRUNG **5098**
GÄRUNGSSTOFF **5097**
GAS **5808**
GAS (GLÜHENDES) **6827**
GAS (KOMPRIMIERTES) **2847**
GAS (VERDICHTETES) **2847**
GAS (VERFLÜSSIGTES) **7637**
GAS (VOLLKOMMENES) **9185**
GAS VERFLÜSSIGEN (EIN) **7640**
GASABZUG **5821**
GASABZUG OHNE VERBRENNUNG MIT WIEDERGEWINNUNG
 5841
GASABZUG UND STAUBABSAUGUNG **4392**
GASANALYSE **5809**
GASANSTALT **5848**
GASÄTHER **9228**
GASBEHÄLTER **5859**
GASBEIZUNG **5832**
GASBELEUCHTUNG **5828**
GASBLASE **7523**
GASBLASE **1353**
GASBLASE **1671**
GASBLASE **5836**

GASBLASE **5825**
GASBRENNER **5811**
GASDICHT **5853**
GASDICHTE **3757**
GASDRUCK **5837**
GASDYNAMO **5850**
GASE ABFÜHREN **10455**
GASEINSCHLUSS **5836**
GASEINSCHLUSS **1353**
GASEINSCHLUSS **5855**
GASEINSCHLUSS **5825**
GASENTWICKLUNG **3818**
GASERZEUGER **5838**
GASEXPANSION **4920**
GASFABRIK **5848**
GASFACKEL **5820**
GASFANG **5842**
GASFEUERUNG **1751**
GASFEUERUNG **5744**
GASFÖRMIGER KÖRPER **5808**
GASGEMISCH **5856**
GASGENERATOR **5838**
GASGLOCKE **5859**
GASGLÜHLICHT **6829**
GASGLÜHLICHTLAMPE **6828**
GASHAHN **5814**
GASHÄRTUNG **5840**
GASKOHLE **5813**
GASKOHLE (ECHTE) **1901**
GASKOKS **5849**
GASKONSTANTE **5815**
GASKRAFTMASCHINE **5819**
GASLAMPE **5826**
GASLEITUNG **5834**
GASLICHT **5827**
GASLÖTEN **5810**
GASLÖTKOLBEN **5824**
GASMESSER **5829**
GASMOTOR **5819**
GASNITRIEREN **5830**
GASÖL **5831**
GASOLIN **9228**
GASOMETER **5859**
GASPORE **9629**
GASPRÜFER **4852**
GASREINIGUNG **5839**
GASREINIGUNG **10018**
GASRILLEN **5823**
GASROHR **5833**
GASROHRGEWINDE **5843**
GASROHRZANGE **5835**

GASSCHWEISSEN **5847**
GASSTROM **5816**
GASTEER **2570**
GASTHERMOMETER **265**
GASTURBINE **5844**
GASUHR **5829**
GASWASCHFLASCHE **5845**
GASWASSER **456**
GASWERK **5848**
GATTERSÄGE **5629**
GATTERSTAB (UNTERER) **7796**
GAUSS'SCHE VERTEILUNG **5887**
GAVALNOPLASTIK **4651**
GAZE **5888**
GEBLÄSE **242**
GEBLÄSE **1361**
GEBLÄSEDÜSE **8685**
GEBLÄSEMASCHINE **1363**
GEBLÄSEWIND **243**
GEBLÄSEWIND **1315**
GEBRAUCHSANWEISUNG **6992**
GEBÜNDELTER STRAHL **5504**
GEBUNDENER KOHLENSTOFF **5289**
GEDIEGEN (MIN.) **5647**
GEDRÜCKTER STAB **2855**
GEFÄLLE (NEIGUNG) **3792**
GEFÄSSBAROMETER **2440**
GEFÄSSE (KOMMUNIZIERENDE) **13553**
GEFÄSSMANOMETER **8827**
GEFLANSCHTE ENDEN **5339**
GEFLECHTÜBERZUG **1526**
GEFRIERPUNKT **5656**
GEFÜGE **12395**
GEFÜGE **12394**
GEFÜGE **12800**
GEFÜGE (BLÄTTRIGES) **7374**
GEFÜGE (FASERIGES) **5141**
GEFÜGE (FEINKÖRNIGES) **5200**
GEFÜGE (GLEICHARTIGES) **6560**
GEFÜGE (HETEROGENES) **6452**
GEFÜGE (HOMOGENES) **6560**
GEFÜGE (INHOMOGENES) **6452**
GEFÜGE (KÖRNIGES) **6053**
GEFÜGE (KRISTALLINISCHES) **3434**
GEFÜGE (UNGLEICHARTIGES) **6452**
GEFÜGE DER LEGIERUNGEN **2058**
GEFÜHLSBOHRMASCHINE **11186**
GEGEN DEN STROM **13407**
GEGEN-EMK **3231**
GEGENBEWEGUNG **3243**
GEGENDRUCK **3232**

GEGENDRUCKBETRIEB **894**
GEGENDRUCKSCHRAUBE **12909**
GEGENDRUCKVENTIL **895**
GEGENDRUCKVENTIL **5523**
GEGENEINANDER VERSETZT **12048**
GEGENFEDER **11974**
GEGENFLANSCH **8054**
GEGENGEWICHT **3251**
GEGENHALTEN **6524**
GEGENHALTER **4115**
GEGENHALTER **1305**
GEGENHALTER **898**
GEGENKEIL **5937**
GEGENKOPPLUNG **10536**
GEGENKRAFT **10248**
GEGENKURBEL **10528**
GEGENLOGARITHMUS **620**
GEGENMUTTER **7702**
GEGENPROE **4384**
GEGENSCHRAUBE **2281**
GEGENSCHRITTSCHWEISSEN **918**
GEGENSTOSS **1918**
GEGENSTREBE **3179**
GEGENSTROM **3230**
GEGENSTROMMESSER IN ÖLBAD **8747**
GEGENWELLE **3233**
GEGOSSEN (IN EINEM STÜCK) **2054**
GEHALT **3011**
GEHALT **10897**
GEHÄMMERT **6220**
GEHÄUSE **1385**
GEHÄUSE **6660**
GEHÄUSE **227**
GEHÄUSE (EINES MASCHINENTEILS) **2026**
GEHÄUSE EINES HAHNES **11334**
GEHÄUSE EINES SCHIEBERS **2229**
GEHÄUSE EINES VENTILS **2229**
GEHRDREIECK **8325**
GEHRFUGE **8324**
GEHRMASS **8325**
GEHRSTOSS **8324**
GEHRUNG **8326**
GEIGER-ZÄHLER **5911**
GEISER **5936**
GEISSFUSS (STEMMEISEN) **13432**
GEKRÄTZ **4331**
GEL **5912**
GELAGERT (AUF EINE SCHNEIDE) **6669**
GELAGERT (AUF EINER SPITZE) **8432**
GELAGERT (AUF ZAPFEN) **9422**
GELÄNDER-EISEN **1086**

GELÄNDEREISEN **6265**
GELÄNDERSTÜTZE **6266**
GELÄNDEWAGEN **3362**
GELATINE **5913**
GELBBLEIERZ **13991**
GELBBRENNE **1608**
GELBBRENNSÄURE **657**
GELBES BLUTLAUGENSALZ **9685**
GELBES CHROMKALI **9682**
GELBES CHROMSAURES KALI **9682**
GELBES ZYANEISENKALIUM **9685**
GELBGUSS **1562**
GELBKUPFER **1562**
GELBÖL **1668**
GELDSTRAFE **5195**
GELENK (MECH.) **7238**
GELENKBAND **6505**
GELENKBOLZEN **7242**
GELENKIG VERBUNDEN **7629**
GELENKKETTE **12002**
GELENKMASSSTAB **5512**
·GELENKPUNKT **5717**
GELENKROHR **5418**
GELENKSPINDELBOHRMASCHINE **12550**
GELENKSTEIN **7625**
GELENKSYSTEM **7628**
GELENKZAPFEN **7242**
GELERNTER ARBEITER **3289**
GEMISCH **8327**
GEMISCH (BRENNBARES) **4938**
GEMISCH (ENTZÜNDLICHES) **4938**
GEMISCH (ZÜNDFÄHIGES) **4938**
GENAUIGKEIT **9763**
GENAUIGKEIT **82**
GENAUIGKEIT DER AUSFÜHRUNG **83**
GENAUIGKEITSGRAD **3721**
GENAUIGKEITSTEILGERÄT **9769**
GENEHMIGUNGSDRUCK **8614**
GENERALNENNER **2789**
GENERATOR (ELEKTRISCHER) **5924**
GENERATOR (ELEKTROSTATISCHER) **4643**
GENERATORGAS **9890**
GENERATRIX **5925**
GENIETETES ROHR **10640**
GEOMETRIE **5931**
GEOMETRISCH **5927**
GEPULVERT **9983**
GERADE **12310**
GERADE LINIE **12310**
GERADER SETZHAMMER **5410**
GERADERICHTEN **12319**

GERADERICHTEN **12318**
GERADFÜHRUNG **12315**
GERADFÜHRUNG (DAMPFM.) **6169**
GERADLINIG **10319**
GERADLINIGE LÄNGSTEIGUNG **7615**
GERADVERZAHNUNG **12311**
GERADVERZAHNUNGSKUPPLUNG **12313**
GERÄT **641**
GERÄUSCH **8608**
GERBSTOFF **5780**
GERIEFT **5484**
GERIFFELT **5484**
GERINNE **2251**
GERINNEN **2556**
GERINNEN **2555**
GERINNSEL **2557**
GERIPPE **5624**
GERIPPE (SELBSTTRAGENDES) **11161**
GERMANIUM **5933**
GERÖLLE **11354**
GERÜST **10965**
GERÜST **5624**
GERÜST (SELBSTTRAGENDES) **11161**
GERÜSTKRAN **5804**
GESAMTBELASTUNG **13062**
GESAMTDRUCK **13065**
GESAMTLÄNGE **8925**
GESAMTLEISTUNG **13064**
GESAMTTOTGANG **13063**
GESAMTWIRKUNGSGRAD **8927**
GESCHÄFT **7229**
GESCHICHTET **7375**
GESCHIEBE **11354**
GESCHMIEDET (AUS DEM VOLLEN) **5552**
GESCHRECKT **4247**
GESCHÜTZBRONZE **6178**
GESCHÜTZBRONZE **197**
GESCHÜTZTER MOTOR **9941**
GESCHWINDIGKEIT **10226**
GESCHWINDIGKEIT **11875**
GESCHWINDIGKEIT **11868**
GESCHWINDIGKEIT (ABNEHMENDE) **3679**
GESCHWINDIGKEIT (KONSTANTE) **13363**
GESCHWINDIGKEIT (KRITISCHE) **3353**
GESCHWINDIGKEIT (LINEARE) **7617**
GESCHWINDIGKEIT (TANGENTIALE) **12623**
GESCHWINDIGKEIT (UNVERÄNDERLICHE) **13363**
GESCHWINDIGKEIT (VERÄNDERLICHE) **13494**
GESCHWINDIGKEIT (VIRTUELLE) **13573**
GESCHWINDIGKEIT (ZUNEHMENDE) **6852**
GESCHWINDIGKEIT ERHÖHEN **6851**

GESCHWINDIGKEIT STEIGERN **6851**
GESCHWINDIGKEIT VERRINGERN **3678**
GESCHWINDIGKEITSABNAHME **3677**
GESCHWINDIGKEITSÄNDERUNG **2245**
GESCHWINDIGKEITSDIAGRAMM **13514**
GESCHWINDIGKEITSERHÖHUNG **10185**
GESCHWINDIGKEITSHÖHE **13515**
GESCHWINDIGKEITSIKURVE **13513**
GESCHWINDIGKEITSMESSER **11877**
GESCHWINDIGKEITSMESSER **11869**
GESCHWINDIGKEITSPARALLELOGRAMM **9072**
GESCHWINDIGKEITSREGLER **11872**
GESCHWINDIGKEITSREGLER **11874**
GESCHWINDIGKEITSSCHREIBER **12586**
GESCHWINDIGKEITSSCHWANKUNGEN **13499**
GESCHWINDIGKEITSSTUFE **13523**
GESCHWINDIGKEITSVERMINDERUNG **10363**
GESCHWINDIGKEITSZAHL **2638**
GESCHWINDIGKEITVERMINDERN **3678**
GESCHWINDIGKEITZUNAHME **6849**
GESENK **3897**
GESENK (OFFENES) **8810**
GESENK (ZWEITEILIGES) **2818**
GESENKFRÄSEN **3892**
GESENKKLOTZ **12531**
GESENKMETALL **3889**
GESENKPLATTE **12531**
GESENKSCHMIEDEN **12532**
GESENKSCHMIEDEN **12534**
GESENKSCHMIEDEN **12059**
GESENKSCHMIEDEN **3894**
GESENKSCHMIEDESTÜCK **4324**
GESENKSTOCK **12531**
GESETZ (BRAGGSCHES) **1525**
GESETZ VON BRAGG **1525**
GESICHTSFELD **5143**
GESICHTSSCHUTZMASKE **5000**
GESICHTSWINKEL **13581**
GESIMSHOBEL **9522**
GESPERRE **7709**
GESPINSTFASER **12799**
GESTALTUNG **3794**
GESTÄNGE **12577**
GESTEHUNGSKOSTEN **3207**
GESTEHUNGSPREIS **3210**
GESTEIN **10651**
GESTEIN (SPALTBARES) **2486**
GESTELL **5624**
GESTELL **12490**
GESTELLSÄGE **9747**
GESTÜTZT (BEIDERSEITIG) **11474**

GETREIDEFÖRDERER UND HEBER **6034**
GETREIDETROCKNER **4258**
GETRIEBE **8122**
GETRIEBE MIT AUSSENVERZAHNUNG **4956**
GETRIEBE MIT INNENVERZAHNUNG **7058**
GETRIEBELEHRE **7306**
GETRIEBENE SCHEIBE **4298**
GEWALZTEN ZUSTAND (IM) **744**
GEWÄSSERTES NATRIUMKARBONAT **11711**
GEWEBE (WASSERDICHTES) **13690**
GEWICHT **13747**
GEWICHT (SPEZIFISCHES) **11850**
GEWICHT (SPEZIFISCHES) **3756**
GEWICHT (TOTES) **3637**
GEWICHT (VERBRAUCHTES) **2992**
GEWICHTSANALYSE **6077**
GEWICHTSARÄOMETER **6723**
GEWICHTSATZ **11219**
GEWICHTSEINHEIT **13378**
GEWICHTSFALLVERSUCH **4330**
GEWICHTSHEBEL **13751**
GEWICHTSHEBELBREMSE **3638**
GEWICHTSPROZENT **9180**
GEWICHTSREGLER **13752**
GEWICHTSSCHALE **10976**
GEWICHTSSICHERHEITSVENTIL **13749**
GEWICHTSSYSTEM **12580**
GEWICHTSTEIL **9180**
GEWICHTSVERLUST **7764**
GEWINDE **11045**
GEWINDE **12857**
GEWINDE (AUSGELEIERTES) **13981**
GEWINDE (FLACHES) **12030**
GEWINDE (FLACHGÄNGIGES) **12030**
GEWINDE (GROBES) **2575**
GEWINDE (HALBIERTES) **1814**
GEWINDE (INTERNATIONALES METRISCHES) **7083**
GEWINDE (KONISCHES) **12666**
GEWINDE (LINKSGÄNGIGES) **7508**
GEWINDE (METRISCHESINTERNATIONALES) **7084**
GEWINDE (RECHTSÄNGIGES) **10585**
GEWINDE (RUNDES) **10802**
GEWINDE (SCHARFES) **545**
GEWINDE (SCHARFGÄNGIGES) **545**
GEWINDE (WHITWORTHSCHES) **13844**
GEWINDE (ZYLINDRISCHES) **9065**
GEWINDE LEHREN **12860**
GEWINDE MIT SCHWACHER STEIGUNG **11647**
GEWINDE MIT STARKER STEIGUNG **10105**
GEWINDE SCHNEIDEN **11046**
GEWINDEBASIS **1028**

GEWINDEBOHREN 12678
GEWINDEBOHRER 11044
GEWINDEBOHRER 12640
GEWINDEBOLZEN 12398
GEWINDEDORN 12867
GEWINDEDREHBANK 12859
GEWINDEDREHBANK 11029
GEWINDEDURCHMESSER (ÄUSSERER) 3852
GEWINDEEISEN 11039
GEWINDEFLANSCH 11064
GEWINDEFLANSCH 12866
GEWINDEFLANSCH 11055
GEWINDEFRÄSMASCHINE 11036
GEWINDEGANG 13279
GEWINDEGANGZAHL 8692
GEWINDEGRUND 1028
GEWINDEKERN 12864
GEWINDEKLUPPE 1121
GEWINDEKONTROLLEHRE 2760
GEWINDEKONTROLLEHRE 11038
GEWINDEKOPF 12455
GEWINDEKUPPLUNG 11026
GEWINDELEHRDORN 12862
GEWINDELEHRDORN 11049
GEWINDELEHRE 11047
GEWINDELEHRMUTTER 11050
GEWINDELOCH 11056
GEWINDEMITTELSCHNEIDER 11102
GEWINDEMUFFE 11061
GEWINDENACHSCHNEIDER 9534
GEWINDEPFROPFEN 11060
GEWINDEROHR 11059
GEWINDESCHNEID-KLUPPE 12266
GEWINDESCHNEIDBACKEN 3896
GEWINDESCHNEIDEN 11030
GEWINDESCHNEIDEN 12868
GEWINDESCHNEIDKOPF (RUNDER) 11781
GEWINDESCHNEIDMASCHINE 11066
GEWINDESCHNEIDVORRICHTUNG 12676
GEWINDESPINDEL 11062
GEWINDESPITZE 12455
GEWINDESPITZE 3333
GEWINDESTAHL 2274
GEWINDESTAHLLEHRE 12865
GEWINDESTEIGUNG 9399
GEWINDESTIFT 6156
GEWINDESTRÄHLER 2274
GEWINDETIEFE 3784
GEWINDEVORSCHNEIDER 5268
GEWINDEWEITEN 11054
GEWINN AN ARBEIT 8094

GEWINNUNG DER KOHLE 13877
GEWINNUNG VON METALLEN 4966
GEWÖLBESTEIN 687
GEWUNDENE KURVE 13307
GEZÄHNELT 6862
GEZAHNTER SEKTOR 13024
GICHT 2076
GICHT 1059
GICHTGAS 1313
GICHTSTAUB 5469
GICHTSTAUBSAMMLER 4390
GIESSBAR 1914
GIESSEN 9715
GIESSEN 5592
GIESSEN 2068
GIESSEN 2030
GIESSEN 2070
GIESSEN 12703
GIESSEN (KONTINUIERLICHES) 3019
GIESSEN (STEIGENDES) 1488
GIESSEN (WIRBELFREIES) 4388
GIESSEN (WIRBELFREIES) (DURVILLE-VERFAHREN) 10109
GIESSEN MIT VERLORENER GIESSFORM 7115
GIESSER 5591
GIESSEREI 5592
GIESSEREI 5593
GIESSEREI 5597
GIESSEREI 2071
GIESSEREIKOKS 2662
GIESSEREIKOKS 5595
GIESSEREIROHEISEN 5596
GIESSHALLE 2071
GIESSHALLE 2077
GIESSHAUS 5597
GIESSKANNE 13687
GIESSKOPF 11506
GIESSKUNST 5592
GIESSPFANNE 7356
GIESSPFANNE (GEMAUERTE) 7618
GIESSPFANNENWAGEN 2072
GIESSRAHMEN 8354
GIESSTRICHTER 9716
GIESSTROMMEL 4336
GIPS 6190
GIPS (GEBRANNTER) 9466
GIPSFORM 6188
GIPSFORM 9465
GIPSMODELL 6189
GITTER 7429
GITTER 6096
GITTER 6071

GITTER (REZIPROKES) **10281**
GITTERAKKUMULATOR **6097**
GITTEREBENE **804**
GITTERKONSTANTE **9075**
GITTERKONSTANTE **7430**
GITTERPAPIER **11530**
GITTERROST **6101**
GITTERROSTSTUFEN **6102**
GITTERSCHIEBER **6100**
GITTERTRÄGER **7431**
GITTERVERZERRUNGSTHEORIE **7434**
GLANZ (METALLISCHER) **8195**
GLANZ (MIN.) **7830**
GLANZ (VERLIEREN DEN) **12685**
GLANZ SCHLEIFEN **9598**
GLÄNZBAD (CHEMISCHES) **1564**
GLANZBRENNE **1609**
GLANZDECKEL **9801**
GLANZKOBALT **2614**
GLANZLEDER **9117**
GLANZLICHTER **6472**
GLANZMITTEL **1615**
GLANZPAPPE **9801**
GLANZPUNKTE **6472**
GLANZSCHLEIFEN **9605**
GLANZVERLUST **4369**
GLANZZIEHEN **4347**
GLANZZUSATZ **1615**
GLAS **5956**
GLAS **11589**
GLAS (BÖHMISCHES) **1400**
GLAS (FARBLOSES) **2746**
GLAS (GEBLASENES) **1365**
GLAS (NICHT SPLITTERNDES) **13894**
GLAS-HALBZELLE **5959**
GLASBAUSTEIN **5957**
GLASERKITT **5971**
GLASFLUSS **4709**
GLASGLANZ **13583**
GLASIEREN **5972**
GLASIEREN **5969**
GLASIG **13582**
GLASKOLBEN **5370**
GLASLEINWAND **5958**
GLASMEHL **5962**
GLASPAPIER **5961**
GLASPLATTE **5963**
GLASRÖHRE **5965**
GLASSCHEIBE **5963**
GLASTAFEL **5963**
GLASUR **5972**

GLASUR **5968**
GLASURSTEIN **5970**
GLASWOLLE **5966**
GLASZIEGEL **5964**
GLÄTTEN **5408**
GLÄTTEN **11686**
GLÄTTEN **11679**
GLÄTTEN **9157**
GLÄTTEN **9454**
GLATTWALZEN **10695**
GLÄTTWALZWERK **9456**
GLÄTTZAHN **1757**
GLAUBERSALZ **5967**
GLEICHACHSIG **2605**
GLEICHFÖRMIGKEIT DER BEWEGUNG **10417**
GLEICHFÖRMIGKEIT DES MATERIALS **13364**
GLEICHFÖRMIGKEITSGRAD **3719**
GLEICHGEWICHT **4793**
GLEICHGEWICHT **4786**
GLEICHGEWICHT (INDIFFERENTES) **6882**
GLEICHGEWICHT (LABILES) **13401**
GLEICHGEWICHT (STABILES) **12045**
GLEICHGEWICHT SEIN (MIT) **1076**
GLEICHGEWICHTSDIAGRAMM **4788**
GLEICHGEWICHTSKONSTANTE **4787**
GLEICHGEWICHTSLAGE **9647**
GLEICHGEWICHTSLIEHRE **12140**
GLEICHGEWICHTSPOTENTIAL EINER ELEKTRODE **4789**
GLEICHGEWICHTSTEMPERATUR **4791**
GLEICHGEWICHTSWERT **4792**
GLEICHGEWICHTSZUSTAND **12132**
GLEICHGEWISCHT (CHEMISCHES) **2307**
GLEICHLÄUFIGES FRÄSEN **13408**
GLEICHLÄUFIGES FRÄSEN **4182**
GLEICHMASSGRENZE **7590**
GLEICHMÄSSIG **6555**
GLEICHRICHTER **10314**
GLEICHRICHTER (ELEKTROCHEMISCHER) **4572**
GLEICHRICHTER (ELEKTRONISCHER) **4634**
GLEICHRICHTERANODE **10315**
GLEICHRICHTERKATODE **10316**
GLEICHRICHTERRÖHRE **10317**
GLEICHRICHTUNG **10312**
GLEICHSCHLAGRAHTSEIL **7390**
GLEICHSTROM **3023**
GLEICHSTROM (VON FLÜSSIGKEITEN) **13360**
GLEICHSTROM-LICHTBOGENSCHWEISSEN **686**
GLEICHSTROMDAMPFMASCHINE **12407**
GLEICHSTROMDYNAMO **3021**
GLEICHSTROMGENERATOR **3021**
GLEICHSTROMMASCHINE **3021**

GRENZLEHRE 3905
GRENZLINIEN 5689
GRENZMASS 7587
GRENZWELLENLÄNGE 8309
GRENZWELLENLÄNGE 10072
GRENZWERT 7594
GRENZWERTVERTEILUNG 4968
GRENZWINKEL (OPT.) 3345
GRENZZUSTAND 3349
GRIFF EINES WERKZEUGS 6263
GRIFFKURBEL 6231
GRIT 6123
GROBBLECH 9477
GROBBLECH 6409
GROBBLECHPRESSE 6411
GROBEN BEARBEITEN (AUS DEM) 10775
GROBFEILE 10774
GROBGEFÜGE 7869
GROBGEWINDE 2575
GROBGEWINDE 2576
GROBKALK 5664
GROBKOHLE 7826
GROBKORN 2573
GROBKORNEISEN 8819
GROBMASCHIG 2578
GROBMÖRTEL 2894
GROBSCHMIED 11671
GROBWALZWERK 6410
GROBZUG 1727
GRÖNTARNSPAT 3423
GRÖSSE (BEKANNTE) 7335
GRÖSSE (GEGEBENE) 5951
GRÖSSE (GESUCHTE) 10469
GRÖSSE (MATH.) 10065
GRÖSSE (UNBEKANNTE) 13391
GRÖSSE (WIRKLICHE) 152
GROSSE ACHSE EINER ELLIPSE 7947
GRÖSSE EINER KRAFT 7939
GRÖSSENORDNUNG 8852
GRÖSSENVERHÄLTNIS 10982
GROSSGASMASCHINE 7410
GROSSLUCKIGES EISEN 8819
GROSSSERIENANFERTIGUNG 7727
GRÖSSTWERT 8069
GROSSWASSERRAUMKESSEL 7412
GRÜBCHEN 9393
GRUBE 8300
GRUBENGAS 8215
GRUBENGUSS 9394
GRUBENHOBELMASCHINE 9395
GRUBENKLEIN (MITTELGROSSES) 8267

GRUBENSAND 9396
GRUBENSCHACHT 9398
GRUBENWASSER 9397
GRÜN (SCHEELESCHES) 10996
GRUND 1028
GRUNDANSTRICH 9864
GRUNDBELASTUNG 8655
GRUNDBÜCHSE 1565
GRUNDFARBE 9847
GRUNDFLÄCHE 1025
GRUNDIERLACK 9855
GRUNDIERUNG 9864
GRUNDKEGEL 9406
GRUNDKREIS 1030
GRUNDLAGENFORSCHUNG 5735
GRUNDLAGENFORSCHUNG 1052
GRUNDLAGENFORSCHUNG 10016
GRUNDLAGER 7941
GRUNDLAST 8655
GRUNDLINIE 1027
GRUNDMASS 1053
GRUNDMASSE 8055
GRUNDMESSING 8056
GRUNDMETALL 8057
GRUNDMETALL 1032
GRUNDMETALL 9079
GRUNDMETALLPROBESTÜCK 1039
GRUNDPLATTE 1026
GRUNDPLATTE 1127
GRUNDPLATTE 1486
GRUNDPLATTE EINER MASCHINE 1037
GRUNDRING 9000
GRUNDRISS 11124
GRUNDRISS 5914
GRUNDSCHICHT 9857
GRUNDSCHWINGUNG 5736
GRUNDSTOFF 9852
GRUNDSTOFF 4654
GRUNDSTOFF (HOMOGENER) 6558
GRUNDWASSER 6147
GRUNDWASSERSPIEGEL 6148
GRUNDWERKSTOFF 909
GRUNDZAHL 9860
GRUNDZAHL 1033
GRÜNFUTTERLADER 7687
GRÜNFUTTERLADER 6091
GRÜNLING 6090
GRÜNSANDBINDER 1445
GRÜNSPAN 220
GRÜNSPAN 219
GRÜNSPAN 13533

GRÜNSPANÄHNLICH 219
GRÜNSTEIN 3828
GRUPPENANTRIEB 6150
GRUPPENVENTIL 8476
GRUPPIERUNG DER NIETE 717
GRUSKOHLE 11557
GUINIER-PRESTON ZONE 6171
GUMMI 10816
GUMMI (ARABISCHES) 6172
GUMMI MIT GEWEBEEINLAGE 6973
GUMMI- (ODER NEOPREN-) ISOLIERUNG 11329
GUMMIGUTT 5793
GUMMIHARZ 6173
GUMMIISOLIER BAND 10822
GUMMILACK 12254
GUMMIMEMBRAN 10819
GUMMIPFROPFEN 10824
GUMMIRIEMEN 10817
GUMMIRING 10823
GUMMIRINGDICHTUNG 10820
GUMMISACKVERFAHREN 923
GUMMISCHLAUCH 10821
GUMMISTOFF 10818
GUMMITRAGANT 6174
GURTFÖRDERER 1164
GUSS 2068
GUSS 9715
GUSS 2034
GUSS (BLASIGER) 6563
GUSS (LUNKERFREIER) 2075
GUSS (SCHMIEDBARER) 7959
GUSS (SPANNUNGSFREIER) 2074
GUSS FÜR DEN MASCHINENBAU 4763
GUSSALUMINIUM 435
GUSSASPHALT 9713
GUSSBETON 8426
GUSSBLASE 1353
GUSSEISEN 2034
GUSSEISEN 9299
GUSSEISEN 7157
GUSSEISEN (KORROSIONSBESTÄNDIGES) 2039
GUSSEISEN (LEGIERTES) 363
GUSSEISEN (LEGIERTES) 2037
GUSSEISEN (MELIERTES) 2044
GUSSEISEN (PERLITISCHES) 2046
GUSSEISEN (VERSCHLEISSFESTES) 2048
GUSSEISEN (WEISSES) 2049
GUSSEISEN MIT KUGELGRAPHIT 2045
GUSSEISEN MIT KUGELGRAPHIT 11895
GUSSEISEN MIT KUGELGRAPHIT 11894
GUSSEISEN MIT KUGELGRAPHIT 2040

GUSSEISEN MIT LAMELLENGRAPHIT 2041
GUSSEISEN MIT STAHLZUSATZ 11177
GUSSEISEN-ROHRLEITUNG 2062
GUSSEISENQUELLUNG 2061
GUSSEISENROHR 2035
GUSSEISENTHERMIT 2036
GUSSFEHLER 5056
GUSSFORM 8422
GUSSFORM (GIESS.) 8423
GUSSHAUT 11538
GUSSKUPFER 2073
GUSSMESSING 2031
GUSSMETALL 2050
GUSSMODELL 9129
GUSSNAHT 5184
GUSSPLATTIERUNG 2032
GUSSPUTZER 4254
GUSSPUTZEREI 5130
GUSSPUTZHAMMER 10986
GUSSRINNE 10844
GUSSSPANNUNG 7075
GUSSSPANNUNG 2081
GUSSSPANNUNG 2078
GUSSSTAHL 2057
GUSSSTÜCK 2069
GUSSSTÜCK 2079
GUSSSTÜCKE (HITZEBESTÄNDIGE) 2084
GUSSTÜCKE 2080
GUSSTÜCKE (KORROSIONSBESTÄNDIGE) 2082
GUSSTÜCKENZUSAMMENSCHWEISSEN 2060
GUSSWERK 5593
GUSSZAPFEN 12004
GUT-UND-AUSSCHUSS-GRENZLEHRE 5991
GUTACHTEN 4931
GUTACHTER 4930
GÜTEPRÜFUNG 10063
GÜTEVERHÄLTNIS 4463
GUTTAPERCHA 6184
H.D. ÖL 6403
HAARRIEMEN 6200
HAARRISS 6202
HAARRISS 4377
HAARRISS 3318
HAARROHR 1921
HAARRÖHRCHENWIRKUNG 1920
HAARSEITE 6038
HAARZIRKEL 6201
HACKFRÄSEN UND BODENFRÄSEN 10753
HAFNIUM 6199
HAFT- 175
HAFTEND 175

HAFTFÄHIGKEITSVERSUCH **172**
HAFTFESTIGKEIT **173**
HAFTREIBUNG **12134**
HAFTREIBUNG **5678**
HAFTVERMÖGEN **176**
HAHN **2617**
HAHN **12639**
HAHN (SELBSTDICHTENDER) **7111**
HAHN MIT ABLAUF **1234**
HAHN MIT GEHÄUSESCHMIERUNG **7802**
HAHNGEHÄUSE **11334**
HAHNKEGEL **9532**
HAHNKÜKEN **9532**
HAHNREIBER **9532**
HAHNSCHLÜSSEL **9532**
HAHNSCHLÜSSEL **2618**
HAHNVENTIL **9535**
HAHNVENTIL **8997**
HAHNVENTIL (GEÖLTES) **7802**
HAHNWIRBEL **9532**
HAKEN **6568**
HAKENKETTE **6570**
HAKENNAGEL **2457**
HAKENNÄGEL **11902**
HAKENSCHLÜSSEL **1822**
HAKENSCHRAUBE **6569**
HALB-V-HAHT MIT LUFTSPALT **8822**
HALB-V-NAHT **1800**
HALB-V-NAHT MIT LUFTSPALT **1797**
HALB-V-NAHT OHNE LUFTSPALT **2528**
HALB-V-NAHT OHNE LUFTSPALT **1789**
HALBACHSE (GEOM.) **11167**
HALBDIESELMOTOR **11171**
HALBFABRIKAT **11174**
HALBHOLZ **6210**
HALBIEREN (MATH.) **1267**
HALBIEREN (MATH.) **1266**
HALBIERUNGSLINIE EINES WINKELS **1268**
HALBKREIS **11179**
HALBKREISFÖRMIG **11169**
HALBKREISTRANSPORTEUR **11180**
HALBKREUZRIEMENTRIEB **6207**
HALBKUGEL **6438**
HALBLEHRENBOHRMASCHINE **11172**
HALBMESSER **10516**
HALBMONDFÖRMIGE SCHWEISSNAHT-VERZERRUNGEN **5212**
HALBRAUPENSCHLEPPER **13093**
HALBRUNDEISEN (HOHLES) **6541**
HALBRUNDEISEN (VOLLES) **11786**
HALBRUNDER MEISSEL **6546**
HALBRUNDFEILE **6208**

HALBRUNDPROFILEISEN **6209**
HALBRUNDSTAHL (GEZOGENER) **4249**
HALBSCHATTEN **9177**
HALBSCHLICHTFEILE **11100**
HALBVERSENKNIET **3249**
HALBWASSERGAS **4194**
HALBWELLENGLEICHRICHTEN **6212**
HALBWERTSCHICHT **6211**
HALBZELLE **6204**
HALBZEUG **11174**
HALLE **1072**
HALO **6213**
HALOGEN **6214**
HALS **8512**
HALSLAGER EINER LIEGENDEN WELLE **6586**
HALSLAGER EINER STEHENDEN WELLE **13547**
HALSSTUTZEN **8514**
HALSZAPFEN **8513**
HALT (WAHLWEISER) **8850**
HALT (WAHLWEISER) **6521**
HALTBARKEIT **4385**
HALTEPUNKT **3347**
HALTER **12490**
HALTERUNG **1522**
HALTESEILE **6186**
HALTESTEIN **9159**
HALTESTIFT **9159**
HALTESTIFT **11220**
HALTESTREIFEN **10516**
HALTEZEIT **6528**
HÄMATIT **11865**
HÄMATIT **6434**
HÄMATITROHEISEN **6435**
HAMMER **6215**
HAMMER MIT GESPALTENER FINNE **2462**
HAMMER MIT KUGELFINNE **960**
HAMMER MIT KUGELFINNE **962**
HAMMER MIT ZWEI BAHNEN **4169**
HAMMERBAHN **4993**
HÄMMERBAR **7960**
HÄMMERBARKEIT **7958**
HAMMERFINNE **9031**
HAMMERKOPF **6317**
HAMMERKOPFSCHRAUBE **12701**
HAMMERLÖTKOLBEN **2350**
HAMMERMÜHLEN **8297**
HÄMMERN **9157**
HÄMMERN **6223**
HÄMMERN **6216**
HAMMERPINNE **9031**
HAMMERSCHLAG **6217**

HAMMERSCHWEISSEN **6218**
HAMMERSTAHL **6222**
HAMMERSTIEL **11244**
HANBREIBAHLE **6247**
HAND (NACHSTELLBARE) **4917**
HAND ANGETRIEBEN **6241**
HAND-STOSSELEKTRODENSCHWEISSEN **10027**
HAND-TEILEPROGRAMM **6242**
HANDANTRIEB **6233**
HANDARBEIT **6237**
HANDBETRIEB **6233**
HANDBETRIEB (FÜR) **6241**
HANDBOHRMASCHINE **6258**
HANDBOHRMASCHINE (ELEKTRISCHE) **9732**
HANDBREMSE **9082**
HANDBREMSE **6227**
HANDELSBEZEICHNUNG **13101**
HANDELSEISEN **8148**
HANDELSGÜTEBRONZE **1657**
HANDELSPROFILE **8149**
HANDELSSCHWEFELSÄURE **8856**
HANDELSÜBLICHE BEZEICHNUNG **13101**
HANDELSZINK **2785**
HANDELSZINK **11879**
HANDFEILE **6234**
HANDGEARBEITET **6240**
HANDGEFERTIGT **6240**
HANDGESCHMIEDET **6261**
HANDGEWINDEBOHRER **6257**
HANDGLANZSCHLEIFEN **6228**
HANDGRIFF **6262**
HANDHABE **6262**
HANDHAMMER **6235**
HANDHEBEL **6239**
HANDKARREN **6230**
HANDKARREN **13231**
HANDKLOBEN **6254**
HANDKURBEL **6231**
HANDLAUF **13035**
HANDLÄUFEREISEN **6265**
HANDLEISTE **13035**
HANDLEISTENEISEN **6245**
HANDLEISTENEISEN **6265**
HANDLICH **6269**
HANDLOCH **6236**
HANDNIETUNG **6248**
HANDPFANNE **6238**
HANDPUMPE **6244**
HANDRAD **6267**
HANDRAD **13810**
HANDRAD **6255**

HANDRADMUTTER **6268**
HANDREIBAHLE (EINSTELLBARE) **184**
HANDSÄGE **6249**
HANDSCHERE **6251**
HANDSCHIRM **6252**
HANDSCHMIEDEN **11669**
HANDSCHNEIDEN **6232**
HANDSCHUHKASTEN **5980**
HANDSTAMPFER **6246**
HANDSTICHTORF **4364**
HANDWERKSZEUG **13015**
HANDWERKZEUG **6253**
HANDWINDE **6256**
HANDZEICHNUNG **5648**
HANF **6439**
HANFDICHTUNG **6445**
HANFLIDERUNG **6445**
HANFÖL **6446**
HANFPACKUNG **6445**
HANFRIEMEN **1907**
HANFSCHLAUCH **1908**
HANFSEELE **6440**
HANFSEIL **6443**
HANFWERG **6444**
HANFZOPF **6441**
HANFZOPF (GEFETTETER) **6442**
HÄNGEBOCK **9166**
HÄNGEBRÜCKE **12528**
HÄNGEDACHTANK **12525**
HÄNGEDECKE **12524**
HÄNGELAGER **6321**
HÄNGELAGER (GESCHLOSSENES) **6270**
HÄNGELAGER (OFFENES) **6271**
HÄNGENBLEIBEN **7210**
HÄNGESITZ **1383**
HARDENIT **8022**
HÄRTBARKEIT **6288**
HÄRTBARKEIT EINES METALLES **6296**
HARTBLEI **6276**
HARTBRANDSTEIN **6273**
HÄRTE **6297**
HÄRTE BEI HOCHTEMPERATUR **6629**
HÄRTE DES WASSERS **6298**
HÄRTE DES WASSERS **9203**
HÄRTE DES WASSERS (SCHWINDENDE) **12735**
HÄRTE DES WASSERS (TEMPORÄRE) **12735**
HÄRTE DES WASSERS (VORÜBERGEHENDE) **12735**
HÄRTE-EINDRUCK **6859**
HÄRTEBAD **10096**
HÄRTEGRAD **3720**
HÄRTEGRAD DES STAHLES **6299**

HÄRTEMASS 3720
HÄRTEN 6219
HÄRTEN (GEBROCHENES) 12249
HÄRTEN (PARTIELLES) 7697
HÄRTEN UND ANLASSEN 10092
HÄRTEN VON METALLEN 6293
HÄRTEÖL 6295
HÄRTEÖL 10099
HÄRTEPROBE 6300
HÄRTEPRÜFER 11366
HÄRTEPRÜFER (ELEKTROMAGNETISCHER) 4614
HÄRTEPRÜFUNG 6300
HÄRTEPULVER 2139
HÄRTERISS 3282
HÄRTERISS 10097
HARTES HOLZ 6282
HÄRTESKALA 10974
HÄRTESKALA (MOHSCHE) 8346
HÄRTESTUFE 3720
HARTGUMMI 2363
HARTGUSS 2335
HARTGUSS 2038
HARTGUSS (UMGEKEHRTER) 7109
HARTHOLZ 6301
HARTHOLZ 6282
HARTLOT 6279
HARTLÖTEN 1576
HARTLÖTEN 1573
HARTLÖTKUPFER 1575
HARTLÖTLEGIERUNG 386
HARTLÖTOFEN 1577
HARTLÖTUNG 1576
HARTLÖTUNGSEINLAGEN 1578
HARTMETALL 6277
HARTMETALL-AUFTRAGSCHWEISSUNG 6283
HARTMETALL-AUFTRAGSLEGIERUNG 6284
HARTMETALLBESTÜCKUNG 6278
HARTMETALLLEGIERUNG 11510
HARTMETALLSCHNEIDWERKZEUG 2144
HARTPARAFFIN 9045
HARTSTAHL 1957
HARTSTAHL 6280
HÄRTUNG 6291
HÄRTUNG (ISOTHERMISCHE) 7195
HÄRTUNGSBAD 6294
HÄRTUNGSFLÜSSIGKEIT 6294
HÄRTUNGSGRAD BEIM KALTZIEHEN 6274
HÄRTUNGSMINDERUNG 4
HÄRTUNGSMITTEL 6290
HARTZINN 9230
HARZ 10478

HARZHALTIG 10479
HARZKITT 10482
HARZÖL 10749
HASPEL 13863
HAUBE 3276
HAUBE 1451
HAUBE 6566
HAUBE EINES SCHIEBERS 3266
HAUBE EINES VENTILS 3266
HAUBENFLANSCH 1452
HAUBENGEWINDEBOLZEN 1453
HAUBENOFEN 1132
HAUBENOFEN 1135
HAUCHBILDUNG 1347
HAUEN (DURCH TRAKTOREN ANGETRIEBEN) 6515
HÄUFEPFLÜGE 9527
HÄUFIGKEITSFUNKTION 5662
HAUPTABMESSUNGEN 7481
HAUPTACHSE 9866
HAUPTACHSE DER HYPERBEL 13143
HAUPTAUSLEGER 7226
HAUPTDÜSE 7942
HAUPTDYNAMO 9867
HAUPTLAGER 7941
HAUPTLAGER 3303
HAUPTLEITUNG (ELEKTR.) 5081
HAUPTMASCHINE 9867
HAUPTMASSE 7481
HAUPTNENNER 2789
HAUPTSCHLUSSDYNAMO 11201
HAUPTSCHLUSSMOTOR 11202
HAUPTSCHNITT 9868
HAUPTSPANNUNG 9869
HAUPTSPEISELEITUNG 7940
HAUPTVERFORMUNG 2000
HAUPTWELLE 7943
HAUPTWELLE 7944
HAUPTZYLINDER 8037
HAUSENBLASE 7174
HAUSTEIN 761
HAUT 5171
HAUT-LEIM 11539
HAUTBILDUNG 11545
HAUTPACKUNG 11540
HAUTRISS 12501
HÄUTUNG 4889
HEAV DUTY ÖL 6403
HEBEBAUM 7725
HEBEBOCK 5945
HEBEBOCK (HYDRAULISCHER) 6688
HEBEDAUMEN 13878

HEBEHAKEN **7558**
HEBEKOLBEN **7555**
HEBEL **7530**
HEBEL (BELASTETER) **13751**
HEBEL (DOPPELARMIGER) **4147**
HEBEL (DREIARMIGER) **4131**
HEBEL (GEGABELTER) **5569**
HEBEL (GERADARMIGER) **12306**
HEBEL (ZWEIARMIGER) **4147**
HEBELADE **7533**
HEBELARM **7531**
HEBELARM **706**
HEBELBREMSE **7532**
HEBELPRESSE **7534**
HEBELÜBERSETZUNG **13132**
HEBELÜBERSETZUNG (VERHÄLTNIS) **7536**
HEBELWERK **12577**
HEBEMASCHINE **6519**
HEBEMASCHINEN (SÄCKEBÜNDEL UND ZUCKE PRÜBEN)
   7688
HEBEN (EINE LAST) **7542**
HEBEN EINER LAST **7559**
HEBER **11517**
HEBERBAROMETER **11515**
HEBERMANOMETER **2534**
HEBESTUTZEN **7206**
HEBEZEUG **6519**
HEBEZEUG **6516**
HEBLING **12672**
HECKENSCHNEIDER **6418**
HECTOWATTSTUNDE **6417**
HEFNERKERZE **6419**
HEFT EINES WERKZEUGES **6263**
HEFTKURBEL **3305**
HEFTNIET **12592**
HEFTNIETUNG **12922**
HEFTSCHWEISSEN **12590**
HEFTSCHWEISSEN **12589**
HEFTSTIFT **4242**
HEFTZWEKE **4242**
HEISSDAMPF **12471**
HEISSDAMPFMASCHINE **4759**
HEISSFERTIGPUTZEN **6647**
HEISSINDLEITUNG **1774**
HEISSLAUFEN **6334**
HEISSLAUFEN DER LAGER **6400**
HEISSLUFTMASCHINE **6615**
HEISSSÄGE **6630**
HEISSWASSERHEIZUNG **6489**
HEISSWINDLEITUNG **6651**
HEIZDAMPF **6397**

HEIZELEMENT **10150**
HEIZEN **5259**
HEIZEN **5235**
HEIZER **12271**
HEIZER **12272**
HEIZFLÄCHE **6398**
HEIZGAS **5709**
HEIZKANAL **5743**
HEIZKÖRPER **10150**
HEIZMANTELSCHIEBER **7205**
HEIZÖL **5714**
HEIZRAUM **12270**
HEIZRÖHRENKESSEL **5249**
HEIZSCHLANGE **12161**
HEIZSTOFF **5705**
HEIZTÜR **5242**
HEIZUNG **5259**
HEIZUNG (ELEKTRISCHE) **4513**
HEIZUNG (ZUSÄTZLICHE) **2904**
HEIZUNG VON GEBÄUDEN **6395**
HEIZUNGS-REGULIERVENTIL **10153**
HEIZUNGSANLAGE **6394**
HEIZUNGSROHR **6396**
HEIZWERT **6378**
HEIZWERT **6392**
HEKTOLITER **6415**
HEKTOWATT **6416**
HELIANTHIN **8223**
HELIARC-BRENNER **6424**
HELIUM **6429**
HELLIGKEIT **1616**
HELLIGKEITSMESSER **9267**
HELLROTGLÜHEND **1614**
HELLROTGLUT **1613**
HELM EINER AXT **6433**
HELM EINES BEILES **6433**
HELMLOCH EINER AXT **11231**
HEMIEDERKRISTALL **6436**
HEMMEN **6926**
HEMMSTOFF **6927**
HERABSETZUNG DER GESCHWINDIGKEIT **10363**
HERAUSGESCHLEPPTE LÖSUNG **4202**
HERD **6331**
HERD **3401**
HERDGLÜHOFEN **10532**
HERDGUSS **8820**
HERDOFEN **6332**
HERDSCHLACKE **2500**
HERDWANNE **9029**
HERMETISCH VERSCHLOSSEN **6447**
HERSTELLEN **7948**

HERSTELLER **7992**
HERSTELLERBESCHREIBUNG **7993**
HERSTELLUNG **4983**
HERSTELLUNG **7989**
HERSTELLUNGSFEHLER **3693**
HERSTELLUNGSHANDBUCH **9885**
HERSTELLUNGSKOSTEN **3207**
HERSTELLUNGSVERFAHREN **8218**
HERUNTERWALZEN **2691**
HERZEXZENTER **6329**
HERZKAUSCHE **6330**
HERZKURVE **2001**
HERZSCHEIBE **6329**
HETEROGEN **6451**
HETEROGENE STRAHLUNG **6450**
HEUBÜNDLER **946**
HEUGABELN UND GREIFER **6310**
HEUMÄHDRESCHBINDER **6311**
HEURAFFER **12541**
HEURECHEN **6312**
HEXAEDER **6463**
HEXAGONAL-DICHTGEPACKTE STRUKTUR **6458**
HEXAGONALER KRISTALL **6459**
HIEB EINER FEILE **3518**
HILFS- **12068**
HILFSARBEITER **8048**
HILFSBELEUCHTUNG **4688**
HILFSFUNKTION **854**
HILFSFUNKTION **8319**
HILFSKOMPRESSOR **12069**
HILFSMASCHINE **853**
HILFSMOTOR **11211**
HILFSVENTIL **1819**
HILFSVERDICHTER **12069**
HILFSWELLE **7049**
HIN- UND HERBIEGEPROBE **12774**
HIN-UND HERBIEGEVERSUCH **10535**
HINDERN **6926**
HINDERN (SICH BEI DER BEWEGUNG GEGENSEITIG) **5587**
HINTERACHSE **10265**
HINTERACHSWELLENRAD **3917**
HINTERDREHBANK **10443**
HINTERDREHBANK **913**
HINTERDREHEN **910**
HINTERDREHEN **893**
HINTEREINANDERSCHALTUNG **11199**
HINTERFRÄSEN **1387**
HINTERFRÄSWINKEL **1388**
HINTERFÜLLUNG **900**
HINTERTÜR **891**
HINZUFÜGEN **169**

HIRNHOLZ **4719**
HIRNSCHNITT **4718**
HIRSCHHORNSALZ **461**
HITZE **6333**
HITZEMESSER **10049**
HOBEL **9432**
HOBEL-UND FRÄSMASCHINE **9453**
HOBELBANK **7234**
HOBELEISEN **9436**
HOBELKASTEN **12262**
HOBELKEIL **13733**
HOBELMASCHINE **9452**
HOBELMASCHINE **9443**
HOBELMASCHINE **9451**
HOBELMESSER **9436**
HOBELN **9434**
HOBELN **9450**
HOBELSPÄNE **13919**
HOBELSTAHL **9436**
HOBELSTICHEL **9436**
HOBLER **9443**
HOCH GESPANNTER DAMPF **6491**
HOCHBAU **2442**
HOCHBEHÄLTER **4492**
HOCHDRUCK **6473**
HOCHDRUCKDAMPF **6491**
HOCHDRUCKDAMPFHEIZUNG **6493**
HOCHDRUCKDAMPFMASCHINE **6492**
HOCHDRUCKKESSEL **6487**
HOCHDRUCLEITUNG **6490**
HOCHDRUCKTURBINE **6494**
HOCHDRUCKWASSERHEIZUNG **6489**
HOCHDRUCKZYLINDER **6488**
HOCHFREQUENZ **6467**
HOCHFREQUENZOFEN **6469**
HOCHFREQUENZSTROM **6468**
HOCHGLANZ **1755**
HOCHGLANZPOLIEREN **1758**
HOCHGLANZPOLIEREN **1756**
HOCHGLÜHEN **5719**
HOCHHUBSICHERHEITSVENTIL **10890**
HOCHKANTRIEMEN **7376**
HOCHLEISTUNGSDUPLEXFRÄSMASCHINE **4379**
HOCHLEISTUNGSMASCHINE **6466**
HOCHOFEN **1311**
HOCHOFEN-ROHEISEN **2657**
HOCHOFENBODEN **9011**
HOCHOFENGAS **1313**
HOCHOFENSCHLACKE **1312**
HOCHPROZENTIG **6483**
HOCHRESERVOIR **4492**

HOCHSCHMELZEND **6308**
HOCHSPANNUNG **6478**
HOCHSPANNUNGSLEITUNG **6501**
HOCHSPANNUNGSMOTOR **6502**
HOCHSPANNUNGSSTROM **6500**
HOCHSPANNUUG (ELEKT.) **6479**
HÖCHSTBELASTUNG **8065**
HÖCHSTGESCHWINDIGKEIT **8067**
HÖCHSTGEWICHT **8070**
HÖCHSTLEISTUNG **8066**
HÖCHSTTEMPERATUR **8068**
HÖCHSTWERT **8069**
HOCHVAKUUM **6477**
HOCHWINDEN **6517**
HOCHWINDEN **6553**
HÖHE **6420**
HÖHE UNTER QUERBALKEN **6423**
HOHLBOHRMASCHINE **13182**
HOHLBOHRMASCHINE **1013**
HOHLEISEN **6010**
HOHLGUSS **6539**
HOHLKEHLE **5161**
HOHLKEHLE **2122**
HOHLKEHLNAHT **2878**
HOHLKEHLNAHT **7561**
HOHLKEIL **6542**
HOHLKOLBEN **1514**
HOHLKÖRPER **6538**
HOHLKUGELWALZE **6549**
HOHLLINSE **4095**
HOHLMASS **8079**
HOHLNAHT (KONCAVE) **1792**
HOHLRAUM **2125**
HOHLRAUMBILDUNG **2123**
HOHLSCHLEIFMASCHINE **7068**
HOHLSOGBILDUNG **2123**
HOHLSPIEGEL **2879**
HOHLWÖLBUNG **2880**
HOHLZAHNRAD **7079**
HOHLZIEGEL **245**
HOHLZIRKEL **6975**
HOHLZYLINDER **6540**
HÖLLENSTEIN **7828**
HOLMIUM **6551**
HOLOEDERKRISTALL **6552**
HOLZ **13913**
HOLZ (ASTFREIES) **12940**
HOLZ (GEBOGENES) **1196**
HOLZ (GELAGERTES) **297**
HOLZ (GESUNDES) **11820**
HOLZ (KERNFAULES) **5610**

HOLZ (LUFTTROCKENES) **297**
HOLZ (MORSCHES) **10768**
HOLZ (ROTFAULES) **5610**
HOLZ (WURMSTICHIGES) **13977**
HOLZ AN DER LUFT TROCKNEN **11094**
HOLZ AUSTROCKNEN **11094**
HOLZALKOHOL **13920**
HOLZART **13503**
HOLZÄTHER. METHYLOXYD **8222**
HOLZBAU **12942**
HOLZBEARBEITUNG **13924**
HOLZBEARBEITUNGSMASCHINE **13925**
HOLZBEARBEITUNGSMASCHINEN **13931**
HOLZDREHBANK **13922**
HOLZEISENRAD **8400**
HOLZESSIG **10046**
HOLZFRÄSMASCHINE **9901**
HOLZFUTTER **13927**
HOLZGEIST **13920**
HOLZHAMMER **7962**
HOLZHAMMER **13926**
HOLZKOHLE **13914**
HOLZKOHLENROHEISEN **2258**
HOLZKONSTRUKTION **12942**
HOLZLATTE **7426**
HOLZMEHL **13916**
HOLZNAGEL **13179**
HOLZNAPHTA **13920**
HOLZROHR **13928**
HOLZRÖHRE **13928**
HOLZSÄGE **13917**
HOLZSÄURE **10046**
HOLZSCHEIT **1243**
HOLZSCHLIFF **8116**
HOLZSCHRAUBE **13918**
HOLZSCHWELLE **12941**
HOLZSPIRITUS **13920**
HOLZSTOFF **8116**
HOLZTEER **13921**
HOLZVERKLEIDUNG **13915**
HOLZVERSCHALUNG **13915**
HOLZWOLLE **13923**
HOLZZELLSTOFF **2316**
HOMOGEN **6555**
HOMOGENISIERUNG **6561**
HOMOGENISIERUNGSMITTEL **3754**
HOMÖOMORPHIE **7191**
HONEN **6564**
HOOKESCHER SCHLÜSSEL **6573**
HOOKESCHES GESETZ **6574**
HÖRBARES ZEICHEN **816**

HORIZONTAL **6578**
HORIZONTALBOHRMASCHINE **6581**
HORIZONTALE **6585**
HORIZONTALKEHLNAHTSCHWEISSEN **6583**
HORIZONTALPROJEKTION **11124**
HORIZONTALSCHUB **6590**
HORN **6592**
HORNBLENDE **6598**
HORNBLENDEASBEST **6599**
HORNEINGUSS **6600**
HORNSPÄNE **6595**
HORNZULAUF **6600**
HÖRSIGNAL **816**
HOSENROHR **1595**
HUB **12387**
HUB **7519**
HUB EINES KOLBENS **9389**
HUBBEGRENZER **12278**
HUBBEGRENZER EINES VENTILS **13461**
HUBHÖHE **7551**
HUBHÖHE EINES KOLBENS **7519·**
HUBKETTE **6518**
HUBKOLBEN **7555**
HUBLÄNGE **7519**
HUBMAGNET **7554**
HUBPUMPE **7545**
HUBVENTIL **7547**
HUBVOLUMEN **13599**
HUBZYLINDER **7550**
HUFEISENMAGNET **6606**
HUFMEHL **6567**
HUFNÄGEL **6601**
HUFSTABEISEN **6605**
HÜLLKURVE **4776**
HÜLSE **1773**
HÜLSE **11576**
HÜLSE **11303**
HÜLSENKEGEL **2931**
HÜLSENKEIL **2931**
HÜLSENKUPPLUNG **11577**
HÜLSENSCHLÜSSEL **509**
HUMUS **6667**
HUMUSSÄURE **6665**
HUPE **6592**
HUTMUTTER **4118**
HÜTTENBLEI **8014**
HÜTTENKOKS **2662**
HÜTTENMÄNNISCH **8206**
HÜTTENSCHROTT **4124**
HÜTTENSCHROTT **6554**
HÜTTENWERK **11668**

HÜTTENWESEN **8208**
HÜTTENZINK **2785**
HYBRID-VERFAHREN **6699**
HYDRANT **6672**
HYDRANTENSTANDROHR **12066**
HYDRAT **6673**
HYDRATWASSER **6675**
HYDRATWASSER ENTZIEHEN (DAS) **3727**
HYDRATWASSER VERLIEREN (DAS) **3727**
HYDRAULIK **6698**
HYDRAULIKPUMPE **8769**
HYDRAULISCH **6676**
HYDRAULISCHE ÜBERTRAGUNG **5472**
HYDRODYNAMIK **6710**
HYDRODYNAMISCH **6708**
HYDROKARBÜR **6705**
HYDROLYSE **6719**
HYDROMECHANIK **6720**
HYDROMETALLURGIE **6721**
HYDROMETRISCHER FLÜGEL **11027**
HYDROPHIL **6724**
HYDROPHOBIERUNGSFILM **13648**
HYDROSTATIK **6728**
HYDROXYD **6673**
HYGROMETER **6729**
HYGROSKOPISCH **6730**
HYGROSKOPISCHE EIGENSCHAFT **6732**
HYGROSKOPIZITÄT **6732**
HYPERBEL **6734**
HYPERBELFUNKTION **6737**
HYPERBELRAD **11534**
HYPERBOLISCH **6735**
HYPERBOLOID (EINSCHALIGES) **6739**
HYPERBOLOID (ZWEISCHALIGES) **6740**
HYPERBOLOIDRAD **11534**
HYPERMANGANSAURES KALI **9695**
HYPOIDGETRIEBE **6745**
HYPOSULFIT **12852**
HYPOTENUSE **6746**
HYPOZYKLOIDE **6743**
HYPOZYKLOIDE (VIERSPITZIGE) **5602**
HYPOZYKLOIDENVERZAHNUNG **6744**
HYSTERESE **6747**
HYSTERESIS **6747**
I-EISEN **6194**
I-NAHT MIT LUFTSPALT **8824**
I-NAHT OHNE LUFTSPALT **2532**
I-NAHTVERBINDUNG (I-STOSS) OHNE LUFTSPALT **1791**
I-STUMPFNAHT **12032**
I-TRÄGER **6748**
ICHTERSCHEINUNG **7566**

JODZINNOBER 8155
JOMINY-PROBE 7254
JOULE 7255
JURAKALK 7267
JUSTIEREN 5280
JUSTIEREN 5276
JUTE 7268
K-NAHT 1793
K-NAHT MIT LUFTSPALT 8811
K-NAHT MIT LUFTSPALT 1796
K-NAHT OHNE LUFTSPALT 1787
KABBER 2526
KABEL 13899
KABEL 1829
KABEL (KUNSTSTOFFISOLIERTES) 9472
KABEL MIT METALLMANTEL 8177
KABELARMATUR 715
KABELBÜNDEL 7452
KABELKRAN 1344
KABELVERSEILUNG 1828
KABRIOLETT 3073
KADMIEREN 1832
KADMIUM 1831
KÄFIG EINES KUGELLAGERS 957
KÄFIG EINES ROLLENLAGERS 12040
KAHLBAUM-EISEN 7270
KALAMIN 1836
KALDOVERFAHREN 7271
KALI (OXALSAURES) 118
KALI (SAURES) 118
KALIALAUN 2787
KALIBER 1873
KALIBER (METR) 5871
KALIBER UND KALIBERRING 3581
KALIBERFOLGE 9097
KALIBERRING 2721
KALIBERSTOPFEN 5883
KALIBRIER-UND GESENKSCHMIEDE-PRESSE 1864
KALIBRIEREN 2651
KALIBRIEREN 1861
KALIBRIEREN 1867
KALIBRIERMASCHINE 1865
KALIBRIERPRESSE 11525
KALIBRIERTE FEDER 1866
KALIBRIERUNG 1867
KALIBRIERUNG 11524
KALICHROMALAUN 2370
KALIHYDRAT 9687
KALIKALKGLAS 1400
KALILAUGE 9671
KALINATRON (WEINSAURES) 11729

KALISALPETER 10903
KALIUM 9672
KALIUMALUMINAT 9674
KALIUMALUMINIUMSULFAT 2787
KALIUMAZETAT 9673
KALIUMBICHROMAT 9677
KALIUMBIKARBONAT 9676
KALIUMBIOXALAT 118
KALIUMBITARTRAT 119
KALIUMBROMID 9678
KALIUMCHLORAT 9680
KALIUMCHLORID 9681
KALIUMDICHROMAT 9677
KALIUMEISENZYANID 9684
KALIUMEISENZYANÜR 9685
KALIUMFLUORID 9686
KALIUMHYDROKARBONAT 9676
KALIUMHYDROXYD 9687
KALIUMHYPOCHLORITLÖSUNG 9688
KALIUMJODID 9689
KALIUMKARBONAT 9679
KALIUMNATRIUMTARTRAT 11729
KALIUMNITRAT 10903
KALIUMNITRIT 9691
KALIUMOXYDHYDRAT 9687
KALIUMPERCHLORAT 9694
KALIUMPERKARBONAT 9693
KALIUMPERMANGANAT 9695
KALIUMPHOSPHAT 9696
KALIUMPOLYSULFID 9697
KALIUMRHODANID 9701
KALIUMSILIKAT 9698
KALIUMSULFAT 9699
KALIUMSULFOZYANAT 9701
KALIUMZYANID 9683
KALIWASSERGLAS 9698
KALK 7582
KALK (FETTER) 5046
KALK (GEBRANNTER) 1854
KALK (GELÖSCHTER) 1851
KALK (HYDRAULISCHER) 6689
KALK (KOHLENSAURER) 1849
KALK (MAGERER) 9623
KALK (OXALSAURER) 1853
KALK (PHOSPHORSAURER) 1855
KALK (WASSERFREIER SCHWEFELSAURER) 549
KALK LÖSCHEN 11568
KALKBASISCHE UMHÜLLUNG 7584
KALKDINAS 5803
KALKHYDRAT 1851
KALKLICHT 8987

KALKMERGEL 2225
KALKMILCH 8274
KALKMÖRTEL 8854
KALKSINTER 1837
KALKSPAT 1845
KALKSTEIN 7586
KALKSTREUER 4089
KALKTUFF 1838
KALKWASSER 7585
KALOMEL 8159
KALOMELELEKTRODE 1876
KALORIE (ENGLISCHE) 1625
KALORIMETER 1877
KALORIMETRIE 1881
KALORIMETRISCH 1878
KALORISCHES ARBEITSÄQUIVALENT 6347
KALORISIERUNG 1882
KALOTTE 11137
KALT GEZOGEN 2677
KALT-ODER WARM-AUSHÄRTUNG 224
KALTABGRATEN 2711
KALTABSPRITZEN 2707
KALTANKÖPFEN 2686
KALTBEARBEITUNG 2713
KALTBEHANDLUNG 2710
KALTBIEGEPROBE 2667
KALTBIEGUNG 2668
KALTBRÜCHIGKEIT 2703
KALTBRÜCHIGKEIT DES EISENS 2704
KALTDURCHBOHRUNG 2689
KÄLTEBESTÄNDIG 8635
KÄLTEERZEUGUNG 9895
KÄLTEISOLIERUNGSMITTEL 2700
KÄLTEMASCHINE 10407
KÄLTEMISCHUNG 5654
KÄLTEMITTEL 3086
KÄLTESCHUTZ 7003
KALTFLIESSEN 2680
KALTFORMUNG 2682
KALTGEWALZT 2716
KALTGEWALZTER STABSTAHL 2695
KALTGUSS 2706
KALTHÄMMERN 2684
KALTHÄMMERN 2685
KALTHÄRTUNG 12326
KALTHÄRTUNG 13715
KALTKAMMER- DRUCKGIESSEN 2671
KALTKAMMER-DRUCKGIESSMASCHINE 2670
KALTLÖTSTELLE 2705
KALTMEISSEL 2717
KALTNACHPRESSEN 2679

KALTNACHWALZUNG 9330
KALTNIETEN 2693
KALTNIETEN 2692
KALTPRESSEN 2673
KALTPRESSEN 2690
KALTPROFILIEREN 2682
KALTPRÜFUNG 2687
KALTRECKEN 830
KALTRISS 2674
KALTSÄGE 2688
KALTSÄGEN 2699
KALTSCHMIEDEN 2681
KALTSCHROTMEISSEL 2701
KALTSCHWEISSE 2705
KALTSCHWEISSEN 2712
KALTSPRITZEN 4970
KALTSTAUCHEN 2686
KALTSTICH 11541
KALTVERFORMUNG 2713
KALTVERSUCH 2709
KALTWALZEN 2698
KALTWALZEN 2694
KALTWALZWERK 2708
KALTZIEHEN 2676
KALTZIEHEN 2675
KALTZIEHEN VON NAHTLOSGEGOSSENEN RÖHREN 2053
KALZINIEREN 1841
KALZINIEREN 1839
KALZINIEREN VON ERZEN 1840
KALZINIEROFEN 1843
KALZINIERUNG 1839
KALZIT 1844
KALZIT 1845
KALZIUM 1846
KALZIUM (DOPPELTKOHLENSAURES) 1847
KALZIUM (KOHLENSAURES) 1849
KALZIUMBIKARBONAT 1847
KALZIUMCHLORID 1850
KALZIUMFLUORID 5479
KALZIUMHYDROXYD 1851
KALZIUMKARBID 1848
KALZIUMKARBONAT 1849
KALZIUMOXALAT 1853
KALZIUMOXYD 1854
KALZIUMPHOSPHAT 1855
KALZIUMSULFAT 6190
KALZIUMSULFAT 1856
KALZIUMSULFID 1852
KAMELHAAR 1890
KAMELHAARRIEMEN 1891
KAMERA 9264

KAMIN 2339
KAMINZUG 4212
KAMM 12904
KÄMMEN VON ZAHNRÄDERN 5910
KAMMER 2776
KAMMEROFEN 10524
KAMMERSÄURE 2228
KAMMERSTROM (IONISATION) 7126
KAMMLAGER 12903
KAMMZAHNRAD 8400
KAMMZAPFEN 12906
KAMPFER 1893
KAMPFERÖL 1894
KANADAFASER 1898
KANAL 2248
KANALBLECH 2252
KANISTER 12961
KANNELKOHLE 1901
KANONENBOHRER 3563
KANONENBOHRER 6177
KANONENGUT 6178
KANONENMETALL 6178
KANTE 4441
KANTE (ABGEFASTE) 2231
KANTE (ABGERUNDETE) 10808
KANTE (GESCHNITTENE) 3506
KANTE (SCHARFE) 11272
KANTE RUNDEN (EINE) 10793
KANTENABSCHRÄGWINKEL 1223
KANTENDETEKTIONSVORRICHTUNG 4444
KANTENPRESSUNG 9826
KANTENRISS 3281
KANTENVORBEREITUNG 4449
KANTHOLZ 12035
KANTVORRICHTUNG 7982
KAOLIN 2340
KAPAZITÄT 1917
KAPAZITÄT (ELEKTR.) 1917
KAPILLARDEPRESSION 1924
KAPILLARELEVATION 1925
KAPILLARITÄT 1920
KAPILLARKONSTANTE 1923
KAPILLARROHR 1921
KAPILLARWIRKUNG 1922
KAPITÄL EINER SÄULE 1926
KAPPE 1910
KAPPE EINES HOBELEISENS 13033
KAPPENKOPF EINER SCHUBSTANGE 12345
KAPPENMUTTER 1512
KAPSELGEBLÄSE 10751
KAPSELGUSS 2335

KAPSELPUMPE 10758
KARABINERHAKEN 11691
KARBID 1933
KARBID 1848
KARBIDAUSSCHEIDUNG 1934
KARBINOL 13920
KARBOLINEUM 1939
KARBOLSÄURE 1938
KARBONAT 1967
KARBONAT 1968
KARBONISATION 1970
KARBONISIEREN 1971
KARBONISIERUNG 1974
KARBONISIERUNGSWÄRME 6355
KARBONITRIEREN 1972
KARBONITRIERUNGSATMOSPHÄRE 1973
KARBONYL 1976
KARBONYLEISEN 1977
KARBONYLNICKEL 1978
KARBONYLPULVER 1979
KARBORUND 1980
KARBORUNDUM 1980
KARBURATOR 1984
KARBURIEROFEN 1991
KARDANGELENK 6573
KARDANGELENK 13385
KARDANISCHES GELENK 6573
KARDANKUPPLUNG 1997
KARDIOIDE 2001
KARENZZEIT 6020
KARMINPAPIER 2002
KAROSSERIE 1385
KAROSSERIE (SELBSTTRAGENDE) 13381
KAROSSERIENAGEL 2554
KAROSSERIEWERKSTATT 1398
KARRE 2013
KARREN 2013
KARREN UND BREITSÄMASCHINEN 11129
KARTOFFELERNTEMASCHINEN 9704
KARTOFFELLEGEMASCHINEN 9705
KARTOFFELPFLÜGE 9707
KARTOFFELRODER 9703
KARTOFFELSORTIERMASCHINEN 9706
KARTOFFELZUCKER 6058
KARTON 1999
KARUSSELLDREHBANK 1469
KARUSSELLDREHMASCHINE 13545
KARUSSELLPRESSE 10757
KASEIN 2025
KASKADE 2018
KASSETTE 2028

KASSITERIT 12959
KASTEN 2798
KASTENGLÜHEN 5371
KASTENGUSS 1508
KASTENKABBER 2526
KASTENRAHMEN 1511
KASTENZEMENTIERUNG 1996
KASTORÖL 2086
KATAKAUSTISCHE LINIE 2089
KATALYSATOR 2091
KATALYSE 2090
KATALYTISCH 2092
KATAPHORESE 4639
KATARAKT 3611
KATHETE 9218
KATHETOMETER 2096
KATHODE 2106
KATHODENNICKEL 4600
KATION 2112
KATION 9657
KATODE 2097
KATODENBEIZUNG 2103
KATODENPOLARISATION 2108
KATODENSPANNUNGSABFALL 2100
KATODENSTRAHLEN 2104
KATODENSTRAHLRÖHRE 2105
KATODENWIRKUNGSGRAD 2101
KATODISCHE REINIGUNG 2098
KATOLYT 2111
KATZENAUGE 2088
KAUFBLEI 8014
KAUSCHE 12848
KAUSTISCHE SODA 11722
KAUTSCHUK 10825
KAUTSCHUK (VULKANISIERTER) 13610
KAUTSCHUKSTOPFEN 10824
KAVITATION 2123
KB-TI-MISCHTYP-ÜBERZUG 7583
KEGEL 2924
KEGEL 4016
KEGEL (ABGESTUMPFTER) 13235
KEGEL (GERADER) 10578
KEGEL (SCHIEFER) 8709
KEGEL (SELBSTLÖSENDER) 11159
KEGELAUSZIEHER 2929
KEGELBREMSE 2926
KEGELDACH 2947
KEGELFEDER 2942
KEGELFEDER MIT RECHTECKIGEM QUERSCHNITT 13605
KEGELFORM 2948
KEGELFRÄSER 543

KEGELHAHN 9530
KEGELHALTERUNG 4008
KEGELHANDREIBAHLE 12650
KEGELIGDREHEN 12668
KEGELKUPPLUNG 2927
KEGELKUPPLUNG 2928
KEGELKUPPLUNG 6839
KEGELLEHRDORN 12656
KEGELLEHRDORN 7956
KEGELLEHRDORN 7076
KEGELLEHRDORN MIT MITNEHMER-LAPPEN 12657
KEGELLEHRHÜLSE (METRISCHE) 8230
KEGELLEHRHÜLSEN 12659
KEGELMUTTER 4009
KEGELPENDELREGLER 2943
KEGELRAD 1226
KEGELRAD 1229
KEGELRADANTRIEB 1230
KEGELRÄDERWENDEGETRIEBE 10545
KEGELRADGETRIEBE 1227
KEGELRADPAAR MIT ÜBERSETZUNGS-VERHÄLTNIS 1:1 8323
KEGELREIBAHLE 12658
KEGELREIBUNGSKUPPLUNG 2930
KEGELROLLE 12660
KEGELSCHAFT 12664
KEGELSCHAFTSPANNDORN 12667
KEGELSCHEIBE 12669
KEGELSCHLICHTBOHRER 5226
KEGELSCHNITT 2940
KEGELSENKKOPF 13044
KEGELSITZ 12649
KEGELSITZVENTIL 2951
KEGELSPINDEL 2009
KEGELSTIFT 12655
KEGELSTUMPF 13235
KEGELSTUMPFFEDER 13236
KEGELTROMMEL 2941
KEGELTROMMEL 12669
KEGELWALZE 12660
KEHLE 8512
KEHLHOBEL 10800
KEHLMASCHINE 8429
KEHLNAHT (UNTERBROCHENE VERSETZTE) 12049
KEHLNAHTDICKE 12886
KEHLNÄHTE (SYMMETRISCH VERSETZTE) 2210
KEHLNAHTSCHWEISSEN 5162
KEHRGETRIEBE 10544
KEHRPFLÜGE UND ANHANGEPFLÜGE 9526
KEIL 13731
KEIL 13739
KEIL 7276

KEIL 5211
KEIL (EINTEILIGER) 11492
KEIL (MASSIVER) 11796
KEIL (VERSENKTER) 12458
KEIL LÖSEN (EINEN) 7750
KEILANSTELLUNG 12653
KEILBEILAGE 5937
KEILBOLZEN 11262
KEILFÖRMIG 13737
KEILLOCH 3214
KEILLOCH EINES HOBELS 8434
KEILNASE 6318
KEILNASE 7286
KEILNUT 7282
KEILNUT 7291
KEILNUTENFLANSCHEN 7289
KEILNUTENFRÄSMASCHINE 7281
KEILNUTENSCHABLONE 7283
KEILNUTENSCHABLONE 7290
KEILPRESSE 13734
KEILRAD 13732
KEILRÄDERGETRIEBE 13736
KEILRIEMEN 2925
KEILRIEMEN 13431
KEILRIEMEN 13511
KEILRILLE 7282
KEILRING 2945
KEILSCHIEBER 12662
KEILSICHERUNG 7704
KEILSTEIN 13738
KEILTREIBER 7278
KEILVERBINDUNG 7285
KEILVERBINDUNG 7288
KEILVERSPANNUNG 12929
KEILWELLE 11935
KEIM 8688
KEIM 4686
KEIMBILDUNG 5934
KEIMBILDUNG 8687
KENNBLATT (TECHNISCHES) 3615
KENNELKOHLE 1901
KENNLINIE 2256
KENNWERT 9075
KENNZAHL 2621
KENNZIFFER EINES LOGARITHMUS 2257
KERAMIK 2193
KERBBIEGEVERSUCH 8555
KERBE 8672
KERBE 8668
KERBEMPFINDLICHKEIT 8670
KERBEN 8554

KERBEN 8676
KERBSCHLAGPROBE 1190
KERBSCHLAGPROBE 8673
KERBSCHLAGVERSUCH 2270
KERBSCHLAGZÄHIGKEIT 6788
KERBSCHLAGZÄHIGKEIT 8671
KERBSPRÖDIGKEIT 8669
KERBSPRÖDIGKEIT 1632
KERBZÄHIGKEIT 8671
KERBZÄHIGKEIT 12738
KERN 13726
KERN 3162
KERN 8688
KERN 3142
KERN EINER SCHRAUBE 1394
KERN EINES QUERSCHNITTES 3156
KERNBILDUNG 8687
KERNBINDER 3145
KERNBLASMASCHINE 3146
KERNDURCHMESSER 8314
KERNDURCHMESSER 10719
KERNDURCHMESSER EINER SCHRAUBE 3850
KERNFORMMASCHINE 3154
KERNFÜHRUNGSBLOCK 3144
KERNGEFÜGE 3165
KERNHAKEN 3152
KERNKASTEN 3147
KERNKLEBEMITTEL 3151
KERNLEDER 1780
KERNMARKE 3161
KERNÖL 3157
KERNSAND 3163
KERNSCHÄLE DES HOLZES 3460
KERNSCHLEIFMASCHINE 3150
KERNSEIFE 3482
KERNSTANGE 3143
KERNSTÜCK 5027
KERNSTÜTZE 2253
KERNTROCKENOFEN 3158
KERRSCHLAGZÄHIGKEIT 10477
KERZE 1899
KERZE (INTERNATIONALE) 3663
KERZENSTÄRKE 1900
KESSEL (LIEGENDER) 6579
KESSEL (STEHENDER) 13544
KESSEL EINMAUERN (EINEN) 11215
KESSELABDECKUNGSMATERIAL 1408
KESSELARMATUR 1411
KESSELAUSRÜSTUNG 1411
KESSELBATTERIE 1069
KESSELBLECH 1416

KESSELBODEN (GEWÖLBTER) 4038
KESSELBODENWINKELEISEN 1490
KESSELBOHRMASCHINE 1409
KESSELDRUCK 1417
KESSELDRUCKPROBE 1421
KESSELEXPLOSION 1410
KESSELFABRIK 1423
KESSELHAUS 1413
KESSELISOLIERMATERIAL 1408
KESSELKONTROLLE 9193
KESSELMANTEL 1419
KESSELRAUM 1418
KESSELROHR 1422
KESSELSCHLACKE; KLINKER 2500
KESSELSCHMIED 1415
KESSELSCHMIEDE 1420
KESSELSCHUSS 3260
KESSELSTEIN 10969
KESSELSTEIN ENTFERNEN 10971
KESSELSTEINENTFERNUNG 10987
KESSELSTEINHAMMER 10986
KESSELSTEINLÖSUNGSMITTEL 4041
KESSELSTEINVERHÜTUNGSMITTEL 10978
KESSELTROMMELBODEN 6316
KESSELUNTERSUCHUNG 1414
KESSELWAGEN 12626
KESSELWÄRTER 12272
KESSELZUBEHÖR 1411
KETTE 2201
KETTE (ADJUSTIERTE) 9412
KETTE (ENDLOSE) 4732
KETTE (GERÄUSCHLOSE) 11432
KETTE (GESCHWEISSTE) 13770
KETTE (KALIBRIERTE) 9412
KETTE (KURZGLIEDRIGE) 11373
KETTE (LANGGLIEDRIGE) 7730
KETTE EINES GEWEBES 13627
KETTEN (KALIBRIERTE) 1862
KETTENANTRIEB 2206
KETTENBAND 7496
KETTENBOLZEN 9320
KETTENBREMSE 2203
KETTENEISENSTÄRKE 3856
KETTENFLASCHENZUG 2213
KETTENGLIED 2211
KETTENGLIED 7627
KETTENKÄSTEN 2204
KETTENLINIE 2095
KETTENMASSSYSTEM 6854
KETTENNIET 2214
KETTENNIETUNG 2215

KETTENNUSS 11662
KETTENRAD 2223
KETTENRADANTRIEB 2208
KETTENRADSCHIEBER 2220
KETTENROLLE 2217
KETTENROST 2209
KETTENSÄGE 2216
KETTENSCHMIERUNG 2212
KETTENSPANNER 2219
KETTENSPANNER 2218
KETTENSTEG 12402
KETTENTEILUNG 6978
KETTENTRIEB 2207
KETTENTROMMEL 2202
KETTENZAHNRAD 9402
KIELBLOCK 7273
KIELKLOTZ 7273
KIENÖL 10862
KIENRUSS 7386
KIES 6073
KIESBODEN 6076
KIESELERDE 11436
KIESELERDE 11434
KIESELFLUORWASSERSTOFFSÄURE 11441
KIESELFLUSSSÄURE 11441
KIESELGALMEI 11438
KIESELGUR 7292
KIESELGUR 3880
KIESELKALK 11451
KIESELSÄURE 11436
KIESELSÄUREANHYDRID 11436
KIESELSAURES KALIUM 9698
KIESELSAURES NATRIUM 11731
KIESELSAURES SALZ 11437
KIESELSTEIN 9152
KIESELZINKERZ 11438
KIESFILTER 6074
KIESROST 6075
KILO 7297
KILODYN 7296
KILOGRAMM 7297
KILOGRAMMETER 7299
KILOGRAMMKALORIE 7298
KILOMETER 7300
KILOVOLT 7301
KILOVOLTAMPERE 7302
KILOWATT 7303
KILOWATTSTUNDE 7304
KINEMATIK 7306
KINEMATISCH 7305
KINETIK 7310

KINETISCH 7307
KINKE 7313
KIP = 1000 PFD = 453,59 KG 7316
KIPPBAR 6835
KIPPEN 4376
KIPPER 1932
KIPPER 4375
KIPPER 12967
KIPPFANNE 12935
KIPPHEBEL 10654
KIPPMOMENT 8966
KIPPMOMENT 12936
KIPPSCHALTUNG (BI-STABILE) 5429
KIPPSCHERE 10658
KIRSCHROTGLÜHEND 2327
KIRSCHROTGLUT 2326
KISTENZEMENTIERUNG 1996
KITT 10037
KITT 8038
KITTEN 2146
KLAFFEN EINER FUGE 5806
KLAMMER 2502
KLAMMER 12105
KLAMMERHAKEN 12105
KLAPPDECKEL 6510
KLAPPE 5353
KLAPPE 2617
KLAPPENVENTIL 5353
KLAPPERN DES VENTILS 2277
KLAPPMASSSTAB 5512
KLAPPSCHRAUBE 6509
KLAPPSITZ 7261
KLAPPVERDECK 2719
KLÄRBOTTICH 11229
KLÄREN 2456
KLÄREN 2455
KLÄRGEFÄSS 11229
KLASSIERUNG 2460
KLAUE 2464
KLAUE EINES HAMMERS 2463
KLAUENFETT 1450
KLAUENHAMMER 2462
KLAUENKUPPLUNG 2461
KLAUENKUPPLUNG 4113
KLAUENMEHL 6567
KLAUENÖL ; KNOCHENÖL 1450
KLAVIERSAITENDRAHT 9287
KLEBÄTHER 2733
KLEBEMITTEL 177
KLEBRIG 13577
KLEBRIGKEIT 12591

KLEBSTOFF 177
KLEESALZ 10907
KLEESÄURE 8969
KLEINEISENWAREN 11660
KLEINGASMASCHINE 11658
KLEINLASTWAGEN 7563
KLEINMOTOR 11661
KLEINSCHLAG 1644
KLEINSTWERT 8307
KLEINWASSERRAUMKESSEL 11663
KLEISTER 12112
KLEMMBACKE 5675
KLEMMBACKENSCHALTGETRIEBE 5676
KLEMME (ELEKTR.) 12762
KLEMMEN 12765
KLEMMEN 7212
KLEMMEN (SICH) 7210
KLEMMEN DES SEILES IN DER RILLE 1278
KLEMMENSPANNUNG 12764
KLEMMGESPERRE 5676
KLEMMKLINKE 5675
KLEMMKUPPLUNG 11164
KLEMMNABE 11938
KLEMMRING 11939
KLEMMSCHELLE 2720
KLEMMSCHRAUBE 2453
KLEMPNER 12964
KLETTERN DES RIEMENS 2499
KLIMAANLAGE 250
KLINGE 1299
KLINGE EINES SCHNEIDWERKZEUGS 1300
KLINGELVORRICHTUNG 1137
KLINGERIT 7318
KLINGWERK 1061
KLINKE 6608
KLINKEN 9138
KLINKENGETRIEBE 9138
KLINKENHEBEL 10223
KLINKENRAD 10225
KLISCHEE 1340
KLOBEN 1333
KLOBENZUG 9973
KLOBSÄGE 13526
KLOPFEN 7330
KLOPFEN (IN DIE FORM) 1482
KLOPPERBODEN 4038
KLOTZBREMSE 1335
KLUPPE 12266
KLUPPENHALTER 3895
KNABBERN 8552
KNAGGE 2485

KNAGGE **1883**
KNAGGENSCHEIBE **13221**
KNALL **7330**
KNALLGAS **8985**
KNALLGASGEBLÄSE **8986**
KNALLQUECKSILBER **8154**
KNALLSÄURE **5731**
KNALLSILBER **11463**
KNARRE **10218**
KNARRENSCHRAUBENSCHLÜSSEL **10224**
KNEBELSCHRAUBE **1930**
KNEIFZANGE **3532**
KNEIPZANGE **3532**
KNETALUMINIUM-LEGIERUNG **433**
KNICK **7313**
KNICKBAND **7314**
KNICKBEANSPRUCHUNG **3339**
KNICKBELASTUNG **3338**
KNICKEN **3343**
KNICKFESTIGKEIT **3340**
KNICKFESTIGKEIT **1682**
KNICKLAST **3338**
KNICKSPANNUNG **3341**
KNICKUNG **3343**
KNICKUNG **1681**
KNICKUNG **7315**
KNICKVERSUCH **3342**
KNIEHEBEL **12979**
KNIEHEBELBREMSE **12978**
KNIEHEBELPRESSE **12980**
KNIEHEBELVERBINDUNG **12979**
KNIEHEBELVERCHLUSS **7336**
KNIEROHR **7320**
KNIESTÜCK **12546**
KNIESTÜCK **7320**
KNIESTÜCK **4493**
KNOCHENGELENK **7336**
KNOCHENKOHLE **554**
KNOCHENLEIM **1449**
KNOCHENMEHL **1448**
KNOCHENSÄURE **8873**
KNODIG **8606**
KNOLLIG **8606**
KNÜPPEL **1243**
KNÜPPEL AUS LEGIERTEN STÄHLEN **370**
KNÜPPEL U. PLATINENSCHERE **1244**
KNÜPPELSCHERE **1246**
KNÜPPELWALZWERK **1245**
KOAGULIEREN **2556**
KOAGULIEREN **2555**
KOAGULUM **2557**

KOALESZENZ **2572**
KOALESZIERTES ELEKTROLYTKUPFER **2571**
KOAXIAL **2605**
KOBALT **2607**
KOBALT (SALPETERSAURER) **2610**
KOBALT-CHROMSTAHL **2613**
KOBALT-KARBONYL **2608**
KOBALTCHLORÜR **2609**
KOBALTGLANZ **2614**
KOBALTNITRAT **2610**
KOBALTOCHLORID **2609**
KOBALTOSULFAT **2615**
KOBALTOXYD **2611**
KOBALTOXYDUL (SCHWEFELSAURES) **2615**
KOBALTOXYDULSULFAT **2615**
KOBALTSTAHL **2612**
KOBALTVITRIOL **2615**
KOCHEN **1401**
KOCHEN **1429**
KOCHER **1406**
KOCHKESSEL **1406**
KOCHSALZ **2790**
KODIERUNG **4716**
KOEFFIZIENT **2621**
KOERZITIV **2640**
KOERZITIVEFELD **2639**
KOERZITIVKRAFT **2641**
KOERZITIVKRAFT **2639**
KOFFERRAUM **1458**
KOFFERRAUM **13239**
KOFFERRAUMDECKEL **13240**
KOHÄRENZ **2644**
KOHÄSION **2644**
KOHÄSTONSKRAFT **2644**
KOHLE **2558**
KOHLE (ANTHRAZITISCHE) **11165**
KOHLE (BACKENDE) **1835**
KOHLE (KURZFLAMMIGE) **11371**
KOHLE (LANGFLAMMIGE) **7724**
KOHLE (MAGERE) **11168**
KOHLE (NICHTBACKENDE) **8626**
KOHLE (TROCKNE) **4344**
KOHLE(LICHTBOGEN)SCHWEISSEN **1943**
KOHLEBÜRSTE **1946**
KOHLEELEKTRODEN-LICHTBOGENOFEN **670**
KOHLEFADENGLÜHLAMPE **1951**
KOHLEHYDRAT **1937**
KOHLELEKTRODE **1950**
KOHLELICHTBOGEN **1941**
KOHLELICHTBOGENSCHNEIDEN **1942**
KOHLENBERGWERK **2566**

KRACKVERFAHREN 3285
KRAFT 5532
KRAFT 9730
KRAFT 12352
KRAFT (ELEKTROMOTORISCHE) 4622
KRAFT (ELEKTROSTATISCHE) 4642
KRAFT (GEGENELEKTROMOTORISCHE) 3231
KRAFT (GLEICHBLEIBENDE) 2973
KRAFT (INNERE) 7064
KRAFT (KONTINUIERLICHE) 2973
KRAFT (MAGNETISCHE) 7904
KRAFT (MAGNETISIERENDE) 7925
KRAFT (MAGNETOMOTORISCHE) 7932
KRÄFT (PARALLELE) 9056
KRAFT (THERMOELEKTRISCHE) 12831
KRAFT (THERMOELEKTROMOTORISCHE) 12807
KRAFTANLAGE 9739
KRAFTANTRIEB 9733
KRAFTARM 706
KRAFTAUFWAND 2876
KRAFTBEDARF 3746
KRAFTBEDARFSKOSTEN 3206
KRÄFTE (ENTGEGENGESETZT GERICHTETE) 5537
KRÄFTE (GLEICHGERICHTETE) 5538
KRÄFTE MIT GEMEINSAMEM ANGRIFFSPUNKT 2903
KRAFTECK 9610
KRÄFTEDREIECK 13186
KRÄFTEPARALLELOGRAMM 9071
KRÄFTEPLAN 3835
KRÄFTEPLAN 5526
KRÄFTEPOLYGON 9610
KRAFTERSPARNIS 10948
KRAFTGAS 4194
KRAFTGAS 9736
KRAFTLEITUNG (ELEKTRISCHE) 4526
KRAFTLINIE 7603
KRAFTLINIENBILD 7902
KRAFTLINIENFLUSS 7903
KRAFTLINIENSTROM 7903
KRAFTMACHINE 9859
KRAFTMESSER 4403
KRAFTNIETUNG 9740
KRAFTRICHTUNG 3990
KRAFTRÖHRE 13253
KRAFTSINN 11182
KRAFTSPIRITUS 1983
KRAFTSTATION 9742
KRAFTSTOFF 9222
KRAFTSTOFF (KLOPFFESTER) 613
KRAFTSTOFFBEHÄLTER 5716
KRAFTSTOFFILTER 5708

KRAFTSTOFFLUFTGEMISCH 8337
KRAFTSTOFFPUMPE 5715
KRAFTSTOFFSPARER 4437
KRAFTSTOFFSTANDMESSER 5710
KRAFTTERZEUGUNG 9737
KRAFTÜBERTRAGUNG 9746
KRAFTÜBERTRAGUNG 13134
KRAFTÜBERTRAGUNG 4294
KRAFTÜBERTRAGUNG (ELEKTRISCHE) 13129
KRAFTÜBERTRAGUNG DURCH HEBEL 13132
KRAFTÜBERTRAGUNG DURCH RÄDER 13131
KRAFTVERBRAUCH 2876
KRAFTVERLUST 7758
KRAFTVERSORGUNG 9744
KRAFTWAGEN 8411
KRAFTWERK 9742
KRAFTWIRKUNG 143
KRAFTWIRKUNGSFIGUR 4845
KRAFTZAHNRAD 6402
KRAFTZENTRALE 9742
KRAGSTÜCK 3140
KRAMPE 12105
KRAMPE 12106
KRAN 3290
KRAN (STANDFESTER) (ORTSFESTER) 5291
KRANAUSLEGER 3291
KRANGIESSPFANNE 3292
KRANGIESSPFANNE 1728
KRANKHEITSERREGER 9126
KRANKHEITSKEIM 9126
KRANPFANNE 1728
KRANSÄGE 10609
KRANSCHIENE 3293
KRANZ 3396
KRANZ EINER RIEMENSCHEIBE 10595
KRANZ EINES ZAHNRADES 10594
KRATER 3317
KRATERBILDUNG 9334
KRATZBÜRSTE 11014
KRÄTZE 4331
KRATZEN 11011
KRATZER 11010
KRATZFESTIGKEIT 11015
KRÄUSELN 11399
KRAUSKOPF 3245
KREIDE 13842
KREIDE 2224
KREIS 2401
KREIS 2404
KREIS (FESTER) 5290
KREIS (UMGESCHRIEBENER) 2435

KURBELWELLE (DREIFACH GEKRÖPFTE) 12880
KURBELWELLE (MEHRFACH GEKRÖPFTE) 8478
KURBELWELLE MIT GEGENGEWICHT 3235
KURBELWELLENDREHBANK 3316
KURBELWELLENLAGER 3303
KURBELWELLENLAGER 3314
KURBELWELLENZAHNRAD 3315
KURBELWELLENZAPFEN 3311
KURBELZAPFEN 3302
KURBELZAPFENLAGER 1242
KURKUMAGELB 3481
KURKUMAPAPIER 13275
KURKUMIN 3481
KURVE 3496
KURVE (CASSINISCHE) 2029
KURVE (EBENE) 9435
KURVE (GESCHLOSSENE) 2520
KURVE (ISODYNAME) 7182
KURVE (ISODYNAMISCHE) 7182
KURVE (POLYTROPISCHE) 9618
KURVE (STEILE) 10102
KURVE (ZYKLISCHE) 3555
KURVE DOPPELTER KRÜMMUNG 13307
KURVENFESTIGKEIT 3180
KURVENLINEAL 5660
KURVENSCHAR 5028
KURVENSCHEIBE 4442
KURVENSCHIENE 5660
KURVENTRIEB 1886
KURVIMETER 3501
KURZSCHLUSS 11372
KURZSCHLUSS 11369
KURZSCHLUSSÜBERTRAGUNG 11370
KÜRZUNG EINES RIEMENS 11377
KÜRZUNG EINES SEILES 11377
KÜSTENNAHE..... 8738
KYANISIEREN DES HOLZES 7340
L-EISEN 514
LABARRAQUESCHE LAUGE 4422
LABIL 7343
LABORATORIUM 7344
LABORGERÄTESATZ 7345
LABYRINTHDICHTUNG 7347
LACK (ALKOHOLISCHER) 7354
LACK IN KÖRNERN 11131
LACK IN STANGENFORM 12254
LACKANSTRICH 2581
LACKIEREN 7353
LACKIERUNG 2581
LACKMUSPAPIER 7668
LACKÜBERZUG 2595

LADEFÄHIGKEIT 9139
LADEKONTROLLAMPE 2267
LADEN (EINEN AKKUMULATOR) 2261
LADEN EINES AKKUMULATORS 2265
LADESTROM 2266
LADUNG 2262
LADUNG 7679
LADUNG (ELEKTR.) 2259
LAFETTE 6175
LAGE 9096
LAGE 10832
LAGE 7382
LAGE 7442
LAGEFLÄCHE 1101
LAGENRISS 7383
LAGENSICHERUNG 7711
LAGEPLAN 9430
LAGER 12263
LAGER 1100
LAGER 6660
LAGER 1118
LAGER 12297
LAGER 12292
LAGER 1117
LAGER (AUSGELAUFENES) 13980
LAGER (EINTEILIGES) 11788
LAGER (EINTEILIGES) 11777
LAGER (GESCHLOSSENES) 11788
LAGER (GETEILTES) 4098
LAGER (SCHIEF GESCHNITTENES) 508
LAGER (SCHRÄG) 508
LAGER (SELBSTSCHMIERENDES) 11158
LAGER (SICH SELBSTEINSTELLENDES 11150
LAGER (UNRUNDGEWORDENES) 13980
LAGER-WEISSMETAL 889
LAGERBESTAND 12263
LAGERBOCK 12064
LAGERBOCK 1103
LAGERBRONZE 1660
LAGERBÜCHSE 1771
LAGERBÜCHSE 1773
LAGERDECKEL 1104
LAGERDECKEL 5323
LAGERDECKEL 1912
LAGERENTFERNUNG 4064
LAGERFUGE 7241
LAGERFUSS 1034
LAGERFUTTER 7624
LAGERHALS EINER WELLE 1107
LAGERHAUS 12297
LAGERHÜLSE 1771

LAST (VERTEILTE) 4078
LAST (WANDERNDE) 10706
LAST SENKEN (EINE) 7793
LASTAUTO 8414
LASTDRUCKBREMSE 7681
LASTENAUFZUG 6007
LASTENHEFT 11855
LASTHAKEN 7552
LASTKAHN 991
LASTKETTE 6518
LASTKRAFTWAGEN 7753
LASTKRAFTWAGEN 8414
LASTKRAFTWAGEN (DREIACHSIG) 12871
LASTKRAFTWAGEN MIT PLANE UND SPRIEGEL 12686
LASTMAGNET 7554
LASTSCHEIBE 5040
LASTVERTEILUNG 4081
LASTVERTEILUNG 7684
LASTWAGEN 7752
LASTZANGE 7557
LASURBLAU 13334
LATERNE 7393
LATERNE EINES SCHIEBERS 7392
LATEX 7425
LAUBSÄGE 5665
LAUF 13181
LAUFBÜCHSE (NASSE) 13804
LAUFBÜCHSE (TROCKENE) 4348
LAUFENDES BAND 1164
LAUFENDES TRUMM EINES FLASCHENZUGES 10851
LÄUFER 10767
LÄUFER 10856
LÄUFER (BAUW.) 12369
LÄUFER EINER WAAGE 13160
LAUFFLÄCHE 1116
LAUFFLÄCHE EINES RADES 13173
LAUFGEWICHT 13160
LAUFGEWICHTSWAAGE 12225
LAUFKATZE 13155
LAUFKRAN 13156
LAUFKRAN 8942
LAUFKRANTRÄGER 13157
LAUFRAD 10854
LAUFRAD 6795
LAUFRAD EINER TURBINE 13823
LAUFRADSCHAUFEL EINER TURBINE 13820
LAUFRILLE 13085
LAUFRING 964
LAUFRING 1111
LAUFROLLE 10844
LAUFSCHIENE 3293

LAUFSEITE EINES RIEMENS 10853
LAUFSITZ 10847
LAUFSTEG 13616
LAUGE 7835
LAUGENSPRÖDIGKEIT 2118
LÄUTERN 10370
LÄUTERN 10377
LÄUTWERK 1137
LAVA 7437
LAVE-VERFAHREN 7438
LEBENDIGE KRAFT 7308
LEBENSDAUER 7541
LEBENSMITTELINDUSTRIE 5517
LEBLANC-SODA 12445
LECH 11878
LECH 8058
LECK 7486
LECK 7488
LECK SEIN 7487
LECKEN 7487
LECKEN 7492
LECKSTELLE 7489
LEDEBURIT 7504
LEDER 7494
LEDER (CHROMGARES) 2377
LEDER (GERÖSTETES) 2272
LEDERABFÄLLE 7497
LEDERDICHTUNG 7502
LEDERGLIEDERRIEMEN 7496
LEDERHOCKANTRIEMEN 7376
LEDERLASCHENKUPPLUNG 7501
LEDERLEIM 11539
LEDERLEIM 2140
LEDERMANSCHETTE 3458
LEDERMEMBRAN 7498
LEDERPACKUNG 7502
LEDERPOLIERSCHEIBE 1684
LEDERRIEMEN 7503
LEDERRIEMEN 7495
LEDERRIEMENKUPPLUNG 7349
LEDERSCHLAUCH 7499
LEDERSTULP 3458
LEDERSTULPDICHTUNG 3459
LEERGANG (EINER WERKZEUGMASCHINE) 10529
LEERLAUF 8598
LEERLAUF 6752
LEERLAUF 6755
LEERLAUF (ERHÖHTER) 5038
LEERLAUF (SCHNELLER) 5038
LEERLAUFARBEIT 8601
LEERLAUFBÜCHSE 7748

LEERLAUFDÜSE 11646
LEERLAUFEN 1081
LEERLAUFSPANNUNG 8808
LEERLAUFVERLUST 8599
LEERLAUFWIDERSTAND 8600
LEERLAUFZET 6753
LEERRAUM 13326
LEERSCHEIBE 7747
LEERSTELLE 13433
LEERZEIT 6753
LEESEITE 7505
LEGENDE 7513
LEGIEREN 361
LEGIERT 379
LEGIERT 371
LEGIERUNG 360
LEGIERUNG (BINÄRE) 1247
LEGIERUNG (FEUERFISTE) 393
LEGIERUNG (HITZE-UND KORROSIONSBESTÄNDIGE) 390
LEGIERUNG (HITZEBESTÄNDIGE) 391
LEGIERUNG (KORROSIONSBESTÄNDIGE) 387
LEGIERUNG (LEICHT SCHMELZBARE) 5757
LEGIERUNG (MAGNETISCHE) 392
LEGIERUNG (NATÜRLICHE) 8498
LEGIERUNG (NICHT-FEUERFESTE) 8622
LEGIERUNG (QUATERNÄRE EUTEKTISCHE) 10085
LEGIERUNG (QUATERNÄRE) 10084
LEGIERUNG (SÄUREBESTÄNDIGE) 384
LEGIERUNG (VERSCHLEISSFESTE) 383
LEGIERUNG GROSSER DICHTE 6481
LEGIERUNG MIT ALUMINIUM ALS GRUNDMETALL 439
LEGIERUNGEN (ABRIED-UND KORROSIONFESTE) 396
LEGIERUNGEN (NIEDRIG SCHMELZENDE) 7773
LEGIERUNGSABSCHEIDUNG 364
LEGIERUNGSKONTAMINATIONVERSEUCHUNG 365
LEGIERUNGSSYSTEM 381
LEGIERUNGSÜBERZUG 364
LEGIERWAAGSCHALE 362
LEHM 2465
LEHM 7691
LEHMBODEN 2466
LEHMFORM 7693
LEHMMÜHLE 9966
LEHMPATZEN 7692
LEHMSTEIN 7692
LEHRBOLZEN 3581
LEHRDORN 9531
LEHRDORN UND LOCHLEHRE 3581
LEHRE 5768
LEHRE 5769
LEHRE 5871

LEHRE (MIKROMETRISCHE) 8251
LEHRE (UNVERSTELLBARE) 8625
LEHRE (ZUSAMMENSTELLBARE) 4048
LEHRENBOHRMASCHINE 7228
LEHRGERÄT 3053
LEHRRING (NORMALE) 9431
LEHRSTÜCKE 5373
LEHRSTÜCKENMONTAGE 5884
LEICHTES KOHLENWASSERSTOFFGAS 8215
LEICHTFLÜSSIG 5471
LEICHTFLÜSSIGKEIT 5474
LEICHTMETALL 7564
LEICHTMETALL 7779
LEICHTMETALLDREHBANK 7565
LEICHTÖL 7570
LEICHTSIEDEND 1426
LEIM 5984
LEIMEN 5987
LEIMEN 5985
LEIMVERBAND 5986
LEIMZWINGE 2451
LEINEN 7619
LEINÖL 7631
LEINÖLFIRNIS 8786
LEINWAND 7619
LEISTEN 5940
LEISTENHOBEL 6544
LEISTUNG 5459
LEISTUNG 4463
LEISTUNG 9730
LEISTUNG 8908
LEISTUNG 8896
LEISTUNG (ABGEGEBENE) 13940
LEISTUNG (INDIZIERTE) 6875
LEISTUNG (TATSÄCHLICHE) 4458
LEISTUNG ABBREMSEN (DIE) 8083
LEISTUNGSFAKTOR 9735
LEISTUNGSFAKTORMESSER 9235
LEISTUNGSGEWICHT (FAHRFERTIG) 3480
LEISTUNGSMESSER 4403
LEISTUNGSREGLER 8909
LEISTUNGSVERLUST 7762
LEITBLECH 921
LEITBLECH 922
LEITEN (ELEKTRIZITÄTWÄRME) 2913
LEITEND (WÄRMEELEKTRIZITÄT) 2918
LEITER (ELEKTR.) 2921
LEITER (ELEKTRISCHER) 4537
LEITER (SPANNUNGSLOSER) 3627
LEITER (STROMLOSER) 3627
LEITER (UNTER SPANNUNG STEHENDER) 7670

LEITFÄHIGKEIT 2919
LEITFÄHIGKEIT (ELEKTRISCHE) 4536
LEITFÄHIGKEIT (THERMO-ELEKTRISCHE) 10044
LEITFÄHIGKEITSMOLARE 8350
LEITKUPFER OHNE SAUERSTOFF 8983
LEITLINIE 3996
LEITRAD 6168
LEITRAD EINER TURBINE 6168
LEITROLLE 6161
LEITSALZE 2915
LEITSCHAUFEL EINER TURBINE 6167
LEITSPINDEL 7471
LEITSPINDEL 6165
LEITSPINDELDREHBANK 7428
LEITSTRAHL 10170
LEITUNG 9339
LEITUNG (ELEKTRISCHE) 4523
LEITUNG (FALLENDE) 3791
LEITUNG (UNTERIRDISCHE) 13353
LEITUNGSDRAHT 2922
LEITUNGSDRUCK 9822
LEITUNGSKABEL 1825
LEITUNGSKANAL 8807
LEITUNGSMAST 7608
LEITUNGSNETZ 4080
LEITUNGSSCHUH 1827
LEITUNGSWASSER 13075
LEITUNGSWIDERSTAND (ELEKTR.) 10491
LEITWIDERSTAND (SPEZIFISCHER) 10503
LEMNISKATE 7514
LENARD-STRAHLEN 7515
LENKGESTÄNGE 12232
LENKGETRIEBE 12229
LENKRAD 12233
LENKSÄULE 12228
LENKSTOCKHEBEL 9414
LENKSTOCKHEBEL 12227
LEONARD-EFFEKT 7523
LESEN 10252
LEUCHTBOMBE 7573
LEUCHTDICHTE 7104
LEUCHTGAS 6775
LEUCHTGASLAMPE 5826
LEUCHTKRAFT 6776
LEUCHTÖL 7275
LEUCHTPETROLEUM 7275
LEUCHTSCHIRM 13568
LEUCHTSCHIRM 5477
LIBELLE 13259
LICHBOGEN(ELEKTRO)OFEN 4495
LICHSTÄRKE 7822

LICHT 7560
LICHT (DIFFUSES) 3927
LICHT (ELEKTRISCHES) 4517
LICHT (KÜNSTLICHES) 734
LICHT (POLARISIERTES) 9585
LICHT (REFLEKTIERTES) 10383
LICHT (ZERSTREUTES) 3927
LICHT (ZURÜCKGEWORFENES) 10383
LICHT BOGEN-WIDERSTANDSOFEN 10485
LICHTBILD 9263
LICHTBOGEN (ELEKTR.) 4494
LICHTBOGEN-ARBEITSSPANNUNG 679
LICHTBOGEN-HARTLÖTEN 666
LICHTBOGEN-SCHWEISSEN (TEILAUTOMATISCHES) 11166
LICHTBOGEN-SCHWEISSMASCHINE 680
LICHTBOGENOFEN 668
LICHTBOGENOFEN (DIREKTER) 671
LICHTBOGENOFEN (DIREKTER) 667
LICHTBOGENOFEN (INDIREKTER) 669
LICHTBOGENSCHWEISSEN 683
LICHTBOGENSCHWEISSEN 682
LICHTBOGENSCHWEISSEN MIT WECHSELSTROM 685
LICHTBOGENSCHWEISSUNG 681
LICHTBOGENSPANNUNG 676
LICHTBRECHEND 10402
LICHTBRECHUNG 10400
LICHTBRECHUNGSVERMÖGEN 10403
LICHTDICHT 7572
LICHTDURCHGANG 9099
LICHTDURCHLÄSSIG 3869
LICHTDURCHLÄSSIGKEIT 3868
LICHTE WEITE EINES ROHRES 1464
LICHTEINHEIT 9270
LICHTGESCHWINDIGKEIT 7569
LICHTGESCHWINDIGKEIT 13517
LICHTHOFBILDUNG 6203
LICHTINTENSITÄT 7822
LICHTISTÄRKEMESSUNG 9271
LICHTKEGEL 2932
LICHTLEITUNG (ELEKTRISCHE) 4519
LICHTMASCHINE 4402
LICHTMASCHINE 5924
LICHTPAUSAPPARAT 9874
LICHTPAUSE 9276
LICHTPAUSE (NEGATIVE) 1375
LICHTPAUSE (POSITIVE) 1373
LICHTPAUSPAPIER 9257
LICHTPHÄNOMEN 7566
LICHTQUELLE 7568
LICHTQUELLE 11826
LICHTSCHWINGUNG 7825

LICHTSPIEGEL **10394**
LICHTSTÄRKEMESSER **9267**
LICHTSTRAHL **7824**
LICHTSTRAHLENBÜNDEL **9162**
LICHTSTRAHLUNG **10147**
LICHTSTROM **7821**
LICHTUNDURCHLÄSSIG **7572**
LICHTWELLE **7571**
LIDERN **8990**
LIDERUNG **9009**
LIDERUNGSRING **9007**
LIEFERMENGE EINER PUMPE **10228**
LIEFERUNGSBEDINGUNGEN **3737**
LIEFERUNGSVERZUG **3731**
LIEFERWAGEN **13474**
LIEFERWALZE **5074**
LIEFERZUSTAND (IM) **741**
LIEGEND **6578**
LIEGENDE FRÄSMASCHINE **6589**
LIGNIN **7577**
LIGNIT **7578**
LIGROIN **7580**
LIMBUS **6026**
LIMONIT **7596**
LIMOUSINE **11521**
LINEAL **10831**
LINEARZEICHNUNG **8905**
LINER (GESCHLITZER) **11639**
LINER (KIESBEDECKTER) **9783**
LINIE **7597**
LINIE (ATMOSPHÄRISCHE) **798**
LINIE (AUSGEZOGENE) **5725**
LINIE (GEBROCHENE) **7166**
LINIE (GESTRICHELTE) **4125**
LINIE (PUNKTIERTE) **4125**
LINIE (STRICHPUNKTIERTE) **2205**
LINIE GLEICHEN DRUCKES **7177**
LINIE GLEICHEN RAUMINHALTES **7187**
LINIEN (LÜDERSCHE) **7815**
LINIEN (NEUMANNSCHE) **8540**
LINIENENDE **4731**
LINIENINTEGRAL **7600**
LINIENNETZ (QUADRATISCHES) **6099**
LINIENSPEKTRUM **7610**
LINKSDREHEND **7365**
LINKSDREHEND **3237**
LINKSDREHUNG **3238**
LINKSGEWICKELT **3239**
LINKSGEWINDE **7508**
LINKSGEWUNDEN **3239**
LINKSSCHWEISSUNG **5539**

LINKSVERSCHIEBUNG **7507**
LINKSWEINSÄURE **7364**
LINOLEUM **7630**
LINSE **7522**
LINSE (BIKONKAVE) **1237**
LINSE (BIKONVEXE) **1238**
LINSE (KOKAVE) **4095**
LINSE (KONKAVKONVEXE) **8147**
LINSE (KONVEXE) **2910**
LINSE (PLANKONKAVE) **9458**
LINSE (PLANKONVEXE) **9459**
LINSENSYSTEM **12576**
LIQUIDUSLINIE **7661**
LITER **7669**
LITHIUM **7664**
LITHIUMCHLORID **7666**
LITHIUMKARBONAT **7665**
LITHLUMFLUORID **7667**
LITZE **12336**
LITZE EINES SEILES **12339**
LIZENZ (PATENTRECHTLICHE) **7539**
LIZENZINHABER **7540**
LOCH **6530**
LOCH (BLINDES) **3630**
LOCH (DURCHGEHENDES) **12893**
LOCH (GEBOHRTES) **4273**
LOCH (GEGOSSENES) **2033**
LOCH (NICHT DURCHGEHENDES) **3630**
LOCH (VORGEBOHRTES) **10782**
LOCHBEITEL **8398**
LOCHEISEN **6545**
LOCHEISENHOHLRÄUME **6514**
LOCHEN **10008**
LOCHEN **10006**
LOCHEN **9999**
LOCHER **9997**
LÖCHER STANZEN **9999**
LOCHFRASS **9418**
LOCHKANTE **4447**
LOCHKARTE **10002**
LOCHKREIS **1433**
LOCHKREISDURCHMESSER **3853**
LOCHKREUZDURCHMESSER **1438**
LOCHLEHRE **6608**
LOCHLEHRE **7066**
LOCHLEIBUNG **13624**
LOCHLINER **9187**
LOCHMASCHINE **10009**
LOCHMASCHINE **9296**
LOCHMASCHINE MIT SCHERE **10007**
LOCHMUTTER **6533**

LOCHNAHTSCHWEISSUNG 9536
LOCHPLATTE 12531
LOCHPRESS 10001
LOCHRAND 4447
LOCHRINGÖFFNUNG 6532
LOCHSÄGE 2802
LOCHSTAHL 6545
LOCHSTANZE 10009
LOCHSTANZE 10001
LOCHSTEIN 245
LOCHSTEMPEL 9997
LOCHSTREIFEN 10005
LOCHSTREIFENEINGABE 12643
LOCHSTREIFENENDE 4725
LOCHTASTER 6975
LOCHVERSUCH 10010
LOCHWAND 13624
LOCHWEITE 3857
LOCHWINKEL 3781
LOCHZANGE 10000
LOCHZIEGEL 245
LOCHZIRKEL 6975
LOCKER WERDEN 11559
LOCKERN (SICH) 11559
LOCKERUNG 11560
LOCKERWERDEN 11560
LÖFFELBOHRER 11957
LOGARITHMENTAFEL 7721
LOGARITHMUS 7718
LOGARITHMUS (BRIGG'SSCHER) 2788
LOGARITHMUS (GEMEINER) 2788
LOGARITHMUS (HYPERBOLISCHER) 8509
LOGARITHMUS (NATÜRLICHER) 8509
LOHE (VERBRAUCHTE) 12635
LOHGARES 8704
LOHKÄSE 12612
LOHKUCHEN 12612
LOHN 13613
LOHNARBEIT 3621
LOHNSKALA (GLEITENDE) 11607
LOKALANGRIFF 7695
LOKALELEMENT 7694
LOKALHEIZUNG 7696
LOKALKORROSION 7695
LOKOMOBILE 9640
LOKOMOBILE (FESTSTEHENDE) 11175
LOKOMOBILE (ORTSFESTE) 11175
LOKOMOBILE (STATIONÄRE) 11175
LOKOMOTIVE 7713
LOKOMOTIVE (ELEKTRISCHE) 4520
LOKOMOTIVKESSEL 7714

LOS 7766
LÖSBARE KUPPLUNG 4033
LÖSCHE 2659
LÖSCHEN DES KALKES 11569
LÖSCHPAPIER 1350
LOSE AUFGESETZT (AUF DIE WELLE) 7749
LÖSEN 4059
LÖSEN (EINE BREMSE) 10436
LÖSEN (SICH) 4061
LÖSEN EINES KEILS 7751
LOSKEILEN 7750
LOSKUPPELN 10034
LÖSLICH 11807
LÖSLICH MACHEN 10456
LÖSLICHKEIT 11806
LOSLÖTEN 13399
LOSLÖTEN 13400
LÖSS 7716
LOSSCHEIBE 7747
LOSSCHRAUBEN 13396
LOSSCHRAUBEN 13395
LÖSUNG 11811
LÖSUNG (ALKOHOLISCHE) 315
LÖSUNG (AMMONIAKALISCHE) 457
LÖSUNG (ANGEREICHERTE) 2883
LÖSUNG (FESTE) 11794
LÖSUNG (GESÄTTIGTE) 10941
LÖSUNG (KONZENTRIERTE) 2883
LÖSUNG (RECHNERISCHE) 8694
LÖSUNG (SCHWACHE) 13706
LÖSUNG (ÜBERSÄTTIGTE) 12478
LÖSUNG (UNGESÄTTIGTE) 13394
LÖSUNG (WÄSSRIGE) 661
LÖSUNG (ZAHLENMÄSSIGE) 8694
LÖSUNG (ZEICHNERISCHE) 6060
LÖSUNG EINES KEILS 7751
LÖSUNGSBREMSE 3638
LÖSUNGSGLÜHEN 11812
LÖSUNGSMITTEL 11814
LÖSUNGSMITTEL (IONISIERENDES) 7130
LÖSUNGSMITTEL FÜR KORROSION 3192
LÖSUNGSSPANNUNG 4609
LÖSUNGSSTOFFE 8061
LÖSUNGSVERMÖGEN 4062
LOT 9539
LOT 11755
LOT 9217
LOT (SAURES) 110
LÖTAUFBRINGEMITTEL 5490
LÖTBRENNER FÜR VERDICHTETEN SAUERSTOFF 2848
LÖTEISEN 11765

MÄHBINDER UND ERNTEMASCHINEN **1251**
MÄHDRESCHER UND ZUSÄTZLICHE ANLAGE **2766**
MÄHHÄCKSLER **5525**
MAHLEN **6103**
MAHLEN **8287**
MAHLEN **6106**
MAHLEN **6117**
MAHLSTEIN **8298**
MÄHMASCHINEN (DURCH TRAKTOREN ANGETRIEBEN) UND
  HALBSATTELMÄHMASCHINEN **8439**
MÄHNENHAAR **7965**
MAKROÄTZUNG **7870**
MAKROGRAPHIE **7871**
MAKROPHOTOGRAPHIE **9266**
MAKROSEIGERUNG **7876**
MAKROSKOPISCH **7872**
MALACHIT **6089**
MALAKKAZINN **12335**
MALPINSEL **9019**
MANGAN **7966**
MANGAN (ELEKTROLYTISCHES) **4599**
MANGAN-SESQUIOXYD **7970**
MANGANBRONZE **7967**
MANGANCHLORÜR **7972**
MANGANDIOXYD **10047**
MANGANEISEN **5110**
MANGANHALTIG **7971**
MANGANHYPEROXYD **10047**
MANGANMONOXYD **7973**
MANGANOCHLORID **7972**
MANGANOSULFAT **7974**
MANGANOXYDUL **7973**
MANGANOXYDULOXYD **13207**
MANGANSTAHL **7968**
MANGANSUPEROXYD **10047**
MÄNGEL **11537**
MANGELRAD **7975**
MANILAHANF **7980**
MANILAHANFSEIL **7981**
MANNESMANNROHR **7983**
MANNLOCH **7963**
MANNLOCH **7997**
MANNLOCHDECKEL **7977**
MANNLOCHRING **7978**
MANNLOCHRING **7998**
MANO-VAKUUMMETER **2830**
MANOMETER **9820**
MANOMETER **9821**
MANOMETER (REGISTRIERENDES) **10292**
MANOSTAT **9835**
MANSCHETTE **11576**

MANSCHETTENDICHTUNG **3459**
MANTEL **11303**
MANTEL **7986**
MANTELBLECH **11335**
MANTELELEKTRODE **3271**
MANTELELEKTRODE **2584**
MANTELENDE EINES ROHRES **11901**
MANTELFLÄCHE EINES KEGELS **2949**
MANTELFLÄCHE EINES ZYLINDERS **3590**
MANTELFRÄSER **13848**
MANTELROHR **7469**
MANTISSE **7985**
MANUSKRIPT **7996**
MARINEKOLBEN **2944**
MARINEKOPF **8007**
MARINELEIM **8006**
MARK DES HOLZES **9413**
MARKIEREN **8008**
MARKIEREN **8015**
MARKSTRAHL **8128**
MARMOR **8001**
MARSEILLER SEIFE **8020**
MARTENSIT **8022**
MARTENSIT **8000**
MARTENSIT (NADELFÖRMIGER) **103**
MARTENSITHÄRTUNG (GESTAFFELTE) **12243**
MARTIN-STAHL **8816**
MASCHE **8169**
MASCHENGRÖSSE **8170**
MASCHENWEITE **11523**
MASCHINABSCHNEIDEN **7842**
MASCHINE **9859**
MASCHINE **7839**
MASCHINE (DIREKTGEKUPPELTE) **3982**
MASCHINE (ELEKTRISCHE) **4522**
MASCHINE (FREISTEHENDE) **11151**
MASCHINE (KURZHÜBIGE) **11374**
MASCHINE (LANGHÜBIGE) **7731**
MASCHINE (LANGSAMLAUFENDE) **7789**
MASCHINE (LIEGENDE) **6582**
MASCHINE (MAGNETELEKTRISCHE) **7930**
MASCHINE (STEHENDE) **13546**
MASCHINE FÜR RIEMENANTRIEB **1145**
MASCHINE FÜR SEILANTRIEB **10736**
MASCHINE FÜR SONDERZWECKE **11486**
MASCHINE MIT EINZELANTRIEB **6863**
MASCHINE MIT ZAHNRADANTRIEB **5893**
MASCHINELL **7850**
MASCHINELL ANGETRIEBEN **8119**
MASCHINEN FÜR MAULWURFSDRÄNUNG **8344**
MASCHINEN HERGESTELLT (MIT) **7850**

MATERIALKOSTEN **8049**
MATERIALPRÜFUNG **12782**
MATERIE **12419**
MATHEMATIK **8053**
MATRIZE **8055**
MATRIZE **3883**
MATRIZENSPUR **3888**
MATTBEIZEN **3628**
MATTBRENNE **8059**
MATTBRENNEN **3628**
MATTGESCHLIFFEN **10938**
MATTGLAS **5701**
MATTIEREN **12685**
MATTSCHEIBE **6145**
MATTVERCHROMUNG **4366**
MAUERANKER **500**
MAUERKASTEN **13617**
MAUERN **1601**
MAUERN **1597**
MAUERPLATTE **13621**
MAUERSTEIN **1596**
MAUERVERBAND **1444**
MAUERWERK **8025**
MAUERZIEGEL **1596**
MAULRINGSCHLÜSSEL **2764**
MAXIMALE OBERFLÄCHENTEMPERATUR **6654**
MAXIMALGEWICHT **8070**
MAXIMALTEMPERATUR **8068**
MAXIMALWERT **8069**
MAXIMUM-UND MINIMUM-THERMOMETER **8063**
MAZERATION **7837**
MAZERIEREN **7836**
MCQUAID-EHN-PRÜFUNG **8071**
MECHANIK **8121**
MECHANIK DER LUFT **9558**
MECHANISCH **8093**
MECHANISCH **8119**
MECHANISCH **7850**
MECHANISCHE WÄRMELEHRE **12829**
MECHANISMUS **8122**
MEDAILLENBRONZE **8183**
MEDIANE **8124**
MEDIUM (DURCHSICHTIGES) **13139**
MEDIUM (LEITENDES) **2914**
MEDIUM (STRAHLENBRECHENDES) **10401**
MEERESHÖHE **11070**
MEERESSAND **11071**
MEERESSPIEGEL **11070**
MEERWASSER **11072**
MEGADYN **8129**
MEGAVOLT **8130**

MEGERG **8131**
MEGOHM **8132**
MEHRARBEIT **4880**
MEHRFACHBUCKELSCHWEISSEN **8468**
MEHRFACHEXPANSIONSMASCHINE **8477**
MEHRFACHFORM **8466**
MEHRFACHRIEMEN **8463**
MEHRFACHSYSTEM **8472**
MEHRLAGENSCHWEISSEN **8455**
MEHRLAGENSCHWEISSUNG **13757**
MEHRLOCHDÜSE **8452**
MEHRPHASENSTROM **8462**
MEHRROLLENPROFILIERMASCHINE **8457**
MEHRSCHICHTMETALL **7379**
MEHRSPINDEL-BOHRMASCHINE **5799**
MEHRSPINDELAUTOMAT **8469**
MEHRSPINDELBOHRKOPF **8459**
MEHRSPINDLIGE BOHRMASCHINE IN REIHENANORDNUNG
    8460
MEHRSTOFFPRESSLING **2828**
MEHRWEGEVENTIL **8467**
MEHRZWECKMACHINE **8456**
MEHRZYLINDERMASCHINE **8453**
MEILE (ENGLISCHE) **8272**
MEISSEL **2349**
MEISSELKLAPPENHALTER **2454**
MEISSELN **2344**
MEISSELN **2347**
MEISSELSATZ **5801**
MEISSELSPAN **2343**
MEISSELWINKEL **12995**
MEISTER **5541**
MELAPHYR **8133**
MELDEVORRICHTUNG **305**
MEMBRAN **3870**
MEMBRAN-REGELVENTIL **3872**
MEMBRAN-VENTIL **3876**
MEMBRANFEDER **3875**
MEMBRANPUMPE **3874**
MEMBRANPUMPE **61**
MEMBRANVENTIL **13453**
MENGENBESTIMMUNG **10064**
MENGENMESSER **5458**
MENGENMESSER **13604**
MENGENREGLER **13600**
MENGENVERHÄLTNIS **3011**
MENNIGEKITT **10332**
MENSCHENKRAFT **9738**
MERGEL **8019**
MERGEL (KIESELIGER) **11439**
MERKURBLENDE **2400**

METALLPACKUNG 8196
METALLPACKUNG (NACHGIEBIGE) 4482
METALLPULVER 8185
METALLPULVERPRESSE 9718
METALLREINIGUNG 2474
METALLSÄGE 8186
METALLSÄGEBLATT 6197
METALLSANDSTRAHLUNG 6124
METALLSCHLAUCH 5421
METALLSCHLAUCH 5419
METALLSCHRAUBE 8187
METALLSPÄNE 2348
METALLSPRITZEN 8190
METALLSPRITZEN 5361
METALLSULFITFLECK 3429
METALLTROPFEN 11929
METALLÜBERGANG (FLÜSSIGER) 7659
METALLÜBERZUG 2596
METALLÜBERZUG 8194
METALLURGIE 8208
METALLURGIE (MECHANISCHE) 8107
METALLURGIE (PHYSIKALISCHE) 9282
METALLURGIE DES EISENS 5123
METALLURGISCH 8206
METAPHOSPHORSÄURE 8210
METASTABIL 8212
METAZENTRUM 8172
METAZINNSÄURE 12101
METEORISCHES WASSER 8213
METER 8225
METER (LAUFENDES) 10849
METERGEWICHT 13748
METERKILOGRAMM 7299
METERMASS 8226
METERMASSSTAB 8226
METERZENTNER 10111
METHAN 8215
METHANOL 13920
METHYLALKOHOL 13920
METHYLÄTHER 8222
METHYLBENZOL 12983
METHYLCHLORID 8221
METHYLENHYDRAT 8222
METHYLORANGE 8223
METRISCHER ZENTNER 10111
MIKANIT 8233
MIKROAMPERE 8238
MIKROAMPEREMETER 8237
MIKROAUFNAHME 8244
MIKROBOHRWERKZEUG 8240
MIKROBRUCH 1263

MIKROFARAD 8242
MIKROGEFÜGE 8262
MIKROGRAPHISCH 8243
MIKROHÄRTE 8245
MIKROHÄRTEPRÜFUNG 8246
MIKROHM 8247
MIKROMETER (METRISCHER) 8228
MIKROMETERHALTER 8250
MIKROMETERLEHRE 1875
MIKROMETERRUNDSKALA 8253
MIKROMETERSCHRAUBE 8255
MIKROMETERSCHRAUBE 13540
MIKROMETRISCHER SCHRAUBENGEWINDE-KOMPARATOR 11048
MIKROMILLIMETER 8257
MIKROPHOTOGRAMM 8258
MIKROPHOTOGRAMM 9272
MIKRORISS 1263
MIKRORISSE 8234
MIKROSCHLIFF 8261
MIKROSEIGERUNG 8235
MIKROSEIGERUNG 3172
MIKROSKOP 8259
MIKROSKOP (BINOKULARES) 1256
MIKROTOM 8263
MIKROVOLT 8264
MILCHGLAS 8273
MILCHSÄURE 7355
MILLIAMPERE 8284
MILLIAMPEREMETER 8283
MILLIGRAMM 8285
MILLIMETER 8286
MILLIMETERTASTER 8256
MILLIONSTEL ZOLL 8248
MILLIVOLT 8294
MILROMETER 8249
MINDESTGEWICHT 8310
MINDESTSTRECKGRENZE 8308
MINDESTWERT 8307
MINERAL 8301
MINERALFARBE 4412
MINERALFETT 13508
MINERALGRÜN 10996
MINERALÖL 8304
MINERALÖL (GEREINIGTES) 2480
MINERALÖL (RAFFINNIERTES) 2480
MINERALÖL (UNGEREINIGTES) 1291
MINERALSÄURE 8302
MINERALSÄURE 9203
MINERALWEISS 1003
MINETTE 8305

MINIMALGEWICHT 8310
MINIMALTEMPERATUR 8306
MINIMALWERT 8307
MINIUMKITT 10332
MINUEND 8316
MINUSKORNGRÖSSE 8317
MINUSPLATTE 8530
MINUSZEICHEN 8532
MIRBANÖL 8587
MISCHBATTERIE FÜR BADEWANNEN 1063
MISCHDÜSE 2769
MISCHEN 8328
MISCHEN 1323
MISCHEN 8336
MISCHEN 1325
MISCHER 8331
MISCHERPFANNE 8333
MISCHFARBE 11103
MISCHGAS 4194
MISCHKOLLERGANG 8451
MISCHKOLLERGANG 8449
MISCHKONDENSATOR 3988
MISCHKRISTALL 11794
MISCHKRISTALLE 8329
MISCHMASCHINE 8334
MISCHMETALL 8320
MISCHMETALL 817
MISCHÖL 2832
MISCHSÄURE 8595
MISCHTROMMEL 8331
MISCHUNG 8336
MISCHUNG 8327
MISCHUNG 1325
MISCHUNG EUTEKTISCHE 4857
MISCHUNGSVERHÄLTNIS 9934
MISCHVENTIL 8335
MITNEHMEN (DURCH REIBUNG) 4304
MITNEHMER 4301
MITNEHMERSTANGENEINSATZ 4288
MITREISSEN VON WASSER IM DAMPF 9863
MITSCHWINGEN 10506
MITTEL 8072
MITTEL (ARITHMETISCHES) 703
MITTEL (GEOMETRISCHES) 5929
MITTEL (WASSERABWEISENDES) 13674
MITTEL-(WALZ-)GERÜST 7098
MITTELAUFLAGER 2152
MITTELDRUCKKESSEL 8127
MITTELHIEBFEILE 8266
MITTELKRAFT 10513
MITTELLAGE 8265

MITTELLINIE 8124
MITTELLINIE 2151
MITTELLINIE 874
MITTELÖL 8125
MITTELPUNKT 2148
MITTELPUNKT (GEOM.) 2162
MITTELPUNKTSUCHER 2173
MITTELSTANGE 7795
MITTELSTELLUNG 8265
MITTELWASSERRAUMKESSEL 8126
MITTELWERT 8073
MITTEN 2163
MITTENKURBEL 6976
MITTNEHMER 1883
MITVERÄNDERUNG 3261
MKG 7299
MODELL 9133
MODELL 13320
MODELL 9128
MODELL (ZELEGBARES) 7744
MODELL (ZWEITEILIGES) 3105
MODELL MIT EINGUSSTRICHTERN 5868
MODELLFORMUNG 3041
MODELLGUSSPLATTE 2052
MODELLHERSTELLUNG 9135
MODELLPLATTE 1710
MODELLPLATTE (ZWEISEITIGE) 8044
MODELLSAND 5006
MODELLSCHREINER 9130
MODELLTISCHLER 9130
MODELLTISCHLEREI (HERSTELLUNG VON MODELLEN) 9131
MODELLTISCHLEREI (WERKSTÄTTE) 9132
MODEM 8338
MODUL 8339
MODULATOR 8338
MOHNÖL 9624
MOL 6044
MOLEKEL 8364
MOLEKÜL 8364
MOLEKULARGEWICHT 8363
MOLEKULARKRAFT 8360
MOLEKULARVOLUMEN 8362
MOLEKÜLSTRUKTUR 8361
MOLKEREIMASCHINEN UND ZUSATZGERÄTE 3595
MOLKMASCHINEN 8275
MOLYBDÄN 8367
MOLYBDÄNBLEISPAT 13991
MOLYBDÄNEISEN 5111
MOLYBDÄNGLANZ 8366
MOLYBDÄNSÄURE 8369
MOLYBDÄNSTAHL 8368

MOMENT (STATISCHES) **12139**
MOMENT EINER KRAFT **8370**
MOMENTANKRAFT **6811**
MOMENTANWERT **6990**
MOMENTANZENTRUM **6989**
MÖNCH **9997**
MÖNCHSKOLBEN **9545**
MONDGAS **8378**
MONELMETALL **8379**
MONOCHLORÄTHAN **4849**
MONOTEKTOIDE REAKTION **8386**
MONOTRON **8387**
MONOTROP **8388**
MONOXYBENZOL **1938**
MONTAGE **4808**
MONTAGE **4815**
MONTAGE **4813**
MONTAGEAUSRÜSTUNG **4810**
MONTAGEGESTELL **4116**
MONTAGEHALLE **773**
MONTAGEHALLE **4811**
MONTAGEHALLE **770**
MONTAGEMATERIAL **4809**
MONTAGEWERKZEUG **4812**
MONTAGEWERKZEUG **4817**
MONTAGEZEICHNUNG **772**
MONTAGEZEICHNUNG **4814**
MONTANWACHS **7579**
MONTEJUS **8389**
MONTEUR **4816**
MONTIEREN **4805**
MONTIERHAMMER **5279**
MONTIERWERKSTATT **4811**
MOORINGSYSTEM (DREHBARES) **10217**
MORSEKEGELLEHRDORN **8392**
MORSEKEGELLEHRE **8393**
MÖRSER **8396**
MÖRSERKEULE **9221**
MÖRSERLAFETTE **8397**
MÖRTEL **8394**
MÖRTEL (ALUMINIUM-SILIKAT-FEUERFESTER) **443**
MÖRTEL (FEUERFESTER) **10406**
MÖRTEL (HYDRAULISCHER) **6690**
MOSAIKSTRUKTUR **8403**
MOTOR **9859**
MOTOR (BELASTETER) **7686**
MOTOR (GEKAPSELTER) **4715**
MOTOR (GESCHLOSSENER) **4715**
MOTOR (HYDRAULISCHER) **6691**
MOTOR (LUFTGEKÜHLTER) **251**
MOTOR (UMKEHRBARER) **10540**

MOTOR (VENTILIERTER) **13529**
MOTOR-BREMSE **4760**
MOTORAUFHÄNGUNG **4754**
MOTOREINSTELLUNG **8419**
MOTOREN (BENZINKEROSIN UND DIESEL) **4766**
MOTORENSPIRITUS **1983**
MOTORGENERATOR **8413**
MOTORHAUBE **1451**
MOTORÖL **8415**
MOTORRAUM **4757**
MOTORSCHIEBER **8416**
MOTORVENTIL **8417**
MUFFE **3258**
MUFFE **11548**
MUFFE **11705**
MUFFE **3257**
MUFFE (KONISCHE) **12647**
MUFFELOFEN **8447**
MUFFENANSCHLUSS **11053**
MUFFENENDE EINES ROHRES **11710**
MUFFENKUPPLUNG **1510**
MUFFENREGLER **11578**
MUFFENROHR **11707**
MUFFENVERBINDUNG **11900**
MÜHLEN (KOMBINIERT) **8295**
MÜHLSTEIN **8298**
MÜHLSTEIN (GEOL.) **8299**
MÜLL **13074**
MULL **2288**
MULTIPLIKAND **8479**
MULTIPLIKATION **8480**
MULTIPLIKATOR **8482**
MULTIPLIZIEREN **8483**
MUNDSTÜCK **12966**
MUNDSTÜCK **196**
MUNITION **474**
MUNTZMETALL **8486**
MÜNZEN **2651**
MÜNZENBRONZE **8183**
MÜNZGOLD **5996**
MUSCHELKALK **8487**
MUSCHELLINIE **2892**
MUSCHELSCHIEBER **3592**
MUSIVGOLD **12099**
MUSTER **10915**
MUSTER **9128**
MUSTERMODELL **8032**
MUTTER **8698**
MUTTER (AUFGESCHNITTENE) **11942**
MUTTER (BEARBEITETE) **1612**
MUTTER (BLANKE) **1612**

NAHTSCHWEISSUNG 11086
NAHTÜBERHÖHUNG 8948
NAHTWURZEL 10726
NAPHTHA 8493
NAPHTHALIN 8494
NAPHTHOL (A-NAPHTHOLSS-NAPHTHOL) 8495
NAPHTHYLWASSERSTOFF 8494
NARBENSEITE DES LEDERS 6038
NASE 2093
NASE 8667
NASENBILDUNG 10892
NASENKEIL 5939
NASENSCHRAUBE 5062
NASSDAMPF 13805
NASSZUG 13803
NATRIUM 11726
NATRIUM 11714
NATRIUM (ESSIGSAURES) 11715
NATRIUM (GERBSAURES) 11735
NATRIUM (OXALSAURES) 11726
NATRIUM (SALPETRIGSAURES) 11725
NATRIUM (SAURES) 11717
NATRIUM (SCHWEFELSAURES) 11717
NATRIUM (WOLFRAMSAURES) 11737
NATRIUM (ZINNSAURES) 11732
NATRIUMALUMINAT 11716
NATRIUMAMMONIUMPHOSPHAT 8241
NATRIUMAZETAT 11715
NATRIUMBIBORAT 1459
NATRIUMBIKARBONAT 126
NATRIUMBISULFAT 11717
NATRIUMBISULFIT 11718
NATRIUMCHLORID 2790
NATRIUMHYDROXYD 11722
NATRIUMHYDROXYDBAD 2117
NATRIUMHYPOSULFIT 11736
NATRIUMKARBONAT 11713
NATRIUMKARBONAT (WASSERFREIES) 551
NATRIUMLINIE 11723
NATRIUMNITRAT 2330
NATRIUMNITRIT 11725
NATRIUMOXALAT 11726
NATRIUMOXYD 11724
NATRIUMOXYDHYDRAT 11722
NATRIUMPEROXYD 11727
NATRIUMPHOSPHAT 11728
NATRIUMPYROPHOSPHAT 11730
NATRIUMSILIKAT 11731
NATRIUMSTANNAT 11732
NATRIUMSULFAT 11733
NATRIUMSULFIT 11734

NATRIUMSUPEROXYD 11727
NATRIUMTANNAT 11735
NATRIUMTHIOSULFAT 11736
NATRIUMWOLFRAMAT 11737
NATRIUMZITRAT 11719
NATRIUMZYANID 11720
NATRON 11724
NATRON (BORSAURES) 1459
NATRON (DOPPELTKOHLENSAURES) 126
NATRON (DOPPELTSCHWEFLIGSAURES) 11718
NATRON (PHOSPHORSAURES) 11728
NATRON (SCHWEFELSAURES) 11733
NATRON (SCHWEFLIGSAURES) 11734
NATRON (SCHWEFLIGSAURES) 11718
NATRON (UNTERSCHWEFLIGSAURES) 11736
NATRON (ZITRONENSAURES) 11719
NATRONHYDRAT 11722
NATRONLAUGE 2120
NATRONSALPETER 2330
NATRONWASSERGLAS 11731
NATRONWEINSTEIN 11729
NATURBIMSSTEIN 9986
NATURGAS 8504
NATÜRLICHER GRÖSSE (IN) 5723
NATURSTEIN 8507
NEBELLAMPE 5506
NEBENACHSE DER HYPERBEL 2953
NEBENEINANDERSCHALTUNG 9051
NEBENERZEUGNIS 1820
NEBENGESTEIN 5802
NEBENLAGER 7044
NEBENPRODUKT 1820
NEBENSCHLUSS 1818
NEBENSCHLUSS 1821
NEBENSCHLUSS 11403
NEBENSCHLUSSDYNAMO 11404
NEBENSCHLUSSMOTOR 11405
NEBENSPANNUNG 11109
NEBENSPEKTRUM. 11108
NEBENWINKEL 180
NEGATIV (PHOTOGRAPHISCHES) 8524
NEGATIVE 8529
NEGATIVE HÄRTUNG 8527
NEGATIVELEKTRISCH 4631
NEIGUNG 6836
NEIGUNG 1065
NEIGUNG 11627
NEIGUNG EINES KEILS 12654
NEIGUNGSWINKEL 521
NENNABMESSUNG 8612
NENNDRUCK (ND) 8613

NENNDURCHMESSER **8611**
NENNER **3752**
NENNER (GEMEINSCHAFTLICHER) **2789**
NENNLEISTUNG **8658**
NENNMASS **8615**
NENNMASS **11857**
NENNSPANNUNG **8659**
NENNWEITE (NW) **8615**
NENNWERT **8616**
NEODYM **8534**
NEODYM **8533**
NEON **8535**
NEONLAMPE **8536**
NETZANSCHLUSS **2782**
NETZEBENE **804**
NETZEBENE **7432**
NETZEBENEN-ABSTAND **7087**
NETZGEFÜGE **8539**
NETZPAPIER **11530**
NETZSTRUKTUR **8539**
NEUGELB **8031**
NEUSEELANDHANF **8550**
NEUSILBER **8569**
NEUSILBER **5932**
NEUTRALE FASER **2151**
NEUTRALES KALIUMCHROMAT **9682**
NEUTRALES KALIUMOXALAT **9692**
NEUTRALISATION **8548**
NEUTRALISIEREN **8549**
NEUWEISS **1003**
NICHT BRENNBAR **6843**
NICHT OXYDIERBAR **6959**
NICHT RECHTWINKELIG **8895**
NICHT SPIEGELGLEICH **778**
NICHT UMKEHRBAR **7169**
NICHT ZUSAMMENDRÜCKBAR **6847**
NICHT-METALLISCH **8639**
NICHTEISEN- **8631**
NICHTEISENLEGIERUNG **8619**
NICHTEISENMATALL **8632**
NICHTLEITEND **3898**
NICHTLEITER **3898**
NICHTUMKEHRBARKEIT **7168**
NICKEL **8556**
NICKELALUMINIUMBRONZE **8572**
NICKELAMMONIUMSULFAT **8558**
NICKELBRONZE **8560**
NICKELCHLORÜR **8562**
NICKELEISEN **5112**
NICKELERZ **8564**
NICKELKARBONYL **8561**

NICKELKOHLENOXYD **8561**
NICKELLEGIERUNG **8557**
NICKELMESSING **8559**
NICKELOXYD **8574**
NICKELOXYDUL **8565**
NICKELPLATTIERUNG **8567**
NICKELSCHROT **8568**
NICKELSTAHL **8570**
NICKELSTEIN **8563**
NICKELSULFAT **8571**
NICKELVITRIOL **8571**
NIEDERDRUCK **7774**
NIEDERDRUCKDAMPF **7785**
NIEDERDRUCKDAMPFHEIZUNG **7787**
NIEDERDRUCKDAMPFMASCHINE **7786**
NIEDERDRUCKKESSEL **7782**
NIEDERDRUCKTURBINE **7788**
NIEDERDRUCKWASSERHEIZUNG **7784**
NIEDERDRUCKZYLINDER **7783**
NIEDERFREQUENZ **7771**
NIEDERFREQUENZOFEN **7772**
NIEDERHALTER **3887**
NIEDERSCHLAG **9759**
NIEDERSCHLAG **9761**
NIEDERSCHLAG (ANGEBRANNTER) **1761**
NIEDERSCHLAG (GALVANISCHER) **4592**
NIEDERSCHLAG (RADIOAKTIVER) **149**
NIEDERSCHLAGEN **2907**
NIEDERSCHLAGSVERTEILUNGSVERHÄLTNIS **3777**
NIEDERSCHLAGSVERTEILUNGSVERHÄLTNIS **8179**
NIEDERSCHLAGWASSER **8213**
NIEDERSCHLAGWASSER **13662**
NIEDERSCHRAUBHAHN **11033**
NIEDERSPANNUNG (EL.) **7778**
NIEDERSPANNUNGSMOTOR **7791**
NIEDERSPANNUNGSSTROM **7777**
NIEDRIG GESPANNTER DAMPF **7785**
NIEDRIGGESPANNTER STROM **7777**
NIEDRIGTEMPERATURVERBINDUNG **7790**
NIET **10621**
NIET MIT BÜNDIGEM KOPF **10635**
NIET MIT DREIECKPROFILKOPF **2946**
NIET MIT FLACHEM KOPF **10634**
NIET MIT GEHÄMMERTEM KOPF **2946**
NIET MIT HALBRUNDKOPF **10636**
NIET MIT HALBVERSENKTEM KOPF **3249**
NIET MIT KORBBOGENKOPF **10633**
NIET MIT RUNDKOPF **10637**
NIET MIT SCHELLKOPF **10637**
NIET MIT TRAPEZPROFILKOPF **9030**
NIET MIT VERSENKTEM KOPF **10635**

NIET MIT ZYLINDRISCHEM KOPF **10632**

NIETBOLZEN **11263**

NIETE **10646**

NIETE (DIE) HERAUSSCHLAGEN **3511**

NIETE STAUCHEN **2489**

NIETEISEN **10627**

NIETEN **10643**

NIETEN EINSETZEN (DIE) **6967**

NIETEN EINZIEHEN (DIE) **6967**

NIETER **10641**

NIETFLANSCH **10638**

NIETHAMMER **10644**

NIETKOPF **10624**

NIETLOCH **10625**

NIETLOCH (GELOCHTES) **10004**

NIETLOCH (GESTANZTES) **10004**

NIETLOCH AUFDORNEN (EIN) **4262**

NIETLOCH GEBOHRTES **4275**

NIETLÖCHER AUFEINANDERPASSEN (DIE) **9969**

NIETLOCHSCHWEISSEN **9537**

NIETMASCHINE **10645**

NIETMASCHINE **10642**

NIETMITTE **2171**

NIETNAHT **11084**

NIETPFANNE **4115**

NIETREIHE **10813**

NIETSCHAFT **11263**

NIETSCHWEISSEN **10631**

NIETSTÄRKE **3859**

NIETTEILUNG **9409**

NIETUNG **10639**

NIETUNG **10643**

NIETUNG (DOPPELTE) **4152**

NIETUNG (DREIREIHIGE) **13177**

NIETUNG (DREISCHNITTIGE) **13178**

NIETUNG (EINFACHE) **11487**

NIETUNG (EINREIHIGE) **11487**

NIETUNG (EINSCHNITTIGE) **11489**

NIETUNG (VERJÜNGTE) **7799**

NIETUNG (VERSETZTE) **14034**

NIETUNG (ZWEIREIHIGE) **4152**

NIETUNG (ZWEISCHNITTIGE) **4155**

NIETVERBINDUNG **10639**

NIETVERTEILUNG **717**

NIETWÄRMEOFEN **10623**

NIETWINDE **11032**

NIETZANGE **10630**

NIGGERÖL **3222**

NIOB **2747**

NIOBIUM **8576**

NIPPEL **8578**

NIROSTAHL **12051**

NISCHE **10279**

NITRAT **8579**

NITRIERATMOSPHÄRE **8583**

NITRIERHÄRTUNG **8582**

NITRIEROFEN **8584**

NITRIERSÄURE **8595**

NITRIERSTAHL **8585**

NITRIERUNG **8589**

NITRIT **8586**

NITROBENZOL **8587**

NITROGLYZERIN **8593**

NITROMETER **8594**

NITROZELLULOSE **6176**

NITROZEMENTIERUNG **5818**

NIVEAU **13259**

NIVEAU (HÖHENLAGE) **7525**

NIVEAU REGLER **7654**

NIVEAUFLÄCHE **7527**

NIVEAUSCHWANKUNGEN **13500**

NIVEAUSTANDSREGLER **7655**

NIVELLIERSCHRAUBE **7529**

NOCKE **1883**

NOCKEN **1883**

NOCKENANLAUFSCHRÄGER **1887**

NOCKENHUB **1884**

NOCKENROLLE **1895**

NOCKENSCHEIBE **13221**

NOCKENWELLE **1896**

NOCKENWELLE (OBENLIEGENDE) **8939**

NOCKENWELLENDREHBANK **1897**

NODULAR **8606**

NONIUS **13536**

NORDHÄUSER VITRIOLÖL **5733**

NORDPOL EINES MAGNETEN **8665**

NORM **12071**

NORMAL (GEOM.) **8652**

NORMAL LEHRDORN **12087**

NORMALBESCHLEUNIGUNG **8653**

NORMALDRUCK **8657**

NORMALE **12071**

NORMALE EINER KURVE **8662**

NORMALFLAMME **8543**

NORMALGEWINDE (SCHWEIZER) **12552**

NORMALGEWINDELEHRDORN **12861**

NORMALGEWINDELEHRRING **12863**

NORMALGLÜHEN **8664**

NORMALISIEREN **12090**

NORMALISIEREN DES STAHLS **581**

NORMALISIERUNG **12091**

NORMALKRÜMMER **10074**

OBERFLÄCHENPRÜFGERÄT **12498**
OBERFLÄCHENPRÜFGERÄT **12509**
OBERFLÄCHENRAUHIGKEIT **12518**
OBERFLÄCHENRISS **12501**
OBERFLÄCHENSCHICHT **2019**
OBERFLÄCHENSPANNUNG **7033**
OBERFLÄCHENSPANNUNG **12519**
OBERFLÄCHENUNEBENHEIT **12503**
OBERFLÄCHENWASSER **12520**
OBERGESENK **13036**
OBERINGENIEUR **4762**
OBERIRDISCHE LEITUNG **212**
OBERSCHALE **13030**
OBERSCHLÄCHTIGES WASSERRAD **8960**
OBERSEE KUPFER **7370**
OBERWASSER **6322**
OBERWASSERKANAL **6319**
OBERWINKELEISEN **13028**
OBJEKTIV **8708**
OBJEKTPUNKT **8706**
OBJEKTTISCH **12047**
OBJEKTTRÄGER **11589**
OBSIDIAN **8716**
OBUS **13225**
OCKER **8720**
ODONTOGRAPH **8726**
OFEN **8920**
OFEN **5739**
OFEN **12300**
OFEN (ELEKTRISCHER) **4507**
OFEN (FESTSTEGENDER) **12142**
OFEN (GASGEHEIZTER) **5851**
OFEN (KONTINUIERLICHER) **3026**
OFEN (METALLURGISCHER) **8207**
OFENBESCHICKUNGSMASCHINE **2268**
OFENFUTTER **5748**
OFENFUTTER (BASISCHES) **1051**
OFENFUTTER (MONOLITHISCHES) **8385**
OFENFUTTER (SAURES) **123**
OFENHARTLÖTEN **5740**
OFENKÜHLUNG **5742**
OFENMANTEL **5749**
OFENPLATTE **6512**
OFENSAU **10896**
OFF-LINE-BETRIEB **8735**
OFF-SHORE **8738**
OFFENE LEHRE **6608**
OFFENER MOTOR **8818**
OFFENER SCHUBSTANGENKOPF **8007**
OFFENESFEUER **8813**
OFFENHALTUNGSVORRICHTUNG **6525**

ÖFFNUNG **6530**
ÖFFNUNG **9633**
ÖFFNUNG **13528**
OHM **8742**
ÖHR EINES HAMMERS **4975**
OKKLUSION (CHEM.) **8719**
OKTAEDER **8722**
OKTANT **8724**
OKTANZAHL **8723**
OKULAR **4977**
OKULARMIKROMETER **8254**
ÖL **8743**
ÖL (ABLAUFENDES) **13636**
ÖL (FETTES) **5294**
ÖL (FLÜCHTIGES) **4836**
ÖL (GELÄUTERTES) **10373**
ÖL (RAFFINIERTES) **10373**
ÖL (RANZIGES) **10197**
ÖL (ROSTLÖSENDES) **9169**
ÖL (TIERISCHES) **555**
ÖL (TROCKNENDES) **4357**
ÖL (VEGETABILISCHES) **13512**
ÖL (VERSPRITZTES) **11928**
ÖL (WASSERLÖSLICHES) **11810**
ÖL-ABSTREIFRING **13879**
ÖLABSCHEIDER **8779**
ÖLABSCHEIDER **8778**
ÖLABSCHEIDUNG **8777**
ÖLABSTREIFRING **8775**
ÖLABSTREIFRING **8750**
ÖLBAD **8745**
ÖLBADLUFTFILTER **8746**
ÖLBILDENDES GAS **4851**
ÖLBREMSE **8755**
ÖLBUNKER **8748**
ÖLBUNKER **5706**
OLD VERFAHREN (OXYGENE-LINZ-DONAWITZ) **8793**
OLDHAMKUPPLUNG **8794**
ÖLDRUCKMESSER **8768**
ÖLDURCHLÄSSIGKEITSMASS **10232**
OLEINSÄURE **8795**
ÖLEN **8791**
ÖLEN **8744**
OLEUM **5733**
ÖLFÄNGER **4283**
ÖLFANGRING **8776**
ÖLFARBE **8766**
ÖLFARBENANSTRICH **2582**
ÖLFEST **8792**
ÖLFEUERUNG **1752**
ÖLFEUERUNG **5745**

ÖLFILTER 8757
ÖLFLÜGELRAD 8763
ÖLGAS 8758
ÖLHÄRTUNG 8760
ÖLHÄRTUNG 8772
OLIVIN 8796
ÖLKANNE 8749
ÖLKITT 8771
ÖLKÜHLER 8751
ÖLKÜHLUNG 8752
ÖLMASCHINE 8756
ÖLMESSSTAB 3979
ÖLMESSSTAB 8764
ÖLNUT 8759
ÖLPAPIER 8788
ÖLPOLSTER (DÄMPFUNG) 8754
ÖLPRÜFMASCHINE 8785
ÖLPUFFER 8755
ÖLPUMPE 8769
ÖLPUMPENSIEB 8770
ÖLRING 8774
ÖLSÄURE 8795
ÖLSCHALE 4283
ÖLSCHIEFER 8780
ÖLSCHIFF 4283
ÖLSCHMIERUNG 8791
ÖLSPEICHER 8773
ÖLSPRITZE 8784
ÖLSTEIN 8781
ÖLSÜSS 5988
ÖLTROPFAPPARAT 4323
ÖLVASE 8753
ÖLVISKOSITÄTSINDEX 8787
ÖLWANNE 8782
ÖLZUFÜHRUNG 8783
ON-LINE-BETRIEB 8799
OOLITHKALK 8801
OPERMENT 728
OPTIK 8849
OPTISCH 8840
ORANGENSCHALENEFFEKT 8851
ORDINATE 8858
ORDINATENACHSE 873
ORIENTIERUNG 8867
ORIENTIERUNG (VORZUGSWEISE) 9771
ORIGINALGRÖSSE (IN) 5723
ORT (GEOMETRISCHER) 7715
ÖRTERSÄGE 5628
ORTHOKIESELSÄURE 8875
ORTHOPHOSPHORSÄURE 8873
ORTSVERÄNDERUNG 7712

ÖSE 4980
ÖSE EINES HAMMERS 4975
ÖSENHAKEN 4974
ÖSENSCHRAUBE 4973
OSMIRIDIUM 8888
OSMIUM 8889
OSMOSE 8890
OSTEOPLASTIKLEGIERUNG 498
OSZILLATION 8884
OSZILLIEREN 8877
OSZILLIERENDE DAMPFMASCHINE 8880
OSZILLOGRAPH 8885
OSZILLOSKOP 8886
OVALDREHBANK 8919
OVALEISEN 8916
OXALAT 8968
OXALSÄURE 8969
OXALSAURES KALI 9692
OXALSAURES SALZ 8968
OXYD 8971
OXYDATION 8970
OXYDATION (CHROMSAURE ANODISCHE) 2380
OXYDATION (ELEKTROCHEMISCHE) 4601
OXYDATION (ODER KORROSION) (TROCKENE) 4350
OXYDATION BEI HOCHTEMPERATUR 6498
OXYDATIONSFLAMME 4883
OXYDATIONSMITTEL 8974
OXYDATIONSWÄRME 6361
OXYDBELAG 2598
OXYDHAUT 5168
OXYDIERBAR 8972
OXYDIEREN 8984
OXYDIEREN 8973
OXYDIERUNG 8970
OXYDUL 5124
OZOKERIT 4413
OZON 8989
OZOZEROTIN 2194
P-DIAMIDODIPHENYL 1201
PAAR 3252
PACKEN 8990
PACKFONG 5932
PACKHAHN 5954
PACKHAHN 8997
PACKLEINWAND 6449
PACKMATERIAL 9004
PACKPAPIER 13984
PACKUNG 9009
PACKUNG 8998
PACKUNG (METALLISCHE) 8196
PACKUNG (WEICHE) 11739

PACKUNG MIT EINLAGE **3159**
PACKUNGSRAUM **9008**
PACKUNGSRING **9007**
PACKUNGSZOPF **5863**
PAKET **5017**
PAKETIEREN **5018**
PAKETIEREN **1772**
PAKETIERSCHROTT **5016**
PAKETWALZEN **8994**
PAKETWÄRMOFEN **9308**
PALETTEN-HUBLASTWAGEN **9025**
PALETTIEREN **9026**
PALLADIUM **9023**
PALMBUTTER **9027**
PALMFETT **9027**
PALMITINSÄURE **9028**
PALMÖL **9027**
PANNE **1581**
PANTOGRAPH **9033**
PANZERBLECH **11343**
PANZERKABEL **713**
PANZERPLATTE **712**
PAPIER (ENDLOSES) **3013**
PAPIER (GEFIRNISSTES) **13506**
PAPIER (KARIERTES) **11530**
PAPIER (LACKIERTES) **13506**
PAPIER (METALLISIERTES) **8199**
PAPIERMACHE **9036**
PAPIERRIEMENSCHEIBE **9034**
PAPIERROHR **6803**
PAPIERSTOFF **9035**
PAPIERZEUG **9035**
PAPINSCHER TOPF **829**
PAPPDECKEL **1381**
PAPPE **1381**
PARABEL **9038**
PARABELINTERPOLATION **9040**
PARABOLSPIEGEL **9041**
PARAFFIN **9043**
PARAFFINGASÖL **1668**
PARAFFINÖL **9044**
PARAGUMMI **9037**
PARAKAUTSCHUK **9037**
PARALLAXE **9047**
PARALLAXENFEHLER **9048**
PARALLEL **9050**
PARALLELANZEIGE **9054**
PARALLELDREHBANK **9428**
PARALLELDREHEN **13164**
PARALLELDREHEN **13162**
PARALLELE **9049**

PARALLELENDMASS **5769**
PARALLELEPIPED **9068**
PARALLELEPIPED (ABGESTUMPFTES) **13237**
PARALLELEPIPED (GERADES) **10581**
PARALLELEPIPED (NORMALES) **10581**
PARALLELEPIPED (RECHTWINKLIGES) **10305**
PARALLELEPIPED (SCHIEFES) **8713**
PARALLELHAMMER **4326**
PARALLELITÄT **9069**
PARALLELLINEAL **9059**
PARALLELMASS **8016**
PARALLELNIETUNG **2215**
PARALLELOGRAMM **9070**
PARALLELOGRAMM DER GESCHWINDIGKEITEN **9072**
PARALLELOGRAMM DER KRÄFTE **9071**
PARALLELOGRAMMGETRIEBE **9057**
PARALLELPROJEKTION **9058**
PARALLELPROJEKTION (KLINOGRAPHISCHE) **2501**
PARALLELPROJEKTION (ORTHOGONALE) **8870**
PARALLELPROJEKTION (ORTHOGRAPHISCHE) **8870**
PARALLELPROJEKTION (RECHTWINKLIGE) **8870**
PARALLELPROJEKTION (SCHIEFE) **2501**
PARALLELREISSER **8016**
PARALLELREISSLINIE **7604**
PARALLELSCHALTUNG **9051**
PARALLELSCHIEBER **9060**
PARALLELSCHIEBER (SELBSTDICHTENDER) **9061**
PARALLELSCHIEBER MIT FEDERVERSCHLUSS **9052**
PARALLELSCHIEBER MIT MECHANISCHEM VERSCHLUSS **9053**
PARALLELSCHRAUBSTOCK **9066**
PARALLELSTRÖMUNG **9055**
PARALLELVERSCHIEBUNG **13123**
PARAMAGNETISCH **9073**
PARAMAGNETISMUS **9074**
PARAMETER **9075**
PARAPHOSPHORSÄURE **10054**
PARISER STIFT **13895**
PARITÄTSPRÜFUNG **9081**
PARTIALTURBINE **9085**
PARTIELLE DIFFERENTIALGLEICHUNG **9086**
PASSBLECH **7621**
PASSBOLZEN **4181**
PASSFLÄCHE **5058**
PASSFLÄCHENKORROSION **5666**
PASSIG DREHEN **13280**
PASSIGDREHEN **8869**
PASSIVIERBAD **9103**
PASSIVIERUNG **9105**
PASSIVIERUNGSMITTEL **9106**
PASSIVIERUNGSSCHICHT **9104**
PASSIVITÄT **9107**

PHYLLIT **9278**
PHYSIK **9285**
PHYSIKALISCH **9279**
PICKELBILDUNG **9314**
PIEZOELEKTRIZITÄT **9297**
PIEZOMETER **12067**
PIKIERMASCHINEN UND SISALPFLANZMASCHINEN **11133**
PIKRINSÄURE **9291**
PILGERSCHRITTSCHWEISSVERFAHREN **901**
PINAKOID **9327**
PINCHEFFEKT **9329**
PINKSALZ **470**
PINNE EINES HAMMERS **9031**
PINSEL **2743**
PINT **9337**
PINZETTE **5536**
PIPETTE **9368**
PISTOLENGRIFFMETALLSÄGE **9369**
PITOTROHR **9415**
PLANDREHBANK **5001**
PLANDREHBANK **5005**
PLANDREHBANK MIT WAAGERECHTER PLANSCHEIBE **1469**
PLANDREHEN **12497**
PLANDREHEN **12522**
PLANETARISCH **9445**
PLANETENBEWEGUNG **9447**
PLANETENGETRIEBE **12456**
PLANETENGETRIEBE **9446**
PLANETENRAD **9444**
PLANFILM **8993**
PLANFLÄCHE (OPTISCHE) **8843**
PLANFRÄSER **4446**
PLANIEREN **9454**
PLANIERHAMMER **9455**
PLANIERKOLBEN **4116**
PLANIMETER **9448**
PLANIMETRIE **9449**
PLANIMETRIEREN **8084**
PLANIMETRIERUNG **8087**
PLANROST **6584**
PLANSCHEIBE **4999**
PLANSCHEIBENGETRIEBE **4005**
PLANSCHEIBENKUPPLUNG **9486**
PLANSCHLEIFEN **12506**
PLANSCHLEIFMASCHINE **12505**
PLANUNG **3794**
PLASTIZITÄT **9474**
PLASTMASSE (FEUERFESTE) **9473**
PLASTSTAMPFER (MECHANISCHER) **1732**
PLATIN **9510**
PLATIN-IRIDIUM **9515**

PLATINCHLORID **9507**
PLATINCHLORÜR **9509**
PLATINDRAHT **9513**
PLATINE **11306**
PLATINE **4174**
PLATINEINHEIT (DER LICHTSTÄRKE) **13569**
PLATINEN AUS LEGIERTEN STÄHLEN **1956**
PLATINEN AUS LEGIERTEN STÄHLEN **378**
PLATINENSCHNITT **1310**
PLATINID **9508**
PLATINMOHR **9511**
PLATINMOHR **9514**
PLATINSCHWAMM **11955**
PLATINSCHWARZ **9511**
PLATINTETRACHLORID **9507**
PLATINTHERMOMETER **10495**
PLATINTIEGEL **9512**
PLÄTTCHEN **9498**
PLATTE **11551**
PLATTE **9477**
PLATTE (LICHTEMPFINDLICHE) **11187**
PLATTE (NEGATIVE) **8530**
PLATTE (PHOTOGRAPHISCHE) **11187**
PLATTE (POSITIVE) **9658**
PLATTENABDICHTUNGSSCHALE **11073**
PLATTENFEDERMANOMETER **3873**
PLATTENFÖRDERER **11570**
PLATTENÜBERLAPPUNG **9501**
PLATTENVENTIL **4011**
PLATTFORM (AUSKRAGENDE) **13875**
PLATTIEREN **9503**
PLATTIEREN **9479**
PLATTIERUNG **2447**
PLATTIERUNG (MECHANISCHE) **8108**
PLATZBEDARF **11829**
PLATZEN **1767**
PLATZERSPARNIS **5774**
PLEUELKOPF **2963**
PLEUELSTANGE **2962**
PLOMBE **7480**
PLOMBE (MIT EINER) VERSEHEN **11077**
PLOMBIEREN **11077**
PLUMBAGO **9542**
PLUNGER **9545**
PLUNGERPUMPE **9546**
PLUNSCHEPUMPE **9546**
PLUNSCHER **9545**
PLUSPLATTE **9658**
PLUSZEICHEN **9661**
PNEUMATISCH **9549**
POCHSTEMPEL **12055**

POCHWERK 12057
POL (ELEKTRISCHER) 9591
POL (GEOM.) 9592
POL (NEGATIVER) 8531
POL (POSITIVER) 9659
POL EINES MAGNETEN 9595
POLABSTAND 9579
POLARDREIECK 9581
POLARE 9577
POLARISATION 9582
POLARISATION 9589
POLARISATION (ELEKTROLYTISCHE) 4604
POLARISATIONSAPPARAT 9584
POLARISATIONSMIKROSKOP 9583
POLARISATIONSPRISMA 9586
POLARISATOR 9586
POLARISKOP 9584
POLARITÄT 9587
POLARKOORDINATEN 9578
POLAROGRAPHIE 9590
POLE (GLEICHNAMIGE) 7581
POLE (UNGLEICHNAMIGE) 13392
POLEN 9597
POLGEHÄUSE 5624
POLIERBAR 9599
POLIEREISEN 1757
POLIEREN 9598
POLIEREN 9605
POLIEREN 9600
POLIEREN (ELEKTROLYTISCHES) 4605
POLIERFASS 9602
POLIERKUGELN 1759
POLIERMASCHINE 1686
POLIERMASCHINE 9603
POLIERMITTEL 11
POLIERMITTEL 22
POLIERMITTEL 9601
POLIERPASTE 2827
POLIERROLLEN (STOFFBEKLEIDETE) 2539
POLIERROT 7148
POLIERSCHEIBE 9604
POLIERSCHEIBE 1693
POLIERSCHIEFER 13220
POLIERSTAHL 1757
POLIERSTICH 11541
POLIERSTICH 11542
POLIERSTOCK 632
POLIERTONNE 9602
POLIERTROMMEL 9602
POLONIUM 9606
POLSCHUHEN 9593

POLSTÄRKE 9596
POLSTRAHL 9580
POLSUCHPAPIER 9594
POLYEDER 9612
POLYGON 9609
POLYGON (EINGESCHRIEBENES) 6964
POLYGON (IRREGULÄRES) 7164
POLYGON (REGELMÄSSIGES) 10416
POLYGON (REGULÄRES) 10416
POLYGON (UMSCHRIEBENES) 2436
POLYGON (UNREGELMÄSSIGES) 7164
POLYGONSEITE 11414
POLYGONZUG 9611
POLYKRISTALLINES METALL 9608
POLYMER 9613
POLYMEREKÖRPER 9613
POLYMERIE 9616
POLYMERISATION 9614
POLYMERISIEREN (SICH) 9615
POLYSULFID DES KALIUMS 9697
POLYTROPE 9618
PONTONSMANNLOCH 9620
PORE 9627
PORENVERMUTTUNGSMITTEL 616
PORIG 9630
PORIGKEIT 9629
PORÖS 9630
POROSITÄT 9629
PORPHYR 9632
PORPHYRTUFF 9631
PORTALFRÄSWERKE 9642
PORTALKRAN 13224
PORTALKRAN 9641
PORTLANDZEMENT 9644
PORZELLAN 9625
PORZELLANERDE 2340
PORZELLANTIEGEL 9626
PORZELLANTON 2340
POSITIONIERZEIT 9651
POSITIONSSTEUERUNG 9650
POSITIVELEKTRISCH 4641
POSITRON 9662
POST-PROZESSOR 9665
POTENTIAL 9708
POTENTIALDIFFERENZ 9709
POTENTIALGEFÄLLE 13594
POTENTIOMETER 9711
POTENZ (MATH.) 9729
POTENZ (VIERTE) 5608
POTENZIEREN 10180
POTTASCHE 9679

POTTASCHE (GEREINIGTE) 9143
PRÄGEN 12056
PRÄGEN 12059
PRÄGEN 2651
PRÄGEPOLIEREN 1758
PRÄGEPRESSE 2653
PRÄGEPRESSE 12060
PRÄGEPROBE 12061
PRÄGESTEMPEL 2652
PRÄGESTEMPEL 3883
PRÄGESTEMPEL 5580
PRÄGUNG 3894
PRÄGWERK 12060
PRÄPARAT 9784
PRÄPARIERSALZ 11732
PRASEODYM 9750
PRÄZESSION 9754
PRÄZIPITAT 9759
PRÄZISIONSGUSS 9765
PRÄZISIONSHÖHENMESSER 9768
PRÄZISIONSLEHRE 9766
PRÄZISIONSMASCHINE 9770
PRÄZISIONSMASSSTAB 9767
PRÄZISIONSSCHLEIFMASCHINE 6113
PRÄZISIONSWAAGE 9764
PRESSASBEST 747
PRESSBACKEN 6120
PRESSBLOCK (IN DER STRANGPRESSE) 4371
PRESSE 9795
PRESSE (GERADSEITIGE MECHANISCHE) 12316
PRESSE (HYDRAULISCHE) 6693
PRESSEN 9806
PRESSEN (MIT DER SCHMIEDEPRESSE) 9800
PRESSEN (MIT DER SCHMIEDEPRESSE) 9807
PRESSKÖRPER 6088
PRESSKÖRPER 2791
PRESSLING 1623
PRESSLING 6088
PRESSLING 2791
PRESSLUFT 2839
PRESSLUFT-AUSWURFVORRICHTUNGEN 9796
PRESSLUFTDÜSE 275
PRESSLUFTHAMMER 9553
PRESSLUFTLEITUNG 2845
PRESSLUFTLOKOMOTIVE 2844
PRESSLUFTMOTOR 2843
PRESSLUFTNIETHAMMER 9555
PRESSLUFTSTAMPFER 286
PRESSLUFTVENTIL 9557
PRESSLUFTWERKZEUG 9556
PRESSMITTEL 1250

PRESSRING 3883
PRESSSCHMIEDEN 9798
PRESSSCHMIERUNG 9818
PRESSSCHWEISSEN 9842
PRESSSCHWEISSEN (ALUMINOTHERMISCHES) 9837
PRESSSITZ 5527
PRESSSPAN 9801
PRESSSTAHL 2849
PRESSTORF 9804
PRESSZUSATZ 1250
PREUSSISCHBLAU 1206
PRIMÄRELEMENT 9846
PRIMÄRSTROMVERHÄLTNIS 9850
PRIMITIVE TRANSLATION 9865
PRIMZAHL 9860
PRISMA 9875
PRISMA (DREISEITIGES) 13192
PRISMA (FÜNFSEITIGES) 9173
PRISMA (SCHIEF ABGESCHNITTENES) 9876
PRISMA (SECHSSEITIGES) 6460
PRISMATISCH 9879
PRISMATOID 9880
PRISMENHAMMER 4326
PRISMENSPEKTRUM 9878
PROBE 12770
PROBE 10915
PROBE 11858
PROBE (HYDRAULISCHE) 6695
PROBE (HYDROTATISCHE) 6727
PROBE(STÜCK) 12788
PROBEBELASTUNG 12781
PROBEBLOCK 12777
PROBEDRUCK 12786
PROBEENTNAHME 10916
PROBEFAHRBAHN 12794
PROBENEHMEN 10916
PROBESTAB 12772
PROBESTAB MIT EINKERBUNG 8675
PROBESTABDREHBANK 12771
PROBESTÜCK 12784
PROBESTÜCK 10915
PROBESTÜCK 12789
PROBIERGLAS 12780
PROBIERHAHN 12776
PROBIERHAHN 5875
PROBIERSTAND 12773
PROBIERVENTIL 5875
PRODUKT 9892
PRODUKT (MATH.) 9891
PRODUKT (SKALARES) 10966
PRODUKT (VEKTORISCHES) 13509

PRODUKTE (ERSTKLASSIGE) 9862
PRODUKTIONSKOSTEN 3207
PRODUKTIVITÄT 9897
PROFIL 9898
PROFILDRAHT 11847
PROFILE 6413
PROFILEISEN 11119
PROFILFRÄSER 9899
PROFILFRÄSMASCHINE 9900
PROFILKALIBER 11123
PROFILSTAHL 11125
PROFILWALZE 6136
PROFILWALZEN 10668
PROGRAMM 9902
PROGRAMMENDE 4724
PROGRAMMIEREN 9904
PROGRAMMIEREN (MASCHINELLES) 850
PROGRAMMIERTER HALT 9903
PROGRESSIVE ALTERUNG 9906
PROGRESSIVINDUKTIONSSCHWEISSEN 9908
PROJEKTION 9912
PROJEKTION (DIMETRISCHE) 3961
PROJEKTION (GNOMONISCHE) 5990
PROJEKTION (ISOMETRISCHE) 7188
PROJEKTION (TRIMETRISCHE) 13208
PROJEKTIONSEBENE 9438
PROJEKTIV 9914
PROJIZIEREN 9909
PROMETHIUM 6771
PROPAN 9927
PROPELLER 11041
PROPORTION 9933
PROPORTIONAL 9935
PROPORTIONALITÄT 9938
PROPORTIONALITÄTSFAKTOR 2631
PROPORTIONALITÄTSGRENZE 9936
PROPORTIONALITÄTSGRENZE 7590
PROPORTIONALZIRKEL 9937
PROPYLEN 9939
PROTAKTINIUM 9940
PROTON 9954
PROZESS 2291
PROZESSOR 9889
PRÜFDEHNGRENZE 9922
PRÜFDRUCK 12786
PRÜFGLAS 12780
PRÜFHAHN 12776
PRÜFLEHRE 12779
PRÜFLEHRE 8035
PRÜFLEHRE 10366
PRÜFPUMPE 12787

PRÜFSTAND 12773
PRÜFSTANDVERSUCH 1170
PRÜFSTEMPEL 3893
PRÜFSTÖPSEL 12785
PRÜFSTRASSE 6986
PRÜFSTÜCK 12784
PRÜFSTÜCK 1039
PRÜFTANK 12790
PRÜFUNG 13534
PRÜFUNG 12770
PRÜFUNG 6987
PRÜFUNG (AKUSTISCHE) 11815
PRÜFUNG (AMTLICHE) 8740
PRÜFUNG (MAGNETOGRAPHISCHE) 7931
PRÜFUNG (MECHANISCHE) 8112
PRÜFUNG (PHYSIKALISCHE) 9284
PRÜFUNG (ZERSTÖRENDE) 3806
PRÜFUNG (ZERSTÖRUNGSFREIE) 8630
PRÜFUNG BEI HOCHTEMPERATUR 6499
PRÜFUNG DES MATERIALS 12782
PRÜFUNG MIT DEM LÜCKENSUCHGERÄT 6535
PRÜFUNGSANALYSE 2280
PRÜFUNGSERGEBNIS 10512
PRÜFUNGSRESULTAT 10512
PRÜFUNGSZEUGNIS 12775
PSE 151
PSI 6873
PSYCHROMETER 9960
PUDDELEISEN 9962
PUDDELLUPPE 1345
PUDDELN 9964
PUDDELOFEN 9965
PUDDELROHEISEN 5544
PUDDELSTAHL 9963
PUFFER 1686
PUFFER-SPEICHER 1689
PUFFERBATTERIE 1687
PUFFERFEDER 1688
PUFFERSPEICHER 12736
PUFFERVORRICHTUNG 3611
PULSATIONSSCHWEISSVERFAHREN 9978
PULSIONSVENTILATION 9517
PULSOMETER 9980
PULVER (DENDRITISCHES) 3749
PULVER (FEINGEMAHLENES) 6789
PULVER (GROBGEMAHLENES) 2580
PULVER (KUGELIGES) 5979
PULVER (LEGIERTES) 367
PULVERBRENNSCHNEIDEN 9719
PULVERFORM (IN) 9728
PULVERFÖRMIG 9728

QUARZSAND **10082**
QUECKSILBER **8168**
QUECKSILBER-SCHALTER **8166**
QUECKSILBERBAROMETER **8150**
QUECKSILBERCHLORID **8153**
QUECKSILBERCHLORIDBAD **1369**
QUECKSILBERCHLORÜR **8159**
QUECKSILBERDAMPF-GLEICHRICHTER **8164**
QUECKSILBERDAMPFLAMPE **8167**
QUECKSILBERJODID **8155**
QUECKSILBERLUFTPUMPE **8163**
QUECKSILBERMANOMETER **8151**
QUECKSILBEROXYD **8157**
QUECKSILBEROXYD (SALPETERSAURES) **8156**
QUECKSILBEROXYDNITRAT **8156**
QUECKSILBEROXYDSULFAT **8158**
QUECKSILBEROXYDUL **8161**
QUECKSILBEROXYDUL (SALPETERSAURES) **8160**
QUECKSILBEROXYDULNITRAT **8160**
QUECKSILBEROXYDULSULFAT **8162**
QUECKSILBERSÄULE **2751**
QUECKSILBERTHERMOMETER **8152**
QUECKSILBERZELLE **8165**
QUELLEN DES HOLZES **12544**
QUELLWASSER **11996**
QUER ZUR FASERRICHTUNG **138**
QUERACHSE **13142**
QUERAUSBAUCHUNG **7421**
QUERBALKEN **3361**
QUERDEHNUNG **7421**
QUERDRUCKLAGER **7257**
QUERFINNE **3369**
QUERGLEITUNG **3380**
QUERHAUPT **3388**
QUERKEIL **3215**
QUERKONSOLE **4726**
QUERKONTRAKTION **7424**
QUERKRAFT **11291**
QUERNIETTEILUNG **13145**
QUERPROFIL **3377**
QUERRIEGEL **1432**
QUERRIEGEL **3361**
QUERRISS **3281**
QUERSÄGE **3365**
QUERSCHLITTEN **3379**
QUERSCHNITT **3378**
QUERSCHNITT **11111**
QUERSCHNITT (DREIECKIGER) **13188**
QUERSCHNITT (EIRUNDER) **8917**
QUERSCHNITT (ELLIPTISCHER) **4674**
QUERSCHNITT (GEDRÜCKTER) **3375**

QUERSCHNITT (GEFÄHRLICHER) **3609**
QUERSCHNITT (GEZOGENER) **3376**
QUERSCHNITT (HALBRUNDER) **11170**
QUERSCHNITT (HOHLRECHTECKIGER) **6550**
QUERSCHNITT (KEILFÖRMIGER) **13735**
QUERSCHNITT (KREISFÖRMIGER) **2410**
QUERSCHNITT (KREISRINGFÖRMIGER) **587**
QUERSCHNITT (MAKROSKOPISCHER) **7875**
QUERSCHNITT (NUTZBARER) **13422**
QUERSCHNITT (OVALER) **8917**
QUERSCHNITT (QUADRATISCHER) **12017**
QUERSCHNITT (RECHTECKIGER) **10302**
QUERSCHNITT (RINGFÖRMIGER) **587**
QUERSCHNITT (STERNFÖRMIGER) **12110**
QUERSCHNITT (TRAPEZFÖRMIGER) **13183**
QUERSCHNITT (WIRKSAMER) **4460**
QUERSCHNITT DES HOLZES **4718**
QUERSCHNITTFLÄCHE **11120**
QUERSCHNITTSVERGRÖSSERUNG **5869**
QUERSCHNITTSVERMINDERUNG **10362**
QUERSCHWINGUNG **13146**
QUERTRÄGER **3367**
QUERVERSCHIEBUNG **7420**
QUERVERSTREBUNG **1516**
QUERWALZEN **1643**
QUERWALZEN **3384**
QUERWALZWERK **1642**
QUERWAND **5692**
QUETSCH-UND SCHROTMÜHLEN **8296**
QUETSCHGRENZE **2873**
QUETSCHHAHN **8345**
QUOTIENT **10114**
RACHENLEHRE **11693**
RACHENLEHRE **6608**
RAD **13810**
RAD (ANGETRIEBENES) **4300**
RAD (KONISCHES) **1229**
RAD (ZYLINDRISCHES) **3591**
RAD(NABEN)KAPPE **6662**
RADACHSE **884**
RADANTRIEB **5186**
RADARM **13811**
RADAUSFLUCHTUNGSKONTROLLE **13086**
RÄDELERZ **1502**
RÄDER (AUSTAUSCHBARE) **7027**
RÄDERANTRIEB **13025**
RÄDERBOHRER **5905**
RÄDERDREHBANK **13816**
RÄDERFRÄSMASCHINE **5894**
RÄDERGETRIEBE **5907**
RÄDERGETRIEBE **13021**

RAUCHFANG 1455
RAUCHFANG 5467
RAUCHGASE 5470
RAUCHGASE 5732
RAUCHGASVORWÄRMER 4437
RAUCHKANAL 5467
RAUCHROHR 5248
RAUCHRÖHRENKESSEL 5249
RAUCHSCHIEBER 3603
RAUCHVERHÜTUNG 11677
RAUCHVERZEHREND 11673
RAUHBANK 7245
RAUHIGKEIT 9011
RAUHIGKEIT 10792
RAUM 13434
RAUM 2798
RAUM (SCHÄDLICHER) 2482
RAUM (TOTER) 2482
RAUMAUSDEHNUNG 13603
RAUMAUSDEHNUNGSZAHL 2625
RAUMBEANSPRUCHUNG 11829
RAUMBEDARF 11829
RAUMEINHEIT 13377
RÄUMEN 1637
RÄUMER 1636
RAUMERSPARNIS 5774
RAUMGITTER 11828
RAUMINHALT 13598
RAUMINHALTSBERECHNUNG 3444
RAUMKOORDINATEN 13203
RAUMKURVE 13307
RÄUMLICHE KURVE 13307
RÄUMMASCHINE (SENKRECHTE) 10190
RAUMPROZENT 9179
RAUMPUNKT 9560
RAUMTEIL 9179
RAUMTEMPERATUR 10713
RAUPE (FLACHE) 5401
RAUPENSCHLEPPER 13094
RAUSCHGELB 728
RAUSCHROT 10262
RAUTE 10568
RAUTENFLÄCHNER 10566
REAGENS 10257
REAGENZGLAS 12791
REAGENZGLAS 12780
REAGENZPAPIER 12783
REAKTANZ 10247
REAKTION 10248
REAKTION (ALKALISCHE) 337
REAKTION (BASISCHE) 337

REAKTION (CHEMISCHE) 2312
REAKTION (IRREVERSIBEL ELEKTROLYTISCHE) 7171
REAKTION (PERITEKTISCHE) 9199
REAKTION (SAURE) 122
REAKTIONSENERGIE 2305
REAKTIONSGRENZE 10249
REAKTIONSTURBINE 10250
REAKTOR (KERN) 10251
REALGAR 10262
REAUMURGRAD 3724
REAUMURSKALA 10267
REAUMURTHERMOMETER 10268
RECHENFEHLER 4824
RECHENMASCHINE 1859
RECHENSCHIEBER 11591
RECHENTABELLE 8052
RECHENTAFEL 8052
RECHNERPROGRAMM 10811
RECHNERSTEUERUNG (DIREKTE) 3981
RECHNERSTEUERUNG (DIREKTE) 2875
RECHNERSTEUERUNG (DIREKTE) 3985
RECHTECK 10299
RECHTECKFEDER 10306
RECHTECKFEDER (GESCHICHTETE) 7380
RECHTECKFEDER (ZUGESCHÄRFTE) 10307
RECHTECKGEWINDE 12030
RECHTECKLAST 10303
RECHTECKMASCHE 10304
RECHTKANT 10305
RECHTSDREHEND 3827
RECHTSDREHEND 2506
RECHTSDREHUNG 2507
RECHTSGEWICKELT 2508
RECHTSGEWINDE 10585
RECHTSGEWUNDEN 2508
RECHTSSCHWEISSUNG 907
RECHTSVERSCHIEBUNG 10583
RECHTSWEINSÄURE 3826
RECHTWINKLIG 10310
RECKALTERUNG 12324
RECKEN 4235
RECKEN 4219
RECKSPANNUNG 12370
REDUKTION (CHEM.) 10356
REDUKTION (ELEKTROLYTISCHE) 4607
REDUKTIONS... 10345
REDUKTIONSATMOSPHÄRE 10347
REDUKTIONSFLAMME 4882
REDUKTIONSFLANSCH 10350
REDUKTIONSGETRIEBE 11873
REDUKTIONSMITTEL 10346

REDUKTIONSMUFFE **3962**
REDUKTIONSROHR **10352**
REDUKTIONSTEMPERATUR **10354**
REDUKTIONSWÄRME **6362**
REDUKTIONSZIRKEL **9937**
REDUNDANZ **10364**
REDUZIER... **10345**
REDUZIER-ELEMENTE (KEGELSTUMPFE) **2950**
REDUZIERANSCHLUSS **13121**
REDUZIEREN (CHEM.) **10342**
REDUZIEREND **10345**
REDUZIERMASCHINE **12533**
REDUZIERMUFFE **10348**
REDUZIERNIPPEL **10351**
REDUZIEROFEN **10359**
REDUZIERSTÜCK **1773**
REDUZIERVENTIL **10355**
REDUZIERWALZSTRASSE **10353**
REDUZIERWALZWERK **11508**
REFBUNGSKUPPLUNG **5672**
REFLEKTIEREN **10382**
REFLEKTION **10387**
REFLEKTOR **10394**
REFLEXION **10387**
REFLEXION (TOTALE) **13060**
REFLEXION DES LICHTES **10390**
REFLEXIONS-WASSERSTANDSANZEIGER **10395**
REFLEXIONSGITTER **10388**
REFLEXIONSGLAS **10396**
REFLEXIONSVERHÄLTNIS **10391**
REFLEXIONSVERLUST **10389**
REFLEXIONSVERMÖGEN **10393**
REFLEXIONSVERMÖGEN **10392**
REFLEXIONSWINKEL **525**
REFRAKTION **10400**
REFRAKTIONSVERMÖGEN **10403**
REFRAKTOMETER **10404**
REGELABWEICHUNG **3819**
REGELANTRIEBSTEUERUNG **13492**
REGELDRUCK **9832**
REGELLEISTUNG **8658**
REGELLOS ORIENTIERT **10199**
REGELN **182**
REGELUNG **195**
REGELVENTIL **12890**
REGELVORRICHTUNG **10418**
REGELVORRICHTUNG **3052**
REGELWIDERSTAND **10563**
REGENERATIVFEUERUNG **10410**
REGENERATOR **10412**
REGENRINNENZANGE **4286**

REGENWASSER **10179**
REGENWASSERFLECKEN **10178**
REGISTRIEREN **10288**
REGISTRIERENDES INSTRUMENT **10289**
REGISTRIERTROMMEL **10290**
REGISTRIERVORRICHTUNG **10294**
REGISTRIERWALZE **10290**
REGLER **10421**
REGLER (ASTATISCHER) **775**
REGLER (ELEKTRISCHER) **4527**
REGLER (ELEKTRONISCHER) **4635**
REGLER (ISOCHRONER) **7180**
REGLER (PSEUDOASTATISCHER) **9957**
REGLER (SELBSTTÄTIGER) **843**
REGLER (STATISCHER) **12135**
REGLER MIT GEKREUZTEN STANGEN **3386**
REGLERARM **6014**
REGLERFEDER **6019**
REGLERGEHÄUSE **6016**
REGLERHÜLSE **6017**
REGLERKUGEL **6015**
REGLERMUFFE **6017**
REGLERSPINDEL **6018**
REGLERWELLE **6018**
REGULATOR **10421**
REGULIEREN **182**
REGULIERUNG **195**
REGULIERWIDERSTAND **10563**
REIBAHLE **10263**
REIBAHLE **1636**
REIBAHLE (GENUTETE) **5485**
REIBAHLE (GERADEGENUTETE) **12305**
REIBAHLE (GERIFFELTE) **5197**
REIBAHLE (HARTMETALLBESTÜCKTE) **1935**
REIBAHLE (KEGELIGE) **12658**
REIBAHLE (KONISCHE) **12658**
REIBAHLE (NACHSTELLBARE) **4923**
REIBAHLE (VERSTELLBARE) **188**
REIBAHLENHALTER **10264**
REIBER **9532**
REIBER **9221**
REIBGETRIEBE **5674**
REIBKEGELGETRIEBE **1224**
REIBKEGELRAD **1225**
REIBRAD **5682**
REIBRAD (ZYLINDRISCHES) **12007**
REIBRÄDERWENDEGETRIEBE **10546**
REIBROLLE **5679**
REIBSCHALE **8395**
REIBSCHEIBE **5673**
REIBUNG **5668**

REIBUNG (GLEITENDE) **11601**
REIBUNG (INNERE) **7065**
REIBUNG (INNERE) **5473**
REIBUNG (MAGNETISCHE) **6747**
REIBUNG (ROLLENDE) **10700**
REIBUNG (RUHENDE) **5678**
REIBUNG DER BEWEGUNG **5677**
REIBUNG DER RUHE **5678**
REIBUNG MITNEHMEN (DURCH) **4291**
REIBUNGERMÜDUNG **2200**
REIBUNGSARBEIT **13942**
REIBUNGSBREMSE **5669**
REIBUNGSDRUCK **9816**
REIBUNGSFLÄCHE **5687**
REIBUNGSGESPERRE **5676**
REIBUNGSGETRIEBE **5684**
REIBUNGSGRENZE **7309**
REIBUNGSKEGEL **5671**
REIBUNGSKLINKE **5675**
REIBUNGSKOEFFIZIENT **2629**
REIBUNGSKOEFFIZIENT FÜR GLEITENDE REIBUNG **2634**
REIBUNGSKOEFFIZIENT FÜR ROLLENDE REIBUNG **2633**
REIBUNGSKRAFT **5683**
REIBUNGSKUPPLUNG **11614**
REIBUNGSLOS **5688**
REIBUNGSMOMENT **8371**
REIBUNGSOXYDATION **5666**
REIBUNGSOXYDATION **5667**
REIBUNGSSÄGEN **5680**
REIBUNGSSCHALTWERK **5676**
REIBUNGSVERLUST **5685**
REIBUNGSVERLUST **7755**
REIBUNGSWÄRME **6343**
REIBUNGSWIDERSTAND **5686**
REIBUNGSWINKEL **527**
REIBUNGSZAHL **5670**
REICHGAS **8758**
REICHWEITE DES SCHALLES **10202**
REIFENLAUFFLÄCHE **13172**
REIFENSCHLAUCH **13246**
REIFKLOBEN **13565**
REIHE **11203**
REIHE (ARITHMETISCHE) **704**
REIHE (BINOMISCHE) **1260**
REIHE (DIVERGENTE) **4094**
REIHE (ENDLICHE) **5232**
REIHE (GEOMETRISCHE) **5930**
REIHE (HOMOLOGE) (CHEM.) **6562**
REIHE (KONVERGENTE) **3065**
REIHE (LOGARITHMISCHE) **7719**
REIHE (UNENDLICHE) **6909**

REIHENANFERTIGUNG **10462**
REIHENSCHALTUNG **11199**
REIHENSCHLUSSDYNAMO **11201**
REIHENSCHLUSSMOTOR **11202**
REINGEHALT EINER LEGIERUNG **5209**
REINIGEN **10370**
REINIGEN **10377**
REINIGER **10020**
REINIGER (ALKALISCHER) **2472**
REINIGUNG (ALKALISCHE) **334**
REINIGUNG (ANODISCHE) **592**
REINIGUNG (ELEKTROLYTISCHE) **4589**
REINIGUNG (MECHANISCHE) **8096**
REINIGUNG DES WASSERS **10019**
REINIGUNGS- UND POLIERMITTEL **2475**
REINIGUNGS- UND SCHNEIDEMASCHINEN FÜR WURZELN UND RÜBEN **10715**
REINIGUNGSMITTEL **10995**
REINIGUNGSMITTEL **2473**
REINIGUNGSMITTEL **2471**
REINIGUNGSMITTEL **3813**
REINIGUNGSTROMMEL **13261**
REINIGUNGSWIRKUNG **10017**
REISEGESCHWINDIGKEIT **3417**
REISEOMNIBUS **8412**
REISSBRETT **4231**
REISSEN **10859**
REISSEN DES HOLZES **11948**
REISSFEDER **4241**
REISSLÄNGE **1584**
REISSMASS **8016**
REISSMODELL **8016**
REISSNADEL **11068**
REISSNAGEL **4242**
REISSSCHIENE **12581**
REISSZEUG **8051**
REISSZWECKE **4242**
REITSTOCK **6328**
REITSTOCK **12599**
REITSTOCKSPITZE (UMLAUFENDE) **10560**
REKALESZENZ **10274**
REKRISTALLISATION **10297**
REKRISTALLISATIONSTEMPERATUR **10298**
REKTIFIKATION (CHEM.) **10311**
REKTIFIZIEREN (CHEM.) **10318**
REKUPERATIVOFEN **10321**
REKUPERATIVOFEN **10410**
REKUPERATOR **10322**
RELATIVE FEUCHTIGKEIT DER LUFT **10433**
RELIEFARBEIT **4684**
RELUKTANZ **10444**

RÖHRENSTREIFEN **11527**
RÖHRENTOUR **9357**
RÖHRENWALZWERK **9358**
ROHRFLANSCH **9351**
ROHRFORMSTÜCK **9350**
ROHRFRÄSER **1765**
ROHRFRÄSER ZUM AUSSENFRÄSEN **4951**
ROHRFRÄSER ZUM INNENFRÄSEN **7059**
ROHRGEWINDE **9360**
ROHRGEWINDE-SCHNEIDEMASCHINE **9361**
ROHRHAKEN **9354**
ROHRKALIBER **2027**
ROHRKOLBEN **9545**
ROHRKONTROLLE DURCH TRENNUNG **4877**
ROHRKRÜMMER **1180**
ROHRLEITUNG **9357**
ROHRLEITUNG (UNTERIRDISCHE) **13354**
ROHRMEISSEL **11704**
ROHRMUFFE **11705**
ROHRMÜHLE **13252**
ROHRNAHT **11085**
ROHRNETZ **12579**
ROHRNORMALIEN **9359**
ROHRPLATTE **13254**
ROHRREINIGER **13247**
ROHRRIPPE **10570**
ROHRSCHELLE **13324**
ROHRSCHELLE **9345**
ROHRSCHELLE **9344**
ROHRSCHLANGE **9346**
ROHRSCHLANGE **2645**
ROHRSCHLÜSSEL **9366**
ROHRSCHRAUBSTOCK **9363**
ROHRSCHRAUBSTOCK **13256**
ROHRSTRANG **9357**
ROHRSTRECKE **9357**
ROHRSTÜCK **9292**
ROHRSTUTZEN **5341**
ROHRSTUTZEN **1557**
ROHRSTUTZEN (EINGELASSENER) **6969**
ROHRSTUTZEN (SELBSTVERSTEIFTER) **9012**
ROHRVENTIL **11579**
ROHRVERBINDUNG **9347**
ROHRVERBINDUNG (STUMPFGESCHWEISSTE) **1803**
ROHRVERKLEIDUNG **7369**
ROHRVERSCHRAUBUNG **13369**
ROHRWEITE **1464**
ROHRWEITEMASCHINE **9352**
ROHRZANGE **9362**
ROHSCHIENE **8441**
ROHSTAHL **3414**

ROHSTAHLBLOCK **12210**
ROHSTEIN **11878**
ROHSTOFF **10239**
ROHWASSER **10240**
ROHZINK **2785**
ROLLBAHN **6082**
ROLLBAHN (FÜR ROLLEITER) **10857**
ROLLBANDMASS **8092**
ROLLBEWEGUNG **10702**
ROLLBÖCKE **13231**
ROLLE **2645**
ROLLE **8435**
ROLLE **10671**
ROLLE **10679**
ROLLE **5287**
ROLLE (ALS HEBEZEUG) **9971**
ROLLE (TONNENFÖRMIGE) **1017**
ROLLE EINES FLASCHENZUGS **11301**
ROLLE EINES ROLLENLAGERS **1112**
ROLLEN **10665**
ROLLEN **10696**
ROLLEN **10695**
ROLLEN(NAHT)SCHWEISSEN **11086**
ROLLENBAHN (EINES LAGERS) **10689**
ROLLENBOHRER **1504**
ROLLENDER KREIS **5921**
ROLLENFÖRDERER **6082**
ROLLENKÄFIG **12040**
ROLLENKETTE **10684**
ROLLENKLOBEN **1333**
ROLLENLAGER **10680**
ROLLENLAGER MIT KEGELROLLEN **10683**
ROLLENLAGER MIT TONNENFÖRMIGEN ROLLEN **10681**
ROLLENLAGER MIT ZYLINDRISCHEN ROLLEN **10682**
ROLLENLAGERNADELN **1108**
ROLLENQUETSCHNAHT **8024**
ROLLENRICHTMASCHINE **10685**
ROLLENSPURLAGER **10691**
ROLLENSTÖSSEL **10692**
ROLLENTRANSPORTER **6082**
ROLLENTROMMELSATTEL **10690**
ROLLENZUG **9973**
ROLLGABELSCHLÜSSEL **2551**
ROLLGANG **6082**
ROLLKREIS **5921**
ROLLKURVE **10697**
ROLLQUESTCHE **4450**
ROLLWERKZEUG **3485**
ROLLWIDERSTAND **10703**
ROMANZEMENT **10707**
RÖNTGEN-FLUORESZENZ-ANALYSE **13999**

RÖNTGEN-REFLEXE **3926**
RÖNTGENANALYSE **13993**
RÖNTGENAPPARAT **13994**
RÖNTGENBEUGUNGSBILD **3441**
RÖNTGENBEUGUNGSDIAGRAMM **13997**
RÖNTGENBILD ODER DIAGRAM **9128**
RÖNTGENKRISTALLOGRAPHIE **13996**
RÖNTGENPHOTOGRAPHIE **10163**
RÖNTGENRÖHRE **14002**
RÖNTGENSPEKTROMETER **14000**
RÖNTGENSPEKTRUM **14001**
RÖNTGENSTRAHLEN **14003**
RÖNTGENSTRAHLENBÜNDEL **13995**
RÖNTGENUNTERSUCHUNG **13998**
ROOTGEBLÄSE **10729**
ROSETTENKUPFER **10748**
ROSOLSÄURE **10750**
ROSSHAAR **6604**
ROSSWERK **1731**
ROST **10863**
ROST **6096**
ROST (EISENOXYD) **10335**
ROSTBILDUNG **10865**
ROSTDICHTUNG **10867**
ROSTEN **10865**
RÖSTEN VON ERZEN **10650**
ROSTFLÄCHE **6068**
ROSTFUGE **278**
ROSTIG **10873**
ROSTKITT **7141**
RÖSTOFEN **10649**
ROSTSCHIEBER **6100**
ROSTSCHUTZ **10871**
ROSTSCHUTZ **9945**
ROSTSCHUTZMITTEL **10870**
ROSTSCHUTZMITTEL **10869**
ROSTSCHUTZMITTEL **10866**
ROSTSCHUTZMITTEL **3189**
ROSTSCHUTZMITTEL **2601**
ROSTSPALT **278**
ROSTSTAB **5237**
ROSTSTABEISEN **6098**
ROSTSTÄHLE **6069**
RÖSTUNG **1839**
ROSTVERHÜTEND **10868**
ROT (PARISER) **3355**
ROTATION **10765**
ROTATIONSDAMPFMASCHINE **10761**
ROTATIONSELLIPSOID **4672**
ROTATIONSFLÄCHE **12513**
ROTATIONSKÖRPER **11790**

ROTATIONSPARABOLOID **9042**
ROTATIONSPUMPE **10758**
ROTBRUCHPROBE **10331**
ROTEISENERZ **11865**
ROTEISENSTEIN **11865**
RÖTEL **10340**
ROTER ARSENIK **10262**
ROTER PHOSPHOR **10334**
ROTES BLEIOXYD **10333**
ROTES BLUTLAUGENSALZ **9684**
ROTES CHROMKALI **9677**
ROTES CHROMSALZ **9677**
ROTES CHROMSAURES KALI **9677**
ROTES JODQUECKSILBER **8155**
ROTES MANGANOXYD **13207**
ROTES PRÄZIPITAT **8157**
ROTES SCHWEFELARSEN **10262**
ROTES ZYANEISENKALIUM **9684**
ROTGARES LEDER **8704**
ROTGLÜHEND **10330**
ROTGLUT **10329**
ROTGUSS **10326**
ROTGUSS **1658**
ROTGUSS **12984**
ROTIEREN **10763**
ROTIERENDE DAMPFMASCHINE **10761**
ROTIERENDE WELLE **10764**
ROTKUPFERERZ **3476**
ROTMESSING **10325**
ROTMESSING **12984**
ROTMESSING **2824**
ROTMETALL **12984**
ROTÖL **1668**
ROTOR **10767**
ROTORPLATTE **3160**
ROTOXID **8157**
ROTWARM **10330**
RUBIDIUM **10829**
RÜBÖL **10204**
RÜCKANSICHT **4717**
RÜCKAUFNAHME LAUEDIAGRAMM **7435**
RÜCKBILDUNG **10554**
RÜCKBLICKSPIEGEL **10266**
RÜCKBOGEN **13323**
RÜCKEN **890**
RÜCKENLEHNE **917**
RÜCKENSÄGE **897**
RÜCKENSTÄUBER UND RÜCKENSPRITZGRERÄTE **11969**
RÜCKENWINKEL **2484**
RÜCKFENSTER **899**
RÜCKFLUSSVENTIL MIT KLAPPE **12545**

SÄGEZAHNKUPPLUNG 10960
SAIGERUNG 7635
SALINOMETER 10901
SALIZYLSÄURE 10899
SALMIAK 462
SALMIAKGEIST 660
SALMIAKSPIRITUS 660
SALPETER 10903
SALPETERGAS 8581
SALPETERSALZSÄURE 658
SALPETERSÄURE 8580
SALPETERSÄUREÄTHER 4850
SALPETERSAURES KALI 10903
SALPETERSAURES NATRON 2330
SALPETERSAURES SALZ 8579
SALPETERSSÄUREANHYDRID 8590
SALPETRIGSÄUREANHYDRID 8592
SALPETRIGSAURES KALI 9691
SALPETRIGSAURES SALZ 8586
SALZ (ALKALISCHES) 338
SALZ (CHEM.) 10904
SALZ (GERBSAURES) 12633
SALZ (KOHLENSAURES) 1968
SALZ (NEUTRALES) 8546
SALZ (UNTERSCHWEFLIGSAURES) 12852
SALZÄTHER 4849
SALZBAD 10905
SALZBADHÄRTUNG 10913
SALZBADHÄRTUNG 7651
SALZBADLÖTEN 10911
SALZBADOFEN 10912
SALZBILDNER 6214
SALZGEHALT 10906
SALZKRUSTE 3421
SALZLÖSUNG 10908
SALZSÄURE 6706
SALZSOLE 1617
SALZSPINDEL 10901
SALZSPRÜHNEBELPRÜFUNG 10909
SALZWAAGE 10901
SALZWASSER 10910
SAMARIUM 10914
SÄMASCHINEN 11132
SÄMISCHLEDER 2234
SAMMELBEHÄLTER 10471
SAMMELBEHÄLTER 12292
SAMMELLEITUNG 7979
SAMMELLINSE 2910
SAMMELLINSE 2909
SAMMELPLATTEN 2726
SAMMELROHR 2725

SAMMELSCHIENE 1770
SAMMLER 76
SAMMLERBATTERIE 77
SAMMLERLOKOMOTIVE 79
SAND 10917
SAND (FEINKÖRNIGER) 5199
SAND (GROBKÖRNIGER) 2574
SAND (TROCKENER) 4255
SANDARAK 10927
SANDBAD 10918
SANDBODEN 10930
SANDEINSCHLUSS IM GUSS 3998
SANDFILTER 10923
SANDFORM 10924
SANDFORMMASCHINE 8429
SANDGUSS 10922
SANDHAKEN 5772
SANDKOHLE 4344
SANDPAPIER 10925
SANDPAPIERMASCHINE 10926
SANDSCHLEUDER 209
SANDSIEB 11628
SANDSTEIN 10928
SANDSTEIN (ROTER) 10336
SANDSTRAHL BEARBEITEN (MIT) 10919
SANDSTRAHLEN 1316
SANDSTRAHLGEBLÄSE 10920
SANDTRAHLUNG 10921
SANDWICHWALZEN 10929
SAPONIT 10935
SATTDAMPF 10942
SATTEL 10876
SATTELBEFESTIGUNG 3288
SATTELENDE 6597
SATTELHALTER 10877
SATTELSCHLEPPER 11178
SATTELZUG 13092
SÄTTIGEN 10939
SÄTTIGUNG 10943
SÄTTIGUNG (MAGNETISCHE) 7916
SÄTTIGUNGSGRAD 3722
SÄTTIGUNGSGRENZE 7591
SÄTTIGUNGSTEMPERATUR 10945
SÄTTIGUNGSDRUCK 10944
SATZ 10226
SATZ 11213
SATZ 1332
SATZ 1059
SATZADRESSEEINGABEFORMAT 1334
SATZFRÄSER 5800
SATZNUMMER 11193

SATZRÄDER 7027
SATZUNTERDRÜCKUNG 1337
SAUERKLEESALZ 10907
SAUERKLEESÄURE 8969
SAUERSTOFF 8980
SAUERSTOFF ENTZIEHEN (DEN) 3763
SAUERSTOFF-AZETYLEN 8979
SAUERSTOFF-WASSERSTOFF-SCHWEISSUNG 8988
SAUERSTOFFENTZIEHUNG 3762
SAUERSTOFFGEBLÄSE 8986
SAUERSTOFFION 556
SAUERSTOFFLANZE 8981
SAUERSTOFFSCHNEID-MASCHINE (AUTOGENE) 833
SAUERSTOFFVERFAHREN 8982
SAUG- UND DRUCKLÜFTER 2768
SAUG- UND DRUCKVENTILATOR 2768
SAUG-DRUCKVENTIL 9843
SAUGEN 12428
SAUGENDER LÜFTER 12429
SAUGGAS 4194
SAUGGASGENERATOR 12430
SAUGHEBER 11517
SAUGHÖHE EINER PUMPE 12432
SAUGHUB 12435
SAUGHUB 7008
SAUGKLAPPE 12436
SAUGKOPF 12332
SAUGKORB 12332
SAUGLEITUNG 12433
SAUGLEITUNG 6948
SAUGLÜFTER 12429
SAUGLÜFTUNG 13450
SAUGPUMPE 12434
SAUGROHR 12433
SAUGROHR 6945
SAUGSIEB 12332
SAUGTRICHTER 9340
SAUGVENTIL 12436
SAUGZUG 6890
SÄULE 2748
SÄULE (VOLTASCHE) 13596
SÄULEFÜHRUNGSGESTELL 3890
SÄULENBOHRMASCHINE 9312
SÄULENFUSS 1035
SÄULENGERÜST 2759
SÄULENKAPITÄL 1926
SÄULENKOPF 2749
SÄULENLAGER 9663
SÄULENSTAMM 11243
SÄURE 104
SÄURE (ARSENIGE) 727

SÄURE (ORGANISCHE) 8864
SÄURE (SALPETRIGE) 8596
SÄURE (SCHWEFLIGE) 12450
SÄURE (UNTERCHLORIGE) 6742
SÄUREÄTHER 4837
SÄUREBESTÄNDIG 124
SÄUREBESTÄNDIG 128
SÄUREBESTÄNDIGKEIT 10496
SÄUREBILDEND 130
SÄUREBILDNER 115
SÄUREDÄMPFE 127
SÄUREFEST 128
SÄUREFEST 124
SÄUREFESTE LEGIERUNGEN 125
SÄUREFESTIGKEIT 10496
SÄUREFREI 5635
SÄUREGEHALTSBESTIMMUNG 131
SÄUREMESSER 89
SÄUREN 132
SÄUREPUMPE 121
SAUSCHWANZHAKEN 9301
SCHABEMASCHINE 11012
SCHABEN 11008
SCHABEN 11011
SCHABER 11009
SCHABLONE 9128
SCHABLONE 12238
SCHABLONE 12734
SCHABLONE (FIXE) 5293
SCHABLONENDREHBANK 3138
SCHACHT 9398
SCHACHT 12046
SCHACHTHYDRANT 6672
SCHACHTOFEN 11241
SCHADHAFT 5057
SCHÄDLICHER WIDERSTAND 10497
SCHAFT 12237
SCHAFT 11260
SCHAFT 11257
SCHAFT DER SCHUBSTANGE 11261
SCHAFTFRÄSER 11259
SCHAFTFRÄSER 11783
SCHAFTFRÄSER 4723
SCHAFWOLLE 13932
SCHAKE 7627
SCHÄKEL 11232
SCHÄKEL 2491
SCHAKENKETTE 8855
SCHALE 11303
SCHALE 9029
SCHALE 9155

SCHEIBENFRÄSERPAAR **12301**
SCHEIBENKOLBEN **11791**
SCHEIBENKRANZ **10595**
SCHEIBENKUPPLUNG **5338**
SCHEIBENKURBEL **4433**
SCHEIBENLAGER **12903**
SCHEIBENMESSER **2417**
SCHEIBENPFLÜGE UND ANBAUPFLÜGE **9523**
SCHEIBENRAD **4012**
SCHEIBENRAD (GERADES) **5379**
SCHEIBENRAD (GEWÖLBTES) **4037**
SCHEIBENVENTIL **4011**
SCHEIBENWASCHER **13870**
SCHEIBENWISCHER **13871**
SCHEIBENWÖLBUNG **3398**
SCHEIBENZAPFEN **12906**
SCHEIDEWAND **3870**
SCHEIDEWAND **4110**
SCHEIDEWASSER **8580**
SCHEIDUNG **9091**
SCHEIDUNG (ELEKTROLYTISCHE) **4602**
SCHEIDUNG (ELEKTROMAGNETISCHE) **4618**
SCHEINBILD **13572**
SCHEINDICHTE **1725**
SCHEINDICHTE **642**
SCHEINWERFER **11091**
SCHEINWERFER **6326**
SCHEINWERFER **6327**
SCHEINWERFER (EINGELASSENER) **5480**
SCHEINWIDERSTAND **6794**
SCHEIT **1243**
SCHEITEL **638**
SCHEITELDRUCK **12908**
SCHEITELTANGENTE **12617**
SCHELF.... **8738**
SCHELLACK **11338**
SCHELLACKFIRNIS **10998**
SCHELLACKPAPIER **13506**
SCHELLEISEN **10626**
SCHELLHAMMER **4979**
SCHENKEL EINES TAUES **12338**
SCHENKEL EINES WINKELEISENS **13728**
SCHENKEL EINES WINKELS **7512**
SCHERBACKEN **11282**
SCHERBEANSPRUCHUNG **11295**
SCHERBLATT **11282**
SCHERE **11297**
SCHERE **1333**
SCHERE **11279**
SCHERE (FLIEGENDE) **5495**
SCHERE (ROTIERENDE) **10759**

SCHERFESTIGKEIT **11285**
SCHERFESTIGKEIT **13330**
SCHERFLÄCHE **11284**
SCHERKLINGE **11282**
SCHERKRAFT **11291**
SCHERMESSER **11282**
SCHERSPANNUNG **11296**
SCHERUNG **11290**
SCHERVERSUCH **11287**
SCHERVERSUCH **11949**
SCHERWINKEL **11281**
SCHEUERTROMMEL **9602**
SCHICHT **5166**
SCHICHT **2590**
SCHICHT **7442**
SCHICHT (ARBEITERGRUPPE) **11348**
SCHICHT (ARBEITSZEIT) **11347**
SCHICHT (DÜNNE) **5171**
SCHICHTARBEIT **11350**
SCHICHTENRISS **7383**
SCHICHTENSTRÖMUNG **9055**
SCHICHTFLÄCHE **7527**
SCHICHTKORROSION **7443**
SCHICHTLINIE **7444**
SCHICHTUNG **7382**
SCHIEBEDECKEL **11604**
SCHIEBEKLINKE **2496**
SCHIEBELEHRE **13541**
SCHIEBER **11592**
SCHIEBER MIT MECHANISCHER ABSPERRUNG **9062**
SCHIEBERÄDERGETRIEBE **11598**
SCHIEBERDIAGRAMM **11593**
SCHIEBERSPIEGEL **4996**
SCHIEBERSTEUERUNG **11594**
SCHIEBESITZ **10025**
SCHIEBLEHRE **13537**
SCHIEBUNG **11288**
SCHIEFER (BAUW.) **11571**
SCHIEFER (GEOL.) **10999**
SCHIEFERBRUCH **11615**
SCHIEFERÖL **11256**
SCHIEFERTEER **11255**
SCHIEMANNSGARN **12005**
SCHIENE **10175**
SCHIENEN- U. LASCHENBOHRMASCHINE **10176**
SCHIENENSCHRAUBE **11043**
SCHIENENSCHRAUBEN **11575**
SCHIENENSTAHL (UMGEWALZTER) **10177**
SCHIESSBAUMWOLLE **6176**
SCHIESSPULVER **6181**
SCHIFFBAU **8510**

SCHIFFHOBEL 2801
SCHIFFSBLECH 11356
SCHIFFSKESSEL 8003
SCHIFFSKETTEN 8004
SCHIFFSKOPF 8007
SCHIFFSMASCHINE 8005
SCHIFFSNAGEL 1382
SCHIFFSPECH 9400
SCHIMMEL 8271
SCHIMMELBILDUNG 5575
SCHIPPE 11385
SCHIRM 11346
SCHIRM 11018
SCHIRM (OPT.) 11019
SCHLABBERVENTIL 8933
SCHLACKE 11561
SCHLACKE 2397
SCHLACKE (EINGESCHLOSSENE) 11567
SCHLACKE (EINGEWALZTE) 11567
SCHLACKE (GEKÖRNTE) 6055
SCHLACKE (GRANULIERTE) 6055
SCHLACKENABZUG 3803
SCHLACKENBETON 11565
SCHLACKENBLECH 3596
SCHLACKENEINSCHLUSS 4772
SCHLACKENFORM 2399
SCHLACKENFORMKÜHLER 8380
SCHLACKENFREI 5638
SCHLACKENHALTIG 11562
SCHLACKENKRANZBILDUNG 1604
SCHLACKENLOCH 2399
SCHLACKENLOCH 4332
SCHLACKENSTEIN 11563
SCHLACKENWOLLE 11566
SCHLACKENZEMENT 11564
SCHLACKENZUSCHLAG 5475
SCHLAFKOJE 1739
SCHLAG 12375
SCHLAG 6783
SCHLAG (ELEKTRISCHER) 4529
SCHLAGBIEGEPROBE 4328
SCHLAGBIEGEPROBE MIT EINGEKERBTEN PROBESTÜCKEN 1190
SCHLÄGEL 7962
SCHLAGEN 6224
SCHLAGEN DER PUMPE 7331
SCHLAGEN DES RIEMENS 13825
SCHLAGFESTIGKEIT 6785
SCHLAGFLIESSPRESSEN 6784
SCHLAGFRÄSER 5494
SCHLAGLOT 6279

SCHLAGNIETHAMMER 9555
SCHLAGPRESSE (MECHANISCHE) 9182
SCHLAGSCHATTEN 13339
SCHLAGSCHWEISSEN 9183
SCHLAGSCHWEISSEN (ELEKTROMAGNETISCHES) 4617
SCHLAGSTÄRKE 12757
SCHLAGSTÖCKCHEN 632
SCHLAGVERSUCH 6786
SCHLAGVERSUCH 4328
SCHLAMM 11609
SCHLAMM 8444
SCHLAMM 11655
SCHLAMM 11654
SCHLÄMMEN 4681
SCHLÄMMEN 2550
SCHLAMMHAHN 1357
SCHLAMMIG 8445
SCHLÄMMKREIDE 13841
SCHLAMMLOCH 8442
SCHLÄMMUNG 4681
SCHLAMMVENTIL 1358
SCHLANGENBOHRER 4160
SCHLANGENROST 3200
SCHLANKHEIT 11580
SCHLANKHEIT (EINES TRÄGERS) 11581
SCHLANKHEITSGRAD 11582
SCHLAUCH 9339
SCHLAUCH 6614
SCHLAUCH (GEPANZERTER) 714
SCHLAUCHGEWEBERIEMEN 5397
SCHLAUCHKLAMMER 6611
SCHLAUCHKLEMME 6611
SCHLAUCHKUPPLUNG 6612
SCHLAUCHSCHELLE 6611
SCHLAUCHSCHELLE 6610
SCHLAUFE 8651
SCHLECHTAUSGELAUFEN 8321
SCHLEIERBILDUNG 2541
SCHLEIERBILDUNG 1347
SCHLEIERBILDUNG 6313
SCHLEIFBAND 18
SCHLEIFE 11633
SCHLEIFE EINER KURVE 7741
SCHLEIFEN 11275
SCHLEIFEN 6106
SCHLEIFEN 9605
SCHLEIFEN (SPITZENLOSES) 2156
SCHLEIFENKURVE 7514
SCHLEIFER 6105
SCHLEIFGRAT 13890
SCHLEIFHÄRTE 14

SCHLEIFKOLBEN 9392
SCHLEIFKORN 21
SCHLEIFKÖRPER 26
SCHLEIFKÜHLMITTEL 6109
SCHLEIFMASCHINE (SPITZENLOSE) 2174
SCHLEIFMASCHINE (ZUM PUTZEN UND SCHLICHTEN) 6113
SCHLEIFMATERIALKORNGRÖSSE 23
SCHLEIFMITTEL 22
SCHLEIFMITTEL 17
SCHLEIFMITTEL 6107
SCHLEIFMITTEL 9601
SCHLEIFMITTEL (MILDES) 8269
SCHLEIFMITTEL (NATÜRLICHES) 8499
SCHLEIFPLATTE 6114
SCHLEIFPULVER 6110
SCHLEIFRING 11618
SCHLEIFSCHEIBE 6118
SCHLEIFSCHEIBE 6115
SCHLEIFSCHEIBE 4010
SCHLEIFSCHEIBE 4006
SCHLEIFSCHEIBEN-AUSGLEICHAPPARATE 6116
SCHLEIFSCHMIERSTOFF 6112
SCHLEIFSCHMIRGELSCHEIBE 20
SCHLEIFSCHMIRGELSCHEIBEN 26
SCHLEIFSPITZEN 24
SCHLEIFSPUREN 15
SCHLEIFSTEIN 19
SCHLEIFSTEIN 13824
SCHLEIFSTEIN 6118
SCHLEIMHARZ 6173
SCHLEPPER 11010
SCHLEPPER 13097
SCHLEPPKETTE 11010
SCHLEPPKURVE 13098
SCHLEPPSEIL 11010
SCHLEUDER 6703
SCHLEUDERBREMSE 2175
SCHLEUDERGEBLÄSE 2179
SCHLEUDERGUSS 2176
SCHLEUDERMÜHLE 4045
SCHLEUDERPUMPE 2184
SCHLEUDERSTRAHLEN 6124
SCHLEUDERTHERMOMETER 13826
SCHLEUSE 7698
SCHLICHTE 3171
SCHLICHTEN 11682
SCHLICHTEN 5214
SCHLICHTFEILE 11680
SCHLICHTHAMMER 9455
SCHLICHTHAMMER 1734
SCHLICHTHOBEL 11687

SCHLICHTSTAHL 5228
SCHLIERE 5061
SCHLIESSKOFF (DEN) BILDEN 2516
SCHLIESSKOLBEN (FORM) 2535
SCHLIESSKOPF 10628
SCHLIFF 6106
SCHLIFF FÜR MIKROSKOPIE 11113
SCHLIFFFLÄCHE 6142
SCHLINGE 11610
SCHLITTEN 10188
SCHLITTEN MIT GLEITPLATTEN 11606
SCHLITTEN MIT ROLLEN 10694
SCHLITTENKREUZBEWEGUNGS-HANDRAD 13163
SCHLITTENLAGER 11596
SCHLITTENWINDE 13165
SCHLITZHEBEL 11638
SCHLITZMUTTER 11942
SCHLITZNAHT 11635
SCHLITZSÄGE 8193
SCHLITZSCHRAUBE 11637
SCHLITZSCHWEISSEN 11636
SCHLITZTROMMEL 3579
SCHLOSSER 5278
SCHLOSSERWERKSTATT 5282
SCHLOSSSCHRAUBE 2553
SCHLOT 2339
SCHLUBKLINKE 2496
SCHLÜPFEN 11612
SCHLÜPFEN DER RÄDER 11621
SCHLÜPFEN DES RIEMENS 11620
SCHLÜPFRIGKEIT EINES SCHMIERSTOFFES 7813
SCHLÜSSEL 9532
SCHLÜSSEL (GEKRÖPFTER) 509
SCHLÜSSELFEILE 13625
SCHLÜSSELFERTIG 13284
SCHLÜSSELLOCHDECKEL 4835
SCHLÜSSELLOCH 7287
SCHLÜSSELWEITE 11832
SCHLUSSLATERNE 12595
SCHLUSSLINIE 2536
SCHMALZÖL 7409
SCHMEIZTIEGEL 3401
SCHMELZ 4709
SCHMELZBAD 8137
SCHMELZBAR 5756
SCHMELZBARKEIT 5755
SCHMELZBEREICH 5761
SCHMELZBEREICH 8143
SCHMELZE 2076
SCHMELZE 6333
SCHMELZE 8134 .

SCHMELZE 8365
SCHMELZEN 5764
SCHMELZEN 8135
SCHMELZEN 5760
SCHMELZEN 8136
SCHMELZEN 11666
SCHMELZEN (KONGRUENTES) 2938
SCHMELZFARBE 4711
SCHMELZFUSS 8365
SCHMELZGESCHWINDIGKEIT 8144
SCHMELZKEGEL 11134
SCHMELZKEGEL 10050
SCHMELZKOKS 2662
SCHMELZLEGIERUNG 5757
SCHMELZLEGIERUNG 389
SCHMELZOFEN 11665
SCHMELZOFEN 11667
SCHMELZOFEN 8138
SCHMELZPFROPFEN 5758
SCHMELZPUNKT 8141
SCHMELZPUNKT 5759
SCHMELZSCHWEISSEN 5762
SCHMELZSCHWEISSEN 8643
SCHMELZTAUCHVERFAHREN 6659
SCHMELZTIEGELZANGE 3405
SCHMELZTOPF 8142
SCHMELZUNG 5764
SCHMELZUNG (EUTEKTISCHE) 4856
SCHMELZVERLUSTE 8140
SCHMELZWÄRME 6360
SCHMELZZONE 5763
SCHMIED 11671
SCHMIEDBAR 7960
SCHMIEDBARKEIT 7958
SCHMIEDBARMACHEN 7961
SCHMIEDE 5547
SCHMIEDEEISEN 13988
SCHMIEDEEISEN (ANORMALES) 9
SCHMIEDEESSE 5548
SCHMIEDEFEUER 5548
SCHMIEDEHAMMER 6235
SCHMIEDEHAMMER 11572
SCHMIEDEHAMMER 9552
SCHMIEDEHAMMER (MECHANISCHER) 5559
SCHMIEDEHERD 5548
SCHMIEDEISEN (PERLITISCHES) 2047
SCHMIEDEKOHLE 5543
SCHMIEDEMASCHINE 5561
SCHMIEDEMESSING 5558
SCHMIEDEMESSING 8486
SCHMIEDEN 5542

SCHMIEDEN 5555
SCHMIEDEN 5557
SCHMIEDEN (IM GESENK) 12530
SCHMIEDEPRESSE 5562
SCHMIEDEPRESSE (HYDRAULISCHE) 6687
SCHMIEDEPROBE 5545
SCHMIEDESCHWEISSUNG 5546
SCHMIEDESINTER 6217
SCHMIEDESPANNUNG 5564
SCHMIEDESTAHL 5554
SCHMIEDESTAHLANSCHLUSS 5549
SCHMIEDESTÜCK 5556
SCHMIEDESTÜCK (SCHWERES) 6406
SCHMIEDESTÜCK (ÜBERENTGRATETES) 8322
SCHMIEDESTÜCKE 5566
SCHMIEDETEMPERATUR 5560
SCHMIEDETEMPERATUR 5565
SCHMIEDEWALZWERK 5563
SCHMIEDEWERKSTATT 5547
SCHMIEDEZANGE 1298
SCHMIEGE 1228
SCHMIEGSAM-STRECKBAR 4361
SCHMIEGUNGSEBENE 8887
SCHMIERBOHRER 8762
SCHMIERBOHRUNG 8761
SCHMIERBÜCHSE 8753
SCHMIERBÜCHSE 6085
SCHMIERDOCHT 7807
SCHMIERE (ALTE) 3308
SCHMIEREN 7808
SCHMIEREN 7801
SCHMIERFETT 7804
SCHMIERGEFÄSS 7812
SCHMIERGEFÄSS (UMLAUFENDES) 2182
SCHMIERKANNE 8749
SCHMIERKELCH 8753
SCHMIERLOCH 8761
SCHMIERMATERIAL 7800
SCHMIERMITTEL 7800
SCHMIERNAPF 8753
SCHMIERNIPPEL 6086
SCHMIERNUT 7810
SCHMIERÖL 7805
SCHMIERÖLPUMPE 8769
SCHMIERPRESSE 8105
SCHMIERRING 8790
SCHMIERRING 8774
SCHMIERSCHICHT 5167
SCHMIERSCHRAUBE 7811
SCHMIERSEIFE 11740
SCHMIERSTELLE 8789

SCHMIERSTOFF **7800**
SCHMIERUNG **7808**
SCHMIERUNG **7803**
SCHMIERUNG **7814**
SCHMIERUNG (BESTÄNDIGE) **3012**
SCHMIERUNG (SELBSTTÄTIGE) **848**
SCHMIERUNG (UNTERBROCHENE) **9195**
SCHMIERVORRICHTUNG **7812**
SCHMIERVORRICHTUNG (SELBSTTÄTIGE) **851**
SCHMIERWERT **7806**
SCHMIRGEL **4692**
SCHMIRGEL **4699**
SCHMIRGEL (GESCHLÄMMTER) **5453**
SCHMIRGELLEINEN **4693**
SCHMIRGELLEINWAND **4693**
SCHMIRGELN **4695**
SCHMIRGELPAPIER **4696**
SCHMIRGELPULVER **4697**
SCHMIRGELSCHEIBE **4698**
SCHMUTZANHAFTUNG **3997**
SCHMUTZFÄNGER **12333**
SCHMUTZFÄNGER **8443**
SCHNAPPSCHALTER **11690**
SCHNARCHVENTIL **11694**
SCHNECKE **6430**
SCHNECKE **13979**
SCHNECKE **13970**
SCHNECKE **13969**
SCHNECKENBOHRER **11023**
SCHNECKENFEDER **5396**
SCHNECKENFLASCHENZUG **13973**
SCHNECKENFRÄSER **13972**
SCHNECKENFRÄSER **13971**
SCHNECKENGEBLÄSE **9929**
SCHNECKENRAD **13975**
SCHNECKENRADGETRIEBE **13974**
SCHNEEKETTEN **11696**
SCHNEELAST **11695**
SCHNEEWEISS **1003**
SCHNEIBRENNERDÜSE **3537**
SCHNEIDBACKEN **3896**
SCHNEIDBRENNER **3216**
SCHNEIDBRENNER **3539**
SCHNEIDBRENNER **5307**
SCHNEIDBRENNER (AUTOGEN) **5312**
SCHNEIDE **3528**
SCHNEIDE (STUMPFE) **4365**
SCHNEIDEARBEIT **1309**
SCHNEIDEINSATZ **3525**
SCHNEIDEISEN **11039**
SCHNEIDEN **9998**

SCHNEIDEN **3523**
SCHNEIDEN **10012**
SCHNEIDEN **3504**
SCHNEIDEN **11290**
SCHNEIDEN (AUTOGENES) **832**
SCHNEIDEN (BLECHE) **3513**
SCHNEIDEN (INNENGEWINDE) **12636**
SCHNEIDEN (SICH) (GEOM.) **7094**
SCHNEIDEN EINES AUSSENGEWINDES **3535**
SCHNEIDEN EINES GEWINDES **11030**
SCHNEIDEN EINES INNENGEWINDES **12678**
SCHNEIDENAUFHÄNGUNG **7326**
SCHNEIDENDE WELLEN (SICH) **7096**
SCHNEIDENLAGER **7325**
SCHNEIDFLÜSSIGKEIT **3530**
SCHNEIDFLÜSSIGKEITEN (CHEMISCHE) **2303**
SCHNEIDKANTE EINES WERKZEUGS **3528**
SCHNEIDKLINGE **11039**
SCHNEIDKLUPPE **12266**
SCHNEIDLIPPEN **7633**
SCHNEIDMASCHINE **833**
SCHNEIDÖL **3534**
SCHNEIDSCHEIBE **3542**
SCHNEIDSTAHL **13011**
SCHNEIDWERKZEUG **3538**
SCHNEIDWERKZEUGE AUS SCHNELLSTAHL **6475**
SCHNEIDWINKEL **3524**
SCHNELLARBEITSSTAHL **6497**
SCHNELLAUFEN **10833**
SCHNELLAUFENDE MASCHINE **6496**
SCHNELLÄUFER **6496**
SCHNELLBOHRER **10206**
SCHNELLBOHRMASCHINE **6495**
SCHNELLDREHSTAHL **6497**
SCHNELLDREHSTAHL **6474**
SCHNELLGANG **10208**
SCHNELLGANG **8930**
SCHNELLÖFFNUNGS-SCHIEBER **10103**
SCHNELLOT **11742**
SCHNELLRÜCKLAUFGETRIEBE **10106**
SCHNELLSCHLUSSVENTIL **10104**
SCHNELLTROCKNER **5360**
SCHNELLWAAGE **12225**
SCHNELLWECHSELEINRICHTUNG **10108**
SCHNITT **9898**
SCHNITT **11112**
SCHNITT **3503**
SCHNITT (GOLDENER) **8123**
SCHNITT (SCHIEFER) **8711**
SCHNITTEBENE **7095**
SCHNITTFLÄCHE **12511**

SCHRAUBENSCHAFT 11260
SCHRAUBENSCHLOSS 13285
SCHRAUBENSCHLÜSSEL 11833
SCHRAUBENSCHLÜSSEL (DOPPELMÄULIGER) 4167
SCHRAUBENSCHLÜSSEL (EINFACHER) 11499
SCHRAUBENSCHLÜSSEL (EINMÄULIGER) 11499
SCHRAUBENSCHLÜSSEL (ENGLISCHER) 11052
SCHRAUBENSCHLÜSSEL (GEWÖHNLICHER) 7284
SCHRAUBENSCHLÜSSEL (VERSTELLBARER) 11052
SCHRAUBENSCHLÜSSEL MIT SCHRÄGEM MAUL 1195
SCHRAUBENSCHLÜSSELWEITE 11832
SCHRAUBENSCHNEIDEN 11028
SCHRAUBENSCHNEIDMASCHINE 11066
SCHRAUBENSICHERUNG 7707
SCHRAUBENSICHERUNGSVORRICHTUNG 8701
SCHRAUBENSPINDEL 11062
SCHRAUBENVERBINDUNG 1442
SCHRAUBENVERSETZUNG 11031
SCHRAUBENVERSPANNUNG 12930
SCHRAUBENWELLE 9930
SCHRAUBENWINDE 11035
SCHRAUBENWINDE MIT RATSCHE 10222
SCHRAUBENZIEHER 13294
SCHRAUBGETRIEBE 13974
SCHRAUBKAPPE 11025
SCHRAUBLEHRE 11034
SCHRAUBMUFFE 11061
SCHRAUBSTAHL 2274
SCHRAUBSTOCK 13579
SCHRAUBSTOCK 13561
SCHRAUBSTOCK (EINFACHER) 9429
SCHRAUBSTOCK (SCHWENKBARER) 12562
SCHRAUBSTOCKBACKE 6122
SCHRAUBSTOCKBACKEN (BEWEGLICHER) 8437
SCHRAUBSTOCKBACKEN (FESTER) 5299
SCHRAUBVERBINDUNG 11058
SCHRAUBVERSHLUSS 11025
SCHRAUBZWINGE 2452
SCHREIBER 10289
SCHREIBSTIFT (EINER REGISTRIERVORRICHTUNG) 9161
SCHREIBVORRICHTUNG 10294
SCHREIBWERK 10294
SCHREIBZEUG 10294
SCHREINER 7233
SCHREINEREI 7235
SCHREINEREI (HANDWERK) 7236
SCHRIFT (CHINESISCHE) 2341
SCHRIFTMETALL 13319
SCHRIFTSTEMPEL 1560
SCHRITT 12239
SCHROBHOBEL 7201

SCHROPPHOBEL 7201
SCHROT 7472
SCHROT 11379
SCHROTHOBEL 7201
SCHROTMEISSEL 11217
SCHROTSÄGE 3365
SCHROTSTRAHLPUTZEN 11380
SCHROTT 11007
SCHRUMPFEN 11397
SCHRUMPFEN DES NIETSCHAFTES 3046
SCHRUMPFMASS 477
SCHRUMPFRING 11391
SCHRUMPFRING 1252
SCHRUMPFRING 7203
SCHRUMPFSITZ 11389
SCHRUMPFUNG 11392
SCHRUMPFUNG 3043
SCHRUPPARBEIT 10786
SCHRUPPDREHBANK 10788
SCHRUPPEN 10786
SCHRUPPEN 10775
SCHRUPPHOBEL 7201
SCHRUPPSTAHL 10791
SCHUB 11279
SCHUB (FESTIGKEITSL.) 11288
SCHUBBEANSPRUCHTWERDEN 1078
SCHUBBEANSPRUCHUNG 11295
SCHUBELASTIZITÄT 13144
SCHUBELASTIZITÄTSMODUL 8342
SCHUBELASTIZITÄTSMODUL 5974
SCHUBFALLE 2496
SCHUBFESTIGKEIT 13330
SCHUBKARREN 13812
SCHUBKRAFT 11291
SCHUBLEHRE 11597
SCHUBLEHRE 13537
SCHUBLEHRE MIT DIREKTER ABLESUNG 3986
SCHUBLEHRE MIT TIEFENMASSSTAB 1871
SCHUBMODUL 3226
SCHUBSPANNUNG 5460
SCHUBSPANNUNG 11296
SCHUBSTANGE 2962
SCHUBSTANGENAUGE 4976
SCHUBSTANGENGABEL 5568
SCHUBSTANGENKOPF 2963
SCHUBSTANGENKOPF (GEGABELTER) 5568
SCHUBSTANGENKOPF (GESCHLOSSENER) 1509
SCHUBVERFORMUNG 11286
SCHUBWINKEL 3781
SCHUH 11360
SCHUH 11587

SCHUH FÜR HEBEVORRICHTUNG **7207**
SCHULTER **11383**
SCHULTER EINER WELLE **11384**
SCHUPPE **10964**
SCHUPPE **10968**
SCHUPPE **288**
SCHUPPEN **11302**
SCHUPPENGRAPHIT **10989**
SCHUPPENGRAPHIT **5305**
SCHÜREISEN **10116**
SCHÜREN **9575**
SCHURFHOBEL **7201**
SCHÜRFHOBEL **7201**
SCHÜRFRAUPE **4414**
SCHÜRZE **11549**
SCHUSS EINES GEWEBES **13743**
SCHUSS EINES KESSELS **3260**
SCHUSTERPECH **9400**
SCHÜTTBETON **9714**
SCHÜTTELRINNE **11253**
SCHÜTTELROST **10655**
SCHÜTTELSIEB **11252**
SCHÜTTRUMPF **6577**
SCHÜTTUNG **971**
SCHÜTTUNG **969**
SCHÜTZ **11652**
SCHUTZ **9943**
SCHUTZ (KATHODISCHER) **2109**
SCHUTZ DES GEWERBLICHEN EIGENTUMS **9946**
SCHUTZANSTRICH **9793**
SCHUTZATMOSPHÄRE **9947**
SCHUTZATMOSPHÄRE **790**
SCHUTZBLECH **11343**
SCHUTZBRILLEN **5992**
SCHUTZDECKEL **9944**
SCHÜTZE **11652**
SCHUTZGAS **6905**
SCHUTZGAS-LICHTBOGEN-SCHWEISSEN (ATOMARES) **801**
SCHUTZGAS-LICHTBOGENSCHWEISSEN **5852**
SCHUTZGAS-SCHWEISSEN **683**
SCHUTZGASLICHTBOGEN **1945**
SCHUTZGASOFEN **3055**
SCHUTZGEHÄUSE **9948**
SCHUTZGITTER **9951**
SCHUTZHAUBE **6432**
SCHUTZHÜLLE **9949**
SCHUTZKAPPE **9944**
SCHUTZMARKE **13100**
SCHUTZMASSE **6997**
SCHUTZMUFFE **9942**
SCHUTZNETZ **6157**

SCHUTZRAHMEN **8354**
SCHUTZSCHICHT **9793**
SCHUTZSCHICHT (EINE) AUFTRAGEN **6791**
SCHUTZSCHICHT AUFBRINGEN **6791**
SCHUTZSCHILD **11342**
SCHUTZSTOFF **9952**
SCHUTZSTREIFEN AUS METALL **3268**
SCHUTZÜBERZUG **9793**
SCHUTZÜBERZÜGE **9792**
SCHUTZVORRICHTUNG **10881**
SCHUTZVORRICHTUNG **11346**
SCHUTZVORRICHTUNG **11342**
SCHUTZWAND **9953**
SCHWABBEL **2538**
SCHWABBELMITTEL **1692**
SCHWABBELN **1691**
SCHWABBELSALZE **1690**
SCHWABBELSCHEIBE **1693**
SCHWABBELSCHEIBE **4116**
SCHWABBELSCHEIBE **10173**
SCHWACHGAS **9890**
SCHWACHSTROM **13705**
SCHWÄCHUNG **13707**
SCHWÄCHUNG DES MATERIALS **13708**
SCHWADENMÄHER **11411**
SCHWADENWENDER **12537**
SCHWALBENSCHWANZ **4175**
SCHWALBENSCHWANZFÖRMIG **4177**
SCHWALBENSCHWANZFRÄSER **4176**
SCHWALBENSCHWANZVERBINDUNG **4178**
SCHWAMMEISEN **11953**
SCHWANENSALZ **11729**
SCHWANKUNG **13495**
SCHWANKUNGEN DES WASSERSTANDES **13500**
SCHWANZENDE EINES ROHRES **11901**
SCHWARZBLECH **1296**
SCHWARZBLECHE **1289**
SCHWARZGLÜHEN **571**
SCHWARZKIEFER **1293**
SCHWARZKUPFER **1287**
SCHWARZROTGLÜHEND **1288**
SCHWARZROTGLUT **1295**
SCHWEBEACHSE **5435**
SCHWEBEMANTELMATRIZE **5437**
SCHWEBESTOFFE **8062**
SCHWEBUNG **1119**
SCHWEDISCHER KOLBEN **10196**
SCHWEDISCHGRÜN **10996**
SCHWEFEL **12449**
SCHWEFEL (AMORPHER) **476**
SCHWEFEL (KRISTALLINISCHER) **3435**

SCHWEFELALKOHOL 1964
SCHWEFELAMMONIUM 472
SCHWEFELÄTHER 8853
SCHWEFELBARIUM 1004
SCHWEFELBESTÄUBUNG 4394
SCHWEFELBLEI 5775
SCHWEFELBLUMEN 5464
SCHWEFELBLÜTE 5464
SCHWEFELCHLORÜR 12451
SCHWEFELDIOXYD 12450
SCHWEFELKALIUM 9700
SCHWEFELKALZIUM (EINFACH) 1852
SCHWEFELKIES 7146
SCHWEFELKOHLENSTOFF 1964
SCHWEFELLEBER 7674
SCHWEFELMONOCHLORID 12451
SCHWEFELSÄURE 12453
SCHWEFELSÄURE (KONZENTRIERTE) 2884
SCHWEFELSÄURE (RAUCHENDE) 5733
SCHWEFELSÄURE (ROHE) 8856
SCHWEFELSÄUREANHYDRID 12452
SCHWEFELSAURER KALK 6190
SCHWEFELSAURES CHROMOXYDKALL 2370
SCHWEFELSAURES EISENOXYDUL 6093
SCHWEFELSAURES KALI 9699
SCHWEFELSAURES NICKELOXYDUL 8571
SCHWEFELSAURES NICKELOXYDULAMMONIAK 8558
SCHWEFELSAURES SILBER 11467
SCHWEFELSTANGEN 12255
SCHWEFELTRIOXYD 12452
SCHWEFELWASSERSTOFF 6717
SCHWEFELZYANKALIUM 9701
SCHWEFLIGSÄUREANHYDRID 12450
SCHWEIFHAAR 12594
SCHWEIFSÄGE 1505
SCHWEINFURTERGRÜN 11000
SCHWEINSLEDER 9300
SCHWEISS- UND BELEUCHTUNGSANLAGEN 4524
SCHWEISSBAD 9961
SCHWEISSBAR 13769
SCHWEISSBARKEIT 13768
SCHWEISSBRENNER 5307
SCHWEISSDRAHT 13789
SCHWEISSDRAHT 13793
SCHWEISSDRAHT 8180
SCHWEISSEISEN 13766
SCHWEISSELEKTRODE 13789
SCHWEISSEN 13779
SCHWEISSEN 13755
SCHWEISSEN (ELEKTRISCHES) 4533
SCHWEISSEN (SPIRALFÖRMIGES) 11921

SCHWEISSEN (WAAGERECHTES) 6591
SCHWEISSEN MIT VERDECKTEN LICHTBOGEN 684
SCHWEISSER 13778
SCHWEISSFEHLER 13760
SCHWEISSFITTING 13781
SCHWEISSFOLGE 13790
SCHWEISSGENERATOR FÜR KONSTANTEN STROM 2970
SCHWEISSGUT 13763
SCHWEISSGUT 13761
SCHWEISSGUT (EINGEBRACHTES) 3776
SCHWEISSGUT (REINES) 347
SCHWEISSHITZE 13784
SCHWEISSKONSTRUKTION 13795
SCHWEISSKOPF 13783
SCHWEISSLINSE 13764
SCHWEISSMASCHINE 13786
SCHWEISSNAHT 13776
SCHWEISSNAHT (EINLAGIGE) 11485
SCHWEISSNAHT (LEICHTE) 1792
SCHWEISSNAHTLÄNGE 7520
SCHWEISSNAHTOBERFLÄCHE 4998
SCHWEISSNAHTSINNBILDER 13791
SCHWEISSNAHTÜBERHÖHUNG 10427
SCHWEISSPAKET 5015
SCHWEISSPAKET 5017
SCHWEISSPARAMETER 13787
SCHWEISSPERLE 11842
SCHWEISSRAUPE 13756
SCHWEISSRAUPE 1082
SCHWEISSRAUPENACHSE 871
SCHWEISSRIPPE 5357
SCHWEISSROLLE 2412
SCHWEISSSPIELZEIT 13788
SCHWEISSSTAB 13793
SCHWEISSSTAHL 13765
SCHWEISSSTROMSTÄRKE 13780
SCHWEISSUNG 13767
SCHWEISSUNG 13779
SCHWEISSUNG (AUTOGENE) 834
SCHWEISSUNG (ELEKTRISCHE) 4533
SCHWEISSUNG (HYDROOXYGENE) 8988
SCHWEISSUNG (STUMPFE) 1802
SCHWEISSUNG (WAAGERECHTE) 4190
SCHWEISSVERBINDUG (SCHLECHTE) 920
SCHWEISSVERBINDUNG 13767
SCHWEISSVERBINDUNG 7237
SCHWEISSVERBINDUNG 13774
SCHWEISSVERBINDUNG 13772
SCHWEISSVERBINDUNG (GUTE) 6005
SCHWEISSVERBINDUNG (ÜBERLAPPTE) 7405
SCHWEISSVERSUCH 13792

SICHERHEITSHAKEN **10888**
SICHERHEITSKOEFFIZIENT **5009**
SICHERHEITSKUPPLUNG **10884**
SICHERHEITSMEMBRANE **10860**
SICHERHEITSMUTTER **7702**
SICHERHEITSVENTIL **10891**
SICHERHEITSVENTIL **1593**
SICHERHEITSVENTIL (FEDERBELASTETES) **11986**
SICHERHEITSVENTIL MIT FEDERBELASTUNG **11998**
SICHERHEITSVENTIL MIT GEWICHTSBELASTUNG **13749**
SICHERHEITSVORRICHTUNG **10881**
SICHERHEITSVORSCHRIFTEN **10889**
SICHERHEITSZUGABE **5009**
SICHERN (EINE MUTTER) **7700**
SICHERN (EINEN KEIL) **7700**
SICHERUNG **5751**
SICHERUNG (ELEKTR.) **10886**
SICHERUNG GEGEN VERSCHIEBEN **7711**
SICHERUNG VON SCHRAUBEN **7707**
SICHERUNGSKASTEN **5752**
SICHERUNGSRING F. KOLBENBOLZEN **2405**
SICHERUNGSSCHEIBE **7703**
SICHERUNGSSCHRAUBE **7708**
SICHERUNGSSCHRAUBENMUTTER **10514**
SICHTSIGNAL **13580**
SICKENHAMMER **3321**
SICKENMASCHINE **3323**
SICKENSTOCK **3322**
SICKERWASSER **9181**
SIDERIT **11421**
SIEB **10572**
SIEB **11422**
SIEB **11018**
SIEBANALYSE **11020**
SIEBEN **11424**
SIEBEN **11423**
SIEBKERN **3164**
SIEBMASCHE **8169**
SIEBRÜCKSTAND **9548**
SIEDEDAMPF **1404**
SIEDEDAMPFABLASSDÜSE **1405**
SIEDEHITZE **1427**
SIEDEN **1429**
SIEDEN **1401**
SIEDEPUNKT **1424**
SIEDESALZ **2790**
SIEDETEMPERATUR **1427**
SIEDEVERZUG **10517**
SIEKENSTOCK **3322**
SIEMENS MARTINSTAHL (SAURER) **116**
SIEMENS-MARTIN-OFEN **8829**

SIEMENS-MARTIN-OFEN **8815**
SIEMENS-MARTIN-STAHL **8816**
SIEMENS-MARTIN-STAHL (BASISCHER) **1048**
SIEMENS-MARTINS-VERFAHREN **8830**
SIEMENS-MARTINSTAHL **8023**
SIEMENSMARTINSTAHL (BASISCHER) **1047**
SIGMA-SCHWEISSEN **11428**
SIGNAL (AKUSTISCHES) **816**
SIGNAL (OPTISCHES) **13580**
SIGNALPFEIFE **307**
SIKKATIV **4256**
SIKKATIV **4257**
SILBER **11458**
SILBER (92/5% FEIN) **12253**
SILBERBROMID **1645**
SILBERCHLORID **11461**
SILBERERZ **11625**
SILBERHARTLÖTEN **11460**
SILBERJODID **11464**
SILBERNITRAT **7828**
SILBERSCHLAGLOT **11466**
SILBERSULFAT **11467**
SILBERZYANID **11462**
SILIERMASCHINEN **3522**
SILIKASTEIN **11435**
SILIKAT **11437**
SILIKON **11456**
SILIKOSPIEGEL **11440**
SILIZIUM **11442**
SILIZIUM-MANGANSTAHL **11450**
SILIZIUMBRONZE **11444**
SILIZIUMDIOXYD **11434**
SILIZIUMDIOXYD **11436**
SILIZIUMEISEN **5113**
SILIZIUMKARBID **1980**
SILIZIUMKARBID **11445**
SILIZIUMKUPFER **11446**
SILIZIUMMANGANEISEN **11440**
SILIZIUMMESSING **11443**
SILIZIUMSTAHL **11449**
SILIZIUMSTAHL **11447**
SIMSHOBEL **9522**
SINKSTOFFE **8062**
SINN EINER KRAFT **11182**
SINTERGRUBE **10977**
SINTERKOHLE **4344**
SINTERN **11511**
SINTERN (KONTINUIERLICHES) **3014**
SINTEROFEN **11512**
SINTERSTAHL **10599**
SINTERSTAHL **10598**

SINTERUNG 13584
SINTERUNG 5690
SINTERUNG 11511
SINUS 11475
SINUS-SCHRAUBSTROCK 535
SINUSKURVE 11476
SINUSOIDE 11513
SINUSSCHWINGUNG 6302
SIPHON 11517
SIPHONPFANNE 12693
SIRENE 11518
SITUATIONSPLAN 9430
SITZ 11098
SITZ (ANGEGOSSENER) 7816
SITZ (SELBSTEINSTELLENDER) 9741
SITZE (PARALLELE) 9063
SITZE (SCHRÄGE) 12663
SITZFLÄCHE 11098
SITZGURT 11096
SITZRÜCKENLEHNE 12012
SKALA 10981
SKALAR 10967
SKALENARÄOMETER 6072
SKALENSCHEIBE 6027
SKALENSCHEIBENMIKROMETER 6877
SKANDIUM 10991
SKELETT-MODELL 11526
SKINPACKUNG 11540
SKIZZE 11532
SKIZZE 11528
SKIZZIEREN 11533
SKIZZIEREN 11529
SKLEROMETER 11002
SKLEROSKOP NACH SHORE 11366
SMITHSONI 1836
SODA 11713
SODA (GEGLÜHTE) (KALZINIERTE) 551
SODALAUGE 11712
SODASALZ (KALZINIERTES) 551
SODASTANNAT 11732
SOHLE 2900
SOHLE (SAURE) 108
SOHLE AUF PFÄHLEN 9305
SOHLE EINES HOBELS 11771
SOHLLEDER 11770
SOHLPLATTE (FÜR LAGER) 11772
SOLARÖL 11753
SOLE 1617
SOLENOID 11773
SOLIDUS 11805
SOLIDUSLINIE 11805

SOLLDURCHMESSER 8611
SOLLWERT 11221
SOLLWERT 1858
SOLVAYSODA 455
SOLWAAGE 10901
SONDE 9882
SONDERERREGUNG 11190
SONDERERZEUGNIS 11849
SONDERGEBIET 11845
SONDERMESSING (SEEWASSERFESTES) 8511
SONDERSTAHL 11848
SONDERSTAHL 368
SONDERZWECKATMOSPHÄRE 791
SONNENRAD 12457
SONNENSPEKTRUM 11754
SORBIT 11817
SORTIER-UND WASCHMASCHINEN FÜR FRÜCHTE UND GEMÜSE 5704
SORTIERUNG 11524
SPACHTEL 11844
SPALT 11629
SPALT 11937
SPALTE 3278
SPALTEN 11622
SPALTEN (CHEM.) 11946
SPALTEN (CHEM.) 11950
SPALTFLÄCHE 2488
SPALTFLÄCHE 2487
SPALTKORROSION 3185
SPALTQUERSCHNITT EINES VENTILS 11121
SPALTROHR 2732
SPALTSÄGE 10609
SPALTSCHIEBER 6100
SPALTUNG 2487
SPALTUNG 4058
SPALTUNGSDESTILLATION 3805
SPAN 2346
SPAN 2343
SPAN 12536
SPANANALYSE 2345
SPÄNE 3543
SPÄNE 11278
SPANHOBELMASCHINE 13525
SPANNBACKE 6121
SPANNDORN 2727
SPANNDRAHT 6187
SPANNFINGER 7280
SPANNFUTTER 2393
SPANNHÜLSE 2931
SPANNKEIL 185
SPANNKLUPPE 13562

SPANNKRAFT **12373**
SPANNKRAFT EINER FEDER **5528**
SPANNKRAFT EINES GASES **4480**
SPANNMASCHINE **12372**
SPANNRING **11990**
SPANNROLLE **12334**
SPANNROLLE **1161**
SPANNROLLE **4910**
SPANNROLLE NACH LENIX **13750**
SPANNSÄGE **5628**
SPANNSCHELLE **2448**
SPANNSCHLITTEN **12756**
SPANNSCHLOSS **13285**
SPANNSCHRAUBE **2453**
SPANNTISCHE **8290**
SPANNUNG **12367**
SPANNUNG **9808**
SPANNUNG **13593**
SPANNUNG **12322**
SPANNUNG **9708**
SPANNUNG **12753**
SPANNUNG (INNERE) **7074**
SPANNUNG (KRITISCHE) **3350**
SPANNUNG (ZULÄSSIGE) **9215**
SPANNUNGS-DEHNUNGSKURVE **12366**
SPANNUNGSABFALL **13594**
SPANNUNGSABFALL AN DER ANODE **595**
SPANNUNGSDIAGRAMM **6881**
SPANNUNGSDIAGRAMM **12366**
SPANNUNGSFELD **4505**
SPANNUNGSFLÄCHE **12368**
SPANNUNGSFREIGLÜHEN **12362**
SPANNUNGSGEFÄLLE **13594**
SPANNUNGSKALTRISS **2674**
SPANNUNGSKONZENTRATION **12360**
SPANNUNGSKORROSION **3195**
SPANNUNGSKORROSION **12361**
SPANNUNGSMESSER **13597**
SPANNUNGSREGLER **13595**
SPANNUNGSREGULATOR **13595**
SPANNUNGSRISS **5304**
SPANNUNGSRISSKORROSION **11093**
SPANNUNGSUNTERSCHIED **9709**
SPANNUNGSVERLUST **13594**
SPANNUNGSZEIGER **13597**
SPANNUNGSZUSTAND (KRITISCHER) **5935**
SPANNUT **5483**
SPANNVORRICHTUNG **3370**
SPANNVORRICHTUNG **13285**
SPANNWAGEN **12756**
SPANNWEITE **11831**

SPANWINKEL **11418**
SPANWINKEL **10186**
SPARBEIZE **10510**
SPATEISENSTEIN **11421**
SPECKÖL **7409**
SPECKSTEIN **11703**
SPECKSTEIN **5659**
SPEICHENRAD **11951**
SPEICHER **76**
SPEICHER **8146**
SPEICHER **12297**
SPEICHERUNG **1705**
SPEICHERUNG **12290**
SPEICHERWÄRME **10411**
SPEISEBEHÄLTER **5075**
SPEISEDÜSE **5084**
SPEISEHAHN **5066**
SPEISELEITUNG **5070**
SPEISELEITUNG **5081**
SPEISELEITUNG **12484**
SPEISEN **5064**
SPEISEPUMPE **5071**
SPEISEREGLER **5072**
SPEISEROHR **5070**
SPEISEVENTIL **5076**
SPEISEVORRICHTUNG **5083**
SPEISEWALZE **5074**
SPEISEWASSER **5077**
SPEISEWASSER-VORWÄRMER **5078**
SPEISUNG **5063**
SPEKTRALANALYSE **11861**
SPEKTRALANALYSE **11864**
SPEKTRALAPPARAT **11862**
SPEKTRALFARBEN **11859**
SPEKTROGRAPH **11860**
SPEKTROGRAPH **11862**
SPEKTROMETER **11862**
SPEKTROSKOP **11862**
SPEKTRUM **11863**
SPEKTRUM (KONTINUIERLICHES) **3036**
SPENGLER **12964**
SPERRAD **10225**
SPERRAD **2495**
SPERRADHEBEL **10223**
SPERRE **7705**
SPERREN **7699**
SPERRFLÜSSIGKEIT **11082**
SPERRHAKEN **2094**
SPERRHORN **1091**
SPERRING **5128**
SPERRKEGEL **2094**

SPLINT 11943
SPLINT 10936
SPLINTHOLZ 10936
SPLINTLOCH 9333
SPLINTLOCH 9332
SPLINTTREIBER 9326
SPLISS 11933
SPLISSEN 11934
SPLISSEN (EIN SEIL) 11931
SPREIZBARE DORNE 4906
SPREIZDORN 4907
SPREIZRING 11990
SPRENGRING 10515
SPRENGRING 11692
SPRENGSTOFF 4937
SPRINGENDES ZÄHLWERK 3559
SPRINGSCHALTER 11690
SPRITZDUSE ZERSTÄUBER 11967
SPRITZEN 11968
SPRITZEN 11842
SPRITZER 11967
SPRITZFLASCHE 13633
SPRITZLACKIEREN 9020
SPRITZLACKIERUNG 11965
SPRITZPISTOLE 11962
SPRITZVERFAHREN 11972
SPRITZVERLUSTE 11843
SPRITZVERZINKEN 14045
SPRITZWAND 5251
SPRITZWASSERSCHUTZ 13720
SPRÖDE 1627
SPRÖDIGKEIT 1628
SPRÖDIGKEIT 4685
SPRÖDIGKEIT DES EISENS 1629
SPRÖDIGWERDEN 4685
SPROSSENEISEN 10937
SPRÜHDÜSE 11964
SPRÜHHÄRTUNG 11966
SPRUNG 3278
SPRUNG-BILDUNG 5274
SPRUNGBILDUNG 2286
SPRUNGGERÜST 7260
SPULE 1384
SPULE 2645
SPULE 4499
SPÜLEN 10607
SPÜLEN 10608
SPULENWICKLUNG 2648
SPÜLSCHMIERUNG 5442
SPÜLVERSATZ 6686
SPUND 12989

SPUNDEN 12988
SPUNDEN 12991
SPUNDHOBEL 8047
SPUNDMASCHINE 12992
SPUR 13085
SPUR 13076
SPURKRANZ EINES DRUCKLAGERS 12904
SPURKRANZ EINES RADES 5331
SPURLAGER 12241
SPURLAGER 5524
SPURPFANNE 12905
SPURPLATTE 12905
SPURRING 12904
SPURSTANGE 12231
SPURSTANGE 12915
SPURSTANGE 12916
SPURWEITE 13085
SPURWEITE (EISENBAHN) 5871
SPURZAPFEN 9421
SS-STAHL 8000
STAB 10660
STAB 981
STAB (EINES FACHWERKES) 986
STAB (GEZOGENER) 12754
STAB MIT SCHMELZKERN 3168
STABEISEN 984
STABILISATOR 12042
STABILISIERUNG 12043
STABILITÄT 12041
STABILITÄTSMOMENT 8375
STABMAGNET 985
STABSTAHL (GEZOGENER) 4248
STABTHERMOMETER 12235
STACHELDRAHT 987
STACHELDRAHTMASCHINE 988
STADTBUS 13421
STADTOMNIBUS 8410
STAHL 9427
STAHL 13292
STAHL 13011
STAHL 12196
STAHL 12223
STAHL (ALUMINIUM-BERUHIGTER) 442
STAHL (ANORMALER) 8
STAHL (AUSHÄRTENDER) 8000
STAHL (AUSTENITISCHER) 826
STAHL (BERUHIGTER UND GEHÄRTETER) 10093
STAHL (BLASENFREIER) 12209
STAHL (DEN) BLAU ANLAUFENLASSEN 1368
STAHL (GEHÄRTETER) 6289
STAHL (GESCHMIEDETER) 5554

STAHL (GEZOGENER) **4251**
STAHL (GROBKÖRNIGER) **2577**
STAHL (HALBBERUHIGTER) **11173**
STAHL (HALBHARTER) **1959**
STAHL (HARTER) **6280**
STAHL (HOCHWERTIGER) **6486**
STAHL (KALTGERECKTER) **2697**
STAHL (KOHLENSTOFFARMER) **8270**
STAHL (KOHLENSTOFFARMER) **7770**
STAHL (KOHLENSTOFFEICHER) **1957**
STAHL (KOMPRIMIERTER) **2849**
STAHL (LEGIERTER) **368**
STAHL (NATURHARTER) **8000**
STAHL (NICHTROSTENDER) **8644**
STAHL (NICKELPLATTIERTER) **8573**
STAHL (NIEDRIGLEGIERTER) **382**
STAHL (PERLITISCHER) **9145**
STAHL (PLATTIERTER) **2446**
STAHL (ROSTFREIER) **12051**
STAHL (ÜBERHITZTER) **8945**
STAHL (UNBERUHIGTER) **4462**
STAHL (UNBERUHIGTER) **10598**
STAHL (UNLEGIERTER) **1954**
STAHL (UNMAGNETISIERBARER) **8638**
STAHL (UNVOLLSTÄNDIG DESOXYDIERTER) **8825**
STAHL (VERBRANNTER) **1763**
STAHL (VERGÜTETER) **10093**
STAHL (VERSCHLEISSFESTER) **13718**
STAHL (WEICHER) **11744**
STAHL (WEICHER) **8270**
STAHL (WEICHER) **7770**
STAHL (WEICHER) **1958**
STAHL (ZWEIMAL GEGÄRBTER) **4156**
STAHL ABSCHRECKEN (DEN) **10091**
STAHL AUSGLÜHEN **560**
STAHL FÜR KÜSTENNAHE ÖLBOHRUNG **8739**
STAHL MIT MANGAN (BERUHIGTER) **7969**
STAHL MIT NADELIGER STRUKTUR **8523**
STAHL NORMALISIEREN **560**
STAHL VERGÜTEN (DEN) **6808**
STAHLBAND **12197**
STAHLBAND **12343**
STAHLBAND (ROSTFREIES) **10875**
STAHLBANDRIEMEN **12200**
STAHLBARREN **12210**
STAHLBAU **12218**
STAHLBILDUNG **145**
STAHLBLECH **12213**
STAHLBLECH **11325**
STAHLBLOCK **12210**
STAHLDEUL **951**

STAHLDRAHT **12221**
STAHLDRAHTSEIL **12215**
STÄHLE (KORROSIONS-UND WÄRMEBESTÄNDIGE) **3184**
STÄHLEN **12205**
STÄHLEN **12206**
STAHLERZEUGUNG **12214**
STAHLFLASCHE **12204**
STAHLFORMGUSS **12202**
STAHLFORMGUSS AUS LEGIERTEN STÄHLEN **372**
STAHLGERIPPE **12208**
STAHLGERÜST **12218**
STAHLGIESSEREI **12207**
STAHLGUSS **2055**
STAHLGUSSRADSTERN **2056**
STAHLHALTER **13010**
STAHLHALTER **13005**
STAHLHALTER-REVOLVERKÖPFE **13004**
STAHLKABEL **12201**
STAHLKONSTRUKTION **12218**
STAHLKONSTRUKTION **12208**
STAHLLINEAL MIT TEILUNG **4100**
STAHLLITZEN **12217**
STAHLROHR **12219**
STAHLROHR **12212**
STAHLSANDBLASEN **11380**
STAHLSORTE **2459**
STAHLSORTE **6022**
STAHLSPITZE **8666**
STAHLUNG **135**
STAHLWERK **12222**
STAHLWERKE **12211**
STAHLWERKSKOKILLE **6923**
STALAGMOMETRIE **12054**
STAMM EINER SÄULE **11243**
STAMPFASPHALT **12610**
STAMPFBETON **10193**
STAMPFEN **10194**
STAMPFEN **10195**
STAMPFWERK **12057**
STANDARDABWEICHUNG **12074**
STANDARDKERZE **3663**
STANDARDMODELL **12073**
STÄNDER **12490**
STÄNDER **12146**
STÄNDER **12062**
STÄNDER **13219**
STÄNDERBOHRMASCHINE **13412**
STÄNDERBOHRMASCHINE **2755**
STÄNDERTRAGWERK **2754**
STANDFESTIGKEIT **2643**
STANDFESTIGKEIT **12041**

STANDGUSS **6079**
STANDPEILER **7526**
STANDROHR **12066**
STANDROHR **12093**
STANDROHR **12067**
STANDSICHERHEIT **12041**
STANDVENTIL **13630**
STANGE **10663**
STANGE **10660**
STANGE EINES SCHIEBERS **12236**
STANGE EINES VENTILS **12236**
STANGEN-SPANNZANGEN **12981**
STANGEN-UND ROHRZIEHMASCHINE **982**
STANGENDICHTUNG **9386**
STANGENDRAHT **13882**
STANGENDREHAUTOMAT **983**
STANGENLACK **12254**
STANGENSCHWEFEL **12255**
STANGENZIRKEL **1096**
STANGENZIRKEL **13105**
STANNAT **12095**
STANNICHLORID **12096**
STANNIOL **12955**
STANNOCHLORID **12100**
STANNOOXYD **12102**
STANZE **12060**
STANZE **10009**
STANZE (AUSSCHNEIDEMASCHINE) **10009**
STANZEINHEIT **10011**
STANZEN **10012**
STANZEN **10006**
STANZEN **9998**
STANZEN **12059**
STANZEN **12056**
STANZEN VON LÖCHERN **10008**
STANZPROBE **12061**
STAPEL **9304**
STÄRKE **12843**
STÄRKE **12111**
STÄRKEGUMMI **3825**
STÄRKEKLEISTER **12112**
STÄRKEZUCKER **6058**
STARKSTROM **6401**
STARKWANDIG **12841**
STARRBOHRMASCHINEN **10590**
STARRFETT **6275**
STARRSCHMIERE **6275**
STARRSCHMIERUNG **6087**
START/STOP DRUCKKNOPF **12115**
STARTERKLAPPE **292**
STARTERKLAPPE **12342**

STATIK **12140**
STATIK (GRAPHISCHE) **6061**
STATISCH **12133**
STATISCHES GLEICHGEWICHTSPOTENTIAL EINER REAKTION
 4790
STATIV **13219**
STATOR (ELEKTR.) **12146**
STATUENBRONZE **12147**
STAU **10182**
STAUANLAGE **13753**
STAUB **4389**
STAUBDICHT **4393**
STAUBFÄNGER **4390**
STAUBKOHLE **2563**
STAUCHALTERUNG **12324**
STAUCHEN **4452**
STAUCHEN **13414**
STAUCHEN DER NIETE **2490**
STAUCHGERÜST **4451**
STAUCHGERÜST **4454**
STAUCHGRAT **5357**
STAUCHKALIBER **4455**
STAUCHMASCHINE **13415**
STAUCHMASCHINE **5561**
STAUCHSCHLITTEN **8436**
STAUCHSTICH **4455**
STAUCHSTREIFEN **3382**
STAUCHUNG **3420**
STAUCHVERSUCH **11949**
STAUCHVERSUCH **4374**
STAUCHVERSUCH **13416**
STAUCHVERSUCH **7262**
STAUCHWALZE **4456**
STAUFFERBÜCHSE **12148**
STAUHÖHE **13515**
STEARIN **12195**
STEARINÖL **8795**
STEARINSÄURE **12194**
STEATIT **11703**
STECHBEITEL **9080**
STECHEN **8515**
STECHKARREN **947**
STECHUHR **12946**
STECKKONTAKT **13622**
STECKSCHLÜSSEL **1515**
STECKSCHLÜSSEL **11709**
STEG **13729**
STEG **13726**
STEG EINER KETTE **12402**
STEG EINES FORMEISENS **13730**
STEGFLANKE **10721**

STEUERKETTE 12950
STEUERLEISTUNG 10233
STEUERN 3049
STEUERRAD 12233
STEUERSCHEIBEN 9497
STEUERSEIL 3051
STEUERUNG 3057
STEUERUNG 3048
STEUERUNG 12226
STEUERUNG 13471
STEUERUNG (ADAPTIVE) 161
STEUERUNG (DAMPFM.) 13460
STEUERUNGS-SYSTEM (NUMERISCHES) 8695
STEUERUNGSEINSTELLUNG 12952
STEUERWELLE 10550
STEUERWELLE 4315
STICH 10613
STICHEL 13011
STICHFLAMME 8975
STICHFOLGE 9097
STICHHÖHE 10613
STICHLOCH 12645
STICHLOCH 7287
STICHMASS 10661
STICHSÄGE 2802
STICKOXYD 8581
STICKSTOFF 8588
STICKSTOFF MONOXYD 8581
STICKSTOFFDIOXYD 8591
STICKSTOFFOXYDUL 8597
STICKSTOFFPENTOXYD 8590
STICKSTOFFPEROXYD 8591
STICKSTOFFSÄURE 8580
STICKSTOFFSESQUIOXYD 8592
STICKSTOFFTETROXYD 8591
STICKSTOFFTRIOXYD 8592
STIEL EINES HAMMERS 11244
STIELPFANNE 6250
STIFT 12397
STIFT 9315
STIFT (KONISCHER) 12655
STIFTLOCH 9332
STIFTNIETUNG 9321
STIFTÖLER 8520
STIFTRAD 9324
STIFTSCHLÜSSEL 9322
STIFTSCHRAUBE 12399
STIFTSCHRAUBE 11220
STIFTSCHRAUBE 12397
STILL SETZEN 11406
STILLEGUNG 11409

STILLSETZEN EINER MASCHINE 11410
STILLSTAND EINER MASCHINE 12094
STILLSTANDZEIT 4191
STIRNABSCHRECKPROBE 7254
STIRNFLÄCHE 4989
STIRNFRÄSER 4722
STIRNFRÄSER 4992
STIRNKURBEL 8947
STIRNLAGER 4720
STIRNRAD 12009
STIRNRADANTRIEB 12010
STIRNRÄDERGETRIEBE 12008
STIRNRÄDERWENDEGETRIEBE 10547
STIRNRADFLASCHENZUG 12011
STIRNSENKUNG 11960
STIRNWAND 4728
STIRNZAPFEN 4721
STÖCHIOMETRIE 12268
STÖCHIOMETRISCH 12267
STÖCKEL 632
STOCKFLECKEN 4843
STOCKGETRIEBE 9325
STOCKLACK 12254
STOCKPUNKT 11804
STOCKSCHERE 12264
STOCKTHERMOMETER 12235
STOFF 12419
STOFF (ORGANISCHER) 8866
STOFFE (FEUERFESTE) 10466
STOFFMANGEL 13348
STOPFBÜCHSE 12405
STOPFBÜCHSE 9001
STOPFBÜCHSE 5953
STOPFBÜCHSENBRILLE 9002
STOPFBÜCHSENFLANSCH 5955
STOPFBÜCHSENHAHN 5954
STOPFBÜCHSENMUTTER 9005
STOPFBÜCHSENPACKUNG 12406
STOPFBÜCHSGEHÄUSE 9008
STOPFBÜCHSRAUM 12405
STOPFBÜCHSTOPF 9008
STOPFEN 9529
STOPFEN 12285
STOPFEN 9011
STOPFEN 1483
STOPFENHAHN 8997
STOPFENPFANNE 1496
STOPFENSTECKER 9529
STOPPUHR 12283
STÖPSEL 12285
STÖPSEL 12639

STÖPSEL (EINGESCHLIFFENER) **6146**
STÖPSELKONTAKT **13622**
STORCHSCHNABEL **9033**
STÖRSCHUTZ **10156**
STÖRUNG **1581**
STOSS **6783**
STOSS **7238**
STOSS (MECH.) **11359**
STOSS (MIT LASCHE) **1782**
STOSS AUF GEHRUNG **8324**
STOSS-UND RÜTTELPRÜFUNG **7249**
STOSSAPPARATE **11642**
STOSSDÄMPFER **11358**
STOSSEISEN **7333**
STÖSSEL **12674**
STÖSSEL **9221**
STÖSSEL **10188**
STÖSSEL **1885**
STÖSSEL **12672**
STÖSSELKOPFSTAHLHALTER **13003**
STÖSSELSPIEL **12673**
STÖSSELSTANGE **10031**
STOSSEN **11630**
STOSSEN **11641**
STOSSFÄNGER **1732**
STOSSFESTIGKEIT **1918**
STOSSFESTIGKEIT **6785**
STOSSHEBER **6694**
STOSSKRAFT **6811**
STOSSLASCHE **1783**
STOSSLASCHENTEIL **11932**
STOSSMASCHINE **11643**
STOSSNAHT **1784**
STOSSOFEN **10030**
STOSSOFEN **10024**
STOSSPLATTE **1783**
STOSSPUNKTER **6180**
STOSSSIEB **1733**
STOSSSTANGENHORN **8956**
STOSSSTELLE **7238**
STOSSSTANGE **10026**
STOSSVERBINDUNG **1778**
STOSSVERBINDUNG **1776**
STOSSVERLUST **7757**
STOSSVERSUCH **6786**
STOSSVORRICHTUNG **10029**
STRAHL (AUFTREFFENDER) **6832**
STRAHL (AUSTRETENDER) **4691**
STRAHL (EINFALLENDER) **6832**
STRAHL (GEBROCHENER) **10397**
STRAHL (REFLEKTIERTER) **10384**

STRAHL (ZUSAMMENHALTENDER) **11787**
STRAHL (ZUÜCKGEWORFENER) **10384**
STRAHLDÜSENBOHREN **7222**
STRAHLEN **10141**
STRAHLEN **10242**
STRAHLEN (ARTINISCHE) **140**
STRAHLEN (INFRAROTE) **7103**
STRAHLEN (ULTRAROTE) **13332**
STRAHLEN (ULTRAVIOLETTE) **13333**
STRAHLEN AUFFANGEN **7024**
STRAHLEN AUSSENDEN **4703**
STRAHLENBRECHUNG **10399**
STRAHLENBÜNDEL **9164**
STRAHLENEMISSION **10143**
STRAHLENKUNDE **10164**
STRAHLENMESSER **142**
STRAHLENMESSER **1431**
STRAHLENRISS **12108**
STRÄHLER **2274**
STRAHLHONVERFAHREN **7653**
STRAHLKONDENSATOR **4470**
STRAHLPUMPE **7223**
STRAHLROHR **8679**
STRAHLSANDBLASEN **2540**
STRAHLUNG (HOMOGENE) **6559**
STRAHLUNG (MONOCHROMATISCHE) **8383**
STRAHLUNGSINTENSITÄT **7019**
STRAHLUNGSKONSTANTE **10144**
STRAHLUNGSKONSTANTE (SCHEINBARE) **643**
STRAHLUNGSVERMÖGEN **4702**
STRAHLUNGSWÄRME **10140**
STRAHLUNGSWÄRMEVERLUST **10145**
STRANG EINES TAUES **12338**
STRANGGIESSEN **3029**
STRANGPRESSEN **4970**
STRANGPRESSENROHLING **4971**
STRASSENÖL **10647**
STREBE **12396**
STREBE **1516**
STREBEBOGEN **688**
STRECKBAR **4361**
STRECKBARKEIT **4363**
STRECKDRAHT **4252**
STRECKE (GEOM.) **11118**
STRECKEN (EISEN) **4219**
STRECKEN DES EISENS **4235**
STRECKENSTEUERUNG **12314**
STRECKFESTIGKEIT **14012**
STRECKGESENK **5726**
STRECKGRENZE **14012**
STRECKGRENZE **14011**

STRECKGRENZE 14013
STRECKMETALL 4904
STRECKPROBE 5409
STRECKRICHTEN 9118
STREICHMASS 8016
STREICHMASS (STEHENDES) 11069
STREICHMODELL 8016
STREIFEN 12378
STREIFEN 11613
STREIFEN 980
STREIFIGKEIT 11453
STRENGFLÜSSIG 13577
STRENGLOT 6279
STREUDÜSE 11963
STREULICHT 3955
STREUSAND 9094
STREUSTROM 7490
STREUSTROM 12349
STREUUNG 10994
STREUUNG 3933
STREUUNG (ELEKTROMAGNETISCHE) 4616
STRICHFOKUS 7599
STRICHNAHT 1085
STRICHRAUPE 12376
STRICHZEICHNUNG 8905
STRIPPERKRAN 12383
STROBOSKOP 12385
STROBOSKOPISCHE METHODE 12386
STROHFASER 12348
STROHFEILE 10774
STROHHÄCKSLER 2199
STROM 3486
STROM (ELEKTRISCHER) 4525
STROM (ELEKTRISCHER) 4500
STROM (HOCHGESPANNTER) 6500
STROM (INDUZIERTER) 6889
STROM (MIT DEM) 4184
STROMABWÄRTS 4184
STROMAUFWÄRTS 13407
STROMAUSBEUTE 3489
STROMBEGRENZUNGSVENTIL 4879
STROMDICHTE 3488
STROMDICHTE 4501
STROMERZEUGER 5924
STROMERZEUGUNG 9894
STROMERZEUGUNGSANLAGE 5920
STROMFÜHRENDER LEITER 7670
STROMKREIS 2406
STROMKREIS (DEN) SCHLIESSEN 2515
STROMKREIS ÖFFNEN (DEN) 8826
STROMLEITUNGSNETZ 4080

STROMLIEFERUNG 12487
STROMLINIE 12351
STROMMESSER 452
STROMPAUSE 8734
STROMQUELLE 11823
STROMREGLER 3491
STROMRICHTUNGSANZEIGER 9588
STROMSCHIENE 1770
STROMSTÄRKE 7021
STROMSTÄRKE 478
STROMSTÄRKEREGLER 7020
STROMUMFORMER 3071
STRÖMUNG (AUFGEZWUNGENE) 5534
STRÖMUNG (FREIE) 5634
STRÖMUNG (GEORDNETE) 9055
STRÖMUNG (LAMINARE) 9055
STRÖMUNG (TURBULENTE) 13274
STRÖMUNG (UNGEORDNETE) 13274
STRÖMUNG (WIRBELIGE) 13274
STRÖMUNGSDRUCK 9825
STRÖMUNGSGESCHWINDIGKEIT 13516
STRÖMUNGSGESCHWINDIGKEIT 5462
STRÖMUNGSMESSER 3490
STRÖMUNGSWEG 9125
STROMUNTERBRECHUNG 7093
STROMVERBRAUCH 2993
STROMVERSORGUNG 12487
STROMVERSORGUNGSNETZ 6096
STROMVERTEILUNGSNETZ 4080
STROMZÄHLER 4553
STROMZEIGER 452
STROMZEIT 6382
STRONTIUM 12389
STRONTIUMFLUORID 12390
STROPHOIDE 7722
STRUKTUR 12395
STRUKTUR 12394
STRUKTUR 12800
STRUKTUR (DENDRITISCHE) 3750
STRUKTUR (KÖRNIGE) 6052
STRUKTURELEMENT 12391
STÜCK 4067
STÜCK (EINGESETZTES) 6970
STÜCKARBEIT 9294
STÜCKKOHLE 7826
STÜCKLISTE 7662
STÜCKLOHN 9295
STÜCKZEICHNUNG 3809
STUFE 6021
STUFE 12239
STUFE 11383

T-NAHT **1801**
T-SPANN-NUTEN FRÄSER **12582**
T-STÜCK **12699**
T-STUCK **1559**
T-TRÄGER **12700**
T-VERSCHRAUBUNG **1559**
T-WULST **1719**
TABELLE **2273**
TABELLE **12584**
TABLETT **13167**
TABLETTABSTAND **13169**
TABLETTENMANNLOCH **13168**
TABLETTENMASCHINE **9310**
TABLETTRINGHALTER **13170**
TACHOGRAPH **12586**
TACHOMETER **12587**
TACHOMETER **11869**
TAFEL **12584**
TAFEL **9032**
TAG DER ANMELDUNG **3616**
TAG DER ERTEILUNG EINES PATENTES **3617**
TAGELOHN **3594**
TAGESLICHT **3622**
TAGSCHICHT **3620**
TAGWASSER **12520**
TALG **12609**
TALK **12608**
TALKPULVER **12607**
TALKSPAT **7880**
TALKUM **12607**
TALKUM **5659**
TALSPERRE **3597**
TANDEMDAMPFMASCHINE **12613**
TANGENS **12615**
TANGENTE **12616**
TANGENTE **12615**
TANGENTENPOLYGON **2436**
TANGENTENSCHRAUBE **12619**
TANGENTIALBESCHLEUNIGUNG **12620**
TANGENTIALKEIL **12618**
TANGENTIALKRAFT **12622**
TANGENTIALSPANNUNG **12624**
TANGENTIALTURBINE **12621**
TANGENTIATLSCHNITT DES HOLZES **1058**
TANK **4161**
TANK **12625**
TANK **3973**
TANK (WÄRMEISOLIERTER) **6995**
TANK FÜR FLÜSSIGES ERDGAS **7678**
TANK MIT GEWÖLBTEM DECKEL **4121**
TANK MIT KONKAVEM BODEN **12631**

TANK MIT KONVEXEM BODEN **12632**
TANK MIT SELBSTTRAGENDEM KONUSDACH **11160**
TANKBODEN **6314**
TANKGRUBE **12629**
TANKLAGER **12627**
TANKUNTERBAU **12628**
TANKWALL **1735**
TANKZUGANGSLEITER **5646**
TANNAT **12633**
TANNENBAUMKRISTALL **3748**
TANNIN **5780**
TANTAL **12634**
TANZEN DES REGLERS **6671**
TANZMEISTER **6975**
TARIEREN **11216**
TARTRAT **12691**
TASTER **1874**
TASTERLEHRE **6608**
TASTZIRKEL **1874**
TAU **1829**
TAUCHAUFTRAG **3970**
TAUCHBADSCHMIERUNG **11925**
TAUCHBEHÄLTER **3973**
TAUCHEN **3975**
TAUCHHARTLÖTUNG **3969**
TAUCHKOLBEN **9545**
TAUCHKOLBENPUMPE **9546**
TAUCHKORB **3976**
TAUCHLÖTEN **2304**
TAUCHROHR **1711**
TAUCHSCHMIERUNG **11925**
TAUCHÜBERZUG **2592**
TAUKLOBEN **10731**
TAUMELSÄGE **4337**
TAUPUNKT **3824**
TECHNETIUM **8040**
TECHNIK **12695**
TECHNISCH **12694**
TEER **12679**
TEEREN **12689**
TEEREN **12680**
TEERFARBSTOFF **2567**
TEERÖL **12681**
TEERPAPPE **10712**
TEESTÜCK **12697**
TEIGZUSTAND **8488**
TEIL **9083**
TEIL (LOSER) **7745**
TEIL (ZENTRALER) **2287**
TEILCHEN **9088**
TEILCHEN (NADELFÖRMIGES) **8517**

TEILCHEN (UMHÜLLTE) **2585**
TEILCHENGRÖSSE **9089**
TEILDRUCK **9087**
TEILEN **4097**
TEILEPROGRAMM **9084**
TEILFERTIGUNG **11960**
TEILFUGE **7241**
TEILFUGE **3888**
TEILGERÄT **4100**
TEILHÄRTUNG **11145**
TEILKEGELWINKEL **9401**
TEILKRAFT **2816**
TEILKRAFT **2817**
TEILKREIS **6026**
TEILKREIS **2423**
TEILKREIS **9404**
TEILKREISDURCHMESSER **9407**
TEILLISTE **9095**
TEILMASCHINE **4102**
TEILMONTAGE-WERKSTATT **5282**
TEILPLATTE **6867**
TEILSCHEIBE **4109**
TEILSCHEIBE **4106**
TEILSCHEIBE **4105**
TEILSCHEIBEN **6870**
TEILSTRICH **8009**
TEILUNG **10981**
TEILUNG **9399**
TEILUNG **4107**
TELEGRAMM **1830**
TELEGRAPH **12707**
TELEGRAPHENDRAHT **12708**
TELEGRAPHIE **12709**
TELEGRAPHIE (DRAHTLOSE) **13907**
TELEPHON **12710**
TELEPHONDRAHT **12711**
TELEPHONIE **12712**
TELEPHONIE (DRAHTLOSE) **13908**
TELESKOP **10386**
TELESKOPLEHRE **12713**
TELESKOPSTEIL **4945**
TELLER (EINES VENTILS) **13463**
TELLERKOLBEN **11791**
TELLERVENTIL **4011**
TELLUR **12716**
TEMPERATUR **12721**
TEMPERATUR (ABNEHMENDE) **5025**
TEMPERATUR (ABSOLUTE) **35**
TEMPERATUR (DER UMGEBENDEN LUFT) **12729**
TEMPERATUR (ERHÖHTE) **4658**
TEMPERATUR (EUTEKTISCHE) **4858**

TEMPERATUR (FALLENDE) **5025**
TEMPERATUR (GERECHNETE) **3798**
TEMPERATUR (GLEICHBLEIBENDE) **2977**
TEMPERATUR (HERABGESETZTE) **10344**
TEMPERATUR (HOHE) **6476**
TEMPERATUR (KONSTANTE) **2977**
TEMPERATUR (KRITISCHE) **3352**
TEMPERATUR (MITTLERE) **855**
TEMPERATUR (NIEDRIGE) **7775**
TEMPERATUR (SINKENDE) **5025**
TEMPERATUR (STEIGENDE) **10618**
TEMPERATUR (TIEFE) **7775**
TEMPERATUR (TIEFSTE) **8306**
TEMPERATUR (ZUNEHMENDE) **10618**
TEMPERATUR-REGELVENTIL **12840**
TEMPERATURABNAHME **5022**
TEMPERATURÄNDERUNG **2239**
TEMPERATURANSTIEG **10231**
TEMPERATURANSTIEG **10612**
TEMPERATURBEREICH **12730**
TEMPERATURDIAGRAMME **12725**
TEMPERATURERHÖHUNG **10612**
TEMPERATURERNIEDRIGUNG **5022**
TEMPERATURFUGE **4918**
TEMPERATURGEFÄLLE **12724**
TEMPERATURGRENZE **7592**
TEMPERATURKURVEN **12723**
TEMPERATURMESSUNG **8086**
TEMPERATURREGLER **12840**
TEMPERATURRÜCKGANG **5022**
TEMPERATURSCHOCK **12812**
TEMPERATURSCHWANKUNG **13498**
TEMPERATURSKALA **12835**
TEMPERATURSTEIGERUNG **10612**
TEMPERATURUNTERSCHIED **3904**
TEMPERATURVERTEILUNG **4083**
TEMPERATURWECHSEL **2239**
TEMPERATURWECHSELBESTÄNDIGKEIT **12813**
TEMPERATURZUNAHME **10612**
TEMPERGUSS **7959**
TEMPERGUSS **2043**
TEMPERKOHLE **13345**
TEMPERKOHLEABSCHEIDUNG **6066**
TEMPERN **4230**
TEMPERN **562**
TEMPERN **568**
TEMPERN **12717**
TEMPERN **13175**
TEMPERN **7952**
TEMPERSTAHLGUSS **7959**
TERBIUM **12760**

TERPENTIN 3415
TERPENTIN (VENEZIANER) 13527
TERPENTINÖL 8765
TERPENTINÖLLACK 13296
TERRAKOTTA 12767
TERRAZZO 12768
TETRACHLORKOHLENSTOFF 1960
TETRAEDER 12797
TETRAGONAL 12795
TETRAJODFLUORESZEIN 4831
TEXEL 207
TEXTILFASER 12799
TEXTILRIEMEN 4981
THALLIUM 12801
THERMISCHE ERWEICHUNG 5319
THERMISCHER WIRKUNGSGRAD 12806
THERMIT 12817
THERMIT-VERFAHREN 431
THERMITPRESSSCHWEISSEN 9837
THERMITSCHWEISSUNG 12818
THERMITSCHWEISSUNG 12819
THERMOCHEMIE 12822
THERMODYNAMIK 12829
THERMODYNAMISCH 12828
THERMOELEKTRISCH 12823
THERMOELEKTRISCHER STROM 12824
THERMOELEKTRIZITÄT 12826
THERMOELEMENT 12827
THERMOELEMENT 12821
THERMOGRAPH 10293
THERMOMETER 12833
THERMOMETER 12726
THERMOMETER (ACHTZIGTEILIGES) 10268
THERMOMETER (AUFZEICHNENDES) 10293
THERMOMETER (HUNDERTTEILIGES) 2186
THERMOMETER REGISTRIERENDES 10293
THERMOMETER SCHREIBENDES 10293
THERMOMETEREINTEILUNG 12835
THERMOMETERKUGEL 1717
THERMOMETERRÖHRE 13255
THERMOMETERSKALA 12835
THERMOSÄULE 12820
THERMOSÄULE 12836
THERMOSKOP 12837
THERMOSTAT 12838
THERMOSTAT 12722
THERMOSTROM 12824
THIOSCHWEFELSÄURE 12853
THIOSULFAT 12852
THIXOTROPIE 12854
THOMAS-VERFAHREN 1043

THOMASROHEISEN 1049
THOMASSCHLACKE 1054
THOMASSTAHL 1044
THOMASSTAHL 12855
THORIUM 12856
THULIUM 12910
THYRISTOR 12913
TIEFBAU 2444
TIEFBEHÄLTER 13047
TIEFBEIZEN 3685
TIEFBEIZPROBE 3684
TIEFBETTFELGE 4322
TIEFBOHRUNG 3780
TIEFBRUNNENPUMPE 3689
TIEFE 3779
TIEFENLEHRE 3781
TIEFENLEHRE 13538
TIEFENLEHRE MIT MIKROMETERSCHRAUBE 13539
TIEFENMASS 8357
TIEFENMASS 3781
TIEFENMIKROMETER 8252
TIEFENSTREUUNG 12899
TIEFLOCH-BOHRMASCHINE 3687
TIEFLOCHBOHRER 3686
TIEFZIEHBANDSTAHL 3688
TIEFZIEHEN 3683
TIEFZIEHEN 7158
TIEFZIEHEN 3469
TIEFZIEHQUALITÄTSMESSING 3690
TIEGEL 3401
TIEGELGUSSSSTAHL 3410
TIEGELGUSSSSTAHL 2057
TIEGELOFEN 4597
TIEGELOFEN 3402
TIEGELSCHERE 3403
TIEGELSCHMELZOFEN 9669
TIEGELSCHMELZVERFAHREN 3404
TIEGELZANGE 7548
TIEGELZANGE 3405
TINTENEINSATZ 9160
TINTENGUMMI 6939
TISCH-QUERBEWEGUNG 12585
TISCHBOHRMASCHINE 1168
TISCHHOBELMASCHINE 2007
TISCHHOBELMASCHINE (EINSEITIG OFFENE) 8821
TISCHLER 7233
TISCHLEREI 7236
TISCHLEREI (WERKSTÄTTE) 7235
TISCHLERSÄGE 5628
TITAN 12970
TITANDIOXYD 12971

TITANEISEN **5114**
TITANSÄUREANHYDRID **12971**
TITANSTAHL **12972**
TITER EINER LÖSUNG **12354**
TITERFLÜSSIGKEIT **12084**
TITRATION **12968**
TITRIEREN **12968**
TITRIEREN **12973**
TITRIERTE LÖSUNG **12084**
TITRIERUNG **12968**
TITRIERUNG **12974**
TOBOGGAN **11914**
TOLERANZ **12982**
TOLERANZ **359**
TOLERANZ **2483**
TOLERANZ **358**
TOLERANZGRENZE **7595**
TOLERANZKALIBER **3905**
TOLERANZLEHRE **3905**
TOLERANZLEHRE **3903**
TOLUOL **12983**
TOMBAK **12984**
TON **2465**
TON (FEUERFESTER) **5241**
TON (GEBRANNTER) **926**
TON (GEBRANNTER) **6125**
TONBANDGERÄT **12644**
TONEISENSTEIN **2467**
TONERDE **413**
TONERDE **412**
TONERDE **438**
TONERDE (ESSIGSAURE) **417**
TONERDE (KIESELSAURE) **425**
TONERDE (SCHWEFELSAURE) **426**
TONERDEHYDRAT **423**
TONERDENATRON **11716**
TONERDESILIKAT **425**
TONMERGEL **699**
TONNE (1000 KG) **12986**
TONNENLAGER **10681**
TONRING **2468**
TONROHR **4416**
TONSCHIEFER **700**
TOPAS **13038**
TOPOCHEMIE **13039**
TOPPRÜCKSTAND **13040**
TORF **9147**
TORFBRIKETT **9149**
TORFKUCHEN **9151**
TORFMEHL **9726**
TORFMULL **9726**

TORFPLATTE **9150**
TORFSODE **9151**
TORFSTEIN **9148**
TORSION **13054**
TORSION **13051**
TORSION **13056**
TORSIONSELASTIZITÄT **4490**
TORSIONSFEDER **11994**
TORSIONSFESTIGKEIT **13057**
TORSIONSKRAFT **13055**
TORSIONSSCHWINGUNG **13557**
TORSIONSSPANNUNG **13058**
TORSIONSWAAGE **13052**
TORSONSBEANSPRUCHUNG **13056**
TORUSRING **8703**
TOTALRÖSTUNG **3636**
TOTER PUNKT **3626**
TOTLAGEKREIS **8604**
TOTPUNKT **3625**
TOTPUNKT **3626**
TOTPUNKT **3635**
TOTPUNKT (OBERER) **13411**
TOTPUNKT (OBERER) **13032**
TOTPUNKT (UNTER) **1491**
TOTPUNKTLAGE **9646**
TOTPUNKTSTELLUNG **9646**
TOTRAUM **2482**
TOTWASSER **12050**
TOUR **10557**
TOURENZAHL **10814**
TOURENZAHL **8690**
TOURENZÄHLER **10556**
TRACHYT **13081**
TRAFOBLECH **3160**
TRAGACHSE **2010**
TRAGANT **6174**
TRAGANTGUMMI **6174**
TRAGBALKEN **13243**
TRAGBELAG **9011**
TRÄGER **12490**
TRÄGER **5946**
TRÄGER **1522**
TRÄGER **1093**
TRÄGER (BAUW.) **1099**
TRÄGER (BEIDERSEITS EINGESPANNTER) **2756**
TRÄGER (DURCHGEHENDER) **3018**
TRÄGER (DURCHLAUFENDER) **3018**
TRÄGER (EINGESPANNTER) **10131**
TRÄGER (EINSEITIG EINGESPANNTER) **1097**
TRÄGER (EISERNER) **12199**
TRÄGER (FREI AUFLIEGENDER) **1098**

TRÄGER (HALBEINGESPANNTER) **1097**
TRÄGERFUSSPLATTE **5947**
TRAGFÄHIGKEIT **7682**
TRAGFÄHIGKEIT **2011**
TRAGFÄHIGKEIT DES BODENS **11751**
TRAGFÄHIGKEIT DES BODENS **11752**
TRAGFEDER **1114**
TRAGFLÄCHE **1115**
TRAGGESTELL **4116**
TRÄGHEIT **6907**
TRÄGHEITSHALBMESSER **10169**
TRÄGHEITSKRAFT **6907**
TRÄGHEITSMOMENT **8372**
TRAGKRAFT **13090**
TRAGLAGER **7257**
TRAGMAGNET **7554**
TRAGSCHERE **3403**
TRAGSTANGE **11365**
TRAGSTÜTZE **1522**
TRAGWEITE **11831**
TRAGZAPFEN **7256**
TRAINEUR **1164**
TRAJEKTORIE **13104**
TRAKTION **13088**
TRAKTOR **13097**
TRAKTOR **9859**
TRAKTOREN (GÄRTNEREI-) **13096**
TRAKTRIX **13098**
TRAN **13103**
TRÄNENBLECH **10181**
TRÄNKEN **6805**
TRÄNKUNG **6805**
TRÄNKUNGSMITTEL **6804**
TRANSDUCER **13108**
TRANSDUKTOR **13108**
TRANSFORMATIONSGESCHWINDIGKEIT **13521**
TRANSFORMATOR **13115**
TRANSFORMATORBLECH **13116**
TRANSFORMATORENBLECH **4544**
TRANSISTOR **13119**
TRANSLATION **13123**
TRANSLATIONSBEWEGUNG **8407**
TRANSLATIONSGITTER **13120**
TRANSMISSION **11248**
TRANSMISSIONSANTRIEB **7609**
TRANSMISSIONSDYNAMOMETER **13127**
TRANSMISSIONSSEIL **4314**
TRANSMISSIONSWELLE **7944**
TRANSPARENT **13125**
TRANSPARENZ **13124**
TRANSPORT **13140**

TRANSPORT IM WERK **8106**
TRANSPORTBAND **1164**
TRANSPORTEUR **9956**
TRANSPORTKARREN **13230**
TRANSPORTKOSTEN **3209**
TRANSPORTPFANNE **13110**
TRANSPORTRAD **6754**
TRANSPORTROHR **10839**
TRANSPORTSCHNECKE **13970**
TRANSVERSALE **8124**
TRANSVERSALKRAFT **11291**
TRANSVERSALMASSSTAB **3833**
TRAPEZ **13148**
TRAPEZFEDER **13151**
TRAPEZFÖRMIGES GEWINDE **1814**
TRAPEZGEWINDE **1814**
TRAPEZLAST **13150**
TRAPEZOID **13149**
TRASS **13152**
TRAUBENZUCKER **6058**
TRAVERTIN **13166**
TRECKSÄGE **3365**
TREFFPUNKT **9565**
TREIBACHSE **4290**
TREIBENDE SCHEIBE **4313**
TREIBENDES RAD **4318**
TREIBKEIL **2485**
TREIBKEIL **12665**
TREIBKETTE **4306**
TREIBMITTEL **8408**
TREIBÖL **5714**
TREIBPROBE **10184**
TREIBRIEMEN **1163**
TREIBSCHRAUBE **11041**
TREIBSEIL **4314**
TREIBSITZ **4307**
TREIBSTANGE **2962**
TREIBSTOFFSYSTEM **9339**
TREIBWELLE **9930**
TRENNLINIE **9093**
TRENNMITTEL **9092**
TRENNSCHEIBE **3542**
TRENNUNG (ELEKTROSTATISCHE) **4646**
TRENNUNG (MAGNETISCHE) **7917**
TRENNUNGSFLÄCHENGEBIET **7034**
TRENNUNGSFUGE EINES LAGERS **7241**
TRENNUNGSWAND **4110**
TRENNUNGSWÄRME **8528**
TRENNWAND **1726**
TREPPE **12053**
TREPPE (GERADE) **5641**

TRUNKKOLBEN 13241
TS-DIAGRAMM 6346
TUCHSCHEIBE 10173
TUFF 13260
TUFF (VULKANISCHER) 13590
TUFFSTEIN 1838
TUNGSTEINSÄURE 8876
TÜPPELUNG 8421
TURBINE 13267
TURBINE (GEMISCHTE) 8330
TURBINENRAD 13823
TURBINENSCHAUFELN 13268
TURBODYNAMO 13271
TURBOGEBLÄSE 13270
TURBOGENERATOR 13271
TURBOKOMPRESSOR 2178
TURBOKOMPRESSOR 13272
TURBOPUMPE 13269
TURBULENZ 13273
TURMALIN 13072
TÜRPFOSTEN 6507
TÜRSCHLOSS 4123
TÜRSTEIN 7211
TUSCHE 6871
TUSCHEINSATZ FÜR ZIRKEL 9160
TUSCHNAPF 10946
TUSCHPINSEL 2743
TUSCHSCHALE 10946
TUSCHZEICHNUNG 12965
TYP 13320
U-EISEN 2248
U-EISEN 2250
U-FUGENNAHT OHNE LUFTSPALT 2530
U-MONTAGETEIL 12388
U-ROHR 13323
U-TRÄGER 2249
U/MIN 10814
ÜBERALTERUNG 8923
ÜBERBEANSPRUCHUNG 8963
ÜBERBELASTEN 8954
ÜBERBELASTUNG 8953
ÜBERBLATTUNGSNIETUNG 7400
ÜBERCHLORSAURES KALI 9694
ÜBERDECKUNG DES SCHIEBERS 7399
ÜBERDICKE 1695
ÜBERDREHEN (EIN GEWINDE) 12379
ÜBERDRUCK 9809
ÜBERDRUCK 5882
ÜBERDRUCK 8955
ÜBERDRUCKTURBINE 10250
ÜBERDRUCKVENTIL 10441

ÜBEREINANDERGREIFEN 8950
ÜBEREINANDERGREIFEN 8951
ÜBEREINANDERLIEGEN 8950
ÜBEREINANDERSTEHEN 8950
ÜBEREUTEKTOID 6733
ÜBERFAHREN 8959
ÜBERFALL 8931
ÜBERFALLKANTE 13754
ÜBERFALLROHR 8932
ÜBERFLÜSSIGE ZUTAT 5766
ÜBERGANG (KONGRUENTER) 2939
ÜBERGANGSFLANSCH 10350
ÜBERGANGSMUFFE 3962
ÜBERGANGSPUNKT 13122
ÜBERGANGSROHEISEN 8733
ÜBERGANGSROHR 10352
ÜBERGLASUNG 2603
ÜBERGRÖSSE 8961
ÜBERHANG 8934
ÜBERHÄNGEND 8936
ÜBERHITZEN 12473
ÜBERHITZEN 12468
ÜBERHITZER 12472
ÜBERHITZUNG 8946
ÜBERHITZUNG 12460
ÜBERHITZUNG 1754
ÜBERHOLEN 8937
ÜBERHOLUNGSLEUCHTE 9102
ÜBERKOHLENSAURES KALIUM 9693
ÜBERKOPFSCHWEISSEN 7113
ÜBERKOPFSCHWEISSNAHT 8944
ÜBERKOPFSCHWEISSUNG 8941
ÜBERKRAGUNG 9913
ÜBERKRIECHEN EINER LÖSUNG 3330
ÜBERLADUNG 8953
ÜBERLAGERUNG 8948
ÜBERLAGERUNG 8952
ÜBERLAGERUNG VON SCHWINGUNGEN 12476
ÜBERLANDBUS 7099
ÜBERLAPPEN 8951
ÜBERLAPPNAHTSCHWEISSEN 7401
ÜBERLAPPT SCHWEISSEN 7404
ÜBERLAPPTE SCHWEISSUNG 7405
ÜBERLAPPTER STOSS 7396
ÜBERLAPPTER STOSS 7397
ÜBERLAPPUNG 7396
ÜBERLAPPUNG 8949
ÜBERLAPPUNG 8948
ÜBERLAPPUNG 8951
ÜBERLAPPUNG DES SCHIEBERS 7399
ÜBERLAPPUNGSFLANSCH 7398

UMHÜLLEN VON ROHREN 7369
UMHÜLLUNG 8999
UMHÜLLUNG 3274
UMHÜLLUNG 2587
UMHÜLLUNGSKURVE 4776
UMKEHRBAR 10538
UMKEHRBARKEIT 10537
UMKEHRDAMPFMASCHINE 10543
UMKEHREN (DIE BEWEGUNG) 10534
UMKEHRUNG 8965
UMKEHRUNG 7110
UMKEHRUNG EINER BEWEGUNG 10552
UMKEHRVORGELEGE 10542
UMKIPPBAR 6835
UMKLEIDEBUDE 2247
UMKREIS 2435
UMLAUF 10557
UMLAUF 2433
UMLAUFBEWEGUNG 10557
UMLAUFEN 10763
UMLAUFEN 2432
UMLAUFENDE WELLE 10764
UMLAUFGESCHWINDIGKEIT 11876
UMLAUFGETRIEBE 12456
UMLAUFKÜHLER 10752
UMLAUFMOTOR 10754
UMLAUFSCHMIERUNG 5442
UMLAUFVENTIL 1819
UMLAUFVORGELEGE 4778
UMLAUFZAHL 8690
UMLAUFZÄHLER 10556
UMLENKKONUS 11924
UMLENKSCHEIBE 10551
UMRECHNUNG 3067
UMRECHNUNGSTAFEL 3069
UMRISS 8907
UMRISSFRÄSEN 3040
UMRISSLINIE 8907
UMROLL-RÜTTELFORM MASCHINE 7251
UMSCHALTEN (ELEKTR.) 2246
UMSCHALTEN (ELEKTR.) 2241
UMSCHALTER (ELEKTR.) 2244
UMSCHALTVENTIL 10553
UMSCHAUFELN 13283
UMSCHLINGUNGSBOGEN 675
UMSCHLINGUNGSWINKEL 517
UMSCHMELZEN 10447
UMSCHMELZEN 10446
UMSCHMELZMETALL 11107
UMSETZUNG 13123
UMSETZUNG (CHEMISCHE) 4136

UMSPANNER 13115
UMSPANNUNGSBOGEN 675
UMSPONNENER DRAHT 3219
UMSTELLVENTIL 12882
UMSTELLZEIT 10533
UMSTEUERGETRIEBE 10548
UMSTEUERHEBEL 10549
UMSTEUERN 10534
UMSTEUERUNG 10543
UMSTEUERUNG EINER MASCHINE 10541
UMSTEUERWELLE 10550
UMSTÜRZEN 8965
UMWANDLER 13115
UMWANDLUNG 13136
UMWANDLUNG (CHEMISCHE) 2296
UMWANDLUNG (ISOTHERMISCHE) 7196
UMWANDLUNG (PHYSISCHE) 9280
UMWANDLUNG VON ENERGIE 3070
UMWANDLUNGSBEREICH 13113
UMWANDLUNGSPUNKT 3347
UMWANDLUNGSPUNKT 718
UMWANDLUNGSTEMPERATUR 13114
UMWANDLUNGSWÄRME 6364
UMWICKLUNG 13629
UNBALANZ 13341
UNBEKANNTE 13391
UNBELASTET 1081
UNDICHT 7493
UNDICHTE STELLE 7486
UNDICHTHEIT 7491
UNDICHTIGKEIT 7486
UNDICHTIGKEIT 7491
UNDICHTIGKEITSVERLUST 7756
UNDURCHLÄSSIG 6798
UNDURCHLÄSSIG FÜR WÄRMESTRAHLEN 786
UNDURCHLÄSSIGKEIT 6797
UNDURCHLÄSSIGKEIT FÜR WÄRMESTRAHLEN 785
UNDURCHSICHTIG 8803
UNDURCHSICHTIGKEIT 8802
UNEBENHEITEN 6464
UNELASTISCH 6904
UNEMPFLINDLICHKEITSGRAD 3723
UNENDLICH GROSS 6910
UNENDLICH KLEIN 6911
UNFALL 75
UNGEBUNDEN 6818
UNGEHÄRTET 13359
UNGENÜGENDE DECKKRAFT 6119
UNGESCHÜTZTER KOHLELICHTBOGEN 1944
UNGLEICHARTIGKEIT 3170
UNGLEICHFÖRMIGKEIT DER BEWEGUNG 7167

UNGLEICHFÖRMIGKEIT DES MATERIALS 7352
UNGLEICHFÖRMIGKEITSGRAD 3719
UNGLEICHGEWICHT 13341
UNHANDLICH 13402
UNIVERSAL-GELENKKUPPLUNG 6572
UNIVERSAL-RUNDSCHLEIFMASCHINE 13383
UNIVERSALEISEN 5375
UNIVERSALFRÄSMASCHINE 13387
UNIVERSALFUTTER 13382
UNIVERSALGELENK 6573
UNIVERSALPRÜFMASCHINE 13388
UNIVERSALSCHLEIFMASCHINE 13384
UNIVERSALSCHRAUBENSCHLÜSSEL 11052
UNIVERSALSCHRAUBSTOCK 13389
UNIVERSALSCHWEISSKOPF 13390
UNIVERSALSTAHLWALZWERK 13386
UNKOMPRIMIERBAR 6847
UNKOSTEN (ALLGEMEINE) 8798
UNKRAUTDISTEL-UND FARNKRAUTSCHNEIDER 13741
UNLÖSLICH 6984
UNLÖSLICHKEIT 6983
UNMAGNETISCH 8637
UNMAGNETISCH WERDEN 1123
UNMITTELBAR WIRKEND 3980
UNORGANISCHER STOFF 6957
UNREGELMÄSSIG 7163
UNREINHEIT 6814
UNREINIGKEITEN 6813
UNRUND DREHEN 13280
UNRUND LAUFEN 10840
UNRUNDDREHEN 8869
UNRUNDE RIEMENSCHEIBE 9974
UNRUNDE SCHEIBE 4442
UNRUNDHEIT 8894
UNRUNDWERDEN EINES LAGERS 13719
UNSCHLITT 12609
UNSCHMELZBAR 6918
UNSCHMELZBARKEIT 6917
UNSTETIGKEIT 11546
UNSYMMETRIE 13341
UNSYMMETRISCH 778
UNTER LUFTZUTRITT ERHITZEN 6351
UNTERBAU 1026
UNTERBAU 5589
UNTERBELASTET LAUFEN 1080
UNTERBELASTUNG 7562
UNTERBRECHERKONTAKT 3009
UNTERBRECHUNG DES ELEKTRISCHEN STROMES 7093
UNTERCHLORIGSÄUREANHYDRID 2355
UNTERDETERMINANTE 8315
UNTERDRUCK 9811

UNTERDRUCK 13434
UNTERDRUCK 3778
UNTERDRUCKBREMSE 13436
UNTERDRUCKMESSER 13442
UNTERDRUCKREGLER 12431
UNTEREUTEKTOID 6741
UNTERFLURHYDRANT 5244
UNTERGESENK 1497
UNTERGESTELL MIT GLEITSCHUH 11606
UNTERHALTUNG 7946
UNTERHALTUNGSKOSTEN 7945
UNTERKASTEN 8678
UNTERKÜHLEN 12465
UNTERKÜHLEN 12463
UNTERKÜHLTE FLÜSSIGKEIT 12464
UNTERKÜHLUNG 13347
UNTERLAGE 12490
UNTERLAGE 12416
UNTERLAGEN 12494
UNTERLAGKEIL 13731
UNTERLAGSCHEIBE 13631
UNTERLAGSCHEIBE (FERERNDE) 7703
UNTERLAST 7562
UNTERLAUFHAHN 1234
UNTERLEGBLECH 11351
UNTERLEGRING (FEDERNDER) 4486
UNTERMASS 477
UNTERMASS 13349
UNTERNAHTRISS 13350
UNTEROXYDSCHICHT 12416
UNTERPLATTE 1040
UNTERPROGRAMM 12415
UNTERPULVER-SCHWEISSEN 684
UNTERPULVERSCHWEISSEN 12414
UNTERROSTUNG 13352
UNTERSALPETERSÄURE 8591
UNTERSATZ 12490
UNTERSCHALE 1487
UNTERSCHICHT 9857
UNTERSCHNITT 13351
UNTERSCHWEFLIGE SÄURE 12853
UNTERSETZUNG 10357
UNTERSETZUNG 5908
UNTERSETZUNGSVERHÄLTNIS 10360
UNTERSTEMPEL 7794
UNTERSTÜTZUNGSPUNKT 9567
UNTERSUCHUNG (MAKROSKOPISCHE) 7873
UNTERSUCHUNG (MIKROSKOPISCHE) 7114
UNTERSUCHUNG (MIKROSKOPISCHE) 8260
UNTERSUCHUNG (ULTRAMIKROSKOPISCHE) 13336
UNTERTAUCHEN 12413

UNTERTEILUNG 12409
UNTERVERBINDUNG 12408
UNTERWASSER 12597
UNTERWASSERKANAL 12596
UNVERBRANNT 13344
UNVERBRENNBAR 6843
UNVERBRENNBARKEIT 6842
UNVERBRENNLICH 6843
UNVERBRENNLICHKEIT 6842
UNWÄGBAR 6800
UNZERBRECHLICH 13343
UNZERBRECHLICHKEIT 13342
UNZUGÄNGLICH 6821
UNZUGÄNGLICHKEIT 6820
UNZUSAMMENDDRÜCKBARKEIT 6846
UNZUSAMMENDRÜCKBAR 6847
URAN 13420
URMUSTER 9955
URSPRUNG DER KOORDINATEN 8868
URSPRUNGSFESTIGKEIT 9854
URTONSCHIEFER 9278
V-FUGENNAHT OHNE LUFTSPALT 2531
V-LUNKER 5271
VAKUMMETALLISIERUNG 13445
VAKUUM 13434
VAKUUM WÄRMEBEHANDLUNG 13443
VAKUUMBREMSE 13436
VAKUUMDESTILLATION 4075
VAKUUMFORMEN 13439
VAKUUMGEHÄUSE 13438
VAKUUMHEIZUNG 8318
VAKUUMKAMMER 13440
VAKUUMKAMMER 13449
VAKUUMKASTEN 13435
VAKUUMKESSEL 13451
VAKUUMMETALLURGIE 13446
VAKUUMMETER 13442
VAKUUMPUMPE 284
VAKUUMRAFFINIERUNG 13447
VAKUUMSCHMELZEN 13444
VAKUUMSINTERUNG 13448
VALENZ 13452
VANADIN 13475
VANADIUM 13475
VANADIUMEISEN 5116
VANADIUMSTAHL 13476
VARIABLE 13488
VARIANTE 410
VARIATION (MATH.) 13496
VASELIN 13508
VATERGEWINDE 7955

VAUCANSONSCHE KETTE 6570
VEBRENNUNGSRÜCKSTAND 10476
VEKTOR 13510
VEKTORPRODUKT 13509
VENTIL 13453
VENTIL 970
VENTIL 2617
VENTIL (AUSBALANCIERTES) 934
VENTIL (DOPPELSITZIGES) 4166
VENTIL (EINFACHES) 11491
VENTIL (ELEKTRISCHES) 4549
VENTIL (ENTLASTETES) 934
VENTIL (GESTEUERTES) 8120
VENTIL (HYDRAULISCH BETÄTIGTES) 6697
VENTIL (MEHRFACHES) 8476
VENTIL (MEHRSTÖCKIGES) 8475
VENTIL (PNEUMATISCH GESTEUERTES) 9557
VENTIL (SELBSTTÄTIGES) 852
VENTIL (STOPFBÜCHSLOSES) 9010
VENTIL (UNGESTEUERTES) 852
VENTIL (VIERSITZIGES) 10060
VENTIL MIT FALTENBALG-ABDICHTUNG 1138
VENTIL MIT HEBELFEDERBELASTUNG 11985
VENTIL MIT HEBELGEWICHTSBELASTUNG 7535
VENTIL MIT UNMITTELBARER FEDERBELASTUNG 3987
VENTIL MIT UNMITTELBARER GEWICHTSBELASTUNG 3632
VENTILATION 13530
VENTILATOR 5033
VENTILATOR 12429
VENTILATOR 9817
VENTILATORFLÜGELBLATT 5030
VENTILATORRIEMEN 5029
VENTILATORRIEMENSCHEIBE 5032
VENTILAUSSCHLAG 7544
VENTILDIAGRAMM 13457
VENTILE (HÄNGENDE) 8943
VENTILEINSTELLUNGTARIEREN 13469
VENTILERÖFFNUNG 8832
VENTILFÄNGER 13461
VENTILFEDER 13470
VENTILFÜHRUNG 13462
VENTILHAHN 13455
VENTILHEBEL 13465
VENTILHUB 7544
VENTILKAMMER 2229
VENTILKASTEN 2229
VENTILKLAPPE 9024
VENTILKLAPPE 13459
VENTILKOLBEN 1673
VENTILKÖRPER 13454
VENTILKÖRPERSITZ 1397

VERDAMPFERKÖRPER 4870
VERDAMPFUNGSKONDENSATOR 4868
VERDAMPFUNGSTEMPERATUR 4865
VERDAMPFUNGSVERLUST 4866
VERDAMPFUNGSWÄRME 6366
VERDAMPFUNGSWÄRME 6359
VERDAMPFUNGSWÄRME (LATENTE) 7416
VERDICHTBARKEIT 2851
VERDICHTEN 2837
VERDICHTER 2874
VERDICHTER (HYDRAULISCHER) 6680
VERDICHTER (NASSER) 13801
VERDICHTER (TROCKENER) 4345
VERDICHTERRAD 6795
VERDICHTUNG 11355
VERDICHTUNG 2854
VERDICHTUNG 3753
VERDICHTUNGSARBEIT 13941
VERDICHTUNGSHUB 2864
VERDICHTUNGSRING 2867
VERDICHTUNGSVERHÄLTNIS 2860
VERDICKER 12842
VERDICKUNG 7675
VERDICKUNG (FASERIGE) 10745
VERDRÄNGERPUMPE 9653
VERDRÄNGUNG EINER FLÜSSIGKEIT 4054
VERDREHUNG 13054
VERDREHUNGSBEANSPRUCHUNG 13056
VERDREHUNGSFESTIGKEIT 13057
VERDREHUNGSPROBE 13306
VERDREHUNGSSCHWINGUNG 13557
VERDREHUNGSVERSUCH 13059
VERDREHUNGSWAAGE 13052
VERDREHUNGSWINKEL 532
VERDRILLUNG 13054
VERDÜNNEN (EIN GAS) 10214
VERDÜNNEN (EINE LÖSUNG) 3952
VERDÜNNEN EINER LÖSUNG 3954
VERDÜNNTE LÖSUNG 3953
VERDÜNNUNG EINER LÖSUNG 3954
VERDÜNNUNG EINES GASES 10211
VERDÜNNUNGSMITTEL 3951
VERDUNSTEN 4863
VERDUNSTUNG 8502
VERDUNSTUNGSHÖHE 1402
VERDUNSTUNGSKONDENSATOR 4868
VERDUNSTUNGSWÄRME 6366
VERDUNSTUNGSWÄRME 6359
VEREDELUNG 10377
VEREDELUNG (ELEKTROLYTISCHE) 4558
VEREDELUNG (THERMISCHE) 5245

VEREDELUNG EINES METALLS 10378
VEREDLUNGSINDUSTRIE 9888
VERENGERUNG 3045
VERENGT 3042
VERFAHREN 9883
VERFAHREN 8216
VERFAHREN (BASISCHES) 1050
VERFAHREN (HANSGIRGSCHE) 6272
VERFAHREN (SAURES) 120
VERFÄLSCHEN 204
VERFÄLSCHUNG 205
VERFÄRBUNG 4026
VERFESTIGUNG 6850
VERFEUERN GASFÖRMIGER BRENNSTOFFE 1751
VERFEUERN VON KOHLEN 2564
VERFEUERN VON KOHLENSTAUB 1753
VERFEUERN VON ÖL 1752
VERFLÜCHTIGEN 13588
VERFLÜCHTIGUNG 13587
VERFLÜSSIGBAR 7639
VERFLÜSSIGER 2908
VERFLÜSSIGUNG EINES GASES 7636
VERFORMBARKEIT 3705
VERFORMUNG 3698
VERFORMUNG 4077
VERFORMUNG 12322
VERFORMUNG (BLEIBENDE) 9202
VERFORMUNG (PLASTISCHE) 9469
VERFORMUNG (STUFENLOSE) 6853
VERFORMUNGGESCHWINDIGKEIT 12330
VERFORMUNGSALTERUNG 12328
VERFORMUNGSALTERUNG 12327
VERFORMUNGSBAND 3706
VERGÄLLUNGSMITTEL 3747
VERGASBAR 5860
VERGASEN 5862
VERGASER 1984
VERGASER 1982
VERGASUNG 5861
VERGASUNG 5865
VERGIESSBARKEIT 2064
VERGIESSBARKEITSVERSUCH 2065
VERGIESSEN 2030
VERGIESSEN 2068
VERGIESSEN (MIT MÖRTELZEMENT) 6153
VERGIESSEN (MIT ZEMENTMÖRTEL) 6154
VERGLASEN 5969
VERGLASEN 5972
VERGLASUNG 13584
VERGLEICH VON ABMESSUNGEN 7445
VERGLEICHER 2795

VIERKANTFEILE 12019
VIERKANTMUTTER 12024
VIERKANTMUTTER (SCHWEISSBARE) 1307
VIERKANTSCHRAUBE 12033
VIERKANTSCHRAUBENKOPF 10309
VIERKANTSEIL 12028
VIERKANTSTAHL 379
VIERKANTWELLE 12029
VIERRADANTRIEB 5601
VIERSEITIGES PRISMA 10055
VIERTAKTMOTOR 5600
VIERTAKTMOTOR 5603
VIERTELHOLZ 10076
VIERTELKREIS 10056
VIERTÜRIGE LIMOUSINE 5599
VIERWEGESTÜCK 3368
VIERWEGHAHN 5605
VIERWEGVENTIL 5606
VIERWERTIG 12798
VISIERSCHEIBE 6145
VISKOS 13577
VISKOSE 13575
VISKOSIMETER 13574
VISKOSITÄT 13576
VITRIOL (BLAUER) 1376
VITRIOL (GRÜNES) 6093
VITRIOL (WEISSER) 14046
VOGELPERSPEKTIVE 1264
VOGELZUNGE 3392
VOLLAST-EINSTELLSCHRAUBE 187
VOLLBELASTET LAUFEN 1079
VOLLDRUCKDAMPFMASCHINE 12165
VOLLDRUCKLINIE 199
VOLLKEHLNAHT 3077
VOLLKOMMENE VERBRENNUNG 9184
VOLLKREISTRANSPORTEUR 2426
VOLLPIPETTE 1718
VOLLSCHEIBE 5040
VOLLSTEIN 11780
VOLLTURBINE 5718
VOLLZIEGEL 11780
VOLT 13591
VOLT-AMPERE 13592
VOLTAMETER 3227
VOLTMETER 13597
VOLTZAHL 13593
VOLUMEN 13598
VOLUMEN (KRITISCHES) 3354
VOLUMEN (SPEZIFISCHES) 11854
VOLUMEN (VERDRÄNGTES) (PHYS.) 4053
VOLUMEN-AUFNAHME 4912

VOLUMENÄNDERUNG 2240
VOLUMENELASTIZITÄT 2852
VOLUMPROZENT 9179
VON GASEN 4833
VOR-EUTEKTISCHER BESTANDTEIL 9848
VOR-UND RÜCKSPRUNG 7957
VORANSCHLAG 4838
VORARBEIT 1583
VORARBEITEN 9787
VORARBEITER 7482
VORBEARBEITEN 10775
VORBEARBEITUNG 10786
VORBEHANDLUNG 9780
VORBELASTUNG 9782
VORBLECH 11306
VORBLOCK 1346
VORBLOCKWALZWERK 1348
VORBOHREN 10771
VORBRENNE 2476
VORDERACHSE 5691
VORDERACHSGEOMETRIE 5696
VORDERANSICHT 5694
VORDERQUERTRÄGER 5693
VORDERRADANTRIEB 5699
VOREILEN 7448
VOREILWINKEL 515
VORFEILE 1057
VORFILTER 5265
VORFILTER 9751
VORFORMLING 9772
VORFORMUNG 9773
VORFORMUNG 1341
VORFRÄSEN 5858
VORGANG 9883
VORGANG 9240
VORGANG (CHEMISCHER) 2291
VORGEKERBTER BLOCK 8674
VORGELEGE 3244
VORGELEGEWELLE 3240
VORGELEGEWELLE 3233
VORGELEGEWELLE 6756
VORGEWALZT 10778
VORGEWALZTER BLOCK 1346
VORHALTEN 6524
VORHALTER 4115
VORHALTHAMMER 6529
VORHÄNGEBILDUNG 3492
VORHÄNGEBILDUNG 10895
VORKRAGUNG 9917
VORLAGE 10277
VORLAUF (DER DESTILLATION) 5267

WALZDRAHT AUS LEGIERTEN STÄHLEN **376**
WALZE **10671**
WALZE (GEOM.) **3561**
WALZE (GERIFFELTE) **6136**
WALZE (RAUHE) **10174**
WALZE EINER REGISTRIERVORRICHTUNG **10290**
WALZEISEN **10676**
WALZEN **10665**
WALZEN **10696**
WALZEN (KONTINUIERLICHES) **3035**
WALZENBIEGEMASCHINE **1186**
WALZENFRÄSER **13848**
WALZENKESSEL **3578**
WALZENLAGER **10680**
WALZENMÜHLE **10687**
WALZENPUMPE **10758**
WALZENSPURLAGER **10691**
WALZENSTIRNFRÄSER **11337**
WALZENSTRASSE **10669**
WALZENSTRASSE **10701**
WALZENSTRASSE **10670**
WALZFERTIGUNG **8278**
WÄLZFLÄCHE **10705**
WALZFLANSCH **10672**
WALZFLANSCH **10699**
WALZGERÜST **10704**
WALZGERÜST **12065**
WALZGOLD **10673**
WALZHAUT **11544**
WÄLZHEBEL **10686**
WÄLZKREIS **5921**
WÄLZLAGER **618**
WALZNARBE **10171**
WALZNARBE **9417**
WALZNUTEN (GESCHLOSSENE) **2527**
WALZNUTZBREITE **13427**
WALZPROFIL **10675**
WALZRICHTUNG **10698**
WALZSCHMIEDEN **10668**
WALZSPLITTER **7383**
WALZSPLITTER **11626**
WALZSTRASSE (KONTINUIERLICHE) **3030**
WALZSTRECKE **10701**
WALZWERK **10701**
WALZZUNDER **8279**
WANDARM **13618**
WANDBETT **13621**
WANDBOHRMASCHINE **13620**
WANDDAMPFMASCHINE **13620**
WANDDICKE EINES ROHRES **12844**
WANDDREHKRAN **13623**

WANDERMUTTER **13159**
WANDERUNG **8268**
WANDHOBEL **11417**
WANDKONSOLE **13618**
WANDLAGER **1523**
WANDLAGERSTUHL **13618**
WANDPLATTE **13621**
WANDSTÄRKE EINES ROHRES **12844**
WANDUNG **11330**
WANDVORGELEGE **13619**
WANGENHOBEL **11417**
WANZE **1327**
WARENAUFZUG **6007**
WARENZEICHEN **13100**
WARM AUFZIEHEN **11390**
WARMAUFZIEHEN **11398**
WARMBADHÄRTUNG **8021**
WARMBEARBEITUNG **6650**
WARMBEHANDLUNG **6377**
WARMBIEGEPROBE **6618**
WARMBIEGUNG **6619**
WARMBRUCH **6621**
WARMBRÜCHIG **6658**
WARMBRÜCHIG **10337**
WARMBRÜCHIGKEIT **6641**
WARMBRÜCHIGKEIT **10339**
WÄRME **6333**
WÄRME (FREIGEWORDENE) **6370**
WÄRME (FÜHLBARE) **11184**
WÄRME (INNERE) **7070**
WÄRME (LATENTE) **7415**
WÄRME (LATENTE) **6365**
WÄRME (MITTLERE SPEZIFISCHE) **8076**
WÄRME (SPEZIFISCHE) **11852**
WÄRME (STRAHLENDE) **10140**
WÄRME ABFÜHREN **4663**
WÄRME ABGEBEN **5950**
WÄRME AUFSPEICHERN **12296**
WÄRME ENTZIEHEN **13910**
WÄRME ZUFÜHREN **12485**
WÄRMEABFUHR **3735**
WÄRMEABGABE **3735**
WÄRMEABLEITUNG **4666**
WÄRMEAKKUMULATOR **6336**
WÄRMEÄQUIVALENT (MECHANISCHES) **8104**
WÄRMEAUFNAHME **51**
WÄRMEAUFNAHMEFÄHIGKEIT **6335**
WÄRMEAUFNAHMEFÄHIGKET **41**
WÄRMEAUFSPEICHERUNG **12291**
WÄRMEAUSDEHNUNG **12808**
WÄRMEAUSDEHNUNG **4948**

WARNPFEIFE 307
WARNZEICHEN 306
WARTUNG 814
WARTUNG 7946
WARZENBLECH 12769
WASCHEN 13632
WASCHLEDER 2234
WASSER 13640
WASSER (ANGESÄUERTES) 134
WASSER (DESTILLIERTES) 4076
WASSER (FLIESSENDES) 5466
WASSER (GEBUNDENES) 2318
WASSER (HARTES) 6281
WASSER (HYGROSKOPISCHES) 6731
WASSER (KALKHALTIGES) 6281
WASSER (KALKHALTIGES) 2227
WASSER (KEIMFREIES) 13650
WASSER (KOCHENDES) 1428
WASSER (SALZHALTIGES) 10910
WASSER (SCHLAMMIGES) 8446
WASSER (SIEDENDES) 1428
WASSER (STAGNIERENDES) 12050
WASSER (STEHENDES) 12050
WASSER (STRÖMENDES) 5466
WASSER (TOTES) 12050
WASSER (ÜBERHITZTES) 12470
WASSER ENTHÄRTEN 11748
WASSER ERHÄRTEN (IM) (ZEMENT) 6287
WASSERABFLUSS 8899
WASSERABSCHEIDER 12188
WASSERABWEISEND 8349
WASSERABWEISEND 13685
WASSERANZIEHEND 6730
WASSERAUFNAHME 52
WASSERAUFSAUGEND 6730
WASSERBAD (CHEM.) 13641
WASSERBAU 6685
WASSERBEHÄLTER 13679
WASSERBESTÄNDIG 13671
WASSERDAMPF 12192
WASSERDICHT 13686
WASSERDICHT 13688
WASSERDICHT MACHEN 13689
WASSERDICHTMACHEN 13691
WASSERDRUCK 13669
WASSERDRUCKPROBE 6695
WASSERDYNAMO 6700
WASSERENTHÄRTUNG 13676
WASSERENTZIEHUNG (CHEM.) 3728
WASSERENTZUNDERUNG 13647
WASSERFARBE 13664

WASSERFEST 13671
WASSERFREI (CHEM.) 550
WASSERFREIE SCHWEFELSÄURE 12452
WASSERFREIES KALZIUMSULFAT 549
WASSERGAS 13651
WASSERGEHALT 13645
WASSERGEKÜHLT 13683
WASSERGIER 6732
WASSERGIERIG 6730
WASSERGLAS 11808
WASSERHAHN 13644
WASSERHALTIG 6674
WASSERHÄRTUNG 13654
WASSERHÄRTUNG 13672
WASSERHAUT 5169
WASSERHEIZUNG 6648
WASSERKALK 6689
WASSERKALORIMETER 13643
WASSERKOLBEN 13667
WASSERKRAFT 13668
WASSERKRAFTANLAGE 6692
WASSERKRAFTMASCHINE 6691
WASSERKÜHLUNG 13646
WASSERLEITUNG 13666
WASSERLEITUNGSROHR 13665
WASSERLÖSLICH 11809
WASSERMANTEL 13655
WASSERMESSER 13661
WASSERMESSER (VENTURISCHER) 13532
WASSERMESSFLÜGEL 11027
WASSERMESSSCHRAUBE 11027
WASSERMÖRTEL 6690
WASSERMOTOR 6691
WASSERRAD 13692
WASSERRAD (MITTELSCHLÄCHTIGES) 7769
WASSERRAD (RÜCKENSCHLÄCHTIGES) 6465
WASSERRAD (UNTERSCHLÄCHTIGES) 13355
WASSERRAUM EINES KESSELS 13677
WASSERREINIGUNG 10019
WASSERROHR (DAMPFKESSEL) 13681
WASSERRÖHRENKESSEL 13682
WASSERSACKROHR 11514
WASSERSAMMLER 12188
WASSERSÄULE 2752
WASSERSÄULENMASCHINE 13670
WASSERSÄULENVERDICHTER 13801
WASSERSCHLAG 13653
WASSERSPIEGEL 13659
WASSERSTAND 13652
WASSERSTAND 13659
WASSERSTANDMARKE 13660

WÜRFEL **3445**
WÜRFELKOHLE **2606**
WURFGITTER **11628**
WÜRGELPUMPE **10758**
WURM **13979**
WURMGETRIEBE **13974**
WURMSCHRAUBE **6156**
WURMSTICHIG **13976**
WURZEL (BIQUADRATISCHE) **5609**
WURZEL (FÜNFTE) **5147**
WURZEL (MATH.) **10714**
WURZEL (VIERTE) **5609**
WURZEL (ZWEITE) **12027**
WURZELFLANKE **10720**
WURZELKREIS **10723**
WURZELSEITIGE GEGENNAHT **904**
WURZELSPALT **10727**
X-ACHSE **872**
X-FUGENNAHT OHNE LUFTSPALT **2525**
XENON **14004**
XEROGRAPHIE **14005**
XYLOL **14006**
Y-ACHSE **873**
YARD **14007**
YTTERBIUM **14019**
YTTRIUM **14020**
Z-EISEN **14023**
ZACKIG **6862**
ZÄH **13067**
ZÄHES EISEN **13068**
ZÄHFLÜSSIG **13577**
ZÄHFLÜSSIGKEIT **13576**
ZÄHIGKEIT **13576**
ZÄHIGKEIT **13071**
ZÄHIGKEIT **12737**
ZÄHIGKEITSMESSER **13574**
ZÄHIGKEITSREIBUNG **5473**
ZAHL **3940**
ZAHL (GANZE) **7012**
ZAHL (GEBROCHENE) **5613**
ZAHL (GERADE) **4872**
ZAHL (IMAGINÄRE) **6781**
ZAHL (IRRATIONALE) **7162**
ZAHL (IRRATIONALE) **10260**
ZAHL (KOMPLEXE) **2813**
ZAHL (NATÜRLICHE) **7012**
ZAHL (RATIONALE) **10235**
ZAHL (UNGERADE) **8725**
ZAHL (ZUSAMMENGESETZTE) **2820**
ZAHLENWERT **8697**
ZÄHLER **3229**

ZÄHLER (FLÜSSIGKEITS-) **8214**
ZÄHLER (MATH.) **8693**
ZÄHLVORRICHTUNG **3229**
ZÄHLWERK MIT SPRINGENDEN ZAHLEN **3559**
ZAHN **13016**
ZAHN (BEARBEITETER) **3514**
ZAHN (UNBEARBEITETER) **2059**
ZAHN UND EINZAHNUNG **3761**
ZAHNBOGEN **13024**
ZAHNBREITE **1579**
ZAHNDICKE **12846**
ZÄHNEZAHL **8691**
ZAHNFLANKE **13018**
ZAHNFLANKE (ÜBER DEM TEILKREIS) **4997**
ZAHNFLANKE (UNTER DEM TEILKREIS) **5351**
ZAHNFLANKENZIRKEL **8726**
ZAHNFORM **13019**
ZAHNFUSS **10725**
ZAHNFUSSHÖHE **3681**
ZAHNFUSSKEGEL **10718**
ZAHNFUSSLÄNGE **3681**
ZAHNFUSSWINKEL **10716**
ZAHNGESPERRE **9138**
ZAHNHOBEL **13027**
ZAHNHÖHE **13017**
ZAHNHÖHE **12846**
ZAHNHÖHE **13845**
ZAHNKETTE **11432**
ZAHNKOPF **9569**
ZAHNKOPFLÄNGE **164**
ZAHNKRANZ **10594**
ZAHNKRANZ **13023**
ZAHNKRONE **9569**
ZAHNKUPPLUNG **2461**
ZAHNLÄNGE **3785**
ZAHNLÜCKE **6548**
ZAHNMESS-SCHIEBLEHRE **5899**
ZAHNPROFIL **13019**
ZAHNPROFIL **13020**
ZAHNRAD **5903**
ZAHNRAD **5889**
ZAHNRAD (BEARBEITETES) **3508**
ZAHNRAD (GESCHNITTENES) **3508**
ZAHNRAD (KLEINERES) **9335**
ZAHNRAD (ROH GEGOSSENES) **10773**
ZAHNRAD MIT BORDSCHEIBEN **11402**
ZAHNRAD-NONIUSSCHUBLEHRE **5900**
ZAHNRADANTRIEB **13025**
ZAHNRADANTRIEB **5892**
ZAHNRÄDERPRÜFMASCHINE **5902**
ZAHNRADFRÄSER **5898**

ZAHNRADGETRIEBE 5907
ZAHNRADGETRIEBE 13021
ZAHNRADPUMPE 5895
ZAHNRADSATZ (VERSCHIEBBARER) 11602
ZAHNRADSCHNEIDEMASCHINE 5891
ZAHNRADSTAHL 5904
ZAHNRADVORGELEGE 7050
ZAHNRÜCKEN 12514
ZAHNSCHEITEL 12514
ZAHNSEKTOR 13024
ZAHNSTANGE 13022
ZAHNSTANGE 10124
ZAHNSTANGE 10119
ZAHNSTANGENGETRIEBE 10120
ZAHNSTANGENSCHIEBER 1359
ZAHNSTANGENTEILVORRICHTUNG 10123
ZAHNSTANGENWINDE 10121
ZAHNSTÄRKE 2365
ZAHNSTÄRKE 12846
ZAHNSTÄRKE (IM ROLLKREIS) 678
ZAHNSTÄRKE IM ROLLKREIS 2431
ZAHNTEILKREIS 9403
ZAHNTEILUNG 9408
ZAHNTEILUNG 2423
ZAHNTIEFE 13845
ZAHNTRIEB 13021
ZAHNWURZEL 1036
ZAHNWURZEL 10724
ZANGE 12987
ZANGENFUTTER 2728
ZANGENFUTTER 2729
ZANGENSPANNFUTTER 11979
ZAPFEN 9420
ZAPFEN 9419
ZAPFEN 9315
ZAPFEN 7256
ZAPFEN (TISCHLEREI) 12740
ZAPFENBOHRER 9316
ZAPFENDREHRINGE 3312
ZAPFENDRUCK 7259
ZAPFENDÜSE 9338
ZAPFENFUGE 7231
ZAPFENGELENK 9318
ZAPFENLAGER 11905
ZAPFENLAGER 1117
ZAPFENLAUFFLÄCHE 1116
ZAPFENLOCH 8401
ZAPFENREIBUNG 7258
ZAPFENSÄGE 12741
ZAPFENSCHNEIDMASCHINE 12742
ZAPFENSTREICHMASS 8399

ZAPFENZAHNRAD 9324
ZAPFHAHN 1234
ZAPFWELLE 9745
ZAPONLACK 14022
ZÄSIUM 1833
ZÄSIUM 2197
ZAUM (PRONYSCHER) 9920
ZAUN 5094
ZECHE 8300
ZECHENKOKS 2662
ZEHNECK 3644
ZEICHEN 3940
ZEICHEN 2255
ZEICHEN 8012
ZEICHEN (CHEMISCHES) 2315
ZEICHEN (SICHTBARES) 13580
ZEICHENBRETT 4231
ZEICHENBÜRO 4238
ZEICHENBÜRO 4198
ZEICHENFEDER 7999
ZEICHENGERÄT 4236
ZEICHENLEINWAND 13083
ZEICHENPAPIER 4240
ZEICHENSAAL 4238
ZEICHENSTIFT 7478
ZEICHENTISCH 4243
ZEICHENUTENSILIEN 4236
ZEICHNEN 4195
ZEICHNEN 4246
ZEICHNER 4213
ZEICHNUNG 4229
ZEICHNUNG 4245
ZEICHNUNG (AUSGEZOGENE) 6941
ZEICHNUNG (EINE) ANLEGEN 2742
ZEICHNUNG (EINE) AUSTUSCHEN 2742
ZEICHNUNG (ENDGÜLTIGE) 5185
ZEICHNUNG (GEOMETRISCHE) 5928
ZEICHNUNG (GETUSCHTE) 12965
ZEICHNUNG (MASSSTÄBLICHE) 10983
ZEICHNUNG (SCHATTIERTE) 11235
ZEICHNUNG (SCHEMATISCHE) 3836
ZEICHNUNG (TECHNISCHE) 8099
ZEICHNUNG IN NATÜRLICHER GRÖSSE 5724
ZEICHNUNG MIT EINGESCHRIEBENEN MASSEN 3959
ZEICHNUNGSMASSSTAB 10973
ZEIGER 13153
ZEIGER 6226
ZEIGER (EINES INSTRUMENTS) 6868
ZEIGERAUSSCHLAG 3699
ZEIGERWERK 3843
ZEILE 10812

BIBLIOGRAPHIE

Annuaire du Syndicat du revêtement et Traitement des Métaux. PARIS

Assembly engineering master catalog.

ASSIDER Repertorio della Industrie Siderurgiche Italiane

BABCOK and WILCOX Dictionnaire de la technique de production de vapeur.
CLASSEN ESSEN 1972.

BADER Dictionnaire de métallurgie EYROLLES PARIS 1962.

BENDIX Technologie des travaux sur métaux EYROLLES PARIS.

BROCKHAUS Der' wissenschaften und der technik. BROCKHAUS WISSBADEN.

BUCKSCH Dictionnaire Bâtiment et travaux publics. Anglais Allemand.

CHAMBERS Technical dictionary. CHAMBERS.

CHESTERS Refractories for iron and steelmaking. METALS STY LONDON.

CLASON Dictionnaire de la métallurgie. ELSEVIER.

Comment ça marche ? BORDAS PARIS.

CROCKER Piping handbook. Mc GRAW HILL N.Y.

C.T.I.F. Dictionnaire international de fonderie. PARIS.

DEPECKER Les organes de transmission mécanique. LE PRAT PARIS.

Dictionnaire polyglotte : technique générale, métallurgie, machines, pétrole.
EDITURA TEHNICA BUCAREST.

DOEHLER Die casting. Mc GRAW HILL N.Y.

ERNST Worterbuch der industriellen technik. Anglais Allemand.
BRANDSTETTER WIESBADEN.

Fer et Acier. MONTAN WIRTSCHAFT FRANKFURT.

Foundry sands book.

GRAHAM Electroplasting engineering handbook. VAN NOSTRAND REINHOLD N.Y.

HEILER Dictionnaire des canalisations à longue distance. EYROLLES PARIS.

HEILER Dictionnaire des outils-coupants. EYROLLES PARIS.

HENDERSON Metallurgical dictionary. REINHOLD N.Y.

HERSCU Rollingmill terminology. ELSEVIER AMSTERDAM.

HOPFINGER GOLDSTEIN Dictionnaire technique de l'automobile. Angl/Franç/All.
DUNOD PARIS.

Machining data handbook. METCUT RESEARCH ASSOCIATES CINCINNATI.

MARKS Mechanical engineer's handbook. Mc GRAW HILL N.Y.

Metals handbooks. A.S.M.

MOUREAU ET ROUGE Dictionnaire du pétrole. TECHNIP PARIS.

PARKINSON Engineering inspection. PITMAN LONDRES.

PECHEUX Précis de métallurgie. BAILLIERE 1931. PARIS.

PHILIPPE Lexique de la machine-outil. DUNOD PARIS.

SCHLOMAN Dictionnaires techniques divers. OLDENBURG.

SCHROCK Assemblage, ajustage et vérification d'éléments de machine. EYROLLES PARIS.

Sheet metal industries.

STELLHORN Das franzosiche fachwort. GIRARDET ESSEN.

Technical economics for engineers. Pump manual. AICHE N.Y.

TUTZAUER Dictionnaire technique de l'automobile. All/Angl/Franç. HEYMANNS KOLN.

Vocabulaires de fonderie. Angl/All. C.T.I.F. PARIS.

WEAVER Process piping drafting. GULF PUBL. C HOUSTON.

Wie funktioniert das ? Bibliographisches Institut. MANNHEIM.

WUSTER Dictionnaire international de la machine-outil. TECHNICAL PRESS LONDON.

Dépôt légal : 4ᵉ Trimestre 1976
Nº d'éditeur 0001/1976
Achevé d'imprimer
sur les presses de
l'imprimerie ALIN - 93220 Gagny - France
le 30 Septembre 1976.